Fundamental Medical Mycology

Fundamental Medical Mycology

Errol Reiss
Mycotic Diseases Branch, Centers for Disease Control and Prevention, Atlanta, Georgia

H. Jean Shadomy
Department of Microbiology and Immunology, Virginia Commonwealth University, School of Medicine, Richmond, Virginia

G. Marshall Lyon, III
Department of Medicine, Division of Infectious Diseases, Emory University, School of Medicine, Atlanta, Georgia

A JOHN WILEY & SONS, INC., PUBLICATION

This book was written by Errol Reiss in his private capacity. No official support or endorsement by the Centers for Disease Control and Prevention, Department of Health and Human Services is intended, nor should be inferred.

Copyright © 2012 by Wiley-Blackwell. All rights reserved

Published by John Wiley & Sons, Inc., Hoboken, New Jersey
Published simultaneously in Canada

No part of this publication may be reproduced, stored in a retrieval system, or transmitted in any form or by any means, electronic, mechanical, photocopying, recording, scanning, or otherwise, except as permitted under Section 107 or 108 of the 1976 United States Copyright Act, without either the prior written permission of the Publisher, or authorization through payment of the appropriate per-copy fee to the Copyright Clearance Center, Inc., 222 Rosewood Drive, Danvers, MA 01923, (978) 750-8400, fax (978) 750-4470, or on the web at www.copyright.com. Requests to the Publisher for permission should be addressed to the Permissions Department, John Wiley & Sons, Inc., 111 River Street, Hoboken, NJ 07030, (201) 748-6011, fax (201) 748-6008, or online at http://www.wiley.com/go/permission.

Limit of Liability/Disclaimer of Warranty: While the publisher and author have used their best efforts in preparing this book, they make no representations or warranties with respect to the accuracy or completeness of the contents of this book and specifically disclaim any implied warranties of merchantability or fitness for a particular purpose. No warranty may be created or extended by sales representatives or written sales materials. The advice and strategies contained herein may not be suitable for your situation. You should consult with a professional where appropriate. Neither the publisher nor author shall be liable for any loss of profit or any other commercial damages, including but not limited to special, incidental, consequential, or other damages.

For general information on our other products and services or for technical support, please contact our Customer Care Department within the United States at (800) 762-2974, outside the United States at (317) 572-3993 or fax (317) 572-4002.

Wiley also publishes its books in a variety of electronic formats. Some content that appears in print may not be available in electronic formats. For more information about Wiley products, visit our web site at www.wiley.com.

Library of Congress Cataloging-in-Publication Data:

Reiss, Errol.
Fundamental medical mycology / Errol Reiss, H. Jean Shadomy, and G. Marshall Lyon III.
p. ; cm.
Includes bibliographical references and index.
ISBN 978-0-470-17791-4 (cloth)
1. Medical mycology. I. Shadomy, H. Jean. II. Lyon, G. Marshall. III. Title.
[DNLM: 1. Mycology–methods. 2. Mycoses–microbiology. 3. Mycoses–therapy. QY 110]
QR245.R45 2012
616.9′6901–dc22

2011009910

Printed in the United States of America

oBook ISBN: 978-1-118-10177-3
ePDF ISBN: 978-1-118-10175-9
ePub ISBN: 978-1-118-10176-6

10 9 8 7 6 5 4 3 2 1

To our spouses, with gratitude:
Cheryl (E. R.), "Shad" (H. J. S.), and Tabitha (G. M. L.)

Contents

Preface

RATIONALE FOR THIS TEXT

Medical mycology is a distinct subspecialty of medical microbiology and infectious disease. The field has progressed along with advances in both disciplines, informed by new knowledge from general mycology, immunology, and molecular biology. This textbook aspires to integrate that knowledge. It is designed to function as a reference work for the clinical microbiology laboratory, a textbook for a course in medical mycology, and for independent reading and reference by physicians and research microbiologists.

Textbooks in medical mycology are few in number and those that exist are, by-and-large, outdated. The text's scope is balanced between medical and microbiologic knowledge of the fungi pathogenic for humans. It is designed to accompany an upper level course in medical mycology, e.g., a six-week elective consisting of twelve 2-hour lectures. The material is sufficiently detailed so that it may also be presented as a semester course. The chapters are organized by disease and contain numerous illustrations and one-to-three case presentations. A series of questions is appended at the end of each chapter to reinforce learning. The text is annotated with an extensive glossary. The bibliography emphasizes selected references. The text assumes no prior knowledge of mycology but assumes a foundation in modern biology and medical microbiology.

SCOPE OF FUNDAMENTAL MEDICAL MYCOLOGY

Three cross-cutting chapters are followed by 19 disease-specific chapters. The introductory chapter is designed to orient the reader to the spectrum of fungal diseases, taxonomy within the fungal kingdom, reproduction of fungi, the composition of the fungal cell, primary and opportunistic pathogens, and determinants of pathogenicity. The second chapter presents a systematic treatment of laboratory diagnostic methods in medical mycology, including morphologic, genetic, and nonculture methods. That chapter is structured and annotated for ease of use. This is followed by a chapter introducing antifungal therapy. The antifungal agents in current use are discussed with regard to their action spectrum and applications in clinical medicine. This is followed by a subchapter on the specialized subject of antifungal susceptibility tests. These chapters set the stage for the disease-specific chapters which focus with greater granularity on the pertinent laboratory diagnostic methods and therapy.

ORGANIZATION OF THE DISEASE-SPECIFIC CHAPTERS

Each disease-specific chapter is aligned to the same format in order to direct the reader or course participant to sections of particular interest as outlined in the following section, **Generic Format for the Mycotic Disease Chapters**. The participant will become knowledgeable about mycotic diseases through case presentations, including the disease definition and differential diagnosis. The etiologic agents are described according to their general properties, taxonomic relationships, and their ecologic niche. Geographic distribution of each mycotic disease is presented along with a contemporary view of the epidemiology including incidence, prevalence, risk groups, risk factors, and disease transmission. Current knowledge of the determinants of pathogenicity is reviewed from the vantage point of both host and microbial factors. The clinical forms of each disease are detailed according to the organ system involved, clinical signs and symptoms, and pathology. All phases of the laboratory detection, recovery and identification are included for each fungal pathogen. Emphasis is placed on the use of direct examination to achieve a rapid diagnosis. Specialized tests and genetic identification are covered for those methods in clinical use.

OBJECTIVES OF FUNDAMENTAL MEDICAL MYCOLOGY

The clinical laboratory scientist will gain knowledge about the etiologic agents, determinants of pathogenicity, laboratory detection, recovery and identification. The physician will refresh and extend knowledge of mycotic diseases through case presentations, epidemiology, clinical forms, and therapy. There is good reason for both professions to cross-train in the areas of the other, in order to gain a more balanced view of the totality of fungal diseases.

Target audiences:

- **a.** Clinical laboratory scientists who are called upon to identify fungi in the clinical laboratory e.g., medical technologists, clinical laboratory supervisors.
- **b.** Physicians who wish to refresh and extend their knowledge with emphasis on antifungal therapy.
- **c.** Microbiologists who wish to become cross-trained in medical mycology.
- **d.** Microbiologists who wish to conduct research in medical mycology.
- **e.** Medical students wishing to take an elective in medical mycology.
- **f.** Students registered in clinical laboratory science programs.

GENERIC FORMAT FOR THE MYCOTIC DISEASE CHAPTERS

This format facilitates reader's comfort with the material because of the treatment given to each category with content tailored for each mycotic disease.

1. Mycosis-at-a-glance
2. Introduction/Disease Definition
3. Case Presentation
   - **a.** Diagnosis
4. Etiologic Agents
5. Geographic Distribution/Ecologic Niche
6. Epidemiology
   - **a.** Incidence and Prevalence
7. Risk Groups/Factors
8. Transmission
9. Determinants of Pathogenicity
10. Clinical Forms
11. Veterinary Forms
12. Therapy
13. Laboratory Detection, Recovery, and Identification
14. Selected References
15. Questions & Answers

Errol Reiss, Ph.D.
H. Jean Shadomy, Ph.D.
G. Marshall Lyon, III, M.D.

Atlanta, Georgia
September 2011

Acknowledgments

We express our gratitude to the following scientists who provided critiques of the book chapters:

Raza Aly, Ph.D., University of California School of Medicine, San Francisco, California

Ruth Ashbee, Ph.D., University of Leeds, Leeds, England

John W. Baddley M.D., M.P.H., School of Medicine, University of Alabama, Birmingham, Alabama

Mônica Bastos de Lima Barros, M.D., Hospital Evandro Chagas-Fiocruz, Rio de Janeiro, Brazil

Andrew M. Borman, Ph.D., Mycology Reference Laboratory, Bristol, United Kingdom

Iracilda Z. Carlos, Ph.D., D. Pharm., São Paulo State University, São Paulo, Brazil

John D. Cleary, Pharm. D., University of Mississippi School of Pharmacy, University, Mississippi

Garry T. Cole, Ph.D., University of Texas, San Antonio, Texas

Chester R. Cooper, Ph.D., Youngstown State University, Youngstown, Ohio

Arthur F. Di Salvo, M.D., University of South Carolina School of Medicine, Columbia, South Carolina

E. López-Romero, Ph.D., Universidad Autonoma de Guanajuato, Mexico

Xiarong Lin, Ph.D., Texas A&M University, College Station, Texas

Ronald E. Garner, Ph.D., Mercer University School of Medicine, Macon, Georgia

Cornelia Lass-Flörl, M.D., Innsbruck Medical University, Innsbruck, Austria.

Paul F. Lehmann Ph.D., The University of Toledo Health Sciences, Toledo, Ohio

Christine J. Morrison Ph.D., U.S. Centers for Disease Control and Prevention, Atlanta, Georgia

Stephen A. Moser Ph.D., School of Medicine, University of Alabama, Birmingham, Alabama

Marcio Nucci, M.D., Universidade Federal do Rio de Janeiro, Rio de Janeiro, Brazil

Flávio Queiroz-Telles, M.D., Universidade Federal do Paraná, Curitiba, Brazil

Wiley A. Schell, M.S., Duke University Medical Center, Durham, North Carolina

Jerry D. Smilack, M.D., Mayo Clinic Hospital, Phoenix, Arizona

Deanna A. Sutton, Ph.D., University of Texas Health Science Center, San Antonio, Texas

Carlos P. Taborda M.D., University of São Paulo, São Paulo, Brazil

Carolina Talhari M.D., Institute of Tropical Medicine Amazonas, Manaus, Brazil

Uma M. Tendolkar, M.D., LTM Medical College, and LTM General Hospital, Mumbai, India

Brian L. Wickes, Ph.D., University of Texas Health Sciences Center, San Antonio, Texas

Peter R. Williamson, M.D., Ph.D., School of Medicine, University of Illinois, Chicago, Illinois

Jon P. Woods, Ph.D., University of Wisconsin School of Medicine, Madison, Wisconsin

Rosely Zancopé-Oliveira, Ph.D., Fundação Oswaldo Cruz, Rio de Janeiro, Brazil

Many laboratory scientists and physician-scientists have contributed individual illustrations for this book and they are acknowledged in the figure legends. A smaller group went to great lengths to supply several illustrations, many previously unpublished, from their collections, and they deserve special mention and gratitude.

George Barron, Ph.D., Ontario Agricultural College, University of Guelph, Ontario, Canada

Edward P. Ewing, Jr. M.D., Public Health Image Library, Centers for Disease Control, Atlanta, Georgia

Jim Gathany, Scientific photographer, the Centers for Disease Control Creative Arts Branch, Atlanta, Georia

Rajesh T. Gandhi, M.D., Editor for Partners in Infectious Disease Images, www.idimages.org

Arvind A. Padhye, Ph.D., Centers for Disease Control, Atlanta, Georgia

Brian J. Harrington, Ph.D., University of Toledo Health Sciences, Toledo, Ohio

Christoph U. Lehmann, M.D., Johns Hopkins University School of Medicine, Baltimore, Maryland and Chief Information Officer of DermAtlas, www.dermatlas.org

Tadahiko Matsumoto, M.D., Yamada Institute of Health and Medicine, Tokyo, Japan

Jorge Musa, Sr. MLT, Life Laboratories, Toronto, Canada

Stephen W. Peterson, Ph.D., National Center for Agricultural Utilization Research, U.S. Department of Agriculture, Peoria, Illinois

Lynne Sigler, M.S., Professor, University of Alberta, Edmonton, Canada

Robert Simmons, Ph.D., Director, Biological Imaging Core Facility, Department of Biology, Georgia State University, Atlanta, Georgia

Uma M. Tendolkar, M.D., LTM Medical College, and LTM General Hospital, Mumbai, India

Robert L. Wick, Ph.D., Plant, Soil and Insect Sciences Department, University of Massachusetts, Amherst, Massachusetts

Calvin O. McCall, Jr., M.D., School of Medicine, Virginia Commonwealth University, Richmond, Virginia, provided expert advice on therapy for dermatophytosis.

Michael M. McNeil, M.D., Centers for Disease Control, Atlanta, Georgia, provided expert advice on therapy for candidiasis.

We express our sincere appreciation to Dr. Karen E. Chambers, Editor, Ms. Lisa Van Horn, Production Editor, and Ms. Anna Ehler, Editorial Assistant, Wiley-Blackwell Publishers.

E. R.
H. J. S.
G. M. L., III

Part One

Introduction to Fundamental Medical Mycology, Laboratory Diagnostic Methods, and Antifungal Therapy

Fundamental Medical Mycology, First Edition. By Errol Reiss, H. Jean Shadomy and G. Marshall Lyon, III.
© 2012 Wiley-Blackwell. Published 2012 by John Wiley & Sons, Inc.

Chapter 1

Introduction to Fundamental Medical Mycology

1.1 TOPICS NOT COVERED, OR RECEIVING SECONDARY EMPHASIS

The Table of Contents is explicit but it is well to advise readers that some topics are either outside the scope of *Fundamental Medical Mycology* or receive secondary emphasis. Although caused by, or associated with, fungi it is not within the scope of this work to discuss mushroom poisoning (ingestion of toxins present in mushrooms), mycotoxicosis (ingestion of a fungal toxin), or allergies, except when encountered as a complication of one of the fungal diseases discussed.

- More information on mushroom poisoning (mycetismus) can be found in Benjamin (1995).
- Allergies caused by fungi are discussed in Kurup and Fink (1993) and Breitenbach et al. (2002). The health effects of exposure to molds, apart from infection, may be found in Storey et al. (2005) and U.S. Environmental Protection Agency publication 402K-01-001 (2001).
- Environmental mycology is discussed as it relates to the ecologic niche of the causative agents of mycoses.
- Veterinary medical mycology is covered in a concise section, "Veterinary Forms," in each disease-specific chapter.

1.2 BIOSAFETY CONSIDERATIONS: BEFORE YOU BEGIN WORK WITH PATHOGENIC FUNGI...

Safety in the laboratory is of prime importance. Clinical laboratory supervisors and principal investigators have the serious responsibility to train all technologists and students in the safe manipulation of clinical specimens and pathogenic fungi. Before working with pathogenic microbes, including fungi, microbiologists should participate in their organization's safety training program, be certified to work with pathogens, and, when questions about biosafety arise, consult the supervisor and the CDC/NIH biosafety manual: *Biosafety in Microbiological and Biomedical Laboratories*, 5th edition (BMBL). The manual is available online at the URL http://www.cdc.gov/biosafety/publications/bmbl5/index.htm. This will ensure a safe work environment where the workers will not be afraid to work with fungi but instead will have confidence that they are observing prudent precautions.

Molds growing on Petri plates can produce far more infectious propagules (conidia or spores) than an environmental exposure! Therefore, mold cultures should be transferred to screw cap- or cotton-stoppered agar slants. Mold cultures on Petri plates should never be opened on the open laboratory bench. All cultures of unknown molds should be handled inside a biological safety cabinet (BSC). Petri plates should be sealed with shrink seals, which are colorless transparent cellulose bands. "Occupational Hazards from Deep Mycoses" is a useful and cautionary article summarizing laboratory infections (Schwarz and Kauffman, 1977; Padhye et al., 1998).

The BMBL should also be consulted for further information about selection of BSCs, and biosafety considerations of work with pathogenic fungi. If further questions arise on matters of fungal biosafety, please contact the State Department of Health in the United States of America or the CDC Mycotic Diseases Branch, which is the World Health Organization Center for Mycoses.

Fundamental Medical Mycology, First Edition. By Errol Reiss, H. Jean Shadomy and G. Marshall Lyon, III.
© 2012 Wiley-Blackwell. Published 2012 by John Wiley & Sons, Inc.

1.2.1 Biological Safety Cabinets (BSC)

What are the characteristics of Class II Biological Safety Cabinets? The Class II BSC is designed with inward airflow velocity (75–100 linear feet/min) and is fitted with high efficiency particulate air (*HEPA*) filters. This design ensures that the workspace in the cabinet receives filtered, downward, vertical *laminar airflow*. These characteristics protect personnel and the microbiologic work conducted in the BSC.

HEPA-filtered-exhaust air ensures protection of the laboratory and the outside environment. All Class II cabinets are designed for work involving microorganisms assigned to biosafety levels 1, 2, and 3.1. Fungi pathogenic for humans are classed in biosafety level 2 and work with them should be conducted in the BSC, and not on the open bench. Certain manipulations of fungal pathogens or environmental samples require biosafety level 3 (please see below).

Class II BSCs provide a microbe-free work environment. Class II BSCs are classified into two types (A and B) based on construction, airflow, and exhaust systems. Type A cabinets are suitable for microbiologic work *in the absence of* volatile or toxic chemicals and radionuclides, since air is recirculated within the cabinet. Type A cabinets may be exhausted into the laboratory or to the outdoors via a special connection to the building exhaust system. Type B cabinets are hard-ducted to the building exhaust system and contain negative pressure plenums to allow work to be done with toxic chemicals or radionuclides. A list of products that meet the standards for Class II BSCs are available from the National Sanitation Foundation International, Ann Arbor, Michigan. It is mission-critical that BSCs be tested and certified in situ at the time of installation, at any time the BSC is moved, and at least annually after that.

1.2.2 Precautions to Take in Handling Etiologic Agents that Cause Systemic Mycoses

The major known reasons for laboratory exposures to pathogenic fungi are dropped cultures, preparing soil suspensions and inoculating animals, opening Petri plates, and aerosols from needles and syringes (Padhye et al., 1998). The portals of entry for the fungi, resulting from the above exposures, are minor skin wounds or the inhalation of fungal conidia.

Pathologists and veterinarians should be mindful that autopsies and necropsies have caused accidental hand wounds, which have become infected. These infections are localized to the wound and have not disseminated. Needle stick injuries have also been the source of laboratory infections, and they too have remained localized. Localized infections require systemic antifungal therapy.

Laboratory exposures to aerosolized conidia (spores) have led to pulmonary infections and, in the case of *Coccidioides* species, to serious or even fatal infections. Some cases of coccidioidomycosis have occurred in laboratories beyond the endemic area and resulted when the laboratory did not suspect the mold they had isolated was *Coccidioides*.

Biosafety level 2 (BSL 2) practices, containment equipment, and facilities are recommended for handling and processing clinical specimens, identifying isolates, and processing animal tissues suspected of containing pathogenic fungi. BSL 2 is also sufficient for mold cultures identified as *Blastomyces dermatitidis, Cryptococcus neoformans*, dermatophytes, *Penicillium marneffei*, and *Sporothrix schenckii*. In addition to these agents, certain melanized molds have caused serious infection in immunocompetent hosts following inhalation or accidental penetrating injuries: *Bipolaris* species, *Cladophialophora bantiana, Wangiella* (*Exophiala*) *dermatitidis, Exserohilum* species, *Fonsecaea pedrosoi, Ochroconis gallopava, Ramichloridium mackenziei*, and *Scedosporium prolificans*.

All manipulations of clinical specimens and culture work are performed inside an annually inspected and certified, well-functioning, laminar flow biological safety cabinet (BSC), equipped with HEPA filtered exhaust. Workers should wear personal protective equipment (PPE).

- Clothing: laboratory coats with fronts fastened and shoes with closed fronts.
- Eye protection: safety glasses, goggles, as recommended by the supervisor.
- Gloves: latex or plastic.
- Respiratory protection: goggles, mask, face shield, or other splatter guards are used when the cultures must be handled outside the BSC. Surgical masks are not respirators and do not provide protection against aerosolized infectious agents. The N95 disposable respirator provides a level of protection. Supervisors should consult the National Institute of Occupational Safety and Health (NIOSH) Publication No. 99–143: TB Respiratory Protection Program in Health Care Facilities to match the respiratory protection to their risk assessment at the URL http://www.cdc.gov/niosh/docs/99-143.
- Sharps jars should be provided for disposal of needles and syringes.
- All waste should be autoclaved before disposal.

1.2.3 Additional Precautions at Biosafety Level 3 (BSL 3)

BSL 3 conditions should be observed when working with mold-form cultures identified as *Coccidioides* species and *Histoplasma capsulatum* according to the following specific situations.

- *Coccidioides immitis* and *C. posadasii*. Once *Coccidioides* species are identified in the clinical laboratory, biosafety level 3 practices, equipment, and facilities are required for manipulating sporulating cultures and for processing soil or other environmental materials known to contain infectious arthroconidia. *Coccidioides* species are subject to the regulations regarding Select Agents: biological agents and toxins that could pose a severe threat to public health and safety. These regulations are discussed in "Appendix F" of the BMBL.
- Clinical laboratory supervisors and principal investigators should be aware of what to do if there is a *Coccidioides* exposure in their laboratory (Stevens et al., 2009).
- BSL 3 practices, containment equipment, and facilities are recommended for propagating sporulating cultures of *H. capsulatum* in the mold form, as well as processing soil or other environmental materials known or likely to contain infectious conidia.

The criteria for BSL 3 practices are numerous and are detailed in the CDC/NIH manual, *Biosafety in Microbiological and Medical Laboratories*. BSL 3 conditions may require facility reconstruction: for example, two doors should separate the laboratory from a public area; the laboratory should be at negative pressure with respect to the entrance; and air that enters the laboratory should be vented through the biological safety cabinet HEPA filter and then vented outside the building. Waste is to be autoclaved before it leaves the BSL 3 laboratory, so that a room adjacent to the level 3 laboratory should be equipped with an autoclave.

1.2.4 Safety Training

The U.S. Centers for Disease Control and Prevention sponsor the International Symposium on Biosafety. Short courses at that conference cover biosafety practices in laboratories and in veterinary practice: for example, "Infection Control, Biosafety in Research and Clinical Settings" and "Moving from BSL 2 to BSL 3." Topics included are engineering controls, personal protective equipment (PPE), hand hygiene, environmental disinfection, and waste disposal.

1.2.5 Disinfectants and Waste Disposal

For information on these topics for the laboratory please see **Chapter 2, Section 2.3.1.1**, Disinfectants and Waste Disposal.

1.3 FUNGI DEFINED: THEIR ECOLOGIC NICHE

What are fungi? Where are they found? The kingdom Fungi is composed of unicellular or multicellular, *eukaryotic*, *heterotrophic* microbes. Each fungal cell contains a full array of organelles and is bound by a rigid cell wall containing chitin, glucan, and/or cellulose (Table 1.1). Please also see Section 1.12, General Composition of the Fungal Cell.

Of the thousands of fungal species that are free-living in nature or are pathogenic for plants, only a small group are known to be pathogenic for humans and animals. It is also true that *any fungus capable of growing at 37°C is a potential pathogen in a debilitated or immunocompromised host*.

Some fungi are primary pathogens (e.g., *Coccidioides* species) and can cause disease in immune-normal persons. Severity of a fungal disease is related to host factors (immune status, general health status) and the number of infectious propagules (conidia or spores) inhaled, ingested, or injected. Persons who are immunocompromised, or otherwise debilitated, are prone to develop more serious disease and to be susceptible to opportunistic fungi against which immune-normal persons have a high level of resistance.

Fungi are ubiquitous in nature, being found in the air, in soil, on plants, and in water, including the oceans, even as a part of lichens growing on rock. There is essentially no part of our earth where fungi are not found. A few fungal species are adapted to live as *commensals* in humans but for most fungal pathogens humans are accidental hosts. Of all the fungi with pathogenic potential most are opportunistic, whereas a select few are able to cause disease in otherwise healthy humans who have intact immune and endocrine systems.

1.4 MEDICAL MYCOLOGY

What is medical mycology? Medical mycology is a distinct discipline of medical microbiology concerned with all aspects of diseases in humans and lower animals caused by pathogenic fungi.

What are the mycoses? Mycoses are diseases of humans and lower animals caused by pathogenic fungi.

Table 1.1 Comparison of Structure/Function of Bacterial and Fungal Organelles

Organelles	Bacteria	Fungi
Cell wall	Murein = peptidoglycan	Glucan, mannan, chitin
Cell membrane	No sterols	Ergosterol
Metabolism	Aerobic or anaerobic	Aerobic
Energy transduction	Cell membrane	Mitochondria
Cytoplasm	Proteins	Glycoproteins, actin, tubulin, mitotic spindle, Golgi
Gene structure	Operons, no introns	Single genes, introns, repetitive DNA
Nuclear material	Prokaryotic: a single chromosome; no nuclear membrane.	Eukaryotic: multiple chromosomes contained in a nucleus bounded by a nuclear membrane; for example *Candida albicans* has 8 chromosomes
Ribosome	30S + 50S = 70S	40S + 60S = 80S
Extrachromosomal DNA	Plasmids	Mitochondrial DNA, dsRNA
Capsular polysaccharide	Several agents of meningitis	One agent of meningitis: *Cryptococcus neoformans*

There is a broad spectrum of mycoses ranging from superficial skin diseases to deep-seated, multisystem disseminated diseases. Please see Section 1.10, Classification of Mycoses Based on the Primary Site of Pathology.

1.5 A BRIEF HISTORY OF MEDICAL MYCOLOGY

Because the fruiting bodies of some fungi are large enough to see without the aid of a microscope, such as mushrooms, they were the first microorganisms known. Centuries later, it was discovered that mushrooms are only the obvious structures of complex fungi with a vast network of fungal cells found beneath the soil, tree bark, and so on.

1.5.1 Ancient Greece

Fungi have caused a variety of maladies affecting our quality of life for millennia. Hippocrates (460–377 B.C.E.), the father of medicine, recognized that persons with oral thrush (due to *Candida albicans*) were already debilitated by other diseases. This thought was echoed in our own time when Professor Graham S. Wilson, Director of the U.K's Public Health Laboratory Service said: "*Candida* is a much better clinician than we are, in its ability to detect abnormalities earlier in the course of development of such abnormalities than we can with all our chemical tests." This comment was made before the advent of AIDS but now it is well known that oral–esophageal candidiasis heralds the onset of that disease.

1.5.2 Middle Ages

In general, the vast majority of fungal infections are not spread from person to person (are not communicable). However, there are significant exceptions, for example, the *dermatophytes*. In the Middle Ages, children in Europe became infected with *favus*, a fungal disease of the scalp, smooth skin, even nails, due to *Trichophyton schoenleinii*. It was devastating to these individuals because they were considered unclean, separated from their peers, and sent to separate schools. There was no specific treatment at that time. Favus was so disfiguring that it was mistaken for leprosy by artists of the Renaissance (Goldman, 1968). A modern case of scalp ringworm is shown in Fig. 1.1.

1.5.3 Twentieth Century

Examples of outbreaks from a single source are not uncommon. In Witwatersrand, South Africa, from 1941 through 1944 nearly 3000 gold mine workers were infected with the subcutaneous fungal pathogen *Sporothrix schenckii*, which they acquired by brushing against mine timbers on which the fungus was growing (Du Toit, 1942). Figure 1.2 depicts the classic appearance of lymphocutaneous sporotrichosis.

1.5.4 Endemic Mycoses in the Americas

A review of thousands of induction center roentgenograms of young men inducted into the armed forces in World War II noted the "incidence of calcified lesions presumed to represent healed tuberculosis corresponded to the now well-known pattern of regional differences in the U.S." (reviewed by Iams, 1950). Mycologic investigations, including large scale skin testing with histoplasmin, established that delayed type hypersensitivity to *Histoplasma capsulatum* was widespread among residents of the major river valleys of the central United States, thus establishing the boundaries of the histoplasmosis endemic area.

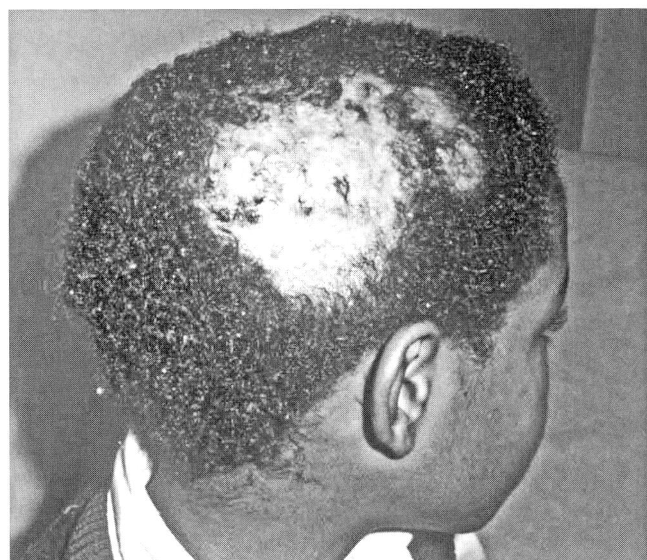

Figure 1.1 Head of a child with tinea capitis (scalp ringworm). A 9-year-old girl complained of cradle cap for over a year before developing itchy red patches on the right parietal scalp. She shed most of the hair in these areas. A culture was positive for *Trichophyton tonsurans*, and she was treated with oral griseofulvin with complete clearing of the tinea and regrowth of hair.
Source: Copyright Bernard A. Cohen, M.D. Used with permission from Dermatlas; www.dermatlas.org.

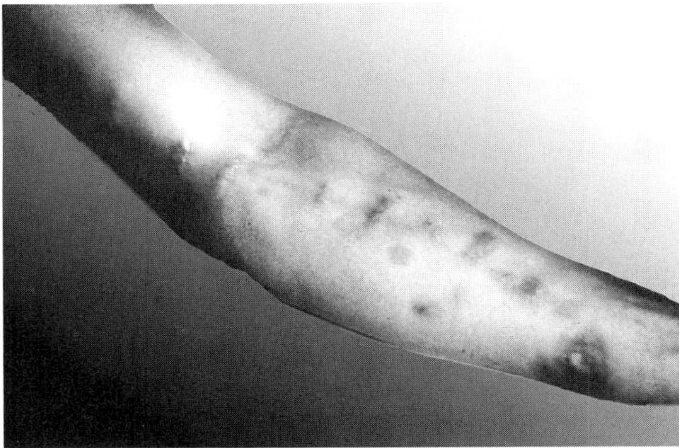

Figure 1.2 Sporotrichosis of the arm. Lesions draining each lymph node, from a primary lesion on the hand.
Source: Wilson and Plunkett (1965), used with permission from the University of California Press.

1.5.5 Era of Immunosuppression in the Treatment of Cancer, Maintenance of Organ Transplants, and Autoimmune Diseases

The era of cancer chemotherapy began when Sidney Farber synthesized a folic acid antagonist, now called methotrexate, and used it to treat childhood leukemia. His report in 1948 in the *New England Journal of Medicine* was greeted with ridicule because, at the time, the medical community held that childhood leukemia was incurable and children so afflicted should be allowed to die in peace. Since then, other researchers discovered drugs that blocked different functions involved in cell growth and replication ushering in the era of chemotherapy. The first cure of metastatic cancer was obtained in 1956 when methotrexate was used to treat a rare tumor called choriocarcinoma.

1.5.6 Opportunistic Mycoses

Cancer chemotherapy using cytotoxic drugs and systemic corticosteroids, by weakening the immune system, created opportunities for yeast and mold disease, principally candidiasis and aspergillosis, but also a long list of fungi formerly regarded as "saprophytes." Members of the genus *Aspergillus*, consisting of common environmental molds, have been known as pathogens since 1842, when one species was detected in the air sac of a bullfinch. Later, *Aspergillus fumigatus* was identified in other birds, and in humans where infections have increased proportional to the use of immunosuppressive therapy. Masses of the fungus have been found behind suspended ceilings in hospitals, in building materials, and outside windows of hospitals, especially during renovation or construction, and are the cause of single cases of aspergillosis as well as outbreaks in hospitalized patients.

1.5.7 HIV/AIDS

The AIDS epidemic in the United States, beginning in 1981, was brought to the attention of infectious disease specialists when two previously rare diseases, Kaposi's sarcoma and pneumocystosis, were encountered in men having sex with men (MMWR, 1981). An increase in *Pneumocystis* pneumonia (PCP) was noticed at the Centers for Disease Control (CDC) in April 1981, by Sandra Ford, a drug technician, who reported a high number of requests for the drug pentamidine, used to treat PCP. According to Ford, "A doctor was treating a gay man in his 20's who had pneumonia. Two weeks later, he called to ask for a refill of a rare drug that I handled. This was unusual, nobody ever asked for a refill. Patients usually were cured in one 10-day treatment or they died." As HIV, the cause of AIDS, and its role as a bloodborne and sexually transmitted pathogen was elucidated, AIDS brought to the forefront those opportunistic infections whose host defense was coupled to T-cell mediated immunity. In addition to pneumocystosis, other fungal opportunistic infections of AIDS were identified in quick succession as oro-esophageal candidiasis, cryptococcosis, *endemic mycoses* in the United States and, in Southeast Asia, penicilliosis.

1.5.8 Twenty-first Century

1.5.8.1 Advances in Clinical Laboratory Mycology

Important milestones in clinical laboratory mycology include (i) more rapid tests to identify *Candida* species in blood cultures, (ii) the wider availability of antifungal susceptibility tests because of new commercial kits, and (iii) the transfer of technology into the clinical laboratory for sequence-based identification of fungi. These are state-of-the-art tests appropriate for well-resourced hospitals. In resource-limited countries, a lack of training, proper reagents, supplies, and equipment impacts their laboratories' ability to identify pathogens and to detect antimicrobial resistance. Beginning in 2005, the American Society for Microbiology (ASM) International Laboratory Capacity Building Program (URL www.labcap.org) began to strengthen and expand clinical microbiology services in those regions.

Rapid Results for Candidemia The *Candida albicans/Candida glabrata* peptide nucleic acid fluorescent in situ hybridization assay (PNA-FISH, AdvanDx, Inc. Woburn, MA) was described in 2002 and is approved by the U.S. Food and Drug Administration (FDA) as a kit to identify yeast directly from a newly positive blood culture (Shepard et al., 2008). Cost savings accrue because, by ruling out *C. glabrata*, unnecessary echinocandin therapy can be avoided. Please see Chapter 11, Section 11.12, Laboratory Detection, Recovery and Identification, for further information.

Wider Availability of Antifungal Susceptibility (AFS) Testing The availability of commercial AFS tests in good agreement with reference methods facilitates rapid test results, thus aiding in clinical treatment decisions. These methods are discussed in Chapter 3B and are referred to here as important milestones in advancing clinical laboratory mycology (Cantón et al., 2009). The tests listed below have good correlation with reference methods standardized by the U.S. Clinical Laboratory and Standards Institute (CLSI).

Microtitration Plates Precoated with Drugs Sensititre® YeastOne™ (Trek Diagnostic Systems Inc., Westlake, OH) is a broth microdilution method. Wells of microtitration plates come precoated with dried dilutions of antifungal agents. Because of that they are stable in storage for prolonged periods, even at room temperature.

YeastOne is approved by the U.S. FDA for testing fluconazole (FLC), itraconazole (ITC), and flucytosine against clinical yeast isolates, and is available for investigational use to test amphotericin B, ketoconazole, voriconazole (VRC), posaconazole (PSC), and caspofungin (CASF). The YeastOne method was also evaluated to test the susceptibility of molds to triazoles and to amphotericin B.

Automated Spectrophotometric Microdilution Susceptibility The VITEK® 2 (bioMérieux, Inc., France) is FDA approved for testing FLC against *Candida* species. It has also been evaluated for testing amphotericin B, VRC, and flucytosine.

Disk Diffusion Neo-Sensitabs® tablet (Rosco, Taastrup, Denmark) is an easy to perform disk diffusion method available in Europe for testing yeasts and molds against polyenes, azoles, and echinocandins.

Drug Gradient Strips Etest® (AB Biodisk, Solna, Sweden) is not new but accommodates newer antifungal agents so it can be viewed as an improved method for AFS testing of yeasts and molds. It is an agar diffusion method with each drug applied to a plastic strip in a gradient of concentrations and printed with a minimum inhibitory concentration scale. The FDA has approved the Etest for testing FLC, ITC, and flucytosine.

Sequence-Based Identification of Fungi Unusual yeasts and molds, fungi that are slow growing, or those that fail to sporulate pose challenges to morphologic identification. Sequence-based identification is moving from the research laboratory to the clinical microbiology laboratories of tertiary care medical centers, aided by the availablility of kits for DNA preparation and purification, and biotechnology core facilities.

As an indication of progress, the CLSI issued a guideline, "Interpretive Criteria for Identification of Bacteria and Fungi by DNA Target Sequencing" (CLSI, 2008) to address sequence analysis in clinical laboratory practice. The guideline provides a standardized approach to identify fungi by DNA sequencing using the most common target, the ITS region of rDNA. Topics covered include primer design, quality control of amplification and sequencing, and reference sequence databases. The progress-to-date and remaining challenges of DNA sequence-based identification of opportunistic molds are discussed by Balajee et al., (2009).

1.5.8.2 Advances in Antifungal Therapy

Licensing of extended spectrum azoles, and a new class of antifungal drugs, the echinocandins, in this century, expand the therapeutic choices to treat invasive mycoses. Extended spectrum triazole antifungal agents (VRC, PSC, and the echinocandins CASF, micafungin, and anidulafungin) have become licensed for use and advance the therapy of serious mycoses (Fera et al., 2009). The echinocandins are the fourth class of antifungal agents available to treat systemic mycoses. The other classes are the polyenes (amphotericin B and its lipid formulations), azoles (ketoconazole, ITC, FLC, VRC, PSC), and pyrimidines (flucytosine). Please see Chapter 3A, for further information.

Echinocandins The echinocandins attack fungi via a novel mode of action by inhibiting ß-(1→3)-D-glucan synthase, a key enzyme in synthesis of ß-glucan,the fibrillar component of the fungal cell wall. CASF was approved by the FDA in 2001, followed by micafungin in 2005, and anidulafungin in 2006. In addition to approval for treating candidiasis, CASF is approved to treat invasive aspergillosis in patients intolerant of or refractory to other therapy.

Extended Spectrum Triazoles The spectrum of activity for VRC and PSC includes *Candida* species, molds, and dimorphic fungi. Their activity extends to both FLC- and ITC-resistant strains of *Candida*. VRC was approved by the U.S. FDA in May 2002 to treat invasive aspergillosis and infections caused by *Scedosporium apiospermum* and *Fusarium* species in cases of intolerance to or failure of other antifungal agents.

PSC was approved by the FDA in 2006 for the prophylaxis of invasive *Aspergillus* and *Candida* infections in adolescent and adult patients who are heavily immunosuppressed for a stem cell transplant or in transplant recipients with graft-versus-host disease or those with hematologic malignancies and prolonged neutropenia. Later, PSC was approved for oropharyngeal candidiasis.

1.5.8.3 Advances in the Molecular Basis of Pathogenesis and the Host Response

Fungal Genome Initiative This is an organized genome sequencing effort aimed at illuminating aspects of fungal biology which are fundamental to an understanding of fungal pathogenesis. To date, 110 assembled genomes for 81 fungi are available in public databases and more sequencing projects are underway. A list of fungal pathogens whose genomes have been sequenced is found in Cuomo and Birren (2010).

Genome-wide Expression Profiling During Pathogenesis Completion of the *Candida albicans* genome sequence, and those of other fungal pathogens, is being followed by annotation of the genes, which is a work in progress. Microarray technology[1] is used to capture a genome-wide portrait of the transcriptome expressed during infection in order to identify the genes, signaling pathways, and transcription factors involved in pathogenesis. The investigator is able to observe pathogen gene expression and to compare it, in the same system, with host gene expression to compile the time-sensitive events during pathogenesis and the host response (Wilson et al., 2009).

As an example, genes expressed by *C. albicans* growing on oral epithelial cells, and the level of gene expression, was revealed during hyphal formation and adherence, reflected at the molecular level with a number of genes encoding adhesins or other hyphal-associated functions (Wilson et al., 2009).

Proteomics Proteomics is defined as the set of proteins expressed in a given type of cell or by an organism at a given time under defined conditions. "Proteopathogen" (http://proteopathogen.dacya.ucm.es) is a protein database focused on the *Candida albicans* – macrophage interaction model (Vialás et al., 2009). There are 66 *C. albicans* proteins and 38 murine macrophage proteins identified in the model. Whereas two-dimensional polyacrylamide gel electrophoresis (2D-PAGE) is the most widely used method to separate complex protein mixtures, there are newer methods that combine high pressure liquid chromatography with mass spectrometry to generate large data sets that have to be stored in a website to enable their retrieval for efficient data mining. The goal of proteomics is to elucidate the proteins of the fungus and of the host, which are produced during different stages of the pathogenic process. The objective of such analysis is to understand pathogenesis at the molecular and cellular levels, to develop better diagnostic tests, and to discover rational methods of antifungal therapy.

1.6 RATIONALE FOR FUNGAL IDENTIFICATION

Why do we need to identify fungi? When dealing with microbes causing disease in humans there are important reasons to identify the causal agent.

1.6.1 Developing the Treatment Plan

Knowledge of the pathogen will increase chances for successful therapy because the pathogenesis of most mycoses

[1] A microarray (or DNA chip) is a series of gene targets immobilized on glass. Hybridization of cDNA probes made from mRNA in the test sample to these targets allows analysis of relative amounts of gene expression (Bryant et al., 2004).

is well studied. That will influence the choice of diagnostic tests, medical and surgical procedures, and antifungal therapy.

1.6.2 Investigating Outbreaks

- *The Hospital Setting.* The source of infection may be in the environment, including construction or renovation near patient wards, faulty HVAC systems affecting airflow in operating rooms or patient wards, contaminated hospital injectable solutions, indwelling medical devices, substandard hand-washing procedures, or other factors implicated in the healthcare environment.
- *The Community Setting.* An outbreak of fungal disease in the community typically requires an investigation of the causal agent: for example, histoplasmosis during spring break in 2001 among college students in Acapulco, Mexico (MMWR, April 13, 2001).

1.6.3 Determining the Susceptibility to Antifungal Agents

Because different fungi are susceptible to different antifungal agents it is important to:

- Identify the causal agent in order to select the most appropriate antifungal agent.
- Determine its in vitro killing effectiveness. Susceptibility in vitro does not uniformly predict clinical success in vivo because host factors play a critical role in determining clinical outcome. Resistance in vitro, however, will often, but not always, correlate with treatment failure. Please see Chapters 3A and 3B.
- Monitor the therapeutic response of the patient. Additional diagnostic tests and/or surgical intervention may be necessary.

1.6.4 Estimating the Significance of Fungi Generally Considered to be Opportunists or Saprobes

The physician should consider the immunocompetence of the patient, among other risk factors, to assess whether a fungus generally considered a saprobe may be the cause of disease.

1.6.5 Types of Vegetative Growth

What are the major forms of microscopic fungi? The microscopic fungi are classified by the type of vegetative growth as either yeasts or molds.

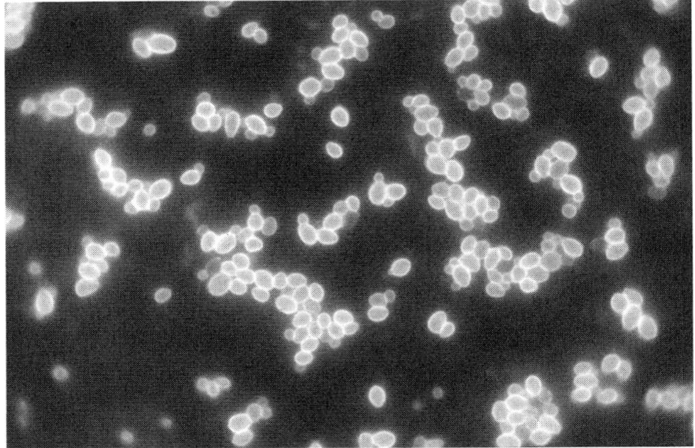

Figure 1.3 Budding yeast cells of *C. albicans*; immunofluorescence stained with an anti-mannan mAb. *Source:* PHIL, CDC, E. Reiss.

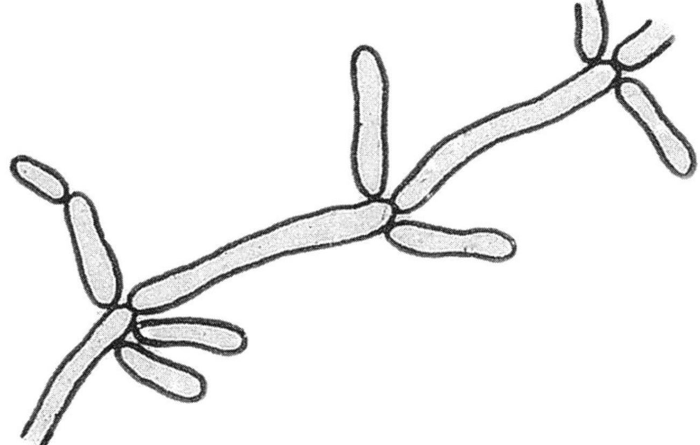

Figure 1.4 *Candida parapsilosis* pseudohyphae. *Source:* Adapted from Haley and Rice (1983).

1.6.5.1 Yeasts

The simplest fungi are the *yeasts* (Fig. 1.3). They are unicellular and reproduce by budding (e.g., *Candida albicans*) or fission (e.g., *Schizosaccharomyces pombe*). Yeasts causing disease in humans produce buds termed *blastoconidia*. An exception is *Trichosporon*, which, in addition to budding, produces hyphae that fragment into *arthroconidia*.

Pseudohyphae Pseudohyphae are seen in a wide variety of yeasts. Pseudohyphae are distinct from yeast forms and true hyphae. When blastoconidia remain attached in a chain of round to elongate cells, often resembling a string of pearls, the entire structure is called *pseudohyphae* (singular: pseudohypha). A mass of pseudohyphae is a pseudomycelium (Fig. 1.4).

The types of budding patterns in pseudohyphae are (i) unipolar, synchronous budding in which the first and later daughter buds are formed at the apex extending the length of the pseudohyphae; (ii) axial budding to form clusters of buds (whorls or *verticils*) behind pseudohyphal junctions; (iii) bipolar budding in which daughter cells are formed at both poles of a pseudohypha; and (iv) budding in which daughter cells are formed from both the distal and proximal ends of adjacent cells within a pseudohypha (Veses and Gow, 2009).

Nuclear division in pseudohyphae occurs at the point where the mother and daughter cells are most constricted. Septum formation also occurs at this point in the neck. Mitosis and septum formation in true hyphae of *C. albicans* are located at some distance within the true hypha. Cell divisions are near synchronous in pseudohyphae but in true hyphae, subapical cells are often arrested in G1 phase for several cell cycles (please see Section 1.12.1, yeast Cell Cycle.

1.6.5.2 Molds

The *molds* are formed by filamentous, cylindrical, often branching cells called *hyphae* (singular: *hypha*) (Fig. 1.5). A mass of hyphae is termed a *mycelium* (Fig. 1.6). The term *thallus* is sometimes used to refer to the entire body of a fungus (Fig. 1.7a). For a comparison, an agar plate with a yeast colony is shown in Fig. 1.7b. Hyphae occur in two different forms, depending on the phylum of fungi involved. The *Mucoromycotina* (please see Section 1.11.2.1) produce hyphae with sparse *crosswalls* or *septa* (singular: septum). Where septa occur, they are not perforated but serve to isolate reproductive structures or vacuolated regions in the mycelium (Fig. 1.8).

All other clinically encountered molds produce hyphae with crosswalls (septa) to separate the nuclei in different cells. Pores within the septa allow exchange

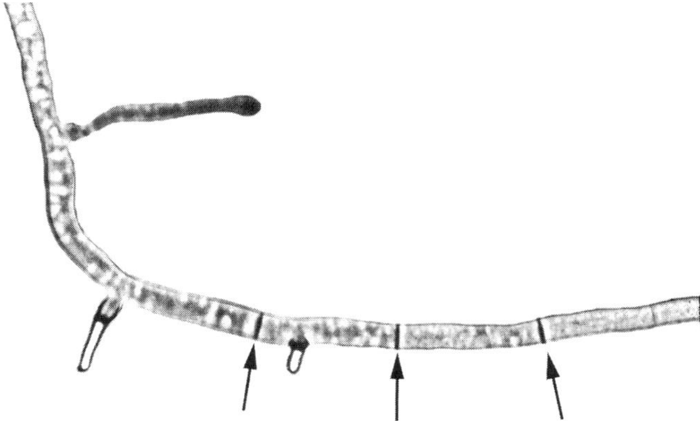

Figure 1.5 Microscopic view of septate hyphae: *arrows* point to septa.
Source: H. J. Shadomy.

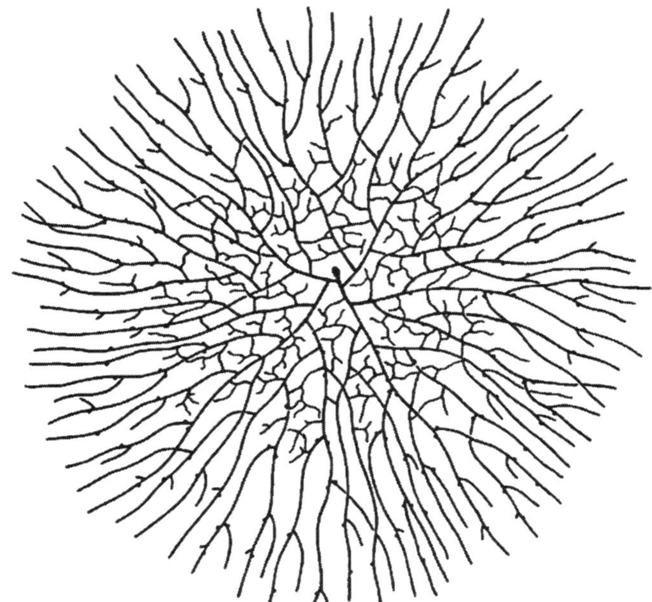

Figure 1.6 Radiating hyphae of a mycelium.
Source: Buller (1931).

of cytoplasm and even nuclei. The types of septa differ physically depending on the phylum of fungi involved. This characteristic is not typically used in the mycology laboratory for identification purposes.

1.7 SPORULATION

What is fungal sporulation and how does it differ among species? Vegetative growth is necessary but is not sufficient to perpetuate fungi and a variety of reproductive propagules are formed for dispersal with the aid of air currents or in water. Fungal propagules are different types of spores, a means of asexual reproduction. The method of sporulation used by fungi is the major character with which clinical laboratory scientists use to identify fungi in the clinical laboratory and, as such, is discussed in Chapter 2, Section 2.3.8.7, Common Types of Asexual Sporulation Seen in the Clinical Laboratory and Generally Termed a Type of Conidium (or Spore).

1.8 DIMORPHISM

Dimorphism (*definition*: two forms) is an important characteristic of certain fungal pathogens. Dimorphism is morphogenesis that allows growth to occur in either the mycelial or yeast forms, (mycelium → yeast, or yeast → mycelium conversion); for example, *Histoplasma capsulatum* is a dimorphic fungus (Please see Chapter 6 for illustrations of this dimorphism.) Fungi causing

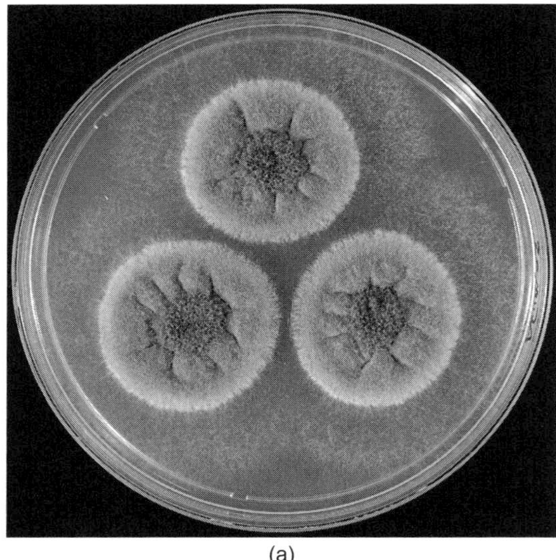

(a)

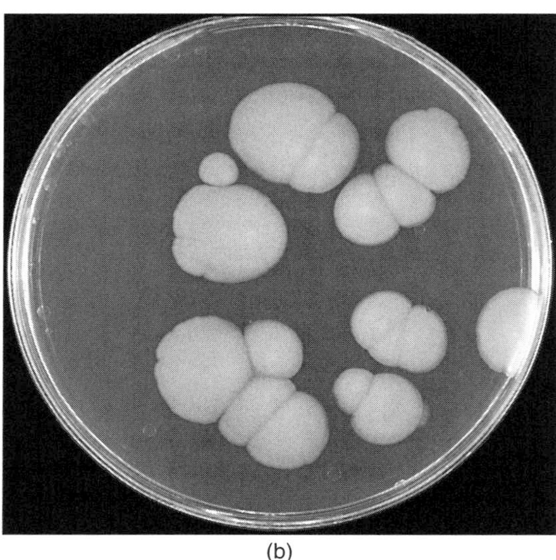

(b)

Figure 1.7 (a) Agar plate with mould colony: *Aspergillus fumigatus*. *Source:* Mr. Jim Gathany, CDC Creative Arts Branch. (b) Agar plate with yeast colony: *Candida albicans* growing on SABHI agar.
Source: Dr. William Kaplan, CDC.

primary systemic infections are typically filamentous soil-dwelling molds. The infectious propagules most frequently are conidia that are inhaled, along with hyphal fragments. Morphogenesis to the yeast form occurs during infection of tissues, usually in the lungs. This conversion is temperature sensitive, with the yeast form developing at 37°C. In the laboratory, growth at 35–37°C on an *enriched medium* may be used to help identify the fungus by this form change also known as "morphogenesis."

There are notable exceptions to the mold-to-yeast dimorphism. The primary systemic pathogens *Coccidioides immitis* and *C. posadasii* grow as a mold form in the environment. The mold form fragments into *arthroconidia*, which are the infectious propagules. Once inhaled, arthroconidia convert to *spherules*, enlarge, and segment into endospores. *Melanized* molds (e.g., *Fonsecaea pedrosoi, Cladophialophora carrionii*), the causative agents of chromoblastomycosis, grow as molds in the environment but in the cutaneous and subcutaneous tissues convert to *muriform* cells—round cells that do not bud but enlarge and divide by internal septation. Growth by enlargement in all directions is called *isotropic*.

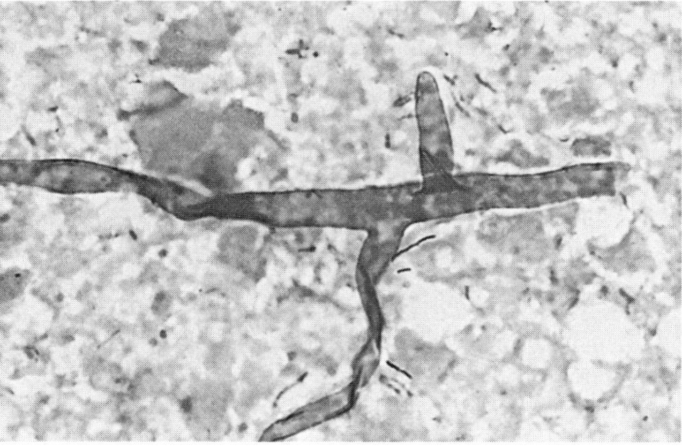

Figure 1.8 Broad aseptate hypha of *Rhizopus oryzae* in the histopathology section of a nasal mucosal biopsy from a case of rhinocerebral mucormycosis (GMS stain, 400×).
Source: Used with permission from Dr. Uma M. Tendolkar, LTM Medical College, Mumbai.

1.8.1 Dimorphism and Pathogenesis

How does dimorphism function in the pathogenesis of mycoses? As an adaptation to the host environment, dimorphism improves a fungus's ability as a pathogen; for example, *Histoplasma capsulatum* yeast forms survive after phagocytosis within alveolar macrophages and travel from the lungs via the bloodstream into the spleen and liver. Spherules produced by during infection by *Coccidioides* species produce many endospores, which spread the infection within the lung and to other body sites.

Although true of the primary systemic fungal pathogens, not all fungi that produce disease in humans are dimorphic. The opportunistic fungi may or may not be dimorphic. Monomorphic yeasts do not exhibit dimorphism and are seen as yeast in culture and in host tissues, for example, *Candida glabrata* and *Cryptococcus neoformans*. However, special studies can demonstrate dimorphism during the basidiomycetous sexual cycle of *C. neoformans* when that yeast produces a filamentous form with clamp connections. Many opportunistic pathogens are monomorphic molds, for example,

Aspergillus species or members of the *Mucorales*. They exhibit only the mold form in diseased tissue.

1.9 SEX IN FUNGI

Do fungi have sex and why is that important? Fungi, in addition to producing asexual spores or conidia, can undergo meiosis. Genetic coupling of nonidentical DNA occurs during meiosis, resulting in progeny with a new combination of the genes that were present in the parental haploid genomes. Diversity is produced by recombination of homologous chromosomes and crossing-over of chromosomal segments. This process results in a new and unique set of chromosomes, which, seen on a large scale, increases the level of genetic diversity in the entire population. With only mitosis, there would be no sharing of genetic information between compatible mating types; only division would be possible. The structures specialized to accomplish meiosis are the foundation used to classify fungi into Orders, Families, Genera, and Species. (Please see Section 1.11, Taxonomy/Classification: Kingdom Fungi.)

1.9.1 Anamorph and Teleomorph Nomenclature

The asexual state of fungi is termed the *anamorphic* state, while the sexual state is termed the *teleomorphic* state: for example, *Histoplasma capsulatum* (anamorph) and *Ajellomyces capsulatus* (teleomorph). Although the fungi are in a separate and unique kingdom, rules of nomenclature (naming genus and species) still follow the "International Code of Botanical Nomenclature." When the sexual state (teleomorph) of a fungus is identified, the genus and species of the teleomorph form takes priority and should be used thereafter. The asexual or anamorph name is subsidiary to the sexual state. In both medicine and clinical laboratory practice, however, the anamorph names persist so as to avoid confusion in understanding the actual causal fungus.

1.10 CLASSIFICATION OF MYCOSES BASED ON THE PRIMARY SITE OF PATHOLOGY

The thrust of medical mycology is to understand fungi as the causative agents of disease in humans and lower animals. This is the major difference between medical and general mycology. This section introduces the major fungal pathogens according to the organ system affected by fungal disease activity. In Section 1.11, Taxonomy/Classification: Kingdom Fungi, we will consider classification based strictly on cladistic analysis, method of sexual reproduction, and phenotypic characters. The following brief listing of the categories of fungal diseases is an opportunity to introduce the etiologic agents, which are covered in depth in the individual chapters.

1.10.1 Superficial Mycoses

Pityriasis versicolor is a mild infection of the nonliving keratinized outer layer of the epidermis caused by lipophilic yeasts, *Malassezia* species, and is mostly a cosmetic issue. More serious bloodstream infections caused by *Malassezia* species do occur, most often in neonates.

1.10.2 Cutaneous Mycoses

Dermatophytes or "ringworm" fungi cause disease of the skin, hair, and nails. These fungi are restricted to grow only on nonliving keratinized tissues. Dermatophytosis agents are, in order of prevalence, *Trichophyton tonsurans* > *T. rubrum* > *T. interdigitale* > *Microsporum canis*. Skin lesions may also be the cutaneous manifestations of deep-seated systemic mycoses: that is, the skin is a frequent site of dissemination for *Blastomyces dermatitidis*.

1.10.3 Systemic Opportunistic Mycoses

Systemic opportunistic mycoses cover a wide range of etiologic agents and clinical forms caused by molds and yeasts including environmental fungi and *endogenous commensal* fungi of the human microbiota (Table 1.2). Persons with normal immune and endocrine functions have normal levels of natural immunity sufficient to prevent these diseases. Factors affecting susceptibility to these fungi are presented in Section 1.15, Opportunistic Fungal Pathogens.

1.10.4 Subcutaneous Mycoses

These will be discussed in Part 5, Mycoses of Implantation. Usually initiated by a puncture with a thorn or splinter, this broad category of mycoses causes subcutaneous disease, in which *melanized* molds and their yeast-like relatives play an important role (Table 1.2).

1.10.5 Endemic Mycoses Caused by Dimorphic Environmental Molds

Endemic mycoses have a restricted geographic distribution as shown in Table 1.3. Most are primary pulmonary pathogens affecting immune-normal as well as immunocompromised persons. Exceptions are the dimorphic pathogens, *Penicillium marneffei* and *Sporothrix schenckii*. *Penicillium marneffei* is an

Table 1.2 Systemic Opportunistic Mycoses and Subcutaneous Mycoses

Category	Mycosis	Examples of etiologic agent(s)
Systemic mycoses caused by opportunistic yeasts	Candidiasis	*Candida albicans*, other *Candida* species
	Cryptococcosis	*Cryptococcus neoformans, Cr. gattii*
Systemic mycoses caused by opportunistic molds.	Aspergillosis	*Aspergillus fumigatus*, other *Aspergillus* species
	Mucormycosis, Entomophthoramycosis	*Mucorales, Entomophthorales*
	Fusarium mycosis	*Fusarium* species
	Pseudallescheria/Scedosporium Mycosis	*Pseudallescheria boydii, Scedosporium apiospermum*
Systemic mycoses caused by other opportunistic fungi	Pneumocystosis	*Pneumocystis jirovecii*
Subcutaneous mycoses of implantation.	Chromoblastomycosis	Dimorphic melanized molds (e.g., *Fonsecaea pedrosoi*)
	Phaeohyphomycosis	Melanized yeasts and their mold relatives (e.g., *Exophiala jeanselmei*)
	Sporotrichosis	*Sporothrix schenckii*
	Eumycetoma	*Pseudallescheria boydii, Madurella mycetomatis*

Table 1.3 Endemic Mycoses Caused by Dimorphic Environmental Molds

Endemic mycosis	Etiologic agent	Major endemic area
Histoplasmosis	*Histoplasma capsulatum*	States bordering major river valleys of central United States
Blastomycosis	*Blastomyces dermatitidis*	Overlaps with *Histoplasma* endemic area
Coccidioidomycosis	*Coccidioides immitis, C. posadasii*	Two major foci: Central Valley of California and Arizona endemic area
Paracoccidioidomycosis	*Paracoccidioides brasiliensis*	Brazil, Colombia
Penicilliosis	*Penicillium marneffei*	Southeast Asia
Sporotrichosis	*Sporothrix schenckii*	Worldwide with areas of high endemicity

opportunistic endemic mycosis, causing pulmonary and disseminated disease in immunocompromised persons, especially in AIDS patients living in or traveling to Southeast Asia. Sporotrichosis is distributed worldwide but has regions of high endemicity. It is most often a subcutaneous mycosis caused by a penetrating injury with a thorn or splinter, but can spread by direct extension to joints and other organs.

1.11 TAXONOMY/ CLASSIFICATION: KINGDOM FUNGI

How are fungi organized in a taxonomic scheme? (See Fig. 1.9.) The taxonomic classification of fungi is based on sexual reproduction (meiosis). The mode of sexual reproduction is an important taxonomic criterion, if it can be demonstrated. In addition to the mode of reproduction, other characteristics useful in classification are morphology—including the structure of crosswalls or septa—life cycle, and physiology. If no sexual reproductive cycle has been observed, the fungi are referred to as *mitosporic* and are further classified by *cladistic* analysis. The ultimate determinant of relationships is a comparison of genetic sequences.

What is the value of knowing the classification? Clinical microbiologists may wonder about the value of studying the classification schemes of fungi. Understanding the "fungal tree of life" gives a holistic view of the subject

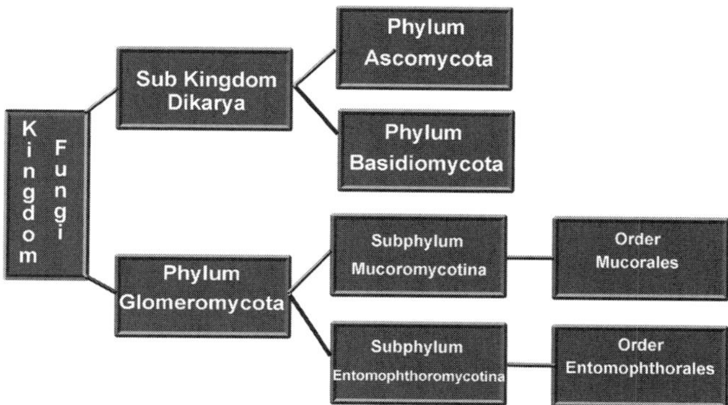

Figure 1.9 A higher level classification of the kingdom Fungi: phyla and subphyla containing pathogenic fungi (Hibbett et al., 2007).

that informs various important aspects, such as how individual species will respond to antifungal agents, the extent of their invasive potential, and their ecologic niche, which affects the mode of transmission.

1.11.1 The Phylogenetic Species Concept for Classification

Definition: A group of individuals with a shared genealogic relationship determined by phylogenetic analysis. Phylogenetic analysis may depend on phenotype but more reliably depends on genetic sequences. Another term for phylogenetic analysis is cladistics. Modern fungal taxonomic classification depends on *cladistic analysis*, which is the method of classifying organisms based on their phylogenetic relationships and evolutionary history. This method hypothesizes relationships among organisms determined through the construction of evolutionary trees. Organisms are classified exclusively on the basis of joint descent from a single ancestral species. The order of descent is represented in a branching diagram (a dendrogram or cladogram). Based on the phylogenetic classification, cladistic analysis produces a nested hierarchy where an organism is assigned a series of names to specifically locate it within the tree. A *monophyletic* group or clade is comprised of a single common ancestor and all the descendants of that ancestor. Another way to express monophyletic classification is that all groups within a phylum are descendants of one ancestor. The major gene targets for cladistic analysis are the DNA sequences in the ribosomal RNA genes (referred to here as rDNA) and also in selected somatic genes (e.g., translation-elongation factor 1α). *Multilocus sequence* analysis also plays a part in cladistic analysis.

The phylogenetic species concept stands in contrast to the older Linnaean system of classification, which depended on assigning an individual to a kingdom, phylum, class, order, family, genus, and species based on phenotype without taking into account the genotype. In that way birds and reptiles were placed in separate lineages, whereas we know from cladistics that birds are descended from reptiles.

Three assumptions of cladistic analysis are: (i) changes in characteristics in organisms occur in lineages over time; (ii) any group of organisms can be related by descent from a common ancestor; and (iii) there is a branching structure to lineage splitting. In this chapter we will rely on the phylogenetic species complex to make associations among fungal pathogens, especially when no sexual stage is known. The construction of phylogenetic trees is a subject in itself (Hall, 2007). A useful resource is TreeBASE at the URL www.treebase.org. Its main function is to store published phylogenetic trees and data matrices. The "how to" method for genetic identification of an unknown clinical isolate is discussed in Chapter 2, Section 2.4, Genetic Identification of Fungi.

Fungi are classified in the clinical microbiology laboratory by genus and species, and generally by the asexual state. The sexual state is rarely formed by cultures in the clinical laboratory. The identification is based on microscopic morphology and other phenotypic characters (e.g., enzymatic activities, presence of a capsule, temperature tolerance). Increasingly, the technology to conduct genetic identification is being used to supplement morphologic and other phenotypic characters.

1.11.2 The Higher Level Classification of Kingdom Fungi

The higher level classification of kingdom Fungi was revised by Hibbett et al. (2007). These revisions take into account *cladistic analysis*. The largest category of fungi pathogenic for humans is the subkingdom *Dikarya*, consisting of two phyla: *Ascomycota* and *Basidiomycota*. (The familiar phylum *Zygomycota* is not considered a valid taxon because it is not *monophyletic*.) Fungal pathogens previously classed in the *Zygomycota* are now found in the phylum *Glomeromycota*, subphylum *Mucoromycotina* and subphylum *Entomophthoromycotina*. In summary, the phyla and subphyla are constructed based on the the result of cladistic analysis and the mode of sexual reproduction: the *Mucoromycotina, Ascomycota*, and *Basidiomycota* (Fig. 1.9). The *Entomophthoromycotina* formerly classed with the *Zygomycota* are now considered separately. Other changes in taxonomy of fungi may be found in Boekhout et al. (2009). The mycotic disease agents in *Fundamental Medical Mycology* are aligned with these changes.

1.11.2.1 Mucoromycotina

The *Mucoromycotina* is considered the more primitive of these phyla and subphyla. Its members are identified by the production of sparsely septate or coenocytic hyphae, and sporangia with sporangiospores. The only septa in the *Mucoromycotina* isolate reproductive structures and wall-off vacuolated regions of the mycelium. Under special conditions a zygospore, the thick-walled sexual spore characteristic of the *Mucoromycotina*, is formed by the fusion of two gametangia (Fig. 1.10a, b). The *Mucoromycotina* are prolific producers of asexual spores formed inside the sporangia. Sporangiospores develop differently from the asexual spores termed conidia, of the *Ascomycota* and *Basidiomycota*. Sprangiospores form by internal cleavage of the sporangial cytoplasm. When mature, the sporangial wall deliquesces (dissolves) with resultant dispersal of the spores in air currents or water. Two orders of the subphyla *Mucoromycotina* and *Entomophthoromycotina*, which harbor species pathogenic for humans, are the *Mucorales* and the *Entomophthorales*.

The subkingdom *Dikarya* is so-named because of the feature held in common by the phyla *Ascomycota* and *Basidiomycota*: their hyphae contain pairs of genetically dissimilar unfused nuclei (dikaryons), which coexist and divide within hyphae before nuclear fusion (karyogamy) occurs. The *Ascomycota* and *Basidiomycota* are structurally similar, and it is believed that the *Ascomycota* probably gave rise to the *Basidiomycota*. The dikaryotic state in the *Basidiomycota* may be long-lasting. Furthermore, there is homology between structures that synchronize mitosis of the dikaryotic nuclei (*Ascomycota* croziers and *Basidiomycota clamp connections*). In the *Ascomycota*, the dikaryotic phase is limited to mycelium within the fruiting body (*ascoma*), but in the *Basidiomycota* growth in the dikaryotic stage lasts for some time before sexual reproduction occurs.

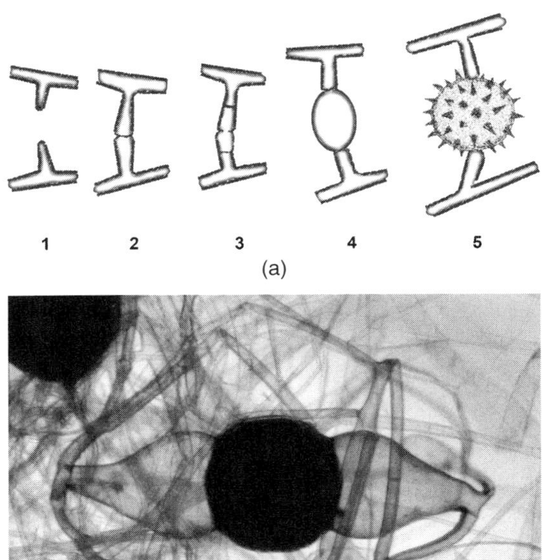

Figure 1.10 (a) Form development of zygospore production: 1, growth and attraction of hyphal branches of two compatible mating types; 2, progametangia—the branches touch and their tips swell; 3, gametangia—a septum forms between the gametangia and their vegetative hyphae; 4, fusion of gametangia occurs with formation of the zygospore; and 5, the mature ornamented zygospore. Genetic events that accompany sexual reproduction are displayed in Fig. 17A.2. (b) Zygospore with suspensor cells of *Syzgyites megalocarpus*.
Source: Used with permission from Dr. Gerald L. Benny, University of Florida, Gainesville. URL: www.zygomycetes.org.

1.11.2.2 Ascomycota

The *Ascomycota*, or sac fungi, are members of a *monophyletic* group that accounts for approximately 75% of all described fungi, including yeasts and molds. The *Ascomycota* reproduce sexually after plasmogamy, a *brief* dikaryotic stage, followed by karyogamy and meiosis within a sac or ascus (Fig. 1.11). One round of mitosis typically follows meiosis to leave 8 nuclei, packaged into 8 ascospores. The asci are formed within fruiting bodies, usually a cleistothecium or perithecium. A *cleistothecium* is a completely enclosed structure formed from specialized hyphae (Fig. 1.12a.). When mature, the cleistothecium ruptures, releasing the asci. A *perithecium* is similar but contains a pore or *osteole* from which the asci are extruded upon maturity (Fig. 1.12b) When ascospores are released, they germinate and develop as a haploid mycelium. Ascomycetous yeasts do not produce fruiting bodies; instead, "naked" asci are formed containing 4 ascospores. After germination, budding yeast forms develop. In that case identification depends on the texture, shape, and color of the ascospores (e.g., hat, walnut, spiny surface, etc.).

The *Ascomycota* also reproduce asexually by means of conidia produced by molds and blastoconidia or budding in yeast. The various genera and species are identified in the clinical laboratory by the method of conidiogenesis, or by genetic identification if they are slow-growing or fail to sporulate. Current updated classification of the phylum *Ascomycota* are published in electronic form (Lumbsch and Huhndorf, 2007) and may be viewed at the Myco-Net, URL http://www.fieldmuseum.org/myconet/outline.asp#Asco.

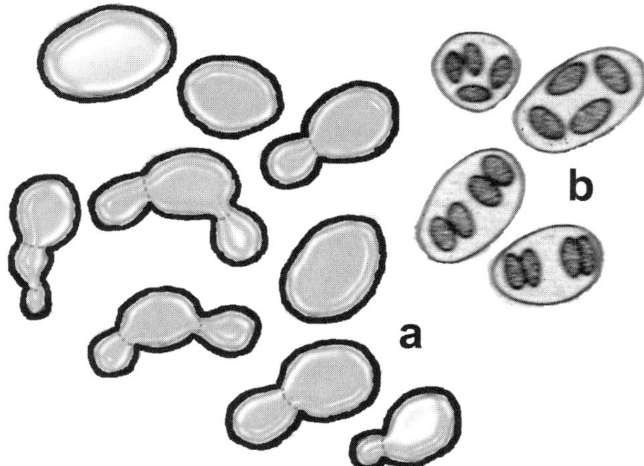

Figure 1.11 Asci with ascospores, yeast in the order *Saccharomycetales*.
Source: Adapted from Wilson and Plunkett (1965).

Subphyla of the *Ascomycota* containing fungi pathogenic for humans and animals are:

1. *Taphrinamycotina.* A basal lineage of *Ascomycota* consists of the class *Pneumocystidomycetes* containing the lung-dwelling opportunist fungus, *Pneumocystis jirovecii*.
2. *Saccharomycotina.* The class *Saccharomycetes* or "true yeasts" are composed of one large order, the *Saccharomycetales*. Pathogenic yeasts are found within three clades (Diezmann et al., 2004):
   - *Clade 1. Candida albicans, C. dubliniensis, C. tropicalis, C. viswanathii, C. parapsilosis, C. orthopsilosis, C. metapsilosis*.
   - *Clade 2. Candida guilliermondii*, and the teleomorph *Pichia guilliermondii, Clavispora* (*Candida*) *lusitaniae, Candida zeylanoides, Pichia* (*Candida*) *norvegensis*.
   - *Clade 3. Candida glabrata* and *Issatchenkia orientalis* (anamorph: *Candida krusei*). *Saccharomyces cerevisiae* is also in Clade 3.
3. *Pezizomycotina.* This subphylum contains over 90% of *Ascomycota*, as mold species, with sporocarp[2]-producing and mitosporic species. A feature characteristic of hyphal *Ascomycota* is the septal pore, which allows nuclei to migrate from cell to cell. The pore is at times blocked by a vesicle called the *Woronin body*, comprised of HEX-1 protein, which functions to respond to cell damage. Important orders in this subphylum are:
   - *Onygenales.* The *Onygenales* is a monophyletic lineage within the *Ascomycota* containing five families. Some members of this order are able to degrade keratin, the principal protein of the outer layer of the epidermis, hair, and nails. Keratinolytic activity defines the *Arthrodermataceae* and *Onygenaceae*, whereas the *Gymnoascaceae* and *Myxotrichaceae*, are nonkeratinolytic and cellulolytic. A fifth family, the *Ajellomycetaceae*, contains medically important genera.

[2]Sporocarp. *Definition*: A multicellular structure in which spores form; a fruiting body (e.g., perithecium, cleistothecium).

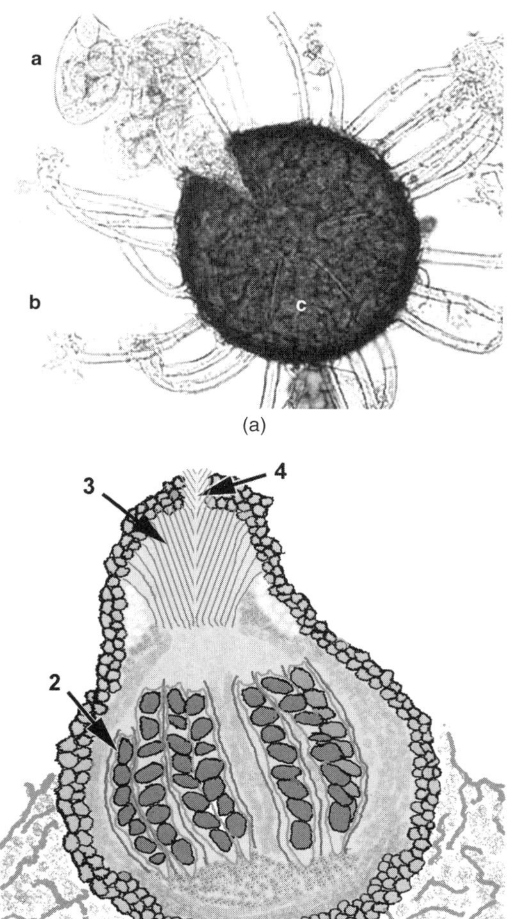

Figure 1.12 (a) Cleistothecium of *Erisiphe* (*Microsphaera*) species, an agent of powdery mildew. The cleistothecium is ruptured showing transparent asci containing oval ascospores: *a*, asci; *b*, appendages; *c*, cleistothecium. *Source:* Used with permission from Dr. Robert L. Wick, Plant, Soil and Insect Sciences Department, University of Massachusetts, Amherst. (b) Perithecium in longitudinal section, *Sordaria* species: *1*, the peridium is the perithecium wall; *2*, asci and paraphyseal hyphae are the fertile layer of the ascoma—the hymenium; *3*, periphyseal hyphae provide a channel for escape of the ascospores through *4*, the pore or ostiole. *Source:* E. Reiss.

- *Teleomorphs*. Members of the *Onygenales* produce ascospores in *gymnothecia* or *cleistothecia*. The ascomata contain round asci with unicelled, hyaline ascospores. The characteristic mode of conidiogenesis of their anamorphs is *thallic*, producing *aleurioconidia* or arthroconidia.

  The major lineages within the *Onygenales* are:

  - The *Ajellomycetaceae* consists of the *Ajellomyces-Paracoccidioides* clade (Untereiner et al., 2004). Important pathogens include *Histoplasma capsulatum* (teleomorph: *Ajellomyces capsulatus*) and *Blastomyces dermatitidis* (teleomorph: *Ajellomyces dermatitidis*). Here too is the anamorph species *Paracoccidioides brasiliensis* and the rodent and other small mammal pathogen *Emmonsia parva* and *E. crescens* (teleomorph: *Ajellomyces crescens*). Species of *Ajellomyces* form round cleistothecia with coiled appendages and small, finely ornamented ascospores. Anamorphs are the primary means of identification in the clinical setting. Conidia are smooth to slightly echinulate or are solitary, tuberculate *aleurioconidia*. Temperature-sensitive mold-to-yeast dimorphism is characteristic of this family, except for *Emmonsia* species, the tissue form of which are large, round, and nonbudding *adiaspores*. None of the members of this clade possess keratinolytic activity.
  - *Coccidioides*-containing clade. The remaining medically important anamorph genus, *Coccidioides*, remains in the *Onygenales* but is not closely related to the *Ajellomycetaceae*. This well-supported group also contains the look-alike arthroconidia-producing *Auxarthron* (teleomorph) and its anamorph, *Malbranchea aurantiaca*.
  - The *Arthrodermataceae* contains the keratin-degrading dermatophyte genera: *Trichophyton, Microsporum*, and their teleomorph *Arthroderma* species. Also present is the third and less common genus with a single species: *Epidermophyton floccosum*. The *Arthrodermataceae* also contains anamorphs assigned to *Chrysosporium*.

- *Hypocreales. Fusarium* species in this order are important plant pathogens and an emerging cause of deep-seated mycosis in immunocompromised hosts. The teleomorphs in the order are characterized by perithecia, which are brightly colored, yellow, orange, or red. Species pathogenic for humans are classed in the genus *Gibberella*.

  The characteristic mode of conidiogenesis is via phialides. Multilocus sequence typing found that Clade 3 of the *Fusarium solani* species complex (FSSC) contains at least 18 species isolated from mycoses of humans and animals (O'Donnell et al., 2008).

  Members of this complex are responsible for approximately two-thirds of *Fusarium* mycoses worldwide. The second most frequent causative agents of *Fusarium* mycosis are members of the *F. oxysporum* complex, where phylogenetic analyses has shown that a recently dispersed, geographically widespread clonal lineage is responsible for over 70% of all clinical isolates (O'Donnell et al., 2004).

- *Eurotiales*. This order is notable for the classification of *Aspergillus* species and their teleomorphs. The characteristic conidiogenous structure of the genus *Aspergillus* is the aspergillum, a stalk terminating in a bulbous vesicle on which a row of phialides are erected, either directly or upon another layer of basal cells or *metulae*. The conidia are produced within the phialides and released in an enteroblastic mode. (Please see Chapters 2 and 13.) Teleomorphs in the order form cleistothecia with *bitunicate*,[3] spherical asci and ornamented, unicellular ascospores (Table 1.4).

  The significant pathogen, *Penicillium marneffei*, like other penicillia, have rDNA sequences that cluster with the *Talaromyces* species. *Penicillium* species are now considered to be the asexual forms of *Talaromyces* within the *Eurotiales*.

- *Microascales*. Important pathogens are in this order, characterized by the cleistothecia-forming teleomorph *Pseudallescheria*, a rare example of a sexual stage that can be observed in the clinical laboratory, partly because the genus is homothallic. Anamorphs in the order include the genus *Scedosporium*, notably *S. apiospermum* and *S. prolificans*.

The following are three orders of melanized fungi:

- *Chaetothyriales*. The order contains the black yeasts and their melanized filamentous relatives and is related to the *Eurotiales* and the *Onygenales*. Melanized molds in this order, which are pathogenic for humans, are in a monophyletic clade, the family *Herpotrichiellaceae*. The teleomorph associated with this clade is *Capronia*. Genera of medical interest include *Exophiala* (yeast+filaments), *Cladophialophora* (filaments with or without yeast forms), *Fonsecaea* (filaments), *Phialophora* (filaments), *Ramichloridium* (filaments with or without yeast forms), and *Rhinocladiella* (filaments and yeast forms).

[3]Bitunicate. *Definition*: Ascus is enclosed in a double wall.

Table 1.4 Sexual Stages Associated with the Genus *Aspergillus*

Genus	Anamorphs of medical importance	Properties of the ascomata, cleistothecia
Neosartorya	*A. fumigatus, N. fischeri, A. lentulus*	Cleistothecial wall composed of flattened hyphae
Emericella	*E. nidulans*	Dark cleistothecia with walls composed of flattened cells; red ascospores
	A. ustus, A. calidoustus, A. versicolor	Cleistothecium surrounded by a stromatic layer of hülle cells
Fennellia	*A. terreus*	Cleistothecial wall composed of thick-walled cells
Petromyces	*A. flavus, A. niger (?)*	Multiple cleistothecia enclosed in a usually dark sclerotial stroma

Source: Adapted from Geiser (2008).

The following two orders are classed in the *Dothideomycetes*:

- *Dothidiales.* Black yeasts of medical importance in this order are *Aureobasidium* (yeast and filaments are produced) and *Hortaea* (yeast and filaments).
- *Pleosporales.* This order of melanized fungi is characterized by hairy, black, rapidly growing colonies. The anamorphs produce poroconidia: large multicelled conidia produced through a minute pore in the wall of the conidiophore or conidiogenous cell. The poroconidia are visible in the dissecting microscope. The pathogenic species are in a single family, the *Pleosporaceae*. Two groups are recognized, one group consisting of *Bipolaris, Curvularia*, and *Exserohilum* species have large-sized conidia, which can lodge in the nasal sinuses causing fungal sinusitis. Their teleomorphs are in the genera *Cochiobolus* and *Setosphaeria*. A second group contains *Alternaria* and *Ulocladium* species whose teleomorph genus is *Lewia*.

  Conidiogenesis in the *Pleosporales* is further characterized by the presence in some species of apical beaks on their poroconidia. Sometimes a new conidium is formed on top of the beak (e.g., *Alternaria*). The multicelled conidia are septate if the crosswall in the septum is continuous with the outer wall, or pseudoseptate if the septations are in the inner wall only, surrounded by an outer wall layer that forms a sac around the entire structure.

Further Reading About the *Ascomycota* Please consult "Myconet" an electronic and print journal specializing in fungal classification at the URL http://www.fieldmuseum.org/myconet/printed.asp. Monographs of fungal genomics are published in the journal *Fungal Biology*.

1.11.2.3 Basidiomycota

These fungi reproduce sexually by means of *basidiospores* produced on the outside of the spore-mother cell called a *basidium* (Fig. 1.13). A small number of species also produce conidia.

Characteristics of the *Basidiomycota*

Dolipore Septum (Fig. 1.14) The septum (crosswall) separates individual cells. Adjacent to the central pore in the septum each end flares out to produce the dolipore swelling. A septal plug is positioned on each side of the central pore, capable of blocking organelles from passing from one cell to the next. The septum is surrounded on each side by a dome shaped structure, which is perforated. This is the parenthesome, seen as crescent-shaped in cross section (Fig. 1.14). The dolipore septum of *Filobasidiella* species (including the perfect state of *Cryptococcus neoformans*) lacks the parenthesome. That structure is replaced by a simpler layer of porous striated material (Rhodes et al., 1981).

Clamp Connections in the Hyphae *Basidiomycota* occur as a binucleate cell type, containing two dissimilar but sexually compatible haploid nuclei that divide synchronously: for example, nuclei *a* and *b* both divide synchronously to give *a, a* and *b, b*. Nucleus *b* moves into the clamp and is sealed there by a new septum. Next, nuclear pair *a* and *b* migrate to the hyphal tip. The remaining nucleus *a* remains behind and a new septum is laid down separating it from the apical tip cell. Nucleus *b* in the clamp then joins nucleus *a* by dissolving the hyphal wall at that point. This process of synchronous binucleate cell division may continue for a long time. Eventually, karyogamy and meiosis occur to form a basidium and basidiospores.

Mushrooms are found here. The diverse human pathogens classed as *Basidiomycota* are illustrated:

- *Schizophyllum commune* is a bracket fungus classed in the order *Agaricales* with other gilled mushrooms.

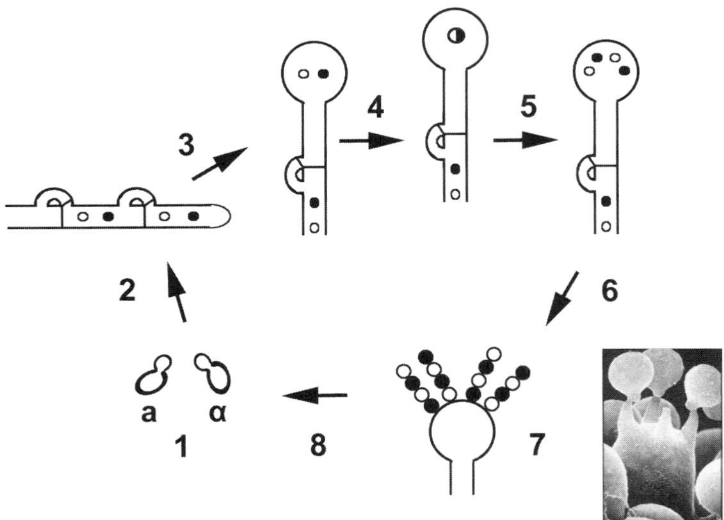

Figure 1.13 Sexual reproduction in a basidiomycete yeast: *Filobasidiella neoformans. Inset*: Scanning EM of a basidium with basidiospores. 1, Compatible mating type haploid yeast cells **a** and α secrete peptide pheromones that stimulate cell fusion (plasmogamy). 2, The resulting dikaryotic cell develops as a filamentous phase maintaining the dikaryon: the two parental nuclei migrate coordinately in the hyphae, divide, and septa separate the cells via clamp connections. 3, Later, the tip of the hypha enlarges into a round basidium. 4, Nuclear fusion (karyogamy) occurs in the basidium. 5, Meiosis occurs producing four haploid nuclei. 6, The haploid nuclei divide (mitosis) as they are packaged into basidiospores. 7, Basidiospores bud from the cell surface forming chains of basidiospores (these are infectious propagules). 8, Basidiospores disperse in air currents, germinate, and produce haploid yeast cells.

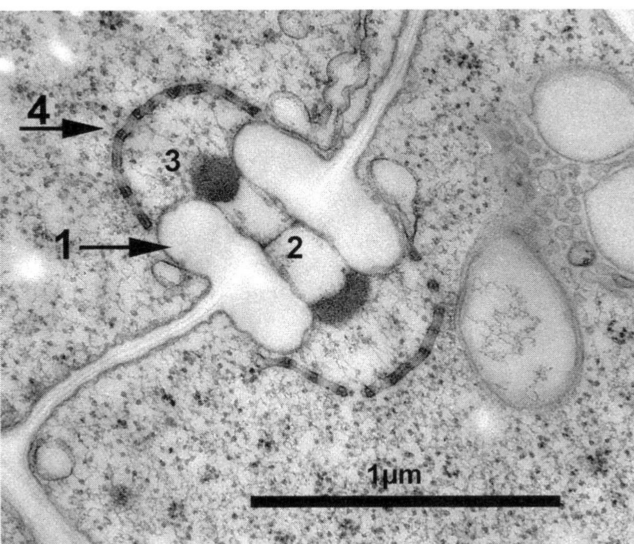

Figure 1.14 Dolipore septum of *Psilocybe cubensis*, present in all basidiomycetes. Transmission EM shows the following key features: 1, dolipore swelling; 2, septal pore; 3, pore plug; 4, dome-shaped perforated parenthesome.
Source: Used with permission from Dr. Robert Simmons, Director, Biological Imaging Core Facility, Department of Biology, Georgia State University, Atlanta, GA.

Schizophyllum commune, as a human pathogen, is found in chronic lung disease, meningitis, sinusitis, and onychomycosis (Sigler et al., 1999). *Schizophyllum commune* is identified in the clinical laboratory by the presence of the fruiting bodies, or by the occurrence of clamp connections in the hyphae, or, if neither is present, by genetic identification of the ITS region of rDNA. (Please see Chapter 2, Section 2.4, Genetic Identification of Fungi.)

- Order *Tremellales*. This order, notable for the jelly fungi found on rotting wood, contains important pathogenic yeasts and yeast-like fungi.
  - *Cryptococcus neoformans* and *C. gattii* are the etiologic agents of cryptococcal meningoencephalitis, an AIDS-associated opportunistic infection, also occurring in other immunosuppressed patients and, infrequently, in persons with no known underlying deficit. The haploid budding yeast has two mating types: α and a. In response to mating pheromones, the two mating partners produce conjugation tubes, and plasmogamy occurs, resulting in a heterokaryon filamentous form with clamp connections. Ultimately, meiosis occurs in a basidium. Spores are then produced on the surface of the basidium.
  - *Trichosporon* Species. Of 23 total species, the medically important species include *T. asahii*, recovered from superficial and deep-seated infections, *T. cutaneum* from superficial infections, and *T. inkin* from white piedra of the genital area (Diaz and Fell, 2004). The morphology of *T. asahii* consists of hyphae that disarticulate into cubic shaped arthroconidia. *Trichosporon inkin* also produces hyphae and arthroconidia, cylindric in shape. Budding is absent in these two species.
- Order *Malasseziales*. At least 14 species are known in the genus *Malassezia*, which are yeasts and members of the microbiota of human skin. *Malasseziales* are taxonomically related to the corn smut fungus, *Ustilago maydis*. No sexual stage for members of this order is known. These yeasts produce a distinct scar by repetitive bud formation from a single location on the mother cell, a feature most useful in their identification in direct examination of skin scrapings. Some

species are etiologic agents of pityriasis versicolor (please see **Chapter 21**) and others are implicated in seborrheic dermatitis. All of the species within the genus, with the exception of *Malassezia pachydermatis*, are lipid dependent due to an inability to synthesize C_{14} or C_{16} fatty acids (reviewed by Ashbee, 2007). Some species produce a mycelium as well as budding yeast cells.

Older texts refer to a form-phylum, the *Deuteromycota* (syn: *Fungi imperfecti*), composed of *mitosporic* fungi that have no known sexual reproductive cycle. This nomenclature allowed two broad categories: *Blastomycetes* containing the yeast *anamorphs.* Molds were described in a second category, *Hyphomycetes*, and were further classed in the form family *Dematiaceae*, with melanin pigment in their cell walls, or *Hyalohyphomycetes*, which lack melanin pigment. This nomenclature has become obsolete because cladistic analysis using multigene sequence comparisons can determine phylogenetic relationships even when the teleomorph has not been observed.

1.12 GENERAL COMPOSITION OF THE FUNGAL CELL

How is the fungal cell organized and how does it differ from the bacterial life form? (Table 1.4). (See Griffin, 1994 and Howard and Gow, 2007.)

Cells that do not have a nucleus are called prokaryotes (e.g., bacteria). Cells that have a nucleus are called *eukaryotes*. *Fungi* are simple eukaryotes whereas plants and animals are higher eukaryotes. Fungi are not plants—they lack chlorophyll and cannot undergo photosynthesis. Although most fungi are multicelled, some are single celled, the *yeasts*. Fungal cells possess a nucleus containing chromosomes and contain a full array of intracellular organelles that allow the cell to function (Fig. 1.15). The yeast cell cycle and hyphal morphogenesis introduce the topic of the organization of the fungal cell.

1.12.1 Yeast Cell Cycle

The cell cycle traces cell growth and division. A visual three-dimensional (3D) representation of the yeast cell cycle is found at the URL http://www.nformationdesign.com/portfolio/portfolio09.php. Each cell division requires the duplication of all essential components of the cell. The most important component is the DNA, organized in chromosomes. DNA must be accurately replicated into two copies and segregated to the mother and daughter cells. The processes of DNA replication and sister chromatid separation occur in distinct timed phases of the cell

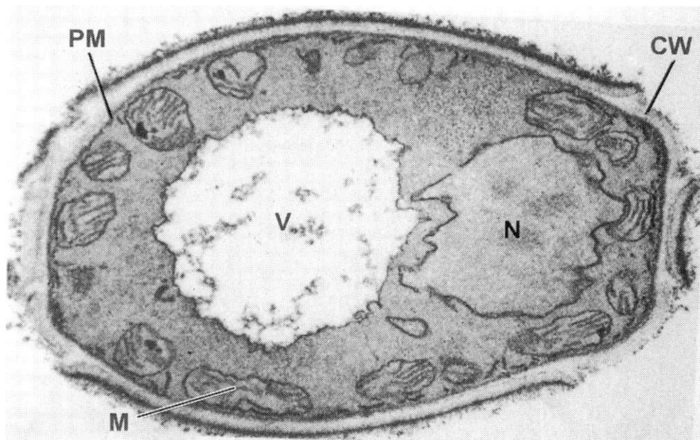

Figure 1.15 Ultrastructural features of the yeast cell visualized in a transmission EM depicting a cross section of yeast cell. CW, cell wall; PM, plasma membrane; N, nucleus; V, vacuole; M, mitochondria.

cycle. In phase G1 the yeast cell grows until it reaches a critical mass before DNA synthesis. Then, at START, the cell is committed to replicate its DNA and progresses into the S phase (DNA synthesis). The yeast bud is clearly visible. Following S phase, cells enter gap-2 (G2). The nucleus and mitochondria are oriented by microtubules. Mitochondria and vacuole inheritance into the daughter bud occurs. Next, M phase (metaphase) marks the stage of mitosis. Chromosomes align on the equator of the microtubule spindle, halfway between its two poles. The nuclear membrane remains intact.

Once the genome replicates, the spindle aligns parallel to the mother-bud axis and elongates. Anaphase is the stage of mitosis when two copies of each chromosome move to opposite poles of the spindle. The spindle elongates and provides the mother and daughter cell each with one nucleus. The nucleus moves through the bud neck, assuming an hourglass shape. Telophase is the last stage of mitosis when a complete set of chromosomes aligns at two poles of the mitotic spindle and the nuclei separate into mother and daughter cells Cytokinesis[4] occurs and the mother and daughter cells separate. Please consult Berman (2006) for the topic of cell cycle regulators including cyclins, cyclin-dependent kinases, and CDC proteins.

1.12.2 Hyphal Morphogenesis

Fungal hyphae originate from a germinating conidium or another hypha during branch formation (Fig. 1.16). An axis of polarity is set and cell surface expansion occurs along the axis to the hyphal tip and linear extension from it. Successive hyphal branching results in a tree-like

[4]Cytokinesis. *Definition*: Following the telophase of mitosis in cell division, the cytoplasm is divided between mother and daughter cells.

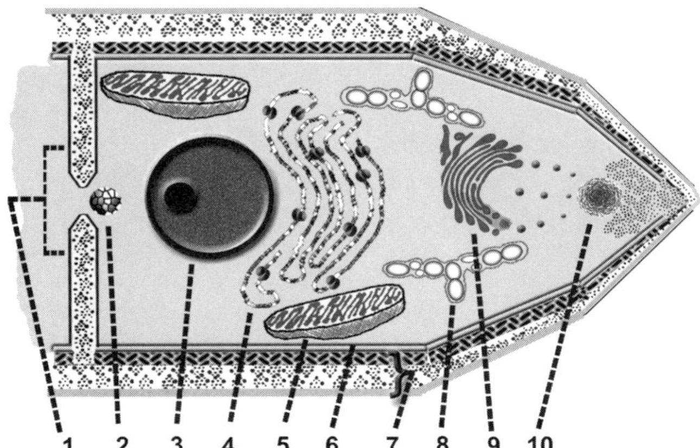

Figure 1.16 Organelles of the hyphal tip: *1*, septum with septal pore; 2, Woronin body; 3, nucleus with nucleolus; 4, rough endoplasmic reticulum; 5, mitochondrion; 6, cell membrane; 7, cell wall in three layers—inner layer (chitin), middle layer (glucan), and outer layer (mannan); 8, tubular vacuole; 9, Golgi; 10, *spitzenkörper*.
Source: E. Reiss.

mycelial network. Nutrients acquired at the colony periphery are distributed in the mycelium. As it grows older parts of the mycelium are recycled to support new growth. In the subkingdom *Dikarya* (including phyla *Ascomycota* and *Basidiomycota*) adjacent branches may fuse (anastomose), forming an interconnected network.

The establishment and maintenance of hyphal polarity require microtubule- and microfilament-based motor proteins, chitin deposition, and both cyclic AMP and mitogen-activated protein kinase signaling (Harris et al., 2005). *Spitzenkörper* are present in growing vegetative hyphal tips and branch points. They are located within the hyphal tip in the direction of hyphal growth. *Spitzenkörper* are complex, multicomponent structures that function to support directional growth by concentrating and delivering secretory vesicles to the hyphal tip. Some of these vesicles are chitosomes, containing chitin synthase. That enzyme is activated when the chitosome fuses with the plasmalemma, initiating new chitin synthesis and polymerization of chitin microfibrils.

The cytoskeleton is very important in hyphal morphogenesis. Microtubules are responsible for the transport of secretory vesicles to the *spitzenkörper*, while actin microfilaments control vesicle organization within the *spitzenkörper* and transport from there to the plasmalemma. The *spitzenkörper* may be viewed as a supply and distribution center for vesicles involved in apical growth.

Pseudohyphae are a feature of *Candida* species. Pseudohyphal cells bud in a unipolar manner, and the cells remain attached after cytokinesis. Pseudohyphae are described in detail in Section 1.6.5.1.

1.12.3 Cell Wall

All fungal cells are bounded by a rigid, laminated, cell wall that imparts protection and firmness to the internal structures. The outer layer consists of readily soluble mannan (or galactomannan) and inner layers of glucan fibrils. The cell walls of molds have an inner microcrystalline sleeve of chitin [poly ß-(1→4)-*N*-acetylglucosamine]. Chitin in yeast cells is also concentrated as disks in birth scars and bud scars. Some fungi have melanin in their cell walls.

*1.12.3.1 Molecular Architecture of the Yeast (*Candida*) Cell Wall*

Overview The outer surface layer of the *Candida* cell wall (Ruiz-Herrera et al., 2006) consists of mannoprotein, which is the major surface antigen, "mannan."[5] The fibrillar polysaccharides of the cell wall are ß-(1→6)-D-glucan (superficial layer), ß-(1→3)-D-glucan (deep layer), and the innermost layer is a microcrystalline sleeve of chitin: poly-ß-(1→4)-*N*-acetylglucosamine. Proteins are embedded in the cell wall either noncovalently or bound to fibrillar cell wall polysaccharides: to ß-(1→6)-D-glucan, to chitin via short chains of ß-(1→6)-D-glucan, or directly to chitin. The types of linkages between proteins and the glucans and chitin are:

1. Phosphodiester linked *g*lycosyl *p*hosphatidyl*i*nositol proteins ("GPI" binding motif).
2. *N*-glycosidic linkages between chitobiose and asparagine of the protein moiety (alkali stable). *This is the major linkage between the "inner core" mannan to protein* (please see "Mannan" for further explanation.
3. Reducing agent-extractable (RAE) proteins linked by disulfide bonds, which, in turn, are linked to ß-(1→3)-D-glucan by *O*-glycosidic bonds to threonine or serine. Several proteins can be extracted from the cell wall with sulfhydryl reagents such as 2-mercaptoethanol. This can occur because they are covalently linked to the cell wall via disulfide bridges or because they are released when the cell wall structure is loosened by reducing disulfide bridges. Among them are phospholipomannan and Pir proteins (please see below for definition and functions).

Mannan Mannoproteins comprise 38–40% of the cell wall mass. The largest mannoprotein constituent is mannan, the antigenic coat of *C. albicans*. Antigenic variations in mannan occur among the *Candida* species (Suzuki,

[5]Mannan is also referred to as "peptidophosphomannan," to distinguish it from glycoprotein enzymes that may also contain mannose-oligosaccharides.

1997). Mannan is of interest because it is the major surface antigen of *Candida* species, and tests to detect it in serum during candidemia and deep-seated candidiasis have been developed. Mannan is comprised of an inner core linked to protein and an antigenic outer chain region. The inner core consists of a linear α-(1→6)-mannan backbone substituted with 12–17 mannose residues and di-*N*-acetylglucosamine (chitobiose) units, which are linked via *N*-glycosidic bonds to asparagine of the protein moiety. Please see Fig. 11.11 for an illustration of mannan structure.

The outer chain region is disposed along the linear α-(1→6)-mannan backbone in two domains. An α-linked domain consists of oligomannosides up to 6 or 7 units linked α-(1→2) and α-(1→3). These are the antigenic epitopes. A separate domain in the outer chain is composed of ß-(1→2)-linked oligomannosides with a degree of polymerization of between 2 and 6. The ß-mannan is appended from the backbone via acid-labile phosphodiester bonds, the *O*-phosphonomannan. Both α- and ß-linked mannose oligosaccharide epitopes contribute to the antigenic mosaic of the *Candida* cell wall. Tests to detect mannan should incorporate antibodies to both the α- and ß-oligomannose epitopes.

The question of the relative importance of α- and ß-mannan domains to pathogenicity has been approached (Hobson et al., 2004). The *MNN4* gene is required for mannosylphosphate transfer, so that the Δ*Mnn4* mutant lacks the ß-mannan domain. The null mutant was unaffected in its growth, form development, mouse virulence, adherence to and uptake by macrophages. Other studies, however, implicated ß-(1→2)-oligomannosides in murine gastrointestinal colonization (reviewed by Masuoka, 2004). During infection mannan is sloughed from the cell wall and circulates in the plasma. It is rapidly cleared from the circulation so that its detection by sandwich EIA is difficult.

PLM PLM is a unique glycolipid shed from the cell when *C. albicans* comes in contact with host cells (Poulain et al., 2002). PLM secretion has been linked to survival of *C. albicans* in macrophages and in promoting macrophage apoptosis. PLM is a member of the mannose-inositol- phosphoceramide family.

PLM Structure: PLM is composed of a long linear chain of ß-(1→2)-mannose (average degree of polymerization of 14) linked via a ß-(1→2) glycosidic bond to the polyol, phosphatidyl inositol. Phosphatidyl inositol is then linked via phosphodiester bonds to ceramide. (*Definition*: Ceramide—a lipid composed of sphingosine linked to a fatty acid via an amide bond. Sphingosine is an 18-carbon amino alcohol with an unsaturated hydrocarbon chain.)

1.12.3.2 Considerations for the Fibrillar Polysaccharides of the Candida Cell Wall

Chitin The yeast form of *C. albicans* cell wall contains 2% chitin, whereas that of the mycelial form is higher, 6%. Twenty-four hundred chains of the poly-ß-(1→4)-*N*-acetylglucosamine polymer are united in antiparallel bundles to form highly insoluble crystalline chitin microfibrils. This microcrystalline sleeve is the base layer to which ß-(1→3)- and ß-(1→6)-D-glucan fibrils are linked. This chitin–glucan complex forms the scaffold of the cell wall onto which mannoproteins and ß-(1→3)-D-glucan are disposed. In addition to forming a microcrystalline sleeve in both yeast and mycelia forms, chitin is concentrated in the bud and birth scars and septa. At first chitin was thought to be a good target for antifungal therapy because it is absent in mammals, but up to now chitin synthase inhibitors, polyoxins and nikkomycins, have showed low in vivo activity.

Four genes, *CHS1, CHS2, CHS3*, and *CHS8*, encode chitin synthase isoenzymes with different biochemical properties and physiological functions (Lenardon et al., 2007). Specific chitin synthases produce chitin microfibrils of different lengths and act cooperatively to generate the pattern of chitin microfibrils in the cell wall. Electron microscopy in *chs8* mutants and *chs3* mutants showed that Chs8 synthesizes long-chitin microfibrils in the septum formed at cytokinesis. Chs3 synthesizes short-chitin rodlets in the septum and is responsible for the synthesis of the majority of chitin in the cell wall of yeast and hyphal cells. Chs1 is essential for growth and viability. Chs2 is the major chitin synthase activity measured in cell membrane preparations.

ß-(1→3)-D-Glucan, ß-(1→6)-D-Glucan The ß-(1→3)-D-glucan strands are helical and can form a tertiary structure of three strands united in a triple helix stabilized by interchain hydrogen bonds between C-2 hydroxyl groups. The glucan fraction comprises 58–60% of the cell wall dry weight. The relative proportions of ß-(1→3) and ß-(1→6) glucans in the cell wall alkali-insoluble fraction are 30–39% and 43–53%, respectively. Although there can be links between the two types of glucan, there are no polysaccharides in which the two glycosidic bond arrangements occur in the same strand.

The enzymes involved in ß-(1→3)-D-glucan synthesis are important because they are the target for the echinocandin class of antifungal agents (e.g., caspofungin, micafungin, and anidulafungin). The biosynthetic glucosyl transferases are encoded by the *FKS* genes. Mutations in *FKS1* result in echinocandin resistance, indicating it is the most important glucan synthase in *C. albicans*. Glucans are an important example of a "pathogen-associated molecular pattern" (PAMP) functioning in the innate immune response by interacting with receptors such as dectin-1 on macrophages.

1.12.3.3 Cell Wall Associated Proteins

GPI Proteins The glycosyl–phosphatidylinositol linkage is the most prevalent means by which proteins are linked to ß-1→6-glucan (90%) or chitin (10%) in the cell wall. Function: Examples of GPI proteins are the adhesins, Hwp1, and the Als family in *C. albicans*. Other functions of GPI proteins are involved as enzymes in cell wall biosynthesis.

Pir Proteins These are a smaller group than GPI proteins and are so-named because they are *p*roteins with *i*nternal *r*epeats. They can be solubilized from the cell wall with dilute alkali. They contain a domain with 2–11 repetitive amino acid sequences and 4 cysteine residues arranged in the motif: (repeat(s)-Cys-6aa-Cys-16aa-Cys-12aa-Cys-COOH). They are attached to ß-(1→3)-D-glucan by alkali-labile bonds. Function: Pir proteins are believed to contribute to structural strength of the cell wall by crosslinking ß-(1→3)-D-glucan strands.

1.12.3.4 Cell Surface Carbohydrates Implicated in Pathogenicity

"Fuzzy Coat" Layer A fuzzy coat external to the cell wall and localized at the hyphal tips of *C. albicans* appears to be enriched in mannoprotein, possibly with *O*-phosphono-linked ß-(1→2)-mannan (Ruiz-Herrera et al., 2006). A role for this coat layer in adherence to and invasion of host tissue is speculated.

Surface Microfibrils and Fimbriae Various investigators have observed microfibrils or fimbriae on the surface of *C. albicans* blastoconidia and hyphae, also indicating they are composed of mannoproteins, and have proposed that these structures are involved in adherence to mammalian cells. The structures are delicate but are visible in freeze-fracture electron microscopy (Hazen and Hazen, 1993).

Extracellular Matrix Material (EMM) The EMM surrounds *Candida* species embedded in *biofilms* adherent to *biomaterials* and contains a high carbohydrate content, but the relative contribution of mannan and glucan to the EMM is not certain (Al-Fattani and Douglas, 2006). *Candida* embedded in biofilms are less susceptible to antifungal agents. Rats with central intravenous (IV) catheters were injected through the catheter with *C. albicans* and developed biofilms (Nett et al., 2007a). ß-(1→3)-D-glucan concentrations in plasma of rats with biofilm-embedded catheters were ten fold greater than rats infected IV without catheters. ß-(1→3)-D-glucan is a major component of *Candida* biofilms and is implicated as the constituent that binds azole antifungal agents, so that they are unable to enter the *Candida* embedded in biofilms (Nett et al., 2007b). Phospholipomannan, an antigenic wall component, secreted through the cell wall may blend with other constituents of the EMM (Poulain et al., 2002).

Hydrophobicity Hydrophobic *C. albicans* cells are more adherent than hydrophilic cells to host tissues and are also more resistant than hydrophilic cells to phagocytic killing. *Candida albicans* can regulate cell surface hydrophobicity by altering cell surface microfibrils composed of mannoproteins with ß-(1→2)-mannose linkages. *Candida albicans* can change the conformation of the microfibrils: on the surface of hydrophobic cells microfibrils are shorter while those of hydrophilic cells are longer and radiating (Masuoka and Hazen, 2004).

1.12.3.5 *Plasma Membrane* (Also Known as the Plasmalemma)

Just inside the cell wall lies the plasma membrane, a phospholipid bilayer. Importantly, ergosterol is a key membrane component. The plasma membrane controls cell semipermeability, resorption, excretion, and secretion, and is capable of enzymatic breakdown of substrates. Ergosterol is the target for amphotericin B and the azole class of antifungal drugs (please see Chapter 3).

1.12.3.6 Cytoplasm

The protoplasm of the fungal cell is organized as a cytoskeleton consisting of microfilaments (actin) and microtubules (tubulin) (Sudbery and Court, 2007).

Cytoskeleton in Yeast Actin cables, actin cortical patches, and a contractile acto-myosin ring occur at the site of septum formation. This contractile ring consists of the septins, a series of five proteins, which function to separate mother and daughter cells. This function is known as cytokinesis. Microtubules, the tubular network in fungi, are contiguous with the yeast vacuole.

Cytoskeleton in Molds Cortical patches of actin are clustered at the sites of hyphal tips in the molds. Long microtubules are oriented along the axis of hyphal growth and are contiguous with tubular lysosomes. Actin cables are oriented toward these sites. A tubular vacuole network is a normal function of the growing zone of the mycelium used for long distance transport of vesicles to the *spitzenkörper*.[6] Actin cables mediate short distance dispersal of vesicles to the cell surface.

[6] *Spitzenkörper*. Definition: Vesicles present at the hyphal tip involved in cell wall biosynthesis.

1.12.3.7 Genome

Genetic material (DNA) within the fungal cell is in the nucleus. It contains chromosomes and is bounded by the nuclear membrane; for example, *Candida albicans* has 8 chromosomes and a genome size of 16 Mb (haploid genome) (*Candida albicans* Physical Map website). The nucleolus is also found within the nucleus and is the locale where ribosomal RNA is transcribed and ribosomal subunits are assembled.

1.12.3.8 Golgi

The Golgi apparatus are a series of concentrically arranged double membranes that receive and often modify molecules made in the endoplasmic reticulum before exporting them to the exterior of the cell or to another location. It is the central conduit of the secretory pathway to deliver secretory products and membrane proteins to the cell surface or for transit into vacuoles.

1.12.3.9 Endoplasmic Reticulum (ER)

The ER is a system of tubules and vesicles within the cytoplasm where most membrane components and proteins are synthesized on ribosomes (polyribosomes). The ER is also the site of Ca^{2+} storage for release into the cytosol when induced by signaling cascades.

1.12.3.10 Mitochondria

Mitochondria are membrane-bound organelles, about the size of a bacterium, within the cytoplasm of eukaryotic cells. Mitochondria are where respiration and glucose catabolism occurs. They power cells by utilizing oxygen and glucose to generate energy (in the form of adenosine triphosphate, (ATP)) by a series of stepwise enzymatic reactions. Oxidative phosphorylation is the process for generating energy (ATP) via the electron transport sequence. The Krebs cycle (citric acid cycle) is the means by which glucose is converted to energy. Fatty acid oxidation also occurs in mitochondria. Mitochondria are self-replicating because they have their own DNA.

1.12.3.11 Nucleus

A compact sphere, the nucleus is the most prominent organelle and the control center of the eukaryotic cell. The nucleus is bounded by the nuclear envelope composed of two parallel membranes separated by a narrow space and perforated with pores. The nucleus contains chromosomes. The nucleolus is a dark area for rRNA synthesis and ribosome assembly (please see also Section 1.12.3.7, Genome).

1.12.3.12 Ribosomes

These cellular organelles are composed of ribosomal RNA and ribosomal proteins. They associate with messenger RNA and tRNA in order to translate the message into protein synthesis; 40S and 60S ribosomal subunits form the 80S ribosome.

1.12.3.13 Vacuoles

The largest organelles of a yeast cell also occur in the cytoplasm of molds. The vacuole is the key organelle involved in intracellular trafficking of proteins. It is part of the cell's intramembranous system analogous to lysosomes of mammalian cells. Vacuoles contain hydrolytic enzymes involved in digesting proteins: endopeptidases, aminopeptidases, and carboxypeptidases. Here are stored basic amino acids, polyphosphates, and certain metal ions. The vacuole regulates cytoplasmic ion concentrations and intracellular osmotic pressure. Vacuoles in yeast are inherited from the mother cell to the daughter cell as a single stream of vesicles that fuse into the vacuole.

Other smaller membrane-bound vesicles are present in the cytoplasm of fungi and perform a variety of tasks related to the transport of materials within the cell, for example, *spitzenkörper* located at the hyphal tips and involved in cell wall biosynthesis.

1.12.3.14 Intracellular Trafficking of Proteins

The ER is the site of protein synthesis and modification. After synthesis on polyribosomes on the surface of the ER membrane, precursor proteins are translocated into the lumen of the ER, where trimming, chaperone-assisted folding, and glycosylation occur. Proteins then are directed by vesicles to the Golgi apparatus where carbohydrate side chains may be added in a process of mannosylation. Proteins processed in the Golgi are transported via secretory vesicles to the vacuole, the plasma membrane, and the periplasmic space.

1.13 PRIMARY PATHOGENS

What are primary fungal pathogens? Where are they found? Primary dimorphic fungal pathogens are those found in specific geographic areas of the world, their endemic areas, and have the capacity to cause infection in any individual (i.e., immune-normal or compromised). They are *Coccidioides* species, *Blastomyces dermatitidis, Histoplasma capsulatum*, and *Paracoccidioides brasiliensis*. Infection is initiated after the infectious *propagules* (conidia) are inhaled when they are aerosolized by a disturbance of the environment. Subcutaneous mycoses are also caused by primary pathogens (Table 1.1). In that case breaching the normally intact anatomic barrier of the skin by a puncture wound may be sufficient to initiate the pathogenic process.

1.13.1 Susceptibility to Primary Pathogens

Who is susceptible to primary pathogens? Immune normal persons are at risk to become exposed to primary pathogens and to develop disease along a spectrum from subclinical to moderate self-limited, to disseminated disease, the latter requiring timely therapeutic intervention. Whether exposure is benign, self-limited, or moderate to severe, and whether or not it will *disseminate* depends on a number of factors, among which are age, sex, race, physical health, immunologic status, and the number of infectious propagules inhaled. Diseases caused by these fungi are, with rare exceptions, not communicable. When an immunocompromised individual is exposed to a primary pathogen the clinical course may be severe.

Up to now, there are no vaccines available for these diseases, although a vaccine for coccidioidomycosis is under investigation (URL: www.valleyfever.com). Primary systemic fungi differ in whether recovery from an infection results in durable immunity. This will be discussed with each fungus and disease.

1.14 ENDEMIC *VERSUS* WORLDWIDE PRESENCE

Primary systemic fungal pathogens are found in geographically restricted (endemic) areas, whereas fungi producing opportunistic fungal disease are generally found worldwide. An exception is the fungal opportunist *Penicillium marneffei*, a dimorphic endemic pathogen restricted to Southeast Asia.

1.15 OPPORTUNISTIC FUNGAL PATHOGENS

What are opportunistic pathogens? Where are they found?

Opportunistic fungal pathogens (Table 1.5) may be common environmental molds (and some yeasts) whose cells and conidia circulate in the *aerospora* (e.g., *Aspergillus* species, *Cryptococcus* species). Otherwise, they may be endogenous commensal fungi such as *Candida albicans*, a yeast that has adapted to an ecologic niche on the oral, intestinal, and vaginal mucosae of warm-blooded animals and especially humans where it lives an inconspicuous existence, all the while probing the mucosal epithelium for signs of decreased immune surveillance or lack of anatomic integrity. Given the opportunity, any fungus with the ability to grow at 37°C may become an opportunistic pathogen. Physicians and laboratory personnel alike must be aware of this when diagnosis is difficult or unusual.

1.15.1 Susceptibility to Opportunistic Fungal Pathogens: Host Factors

1.15.1.1 Immunocompromised Status

Who is susceptible to opportunistic fungi? Persons may become susceptible to opportunistic fungal pathogens because of immunodeficiency disease (either inborn or acquired, e.g., HIV infection); deliberate immunosuppressive therapy to treat cancer, *collagen vascular disease*, or for maintenance of stem cell or solid organ transplants.

Table 1.5 Some Differences Between Systemic Mycoses Caused by Primary Dimorphic Environmental Molds and Opportunistic Fungal Pathogens

	Type of fungal pathogen	
	Primary systemic dimorphic	Opportunistic
Host resistance	Immunocompetent or immunocompromised	Immunocompetent but debilitated, or immunocompromised
Portal of entry	Lungs or subcutaneous penetrating injury	Lungs, gastrointestinal tract, intravascular catheter
Prognosis	±95% cases resolve spontaneously	Depends on severity of impairment of host defenses, timing of diagnosis
Morphology	Dimorphic, yeast forms or spherules in tissue	Either filamentous or yeast-like; no change between cultural and tissue forms generally seen
Distribution	Geographically restricted (endemic mycoses)	Some endogenous *microbiota*; others ubiquitous in nature; worldwide; one endemic mycosis: penicilliosis

1.15.1.2 Immune-Normal Host

Some of the host factors that allow immune-normal persons to become susceptible to systemic mycoses due to opportunistic pathogens are:

- Age (low birthweight–premature infants; the elderly)
- Burns
- Chronic respiratory disease
- Debilitating illness
- Dialysis, whether hemodialysis or peritoneal
- Endocrine disorders (e.g., diabetes mellitus)
- Intensive care requiring parenteral nutrition, high *APACHE II score*
- Surgery (e.g., cardiothoracic or abdominal)
- Traumatic injury

In many of the above situations the normal anatomic barriers are disrupted, enabling entry of the opportunist fungus, or, as in the case of diabetes mellitus, there is *functional neutropenia*.

1.16 DETERMINANTS OF PATHOGENICITY

Why are fungi pathogenic for humans? As eukaryotes, fungi use various stratagems to evade host defenses. The list below is a summary of microbial factors that have been shown to influence pathogenicity. Further information specific to each pathogen is discussed in the disease chapters under the section heading "Determinants of Pathogenicity."

- Thermotolerance. Fungi that can grow at 37°C are potential pathogens in a suitably susceptible host.
- Adaptation to a parasitic lifestyle, sometimes in an intracellular environment. The traditional assumption is that most primary and opportunistic fungal pathogens are free-living saprobes in nature.
  - Evidence is accumulating that the ecologic niche of *Cryptococcus neoformans* may include intracellular survival within soil amebae. In that case *C. neoformans* "learned how to become a pathogen." This theory may also apply to other environmental fungi and may help explain how some fungi have become adapted to intracellular survival within phagocytes (Steenbergen et al., 2001).
  - Environmental fungi can infect other mammalian hosts, including small rodents, and have adapted to a parasitic lifestyle.
- Adhesins. Pathogenesis of microbial disease proceeds via adherence to host tissues, a process of receptor–ligand interaction: for example, BAD-1 adhesin of *Blastomyces dermatitidis*, the Als family of surface adhesins of *Candida albicans*.
- Attack on host tissues using invasion promoting enzymes:
  - Secreted enzymes that damage host tissues: for example, aspartyl proteinases and phospholipases of *C. albicans*.
  - Production of catalase that decomposes hydrogen peroxide, thus interrupting the oxidative microbicidal pathway of polymorphonuclear neutrophilic granulocytes: for example, catalase of *Aspergillus fumigatus* and *Histoplasma capsulatum*.
- Dimorphism. Morphogenesis to distinct tissue forms confers an advantage to the pathogen. For example, *H. capsulatum* yeast forms are translocated intracellularly within monocytes from the lung to the spleen and liver; spherules and endospores of *Coccidioides* species spread the infection; yeast forms of *B. dermatitidis* are too large for endophagocytosis.
- Evasion of host immune defenses. For example, *Histoplasma capsulatum* survival in the phagosome is linked to preventing phago-lysosome fusion and by being a resourceful scavenger of iron from the host through secretion of siderophores, ferric reductase, and directly from host transferrin.
- Cell wall molecules are barriers that resist lysis by phagocytes and antifungal agents: for example, cell wall polymers, including α-(1→3)-D-glucan, melanin, and the glucuronoxylomannan polysaccharide capsule of *Cryptococcus neoformans*. To this list we add ß-(1→3)-D-glucan of *Candida* species, in shielding the yeast from antifungal agents by functioning as a major component of extracellular matrix material of biofilms embedded on intravascular catheters.

GENERAL REFERENCES IN MEDICAL MYCOLOGY

These texts provide depth in the form of monographs and reviews from the viewpoints of infectious disease specialists and basic scientists. Other texts, which are primarily laboratory manuals, are listed in Chapter 2.

Baker RD (ed.), 1971. *Human Infection with Fungi, Actinomycetes, and Algae*. Springer Verlag, New York.

Bennett JE, 2005. Section G, Mycoses, *in:* Mandell GL, Douglas RG, Bennett JE (eds.), *Principles and Practice of Infectious Diseases* 6th ed., 2 vols. Elsevier/Churchill Livingstone, New York.

Breitenbach M, Crameri R, Lehrer SB (eds.), 2002. Fungal Allergy and Pathogenicity, *in*: *Chemical Immunology*, Vol. 81. Karger, Basel.

Bulmer GS, 1995. *Fungus Diseases in the Orient*, 3rd ed. Rex Book Stores, Manila, Philippines.

Casadevall A, Perfect JR, 1998. *Cryptococcus neoformans*. ASM Press, Washington, DC.

Chandler FW, Watts JC, 1987. *Pathologic Diagnosis of Fungal Infections*. ASCP Press, Chicago.

DIAMOND RD, MANDELL G, 2000. *Atlas of Infectious Diseases: Fungal Infections*. Current Medicine, Philadelphia.

DISMUKES WE, PAPPAS PG, SOBEL JD (eds.), 2003. *Clinical Mycology*. Oxford University Press, Oxford, UK.

FIDEL PL, HUFFNAGLE GB (eds.), 2005. *Fungal Immunology: From an Organ Perspective*. Springer Science+Business Media, Inc., New York.

HEITMAN J, KRONSTAD JW, TAYLOR JW, CASSELTON LA (eds.), 2007. *Sex in Fungi: Molecular Determination and Evolutionary Implications*. ASM Press, Washington, DC.

HEITMAN J, FILLER SG, EDWARDS JE Jr, MITCHELL AP (eds.), 2006. *Molecular Principles of Fungal Pathogenesis*. ASM Press, Washington, DC.

HOSPENTHAL DR, RINALDI MG (eds.), 2008. *Diagnosis and Treatment of Human Mycoses*. Humana Press, Totowa, NJ.

JANEWAY, CA Jr, TRAVERS P, WALPORT M, SHLOMCHIK MJ, 2001. *Immunobiology, The Immune System in Health and Disease*, 5th ed. Garland Science, New York. Available on the internet at http://www.ncbi.nlm.nih.gov/sites/entrez?db=books.

KIBBLER CC, MACKENZIE DWR, ODDS FC, 1996. *Principles and Practice of Clinical Mycology*. Wiley, Chichester, UK.

KURUP VP, FINK JN, 1993. Fungal Allergy, *in*: MURPHY JW, FRIEDMAN H, BENDINELLI M (eds.), *Fungal Infections and Immune Responses*. Springer Publishing, New York.

KWON-CHUNG KJ, BENNETT JE, 1992. *Medical Mycology*. Lea & Febiger, Philadelphia.

LATGÉ J-P, STEINBACH WJ (eds.), 2008. *Aspergillus fumigatus and Aspergillosis*. ASM Press, Washington, DC.

LARONE D, 2000. *Medically Important Fungi: A Guide to Identification*, 4th ed. ASM Press, Washington, DC.

MURPHY JW, FRIEDMAN H, BENDINELLI, M (eds.), 1993. *Fungal Infections and Immune Responses*. Plenum Press, New York.

ODDS FC, 1988. *Candida and Candidosis*. Baillière Tindall, London, UK.

Proceedings of the 2005 ASM Annual General Meeting Symposium, "Sequence-based identification of mycotic pathogens."

REISS E, 1986. *Molecular Immunology of Mycotic and Actinomycotic Infections*. Elsevier, New York.

RICHARDSON MD, WARNOCK DW, 2003. *Fungal Infection—Diagnosis and Management*, 3rd ed. Blackwell Science, Malden, MA.

ST GERMAIN G, SUMMERBELL R, 1996. *Identifying Filamentous Fungi: A Clinical Laboratory Handbook*. Star Publishing, Belmont, CA.

Topley and Wilson's Microbiology and Microbial Infections: Medical Mycology 2007. MERZ WG, HAY RJ (eds.), 10th ed. *(rev)*. Hodder Education, London, UK.

SELECTED REFERENCES FOR INTRODUCTION TO FUNDAMENTAL MEDICAL MYCOLOGY

AL-FATTANI MA, DOUGLAS LJ, 2006. Biofilm matrix of *Candida albicans* and *Candida tropicalis*: chemical composition and role in drug resistance. *J Med Microbiol* 55: 999–1008.

ASHBEE HR, 2007. Update on the genus *Malassezia*. *Med Mycol* 45: 287–303.

BALAJEE SA, BORMAN AM, BRANDT ME, CANO J, CUENCA-ESTRELLA M, DANNAOUI E, GUARRO J, HAASE G, KIBBLER CC, MEYER W, O'DONNELL K, PETTI CA, RODRIGUEZ-TUDELA JL, SUTTON D, VELEGRAKI A, WICKES BL, 2009. Sequence-based identification of *Aspergillus, Fusarium*, and *Mucorales* species in the clinical mycology laboratory: Where are we and where should we go from here? *J Clin Microbiol* 47: 877–884.

BENJAMIN DR, 1995. *Mushrooms: Poisons and Panaceas: A Handbook for Naturalists, Mycologists and Physicians*. WH Freeman, New York.

BERMAN J, 2006. Morphogenesis and cell cycle progression in *Candida albicans*. *Curr Opin Microbiol* 9: 595–601.

Biosafety in Microbiological and Biomedical Laboratories (BMBL) 2009. Centers for Disease Control and Prevention and National Institutes of Health, 5th ed. U.S. Government Printing Office. HHS Publication No. (CDC) 21–1112, Revised December 2009. Please also see "Websites Cited."

BOEKHOUT T, GUEIDAN C, DE HOOG S, SAMSON R, VARGA J, WALTHER G, 2009. Fungal taxonomy: New developments in medically important fungi. *Curr Fungal Infection Rep* 3: 170–178.

BREITENBACH M, CRAMERI R, LEHRER SB (eds.), 2002. Fungal Allergy and Pathogenicity, *in*: *Chemical Immunology*, Vol. 81. Karger, Basel.

BRYANT PA, VENTER D, ROBINS-BROWNE R, CURTIS N, 2004. Chips with everything: DNA microarrays in infectious diseases. *Lancet Infect Dis* 4: 100–111.

BULLER AHR, 1931. *Researches on Fungi*, Vol. 4. Longman's Green and Co., London, UK.

CANTÓN E, ESPINEL-INGROFF A, PEMÁN J, 2009. Trends in antifungal susceptibility testing using CLSI reference and commercial methods. *Expert Rev Anti Infect Ther* 7: 107–119.

CLSI, 2008. Interpretive criteria for identification of bacteria and fungi by DNA target sequencing: Guideline, CLSI document MM18-A. Clinical and Laboratory Standards Institute, Wayne, PA.

CUOMO CA, BIRREN BW, 2010. The fungal genome initiative and lessons learned from genome sequencing. *Methods Enzymol* 470: 833–855.

DIAZ MR, FELL JW, 2004. High-throughput detection of pathogenic yeasts of the genus *Trichosporon*. *J Clin Microbiol* 42: 3696–3706.

DIEZMANN S, COX CJ, SCHÖNIAN G, VILGALYS RJ, MITCHELL TG, 2004. Phylogeny and evolution of medical species of *Candida* and related taxa: A multigenic analysis. *J Clin Microbiol* 42: 5624–5235.

DU TOIT CJ, 1942. Sporotrichosis on the Witwatersrand. *Proc Mine Med Officers Assoc* 22: 111–127.

FERA MT, LA CAMERA E, DE SARRO A, 2009. New triazoles and echinocandins: Mode of action, in vitro activity and mechanisms of resistance. *Expert Rev Anti Infect Ther* 7: 981–998.

GEISER DM, 2008. Sexual structures in *Aspergillus*: Morphology, importance and genomics. *Med Mycol* 47 (Suppl 1): S1–S26.

GOLDMAN L, 1968. Favus mistaken for leprosy by artists of the renaissance. *Arch Dermatol* 98: 660–661.

GRIFFIN DH, 1994. *Fungal Physiology*, 2nd ed. Wiley-Liss, New York.

HALEY LD, RICE EH, 1983. Basic terminology used in identification of some imperfect fungi. U.S. Department of Health and Human Services, Centers for Disease Control, Atlanta, GA.

HALL BG, 2007. *Phylogenetic Trees Made Easy: A How-to Manual*, 3rd ed. Sinauer Assoc., Sunderland, MA.

HARRIS SD, READ ND, ROBERSON RW, SHAW B, SEILER S, PLAMANN M, MOMANY M, 2005. Polarisome meets spitzenkörper: Microscopy, genetics, and genomics converge. *Eukaryot Cell* 4: 225–229.

HAZEN KC, HAZEN BW, 1993. Surface hydrophobic and hydrophilic protein alterations in *Candida albicans*. *FEMS Microbiol Lett* 107: 83–87.

HIBBETT DS, BINDER M, BISCHOFF JF, BLACKWELL M, CANNON PF, ERIKSSON OE, HUHNDORF S, JAMES T, KIRK PM, LÜCKING R, et al., 2007. A higher-level phylogenetic classification of the Fungi. *Mycol Res* 111(Pt 5): 509–547.

HOBSON RP, MUNRO CA, BATES S, MACCALLUM DM, CUTLER JE, HEINSBROEK SEM, BROWN GD, ODDS FC, GOW NAR, 2004. Loss of cell wall mannosylphosphate in *Candida albicans* does not influence macrophage recognition. *J Biol Chem* 279: 39628–39635.

HOWARD RJ, GOW NAR (volume eds.), 2007. *Biology of the Fungal Cell (The Mycota)*, 2nd ed. Springer, Heidelberg.

IAMS AM, 1950. Histoplasmin skin test. *Ann NY Acad Sci* 50: 1380–1387.

KURUP VP, FINK JN, 1993. Fungal Allergy, *in*: MURPHY JW, FRIEDMAN H, BENDINELLI M (eds.), *Fungal Infections and Immune Responses*. Springer Publishing Company, New York.

LENARDON MD, WHITTON RK, MUNRO CA, MARSHALL D, GOW NA, 2007. Individual chitin synthase enzymes synthesize microfibrils of differing structure at specific locations in the *Candida albicans* cell wall. *Mol Microbiol* 66: 1164–1173.

LUMBSCH HT, HUHNDORF SM (eds.), 2007. Outline of Ascomycota–2007. *Myconet* 13: 1–58.

MASUOKA J, 2004. Surface glycans of *Candida albicans* and other pathogenic fungi: Physiological roles, clinical uses, and experimental challenges. *Clin Microbiol Rev* 17: 281–310.

MASUOKA J, HAZEN KC, 2004. Cell wall mannan and cell surface hydrophobicity in *Candida albicans* serotype A and B strains. *Infect Immun* 72: 6230–6236.

MMWR, 1981. Kaposi's sarcoma and *Pneumocystis* pneumonia among homosexual men–New York City and California. *MMWR Morb Mortal Wkly Rep* 30: 305–308 (July 4).

MMWR, 2001. Outbreak of acute respiratory febrile illness among college students. *Morb Mortal Wkly Rep* 50: 261–262 (April 13).

NETT J, LINCOLN L, MARCHILLO K, ANDES D, 2007a. Beta-1,3 glucan as a test for central venous catheter biofilm infection. *J Infect Dis* 11: 1705–1712.

NETT J, LINCOLN L, MARCHILLO K, MASSEY R, HOLOYDA K, HOFF B, VAN HANDEL M, ANDES D, 2007b. Putative role of beta-1,3 glucans in *Candida albicans* biofilm resistance. *Antimicrob Agents Chemother* 51: 510–520.

O'DONNELL K, SUTTON DA, FOTHERGILL A, MCCARTHY D, RINALDI MG, BRANDT ME, ZHANG N, GEISER DM, 2008. Molecular phylogenetic diversity, multilocus haplotype nomenclature, and *in vitro* antifungal resistance within the *Fusarium solani* species complex. *J Clin Microbiol* 46: 2477–2490.

O'DONNELL K, SUTTON DA, RINALDI MG, MAGNON KC, COX PA, REVANKAR SG, SANCHE S, GEISER DM, JUBA JH, VAN BURIK J-AH, PADHYE A, ANAISSIE EJ, FRANCESCONI A, WALSH TJ, ROBINSON JS, 2004. Genetic diversity of human pathogenic members of the *Fusarium oxysporum* complex inferred from multilocus DNA sequence data and amplified fragment length polymorphism analyses: Evidence for the recent dispersion of a geographically widespread clonal lineage and nosocomial origin. *J Clin Microbiol* 42: 5109–5120.

PADHYE AA, BENNETT JE, MCGINNIS MR, SIGLER L, FLISS A, SALKIN IF, 1998. Biosafety considerations in handling medically important fungi. *Med Mycol* 36(Suppl 1): 258–265.

POULAIN D, SLOMIANNY C, JOUAULT T, GOMEZ JM, TRINEL PA, 2002. Contribution of phospholipomannan to the surface expression of ß-1,2-oligomannosides in *Candida albicans* and its presence in cell wall extracts. *Infect Immun* 70: 4323–4328.

RHODES JC, KWON-CHUNG KJ, POPKIN TJ, 1981. Ultrastructure of the septal complex in hyphae of *Cryptococcus laurentii*. *J Bacteriol* 145: 1410–1412.

RUIZ-HERRERA J, ELORZA MV, VALENTIN E, SENTANDREU R, 2006. Molecular organization of the cell wall of *Candida albicans* and its relation to pathogenicity. *FEMS Yeast Res* 6: 14–29.

SCHWARZ J, KAUFFMAN CA, 1977. Occupational hazards from deep mycoses. *Arch Dermatol* 113: 1270–1275.

SHEPARD JR, ADDISON RM, ALEXANDER BD, DELLA-LATTA P, GHERNA M, HAASE G, HALL G, JOHNSON JK, MERZ WG, PELTROCHE-LLACSAHUANGA H, STENDER H, VENEZIA RA, WILSON D, PROCOP GW, WU F, FIANDACA MJ, 2008. Multicenter evaluation of the *Candida albicans/Candida glabrata* peptide nucleic acid fluorescent *in situ* hybridization method for simultaneous dual-color identification of *C. albicans* and *C. glabrata* directly from blood culture bottles. *J Clin Microbiol* 46: 50–55.

SIGLER L, BARTLEY JR, PARR DH, MORRIS AJ, 1999. Maxillary sinusitis caused by medusoid form of *Schizophyllum commune*. *J Clin Microbiol* 37: 3395–3398.

STEENBERGEN JN, SHUMAN HA, CASADEVALL A, 2001. *Cryptococcus neoformans* interactions with amoebae suggest an explanation for its virulence and intracellular pathogenic strategy in macrophages. *Proc Natl Acad Sci USA* 98: 15235–15250.

STEVENS DA, CLEMONS KV, LEVINE HB, PAPPAGIANIS D, BARON EJ, HAMILTON JR, DERESINSKI SC, JOHNSON N, 2009. Expert opinion: What to do when there is *Coccidioides* exposure in a laboratory. *Clin Infect Dis* 49: 919–923.

STOREY E, DANGMAN KH, SCHENCK P, DE BERNARDO RL, YANG CS, BRACKER A, HODGSON MJ, 2005. Guidance for clinicians on the recognition and management of health effects related to mold exposure and moisture indoors. University of Connecticut Health Center, Farmington, CT.

SUDBERY R, COURT H, 2007. Polarized Growth in Fungi, pp. 137–166 *in*: HOWARD RJ, GOW NAR (volume eds.), *Biology of the Fungal Cell* (*The Mycota*), 2nd *ed*. Springer, Heidelberg.

SUZUKI S, 1997. Immunochemical study on mannans of genus *Candida*. I. Structural investigation of antigenic factors 1, 4, 5, 6, 8, 9, 11, 13, 13b and 34. *Curr Topics Med Mycol* 8: 57–70.

UNTEREINER WA, SCOTT JA, NAVEAU FA, SIGLER L, BACHEWICH J, ANGUS A, 2004. The Ajellomycetaceae, a new family of vertebrate-associated Onygenales. *Mycologia* 96: 812–821.

US Environmental Protection Agency, 2001. Mold remediation in schools and commercial buildings. EPA publication 402-LK-01-001, 48 pp.

VESES V, GOW NA, 2009. Pseudohypha budding patterns of *Candida albicans*. *Med Mycol* 47: 268–275.

VIALÁS V, NOGALES-CADENAS R, NOMBELA C, PASCUAL-MONTANO A, GIL C, 2009. Proteopathogen, a protein database for studying *Candida albicans*–host interaction. *Proteomics* 9: 4664–4668.

WILSON D, THEWES S, ZAKIKHANY K, FRADIN C, ALBRECHT A, ALMEIDA R, BRUNKE S, GROSSE K, MARTIN R, MAYER F, LEONHARDT I, SCHILD L, SEIDER K, SKIBBE M, SLESIONA S, WAECHTLER B, JACOBSEN I, HUBE B, 2009. Identifying infection-associated genes of *Candida albicans* in the postgenomic era. *FEMS Yeast Res* 9: 688–700.

WILSON JWW, PLUNKETT OA, 1965. *The Fungous Diseases of Man*. University of California Press: Berkeley and Los Angeles, 428 p.

WEBSITES CITED

The current status of genome sequencing, assembly, and annotation for pathogenic fungi may be found by visiting the National Center for Biotechnology Information homepage and then entering the genus and species into the search Genome Project page. The URL for the NCBI Genome Project is http://www.ncbi.nlm.nih.gov/genomeprj.

American Society for Microbiology (ASM) International Laboratory Capacity Building Program. www.labcap.org.

Biosafety in Microbiological and Biomedical Laboratories 5th ed. (BMBL). http://www.cdc.gov/OD/ohs/biosfty/bmbl5/bmbl5toc.htm.

Candida albicans Physical Map. http://albicansmap.ahc.umn.edu/.

Dr. Fungus. http://www.doctorfungus.org/.

Lumbsch, HT, Huhndorf SM (eds.), 2007. Outline of *Ascomycota*–2007. *Myconet* 13: 1–58. Available at http://www.fieldmuseum.org/myconet/outline.asp#subclassPleo.

The Fifth Kingdom, Bryce Kendrick. http://www.mycolog.com.

Myconet. Current updated classification of the Phylum *Ascomycota* are published in electronic form (Lumbsch and Huhndorf, 2007). http://www.fieldmuseum.org/myconet/outline.asp#Asco

N Formation Design at the University of Michigan has visual representation and commentary about the yeast cell cycle at http://www.nformationdesign.com/portfolio/portfolio09.php

NIOSH Publication No. 99–143: TB Respiratory Protection Program in Health Care Facilities to match the respiratory protection to their risk assessment. http://www.cdc.gov/niosh/docs/99-143

Proteopathogen, a proteomics site. http://proteopathogen.dacya.ucm.es

TreeBASE: Its main function is to store published phylogenetic trees and data matrices. www.treebase.org

Valley Fever Vaccine Project. A vaccine for coccidioidomycosis is under investigation. www.valleyfever.com.

QUESTIONS

The Answer Key to multiple choice questions may be found at the end of the book.

1. Which of the following fungal diseases are communicable?
   A. Aspergillosis
   B. Dermatophytosis
   C. Histoplasmosis
   D. Sporotrichosis
   E. None of the above
2. Which of the following is not true of fungi?
   A. Eukaryotic
   B. Heterotrophic
   C. Mitochondria present
   D. Photosynthetic
3. Any fungus capable of growing at 37°C
   A. Could convert to the yeast form in tissue
   B. Has the capacity to cause disease in an immunocompromised or debilitated person
   C. Is a necessary but not sufficient property to cause disease in humans
   D. Is pathogenic for immune-normal persons
4. This soil dwelling mold produces conidia, which when inhaled germinate and grow in the lung as a budding yeast. The fungus is said to be
   A. Allergenic
   B. Dimorphic
   C. Opportunistic
   D. Thermophilic
5. The higher level classification of fungi pathogenic for humans is:
   A. *Ascomycota, Basidiomycota, Zygomycota, Pneumocystidales*
   B. Coelomycetes, Fungi imperfecti, Hyphomycetes, *Saccharomycetales*
   C. Dematiaceae, Hyalohyphomycetes, Dimorphics, Dermatophytes
   D. Subkingdom *Dikarya*, phylum *Ascomycota*, phylum *Basidiomycota*, subphylum *Mucoromycotina*, subphylum *Entomophthoromycotina*
6. Is vegetative growth sufficient to perpetuate fungi in the environment? What is meant by a fungal propagule? Besides direct extension, how do fungi spread in the environment?
7. Discuss the phylogenetic species concept. Compare it to the Linnean System. What role does cladistic analysis play in determining the relationship of species? Describe how a graphic method is used to illustrate a monophyletic group of related fungi.
8. To what taxonomic group (phylum, subphylum, order) do the *Candida* species belong? Describe the *Candida* species that assort into three clades.
9. Why is the *Onygenales* an important order for fungal pathogens? Indicate the major pathogens and how they associate within the *Onygenales*.
10. Compare the internal organization of the fungal cell with that of bacteria. How is the system of microtubules organized in the yeasts and in molds? What role does actin play in the cytoskeleton? What is the function of the *spitzenkörper*?

Chapter 2

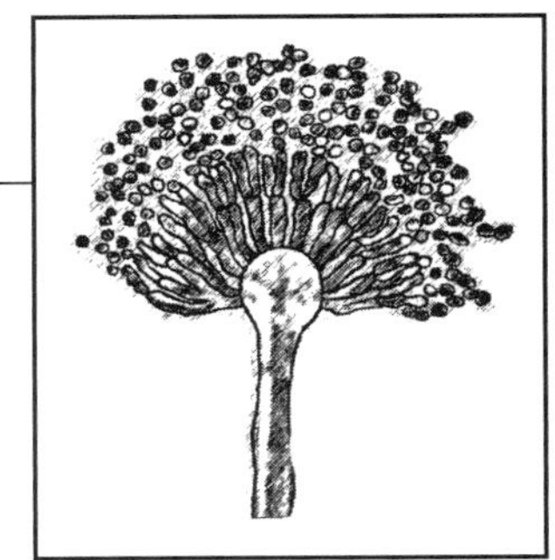

Laboratory Diagnostic Methods in Medical Mycology

2.1 WHO IS RESPONSIBLE FOR IDENTIFYING PATHOGENIC FUNGI?

2.1.1 Role of the Clinical Laboratorian

The clinical laboratorian is responsible for identifying any fungal pathogen, but the physician should advise the laboratory as to what may be the cause of disease, particularly if a systemic pathogen is suspected.

2.1.2 Role of the Physician

Background patient information should help the physician in this determination, and physicians specializing in infectious disease medicine are trained in this kind of clinical observation. Complete and proper questioning of the patient and study of the signs and symptoms are used by the physician to help identify the suspected presence of a fungal pathogen.

Signs and symptoms may vary according to the fungus causing disease, however, relevant questions remain the same: past history including where the patient has lived and traveled, occupation, avocation and recreation, and other medical conditions (e.g., tuberculosis or Hodgkin's disease). A history of residence or travel in particular areas of the world or the United States may help in diagnosis. Living or working near building construction, renovation, or demolition sites may lead to exposure to an *endemic* or opportunistic fungal agent.

In summary, the physician's examination of the patient will reveal:

- Physical diagnosis: signs and symptoms
- Recent history of travel
- Past history of travel and residence
- Occupation
- Recreation
- Other medical conditions

Pathologists may identify or tentatively identify the causal fungus from stained histologic tissue sections. This is preferably followed by culture of the fungus at the mycology bench in the clinical microbiology laboratory for the definitive identification and, when available, determination of antifungal susceptibility. As with most infectious diseases, diagnostic imaging is another key modality.

2.2 WHAT METHODS ARE USED TO IDENTIFY PATHOGENIC FUNGI?

The major methods are direct examination of clinical specimens, histopathology, and culture. Identification is made by morphology, biochemistry, genetics, and serology.

2.2.1 Culture and Identification

2.2.1.1 Morphology

The yeasts and molds are broad categories of the fungal life form. For further information please see Chapter 1, Section 1.6.5, Types of Vegetative Growth.

The most basic element of laboratory diagnosis consists of the microscopic examination of yeasts and molds and how they differ among the various fungal pathogens. This is accomplished by direct examination of exudates, scrapings, smears, and body fluids in the microbiology laboratory, and of biopsied tissue in the pathology laboratory. After recovery of the fungus in pure culture, the

Fundamental Medical Mycology, First Edition. By Errol Reiss, H. Jean Shadomy and G. Marshall Lyon, III.
© 2012 Wiley-Blackwell. Published 2012 by John Wiley & Sons, Inc.

means to identify the causative agent is to observe the shape, size, and pattern of budding in yeasts and of sporulation in molds (referred to as *conidiogenesis*).

2.2.1.2 Biochemistry

Metabolic activities of fungi are measured in biochemical reactions once the fungus is isolated in pure culture: including individual reactions and a pattern of reactivity with respect to a panel of carbohydrate assimilation tests and enzyme activities. Surrogate markers of fungal disease are measured as minute amounts of fungal metabolites directly detected in human body fluids.

2.2.1.3 Genetics

PCR and DNA sequencing are the common language of modern biology in all fields and that includes medical mycology! Once isolated in pure culture, genetic identification of fungi is indicated when growth is slow and sporulation is either lacking or does not fit a profile in a dichotomous key (St. Germain and Summerbell, 1996) or a well-illustrated laboratory manual (Larone, 2002). Another indication is when the patient has a rapidly progressing disease and time is of the essence. In such instances, the use of universal fungal primers for PCR directed against subregions of the ribosomal RNA genes ("rDNA") provide an expedited means to an identification, as developed in this chapter (please see Section 2.4, Genetic Identification of Fungi). Discussion of direct detection of fungi in body fluids and tissues by molecular means is confined to those test kits that are commercially available. When a fungus cannot be identified in a timely manner and PCR-sequencing is unavailable, the culture is referred for further testing to the State Department of Health in the United States or to a regional or national company that performs laboratory tests, including Labcorp, Quest Diagnostics, ARUP, and the Mayo Clinic.

2.2.1.4 Serology

Serodiagnosis is minimally invasive and can be a rewarding diagnostic adjunct. An array of such tests is offered by specialized companies (e.g., Immunomycologics, Inc., MiraVista Laboratories), through the state public health laboratories and the CDC, and also is available through the national and regional diagnostic companies. Some caveats concerning immunodiagnosis are:

- *Latent Period for Antibody Formation.* The adaptive immune system requires time to recruit and expand B-lymphocyte clones and to produce antibody secreting plasma cells. This process may take 21 days after the first symptoms appear, before an antibody titer can be demonstrated.
- *Antibody Production in Immunosuppressed Patients.* Mycoses that afflict immunosuppressed patients are typically due to common environmental molds against which immune-normal individuals have normal levels of natural antibodies and are capable of responding effectively to prevent infection. The same holds true for endogenous commensal members of the human microbiota, especially *Candida albicans* and related *Candida* species. Fluctuation in antibody titers in immunosuppressed patients more often reflects fluctuations in their immunocompetence rather than being proportional to fungal disease activity.
- *Crossreactivity.* Common fungal antigens, especially galactomannans, are found in antigenic *gemisch*[1] preparations such as whole fungal cell lysates. Unless extensive purifications are conducted by the test designers, primary binding assays, such as complement fixation and enzyme immunoassays (EIAs) used to detect circulating antibodies, will tend to have crossreactions. This tendency is minimized when precipitin reactions in agar gel are used because of the ability of immunodiffusion tests to differentiate identity with immunoprecipitin arcs formed between patient sera and positive control serum from nonidentity or crossreactivity.
- *Antigenemia and Antigenuria Detection.* The detection of cryptococcal capsular polysaccharide in cerebrospinal fluid using antibody coated latex particles is the ideal test because it is diagnostic, prognostic, and inexpensive. Detection of antigenemia and antigenuria in other mycoses, usually in an EIA format, is more problematic because the tests are operating near their lower limit of detection. This is because mannan (*Candida*) and galactomannan (*Aspergillus*) antigens, unlike cryptococcal polysaccharide, are rapidly cleared from the circulation, requiring frequent monitoring during the period of high risk. Test kits for antigen detection in aspergillosis and candidiasis are expensive; in the case of histoplasmosis, antigenuria detection testing has limited availability outside the United States.
- Taking these issues into account, immunodiagnostic methods are commercially available and are a useful diagnostic modality. Please see Section 2.3.11, Fungal Serology and Biochemical Markers of Infection.

[1] *Gemisch* (German). *Definition*: Mixture, jumble, medley, potpourri, blend.

2.3 LABORATORY DETECTION, RECOVERY, AND IDENTIFICATION OF FUNGI IN THE CLINICAL MICROBIOLOGY LABORATORY

This section lists and describes tests and media in use in the mycology laboratory. Each disease-specific chapter in the text has a section entitled "Laboratory Detection, Recovery, and Identification," which gives specific guidance to differentiate fungal genera and species.

2.3.1 The Laboratory Manual

Each clinical laboratory must maintain a file containing, in detail, all standard procedures performed within that laboratory, and verified by inspection through participation in a proficiency testing program. Laboratory testing in the United States is regulated by the Clinical Laboratory Improvement Act (CLIA) of 1988 as amended. It was undertaken as a result of false-negative Pap smears, creating a public outcry for more regulation of laboratory testing. The Act requires that all laboratories conducting testing of human specimens participate in a proficiency testing program from an accredited state or national organization approved by the U.S. Centers for Medicare & Medicaid Services, that is, the College of American Pathologists or the New York State Department of Health. A list of such proficiency testing providers is found at the URL http://www.cms.hhs.gov/CLIA/downloads/ptlist.pdf.

Requirements of CLIA include certification and recertification, frequent proficiency testing, a quality control program, employment of properly accredited personnel, submission to on-site inspections, procedure manuals, and extensive documentation. CLIA is enforced in the United States by the U.S. Centers for Medicare & Medicaid Services. Further information about CLIA may be obtained at the website with the URL http://www.cms.hhs.gov/CLIA/05_CLIA_Brochures.asp#TopOfPage.

One brochure details the Laboratory Director's responsibilities including proficiency testing, the procedure manual, training of personnel, and supervision of tests. This is done for quality assurance and quality control. The methods discussed in this chapter are included to help in understanding procedures generally used in the medical mycology section of a clinical laboratory. The information presented here does not supersede any procedures contained in the laboratory manual.

Although similar to isolation and identification in the bacteriology laboratory, certain differences are important and are discussed in this chapter. Important similarities include methods of isolation and inoculation of specimens onto media and the necessity for sterile technique to avoid contamination of the clinical specimens and of the technologist.

2.3.1.1 Disinfectants and Waste Disposal

Disinfectants for the Mycology Laboratory The U.S. Environmental Protection Agency registers disinfectants in the United States and requires manufacturers to specify the activity level of each compound at the working dilution. They may be found at the URL http://www.epa.gov/oppad001/chemregindex.htm.

The following microbicides are to be used on work surfaces or in containers for septic waste. They are not to be used for hand-washing. Avoid contact with skin and wear eye protection when using these chemicals. Here is a short list of the microbicides certified as effective against fungi:

- Phenol Derivatives. Amphyl (Reckitt Benckiser Inc, EPA Reg #: 675-43) is used at a 1:200 dilution in water. Vesphene II is used at a 1:128 dilution in water (Steris Corp., EPA Reg#: 1043-87). These microbicides are used to fill jars that receive disposable Pasteur pipettes, and microliter pipet tips. They are also in the collection of laboratory waste as a layer in discard pans lined with plastic bags.
- Ethyl Alcohol, 70% in water. Used to disinfect surfaces, such as surfaces of biological safety cabinets.
- Bleach, a 5.24% aqueous solution of sodium hypochlorite is diluted 1:10 for use in the laboratory to decontaminate work surfaces.

Waste Disposal Regulated medical waste is a legal term that includes cultures, primary specimens, mail containers, sharps, and all other septic waste from the clinical or research laboratory. State and federal laws in the United States and in many countries govern how this waste is to be collected, packaged, labeled, and disposed. In brief, waste is to be packaged in puncture-proof containers with tight-fitting lids, lined with one or two layers of plastic bags, and labeled with the international biohazard symbol (Padhye et al., 1998). Once packaged, waste can be transported within the clinical facility or to an off-site facility for treatment.

2.3.2 Specimen Collection

Fresh early morning specimens of urine or sputum are optimal for submission to the clinical laboratory. Specimens of all types should be sent to the laboratory for processing as quickly as possible, *preferably within 2 h of collection*, to minimize bacterial growth. (This important topic is covered in detail by Sutton, 2007). Overnight or 24 h specimens should *not* be used. This rule applies generally but specifically to urine specimens. Exceptions to this rule are that hair, nails, and skin scrapings may be stored up to 72 h at room temperature before culture, and they may be shipped by mail.

Specimens received for direct examination and culture vary according to the healthcare facility in which the laboratory is situated.

Appropriate specimens submitted for direct examination and culture are:

- Aspirates, especially from subcutaneous lesions (including fine needle aspirates)
- Blood
- Bone marrow
- Cerebrospinal fluid (CSF)
- Mycetoma grains
- Hair, nails, or skin scrapings
- Mucocutaneous lesions and wound swabs
- Oral or vaginal swabs
- Prostatic secretions
- Pus
- Respiratory tract secretions (sputum, bronchoalveolar lavage fluid)
- Tissue biopsies
    - Preparation of biopsied tissue for culture. Three options are available: mincing, grinding, or use of the Seward Stomacher® (Brinkmann Instruments distributor). Tissue grinds risk disruption of delicate fungal hyphae such as those of the *Mucorales*, with resultant loss of viability. Mincing or use of the Stomacher is recommended. The Stomacher is a closed system in which the sample is placed in a sealed plastic bag. Then reciprocating stainless steel paddles express the cell contents.
- Urine
- Wound drainage with or without grains (please see Chapter 20 for a further description of grains)
- Other specimens as submitted (e.g., catheter tips, corneal scrapings)

2.3.3 Direct Examination

Why is direct examination important? Because it can lead to an early presumptive diagnosis. If sufficient specimen material is not submitted, culturing is preferable to microscopic examination, as it is more sensitive. A presumptive diagnosis is then confirmed by recovery of the pathogen, if it is culturable, and its identification.

2.3.3.1 Wet Mount

Sputum, abscess material, caseous material, aspirates, sinus tract drainage, and similar specimens may first be examined with a dissecting microscope to identify appropriate material for culture (e.g., mycetoma grains). Next, a primary microscopic examination using the 40× objective can give important clues about the type of fungus present (e.g., yeast, mold, pseudohyphae, septate or nonseptate hyphae). These preliminary findings help in selecting the type of media and temperature of incubation to use for culture. However, standard operating procedures require specimens to be processed and plated and, if a microscopic exam is ordered, it is examined after the plating is done.

KOH Prep The 10% KOH wet prep is the standard method used by dermatologists and clinical laboratory scientists to examine skin scrapings, hair, nails, corneal scrapings, and exudates obtained from lesions at other body sites (Monod et al., 1989). KOH acts as a clearing agent in that it dissolves cellular debris without affecting the fibrillar glucan and chitin structure of the fungal cell wall.

To prepare a 10% KOH reagent:

Potassium hydroxide (KOH)	10 g
Glycerol	10 mL
Distilled H_2O	90 mL

Dissolve potassium hydroxide in water as follows:

- *Note*: Alkali is caustic, so protect your eyes. This reaction generates heat. Wear eye protection, laboratory coat, and gloves.
- Add 80 mL of water to a 250 mL Erlenmeyer flask. Chill the water in an ice bath. Then add the weighed amount of KOH pellets. Gently swirl to dissolve the pellets. Do not shake or splash the contents.
- Add 10 mL glycerol. Mix by swirling.
- *Adjust the final volume to 100 mL with water*. Transfer to a screw cap bottle, not one with a glass stopper (alkali fuses glass). The glycerol prevents crystallization of the reagent and allows the preparation to be kept for several days before drying.
    - Alternative formulations include adding Parker ink to the KOH prep or to use DMSO instead of glycerol. These variations may be viewed at Prof. David Ellis's website: www.mycology.adelaide.edu.au/Laboratory_Methods

Procedure: Specimens that are viscous, purulent, or consist of skin scrapings, hair, or nails are relevant for a KOH prep. The specimen is placed on a glass microscope slide; a drop or two of 10% KOH is added. If necessary, the specimen is teased apart with inoculating needles. Then it is coverslipped and incubated at room temperature for 5–15 min, or less if incubated at 37°C. Warming briefly over a Bunsen flame is also used but open flames are being discouraged in many laboratories. Warming over a flame may also result in overheating.

The specimen may be studied microscopically and a decision made about the media on which the original specimen should be planted. At times it is helpful to use 10% KOH with *Calcofluor white*, a fluorescent brightener, which enhances the visualization of fungi when viewed with a fluorescence microscope. When the specimen is not abnormally viscous (e.g., urine, CSF), the sediment can be observed directly after centrifugation.

How soon after a KOH prep is made should it be examined in the microscope? Immediately is preferred, but not later than 2 h.

Digestion of Sputum Bloody or mucoid specimens may first be treated with Mucolyse®—dithiothreitol in phosphate buffer (Pro-Lab Diagnostics, Richmond Hill, Canada).

Fluorescent Brighteners Used with or Without KOH Fluorescent brighteners increase the sensitivity and specificity of detecting fungi in KOH mounts of clinical specimens but require the use of a fluorescence microscope (Harrington, 2008; Hamer et al., 2006). Fungal elements produce bright fluorescence; conidia and yeasts are also stained and can be distinguished from artifacts. The use of Calcofluor-KOH wet mounts containing fluorescent brighteners allow specimens to be screened more easily, rapidly, and with a low power objective (10×). Two products can be used: Calcofluor white and Blankophor. Both are inexpensive chemical compounds that bind to chitin and cellulose and fluoresce when exposed to UV light.

Calcofluor is a stilbene-based dye that fluoresces (peak emission wavelength 450 nm) when excited with UV or near-UV light (optimum excitation wavelength 347 nm). The colorless dye is an "optical brightener," used to brighten paper and textiles and, added to laundry detergent, used to maintain brightness in clothing. The biologic application of Calcofluor white was first reported by Darken (1961) as a dye useful for viewing cell walls of fungi and bacteria. The specific wall component stained has not been resolved with certainty, although it clearly binds to chitin, cellulose, and other ß-1,4-linked polysaccharides (Hoch et al., 2005).

Calcofluor white is sold in the United States as the Fungi-Fluor® Kit (# 17442, Polysciences Inc., Warrington, PA). Detailed instructions on fixation, staining, and fluorescence microscopy are found in Polysciences application note # 316. The Fungi-Fluor kit consists of Cellofluor (another name for Calcofluor), 0.5%, including KOH. A second vial contains Evans Blue dye to reduce background fluorescence.

Specimens are viewed using a UV source with excitation at 340–380 nm, and a suppression filter at 430 nm. Fungal elements are blue against a red background if Evans Blue counterstain is used. Alternative excitation is in the H3 Violet plus Blue, Wide Band FITC filter, with excitation at 420–490 nm and a suppression filter at 520 nm.[2] Fungal elements appear green with a red background with Evans Blue counterstain.

Fungi-Fluor for *Pneumocystis* (#22363, Polysciences, Inc.) is a separate product and contains a positive control slide. Instructions are provided in Polysciences application note #482. Smears are fixed in methanol and then stained. No counterstain is required. Tissue sections are also fixed in methanol and then stained. Filters for excitation and emission are the same choices as for the Fungi-Fluor kit #17442. Other sources for Calcofluor are discussed as well as critiqued in Harrington (2008).

Blankophor is a yellowish liquid (A.G. Scientific, Inc., San Diego, CA) that is dissolved directly in 10% KOH to give a final solution of 0.001% vol/vol (i.e., 1 μL/100 mL of 10% KOH). Thus, Blankophor is a one-step process, whereas Calcofluor, because it is not stable in 10% KOH, requires two steps.

2.3.3.2 *Stained Smears*

Gram-stained or other stained smears usually are submitted from the bacteriology laboratory to confirm the presence of a fungus. Gram and Wright stains also are useful for certain specimens; for example, vaginal smears when looking for *Candida* species and BAL cytocentrifuge smears.

2.3.3.3 *Cerebrospinal Fluid (CSF)*

CSF specimens should be centrifuged, and the sediment studied microscopically before being cultured. Fungi can be detected in the CSF sediment with various staining methods:

- Gram-stained smears, or a wet preps with KOH-Calcofluor are also relevant for detecting fungi in CSF but do not stain the cryptococcal capsule.
- The May–Grünwald–Giemsa stain is effective in demonstrating cryptococci in CSF sediment (Sato et al., 1999).

[2]For excitation and emission maxima of fluorescent brightener products please see the Olympus Microscope web page: http://www.olympusmicro.com/primer/techniques/fluorescence/fluorotable2.html. This also offers a video tutorial on selection of excitation and barrier filters.

- *India ink*, a negative stain, enables detection of the *Cryptococcus* and its capsule.

  India Ink Preparation: India ink is a suspension of fine carbon black particles for use as a negative stain. Place *a drop of sediment on a drop of India ink on a slide*, coverslip, and observe in the microscope for the presence of a capsule, seen as a halo surrounding the yeast cell. If encapsulated yeast cells are seen, a preliminary result of *Cryptococcus* may be noted prior to culturing, and the physician alerted. The same technique may be used in observing yeast colonies. *Note:* If India ink from a dropper bottle is used, *do not add the India ink directly to a drop of the specimen*, because (as has happened) one can touch the specimen itself, and return a contaminated dropper to an India ink bottle.

2.3.4 Histopathology

Histopathology is an important part of direct examination. The specimens are usually obtained by minimally invasive methods such as:

- Abrasive technique—endoscopic brushings
- Collecting exfoliated cells in fluids—effusions, urine, sputum, CSF
- Fine needle aspiration (FNA)

All three methods may at times be used on respiratory and gastrointestinal systems. The pathologist will most often review and report the presence of fungal elements in a biopsy specimen. The microbiology laboratory rarely receives a stained histologic tissue section to study for fungi, but the laboratorian should be versed in the appearance of the various pathogens in tissue (Chandler et al., 1980; Chandler and Watts, 1987). Key features of the fungal elements in histopathologic sections are discussed for each mycotic disease in these texts and in *Fundamental Medical Mycology*.

Histopathologic studies with stains amenable to fungal structures are important to identify fungi in tissue. Although hematoxylin and eosin (H&E) is the most often used stain in the pathology laboratory, it is not the best stain for fungi. A specific indication for H&E is when the fungal elements are naturally pigmented (please see Chapters 18 and 19).

Excellent fungal stains are the periodic acid-Schiff (*PAS*) and the Gomori–methenamine silver stain (*GMS*). The mucicarmine stain is able to stain the cryptococcal capsule a bright red. Formulas for special fungal stains may be found in Baker (1971). A method for preparing GMS stain is given in Larone (2002). A resource for stains is Sigma-Aldrich Company Stain Expert, which lists all reagents and formulas for stains including fungal stains at the URL http://www.sigmaaldrich.com/life-science/cell-biology/hematology-and-histology/learning-center/stain-expert.html.

GMS Gomori–methenamine silver is a histologic stain that deposits silver into the cell walls of fungi resulting in a dark brown to black color outlining the fungal elements. It is a good stain in screening for fungi in tissue because it provides good contrast between fungal elements and the background. Extracellular capsule and intracellular details are not visible using this method. The histologic picture of human and animal tissues is not visible in the standard method, rather the background is uniformly stained with fast green counterstain. If hematoxylin and eosin is used as a counterstain, the tissue reaction is made visible. *Principle:* Chromic acid oxidizes the hydroxyl groups in the cell wall. Then silver nitrate solution is added; the silver ions are reduced by the aldehydes in the cell wall depositing elemental silver on the wall resulting in the brown-black color.

Combined Hematoxylin and Eosin/Methenamine Silver Stain This variation permits both the fungal elements to be demarcated by silver deposition and the tissue reaction to be rendered as well. For the method of procedure please consult the Rowley Biochemical Co. site at the URL http://www.rowleybio.com/pdfs/A-123%20Procedure.pdf.

PAS PAS stains the fungal cell wall pink-red. *Principle*: The cell walls of fungi are oxidized producing a Schiff base, which then reacts with basic fuchsin depositing a pink-red color in the cell walls. The tissue reaction is also visible when hematoxylin is used as a counterstain.

Mucicarmine Mayer's mucicarmine stains acidic capsular polysaccharides red, nuclei black, and the tissue elements, yellow.

Alcian Blue This is a stain for polysaccharides but is most often used in tissue sections, where, in combination with PAS stain, the Alcian blue component reacts with the cryptococcal capsule and PAS with the cell wall and cytoplasm (Lazcano et al., 1993).

Diff-Quik® (Dade Behring Inc., Newark, NJ.) This is a two-step modification of Wright–Giemsa with the advantage of reducing the preparation time to less than 1 min. This stain is one of the most common stains in histopathology. It is used on air-dried smears including cytocentrifuge smears and fine needle aspirates. This is not a special fungal stain, but it is an easy screening method. Results with this method include the following:

- *Candida* species stain well.

- *Sporothrix schenckii* yeasts stain pale to medium blue with an off-center pink or purple nucleus.
- *Histoplasma capsulatum* yeast forms stain pale to medium blue, with a pink to purple crescent-shaped nucleus. A thin clear halo surrounds the yeast.
- *Cryptococcus neoformans* stains pink to blue-purple and may be slightly granular. The capsule is an unstained clear zone.
- *Pneumocystis jirovecii*. Alveolar casts are stained. Trophic forms and ascospores are stained blue, not the ascus wall.
- Many molds stain with Diff-Quik with a blue-purple stain of the hyphal wall that is not clearly differentiated from the surrounding tissue.

Papanicolaou Stain This is capable of staining *Aspergillus* hyphae. *Histoplasma* yeast forms are seen in dark outline.

Other more specialized stains are useful, for example, Fontana–Masson to demonstrate latent melanin pigment.

Immunohistochemical Tests While these tests are useful they are not generally available, the Pathology Branch, CDC, provides this kind of consultative testing. A commercially available monoclonal antibody test is used in fluorescent antibody detection of *Pneumocystis jirovecii* in clinical specimens (please see Section 2.3.11, Fungal Serology and Biochemical Markers of Infection). Immunohistochemical tests can add a level of species specificity that is not possible with histologic stains alone.

Immunohistochemical methods for the mycoses were reviewed by Jensen et al. (1996, 1997).

2.3.5 Culture

2.3.5.1 Additional Resources

Clinical mycologists have produced well-illustrated laboratory manuals and web-based instruction to guide laboratory diagnosis. Some of these are listed here and others are listed in the General References for Laboratory Diagnostic Methods in Medical Mycology.

- As an introduction to the growth of molds, a useful online resource is "Moulds, Isolation Cultivation and Identification" by David Malloch of the University of Toronto which may be viewed at the URL http://website.nbm-mnb.ca/mycologywebpages/mycologywebpages.html.
- *Medically Important Fungi*, *A Guide to Identification* by Larone (2002) is well organized and illustrated and has been continuously improved through four editions.
- *Identifying Filamentous Fungi*: *A Clinical Laboratory Handbook* by St. Germain and Summerbell (1996) is well illustrated with color plates and has a dichotomous key. These two manuals are well-suited to clinical laboratories at all levels.
- The *Atlas of Clinical Fungi*, 2nd ed. by de Hoog et al. (CD version 2004.11) is a comprehensive keyed and well-illustrated reference work of the well-known and emerging fungal pathogens. It is encyclopedic in coverage and is well suited to Reference Laboratories.
- A novel approach to identify molds is the online interactive illustrated key at Prof. David Ellis' website: "Mycology Online" at the URL http://www.mycology.adelaide.edu.au/Fungal_Jungle/Identification_Keys.html.
- The "Dr. Fungus" website contains information for the general public about mycoses, on sick building syndrome, pictures and descriptions of fungi, antifungal agents, laboratory procedures, and a database of minimum inhibitory concentrations at the URL www.doctorfungus.org.
- Please visit the American Society for Microbiology (ASM) Microbe Library at the URL www.microbelibrary.org. William Mayberry has posted a PowerPoint presentation entitled "Introduction to Mycology," which contains useful illustrations relevant to the material in this chapter.
- Two online sources of formulas for mycologic media are:
  - The Centraalbureau voor Schimmelcultures in the Netherlands at the URL http://www.cbs.knaw.nl/pdf/mediaCBS.pdf.
  - Alberta Microfungus Collection and Herbarium, Prof. Lynne Sigler, Principal Investigator, at the URL http://www.devonian2.ualberta.ca/uamh/culture_receipt.htm.
- Other excellent handbooks to identify fungi are listed in the "General Reference for Laboratory Diagnostic Methods in Medical Mycology" followed by "Selected References for Laboratory Diagnostic Methods in Medical Mycology."

2.3.5.2 Media for Recovery of Fungi from Clinical Specimens

Primary media for the isolation of fungi from clinical specimens are listed below and their contents are given in Table 2.1. There are other more specialized media for specific purposes that will be defined and discussed in the text.

Composition and preparation of the various media may be found in standard texts such as St. Germain

Table 2.1 Common Media for Cultivation of Fungi in the Clinical Laboratory

Medium	Purpose	Composition	Comment
Sabouraud–dextrose agar (SDA)	General purpose growth of fungi	Glucose, 4% Peptone, 1% (enzymatically hydrolyzed casein: enzymatically hydrolyzed animal tissue, at a ratio of 1:1), pH 5.6	For this medium and others, 1.5% agar is used for solid medium.
SDA-Emmons modification	General purpose growth of fungi	Glucose, 2%, Peptone, 1% pH 6.9	A higher pH and half the amount of glucose stimulates sporulation.
Potato dextrose agar, PDA	Encourages sporulation	Potato starch, 0.4% Glucose, 1% pH 5.6	
Mycosel™ (Becton, Dickinson, Co., Sparks, Maryland)	Inhibition of bacteria and saprobic fungi	Papaic digest of soybeans, 1% Glucose, 1% Chloramphenicol, 50 μg/mL Cycloheximide, 400 μg/mL pH 6.9	
Mycobiotic agar (Delasco, Council Bluffs, Iowa)	Inhibits bacteria and saprobic fungi	Enzymatic digest of soybean meal, 1% Glucose, 1% Chloramphenicol, 50 μg/mL Cycloheximide, 500 μg/mL pH 6.5	This formulation is similar to Mycosel.
Inhibitory Mold agar	Isolation of fungi from potentially contaminated specimens	Pancreatic digest of casein, 0.5% Peptic digest of animal tissue, 0.2% Yeast extract, 0.5% Glucose, 0.5% Starch, 0.3% Dextrin, 0.1% Chloramphenicol, 125 μg/mL Sodium phosphate, 0.2% Magnesium sulfate, 0.08% Ferrous sulfate, 4 mg% Sodium chloride 4 mg% Manganese sulfate, 1.6 mg% pH 6.7	
Brain–Heart Infusion agar, ±5% sheep blood	This enriched medium is used to stimulate growth in some fungi (e.g., *Histoplasma capsulatum*)	Calf brains, infusion from 200 g, 7.7 g/L Beef heart, infusion from 250 g, 9.8 g/L Proteose peptone, 1% Glucose, 0.2% Sodium chloride, 0.5% Disodium phosphate, 0.25% pH 7.4	This medium is not supplemented with antibiotics unless special-ordered. There are different formulations; please consult http://www.bd.com/ds/technicalCenter/.
SABHI agar (Sabouraud–Brain–Heart Infusion) (Difco®)	The combined BHI and SDA may improve recovery of fungi. Addition of defibrinated sheep blood is used to increase recovery of dimorphic fungi.	Brain–heart digest, 9.25 g Proteose peptone, 5.0 g Enzymatic digest of casein, 5.0 g Dextrose, 21.0 g Sodium chloride, 2.5 g Disodium phosphate, 1.25 g 59 g of dehydrated medium/L of water Final pH 7.0	Formulated with or without cycloheximide and chloramphenicol or supplemented with sheep blood. Please see http://www.bd.com/ds/technicalCenter/inserts/difcoBblManual.asp.

and Summerbell (1996) and Larone (2002), and in technical bulletins of BD Biosciences at the URL http://www.bd.com/ds/technicalCenter/.

Limiting the primary isolation media to the following is advantageous to keep costs low.

- Sabouraud Dextrose Agar, Emmons modified (*SDA-Emmons*) (contains no inhibitors but is not enriched for fastidious fungi)
- *Mycosel*™ or *Mycobiotic* agar. Contains cycloheximide and chloramphenicol to inhibit both bacteria and saprobic fungi.
- *Inhibitory Mold Agar* contains chloramphenicol to inhibit bacteria. A medium containing an antibiotic to inhibit bacteria is often necessary. This may be accomplished by using SDA-Emmons with chloramphenicol or by using Inhibitory Mold Agar.
- Brain–Heart Infusion *(BHI)* with antibiotics to inhibit bacteria, with or without 5% sheep blood, is an enriched medium favoring the growth of dimorphic fungal pathogens (e.g., *Blastomyces dermatitidis* and *Histoplasma capsulatum*).

If a medium containing cycloheximide is used to inhibit saprobic fungi, *a similar medium without that drug should be used to permit the growth of some Candida* species, *Trichosporon* species, *Cryptococcus neoformans*, as well as some molds including *Aspergillus* species and *Pseudallescheria boydii*. The yeast forms of *Histoplasma capsulatum* and *Blastomyces dermatitidis* are also inhibited by cycloheximide.

How long should fungal cultures be incubated and when should they be observed for growth? Each chapter has information about the conditions for "Laboratory Detection, Recovery, and Identification" of each group of etiologic agents. In general, fungal cultures are observed for growth on the first and second day after seeding with the specimen; thereafter, every other day for 1 week, and then every 5–7 days until 4–6 weeks of incubation.

2.3.5.3 Common Media for Cultivation of Fungi

These media are used at the mycology bench of the clinical microbiology laboratory (Table 2.1).

Tube Media The use of tubes rather than Petri plates is preferable when specimens are obtained from sterile sites and contamination is not expected. When tube media are used, whether with cotton plugs or screw cap closures, allow for the exchange of air, since all fungi are aerobic. Tubes should have a relatively large diameter, such as 18 × 150 mm, and contain a thick butt of medium that will not readily become dry.

Petri Plate Media Petri plate media are used when separation of different organisms is required due to the nature of the specimen (e.g., sputum, purulent material, or skin scrapings), where different fungi and bacteria may be found. Solid medium typically contains 1.5% agar. Types of isolation media used, temperature of incubation, and the time specimens must be kept before discarding cultures vary with the specimen. Primary isolation media (SDA, SDA-Emmons) are typically used for most specimens, while different tests and media may be required, particularly on subculture, to differentiate various genera and species (Tables 2.1 and 2.2).

Standard Petri plates (15 × 100 mm) are used and contain 15–20 mL of medium poured/plate. When specimens are planted on Petri plates, the plates should be sealed with material such as Parafilm®, which "breathes." Mycoseals® are shrink seals for Petri plates composed of cellulose and are gas-permeable (Hardy Diagnostics, Santa Maria, CA). They provide an additional amount of biosafety and keep cultures from becoming dehydrated. *Petri plates are incubated upside down* so that water vapor condenses on the plate top, and not on the growing fungus culture. Information including specimen number and date is noted on the bottom of the plate.

2.3.5.4 Blood Cultures

Detection of fungemia is an essential part of laboratory identification. Fungal sepsis is an indication of invasive disease with potential multisystem dissemination. Invasive fungal infections, particularly yeast infections, are often first detected by a positive blood culture. A positive blood culture is a cause for an immediate report to the physician of a presumptive fungal infection, with the specific identification to follow.

The laboratory approach to recover fungi from blood cultures has changed according to the increased variety of fungi seen in the clinical laboratory and, above all, because of technical improvements in commercial blood culture products and systems (Reimer et al., 1997). Originally, fungal blood cultures were constituted in home-brew Castaneda biphasic bottles consisting of a BHI agar layer overlaid with BHI broth (Geha and Roberts, 1994).

Factors Affecting Yield of Fungi from Blood Cultures

Oxygen Tension Venting bottles improves oxygenation and growth of fungi, shortening the time to detect yeasts from 5.2 to 2.6 days.

Incubation Temperature for Blood Cultures The incubation temperature depends on the fungus suspected. If a yeast is suspected the temperature of incubation is 35°C, and in the case of a mold, 30°C or room temperature. If a dimorphic fungus is suspected, the blood sample

Table 2.2 Special Media and Tests for Mold Identification[a,b]

Aspergillus Differential Agar™ BBL Cat. No. 297244	Detects yellow-orange pigment produced by colonies of *Aspergillus flavus*. This differentiates *A. flavus* from most other *Aspergillus* species.
Benomyl selective medium	Several important fungal pathogens will grow on SDA with 10 μg/mL of Benomyl. (Please see Table 2.6.)
Cornmeal agar (CMA)	Enhances pigment production in some molds. Induces sporulation better than SDA. Used in slide culture production (see below).
Czapek–Dox agar	A reference medium to identify the *Aspergillus* species based on their conidiogenesis.
Dermatophyte testing medium (DTM)	Identification of dermatophytes by red color production. *Note:* Occasionally saprobe fungi produce this color.
In vitro hair perforation test	Differentiates *Trichophyton rubrum* (−) from *T. interdigitale* (+).
KNO_3 assimilation	Differentiates *Exophiala jeanselmei* (+), from *Wangiella (Exophiala) dermatitidis* (−).

[a]General media for identification of molds with their formulations are in Table 2.1.

[b]Formulations for these media may be found at BD Biosciences at the URL http://www.bd.com/ds/technicalCenter/.

is processed in an Isolator® tube (please see below). Incubation of blood cultures at 35°C and 30°C ensures maximum coverage, when there is no prediction as to what fungus is suspected.

Subcultures incubated at 35°C are used if there is a question about the isolate's pathogenic potential. Even higher temperatures are required to identify certain thermotolerant genera or species, and these are discussed under each etiologic agent. Molds prefer 27–30°C, whereas yeasts grow best at 37°C (Table 2.3).

Duration Yeasts can be recovered within 2–3 days (except please see below regarding *Candida glabrata*) whereas dimorphic fungi may require 3–6 weeks. Terminal blind subcultures are useful to improve yield but that is not a standard practice when sensor-reporting systems are used such as BactT/ALERT®.

Volume At least 10 mL of blood in a 1:10 ratio of blood to broth facilitates recovery.

False-positives Processes must be controlled from the point of drawing the blood through all laboratory steps in the isolation and detection process to minimize contamination of blood cultures with environmental fungi or skin-dwelling commensals.

Table 2.3 Temperature Tolerance Test

Species[a]	Maximum growth temperature (°C)
Aspergillus fumigatus	48
Cladophialophora bantiana	42–43
Lichtheimia corymbifera	48–52
Apophysomyces elegans	Up to 43
Cunninghamella bertholletiae	45
Rhizomucor pusillus	54–58
R. microsporus var. rhizopodiformis	50–52
Rhizopus oryzae	40–45

[a]Shaded blocks indicate the fungi are all classed in the *Mucorales* in *Mucoromycotina*.

Negative Yield Despite improvements in methods, some fungi fail to grow in blood cultures even though a systemic or disseminated infection may be present. Reasons for failure to grow are several:

- Requirements for a Minimum Population Size. *Histoplasma capsulatum* acquires iron by secreting siderophores. These growth factors are present in sufficient quantity when there is at least a minimum population size in the clinical specimen.
- Adhesins. *Candida albicans* produces adhesins that enable it to bind to endothelial cells and remove it from the peripheral blood. An example is ALS, a family of surface adhesins of *C. albicans* (Filler et al., 2006). Additionally, *Candida* species are removed from the peripheral blood at the splenic marginal zone and in the liver because the outer mannan layer of cell walls bind to mannose receptors on macrophages (Masuoka, 2004).
- Unknown Factors. *Aspergillus fumigatus* is a low-yielding fungus in blood cultures even though it is angioinvasive and has a predilection for the elastic lamina of blood vessels.

2.3.5.5 Types of Blood Culture Methods

Manual Systems Manual blood cultures are conducted in either broth, broth with an agar layer, or on solid agar medium. Broth blood cultures are maintained *for at least 7 days* and visually examined daily for signs of turbidity. They are also "blindly" subcultured after 5 days and terminally subcultured even though there is no visually detected turbidity. Fungal blood cultures are optimally held for 28 days.

Biphasic Broth/Agar Combination System The prototype for this is the Castaneda bottle containing an agar layer overlaid with broth, usually BHI. A newer version is the paddle device marketed as Septi-Chek® (BD Diagnostic Systems, Sparks, MD). Septi-Chek uses soybean-casein digest broth and slide agars, which coat different faces of the paddle. Subcultures are inoculated automatically onto different slide agars by inverting the bottle when it is checked for turbidity. The slide agars used are chocolate agar, MacConkey, and malt agars. This arrangement provides an alternative method of detecting growth since isolated colonies are accessible on the paddle to be directly picked. The biphasic BHI and Septi-Chek systems recover a majority of fungal isolates. Improved performance is linked to lysis of blood cells, aeration by venting, and agitated incubation.

The Isolator® Solid Medium System The Isolator® solid medium system (Inverness Medical Professional Diagnostics, Princeton, NJ) approach to blood cultures uses a lysis-centrifugation step preparatory to planting to agar medium. The Isolator tube, available in adult and pediatric size tubes, contains EDTA anticoagulant and saponin to lyse blood cells. Lysis of WBCs releases phagocytosed yeasts and yeast forms of dimorphic fungi.

Whole blood is introduced, the formed elements of the blood are lysed, and the sediment is recovered by centrifugation through a dense fluorocarbon phase. This sediment is then planted on agar medium or it may be injected into broth blood culture bottles. BHI, Inhibitory Mold agar, and SDA are media on which the sediment may be planted. The lysis and centrifugation steps concentrate the specimen and increase chances for growth of yeasts.

In practice, the Isolator appears to favor recovery from blood of dimorphic pathogens. The mean time to recover *H. capsulatum* was reduced from 24.1 days with biphasic bottles to 8 days when the Isolator system was used (reviewed in Geha and Roberts 1994). Candidemia may also be recovered at a higher rate than in continuously monitored broth systems albeit not significantly so (Cockerill et al., 1997). The importance of the Isolator as a modality is underscored in endemic areas of the primary mycoses (e.g., *Coccidioides* species, *H. capsulatum*) and also in tertiary care centers where immunocompromised patients may be vulnerable to infection with less commonly encountered molds.

Automated Systems Automated systems with continuous monitoring have improved the recovery of yeast from broth blood cultures. Fungemia is detected earlier because of continuous monitoring using a fluorescent sensor to detect CO_2 released by microbial growth. Fungemia usually is detected within 5 days from the time when the blood cultures are constituted.

Mycosis IC/F Medium and the BACTEC® System An additional advance was the availability of selective media to promote the growth of fungi (Meyer et al., 2004). The Mycosis IC/F medium introduced for the BACTEC system (BD Diagnostic Systems, Sparks, MD) is based on BHI broth augmented with sucrose and containing the antibiotics chloramphenicol and tobramycin. Saponin is added to lyse blood cells. This medium led to higher rates of recovery when compared with a standard blood culture broth medium (Plus Aerobic/F medium, BACTEC) designed to recover aerobic bacteria. More cases of fungemia were recovered using the selective medium and this was especially relevant for *Candida glabrata* fungemia, where in one study the mean detection time decreased from 61.5 to 17.8 h (Meyer et al., 2004). Both media support the growth of fungi but the selective Mycosis IC/F medium recovered some yeasts significantly sooner: *C. albicans* and *C. glabrata*. This finding is noteworthy because of the high proportion of fungemia attributed to these two species. This increased rate of recovery and total recovery seemed to depend on the ability of the selective medium to inhibit competing bacterial flora.

BacT/ALERT 3D® BacT/ALERT 3D (BioMérieux Inc., Hazelwood, MD) is another continuously automated blood culture system. A sensor in the base of the culture bottle turns yellow when CO_2 is released by metabolism of the growing microbe(s). The BacT/ALERT 3D system uses a broth that permits the growth of both bacteria and fungi. In a head-to-head comparison between that system and the BACTEC system the nonselective medium was capable of recovering common yeasts and bacteria simultaneously, resulting in cost savings. The BACTEC system with selective medium for fungi was capable of detecting more fungal clinical isolates from blood (McDonald et al., 2001).

2.3.6 Storage and Cryopreservation of Cultures for QA and QC in the Clinical Mycology Laboratory

Maintaining a culture collection is a time-consuming activity and thus cultures maintained as positive controls should be limited in number. Short-term storage of reference cultures in agar slants at 4°C for periods of 1–6 months is feasible. Such cultures should be backed up by more stable frozen storage either at −20°C or preferably at −70°C. Before setting up a culture collection or taking responsibility for an existing one, please consult the American Type Culture Collection website www.ATCC.org and the Cryopreservation Technical Manual (Simione, 2006). In it they discuss freezing at −80°C (please see below) and lyophilization.

Safety Note: The present authors do **not** recommend lyophilization (freeze-drying) of pathogenic fungi because of the potential to create an aerosol biohazard.

The ATCC recommends sterile 10% glycerol in fresh growth medium as the cryoprotective agent and storage at −80°C in a low temperature freezer.

Safety Note: Immersion of cultures in cryovials in liquid nitrogen is **not recommended** because when thawed, conversion of liquid nitrogen to nitrogen gas can cause vials to explode!

Recovery is dependent on the initial concentration of fungal cells. For best recovery $\geq 1 \times 10^7$ yeast cells is recommended. There is no particular concentration recommendation for mold spores and hyphae, approximately 10% of the total volume of 0.5 mL. The authors use, as storage vials, 2 mL Sarstedt screw cap polypropylene cryovials with rubber O-rings in the cap to prevent leakage (CryoPure® tube, No. 72.694.006, Sarstedt AG and Co., Newton, NC). A quantity of 0.5 mL of glycerol-medium-cell suspension is dispensed/vial.

What can I do if I do not have access to a low temperature freezer? Larone (2002) recommends sterile water storage at room temperature or freezing agar slants at −20°C.

2.3.7 Media and Tests for Yeast Identification

Once a fungal colony has grown on an agar medium it is examined microscopically, subcultured to assure purity, and subjected to rapid tests and other identification methods (Table 2.1).

2.3.7.1 Yeast Colonies

What does a yeast colony look like? Yeast colonies look similar to bacterial colonies. Generally, they are raised, smooth, and white to cream colored. *Cryptococcus neoformans* colonies may be mucoid. Some yeast colonies are pink-colored, for example, *Rhodotorula glutinis*, an opportunistic pathogen in the immunocompromised host.

Microscopic Appearance of a Yeast Using sterile technique, place a small loopful from the edge of the yeast colony on a microscope slide to a drop of water or mounting medium and coverslip it. Mounting medium may be lactophenol, *lactophenol cotton blue* (Porrier's blue), or *lactofuchsin*. Observe the yeast cells, especially budding cells. Note the size of the yeast cell, manner of bud attachment, size of typical daughter cell, and thickness of the cell wall. Is there a broad- or narrow-base *isthmus* between mother and daughter yeast cells?

Definitive yeast identification requires additional tests, and these are discussed below.

When Is It Necessary to Determine the Species of a Clinical Yeast Isolate? Smaller resource-limited laboratories will do a germ tube test and a urease test to differentiate *Candida albicans* and *Cryptococcus neoformans*. Informed therapy decisions also need to know if *C. glabrata* is present. Germ tube negative and urease negative yeasts should be tested at least to identify *C. glabrata* (please see below for rapid trehalase test). To ensure broad coverage a panel such as the API 20C® is advisable if resources permit (bioMérieux Inc., Durham, NC).

2.3.7.2 Rapid Tests for Yeast Identification

Once a yeast is isolated, the emphasis here is on rapid tests to aid in making the diagnosis. Rapid tests are presumptive and should be confirmed as indicated below.

Germ Tube Test Time required: 3–4 h. *Principle:* Only *C. albicans* and its genetic sister species *C. dubliniensis* produce germ tubes when incubated in serum at 37°C.

A tube containing 0.5 mL of fetal calf serum or specially designed medium is seeded with a loopful of a yeast suspected of being *C. albicans*. (Germ Tube Solution, R21066, Remel Laboratories, Lenexa, KS; Germ Tube Cryo #Z217, Hardy Diagnostics, Santa Maria, CA).

Germ tubes are produced by *C. albicans* within 3 h at 37°C. They should be read by this time and not held for a longer period. *Candida tropicalis* is operated as a negative control in order to tell the difference between germ tubes and peudohyphae: a germ tube has no constriction at the point where it emerges from the yeast cell. In contrast, a pseudohypha has a definite constriction at that site.

BactiCard *Candida*® (Remel Laboratories, Lenexa KN.) Time required: 5 min. *Principle:* This test for *C. albicans* measures *N*-acetylgalactosaminidase and L-proline aminopeptidase activity using chromogenic substrates. *Candida albicans* produces both enzymes and the colorimetric reactions are read visually. Other vendors sell similar tests.

Rapid Trehalase Test for *Candida glabrata* Time required: 20 min. *Principle:* Although trehalase is present in other yeast species, this test is based on the speed with which the *C. glabrata* enzyme hydrolyzes trehalose.

A commercial source is "*Glabrata* RTT" from Fumouze Diagnostics, Levallois Perret, France.

Twenty-five microliters of the yeast suspended in distilled water are placed in three wells containing trehalose, maltose, and sugar-free base medium as a control. After 10 min at room temperature, 25 μL of a revealing reagent containing glucose oxidase, peroxidase, and a chromogenic substrate is added to each well. Results are read after 5–10 min. An orange color in the well containing trehalose indicates the yeast is *C. glabrata*. Positive reactions in other wells indicate a yeast, not *C. glabrata* (Willinger et al., 2005).

Rapid Urease Test for *Cryptococcus neoformans* Time required: 15 min. *Principle*: Urease produced by the fungus splits urea, producing ammonia, which causes a pH increase. The pH indicator changes from pale yellow to bright magenta. Cotton-tipped medicine applicator sticks soaked in 5× Christensen's urea agar base are frozen and lyophilized. Then they are brushed over 2–3 colonies of the yeast suspected of being *Cryptococcus neoformans*. The applicator stick then is transferred to a test tube containing a few drops of 1% benzalkonium chloride (aids in releasing urease from the yeast cell) previously adjusted to pH 4.86. Swirling the applicator stick into the broth transfers the yeast growth. Then the cotton swab is removed, the tube is capped and incubated at 45°C. Ninety-nine percent of *C. neoformans* cultures tested produced a color change from colorless to purple within 15 min incubation (Zimmer and Roberts, 1979).

Christensen's Urea Agar Time required 18–24 h. This is constituted from commercial powder and may be used as broth or with 1.5% agar (Remel Laboratories, Lenexa, KS). Once seeded with unknown yeast, urea agar takes longer than the rapid urease test to indicate a color change. *Trichosporon* species and *Rhodotorula* species also produce the magenta color and are differentiated by microscopic morphology and/or colony color on standard laboratory media. (Urea agar is also used to differentiate molds such as the dermatophytes *Trichophyton rubrum* and *T. interdigitale*.)

Birdseed (Niger Seed) Agar Test for *Cryptococcus neoformans* Time required 2–24 h. *Principle*: Niger seed (*Guizotia abyssinica*) contains melanin precursors. When planted on birdseed agar *C. neoformans*, but not *Candida* species or *Cryptococcus albidus*, will produce melanin pigment causing the "brown colony effect." Other yeast species colonies produce a cream to beige color.

Birdseed agar is useful for presumptive identification of *C. neoformans* isolated from blood or CSF, and to detect a few colonies of *C. neoformans* among a larger number of *Candida* species colonies, as is sometimes observed in urine of persons living with AIDS and cryptococcosis. The time necessary to produce detectable brown color can vary from 2 to 24 h at 25°C. Birdseed agar can be obtained from Remel Laboratories, Lenexa, Kansas. The medium is simple to make.

> ***Application Note:*** Birdseed (Niger seed) medium (Paliwal and Randhawa, 1978) is prepared as follows. Seventy grams of *Guizotia abyssinica* seeds are pulverized in a Waring blender. Then they are added to 500 mL of distilled water, heated to boiling for 50 min with magnetic stirring, and filtered through cotton gauze. Chloramphenicol, final concentration of 50 μg/mL, is added from a 5 mg/mL stock solution in ethanol. The soluble filtrate is combined with 2% Bacto agar, and autoclaved for 15 min at 121°C. After cooling to 60°C, the medium is dispensed in Petri plates.

Glucose-salts media containing caffeic acid or dihidroxyphenylalanine (1 mM, D,L-DOPA) have been substituted for the birdseed. Of these the first is preferred because it is stable to ordinary daylight.

Glucose-Salts-DOPA Medium The medium contains 15 mM glucose, 10 mM $MgSO_4$, 29.4 mM KH_2PO_4, 13 mM glycine, 1 mM D,L -DOPA (Sigma-Aldrich), 3 μM thiamine, and 1.5% Bacto agar, pH 5.5. Yeasts suspected of being *Cryptococcus* species are streaked for isolation, the Petri plate is incubated in the dark at 30°C, and inspected daily for melanin pigment production (Eisenman et al., 2007).

Caffeic Acid Test Disks (Remel Diagnostics; Hardy Diagnostics, Inc.) Disks rehydrated on a microscope slide or on a nonglucose containing agar are seeded with a colony or two of a culture suspected of being *C. neoformans* or *C. gattii*. After incubation at 37°C for between 30 min and 4 h, the disk will turn brown in the case of a positive reaction.

Caffeic Acid–Cornmeal Agar *Cryptococcus* species colonies turn brown on this medium after 18–24 h incubation at 25°C, but not *Candida, Rhodotorula, Geotrichum*, or *Trichosporon* species (Kaufmann and Merz, 1982). This combination is made by mixing 900 mL of autoclaved, molten cornmeal agar (pH 6.0) (BD Diagnostic Systems, Sparks, MD) with 100 mL of 0.3% aqueous caffeic acid (autoclaved separately).

2.3.7.3 Additional and Confirmatory Tests for Yeast Identification

CHROMagarCandida® (BD Diagnostic Systems, Sparks, MD.) Time required: 24–48 h. The medium is used as both a primary culture medium and a differential and selective medium (Odds and Bernaerts, 1994). A caveat is that, used as a primary medium, recoveries may not be as complete as may be obtained with SDA-Emmons. A major advantage of this medium is its ability to demonstrate the presence of a mixed yeast infection.

How reliable is CHROMagarCandida for definite identification of Candida albicans? It is nearly 100% sensitive and specific for *C. albicans*. Exceptionally, a *C. albicans* isolate will not turn green. For that reason a more definitive identification includes a germ tube test.

Principle: Chromogenic substrates in the agar change from colorless to colored indicative of two enzyme activities: ß-*N*-acetylhexosaminidase and phosphatase: for example, colonies of *C. albicans* are green, *C. tropicalis* is blue, *C. glabrata* is dark purple (after 3–4 days). CHROMagar*Candida* is very useful in identifying mixed cultures of yeast in specimens from HIV-positive patients and other patients. Table 2.4 is a summary of colony colors observed for *Candida* species grown on CHROMagar*Candida* (Hospenthal et al., 2002).

Canavanine–Glycine–Bromothymol Blue (CGB) Agar This is a specialized agar (Kwon-Chung et al., 1982) to separate *Cryptococcus neoformans* (no growth, no color change) from *Cryptococcus gattii* (growth, color change from yellow to blue). *Principle*: *Cryptococcus gattii* (serotypes B and C) rapidly detoxifies L-canavanine and can grow in media containing this compound. Glycine (1%), the sole source of carbon and nitrogen, is utilized by *C. gattii*, increasing the pH and causing a color change.

Preparation of CGB Medium: Solution A Glycine-L-canavanine solution contains 10 g of glycine, 1 g of KH_2PO_4, 1 g of $MgSO_4$, 1 mg of thiamine-HCl, and 30 mg of L-canavanine sulfate in 100 mL of distilled water. The pH is adjusted to 5.6, and the solution is filter sterilized.

Table 2.4 Colony Color of Some *Candida* Species Grown on CHROMagar *Candida*®[a,b]

C. albicans	Green
C. glabrata	Dark violet
C. guilliermondii	Pink-lavender
C. kefyr	Pink-lavender
C. krusei	Pink, pale border
C. lusitaniae	Pink
C. parapsilosis	Pink-lavender
C. rugosa	Light blue-green, pale border
C. tropicalis	Steel blue to blue-gray, purple diffusion

[a]Data are from Hospenthal et al. (2002).

[b]Grown for 3–4 days.

Solution B This is a 0.4% aqueous solution of bromothymol blue. To prepare a liter of the medium mix 880 mL of water, 20 mL of solution B, and 20 g of agar and autoclave. Then 100 mL of solution A is added to the molten agar, cooled to 65°C, swirled, and dispensed in Petri plates or slants. A positive test with CGB medium indicates a change in pH from 5.8 (greenish yellow) to at least 7.0 (cobalt blue).

Commercial Yeast Identification Panel Kits

API 20C® (bioMérieux, Inc., Durham, NC.) *Principle:* A panel of "cupules" containing 19 carbon assimilation substrates is filled with a diluted suspension of the unknown yeast, incubated at 30°C, and then is read manually for turbidity after 24 and 48–72 h. *Candida* species, cryptococci, and other yeast are identified based on their assimilation patterns with reference to an extensive key provided by the manufacturer.

MicroScan Rapid Yeast Identification Panel® (Siemens Healthcare Diagnostics, Deerfield, IL.) *Principle*: Substrates for this test include those for 13 aminopeptidases, 3 carbohydrates, phosphatase, and urease. The panel is provided in a microtitration plate. Wells are seeded with the suspension of unknown yeast and incubated for 4 h at 37°C. After adding reagents for the aminopeptidases, reactions are read either manually or in a spectrophotometer. The accuracy of this panel was 99.5% with common yeast species and 92.1% with less common species. Accuracy is improved when the Rapid Yeast Identification panel is used in parallel with microscopic morphology on cornmeal agar (St. Germain and Beauchesne, 1991). No more recent reports were found which evaluated this system.

Peptide Nucleic Acid-Fluorescent In Situ Hybridization (PNA-FISH) This commercial test for detecting yeast in smears from blood culture bottles is discussed in Chapter 11, Section 11.12, Laboratory Detection, Recovery, and Identification.

VITEK 2 YST® (bioMérieux, Inc., Durham, NC.) *Principle*: The format is a plastic disposable card containing biochemical test substrates including 4 aminopeptidases, 25 carbohydrates, esculin, 3 glycosidases, KNO_3, 2 nitrogen, 9 organic acid, and a urease test. After inoculation with a yeast suspension the card is incubated for 18 h at 35°C. The colorimetric reactions are read in a spectrophotometer.

Tests with 24 yeast species of 6 genera were evaluated with respect to sequence analysis of the ITS2 region sequence of rDNA serving as the standard. A total of 98.2% of the panel of 750 clinical yeast isolates were correctly identified (Sanguinetti et al., 2007). VITEK 2

YST had difficulty with 2 of 258 *C. albicans* isolates, and 1 *C. glabrata* isolate.

A multicenter study tested the VITEK 2 YST card against the API20C AUX kit as the reference method (both manufactured by bioMérieux) (Hata et al., 2007). Out of 623 yeast clinical isolates the VITEK 2 YST card correctly identified 98.5% of the isolates within 18 h; 1% were incorrectly identified and 0.5% had no identification. Fewer misidentified and unidentified yeast and faster turnaround time favored the VITEK 2 YST card.

RapID Yeast Plus® (Remel Laboratories, Lenexa, KS.) *Principle*: The test format consists of 18 wells in a plastic strip. It is a 4 h identification of yeast based on enzyme technology. Included are five carbohydrates, and substrates for lipase, phosphatase, phosphorylcholine esterase, urease, histidine, leucyl-glycine aminopeptidase, ß-fucosidase, α-galactosidase α-glucosidase, ß-glucosidase, and *N*- acetyl–ß-galactosaminidase. Unknown yeasts are diluted in the provided inoculation fluid and a standard O.D. suspension is transferred into the reaction panel and incubated at 30°C for 4 h. Some wells require additional reagents upon completion of the incubation period. Using the instructions provided in the RapID Yeast Plus Code book, six-digit microcodes are constructed and used to make a species identification.

The RapID Yeast Plus system was evaluated side-by-side with the VITEK 2 YST card with respect to the ITS2 region of rDNA serving as the reference standard (see above and Sanguinetti et al., 2007). A total of 95.5% of 750 clinical isolates were correctly identified. The RapID Yeast Plus kit could not differentiate *C. dubliniensis* from *C. albicans*, and failed to identify *Trichosporon asahii* and *C. norvegensis*. The percentage of misidentifications was higher with the RapID Yeast Plus system than for the VITEK 2 system.

Blood–Glucose–Cysteine Agar This promotes mold to yeast conversion of *Histoplasma capsulatum, Blastomyces dermatitidis, Paracoccidioides brasiliensis*, and *Sporothrix schenckii*.

Cornmeal Agar (CMA) This is used for the microscopic morphology of yeasts, including production of pseudohyphae by *Candida* species and chlamydospores by *C. albicans*. Addition of Tween 80 improves chlamydospore production.

Dalmau Plate Culture A #-shaped pattern is cut into, but not through, a CMA–Tween 80 agar plate using a scant loopful of culture. A coverslip is added and microscopic morphology is observed over a 48–72 h period for the production of yeasts, *verticils* of yeast, pseudohyphae and/or chlamydospores by *Candida albicans* and *C. dubliniensis* but not by other *Candida* species (Fig. 2.1).

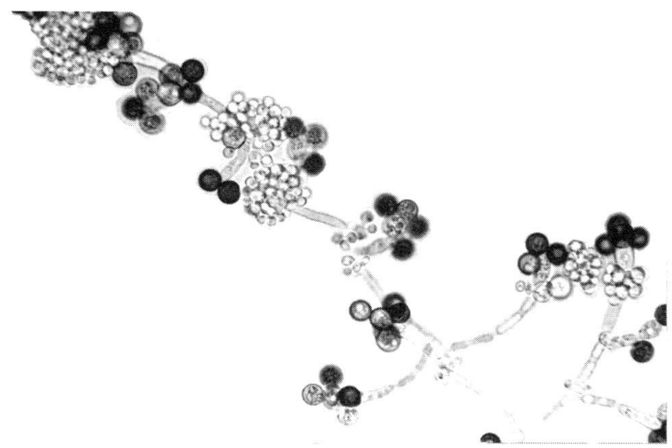

Figure 2.1 *Candida albicans* growth on cornmeal agar, 475×, showing pseudohyphae, verticils of yeast cells, and terminal chlamydospores.
Source: Dr. Libero Ajello, CDC.

The plate is observed by removing the Petri plate cover and placing the plate on the microscope stage. First starting with low power, find the field that may be at or near an edge of the coverslip. Then gradually increase the power until the morphology is in focus with the high-dry (40×) objective. The Dalmau plate culture is a necessary adjunct to the "Commercial Yeast Identification Panel Kits," please see above).

Yeast Morphology Agar (BD Diagnostic Systems, Inc.) This is a chemically defined medium with ammonium sulfate and asparagine as the source of nitrogen and glucose as the carbon source. The complete formulation is found in the Difco Manual at the URL http://www.bd.com/ds/technicalCenter/inserts/Yeast_Media.pdf. Yeast morphology agar is used in the Dalmau plate test for microscopic morphology of yeast and for special studies to stimulate asci and ascospore production.

2.3.8 Methods Useful for Mold Identification

Depending on the laboratory's capabilities, a variety of methods may be used in fungal identification. Further tests also may be performed:

- Scotch® Tape mount method to study structures, conidial attachment, and so on.
- Slide culture to study the method of conidial development. The usual media for this purpose are CMA or PDA.
- Special media and tests for mold identification (Table 2.2).
- Thermotolerance tests will differentiate fungi by their ability to grow at temperatures above 37°C (Table 2.1).

- *Trichophyton* agars 1–7 differentiate various dermatophytes by their ability to grow with only certain growth factors as substrate. (Please see Chapter 21.)

2.3.8.1 Colony Morphology

Descriptions of colony morphology are made from 1–3 week old cultures grown on SDA or PDA. SDA-Emmons is used in many laboratories because it encourages sporulation. Gross observation of colonies includes pigmentation (both top and reverse), rate of growth, texture of colony including topography (is the periphery regular or irregular, flat, heaped, or folded), and texture (please see list below). If the colony is granular, even if only in a small area, this area is to be studied microscopically because of the usual presence of conidia. Record observations of the colony and microscopic morphology.

The following terms are used to describe colony texture:

- Smooth (syn: glabrous)—yeast-like, pasty, creamy
- Downy—covered with short fine hyphae
- Velvety—resembling velvet
- Cottony (syn: floccose)
- Powdery
- Granular
- Woolly (syn: lanate)

2.3.8.2 Microscopic Morphology

Yeast Colonies Please see Section 2.37, Media and Tests for Yeast Identification. If isolated at 37°C, the yeast form of a dimorphic pathogen may be involved. In that case subculture at room temperature should reveal the mold form.

Mold Colonies Direct microscopy, culture, and, if necessary, slide cultures are standard methods usually sufficient to identify the causal mold when the fungal isolate demonstrates typical conidia and conidiophores. Importantly, mold colonies are observed early in their development and, if necessary, after more growth has occurred and the colony has further matured.

2.3.8.3 Tease Mount

Safety Note: Conduct tease mounts, Scotch tape mounts, and slide cultures in an accredited and well-maintained biological safety cabinet.

The primary method of investigating mold colonies is the tease mount. A small drop of *lactophenol cotton blue* (Porrier's blue) or *lactofuchsin* mounting medium is placed on a microscope slide. Wear latex or other type of protective gloves and, using sterile technique, remove a small amount of mycelium from the culture with an inoculating needle and place it in the mounting medium. Gently tease the mycelium apart with inoculating needles. In a well-developed colony a tease preparation should be made from both the edge and an area closer to the mature center, where typical conidia may be found. Coverslip the prep and observe in a microscope.

Note whether the hyphae are hyaline or brown-colored. Are the hyphae septate or sparsely septate (= "pauciseptate") and are conidia present? If so, what is their size, shape, method of attachment? Are they borne singly, in chains, clusters, or clumps? (See Fig.2.2.) Conidiophores are usually single and isolated but they may form columnar aggregates called *synnemata* or *coremia* (e.g., *Graphium* form of *Pseudallescheria boydii*) or macroscopic cushion-shaped masses called *sporodochia* (e.g., *Fusarium* species). These latter structures are uncommonly seen in the clinical laboratory.

If no conidiation is seen, cultures should be subcultured to conidia-enhancing media such as CMA, PDA, or Czapek–Dox Agar. Please see Section 2.3.8.9, Non-sporulating Cultures. If conidia are present but do not help in identification of the fungus, special tests or media may be necessary and will be discussed in each disease chapter. Other aids in morphologic identification are listed in this chapter in Section 2.3.5.1, Additional Resources. When the clinical laboratory is unable to resolve the identification, the problematic isolate should be referred to a reference laboratory for morphologic or genetic identification.

2.3.8.4 Scotch Tape Mount (or Other Adhesive Cellophane Tape) Method

Two things must be remembered: (i) the final slide will last only a short time (hours to days) before drying and wrinkling, and (ii) not all cellophane tape is useful for this purpose. New rolls of tape must be tested for usability.

This method may be used on Petri plates where colonies are **known not to be** *H. capsulatum, B. dermatitidis, P. brasiliensis, Coccidioides* species, or *Penicillium marneffei*.

If the colony to be studied is dark-pigmented (i.e., melanized), a clear lactophenol solution is used. If the colony is *hyaline*, a drop of lactophenol-cotton blue or lactofuchsin is placed on a microscope slide. Wearing latex or other type gloves, hold an approximately 3 cm length of tape between two fingers with the sticky side out, and without touching the colony with the gloved fingers, gently place the tape on the top of the colony. A portion of colony will adhere to the tape. Place the tape, sticky side down, on the slide, covering the drop of mounting medium. Carefully smooth out the tape and press both ends of the tape to the slide.

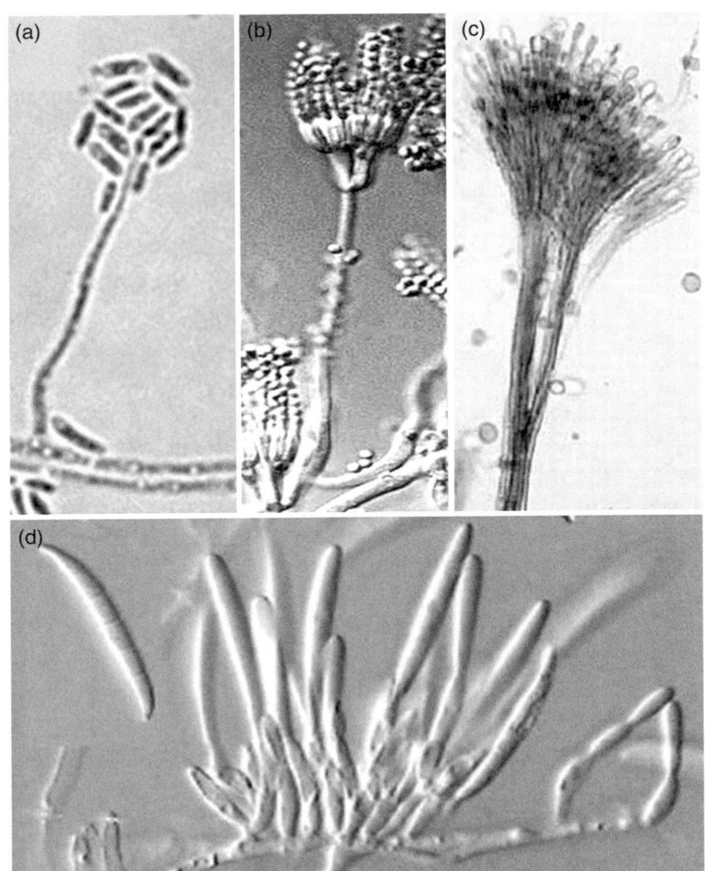

Figure 2.2 Types of conidiophores and conidiogenous cells: simple, complex, synnema, sporodochium. (*Panel a*) Simple, *Acremonium* species. *Source:* E. Reiss. (*Panel b*) Complex, *Penicillium* species. *Source:* Used with permission from Dr. George Barron, Department of Environmental Biology, Ontario Agricultural College, University of Guelph, Ontario, Canada. (*Panel c*) Synnema. Note the bundle of hyphae. *Graphium* synanamorph of *Scedosporium apiospermum*. *Source:* Used with permission from Subhash Mohan, MLT, Mycology Laboratory University Health Network and Mt. Sinai Hospital, Toronto, Ontario, Canada. (*Panel d*) Sporodochium of *Fusarium graminearum*.
Source: Used with permission from Dr. Keith Seifert, Department of Agriculture and Agri-Food, Government of Canada, © 2003, Minister of Public Works and Government Services Canada.

For a permanent slide, the method of Rodriguez-Tudela and Aviles (1991) may be used. A device (Pelifix® adhesive roller, Pelikan International Corp.) dispenses a thin layer of adhesive to the surface of a coverslip. This may be pressed against a colony, then placed on a small drop of lactophenol on a slide. Clear nail lacquer is placed on two edges of the coverslip to hold it firmly. When the lacquer is dry, the slide may be studied.

2.3.8.5 Slide Cultures

At times it is necessary to make a slide culture to demonstrate the relationship between conidium and conidiophore when this cannot be demonstrated by a tease mount. A slide culture clarifies the details of conidia formation. Slide cultures should **never** be made before isolation of the fungus and its study using a tease mount. This will prevent the slide culture preparation of primary systemic fungal pathogens and reduce the risk of exposure to *Blastomyces dermatitidis, Histoplasma capsulatum, Coccidioides* species, *Paracoccidioides brasiliensis*, or *Penicillium marneffei*.

Reagents for Slide Cultures

Lactophenol ***Principle***: The phenol serves to kill the fungus. Glycerol prevents rapid drying of the specimen.

Preparation: Phenol crystals, 20 g (or concentrated liquefied phenol, 20 mL); lactic acid, 20 mL; glycerol, 40 mL, distilled water, 20 mL. Mix phenol crystals or liquefied phenol in water and add glycerol, lactic acid. Stir to dissolve. Use gentle heat (i.e., 37°C) if necessary.

> ***Safety Note:*** **Protect eyes and skin: phenol will burn skin**. Store away from direct light.

Lactophenol Cotton Blue (LPCB) Add 50 mg of cotton blue (syn: Porrier's blue, aniline blue) to the lactophenol reagent, mix, and let stand to dissolve.

Lactophenol Cotton Blue with Polyvinyl Alcohol Please consult Larone (2002) for its preparation. This mounting medium gives a permanent mount, whereas LPCB gradually dries.

Lactofuchsin This mounting medium consists of acid fuchsin, 100 mg dissolved in 100 mL of lactic acid.

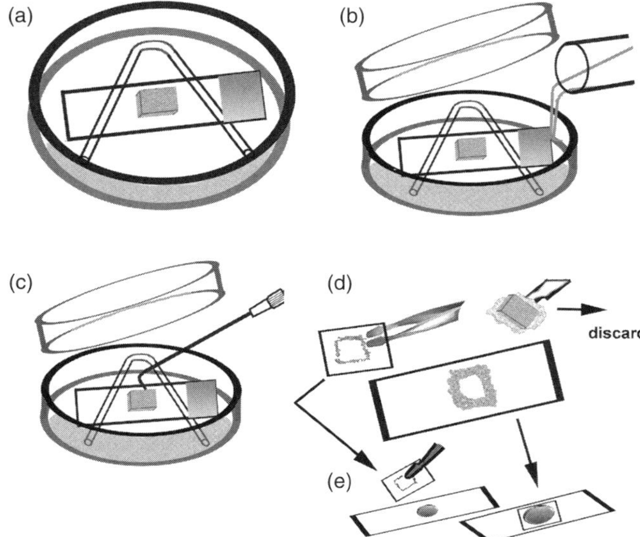

Figure 2.3 The slide culture method. Please see text for step-by-step instructions.
Source: E. Reiss.

Preparation of a Slide Culture Slide cultures should be made using CMA or PDA to enhance the early production of conidia. These media are prepared in Petri plates. All glassware is presterilized (Fig. 2.3).

Method A A sterile Petri plate is prepared with a V-curved piece of sterile glass rod or tube on the bottom. A microscope slide is placed on the glass rod or glass tube (Fig. 2.3A); 22 mm^2 sized coverslips are sterilized at the same time, separating each with a piece of paper.

- Add 8–10 mL sterile water to the Petri plate (Fig. 2.3B).
- Using a sterile scalpel, cut 1 cm squares of agar medium. Place a square near the center of the microscope slide.
- Inoculate the center of all sides of the agar square with an inoculating needle. Add a sterile coverslip, gently tapping the top to ensure contact with the medium. Replace the lid of the Petri plate (Fig. 2.3C) and incubate at 30°C.
- When growth appears at the upper edge of the square, remove the coverslip. Now you have a coverslip with the outline of growth on it. A fresh coverslip may be added if more time is needed to achieve sporulation. If not, discard the agar block in the septic waste container (Fig. 2.3D).
- Add a drop of lactophenol, LPCB, or lactofuchsin to a new microscope slide. (Optionally, add a drop of absolute ethanol to the specimen on the coverslip before placing it on the slide. As the ethanol evaporates, the lactophenol will replace it, and few or no bubbles will form.)
- Place the coverslip on the drop of mounting medium with a forceps, growth side down (Fig. 2.3E). Position one edge of the coverslip against the stain and carefully lower it over the slide. Gently press down. This will minimize air bubbles in the specimen.
- Remove excess mounting fluid. A small drop of clear nail lacquer added to two sides of the coverslip will keep it from moving and will allow space to add more fluid to the specimen later as it dries.
- Optionally, the microscope slide used to make the slide culture may also be preserved by adding a drop of mounting fluid and coverslipping it (Fig. 2.3E).
- Study the slide culture preps microscopically for conidia and conidiophores.

Method B This method utilizes a Petri plate with a base of sterile medium on which the slide culture will be performed. The base medium may be any standard medium, for example, SDA, CMA, or PDA. The medium used for the slide culture itself is either CMA or PDA.

- Using a sterile scalpel, cut 1 cm squares of medium for the slide culture.
- Place one or several squares of medium on the agar base, far enough from the edge of the Petri plate to allow microscopic observation. Inoculate all sides of each agar square, add a coverslip, and incubate at 30°C.
- When conidia appear on the upper edge of a block, carefully remove the coverslip and treat it above as in Method A. A fresh coverslip may be added if more time is needed to achieve sporulation.

Direct microscopy, culture, and, if necessary, slide cultures are standard methods usually sufficient to identify the causal mold when the fungal isolate demonstrates typical conidia and conidiophores. Definitive yeast identification requires additional assimilation tests; please see Section 2.3.7, Media and Tests for Yeast Identification.

Additional tests are brought to bear when microscopic morphology is not revealing and/or the assimilation test results are ambiguous (please see Section 2.4, Genetic Identification of Fungi). Another situation calling for genetic identification is when the fungus is slow growing and/or the clinical situation is acute.

2.3.8.6 Sporulation in fungi: A Major Characteristic Used to Identify Species in the Clinical Laboratory

Fungal propagules are different types of spores, a means of asexual reproduction. Mycologists refer to asexual spores

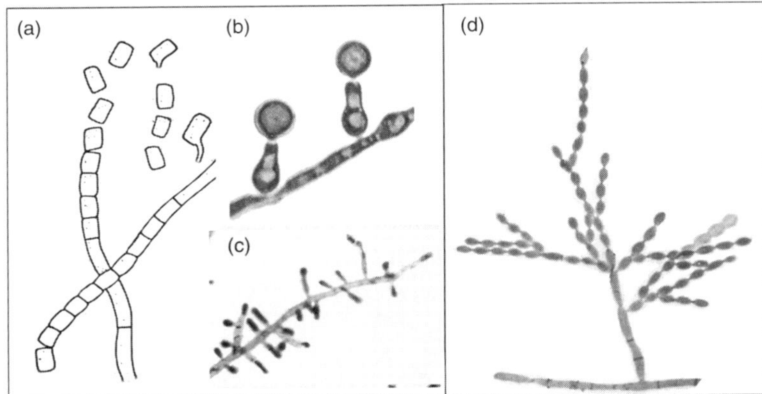

Figure 2.4 Thallic and holoblastic conidiogenesis. (A–C) Thallic conidiogenesis. (A) Arthroconidia (*Geotrichum* species). *Source:* H. J. Shadomy. (B) Chlamydospores (chlamydoconidia). These may be terminal as above in *Candida albicans*, or *intercalary*. *Source:* E. Reiss. (C) Aleurioconidia from *Trichophyton tonsurans*, some display the "matchstick" formation. *Source:* H. J. Shadomy. (D) Holoblastic conidiogenesis, *Cladophialophora carrionii*. Conidia are produced in an acropetal manner (with the youngest conidium at the tip and the oldest at the base). *Source:* H. J. Shadomy.

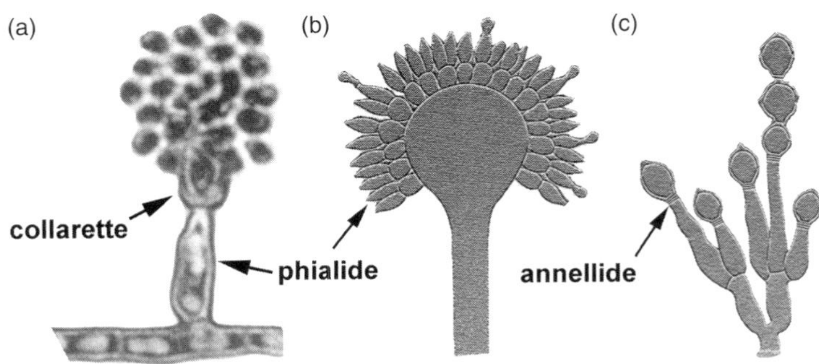

Figure 2.5 Enteroblastic conidiogenesis. Conidia are produced within the conidiogenous cell and released singly in a basipetal manner so that the youngest cell is nearest the conidiogenous cell. (A) *Phialophora verrucosa*. Note the conidium is emergent within the phialide. (B) *Aspergillus* species. Here the phialides are borne directly on the bulbous vesicle, or are joined to it by an intermediary cell, the metula. (C) *Scopulariopsis* species. Annellides are characterized by their method of synthesis which leaves a fragment of the conidiogenous cell wall at the tip each time a new conidium is released to the exterior. The accumulated cell wall fragments result in a narrowing of the conidiogenous cell and an accumulation of ringlets. *Source:* H. J. Shadomy.

of *Ascomycota* as conidia whereas those of *Mucoromycotina* are properly referred to as spores. Sporulation can be induced in the laboratory on mycologic agar media. Their characteristics are used to identify fungi, along with colony morphology, segmentation and pigmentation of hyphae, and biochemical and other tests. (Refer to Safety Guidelines in the BMBL[3] when working with sporulating fungi.) The name given to each type of conidium is determined by the manner in which the propagule is produced (conidiogenesis).

2.3.8.7 Common Types of Asexual Sporulation Seen in the Clinical Laboratory and Generally Termed a Type of Conidium (or Spore)

Conidia are termed either (i) *blastic*, that is, formed by enlargement of the parent cell, division by a septum to form a daughter cell, and its initial development prior to separation from the parent cell or (ii) *thallic*, that is, division of a hyphal strand by septal formation into single cell units that then become the conidia (e.g., *Coccidioides immitis*) (Fig. 2.4).

Holoblastic The cell wall of the new conidium is contiguous with the wall of the conidiophore. The new conidium breaks through to the exterior of the cell. This conidium produces the second, and that the third, and so on, making the farthest from the conidiophore the youngest (*acropetal*): for example, budding in yeast; sporulation in *Cladophialophora carrionii*, an agent of chromoblastomycosis (Fig. 2.4D).

Enteroblastic The new conidium is formed within the conidiogenous cell, which may be an annellide (Fig. 2.5C) or a *phialide* (Fig. 2.5A, 2.5B). The cell wall of the new conidium is independent of the conidiogenous cell wall. New conidia are produced sequentially and are extruded to the exterior through an opening in the conidiogenous

[3] *Biosafety in Microbiological and Biomedical Laboratories* (BMBL), 5th edition. Available at the URL http://www.cdc.gov/od/ohs/biosfty/bmbl5/bmbl5toc.htm.

cell, making the closest to the conidiophores the youngest (*basipetal*): for example, phialidic—*Aspergillus fumigatus, Phialophora verrucosa*; annellidic—*Scedosporium* species.

2.3.8.8 Microscopic Identification of Molds Based on the Specific Conidia-Forming Structures

Identification of fungi can seem difficult because of the many different conidia-forming structures and the terminology used to define them. Here, we list and describe these structures as they occur in a mature and developing colony.

How can I use these terms to better identify fungi in the clinical laboratory? In this text the microscopic morphology of the major and emerging fungal pathogens is illustrated in each disease-specific chapter. This information is provided in the sections "Etiologic Agent(s)" and "Laboratory Detection, Recovery, and Identification." There, along with an illustration of the common pathogens, will be found (i) the type of spore-forming structure as described below and, (ii) the position of the fungus in the overall classification of fungi.

Aleurioconidia An aleurioconidium is a simple terminal or lateral conidium that develops as an expanded end of an undifferentiated hypha, or on a short stalk (pedicel). It is a form of thallic conidiogenesis. The aleurioconidium detaches by lysis or fracture of the wall of the supporting cell. It usually is recognized by its truncate base: for example, macroconidia and microconidia of dermatophytes (Fig. 2.4C).

Annelloconidia This type of conidium arises from an annellide, either a vase-shaped or tube-shaped cell. The first conidium develops directly from the end of the annellide, bursting through the tip (*holoblastic*). Subsequent conidia are formed through the opening in a *basipetal* manner, forming their own cell walls (*enteroblastic*), and leaving an annular ring at the tip of the annellide as they emerge (Fig. 2.5C). The annellide becomes more slender with each conidium produced, leaving a series of annular rings. The oil immersion lens and phase-contrast microscopy are most often necessary to see the annular rings, which are difficult to observe using transmitted light microscopy and the 40× dry objective.

Arthroconidia These are a *thallic* form of conidiogenesis in which conidia are produced by the fragmentation of a hyphal strand by means of double septal formation along the hyphae, and lysis of the material between the septa. The arthroconidia may be either adjacent to each other or may have empty hyphal cells between each or several conidia (i.e., intercalary arthroconidia). The conidia may be either rectangular or barrel-shaped (Fig. 2.4A).

Blastoconidia This is a synonym for yeasts, that is, vegetative reproduction by budding. Small round *holoblastic* projections from the parent cell are produced during mitosis resulting in a daughter cell termed a bud or *blastoconidium*. The parent cell may produce single or multiple buds. Buds may be either solitary or in *acrogenous* chains. Blastoconidia may occur along pseudohyphae or hyphae (Figs. 1.3 and 2.1).

Chlamydospore (syn: Chlamydoconidium) This type of conidium is an example of thallic development: an enlarged, thick walled cell found either terminally or on a side branch of a hypha, or intercalary (within the hypha). It is believed to be a survival propagule but its precise function is unknown (Figs. 2.1 and 2.4B).

Macroconidium The larger of two asexual propagules of different sizes produced in the same manner, in the same colony, by a single fungus is termed a macroconidium. They may be single-celled or, more often, multicellular, thick- or thin-walled, smooth or rough, single or in clusters, and club, spindle, or oval in shape (Fig. 2.6).

Microconidium Small asexually produced conidia formed on aerial hyphae, or the smaller of two asexual propagules, are termed microconidia.

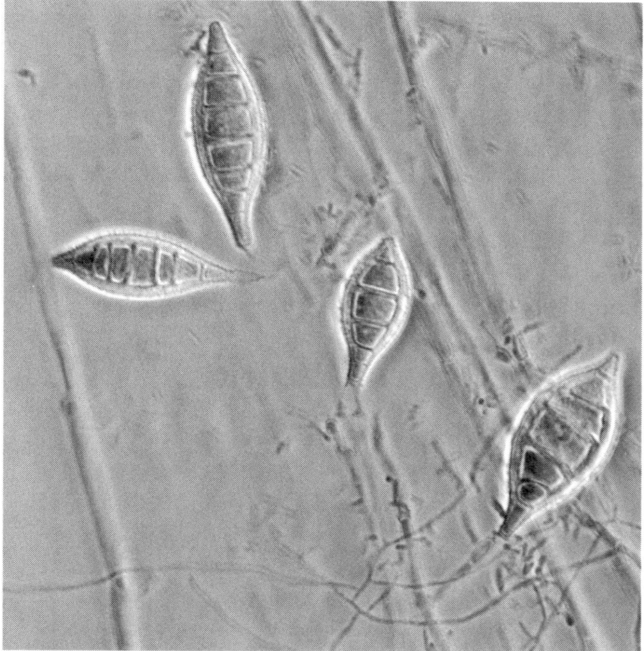

Figure 2.6 Macroconidia of *Microsporum canis*, 400× phase. *Source:* E. Reiss, CDC.

Phialoconidia Conidia produced within a phialide (a tube, or more often, a vase-shaped cell). The first conidium is *holoblastic*, the subsequent conidia are *enteroblastic* and form below the previous conidium (*basipetal*). The cell wall arises de novo rather than from the phialide cell wall. As a result the phialides does not lengthen as the number of conidia produced increases, and there are no annellations in the conidiogenous cell. Conidia often adhere to the phialide as balls held together by polysaccharide (*Phialophora verrucosa*), or they may form easily dispersed dry chains (*Aspergillus fumigatus*) (Fig. 2.5A, 2.5B).

Poroconidia Conidia produced through a minute pore in the wall of the conidiophore or conidiogenous cell wall are poroconidia. Often seen in *Alternaria alternata, Bipolaris* species, *Curvularia lunata*, and *Exserohilum* species, all are members of the *Pleosporaceae* characterized by their production of large multicelled *poroconidia* (Fig. 2.7).

Sporangiospore Asexual propagules of the subphylum *Mucoromycotina* are produced within a specialized sac, the *sporangium*, and are termed sporangiospores. They are freed by *deliquescence* of the sporangium. The sporangium is formed at the apex of a stalk-like structure termed a *sporangiophore* (Fig. 2.8).

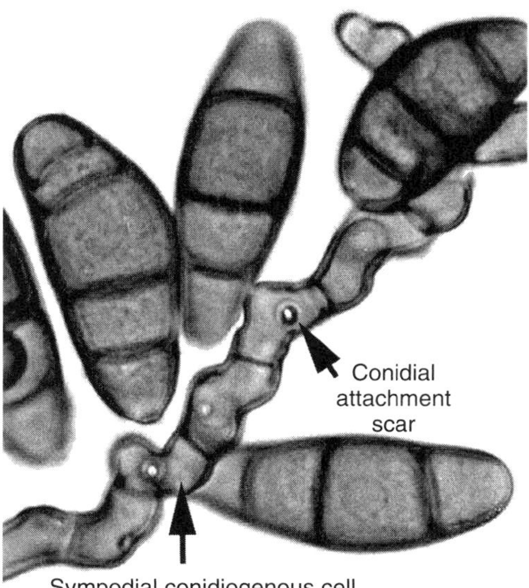

Figure 2.7 Poroconidia of *Curvularia lunata*. Note the geniculate ("knee"-like) conidiophore, and the pore left behind after the poroconidium is released. Each time a conidium is produced, the conidiophore grows around it in a *sympodial* pattern. In this species the central conidium is larger and may be darker than those adjacent to it.
Source: Used with permission from Dr. George Barron, Department of Environmental Biology, Ontario Agricultural College, University of Guelph, Ontario, Canada.

Some Other Terms Associated with Asexual Reproductive Structures The terms in Table 2.5 are defined in Part 7, Glossary.

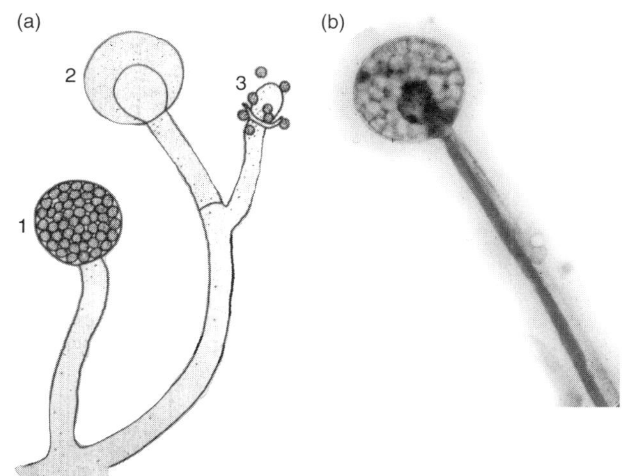

Figure 2.8 *Mucorales* sporangium with sporangiospores. (A) Line drawing shows (1) sporangium with sporangiospores, (2) cross section of sporangium shows columella; and (3) sporangium has dissolved releasing sporangiospores and leaving a collarette remnant of the sporangial wall. (B) Photomicrograph of an intact *Mucorales* sporangium (Wilson and Plunkett, 1965).

Table 2.5 Other Terms Associated with Asexual Reproduction in Fungi[a]

Acropetal
Annellide
Ballistospore
Basipetal
Biseriate
Collarettes
Columella
Conidiophore
Denticle
Disjunctor
Geniculate
Metula
Muriform cell
Phialide
Sporangiophore
Stipe
Sympodial
Vesicle

[a]Definitions of these terms are found in Part 7, Glossary.

2.3.8.9 Nonsporulating Cultures

Occasionally molds will fail to sporulate on the common laboratory media. PDA and cornmeal agar are generally available and sporulation should be attempted on these media. If there is a delay in obtaining sporulation the situation may call for expediting PCR-sequencing and genetic identification (please see Section 2.4, Genetic Identification of Fungi).

Media Specialized for Promoting Sporulation Regional reference laboratories may desire to encourage sporulation through the use of "homebrew" media that have been successful for this purpose. Formulas for these media are available from the website of the University of Alberta Microfungus Collection and Herbarium, Prof. Lynne Sigler, at the URL http://www.devonian2.ualberta.ca/uamh/culture_receipt.htm.

The following media are recommended to promote sporulation (specific indications are noted):

- Cereal agar: many fungi
- Modified Leonian's agar: Ascomycetes
- Oatmeal-salts agar: sexual reproduction in the *Onygenales*
- Phytone yeast extract agar: dermatophytes
- Takashio's agar: *Onygenales*.
- V8 Juice agar: melanized molds, yeasts, and Ascomycetes

Media Containing Plant Material A Petri plate is lined with sterile filter paper moistened with sterile tap water (a good source of trace elements). Because many environmental or "contaminant" fungi grow on plant material, add sterile wood medicine applicator sticks, or slices of carrot or potato, plant leaves, or flower petals. Inoculate and incubate at ambient temperature until growth appears. Carnation-leaf medium is recommended for induction of sporulation in *Fusarium* species. Carnation-leaf pieces, sterilized with gamma irradiation, in a water agar medium (carnation-leaf agar) promote good growth, sporulation, and maintenance of *Fusarium* species (Nelson et al., 1983).

Photoinduction of Sporulation Young colonies (3–4 days old) grown in glass Petri plates are exposed to full sunlight or light from a cool-white daylight fluorescent lamp, then replaced in dark. (Cool-white emits in the near UV between 310 and 410 nm). The period of exposure to fluorescent light is a variable from 15 to 20 min or longer) (McGinnis, 1980; Dahlberg and Etten, 1982).

Septum Formation It may seem elementary, but basic information can be obtained from microscopic examination of hyphae: lack of septum formation indicates a member of the subphylum *Mucoromycotina*. Septate hyphae with clamp connections indicate a basidiomycete (e.g., *Schizophyllum commune*). This fungus occurs in dikaryotic form with clamp connections and is capable of producing fruiting bodies (basidiocarps). The monokaryotic form may lack clamp connections and is unable to form fruiting bodies unless mated with single basidospores from a standard tester strain (Sigler et al., 1999).

2.3.8.10 Specialized Tests for Molds

(Please also see Table 2.2.)

Temperature tolerance tests These tests are useful for species shown in Table 2.3.

Benomyl Selective Medium Benomyl (Summerbell 1993), an agricultural fungicide, may be used in SDA or modified Leonian's agar as a selective medium that will facilitate growth of certain fungi: *these fungi are resistant to Benomyl at 10 μg/mL*. Isolates of *A. fumigatus* and *Fusarium* species show a range of susceptibility from susceptible, partially inhibited, to resistant (Table 2.6). The formula for modified Leonian's agar may be found at the URL http://www.devonian2.ualberta.ca/uamh/culture_receipt.htm.

Fungi Inhibited by Cycloheximide Addition of cycloheximide to mycologic media such as SDA (with cycloheximide), Mycosel™, and Mycobiotic agars is designed to inhibit saprobic fungi and allow the growth

Table 2.6 Benomyl Selective Medium

Phylum or subphylum	Genus/species resistant to Benomyl at 10 μg/mL
Mucoromycotina	Most members of the *Mucorales*
Ascomycota	Members of the *Pleosporaceae*: *Alternaria, Bipolaris, Curvularia, Exserohilum*
	Members of the *Microascaceae: Pseudallescheria boydii, Scedosporium* species
	Scopulariopsis
	Candida species
Basidiomycota	*Cryptococcus neoformans*.
	Molds with sterile mycelia where there may be no clamp connections and arthroconidia may grow

Table 2.7 Fungi Inhibited by Cycloheximide

Inhibited by cycloheximide	Examples of cycloheximide sensitive species	Comment
Aspergillus species	Especially *A. fumigatus*	
Cryptococci	*Cryptococcus neoformans* and other *Cryptococcus* species	
Mucorales	*Rhizopus oryzae*	
Most *Candida* species	*C. glabrata, C. krusei, C. tropicalis, C. parapsilosis*[a]	*C. albicans* and *C. dubliniensis* are resistant
Penicillium species	*P. marneffei*	
Scedosporium apiospermum		
Scopulariopsis brevicaulis		Many fungal opportunists are inhibited by cycloheximide
Yeast forms (spherule form) of dimorphic fungi	*Blastomyces dermatitidis, Coccidioides* species, *Histoplasma capsulatum, Paracoccidioides brasiliensis*	No inhibition of mold forms growing at 25–30°C

[a]A more complete list of *Candida* species inhibited by cycloheximide may be found in Larone (2002).

of many fungal pathogens that are resistant to cyclohex-imide, especially those discussed in Part two. Table 2.7 shows a selection of the opportunistic fungal pathogens that are susceptible to cycloheximide and will not grow on media containing it.

2.3.8.11 Saprobic Fungi and Their Pathogenic Fungal Look-alikes in Culture

At times, fungi isolated in the clinical laboratory will resemble known pathogens. These may be saprobes or opportunists but must be differentiated from the frank pathogen. Until their identification is known with certainty the look-alikes should be handled with the precaurtions accorded to the pathogen they resemble. (See Table 2.8.)

2.3.8.12 Common Opportunistic Environmental Molds and Saprobic Fungi

These fungi are occasionally present as contaminants in clinical specimens. The list is endless and Table 2.9 is but a glimpse at the variety of what may be encountered. Several handbooks, atlases, and guides are listed in Chapter 1, "General References in Medical Mycology." These will provide keys and images for morphologic identification of the less common species. The current text provides images of the more frequently encountered molds, and the tools and techniques with which to identify most any unknown fungus. Any fungus capable of growth at 37°C is a potential pathogen in the compromised or debilitated host.

2.3.8.13 Rhinosporidium seeberi — A Protist Previously Classed in the Kingdom Fungi

Rhinosporidium seeberi is a noncultivatable protistan parasite classed in the *Mesomycetozoea*, positioned between the fungi and the metazoa (Mendoza et al., 2002). It is the causative agent of rhinosporidiosis of humans and lower animals characterized by granulomatous polyps on the mucous membranes in the nose, the conjunctiva, and less commonly on other body sites. Rhinosporidiosis is endemic in India and Sri Lanka, and Southeast Asia and may occur in the tropical regions of the Americas and in Africa. Sporadic cases are reported from Europe and other temperate regions. The tissue form of *R. seeberi* consists of endosporulating sporangia (Chandler and Watts, 1987). Fungi that may be considered in the differential histopathologic diagnosis of *R. seeberi* are *Emmonsia crescens* and *Coccidioides* species. The adiaspore walls of the *Emmonsia crescens* are much thicker than those of *R. seeberi* and do not contain endospores. Differentiation from *Coccidioides* species is made by the size of the sporangia: 20–80 μm for *Coccidioides* spherules versus 50–1000 μm for *R. seeberi* sporangia. The endospores of *Coccidioides* are 2–4 μm in diameter. Those of *R. seeberi* are larger (5–10 μm diameter) and more numerous.

2.3.9 Microscopy Basics

A microscope with a 400× magnification resolves the diagnostic features of medically important fungi (40x high-dry objective and 10x ocular). When specimens are nonpigmented, the specimen is mounted in lactophenol-cotton blue (LPCB) or similar mounting agent such as lactofuchsin. Occasionally, they may be mounted in water to observe certain structures. When specimens are pigmented they may be mounted in lactophenol without added stain. Certain small or delicate structures (e.g., annellations, phialides, or phialide collarettes) often require observation under oil immersion and/or the use of phase contrast microscopy or differential

Table 2.8 Saprobic Fungi and Their Pathogenic Fungal Look-alikes in Culture

Pathogen	Saprobic look-alike	Microscopic appearance of saprobe, 30°C	Differential
Blastomyces dermatitidis[a]	*Chrysosporium* species[b]	Single celled, smooth to rough *aleurioconidia* are *sessile* or on tips or sides of short or long stalks that may be branched. Conidia are *pyriform*, *clavate*, or *obovoid*. Intercalary arthroconidia in a barrel shape.	Gen-Probe test is diagnostic; *B. dermatitidis* is dimorphic.
			Microconidia (aleurioconidia) of *B. dermatitidis* are pyriform, or nearly round (subglobose). Some isolates produce no conidia.
Histoplasma capsulatum[a]	*Chrysosporium* species[b]	See entry under *B. dermatitidis* (above). *Chrysosporium* anamorph of *Renispora flavissima* produces *sessile* tuberculate conidia.	*Histoplasma capsulatum* produces microconidia *and* tuberculate macroconidia; *H. capsulatum* is *dimorphic*. The AccuProbe® (Gen-Probe Inc.) test is diagnostic for *H. capsulatum*.
	Sepedonium species	Growth is rapid, wooly, white to golden yellow. Large tuberculate macroconidia are produced. Some species produce alternating arthroconidia.	
	Myceliophthora thermophila[b]	Obovoid or pyriform conidia with rough walls borne sessile or on short swollen cells. Dimensions of about 5 μm × 4 μm. Grows up to 50°C.	Smaller than macroconidia of *H. capsulatum*, but roughened and thick walls may be mistaken for *H. capsulatum* macroconidia.
Fonsecaea pedrosoi	*Cladosporium*	Colony: growth slow to moderate with a velvety surface. Microscopic: Erect, melanized, branching conidiophores ending in shield-shaped cells that branch and develop *acropetal* chains of lemon-shaped conidia. Saprobic species conidia dislodge easily.	*Fonsecaea pedrosoi* produces three types of conidia: *Cladosporium* type, *Rhinocladiella* type, and *Phialophora* type[c]
	Rhinocladiella	Colony: rapid growing velvety surface. Microscopic: conidiophres are cylindrical, brown, unbranched. Conidia are ellipsoid, club-shaped borne on denticles closely spaced under the apex of conidiophore.	

Coccidioides species	*Malbranchea aurantiaca*	Growth is relatively rapid with undifferentiated hyphae producing arthroconidia with alternating disjunctor cells resembling *Coccidioides* species.	*Malbranchea* does not give a positive reaction in the *Coccidioides* Gen-Probe nucleic acid probe test.
	Geotrichum candidum	Growth is relatively rapid with yeast-like colony producing arthroconidia.	The arthroconidia of *Geotrichum* do not have alternating disjunctor cells.
	Neoscytalidium dimidiatum	Not a common saprobe, but may cause nail or skin infections and subcutaneous abscesses. Growth is rapid, producing narrow hyaline arthroconidia, and wider melanized arthroconidia, rectangular or barrel shaped.	The arthroconidia are melanized, and unlike *Coccidioides*, there are no *disjunctor* cells. *Pycnidia* are also produced. Hyaline mutant strains occur. See Fig. 22.4.
	Trichsporon asahii and related spp.	The cream-colored yeast-like colony becomes wrinkled and folded. The growth consists of arthroconidia. Budding cells may be present; pseudohyphae and true hyphae.	No disjunctor cells in *Trichosporon* spp.; *Coccididoides* does not produce budding cells.
Sporothrix schenckii	*Ophiostoma* (*Ceratocystis*) *stenoceras*	A relatively rapid growing hyaline mold producing similar oval conidia on small, cylindrical *denticles* in a *sympodial* arrangement that is flower-like.	*Ophiostoma* does not convert to a yeast form at 37°C. It does not produce melanized conidia. It is homothallic and may produce long necked perithecia after 2–3 weeks incubation.
Penicillium marneffei	Saprobic *Penicillium* species	Rapid growing molds with colony and microscopic morphology similar to the pathogen.	Saprobic *Penicillium* do not convert to fission arthroconidia at 37°C.
Trichophyton species isolates producing only microconidia	Please see Chapter 22, Section 22.4, *Chrysosporium* and Other Nonpathogenic or Opportunistic Fungi Isolated from Skin and Resembling Dermatophytes in Culture.		

[a] *Blastomyces dermatitidis* may also appear similar to *H. capsulatum* strains, which produce only microconidia.

[b] Sigler (1997).

[c] Please see Chapters 18 and 19 for illustrative examples of melanized molds.

Table 2.9 Common Environmental Opportunistic Molds and Saprobic Fungi Occasionally Present as Contaminants in Clinical Specimens

Species	Colony morphology[a] (SDA, 25–30°C)	Microscopic morphology (SDA, 30°C)	Microscopic morphology
		Hyaline molds	
Acremonium	Smooth to floccose, gray to pink to orange	Balls of ellipsoid, closely packed conidia borne at ends of long tapering phialides.	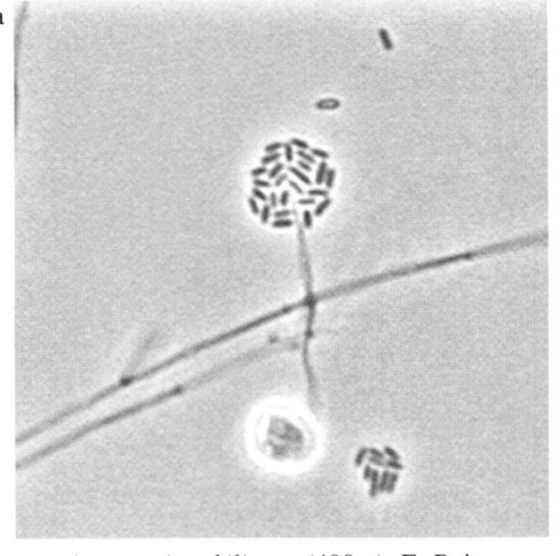 *Acremonium kiliense* (400×). E. Reiss.
Aspergillus	*Rugose*, rapid growing, variously colored depending on species[b] (*Aspergillus fumigatus* is blue-green with a white border)	Rough or smooth conidiophore with a foot cell at its base, terminating in a vesicle covered in part or completely by phialides producing phialoconidia. May be *monoseriate* or *biseriate* depending on the species.	Please see illustrations in Chapter 14.
Fusarium	Wooly to cottony, moderate–rapid growing, white, later may become yellow or lavender	Conidiophores single or branched and terminating in tapering phialides bearing characteristic sickle-shaped two- to three-celled macroconidia. One-celled microconidia in slimy heads or chains.	Please see illustrations in Chapter 15.
Paecilomyces	Powdery to cottony, white becoming lavender to olivaceous	Irregularly branched, elongated, and beaked conidiophores bearing chains of oval, hyaline to pigmented conidia.	 *Paecilomyces lilacinus* (400×). E. Reiss.

Table 2.9 (*Continued*)

Species	Colony morphology[a] (SDA, 25–30°C)	Microscopic morphology (SDA, 30°C)	Microscopic morphology
Penicillium	Powdery, rapid growing, blue-green with white border. Color varies with medium.	Conidiophores ending in brush-shaped penicillia consisting of up to five broad *metulae* with phialides bearing chains of conidia.	An illustration of a *Penicillium* conidiophore and penicillus is given in Chapter 14.
Rhizopus	White cottony hyphae rapidly reaching top of Petri plate dotted with brown-black sporangia	Aseptate hyphae with rhizoids opposite sporangiophores that end in columellae bearing a sporangium containing sporangiospores.	*Rhizopus* morphology is illustrated in Chapter 17A.
Scopulariopsis	Rapid growing, velvety turning granular to loosely floccose; characteristic sandy brown color	Short, branched heads ending in annellides bearing chains of rough or spiny walled annelloconidia with a characteristic neck-like flattened base.	*Scopulariopsis brevicaulis* (400×). E. Reiss.
		Melanized molds	
Alternaria	Rapid growing, downy to wooly; gray to olive-brown. Mature colonies often are covered with loose, light colored aerial hyphae.	Brown, septate hyphae, conidiophores simple with *acropetal* olivaceous to brown ovoid to inverted club-shaped large *muriform* conidia with elongated beak-like apical cells. Conidia may be single or in chains.	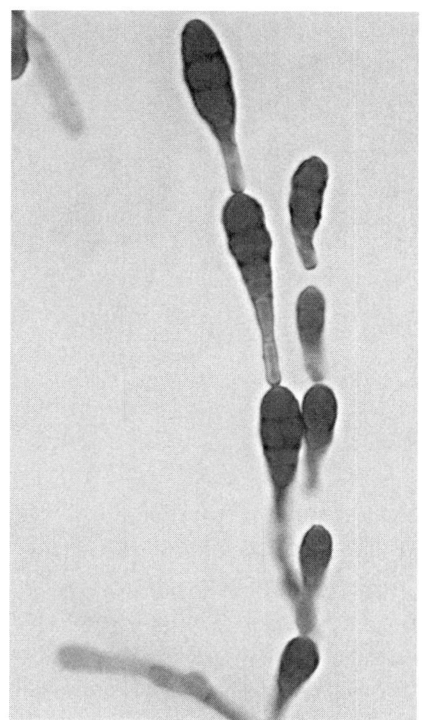 *Alternaria alternata* (400×). E. Reiss.

(*Continued*)

Table 2.9 (*Continued*)

Species	Colony morphology[a] (SDA, 25–30°C)	Microscopic morphology (SDA, 30°C)	Microscopic morphology
Aureobasidium pullulans	Initially white, yeast-like becoming black and leathery with a white edge with age	Brown irregular thick-walled hyphae with undifferentiated denticles bearing single or clusters of hyaline blastoconidia. These may produce hyaline blastoconidia in clusters or tufts.	*Aureobasidium pullulans*. E. Reiss, redrawn from Wilson and Plunkett (1965).
Cladosporium[b]	Slow to moderate growth, powdery to velvety, gray-green to brown	Conidiophores ending in shield-shaped cells, bearing acropetal chains of lemon-shaped conidia. Dark scars of attachment of shield cells and conidia are apparent even when detached.	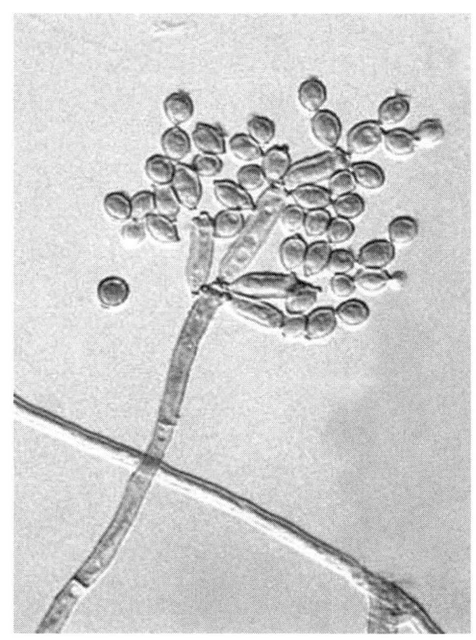 *Cladosporium cladosporoides*. Used with permission from Dr. George Barron, University of Guelph.

[a] Surface texture, color.

[b] Please see illustrations in Chapter 14.

[c] Pathogenic species are discussed in Chapters 18 and 19.

interference microscopy. A good website to aid the clinical laboratorian in the use of the microscope is www.olympusmicro.com. The Nikon microscopy website includes an interactive tutorial on microscope alignment for Köhler illumination, and how to adjust phase contrast microscopy. The URL is http://www.microscopyu.com/tutorials/java.

Fluorescent brighteners such as Calcofluor white in combination with fluorescence microscopy may help in distinguishing specific structures (please see Section 2.3.3.1, Fluorescent Brighteners).

2.3.9.1 Sizing Microscopic Forms

Microscopic measurements are made with the aid of an eyepiece micrometer and a stage micrometer. The graduated scales on the eyepiece micrometer are calibrated against a stage micrometer for each objective or magnification to be used. Details of the method can be found at the URL http://web.uvic.ca/ail/techniques/measuring.html.

Often a photograph of the stage micrometer is made at the same magnification as the micrograph and it can then be used as a direct "ruler" for measurements.

2.3.10 Use of Reference Laboratories

When cultures are requested by the physician but the resulting fungal growth cannot be identified by morphologic means in a reasonable time, a young actively growing subculture should be submitted to a reference laboratory. (please see Section 2.2.1.3, Genetics) Agar slants (slopes) rather than Petri plates must be used, and specific requirements on the *Shipment of Etiologic Agents* code must be followed. The best way to become informed of the shipping requirements is the brochure "Transporting Infectious Substances Safely." It is readable, well illustrated, and detailed. Please see the "General References" section for the URL for this publication.

Clinical laboratories with resources to perform PCR-sequencing studies on cultures may use this option when cultures cannot be identified by morphologic means. A positive sequence result submitted to a recognized database may provide an identification without the need for further bench work (Please see Section 2.4, Genetic Identification of Fungi).

2.3.11 Fungal Serology and Biochemical Markers of Infection

What can be done if morphologic identification is not yielding timely results? Additional tests are brought to bear when microscopic morphology is not revealing or, in the case of yeasts, the assimilation test results are ambiguous. Other situations calling for additional tests are when the fungus is slow-growing or the clinical situation is acute. Alternatives to morphologic identification include serologic tests, biochemical tests, and genetic identification of fungi. Often these tests will be carried out in parallel with the standard method of morphologic identification. These categories are discussed below.

2.3.11.1 Fungal Serology — Detection of Antibodies, Antigenemia, Antigenuria

This section lists and describes serologic tests for antigen and antibody diagnosis of mycoses with further information found in disease-specific chapters. The common formats for fungal serology are double immunodiffusion in agar gel (ID), complement fixation (CF), latex agglutination (LA) for antibodies or antigen detection, and enzyme immunoassay (EIA). For consistency the term EIA will be used in preference to enzyme linked immunosorbent assay (ELISA). Lindsley et al. (2006) should be consulted for a complete review of fungal immunodiagnosis, also including immunohistochemistry. A list and description of commercially available fungal immunoassays is given in Table 2.10.

Detailed standard operating procedures for CF, LA (coccidioidomycosis, histoplasmosis, and sporotrichosis), and ID tests and reagents (aspergillosis, blastomycosis, coccidioidomycosis, histoplasmosis) can be viewed and printed from the Immunomycologics, Inc. website.

Major Commercially Available Serologic Tests for Mycoses The tests discussed below are a partial list of the most useful commercial antibody and antigen tests for mycoses. Vendors of these tests include Bio-Rad Laboratories, Hercules, CA; Gibson Laboratories, Lexington, KY; Immunomycologics Inc., Norman, OK; Meridian Biosciences, Inc., Cincinnati, OH; and MiraVista Diagnostics, Indianapolis, IN. Their websites should be consulted for particular product lines. If no immunodiagnostic test is mentioned for a particular mycosis it is because it is still experimental and is not commercially available.

Antigen Detection Antigen detection is desirable because it can be a direct measure of current disease activity in patients without an intact immune system. Practical antigen detection methods have been achieved for invasive aspergillosis, cryptococcosis, moderate to severe acute pulmonary and disseminated histoplasmosis, and immunohistochemical detection of pneumocystosis. Antigen detection in invasive candidiasis is less satisfactory but remains promising.

Antibody Titer A fourfold rise in the antibody titer in successive specimens is an indication of active infection. Titers are expressed as the reciprocal of the endpoint serum dilution. EIA test results can be similarly expressed or as a ratio of the absorbance of the unknown serum divided by the absorbance of the mean of the normal amount of antibodies present in the asymptomatic population. In the instance of ID tests, seroconversion or the presence of one or more immunoprecipitin arcs are considered presumptive evidence for infection.

Aspergillosis

Tests for Antibodies

- *Immunodiffusion (ID)*. The antigenic complex is a culture filtrate. A single immunoprecipitin arc suggests either a fungus ball (aspergilloma) or *ABPA*. Three or more arcs increases the probability of an

Table 2.10 Current Commercially Available Serologic Immunoassays for the Mycoses

Disease/test	Antigen	Interpretation	Commercial source
Aspergillosis			
Immunodiffusion (ID)	Culture filtrate	A single immunoprecipitin arc suggests aspergilloma or ABPA; more than three lines is evidence pointing to aspergilloma	Gibson, IMMY, Meridian Bioscience
Enzyme immunoassay(EIA)	Galactomannan (antigenemia)	Two consecutive positive results required for reporting as a true-positive	Bio-Rad
Blastomycosis			
Complement fixation (CF)	A-antigen	Antibody titer ≥1:8 = presumptive evidence of infection; ≥1:32 = strong presumptive evidence; four-fold increase in titer indicates recent infection	Gibson, IMMY, Meridian Bioscience
ID	A-antigen	A-immunoprecipitate corresponds to recent infection	IMMY
Candidiasis			
ID	Culture filtrate + cell lysate of *C. albicans* yeast	Seroconversion or increase in number of immunoprecipitates suggests systemic infection in immune-normal persons	IMMY, Gibson
EIA	Cell wall mannan (antigenemia)	Cutoff for positive test is >0.5 ng mannan/mL serum	Bio-Rad
Coccidioidomycosis			
CF	Coccidioidin	Antibody titer ≥1:2 implies acute infection; ≥1:16 suggestive of extrapulmonary dissemination	IMMY, Meridian Bioscience
ID	Coccidioidin	Precipitins include TP (IgM) and CF (IgG)	IMMY, Meridian Bioscience
EIA	Mix of TP and CF antigens	Detects IgG (CF antigen) and IgM (TP antigen) O.D. <0.1.5 = negative; 0.15–0.2 O.D. = intermediate; ≥0.2 O.D. = positive	Meridian Bioscience
EIA	Con A purified polysaccharide (antigenuria)	Cutoff for a positive test is 0.05 ng/mL of *Coccidioides* polysaccharide antigen	MiraVista Laboratories
Cryptococcosis			
Latex agglutination (LA)	Capsular polysaccharide (antigenemia, antigen in CSF)	Serum or CSF antigen titer of 1:4 is suggestive of cryptococcal infection; titers of ≥1:8 usually indicate active cryptococcosis	IMMY, Bio-Rad, Gibson, Inverness, Meridian Bioscience
EIA	Antigenemia and antigen in CSF	O.D. ≥0.15 = positive test	Meridian Bioscience
Histoplasmosis			
CF	Histoplasmin, formalin-killed yeast forms	CF antibody titer of ≥1:32, or fourfold rise indicates active infection; titers of 1:8 or 1:16, with either antigen, are suggestive of histoplasmosis	IMMY, Meridian Bioscience
ID	Histoplasmin	The m-immunoprecipitate arises first and is present 4–6 weeks after symptoms appear; h-precipitin indicates active infection but is less commonly present	Gibson, IMMY, Meridian Bioscience
LA	Histoplasmin	Antibody titers of 1:16–1:32 are considered positive; titers of ≥1:32 are evidence for active or recent disease	IMMY

Table 2.10 (*Continued*)

Disease/test	Antigen	Interpretation	Commercial source
EIA	Inactivated *Histoplasma* antigen	Ratio of O.D. with patient serum/O.D. of negative calibrator: ratio >1.10 is a positive test for antibodies	Focus Diagnostics
EIA	Polysaccharide (antigenuria)	Cutoff value for positive test is ratio of O.D. patient specimen divided by mean of negative control values to generate EU, and EU of ≥1.0 is considered positive	MiraVista Laboratories
Pneumocystosis			
Fluorescent antibody	Antigen detection (induced sputum or BAL)	Stains cysts (ascus) and trophic forms of *P. jirovecii*	Bio-Rad, Meridian Bioscience, Millipore
Sporotrichosis			
LA	*S. schenckii* culture filtrate.	Antibody titers of ≥1:8 indicate active infection	IMMY

Source: Table modified from Lindsley et al. (2006), who should be consulted for a more complete list.

aspergilloma. The reagents include antigens from *A. flavus, A. fumigatus, A. niger*, and *A. terreus*.

- *Antigen Detection*. The Platelia® *Aspergillus* EIA (Bio-Rad Laboratories) is a double antibody sandwich EIA that detects serum galactomannan antigenemia during invasive pulmonary aspergillosis in the immunocompromised host.

Blastomycosis

Tests for Antibodies

- *Complement Fixation* (*CF*). The A-antigen is the immunodominant surface adhesin, also known as BAD-1 for "*Blastomyces* adhesin." Presumptive evidence of infection is a titer of ≥1:8; titers ≥1:32 are strong presumptive evidence of recent infection.
- *ID*. The A-immunoprecitin arc corresponds to active infection.

Candidiasis

Tests for Antibodies

These tests for candidiasis are relevant in the host with a reasonably intact immune system. Antibody titers in immunocompromised patients are not reliable and rather reflect fluctuations in their immunocompetence.

- *ID*. The antigenic complex consists of culture filtrate and yeast lysate. One or multiple immunoprecipitin arcs reflect active infection and can be informative in cases of *Candida* endocarditis.
- *Antigen Detection*. The Platelia® *Candida* Ag EIA (Bio-Rad Laboratories) is a double antibody sandwich EIA to detect cell wall mannan in serum from immunocompromised patients with invasive candidiasis. There are a multiplicity of epitopes in *Candida* species cell wall mannan, affecting the test's sensitivity and specificity. Research suggests that both α-linked mannosyl determinants and ß-linked mannosyl determinants are needed to achieve broad enough specificity to be clinically useful (Ellepola and Morrison, 2005). In addition to issues about the epitope specificity of the capture and indicator antibodies the ability to detect mannanemia is influenced by the frequency of sampling and underlying disease. Until all of these parameters are optimized and evaluated in a multicenter study, this test, while promising, is still a matter for clinical research.

Coccidioidomycosis

Tests for Antibodies

Serologic tests for antibodies are valuable in the diagnosis and prognosis of coccidioidomycosis (Pappagianis and Zimmer, 1990). Detection of coccidioidal antibody in a single test can be diagnostically significant; testing sequential specimens can aid in following the course of disease. The most comprehensive testing is conducted by the Coccidioidomycosis Serology Laboratory at the University of California at Davis, directed by Prof. Demosthenes Pappagianis (http://som.ucdavis.edu/departments/microbiology/cocci/).

- *TP*. This tube precipitin test measures IgM antibodies against the heat-stable antigen in coccidioidin. The active component is a 3-*O*-methylmannose containing polysaccharide or glycoprotein. For use, the kit manufacturer heats coccidioidin at 60°C for 30 min. TP antibodies are first to arise in acute coccidioidomycosis often within 1–2 weeks after the first appearance of symptoms.

- *CF*. The CF antibodies are slower to appear but are more persistent and rise in proportion to disease severity. The antigenic complex used is coccidioidin, a toluene-induced lysate of mycelial form broth cultures. Titers of 1:2 to 1:8 may represent antibody crossreacting with a heterologous antigen and should be verified by the ID test. Titers >1:16 create apprehension about extrapulmonary dissemination.

ID variations have been designed to measure patient antibodies comparable to their TP and CF counterpart tests. The Coccidioidomycosis Serology Laboratory at the University of California at Davis has a qualitative ID test that simultaneously measures patient antibodies against both the CF and TP antigens with respect to lines of identity with reference antiserum known to contain either the CF or TP antibodies.

- *IDCF*. Using coccidioidin, this ID test measures the same antibody detected by the CF test. The active antigen of coccidioidin measured in this test is a chitinase. It can be performed with a single dilution (qualitative), or in the quantitative ID version wherein an endpoint titer can be determined.
- *IDTP*. Heat-stable coccidioidin is the antigen. A positive precipitin arc is interpreted as an earlier diagnostic marker of acute coccidioidomycosis than the IDCF test.
- *LA Test*. Latex particles coated with heat-treated coccidioidin constitute a sensitive test for antibodies, but one prone to crossreactions and where a confirmatory test (IDCF) is required. It is valuable in acute pulmonary coccidioidomycosis.
- *Antigenuria EIA*. A double antibody sandwich EIA detects a heat-stable antigen in the urine of immunosuppressed patients with severe coccidioidomycosis (MVista *Coccidioides* antigen EIA, MiraVista Diagnostics, Indianapolis, IN) (Durkin et al., 2008). Patients with positive antigenuria tests were a group in which nearly one-half had disseminated disease, almost one-third had fatal cases, and 87% were culture-positive for *Coccidioides* species. In a follow-up study, antigenuria was detected in 50% of 16 proven cases of coccidioidomycosis (Durkin et al., 2009). Using heat-EDTA to dissociate immune complexes in *serum* specimens, crossreactions were observed in 22.2% of patients with histoplasmosis or blastomycosis.

Cryptococcosis

- *LA Test for Antigen Detection*. The fungal antigen detection test that is renowned for its diagnostic and prognostic ability is the reverse passive latex agglutination (LA) test to detect capsular polysaccharide of *Cryptococcus neoformans* in CSF and in serum. The capsular polysaccharide is an acidic, viscous, high molecular mass glucuronoxylomannan (GXM) that sloughs off the encapsulated yeast cells during CNS infection. GXM persists in CSF and serum because it is resistant to digestion by host phagocytes and their lysosomal hydrolases.
- *Principle:* Rabbit antibodies to whole killed cryptococci are adsorbed onto polystyrene beads, which agglutinate in the presence of cryptococcal polysaccharide in CSF or serum. Proteinase treatment of serum removes nonspecific agglutination from rheumatoid factor.

  Sources of commercial tests using the LA format are Latex Crypto Antigen Detection System, Immunomycologics Inc.; Crypto-LA® Test, Inverness Medical Professional Diagnostics; Cryptococcal Antigen Latex Agglutination System (CALAS®), Meridian Bioscience Inc.; and Pastorex® Crypto Plus, Bio-Rad Laboratories.

  The latex agglutination test to detect cryptococcal polysaccharide has the required sensitivity to detect cryptococcal meningitis. A further development is an EIA to detect cryptococcal antigen, the Premier® Cryptococcal Antigen EIA, offered by Meridian Bioscience.

Histoplasmosis

Tests for Antibodies

These antibody tests can be applied to serum or CSF specimens. CF antibodies arise in acute pulmonary histoplasmosis and over 90% of the culturally proven cases may be positive if serum samples are collected at 2–3 week intervals.

- *CF*. Two antigenic preparations are used: one test uses histoplasmin, a culture filtrate of mycelial form growth; and another preparation consists of killed yeast-form cells of *H. capsulatum*. The latter test is more sensitive but prone to crossreactivity such that confirmation with the ID test is recommended. Titers of >1:8 are presumptive evidence of infection, and the probability of active disease increases in proportion to the titer. CSF also is a suitable specimen for this test.
- *ID*. Histoplasmin contains two dominant, species-specific and well-studied antigens. The "m" antigen is a catalase, whereas the "h" antigen is a glucosidase. The m-immunoprecipitin arc arises first and is present 4–6 weeks after the first appearance of symptoms. The serum from about 70% of patients with proven histoplasmosis contains m-precipitins,

whereas only 10% of the sera demonstrate both the h and m precipitins.

- *LA*. Latex beads coated with histoplasmin is the format for this test. The LA test is an excellent presumptive test for acute histoplasmosis. Antibody titers of 1:16 to 1:32 are considered to be positive. Titers of 1:32 or greater are strong evidence for active or recent disease. A positive LA test result can be demonstrated as early as 2–3 weeks after infection. Because of the tendency for crossreactions with other systemic mycoses, results should be confirmed with the ID test.
- *Antigenuria Detection*. The HPA test or "*Histoplasma* Polysaccharide Antigen" test is a double antibody sandwich EIA to detect antigenuria in moderate to severe acute pulmonary histoplasmosis (80%) (Miravista Laboratories, Indianapolis, IN). The test is more sensitive in extrapulmonary disseminated histoplasmosis (82%), and most sensitive in AIDS-histoplasmosis (95%). The test is well studied and the titer is directly proportional to disease activity and inversely to response to therapy. The sensitivity of the test in serum specimens is less than in urine specimens. Crossreactions occur in paracoccidioidomycosis, blastomycosis, African histoplasmosis, and *Penicillium marneffei* infections (Wheat, 2003).

Paracoccidioidomycosis

Test for Antibodies

- *ID*. The ID reaction with concentrated yeast-form filtrate as the antigen source and reference sera is entirely specific and has a sensitivity of 94% with sera from patients with paracoccidioidomycosis. The main immunoprecipitin arc shows identity with gp43, the immunodominant antigen for this systemic fungal pathogen.

Pneumocystosis

- *Antigen Detection*. Direct fluorescent antibody staining is a sensitive and specific diagnostic method for pneumocystosis. Tests containing monoclonal antibody "2G2" detect both cysts and trophic forms. At the NIH Clinical Center, the diagnostic approach is to obtain an immunofluorescent-stained smear of induced sputum and, if that smear is negative, to proceed to BAL within 24 h, should that be necessary (Kovacs et al., 2001). Indirect and direct immunofluorescence tests using monoclonal antibodies to detect *Pneumocystis jirovecii* are available (Bio-Rad Laboratories).

Sporotrichosis

Test for Antibodies

- *LA*. The test format consists of latex particles coated with a yeast form *Sporothrix schenckii* culture filtrate antigen. Titers of ≥1:8 are indicative of active infection. The test is highly specific owing to the presence of the immunodominant antigen, a peptido-L-rhamno-D-mannan, originating in the outer layer of the cell wall.

2.3.11.2 Biochemical Tests for Surrogate Markers of Fungal Infection

Pan-Specific Test for Mycoses: Serum ß-(1→3)-D-Glucan Fungitell® (Associates of Cape Cod Inc., East Falmouth, MA) is a rapid, presumptive screen for invasive fungal infections and measures serum levels of ß-(1→3)-D-glucan, a fungal cell wall component.

Principle The hemolymph of the horseshoe crab, *Limulus polyphemus*, undergoes a clotting cascade in the presence of either lipopolysaccharide endotoxin or ß-(1→3)-D-glucan. The test is designed to separate these two activities so that only the ß-(1→3)-D-glucan activity is measured. In the presence of ß-(1→3)-D-glucan, Factor G, a serine proteinase is activated and cleaves the artificial chromogenic substrate: an oligopeptide conjugated to the chromophore pNA (*para*-nitroaniline). pNA is released and detected in a kinetic assay (rate increase in absorbance at A_{405} nm). The test is conducted in a microtitration plate as a homogeneous test; that is, no washing steps are needed. Beacon Diagnostics Laboratory (East Falmouth, MA) conducts the Fungitell assay as a reference diagnostic service.

Clinical Significance Fungal pathogens release ß-(1→3)-D-glucan into the plasma during infection and this circulating metabolite can be detected in *Aspergillus* spp. *Blastomyces dermatitidis, Candida* spp., *Coccidioides* spp., *Fusarium* spp., *Histoplasma capsulatum, Pneumocystis jirovecii, Saccharomyces cerevisiae, Sporothrix schenckii*, and *Trichosporon* spp. Normal amounts of ß-(1→3)-D-glucan, 10–40 pg/mL, are present in human serum. Concentrations exceeding 80 pg/mL are reported as a positive test. This test is therefore a surrogate for invasive fungal infection but cannot discriminate among fungal species. Fungitell is a rapid presumptive screen for invasive fungal infection which is followed up with additional more targeted diagnostic tests.

Factors Limiting Test Performance Lipemic or hemolyzed sera interfere with the test sensitivity. Bacteremic patient sera can result in false-positive reactions and there are various other sources of false-positive reactions (Pickering et al., 2005). Certain fungi do not produce sufficient ß-(1→3)-D-glucan to be detected or are nonproducers, including *Cryptococcus neoformans* and members of the *Mucorales* (e.g., *Lichtheimia, Mucor*, and *Rhizopus* species).

Serum D-Arabinitol/Creatinine Ratio in Candida Sepsis *Candida albicans, C. parapsilosis*, and *C. tropicalis* but not *C. glabrata* produce D-arabinitol during fungemia and this metabolite can serve as a surrogate marker of *Candida* sepsis due to those yeasts. A correction is necessary for kidney function, the arabinitol/creatinine ratio. *Candida* sepsis is indicated when the arabinitol/creatinine ratio is $\geq$3.9. (The units of the ratio are μM D-arabinitol/mg/dL creatinine). That ratio corresponded to 3 standard deviations above the mean of the normal amount of D-arabinitol in healthy adults, and correlated with *Candida* sepsis (Yeo et al., 2006). There is no kit for the arabinitol determination. The major consumable item is a recombinant enzyme D-arabinitol dehydrogenase, in an NAD-linked enzyme reaction, developed by Yeo et al., (2006). The test method uses a clinical chemistry instrument: the Cobas Fara II centrifugal analyzer (Roche Diagnostics, Basel, Switzerland).

2.4 GENETIC IDENTIFICATION OF FUNGI

2.4.1 Commercial Test

This test is produced by Gen-Probe, Inc. (San Diego, CA). When growth occurs on an agar slant a definitive identification can be made using the Accuprobe® nucleic acid probe test for *Coccidioides immitis, Histoplasma capsulatum*, and *Blastomyces dermatitidis*.

Principle: The Hybridization Protection Assay

- A specific DNA probe, labeled with an acridinium ester detector molecule hybridizes to target rRNA of the mold to be tested.
- Nonhybridized probe is inactivated.
- A chemiluminescent signal is emitted and detected in a luminometer (sold by Gen-Probe, Inc.).
- Results are expressed in Relative Light Units—RLU.

2.4.2 Peptide Nucleic Acid–Fluorescent In Situ Hybridization (PNA-FISH)

This commercial test for detecting yeast in smears from blood culture bottles is discussed in Chapter 11, Section 11.12, Laboratory Detection, Recovery, and Identification.

2.4.3 PCR-Sequencing Method

Pathogenic fungi can be identified by PCR, followed by sequencing of target genes that have sufficient diversity to be useful in species identification. Identification is established by alignment of the sequence in question with ones deposited in *GenBank* or another sequence database (please see Section 2.4.13, Other Sequence Databases). A general reference for molecular biology is Sambrook and Russell (2001).

GenBank is an annotated database of all publicly available nucleotide sequences and their protein translations. The database is produced at the U.S. National Library of Medicine's National Center for Biotechnology Information (NCBI). International collaboration is via the European Molecular Biology Laboratory (EMBL) Data Library of the European Bioinformatics Institute, (EBI) and the DNA Data Bank of Japan (DDBJ).

2.4.4 Nuclear rDNA Complex

The ribosomal RNA gene complex contains highly conserved domains interspersed with more variable domains, ones that contain species-specific "signatures." For simplicity the subunits of the complex are referred to as rDNA with the understanding that the genes encode rRNA molecules. The rDNA complex is a target of choice for genetic identification of fungi (Fig. 2.9).

2.4.4.1 Physical Location and Biology

Within the rDNA complex is a tandem array of multiple copies of the repeat unit consisting, from the 5′→3′ direction, of 18S ITS1 5.8S ITS2 28S gene segments. ITS1 and ITS2 are variable internal transcribed spacer regions. Flanking the 28S gene are two variable nontranscribed

Figure 2.9 Organization of fungal rRNA genes. The fungal nuclear-encoded ribosomal RNA genes (rDNA) occur as a multicopy gene family arranged in a tandem array (head-to-toe). Each repeat unit has coding regions for a single ribosome, separated by nontranscribed intergenic spacer (IGS) regions. The 18S small subunit, or SSU, is useful for taxonomic studies. Please see text for examples. The taxonomically useful regions within the 28S large subunit, or LSU, are the first 600–900 bases including three domains (D1, D2, D3), which contain the most variable regions. The internal transcribed spacer, ITS region (ITS1 and ITS2), is widely sequenced for molecular identification of species. In some fungal groups (basidiomycetes and ascomycetous yeasts), each repeat has a separately transcribed coding region for 5S RNA. Most filamentous ascomycetes have a single uninterrupted IGS region (between the end of the LSU and start of the next SSU sequence). *Source:* E. Reiss.

spacer regions: IGS1–5S–IGS2. In *Saccharomyces cerevisiae* there are approximately 140 repeat units. The precursor RNA transcript undergoes enzymatic removal of the intergenic transcribed spacer (ITS) units and, after further modifications, becomes 18S, 5.8S, and 28S mature rRNAs, which combine with proteins to become functional ribosomes. The biological role of the ITS units is involved in early transcription during rRNA processing. Thus, they are noncoding regions of the genome.

Universal Primers They are primer pairs that amplify the rDNA gene subunits 18S, 5.8S, or within the 28S region. (For further information please see "Universal Fungal Primers" below.)

2.4.4.2 Ribosomal RNA Genes and Subregions Useful for Species Identification

18S rDNA The nuclear "small" subunit (SSU) of rDNA is the 18S gene and is particularly useful in taxonomic studies. The SSU is about 1800 bp in size and contains conserved and variable domains.

- Here is an example of the utility of the SSU sequence for taxonomic studies. Phylogenetic analysis of 55 species of black yeasts and related fungi were investigated, including the pathogenic genera: *Cladophialophora, Exophiala, Fonsecaea, Hortaea, Phialophora, Ramichloridium*, and *Rhinocladiella* (Haase et al., 1999). The SSU from position 62 to 1800, the 3′ end of the gene, was amplified with the NS1 and NS8a primers. Primer sequences are listed in the Fungal Tree of Life website, at the URL http://aftol.org/resources.php.
- The yeasts and yeast-like black fungi were classed in the ascomycete family *Herpotrichiellaceae*. The pathogenic species formed groups clustered around *Wangiella* (*Exophiala*) *dermatitidis*, another complex consisting of *Fonsecaea* and *Cladophialophora*, and a cluster containing *Exophiala jeanselmei*.

5.8S rDNA The 5.8S region is about 160 bp long, is highly conserved, and is not considered useful to classify fungi or for phylogenetic analysis. An exception to this assertion is that the alignment of a 143 bp region of 5.8S rDNA resolved the taxonomic placement of basidiomycete yeasts including the medically important genus *Filobasidiella* (anamorphs = *Cryptococcus neoformans* and *C. gattii*) (reviewed by Valente et al., 1999). The 5.8S region is useful as an attachment site for universal primers to PCR-amplify the flanking ITS regions.

28S rDNA The 28S rDNA, ~2900 bp, large subunit (LSU, sometimes described as 26S) is conserved but the first 600 bp, called the D1/D2 subregion, is used to identify yeasts to the species level (Kurtzman, 2000). The D1/D2 sequences of all ascomycetous and basidiomycetous yeast species are known and deposited in GenBank. Most yeast species can be identified by their sequence divergence in the D1/D2 domain. The D1/D2 sequences were utilized to develop probes adherent to microbeads for use in a flow cytometric assay to rapidly identify eight *Candida* species by their D1/D2 amplicons (Page and Kurtzman, 2005).

Sequence-based identification of molds can be accomplished with reference to the MicroSeq D2 LSU rDNA fungal kit and its associated database (Applied Biosystems, Foster City, CA) (Hall et al., 2004). Of 234 clinical mold isolates, 158 were correctly identified to genus or genus and species by DNA sequencing. Sequences for 70 (29.9%) of the isolates (27 genera) were not included in the MicroSeq library, which needs to be expanded.

ITS1 and ITS2 These regions flanking the 5.8S rDNA gene show extensive sequence diversity. Each ITS region is ≤300 bp in length. They are very useful for species identification of both yeasts and molds (Table 2.11) (Iwen et al. 2002). Thirteen known pathogens, also including potentially invasive *Aspergillus* species, could be identified by comparative analysis of their ITS1 and ITS2 sequences. Both spacer regions showed better species differentiation than a single spacer region alone (Hinrikson et al., 2005).

Table 2.11 Species Identification by Direct Analysis of Internal Transcribed Spacer (ITS) Regions of rDNA

ITS1	*Wangiella (Exophiala) dermatitidis*
	Epidermophyton floccosum
	Microsoporum (4 species)
	Trichophyton (6 species)
	Malassezia (6 species)
ITS2	*C. albicans, C. parapsilosis*, 13 other *Candida* species
	Cryptococcus (5 species)
	Rhodotorula
ITS1 and ITS2	*Aspergillus* (13 species) (Hinrikson et al., 2005)
	Cryptococcus neoformans
	Fonsecaea compacta, F. pedrosoi
	Histoplasma capsulatum
	Paracoccidioides brasiliensis
	Pseudallescheria boydii

Source: Adapted from Iwen et al. (2002).

IGS1 This nontranscribed spacer region between the 26S and 5S rDNA of yeast is suitable to delineate *Trichosporon* species (Sugita et al., 2002).

2.4.5 Genetic Tools for Species Identification

- The D1/D2 subregion of the LSU rDNA is suitable for identification of many commonly encountered clinical yeasts and molds.
- The SSU rDNA has been successfully used for speciation of molds.
- The ITS region remains the best tool for identification of any fungus to the species level because of its universality, ease of use, and range of available data. Comprised of the ITS1–5.8S–ITS2, it is approximately 500–1000 bp long and can be amplified in a single PCR. The ITS region, however, can fail to resolve very closely related species, for example, *Fusarium* species. Then protein encoding regions are used because they contain introns that evolve at a rate higher than ITS (e.g., translation elongation factor 1-α).
- *Sequence Databases*. The GenBank database has a large number of D1/D2 28S rDNA, 18S rDNA, and ITS 1, ITS 2 sequences for many fungi; but since it is not curated, improvement is needed in its quality and accuracy.

2.4.6 How Is the Genetic Identification of an Unknown Fungus Accomplished?

Outline of the Procedure for Genetic Identification of an Unknown Fungus

- Grow the fungus.
- Extract and purify the DNA.
- Select the DNA target sequence according to whether the fungus is a yeast or mold.
- Select and obtain the primer pair for PCR.
- Conduct the PCR.
- Clean up the PCR products.
- Set up and run the PCR Cycle Sequencing reaction.
- Clean up products.
- Submit the products to the Biotechnology Core Facility for automated DNA sequencing.
- Import the chromatograms containing forward and reverse DNA sequence into Sequencher for editing, trimming, and assembly of the positive strand sequence.
- Perform a BLAST search and query the GenBank database or submit the sequence to a specialized and curated database.
- Determine the correct identification according to the match with a deposited sequence.

2.4.7 Growth of the Fungus in Pure Culture, Extraction and Purification of DNA

2.4.7.1 *Yeast — DNA Extraction and Purification*

1. A general medium for growth is yeast-extract–peptone–glucose broth (YPG), 10 mL in a 50 mL Erlenmeyer flask will provide ample growth for DNA extraction.
2. Extraction and purification may be accomplished with Fast-Prep tubes, held in a vortex mixer Vortex-Genie® with a circular platform containing the Microtube Foam Insert part No. 504 -0234-00 Scientific Industries, Inc. Bohemia, NY.
3. DNeasy Kit from Qiagen Inc. (Valencia, CA). See the URL http://www1.qiagen.com/Products/GenomicDnaStabilizationPurification/DNeasyTissueSystem/DNeasyBloodTissueKit.aspx.

Procedure

- A single yeast colony (3–4 days old) → YPG broth, 10 mL/50 mL Erlenmeyer flask. (*Note*: for *Cryptococcus* species use YPG broth containing 0.5 M NaCl to suppress capsule formation.)
- Incubate 21–24 h, 35°C, 180 rpm.
- Harvest into a 15 mL screw cap polystyrene centrifuge tube.
    - Centrifuge 1800 rpm, 10 min, swinging bucket rotor at 4°C.
- Wash 1× by centrifugation in 10 mL deionized H_2O (dH_2O).
- Resuspend 1 vol packed cells +5.5 vols of dH_2O.
- Transfer 2.0 mL of this suspension to a new 15 mL centrifuge tube.
    - Centrifuge 1800 rpm, 10 min as above.
- (Warm ATL solution from the Qiagen Kit at 37°C to dissolve. It will be needed in the next step)
- Resuspend pellet in 700 μL of ATL solution. This will give a packed cell to buffer ratio of 1:2 (0.35 mL:0.7 mL).
- Transfer to red top Fast Prep tube. "Lysing Matrix C", 2 mL capacity (#6912-050, MP Biomedicals, Inc., Solon, OH).

- ◦ Heat 65°C, 10 min; cool to room temperature, 10 min.
- Disrupt on Vortex Genie (operated at full speed). Disrupt for 5 min. Rest 5 min, then disrupt for 2.5 min (hold at room temperature between disruptions).
  - Centrifuge, 2 min, 8000 rpm in a benchtop centrifuge (e.g., Eppendorf Microcentrifuge).
  - Aspirate supernatant; expect vol = 1 mL into 1.5 mL snap cap tube.
  - Recentrifuge to clarify (12,000 rpm, 2 min).
- Add 10 μL proteinase K (PrK) to give a final concentration of 50 μg/mL (5 mg PrK/mL stock solution = 100×. (Buffer is 0.05 M Tris pH 8.0 + 1 mM $CaCl_2$.) That is, add 10 μL.
  - Heat at 55°C, 1 h at 60 rpm.
- Cool to room temperature. Add 5 μL of heat-treated RNase A (stock concentration = 10 mg/mL is 200×). The final concentration is 50 μg RNase A/mL.
  - Incubate 15 min at 37° C.
  - Heat at 70°C for 10 min. (Can stop here and store at 4°C.)
- Next, continue by following instructions in Qiagen DNeasy Kit brochure.
- Expected yield is 10–25 μg DNA. Store DNA at −70°C. For PCR, dilute DNA 1/10 in 0.01 M Tris-EDTA buffer pH 8.0, then add 2 μL to 18 μL of PCR reaction mix.

2.4.7.2 Molds — DNA Extraction and Purification

1. *Petri Plate Isolation* (Hurst, 2008). Apparatus needed:
   - Omni Mixer model TH (Omni International, Marietta, GA)
   - Omni Tips, disposable-rotor-stator generator probes
   - Tubes, polypropylene, round bottom, 5 mL
   - This method will yield sufficient DNA for PCR. If the mold is growing in pure culture, remove a 1 cm diameter circle of mycelium with as little agar as possible. Some fungi grow into the agar so it may be necessary to include some agar in the removed sample. Try to do this at a time point before conidia formation occurs. Old plate cultures will not yield useful DNA.
   - Transfer mycelium to a 5 mL polypropylene tube and add 900 μL of ATL buffer from the Qiagen DNeasy kit (see above for Qiagen URL). Add PrK to give a final concentration of 50 μg/mL.
   - Omni-mix for ~30 s at slow speed. Ensure that sample material is broken into small <1 mm particles. Then homogenize at high speed, stopping to avoid excessive foaming, for a total contact time of 15–20 s.
   - (After use, separate parts of the probe, clean and disinfect in detergent solution with 2% Clorox.)
   - Add 10 μL proteinase K (PrK) to give a final concentration of 50 μg/mL.
     - Incubate sample at 55°C, 1 h (digestion of proteins).
     - Cool to room temperature; add heat treated RNase to give a final concentration of 50 μg/mL and incubate at 37°C, 15 min.
   - Follow the procedure in the Qiagen DNeasy Kit. Yield, 5–30 μg DNA. Store at −70°C.

2.4.8 PCR of the Target Sequence

Up to this stage the DNA has been extracted, purified, and stored at −70 to 80°C. Next, the microbiologist selects the DNA target sequence according to whether the fungus is a yeast or mold; selects and obtains the primer pair for PCR; and then conducts PCR.

1. *Universal Fungal Primers*. Use primers provided at the Fungal Tree of Life website URL, http://aftol.org/resources.php, to amplify the target sequence.
   - PCR protocols may be found in Hinrikson et al., 2005.
   - *ITS*. Primer pair consisting of ITS1 and ITS4 amplifies across the ITS1-5.8S-ITS2 region.
   - *28S*. Primer pair 5.8SR and LR7 amplify the first 900 bases of the 28S LSU subunit rDNA which encompasses the D1, D2 and D3 regions.
   - *18S*. Primer pairs of the “NS” and “SR” type amplify particular subregions of the SSU rDNA (please see the Fungal Tree of Life website for details).
2. *Check to see if PCR amplification has occurred*.
   - A small scale agarose gel is used to ensure that amplification has occurred.
   - Cast 90 mL of a 2% agarose gel in 1× TBE buffer in a 7 cm × 15 cm tray. Electrophorese samples at 75 volts until the tracking dye reaches 1.5 cm from the anode end of the gel (~1.5 h). Then the gel is stained with ethidium bromide solution and detected with a UV transilluminator.
3. *Purify PCR products before sequencing*.
   - This is done to remove nucleotide triphosphate monomers and excess PCR primer DNA. One method is to use the ExoSAP-IT® kit. Please refer to the package insert. http://www.usbweb.com/brief_proto/78200b.pdf.

2.4.9 PCR Cycle Sequencing

1. Set up the BigDye Terminator cycle sequencing reaction from the kit (Applied Biosystems). Please refer to the downloadable. pdf file for the manual title: BigDye® Terminator v3.1 Cycle Sequencing Kit: Protocol: Rev A.
2. Two reactions are needed for each unknown fungus: one reaction well containing the forward primer and another well with the reverse primer.
3. *Clean up before submitting to the Core Lab for sequencing*. The cycle sequencing reactions are cleaned-up to remove unincorporated dyes, nucleotides, and salts before submitting for capillary sequencing (*e.g*. on the ABI 3730).
4. Clean-up is accomplished by binding the DNA sequence to the magnetic beads, washing free of contaminants, and elution with water or dilute EDTA solution: for example, Clean-Seq® (Agencourt Bioscience Corp., Beverly, MA). http://www.agencourt.com/documents/products/cleanseq/Agencourt_Clean SEQ_Protocol.pdf.

2.4.10 Assemble the DNA Sequence

Sequence editing software is used to edit, trim, and assemble the sequence: for example, Sequencher® (Gene Codes Corp., Ann Arbor, MI).

1. *Tutorials for Sequencher* are available at the URL http://www.sequencher.com/sequencher/tutorials.html.
2. *Import* the chromatograms containing sequence data obtained by the automated DNA sequencer.
3. *Trim*. Automated DNA sequencers occasionally produce poor quality reads, particularly near the sequencing primer site and toward the end of longer sequences. "Trim Ends" removes misleading data from the ends of sequencing fragments.
4. *Assemble*. The Sequencher option: "Assemble by Name" sorts the sequences in your project, assembles the sequences, and automatically names the contig. (*Definition*: A contig is a contiguous stretch of DNA sequence without gaps that has been assembled solely based on direct sequencing information.)
5. *Check for base call disagreements*. Open the contigs for forward and reverse sequences and where there is disagreement, select "Show chromatogram." Decide the correct call and type it in. When all the bases are edited to remove mismatched or ambiguous results the contig is now complete.
6. *Export contig* into the Consensus folder. Save the file as a Contig Assembly in the run folder.

2.4.11 Perform a BLAST Search

BLAST is the acronym for Basic Local Alignment Search Tool of the sequence being queried in the GenBank database.

1. *More about BLAST*. Clinical laboratory scientists just starting to use genetic identification of fungi should take the NCBI BLAST Course. It is without cost and is located at the URL http://www.ncbi.nlm.nih.gov/Class/BLAST/blast_course.short.html.
2. A useful guide preparatory to taking the NCBI BLAST course is the video "BLAST for Beginners," available without cost in the Microbe Library of the American Society for Microbiology (www.microbelibrary.org) uploaded by Sandra Porter of Geospiza, Inc. Seattle, Washington (http://www.geospiza.com/education/index.html).

2.4.12 The MicroSeq System®

This system (Applied Biosystems, Inc., Foster City, CA, a division of Life Technologies, Inc.) is an alternative to the above method in that it combines all the instruments, reagents (including sample preparation), sequence libraries, and software required for automated yeast and mold DNA sequencing and identification. The database is MicroSeq D2 LSU rDNA Identification System and contains 1072 sequences from 900 yeasts and molds.

2.4.13 Other Sequence Databases

In addition to GenBank the following databases are curated and are specialized for fungal identification.

1. *MycoBank – Yeasts*. http://www.cbs.knaw.nl/yeast/(f3rkvyblldafnu55ugsca445)/BioloMICS.aspx. This is a polyphasic database for identifying unknown yeasts based on morphology, biochemistry, and/or sequence analysis. Paste in the DNA sequence being queried and receive the hits from its database. It is part of the BioloMICS program of Dr. Vincent Robert of the Centraalbureau voor Schimmelcultures (CBS) in the Netherlands.
2. *SmartGene®* (Zug, Switzerland). This is a commercial subscription fee-based Integrated Database Network System (IDNS). SmartGene–Fungi is the module dedicated to fungal identification based on DNA/RNA sequence analysis.
3. *Phylogenetic Tree Database*. TreeBASE (www.treebase.org) is a public access depository of DNA sequence alignments from phylogenetic publications. Its main function is to store published phylogenetic trees and data matrices. It also includes bibliographic

information on phylogenetic studies. TreeBASE allows retrieval of trees and data from different studies, and it can be explored interactively using trees included in the database. TreeBASE provides a means of assessing and synthesizing phylogenetic knowledge. For example, "*Coccidioides*" was entered as a search term and it brought up a tree showing the relationship of *C. immitis* and *C. posadasii* from a 2001 article from the Taylor laboratory at University of California, Berkeley.

GENERAL REFERENCES FOR LABORATORY DIAGNOSTIC METHODS IN MEDICAL MYCOLOGY

Biosafety in Microbiological and Biomedical Laboratories (BMBL), 5th ed. U.S. Department of Health and Human Services, Centers for Disease Control and Prevention and National Institutes of Health, December 2009. U.S. Government Printing Office, Washington, DC. The manual is available at the URL http://www.cdc.gov/od/ohs/biosfty/bmbl5/bmbl5toc.htm.

Bulmer GS, 1995. *Fungus Diseases in the Orient*, 3rd ed. Rex Book Stores, Manila, Philippines.

Campbell CK, Johnson EM, Philpot CM, Warnock DW, 1996. *Identification of Pathogenic Fungi*. Public Health Laboratory Service, London.

de Hoog GS, Guarro J, Gené J, Figueras MJ, 2004. *Atlas of Clinical Fungi*, 2nd ed. American Society for Microbiology, Washington, DC. (CD version 2004.11.)

Ellis D, Davis S, Alexiou H, Handke R, Bartley R, 2007. *Descriptions of Medical Fungi*, 2nd ed. Mycology Unit Women's and Children's Hospital, Nexus Print Solutions, South Australia.

Kwon-Chung KJ, Bennett JE, 1992. *Medical Mycology*. Lea & Febiger, Philadelphia.

Murray PR, Baron EJ, Jorgenson JH, Landry ML, Pfaller MA (eds.), 2007. *Methods in Clinical Microbiology*, 9th ed., vol 2, pp 1723–2256, Pfaller MA (vol ed.), Section VIII, Mycology, Warnock DW (sect ed.), Section IX, Antifungal Agents and Susceptibility Test Methods, Turnidge JD (sect ed.). ASM Press,Washington, DC.

Sutton DA, Fothergill AW, Rinaldi MG, 1998. *Guide to Clinically Significant Fungi*. Williams and Wilkins, Baltimore, MD. This is a comprehensive and authoritative handbook containing first-quality photomicrographs and explanatory notes of the major and minor fungal pathogens.

Transporting Infectious Substances Safely, a publication of the U.S. Department of Transportation Pipeline and Hazardous Materials Safety Administration. URL http://www.phmsa.dot.gov/hazmat/library Write in the name of this publication in the search window.

SELECTED REFERENCES FOR LABORATORY DIAGNOSTIC METHODS IN MEDICAL MYCOLOGY

Baker RD, 1971. *In*: *Human Infection with Fungi, Actinomycetes and Algae*. Springer Verlag; New York, pp. 6–9.

Chandler FW, Kaplan W, Ajello L, 1980. *A Colour Atlas and Textbook of the Histopathology of Mycotic Diseases*. Wolfe Medical Publications, Ltd., London, UK.

Chandler FW, Watts JC, 1987. *Pathologic Diagnosis of Fungal Infections*. ASCP Press, Chicago.

Cockerill FR 3rd, Reed GS, Hughes JG, Torgerson CA, Vetter EA, Harmsen WS, Dale JC, Roberts GD, Ilstrup DM, Henry NK, 1997. Clinical comparison of BACTEC 9240 plus aerobic/F resin bottles and the isolator aerobic culture system for detection of bloodstream infections. *J Clin Microbiol* 35: 1469–1472.

Dahlberg KR, Etten JLV, 1982. Physiology and biochemistry of fungal sporulation. *Annu Rev Phytopathol* 20:281–301.

Darken MA, 1961. Applications of fluorescent brighteners in biological techniques. *Science* 133: 1704–1705.

Durkin M, Connolly P, Kuberski T, Myers R, Kubak BM, Bruckner D, Pegues D, Wheat LJ, 2008. Diagnosis of coccidioidomycosis with use of the *Coccidioides* antigen enzyme immunoassay. *Clin Infect Dis* 47: e69–e73.

Durkin M, Estok L, Hospenthal D, Crum-Cianflone N, Swartzentruber S, Hackett E, Wheat LJ, 2009. Detection of *Coccidioides* antigenemia following dissociation of immune complexes. *Clin Vaccine Immunol* 16:1453–1456.

Ellepola ANB, Morrison CJ, 2005. Laboratory diagnosis of invasive candidiasis. *J Microbiol* (*Korea*) 43 (special issue No. 5): 65–84.

Eisenman HC, Mues M, Weber SE, Frases S, Chaskes S, Gerfen G, Casadevall A, 2007. *Cryptococcus neoformans* laccase catalyses melanin synthesis from both D- and L-DOPA. *Microbiology* 153:3954–3962.

Filler SG, Sheppard DC, Edwards, JE Jr, 2006. Molecular Basis of Fungal Adherence to Endothelial and Epithelial Cells, pp. 187–196, *in*: Heitman J, Filler SG, Edwards JE Jr, Mitchell AP (eds.), *Molecular Principles of Fungal Pathogenesis*. ASM Press, Washington, DC.

Geha DJ, Roberts GD, 1994. Laboratory detection of fungemia. *Clin Lab Med* 14:83–97.

Haase G, Sonntag L, Melzer-Krick B, de Hoog GS, 1999. Phylogenetic inference by SSU-gene analysis of members of the *Herpotrichiellaceae* with special reference to human pathogenic species. *Studies Mycol* 43:80–97.

Hall L, Wohlfiel S, Roberts GD, 2004. Experience with the MicroSeq D2 Large-Subunit Ribosomal DNA Sequencing Kit for identification of filamentous fungi encountered in the clinical laboratory. *J Clin Microbiol* 42:622–626.

Hamer EC, Moore CB, Denning DW, 2006. Comparison of two fluorescent whiteners, Calcofluor and Blankophor, for the detection of fungal elements in clinical specimens in the diagnostic laboratory. *Clin Microbiol Infect* 12:181–184.

Harrington B, 2008. Staining of cysts of *Pneumocystis jiroveci* (*P. carinii*) with the fluorescent brighteners Calcofluor White and Uvitex 2B: A review. *Lab Med*. 39:731–735.

Hata DJ, Hall L, Fothergill AW, Larone DH, Wengenack NL, 2007. Multicenter evaluation of the new VITEK 2 Advanced Colorimetric Yeast Identification Card. *J Clin Microbiol* 45:1087–1092.

Hinrikson HP, Hurst SF, Lott TJ, Warnock DW, Morrison CJ, 2005. Assessment of ribosomal large-subunit D1-D2, internal transcribed spacer 1, and internal transcribed spacer 2 regions as targets for molecular identification of medically important *Aspergillus* species. *J Clin Microbiol* 43:2092–2103.

Hoch HC, Galvani CD, Szarowski DH, Turner JN, 2005. Two new fluorescent dyes applicable for visualization of fungal cell walls. *Mycologia* 97:580–588.

Hospenthal DR, Murray CK, Beckius, ML, Green JA, Dooley DP, 2002. Persistence of pigment production by yeast isolates grown on CHROMagar *Candida* medium. *J Clin Microbiol* 40:4768–4770.

Hurst SF, 2008. Personal communication.

Iwen PC, Hinrichs SH, Rupp ME, 2002. Utilization of the internal transcribed spacer regions as molecular targets to detect and identify human fungal pathogens. *Med Mycol* 40:87–109.

Jensen HE, Schonheyder HC, Hotchi M, Kaufman L, 1996. Diagnosis of systemic mycoses by specific immunohistochemical tests. *APMIS* 104:242–258.

Jensen HE, Salonen J, Ekfors TO, 1997. The use of immunohistochemistry to improve sensitivity and specificity in the diagnosis of systemic mycoses in patients with haematological malignancies. *J Pathol* 181:100–105.

Kaufmann CS, Merz WG, 1982. Two rapid pigmentation tests for identification of *Cryptococcus neoformans*. *J Clin Microbiol* 15:339–341.

Kurtzman CP, 2000. Systematics and taxonomy of yeasts. *Contrib Microbiol* 5:1–14.

Kovacs JA, Gill VJ, Meshnick S, Masur H, 2001. New insights into transmission, diagnosis, and drug treatment of *Pneumocystis carinii* pneumonia. *J Am Med Assoc* 286:2450–2460.

Kwon-Chung KJ, Polacheck I, Bennett JE, 1982. Improved diagnostic medium for separation of *Cryptococcus neoformans* var. *neoformans* (serotypes A and D) and *Cryptococcus neoformans* var. *gattii* (serotypes B and C). *J Clin Microbiol* 15:535–537.

Larone DH, 2002. *Medically Important Fungi, A Guide to Identification*, 4th ed. ASM Press, Washington, DC.

Lazcano O, Speights VO Jr, Strickler JG, Bilbao JE, Becker J, Diaz J, 1993. Combined histochemical stains in the differential diagnosis of *Cryptococcus neoformans*. *Mod Pathol* 6:80–84.

Lindsley MD, Warnock DW, Morrison CJ, 2006. Serological and Molecular Diagnosis of Fungal Infections, Chapter 66, pp. 569–608, *in*: Detrick B, Hamilton RG, Folds JD (eds.), *Manual of Molecular and Clinical Laboratory Immunology*, 7th ed. ASM Press, Washington, DC.

Masuoka J, 2004. Surface glycans of *Candida albicans* and other pathogenic fungi: Physiological roles, clinical uses, and experimental challenges. *Clin Microbiol Rev* 17:281–310.

McDonald LC, Weinstein MP, Fune J, Mirrett S, Reimer LG, Reller LB, 2001. Controlled comparison of BacT/ALERT FAN aerobic medium and BACTEC fungal blood culture medium for detection of fungemia. *J Clin Microbiol* 39:622–624.

McGinnis MR, 1980. *Laboratory Handbook of Medical Mycology*. Academic Press, New York.

Mendoza L, Taylor JW, Ajello L, 2002. The class mesomycetozoea: A heterogeneous group of microorganisms at the animal-fungal boundary. *Annu Rev Microbiol* 56:315–344.

Meyer M-H, Letscher-Bru V, Jaulhac B, Waller J, Candolfi E, 2004. Comparison of Mycosis IC/F and Plus Aerobic/F media for diagnosis of fungemia by the Bactec 9240 system. *J Clin Microbiol* 42:773–777.

Monod M, Baudraz-Rousselet F, Ramelet AA, Frenk E, 1989. Direct mycological examination in dermatology: A comparison of different methods. *Dermatologica* 179:183–186.

Nelson PE, Toussoun TA, Marasas WFO, 1983. *Fusarium Species: An Illustrated Manual for Identification*. Pennsylvania State University, University Park, PA.

Odds FC, Bernaerts R, 1994. CHROMagar *Candida*, a new differential isolation medium for presumptive identification of clinically important *Candida* species. *J Clin Microbiol* 32:1923–1929.

Padhye AA, Bennett JE, McGinnis MR, Sigler L, Fliss A, Salkin IF, 1998. Biosafety considerations in handling medically important fungi. *Med Mycol* 36(Suppl 1): 258–265.

Page BT, Kurtzman CP, 2005. Rapid identification of *Candida* species and other clinically important yeast species by flow cytometry. *J Clin Microbiol* 43:4507–4514.

Paliwal DK, Randhawa HS, 1978. Evaluation of a simplified *Guizotia abyssinica* seed medium for differentiation of *Cryptococcus neoformans*. *J Clin Microbiol* 7:346–348.

Pappagianis D, Zimmer BL, 1990. Serology of coccidioidomycosis. *Clin Microbiol Rev* 3:247–268.

Pickering JW, Sant HW, Bowles CAP, Roberts WL, Woods GL, 2005. Evaluation of a $(1\rightarrow3)$-ß-d-glucan assay for diagnosis of invasive fungal infections. *J Clin Microbiol* 43:5957–5962.

Reimer LG, Wilson ML, Weinstein MP, 1997. Update on detection of bacteremia and fungemia. *Clin Microbiol Rev* 10:444–465.

Rodriguez-Tudela JL, Aviles P, 1991. Improved adhesive method for microscopic examination of fungi in culture. *J Clin Microbiol* 29:2604–2605.

Sambrook J, Russell D, 2001. *Molecular Cloning: A Laboratory Manual*, 3 vol set, 3rd ed. Cold Spring Harbor Press, New York.

Sanguinetti M, Porta R, Sali M, La Sorda M, Pecorini G, Fadda G, Posteraro B, 2007. Evaluation of VITEK 2 and RapID Yeast Plus Systems for yeast species identification: Experience at a large clinical microbiology laboratory. *J Clin Microbiol* 45:1343–1346.

Sato Y, Osabe S, Kuno H, Kaji M, Oizumi K, 1999. Rapid diagnosis of cryptococcal meningitis by microscopic examination of centrifuged cerebrospinal fluid sediment. *J Neurol Sci* 164:72–75.

Sigler L, 1997. Chapter 9, *Chrysosporium* and Molds Resembling Dermatophytes, pp. 261–311, *in*: Kane J, Summerbell R, Sigler L, Krajden S, Land G (eds.), *Laboratory Handbook of Dermatophytes*. Star Publishing Co., Belmont, CA.

Sigler L, Bartley JR, Parr DH, Morris AJ, 1999. Maxillary sinusitis caused by medusoid form of *Schizophyllum commune*. *J Clin Microbiol* 37:3395–3398.

Simione FP, 2006. *Cryopreservation Technical Manual*. Nalge-Nunc International, Rochester, NY.

St Germain G, Beauchesne D, 1991. Evaluation of the MicroScan Rapid Yeast Identification panel. *J Clin Microbiol* 29:2296–2299.

St Germain G, Summerbell R, 1996. *Identifying Filamentous Fungi: A Clinical Laboratory Handbook*. Star Publishing Co., Belmont, CA.

Sugita T, Nakajima M, Ikeda R, Matsushima T, Shinoda T, 2002. Sequence analysis of the ribosomal DNA intergenic spacer 1 regions of *Trichosporon* species. *J Clin Microbiol* 40:1826–1830.

Summerbell RC, 1993. The Benomyl test as a fundamental diagnostic method for medical mycology. *J Clin Microbiol* 31:72–77.

Sutton DA, 2007. Specimen Collection, Transport, and Processing: Mycology, pp. 1728–1736, *in*: Murray PR, Baron EJ, Jorgenson JH, Landry ML, Pfaller MA (eds.), *Manual of Clinical Microbiology*, 9th ed., Vol 2. ASM Press, Washington, DC.

Valente P, Ramos JP, Leoncini O, 1999. Sequencing as a tool in yeast molecular taxonomy. *Can J Microbiol* 45:949–958.

Wheat LJ, 2003. Current diagnosis of histoplasmosis. *Trends Microbiol* 11:488–494.

White PL, Shetty A, Barnes RA, 2003. Detection of seven *Candida* species using the Light-Cycler system. *J Med Microbiol* 52:229–238.

Willinger B, Wein S, Hirschl AM, Rotter ML, Manafi M, 2005. Comparison of a new commercial test, GLABRATA RTT, with a dipstick test for rapid identification of *Candida glabrata*. *J Clin Microbiol* 43:499–501.

Wilson JWW, Plunkett OA, 1965. *The Fungous Diseases of Man*. University of California Press, Berkeley and Los Angeles.

Yeo S-F, Huie S, Sofair AN, Campbell S, Durante A, Wong B, 2006. Measurement of serum d-arabinitol/creatinine ratios for initial diagnosis and for predicting outcome in an unselected, population-based sample of patients with *Candida* fungemia. *J Clin Microbiol* 44:3894–3899.

Zimmer BL, Roberts GD, 1979. Rapid selective urease test for presumptive identification of *Cryptococcus neoformans*. *J Clin Microbiol* 10:380–381.

WEBSITES CITED

American Society for Microbiology (ASM) Microbe Library at www.microbelibrary.org. William Mayberry has posted a PowerPoint presentation entitled "Introduction to mycology" which contains useful illustrations relevant to the material in this chapter.

BigDye Terminator cycle sequencing manual: BigDye® Terminator v3.1 Cycle Sequencing Kit: Protocol: Rev A.

Biosafety in Microbiological and Biomedical Laboratories (BMBL), 5th ed. Available at http://www.cdc.gov/od/ohs/biosfty/bmbl5/bmbl5toc.htm.

Coccidioidomycosis Serology Laboratory at University of California Davis, directed by Prof. Demosthenes Pappagianis. http://som.ucdavis.edu/departments/microbiology/cocci/.

Combined Hematoxylin and Eosin/Methenamine Silver Stain. The method of procedure is given at the Rowley Biochemical Co. site at http://www.rowleybio.com/pdfs/A-123%20Procedure.pdf.

Difco Manual of microbiologic media at the URL http://www.bd.com/ds/technicalCenter/inserts/Yeast_Media.pdf.

DNeasy Kit from Qiagen Inc. (Valencia, CA). http://www1.qiagen.com/Products/GenomicDnaStabilizationPurification/DNeasyTissueSystem/DNeasyBloodTissueKit.aspx.

Dr. Fungus. http://www.doctorfungus.org/. "Dr. Fungus" website contains information on sick building syndrome; pictures and descriptions of fungi; antifungal agents; laboratory procedures; and a database of minimum inhibitory concentrations at www.doctorfungus.org.

Fungal Tree of Life, primers for PCR. http://aftol.org/.

The Fifth Kingdom, by Bryce Kendrick. Available at http://www.mycolog.com.

Formulas of mycologic media:

- BD Biosciences at the URL http://www.bd.com/ds/technicalCenter/.
- Difco Manual at the URL http://www.bd.com/ds/technicalCenter/inserts/Yeast_Media.pdf.
- Centraalbureau voor Schimmelcultures in the Netherlands at the URL http://www.cbs.knaw.nl/pdf/mediaCBS.pdf.
- Alberta Microfungus Collection and Herbarium, Dr. Lynne Sigler, Principal Investigator, at the URL http://www.devonian2.ualberta.ca/uamh/culture_receipt.htm.

Identify molds using the online interactive illustrated key at Prof. David Ellis' website: "Mycology Online" at the URL http://www.mycology.adelaide.edu.au/Fungal_Jungle/Identification_Keys.html.

KOH prep alternative formulations may be viewed at Prof. David Ellis' website at the URL www.mycology.adelaide.edu.au/Laboratory_Methods.

Modified Leonian's agar formula may be found at the URL http://www.devonian2.ualberta.ca/uamh/culture_receipt.htm.

"Molds, Isolation Cultivation and Identification," by David Malloch of the University of Toronto may be viewed at the URL http://labs.csb.utoronto.ca/moncalvo/malloch/Moulds/Identification.html.

MycoBank–Yeasts, a polyphasic database for identifying unknown yeasts. http://www.cbs.knaw.nl/yeast/(f3rkvyblldafnu55ugsca445)/BioloMICS.aspx.

Nikon microscopy website interactive tutorial on Köhler illumination, phase contrast microscopy. http://www.microscopyu.com/tutorials/java.

Olympus Microscope web page. http://www.olympusmicro.com/primer/techniques/fluorescence/fluorotable2.html. Also offering a video tutorial on selection of excitation and barrier filters.

Proficiency Testing Providers are located at the URL http://www.cms.hhs.gov/CLIA/downloads/ptlist.pdf.

Sequencher Tutorial. http://www.sequencher.com/sequencher/tutorials.html.

Sigma-Aldrich Stain Expert at the URL http://www.sigmaaldrich.com/life-science/cell-biology/hematology-and-histology/learning-center/stain-expert.html.

Sizing microscopic forms. http://web.uvic.ca/ail/techniques/measuring.html.

University of Alberta Microfungus Collection and Herbarium, Dr. Lynne Sigler, Principal Investigator at the URL http://www.devonian2.ualberta.ca/uamh/culture_receipt.htm.

U.S. Environmental Protection Agency Disinfectant Registry at the URL http://www.epa.gov/oppad001/chemregindex.htm.

COMMERCIAL MANUFACTURERS AND SUPPLIERS OF FUNGAL MEDIA, STAINS, AND REAGENTS

Associates of Cape Cod
124 Bernard E. Saint Jean Drive
E Falmouth MA
1-800- 568-0058
www.acciusa.com/clinical

BD Diagnostic Systems
7 Loveton Circle
Sparks, MD 21152
1-800-675-0908
www.bd.com

bioMérieux, Inc.
595 Anglum Rd
Hazlewood, MD 63042-2302
1-800-638-4835
www.biomérieux.usa.com

Fisher Scientific
Hampton, NH 03842-1819
1-800-640-0640
www.fishersci.com

Hardy Diagnostics
1430 West McCoy Lane
Santa Maria, CA 933455
1-800-266-2222
www.hardydiagnostics.com

Immunomycologics Inc.
2700 Technology Pl
Norman, OK 73071
1-800-654-3639
www.immy.com

Meridian Bioscience
3471 River Hills Drive
Cincinnati, OH 45244
1-800-543-1980
www.meridianbioscience.com

Remel
12076 Santa Fe Drive
Lenexa, KS 66215
1-800-255-6730
www.remel.com

PACKING AND SHIPPING OF INFECTIOUS AGENTS AND CLINICAL SPECIMENS

Please refer to BMBL, Appendix C, "Transportation and Transfer of Biological Agents", "Packing and Labeling of Infectious Substances," and "Packing and Labeling of Clinical Specimens."

The International Air Transport Association (IATA, www.IATA.org) regulates packing and shipping of infectious agents and clinical specimens by air according to their "Dangerous Goods Regulations Manual." Ordering information for the manual can be found on the IATA website.

QUESTIONS

The Answer Key to multiple choice questions may be found at the end of the book.

1. A sputum specimen comes to the mycology bench. What steps are taken to detect, recover, and identify a fungus in the specimen?
1. Culture on Mycosel agar.
2. Culture on SABHI +5% sheep blood and cycloheximide and chloramphenicol.
3. Culture on SDA-Emmons with chloramphenicol.
4. Direct examination in a KOH prep.

which are correct?
A. 1 + 2 are correct.
B. 1 + 4 are correct.
C. 2 + 3 are correct.
D. 2 + 4 are correct.
E. 3 + 4 are correct.

2. A CSF specimen arrives in the laboratory and a portion is sent to the mycology bench. How would you propose to detect, recover, and identify a fungus in the specimen?
A. Centrifuge and place a drop of sediment in an india ink prep.
B. Centrifuge and plant the sediment on SDA-Emmons agar.
C. Plant a drop of the sediment on birdseed agar.
D. Reserve the supernatant after centrifugation for fungal serology.
E. All are correct.

3. In the previous question what other stain(s) would be appropriate for the centrifuged CSF sediment?
A. Gram stain
B. May–Grünwald Giemsa
C. Calcofluor (FungiFluor®)
D. Alcian Blue
E. All the above

4. An immune-normal young adult has been admitted with community acquired pneumonia. Blood cultures are being prepared. What blood culture system would you recommend for the recovery of a dimorphic environmental mold?
A. BacT/ALERT 3D® system
B. Biphasic broth Septi-Chek® system
C. Isolator® solid medium system
D. Mycosis IC/F medium and the BACTEC® system

5. A mold growing on a blood agar plate is sent to the mycology bench from bacteriology. What are the initial steps to take in the work-up of this culture?
1. Make a tease mount.
2. Set up a slide culture.
3. Transfer to Mycosel agar.
4. Transfer to SDA-Emmons agar.

which are correct?
A. 1 + 3 + 4 are correct.
B. 1 + 2 + 3 are correct.
C. 2 + 3 + 4 are correct.
D. All are correct.

6. A yeast is growing in a blood culture bottle. Which of the following direct examination methods would be appropriate to aid in identifying it?
1. Germ tube test
2. Gram stain
3. KOH prep
4. Peptide nucleic acid test "Yeast traffic light"

Which are correct?
A. 1 + 3
B. 2 + 4
C. 2 + 3
D. 2 + 3
E. 1 + 4

7. A yeast is growing in a blood culture bottle. Which of the following culture methods would lead directly to a presumptive identification?
A. BHI agar
B. CHROMagar *Candida*®
C. SDA-Emmons agar
D. SDA-Emmons agar with cycloheximide

8. Match the serology test to the disease that best fits.
A. Antigenuria–HPA test
B. Antigen in CSF–Latex agglutination
C. VF antibody titer–Complement fixation
D. Antigenemia GM-EIA

1. Aspergillosis
2. Coccidioidomycosis
3. Cryptococcosis
4. Histoplasmosis

9. Compare and contrast the three types of conidiogenesis in fungi: thallic, enteroblastic, and holoblastic, giving examples of each. In your discussion use the terms acropetal and basipetal.

10. A mold has been growing slowly on SDA-Emmons for 5 days with no evidence of sporulation. In the next 2 days growth continued but without sporulation. The physician believes this sterile site isolate is important and is keen to know its identification. Discuss the additional steps to take to expedite this process. Include the different media known to stimulate sporulation including the addition of plant material; the effect of light; septum formation; growth in the presence of inhibitors—cycloheximide and Benomyl; and preparation for genetic identification.

11. Describe the three main subregions of rDNA useful as targets for PCR-sequencing. Give examples of groups of fungi that are capable of being identified to species level with each of the target regions.

12. Make an outline of the steps necessary to use PCR-sequencing as a means to identify an unknown fungus to the species level.

Chapter 3A

Antifungal Agents and Therapy

3A.1 INTRODUCTION

Antifungal agents can be used in four ways: prophylactic, preemptive, empiric therapy, and therapy of a definitively diagnosed fungal infection. Prophylaxis is the use of antifungal agents to prevent fungal infections in high-risk patients. These patients include hematopoietic stem-cell transplant (*SCT*), solid organ transplant (*SOT*) recipients, and other potentially immunocompromised patients. Preemptive therapy is the use of antifungal agents in patients who have positive diagnostic laboratory tests but do not yet show signs of fungal disease. Preemptive therapy using antifungal agents is not well defined. However, as the use of improved nonculture-based diagnostic tests becomes more commonplace, preemptive therapy will likely be used more widely. Empiric therapy is used in patients who have signs and symptoms of a fungal infection but do not yet have a definitive diagnosis of a fungal infection. This category includes patients with fever and neutropenia, or other high risk patients not responding to antibacterial agents. Definitive therapy is defined as the use of an antifungal agent in a patient with a proven fungal infection.

Successful antifungal therapy depends on an informed choice of therapeutic agents, dosing, and the delivery regimen. Factors influencing clinical treatment decisions include:

- Prompt and correct identification of the fungus, if possible
- Matching the drug choice and regimen to the causative agent and clinical syndrome
  - Susceptibility testing of positive cultures
- Controlling or correcting underlying medical or immunosuppressive condition(s)
- Understanding potential drug interactions and pharmacokinetics
- Guarding against allergic reactions
- Awareness of adverse events
- Following primary therapy with maintenance therapy
- Follow-up for relapse after treatment

The diagnosis and treatment of invasive mycoses can be more difficult than that of bacterial infections. The threat of higher attributable mortality adds urgency to establishing an early diagnosis, selecting the appropriate antifungal agent, or combination of agents, and prompt initiation of treatment. Often clinical treatment decisions are made before the etiologic agent is identified. This is done to avoid delay, which may lead to a poor outcome. Two important factors influencing successful clinical management are:

1. *Rapid Diagnosis.* Challenges to a rapid diagnosis of invasive fungal infections are the low yield of blood cultures in candidiasis and the nonutility of blood cultures to detect most *Aspergillus* species. Detection of fungal DNA in blood cultures is hindered by the difficulty in lysing the fungal cell wall and the fungal load in blood, which may be lower than 10 CFU/mL (White et al., 2003). The laboratory can provide presumptive identification by means of rapid diagnostic tests such as direct examination of clinical specimens, histopathology, serology, or other nonculture methods, preferably followed by culture confirmation and, where indicated, susceptibility testing. Radiography is another important diagnostic adjunct.
2. *Immune Status of the Patient.* Prolonged and profound neutropenia due to immunosuppressive therapy increases susceptibility to fungal infections. The incidence of mycoses in leukemia patients, and after an allogeneic SCT, is estimated at 10–25% (Groll and Tragiannidis, 2009). The case-fatality rate continues to exceed 50% and approaches 90%

Fundamental Medical Mycology, First Edition. By Errol Reiss, H. Jean Shadomy and G. Marshall Lyon, III.
© 2012 Wiley-Blackwell. Published 2012 by John Wiley & Sons, Inc.

in patients when infection is disseminated, if there is CNS involvement, or when neutropenia persists. *Candida* and *Aspergillus* species account for the majority of infections. Current trends indicate a shift toward non-*albicans Candida* species and to previously uncommon fungi, which pose challenges in diagnosis and therapy.

3A.1.1 Major Antifungal Agents Approved for Clinical Use

This chapter describes the major antifungal agents approved for clinical use at this time in the United States (Table 3A.1) (Patterson, 2006). Each disease-specific chapter contains a section, "Therapy," providing further targeted information. Fungi contain an array of intracellular organelles similar to mammalian cells, complicating the development of drugs that selectively target fungi. Two broad categories of successful antifungal agents are natural products and synthetic compounds.

3A.1.1.1 Natural Products

The discovery of natural products, produced by competing soil microbes, yielded several useful antifungal agents: for example, amphotericin B, nystatin, griseofulvin, and the echinocandin derivatives (e.g., caspofungin).

3A.1.1.2 Chemically Synthesized Compounds

Azoles Pure chemical synthesis successfully developed increasingly effective compounds based on the imidazole clotrimazole. Miconazole and ketoconazole are imidazoles, and the latter was the first orally administered azole capable of treating endemic mycoses, such as histoplasmosis. First generation triazoles then appeared, including fluconazole and itraconazole, which are in current use. Newer, second generation triazoles, principally voriconazole (Pfizer), but also posaconazole (Schering-Plough/Merck & Co.) and ravuconazole (experimental, Bristol-Myers-Squibb) have been developed to have a broader spectrum of activity, especially against mold infections.

3A.1.1.3 Special Topic — Triazoles

Triazole refers to either one of a pair of isomers of a five-membered ring structure consisting of two carbon atoms and three nitrogen atoms, with the molecular formula $C_2H_3N_3$. The first generation of triazole antifungals includes fluconazole (FLC) and itraconazole (ITC). FLC is indicated for yeast infections but has limited effectiveness against *Candida krusei* and *Candida glabrata*. ITC is indicated for mold infections, but there are concerns about unpredictable bioavailability. Three newer extended spectrum triazoles act by the same mechanism: inhibition of fungal cytochrome P450 (CYP)-dependent 14-α-demethylase, which is critical for synthesis of ergosterol, an important constituent of the fungal cell membrane. The result is accumulation of ergosterol precursors in the fungal membrane and inhibition of fungal growth. Voriconazole (VRC) is derived from FLC. Ravuconazole is similar in structure to FLC, containing a thiazole instead of a second triazole. Posaconazole (PSC) is structurally similar to ITC but with fluorine substituents in place of chlorine and a furan ring in place of the dioxolane ring.

The subject of this chapter is anti-infectives and their therapeutic use in mycoses. Five classes of antifungal agents are used orally or intravenously for this purpose: polyenes, pyrimidine analogs, allylamines, azoles, and echinocandins. Another natural product, griseofulvin, has a specialized use in treating dermatophytosis. In addition to the present text, clinical laboratory scientists and physicians are encouraged to consult other resources to continually increase their knowledge of these agents and their use.

3A.1.1.4 Additional Resources

- An excellent PowerPoint presentation is "Introduction to Antifungal Pharmacology," by Dr. Russell E. Lewis, University of Houston College of Pharmacy, Texas. It can be viewed at the URL www.aspergillus.org.uk/education/introductionAntifung_rev.ppt.
- A pertinent review article is "Pharmacology of systemic antifungal agents," Dodds Ashley et al. (2006).
- Structure diagrams of antifungal agents may be found at the U.S.A. National Center for Biotechnology Information (NCBI) website, PubChem-substance, at the URL http://pubchem.ncbi.nlm.nih.gov/, or at the National Institute of Allergy and Infectious Diseases (NIAID) website at the URL http://chemdb2.niaid.nih.gov/struct_search/.
- A full listing of signs of adverse effects and drugs that interact with antifungal agents may be found at the U.S.A. National Library of Medicine website—"Medline-Plus" at the URL www.medlineplus.gov.
- Drugs.com (professional category) has useful information on clinical pharmacology.
- The Johns-Hopkins Poc-It ABX guide includes indications, cost, dosing regimens, and adverse reactions at the URL http://Hopkins-abxguide.org.

Table 3A.1 Antifungal Agents in Current Use for Invasive Mycoses

Agent	Class	Dosage	Major toxicities, dose adjustments for organ dysfunction	Clinical efficacy
AmB	Polyene	0.6–1.5 mg/kg/day, IV	Dose-limiting renal toxicity; infusion related reactions	Efficacy limited in severely immunosuppressed patients; higher mortality in randomized aspergillosis trial
Lipid formulations of AmB: L-AmB	Polyene	3–6 mg/kg/ day, IV	Less toxic than AmB; systemic toxicities: ABCD>ABLC>L-AmB; high doses ($\geq$10 mg/kg/day) associated with increased renal and liver toxicity	Anecdotal outcomes better with higher doses, but not proved; expensive
Fluconazole	Triazole	400–800 mg/day (IV or orally)	Well tolerated; dose adjustment for renal insufficiency, although toxicity not clearly increased	Active against *C. albicans*, variable activity versus other yeasts, especially *C. glabrata*; poorer outcomes versus echinocandin in one candidemia trial
Itraconazole	Triazole	400–600 mg/day (oral); 200 mg/day, IV	Liver and renal toxicity with some chemotherapeutics; gastrointestinal toxicity; avoid IV with creatinine clearance <30 mL/min	Second line aspergillosis therapy; suspension improves bioavailability; limited efficacy data for IV form
Voriconazole	Extended spectrum triazole (similar in structure to FLC)	6 mg/kg every 12 h × 2 as a loading dose; then 4 mg/kg every 12 h, IV; 200 mg orally twice daily	Visual hepatic and skin toxicity; avoid IV with creatinine clearance <50 mL/min; decrease dose with mild or moderate hepatic toxicity	Indicated as primary therapy for invasive aspergillosis in most patients due to survival advantage versus AmB; effective in *Fusarium* species, *Scedosporium*, and other molds, but not *Mucorales*; azole cross resistance in *Candida* can occur
Posaconazole	Extended spectrum triazole	200 mg orally 4 times daily then 400 mg orally twice daily (investigational in United States)	Oral suspension; well tolerated in clinical trials	Salvage therapy for invasive mycoses, including aspergillosis and mucormycosis; efficacy and survival advantage in prophylaxis of invasive fungal infection in high-risk patients (GvHD and acute leukemia); only oral formulation is clinically available

(*continued*)

Table 3A.1 *(continued)*

Agent	Class	Dosage	Major toxicities, dose adjustments for organ dysfunction	Clinical efficacy
Caspofungin	Echinocandin	A 70 mg IV loading dose then 50 mg/day	Well tolerated; potential cyclosporine drug interaction; uncommon hepatic toxicity; dose reduction in hepatic dysfunction	Less toxic with similar efficacy versus AmB in invasive candidiasis; indicated as salvage therapy against aspergillosis; activity against other molds not established; no clinical activity demonstrated versus *Mucorales* or *Cryptococcus*; well tolerated; only IV; expensive
Micafungin	Echinocandin	50–150 mg/day, IV	Well tolerated in clinical trials	Regulatory approval for prophylaxis in high-risk patients and esophageal candidiasis; investigational for candidemia; only IV; expensive
Anidulafungin	Echinocandin	200 mg IV load followed by 100 mg/day, IV	Well tolerated in clinical trials	Superior efficacy versus fluconazole in one candidemia trial; efficacy in *Candida* esophagitis; spectrum similar to other echinocandins; only IV; expensive

Abbreviations: AmB, amphotericin B deoxycholate; IV, intravenous; L-AmB, liposomal amphotericin B; ABLC, amphotericin B lipid complex; ABCD, amphotericin B colloidal dispersion; FLC, fluconazole; GvHD, graft-versus-host disease.

Source: Patterson (2006). Fungal infection in the immunocompromised patient: Risk assessment and the role of antifungal agents. Table reprinted with permission from Medscape, medscape.com, 2010.

3A.1.1.5 Special Topic: Basic Definitions Relevant to Pharmacology of Antifungal Drugs

Understanding the pharmacology of antifungal agents requires knowledge of basic definitions. Figure 3A.1 depicts the kinetics of a drug concentration in the plasma following its administration, with the relevant terms described below.

- *Pharmacokinetics (PK). Definition:* The time course of antimicrobial concentrations in the body.
- *Pharmacodyamics (PD). Definition:* The relationship between the change in antimicrobial concentration in the body over time and its antimicrobial effect.
- *Elimination Half-life* ($t_{1/2}$). *Definition*: The time taken for plasma concentration of the drug to reduce by 50%. After 4 half-lives, elimination is 94% complete.
- *Cmax. Definition:* Peak plasma concentration.
- *Cmin. Definition:* The trough, or minimum plasma concentration.
- *Area Under the Curve (AUC).* The change over time in the amount of drug available in the plasma, the graphic representation of the plasma drug

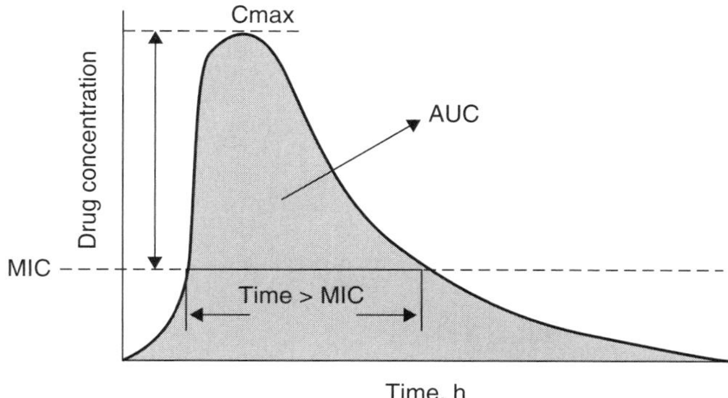

Figure 3A.1 Pharmacokinetic determinants of drug efficacy. The time course of antifungal drug activity is determined by pharmacokinetic (PK) parameters. The peak plasma level is the Cmax, which declines over time to the Cmin, or trough concentration. The curve shows how this change occurs over time with the plasma concentration determined as the area under the curve (AUC). The important ratios that determine drug efficacy are the Peak/MIC ratio—the peak plasma level or Cmax, divided by the MIC. Next, the T > MIC is the time above MIC, the percentage of the time between doses when the plasma level is greater than the MIC. Another ratio, the 24-h AUC/MIC, is the amount of time in a day when the plasma drug concentration is greater than the MIC, determined by dividing the 24-h AUC by the MIC.

concentration over time. The pharmacokinetic index that accounts for the drug exposure over time is the ratio of the 24 h AUC to the MIC. A free-drug 24 h AUC: MIC near 25:1 is predictive of a good clinical outcome for triazoles to treat invasive candidiasis but is less clear for other fungi–drug class combinations.

- *Volume of Distribution (Vd).* This is the amount of drug in the body divided by the concentration in the blood. More precisely it is the *theoretical* volume of fluid into which the total drug dose would have to be diluted to produce the concentration in plasma; for example, if the drug dose is 1000 mg and the plasma concentration achieved is 10 mg/L, that 1000 mg *seems* to be distributed in 100 L (dose/volume = concentration; 1000 mg/x L = 10 mg/L; therefore, x = 1000 mg/10 mg/L = 100 L). The Vd is usually expressed as mL/kg body weight.
  - Vd has nothing to do with the actual volume of the body or its fluid compartments but involves the distribution of the drug within the body. Drugs that are highly lipid soluble, or highly tissue bound, have a low plasma concentration and have a very high Vd (i.e., 500 L). Drugs that are water soluble remain in the blood and have a low Vd. Vd provides a reference for the expected plasma concentration but gives little information about the specific pattern of distribution. Each drug is uniquely distributed in the body.
  - Very few drugs distribute immediately throughout the body. In the two-compartment model, the drug equilibrates with a peripheral compartment slowly over time: (i) the drug is rapidly distributed into a central compartment consisting of blood and well-perfused organs—the liver and kidneys; the central compartment is referred to as "plasma." (ii) The peripheral compartment consists of muscle, lean tissue, and fat and is referred to as "tissue."

  An iv bolus injection of drug is followed by: instantaneous absorption and distribution of drug in the central compartment (plasma) → elimination from central compartment → distribution from central compartment to peripheral compartment (tissues) → distribution continues until the concentration in the central compartment (plasma) equals concentration in peripheral compartment (tissue). This is the distribution at steady state: Vd at steady state (Vdss). It is the sum of central and peripheral volume terms (at equilibrium).
- *Clearance (Cl).* Clearance of a drug is the volume of plasma from which the drug is completely removed per unit time. The amount eliminated is proportional to the concentration of the drug in the blood.
- *Bioavailability. Definition:* This is the fraction of the administered dose that reaches the systemic circulation. Bioavailability is 100% for IV injection. Bioavailability varies for other routes depending on the extent of absorption.
- *Dosing Regimen.* The maintenance dose is equal to the rate of elimination at steady state: that is, at steady state, rate of elimination = rate of administration. Dosing rate = clearance × desired plasma concentration.

3A.1.2 Comparison of Antibacterial and Antifungal Agents According to Their Intracellular Targets

Antibiotics developed to combat bacterial infections do not work against fungi, with the exception that antimetabolite drugs targeting folic acid biosynthetic pathways—trimethoprim-sulfamethoxazole—are active against *Pneumocystis jirovecii* and *Paracoccidioides brasiliensis*. The cell wall of bacteria contains peptidoglycan, which is absent in fungi, so that penicillins and cephalosporins are inactive against fungi. Conversely, antifungal drugs that target ergosterol do not work against bacteria because they lack this sterol in their cell membranes. The class of compounds known as echinocandin derivatives (e.g., caspofungin, micafungin, anidulafungin) target synthesis of ß-(1→3)-D-glucan, a fibrillar constituent of the cell walls of some yeasts and molds. That drug could be said to be analogous to penicillin. Antifungal drugs that target the fungal nucleus are 5-fluorocytosine (flucytosine) and griseofulvin. Flucytosine is a synthetic compound. Griseofulvin interferes with the mitotic spindle apparatus arresting fungal cells in metaphase.

The intracellular targets vulnerable to antifungal therapy are summarized in Fig. 3A.2.

3A.1.2.1 Primary Therapy

The antifungal agents in the current armamentarium for treating invasive fungal infections are shown in Table 3A.1. Primary therapy keyed to the etiologic agent is shown in Table 3A.2. The individual classes of antifungal compounds are considered below.

3A.2 AMPHOTERICIN B (AmB-DEOXYCHOLATE) (FUNGIZONE®, APOTHECON SUBSIDIARY OF BRISTOL-MYERS-SQUIBB)

3A.2.1 Structure

The structure of AmB is depicted in Fig. 3A.3, with the mode of action shown in Fig. 3A.4. AmB is a polyene antifungal agent and the natural product of *Streptomyces nodosus*. Chemically it is a heptaene—containing seven conjugated double bonds. That face of the molecule is the lipophilic face. The hydrophilic face contains six hydroxyl groups.

3A.2.2 Mode of Action

By intercalating with ergosterol in the fungal cell membrane, AmB creates a channel through which K^+ and other

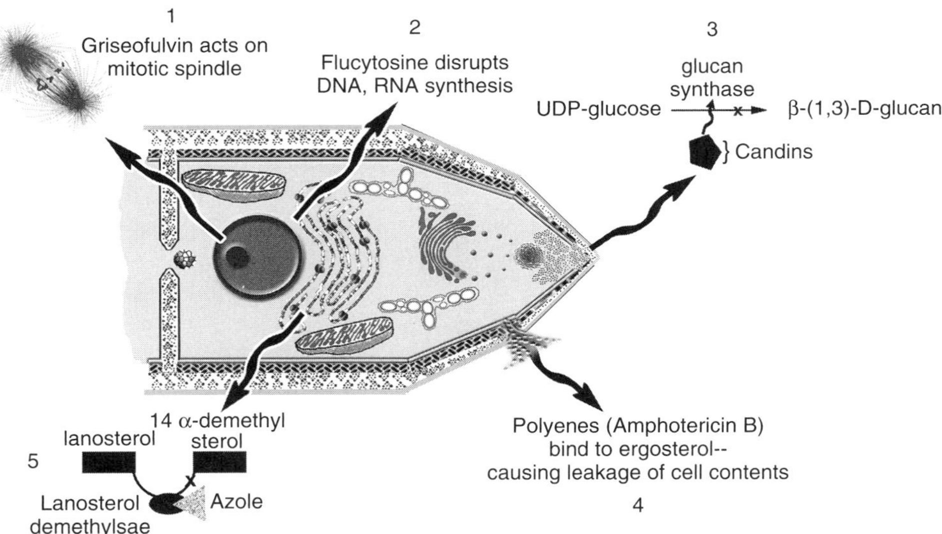

Figure 3A.2 Summary of targets for systemic antifungal agents. The fungal cell shown is the hyphal tip. (1) The mitotic spindle apparatus. The antifungal agent, griseofulvin, decreases the expression of ß-tubulin, thus inhibiting microtubules which are structural components of the spindle. The result is interruption of mitosis and arrest of cell division. (2) DNA is the cell target of the fluoropyrimidine drug, flucytosine. Once inside the cell, flucytosine is converted to 5-fluorouracil. Further metabolism occurs to 5-fluorodeoxyuridinemonophosphate (5-F-dUMP) and 5-fluorouridinetriphosphate (5-F-UTP). The former inhibits DNA synthesis while the latter inhibits RNA synthesis. The result is inhibition of DNA replication and protein synthesis, respectively. (3) Echinocandins (candins) are based on natural products produced by soil fungi. The drug target is glucan synthase, resulting in downregulation of ß-(1→3)-D-glucan, a fibrillar cell wall component of many fungi. ß-(1→3)-D-glucan is synthesized by adding the universal donor, UDP-glucose, to the growing glucan chain. Examples of echinocandins are caspofungin, anidulafungin, and micafungin. The pentagon shape represents an echinocandin interfering with ß-glucan synthase. (4) Polyene antifungal agents (e.g., amphotericin B) bind to ergosterol in the plasma membrane. The hydrophobic face of the polyene intercolates with ergosterol while the hydrophilic face creates a channel through which cell contents (*arrow*) leak out of the fungal cell. (5) Azole antifungal agents inhibit ergosterol biosynthesis by binding to lanosterol demethylase, a product of the *ERG11* gene.

Table 3A.2 Primary Therapy Keyed to the Etiologic Agent

	Mycosis	Antifungal agent for primary therapy	Second line or salvage antifungals
		Yeast infections	
	Candidiasis (systemic)	Fluconazole, echinocandin (CASF, MCF, ANF)	AmB, Voriconazole (VRC)
	Candidiasis (oroesophageal)	FLC (oral), Nystatin (lozenges), ITC (oral), VRC (oral)	AmB, posaconazole (PSC), CASF, MCF, ANF, VRC IV
	Candidiasis (vulvovaginal)	Clotrimazole topical cream, other topical creams (please see Chapter 11) FLC (oral), ITC (oral alternative)	
	Cryptococcosis	AmB + flucytosine; FLC for maintenance	FLC for nonmeningitis
Pneumocystosis		Trimethoprim-sulfamethoxazole (Bactrim)	Pentamidine; Clindamycin + Primaquine; Atovaquone
		Mold infections	
Hyalohyphomycetes	Aspergillosis	VRC	PSC, MCF, CASF, AmB VRC + echinocandin, AmB + echinocandin
	Mucorales	AmB	PSC; AmB + echinocandin
	Fusarium species infections	VRC, AmB	
Endemic mycoses	Blastomycosis	ITC	AmB reserved for treatment failures or meningitis
	Coccidioidomycosis	FLC	AmB, VRC, PSC
	Histoplasmosis	ITC	AmB for treatment failures and meningitis
	Paracoccidioidomycosis	ITC	AmB for treatment failures and meningitis
	Penicilliosis	AmB	
Dark pigmented molds causing phaeohyphomycosis[a]	No standard therapy because: there are a large number of fungal species involved, case series are difficult to establish, the sporadic nature, and different clinical forms of infections. Recovery from neutropenia is critical to success of antifungal therapy.	Suggested therapy for: • Subcutaneous nodules—surgery with or without ITC. • Cerebral abscess—surgery and high dose VRC or PSC; lipid AmB with or without 5FC. • Disseminated disease—*Scedosporium prolificans* has responded to a combination of VRC and terbinafine. Alternate is high dose lipid AmB with or without an echinocandin.	

(*continued*)

Table 3A.2 *(continued)*

	Mycosis	Antifungal agent for primary therapy	Second line or salvage antifungals
Mycotic keratitis	*Fusarium* species; some dark pigmented molds—*Curvularia, Bipolaris*	Topical natamycin with or without topical AmB	Oral VRC
Eumycotic mycetoma		ITC	PSC
Allergic fungal sinusitis	*Bipolaris, Curvularia*	corticosteroids	ITC, although not routinely recommended, may reduce requirement for steroids.
Dermatophytes	*Trichophyton tonsurans, T. rubrum, T. mentagrophytes; Microsporum canis*	Tinea capitis—oral ITC, oral griseofulvin; tinea cruris—terbinafine cream; tinea corporis—terbinafine cream, oral itraconazole; tinea unguium—oral terbinafine or oral itraconazole	

[a]Based on information in Revankar (2005).

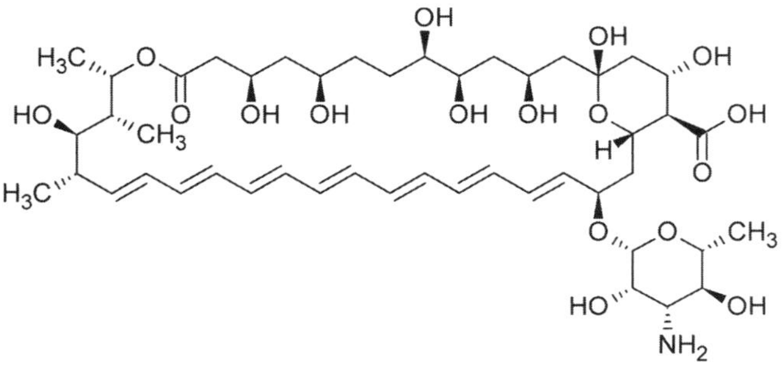

Figure 3A.3 Amphotericin B, chemical structure.
Source: PubChem at the URL http://pubchem.ncbi.nlm.nih.gov, a service of the National Center for Biotechnology Information and the National Library of Medicine, U.S.A.

cell constituents leak out, with resultant death of the fungal cell. Figure 3A.4 is a model of AmB-sterol pores, which depicts the mode of action (Andreoli, 1973). AmB has much higher affinity for ergosterol than for cholesterol and that is the basis of its selective toxicity in both yeasts and molds.

3A.2.3 Indications

AmB has reliable activity against most systemic mycoses. First discovered in Squibb Laboratories by Gold et al. (1956), AmB has played a major role since it first was approved to treat systemic fungal infections in 1959. However, dosages often have to be reduced to forestall renal toxicity. The advent of extended spectrum triazoles and echinocandin derivatives has lessened reliance on AmB, and now it is less often referred to as the "gold standard."

3A.2.4 Formulation

AmB is poorly absorbed from the gastrointestinal tract and must be administered via the intravascular (IV) route to treat systemic mycoses. AmB is insoluble in water and is formulated for use as a colloidal dispersion with sodium deoxycholate and phosphate buffer, "Fungizone." In this discussion the term AmB will apply to the deoxycholate dispersion formula.

3A.2.5 Spectrum of Activity

AmB is active against all sterol-containing microbes including yeasts, algae, protozoa, flatworms, and molds. AmB is a fungicidal drug against most yeasts and molds.

3A.2.6 Clinical Uses

3A.2.6.1 CNS Mycoses

AmB combined with flucytosine is the preferred primary therapy for cryptococcal meningitis. AmB remains the primary therapy for meningitis resulting from infection with primary dimorphic fungal pathogens, *B. dermatitidis, C. immitis, C. posadasii*, and *H. capsulatum*, and in treating rhino-orbito-cerebral abscess resulting from members of the *Mucorales* in the subphylum *Mucoromycotina*.

3A.2.6.2 Fever in Neutropenic Cancer Patients

AmB (as a lipid formulation) is recommended for empiric therapy in this patient group in the setting of profound

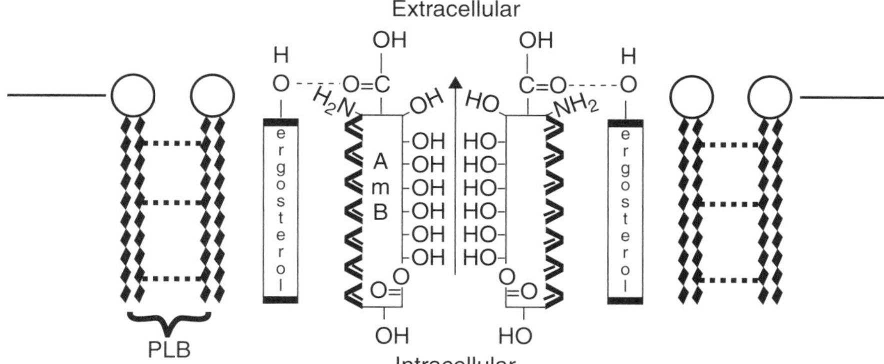

Figure 3A.4 A model for amphotericin B-sterol pores depicts the mode of action of AmB. By intercalating with ergosterol in the cell membrane, a channel is created through which cell constituents leak out *(arrow)* with resultant death of the fungal cell. The dotted lines between phospholipid hydrocarbon chains indicate van der Waals forces. PLB, phospholipid bilayer.
Source: Andreoli et al. (1973). Adapted by permission from Macmillan Publishers Ltd., © 1973.

neutropenia and fever unresponsive to 5 days of antibacterial therapy (Pappas et al., 2009). Caspofungin (CASF) or voriconazole (VRC) are also recommended, though VRC does not have an FDA indication for this purpose. Patients receiving prophylaxis with fluconazole (FLC) whose fever is unresponsive to that drug or to antibiotics are candidates for AmB. AmB-deoxycholate is an effective alternative but carries a higher risk of toxicity.

A lipid formulation of AmB is also indicated as a choice for primary therapy for *Candida* sepsis in neutropenic cancer patients, but is being replaced as the first choice by an echinocandin. Echinocandins (e.g., CASF) are particularly indicated where there is severe sepsis and shock (please see Section 3A.9, Caspofungin).

3A.2.6.3 Pulmonary Aspergillosis

Any formulation of AmB is no longer the first choice for primary therapy, because VRC showed superior survival when compared to AmB (Herbrecht et al., 2002). Liposomal AmB is recommended for salvage therapy for refractory aspergillosis.

3A.2.6.4 Primary Endemic Dimorphic Fungal Pathogens

The role for AmB is in patients with severe or rapidly progressing disease and/or when there is no response to ITC. AmB is primary therapy for penicilliosis, because the host's immune response is often compromised by HIV infection. AmB, or lipid AmB, is recommended for histoplasmosis in AIDS patients.

3A.2.7 Lipid Formulations of AmB

Formulations of AmB compounded with lipids were developed to relieve the potentially harmful side effects of AmB-deoxycholate. Lipid formulations are less nephrotoxic but more expensive than AmB-deoxycholate. Their advantages are increased stability, solubility, and absorption of AmB. Lipid formulations achieve higher concentrations in the lungs, liver, and spleen. They achieve an increased daily dose of AmB, and a decrease in infusion-associated side effects (especially liposomal AmB) and a marked decrease in nephrotoxicity. The different lipid formulations of AmB include:

- Amphotec® (Three Rivers Pharmaceuticals, Cranberry Township, PA). AmB in a colloidal dispersion with cholesteryl sulfate ("ABCD"). This formulation has, at times, been marketed as Amphocil. ABCD has a toxicity profile that is similar to AmB-deoxycholate. As such, it is seldom used.
- Abelcet® (Enzon, Inc, Bridgewater, NJ). An AmB lipid complex, Abelcet consists of a 1:1 ratio of AmB in combination with two phospholipids (a 7:3 ratio of dimyristoyl phosphatidylcholine to dimyristoyl phosphatidylglycerol).
- AmBisome® (Astellas, Inc.). This preparation consists of unilamellar bilayer liposomes with AmB intercalated within the liposome membrane. The liposomes are less than 100 nm in diameter. A schematic of the liposome is presented in Fig. 3A.5.

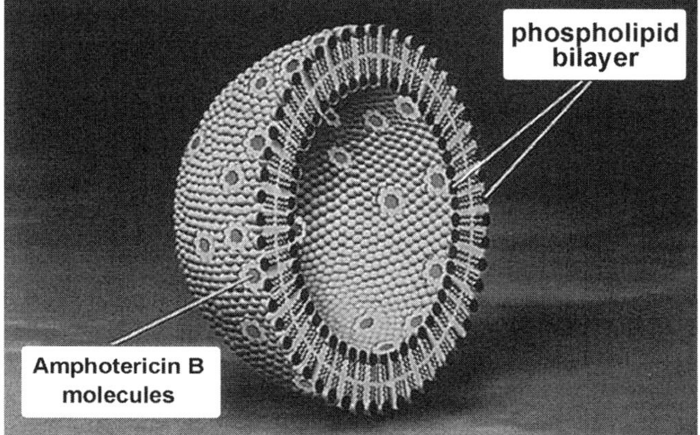

Figure 3A.5 Cross section of liposomal AmB, AmBisome® (Astellas, Inc.) AmBisome is a trademark of Gilead Sciences, Inc.

3A.2.7.1 Indications for Lipid-AmB Formulations

The FDA approved lipid-AmB drugs for systemic mycoses, particularly invasive aspergillosis, which is refractory to AmB-deoxycholate or when there are severe or persistent infusion-related unfavorable events, renal impairment as a result of therapy (i.e., when the serum creatinine >2.5 mg/dL), hypokalemia, hypomagnesemia, tubular acidosis, or polyuria, or when there is disease progression even after a cumulative AmB-deoxycholate dosage of 500 mg (Dismukes, 2000). Liposomal AmB is indicated for the febrile neutropenic patient unresponsive to antibacterial antibiotics.

3A.2.8 Pharmacokinetics

3A.2.8.1 Dosage

The AmB peak plasma concentration of 0.5–2 μg/mL is attained after repeated doses of 0.5 mg/kg/day. The usual dosing regimen for adults is a daily IV infusion, over several hours, of 0.3–0.6 mg/kg of AmB. Alternate day dosing with 60 mg is as effective. Slow infusion is especially necessary in patients with renal insufficiency to prevent severe hyperkalemia resulting in potentially fatal arrhythmia.

3A.2.8.2 Distribution

The distribution of AmB is multicompartmental. The volume of distribution (Vd) is 3.8–4.1 L/kg in adults. Only 10% of the AmB dose can be accounted for in plasma. AmB concentrations in the pleura, peritoneum, and synovium are about 60% of plasma concentrations. AmB is also distributed into other body fluids: pleural, pericardial, peritoneal, synovial fluid, and vitreous gel. AmB crosses the placenta, attaining low concentrations in amniotic fluid.

CSF concentrations of AmB are about 3% of the serum concentrations. Higher CSF concentrations (0.5–0.8 μg/mL) to treat recalcitrant cryptococcal meningitis and coccidioidal meningitis are attained by intrathecal administration into the lumbar, cisternal, or ventricular CSF (Stevens and Shatsky, 2001). Intraventricular administration is made through a subcutaneously implanted reservoir (*Ommaya reservoir*). AmB appears to be stored in the extracellular compartment of the brain, which may itself act as a reservoir.

3A.2.8.3 Metabolism (Elimination Route)

AmB is eliminated very slowly, over weeks to months through the kidneys. Over a 7 day period, about 40% of the administered drug is excreted in urine. After cessation of therapy, AmB can still be detected in blood for up to 4 weeks, and longer in the urine. Protein binding and binding to the tissues is very high. The highest concentrations of AmB are in the liver, spleen, and kidney, where the drug is still detectable up to 1 year after completion of therapy.

In children aged 4 months to 14 years AmB infused in 2–4.5 h at a dose range of 0.25–1.5 mg/kg/day resulted in a mean Cmax of 2.9 μg/mL. The volume of distribution (Vd) and elimination $t_{1/2}$ were 0.75 L/kg and 18.1 h, respectively. Total clearance decreased with age. In children aged 8 months to 9 years, the mean total clearance was 0.57 mL/min/kg, and in children older than 9 years, it was 0.24 mL/min/kg. Interpatient variations in the clearance and volume of distribution of AmB were greater than three fold and greater than eight fold, respectively (Benson and Nahata, 1989).

3A.2.8.4 $t_{1/2}$

Following IV administration of AmB, the initial plasma half-life is approximately 24 h. Thereafter, the elimination rate decreases to a terminal $t_{1.2}$ of approximately 152 h.

3A.2.8.5 Liposomal AmB, Different Pharmacokinetics

Liposomal encapsulation alters the pharmacokinetics of AmB (Bekersky et al., 2002). Liposomal AmB (AmBisome) produces higher plasma concentrations, lower volume of distribution, and greatly reduced renal and fecal clearances, so that it constitutes a stable intravenous drug delivery system that remains in the circulation for a period of time. Urinary and fecal clearances of liposomal AmB are tenfold lower than those of AmB deoxycholate. Liposomal AmB has a long terminal half-life ($t_{1/2\gamma}$, 152 h) similar to the terminal half-life of AmB deoxycholate ($t_{1/2\gamma}$, 127 h). The rate of AmB elimination is similar for the two formulations. However, the terminal phase contained over 80% of the total AUC for AmB-deoxycholate but <50% of the total AUC for liposomal AmB. Earlier phases of liposomal AmB disposition contribute to its pharmacokinetic profile and may explain the relatively minor accumulation observed in plasma after repeated daily administration of liposomal AmB. The low initial Vd for liposomal AmB is because liposomes sequester drug in the plasma compartment, contributing to a high Cmax. For liposomal AmB the initial Vd is 50.1 mL/kg and shifted to at steady state (Vdss) of 774 mL/kg, showing that a substantial fraction of liposomal AmB eventually enters the tissue compartment. Reduced excretion of unchanged drug results from the liposome encapsulation, which prevents renal filtration or biliary secretion. The mononuclear phagocytic system plays a role in the disposition of liposomes, contributing to tissue uptake, retention, and possible metabolism of liposomal AmB. Liposome encapsulation of AmB achieves more extensive and prolonged tissue distribution than AmB-deoxycholate.

3A.2.9 Interactions

3A.2.9.1 Effect of Other Drugs on AmB

Antineoplastic agents, cisplatin, and nitrogen mustard compounds may enhance renal toxicity of AmB.

3A.2.9.2 Effect of AmB on Other Drugs

AmB can enhance the renal toxicity of cyclosporine and aminoglycoside antibiotics. Because of AmB's effect on potassium and magnesium, other drugs that rely on appropriate concentrations of these ions may be adversely affected. Additionally, because of the decrease in these ions, there may be an increase in the potential of other drugs to cause an abnormal cardiac rhythm.

3A.2.10 Adverse Reactions

Both acute and chronic adverse effects are associated with AmB.

3A.2.10.1 Acute Adverse Effects

Within 30–45 min after the first few infusions there are chills, fever, and tachypnea. Headache, nausea, anorexia, and vomiting may ensue. These acute reactions peak in 15–30 min and subside over the next 2–4 h. When such adverse effects occur, premedication with acetaminophen and diphenhydramine (Benadryl™) have been used. If rigors/chills occur, meperidine (Demerol®) may be used, or premedication with hydrocortisone is also effective. Less commonly there is irregular heartbeat, muscle cramps, or pain.

Phlebitis is common if peripheral vein catheters are used. The addition of 500–1000 units of heparin to the AmB infusion, the use of a pediatric scalp-vein catheter, or alternate-day therapy may decrease the incidence of thrombophlebitis.

Acute Adverse Effects with AmB Lipid Preparations A combination of infusion related toxicity may occur including (i) pulmonary symptoms (chest pain, dyspnea, and hypoxia); (ii) abdominal, flank, or leg pain; and (iii) flushing and urticaria (Dodds Ashley et al., 2006).

3A.2.10.2 Chronic Adverse Effects

Nephrotoxicity AmB nephrotoxicity is common, is dose-dependent, and can be severe. Dose-dependent decrease in glomerular filtration accompanies AmB therapy. The mechanism of AmB nephrotoxicity includes a direct vasoconstrictive effect on renal arterioles, reducing glomerular and renal tubular blood flow and the lytic action on cholesterol-rich lysosomal membranes of renal tubular cells.

Adults with no preexisting renal disease will have average serum creatinine of 2–3 mg/dL at therapeutic doses. Therapy should not be withheld unless the serum creatinine rises above 3 mg/dL. Attempting to give AmB to adults without causing acute kidney injury usually leads to inadequate therapy.

Persons at higher risk for permanent kidney injury include those with renal insufficiency and also in SCT recipients. Most patients receiving AmB experience increased BUN and serum creatinine, decreased creatinine clearance, glomerular filtration rate, and renal plasma flow. Many patients also develop hypokalemia and hypomagnesemia. Uric acid excretion is increased and nephrocalcinosis can occur. Renal tubular acidosis may be present. Supplemental alkali therapy may decrease complications related to renal tubular acidosis.

Prehydration with IV normal saline along with avoiding dehydration appears to be effective in reducing the risk of AmB induced nephrotoxicity in adults and children, including low birthweight babies (Holler et al., 2004).

Anemia Anemia is a side effect in 24% of total patients treated with AmB, reaching up to 95% of patients requiring prolonged therapy with the drug (Brandriss et al., 1964; MacGregor et al., 1978). The anemia is normocytic and normochromic, and bone marrow examination, reticulocyte counts, and ferrokinetic studies do not reveal a compensatory erythropoietic response to the anemia. Measurements in patients indicate a suppression of the stimulating hormone, erythropoietin, rather than the increased production seen in hypoplastic anemia (Lin et al., 1990). Decreased erythropoietin concentration occurs 3 weeks after start of AmB therapy and is independent of the total dosage.

Experimental evidence, in rat kidneys and tissue cultures, implicates the transcription factor, HIF-1, regulating erythropoietin production as the site of AmB suppression (Yeo et al., 2006). Specifically, AmB represses the C-terminal transactivation domain (CAD) of HIF-1α by stimulating the binding of the "factor inhibiting HIF-1" to the CAD.

3A.3 FLUCONAZOLE (FLC) (DIFLUCAN®, PFIZER)

Fluconazole is the first broadly effective azole without significant side effects.

3A.3.1 Structure and Mode of Action

FLC inhibits the synthesis of ergosterol by inhibiting the cytochrome P450 enzyme lanosterol-14-α-demethylase, a product of the *ERG11* gene (in *Candida albicans*); thus inhibiting ergosterol biosynthesis. FLC is fungistatic, *requiring a functioning immune system* to achieve a maximum therapeutic effect. See Figs. 3A.6 and 3A.7.

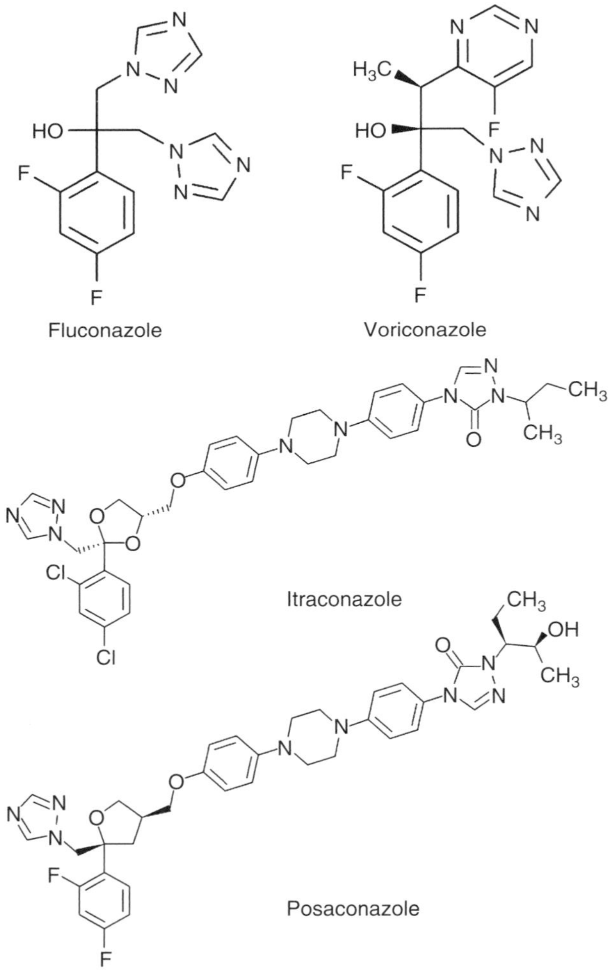

Figure 3A.6 Chemical structures of triazole antifungal agents. *Source:* PubChem at the URL http://pubchem.ncbi.nlm.nih.gov, a service of the National Center for Biotechnology Information and the National Library of Medicine, U.S.A.

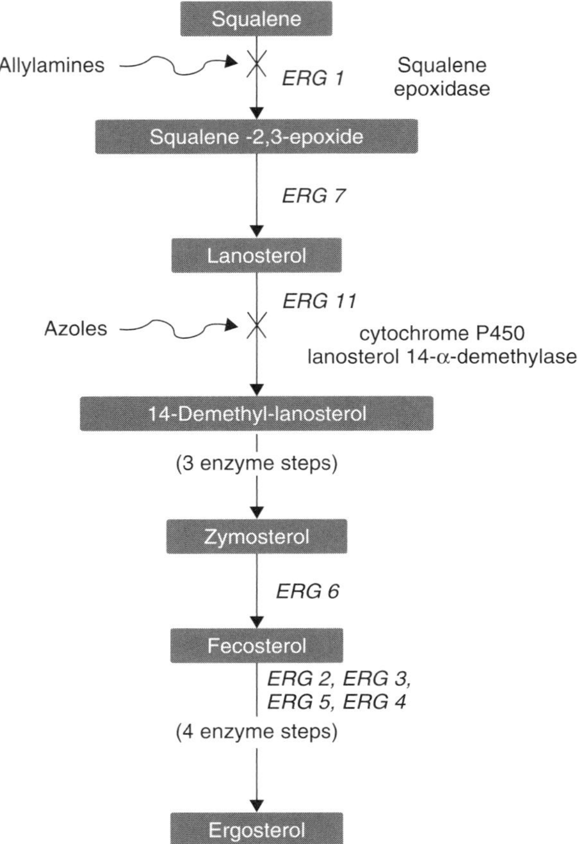

Figure 3A.7 Key steps in the biosynthesis of ergosterol and the enzymes inhibited by allylamine and azole antifungal agents. The ergosterol biosynthetic pathway is depicted for *Candida albicans*. A similar pathway occurs in molds. The steps in the pathway inhibited by allylamines (e.g., terbinafine) and by azole antifungal agents are shown. Genes encoding the enzymes are indicated in the *ERG* series. A complete pathway with descriptions of the intermediary steps may be found at "steroid biosynthesis" at the website http://ncbi.nlm.nih.gov/biosystems.

3A.3.2 Indications

FLC is primarily a drug for treating yeast infections. Oral, once daily FLC is a therapeutic choice, depending on the severity of illness, to treat infections with *Candida* species and *Cryptococcus neoformans* in immunosuppressed patients (e.g., those living with AIDS) and for treatment or primary prophylaxis in neutropenia resulting from cancer chemotherapy or in SCT recipients.

3A.3.2.1 Mucosal Candidiasis

FLC is effective against *Candida albicans* and other *Candida* species as therapy for oropharyngeal and esophageal candidiasis. FLC therapy for oropharyngeal candidiasis (OPC) is more effective than clotrimazole troches or nystatin mouthwash. *Candida glabrata* and *C. krusei* are innately less susceptible to FLC. In vulvovaginal candidiasis, clinical cure can be achieved in >90% of patients after a single 150 mg oral dose of FLC and further doses of 150 mg/month are effective against recurrences.

Dosage The usual dose in mucosal candidiasis is a loading dose of 200 mg, followed by a daily dose of 100 mg for at least 2 weeks. Higher doses can be used for clinically resistant OPC. After a single oral dose, blood concentrations (AUC) correlate well with the administered dose. Blood concentrations show little variation after multiple oral doses. The accumulation factor (defined as the ratio of concentrations after a dose at steady state to that after the first dose) is between 2.1 and 2.8.

Renal Insufficiency The $t_{1/2}$ in renal insufficiency is longer, so that dosages are adjusted in relation to creatinine clearance. In CAPD, a 150 mg dose in a 2 L dialysis solution every 2 days has been proposed. In hemodialysis, a dose of 100–200 mg is recommended at the end of each dialysis session.

3A.3.2.2 Invasive Candidiasis Including Candidemia

Patients who are not neutropenic respond as well to FLC as they do to AmB. Removal of a central IV catheter is important in treating candidemia (Pappas et al., 2009). Patients who are not neutropenic may be treated with a FLC dose of 6 mg/kg/day (~400 mg). The choice of FLC or an echinocandin depends on whether the patient is less critically ill without a history of azole exposure. Patients started on an echinocandin can be switched to FLC when they become clinically stable and if the causative species is susceptible to the drug. Therapy is continued for 2 weeks after the resolution of candidemia and resolution of symptoms. The recommendation for neutropenic patients with candidemia are liposomal AmB or an echinocandin. FLC is an alternative at 6 mg/kg/day (~400 mg).

3A.3.2.3 Cryptococcal Meningitis

FLC is used for maintenance therapy for 2–6 months following AmB (preferably with flucytosine) induction therapy for the first 2 weeks. FLC crosses the meninges easily and is present in the brain and CSF at concentrations inhibitory to *C. neoformans*. A favorable clinical response in cryptococcal meningitis is achieved by FLC at doses of ≥6 mg/kg/day. Because of the relatively high MICs of *C. neoformans*, dosing in cryptococcosis may require 200–400 mg/day.

FLC resolves symptoms in up to 60% of patients with cryptococcal meningitis and AIDS, but is not as effective as AmB. Symptoms of cryptococcosis subside more slowly during FLC therapy as compared with AmB, so it is more often used as maintenance therapy after AmB induction therapy. Cryptococcosis in persons living with AIDS requires long-term lifetime maintenance with FLC, or until the viral load is minimized by availability and compliance with ART. A dose of 200 mg/day is effective in preventing relapse in AIDS patients.

3A.3.2.4 Coccidioidal Meningitis and Disseminated Nonmeningeal Coccidioidomycosis

FLC is indicated for meningitis due to *Coccidioides immitis* because of its excellent penetration into CSF. Long-term treatment with 6 mg/kg/day of FLC is effective in meningeal, pulmonary, or disseminated coccidioidomycosis. When used to treat coccidioidal meningitis at a dose of 6 mg/kg/day the FLC concentration in CSF is 2.3–3.4 μg/mL and the ratio of FLC concentration in CSF versus blood is 0.7–0.81.

3A.3.2.5 Other Indications for FLC

Cutaneous sporotrichosis, ringworm, histoplasmosis, and blastomycosis may respond to FLC but it is *inferior to ITC*. FLC is not effective against mold infections.

3A.3.3 Fluconazole Pharmacokinetics

See Debruyne and Ryckelynck (1993), Goa and Barradell (1995), and Wildfeuer et al. (1997).

3A.3.3.1 Metabolism

FLC circulates as the free active drug and plasma protein binding is low (11–12%). No metabolites represent more than 4% of the dose. After gastrointestinal absorption very little metabolism occurs after the first pass through the liver.

3A.3.3.2 Distribution

Oral dosing results in very reliable gastrointestinal absorption, with or without food, and is unaffected by gastric pH. FLC oral capsules or solution are rapidly absorbed, achieving nearly complete (>90%) bioavailability with both IV and oral formulations. Rapid absorption gives a time to Cmax of 1–3 h. Absorption is not affected by ingestion with food except that the time to Cmax is increased. Persistence of FLC in the kidneys and spleen is longer than in blood.

A 100 mg daily dose results in plasma concentration of 1 μg/mL after the first dose, and between 0.78 and 3.1 μg/mL 24 h after receiving multiple doses. One 3 μg/mL dose is the minimal efficacious concentration for good primary and maintenance therapy. In cases of serious infections the daily dose can be increased up to 400 mg. The Cmax at steady state after a 400 mg/day regimen is 20 μg/mL.

The Vd for FLC it is 50–60 L (equivalent to 0.7–0.8 L/kg), which approximates the volume of body water. Concentrations of FLC in body fluids such as CSF and vitreous gel approach 80% of that in blood. FLC is distributed in many body fluids, especially in CSF and dialysis fluid. This property is significant for treating coccidioidal meningitis and fungal peritonitis. FLC concentrations in the CSF, vaginal secretions, blister fluid, sputum, saliva, joint fluid, and dialysate are similar to that in blood (ratio of body fluid concentration: blood concentration = 0.42–1.36). The ratio of urine:blood concentration of 10 for FLC is suitable for treating urinary tract fungal infections in patients with bladder catheters. Removal of bladder catheters may be sufficient. Symptomatic cystitis and pyelonephritis due to FLC-susceptible *Candida* species may be treated with oral doses of 3–6 mg/kg/day (Pappas et al., 2009). Efficacious anti-*Candida* concentrations are attained in vitreous gel, liver, spleen, and vaginal secretions. In urine the FLC concentration is 10- to 20-fold higher than in blood, a profile that is important in treating urinary tract yeast infections.

3A.3.3.3 $t_{1/2}$

The average elimination $t_{1/2}$ is long: 31.6 ± 4.9 h from plasma, and a similar $t_{1/2}$ for elimination from CSF. For that reason a single daily dose is recommended with a loading dose of twice the daily dose because a minimum of 6 days is needed to reach steady state. The $t_{1/2}$ is not appreciably different in the elderly but is shorter in children; however, in neonates the $t_{1/2}$ is prolonged up to 88 h. The long $t_{1/2}$ combined with multiple dosing results in an accumulation of a high concentration in various body compartments.

3A.3.3.4 Elimination Route

The main route of elimination is renal and the production of metabolites is not significant, 65–90% of the dose is excreted as unchanged parent drug. Renal elimination is influenced by two phenomena: high renal filtration because of low plasma protein binding, and extensive tubular reabsorption, accounting for the long $t_{1/2}$. Total clearance rate is 19.5±4.7 mL/min. Renal clearance is 14.7±2.7 mL/min.

3A.3.4 Efficacy

Except for *C. krusei* and *C. guilliermondii*, the MICs of FLC are low for pathogenic *Candida* species and also for *Cryptococcus neoformans*. MICs are variable for *C. glabrata*. For *C. albicans* the range of MICs is 0.125–0.5 μg/mL. MICs for *C. neoformans* are 0.39–6.25 μg/mL. FLC is not effective against aspergilli, or other molds, and *Aspergillus* species infections have developed during FLC prophylaxis in immunocompromised hosts. Secondary resistance to FLC is uncommon and is associated with the cumulative dose resulting from long-term maintenance therapy in persons living with AIDS. In that case *C. albicans* is seen to develop reduced susceptibility to FLC. Resistant *C. albicans* strains arise in approximately 5% of that patient group.

Treatment failure of FLC therapy has been explained by host and microbial factors: infection with an innately more resistant *Candida* species, prolonged neutropenia, intravascular catheters, and long-term exposure to low dosage FLC (<200 mg/day). The rapid clinical improvement of persons living with AIDS and mucocutaneous candidiasis is not always accompanied by a significant reduction in colonization, so that without maintenance or suppressive therapy, recurrence may occur rapidly.

3A.3.5 Formulations

FLC is supplied in 50, 100, 150, and 200 mg tablets and as 50 mg/5 mL and 200 mg/5 mL suspensions. FLC for injection is an iso-osmotic solution of FLC in a sodium chloride or dextrose diluent. Each milliliter contains 2 mg of FLC and 9 mg of sodium chloride or 56 mg of dextrose.

3A.3.6 Interactions

3A.3.6.1 Effect of Other Drugs on FLC

Oral cimetidine decreases the Cmax and AUC for FLC. Rifampicin accelerates elimination of FLC and can cause clinical failure of the antifungal therapy: that is, rifampicin decreases the AUC and $t_{1/2}$ when coadministered with FLC.

3A.3.6.2 Interactions of FLC with Other Drugs

FLC increases blood concentrations of phenytoin[1] and can also increase blood concentrations of cyclosporine.

3A.3.7 Adverse Reactions

Generally well tolerated, FLC is associated with mild disturbances, for example, headache, nausea, vomiting, abdominal discomfort, and elevated liver function tests. Adverse reactions are more often encountered in persons living with AIDS.

[1] *Definition*: Phenytoin is the generic form of Dilantin, a seizure medication. Phenytoin is indicated for use as an anticonvulsant drug in people of all ages. Developed in 1938, phenytoin remains a first-line medication for epilepsy.

3A.4 ITRACONAZOLE (ITC) (SPORANOX®, JANSSEN PHARMACEUTICA DIVISION OF JOHNSON & JOHNSON)

ITC is a further refinement of a triazole antifungal agent and has structural motifs in common with the earlier imidazole, ketoconazole (Fig. 3A.6).

3A.4.1 Action Spectrum

ITC is effective against a variety of opportunistic and endemic fungi and is the drug of choice for treating moderate cases of blastomycosis, histoplasmosis, coccidioidomycosis, paracoccidioidomycosis, and sporotrichosis. ITC prevents relapsing histoplasmosis in AIDS. ITC is effective against aspergillosis, although less so than VRC. ITC is effective as systemic therapy for ringworm including nail infections and pityriasis versicolor. ITC is effective in treating oral/esophageal candidiasis. At doses of 5–10 mg/kg ITC may be superior to AmB to treat infections with *Aspergillus* species and a range of other molds.

3A.4.2 ITC: Uncertain Bioavailability

Compared with other azole antifungal agents, there have been problems with the formulation and delivery of ITC and uncertainty about effective bioavailability (Prentice and Glasmacher, 2005). It is practically insoluble in H_2O, is lipophilic, and is strongly protein-bound (>99%). Tissue penetration is high and sustained. Oral absorption, and thus bioavailability, is improved with food or acidification of the stomach. The mean absolute bioavailability is 55%. As an oral dose is increased from 50 to 200 mg, increases in the AUC and Cmax are nonlinear, implying saturation of the first pass sites for metabolism by hydroxylation in the gut mucosa and liver.

Very variable plasma concentrations of ITC occur in neutropenic patients and those with hematologic malignancies, with higher doses required to reach effective antifungal concentrations than in normal subjects. Also, the pharmacokinetics of ITC will vary because of genetic variation in cytochrome P450 isozymes among patients, as is the case with all azole antifungal agents. Because of interpatient variation, weekly measurements of serum concentrations may be needed to assure adequate therapeutic levels.

3A.4.3 Properties

Aqueous solutions of ITC can be achieved with the addition of 5% hydroxypropyl-ß-cyclodextrin. Cyclodextrin forms an inclusion complex with ITC. The cyclodextrin ring traps the hydrophobic, water-insoluble drug and releases it at the lipid-rich gut lining for absorption. Janssen Pharmaceutica, in 1997, received approval to market an oral ITC solution containing 40% hydroxypropyl ß-cyclodextrin (HPB-Sporanox®) in the United States.

The oral solution of cyclodextrin-ITC has improved predictable plasma concentrations. Cyclodextrin solution with ITC provides greater Cmax and AUC concentrations than ITC alone in normal volunteers and also in patients receiving chemotherapy for leukemia, in SCT recipients, and children with hematologic malignancies. Cyclodextrin oral solution increases bioavailability by 30–37% after a single dose and by 23–31% at steady state after multiple dosings. Since the first pass metabolite, hydroxyl-ITC, has the same spectrum of activity as the parent drug, the combined bioavailability can reach 80%. The existence of severe chemotherapy-related mucositis or concurrent use of H2-antagonists did not significantly reduce bioavailability in adult chemotherapy or transplant patients. A drawback of the ITC oral solution is that some patients will not tolerate the taste.

The IV formulation of ITC-cyclodextrin avoids the bioavailability and compliance issues. IV ITC-cyclodextrin solution has rapid loading. It is now easy to obtain peak concentrations of 2–4 μg/mL and troughs of ≥1 μg/mL. The IV solution can be used to load patients quickly at induction of neutropenia before switching to the oral solution. The IV formulation of ITC should be used with caution in patients with diminished renal function. The cyclodextrin component can accumulate and worsen the renal impairment. At present there is no source for this formulation in the United States.

3A.4.4 Pharmacokinetics

3A.4.4.1 Distribution

Tissue penetration of ITC is high and sustained in the normal population. Animal studies have shown that tissue-to-plasma ratios are ~3:1 in lung, 10:1 in liver, and 25:1 in fat. Therefore, ITC has continuous bioavailability at the sites of infection. Low ITC concentrations are found in aqueous body fluids: plasma, tears, saliva, and CSF. Good penetration occurs in sputum, bronchial aspirates, and even pus. ITC displays extensive binding to hair and in skin, which is noteworthy for treatment of superficial and cutaneous mycoses.

3A.4.4.2 Dosage

The high Vd ss of ITC (11 L/kg) suggests that loading doses are needed to reach clinically active concentrations quickly. A combination of capsules (800 mg/day, from day 1 to day 7) and oral solution (400 mg/day starting on day 1 until the end of therapy) or by combining the same oral solution with IV solution (400 mg/day on day 1 and 2) can achieve trough concentrations of ITC above 1 μg/mL by the end of the first week. The 1 μg/mL plasma concentration is the established breakpoint for in vitro susceptibility of ITC for *Candida* species. This breakpoint does not include the first pass metabolite hydroxy-itraconazole, which accumulates twice as fast as the parent compound.

3A.4.4.3 Metabolism

First pass metabolism of the drug occurs via hydroxylation in the gut mucosa and liver. The major, and active, metabolite is hydroxy-ITC. The $t_{1/2}$ in steady state is 34 ± 9 h. The elimination route for ITC is via the liver. Very little of the drug is excreted unchanged in the urine.

3A.4.5 Interactions

3A.4.5.1 Effect of Other Drugs on ITC

Absorption from oral capsules is reduced by a mean of 20% by the simultaneous use of H2-antagonists (e.g., cimetidine, rantidine). Increased metabolism of ITC is induced by carbamazepine, isoniazid, rifampicin, phenobarbital, and phenytoin.

3A.4.5.2 Interactions of ITC on Other Drugs

ITC inhibits human cytochrome P450, reducing the metabolism and, as a result, increasing the plasma concentrations of cyclosporine, tacrolimus, and digoxin, reaching potentially toxic concentrations of those drugs. ITC should be avoided in patients receiving calcium channel blockers to treat angina pectoris and high blood pressure because ITC can affect a key interval in the heart rhythm. ITC in combination with vincristine or vinblastine can result in severe neurotoxicity. This effect is due to decreased clearance of vincristine and vinblastine.

3A.4.6 Adverse Reactions

ITC has been reported to result in renal and hepatic toxicity in bone marrow transplant patients, especially those patients receiving cyclophosphamide. ITC has also been shown to cause prolongation in the QT interval of the heart. This can be associated with a potentially fatal arrhythmia known as *torsades de pointe*.

3A.5 VORICONAZOLE (VRC) (VFEND®, PFIZER)

Voriconazole (VRC) has structural similarity to FLC, as shown in Fig. 3A.6. Importantly, VRC has an expanded spectrum of activity because of its tighter binding to lanosterol demethylase and because it inhibits a second enzyme in the pathway to ergosterol. Dosage is administered either orally or IV.

3A.5.1 Action Spectrum

VRC has fungicidal activity against *Aspergillus*, on a par with AmB, and better than ITC. VRC is active against difficult to treat opportunistic molds:

Fusarium spp., *Penicillium marneffei, Pseudallescheria boydii, Scedosporium apiospermum*, and, according to some reports, has success against *Scedosporium prolificans*. VRC has no activity against *Mucorales*, including *Rhizopus* species, and *Mucor* species, as well as *Sporothrix schenckii*. There are instances of mucormycosis breaking through VRC prophylaxis (reviewed by Chayakulkeeree et al., 2006; Kontoyiannis et al., 2005).

A microtitration-based susceptibility test performed according to the M38-A guideline (National Committee for Clinical Laboratory Standards, 2002) of the Clinical Laboratory Standards Institute, CLSI, assessed the susceptibility of 19 strains of *S. schenckii* against a range of antifungal agents including VRC (Alvarado-Ramirez and Torres-Rodriguez, 2007). A wide range of MICs was found with VRC at 4–16 μg/mL and a geometric mean titer of 9.3 μg/mL, indicating reduced susceptibility.

VRC is licensed in the United States for treatment of invasive aspergillosis, invasive candidiasis, including candidemia, mucosal candidiasis, and for serious *S. apiospermum* and *Fusarium* species infections.

3A.5.2 Pharmacokinetics

VRC is well absorbed after ingestion, achieving maximum plasma concentrations in 1–2 h (Donnelly and de Pauw, 2004). The drug displays nonlinear kinetics such that increasing the oral dose from 200 mg to 300 mg twice daily leads to a 2.5-fold increase in plasma concentration (AUC). Interindividual variability in Cmax is high. Absorption is not affected by gastric pH. VRC absorption is inhibited by high-fat food so it should be given at least 1 h before or after a meal. Plasma protein binding is estimated at 58%. Voriconazole is detectable in the CSF.

3A.5.2.1 Dosage

An IV loading dose of 6 mg VRC/kg at 12 h intervals is given on the first day followed by a maintenance dose of

4 mg/kg at 12 h intervals. Maintenance can be switched to the oral form at 200 mg every 12 h.

3A.5.2.2 Metabolism

Metabolism of VRC is via the hepatic P450 isozymes. Metabolism to derivatives and degradation products occurs and approximately 80% of a single dose is excreted in the urine as metabolites without antifungal activity. Because of genetic polymorphism in isozymes, 15–20% of Asian persons are slow metabolizers, and they may have a four fold higher AUC than homozygous extensive metabolizers. The major metabolite is the N-oxide, which lacks antifungal activity.

3A.5.3 Drug Interactions

3A.5.3.1 Effect of Other Drugs on VRC

Other drugs metabolized by cytochrome P450 isozymes interfere with VRC: for example, rifampicin, cimetidine, carbamazepine, and ritonavir all lower the Cmax of VRC. It is not recommended to concurrently use VRC and ritonavir.

3A.5.3.2 Effect of VRC on Other Drugs

VRC inhibits cytochrome P450 enzymes, which can increase plasma concentrations of drugs cleared through this mechanism. This includes immunosuppressants often used in hematopoietic SCT recipients and solid organ transplant recipients, the very patients prone to develop an invasive fungal infection and to need VRC. Because of this inhibition it is recommended that the dose of tacrolimus, cyclosporine, or sirolimus be reduced to 1/3 or 1/2 of the dose prior to initiation of VRC. Other drugs that may interact with VRC include the nonsedating antihistamines terfenadine and andastemizole, the promotility gastrointestinal agent cisapride, pimozide, ergot alkaloids, and quinidine.

3A.5.3.3 Two-Way Drug Interactions

Several drugs have a two-way interaction with VRC; for example, VRC decreases metabolism of the other drug and the other drug decreases or increases concentrations of VRC. These drugs include efavirenz, omeprazole, rifabutin, phenytoin, oral contraceptives, indinavir, other HIV protease inhibitors, and other non-nucleoside reverse transcriptase inhibitors of HIV.

3A.5.4 Adverse Reactions

The major adverse reaction is visual disturbance. Approximately 30% of patients experience blurred vision, color vision change, or photophobia occurring about 30 min after administration and resolving within an hour. Prolonged use (>4 weeks) of therapy can alter visual acuity. In that case visual function should be monitored. Mild to moderate skin reactions have been reported. Liver function abnormalities occurred in 13% of patients, which resolve after the dosage is lowered or after completion of treatment. In clinical trials, 2.4% of patients taking VRC reported hallucinations.

3A.6 POSACONAZOLE (PSC) (NOXAFIL®, SCHERING-PLOUGH/ MERCK & CO.)

See Fig. 3A.6

3A.6.1 Action Spectrum

PSC is an extended spectrum triazole antifungal agent noteworthy for its activity against difficult to treat invasive mold infections including those caused by *Mucorales* (e.g.: *Rhizopus* species) and another opportunistic mold, *Scedosporium apiospermum*. PSC has been successfully used to resolve or improve the clinical status of patients with eumycetoma and chromoblastomycosis (Negroni et al., 2005). Case reports of fungal osteomyelitis suggest that PSC is able to penetrate bone (Negroni et al., 2005).

Currently, PSC is approved in the United States for the prevention of fungal infections in neutropenic cancer patients, as prophylaxis to prevent fungal infections in patients who develop severe graft-versus-host disease following a SCT, and for the treatment of oropharyngeal candidiasis refractory to ITC or FLC. PSC may also be useful in the treatment of aspergillosis (Walsh et al., 2007).

3A.6.2 Pharmacokinetics

PSC is well absorbed from the gastrointestinal tract. When PSC is taken with a high-fat meal, the Cmax is improved by 300–500% over the fasting state. Cmax is achieved in 3 to 5 h after the dose. Potentially therapeutic concentrations are achieved in 2–4 days. Steady state concentrations are not achieved for 7–10 days. The Vd is 1774 L or 25 L/kg. The $t_{1/2}$ of PSC is 35 h. PSC is highly protein bound (>98%).

3A.6.2.1 Elimination

The majority of the PSC dose is eliminated as unchanged drug through the feces. Renal excretion accounts for 13% of the elimination, but <0.2% is unchanged drug. There is no effect on the elimination of PSC by hepatic or renal insufficiency.

3A.6.3 Drug Interactions

PSC has numerous drug interactions because of its inhibitory effect on the cytochrome P450 enzyme, CYP3A4. Consequently, any medication that is metabolized through CYP3A4 will have decreased clearance when coadministered with PSC. Examples of drugs that may have increased concentrations when coadministered with PSC include, but are not limited to, cyclosporine A, tacrolimus, and sirolimus; it is recommended that the concentration of these drugs be closely monitored while being coadministered with PSC. Other medications whose serum concentrations may be increased include vinca alkaloids (vincristine, vinblastine), calcium channel blockers, HMG-CoA reductase inhibitors (statins), and midazolam. Monitoring for side effects is important when these drugs are used with PSC. Because of increased drug concentrations with the following drugs, PSC should not be coadministered with ergot alkaloids, terfenadine, astemizole, cisapride, pimazide, or quinidine. There is a two-drug interaction with rifamycins (rifampin, rifabutin) which may increase side effects of the rifamycin. PSC serum concentrations may be reduced when coadministered with phenytoin and cimetidine.

3A.6.4 Adverse Reactions

In clinical trials, the most common side effects reported in patients taking PSC were nausea, vomiting, diarrhea, elevation in liver enzymes, and creatinine increase.

3A.7 AZOLE RESISTANCE MECHANISMS

3A.7.1 Alteration of Target Enzyme (Lanosterol Demethylase)

Mutations in *ERG11*, the gene encoding the target enzyme, lanosterol C14α-demethylase, prevents binding of azoles to the active site (Kanafani and Perfect, 2008). This type is associated with innate resistance and irreversible acquired resistance; for example, *C. krusei* is innately resistant to FLC because of decreased affinity of Erg11 to the drug. More than 80 amino acid substitutions in Erg11 have been detected. Different mutations can coexist in the same gene with additive effects.

3A.7.2 Overexpression of Target Enzyme

Some *Candida* isolates with reduced azole susceptibility have increased intracellular concentrations of Erg11. In that case, regular therapeutic drug concentrations cannot effectively inhibit ergosterol biosynthesis. Upregulation of the lanosterol demethylase is achieved by gene amplification, increased transcription, or decreased degradation of the enzyme. This mechanism may not contribute much to overall resistance in *Candida* species, because only small increases in enzyme concentrations have been observed. The resistance may be reversible. The mechanism is usually found combined with another means of resistance.

3A.7.3 Increased Efflux of Drug, CDR Efflux Pumps

The activity of efflux pumps results in decreased intracellular drug concentrations. Two gene families of transporters function as efflux pumps in *Candida* species. The CDR genes of the *A*TP-*b*inding *c*assette (ABC) superfamily, and the MDR genes of the major facilitators class. The CDR efflux pumps multiple azole drugs out of *Candida*. Several *Candida* species contain CDR homologs. Overexpression of efflux pumps may be transient and reversible. The MDR efflux pumps are specific for FLC. As with CDR, MDR overexpression may be transient and reversible.

3A.7.4 Bypass Pathways

Exposure to azoles depletes ergosterol in the cell membranes and the toxic product 14α-methyl-3,6-diol accumulates, leading to growth arrest. Mutation in *ERG3* prevents the formation of 14α-methyl-3,6-diol. The resulting sterol 14α-methylfecosterol leads to functional membranes. The action of azoles is thus minimized. *Candida* strains with *ERG3* mutation are also resistant to polyenes, because of the absence of ergosterol.

3A.7.5 Loss of Heterozygosity in Chromosome 5 and Azole Resistance

Loss of heterozygosity at the *C. albicans* mating-type-like (MTL) locus is linked to resistance to azole antifungals (Rustad et al., 2002). Both *MTLa* and *MTLalpha* are present in most *C. albicans* strains. The loss of either *MTLa1* and *MTLalpha1* genes corresponded to loss of all of the loci-specific genes, resulting in homozygosity at the MTL locus.

TAC1 (for transcriptional activator of CDR genes) is critical to upregulate ABC transporters CDR1 and CDR2, which mediate azole resistance in *C. albicans* (Coste et al., 2007). High azole resistance levels are achieved when *C. albicans* carries hyperactive alleles, which occur only as a consequence of loss of heterozygosity at the TAC1 locus on chromosome 5, which is linked to the mating-type-like (MTL) locus. Both are located on the chromosome 5 left arm along with *ERG11* (target of azoles).

Azole-resistant *C. albicans* acquire *TAC1* hyperactive alleles and *ERG11* mutant alleles by loss of heterozygosity. This may occur via mitotic recombination of the left arm of chromosome 5, or the loss and reduplication of the entire chromosome 5. In one case, two independent *TAC1* hyperactive alleles were acquired. In some cases a duplication of an entire isochromosome consisting of the left arm of chromosome 5 increases the copy number of the azole resistance genes located in that arm including *TAC1* and *ERG11*. In summary, azole resistance is due not only to specific mutations in azole resistance genes (such as *ERG11* and *TAC1*) but also to their increase in copy number by loss of heterozygosity and the addition of extra chromosome 5 copies.

3A.7.6 Azole Resistance in Aspergillus Species

Aspergilli are innately resistant to FLC. Resistance to ITC is uncommon but has been characterized by (i) expression of efflux pumps and (ii) a more prevalent mechanism relies on modification of the lanosterol C14α-demethylase, in which amino acid substitutions at the M220 position are associated with a increased MICs to all azoles, whereas amino acid substitutions at G54 result in cross-resistance to itraconazole and posaconazole.

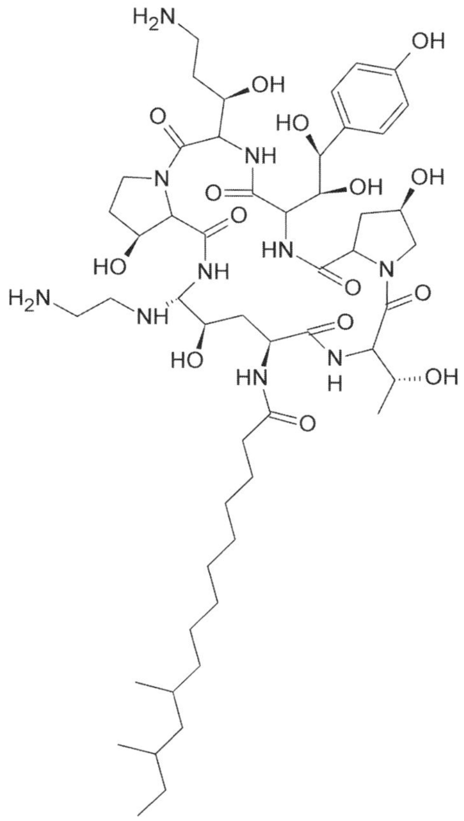

3A.8 ECHINOCANDINS

This class of parenteral antifungal agents inhibits a novel target: blocking the synthesis of fungal cell-wall fibrils composed of ß-(1→3)-D-glucan (Fig. 3A.2). The result is that fungal cells cannot maintain their shape and become osmotically fragile. ß-(1→3)-D-glucan is not present in mammalian cells; consequently, its mechanism is analogous to the role of penicillin in blocking the synthesis of bacterial peptidoglycan.

3A.8.1 Mode of Action

The echinocandins bind to ß-(1→3)-D-glucan synthase, blocking the synthesis of ß-(1→3)-D-glucan, a fibrillar structural component of the cell walls of yeasts and many molds (Cappelletty and Eiselstein-McKitrick, 2007). The treated fungal cell becomes fragile and osmotic lysis of yeast occurs. The effect in molds is to cause aberrant growth at the hyphal tips. Acquired resistance, when it occurs, is caused by mutations in the *FKS1* gene for glucan synthase.

Echinocandins are lipopeptides, that is, cyclic hexapeptides N-acylated with an aliphatic chain of differing lengths. Echinocandins differ in having different substituents in the hexapeptide ring or a distinct fatty acid chain (Fig. 3A.8). The first echinocandin was discovered in the early 1970s; since then many members of the family have been discovered, the natural products of several different fungal species. CASF is a natural product of the fungus *Glarea lozoyensis*. Micafungin (MCF) and anidulafungin (ANF) are semisynthetic lipopeptides modified from fermentation products of *Aspergillus* species. Many experts consider the echinocandins as equivalent in efficacy. Cost, side effects, and drug interactions are similar and minor differences in these are what differentiate the drugs.

Figure 3A.8 Caspofungin (Cancidas®, Merck).
Source: PubChem at the URL http://pubchem.ncbi.nlm.nih.gov, a service of the National Center for Biotechnology Information and the National Library of Medicine, U.S.A.

3A.8.2 Spectrum of Activity

Echinocandins have good efficacy in treating candidemia and invasive candidiasis. A prospective, randomized double blind study of CASF compared with AmB found that the clinical response to CASF was superior to AmB, with fewer adverse events (Mora-Duarte et al., 2002). The result of this study has contributed to a "paradigm shift" in the treatment of candidemia and invasive candidiasis away from AmB to echinocandins as primary therapy. Echinocandins have fungistatic activity against *Aspergillus* species. CASF is approved by the U.S. Food and Drug Administration for invasive aspergillosis refractory to or intolerant of other therapy.

Several important fungal pathogens are not susceptible to echinocandins: *C. neoformans*, and *Trichosporon* species among the yeasts. In vitro data support the activity against the mold forms of primary dimorphic fungal pathogens but echinocandins are *not active against the yeast forms* of *Histoplasma capsulatum, Blastomyces dermatitidis*, nor the spherule form of *Coccidioides* species. Investigators thought that *Mucorales* were not susceptible to echinocandins but evidence reviewed by Spellberg et al. (2009) from experimental murine infections and a series of human mucormycosis patients indicates *some* therapeutic potential. *Rhizopus oryzae*, the most frequent *Mucorales* pathogen, expresses the target enzyme for echinocandins: ß-glucan synthase. Combination therapy of an echinocandin with lipid-AmB in diabetic patients with rhinocerebral mucormycosis resulted in a significantly improved outcome compared with polyene monotherapy. Experimental infection with *R. oryzae* in mice rendered either neutropenic or with diabetic ketoacidiosis had improved survival if treated with micafungin or anidulafungin and lipid-AmB compared with AmB monotherapy. Molds that are not susceptible include the *Entomophthorales* and *Fusarium* species. Echinocandins have variable activity against *Scedosporium* species.

The novel mode of action of echinocandins against a target that is absent in mammalian cells, and minimal drug interactions, provide impetus for trials of combination therapy for invasive mold infections. Experimental aspergillosis in mice treated with combinations of echinocandins with either AmB or triazoles suggests that these combinations have synergistic effects in prolonging survival (reviewed by Kim et al., 2007).

3A.9 CASPOFUNGIN (CASF) (CANCIDAS®, MERCK)

Caspofungin (CASF) was the first echinocandin derivative licensed for human use, Fig. 3A.8. Currently, CASF is indicated for empiric therapy in febrile neutropenia, for *Candida* esophagitis, candidemia, invasive candidiasis, and salvage therapy for aspergillosis. Because of its efficacy against *Candida* species and excellent safety profile it is recommended as a first line drug to treat yeast bloodstream infections complicated by severe sepsis or septic shock (Flückiger et al., 2006). Among mold infections CASF is FDA approved for salvage therapy of refractory invasive aspergillosis.

3A.9.1 Action Spectrum

CASF is regarded as fungicidal for many yeasts and fungistatic for molds. CASF is highly active against all *Candida* species except *C. parapsilosis, C. guilliermondii*, and *C. lusitaniae*, against which it is moderately active. FLC-resistant *C. glabrata* and *C. krusei* infections can be treated with CASF.

3A.9.1.1 Yeasts

Susceptible *Candida albicans, C. glabrata, C. krusei*, and *C. tropicalis. Candida parapsilosis* is a special case because based on in vitro testing, *C. parapsilosis* isolates are susceptible but the MIC_{90} is close to the breakpoint. In clinical trials there is little difference in efficacy between these species.

Reduced Susceptibility *Candida guilliermondii* and *C. lusitaniae* have reduced susceptibility to all echinocandins.

Not Susceptible. *Cryptococcus neoformans* and *Trichosporon* species are not susceptible.

3A.9.1.2 Pneumocystis jirovecii

Good activity of CASF against the cyst (ascus) form of *P. jirovecii* occurs because ß-(1→3)-D-glucan is a major constituent of the cyst (ascus) wall.

3A.9.1.3 Febrile Neutropenia

"Caspofungin is as effective as and generally better tolerated than liposomal amphotericin B when given as empirical antifungal therapy in patients with persistent fever and neutropenia" (Walsh et al., 2004). In that study of a large group of patients with fever and neutropenia, the overall clinical success rates were 33.9% for 556 patients treated with CASF and 33.7% for 539 patients treated with liposomal AmB.

3A.9.1.4 Emerging Resistance Linked to Glucan Synthase Mutations

Resistance of clinical isolates of *C. albicans* to CASF is slowly emerging and is linked to mutations in short conserved regions in the *FKS1* gene (Balashov et al., 2006). Changes occurred at the serine 645 position in glucan synthase with substitutions of proline, tyrosine, and phenylalanine.

Mutations altering serine 645 occurred in both heterozygous and homozygous states (recall that *C. albicans* is diploid). Spontaneous mutants of *C. albicans* strains with MICs >16 μg/mL to CASF all had mutations in *FKS1*; 93% showed changes at Ser645.

3A.9.1.5 Molds

Susceptible *Aspergillus* species. Complete or partial responses to CASF therapy were observed in 5% and 40%, respectively, of 83 patients with invasive aspergillosis who did not respond or were intolerant to conventional antifungal therapy (Maertens et al., 2004). Patients who did not respond to primary therapy with AmB formulations had better clinical outcomes when treated with the combination of VRC and CASF compared with VRC alone (Marr et al., 2004).

Nonsusceptible *Blastomyces dermatitidis, Fusarium* species, *Histoplasma capsulatum, Mucorales, Scedosporium* species, *Paracoccidioides brasiliensis*, and *Pseudallescheria boydii* are all resistant to each of the echinocandins. Members of the *Mucorales* are generally regarded as not susceptible to echinocandins. The in vitro MIC of CASF against *Rhizopus oryzae* was >8 μg/mL compared to an MIC of 0.06–0.12 μg/mL against *Aspergillus fumigatus* (Pfaller et al., 1998).

There is apprehension that a trend toward breakthrough infections with mucoraceous fungi may be developing as either VRC or CASF or combinations of the two are used more often to treat aspergillosis. A case report illustrated breakthrough mucormycosis in a lymphoma patient receiving long-term VRC and CASF therapy (Blin et al., 2004).

Despite the general view that mucormycosis is unresponsive to echinocandins, therapy reviewed for 41 patients with biopsy-proven rhino-orbital-cerebral mucormycosis over a 12 year period at one California hospital suggested otherwise (Reed et al., 2008). Patients treated with polyene–caspofungin combination therapy (6 evaluable patients) had superior success (100% vs. 45%) and Kaplan–Meier survival time compared with patients treated with polyene monotherapy.

The puzzle of susceptibility of the *Mucorales* to echinocandins was investigated by Lamaris et al. (2008). They preincubated *Rhizopus oryzae* with CASF (32 μg/mL) and found that there was increased exposure to ß-glucan sites in the cell wall. PMN-induced damage increased after CASF exposure and was augmented by the addition of anti–ß-glucan antibody. Echinocandins appear to unmask ß-(1→3)-D-glucan in the cell wall of *R. oryzae*, which promotes PMN activity against the fungal hyphae via the PMN glucan receptor, dectin-1.

3A.9.2 Dosage

Intravenous administration of a 70 mg loading dose of CASF followed by 50 mg daily dose is the recommended regimen (Theuretzbacher, 2004).

3A.9.3 Pharmacokinetics

CASF is well tolerated because it lacks mechanism-based toxicity. It does not interfere with drugs that are metabolized by cytochrome P450 isozymes. The $t_{1/2}$ is 8–13 h and CASF is 80% to >99% protein-bound. There is marked inter-individual variability in the AUC, Cmax, and Cmin in normal volunteers and patients. CASF is metabolized slowly and undergoes spontaneous degradation to an inactive open ring compound in the liver.

3A.9.4 Drug Interactions

There is a drug interaction with cyclosporine. In a limited number of patients there has been an increase in hepatic enzymes when cyclosporine and CASF have been used together. This occurred in up to 35% of transplant patients.

3A.9.5 Adverse Reactions

The most common side effects associated with CASF are gastrointestinal associated, elevations in hepatic enzymes, and occasional increases in serum creatinine.

3A.10 MICAFUNGIN (MCF) (MYCAMINE®, ASTELLAS PHARMA, INC.)

3A.10.1 Indications

Currently, micafungin (MCF) is approved in the United States for treating invasive candidiasis including candidemia, esophageal candidiasis, and to prevent fungal infections following a hematopoietic stem cell transplant (SCT).

3A.10.2 Dosage

MCF has different doses for different indications:

- For prophylaxis against fungal infections in SCT patients, the dose is 50 mg IV/day.
- For esophageal candidiasis the dose is 150 mg IV/day.
- For invasive candidiasis and candidemia, the dose is 100 mg IV/day.
- For salvage therapy of aspergillosis or as adjunctive therapy to VRC or AmB, the dose is 150 mg IV/day, which may be escalated up to 300 mg IV/day.
- Doses up to 800 mg/day have been well tolerated in small studies. No dosage adjustments are needed for renal or hepatic insufficiency.

3A.10.3 Metabolism

The $t_{1/2}$ is 11–15 h. Plasma protein binding is >99%. MCF is metabolized by arylsulfatase and then undergoes further hydroxylation in the liver.

3A.10.4 Drug Interactions

There are minimal drug interactions between MCF and other medications. Those that have been studied and no interactions found include AmB, mycophenolate mofetil, cyclosporine, tacrolimus, prednisolone, FLC, VRC, ritonavir, and rifampin. When coadministered with sirolimus, MCF did increase the AUC of sirolimus by 21%, but did not significantly alter the Cmax. Similarly, MCF did increase the AUC and Cmax of nifedipine and ITC. Thus, when MCF is administered with these three drugs, it is recommended to monitor for adverse events from the sirolimus, nifedipine, or ITC. Highlights of prescribing information for MCF may be found at the URL http://www.astellas.us/docs/mycamine.pdf.

3A.11 ANIDULAFUNGIN (ANF) (ERAXIS®, PFIZER)

ANF is a semisynthetic product of echinocandin B, itself a fermentation product of *Aspergillus nidulans*. It was developed by Eli Lilly, underwent preclinical and clinical development at Vicuron Pharmaceuticals, and was sold to Pfizer. It received approval from the U.S. Food and Drug Administration in February 2006.

3A.11.1 Indications

ANF is approved in the United States for the treatment of invasive candidiasis, candidemia, and esophageal candidiasis. In vitro susceptibility data show that *C. albicans, C. glabrata, C. krusei*, and *C. tropicalis* are very susceptible to ANF with MICs of between 0.03 and 0.13 μg/mL. *Candida parapsilosis* and *C. lusitaniae* are less susceptible. *Cryptococcus neoformans* is not susceptible to ANF or to other echinocandins. *Aspergillus fumigatus, A. flavus*, and *A. niger* are very susceptible to ANF with an MIC of 0.03 μg/mL. The MIC for *Pseudallescheria boydii* is 2.5 μg/mL. Of the molds tested so far, *Rhizopus* species are not susceptible to ANF or to other echinocandins (Sabol and Gumbo, 2008) but combination therapy with an echinocandin and AmB was superior to AmB alone in treating a small group of mucormycosis patients (Spellberg et al., 2009).

3A.11.2 Invasive Candidiasis

The efficacy of FLC was compared with that of ANF in treating invasive candidiasis (Reboli et al, 2007). In this double-blind study, 261 patients were stratified by APACHE score and presence of neutropenia. Most patients, 97%, did *not* have neutropenia. Patients were randomized to treatment with ANF (200 mg loading dose then 100 mg/day) or IV FLC (800 mg loading dose then 400 mg/day). Therapy was continued for 14 days after a negative blood culture and absence of signs and symptoms. The most common isolates were *C. albicans* (61.6%), *C. glabrata* (20.4%), and *C. parapsilosis* (11.8%). Successful outcomes were achieved in 75.6% treated with ANF, compared to 60.2% treated with FLC (15.4% difference). Thus, in this study ANF was superior to standard therapy. Patients receiving ANF had successful responses for all *Candida* isolates except *C. parapsilosis*.

ANF may be considered as first-line or primary therapy for candidemia and invasive candidiasis in non-neutropenic patients. Based on safety, compared with AmB, and efficacy, it may be superior to AmB and FLC. There is no clear superiority for ANF compared with the other echinocandins, CASF and MCF, in this application. ANF is fungicidal but as yet there are no data to support its use as primary therapy in neutropenic patients with invasive candidiasis.

3A.11.3 Molds

Data are lacking on the use of ANF therapy for invasive aspergillosis. An open label, noncomparative study of the safety and efficacy of ANF plus liposomal AmB for the treatment of aspergillosis has been completed, but results have not yet been published in full. Details of this trial and the partial results can be found at Clinical Trials.gov (NCT00037206). Thus, the clinical utility of ANF for invasive aspergillosis remains to be determined.

3A.11.4 Dosage

The drug regimen is a 200 mg loading dose and a 100 mg/day for invasive candidiasis and 100 mg loading dose followed by 50 mg/day for esophageal candidiasis.

3A.11.5 Metabolism

ANF achieves lower mean peak plasma concentration than CASF or MCF. Over a range of doses ANF displayed linear pharmacokinetics, with dose-independent plasma clearance and dose proportional increase in the AUC. Mean plasma concentrations range from 1.7 μg/mL to 3.82 μg/mL. At the maintenance dose of 100 mg/day IV, in patients with fungal infections, the steady state $t_{1/2}$ is 26.5 h, Cmax is 7.2 μg/mL, Cmin of 3.3 μg/mL, and AUCss is 110.3 mg·h/L. The Vd is 20–50 L, similar to total body volume. ANF is eliminated via fecal excretion, mostly as metabolites. There is no renal clearance.

All three echinocandin derivatives are highly protein bound in plasma. The effect of binding to plasma proteins on activity of echinocandins in vivo is not known. According to the "free drug hypothesis," protein binding should result in reduced activity of the antimicrobial agent because less free drug is able to enter the tissues (Theuretzbacher, 2004).

3A.11.6 Drug Interactions

No significant interactions were found when ANF was tested with liposomal AmB, tacrolimus, cyclosporine, VRC, and rifampin.

3A.11.7 Adverse Reactions

The most common adverse events during clinical trials were diarrhea and increases in ALT, AST, alkaline phosphatase, and hypokalemia.

3A.12 TERBINAFINE (TRB) (LAMISIL®, NOVARTIS)

Terbinafine (TRB) is an oral and topical agent in the allylamine class of synthetic compounds (Fig. 3A.9).

3A.12.1 Mode of Action

TRB inhibits an early step in the ergosterol biosynthetic pathway: the squalene epoxidase enzyme. Treated fungi accumulate squalene and become deficient in ergosterol, damaging the cell membrane.

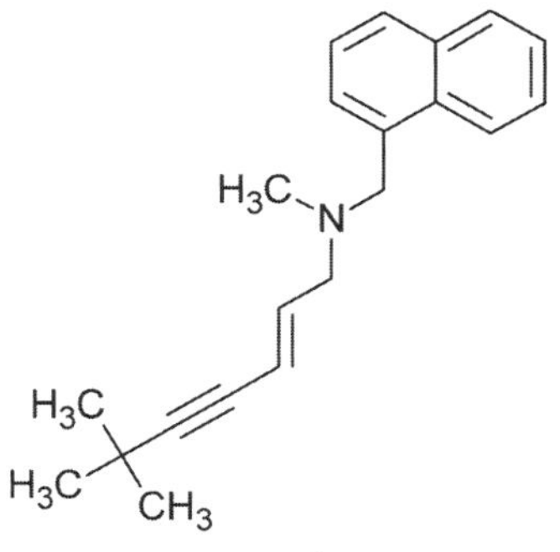

H_3C N H_3C H_3C CH_3

Figure 3A.9 Terbinafine (Lamisil®).

Source: PubChem at the URL http://pubchem.ncbi.nlm.nih.gov, a service of the National Center for Biotechnology Information and the National Library of Medicine, U.S.A.

3A.12.2 Action Spectrum

TRB accumulates in skin, nails, and fat. This drug is effective in the treatment of tinea infections, including nail infections, and is superior to griseofulvin.

3A.12.2.1 Onychomycosis

The dosage regimen for nail infections is TRB, 250 mg/day (oral) for 3 months. TRB is well absorbed from the gastrointestinal tract, and the time to reach effective concentrations in nails is 1–2 weeks, with persistence in nails for longer than 252 days.

3A.12.2.2 Refractory Mycoses

Oral TRB is used in primary therapy, combination therapy, and alternative therapy for refractory subcutaneous mycoses: chromoblastomycosis, eumycetoma, phaeohyphomycotic cysts, and sporotrichosis. Please see the disease chapters mentioned for specific examples.

3A.12.3 Drug Synergy

In vitro evaluations of drug synergy have found that TRB is synergistic with azoles, AmB, and echinocandins (Antoniadou and Kontoyiannis 2003). Examples of such synergy are in combination with:

- Azoles versus susceptible-dose-dependent *C. glabrata*
- VRC versus azole-resistant *C. albicans*
- VRC and MCF versus *Scedosporium prolificans*
- VRC or ITC or AmB versus *Mucorales*

The synergy is facilitated by the different drug target of TRB, squalene epoxidase, in the ergosterol biosynthetic pathway.

3A.12.4 Metabolism

TRB is widely distributed in tissues. It accumulates in the skin and in adipose tissue, from which it is slowly released (Debruyne and Coquerel, 2001). The Vd is very large (948 L) at steady state. It is widely metabolized in the liver into 15 inactive metabolites. Peak plasma TRB concentrations, 860–1340 μg/L, are reached in 2 h after a single oral dose of 250 mg. The plasma concentration of TRB declines in three phases, with apparent half-lives of 1.1, 16–22, and 100 h. The very long $t_{1/2}\beta$ is because of its slow release from the skin and adipose tissue. Steady-state concentrations are reached within 10–14 days of starting therapy. Eighty percent of TRB is excreted as metabolites in the urine, and 20% in the feces. Very little unchanged drug is detected in urine.

3A.12.5 Adverse Reactions

TRB is generally well tolerated. The most common complaints are gastrointestinal disturbances and cutaneous reactions (please see below). Other adverse reactions include elevated liver enzyme levels, visual and taste disturbances, rare severe neutropenia, pancytopenia, and allergic reactions. Adverse reactions may occur in more than 7% of patients including a prevalence of about 2% of various skin manifestations including severe cutaneous reactions, for example, acute generalized exanthematous pustulosis (Beltraminelli et al., 2005), Stevens–Johnson syndrome, toxic epidermal necrolysis, erythroderma, and exacerbation or induction of psoriasis.

3A.13 5-FLUOROCYTOSINE (FLUCYTOSINE, 5FC) (ANCOBON®, VALEANT PHARMACEUTICALS)

Flucytosine (Fig. 3A.10), a fluorinated pyrimidine, inhibits both DNA and RNA synthesis. Flucytosine acts directly on fungi by competitive inhibition of purine and pyrimidine uptake and indirectly by intracellular metabolism to 5-fluorouracil. Flucytosine enters the fungal cell via cytosine permease and is metabolized to 5-fluorouracil by cytosine deaminase. Humans and other mammals lack this enzyme so that flucytosine does not inhibit DNA and RNA synthesis in mammalian cells. In fungi, 5-fluorouracil is then converted into two different nucleotides: 5-fluorouridine triphosphate inhibits RNA processing, and 5-fluorodeoxyuridine monophosphate inhibits thymidylate synthase needed for DNA synthesis. The result is unbalanced growth and death of the target cell.

3A.13.1 Indications

Flucytosine is not the drug of choice for any infection because it is inferior to AmB and because secondary resistance is commonly and rapidly acquired during therapy, limiting flucytosine's usefulness as monotherapy in candidiasis and cryptococcosis. The most common cause of resistance appears to be loss, by the fungus being treated, of cytosine deaminase or UMP pyrophosphorylase activity. Flucytosine is recommended alternative monotherapy for patients with symptomatic candiduria and cystitis due to a fluconazole-resistant *Candida* species, (but not *C. krusei*) (Pappas et al., 2009).

Flucytosine may have some effectiveness against dematiaceous fungi. Fungi that lack cytosine deaminase are not susceptibile to flucytosine. This drug is not effective against *Blastomyces dermatitidis, Coccidioides* species, *Histoplasma capsulatum*, or *Sporothrix schenckii* (Vermes et al., 2000).

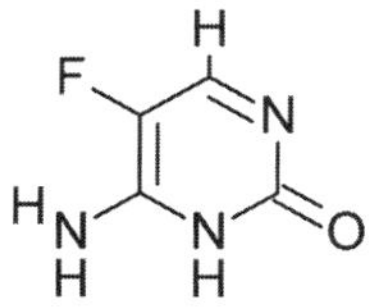

Figure 3A.10 5-Fluorocytosine (flucytosine).
Source: PubChem at the URL http://pubchem.ncbi.nlm.nih.gov, a service of the National Center for Biotechnology Information and the National Library of Medicine, U.S.A.

3A.13.2 Combination Therapy

The mixture of flucytosine in combination with AmB is at least additive, especially in cryptococcal meningitis where their combined use suppresses development of flucytosine resistance, and reduces the required dose of AmB. Most *Cryptococcus neoformans* strains isolated from clinical material have flucytosine MICs of 0.46–7.8 μg/mL. Isolates with an MIC greater than 12.5 μg/mL are considered resistant. Clinical cultures should be referred for susceptibility testing initially and at weekly intervals during therapy, to check for the development of flucytosine resistance.

3A.13.3 Metabolism

Flucytosine is rapidly and almost completely absorbed following oral administration. It is water soluble and penetrates into tissues. The Vd is 0.7–1 L/kg. Flucytosine readily penetrates the blood–brain barrier, achieving clinically significant concentrations in CSF. Bioavailability estimated by comparing the area under the curve of serum concentrations after oral and intravenous administration showed 78–89% absorption of the oral dose.

Mean blood concentrations are approximately 70–80 μg/mL, 1–2 h after a dose, in patients with normal renal function who received a 6 week regimen of flucytosine (150 mg/kg/day given in divided doses every 6 h) in combination with AmB. Flucytosine is excreted intact via the kidneys by means of glomerular filtration without significant tubular reabsorption. The $t_{1/2} = 3–6$ h is prolonged in patients with renal insufficiency and should be used with extreme care in patients with impaired renal function.

3A.13.4 Adverse Reactions

There is a relationship between flucytosine plasma concentrations and bone marrow toxicity which may present as bone marrow suppression. Neutropenia and thrombocytopenia may occur, but these are usually mild and reversible. It is thought that intestinal bacteria convert flucytosine to 5-fluorouracil and that is the source of the metabolite which toxic to mammalian cells. In addition, to the conversion of flucytosine to 5-fluorouracil by intestinal microbiota, 5-fluorouracil may be present in preparations of flucytosine as a result of sterilization and upon storage (Vermes et al., 2000).

Close monitoring of hematologic, renal, and hepatic status of all patients is essential. Gastrointestinal disturbances may occur. In combination therapy the renal toxicity of AmB may increase plasma flucytosine concentration to potentially toxic levels. Flucytosine is contraindicated in pregnancy because of fetal abnormalities observed in animals (Pappas et al, 2009).

3A.14 GRISEOFULVIN (GRIFULVIN V, ORTHO PHARMACEUTICAL CORP.)

Griseofulvin (Fig. 3A.11) is a spiro-benzo[*b*]furan natural product, produced by *Penicillium griseofulvum*. Originally isolated in 1939 as a natural product of *Penicillium* species, it was not until 1947 that it was observed to distort hyphal growth of certain fungi. Before its identity as griseofulvin was firmly established it was named "curling factor." In 1958 it was found to cure experimental ringworm in guinea pigs. Soon thereafter it was successfully used to treat human dermatophyte infections.

3A.14.1 Mode of Action

Griseofulvin has been used for many years to treat tinea capitis and other dermatophytoses. It accumulates in the stratum corneum, hair, and nails. The mechanism of action was thought to involve the inhibition of fungal cell mitosis, by interfering with the spindle microtubule function, arresting fungal cells in metaphase. The antimitotic effect of griseofulvin in mammalian cells is very weak, accounting for its selective toxicity for dermatophytes. Now it appears that griseofulvin acts indirectly on microtubules by inhibiting gene expression. This was discovered when the cloned β-tubulin gene from *T. rubrum* was recombined with yellow fluorescent protein (Zomorodian et al., 2007). Griseofulvin treatment of *T. rubrum* fungal cells decreases the expression of ß-tubulin in a dose-dependent manner.

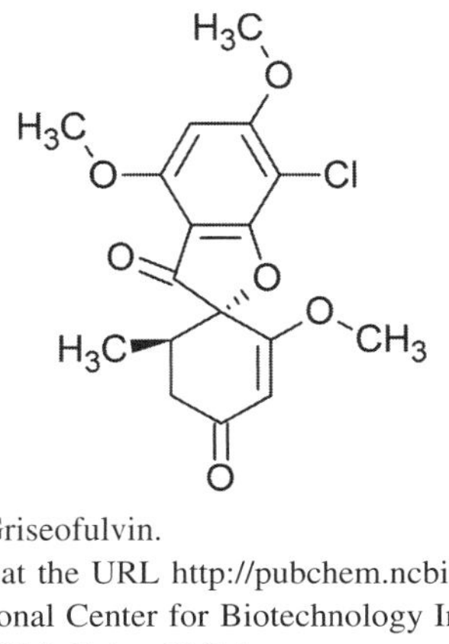

H_3C O H_3C O Cl O O O CH_3 H_3C O

Figure 3A.11 Griseofulvin.

Source: PubChem at the URL http://pubchem.ncbi.nlm.nih.gov, a service of the National Center for Biotechnology Information and the National Library of Medicine, U.S.A.

3A.14.2 Action Spectrum

Griseofulvin is effective to treat ringworm of the scalp, beard, soles, palms, tinea corporis, or tinea cruris. Griseofulvin has no activity against yeasts; thus, it is ineffective in treating pityriasis versicolor or *Candida* species infections. ITC and TRB have supplanted griseofulvin as the treatment of choice for tinea corporis, onychomycosis, and tinea capitis.

3A.14.3 Indications

Griseofulvin inhibits the growth of dermatophyte genera: *Trichophyton* species and *Microsporum* species. Major indications are for tinea capitis (scalp ringworm) and tinea unguium (onychomycosis) and also for tinea pedis, tinea cruris, and tinea corporis.

3A.14.4 Dosage Regimen

After oral administration the drug reaches skin and hair. Topical applications of griseofulvin are *not* effective. The usual daily dose for adults is 500, 750, or 1000 mg for microcrystalline griseofulvin. A daily dose of 500 mg will give a satisfactory response in most patients with tinea corporis, tinea cruris, and tinea capitis. An increased daily dose of 1000 mg is used for tinea unguium. For children weighing 30–50 pounds the daily dose is 125–250 mg; and for children over 50 pounds, 250–500 mg/day. The ultramicrocrystalline form is given in a 330 mg daily dose (divide dose in half, give 2×/day). The two forms have comparable efficacy. Representative treatment periods are tinea capitis, 4–6 weeks; tinea corporis, 2–4 weeks; tinea pedis, 4–8 weeks.

3A.14.5 Metabolism

Griseofulvin absorption from the gastrointestinal tract varies among individuals, mainly because it is insoluble in the aqueous milieu of the upper gastrointestinal tract. Peak plasma concentration after a 500 mg oral dose occurs after 4 h, reaching 0.5–2.0 μg/mL. The blood $t_{1/2}$ is 24–36 h. Some individuals are poor absorbers and attain lower blood concentrations, which may explain unsatisfactory results. Improved blood concentrations can be reached if the drug is taken after a meal with a high fat content. Griseofulvin is metabolized in the liver.

3A.14.6 Adverse Reactions

There is a low incidence of side effects. It is teratogenic in mice, so it is not indicated in pregnancy. Adverse reactions are commonly hypersensitivity reactions, for example, skin rashes, urticaria, and, rarely, angioneurotic edema. Rarely, paresthesias of the hands and feet have been reported after extended therapy. Other occasional side effects are nausea, vomiting, epigastric distress, diarrhea, headache, fatigue, dizziness, insomnia, mental confusion, and impairment of performance of routine activities. Rarely, proteinuria and leukopenia have been reported. Administration of the drug should be discontinued if granulocytopenia occurs. When rare, serious reactions occur with griseofulvin, they are usually associated with high dosages and/or long periods of therapy.

Because of the potential for adverse effects on the human fetus, contraceptive precautions should be taken during treatment with griseofulvin and for a month after termination of treatment. Griseofulvin should not be prescribed for pregnant women or women intending to become pregnant within 1 month following cessation of therapy. Since griseofulvin has demonstrated harmful effects in vitro on the genotype in microbes and plants, males should wait at least 6 months after completing griseofulvin therapy before fathering a child.

This concludes the section on the antifungal agents currently in clinical use. The following section considers how these drugs are deployed in combination, suppressive or maintenance, prophylactic, and empiric therapy of mycoses.

3A.15 COMBINATION THERAPY

Poor outcomes of treating mold infections and a larger selection of antifungal agents have increased interest in combination therapy. Animal models have shown that echinocandins reduce tissue burden and sterilize tissues when used in conjunction with extended spectrum triazoles (Patterson, 2006). One such combination is VRC plus CASF for the treatment of invasive aspergillosis in solid organ transplant recipients. This study showed a statistically significant improvement in the 90 day survival when compared to historical controls (Singh et al., 2006). At present, the safety of this combination remains undetermined. Other studies in hematopoietic stem cell recipients have shown similar results whether the combination is CASF and VRC or CASF with AmB. All studies have shown a survival benefit. There was a 50% reduction in mortality when CASF was combined with VRC compared to VRC alone (Marr et al., 2004). Two other studies have shown a similar benefit when CASF was combined with AmB (Aliff et al., 2003; Kontoyiannis et al., 2003). Another combination that has proved effective in case reports is TRB combined with VRC to treat infections caused by *Scedosporium prolificans*, even in dissemintated infection after an SCT, in a long-term kidney transplant recipient, and in a brain abscess in a patient with a hematologic malignancy (Li et al., 2008; reviewed by Vazquez, 2008). *Scedosporium prolificans* is a mold that is innately resistant to most antifungal agents.

The combination of AmB and flucytosine is standard primary therapy for CNS cryptococcosis. AmB penetrates slowly into the CNS, whereas flucytosine rapidly achieves an effective fungicidal concentration there. The regimen consists of 0.7 mg/kg/day of AmB, and 100 mg/kg/day of flucytosine (the latter drug divided into 4 daily doses) for a total of 2 weeks. Then, pending a negative CSF culture, patients are switched to maintenance therapy.

3A.16 SUPPRESSIVE OR MAINTENANCE THERAPY

Relapse following primary therapy is a formidable problem in certain risk groups. FLC is effective maintenance therapy in cryptococcosis. Mucosal candidiasis in AIDS patients who do not have access to HAART is apt to recur rapidly, thus secondary prophylaxis with 100 mg FLC/day is a recognized part of managing patients with severe or recurring candidiasis.

Cryptococcosis in persons living with AIDS is a prime indication for FLC maintenance therapy. Using Kaplan–Meier curves[2] the probability of remaining relapse-free at 1 year was 95–97% for FLC versus 78% for AmB, and 0% for placebo.

3A.17 PROPHYLACTIC THERAPY

Patients undergoing cytoreductive therapy for stem cell or solid organ transplantation or who are about to receive chemotherapy for cancer are at increased risk of fungal infections and could benefit from antifungal prophylaxis. FLC, AmB, ITC, MCF, and PSC have all been studied for prophylaxis and found beneficial.

[2]Kaplan–Meier. *Definition*: In clinical trials the investigator is interested in the time until participants present a specific event or endpoint, usually a clinical outcome such as death, disappearance of a tumor, and so on. The participants are followed beginning at a certain starting point, and a record is kept of the time necessary for the event of interest to occur. These data can be analyzed by means of a life-table, or Kaplan–Meier curve, which is the most common method to describe survival characteristics. The equation for probability curves and the Kaplan–Meier curve can be found at www.sas.com.

3A.17.1 Bimodal Period of Risk

There is a *bimodal period of risk* for invasive fungal infections in hematopoietic SCT recipients. The neutropenic period before engraftment may expose patients to their endogenous microbiota, especially to *Candida* species, and there is also risk of exposure to airborne conidia of molds in the hospital environment. Invasive mold infections have a bimodal period of risk: a small incidence during the neutropenic period before engraftment, and an increased risk after engraftment because of chronic GVHD, and the need for associated immunosuppressive therapy (Fig. 3A.12) (Dykewicz et al., 2000).

Other transplant recipients are at risk including those receiving liver or pancreas transplants (van Burik, 2005). Solid organ transplants were made possible by the advent of immunosuppressive therapy to prevent rejection (Post et al., 2005). The case of liver transplantation is illustrative of the effect of this therapy on the immune system, leaving the patient vulnerable to fungal infection. In a randomized, placebo-controlled trial investigating the utility of FLC for preventing fungal infections, Winston et al. (1999) found that FLC reduced overall fungal infections from 43% to 9% ($p<0.001$) and invasive fungal infections from 23% to 6% ($p<0.001$). However, other investigators commented that the fungal infection rate of 6% in the FLC treated group was similar to their fungal infection rates without any prophylaxis (Singh and Vu, 2000).

3A.17.2 Fluconazole and Alternatives for Primary Prophylaxis

The CDC guideline recommends FLC at 400 mg/day for patients undergoing SCT as primary prophylaxis against yeast infections (Dykewicz et al., 2000). When prophylaxis is started at conditioning and continued until 75 days post-transplantation, systemic fungal infections decreased from 18% to 7% with overall survival benefit (reviewed by Strasfeld and Weinstock, 2006).

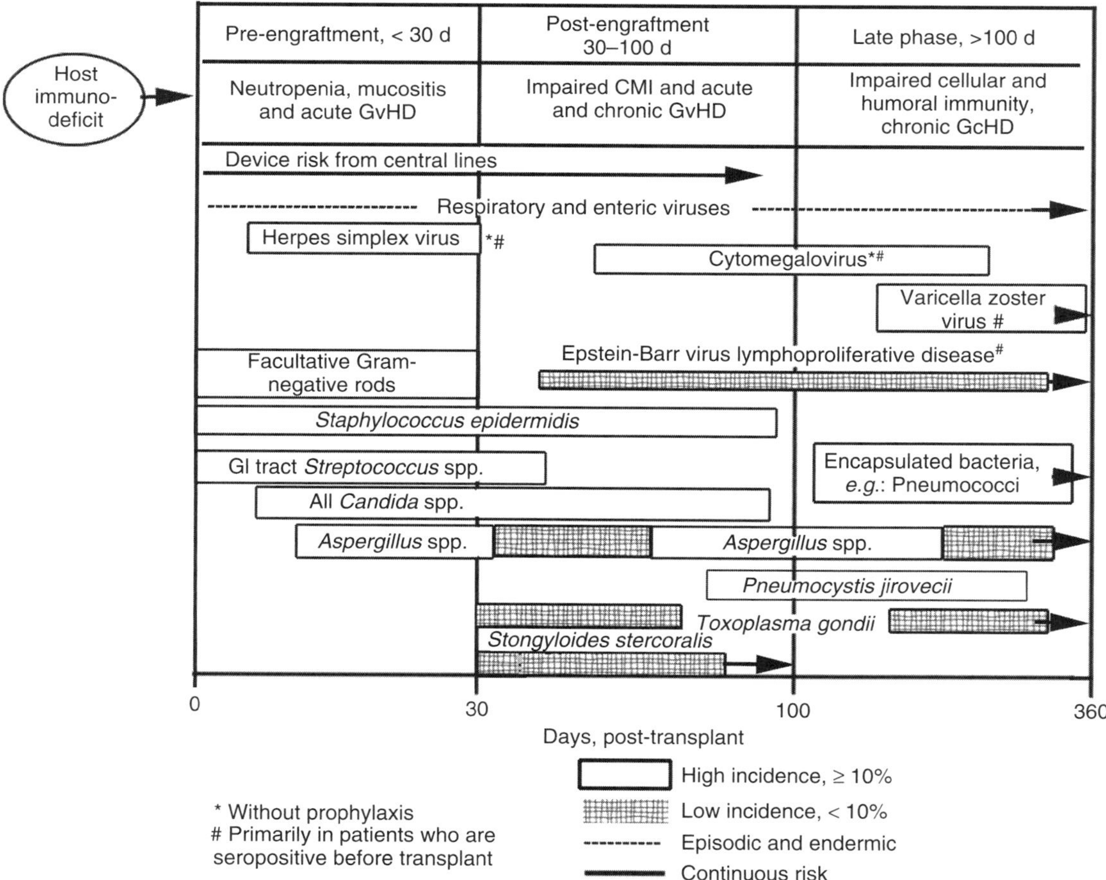

Figure 3A.12 Phases of opportunistic infections among allogeneic SCT recipients; GvHD, graft-versus-host disease; d, days. *Source:* Adapted from Dykewicz et al. (2000).

The advent of extended spectrum triazoles and echinocandin derivatives has resulted in additional choices for prophylaxis. MCF, an echinocandin FDA approved in 2005 for prophylaxis of SCT patients, is fungicidal for *Candida* species and fungistatic for molds. In addition, echinocandins are highly active against *Pneumocystis jirovecii*. In comparison with FLC, patients treated with MCF were better protected against invasive fungal infections, with notably fewer cases of aspergillosis.

In the pre-engraftment period for SCT patients VRC, structurally derived from FLC, is an important agent because of its activity against *Aspergillus* species and excellent activity against FLC-resistant *Candida* species. Posaconazole (PSC) is very promising for prophylaxis because of its activity against both aspergilli and *Mucorales*. In patients with prolonged neutropenia, PSC was superior to FLC or ITC in preventing invasive fungal infections, 2% versus 8% ($p<0.001$) (Cornely et al., 2007).

Alternatives to triazoles, with activity against invasive mold infections, include AmB, AmB–lipid complexes, and the echinocandins. Both VRC and echinocandins lack activity against *Mucorales* explaining, at least partially, the increased incidence of infections with this group of environmental molds.

3A.17.3 Prophylaxis in Patients During the Pre-engraftment Period with a History of Invasive Mold Infections

Stem cell transplant recipients are vulnerable during the neutropenic period of approximately 100 days after transplantation and prior to engraftment. During conditioning chemotherapy SCT patients could benefit from secondary prophylaxis with a mold-active antifungal agent to prevent a flare or recurrence of mold infection. Because some of these agents have only recently been approved, there are insufficient data to support the value of this approach. AmB or lipid derivatives of AmB are relevant for this purpose. Of the extended spectrum azoles, VRC in either oral or IV formulations has excellent bioavailability and both anti-*Candida* and anti-*Aspergillus* activity. PSC, while less effective against yeasts than FLC, can be orally administered and is active against both aspergilli and *Mucorales*. MCF is effective against both *Candida* species and *Aspergillus* species.

3A.17.4 Prophylaxis in the Post-engraftment Period

Antifungal prophylaxis in this period is designed to prevent fungal infections during recovery after allogeneic transplantation. FLC has a role and is efficient in reducing *Candida* species infections, although it is less effective against *C. glabrata* and *C. krusei* and lacks antimold activity. AmB, in low-dose form, has been used but is not preferred because of well-known infusion-related side effects and drug-related toxicity. ITC has the capacity to prevent opportunistic yeast infections and, to a lesser extent, aspergillosis in the post-engraftment period, but adverse reactions can result in its discontinuation. Extended spectrum azoles, VRC and PSC, are effective in preventing yeast and mold infections in the post-engraftment period in patients contending with GVHD. Other alternatives for prophylaxis in this period include the echinocandins, either CASF, MCF, or ANF. They have potent anti-*Candida* and anti-*Aspergillus* activity but only IV formulations are available. PSC was compared to FLC for prophylaxis of fungal infections in patients who had developed GVHD. In a randomized, blinded study of 600 patients, PSC was as effective as FLC in preventing all invasive fungal infections (5.3% and 9.0%, respectively) and was superior to fluconazole in preventing proven or probable invasive aspergillosis (2.3% vs. 7.0%, $p = 0.006$) (Ullman et al., 2007).

3A.18 EMPIRIC THERAPY

When patients with profound neutropenia (<100 wbc/mm^3) have a fever of unknown origin which persists for $\geq$5 days and fail to respond to antibacterial therapy, they are candidates for empiric antifungal therapy (Bodey et al., 2002). Febrile neutropenia may have various causes, one of which is fungal disease. Others include slowly resolving bacterial infection, antimicrobial resistant bacterial infection, and drug or tumor fever. Among the choices for empiric antifungal therapy are low-dose (3 mg/kg) liposomal AmB, FLC, ITC, VRC, or an echinocandin: CASF or MCF. Of these, FLC may be the poorest choice because it is inactive against invasive mold infections; it may be used in institutions where aspergillosis occurs infrequently.

Patients who have received FLC prophylaxis and then develop fever should be treated with an echinocandin or AmB formulation, because of the probability of an azole-resistant *Candida* species or a mold infection. In a randomized, controlled trial, CASF was as effective as liposomal AmB in the empiric treatment of fever and neutropenia but was better tolerated (Walsh et al., 2004). Empiric therapy is tailored to the patient and usually lasts for 10 days, weighing such factors as whether a fungal infection is identified and when recovery from neutropenia occurs. If a particular organ is certainly involved treatment is continued for at least 2 weeks after organ involvement is resolved (Bodey et al., 2002). When febrile neutropenia persists, treatment is continued until resolution of fever or neutropenia, and if not >3 weeks.

3A.19 INNATELY RESISTANT FUNGI

Despite the increased armamentarium of antifungal agents available at present, there are still clinically important fungi for which there is no satisfactory chemotherapeutic approach. Some individual species and classes of fungal pathogens demonstrate unusual patterns of resistance to primary therapy and constitute a "rogue's gallery" of dangerous opportunists.

3A.19.1 Innately Resistant Molds

Rhizopus, Rhizomucor, and *Lichtheimia* species, members of the order *Mucorales*, subphylum *Mucoromycotina* are generally regarded as not susceptible to echinocandins because they do not appear to contain ß-(1→3)-D-glucans in their cell walls. However, *R. oryzae* has an *FKS* (glucan synthase) gene and CASF inhibits glucan synthase activity in *R. oryzae* membrane preparations. As well, CASF demonstrates efficacy in vivo during disseminated *R. oryzae* infection in diabetic ketoacidotic mice (Ibrahim et al., 2005). *Aspergillus terreus* is resistant in vitro and in vivo to AmB.

Fusarium species invasive mold mycoses are relatively resistant to treatment with antifungal agents. For example, ITC has no activity against *Fusarium* species. The preferred primary therapy is with high-dose AmB or, as an alternative, VRC. *Fusarium* species are resistant in vitro to echinocandins, but successful therapy of disseminated *Fusarium* infection was reported in a patient with acute leukemia who failed AmB treatment but became afebrile and fungus culture-negative on CASF (Apostolidis et al, 2003).

Pseudallescheria boydii and *Scedosporium apiospermum* are resistant to AmB, which, even at high doses, evokes a very poor clinical response. The clinical forms involved are eumycotic mycetoma, and pulmonary and pulmonary-disseminated mycosis in the immunocompromised host. The darkly pigmented mold *Scedosporium prolificans* demonstrates high MICs in vitro to all known classes of polyenes, azoles, and echinocandin derivatives. Combination therapy, as mentioned above, has had success in treating infections with this recalcitrant mold. Eumycetoma caused by *Madurella mycetomatis* sometimes responds poorly even when antifungal agents are given in combination at high doses.

3A.19.2 Innately Resistant Yeasts

The susceptibility profile of *Candida krusei* is that of a multidrug-resistant pathogen with decreased susceptibility to FLC (2.9% of isolates are susceptible), AmB (the MIC was $\leq$1 μg/mL for 8.0% of isolates), and flucytosine (4.0% susceptible) (Pfaller and biekema, 2004).

Candida glabrata is innately less susceptible to FLC and AmB than other species of *Candida*, rapidly developing azole resistance by upregulating efflux pumps. The clinical relevance is that suboptimal concentrations of FLC may reduce susceptibility, and patients receiving FLC prophylaxis may experience increased rates of colonization and infection with *C. glabrata. Cryptococcus neoformans* is innately resistant to echinocandin derivatives.

Trichosporon asahii and *T. mucoides* may cause fungemia and hematogenous dissemination in neutropenic patients resulting in a high case/fatality ratio. AmB lacks fungicidal activity against *Trichosporon* and clinical failures with AmB, FLC, and combinations of the two have been reported. The multiresistant nature of *T. asahii* makes it a threat for nongranulocytopenic patients as well.

GENERAL REFERENCE FOR ANTIFUNGAL AGENTS AND THERAPY

Tkacz J S, Lange L (eds.), 2004. *Advances in Fungal Biotechnology for Industry, Agriculture, and Medicine*. Springer, New York.

SELECTED REFERENCES FOR ANTIFUNGAL AGENTS AND THERAPY

Aliff TB, Maslak PG, Jurcic JG, Heaney ML, Cathcart KN, Sepkowitz KA, Weiss MA, 2003. Refractory *Aspergillus* pneumonia in patients with acute leukemia. *Cancer* 97: 1025–1032.

Alvarado-Ramírez E, Torres-Rodríguez JM, 2007. In vitro susceptibility of *Sporothrix schenckii* to six antifungal agents determined using three different methods. *Antimicrob Agents Chemother* 51: 2420–2423.

Andreoli TE, 1973. On the anatomy of amphotericin B–cholesterol pores in lipid bilayer membranes. *Kidney Intern* 4: 337–345.

Antoniadou A, Kontoyiannis DP, 2003. Status of combination therapy for refractory mycoses. *Curr Opin Infect Dis* 16: 539–545.

Apostolidis J, Bouzani M, Platsouka E, Belasiotou H, Stamouli M, Harhalakis N, Boutati EI, Paniara O, Nikiforakis E, 2003. Resolution of fungemia due to *Fusarium* species in a patient with acute leukemia treated with caspofungin. *Clin Infect Dis* 36: 1349–1350.

Balashov SV, Park S, Perlin DS, 2006. Assessing resistance to the echinocandin antifungal drug caspofungin in *Candida albicans* by profiling mutations in FKS1. *Antimicrob Agents Chemother* 50: 2058–2063.

Benson JM, Nahata MC, 1989. Pharmacokinetics of amphotericin B in children. *Antimicrob Agents Chemother* 33: 1989–1993.

Beltraminelli HS, Lerch M, Arnold A, Bircher AJ, Haeusermann P, 2005. Acute generalized exanthematous pustulosis induced by the antifungal terbinafine: Case report and review of the literature. *Br J Dermatol* 152: 780–783.

Bekersky I, Fielding RM, Dressler DE, Lee JW, Buell DN, Walsh TJ, 2002. Pharmacokinetics, excretion, and mass balance of liposomal amphotericin B (AmBisome) and amphotericin B deoxycholate in humans. *Antimicrob Agents Chemother* 46: 828–333.

Blin N, Morineau N, Gaillard F, Morin O, Milpied N, Harousseau J-L, Moreau P, 2004. Disseminated mucormycosis associated with invasive pulmonary aspergillosis in a patient treated for post-transplant high-grade non-Hodgkin's lymphoma. *Leuk Lymphoma* 45: 2161–2163.

Bodey GP, Kontoyiannis DP, Lewis RE, 2002. Empiric antifungal therapy for persistently febrile neutropenic patients. *Curr Treatment Opt Infect Dis* 4: 521–532.

Brandriss MW, Wolff SM, Moores R, Stohlman F Jr, 1964. Anemia induced by Amphotericin B. *J Am Med Assoc* 189: 663–666.

Cappelletty D, Eiselstein-McKitrick K, 2007. The echinocandins. *Pharmacotherapy* 27: 369–388.

Chayakulkeeree M, Ghannoum MA, Perfect JR, 2006. Zygomycosis: The re-emerging fungal infection. *Eur J Clin Microbiol Infect Dis* 25: 215–229.

Cornely OA, Maertens J, Winston DJ, Perfect J, Ullmann AJ, Walsh TJ, Helfgott D, Holowiecki J, Stockelberg D, Goh YT, Petrini M, Hardalo C, Suresh R, Angulo-Gonzalez D, 2007. Posaconazole vs. fluconazole or itraconazole prophylaxis in patients with neutropenia. *N Engl J Med* 356: 348–359.

Coste A, Selmecki A, Forche A, Diogo D, Bougnoux ME, d'Enfert C, Berman J, Sanglard D, 2007. Genotypic evolution of azole resistance mechanisms in sequential *Candida albicans* isolates. *Eukaryot Cell* 6: 1889–1904.

Debruyne D, Coquerel A, 2001. Pharmacokinetics of antifungal agents in onychomycoses. *Clin Pharmacokinet* 40: 441–472.

Debruyne D, Ryckelynck J-P, 1993. Clinical pharmacokinetics of fluconazole. *Clin Pharmacokinet* 24: 10–27.

Dismukes WE, 2000. Introduction to antifungal drugs. *Clin Infect Dis* 30: 653–657.

Dodds Ashley ES, Lewis R, Lewis JS, Martin C, Andes D, 2006. Pharmacology of systemic antifungal agents. *Clin Infect Dis* 43 (S1): S28–S39.

Donnelly JP, De Pauw BE, 2004. Voriconazole—a new therapeutic agent with an extended spectrum of antifungal activity. *Clin Microbiol Infect* 10 (Suppl 1): 107–117.

Dykewicz CA, Jaffe HW, Kaplan JE, et al, 2000. Guidelines for preventing opportunistic infections among hematopoietic stem cell transplant recipients: Recommendations of CDC, the Infectious Disease Society of America, and the American Society of Blood and Marrow Transplantation. *MMWR Morb Mortal Wkly Rep* October 20, 2000/49(RR10): 1–128.

Flückiger U, Marchetti O, Bille J, Eggimann P, Zimmerli S, Imhof A, Garbino J, Ruef C, Pittet D, Täuber M, Glauser M, Calandra T, 2006. Treatment options of invasive fungal infections in adults. *Swiss Med Wkly* 136: 447–463.

Goa KL, Barradell LB, 1995. Fluconazole. An update on its pharmacodynamic and pharmacokinetic properties and therapeutic use in major superficial and systemic mycoses in immunocompromised patients. *Drugs* 50: 658–690.

Gold W, Stout HA, Pacano JS, Donovick R, 1956. Amphotericins A and B: Antifungal Antibiotics Produced by a Streptomycete, In Vitro Studies, *in*: *Antibiotics Annual 1955–1956*. Medical Encyclopedia, Inc., New York.

Groll AH, Tragiannidis A, 2009. Recent advances in antifungal prevention and treatment. *Semin Hematol* 46: 212–229.

Herbrecht R, Denning DW, Patterson TF, Bennett JE, Greene RE, Oestmann JW, Kern WV, Marr KA, Ribaud P, Lortholary O, Sylvester R, Rubin RH, Wingard JR, Stark P, Durand C, Caillot D, Thiel E, Chandrasekar PH, Hodges MR, Schlamm HT, Troke PF, De Pauw B, 2002. Voriconazole versus amphotericin B for primary therapy of invasive aspergillosis.. *N Engl J Med* 347: 408–415.

Holler B, Omar SA, Farid MD, Patterson MJ, 2004. Effects of fluid and electrolyte management on amphotericin B-induced nephrotoxicity among extremely low birth weight infants. *Pediatrics* 113: e608–e616.

Ibrahim AS, Bowman JC, Avanessian V, Brown K, Spellberg B, Edwards JE Jr, Douglas CM, 2005. Caspofungin inhibits *Rhizopus oryzae* 1,3-beta-D-glucan synthase, lowers burden in brain measured by quantitative PCR, and improves survival at a low but not a high dose during murine disseminated zygomycosis. *Antimicrob Agents Chemother* 49: 721–727.

Kanafani ZA, Perfect JR, 2008. Antimicrobial resistance: Resistance to antifungal agents—mechanisms and clinical impact. *Clin Infect Dis* 46: 120–128.

Kim R, Khachikian D, Reboli AC, 2007. A comparative evaluation of properties and clinical efficacy of the echinocandins. *Expert Opin Pharmacother* 8: 1479–1492.

Kontoyiannis DP, Lionakis MS, Lewis RE, Chamilos G, Healy M, Perego C, Safdar A, Kantarjian H, Champlin R, Walsh TJ, Raad II, 2005. Zygomycosis in a tertiary-care cancer center in the era of *Aspergillus*-active antifungal therapy: A case–control observational study of 27 recent cases. *J Infect Dis* 191: 1350–1360.

Kontoyiannis DP, Hachem R, Lewis RE, Rivero GA, Torres HA, Thornby J, Champlin R, Kantarjian H, Bodey GP, Raad II, 2003. Efficacy and toxicity of caspofungin in combination with liposomal amphotericin B as primary or salvage treatment of invasive aspergillosis in patients with hematologic malignancies. *Cancer* 98: 292–299.

Lamaris GA, Lewis RE, Chamilos G, May GS, Safdar A, Walsh TJ, Raad II, Kontoyiannis DP, 2008. Caspofungin-mediated beta-glucan unmasking and enhancement of human polymorphonuclear neutrophil activity against *Aspergillus* and non-*Aspergillus* hyphae. *J Infect Dis* 198: 186–192.

Li JY, Yong TY, Grove DI, Coates PT, 2008. Successful control of *Scedosporium prolificans* septic arthritis and probable osteomyelitis without radical surgery in a long-term renal transplant recipient. *Transpl Infect Dis* 10: 63–65.

Lin AC, Goldwasser E, Bernard EM, Chapman SW, 1990. Amphotericin B blunts erythropoietin response to anemia. *J Infect Dis* 161: 348–351.

MacGregor RR, Bennett JE, Erslev AJ, 1978. Erythropoietin concentration in amphotericin B-induced anemia. *Antimicrob Agents Chemother* 14: 270–273.

Maertens J, Raad I, Petrikkos G, Boogaerts M, Selleslag D, Petersen FB, Sable CA, Kartsonis NA, Ngai A, Taylor A, Patterson TF, Denning DW, Walsh TJ, 2004. Efficacy and safety of caspofungin for treatment of invasive aspergillosis in patients refractory to or intolerant of conventional antifungal therapy. *Clin Infect Dis* 39: 1563–1571.

Marr KA, Boeckh M, Carter RA, Kim HW, Corey L, 2004. Combination antifungal therapy for invasive aspergillosis. *Clin Infect Dis* 39: 797–802.

Mora-Duarte J, Betts R, Rotstein C, Colombo AL, Thompson-Moya L, Smietana J, Lupinacci R, Sable C, Kartsonis N, Perfect J, 2002. Comparison of caspofungin and amphotericin B for invasive candidiasis. *N Engl J Med* 347: 2020–2029.

National Committee for Clinical Laboratory Standards, 2002. Reference methods for broth dilution antifungal susceptibility testing of filamentous fungi. Approved guideline M38-A. National Committee for Clinical Laboratory Standards, Wayne, PA.

Negroni R, Tobón A, Bustamante B, Shikanai-Yasuda MA, Patino H, Restrepo A, 2005. Posaconazole treatment of refractory eumycetoma and chromoblastomycosis. *Rev Inst Med Trop San Paulo* 47: 339–346.

Pappas PG, Kauffman CA, Andes D, Benjamin DK Jr, Calandra TF, Edwards JE Jr, Filler SG, Fisher JF, Kullberg BJ, Ostrosky-Zeichner L, Reboli AC, Rex JH, Walsh TJ, Sobel JD, 2009. Clinical practice guidelines for the management of candidiasis: 2009 update by the Infectious Diseases Society of America. *Clin Infect Dis* 48: 503–535.

Patterson TF, 2006. Fungal infection in the immunocompromised patient: Risk assessment and the role of antifungal agents. www.medscape.com. Release date: July 31, 2006.

Pfaller MA, Diekema DJ, 2004. Rare and emerging opportunistic fungal pathogens: Concern for resistance beyond *Candida albicans* and *Aspergillus fumigatus*. *J Clin Microbiol* 42: 4419–4431.

Pfaller MA, Marco F, Messer SA, Jones RN, 1998. In vitro activity of two echinocandin derivatives, LY303366 and MK-0991 (L-743,792), against clinical isolates of *Aspergillus*, *Fusarium*, *Rhizopus*, and other filamentous fungi. *Diagn Microbiol Infect Dis* 30: 251–255.

Post DJ, Douglas DD, Mulligan DC, 2005. Immunosuppression in liver transplantation. *Liver Transpl* 11: 1307–1314.

Prentice AG, Glasmacher A, 2005. Making sense of itraconazole pharmacokinetics. *J Antimicrobial Chemother* 56 (Suppl 1): i17–i22.

Reboli AC, Rotstein C, Pappas PG, Chapman SW, Kett DH, Kumar D, Betts R, Wible M, Goldstein BP, Schranz J, Krause DS, Walsh TJ, 2007. Anidulafungin versus fluconazole for invasive candidiasis. *N Engl J Med* 356: 2472–2482.

Reed C, Bryant R, Ibrahim AS, Edwards J Jr, Filler SG, Goldberg R, Spellberg B, 2008. Combination polyene–caspofungin treatment of rhino-orbital-cerebral mucormycosis. *Clin Infect Dis* 47: 364–371.

Revankar SG, 2005. Therapy of infections caused by dematiaceous fungi. *Expert Rev Anti Infect Ther* 3: 601–612.

Rustad TR, Stevens DA, Pfaller MA, White TC, 2002. Homozygosity at the *Candida albicans* MTL locus associated with azole resistance. *Microbiology* 148: 1061–1072.

Sabol K, Gumbo T, 2008. Anidulafungin in the treatment of invasive fungal infections. *Ther Clin Risk Manag* 4: 71–78.

Singh N, Limaye AP, Forrest G, Safdar N, Muñoz P, Pursell K, Houston S, Rosso F, Montoya JG, Patton P, Del Busto R, Aguado JM, Fisher RA, Klintmalm GB, Miller R, Wagener MM, Lewis RE, Kontoyiannis DP, Husain S, 2006. Combination of voriconazole and caspofungin as primary therapy for invasive aspergillosis in solid organ transplant recipients: A prospective, multicenter, observational study. *Transplantation* 81: 320–326.

Singh N, Yu VL, 2000. Prophylactic fluconazole in liver transplant recipients. *Ann Intern Med* 132: 843–844.

Spellberg B, Walsh TJ, Kontoyiannis DP, Edwards J Jr, Ibrahim AS, 2009. Recent advances in the management of mucormycosis: From bench to bedside. *Clin Infect Dis* 48: 1743–1751.

Strasfeld L, Weinstock DM, 2006. Antifungal prophylaxis among allogeneic hematopoietic stem cell transplant recipients: Current issues and new agents. *Expert Rev Anti Infect Ther* 4: 457–468.

Stevens DA, Shatsky SA, 2001. Intrathecal amphotericin in the management of coccidioidal meningitis. *Semin Respir Infect* 16: 263–269.

Theuretzbacher U, 2004. Pharmacokinetics/pharmacodynamics of echinocandins. *Eur J Clin Microbiol Infect Dis* 23: 805–812.

Ullmann AJ, Lipton JH, Vesole DH, Chandrasekar P, Langston A, Tarantolo SR, Greinix H, Morais de Azevedo W, Reddy V, Boparai N, Pedicone L, Patino H, Durrant S, 2007. Posaconazole or fluconazole for prophylaxis in severe graft-versus-host disease. *N Engl J Med* 356: 335–347.

Van Burik JA, 2005. Role of new antifungal agents in prophylaxis of mycoses in high risk patients. *Curr Opin Infect Dis* 18: 479–483.

Vazquez JA, 2008. Combination antifungal therapy for mold infections: Much ado about nothing? *Clin Infect Dis* 46: 1889–1901.

Vermes A, Guchelaar HJ, Dankert J, 2000. Flucytosine: A review of its pharmacology, clinical indications, pharmacokinetics, toxicity and drug interactions. *J Antimicrob Chemother* 46: 171–179.

Walsh TJ, Raad I, Patterson TF, Chandrasekar P, Donowitz GR, Graybill R, Greene RE, Hachem R, Hadley S, Herbrecht R, Langston A, Louie A, Ribaud P, Segal BH, Stevens DA, van Burik JA, White CS, Corcoran G, Gogate J, Krishna G, Pedicone L, Hardalo C, Perfect JR, 2007. Treatment of invasive aspergillosis with posaconazole in patients who are refractory to or intolerant of conventional therapy: An externally controlled trial. *Clin Infect Dis* 44: 2–12.

Walsh TJ, Teppler H, Donowitz GR, Maertens JA, Baden LR, Dmoszynska A, Cornely OA, Bourque MR, Lupinacci RJ, Sable CA, dePauw BE, 2004. Caspofungin versus liposomal amphotericin B for empirical antifungal therapy in patients with persistent fever and neutropenia. *N Engl J Med* 351: 1391–1402.

White PL, Shetty A, Barnes RA, 2003. Detection of seven *Candida* species using the Light-Cycler system. *J Med Microbiol* 52: 229–238.

Wildfeuer A, Laufen H, Schmalreck AF, Yeates RA, Zimmermann T, 1997. Fluconazole: Comparison of pharmacokinetics, therapy and in vitro susceptibility. *Mycoses* 40: 259–265.

Winston DJ, Pakrasi A, Busuttil RW, 1999. Prophylactic fluconazole in liver transplant recipients. A randomized, double-blind, placebo-controlled trial. *Ann Intern Med* 131: 729–737.

Yeo EJ, Ryu JH, Cho YS, Chun YS, Huang LE, Kim MS, Park JW, 2006. Amphotericin B blunts erythropoietin response to hypoxia by reinforcing FIH-mediated repression of HIF-1. *Blood* 107: 916–923.

Zomorodian K, Uthman U, Tarazooie B, Rezaie S, 2007. The effect of griseofulvin on the gene regulation of beta-tubulin in the dermatophyte pathogen *Trichophyton rubrum*. *J Infect Chemother* 13: 373–379.

WEBSITES CITED

Drugs.com (professional category) has useful information on clinical pharmacology at www.drugs.com.

"Introduction to Antifungal Pharmacology", by Dr. Russell E. Lewis, University of Houston College of Pharmacy, TX. The PowerPoint can be viewed at the URL www.aspergillus.org.uk/education/introductionAntifung_rev.ppt.

Johns-Hopkins Poc-It ABX guide includes indications, cost, dosing regimens, adverse reactions: http://Hopkins-abxguide.org.

Medline-Plus contains adverse effects and drugs which interact with antifungal agents at the URL www.medlineplus.gov.

Structure diagrams of antifungal agents may be found at:

U.S.A. National Center for Biotechnology Information (NCBI) website, PubChem-substance, at the URL http://pubchem.ncbi.nlm.nih.gov/.

National Institute of Allergy and Infectious Diseases (NIAID) at the URL http://chemdb2.niaid.nih.gov/struct_search/.

QUESTIONS

The Answer Key to multiple choice questions may be found at the end of the book.

1. The azole antifungal drug that has had the most success in treating *Candida* bloodstream infections is
   A. Clotrimazole
   B. Fluconazole

C. Itraconazole
D. Miconazole

2. The mode of action of azole antifungal drugs is that they
A. Combine with ergosterol in the cell membrane resulting in cell lysis
B. Inhibit lanosterol demethylase in the ergosterol synthesis pathway
C. Interfere with cell wall synthesis
D. Substitute for uracil and inhibit RNA synthesis

3. Which is/are true about amphotericin B?
A. Acute adverse effects may include include chills, fever, rapid breathing, headache, nausea, loss of appetite.
B. Chronic adverse effects include kidney toxicity.
C. It is a fungicidal drug that interferes with ergosterol in the cell membrane.
D. It is formulated with deoxycholate to improve solubility.
E. All of the above.

4. Caspofungin, micafungin, and anidulafungin
A. Are good for oral antifungal therapy for systemic yeast infections
B. Can be used in coccidioidomycosis
C. Are suitable to treat cryptococcosis
D. Target the synthesis of cell wall glucan

5. Flucytosine combination therapy is the preferred primary therapy with
A. Amphotericin B to treat cryptococcosis
B. Fluconazole to treat systemic candidiasis
C. Itraconazole to treat aspergillosis
D. Micafungin to treat mucormycosis

6. For which disease is oral terbinafine therapy recommended?
A. Black piedra
B. Oral thrush
C. Pityriasis versicolor
D. Toenail dermatophyte infections (tinea unguium)

7. List two clinically important *Candida* species that are resistant to fluconazole and discuss the use of other systemic antifungal agents suitable to treat bloodstream infections with those two *Candida* species.

8. Even though *Pneumocystis jirovecii* is accepted as a fungus classed in the phylum *Ascomycota*, it is not susceptible to amphotericin B or to the azole class of antifungal agents. Discuss why this is the case and how other classes of antifungal and antibacterial agents may be used to treat PCP.

9. Discuss the issue of uncertain bioavailability associated with the use of oral itraconazole. How can the oral route be used to achieve better bioavailability? How else can it be administered to avoid this issue?

10. Select one drug from each of the following classes of antifungal agents: azoles, polyenes, echinocandins. For that drug describe how it is metabolized, its route of elimination, the elimination half-life, $t_{1/2}$; its peak plasma concentration, Cmax; minimum plasma concentration, Cmin; the change over time in the amount of drug available in the plasma—area under the Curve (AUC); its volume of distribution (Vd) = the amount of drug in the body divided by the concentration in the blood.

Chapter 3B

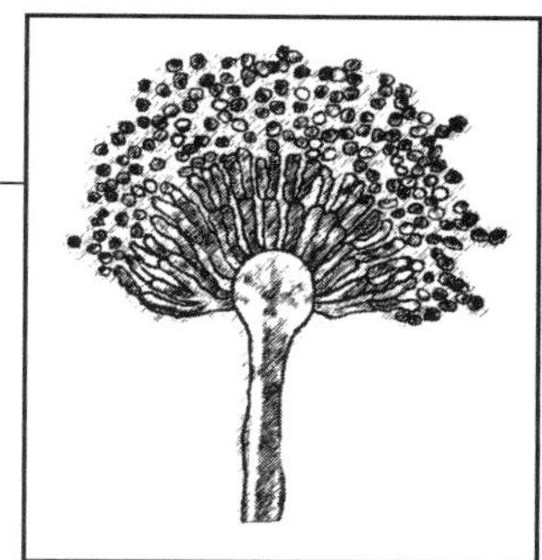

Antifungal Susceptibility Tests

3B.1 ANTIFUNGAL SUSCEPTIBILITY TESTS DEFINED

An antifungal susceptibility (AFS) test is an in vitro laboratory test designed to determine the minimum inhibitory concentration (MIC) of an antifungal drug against a clinical isolate of a pathogenic fungus. The susceptibility profile of an individual clinical isolate, yeast or mold, consists of the array of MICs from a panel of from three to nine antifungal agents.

3B.2 NATIONAL AND INTERNATIONAL STANDARDS FOR AFS TESTS

Where can standard operating procedures for the conduct of specific AFS tests be found? This chapter introduces the terminology and compares the attributes of AFS methods in use and in development but does not provide stepwise standard operating procedures for the conduct of such tests. Those methods of procedure may be found in guidelines of national and international committees. The Subcommittee for Antifungal Testing of the Clinical Laboratory Standards Institute (CLSI, www.clsi.org) is a working group in the United States which develops standard operating procedures for AFS tests. At present, these consist of microtitration plate-based and disk diffusion tests for yeasts and molds. The European Committee on Antimicrobial Susceptibility Testing (EUCAST) performs similar functions abroad.

The procedure methods for commercial tests discussed here may be found in the package inserts for such tests. A detailed review with specific parameters for performance of the CLSI methods for yeasts and molds, a graphic for additive dilutions of antifungal agents, minimum inhibitory concentration ranges for various drugs versus standard tester strains of yeasts and molds, and optical density versus colony forming units (CFU) for mold conidia may be found in Cantón et al. (2009).

3B.3 OBJECTIVE OF AFS TESTS

The objective of AFS testing is to predict the clinical efficacy of drug therapy for a patient's mycotic disease.

3B.4 MINIMUM INHIBITORY CONCENTRATION (MIC) OF AN ANTIFUNGAL DRUG DEFINED

The MIC is the minimum concentration of an antifungal agent (μg/mL) required to prevent the in vitro growth of a yeast or mold. The MIC is determined using an in vitro AFS test.

For yeasts and amphotericin B (AmB), the MIC is the lowest drug concentration (in μg/mL) that results in the absence of growth after 48 h incubation; for yeasts and azole antifungal agents it is the lowest drug concentration resulting in a $\geq$50% reduction in growth, when tested by the broth microdilution method. When azoles are tested by the macrobroth dilution method, using individual tubes instead of wells in a microtitration plate, it is the lowest drug concentration resulting in an 80% reduction in growth.

The difference in computing the MIC for AmB and azole antifungal agents is because AmB is a fungicidal drug, whereas azole drugs are considered to be fungistatic.

Fundamental Medical Mycology, First Edition. By Errol Reiss, H. Jean Shadomy and G. Marshall Lyon, III.
© 2012 Wiley-Blackwell. Published 2012 by John Wiley & Sons, Inc.

3B.4.1 MIC_{50} and MIC_{90}

These terms refer to the *population* of clinical isolates of a fungal species such that MIC_{50} is the lowest concentration of an antifungal agent capable of inhibiting the growth of 50% of the population of clinical isolates of that species. MIC_{90} is the lowest drug concentration capable of inhibiting growth of 90% of that population.

3B.5 BROTH MICRODILUTION (BMD) METHOD

Antifungal susceptibility tests developed by the CLSI are capable of testing a range of drug concentrations. These are broth dilution methods which come in two variations: (i) the broth microdilution method (BMD) conducted in a 96 well microtitration plate and, (ii) broth macrodilution tests conducted in test tubes. The latter method has fallen out of favor because it is slow and awkward compared with the BMD method.

3B.6 CLINICAL INDICATIONS FOR AFS TESTING

AFS testing is important when:

- The isolated yeast or mold is from a sterile site, such as the bloodstream.
  - For yeasts, AFS should be determined after the isolate has been identified to species. In the case of molds, slow growth may delay identification and such a delay in treatment may result in poor outcome—AFS may be determined before the identification is made with certainty.[1]
- The patient has failed to respond to primary therapy or has become intolerant to the drug of choice for the mycosis in question.
- The fungus involved is known to have a wide range of susceptibilities to the antifungal agent(s) in question.
- The fungal species is known to be refractory to a range of antifungal agents.
- A new class of antifungal agents is being considered for use.
- A new member of an existing class of antifungal agents is being considered for use.
- An outbreak of a fungal disease requires treatment of a group of patients.

3B.7 CORRELATION BETWEEN THE IN VITRO DETERMINED MIC AND THE CLINICAL EFFICACY OF DRUG THERAPY

The clinical efficacy of drug therapy depends on several host factors. Those factors that influence clinical outcome and correlation with AFS test results include:

- Contaminated indwelling medical devices (e.g., catheters). This is because yeast embedded in biofilms coating indwelling medical devices are less susceptible to killing than planktonic[2] yeast.
- Dosage regimen.
- Drug interactions between the antifungal drug and the patient's other medications may alter the availability of antifungal therapy (please see Chapter 3A).
- Immune and endocrine functions. Immunosuppressive and/or cytotoxic therapy for an underlying disease
  - Persistent immunosuppression
  - Persistent neutropenia
- Other intercurrent medical problems/poor clinical status of the patient (e.g., high APACHE II score[3]).
- Patient's age—very young (neonates) and elderly are at higher risk.
- Poor drug absorption—measurement of the itraconazole blood concentration is important to test for proper absorption.
- Poor patient compliance
- Timeliness of therapeutic intervention; for example, for a high fungal burden—prompt diagnosis is key.

Microbial factors also influence the clinical course because of differences in the pathogenic potential within a fungal species, although that aspect is difficult to gauge with present technology.

There is a disparity between in vitro AFS tests and in vivo results of antifungal therapy. The "90/60 rule" was proposed as a yardstick to predict clinical outcome

[1]***Safety Note:*** Because of the possibility that aerosols may be generated in setting up and reading AFS tests, it is of prime importance for technologists to work in a well-functioning laminar flow BSC, wearing PPE, trained and supervised by a certified clinical laboratory scientist and according to the CLIA-approved standard operating procedures manual for the laboratory. When in doubt, consult the manual: *CDC\NIH Biosafety in Microbiological and Biomedical Laboratories* (BMBL), 5th ed. Available at the URL http://www.cdc.gov/OD/ohs/biosfty/bmbl5/bmbl5toc.htm.

[2]Planktonic. *Definition*: Free-floating microbes dispersed in the fluid phase.

[3]APACHE II score. *Definition*: "Acute Physiology and Chronic Health Evaluation II" is a severity of disease classification. The score from 0 to 71 is based on several measurements for patients admitted to the intensive care unit; higher scores imply severe disease and risk of death.

(Rex and Pfaller, 2002). According to this theory, (i) infections caused by isolates with MICs considered *susceptible* respond favorably to therapy approximately 90% of the time. That is because 10% of patients may be too sick for AFS test results to impact outcome. (ii) Infections caused by isolates with MICs considered *resistant* should respond favorably approximately 60% of the time. Another 30% could benefit from better drug selection based on determining the AFS profile of the isolate in question.

This rule has support for susceptible *Candida* species isolates, but for those which have a MIC ≥64 μg/mL, *only 42% of the patient-episode-isolate events* were successfully treated with fluconazole (FLC). This was true for both mucosal or invasive candidiasis analyzed using clinical correlations of MICs from 11 different studies (Pfaller et al., 2006). This evidence strongly supports the usefulness of the "R" (for resistant) category.

Clinical correlations of AFS tests for molds have also been problematic but, even with immunocompromised patients, there are examples of successful correlations. Patients (n = 19) with hematologic disease treated by myeloablative chemotherapy and with stem cell transplants later developed invasive aspergillosis (Lass-Flörl et al., 1998). The species distribution among patients was *Aspergillus flavus* in 12 patients (41%), *A. terreus* in 9 (31%), and *A. fumigatus* in 8 (28%). Macrobroth dilution AFS tests with AmB were performed on these clinical isolates. *Aspergillus terreus* isolates had consistently high MICs, whereas variation in MICs occurred among *A. fumigatus* and *A. flavus* isolates. In vitro resistance was the only parameter correlating with clinical outcome in univariate analysis and the only prognostic value in a multivariate analysis, considering known risk factors. All six patients whose isolates had MICs <2 μg/mL survived, whereas most (22/23) with MICs ≥2 μg/mL died. Infections among the 6 survivors were caused by AmB-susceptible *A. fumigatus* and *A. flavus*.

3B.7.1 How Are the Conditions of "Susceptible" or "Resistant" Determined?

They are determined as interpretive "breakpoints" or drug concentrations considered by committees such as the CLSI and EUCAST to correspond to performance of the drug in a population of patients. For example, breakpoints for *Candida* species were determined after consideration of >10,000 AFS tests of *Candida* species and 1295 patient-episode-isolate events (Pfaller et al., 2006).

3B.7.2 What Are Breakpoints?

Breakpoints are discriminatory antimicrobial concentrations (μg/mL) used to interpret AFS test results in order to define isolates as "susceptible," "intermediate" (susceptible-dose-dependent, S-DD), or "resistant." The purpose of the S-DD category is to underline the importance of reaching maximal antifungal drug blood and tissue concentrations for isolates with higher MICs: for example, the maximal FLC dose is ≥400 mg/day in a 70 kg adult with normal renal function.

Table 3B.1 Interpretive MIC Breakpoints for Yeast Susceptibility Tests (μg/mL)[a]

Antifungal agent	Susceptible, S	Susceptible-dose-dependent, S-DD	Resistant, R
Fluconazole[b]	≤2	4	≥8
Itraconazole	≤0.125	0.25–0.5	≥1
Flucytosine	≤4	8–16	≥32
Amphotericin B	≤1		≥2
Voriconazole	≤1	2	≥4
Echinocandins	≤0.25[c]	0.5	≥1

[a] As recommended by the CLSI (Pfaller et al., 2010).

[b] for *C. albicans*, *C. parapsilosis*, *C. tropicalis*; for *C. glabrata*: SDD, ≤32 μg/mL; R ≥64 μg/mL.

[c] for *C. albicans*, *C. tropicalis*, and *C. krusei* (in μg/mL): ≤2 (S), 4 (S-DD), and ≥8 (R) *C. parapsilosis*. Breakpoints for caspofungin and *C. glabrata* are ≤0.12 (S), 0.25 (I), and ≥0.5 l (R) (Pfaller et al., 2011).

The breakpoints of AFS test for yeasts are established for a number of antifungal agents (Table 3B.1). A good clinical response to infection is expected with a susceptible yeast isolate at a typical dosage regimen for the antifungal agent. If the MIC is susceptible dose-dependent a higher dose is likely to result in a good response, but with a MIC in the resistant range a good response is unlikely at any dose. Clinical success may be expected 58% of the time when invasive candidiasis is treated with FLC caused by a *Candida* species isolate with a MIC ≥64 μg/mL, compared with 71% when the MIC is in the susceptible range.

3B.7.3 Minimum Effective Concentration

Echinocandins (e.g., caspofungin) are active against *Aspergillus* species both in vitro and in vivo but estimation of the in vitro activity requires an inverted microscope to determine the drug dilution at which there is an effect on the growing tips of hyphae—producing small rounded compact colonies. Seen at higher magnification, the hyphae are highly branched, short, and stubby. The comparison is made against growth in a control well (no drug). That concentration is referred to as the minimum *effective* concentration (MEC). The MEC is stable across time intervals, varying less than

MIC values at periods of measurement beyond 24 h (at 48 h) (Arikan et al., 2001). Readings are made at 24 h for *Aspergillus* spp. and *Paecilomyces variotii* (used as a control species), 48 h for *Fusarium* spp., and 72 h for *Scedosporium* spp. (Cantón et al., 2009). The MEC is clearly a more labor-intensive measurement, requiring more observer experience.

3B.8 AFS METHODS CURRENTLY AVAILABLE FOR USE IN THE CLINICAL LABORATORY

Three distinctive methods in clinical use are listed below and are discussed in more detail later in this chapter.

3B.8.1 Broth Microdilution (BMD) Method

Microtitration plates are precoated with a panel of drugs in dilutions covering a broad range, for example, YeastOne® Clinical Sensititre® (Trek Diagnostic Systems Division of Magellan Biosciences, Inc., Cleveland, OH).

3B.8.2 Etest®

The Etest (bioMérieux, Marcy l'Etoile, France) consists of plastic strips each embedded with a drug gradient. The bioMérieux Company, in June 2008, acquired AB BIODISK, Sweden, who originated the Etest.

3B.8.3 Disk Diffusion Method

This format is similar to the Kirby–Bauer method in use for many years in antibacterial drug susceptibility tests.

3B.9 WHICH LABORATORIES CONDUCT AFS TESTS?

A survey was conducted of microbiology departments of 171 teaching hospitals in the United States to understand the scope of AFS testing (Pai and Pendland, 2003). An estimated 137,088 *Candida* isolates or 8.5/1000 inpatient days were reported for 2000. Approximately 1% (1300) of the isolates, most from blood specimens, underwent AFS tests. One-hundred-fifteen hospitals or 67.2% of the total conducted AFS tests. On-site testing occurred at 27 hospitals and off-site for 88 hospitals. As of 2006 more than 100 laboratories in the United States were enrolled in the College of American Pathologists (CAP) antifungal proficiency testing program. Commercial AFS tests, either YeastOne® or Etest®, are most popular with these laboratories (Pfaller et al., 2006).

Off-site or reference laboratory AFS testing in the United States is offered by the following:

Focus Diagnostics, Inc., Cypress, CA

Fungus Testing Laboratory, The University of Texas Health Science Center, San Antonio, TX

LabCorp, Inc., Burlington, NC

Mayo Medical Laboratories, Rochester, MN

Quest Diagnostics Nichols Institute, Chantilly, VA

3B.10 PRINCIPLES OF AFS TESTS

The attributes of an ideal AFS test are:

- Accuracy and precision
- Ease in interpretation of endpoints to differentiate susceptible and resistant isolates
- Simplicity of operation
- Speed of performance
- Standard conditions that facilitate interlaboratory comparisons
  - Internal standard tester strains for quality control
  - Standard operating procedure

The CLSI subcommittee for antifungal testing addresses the development and periodic revision of guidelines and standard operating procedures. The current standards are:

- M27-A3: Broth Dilution for Yeasts and M27S3 (Third Informational Supplement)
- M38-A2: Broth dilution for Molds
- M44-A: Disk Diffusion for Yeasts and M44-S2 (Informational Supplement)
- M51-P: Disk Diffusion for Molds

3B.10.1 Standard Method for AFS Testing of Yeasts

This broth microdilution method (CLSI document M27-A3) is approved for testing *Candida* species and *Cryptococcus neoformans*. The test parameters include:

- Medium: RPMI 1640 medium, adjusted to pH 7.0.
- Inoculum: Standardized in a spectrophotometer.
- Incubation in microtitration plates: 35°C for 24–48 h (*Candida* species). The 24 h time point is a means of excluding errors resulting from trailing growth in the presence of azoles, discussed below. *Cryptococcus* spp. are incubated at 35° C for 72 h.
- Endpoint determination: Visual detection of 50% growth inhibition for azoles and flucytosine, 100% inhibition for AmB.

- Quality control strains of defined MIC, so that susceptibility and resistance determinations will have internal standards.

What is a MFC? This is the minimum concentration of a substance required to kill the target fungal isolate, determined by subculture.

3B.10.1.1 Limitations of the M27-A Susceptibility Test Method for Yeasts

Clinical Correlations Clinical correlation does not always match the in vitro susceptibility of the isolate because clinical success is multifactorial (Please see Section 3B.7, Correlation Between the In Vitro Determined MIC and the Clinical Efficacy of Drug Therapy, earlier in this chapter). Despite the observed clinical resistance, the MIC_{50} and MIC_{90} for most yeast species fall in the "susceptible" range. Exceptions are *Candida glabrata* with respect to azoles, *C. krusei* against FLC, and *Cryptococcus neoformans* against CASF. All *C. krusei* strains are innately resistant to FLC; the mechanism of innate resistance of *C. neoformans* to CASF has not been discovered.

Trailing Growth Trailing growth of *Candida* species isolates exposed to azole agents complicates endpoint determination. "Heavy trailers" are *C. albicans* clinical isolates and *C. tropicalis* isolates with a low FLC MIC after 24 h and a high MIC (>64 μg/mL) after 48 h (Agrawal et al., 2007) (Fig. 3B.1).

Trailers are often misinterpreted as resistant in vitro. Experimental infections in mice have shown that heavy trailers are susceptible in vivo to FLC (Rex et al., 1998). A truly resistant isolate MIC will be high after both 24 and 48 h incubation. When the M27-A guidelines are followed (48 h incubation and 50% inhibition) these isolates are erroneously interpreted as resistant.

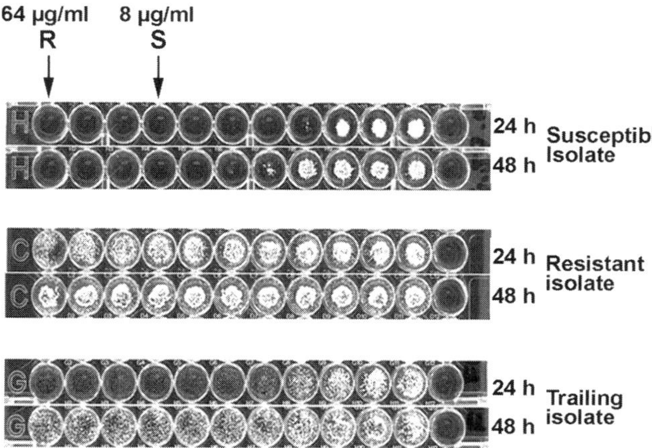

Figure 3B.1 Trailing growth of yeasts. *Candida albicans* versus fluconazole. *Note*: Current breakpoints are shown in Table 3b.1 *Source:* Used with permission from Dr. Beth A. Arthington-Skaggs, CDC.

Approximately 18% of *C. albicans* isolates and 59% of *C. tropicalis* isolates demonstrate the trailing growth phenotype in AFS tests with respect to FLC (Arthington-Skaggs et al., 2002). Trailing growth also may appear in the presence of ITC.

How can the laboratory correct a MIC for fluconazole in a clinical isolate displaying the heavy trailer phenotype? The MIC after 24 h incubation is taken as the true value.

What is the mechanism of the heavy trailer phenotype? Mechanisms proposed to explain trailing growth include the following.

Activation of Calcineurin Calcineurin, a protein phosphatase, is a key member of the calcium-dependent calcineurin signaling pathway, important for survival of yeast under stress conditions (Sanglard et al., 2003). The state of calcineurin activation can increase survival in drug concentrations above the MIC, leading to residual growth of *C. albicans* exposed to high drug concentrations.

Upregulation of *ERG11* This gene encodes lanosterol demethylase, the target of azole antifungal agents, resulting in inhibition of ergosterol synthesis. Quantitation of *ERG11* expression reveals that both trailing, susceptible-dose-dependent, and resistant *C. albicans* isolates are capable of *ERG11* upregulation in response to FLC (Lee et al., 2004).

Incubation Temperature Inhibits Trailing Growth *Candida albicans* isolates exhibiting trailing at 35°C lose the phenotype (appear susceptible) if incubated instead at 25°C and 42°C (Agrawal et al., 2007). No differences were observed between results at 25°C and 42°C.

Genotype Clustering and Resistant or Trailing Growth Phenotype Among 406 *C. tropicalis* clinical isolates collected between 1999 and 2002 in Taiwan hospitals, 23 isolates with FLC MICs ≥64 μg/mL were isolated in a single year, 1999 (Chou et al., 2007). Only 4 had MICs ≥64 μg/mL at 24 h so that 19 of these isolates fit the trailing growth phenotype. Multilocus sequence typing showed that one genotype (DST 140) contained half of the resistant and trailing isolates. This genotype is suspected of causing an outbreak in 1999 at one hospital. The association of a particular genotype with both resistant and trailing growth isolates suggests that trailing growth has some relation to decreased susceptibility, even if that has not been shown to occur in mouse infection studies.

Short of experimental infection of mice, how can the C. albicans displaying the heavy trailer phenotype be tested for reduction in ergosterol synthesis?

Sterol Quantitation Method (SQM) This research tool measures ergosterol content by growing an isolate in Sabouraud broth in the presence of 0, 4, 16, and 64 μg/mL of FLC (Arthington-Skaggs et al., 1999). The sterol is extracted from yeast cells with heat, KOH, and heptane. Sterols are quantitated in a spectrophotometer and the amount in drug-treated yeast cells is compared to the drug-free control. The MIC of FLC, using the SQM, is the lowest concentration of drug resulting in ≥80% inhibition of ergosterol biosynthesis, compared to drug-free control. There is excellent agreement between determination of MIC by the SQM and BMD methods. The heavy trailer phenotype, measured by the SQM method, will contain reduced ergosterol, corresponding to a susceptible MIC.

Paradoxical Growth The ß-(1→3)-D-glucan synthase inhibitors of the echinocandin class of antifungal agents are a choice for primary therapy of invasive candidiasis: MCF, CASF, and ANF. Paradoxical growth is an in vitro AFS test result defined as a resurgence of growth (>50% of that in the drug-free growth control) at drug concentrations above the MIC. Paradoxical in vitro growth is more frequent with CASF than MCF or ANF and is encountered in *C. parapsilosis, C. tropicalis, C. krusei*, but not in *C. glabrata* isolates (Chamilos et al., 2007).

CASF was tested against *Candida* species in AFS tests using the BMD format with both planktonic and biofilm forms of yeast growth. MIC endpoints were determined as the lowest CASF drug concentration resulting in a prominent decrease in growth (planktonic) and 50% reduction in metabolic activity (biofilm) as measured in a tetrazolium dye reduction assay (Melo et al., 2007). Paradoxical growth is more evident when yeast grow as biofilms. Both planktonic and biofilm cells become enlarged and globose, indicative of faulty cell wall development. The proposed use of high concentrations of echinocandins as catheter lock therapy to prevent and treat catheter-associated candidemia may be hindered by the stimulation of *Candida* biofilm growth at CASF concentrations above the MIC.

There is no evidence yet to support the view that paradoxical growth observed in vitro has direct bearing on the clinical efficacy of echinocandins. Invasive *Candida* species isolates ($n = 5346$) collected worldwide between 2001 and 2006 were tested by the M27 BMD method (Pfaller et al., 2008). All three echinocandins are very active against *Candida* species: ANF (MIC_{50}, 0.06 μg/mL; MIC_{90}, 2 μg/mL), CASF (MIC_{50}, 0.03 μg/mL; MIC_{90}, 0.25 μg/mL), and MCF (MIC_{50}, 0.015 μg/mL; MIC_{90}, 1 μg/mL). More than 99% of isolates were inhibited by ≤2 μg/mL of all three agents. There was no significant change in the activities of the three echinocandins over the 6 year study period and no difference in activity by geographic region. The CLSI Antifungal Susceptibility Subcommittee recommends a "susceptible only" breakpoint MIC of ≤2 μg/mL for all three echinocandins because there is no detectable resistance in the population of *Candida* isolates up to 2008. Isolates for which MICs exceed 2 μg/mL should be designated "nonsusceptible" (NS).

Candida parapsilosis is innately less susceptible to echinocandins. *Candida parapsilosis* shows elevated MICs for CASF, MCF, and ANF. This decreased in vitro susceptibility to echinocandins is the result of a naturally occurring Pro-to-Ala substitution at P660A, adjacent to the conserved hot spot 1 region of the ß-(1→3)-D-glucan synthase, Fks1 (Garcia-Effron et al., 2008). This enzyme from *C. parapsilosis* is 10–100× less sensitive to echinocandin drugs than the *C. albicans* enzyme.

The Difficulty in the Detection of AmB Resistant Isolates There is a narrow range of MICs when measured in the M27-A BMD method because the standard test medium, RPMI 1640, is not optimal for use with this drug. *Candida* bloodstream isolates from 66 patients who received AmB showed no correlation of MIC with outcome (Rex et al., 1995). An explanation for this finding is that BMD testing with RPMI 1640 medium generates a narrow range of only four different MICs, from 0.125 to 1 μg/mL. The narrow range of values suggests that AmB susceptibility testing with this medium has limited clinical usefulness (Park et al., 2006). Another medium, Antibiotic Medium 3, produces a broader distribution of MICs, from 0.015 to 0.25 μg/mL.

Bloodstream *Candida* species isolates from 107 patients showed no correlation between in vitro susceptibility to AmB with outcome, despite the fact that both RPMI 1640 and Antibiotic Medium 3 were both used. Patients with AmB resistant isolates did not experience clinical treatment failure more frequently than those with susceptible isolates, even after multivariable analysis considering age, severity of illness, catheter removal, the presence of a subsequent bacteremia, and the type of *Candida* species.

3B.10.2 Modifications Suggested to Improve Performance of the BMD Method for Yeast

Supplementing RPMI 1640 medium with 18 g of glucose/L changes the glucose concentration from 0.2% to 2.0% with resulting increased growth without changing the MICs of AmB, flucytosine, ketoconazole, and FLC.

Agitation of microdilution plates is necessary for spectrophotometric readings and simplifies visual reading of endpoints, especially for *Cryptococcus neoformans*. That yeast aggregates into a button in the wells (Rex et al., 2001). Shortening the incubation time to 24 h and reducing the pH below 5 clarifies trailing endpoints for *Candida* species.

Other modifications have not met with wide acceptance but have certain advantages—reducing the molarity of the morpholinepropanesulfonic acid (MOPs) buffer from 0.165 to 0.0165 M lowers the likelihood of overcalling resistance in *C. albicans*.

3B.10.3 Commercial BMD Method with Precoated Drug Panels: A CLSI-Approved Method for AFS Testing of Yeasts and Molds

3B.10.3.1 YeastOne®

Clinical Sensititre® (Trek Diagnostic Systems Division of Magellan Biosciences, Inc., Cleveland OH).

This commercial test consists of user-defined custom drug-coated plates, manufactured to CLSI guidelines, with or without alamarBlue®,[4] for full range MIC and breakpoint test procedures. The prevalent format is a colorimetric BMD plate test, which includes up to nine antifungal drugs covering a range of concentrations precoated at the factory. Plates can be stored at room temperature.

The colorimetric agent is alamarBlue, and when fungi grow in a particular drug well, the color changes from blue to red. The endpoint is read after 24 h of incubation with a standard inoculum. This visual read-out improves endpoint determination compared with turbidity (Fig. 3B.2).

Antifungal Agents Custom coated panels include between three and nine different antifungal agents. The prevalent three-drug panel contains FLC, ITC, and flucytosine provided at concentrations of 0.125–256 μg/mL, 0.008–16 μg/mL, and 0.03–64 μg/mL, respectively. Recent additions to the precoated drug panels include ANF and MCF, in the echinocandin family of glucan synthase inhibitors. A total of up to nine antifungal agents may be included in one YeastOne BMD plate (Table 3B.2).

There is good agreement between the YeastOne and M27-A BMD method of antifungal susceptibility tests after 24 h incubation (Table 3B.3) (Espinel-Ingroff et al., 1999). In this study, agreement between the YeastOne and M27A BMD methods of antifungal susceptibility testing for 465 clinical isolates of *Candida albicans* versus FLC depended on the time point for recording the results. When both tests were read after 24 h incubation the agreement between the two methods was 87%. When both tests were read after 48 h their agreement was only 60%, and when the M27A BMD test was read at 48 h versus 24 h for the YeastOne test, agreement was 71%. For concordance of results and to minimize the effects of "heavy trailing growth," results with both tests should be recorded after a 24 h incubation period.

[4]alamarBlue®. *Definition*: This cell viability indicator uses the reducing power of live cells to convert resazurin to the fluorescent molecule, resorufin. The active ingredient of alamarBlue (resazurin) is a nontoxic, cell permeable compound that is blue in color and virtually nonfluorescent. Upon entering cells, resazurin is reduced to resorufin, which produces very bright red fluorescence.

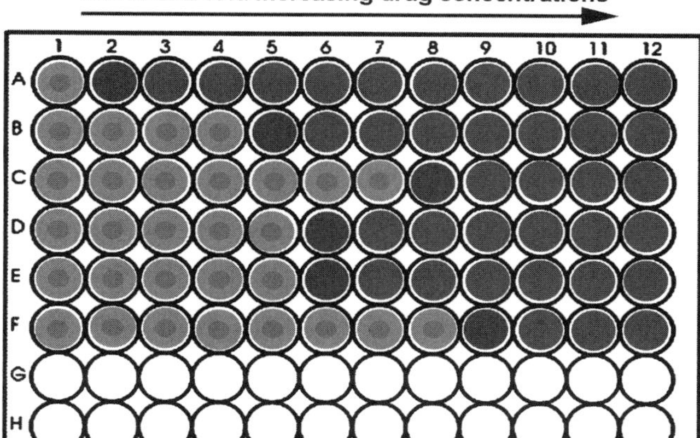

Figure 3B.2 Schematic of YeastOne® Sensititre® broth microdilution plate with alamarBlue® detection reagent. Wells are precoated with a series of 12 twofold increasing fluconazole concentrations. In this diagram each row contains a standard inoculum of a different *Candida albicans* isolate. Incubation is for 24 h at 35°C. The red color indicates growth, and blue, no growth. The endpoint, or MIC, is read as the first well to indicate a color change, expressed as the μg/mL of drug in that well. A susceptible control culture is in row A, *C. parapsilosis*, and a resistant control is in row F, *C. krusei*. See insert for color representation of this figure.

Table 3B.2 YeastOne Sensititre BMD Nine-Drug Panel (TREK Diagnostic Systems, Division of Magellan Biosciences Inc.)

Precoated drug	Dose range (μg/mL)
Amphotericin B	0.12–8
5-Flucytosine	0.06–64
Anidulafungin	0.015–8
Caspofungin	0.008–8
Micafungin	0.008–8
Fluconazole	0.12–256
Itraconazole	0.015–16
Posaconazole	0.008–8
Voriconazole	0.008–8

Table 3B.3 Agreement (%) Between YeastOne and M27A BMD Method of Antifungal Susceptibility Testing for 465 Clinical Isolates of *Candida albicans*[a]

M27 A BMD	24 h	48 h	48 h
YeastOne	24 h	48 h	24 h
Fluconazole	87	60	71
Flucytosine	87	56	89
Itraconazole	89	54	76

[a]Data from Table 1, Espinel-Ingroff et al. (1999). Adapted with permission from the American Society for Microbiology.

Trailing growth issues present in YeastOne test (similar to the M27-A method) are minimized or eliminated by taking 24 h readings. YeastOne colorimetric MICs were consistently more than two dilutions higher after 48 h of incubation than reference BMD MICs, especially for isolates of *C. albicans* recovered from patients with oropharyngeal infections. Patients with oropharyngeal infections may have taken FLC for a prolonged period, increasing drug pressure that could contribute to decreased susceptibility, but for most *C. albicans* 24 h readings of the YeastOne test are more clinically significant.

3B.10.4 Standardization of AFS Tests for Molds, M38-A2: Broth Microdilution for Molds

The M38-A2 protocol developed by the CLSI is a BMD test similar to the M27-A3 test for yeast and is recommended for fast-growing mold species, for example, *Aspergillus, Fusarium, Pseudallescheria boydii, Rhizopus, Sporothrix schenckii* (mold form), and *Scedosporium*. The inoculum consists of conidia or sporangiospores. To induce sporulation these and other molds and are grown for 7 days at 35°C on potato dextrose agar (PDA) and then the conidial suspension is adjusted for use by its optical density (Cantón et al., 2009). Care in preparing the inoculum is important because a conidial concentration above the specified range will result in an elevated MIC to most antifungal agents. Some fungi, in order to induce sporulation, require altering the above conditions, that is, *Fusarium* species sporulate better at 25°C. Nongerminated conidia are used since, for *Aspergillus* species, similar results are obtained for germinated and nongerminated conidia.

After incubation at 35°C for 24–72 h, depending on the growth rate of the species, endpoints are recorded. Incubation times are 24 h for *Rhizopus oryzae* and 48 h for *Aspergillus, Bipolaris, Fusarium, Paecilomyces lilacinus, Sporothrix schenckii*, and *Wangiella dermatitidis* as well as for other *Mucorales*. A 72 h incubation is required for *Pseudallescheria boydii, Scedosporium apiospermum, S. prolificans*, and *Cladophialophora bantiana*. Dimorphic fungi (e.g., *Blastomyces dermatitidis, Histoplasma capsulatum*) may require 5–7 days of incubation. *Note*: AFS testing for *Coccidioides* species is not recommended because of safety concerns to laboratory workers.

Breakpoints are determined for molds in the broth microdilution method as follows: the MIC (or MEC in the case of echinocandins) for AmB, azoles, and CASF: S ≤ 1 μg/mL; intermediate = 2 μg/mL; and R ≥ 4 μg/mL. Reading the endpoint at 100% inhibition improves the detection of resistant *Aspergillus* species isolates to ITC and extended spectrum triazoles, VRC, and PSC (Rex et al., 2001).

3B.10.5 Etest® (bioMérieux, Marcy l'Etoile, France)

Etest is a plastic strip embedded with a gradient of the drug to be tested. Numbers on the scale correspond to drug concentrations on the strip (in μg/mL). The isolate is plated as a "lawn" on agar medium in a Petri plate. The strip is then placed on the agar surface. Both yeasts and molds may be tested using Etest strips. The Etest is particularly useful to assess susceptibility to AmB because it produces results over a greater concentration range that more readily distinguishes susceptible and resistant strains.

Test Conditions The test conditions recommended by the manufacturer can be found in the online manual "Etest® Technical Manual" (AB bioMérieux, www.abbiodisk.com), which is a detailed standard operating procedure and should carefully be followed. The Etest® Technical Manual shows photographs of typical and atypical Etest reactions for yeasts and azoles, AmB, flucytosine, and echinocandins. A summary of Etest results are given in Figs. 3B.3, and 3B.4. The manual includes preparation of the agar medium; how to swab plates; the MIC values for various antifungal agents with standard tester strains of *Candida* species; breakpoints for azoles; and susceptible MICs for AmB and echinocandins. Some key points are presented here.

3B.10.5.1 Etest for Yeasts — Candida *Species and* Cryptococcus neoformans

Medium RPMI + 2% glucose + MOPS + 1.5% Bacto agar.

Inoculum Yeast suspension in saline used to swab plates: 0.5 McFarland standard for *Candida* spp., 1 McFarland for *C. neoformans*.

Incubation Incubation is carried out at 35°C 24–48 h for *Candida* species and 48–72 h for *C. neoformans*.

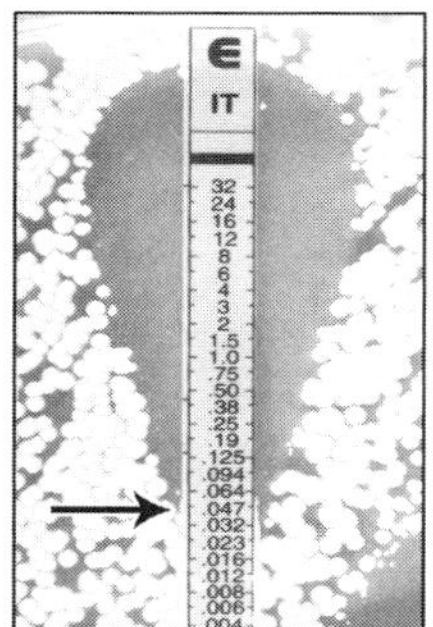

C. albicans vs. Itraconazole. Clear endpoint, 0.047 μg/ml

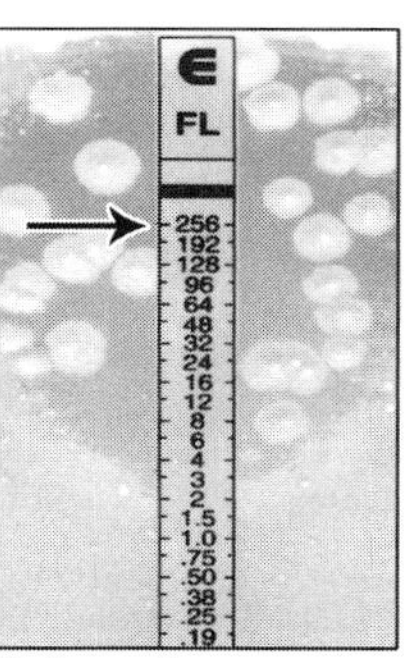

C. krusei vs. Fluconazole. Macrocolonies in the "ellipse". Endpoint, 256 μg/ml

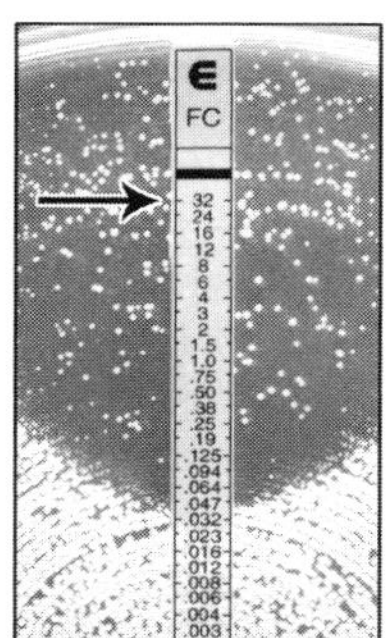

C. albicans vs. Flucytosine. Macrocolonies in the ellipse. Endpoint, >32 μg/ml

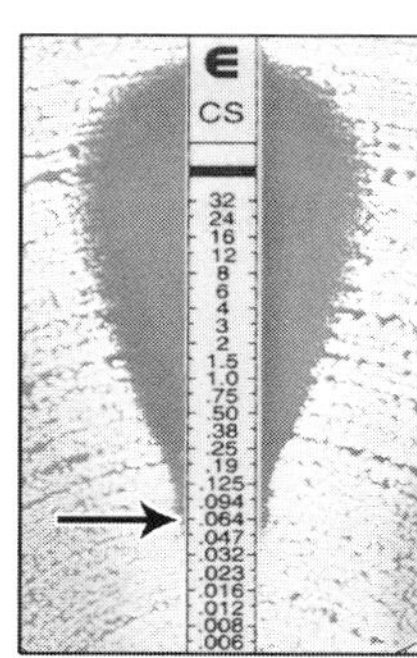

C. albicans vs. Caspofungin. "Dip effect" Endpoint, 0.064 μg/ml

Figure 3B.3 Etest® (Yeast). Results depict various ellipses, which are the areas of effective drug concentrations. The panels should not be taken as "typical" for each drug–yeast pair, but are shown to indicate how different results may be interpreted. Please see text for description of the Etest. *Source:* AB bioMérieux, www.abbiodisk.com.

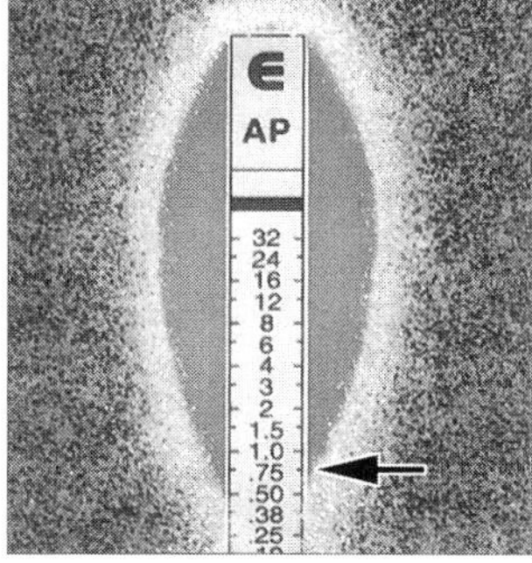

Aspergillus niger, **reading at 48 hr., MIC, 0.75 μg/ml**

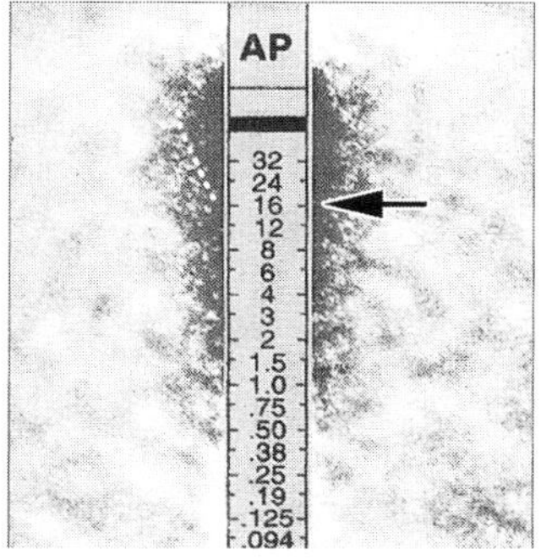

Fusarium **species, reading at 24 hr., MIC, 16 μg/ml**

Figure 3B.4 Etest® (Mold). Amphotericin B versus a susceptible mold isolate of *A. flavus* and, for comparison, a resistant mold isolate of *Fusarium* species.
Source: AB bioMérieux, www.abbiodisk.com.

Reading the MIC Endpoint

1. AmB: Read at the drug concentration resulting in complete inhibition of growth.
2. Flucytosine: Read at almost complete (90%) growth inhibition.
3. Azoles: Read at the first point of significant inhibition which is a marked decrease in growth density. The 80% inhibition principle applies here.
4. Echinocandins: Read endpoint at the first visual point of significant inhibition—80% inhibition.

Quality control MIC limits for M27-A standard tester isolates (*C. krusei* ATCC 6258 and *C. parapsilosis* ATCC 22019) against AmB, flucytosine, azoles, and echinocandins are described in the Etest Technical Manual. Figure 3B.3 shows the reaction of a *C. albicans* isolate with an Etest strip containing ITC. After incubation for 24–48 h the MIC is determined as the point of intersection of a growth inhibition zone with the calibrated strip.

3B.10.5.2 Etest for Molds

See Fig. 3B.4

Medium RPMI + 2% glucose + MOPS agar. Another medium, modified casitone agar, is an alternative for azoles. A chemically defined medium is preferable for quality control purposes.

Inoculum Suspension of conidia and hyphae (mature growth, 5–7 days) in saline with Tween 80. The suspension is adjusted to 0.5 McFarland standard for *Aspergillus* species and 1 McFarland for *Fusarium* and *Rhizopus* species.

Incubation Plates are incubated at 35°C, moist (in loosely folded plastic bags) for 24–72 h, depending on the genus—*Aspergillus* species for 16–24 h; *Fusarium* species at 35°C for 24–48 h, followed by room temperature for another 24–48 h; *Rhizopus* species are incubated for 16–24 h. For other mold species, extend the incubation time as needed and inspect plates daily for growth and the presence of an inhibition ellipse.

Quality control Etest MIC limits are described in the Etest Technical Manual for M38-A standard tester isolates *A. flavus* ATCC 204304 and *A. fumigatus* ATCC 2043051 tested against AmB, azoles, and flucytosine. Nonuniform growth of the fungus and a feathered or trailing growth edge can complicate reading endpoints. For molds, including *Aspergillus* species, good correlations are reported with AmB and ITC comparing Etest and MICs determined using the M38 BMD method.

3B.10.5.3 Agreement Between Etest and BMD Susceptibility Tests for Yeast and Mold Species

See Table 3B.4.

Table 3B.4 Agreement (%) for *Candida albicans* and *C. glabrata* of Etest and YeastOne Compared with the CLSI M27-A Broth Microdilution Reference Method[a]

Test method	Yeast species	No. BSI	5FC	FLC	ITC	VRC	AmB
Etest	*C. albicans*	94	93	98	97	93	99
	C. glabrata	38	100	82	89	89	100
YeastOne	*C. albicans*	94	96	96	100	82	100
	C. glabrata	38	97	66	89	89	100

[a]Data are from Table 5, Alexander et al. (2007). BSI, bloodstream isolates. Percentage values shown are agreement between the commercial test and the reference method (when the test result was within 2 dilutions of the reference value). Adapted with permission from the American Society for Microbiology.

Yeasts Compared to the CLSI M27-A2 reference BMD method of testing, the overall essential agreement was 95% for Etest and 91% for YeastOne (Alexander et al., 2007). *Candida glabrata* had low concordance for FLC and ITC compared to other species. For discrepant results for *C. glabrata* and an azole, the MIC was higher by both test systems compared to the reference method MIC. *Candida glabrata* may be associated with azole resistance detectable by the commercial test systems but undetectable by the reference method. With experience, the correlation between Etest and the BMD method is acceptable for most *Candida* species and azole antifungal agents. MIC determinations, using the Etest method, for azoles versus yeast isolates is more difficult than azoles versus molds because of trailing growth of yeasts.

Molds The performance of the Etest for VRC and for ITC susceptibility testing of 376 isolates of *Aspergillus* species was compared with the standard BMD method (Pfaller et al., 2003). The isolates included *A. fumigatus, A. flavus, A. niger, A. terreus*, and less common *Aspergillus* species. Overall agreement between the Etest and microdilution MICs was 97.6% for VRC and 95.8% for ITC. The Etest method using RPMI agar is a useful method for determining the VRC and ITC susceptibilities of *Aspergillus* species. *Aspergillus* species isolates tested against Etest strips containing echinocandins display small intrazonal colonies demonstrating short and stubby branches and a star-like morphology within the zone of inhibition. This phenomenon does not reflect a heterogeneous population in the isolate, but instead reflects the mode of action of echinocandins, which act only on newly synthesized cell wall.

3B.10.5.4 Clinical Correlation of Etest Compared to BMD Tests for Yeast and Molds

Yeasts Fifty-six *Candida albicans* isolates were obtained from mouth swabs of 41 HIV-positive patients (Ruhnke et al., 1996). Responders were those exposed to FLC for the first and/or second time (<14 days of therapy for oropharyngeal candidiasis) and nonresponders were those repeatedly exposed to FLC for at least 12 months. Clinical success was considered when no clinical signs or symptoms of pseudomembranous candidiasis were present. Overall clinical correlation of the two tests was high; the M27-A BMD test was better at predicting a positive clinical response whereas Etest was better able to predict clinical failure (Table 3B.5).

Molds The correlation between susceptibility tests comparing the BMD and Etest methods was evaluated in 18 cancer patients with invasive aspergillosis receiving monotherapy with AmB lipid complex (Lionakis et al., 2005). Of these patients 16 were treated for hematologic malignancies, and 2 for solid tumors. The response rate to AmB therapy was 22% (4 of 18 patients). Six patients were infected with *A. terreus*, which has high MICs to AmB, all of whom died. The BMD method found that 3 of the 18 patient mold isolates had MICs ≥2 μg/mL; all failed therapy. Mold isolates from 15 patients had MICS <2 μg/mL; of these 4 patients survived (27%).

The Etest found 9 patient isolates had MICs ≥2 μg/mL whereas an equal number of patients had MICs <2 μg/mL. Two patients from each group survived. The small number of patients in the cohort and the relatively high number with refractory *A. terreus* aspergillosis complicated the clinical correlation of MIC results. A larger number of patients at the same hospital could not be evaluated because they were receiving AmB in combination therapy. The only factor in univariate analysis that significantly predicted failure of AmB was admission to the intensive care unit. At this time, there are too few studies with such correlations to see how Etest MICs or BMD MICs, for that matter, correlate with clinical efficacy for mold mycoses.

The Etest, but not the M38 BMD test, indicated that an *A. fumigatus* isolate from a patient who failed CASF therapy was resistant to the drug (MIC >32 μg/mL) (Arendrup et al., 2008). This isolate showed reduced susceptibility to CASF in vivo in a mouse model. This finding provides insight into a possible advantage of the Etest, at least for molds, with respect to the echinocandins. The BMD test, it will be recalled, utilizes

Table 3B.5 Comparison of M27-A Broth Microdilution (BMD) and Etest Susceptibility Tests of Fluconazole in HIV-Positive Patients with Oropharyngeal Candidiasis: Correlation with In Vivo Outcome[a]

56 *Candida albicans* isolates from the oral cavities of 41 HIV + patients	Positive responders (n = 34)		Nonresponders (n = 22) (treatment failure)	
	M27 A BMD	Etest	M27 A BMD	Etest
Agreement with therapy	91%	85%	86%	100%

[a]Data from Table 3, Ruhnke et al. (1996). Adapted with permission from the American Society for Microbiology.

a microscopic examination of the tips of growing hyphae to reveal the MEC for that drug class.

3B.10.6 Disk Diffusion AFS Tests

Disk diffusion is an attractive format because of its simplicity and low cost.

Principle: A standard suspension of yeast or mold is swabbed onto a Petri plate containing Mueller–Hinton–glucose agar medium. Then paper disks or tablets impregnated with known drug concentrations are placed on the agar surface. Plates are incubated at 35°C for between 16 and 72 h (depending on the species rate of growth—18–24 h (*Candida*) or 40–48 h (*Cryptococcus*)). Inhibition is read as the zone diameter to the nearest 1 mm at the point where there is prominent reduction in growth (80%) (Fig. 3B.5).

A determination of relative susceptibility is made according to breakpoints established by the CLSI. For further reading please see the standard methods of procedure available by purchase from the CLSI.

3B.10.6.1 Disk Diffusion AFS Tests for Yeasts: Method for Antifungal Disk Diffusion Susceptibility Testing of Yeasts (M44A2E)

Disk diffusion tests for yeasts are now an accepted standard method. The interpretive criteria for the FLC, VRC, and CASF disk diffusion tests according to the CLSI are shown in the Table 3B.6.

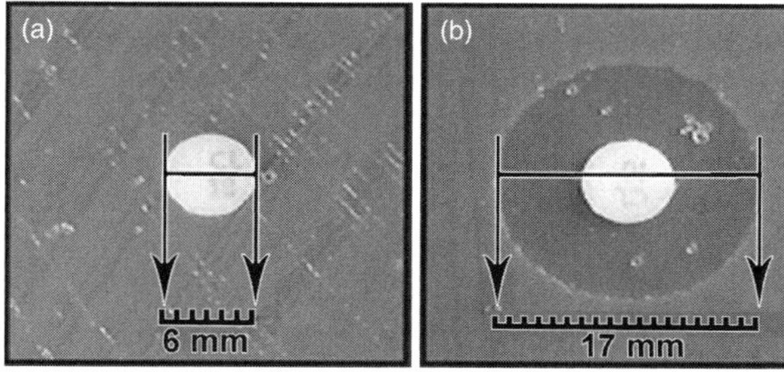

Figure 3B.5 Measuring the zone diameter in the disk diffusion method of susceptibility testing. (a) Disk diameter. (b) Zone of inhibition diameter. The diameter of the entire zone of inhibition is measured (from one edge of inhibition to the opposite edge of inhibition) rather than the radius (from the disk/edge center to the edge of inhibition).
Source: CDC.

Global surveillance for changes in susceptibility of *Candida* species was conducted using disk diffusion tests (Pfaller et al., 2007a). Of 89,750 *C. albicans* isolates tested spanning the period 2001–2005, FLC resistance increased marginally from 0.9% to 1.6%. During the same period resistance to FLC among 16,152 *C. glabrata* clinical isolates declined modestly from 19.2% to 15.2%.

AFS tests were performed on 541 *Candida* species clinical isolates using the BMD and M44-A disk diffusion methods (Noake et al., 2007). Susceptibility determined by disk diffusion and the BMD methods were in concordance for all species tested in the genus *Candida*. Overall agreement between these methods was 94.7% for FLC and 96.7% for VRC. The strong agreement of the simple disk diffusion method with the more standard BMD method supports the use of disk diffusion in routine diagnostic mycology.

In the case of *Cryptococcus neoformans* and *C. gattii*, thousands of isolates collected between 2001 and 2007 were tested for FLC susceptibility with the M44-A disk diffusion method (Pfaller et al., 2009). The percentage of *C. neoformans* isolates resistant to FLC increased from 7.3% to 11.7% over the period, probably as a result of drug pressure; see Table 3B.7. AFS tests may help to optimize therapy when there has been a suboptimal clinical response to an adequate FLC regimen. Then, the flexibility of the disk diffusion method may be advantageous.

Are fluconazole disks available from a commercial source in the United States? Disks (25 μg FLC) are available on special order from BD Diagnostic Systems (Sparks, MD).

3B.10.6.2 Disk Diffusion AFS Tests for Molds

The CLSI guideline for the method for disk diffusion susceptibility testing of nondermatophyte molds is M51AE (CLSI, 2008). Molds of various species (Espinel-Ingroff and Canton, 2008) were tested by both BMD and disk diffusion including *Aspergillus* (5 species), *Mucorales* (3), *Fusarium* (2), *Scedosporium* (2), and one each of *Alternaria, Bipolaris*, and *Paecilomyces* species.

Table 3B.6 Inhibitory Zone Diameter Interpretive Standards for *Candida* Species Tested in the M44-A Disk Diffusion Method Against Fluconazole, Voriconazole, and Caspofungin[a]

Antifungal (μg/disk)	Disk zone diameter (nearest mm)				Equivalent MIC breakpoints (μg/mL)			
	NS	R	S-DD	S	NS	R	S-DD	S
Fluconazole, 25 μg	—	≤14	15–18	≥19	—	≥64	16–32	≤8
Voriconazole, 1 μg	—	≤13	14–16	≥17	—	≥4	2	≤1
Caspofungin, 5 μg	≤10	—	—	≥11	>2	—	—	≤2

Note: Mueller–Hinton agar incubated at 35–37°C and read at 18–24 h. Zone diameter endpoints are read at 80% growth inhibition.

[a]CLSI (2008). NS, not susceptible; R, resistant; S-DD, susceptible-dose-dependent; S, susceptible.

Table 3B.7 In Vitro Susceptibilities of Two *Cryptococcus* Species to Fluconazole (FLC) and Voriconazole (VRC) Determined by CLSI Disk Diffusion Testing[a,b]

Species	FLC[c]			VRC[c]		
	Number of isolates tested	% S	% R	Number of isolates tested	%S	%R
C. neoformans	2,824	77.1	11.2	2,804	97.0	1.7
C. gattii	32	62.5	9.4	32	96.9	3.1

[a]Data from Table 2, Pfaller et al. (2009).

[b]Isolates were obtained from 133 institutions, 2001–2007.

[c]FLC and VRC disk diffusion testing was performed according to CLSI document M44-A. The interpretive breakpoints (zone diameters) were: Susceptible (S), >19 mm (FLC) and >17 mm (VRC); resistant (R), <14 mm (FLC) and <13 mm (VRC).

Readings for *Mucorales*, fast growers, are made after overnight incubation, 16–24 h; at 48 h for most other species and after 72 h for *Scedosporium* species.

Disks Disks containing PSC, VRC, and CASF are supplied by Becton-Dickinson and Co. (Sparks, MD). AmB and ITC disks are sold by Abtek Biologicals, Ltd.

Tablets The manufacturer of tablets (Neo-sensitabs®) is Rosco Diagnostica A/S (Taastrup, Denmark). That supplier has an inventory including AmB, CASF, ITC, PSC, and VRC.

Tentative In Vitro Breakpoints for Disk Diffusion Tests for Molds See Cantón et al. (2009). For triazoles and caspofungin, inhibitory diameters and corresponding breakpoints are: R ≤13 mm diam. equivalent to a MIC (or MEC) of ≥4 μg/mL; intermediate susceptibility, 14–16 mm diam. equivalent to a MIC (or MEC) of 2 μg/mL; and S ≥17 mm diam. equivalent to a MIC (or MEC) of ≤1 μg/mL. For AmB: R ≤12 mm diam. equivalent to a MIC of ≥4 μg/mL; intermediate susceptibility 13–14 mm diam. equivalent to 2 μg/mL; S ≥15 mm diam. equivalent to ≤1 μg/mL. Correlations comparing the M38-A standard BMD test and the M51-P disk diffusion test were determined with 183 mold isolates (Espinel-Ingroff and Canton, 2008). Zone diameters are reproducible and there is good correlation between zone diameters using either disks or tablets with MICs or MECs (the latter for CASF). The Neo-Sensitab assay for CASF, VRC, and PSC is considered valid for all mold isolates tested. ITC tablets are considered valid for all molds except *Mucorales*. AmB tablets are recommended for *Mucorales* only.

3B.10.7 VITEK 2® System for AFS (bioMérieux, Marcy l'Etoile, France)

This automated system both identifies microbes and conducts susceptibility tests. The system is approved for AFS with FLC. The VITEK 2 system also determines MICs for AmB, flucytosine, and VRC tested against *Candida* species (Pfaller et al., 2007b). Results were highly reproducible and reliably predicted MICs as determined by the reference M27-A BMD. The VITEK 2 system is rapid, with a mean time to results of 12–15 h. The use of spectrophotometry to determine the MIC endpoint eliminates observer subjectivity.

3B.10.8 Flow Cytometry AFS Test

Looking ahead, methods that measure killing of the target fungal cell may offer further advantages in improved clinical correlations. Flow cytometry using dyes which differentiate and enumerate live and dead cells could achieve this goal (Vale-Silva and Buchta, 2006). Flow cytometry lends itself to yeast cells but has also been applied to *Aspergillus fumigatus*, as a model mold. There is a detailed standard operating procedure for flow cytometric AFS testing of *Candida* and *Cryptococcus* spp. (Pina-Vaz and Rodrigues, 2010).

Principle: A yeast cell suspension (1×10^6 cells) in RPMI medium is treated with serial twofold dilutions of the antifungal agent. This mixture is allowed to incubate for between 30 min and 4 h. Then the cells are washed and resuspended with a fluorophore, such as FUN-1 (Molecular Probes Division of Invitrogen, Inc., Carlsbad, CA) or propidium iodide, and incubated for 30 min. There is no cell growth during the incubation with dye. This mixture is then passed through the flow cytometer, which measures scattered light and fluorescence of individual cells. The activity of the antifungal agent is detected by dose–response effects measured as changes in cell fluorescence for all cells in the population. When a new probe is introduced, or when results are inconsistent or unexpected, epifluorescence microscopy can assist in the interpretation of flow cytometric results.

For propidium iodide (PI), results are expressed as the % of positive cells showing high fluorescence (PI stains dead or dying cells a red color); for FUN-1, results are expressed as a staining index (SI), defined as the ratio between the mean green-yellow fluorescence (metabolically disturbed cells) in drug-treated cell suspensions divided by the fluorescence of the nondrug treated control cells (orange-red or yellow-orange fluorescent intravacuolar structures). Susceptibility is defined as the minimum concentration of the antifungal agent resulting in a SI above the control, that is, a SI of >1, after drug treatment.

One-hundred clinical isolates of *Candida albicans* were tested for AmB and FLC susceptibilities by the M27-A BMD test and flow cytometry methods (Chaturvedi et al., 2004). Agreement between the M27-A and flow cytometry methods ranged from 96% to 99%. Four isolates showed poor correlation for FLC MICs, possibly due to trailing growth, which is observed in the BMD method after 48 h incubation. (A newer advisory is to measure endpoints at 24 h, reducing the impact of trailers.) Trailing growth is not seen in the FC method because of the short incubation time.

Although fungal cells treated with AmB may be dead they do not stain with either PI or FUN-1. The large size of AmB inserted into the fungal cell membrane may impede permeability of PI. Chaturvedi et al. (2004), in their study, used a detergent, sodium deoxycholate, to render the cells permeable to PI, thus allowing its use to study of the effect of AmB.

Aspergillus fumigatus susceptibility to AmB, ITC, and VRC was tested by flow cytometry with PI dye as the fluorophore and compared with the M38-A BMD reference method (Ramani et al., 2003) (Table 3B.8). The overall agreement, ±1 dilution, between the M38-A and the flow cytometry method for all three drugs was 93%. All PI-positive cells collected by sorter were negative for cell viability. Interestingly, the measurement of minimum fungicidal concentration (MFC) by flow cytometry was comparable to the CFU Petri plate method. VRC exhibited fungicidal activity, to a lesser extent than AmB, but greater than ITC.

Flow cytometry is faster than the AFS tests in current use. The ability to detect and quantitate dead cells may have a closer clinical correlation than determination of MICs. If the isolate population is heterogeneous with respect to susceptibility that event can be observed in the flow cytometer data displayed as a histogram or dot plot.

3B.11 SUMMARY OF THE CURRENT STATUS OF ANTIFUNGAL SUSCEPTIBILITY TESTING

The availability of standardized commercial and simple methods for AFS has led to their more widespread use: YeastOne with alamarBlue colorimetric readout, the Etest, and disk diffusion methods. As these methods come into

Table 3B.8 Minimum Fungicidal Concentrations (MFC) of Three Antifungal Agents Versus *Aspergillus fumigatus* Determined by Flow Cytometry (FC): Comparison of Live–Dead Sorting with Propidium Iodide with Viability of Drug-Treated Cells Measured by CFU[a,b]

Aspergillus fumigatus strain	MFC (μg/mL)					
	Amphotericin B		Itraconazole		Voriconazole	
	FC	CFU	FC	CFU	FC	CFU
817	4	2	>16	>16	4	4
1122	2	1	8	16	8	8

[a]Colony-forming units after 72 h incubation.

[b]Data from Table 2, Ramani et al. (2003).

more common use in hospitals and the turnaround time is shortened, physicians will have more motivation to request AFS tests.

Obtaining a wide enough range of dose–response in the AmB susceptibility tests remains an obstacle. The necessity of using minimum effective concentration for determining susceptibility to echinocandins is also a drawback. Flow cytometry is coming into focus as a promising new technology, albeit its requirement for an expensive apparatus.

The correlation of in vitro susceptibility and clinical efficacy remains an open question. Fungistatic activity, which is what most AFS tests measure, underlines that invasive fungal infections probably will not resolve until a patient can mount a successful immune response. A major function of antifungal therapy is to ensure survival of the patient until any underlying immunosuppression improves.

SELECTED REFERENCES FOR ANTIFUNGAL SUSCEPTIBILITY TESTING

Alexander BD, Byrne TC, Smith KL, Hanson KE, Anstrom KJ, Perfect JR, Reller LB, 2007. Comparative evaluation of Etest and Sensititre YeastOne panels against the Clinical and Laboratory Standards Institute M27-A2 reference broth microdilution method for testing *Candida* susceptibility to seven antifungal agents. *J Clin Microbiol* 45: 698–706.

Agrawal D, Patterson TF, Rinaldi MG, Revankar SG, 2007. Trailing endpoint phenotype of *Candida* spp. in antifungal susceptibility testing to fluconazole is eliminated by altering incubation temperature. *J Med Microbiol* 56: 1003–1004.

Arendrup MC, Perkhofer S, Howard SJ, Garcia-Effron G, Vishukumar A, Perlin D, Lass-Flörl C, 2008. Establishing *in vitro–in vivo* correlations for *Aspergillus fumigatus*: The challenge of azoles versus echinocandins. *Antimicrob Agents Chemother* 52: 3504–3511.

Arikan S, Lozano-Chiu M, Paetznick V, Rex JH, 2001. In vitro susceptibility testing methods for caspofungin against *Aspergillus* and *Fusarium* isolates. *Antimicrob Agents Chemother* 45: 327–330.

Arthington-Skaggs BA, Lee-Yang W, Ciblak MA, Frade JP, Brandt ME, Hajjeh RA, Harrison LH, Sofair AN, Warnock DW, 2002. Comparison of visual and spectrophotometric methods of broth microdilution MIC end point determination and evaluation of a sterol quantitation method for in vitro susceptibility testing of fluconazole and itraconazole against trailing and nontrailing *Candida* isolates. *Antimicrob Agents Chemother* 46: 2477–2481.

Arthington-Skaggs BA, Jradi H, Desai T, Morrison CJ, 1999. Quantitation of ergosterol content: Novel method for determination of fluconazole susceptibility of *Candida albicans*. *J Clin Microbiol* 37: 3332–3337.

Cantón E, Espinel-Ingroff A, Pemán J, 2009. Trends in antifungal susceptibility testing using CLSI reference and commercial methods. *Expert Rev Anti Infect Ther* 7: 107–119.

Chamilos G, Lewis RE, Albert N, Kontoyiannis DP, 2007. Paradoxical effect of echinocandins across *Candida* species in vitro: Evidence for echinocandin-specific and *Candida* species-related differences. *Antimicrob Agents Chemother* 51: 2257–2259.

Chaturvedi V, Ramani R, Pfaller MA, 2004. Collaborative study of the NCCLS and flow cytometry methods for antifungal susceptibility testing of *Candida albicans*. *J Clin Microbiol* 42: 2249–2251.

Chou H-H, Lo H-J, Chen K-W, Liao M-H, Li S-Y, 2007. Multilocus sequence typing of *Candida tropicalis* shows clonal cluster enriched in isolates with resistance or trailing growth of fluconazole. *Diagn Microbiol Infect Dis* 58: 427–433.

CLSI, 2008. Zone diameter interpretive standards, corresponding minimal inhibitory concentration (MIC) interpretive breakpoints, and quality control limits for antifungal disk diffusion susceptibility testing of yeasts (M44S3E), Third informational supplement. CLSI, Wayne, PA.

Espinel-Ingroff A, Canton E, 2008. Comparison of Neo-Sensitabs tablet diffusion assay with CLSI broth microdilution M38-A and disk diffusion methods for testing susceptibility of filamentous fungi with amphotericin B, caspofungin, itraconazole, posaconazole, and voriconazole. *J Clin Microbiol* 46: 1793–1803.

Espinel-Ingroff A, Pfaller M, Messer SA, Knapp CC, Killian S, Norris HA, Ghannoum MA, 1999. Multicenter comparison of the Sensititre YeastOne colorimetric antifungal panel with the National Committee for Clinical Laboratory Standards M27-A reference method for testing clinical isolates of common and emerging *Candida* spp., *Cryptococcus* spp., and other yeasts and yeastlike organisms. *J Clin Microbiol* 37: 591–595.

Garcia-Effron G, Katiyar SK, Park S, Edlind TD, Perlin DS, 2008. A naturally occurring proline-to-alanine amino acid change in Fks1p in *Candida parapsilosis, Candida orthopsilosis*, and *Candida metapsilosis* accounts for reduced echinocandin susceptibility. *Antimicrob Agents Chemother* 52: 2305–2312.

Lass-Flörl C, Kofler G, Kropshofer G, Hermans J, Kreczy A, Dierich MP, Niederwieser D, 1998. *In-vitro* testing of susceptibility to amphotericin B is a reliable predictor of clinical outcome in invasive aspergillosis. *J Antimicrob Chemother* 42: 497–502.

Lee M-K, Williams LE, Warnock DW, Arthington-Skaggs BA, 2004. Drug resistance genes and trailing growth in *Candida albicans* isolates. *J Antimicrob Chemother* 53: 217–224.

Lionakis MS, Lewis RE, Chamilos G, Kontoyiannis DP, 2005. *Aspergillus* susceptibility testing in patients with cancer and invasive aspergillosis: difficulties in establishing correlation between in vitro susceptibility data and the outcome of initial amphotericin B therapy. Pharmacotherapy 25: 1174–1180.

Melo AS, Colombo AL, Arthington-Skaggs BA, 2007. Paradoxical growth effect of caspofungin observed on biofilms and planktonic cells of five different *Candida* species. *Antimicrob Agents Chemother* 51: 3081–3018.

Noake T, Kuriyama T, White PL, Potts AJ, Lewis MA, Williams DW, Barnes RA, 2007. Antifungal susceptibility of *Candida* species using the Clinical and Laboratory Standards Institute disk diffusion and broth microdilution methods. *J Chemother* 19: 283–287.

Pai MP, Pendland SL, 2003. Antifungal susceptibility testing in teaching hospitals. *Ann Pharmacother* 37: 192–196.

Park BJ, Arthington-Skaggs BA, Hajjeh RA, Iqbal N, Ciblak MA, Lee-Yang W, Hairston MD, Phelan M, Plikaytis BD, Sofair AN, Harrison LH, Fridkin SK, Warnock DW, 2006. Evaluation of amphotericin B interpretive breakpoints for *Candida* bloodstream isolates by correlation with therapeutic outcome. *Antimicrob Agents Chemother* 50: 1287–1292.

Pfaller MA, Andes D, Diekema DJ, Espinel-Ingroff A, Sheehan D; CLSI Subcommittee for Antifungal Susceptibility Testing, 2010. Wild-type MIC distributions, epidemiological cutoff values and species-specific clinical breakpoints for fluconazole and *Candida*: time for harmonization of CLSI and EUCAST broth microdilution methods. *Drug Resist Updat* 13: 180–195.

Pfaller MA, Diekema DJ, Andes D, Arendrup MC, Brown SD, Lockhart SR, Motyl M, Perlin DS; the CLSI Subcommittee for Antifungal Testing, 2011. Clinical breakpoints for the echinocandins and *Candida* revisited: Integration of molecular, clinical, and microbiological data to arrive at species-specific interpretive criteria. *Drug Resist Updat* 14: 164–176.

Pfaller MA, Diekema DJ, Gibbs DL, Newell VA, Bijie H, Dzierzanowska D, Klimko NN, Letscher-Bru V, Lisalova M, Muehlethaler K, Rennison C, Zaidi M, 2009. Results from the ARTEMIS DISK Global antifungal surveillance study, 1997 to 2007: A 10.5-year analysis of susceptibilities of noncandidal yeast species to fluconazole and voriconazole determined by CLSI standardized disk diffusion testing. *J Clin Microbiol* 47: 117–123.

Pfaller MA, Boyken L, Hollis RJ, Kroeger J, Messer SA, Tendolkar S, Diekema DJ, 2008. In vitro susceptibility of invasive isolates of *Candida* spp. to anidulafungin, caspofungin, and micafungin: Six years of global surveillance. *J Clin Microbiol* 46: 150–156.

Pfaller MA, Diekema DJ, Gibbs DL, Newell VA, Meis JF, Gould IM, Fu W, Colombo AL, Rodriguez-Noriega E, 2007a. Results from the ARTEMIS DISK Global Antifungal Surveillance study, 1997 to 2005: An 8.5-year analysis of susceptibilities of *Candida* species and other yeast species to fluconazole and voriconazole determined by CLSI standardized disk diffusion testing. *J Clin Microbiol* 45: 1735–1745.

Pfaller MA, Diekema DJ, Procop GW, Rinaldi MG, 2007b. Multicenter comparison of the VITEK 2 antifungal susceptibility test with the CLSI broth microdilution reference method for testing amphotericin B, flucytosine, and voriconazole against *Candida* spp. *J Clin Microbiol* 45: 3522–3528.

Pfaller MA, Diekema DJ, Sheehan DJ, 2006. Interpretive breakpoints for fluconazole and *Candida* revisited: A blueprint for the future of antifungal susceptibility testing. *Clin Microbiol Rev* 19: 435–447.

Pfaller JB, Messer SA, Hollis RJ, Diekema DJ, Pfaller MA, 2003. In vitro susceptibility testing of *Aspergillus* spp.: Comparison of Etest and reference microdilution methods for determining voriconazole and itraconazole MICs. *J Clin Microbiol* 41: 1126–1129.

Pina-Vaz C, Rodrigues AG, 2010. Evaluation of antifungal susceptibility using flow cytometry. *Methods Mol Biol* 638: 281–289.

Ramani R, Gangwar M, Chaturvedi V, 2003. Flow cytometry antifungal susceptibility testing of *Aspergillus fumigatus* and comparison of mode of action of voriconazole vis-à-vis amphotericin B and itraconazole. *Antimicrob Agents Chemother* 47: 3627–3629.

Rex JH, Pfaller MA, 2002. Has antifungal susceptibility testing come of age? *Clin Infect Dis* 35: 982–989.

Rex JH, Pfaller MA, Walsh TJ, Chaturvedi V, Espinel-Ingroff A, Ghannoum MA, Gosey LL, Odds FC, Rinaldi MG, Sheehan DJ, Warnock DW, 2001. Antifungal susceptibility testing: Practical aspects and current challenges. *Clin Microbiol Rev* 14: 643–658.

Rex JH, Nelson PW, Paetznick VL, Lozano-Chiu M, Espinel-Ingroff A, Anaissie EJ, 1998. Optimizing the correlation between results of testing in vitro and therapeutic outcome in vivo for fluconazole by testing critical isolates in a murine model of invasive candidiasis. *Antimicrob Agents Chemother* 42: 129–134.

Rex JH, Cooper CR Jr, Merz WG, Galgiani JN, Anaissie EJ, 1995. Detection of amphotericin B-resistant *Candida* isolates in a broth-based system. *Antimicrob Agents Chemother* 39: 906–909.

Ruhnke M, Schmidt-Westhausen A, Engelmann E, Trautmann M, 1996. Comparative evaluation of three antifungal susceptibility test methods for *Candida albicans* isolates and correlation with response to fluconazole therapy. *J Clin Microbiol* 34: 3208–3211.

Sanglard D, Ischer F, Marchetti O, Entenza J, Bille J, 2003. Calcineurin A of *Candida albicans*: Involvement in antifungal tolerance, cell morphogenesis and virulence. *Mol Microbiol* 48: 959–976.

Vale-Silva LA, Buchta V, 2006. Antifungal susceptibility testing by flow cytometry: Is it the future? *Mycoses* 49: 261–273.

QUESTIONS

The Answer Key to multiple choice questions may be found at the end of the book.

1. You are asked to recommend a fungal susceptibility test method for yeast in your laboratory. Which method is recommended for this purpose for the small-to-medium size laboratory?

1. CLSI broth microdilution method according to M27-A3 procedure
2. Commercial broth microdilution method with precoated drug panel, YeastOne (Trek Diagnostic Systems)
3. Disk diffusion test CLSI standard method M44-A
4. Etest (bioMérieux)

Which are correct?

A. 1 + 3
B. 2 + 4
C. 1 + 4
D. 2 + 3

2. Antimicrobial concentrations (μg/mL) used to interpret antifungal susceptibility test results define isolates as "susceptible," "intermediate" (susceptible-dose-dependent, S-DD), or "resistant." This is the definition of

A. Breakpoints
B. MFC
C. MIC
D. MIC_{90}

3. The lowest concentration of an antifungal agent capable of inhibiting the growth of 50% of the population of clinical isolates of that species is called the

A. Breakpoint
B. MEC
C. MIC
D. MIC_{50}

4. How can the laboratory correct an MIC for fluconazole in a *Candida albicans* clinical isolate displaying the heavy trailer phenotype? The truer value is

A. The MIC after 24 h incubation
B. The MIC after 48 h incubation
C. Only possible after using the sterol quantitation method
D. This determination is not necessary since heavy trailers are susceptible to treatment with fluconazole

5. The conditions for the Etest for *Candida* compared to *Cryptococcus* emphasize that

1. The cell suspension should be 0.5 McFarland units for *Candida* and 1.0 for *Cryptococcus*.
2. The cell suspension should be 1.0 McFarland units for *Candida* and 0.5 for *Cryptococcus*.
3. The incubation at 35°C should be 24–48 h for *Candida* and 48–72 h for *Cryptococcus*.

4. The incubation at 35°C should be 48–72 h for *Candida* and 24–48 h for *Cryptococcus*.

Which are correct?

A. 1 + 3

B. 2 + 4

6. The CLSI-Method M38-A2 for susceptibility testing of molds applied to *Aspergillus* isolates recommends that:

A. To be considered susceptible, the MIC for azoles and amphotericin B is ≤1 μg/mL.

B. Endpoints with itraconazole should be read at 100% inhibition.

C. The inoculum consists of conidia, not germinated conidia.

D. All are correct.

7. For clinical yeasts and azole antifungal agents tested in the broth microdilution method, the MIC is the lowest drug concentration resulting in:

A. 50% reduction in growth

B. 80% reduction in growth

C. 100% reduction in growth

D. Where there is the slightest visual evidence of reduction in growth

8. Explain how the medium affects the results of broth microdilution method of AFS tests of yeasts with amphotericin B?

9. Explain the disparity between in vitro AFS tests for *Candida* species and the results of antifungal therapy, using the "90/60 rule" as a way to predict clinical outcome. What evidence is there to support the use of the "R for resistant category" as it applies to *Candida* species isolates?

10. How is susceptibility estimated in the disk diffusion method of AFS tests? With respect to testing of molds in this assay against itraconazole, posaconazole, and voriconazole what was the correlation between zone diameter and MIC determined by the broth microdilution method?

Part Two

Systemic Mycoses Caused by Dimorphic Environmental Molds (Endemic Mycoses)

Introduction to Endemic Mycoses

The major pathogens included in this group are *Blastomyces dermatitidis*, *Coccidioides* species, *Histoplasma capsulatum*, and *Paracoccidioides brasiliensis*.

The characteristics shared by these fungi are several:

- An endemic area in which they are found in the environment in the soil.
- Production of airborne conidia which are inhaled when the environment is disturbed by human activity or weather patterns.
- The ability to infect immune normal as well as immunocompromised human and lower animal hosts.
- Exposure is believed to be common but disease is usually subclinical or self-limited. More serious disease is related to the infectious dose and host factors.
- The property of dimorphism by which they undergo morphogenesis from the mold form in the environment to a distinctive tissue form during infection.
- They are classed in the family *Ajellomycetaceae* within the *Onygenales*.

Also within the *Ajellomycetaceae* are less frequent and more enigmatic pathogens: *Lacazia loboi* which is at present not cultivatable in the laboratory, so isolations from the environment have not been possible; and *Emmonsia* species, which produce a distinct tissue form consisting of non-replicating isotropic growth in vivo.

Two additional dimorphic environmental fungi included in the group are *Sporothrix schenckii*, phylogenetically classed in the *Ophiostomatales*, and presenting as a subcutaneous mycosis, and *Penicillium marneffei*, a dimorphic respiratory pathogen, encountered most often in HIV/AIDS or other immunocompromised persons.

Despite their common characteristics each of these pathogens has a unique pathogenic profile that continues to present challenges to an understanding of its pathogenesis and, as well, to diagnosis and therapy.

Fundamental Medical Mycology, First Edition. By Errol Reiss, H. Jean Shadomy and G. Marshall Lyon, III.
© 2012 Wiley-Blackwell. Published 2012 by John Wiley & Sons, Inc.

Chapter 4

Blastomycosis

4.1 BLASTOMYCOSIS-AT-A-GLANCE

- *Introduction/Disease Definition.* A chronic, suppurative, and granulomatous pulmonary mycosis initiated by inhaling conidia of the soil-dwelling mold, *Blastomyces dermatitidis*. Dogs also are affected and are a sentinel species. *Blastomyces dermatitidis* converts to yeast forms in the lung, causing lesions. Skin is a frequent site of dissemination.
- *Geographic Distribution.* Ohio–Mississippi River Valley area, Southeast and Midwestern states of the United States and Southern Canada. Small foci in Africa and India.
- *Ecologic Niche.* Blastomycosis is mostly rural but the mold is difficult to isolate from nature.
- *Epidemiology.* The extent of subclinical exposure is unknown. Blastomycosis is notable for outbreaks.
- *Risk Groups/Factors.* Farmers, hunters, campers, and other outdoor vocations or avocations.
- *Transmission.* Respiratory—inhaling conidia from the environment.
- *Determinants of Pathogenicity.* Mold to yeast dimorphism; adhesin "BAD1" is an immunodominant antigen; cell wall α-(1$\rightarrow$3)-glucan.
- *Clinical Forms.* Pulmonary: infiltrates, nodules, cavities, pleuritis, and pneumonia. Chronic pulmonary; Disease may disseminate especially to skin, bones, prostate, and CNS.
- *Therapy.* Itraconazole (ITC), with amphotericin B (AmB) reserved for treatment failures or rapid progression. Voriconazole (VRC) is promising for CNS disease.
- *Laboratory Detection, Recovery, and Identification.* Direct exam and culture from early morning sputum and from scrapings or aspirates of skin lesions: direct smear, histopathology, culture. Identify culture with gene probe "AccuProbe." Serology—test for antibodies.

4.2 INTRODUCTION/DISEASE DEFINITION

Blastomycosis is a primary pulmonary mycosis of humans and lower animals, especially dogs, caused by the dimorphic soil-dwelling mold, *Blastomyces dermatitidis*. Blastomycosis is endemic in the Mississippi and Ohio river basins, extending into Canadian provinces bordering the Great Lakes and in northern Ontario, with smaller foci in Africa and India. Most often blastomycosis is a rural disease occurring sporadically or in outbreaks involving multiple individuals exposed, to the same point source. Blastomycosis is not contagious.

Conidia are inhaled from the environment into the lungs, causing disease that may be asymptomatic, or active disease that may disseminate. The pulmonary infection is either acute, subacute, or frequently, blastomycosis presents in the chronic granulomatous form. The pulmonary infiltrate contains a PMN suppurative component and for that reason blastomycosis is referred to as a *pyogranulomatous* infection. If not contained in the lungs, extrapulmonary dissemination may occur to the skin, bones, genitourinary tract, or central nervous system.

Rarely, primary cutaneous forms may follow accidental laboratory inoculation of infected material or from accidents during work outdoors (Gray and Baddour, 2002). Blastomycosis remains an enigmatic disease because the:

- Ecologic niche is uncertain.

Fundamental Medical Mycology, First Edition. By Errol Reiss, H. Jean Shadomy and G. Marshall Lyon, III.
© 2012 Wiley-Blackwell. Published 2012 by John Wiley & Sons, Inc.

- Extent of subclinical exposure in the community is not known.
- Fungus has a curious predilection for skin dissemination.

The mold form exists in nature in the soil, producing conidia which, when inhaled, convert to the pathogenic yeast form in the lung. This conversion can be demonstrated in the laboratory at 37°C under certain culture conditions. *Blastomyces dermatitidis* is an ascomycete with the *teleomorph* form: *Ajellomyces dermatitidis. Blastomyces dermatitidis* is identified in tissue and body fluids by the particular appearance of the budding cells: a large yeast form with a thick, refractile cell wall and a single, broad-based bud.

4.3 CASE PRESENTATIONS

Case Presentation 1. Outbreak of Blastomycosis in Children Visiting an Environmental Camp, Eagle River, WI (Klein et al., 1986)

History

Between July and August 1984, nine culture-confirmed cases of blastomycosis occurred in 95 members of two school groups who visited an environmental camp in dense forest along the Eagle River, in Vilas County, northern Wisconsin. The camp is a popular destination for outdoor lectures and ecology projects. Exposure was linked to visiting a beaver pond in a wooded area 8 km from the camp and close to the Wisconsin River. Their destination was in a swamp: an abandoned beaver lodge with a mound of rich soil and decaying vegetation 1.5 m above water level.

Forty-eight of the participants in the field trip became ill between 21 and 106 days after exposure to the beaver lodge. Nine cases met the definition of culture-positive, whereas 39 were probable cases. All but 2 patients visited and stood on the beaver lodge, including all definite cases. Forty-six of the patients were children from the 4th to 6th grade, and 32 were female.

Physical and Laboratory Findings

Symptoms, in descending order of occurrence, were cough, headache, chest pain, weight loss, fever, abdominal pain, night sweats, anorexia, and chills. Thirty-one patients had abnormal chest films and, of these, 20 had lower respiratory tract symptoms. A proven case was one in which *Blastomyces dermatitidis* was isolated from sputum (9 cases), whereas a probable case had radiographic evidence of pulmonary disease, and either a positive serum antibody test to the A-antigen (now called BAD1), or a positive lymphocyte blastogenesis assay. Of the culture-confirmed and probable cases the immunodiffusion test was positive in 13 patients or 27%, whereas the enzyme immunoassay was positive in 37 patients, or 77%.

Environmental Sampling

Soil, rotting organic material, wood chips, and animal dung were collected (47 samples) and 16 air samples were taken from sites visited by the two school groups. The supernatants of soil suspensions were injected into mice, and after incubation, the lungs, liver, and spleens were cultured on laboratory media. Two soil samples from the beaver lodge yielded *B. dermatitidis* and one rotting wood sample from the adjacent river bank was positive for the fungus.

Diagnosis

The diagnosis was acute pulmonary blastomycosis after exposure to contaminated soil of a beaver lodge.

Therapy and Clinical Course

Ten patients were hospitalized, nine received AmB. No other patients received antifungal therapy. All recovered and there were no deaths, complications, or recurrences of disease.

Comments

This outbreak was noteworthy because there were no cases of extrapulmonary dissemination, and a high proportion of the afflicted children were females. Exposure was most likely due to inhalation of conidia and/or mycelial fragments stirred up by disturbing the soil mound on the beaver lodge or on the adjacent river bank. Thirty-nine probable cases recovered without specific antifungal therapy. This finding lends support for the idea that many cases of blastomycosis are subclinical or self-limited. The isolation of *B. dermatitidis* is a major coup, because a point source, or any environmental recovery of this fungus, is rare. Clues to the environmental niche of *B. dermatitidis* are its occurrence in moist soil with high organic content (60–70% versus 3–5% organic content in typical soil).

Case Presentation 2. A Young Man Who Could Not Walk (Lee et al., 2006)

History

A 24-yr-old man first appeared in the emergency department with weakness in his right leg, which was diagnosed as right knee arthritis and he was sent home. On this admission he presented with a 1 month history of worsening bilateral leg weakness. He could not walk or get out of bed. He denied fever, chills, or cough but reported night sweats. He lived in Milwaukee and worked as a grocery store clerk.

Physical and Laboratory Findings

For several months red, annular, raised, and crusted skin lesions appeared on his head, torso, and legs, which did not heal with topical treatments. His body temperature was normal, with a blood pressure of 108/64, and a heart rate of 80. Motor strength was absent in the right leg and barely detectable in the left leg.

In the laboratory tests and radiology, the WBC count was 9500/ μL with a normal differential. An MRI of the spine showed partial marrow replacement of the T8 vertebral body and near complete marrow replacement of the T7, L2, and L3 vertebral bodies. An adjacent epidural mass was 6.8 cm long and 10 mm thick. No abscess was noted. Chest radiographs showed an infiltrate in the upper left lobe.

The presumptive diagnosis was neoplasm. A lung biopsy showed granulomatous disease. Fungal elements, stained with PAS, were observed in lung and skin biopsies. Skin scrapings incubated in Mycosel medium at 37°C showed a broad-based budding yeast consistent with *Blastomyces dermatitidis*. The immunodiffusion test was positive for *B. dermatitidis* antibodies.

Diagnosis

The diagnosis was pulmonary blastomycosis with skin and CNS dissemination.

Therapy and Clinical Course

The patient received lipid complex AmB, 6 g over 2 months. After several days he could move his toes and legs. After 1 week he could walk with a walker. The skin lesions resolved over several weeks. MRI of the spine after 2 weeks found the epidural mass decreased to 5 mm thickness, and cord compression decreased. A repeated chest radiograph showed residual scarring in the left upper lobe. After 2 months, therapy was stepped-down to itraconazole (ITC), 200 mg/2×/day. As of 5 mo on ITC, he was ambulatory, but with some night sweats and chills. There was fatty replacement of thoracic vertebral bodies in the area of the previous fungal mass. Some edema of the thoracic cord persisted but with no evidence of spinal cord compression. His chest X-ray revealed slight improvement in the left upper lobe.

Comment

Direct examination of the skin lesions disclosed the causative agent. The pulmonary phase of this case appears to have been mild, without symptomatic pneumonia. Joints and bones are frequent sites of dissemination and the CNS may be involved in 5–10% of cases of disseminated blastomycosis. The therapy for this patient is consistent with the current standard of care for CNS blastomycosis: primary therapy with lipid AmB for 6 weeks followed by step-down to an azole for up to 1 year (Chapman et al., 2008). VRC has better penetration into the CNS than ITC.

4.4 DIAGNOSIS

Diagnosis is made by isolation of the fungus from sputum, or by histopathology and culture of biopsy material from lung or skin lesions, the latter is a frequent site of hematogenous dissemination. Serologic findings of positive precipitins in immunodiffusion are presumptive evidence of blastomycosis.

Differential Diagnosis

Lung involvement may mimic acute bacterial pneumonia or other pulmonary mycoses, whereas a chronic presentation resembles lung cancer or tuberculosis.

4.5 ETIOLOGIC AGENT

The single cause of blastomycosis is the *dimorphic* fungus *B. dermatitidis*.

Blastomyces dermatitidis Gilchrist & W. R. Stokes, *J Exp Med* 3:76 (1898). Position in classification: *Ajellomycetaceae, Onygenales, Ascomycota*. Teleomorph: *Ajellomyces dermatitidis*.

North American cultures produce the characteristic antigen termed “A,” now known as “BAD1.” African isolates of *B. dermatitidis* appear to lack this antigen. When cultured on standard laboratory media and incubated at room temperature, *B. dermatitidis* grows as a mold colony, whereas in human or animal tissue or on enriched media grown at 35–37°C it forms large, thick-walled multinucleate yeast cells that bud with a wide pore or *isthmus* between the mother and daughter cell. The two cells remain attached until the thin-walled daughter cell becomes nearly the size of the parent cell before being released (dehiscence). The cell wall then thickens.

The fungus is *heterothallic*, and genetic studies have demonstrated that both the (+) and (−) mating types produce apparently identical disease. When crossed, they produce fertile *cleistothecia* and the teleomorph, *Ajellomyces dermatitidis*. The *Ajellomyces* genus contains the teleomorph forms of *B. dermatitidis, Histoplasma capsulatum*, and *Emmonsia crescens* (Guarro and Cano, 2002). Phylogenetic analysis targeting the 18S (SSU) rDNA reveals that the endemic mycoses fall into two clades within the *Onygenales: B. dermatitidis* is closely related to *Paracoccidioides brasiliensis* and is related to *Histoplasma* species (Bialek et al., 2000). These species including *Emmonsia crescens* make up the *Ajellomycetaceae. Coccidioides* species, causative agents of the other major dimorphic endemic mycosis, coccidioidomycosis, are in a separate clade within the *Onygenale*s.

4.6 GEOGRAPHIC DISTRIBUTION/ECOLOGIC NICHE

The major endemic region in the United States is generally localized to the Southeast and south-central states bordering the Mississippi and Ohio River basins. This large area overlaps the endemic area for histoplasmosis (Fig. 4.1). Most cases are reported from Arkansas, Kentucky, Mississippi, Tennessee, and Wisconsin, extending into the Canadian provinces bordering the Great Lakes, and to a small region of upper New York and Canada along the St. Lawrence River.

Blastomycosis was long-believed to be a disease restricted to North America and, by far, most cases have been reported from the eastern half of North America, where it is associated with outdoor occupations and recreational activities in rural areas. The disease is now known to be endemic in at least 17 African countries (Bally et al., 1991) and India.

Ecologic Niche

Although it has not been proved, epidemiologic and clinical data support the concept that *B. dermatitidis* lives as a *saprobe* in soil, probably enhanced by decaying wood and/or decaying vegetation. The fungus has been isolated from nature in soil, in a tobacco barn, and in soil samples from a beaver lodge and a soil sample of nearby decomposed wood (please see Section 4.3, Case Presentation 1).

Intravenous inoculation of supernatants of soil suspensions in the tail veins of mice was successful in recovering *B. dermatitidis* from soil obtained from a tobacco barn in Lexington, Kentucky. That barn had sheltered a dog that died of blastomycosis 2 years before the sample was collected (Denton et al., 1961).

The relationship of moisture or nearness to water seems to indicate the fungus prefers a moist environment, enriched with organic matter. In a careful and systematic soil screening, *B. dermatitidis* was recovered from 10 of 356 samples collected in an endemic area at Augusta, Georgia, along the flood plain of the Savannah River, where the soil is a fine alluvial silt (Denton and DiSalvo,

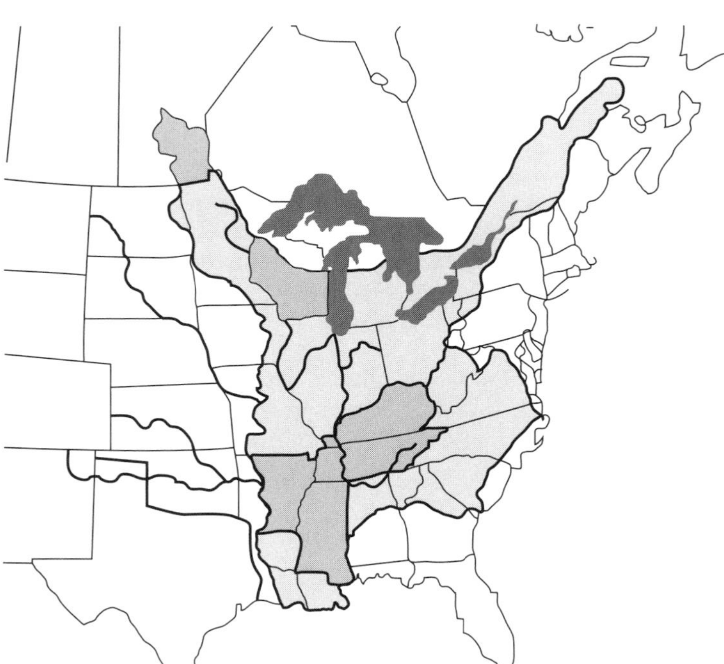

Figure 4.1 Endemic area in North America for blastomycosis indicated by the location of cases. Areas of higher disease prevalence are a darker shade.

1964). Structures from which positive samples were collected include chicken houses, a rabbit pen, a mule stall, a cattle-loading ramp, and an abandoned kitchen. All 10 strains of the fungus isolated are typical of the species and were virulent for mice. Repeated isolations of *B. dermatitidis* from soil, however, have proved fruitless, and the certain ecologic niche remains to be determined.

4.7 EPIDEMIOLOGY

The true incidence of blastomycosis as well as its specific endemic areas are uncertain because:

- Cases of blastomycosis are not reported through an organized system.
- The number of subclinical or asymptomatic benign cases cannot be estimated because there is no skin test for exposure. Although *B. dermatitidis* appears from outbreaks to be highly infectious, symptoms develop in a fraction of those who are exposed, suggesting there may be a significant number of asymptomatic exposed persons.

The endemic areas are delineated by known human and animal (canine) disease. Possibly, blastomycosis is a subclinical disease which at times may resolve spontaneously, as seen with other fungal infections. Blastomycosis is seen in men more often than women, but the numbers vary considerably depending on the study group. Ratios of men to women vary from 5:1 to 15:1. Males are more likely to have an environmental exposure because of greater outdoor vocations or avocations. Other studies indicate that the numbers of females to males is much closer. In epidemics, men, women, and children are all infected at a similar rate. The primary lower animal to be infected is the dog, regarded as a sentinel species indicating where *B. dermatitidis* may occur in the environment.

Between 1896 and 1968, 1476 cases were compiled in the United States (Bradsher et al., 2003). A national database of hospital inpatient stays recorded, in 2002, that 68 children and 703 adults were diagnosed with blastomycosis (Chu et al., 2006). Some areas are considered "hyperendemic." Mississippi is a state of high prevalence (1.3 cases/100,000 population). The University of Mississippi Medical Center reported that 123 patients were treated in their hospital for blastomycosis between 1980 and 2000 (Lemos et al., 2002). Incidence in Wisconsin is 1.4 cases/100,000 population. The Wisconsin Department of Health reported 670 cases in that state between 1986 and 1995 (Proctor and Davis, 1996).

Blastomycosis in Canada

Cases have been reported from Alberta, Ontario, Manitoba, New Brunswick, Nova Scotia, Quebec, and Saskatchewan provinces. Blastomycosis is a reportable disease in some provinces of Canada. Ontario has the highest incidence of blastomycosis, 0.30 cases/100,000 population. A total of 309 cases of blastomycosis were seen in Ontario during a 10 year period, 1994–2003. The mean number of cases diagnosed per year was 33.7; 57% of which occurred from 2001 to 2003. Most patients were from northern Ontario ($n = 188$), where the incidence was 2.44 cases/100,000 population.

Hyperendemicity in the region surrounding Kenora (near Lake-of-the-Woods, Ontario) was estimated to have an annual incidence rate of 117.2 cases/100,000 population, exceeding the next highest rate reported in North America of 100 cases/100,000 population in the Eagle River area in Vilas County, Wisconsin (reviewed in Morris et al., 2006).

During a 12 year period between 1988 and 1999, there were 143 confirmed cases of blastomycosis treated in Manitoba hospitals, 11.9 cases/year (Crampton et al., 2002). The localities of highest endemicity were the census districts of Kenora, Rainy River (on the Ontario–Minnesota border), and Thunder Bay, Ontario (on Lake Superior).

4.8 RISK GROUPS/FACTORS

Persons at risk include those with normal immune responses who become exposed during their recreation or occupation in rural areas by disturbing the environment where the fungus is growing in the soil. Males more frequently present with blastomycosis than females, perhaps due to their outdoor activities. In outbreaks there is a more even ratio of affected males and females (please see Section 4.3, Case Presentation 1). Men appear to be at greater risk for extrapulmonary blastomycosis. Immunosuppresive conditions are encountered in a significant minority of blastomycosis patients. In one series of 123 patients, 18 (14.6%) were fully immunosuppressed and a further 13 (10.6%) were partially immunosuppressed (Lemos et al., 2002). Additionally, diabetes mellitus was an underlying condition in 10 (16.3%) in that cohort.

4.9 TRANSMISSION

Except for exceptional instances, blastomycosis is not communicable between humans, or between animals and humans. Conidia of *B. dermatitidis* aerosolized during disturbance of the environment are inhaled into the lungs resulting in exposure, which may be asymptomatic or may produce active pulmonary disease, which may then disseminate. Zoonotic transmission of *B. dermatitidis* is uncommon; but transmission to veterinarians via accidental injection of a contaminated pulmonary aspirate, accidental laceration during necropsy, and following a bite from an infected dog have been reported (reviewed in Weese et al., 2002). In a rare case, a dog was reported to transmit blastomycosis to a human following a dog bite (Gnann et al., 1983). Another exceptionally rare report was of sexual transmission from a man to his wife (Craig et al., 1970). Primary cutaneous forms may follow inoculation of infected material or cultures (Gray and Baddour, 2002).

4.10 DETERMINANTS OF PATHOGENICITY

4.10.1 Pathogenesis

Alveolar macrophages are the first line of defense against *B. dermatitidis*, but those from normal persons are only modestly capable of phagocytosis and killing of *B. dermatitidis* conidia or yeast forms. In a natural infection, conidia induce an influx of primarily PMN and, to a lesser extent, mononuclear phagocytes. This gives rise to the pyogranulomatous reaction, which is a hallmark of acute pulmonary blastomycosis. PMN phygocytose conidia rapidly and efficiently but killing, due to reactive oxygen species, proceeds slowly and is incomplete. Nonetheless, the susceptibility of conidia to the range of nonspecific defense mechanisms perhaps explains the relative rarity of blastomycosis as a clinical problem (Deepe et al., 2005). Yeast forms are much more difficult for PMN to phagocytose and kill than conidia, even in immune normal hosts: they are 50× more resistant than conidia to H_2O_2 killing. Thus, germination of conidia and their conversion to yeast forms confers a survival advantage.

The typical granulomatous lesion in blastomycosis consists of an outer collar of lymphocytes, an intermediate concentric ring of epithelioid cells and macrophages, and an inner zone containing multinucleate giant cells with phagocytosed yeast forms (Chick, 1971). PMNs often are present in this inner zone. Necrosis is common; fibrosis is found in most every case and is encountered in the tissue response both in pulmonary lesions and in almost any chronically infected tissue. Areas of new pneumonic involvement often are exudative or suppurative with alveolar spaces filled with a myriad of PMN and distinctly fewer chronic inflammatory cells. Yeast forms are more abundant in the suppurative type lesions than in granulomatous ones.

Developing protective immunity to *B. dermatitidis* in a naïve host depends on activation of alveolar macrophages. Interleukin (IL)-12 production is a principal stimulus for inducing a Th1 response, and it does so by inducing the production of IFN-γ. The key cytokines are IL-12, IFN-γ, and tumor necrosis factor (TNF-α). TNF-α (produced by macrophages) synergizes with IL-12 to stimulate IFN-γ. In their absence, animals succumb to blastomycosis. Thus, the IL-12–IFN-γ axis is a critical feature of protective immunity to *B. dermatitidis*.

The *B. dermatitidis* surface adhesin and antigen, BAD-1, induces transforming growth factor-(TGF)-ß, which *blocks* production in mononuclear phagocytes of the key proinflammatory cytokine, TNF-α. The countervailing forces of high TGF-ß and *dysfunctional* macrophages, and high TNF-α and functional macrophages provide an understanding of the influence of TNF-α in the protective immune response (Wüthrich and Klein, 2005).

How do *B. dermatitidis* yeast forms spread to other organs from lesions in the lungs? Logically, hematogenous dissemination is the means of spread. It is thus curious that the first report of a positive blood culture in blastomycosis was by Musial et al. (1987), using the Isolator® system. A literature search at that time revealed no previous reports of *B. dermatitidis* fungemia. The yeast forms are too large for endophagocytosis, and it may be postulated that free yeast forms circulate in the blood.

4.10.2 Host Factors

Blastomyces dermatitidis and its major pathogenic determinant, the adhesin BAD1, block production of TNF-α by host macrophages (Wüthrich and Klein, 2005). Lung surfactant protein D (SP-D) is capable of binding to ß-(1→3)-D-glucan on *B. dermatitidis* yeast forms, with the result that it too inhibits TNF-α production by alveolar macrophages (Lekkala et al., 2006). The mechanism at work here is that the receptor–ligand reaction between ß-(1→3)-D-glucan and Dectin-1 on macrophages normally induces TNF-α. A blockade of ß-(1→3)-D-glucan by SP-D interrupts this signaling and inhibits TNF-α.

T-cell-mediated immunity is critical in host defense against *B. dermatitidis*. The principal lymphokine in conferring benefit is IFN-γ and its activation of macrophages. Production by $CD4^+$ T lymphocytes of Th1 lymphokines, IFN-γ, and, also, GM-CSF is critical to host defense. In CD4-depleted mice, $CD8^+$ T lymphocytes confer protection against *B. dermatitidis* via their production of IFN-γ, GM-CSF, and TNF-α (Wüthrich and Klein, 2005). Furthermore, $CD8^+$ T lymphocytes require MHC class I compatibility for antigen presentation and protection. Dendritic cells can degrade *B. dermatitidis* yeast forms and present them with MHC class I to $CD8^+$ T lymphocytes.

4.10.3 Microbial Factors

Adhesin

BAD1, a 120 kDa protein, is a major surface antigen and also functions as an adhesin. Adherence to host tissue is a primary requirement for a pathogen, and without this ability, *B. dermatitidis* may not be able to establish an infection in the lower respiratory tract. Gene knock-out experiments have shown that the *null* mutant lost the ability to bind to lung tissue and was avirulent for mice (Brandhorst et al., 1999).

The adhesin activity of BAD1 is mediated by a 24 amino acid tandem repeat that is 90% homologous with the *Yersinia* adhesin, called invasin. An engineered mutant that secretes BAD1 but does not "decorate" the cell wall is still pathogenic for mice. This second mechanism of pathogenicity appears to influence "immune deviation" away from protective T-cell responses (Wüthrich et al., 2006). Deletion of the C-terminal epidermal growth factor-like domain from BAD1 results in a BAD1 incapable of binding to yeast forms and binds poorly to macrophages. Even in this form BAD1 retained the ability to inhibit production of TNF-α by phagocytes (Wüthrich and Klein, 2005).

Cell Wall α-(1→3)-D-Glucan

Blastomyces dermatitidis strains that contain less α-(1→3)-D-glucan in their cell walls also have reduced virulence (Hogan and Klein, 1994).

4.11 CLINICAL FORMS

Subclinical Blastomycosis

Unlike the other systemic mycoses, the component of subclinical blastomycosis is largely unknown. There is no available immunologic tool with which to conduct a survey in the endemic areas where a relatively large number of cases have been recorded. The inference from outbreak investigations is that a significant proportion of exposed persons do not develop symptoms. Whether subclinical infection confers immunity to reinfection is unknown.

Pulmonary Blastomycosis

Acute

The median incubation period is 30–45 days (determined from outbreaks such as described in Section 4.3, Case Presentation 1). Acute pulmonary blastomycosis takes different clinical forms. There may be mild pulmonary infiltrates and few or no clinical symptoms. Self-limited resolution of acute blastomycosis pneumonia is reported to occur from investigations of outbreaks, but the proportion of cases with this favorable outcome is unknown. Some patients who present with pulmonary infiltrates often have no pulmonary symptoms, and the diagnosis is suspected, even in the absence of pulmonary complaints, by findings during routine chest radiographs.

Blastomycosis often presents as an influenza-like illness with arthralgia and myalgia. The fungal pneumonia may resolve or progress to localized pulmonary involvement or extrapulmonary disease. As pulmonary lesions become more extensive the patient may develop fever, weight loss, and cough with or without mucopurulent sputum. Pleural effusions were reported in 21–42% in two series (Kinasewitz et al., 1984; Sheflin et al., 1990) but a cooperative study of 198 patients in U.S. Veterans' Administration hospitals described only 4 pleural effusions (Kaplan and Clifford, 1964).

Some patients progress to chronic disease and pulmonary cavitation. Cavitary lesions are uncommon in blastomycosis; their prevalence varies widely, from 6% to 37% of cases of chronic blastomycosis (reviewed by Gadkowski and Stout, 2008). Cavities may be associated with both acute and chronic symptoms (Fang et al., 2007).

Other patients experience an abrupt onset of pleuritic chest pain without fever or other symptoms. Another type is severe illness requiring hospitalization, with diffuse pulmonary infiltrates, severe hypoxemia, and *ARDS*.

Differential Diagnosis Treatment for disease other than blastomycosis may have been started for weeks or even months for blastomycosis, especially when the pulmonary form is not accompanied by skin or other organ involvement. Bacterial pneumonia, tuberculosis, lung cancer, and other systemic mycoses (e.g., coccidioidomycosis or histoplasmosis) may be suspected. In one series (Lemos et al., 2002) the correct initial diagnosis of blastomycosis was made in only 5% of cases (4 of 78).

Chronic

Chronic pulmonary blastomycosis may present with fever, night sweats, weight loss, and a cough with sputum production. Radiographic findings include:

- Air-space consolidation and alveolar disease is the most common finding.
- Interstitial disease including fibronodular or reticulonodular findings, and mass lesions, which may be well circumscribed with homogeneous opacities.
- Cavitary disease.
- Miliary disease. Uncommon and associated with underlying immunosuppression and carrying a poor prognosis.
- ARDS. Diffuse bilateral interstitial–alveolar infiltrates accompanied by a large number of yeasts seen in sputum.
- Pleural disease.
- Post-infectious calcification of lymph nodes or pulmonary parenchyma is rare, in contrast to histoplasmosis.

Differential Diagnosis The diagnosis can be bacterial pneumonia, tuberculosis, histoplasmosis, ARDS, or bronchogenic carcinoma.

Disseminated Blastomycosis

Although the primary focus of disease is the lung, disseminated blastomycosis may involve any body organ, but cutaneous, genitourinary, and osseous blastomycosis most frequently are seen.

Cutaneous Blastomycosis

When present, skin lesions are quite typical and should alert the physician to study the patient further because they are indicative of multi-organ dissemination. Skin lesions are present in percentages that vary according to the series from 20%, to one-third, or $\geq$40% of patients. Lesions may be dry and scaly, but more often are *verrucous* with microabscess formation at the periphery that may ooze small beads of pus when pressure is applied (Fig. 4.2). A purple margin often surrounds the lesions. These lesions are described as "pseudoepitheliomatous hyperplasia," which clinically and histologically resemble skin cancer. A second type of lesion is ulcerative and bleeds easily.

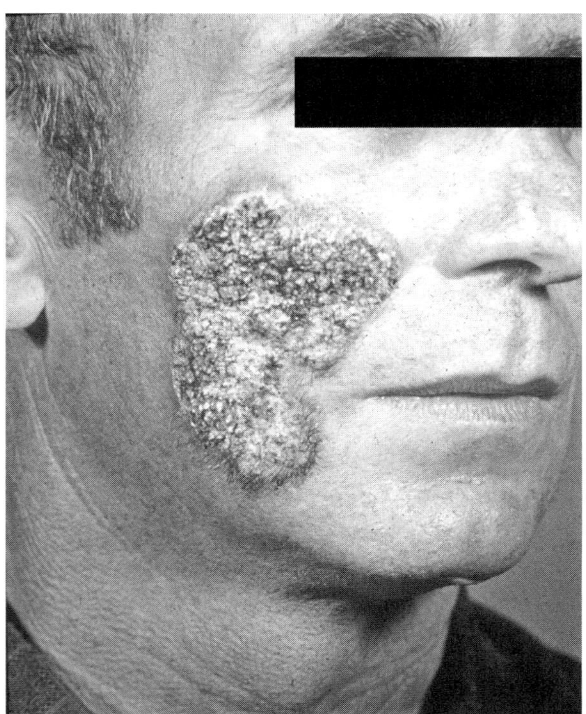

Figure 4.2 Blastomycotic skin lesion disseminated from a lung infection.
Source: H. J. Shadomy.

As the lesions progress, clearing and scarring occur in the center, and the yeast form may not be cultured from there. However, the fungus is readily seen in KOH preps and cultured from material taken from the vesicles and pus. Upon healing, the lesions are scarred, and hypopigmented. Occasionally, lesions appear below the patient's nose, and without microscopic examination have been mistaken for the dermatophyte *Trichophyton rubrum*.

Differential diagnosis of skin lesions includes bromoderma, pyoderma granulosum, Majocchi's granuloma, leishmaniasis, *Mycobacterium marinum*, giant keratocanthoma, and squamous cell carcinoma (Chapman, 2005).

Osseous Blastomycosis

Bone involvement is reported in up to one-quarter of extrapulmonary cases. Although any bones may become infected via hematogenous spread, the most often involved are ribs, vertebrae, sacrum, long bones, and skull. Sharply defined osteolytic lesions are typical. Most bone lesions resolve with antifungal therapy alone, without surgical debridement.

Differential Diagnosis The radiologic appearance of blastomycosis in bone is not specific and cannot be discriminated from other fungal, bacterial, or neoplastic diseases.

Genitourinary Blastomycosis

This clinical form follows pulmonary, cutaneous, and osseous blastomycosis in frequency. Between 10% and 30% of patients with blastomycosis may have genitourinary involvement. The fungus may be recovered from centrifuged urine of patients with kidney involvement. Most often seen in men as prostatitis and epididymo-orchitis, yeasts are observed microscopically in pus or urine after prostatic massage.

CNS Blastomycosis

CNS involvement is reported to occur in 5–10% of cases of disseminated blastomycosis with greater frequency in immuncompromised patients (e.g., those living with AIDS). CNS blastomycosis occurs most often as epidural or cerebral abscesses or granulomatous masses (blastomycomas). MRI or CT guided biopsy is appropriate. Craniotomy and surgical debridement usually is necessary, if eloquent areas of the brain are not endangered. The infection is difficult to identify; however, material from abscesses or masses may demonstrate the fungus. Meningitis usually is a late stage complication of multi-organ involvement and represents severe, usually fatal dissemination. Therapy of CNS blastomycosis has relied on AmB or ITC or AmB induction therapy and ITC step-down therapy. Cases refractory to these therapies have responded to VRC (please see Section 4.13, Therapy).

Blastomycosis and ARDS (Adult Respiratory Distress Syndrome)

In both the immunocompetent and immunocompromised patient, ARDS may be the presenting symptom. Some of the patients with blastomycosis and ARDS have been described as immunocompromised by treatment with corticosteroids as well as in patients living with AIDS.

Blastomycosis in Africa

Pulmonary involvement is the commonest feature as it is in North America. Distinguishing features of African blastomycosis are (i) frequent bone involvement, with vertebral disease, and paraplegia prominent and (ii) subcutaneous suppuration and draining sinuses on the chest wall or the legs (Bally et al., 1991).

4.12 VETERINARY FORMS

Domestic dogs have been the most commonly infected domestic animals seen with blastomycosis. Hunters and their dogs have been infected during hunting season. Four hunters and four of their hounds developed blastomycosis after hunting raccoons in a wooded swampy area of Southeast Virginia (Armstrong et al., 1987). Asymptomatic hunting dogs belonging to the same family as a hunter-patient had positive precipitins for *B. dermatitidis*. Because both dogs and patients were probably infected at the same place, canine blastomycosis is an important epidemiologic marker, and a harbinger of human disease (Sarosi et al., 1979). Cats occasionally become infected, but horses, cattle, and other domesticated animals do not seem susceptible. Wild animals, particularly captive sea mammals (e.g., sea lions and dolphins) may become infected when in an endemic area. Isolated reports of other animals—wolf, African lion, polar bear, and ferret—are found in the literature.

4.13 THERAPY

Acute pulmonary blastomycosis (Chapman et al., 2008) in an immune-normal person can be mild and self-limited, resolving without antifungal therapy, but all infected persons should be considered for treatment in order to prevent extrapulmonary dissemination. Antifungal therapy is necessary for any individual with moderate to severe pneumonia, an extrapulmonary site of infection, or for those in a state of immune compromise. Untreated blastomycosis can be associated with mortality rates approaching 60%. Table 4.1 reviews the treatment recommendations of the *IDSA*, including summary dosage regimens.

Table 4.1 Therapy for Blastomycosis (Recommendations of the IDSA)[a]

Clinical form of disease (or special patient group, as indicated)		Therapy
Pulmonary	Mild–moderate	ITC 200 mg 2×/day for 6–12 months
	Moderately severe to severe	Lipid AmB, 3–5 mg/kg/day or AmB-deoxycholate, 0.7–1.0 mg/kg/day for 1–2 weeks then ITC 200 mg 2×/day for 6–12 months
CNS/meningeal		Lipid AmB, 5 mg/kg/day for 4–6 weeks; step-down to: FLC, 800 mg/day; ITC, 200 mg 2–3×/day; or VRC, 200–400 mg/day, 2×/day
Disseminated non-CNS disease	Life-threatening	Lipid AmB, 3–5 mg/kg/day or AmB-deoxycholate 0.7–1 mg/kg/day for 1–2 weeks; step-down to ITC 200 mg 2×/day for 12 months
	Mild–moderate	ITC, 200 mg once or 2×/day for 6–12 months
Immunocompromised		Lipid AmB, 3–5 mg/kg/day or AmB-deoxycholate, 0.7–1 mg/kg/day for 1–2 weeks then suppressive therapy 200 mg 2×/day for 12 months or until immune functions are restored
Pregnant women		Lipid AmB, 3–5 mg/kg/day
Children	Moderate–severe	AmB-deoxycholate, 0.7–1 mg/kg/day; or lipid-AmB, 3–5 mg/kg/day for 1–2 weeks; step-down to ITC, 10 mg/kg/day for 12 months
	Mild–moderate	ITC, 10 mg/kg/day for 6–12 months

[a]Data from Table 2, Chapman et al. (2008). Used with permission from the University of Chicago Press, © 2008.

Acute Pulmonary Blastomycosis

There are no firm guidelines on how to gauge the severity of illness, which should be determined by clinical judgment. The preferred agent to treat mild to moderate disease is ITC for a minimum of 6–12 months, at a dosage of 200–400 mg/day. The duration of therapy past 6 months depends on the resolution of radiographic findings and clinical symptoms. Of 40 patients with blastomycosis treated with 200–400 mg/day of ITC for >2 months (median 6 months, range 3–24 months), 95% were considered cured (Dismukes et al., 1992). Patients intolerant of ITC may be switched to ketoconazole or fluconazole (FLC), agents that are less effective and, in the case of ketoconazole, more potentially toxic. Extended spectrum triazoles, VRC and posaconazole (PSC), offer other alternatives but the record of VRC up to now has had mixed results (Freifeld et al., 2009), and cases treated with PSC are too few to evaluate.

Patients severely ill with blastomycosis are treated with AmB. AmB-deoxycholate (total dose, 11 g) cures without relapse in 70–91% of cases (reviewed by Chapman et al., 2008). When the patient is stabilized on AmB therapy the preference is to step-down to ITC.

Increased mortality in blastomycosis patients is associated with the elderly, *COPD*, and cancer. The most common cause of death is overwhelming pulmonary disease and respiratory failure due to ARDS, often during the early days of therapy. Corticosteroids may be helpful in patients with ARDS. Higher dosing regimens made possible through the use of lipid AmB result in improved outcome.

Disseminated, Non-CNS Blastomycosis

Mild to moderate extrapulmonary disseminated disease (non-CNS) is treated with ITC for 6–12 months with the endpoint after 6 months determined by the resolution of skin lesions, other foci of infection, and subsidence of clinical symptoms, with therapy continuing beyond that point for an additional few months. If disease progresses, or intolerance to ITC occurs, switching to AmB-deoxycholate or lipid AmB is appropriate.

Osteoarticular Blastomycosis

This clinical form is more difficult to treat requiring at least 1 year of therapy with ITC.

CNS Blastomycosis

Rates of dissemination to the CNS, in immune-normal patients, are ~5% but are much higher in persons living with AIDS. Lipid AmB is preferred because of the length of therapy, up to 6 weeks, and is followed by step-down to an extended spectrum azole for 1 year. VRC has excellent penetration into the CNS. FLC at 800 mg/day is an option. Surgery may be important in the treatment plan, for example, draining of an epidural abscess or of other lesions. Two patients who suffered exacerbations of disseminated blastomycosis after completion of AmB and ITC therapy were clinically cured after switching to oral VRC at 200–300 mg 2×/day for a total of 13 months of therapy (Borgia et al., 2006). A third patient with CNS blastomycosis had multiple cerebral abscesses. One lesion was resected and the patient was treated with AmB for 14 days and then was switched to VRC for a total of 13 months with complete resolution of symptoms and improved neuroimaging. VRC has excellent penetration across the blood–brain barrier, unlike ITC which is largely plasma-protein bound. PSC may have some utility but has not been adequately studied. The echinocandin class of antifungals does not have activity against *B. dermatitidis*.

Immunosuppressed Patients

Primary therapy with lipid AmB is recommended for 1–2 weeks, with step-down to ITC after that. Maintenance on ITC is to be indefinite or for the duration of immunosuppression.

In Pregnancy or in Children

Pregnant women with blastomycosis should be treated with lipid AmB, avoiding azoles owing to their potential teratogenicity. Children with severe blastomycosis should receive AmB or lipid AmB as primary therapy and upon a positive clinical response switched to ITC for 1 year. Mild to moderate blastomycosis in children may be treated with ITC.

Determination of Serum Concentrations of Itraconazole

Because of its uncertain bioavailablity, measurements should be taken after a 2 week course of ITC therapy to assure that adequate blood levels are being maintained. Please see Chapter 3A, Section 3A.4.2, ITC: Uncertain Bioavailability for further information on its bioavailability.

4.14 LABORATORY DETECTION, RECOVERY, AND IDENTIFICATION

Direct Examination

Specimens include early morning sputum, aspirates, scrapings from skin lesions, and biopsy specimens. Bronchoscopy may be needed where patients do not produce sputum.

Skin Lesions The appearance is one of central healing with microabscesses at the periphery. A pus specimen may be obtained by nicking the top of a microabscess with a scalpel. The purulent material is aspirated and examined in a KOH prep.

Sputum Specimens A purulent inflammatory component is typical of *B. dermatitidis* lung infection. Bloody or mucoid specimens may be first treated with Mucolyse®—dithiothreitol in phosphate buffer (Pro-Lab Diagnostics, Richmond Hill, Canada).

When available, sputum specimens should be smeared and cultured on more than one occasion to increase yield.

Wet Mount A KOH prep of patient material viewed microscopically shows *B. dermatitidis* as a large (10–15 μm diam) thick-walled yeast with a single bud and a wide isthmus between mother and daughter cells. Addition of Calcofluor white to KOH increases the chances of detection. This observation usually is sufficient for the physician to begin treatment; however, culture of the specimen is necessary for definitive identification. The yield of positive results of KOH preps of a single sputum specimen was 25% (Martynowicz and Prakash, 2002). Yield can be improved by direct examination of tracheal secretions.

Histopathology

Cytologic preparations of sputum, or bronchial washings of patients with pulmonary nodules can be yield a positive result in 70–93% of cases (reviewed by Martynowicz and Prakash, 2002). Such samples may be requested to rule out malignancy. Yeast forms of *B. dermatitidis* are multinucleate with a thick "doubly contoured" refractile wall, best seen in H&E stain (Fig. 4.3). Figure 4.4 illustrates the broad-based bud of *B. dermatitidis* yeast forms compared with other monopolar budding yeasts.

Culture

Growth of *B. dermatitidis* in culture of sputum specimens takes an average of 5–30 days of incubation, so culture tubes should be kept up to 5 weeks. The yield of positive results for sputum cultures varies; in Case Presentation 1, the Eagle River outbreak, sputum specimens for 9 of 48 (19%) of proven or probable cases were culture positive; whereas, 25 of 31 (81%) of patients with confirmed pulmonary blastomycosis were positive in the review of Martynowicz and Prakash (2002). This range underlines the need to attempt to culture the fungus from more than one sputum specimen from each patient.

Culture medium for dimorphic pathogens should contain an antibiotic to inhibit bacteria and cycloheximide to

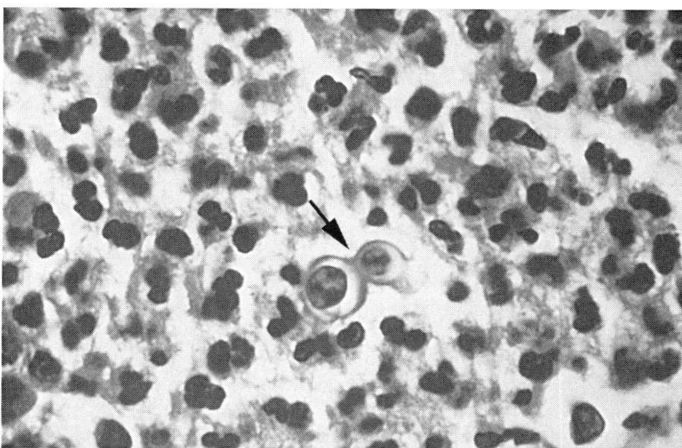

Figure 4.3 Histopathology of a blastomycosis skin lesion. Budding yeast of *Blastomyces dermatitidis* (*arrow*) surrounded by neutrophils. Multiple nuclei are visible.
Source: Dr. Edwin P. Ewing, Jr., CDC Public Health Image Library (PHIL).

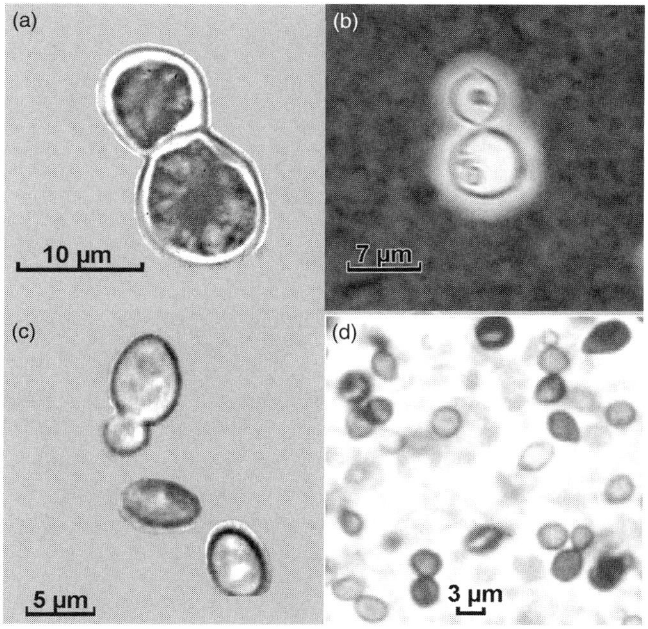

Figure 4.4 *Blastomyces dermatitidis* yeast forms compared with other monopolar budding yeasts. (a) *Blastomyces dermatitidis* yeast forms have a broad-based bud. The yeast size ranges from 10 to 15 μm in diameter. (b) *Cryptococcus neoformans* yeast are round or ellipsoid. The entire cell diameter including capsule may be 15–20 μm with the cell diameter of 7 μm. Yeast cells vary considerably in both size and shape. The buds appear "pinched off" from the mother cell. (c) *Candida albicans* yeast are 3–5 μm in diameter with single buds. (d) *Histoplasma capsulatum* yeast forms are ovoid and very small, with dimensions of 2–4 μm in diameter. Budding cells have a narrow base.
Source: E. Reiss.

inhibit rapid-growing saprobic fungi. Addition of cycloheximide applies to cultures incubated at 30°C, because the yeast form that grows at 35–37°C is inhibited by cycloheximide. A separate Petri plate, without cycloheximide, also is desired in order to recover any opportunist that could conceivably cause disease. SDA-Emmons is an appropriate primary medium for culturing clinical specimens for *B. dermatitidis* growing in the mold form at 30°C. Enriched media (e.g., BHI or SABHI) are recommended primary media to improve recovery of *B. dermatitidis* and *H. capsulatum*, also containing an antibiotic and cycloheximide (when incubated at 30°C). Please see Table 2.2, in Chapter 2 for information about these media. When the culture specimen is obtained from a normally sterile site such as blood or bone marrow, it can be plated onto media without antibiotics.

Safety Note: Petri plate cultures suspected of containing *B. dermatitidis* or other primary mold pathogens should be manipulated only in a well-ventilated and certified biological safety cabinet. Petri plates should be sealed with gas-permeable "shrink-seals" (Myco-seals®, Hardy Diagnostics Inc.) to prevent the drying out of agar during the prolonged incubation period.

Direct smear, cytology, and serology, along with culture facilitate the diagnosis of blastomycosis. Patient management can be affected by the long periods necessary for culture results and less than ideal yields of individual direct examination methods. This is especially true where there is lack of skin lesions and the differential diagnosis includes tuberculosis, bacterial pneumonia, and bronchogenic carcinoma. The interventions for the former, systemic antibiotics, and for the latter, surgical resection, could be spared if there is increased index of suspicion of blastomycosis coupled with appropriate laboratory work-up.

Serologic testing also has a long latent period and less than ideal sensitivity (please see below, "Serology." For these reasons it is prudent to conduct multiple samplings of sputum, other respiratory specimens, or exudates from cutaneous lesions for direct examination and culture.

Colony Morphology

Blastomyces dermatitidis is a *dimorphic* fungus that grows in the mycelial (mold) form at room temperature and as yeast when incubated at 37°C on enriched medium. The mycelial form grows slowly; a colony becoming visible in 5–10 days, but may take up to 30 days. The colony usually is white and cottony, but may become light tan with age. Tan colonies have more abundant sporulation than the white colonies.

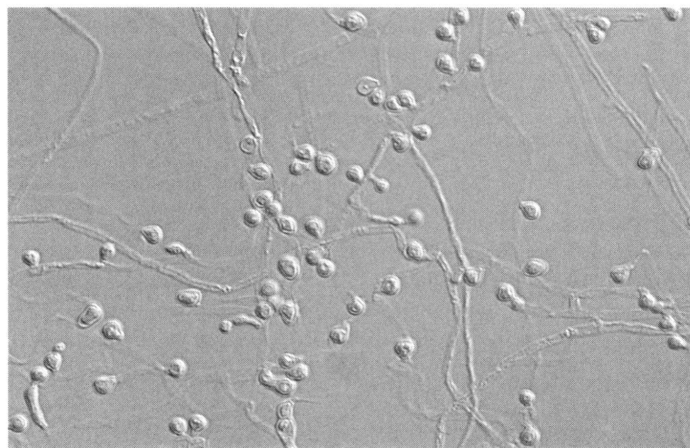

Figure 4.5 Aleurioconidia of *Blastomyces dermatitidis* mold form arise directly from the hyphae (holoblastic conidiogenesis). SDA at 25°C (DIC).
Source: Used with permission from Dr. Arvind A. Padhye, CDC.

Microscopic Morphology

The mycelial form of *B. dermatitidis* produces small, 10 μm diameter, smooth walled or slightly echinulate pear-shaped conidia (solitary *aleurioconidia*). They are attached directly to the supporting hyphae (sessile) or on the sides or tips of narrow conidiophores (Fig. 4.5). These morphologic structures are not unique to *B. dermatitidis* and the fungus must be converted to the yeast form for verification. About 25% of isolates will not sporulate.

Dimorphism

The conversion from mold to yeast form is a practical laboratory diagnostic method for *B. dermatitidis*. Conversion is effected by transferring hyphae from the mycelial colony to an enriched medium such as *BA* or Middlebrook's 7H10 Agar and incubation at 37°C. Within 3–4 days wrinkled yeast-like colonies develop that are a dirty white to brown color. Serial subculture may be necessary for production of the yeast-like colony. Yeast forms are large, with a refractile cell wall, and are multinucleate. Budding results in a single broad-based bud. Specialists working with *B. dermatitidis* have commented that medium with chloramphenicol (added as a bacteriostat) inhibits the growth of yeast form *B. dermatitidis* (Kane, 1984). Moreover, BHI, a common enriched medium, does not provide good nutritional support for conversion to the yeast form. Kane (1984) developed medium "KT," which stimulates conversion to the yeast form at 37°C within 72 h, and permits partial conversion to the yeast form at 26°C. KT medium is a semisynthetic medium.

Nucleic Acid Probe Test

Identify the young mycelial colony with the gene probe "AccuProbe" (Gen-Probe Inc.) but note that this test cross-hybridizes with *P. brasiliensis*.

Serology

The immunodiffusion (ID) test is the most reliable serologic test for detecting antibodies in blastomycosis. The ID test with positive antiserum controls is available from Immunomycologics, Inc. The latent period before precipitins become obvious is 4–6 weeks after the first appearance of symptoms, and the percent positive serology in blastomycosis in one series of proven cases was 40% (Martynowicz and Pradash, 2002). The immunodominant antigen is the "BAD1," a surface-located adhesin. *Blastomyces dermatitidis* sera react with common fungal galactomannan antigens in several other fungal pathogens, complicating serologic assessment. The BAD1 antigen, however, appears to be specific and primary binding assays (i.e., EIA tests, to measure antibodies against BAD1 are needed).

Antigen Detection

This is accomplished with an EIA using urine specimens, but may also be used with CSF, BAL, and serum (Durkin et al., 2004). The antigen being detected is heat stable and is most likely a galactomannan polysaccharide, because this test is cross reactive with clinical specimens from histoplasmosis patients. Histoplasmosis can be diagnosed with the HPA (*Histoplasma* polysaccharide antigen) EIA in similar types of clinical specimens. The overlap in the histoplasmosis and blastomycosis endemic areas is another possible factor to consider in evaluating a positive antigenemia or antigenuria result.

Diagnosis of blastomycosis, using the antigenuria EIA, was assessed in 42 patients with blastomycosis and in 39 healthy volunteers (Durkin et al., 2004). Antigen was detected in the urine of 93% of blastomycosis patients including 25 of 28 with disseminated blastomycosis (89%) and all 14 cases of pulmonary blastomycosis. No positive reactions were seen in urine of healthy volunteers. Crossreactions occurred in 96% of patients with histoplasmosis, 100% of patients with paracoccidioidomycosis, 70% of patients with penicilliosis due to *P. marneffei*, 3% of patients with cryptococcosis, and 1% of patients with aspergillosis. This test has utility in the differential diagnosis, considering that blastomycosis is one of the last etiologic agents to be considered in community acquired pneumonia or in one of the many possible sites of *B. dermatitidis* extrapulmonary dissemination.

GENERAL REFERENCE FOR BLASTOMYCOSIS

Al-Doory Y, DiSalvo AF (eds.), 1992. *Blastomycosis*. Plenum Medical Book Co., New York.

SELECTED REFERENCES FOR BLASTOMYCOSIS

Armstrong CW, Jenkins SR, Kaufman L, Kerkering TM, Rouse BS, Miller GB Jr, 1987. Common-source outbreak of blastomycosis in hunters and their dogs. *J Infect Dis* 155: 568–570.

Bally GG, Robertson VJ, Neill P, Garrido P, Levy LF, 1991. Blastomycosis in Africa: Clinical features, diagnosis and treatment. *Clin Infect Dis* 13: 1005–1008.

Bialek R, Ibricevic A, Fothergill A, Begerow D, 2000. Small subunit ribosomal DNA sequence shows *Paracoccidioides brasiliensis* closely related to *Blastomyces dermatitidis*. *J Clin Microbiol* 38: 3190–3193.

Borgia SM, Fuller JD, Sarabia A, El-Helou P, 2006. Cerebral blastomycosis: A case series incorporating voriconazole in the treatment regimen. *Med Mycol* 44: 659–664.

Bradsher RW, Chapman SW, Pappas PG, 2003. Blastomycosis. *Infect Dis Clin North Am* 17: 21–40.

Brandhorst TT, Wuthrich M, Warner T, Klein B, 1999. Targeted gene disruption reveals an adhesin indispensable for pathogenicity of *Blastomyces dermatitidis*. *J Exp Med* 189: 1207–1216.

Chapman SW, 2005. Chapter 263, *Blastomyces dermatitidis*, pp. 3026–3040, *in:* Mandell GL, Bennett JE, Dolin R (eds.), *Mandell, Bennett, & Dolin's Principles and Practice of Infectious Diseases*, 6th ed., Vol. 2. Churchill Livingstone, An Imprint of Elsevier, Philadelphia, PA.

Chapman SW, Dismukes WE, Proia LA, Bradsher RW, Pappas PG, Threlkeld MG, Kauffman CA, 2008. Clinical practice guidelines for the management of blastomycosis: 2008 update by the Infectious Diseases Society of America. *Clin Infect Dis* 46: 1801–1812.

Chick EW, 1971. North American Blastomycosis, pp. 465–506, *in*: Baker RD (ed.), *Human Infection with Fungi, Actinomycetes and Algae*. Springer Verlag, New York.

Chu JH, Feudtner C, Heydon K, Walsh TJ, Zaoutis TE, 2006. Hospitalizations for endemic mycoses: A population-based national study. *Clin Infect Dis* 42: 822–825.

Craig MW, Davey WN, Green RA, 1970. Conjugal blastomycosis. *Am Rev Respir Dis* 102: 86–90.

Crampton TL, Light RB, Berg GM, Meyers MP, Schroeder GC, Hershfield ES, Embil JM, 2002. Epidemiology and clinical spectrum of blastomycosis diagnosed in Manitoba Hospitals. *Clin Infect Dis* 34: 1310–1316.

Deepe GS Jr, Wüthrich M, Klein BS, 2005. Progress in vaccination for histoplasmosis and blastomycosis: Coping with cellular immunity. *Med Mycol* 43: 381–389.

Denton JF, Di Salvo AF, 1964. Isolation of *Blastomyces dermatitidis* from natural sites at Augusta, Georgia. *Am J Trop Med Hyg* 13: 716–722.

Denton JF, McDonough ES, Ajello L, Ausherman RJ, 1961. Isolation of *Blastomyces dermatitidis* from soil. *Science* 133: 1126–1127.

Dismukes WE, Bradsher RW, Cloud GC, Kauffman CA, Chapman SW, George RB, Stevens DA, Girard WM, Saag MS, Bowles-Patton C, 1992. Itraconazole therapy for blastomycosis and histoplasmosis. NIAID Mycoses Study Group. *Am J Med* 93: 489–497.

Durkin M, Witt J, Lemonte A, Wheat B, Connolly P, 2004. Antigen assay with the potential to aid in diagnosis of blastomycosis. *J Clin Microbiol* 42: 4873–4875.

Fang W, Washington L, Kumar N, 2007. Imaging manifestations of blastomycosis: A pulmonary infection with potential dissemination. *Radiographics* 27: 641–655.

Freifeld A, Proia L, Andes D, Baddour LM, Blair J, Spellberg B, Arnold S, Lentnek A, Wheat LJ, 2009. Voriconazole use for endemic fungal infections. *Antimicrob Agents Chemother* 53: 1648–1651.

Gadkowski LB, Stout JE, 2008. Cavitary pulmonary disease. *Clin Microbiol Rev* 21: 305–333.

Gnann JW Jr, Bressler GS, Bodet CA III, Avent CK, 1983. Conjugal blastomycosis. *Ann Intern Med* 98: 48–49.

Gray NA, Baddour LM, 2002. Cutaneous inoculation blastomycosis. *Clin Infect Dis* 34: e44–e49.

Guarro J, Cano J, 2002. Phylogeny of Onygenalean fungi of medical interest. *Stud Mycol* 47: 1–4.

Hogan LH, Klein BS, 1994. Altered expression of surface α-1,3-glucan in genetically related strains of *Blastomyces dermatitidis* that differ in virulence. *Infect Immun* 62: 3543–3546.

Kane J, 1984. Conversion of *Blastomyces dermatitidis* to the yeast form at 37°C and 26°C. *J Clin Microbiol* 20: 594–596.

Kaplan W, Clifford MK, 1964. Blastomycosis I. A review of 198 collected cases in Veterans Administration hospitals. *Am Rev Respir Dis* 89: 659–672.

Kinasewitz GT, Penn RL, George RB, 1984. The spectrum and significance of pleural disease in blastomycosis. *Chest* 86: 580–584.

Klein BS, Vergeront JM, Weeks RJ, Kumar UN, Mathai G, Varkey B, Kaufman L, Bradsher RW, Stoebig JF, Davis JP, et al., 1986. Isolation of *Blastomyces dermatitidis* in soil associated with a large outbreak of blastomycosis in Wisconsin. *N Engl J Med* 314: 529–534.

Lekkala M, LeVine AM, Linke MJ, Crouch EC, Linders B, Brummer E, Stevens DA, 2006. Effect of lung surfactant collectins on bronchoalveolar macrophage interaction with *Blastomyces dermatitidis*: Inhibition of tumor necrosis factor alpha production by surfactant protein D. *Infect Immun* 74: 4549–4556.

Lee D, Eapen S, Van Buren J, Jones P, Baumgardner DJ, 2006. A young man who could not walk. *Wisconsin Med J* 105: 58–61.

Lemos LB, Baliga M, Guo M, 2002. Blastomycosis: The great pretender can also be an opportunist. Initial clinical diagnosis and underlying diseases in 123 patients. *Ann Diagn Pathol* 6: 194–203.

Martynowicz MA, Prakash UBS, 2002. Pulmonary blastomycosis: An appraisal of diagnostic techniques. *Chest* 121: 768–773.

Morris SK, Brophy J, Richardson SE, Summerbell R, Parkin PC, Jamieson F, Limerick B, Wiebe L, Ford-Jones EL, 2006. Blastomycosis in Ontario, 1994–2003. *Emerg Infect Dis* 12: 274–279.

Musial CE, Wilson WR, Sinkeldam IR, Roberts GD, 1987. Recovery of *Blastomyces dermatitidis* from blood of a patient with disseminated blastomycosis. *J Clin Microbiol* 25: 1421–1423.

Proctor ME, Davis JP, 1996. Blastomycosis—Wisconsin, 1986–1995. *MMWR Morb Mortal Wkly Rep* 45: 601–603.

Sarosi GA, Eckman MR, Davies SF, Laskey WK, 1979. Canine blastomycosis as a harbinger of human disease. *Ann Intern Med* 91: 733–775.

Sheflin JR, Campbell JA, Thompson GP, 1990. Pulmonary blastomycosis: Findings on chest radiographs in 63 patients. *AJR Am J Roentgenol* 154: 1177–1180.

Weese JS, Peregrine AS, Armstrong J, 2002. Occupational health and safety in small animal veterinary practice: Part I—nonparasitic zoonotic diseases. *Can Vet J* 43: 631–636.

Wüthrich M, Finkel-Jimenez B, Brandhorst TT, Filutowicz HI, Warner T, Klein BS, 2006. Analysis of non-adhesive pathogenic mechanisms of BAD1 on *Blastomyces dermatitidis*. *Med Mycol* 44: 41–45.

Wüthrich M, Klein B, 2005. Lung Immunity to *Blastomyces dermatitidis* Infection, pp. 113–133, *in*: Fidel PL Jr, Huffnagle GB (eds.), *Fungal Immunology: From an Organ Perspective*. Springer Science and Business Media, New York.

QUESTIONS

The Answer Key to multiple choice questions may be found at the end of the book.

1. A sputum specimen comes in the laboratory from a suspected case of fungal pneumonia. Which of the following is true in setting up the specimen for direct examination to include *B. dermatitidis* in the diagnosis?
   A. A portion of the specimen should be reserved for culture.
   B. A single sputum specimen is usually sufficient for diagnosis.
   C. Calcofluor is not helpful because the yeast forms are very large.
   D. Thick bloody sputum specimens can be used directly.

2. The microscopic morphology (tease mount) of *B. dermatitidis* growing at 30°C will reveal :
   1. Approximately 25% of isolates will not sporulate.
   2. It is not necessary to convert the isolate to the yeast form to confirm the identification.
   3. Solitary pear-shaped aleurioconidia are attached directly to hyphae or at the sides or tips of narrow conidiophores.
   4. The aleurioconidia are diagnostic for *B. dermatitidis* and are not found in other molds.

   Which are correct?
   A. 1 + 2
   B. 1 + 3
   C. 2 + 3
   D. 2 + 4

3. A hyaline mold is isolated from sputum but failed to sporulate on SDA-Emmons agar medium. Which of the following would be an appropriate test to identify it?
   A. Attempt conversion to the yeast form on enriched medium
   B. Gen-Probe
   C. PCR-sequencing
   D. All are correct

4. The characteristic(s) of the yeast form of *B. dermatitidis* in tissue is (or are):
   A. Broad-based bud
   B. Multipolar budding
   C. One nucleus per cell
   D. Polysaccharide capsule
   E. All are correct

5. What lower animal is the sentinel species for where *Blastomyces dermatitidis* might occur in the environment?
   A. Armadillo
   B. Bat
   C. Cat
   D. Dog

6. In Case Presentation 1, the Eagle River outbreak, which of the following was the most sensitive diagnostic indicator of blastomycosis?

A. Histopathology

B. Positive enzyme immunoassay for antibodies

C. Positive immunodiffusion test for antibodies

D. Positive sputum culture

7. This disease of humans and dogs is caused by a dimorphic primary systemic fungal pathogen. The endemic area is in the central United States. It is rarely isolated from soil. The extent of clinical exposure in the community is not known because there has never been a skin test survey. The disease has a predilection to disseminate from the lungs to the skin. KOH preparations from skin exudates or other infected sites show a large yeast form with a broad-based bud. The yeast forms are usually too large for endophagocytosis by macrophages. The etiologic agent is:

A. *Blastomyces dermatitidis*

B. *Coccidioides immitis*

C. *Histoplasma capsulatum*

D. *Paracoccidioides brasiliensis*

8. Compare the therapy for pulmonary blastomycosis and central nervous system blastomycosis. What therapy is recommended for moderate acute, chronic pulmonary, or severe acute pulmonary blastomycosis?

9. How are the boundaries of the endemic area for blastomycosis estimated? In this regard, discuss the role of environmental sampling, human and animal cases, and skin testing.

10. How does knowledge of the teleomorph form of *Blastomyces dermatitidis* help to explain its position in classification? Discuss the relatedness of *B. dermatitidis* to other dimorphic environmental molds with endemic areas in North America.

Chapter 5

Coccidioidomycosis

5.1 COCCIDIOIDOMYCOSIS-AT-A-GLANCE

- *Introduction/Disease Definition.* A community acquired pneumonia that is acute, self-limited, or progressive. Also known as "cocci," "Valley Fever," and desert rheumatism.
- *Etiologic Agent.* Soil-dwelling molds, *Coccidioides immitis, C. posadasii*.
- *Geographic Distribution.* Major area for *C. immitis*—Central Valley of California; *C. posadasii* occurs in Arizona near Phoenix, in Mexico, and in smaller foci in Central America and South America.
- *Ecologic Niche.* Lower Sonoran life zone, semiarid, very hot summers, little rain, few freezes.
- *Epidemiology.* Majority of infections are subclinical, 40% have symptoms ("flu," "desert rheumatism"), 2% develop chronic pulmonary forms, and 0.5–1% develop extrapulmonary disseminated forms.
- *Risk Groups/Factors.* Persons with AIDS, archeologists, diabetics, elderly retirees, farmers, the immunosuppressed, military personnel on maneuvers, pregnant women.
- *Transmission.* Disturbance of desert soil or dust storms; inhaled arthroconidia in dusts; handling materials shipped from endemic areas.
- *Determinants of Pathogenicity.* Dimorphism—the tissue form consists of spherules+endospores. The spherule outer wall lipids and glycoprotein are pathogenic factors.
- *Clinical Forms.* Acute or chronic pulmonary; extrapulmonary—skin, bones, meninges.
- *Therapy.* Most recover without therapy. Fluconazole (FLC), 400 mg/day; amphotericin B (AmB) for treatment failures or rapid progression. Meningitis requires higher doses, intrathecal AmB, and consideration of combined therapy with an azole and AmB.
- *Laboratory Detection, Recovery, and Identification.* Culture early morning sputum + genetic identification with AccuProbe®; serology—CF test (titer is prognostic = "VF titer"); ID test—diagnostic.

5.2 INTRODUCTION/DISEASE DEFINITION

Coccidioidomycosis is an acute, self-limited, or progressive community acquired pneumonia caused by the ascomycetous, soil-dwelling molds, *Coccidioides immitis* and *C. posadasii. Coccidioides* species produce easily dispersed arthroconidia, which, once inhaled, undergo dimorphic change in the host to form large, spherical stuctures (approximately 10–80 μm diameter) called spherules containing many endospores. Disease heals without antifungal therapy in the great majority of patients. In a small number of patients chronic fibrocavitary disease develops, requiring treatment. Extrapulmonary dissemination is uncommon and sites are skin, meninges, bones, and joints. Coccidioidal meningitis is the third most frequent cause of granulomatous meningitis behind tuberculosis and cryptococcosis.

- Ninety-five percent of initial pulmonary infections resolve without specific therapy. Of all infected individuals, 60% have subclinical disease with conversion of the *coccidioidin* skin test as the only clinical sign.
- Approximately 40% of those infected with *Coccidioides* species develop a mild febrile to moderately severe respiratory disease. Most disease will resolve

Fundamental Medical Mycology, First Edition. By Errol Reiss, H. Jean Shadomy and G. Marshall Lyon, III.
© 2012 Wiley-Blackwell. Published 2012 by John Wiley & Sons, Inc.

without specific therapy but these patients should be followed for a year or more.

- Approximately 5–10% of infections result in residual pulmonary nodules or thin-walled cavities.
- Approximately 0.5–1.0% of infections result in chronic pulmonary disease or extrapulmonary dissemination.

Recovery from a primary infection with *Coccidioides* species confers immunity to a new primary infection, but if there is a persistent nidus of infection, such as may occur in a pulmonary nodule or cavity, *there is risk of endogenous reactivation should the person become immunosuppressed*.

Coccidioidomycosis is a New World mycosis endemic to North America with smaller endemic areas in Central and South America. *Coccidioides immitis* is restricted in the United States to Southern California; *C. posadasii* is the etiologic agent east and south of the Sierra Nevada range. The spectrum of disease caused by the two species is the same. Persons become infected as a result of (i) living and working, (ii) visiting, or (iii) exposure to artifacts or fomites contaminated with infected soil from the endemic areas. Coccidioidomycosis is termed "Valley Fever", referring to the Central Valley of California, the San Joaquin Valley, which is the center of the endemic area in that state. Most reported cases, however, occur in the vicinity of Phoenix, Arizona. Coccidioidomycosis also is known as *desert rheumatism* because of the combination of fever, joint pain, and a skin rash that accompanies acute infection (please see Section 5.10, Determinants of Pathogenicity and Pathogenesis—Allergic Findings).

5.3 CASE PRESENTATIONS

Case Presentation 1. Lung Infection in a 57-yr-old Male Filipino Farm Worker (Bayer et al., 1979)

History

A 57-yr-old male Filipino farm worker was admitted to Harbor General Hospital, Torrance, California (near Los Angeles) with a 3 week history of fever, chills, chest pain, and 4.5 kg weight loss.

Clinical and Laboratory Findings

The physical exam showed fever, 40°C, with a pulse of 120 beats/min. There were rales, diminished breath sounds, absent vocal, and tactile fremitus at base of the left lung. The chest radiograph showed an extensive left lower lobe infiltrate and a large left pleural effusion. The fungal skin test was negative, but was positive with PPD. Skin tests with *Candida* and mumps antigens were also positive. On admission the WBC count was 14,000/ mm^3, with 80% PMNs. *Coccidioides* species was cultured from sputum. The coccidioidomycosis complement fixation (CF) antibody titer was 1:8. Biopsy examination of liver, bone marrow, and cultures taken from these sites were negative. The lumbar puncture results were normal.

During the next 3 weeks, fever and weight loss continued; sputum cultures remained positive; and chest X-rays showed no improvement. The serum CF titer rose to 1:64.

Diagnosis

The diagnosis was pulmonary coccidioidomycosis caused by *Coccidioides immitis*.

Therapy and Clinical Course

Amphotericin B therapy was begun and he received 2.4 g in a 4 month period. Sputum cultures became negative, the fungal skin test converted to positive, and 1 year after discharge the CF titer subsided to 1:4.

Comments

This case differed from typical acute pulmonary coccidioidomycosis because of a longer duration, >6 weeks, and the high-titered CF antibodies. Titers $\geq$ 1:16 create apprehension about extrapulmonary dissemination. This patient's extensive unresolving pulmonary infiltrates also are not typical. Clinicians should consider the diagnosis of coccidioidomycosis in persons with undiagnosed respiratory illness, community-acquired pneumonia, or syndromes that could represent disseminated coccidioidomycosis for those who reside in, or have traveled to, areas where the disease is endemic, or who have had occupational exposure to *Coccidioides* species. Laboratory personnel should be reminded of necessary safety precautions when handling cultures suspected to contain *Coccidioides* species (please see Section 5.14. Laboratory Detection, Recovery, and Identification).

Case Presentation 2. Outbreak of Coccidioidomycosis Among Archeology Students in Northern California (Werner et al., 1972)

History

During the summer of 1970, 103 archeology students were excavating Indian ruins of the Northern Maidu civilization dating from A.D. 1400 near Chico, California, 113 km north of any previously recognized endemic area. The site was in Butte County California, 15 km northeast of Chico in the foothills of the Sierra range, elevation 75 m. Only two cases of coccidioidomycosis were reported from Butte County in the previous decade.

Students were attending a 5 week archeology summer course beginning on June 15, cosponsored by Queens College, City University of New York, and Chico State College. They excavated an area 100 m × 150 m to a depth of up to 1.5 m or until they reached soil devoid of beads, bones, and arrowheads. They sifted dry dirt through screens to filter artifacts. The resulting dust caused coughing spells, but no precautions were taken and masks were not worn.

Epidemic Curve, Clinical and Laboratory Findings

On the basis of chest pains and cough, 61 of 94 (65%) of the archeology students who responded to the questionnaire became ill during the period June 15 to August 31. The peak time of onset of illness was 3.5 weeks from the first soil exposure. Seventy-seven percent of the cases were symptomatic and slightly more than half had skin lesions. Eighty-two students were from New York and most had not previously been in an endemic area.

The first illness was reported on June 28. After that, increasing numbers of students were seen at the Chico State College Student Health Center. Coccidioidomycosis was first suspected on July 6th after a chest radiograph from a 19-yr-old woman showed extensive bilateral pneumonitis with hilar adenopathy. She presented with fever, shaking chills, headache, malaise, myalgia, cough, chest pain, and a generalized *maculopapular* rash. The WBC count was between 6000 and 11,000/mm^3. A coccidioidin skin test applied on July 6 was negative, but when repeated on July 16 converted to positive.

A rash similar to that in the index case appeared at the outset of disease in slightly more than half of the participants with confirmed coccidioidomycosis. Ten patients reported symptoms compatible with *erythema nodosum* (please see Section 5.10, Determinants of Pathogenicity and Pathogenesis—Allergic Findings). The diagnosis was made by coccidioidin skin tests or serologic tests. Of 61 persons reporting illness, serologic tests were positive in 20 of 51 studied (39%). Coccidioidin skin tests were positive in 24 of 53 (45%), indicating that skin testing is at least as sensitive as serologic tests.

Clinical Course

At least four patients were hospitalized, one for almost 4 weeks, but neither surgery nor amphotericin B was required.

Environmental Sampling

Soil samples obtained from the excavation site and nearby rodent burrows were planted on SDA. After 14 days at room temperature a single positive sample grew from a rodent burrow source. The identification was confirmed after intraperitoneal injection of a supernatant of soil suspension in mice and recovery of endosporulating spherules from lesions in the thoracic and abdominal cavities. (*Note*: The use of mouse passage is not a routine method and is rarely used in the present day).

Comments

The appearance of a generalized pruritic maculopapular rash in patients with influenza-like symptoms coming from endemic areas should alert physicians to the possibility of coccidioidomycosis. A majority of the college students who fell ill developed skin rashes.

How would C. immitis be identified in laboratory in the current day? (Please see Section 5.14. Laboratory Detection, Recovery, and Identification)

5.4 DIAGNOSIS

Community acquired pneumonia, with influenza-like symptoms, with or without a skin rash and joint pain in a person relating a history of travel or residence in the endemic area, or handling materials from those areas, increases suspicion for coccidioidomycosis. Chest radiographs are not specific. Laboratory tests consist of direct examination and culture of sputum, and serology. The complement fixation titer also known as the Valley Fever titer is diagnostic and prognostic.

Differential Diagnosis

Community-acquired bacterial pneumonia, tuberculosis, or other pulmonary mycoses are possibilities.

5.5 ETIOLOGIC AGENTS

Coccidioides immitis GW Stiles, *in* Rixford & Gilchrist, *Rep. Johns Hopkins Hosp* 1: 243 (1896). Position in classification: *Onygenaceae, Onygenales, Ascomycota* (an anamorphic species).

Coccidioides posadasii MC Fisher, GL Koenig, TJ White & JW Taylor (2002). Position in classification: *Onygenaceae, Onygenales, Ascomycota* (an anamorphic species).

The two causative agents, *C. immitis* and *C. posadasii*, are phenotypically identical or very similar and the spectrum of disease they cause appears to be the same. (Please see Section 5.6. Geographic Distibution/Ecologic Niche—The Two *Coccidioides* Species). *Coccidioides* species are filamentous, usually white, fluffy, rapid-growing molds. The hyphae fragment into intercalary barrel-shaped arthroconidia. The hydrophobic cell wall remnants from empty cells produce flange-shaped structures which aid in aerial dispersion. *These arthroconidia are highly infectious propagules*.

Phenotype and genotype analyses were conducted of *C. immitis* compared with the *Malbranchea* states of *Uncinocarpus reesei, Auxarthron zuffianum*, and four additional *Malbranchea* species (*M. albolutea, M. dendritica, M. filamentosa, M. gypsea*) (Pan et al., 1994). *Malbranchea* species are the anamorph forms of which *Uncinocarpus* and *Auxarthron* species are the teleomorphs. *Coccidioides immitis* taxonomic classification was clarified by alignments of chitin synthase, *CHS1*, and 18S (small subunit or SSU) rDNA. The phylogenetic tree places *C. immitis* among the *Onygenales*, a large order of ascomycetes, in the family *Onygenaceae*. Its closest neighbor is *U. reesei*. Also nearby are *Auxarthron* and *Malbranchea* species.

Both of these fungi sporulate via production of intercalary arthroconidia, similar to *Coccidioides* species. Other phenotypic characteristics were considered, only *C. immitis* and *U. reesei* release amber colored exudates into the culture medium, which are identical upon UV/visible absorption spectral analysis. Human sera from coccidioidomycosis patients crossreact with antigens from *U. reesei* and *Malbranchea* species. Especially noteworthy, a 120 kDa ß-glucanase antigen is produced abundantly in extracts of *C. immitis* and *U. reesei*.

The other major primary dimorphic fungal pathogens, *Histoplasma capsulatum, Blastomyces dermatitidis*, and *Paracoccidioides brasiliensis*, are classed in a separate smaller, highly related family, the *Ajellomycetaeceae*, also part of the *Onygenales* (Untereiner et al., 2004).

Dimorphism

Once inhaled, the propagules undergo *morphogenesis*, usually in the lungs, producing *spherules* (*definition:* round, approximately 10–80 μm diameter sporangia) inside of which numerous endospores develop.

Development of the Spherule Form

See Fig. 5.1. Arthroconidia are multinucleate, usually as dikaryons. In a suitable environment, such as the alveoli, an individual arthroconidium swells and becomes round and uninucleate. The single nucleus is 2–3× larger than before, but it is unknown whether this occurs by loss of all but one nucleus, or by karyogamy. Later, the round cell enlarges and undergoes multiple nuclear divisions (I–P in Fig. 5.1). These are synchronous judging from mitotic figures present in all nuclei, as visualized by transmission EM (Huppert et al., 1982). At some point between "O" and "Q" (in Fig. 5.1) nuclear replication stops and segmentation of protoplasm begins. Segmentation begins with invagination of the two innermost cell wall layers, and progresses from a primary plane to secondary and successive planes (Q, R) resulting in the production of endospores. Before the spherule ruptures (S), endospores are uninucleate and are clustered in packets held by a membrane. Endospores escape upon rupture of the spherule. The sac-like membrane supporting clusters stretches and twists into fibrils between endospores as they begin to separate (not shown in Fig. 5.1). Endospores often are carried to other areas of the body and, once there, begin to form new spherules (Fig. 5.1).

5.6 GEOGRAPHIC DISTRIBUTION/ECOLOGIC NICHE

North American Endemic Areas

The map in Fig. 5.2 depicts the endemic areas in California, Arizona, Texas, and Mexico (Mojave, Sonoran, and Chihuahuan deserts), extending into parts of Central and South America. The California epicenter is the San Joaquin Valley in King and Tulare counties. In Arizona the highly endemic areas include Tucson and Phoenix.

Habitat

The habitat is the lower Sonoran life zone, semiarid with alkaline and sandy soil. The weather pattern is one of extremely hot summers, few winter freezes, and 10–50 cm of rainfall/year. Vegetation in the ecosystem is sparse, consisting of a desert scrub community including cholla, creosote bush, prickly pear cactus, saguaro, yucca, and other desert shrubs and succulents, occurring at elevations from 30 m to 1050–1200 m. Cedar, cottonwood, and olive trees are found there, too. Small mammals inhabiting the ecosystem include ground squirrels, the kangaroo rat, and the white-footed mouse.

Expansion of the Endemic Area

The 2001 outbreak of coccidioidomycosis in Dinosaur National Monument in northeast Utah at a Native American archeologic site extended the endemic area 200 miles north of the previously known area.[1] The range of the

[1] To find out more about this outbreak refer to "Coccidioidomycosis in workers at an archeologic site—Dinosaur National Monument, Utah, June–July 2001." *MMWR Morb Mortal Wkly Rep* 2001;50:1005–1008 The URL is www.cdc.gov/mmwr.

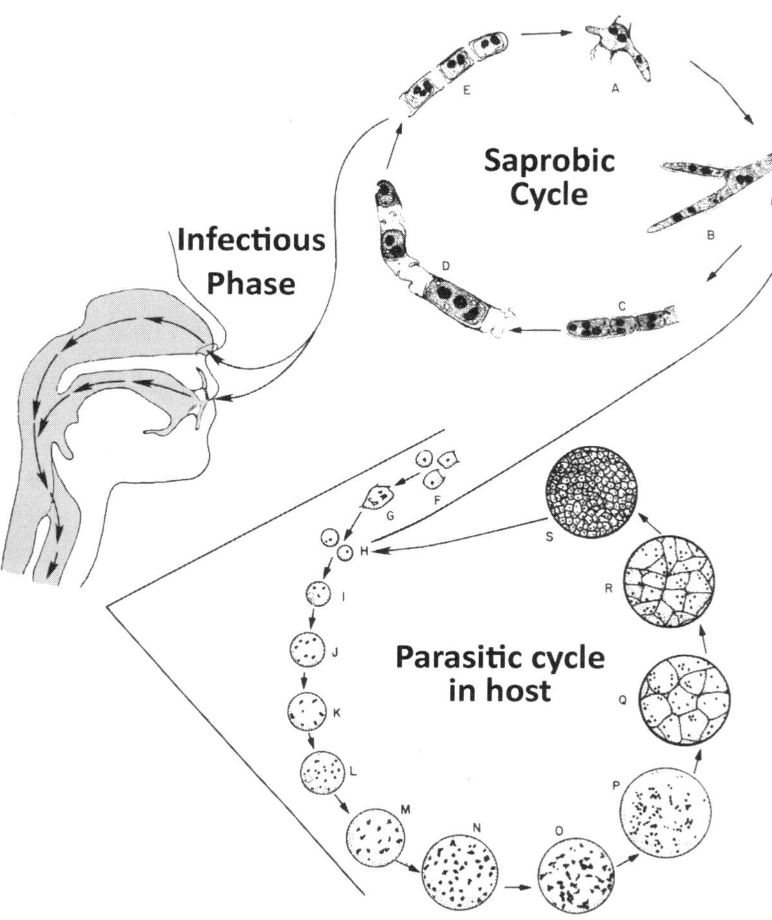

Figure 5.1 The dimorphic life cycle of *Coccidioides* species illustrating the saprobic, parasitic, and nuclear cycles. Arthroconidia convert to spherules (F–H), which become multinucleate (I–P) and then undergo segmentation (Q, R) and endosporulation (S–H) within host tissue. The endospores may convert back to the saprobic phase if the host dies. Nuclei in F–S are represented by darkly stippled regions.
Source: Cole and Sun (1985). Used with permissions from Dr. Garry T. Cole, University of Texas, San Antonio, and from Springer Science and Business media, © 1985.

Coccidioides endemic area may have been much larger at an earlier time. Examination of early Holocene bison mandibles (*Bison antiquus*) from Nebraska, ~8500 years ago, revealed moderately severe, extensive, mandibular osteomyelitis containing spherules consistent with the genus *Coccidioides* (Morrow, 2006). This finding implies that the range of *Coccidioides* was larger then, or that the animals migrated from endemic to nonendemic areas in the early Holocene.

Mexico

Coccidioidomycosis is found in three endemic areas: northern, Pacific Coast, and central parts of the country. Skin test surveys in Tijuana indicated a 10% infection rate and higher rates in Coahuila State.

Endemic Areas Outside the United States

Data for Central America are scarce because the disease is not reportable through an organized system. The first case of coccidioidomycosis was reported in Argentina in 1892 by Posadas in his patient Mr. Silveira. The *Silveira* strain of *Coccidioides posadasii* is still maintained in some culture collections. Coccidioidomycosis is endemic to semidesert areas from Puna to Patagonia. Endemic areas in Brazil occur in the northeastern states of Bahia, Ceará, Piaúi, and Maranhão.

The Two *Coccidioides* Species

Clinical and environmental isolates of *Coccidioides* sampled throughout its New World range were studied by DNA polymorphisms resulting in a division of *C. immitis* into two species (Fisher et al., 2001). One, *C. posadasii*, is the non-California species found in the Western Texas and Arizona endemic areas and in Central and South America. *Coccidioides immitis* is restricted geographically to the California endemic area by the Sierra Nevada range.

The Arizona *Coccidioides* isolates clustered independent of ones from Mexico. Significantly, the South American isolates grouped with those from Texas. No genetic

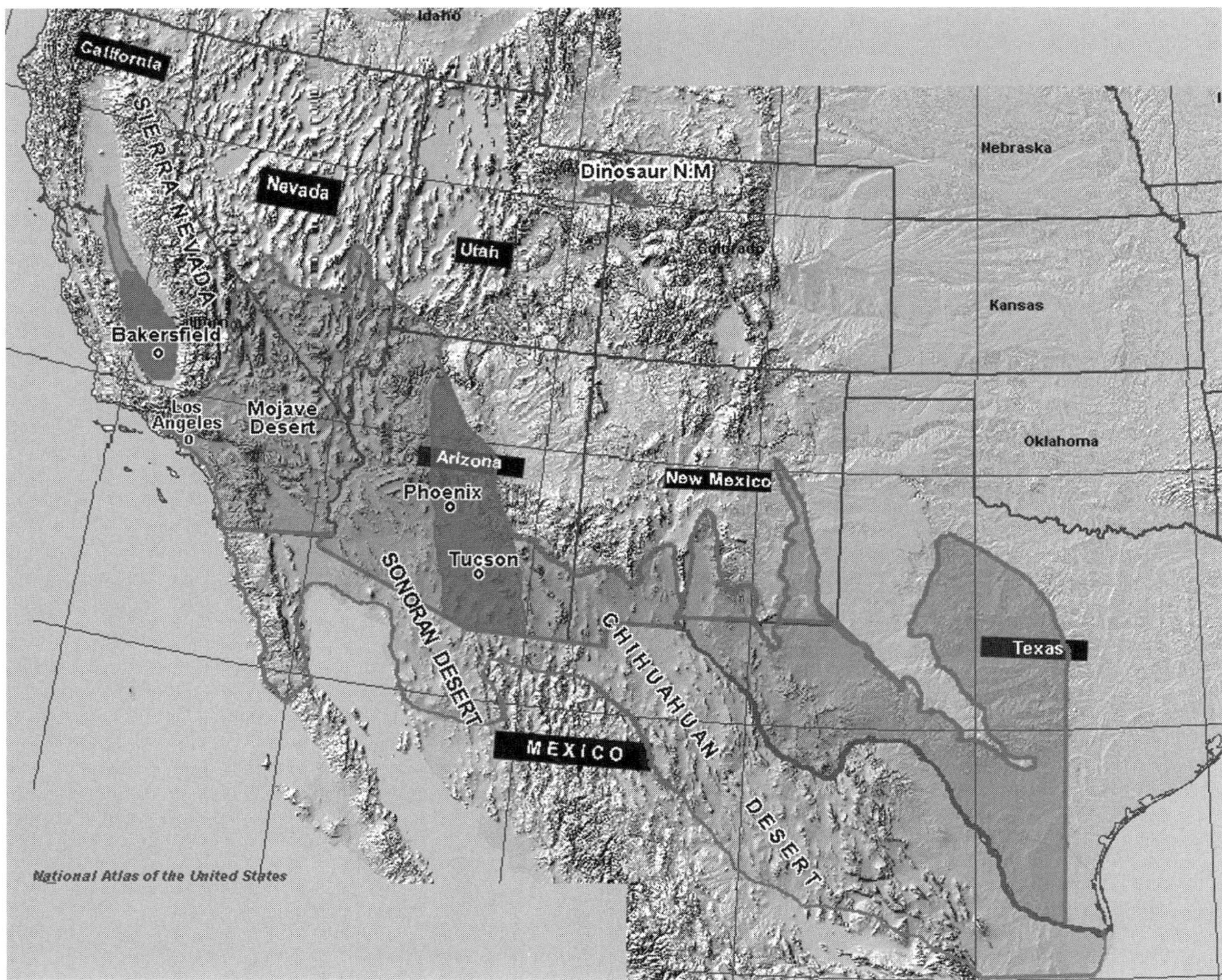

Figure 5.2 Map depicts the coccidioidomycosis endemic areas in California, Arizona, Texas, and Mexico (Mojave, Sonoran, and Chihuahuan deserts). The intense colored areas are highly endemic, moderate colored areas are established as endemic, and the lightly colored area is suspected as endemic. Endemic areas in Texas and Mexico are shown as those matching the Lower Sonoran Life Zone. Focus in northern Utah and Colorado indicates Dinosaur National Monument, site of an outbreak at a Native American archeologic site. Smaller endemic areas extend into parts of Central and South America. Please see text for further details. See insert for color representation of this figure.
Source: Used with permission from Dr. Richard F. Hector, University of California, Berkeley.

variation was found among the South American isolates. This suggested that the South American isolates were descended from a "bottlenecked" population originating in Texas and that too little time had passed for additional genetic variation to appear in the South American *Coccidioides* species.

The strong population structure of the California isolates implies that *Coccidioides* was *not carried southward by air currents, but instead was carried as encysted spherules in mammalian, probably human, hosts*. This interpretation is supported by finding *Coccidioides* in and around ancient Native American middens, which suggests the fungus was prevalent among the local tribes. The arrival of Native Americans in Southern Chile at least 12,500 years ago coincides with the time-frame of southward migration of *Coccidioides*. Human migration the length of South America appears to have taken place in an 800 year time span. This wave of human migration acted as a rapidly dispersing vector of *Coccidioides*. The fungus would have to set up a cycle of infection between migrating tribes and their environment. If true, the *arrival of Coccidioides in South America along with human migration will be the earliest example of a disease spread by human migration*. In addition, several small mammal species capable of carrying *Coccidioides* arrived in South America from the north, including especially kangaroo rats, which have been implicated as reservoirs for *Coccidioides* species.

5.7 EPIDEMIOLOGY

5.7.1 Incidence and Prevalence

An estimated 30 million persons, or 10% of the U.S. population, who reside in central and Southern California, most of Arizona, and parts of Nevada, New Mexico, Texas, and Utah are at risk of infection with *Coccidioides* species (Fig. 5.3). Five million persons are residents of *highly* endemic areas for coccidioidomycosis. The burden of coccidioidomycosis in the United States was estimated at 130,000–150,000 infections/year, 40,000 cases of primary disease (5000 reported), with 3500 cases of disseminated disease and 200 deaths/year (Flaherman et al., 2007; Hector and Rutherford, 2007).

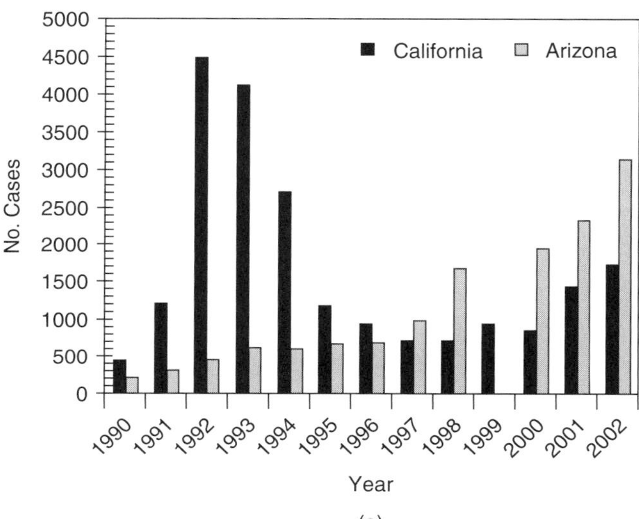

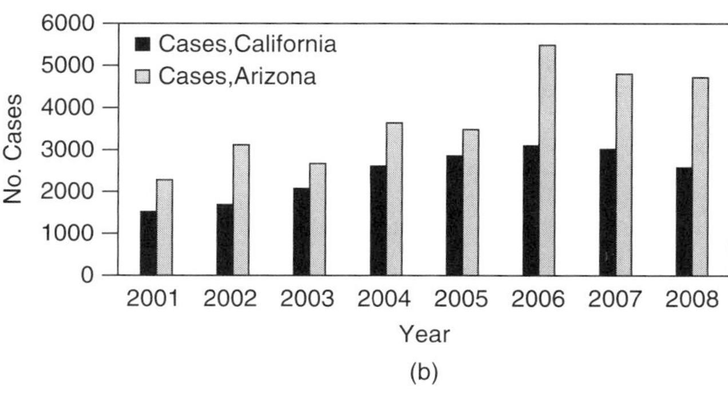

Figure 5.3 (a) Reported cases of coccidioidomycosis, California and Arizona, 1990–2002. *Source:* Used with permission from Dr. George W. Rutherford, School of Medicine, University of California, San Francisco (b) Reported cases of coccidioidomycosis, California and Arizona, 2001–2008.
Source: Collated from annual summaries of notifiable diseases in the United States, *MMWR Morbidity and Mortality Weekly Report*.

Arizona Endemic Area

Approximately 60% of coccidioidomycosis disease is reported from the Arizona endemic area. Over the decade from 1997 to 2007 the incidence in Arizona rose from 21 cases/100,000 population (958 cases) in 1997 to 91 cases/100,000 (5506 cases) in 2006 with case rates doubling since 2000 for the population >65 years old (Sunenshine et al., 2007). Mandatory reporting by physicians, begun in 1997, led to an increase in cases but the cause of further increases since 1999 is unclear. In the most densely populated areas in Maricopa, Pima, and Pinal counties Arizona, the incidence rates were 67.5, 78.3, and 63.7 cases/100,000, respectively.

In the Arizona endemic area cases follow a bimodal annual curve with peak numbers reported in the winter between October and January and a smaller summer peak in May to August. Because of the increased incidence of coccidioidomycosis, the Arizona Department of Health has recommended that all persons presenting with community-acquired pneumonia be tested for coccidioidomycosis.

California Endemic Area

Coccidioidomycosis is a reportable disease in California. The annual total cases reported by the California Department of Health Services were 2641 in 2004, 2885 in 2005, and 2828 in 2006.[2] The most sensational epidemic of coccidioidomycosis occurred in 1992. Cases in Kern County rose from 991 in 1991 to 3027 cases in 1992, spurring formation of the Valley Fever vaccine initiative. (Please see Section 5.10, Determinants of Pathogenicity and Pathogenesis—Vaccine Development). This large increase is explained by drought in the preceding 5 years followed by heavy rains in March 1991 and February–March 1992. Hospitalizations for coccidioidomycosis in California over the period 1997–2002 were 7457 with an average 1258/ year and an incidence of 3.7/100,000/ year.[3] Los Angeles and San Diego county hospitals accounted for 47% of these cases. Coccidioidal meningitis accounted for 19% of all hospitalizations. In terms of mortality, 417 died, for an average of 70 deaths/year or a fatal outcome for 8.9% of hospitalizations.

[2]California Department of Health Services, Division of Communicable Disease Control, Surveillance and Statistics Section, California Selected Reportable Diseases December (week 49–52), 2006.

[3]According to the Inpatient Hospital Discharge Data set, California Office of Statewide Health Planning and Development.

Within Kern County, California, of those who developed symptomatic disease, 85% had a benign course resembling ordinary influenza; 8% developed severe acute pulmonary disease requiring hospitalization; and an additional 7% developed extrapulmonary disseminated coccidioidomycosis including meningitis, lesions of the skin, and skeletal system.

Coccidioidomycosis Outside the Endemic Areas

Infection with *Coccidioides* species in persons residing outside the endemic areas may occur as a result of travel in these areas, laboratory exposure, or inhalation of material from contaminated fomites (e.g., soil, cotton, packing material, or museum artifacts) taken from areas with endemic coccidioidomycosis.

Coccidioidomycosis is a disease with implications for travel medicine. In 1996 members of a church group from Washington State developed influenza-like symptoms and rash after returning from Tecate, Mexico, to assist in the construction of an orphanage. Twenty-one cases were serologically confirmed as coccidioidomycosis.

The New York State Department of Health reported that, in the 5 year period 1992–1997, 161 New York residents had a hospital discharge diagnosis of coccidioidomycosis. They had a history of travel to the Southwest especially to Arizona.

5.7.2 Effect of Weather on Annual Fluctuations in Prevalence of Coccidioidomycosis

The influence of climate on seasonality of coccidioidomycosis is based on the premise that *Coccidioides* hyphae in dry soil need increased soil moisture (i.e., rainfall) to germinate and grow, followed by a dry period in which hyphae desiccate and arthroconidia are produced. Next, wind or other disturbance is required to disperse the arthroconidia, which are then inhaled by the human or animal hosts.

Kern County, San Joaquin Valley, California

Two periods of rain occur: the majority of rains occur in the winter (average rainfall 16.5 cm) and the other in late summer. The yearly cycle of coccidioidomycosis cases reaches a peak toward late fall, decreasing in winter, reaching a minimum in spring and summer. During the 1991–1993 period most (70%) cases in California were from Kern County. Prolonged drought followed by occasional heavy rains stimulated the growth and spread of *C. immitis* (Jinadu et al., 1994).

Arizona

The Southern Arizona endemic area has a bimodal precipitation pattern consisting of late summer monsoon rains and frontal systems in winter. The winter peak precipitation occurs between December and March, followed by the arid season, April through June. The monsoon, a distinctive climate feature in the region, brings periodic rains in July, August, and September. A brief fall dry season follows in October and November.

In Pima County, including Tucson, Arizona, coccidioidomycosis seasons consist of a winter decrease in cases in January to April, a foresummer peak in May through July, a monsoon decrease that takes place in August and September, and a fall peak in October through December. Peak exposure to *Coccidioides* arthroconidia occurs in June and July and in October and November, paralleling the dry and dusty months of the year. Fewer exposures occur in February and March and August and September, consistent with the wetter and less dusty months (Talamantes et al., 2007). Although difficult at any time of year, the fungus may be recovered from soil at the end of winter rains (Galgiani, 2005).

Weather Disturbances

Dust Storms On December 20, 1977 a very strong ground-level windstorm blew through Kern County in the San Joaquin Valley, carrying large amounts of dust into the upper atmosphere and depositing it over large areas of northern California including the San Francisco Bay area, Sacramento, Santa Clara and Monterey counties, and cities in the North Sacramento Valley (Pappagianis and Einstein, 1978; Flynn et al., 1979). Winds gusting to 160 km/h in the Southern San Joaquin Valley scoured the topsoil to a depth of 15 cm, raising it in a huge dust cloud reaching an elevation of approximately 1500 m. Persons in the affected area began to notice symptoms in late December and, in the following 12 weeks, 379 serologically confirmed cases were reported, most in nonendemic areas of Northern California.

Earthquakes An outbreak of coccidioidomycosis in January 1994 followed the earthquake in Northridge in the San Fernando Valley, Ventura County, California, not a known highly endemic area. Landslides in the Santa Susanna Mountains overlooking Simi Valley caused by the Northridge earthquake raised plumes of dust carrying *C. immitis* arthroconidia, which resulted in 170 cases in a 7 week period.[4]

[4]For further information on this outbreak please see the *MMWR Morb Mortal Wkly Rep* issue "Emerging Infectious Diseases: Coccidioidomycosis Following the Northridge Earthquake—California, 1994," at the URL http://www.cdc.gov/mmwr/preview/mmwrhtml/00025779.htm.

5.8 RISK GROUPS/FACTORS

Risk Factors for Contracting Coccidioidomycosis

Opportunities for increased exposure result from occupational or recreational activities in the endemic area including:

- Archeology students on digs (please see Section 5.3, Case Presentation 2)
- Camping
- Construction
- Population migration into endemic areas. The Arizona endemic area has experienced rapid population growth, especially relevant for *a growing demographic of unusually susceptible persons*: elderly retirees settling in the retirement communities built in the desert. The population in Maricopa County (Phoenix) has increased from 100,000 in 1950 to 3.6 million in 2005. Similarly, the 1950 population in Pima County (Tucson) was 100,000 and stood at 924,000 in 2005. Population in-migration has also occurred into the central California and West Texas endemic areas. This migration is reflected in an increase in cases of coccidioidomycosis. (Please see Section 5.7.1, Incidence and Prevalence).
- Farm workers (please see Section 5.3, Case Presentation 1)
- Military on maneuvers. Ten (45%) of a contingent of 23 Navy SEALs who conducted training in Coalinga, California, in the San Joaquin Valley, became symptomatic and had abnormal chest radiographs. The diagnosis of coccidioidomycosis was serologically confirmed (Crum et al., 2002). Skin test surveys in soldiers training at the National Tank Training Center at Fort Irwin, California, in the Mojave Desert indicated a 4.9% annual rate of infection.

Risk Factors for Acquiring Serious or Extrapulmonary Coccidioidomycosis

- Age
- Pregnancy
- Immunosuppressive conditions, including AIDS
- Influence of human genetics on coccidioidomycosis severity

Age

Age is a risk factor for hospitalization with coccidioidomycosis. The rate of hospitalization of <1 case/100,000 population in the 0–14-yr-old group increases gradually to 7.7/100,000 population for those aged 50 years and older (Hector and Rutherford, 2007).

Pregnancy

Pregnant women with respiratory symptoms of pleuritic pain and productive cough should undergo evaluation for coccidioidomycosis commonly associated with travel or residence in an endemic area but it also may result from exposure to infected fomites or occupational exposure to cultures (laboratory workers) (Peterson et al., 1993). Assessment for toxic erythema, erythema nodosum, or erythema multiforme also provides clues. The risk of dissemination, highest in the second and third trimesters, can be estimated by a complement fixation titer >1:16. In disseminated cases aggressive treatment with amphotericin B has improved the otherwise high maternal and neonatal mortality rate.

Women who acquire primary coccidioidal disease in the third trimester or postpartum have more severe illness, dissemination, and death. An analysis of patients from a 1993–1994 epidemic in Kern County, California, found 68 pregnant women with coccidioidomycosis (Einstein et al., 1995). Disease occurred most often in the third trimester (40%), less often in the second trimester (21%), and least in the first trimester (16%). Fifty-seven percent of symptomatic cases did well without treatment. Of the 29 treated cases, there was a trend toward decreased dissemination and the need for chronic treatment when therapy was given during pregnancy as compared with postpartum. Dissemination occurred in 16% of symptomatic cases. Meningeal dissemination was more frequent in pregnancy than in nonpregnant women.

Effect of Sex Hormones on Stimulation of Coccidioides *Spherules*

A potential explanation for the increased susceptibility of pregnant women and also of the male sex to serious or disseminated coccidioidomycosis was the finding that 17-ß-estradiol, progesterone, and testosterone were all stimulatory to the rate of spherule maturation and endospore release (Drutz et al., 1981).

Immunosuppressive Conditions, Including AIDS

AIDS In the late 1980s, 25% of the cohort of HIV-positive persons living in the endemic area developed symptomatic cases of coccidioidomycosis within 2.5 years of follow-up when their $CD4^{+}$ T lymphocytes fell below 250 cells/μL and a diagnosis of AIDS was made (Fish et al., 1990). The type of disease was characterized by overwhelming diffuse bilateral pulmonary infiltrates, thought to be the result of fungemia seeding the lung in many sites. This clinical form often is fulminant, mimicking septic shock or a bacterial infection, and despite treatment, mortality is high. With the advent of combination *ART*, there has been a dramatic decline in AIDS–coccidioidomycosis.

Solid Organ Transplant Recipients Coccidioidomycosis is the most common endemic mycosis in solid organ transplant recipients in North America. The risk of dissemination is much greater in these patients because of antirejection therapy. Another factor contributing to increased risk is *activation of immunomodulating viruses*, for example, cytomegalovirus infection. Half of the coccidioidomycosis cases in the solid organ transplant group were reactivation of previously acquired coccidioidomycosis during the first post-transplant year.

Influence of Human Genetics on Coccidioidomycosis Severity

All races have the same risk of acquiring disease, but there is believed to be a genetic basis for a higher tendency toward disseminated infection among persons of Filipino and African-American ethnicity. The magnitude of the relative risk is not well delineated. The following epidemiologic studies have addressed this aspect:

- A study of patients hospitalized in the Naval Medical Center, San Diego, California, during 1994–2002 revealed that the percentage of patients with disseminated coccidioidomycosis was: Caucasians (14%), African-Americans (44%), and Filipinos (28%) (Crum et al., 2004). The relative risk for disseminated disease was calculated to be 41.9 for African-Americans and 9.6 for Filipinos. Please also see Section 5.3, Case Presentation 1.
- Hospitalizations for coccidioidomycosis in California over the period 1997–2002 showed that African-American race and male gender had an increased rate of hospitalization compared with Native American and Asian-Pacific islanders who had lower rates than Caucasians.
- The 1977 dust storm in the Central Valley of California caused a wind-borne outbreak in a nonendemic area in the vicinity of Sacramento, California. The rate of dissemination was 23.8 disseminated cases/100,000 population for African-American men compared to 2.5 disseminated cases/100,000 for Caucasian men (ratio 9.1–1). In highly endemic Kern County, California, the odds ratio for disseminated coccidioidomycosis is 28× higher in African-American men than for other ethnic groups.
- Risk factors for severe pulmonary and disseminated coccidioidomycosis were evaluated in Kern County, California (Rosenstein et al., 2001). During the period January 1995 to December 1996, there were 905 cases identified. Risk factors for disseminated coccidioidomycosis were African-American race, low income (<$15,000/year), and pregnancy.
- An ABO blood group and histocompatibility gene-linked predisposition to disseminated disease was characterized in a group of African-American, Caucasian, and Hispanic persons with mild or severe disseminated coccidioidomycosis from Kern County, California (Louie et al., 1999). Among Hispanics, predisposition to symptomatic disease and severe disseminated disease was associated with blood types A and B, respectively. The HLA class II DRB1*1301 allele marked a tendency to severe disseminated disease in each of the three groups. Host genes, in particular, HLA class II and the ABO blood group, influence susceptibility to severe coccidioidomycosis.

5.9 TRANSMISSION

Coccidioidomycosis is not a communicable disease. Infection occurs after inhalation of dusts containing *Coccidioides* arthroconidia. Exposure may not require an individual to deliberately disturb the environment but, usually, dusts aerosolized by human activities create opportunities for exposure and infection. Examples include archeological digs (especially those near Native American middens), agricultural labor, military training in desert warfare at the U.S. National Tank Training Center, Fort Irwin, California, and gatherings in the desert (e.g., model airplane flying contests) (MMWR, 2001). That international event occurred in Lost Hills, Kern County, California, in October 2001. Some of the participants developed symptomatic coccidioidomycosis after returning to their homes in Australia, England, Finland, and New Zealand.

The handling of fomites such as cotton packing material or archeological artifacts prepared and shipped from endemic areas to nonendemic areas also may result in infection and clinical illness. Handling of cultures in the laboratory by personnel not following the proper guidelines may cause incidents of exposure (please see Section 5.14, Laboratory Detection, Recovery, and Identification—Safety Considerations). Laboratory exposures may involve large numbers of aerosolized arthroconidia resulting in symptomatic coccidioidomycosis.

5.10 DETERMINANTS OF PATHOGENICITY AND PATHOGENESIS

5.10.1 Allergic Findings

Skin manifestations develop as part of the primary illness (reviewed in di Caudo, 2006).

Toxic Erythema

Most frequent and easily missed is a nonpruritic fine papular rash that occurs within the first 48 h after a patient's first symptoms and then subsides. This is the acute exanthem.

Erythema Multiforme

Erythema multiforme consists of target-like lesions, oral involvement, pruritus, and palmar desquamation. The clinical signs of this skin reaction overlap with those of the acute exanthem early in the acute stage of infection. This combination of allergic skin findings is termed toxic erythema.

Erythema Nodosum

One to 3 weeks into the illness, a subset of patients develop a rash consisting of *tender subcutaneous red nodules, typically on the anterior part of legs* (shins) (Fig. 5.4). Locals in the endemic area refer to this symptom as "the bumps." Ethnicity and gender appear to influence the occurrence of erythema nodosum in coccidioidomycosis. In a large study of military personnel from the early 1940s, erythema nodosum developed in 50% of white women and in 18% of white men with coccidioidomycosis, but appears to be rare in African-American men with the infection.

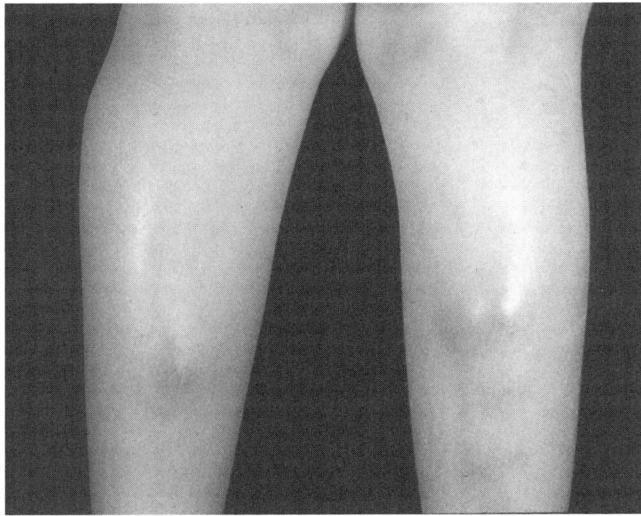

Figure 5.4 Erythema nodosum on the anterior portion of shins, Yale Street Clinic, Los Angeles CA, February 16, 1940.
Source: H. J. Shadomy.

Erythema nodosum may reflect a vigorous and protective CMI response, because the eruptions coincide with the appearance of delayed hypersensitivity, determined in coccidioidin skin tests. In a report of 61 pregnant women with coccidioidomycosis, erythema nodosum was associated with a favorable outcome (reviewed in di Caudo, 2006).

Migratory Arthralgias

Arthralgias (joint pain) also are common complaints. The triad of fever, erythema nodosum, and arthralgias is referred to by the local population as "desert rheumatism."

5.10.2 Pathology

The inflammatory reaction is either suppurative with PMNs attracted when spherules rupture releasing endospores; or granulomatous when macrophages encircle growing spherules; or the two types may be mixed. *Coccidioides immitis* evokes a chronic granulomatous reaction in host tissue, often with caseation necrosis. Necrotizing granulomas are a hallmark of the immune response in coccidioidomycosis.

Chest Radiographic Findings

These are nonspecific in most acute cases. In chronic pulmonary or chronic progressive pulmonary coccidioidomycosis, there often are 2–3 cm diameter nodules in the lung parenchyma. These may resolve or form thin-walled cavities, which either fibrose or calcify. Large inactive cavities can remain for years; about half of these contain spherules. The rest appear devoid of fungi.

Recovery Confers Immunity to Reinfection

Subclinical exposure, sufficient to convert the coccidioidin skin test, or recovery from acute pulmonary coccidioidomycosis is reported to result in durable immunity to reinfection.

5.10.3 Host Factors

See Cox and Magee (2004).

Innate Immunity

Polymorphonuclear Leukocytes (PMNs) PMNs are first in the cellular influx against inhaled arthroconidia.

Phagocytosis is enhanced by immune serum, but despite a respiratory burst and defensins produced by PMNs, fewer than 20% of the arthroconidia are killed. Spherules are too large for endophagocytosis. When they rupture, endospores are surrounded by a fibrillar matrix. This wears away and endospores then are phagocytosed by PMNs, but intracellular killing is inadequate, similar to the interaction with arthroconidia.

Monocytes/Macrophages Nonimmune alveolar macrophages are inadequate at killing arthroconidia and endospores but newer assays suggest that normal blood monocytes can inhibit arthroconidia. Coincubation of monocytes/macrophages with immune T-lymphocytes or IFN-γ significantly improves killing of arthroconidia and endospores.

Cell wall polysaccharides and carbohydrate domains of glycoproteins on the surface of *Coccidioides* spherules and endospores bind to pattern recognition receptors. There are two types: lectins, termed C-lectins, and Toll-like receptors (TLR), both of which recognize structurally conserved pathogen-associated microbial products (PAMPs). Pattern recognition receptors of the C-lectin type include lectins such as the mannose receptor (MR), soluble mannose binding lectin (MBL), and ß-(1→3)-glucan receptor, dectin-1. Peripheral blood monocytes from *Coccidioides*-immune donors are activated by a *Coccidioides* antigen complex "T27K" to produce INF-γ and IL-2. Preliminary evidence implies a role of the MR on monocytes in this activation (Ampel, 2007).

Monocytes from *Coccidioides*-immune donors activated by T27K *Coccidioides* antigenic complex produce proinflammatory cytokines. Tumor necrosis factor alpha (TNF-α). TLR2, and, possibly, TLR4 are implicated in this activation (Ampel, 2007).

Natural Killer (NK) Cells NK cells can migrate to sites of infection and secrete cytokines, notably IFN-γ, and chemokines that induce inflammatory responses and control the growth of monocytes and granulocytes. Before adaptive immunity has fully developed, NK cells are the main source of IFN-γ. Direct cytotoxicity by NK cells occurs on *Coccidioides* spherules.

Dendritic Cells Dendritic cells are potent antigen-presenting cells playing a pivotal role in innate and adaptive immunity. On initial infection, precursor dendritic cells are recruited from the blood to inflammatory sites. *Coccidioides* antigen-pulsed dendritic cells stimulate mononuclear cells to produce IFN-γ. The pathogen binds to TLR and C-type lectins on monocytes and dendritic cells. Phagocytosis of *Coccidioides* spherules by human dendritic cells is mediated by mannose receptors on the dendritic cell surface (Ampel, 2007).

Adaptive Immunity

A profile of adaptive immune responses in persons with different clinical forms of coccidioidomycosis has emerged. Those with primary, asymptomatic, or benign disease have strong skin test reactivity to coccidioidin and low or no levels of anti-*Coccidioides* complement fixing (CF) antibodies. Patients with severe, chronic pulmonary, or disseminated coccidioidomycosis may demonstrate cutaneous anergy and have high levels of anti-*Coccidioides* IgG antibody to the CF antigen. Recovery from active disease is accompanied by reacquisition of T-cell reactivity to *Coccidioides* antigens and subsidence of CF antibody titers. Responses of patients with inactive disease tend to be intermediate between those of the healed primary pulmonary coccidioidomycosis and those with active disseminated disease.

Cellular Immunity

Cutaneous Delayed-type Hypersensitivity Coccidioidin is the classic antigenic preparation used in skin tests and serologic studies. This antigen is the soluble portion of a toluene-induced lysate of mycelia grown in a synthetic asparagine–glycerol–salts medium (Pappagianis and Zimmer, 1990). Coccidioidin skin test reactivity persists in most persons who recover from primary infection. Such individuals maintain immunity to exogenous reinfection. Upwards of 80% of persons who develop solitary pulmonary lesions have positive skin tests to coccidioidin. Less than one-third of those who develop progressive or chronic pulmonary disease are skin test reactors. Approximately 70% of patients who have a single extrapulmonary site of involvement show reactivity to coccidioidin, whereas fewer than 30% of patients with multifocal disease are reactive to coccidioidin or to spherulin, an extract of the spherule phase. Low skin test reactivity or cutaneous anergy in patients with extrapulmonary coccidioidomycosis denotes a poor prognosis. Seventy-five percent of patients who were skin test reactors to coccidioidin recovered compared with only 17% recovery in skin test negative patients.

Cytokine Production TNF-α is a cytokine produced by macrophages, dendritic cells, $CD4^+$ and $CD8^+$ T lymphocytes, and B lymphocytes. TNF-α is responsible for many effects on acute infections, immunologic reactions, and tissue injury. TNF-α can activate neutrophils, enhance the activity of macrophages and NK cells, stimulate proliferation of T and B lymphocytes, and modulate endothelial surface antigens. TNF-α is required to control acute infection and in the formation of granulomas. Formalin-killed and live spherules induce TNF-α and other cytokines from monocytes and macrophages of healthy persons and coccidioidomycosis patients.

Th1 Lymphokines Skin test-positive but not skin test-negative donors secrete both IL-2 and IFN-γ in response to in vitro stimulation with spherulin. IFN-γ production is significantly lower in lymphocytes from patients with disseminated disease than in those from healthy, skin test-positive persons.

Cytokine Activation of Monocytes The lack of killing by normal blood monocytes of *Coccidioides* arthroconidia and endospores is attributed to a lack of phagosome–lysosome fusion. In the presence of lymphocytes from immune persons, both phagosome–lysosome fusion and killing of the fungi are augmented. Both IFN-γ and TNF-α are capable of this augmentation. IFN-γ and TNF-α activate macrophages to generate nitric oxide and related reactive nitrogen intermediates.

Humoral Immunity: Antibodies

Chronic or progressive coccidioidomycosis is associated with a polyclonal B-lymphocyte activation, with elevated plasma levels of IgG, IgA, and IgE. Serum IgG levels are directly proportional to disease activity, highest in patients with multifocal disease. The serum IgA level is elevated in about 20% of patients, particularly in patients with chronic pulmonary disease. Hyperproduction of IgE occurs in approximately 23% of patients with active disease, highest in patients with disseminated disease. Possibly, atopic persons are at greater risk of developing symptomatic coccidioidomycosis.

5.10.4 Pathogenesis

Solving the riddle of *Coccidioides* pathogenesis requires a combined morphologic, biochemical, immunologic, molecular biologic, and genomic approach. The following section owes much to the cadre of dedicated coccidioidomycosis researchers who pursue this combined approach (reviewed by Cole et al., 2006). Much knowledge has been gained and yet mysteries remain.

Arthroconidia are small enough to reach the alveoli. Once there, the process of morphogenesis to the tissue form begins. To survive the initial host response, the arthroconidia have a coat of hydrophobic rodlet proteins. They also contain superoxide dismutase, catalase, and other antioxidants, which enable them to survive oxidative microbicidal processes of the PMNs and naive alveolar macrophages. *What triggers the morphogenesis from arthroconidia to spherules?* Elevated partial pressure of CO_2 present in host tissues appears to be the trigger.

As the spherule initially enlarges, a lipid-rich outer wall layer, the spherule outer wall (SOW), becomes evident. An immunodominant glycoprotein is associated with this lipid layer and is referred to as the SOWgp component. Growth of the spherule to its maximum size of between 40 and 80 μm diameter occurs within 36–96 h. Its nucleus is dividing during this time, the spherule becomes multinucleate, and then the cytoplasm is segmented by invagination of the plasma membrane and new cell wall synthesis. This step is accompanied by upregulation of ornithine decarboxylase. The result of its activity is to increase the pool of polyamines that act as important growth regulators.

What role does chitinase play in spherule maturation? Excess cell wall material in the segments is digested by chitinase. This step is essential for spherule maturation and chitinase knock-out mutants cannot develop further. During this stage excess cytoplasm is squeezed out, resulting in individual endospores, each initially with a single nucleus and a thin cell wall. (The excess cytoplasmic material will be released when the spherule ruptures and will evoke inflammation and tissue damage.) A central vacuole in the maturing spherule disappears, which had maintained its shape.

How do newly released endospores avoid destruction by host phagocytes? The growth and maturation of endospores eventually ruptures the spherule outer wall releasing many endospores (approximately 200–300 per spherule), which seed new sites of infection. Curiously, the endospores are not subject to much phagocytosis. This is the work of an ingenious stratagem of the fungus. The amorphous SOW layer is released when spherules rupture and is the object of phagocytic activity. The newly released endospores, meanwhile, lack the SOWgp component of this outer wall layer and evade phagocytosis because they have secreted a proteinase, in the metzincin family, which digests the SOW glycoprotein.

Why have the spherules been able to mature and proliferate with less than effective control exerted by the host? Complex lipids, phospholipids and glycosphingolipids, and SOWgp in the SOW layer appear to both stimulate cytokine production that misdirects the host response toward a nonprotective immune pathway, and inhibit specific cytokine production especially the key macrophage secreted cytokine, TNF-α. Inhibition of the release of this key inflammatory effector molecule by the innate immune system suppresses T-cell-mediated immunity.

What role does antibody-mediated immunity have in disrupting the process of spherule maturation? Clinical studies show that the titer of complement-fixing antibodies[5] rises in direct proportion to disease severity. Titers rising above 1:16 create apprehension about extrapulmonary dissemination. This observation is based on many thousands of tests conducted at the Coccidioidomycosis Serology Laboratory at University of California, Davis (Smith et al., 1956; Pappagianis and Zimmer, 1990). Subsidence

[5] Also known as the "VF titer"; *definition*: Valley fever titer.

of the titer is a sign of favorable response to therapy. At present, no protective role is known for antibody-mediated immunity in coccidioidomycosis.

Why is the antibody-mediated pathway activated in the host if it has no active role in protection? The SOW glycoprotein is an immunodominant antigen of *Coccidioides* species. Its molecular mass is ~60 kDa but varies from strain to strain because the immunodominant epitope is a 47 amino acid segment which is tandemly repeated 4–6 times. SOWgp synthesis peaks during spherule maturation and falls during endospore release. SOWgp stimulates a Th2 pathway favoring antibody production and upregulation of IL-6, an inflammatory cytokine, and IL-5, an eosinophil differentiation factor. Clinical observations have reported eosinophilia in approximately 25% of coccidioidomycosis patients corresponding to active disease. Thus the role of SOWgp appears to be a means of deviating the host immune response along a pathway designed to have little effect on the pathogen.

5.10.5 Vaccine Development

A consortium of researchers (Pappagianis, 2001; Hector and Rutherford, 2007). investigates the discovery phase and animal protection experiments in the development of a Valley Fever vaccine suitable for human use. Rotary Clubs in California established the nonprofit Valley Fever Americas Foundation in 1995 to fund development of the Valley Fever vaccine. The project is in progress and some of the milestones will be reviewed here. News about progress of the vaccine effort is published in the quarterly newsletter of the Valley Fever Americas Foundation at the URL www.valleyfever.com.

CSA

The *CSA* gene encodes a *Coccidioides*-specific antigen, Csa, which is a secreted protein closely related to fungal ceratoplatanin phytotoxic proteins. This family of fungal proteins of low molecular mass contains a high percentage of hydrophobic amino acids and they self-assemble into rodlets. The Csa antigen is a 19 kDa heat-stable protein present in coccidioidin, and is thus produced by the hyphal form. By itself, immunization with recombinant Csa in a monophosphoryl–lipid A adjuvant protected mice against intranasal challenge with 30% survival at 60 days postinfection. Coimmunization of mice with 1 μg of rCsa and rAg2/PRA increased survival to 90% (reviewed by Cox and Magee 2004).

AG2/PRA

The sequences of *AG2* and *PRA* are identical, encoding a proline-rich cell wall protein, thus the terminology Ag2/Pra. Ag2 was cloned and sequenced and encodes a 19.5 kDa protein with 10 tandem repeats of the tetrapeptide: Thr-Ala-Glu-Pro (Zhu et al., 1996; Dugger et al., 1996). Spherule-derived deglycosylated Pra has a molecular mass of 33 kDa. By itself, immunization with Ag2/PrA provides only partial protection in experimental mouse models (reviewed by Cox and Magee, 2004).

The two promising proteins Ag2/Pra and Csa were developed as a single chimeric recombinant fusion protein. The recombinant fusion protein vaccine (Ag2/Pra and Csa) emulsified in an ISS-Montanide water-in-oil adjuvant was tested for safety and efficacy in female Cynomolgus monkeys (*Macaca fasiculata*) (Johnson et al., 2007). Vaccination with 50 μg of protein antigen in this adjuvant resulted in pronounced release of IFN-γ from mononuclear cells, indicative of the desired Th1 response. The vaccine induced sensitization and reduced disease burden compared to adjuvant alone, but did not prevent pulmonary disease. Present efforts are directed to linking the recombinant protein antigens to a proprietary delivery system to avoid the need for adjuvant.

Aspartyl Proteinase

Working on the hypothesis that antigens eliciting a Th1 response are more likely to be protective, an aspartyl proteinase cloned from *C. posadasii* was predicted to contain epitopes that bind to human major histocompatibility complex class II molecules (Tarcha et al., 2006). Synthetic peptides corresponding to the predicted T-cell epitopes induced IFN-γ production by immune T lymphocytes. This T-lymphocyte-reactive secreted aspartyl proteinase protected mice against pulmonary challenge with *Coccidioides posadasii*.

T27K

T27K is a soluble subcellular antigenic complex derived from *C. immitis* spherules, which, when delivered with an alum adjuvant, is *strongly protective* in experiments where animals are challenged intranasally with *C. immitis* arthroconidia (Pappagianis, 2001). Dissecting the protective components within the T27K complex has resulted in cloning and expression of a Cu, Zn superoxide dismutase, α-(1→2)-mannosidase, and an aspartyl proteinase.

5.10.6 Attenuated Live Chitinase Mutant *C. posadasii*

An alternative approach to vaccine development involves a live attenuated chitinase mutant of *C. posadasii*. Of eight deduced *C. posadasii* chitinase genes, only *CTS2* and *CTS3* revealed increased expression during the transition in spherule development wherein the septal wall complex undergoes digestion and endospore differentiation is initiated (Xue et al., 2009). Disruption of either the *CTS2* gene alone or *CTS3* genes had no significant effect on endosporulation. Parasitic-phase cultures of the *cts2/cts3* double mutant produced only sterile spherules, unable to form endospores in vitro. The attenuated strain remained viable after intranasal challenge in mice but converted to sterile spherules and the mice survived for a 40 day period. Hence, the *cts2/cts3* double mutant lost its disease-causing capacity after disruption of the two chitinase genes. Mouse protection experiments showed that mice primed and boosted with 7000 arthroconidia of the double mutant and challenged with 50–70 arthroconidia of the virulent parental strain survived to 75 days post infection. Unimmunized mice died in 16 days. Culture of organs of immunized, challenged, and surviving mice recovered 1000 colonies/lung but no dissemination occurred to the spleen. Current efforts are directed to induce additional mutations to increase the margins of safety.

5.11 CLINICAL FORMS

Acute Primary Pulmonary Coccidioidomycosis

Symptomatic primary pulmonary coccidioidomycosis resembles community-acquired pneumonia or influenza: a self-limited disease usually of 4–6 weeks duration accompanied by fever, cough, and pulmonary infiltrates. Allergic findings include a positive skin test, joint pain, toxic erythema, erythema nodosum, erythema multiforme, and peripheral eosinophilia.

Pleural effusion may be the only sign of primary infection. Fungi can be recovered from a pleural biopsy and when this occurs in primary coccidioidomycosis, azole therapy is indicated (please see Section 5.13, Therapy). Azole therapy should be used in groups at high risk for dissemination, such as African-Americans, Filipinos, pregnant women, and Hispanics.

Sometimes an area of pneumonitis may heal by forming a solitary pulmonary nodule termed a *coccidioidoma*, up to 5 cm in diameter. These may clear over time or calcify. Such nodules are sometimes diagnosed at surgery to remove a potential lung cancer. Other lung and hilar node lesions may show calcifications. Immunosuppressed patients are apt to develop progressive disease and dissemination and warrant specific therapy. (Please see Section 5.8, Risk Groups/Factors.)

Progressive or Persistent Pneumonia

Diffuse pulmonary or miliary disease may occur after a large dose of inhaled arthroconidia, sometimes leading to acute respiratory failure in immuosuppressed persons. Persistent pneumonia lasts for more than 2 months with extensive infiltrates, sometimes with formation of one or more thin-walled cavities.

Chronic Fibrocavitary Pneumonia

Approximately 5% of persons who develop symptomatic coccidioidomycosis will develop chronic pneumonia with pulmonary infiltrates and thin-walled cavities, suffering flares and remissions over a period of years (Fig. 5.5). This clinical form is associated with type 2 diabetes or with pulmonary fibrosis related to smoking or other causes. Involvement of more than one lobe is more common, and such lesions may cause systemic symptoms such as night sweats and weight loss (Galgiani, 2005). Thin-walled cavities smaller than 2.5 cm diameter tend to resolve over a year, but larger cavities persist. They may be stable for years or become secondarily infected with an *Aspergillus* fungus ball. Chronic pulmonary coccidioidomycosis may damage the lungs progressively.

Extrapulmonary Coccidioidomycosis

An uncommon (0.5% of those infected) but potentially dire complication of coccidioidomycosis is dissemination beyond the lung and hilar lymph nodes to include skin, bones, joints, and meninges. It is noteworthy that some

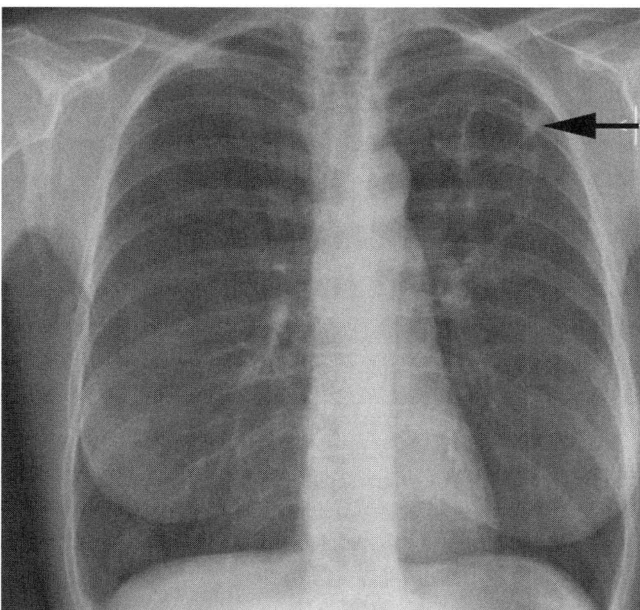

Figure 5.5 Radiograph of chronic fibrocavitary pneumonia in coccidioidomycosis (*arrow* points to cavity).
Source: Used with permission from Dr. Neil M. Ampel, College of Medicine, University of Arizona, Tucson, Arizona.

persons with extrapulmonary coccidioidomycosis have normal chest radiographs. Dissemination occurs within a few months of infection and may involve a single remote lesion or multifocal organ involvement.

Chronic Cutaneous Coccidioidomycosis

The most common site of dissemination is the skin. Lesions range from superficial maculopapular lesions to keratotic and verrucous ulcers, to subcutaneous fluctuant abscesses (Galgiani, 2005). Verrucous skin lesions may mimic squamous or basal cell carcinoma. Scrofulous-like lesions in untreated cases resemble cutaneous tuberculosis and also blastomycosis: "pseudoepitheliomatous hyperplasia."

Osteoarticular Disease

Joints and bones are common sites of dissemination. Joint infections are associated with prominent synovitis and effusion. Weight-bearing joints are more vulnerable, including vertebral disease. Joint infections may extend to underlying bone. Bones may be involved first with extension to the joint (Galgiani, 2005).

Meningitis

Meningitis with granuloma formation or coccidioidoma usually develops relatively soon after the initial infection (within 6 months). The main areas of involvement are the basal meninges. Common presenting symptoms are headache, vomiting, and altered mental status. The differential diagnosis includes cryptococcal meningitis. For further discussion of clinical forms please see Anstead and Graybill (2006) and Galgiani (2005).

5.12 VETERINARY FORMS

Fact sheets on canine coccidioidomycosis are found at the URL http://www.vfce.arizona.edu/ValleyFeverInPets/Default.aspx.

Although dogs may be the most commonly infected lower animals, *C. immitis* infection also has been reported in cats, horses, cattle, swine, llamas, sheep, tigers, coyotes, nonhuman primates, rodents, sea otters, and snakes, but not birds. Treatment for animals consists of itraconazole (ITC), ketoconazole, or fluconazole (FLC), with AmB reserved for severe cases.

5.13 THERAPY

Indications for specific antifungal treatment include weight loss of >10%, intense night sweats persisting 1–2 weeks, pleural effusions, infiltrates involving more than one-half of a lung or portions of both lungs, prominent or persistent hilar adenopathy, a CF antibody titer >1:16, failure to develop a positive skin test (anergy of infection), inability to work, symptoms persisting for >2 months, ethnicity, or medical condition that predisposes to dissemination.

Therapy Options

This section lists the antifungal agents indicated for coccidioidomycosis, followed by the indications for specific clinical forms of disease (Galgiani et al., 2005). Standard therapy options consist of AmB or azoles.

AmB

AmB is indicated for rapid disease progression. AmB-deoxycholate is used for patients with low risk of developing nephrotoxicity. Dosages consist of 0.5–1.5 mg/kg IV, daily or on alternate days. AmB-lipid formulations are preferred for patients at increased risk for developing nephrotoxicity and are given IV at 2–5 mg/kg/day or at higher doses.

Azoles

Azoles are indicated for subacute or chronic disease with dosages consisting of the following:

- Ketoconazole: 400 mg/day oral
- Fluconazole: 400–800 mg/day oral or IV
- Itraconazole: 200 mg/2–3 times/day oral. Measurements of blood concentration of drug are needed to verify that absorption occurs. Cyclodextrin suspensions of ITC increase absorption.

Extended Spectrum Azoles

Newer therapy options include:

- Voriconazole (not FDA-approved at this time). Case reports successful clinical responses.
- Posaconazole. Effective in a small clinical trial and in patients with refractory disease.

ß-(1→3) Glucan Synthase Inhibitors: Caspofungin

These are effective in experimental murine coccidioidomycosis but in vitro susceptibility of isolates varies widely.

Combination Therapy

AmB and an azole may hasten the clinical response, depending on the individual patient. If clinical response is improved, the AmB can be tapered, maintaining the azole dosage.

Outcomes

These include resolution of signs and symptoms of disease, reduced complement fixing antibody titer, and return of organ functionality. Current drug regimens cannot ensure against a relapse of symptoms after cessation of therapy.

Uncomplicated Acute Coccidioidal Pneumonia

Supportive therapy usually is sufficient along with periodic reassessment and radiographic findings to assure resolution of symptoms. Patients undergoing immunosuppressive therapy or who are HIV-positive should be treated. Other categories of patients who should be treated include pregnant women, persons with diabetes mellitus, those with preexisting cardiopulmonary disease, Filipinos, or African-Americans. FLC (400 mg/day) and ITC (200 mg twice/day), were within 20% of each other in producing responses. No studies demonstrate superiority of one azole drug over another.

Patients during therapy and following remission of symptoms should be reassessed at 1–3 month intervals for 1 year or longer. Criteria for follow-up are resolution of pulmonary infiltrates, and to detect any early signs of extrapulmonary dissemination. Follow-up includes patient interviews, physical examination, serologic tests, and radiography.

Diffuse Pneumonia

Bilateral reticulonodular or miliary infiltrates caused by *Coccidioides* species imply an underlying immunodeficiency or a large inhaled dose of arthroconidia. AmB or high-dose FLC is primary therapy, with AmB indicated when hypoxia is present or for rapid disease progression. Therapy should be continued for at least 1 year. Diffuse pneumonia is suggestive of fungemia, and sites of potential dissemination should be studied.

Solitary Pulmonary Nodule, Asymptomatic

If analysis of a fine needle aspirate shows the nodule to be of *Coccidioides* species origin, specific therapy or surgery is not necessary for immune-normal individuals. Periodic chest radiography over the next 2 years should determine whether the nodule is stable.

Pulmonary Cavity, Asymptomatic

Many cavities due to *Coccidioides* species are benign and no intervention is required. Cavities can contain live fungus and sputum cultures commonly yield colonies. Some cavities disappear with time. Cavities that persist for >2 years are indicated for surgical resection to avoid future complications.

Pulmonary Cavity, Symptomatic

Complications include discomfort, superinfection, or hemoptysis. Oral azole therapy may bring improvement but symptoms can recur in some patients after completion of therapy. Surgical resection is recommended as an alternative to chronic or periodic therapy.

Pulmonary Cavity, Ruptured

Rupture of a cavity into the pleural space is an uncommon but well-recognized complication, resulting in pyopneumothorax. Surgical closure by lobectomy with decortication is the preferred management for young, otherwise healthy patients. Antifungal therapy is recommended, particularly in acute cases with active disease. If the diagnosis was delayed for 1 or more weeks, a more conservative management includes antifungal therapy with AmB or FLC before surgery, or chest tube drainage without therapy.

Chronic Fibrocavitary Pneumonia

Oral azole therapy for at least 1 year is recommended, with the option of increasing the daily dose of FLC, switching to an alternative azole, or switching to AmB. Surgical resection is an option.

Extrapulmonary Disseminated Coccidioidomycosis

Primary therapy with an oral azole is recommended: either FLC (up to 2000 mg/day) or ITC (up to 800 mg/day). Alternative therapy with AmB is indicated for rapid disease progression, or if lesions are in critical locations such as the vertebral column. Lipid formulations of AmB are effective and preferred where patients are intolerant of AmB-deoxycholate. Combination therapy with AmB and an azole can be considered but clinical response is variable among patients. Surgical debridement is an occasionally important if not critical option.

Meningitis

The preferred primary therapy is oral FLC at dosages ranging from 400 to 1000 mg/day. A favorable response to FLC serves as a basis for *indefinite therapy with that drug*. This is because it is fungistatic. Intrathecal AmB is sometimes preferred as primary therapy in addition to an azole, expecting a more prompt clinical response. Intra-CSF injections increase survival and cure is obtained in one-third of patients. Intrathecal instillation of AmB via lumbar tap is an effective route for drug delivery, because it has fewer associated risks and fewer complications than other CSF routes. Repeated lumbar injections are feasible in patients who respond quickly but if frequent or prolonged treatments are required a reservoir may be considered. A recommended method is a reservoir implanted subcutaneously and connected to a catheter placed in the intrathecal lumbar space (Stevens and Shatsky, 2001).

5.14 LABORATORY DETECTION, RECOVERY, AND IDENTIFICATION

To confirm a diagnosis, a wet mount, sputum culture, and serologic tests should be performed.

Direct Examination

Rapid diagnosis may be achieved by direct microscopic examination of sputum, bronchoalveolar lavage fluid, CSF, aspirates, exudates, tissue biopsy, or skin scrapings.

Wet Mount, KOH Prep

In KOH preps of clinical specimens the *Coccidioides* tissue form consists of thick-walled spherules, varying in size up to 80 μm diameter, filled with endospores (2–4 μm diameter). Phase contrast optics can help delineate the spherules. Spherules with endospores are diagnostic. Spherules may rupture, discharging their contents, so it is possible to see various sizes and developmental stages within any specimen.

When few or no endospores are present, spherules may be mistaken for phagocytes, or yeast forms of *Blastomyces dermatitidis*, or cryptococci. (*Note*: Phagocytes will be lysed in KOH preps.) Free endospores may resemble yeast forms of *Histoplasma capsulatum* or *Candida glabrata* except that endospores do not bud.

Stains helpful in revealing the spherule form of *Coccidioides* species include Gram, KOH-Calcofluor, and Papanicolaou methods.

Tips and Tricks: To rule out artifacts, place a small amount of a fresh specimen and a drop of sterile saline on a microscope slide, coverslip it, and seal with molten Vaspar. Incubate for 2–3 days at room temperature and observe this microculture for the outgrowth of hyphae from the spherule(s) (Sutton, 2007). This maneuver should only be done by highly experienced clinical microbiologists because of anecdotes of growth of *Coccidioides* from under microscope slide mounts in lactophenol cotton blue. Please see "Safety Considerations" below.

Calcofluor stain for spherules has been successfully applied to diagnose coccidioidomycosis in cytocentrifuge smears of BAL fluid (de la Maza, 1993).

Histopathology

Spherules and endospores (Fig. 5.6) may be viewed after staining with H&E, PAS, GMS, or Papanicolaou stains. Empty spherules and isolated endospores are more difficult to stain but GMS stain is most effective. Isolated endospores look similar to *Cryptococcus neoformans* or *Histoplasma capsulatum*. They stain with GMS or, if available, fluorescent antibodies (immature spherule walls stain and endospores also are positive). The differential diagnosis for spherules includes adiaspores of *Emmonsia crescens* (these enlarged spores have mostly empty interiors). Immature spherules may resemble cryptococci, *Blastomyces dermatitidis*, or *Paracoccidioides brasiliensis* when these yeast forms are not budding. The bodies seen in the rare condition, myospherulosis,[6] may appear strikingly like *Coccidioides* spherules, but the former are pigmented (please see Fig. 21 in Chandler and Watts, 1987). Pollen grains also may appear similar to spherules, especially after KOH digestion (Berman and Yost, 1989).

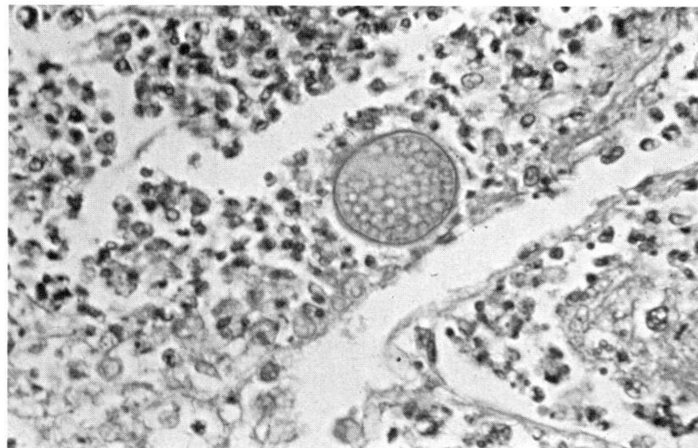

Figure 5.6 A *Coccidioides* spherule in tissue; the spherule wall is intact, the contents are endospores. Histopathology, H&E stain, 375 ×.
Source: CDC.

[6]Myospherulosis. *Definition*.: A chronic granulomatous reaction to undetermined spherical structures frequently contained within a microscopic cyst; first reported in cystic lesions in skeletal muscle from eastern Africa and later in nasal infections in the United States. *Source: Stedman's Medical Dictionary*.

Clusters of endospores in GMS stain may appear to be yeast forms of *Histoplasma capsulatum* (Chandler and Watts, 1987).

Culture

Safety Considerations

Caution! Never work with cultures suspected of being *Coccidioides* on the laboratory bench outside of a well-ventilated and certified laminar flow HEPA biological safety cabinet. Conditions suitable for working with *Coccidioides* species are detailed in the manual *CDC/NIH Biosafety in Microbiological and Biomedical Laboratories* (BMBL), 5th edition; available at the URL http://www.cdc.gov/od/ohs/biosfty/bmbl5/bmbl5toc.htm.

Selected points from the manual and related safety information are summarized below but the original guideline should be consulted for specific information and definition of the terms "BSL 2, BSL 3".

- An important guideline is that laboratory personnel are required to have specific training in handling etiologic agents, and that they be supervised by qualified clinical laboratory scientists.
- Laboratory cultures of *Coccidioides* species rapidly develop as the highly infectious mycelial form and pose a hazard to laboratory workers if arthroconidia from cultures are accidentally aerosolized and inhaled.
- **Handling and processing clinical specimens, identifying isolates, and processing animal tissues are to be conducted using BSL 2 practices, containment equipment, facilities, and personal protective equipment**.
- **BSL 3 practices, containment equipment, and facilities are to be used when propagating and manipulating sporulating cultures identified as *Coccidioides* species and for processing soil or other environmental materials known to contain infectious arthroconidia**.
- Subculturing arthroconidia and opening tubes or plates containing cultures of *Coccidioides* species should be performed only in an well-ventilated, currently certified, laminar flow HEPA biological safety cabinet. Agar slants or bottles should be used in preference to Petri plates for isolating *Coccidioides* species.
- Under **no** circumstances are mold cultures sniffed, a technique that bacteriologists have used to identify bacteria.
- Laboratory workers should be aware that *Coccidioides* species will grow on bacteriologic media and thus may be referred from the bacteriology bench. All laboratory personnel should be wary of mold colonies that appear on bacteriologic media and **never** open such cultures outside the biological safety cabinet.
- If a plate culture is prepared, the Petri plate should be sealed with gas permeable shrink seals[7] and stored in a well-ventilated, certified, laminar flow HEPA biological safety cabinet. The culture plate should be destroyed after 10–14 days.
- Packaging and labeling of *Coccidioides* species cultures also is subject to the regulations concerning interstate shipment of etiologic agents.[8]

Select Agent

Coccidioides immitis and *C. posadasii* are on the Select Agent list and their distribution is restricted to registered laboratories. The regulation and additional information may be found at the URL http://www.cdc.gov/od/sap/. That website will help you to navigate to the list of select agents. Laboratories are required to register with CDC prior to transfer of select agents.

Any diagnostic or CLIA lab that does diagnostic testing, verification, or proficiency testing is exempt from the Select Agent regulation. ***Note*****: Retention of any select agent as a positive control or reference sample is no longer exempt for any reason**.

What must a diagnostic laboratory do when a culture of *Coccidioides immitis* of *C. posadasii* is identified?

After diagnosis, verification, or proficiency testing, the laboratory either transfers the specimens or isolates to a registered facility or destroys them on-site by an appropriate method.

- Select agents used for diagnosis or verification testing are transferred or destroyed within 7 days after identification, unless directed otherwise by the FBI or other law enforcement agency after consultation with the HHS Secretary.
- Select agents used for proficiency testing are transferred or destroyed within 90 days after receipt. The usual practice, for proficiency testing purposes, with dangerous microbes such as *Coccidioides* species, is for them to be killed before shipping.

[7]Shrink seals are colorless, transparent cellulosic bands. The vendor is Scientific Device Laboratory, Des Plaines, Ilinois.

[8]June 7, 2006—Final Rule, Department of Transportation Pipeline and Hazardous Materials Safety Administration, Hazardous Materials: Infectious Substances; Harmonization with the United Nations Recommendations. A summary is provided at the URL http://www.asm.org/Policy/index.asp?bid=43417.

- The laboratory makes a written record of the identification and transfer or destruction on CDC Form 0.1318, and then submits the form to the HHS Secretary (within 7 days after identification or 90 days after receipt for proficiency testing).
- The laboratory maintains a copy of the record for 3 years.

Mycologic Characteristics of the Mold Form of Coccidioides *Species*

Coccidioides immitis and *C. posadasii* are dimorphic fungal pathogens (Fig. 5.1). At room temperature the fungus most often grows rapidly, producing a "dirty" gray-white colony of fine, nonpigmented hyphae. Colony color can vary in atypical isolates. At maturity, branches of the hyphae develop "intercalary arthroconidia." This is a form of "thallic conidiogenesis" in which alternate cells lose their cytoplasm. These vacant cells are sometimes called "disjunctor cells." The remaining viable cells swell, becoming barrel-shaped. The hyphae then break easily into separated barrel-shaped arthroconidia each bearing remnants of the disjunctor cells. The remnants are termed "annular frills." This breakage or dehiscence is called rhexolytic cleavage (rhexolytic, *definition*: dehiscence in which the outer wall of a cell between conidia breaks down). Arthroconidia become easily airborne and are spread by air currents.

Because of the easy separation and dissemination of arthroconidia in the environment, laboratory personnel must consider mature colonies **very hazardous** unless handled in a properly operating biological safety cabinet (please see Safety Considerations earlier in this section).

General Considerations in Preparing Specimens for Culture

Respiratory specimens are planted into agar slants, not Petri plates. Tissues are minced with a scalpel, not ground.

Culture Media

Coccidioides species are not fastidious and will grow in 3–5 days at room temperature or 30°C on BHI, SDA, or PDA. Addition of cycloheximide will inhibit saprobe fungi but will allow growth of primary dimorphic fungi including *Coccidioides*. Mycosel or other similar medium containing, in addition to cycloheximide, antibiotics chloramphenicol, gentamicin, penicllin, or streptomycin will inhibit bacteria and are advised when cultures originate from nonsterile sites. Yeast-extract–glucose chloramphenicol agar (DIFCO, product #219001, BD Diagnostic Systems) also supports good growth of the mold form of *Coccidioides*.

Unopened tubes should be inspected daily for growth. After 3 weeks, tubes showing no growth can be discarded, except when they originated from patients who received antifungal therapy.

Colony Morphology

Colonies begin as white, cream, glistening, glabrous growth (Fig. 5.7a). Very soon they develop cottony

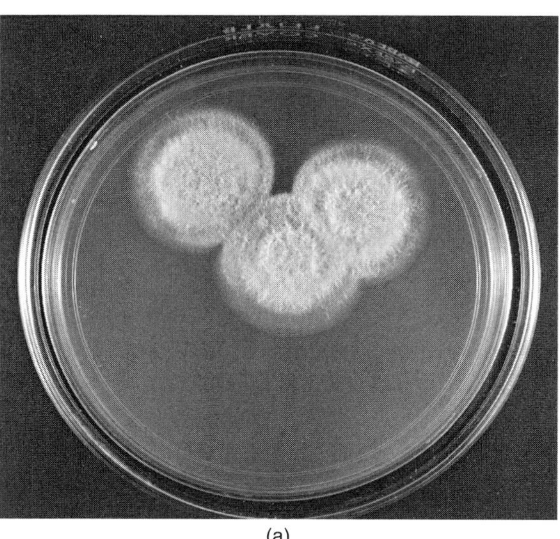

(a)

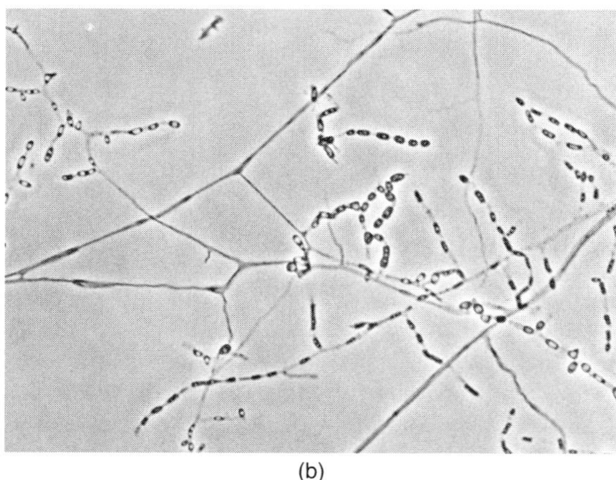

(b)

Figure 5.7 (a) Colonies of the mold form of *Coccidioides* species, a primary culture from Arizona soil. Cultures of *Coccidioides* should be transferred in a BSC from Petri plates to agar slants for safety reasons. The surface appearance of a *Coccidioides* colony is that of a white, smooth to wooly mold. The surface color may also be beige, pink, cinnamon, yellow or brown. *Source:* Used with permission from Dr. Arvind A. Padhye, CDC (b) Intercalary arthroconidia of the mold form of *Coccidioides* species. Microscopic morphology of a tease mount, 400 ×, phase contrast. Saprobic molds display similar arthroconidia (Fig. 5.8), but *any* mold demonstrating this type of sporulation should be regarded as a frank pathogen until proven otherwise.
Source: E. Reiss, CDC.

aerial hyphae, which are the source of the arthroconidia. Colony color turns to dirty-white/gray or tan and later becomes brown with age. Other colony colors have been observed varying from tan to yellow, rose to lavender, and pale brown to gray-brown. Colony morphology is not a species-specific characteristic. All such molds should be handled in a certified laminar flow BSL 2 cabinet, and treated as potential respiratory pathogens.

Microscopic Morphology

Tease mounts are made in lactophenol cotton blue or lacto-fuchsin mounting fluids. It is not necessary or advisable to prepare slide cultures of *Coccidioides* and *that method should be avoided*. Hyphae are thin, hyaline, and septate. Thicker side branches undergo lysis of alternate disjunctor cells, releasing the barrel-shaped arthroconidia (Fig. 5.8). Remnants of the disjunctor cells, the "annular frills," cling to the arthroconidia. Once finished observing the tease mounts, they are immersed in disinfectant solution and autoclaved in laboratory waste.

Look-alike Molds

Some soil-dwelling saprobe fungi also produce arthro-conidia but are incapable of dimorphism and do not produce spherules. They are not pathogenic for humans or animals. They are, however, genetically related to *Coccidioides* (Bowman et al., 1996), The arthroconidia of *Coccidioides* are similar to those of *Malbranchea* species. *Malbranchea* is the asexual stage of *Uncinocarpus reesei* and *Auxarthron zuffianum*. They have also been collected from the lungs of rodents but seem to be only transient colonizers and cannot produce a dimorphic tissue form. They have well-defined sexual stages. These two species are classed in the *Onygenaceae*, the same family as *Coccidioides*. Furthermore, *Coccidioides* species are believed to have descended from *Uncinocarpus* species.

The *Malbranchea* species microscopic morphology is similar to that of *Coccidioides*—hyaline, one-celled, cylindrical, truncate, and alternate arthroconidia produced in terminal portions of the hyphae. Arthroconidia are released by lysis of the disjunctor cells (Fig. 5.8). Table 2.8 in Chapter 2 compares arthroconidia-producing fungi that must be differentiated from *Coccidioides*.

Conversion to the *spherule* form is not done. Instead, identify cultures using AccuProbe genetic probe method.

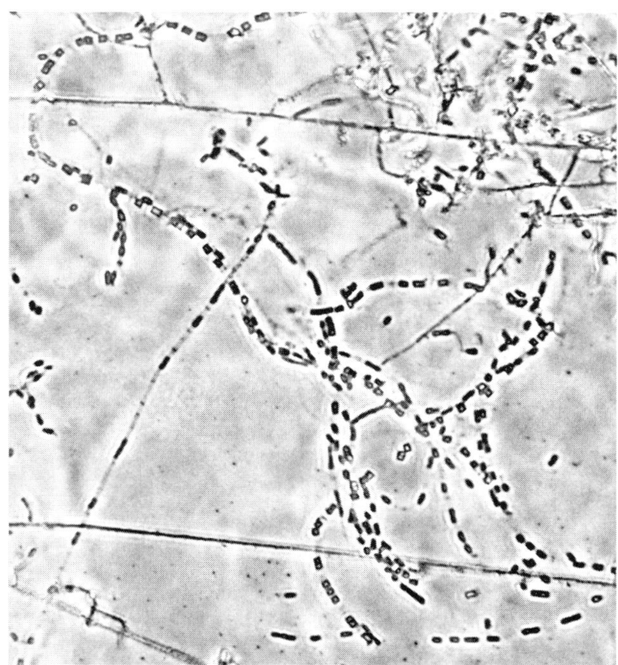

Figure 5.8 Arthroconidia of *Malbranchea* species, microscopic morphology, 400 ×, phase contrast, lactofuchsin stain.
Source: E. Reiss, CDC.

Genetic Identification of *Coccidioides* Species

Nucleic Acid Probe Test

When growth occurs in an agar slant a definitive identification can be made using the AccuProbe® test (Gen-Probe Inc., San Diego, CA). The principle of the test is the hybridization protection assay: a specific DNA probe, labeled with an acridinium ester detector molecule, hybridizes to target rRNA of *Coccidioides* species. Next, nonhybridized probe is inactivated. Then a chemiluminescent signal is emitted and detected in a luminometer (sold by Gen-Probe, Inc.). The result is expressed in relative light units—RLU.

Species-specific identification of *C. immitis* and *C. posadasii* is accomplished by PCR-sequencing of the ITS2 region of rDNA genes (Tintelnot et al., 2007).

Serology

Tube Precipitin ("TP" Antigen) An IgM antibody that arises early in the acute stage of infection, the TP antigen of *C. immitis* is a heat-stable, glycosylated 120 kDa protein containing 3-*O*-methylmannose. That sugar is probably unique to *C. immitis* among the systemic fungal pathogens. The TP antigen is localized within cytoplasmic vesicles and the wall of *spherules*. The TP reaction is not entirely specific for *C. immitis* and is shared by related look-alike but *monomorphic arthroconidial* producing saprobes of the genera *Malbranchea, Auxarthron*, and *Uncinocarpus*. The TP reaction has been modified so it can conveniently be done in immunodiffusion. Reagents are available from Immunomycologics Inc.; see the URL www.immy.com.

Complement Fixation ("F" Antigen) The antigen detected is chitinase with a mass, under nonreducing conditions, of 110 kDa and, when reduced, to a monomer of 48 kDa (Pappagianis, 2001). This test is both diagnostic and prognostic. Performance of CF tests requires considerable experience and they are best performed in reference laboratories, for example, the Coccidioidomycosis Serology Laboratory of the University of California at Davis.[9]

> *Any CF titer with coccidioidin should be considered presumptive evidence for* C. immitis *infection. Titers of 1:2 and 1:4 in the CF test usually indicate early, residual, or meningeal coccidioidomycosis. However, such low titers have also been obtained in sera from patients not known to have coccidioidomycosis. The parallel use of CF and IDCF tests is effective for the specific diagnosis of coccidioidomycosis in patients with low titers of CF antibodies. Sera positive in the CF test at the 1:2 to 1:8 range and also positive in the IDCF test reflect currently active, recent, or remote* C. immitis *infections. When low titers are obtained, a diagnosis of coccidioidomycosis must be based on further serologic tests and on clinical and mycologic studies. Titers >1:16 create apprehension about extrapulmonary dissemination. (Reiss* et al., *2002)*

Note: "IDCF" refers to an immunodiffusion test to detect the "F" antigen.

This reaction is well studied, diagnostic, and prognostic—39,000 tests in one report by coccidioidomycosis pioneer C. E. Smith (Smith et al., 1956). Titers rise in direct proportion to severity. Some laboratories use an easier version of this test, modified for immunodiffusion—the IDCF test—which can be run by diluting patient serum to provide a quantitative titer that compares favorably with the classical CF test (for details please see Pappagianis and Zimmer, 1990). Recombinant chitinase has serologic activity in both the CF, IDCF, and EIA tests comparable to the unpurified coccidioidal antigen (Johnson et al., 1996).

The CF test was used advantageously in Section 5.3, Case Presentation 1 of the 57-yr-old farm worker to support the clinical treatment decision.

EIAs have been developed but will have to be evaluated in parallel with the CF test in large collections of acute and convalescent serum samples from coccidioidomycosis patients.

Skin Test At present, there is no commercial source of the skin test antigen, coccidioidin.[10] Symptomatic pulmonary infection results in a strong skin test reaction—delayed type hypersensitivity (DTH)—and implies immunity to infection. The skin test is a good indication of exposure and usually is among the first tests to become positive after exposure or infection. Asymptomatic exposure proportional to the duration of residence in an endemic area also may convert the skin test to give a positive reaction. A sizable proportion, one-third of the residents in the California and Arizona endemic areas, are skin test positive, whether or not they have been diagnosed with coccidioidomycosis.

Cutaneous Anergy The case of the 57-yr-old farm worker provided a good example of the "anergy of infection." When active pulmonary disease and progressive fibrocaseation occur, *C. immitis*-specific anergy is noticed in at least half of the cases. Successful response to therapy is accompanied by a return of DTH to coccidioidin.

SELECTED REFERENCES FOR COCCIDIOIDOMYCOSIS

Ampel NM, 2007. The complex immunology of human coccidioidomycosis. *Ann NY Acad Sci* 1111: 245–258.

Anstead GM, Graybill JR, 2006. Coccidioidomycosis. *Infect Dis Clin North Am* 20: 621–643.

Bayer AS, Yoshikawa TT, Guze LB, 1979. Chronic progressive coccidioidal pneumonitis. Report of six cases with clinical, roentgenographic, serologic, and therapeutic features. *Arch Intern Med* 139: 536–540.

Berman JJ, Yost B, 1989. Coccidioidomycosis versus pollen. *Southern Med J* 82: 277.

Bowman BH, White TJ, Taylor JW, 1996. Human pathogenic fungi and their close nonpathogenic relatives. *Mol Phylogenetics Evol* 6: 89–96.

Chandler FW, Watts JC, 1987. *Pathologic Diagnosis of Fungal Infections*. ASCP Press, Chicago.

Cole GT, Xue J, Seshan K, Borra P, Borra R, Tarcha E, Schaller R, Yu J-J, Hung C-Y, 2006. Virulence Mechanisms in *Coccidioides*, pp. 363–391, *in*: Heitman J, Filler SG, Edwards JE Jr, Mitchell A (eds.), *Molecular Principles of Fungal Pathogenesis*. ASM Press, Washington, DC.

Cole GT, Sun SH, 1985. Arthroconidium–Spherule–Endospore Transformation in *Coccidioides immitis*, pp. 281–333, *in*: Staniszlo PJ, Harris JL (eds.), *Fungal Dimorphism with Emphasis on Fungi Pathogenic for Humans*. Plenum Press, New York.

Cox RA, Magee DM, 2004. Coccidioidomycosis: Host response and vaccine development. *Clin Microbiol Rev* 17: 804–839.

Crum N, Lamb C, Utz G, Amundson D, Wallace M, 2002. Coccidioidomycosis outbreak among United States Navy SEALs training in a *Coccidioides immitis* endemic area Coalinga, California. *J Infect Dis* 186: 865–868.

Crum NF, Lederman ER, Stafford CM, Parrish JS, Wallace MR, 2004. Coccidioidomycosis: A descriptive survey of a reemerging disease. Clinical characteristics and current controversies. *Medicine* 83: 149–175.

De la Maza LM, 1993. Images in clinical medicine. *Coccidioides immitis*. *N Engl J Med* 329: 1935.

[9]http://www.ucdmc.ucdavis.edu/medmicro/cocci.html. Coccidioidomycosis Serology Laboratory, School of Medicine, University of California, Davis, California, 95616, Demosthenes Pappagianis, M.D., Ph.D., Director.

[10]A reserve supply of coccidioidin is maintained by Professor George W. Rutherford, Director, Institute for Global Health, University of California, San Francisco, CA.

Di Caudo DJ, 2006. Coccidioidomycosis: A review and update. *J Am Acad Dermatol* 55: 929–942.

Dugger KO, Villareal KM, Ngyuen A, Zimmermann CR, Law JH, Galgiani JN, 1996. Cloning and sequence analysis of the cDNA for a protein from *Coccidioides immitis* with immunogenic potential. *Biochem Biophys Res Commun* 218: 485–489.

Drutz DJ, Huppert M, Sun SH, McGuire WL, 1981. Human sex hormones stimulate the growth and maturation of *Coccidioides immitis*. *Infect Immun* 32: 897–907.

Einstein H, Johnson R, Caldwell J, Antonescu A, Reddy V, 1995. Coccidioidomycosis and Pregnancy: The Kern County Experience, p. 324, *in*: *Abstracts Interscience Conference on Antimicrobial Agents and Chemotherapy*, September 17–20, 1995. American Society for Microbiology, Washington, DC.

Fish DG, Ampel NM, Galgiani JN, Dols CL, Kelly PC, Johnson CH, Pappagianis D, Edwards JE, Wasserman RB, Clark RJ, et al., 1990. Coccidioidomycosis during human immunodeficiency virus infection. *Medicine (Baltimore)* 69: 384–391.

Fisher MC, Koenig GL, White TJ, San-Blas G, Negroni R, Alvarez IG, Wanke B, Taylor JW, 2001. Biogeographic range expansion into South America by *Coccidioides immitis* mirrors New World patterns of human migration. *Proc Natl Acad Sci* 98: 4558–4562.

Flaherman VJ, Hector R, Rutherford GW, 2007. Coccidioidomycosis in California: Estimating burden of severe disease using hospital discharge data. *Emerging Infect Dis* 13: 1087–1090.

Flynn NM, Hoeprich PD, Kawachi MM, Lee KK, Lawrence RM, Goldstein E, Jordan GW, Kundargi RS, Wong GA, 1979. An unusual outbreak of windborne coccidioidomycosis. *N Engl J Med* 301: 358–361.

Galgiani J, 2005. Coccidioidomycosis, Chapter 264, *in:* Mandell GL, Bennett JE, Dolin R (eds.), *Mandell, Douglas, and Bennett's Principles and Practice of Infectious Diseases*, 6th ed., Vol. 2. Elsevier-Churchill Livingstone, Philadelphia.

Galgiani JN, Ampel NM, Blair JE, Catanzaro A, Johnson RH, Stevens DA, Williams PL, 2005. Infectious Diseases Society of America 2005. Coccidioidomycosis. *Clin Infect Dis* 41: 1217–1223.

Hector R, Rutherford GW, 2007. The public health need and present status of a vaccine for the prevention of coccidioidomycosis. *Ann NY Acad Sci* 1111: 259–268.

Huppert M, Sun SH, Harrison JL, 1982. Morphogenesis throughout saprobic and parasitic cycles of *Coccidioides immitis*. *Mycopathologia* 78: 107–122.

Jinadu BA, Welch G, Talbot R, Duffey P, Rutherford GW III, 1994. Update: Coccidioidomycosis—California, 1991–1993. *MMWR Morb Mortal Wkly Rep* 43: 421–423.

Johnson SC, Zimmermann CR, Pappagianis D, 1996. Use of a recombinant *Coccidioides immitis* complement fixation antigen/chitinase in conventional serologic assays. *J Clin Microbiol* 34: 3160–3164.

Johnson SM, Lerche NW, Pappagianis D, Yee JL, Galgiani JN, Hector RF, 2007. Safety, antigenicity, and efficacy of a recombinant coccidioidomycosis vaccine in Cynomolgus macaques (*Macaca fascicularis*). *Ann NY Acad Sci* 1111: 290–300.

Kwon-Chung KJ, Bennett JE, 1992. *In*: *Medical Mycology*. Lea and Febiger, Philadelphia, pp. 380, 400.

Louie L, Ng S, Hajjeh R, Johnson R, Vugia D, Werner SB, Talbot R, Klitz W, 1999. Influence of host genetics on the severity of coccidioidomycosis. *Emerging Infect Dis* 5: 672–680.

MMWR, 2001. Public Health Dispatch: Coccidioidomycosis among persons attending the World Championship of Model Airplane Flying–Kern County, California, October 2001. *MMWR Morb Mortal Wkly Rep* 50: 1106–1107 (December 14).

Morrow W, 2006. Holocene coccidioidomycosis: Valley Fever in early Holocene bison (*Bison antiquus*). *Mycologia* 98: 669–677.

Pan S, Sigler L, Cole GT, 1994. Evidence for a phylogenetic connection between *Coccidioides immitis* and *Uncinocarpus reesei* (Onygenaceae). *Microbiology* 140: 1481–1494.

Pappagianis D, 2001. Seeking a vaccine against *Coccidioides immitis* and serologic studies: Expectations and realities. *Fungal Genet Biol* 32: 1–9.

Pappagianis D, Einstein H, 1978. Tempest from Tehachapi takes toll or *Coccidioides* carried aloft and afar. *Western J Med* 129: 527–530.

Pappagianis D, Zimmer BL, 1990. Serology of coccidioidomycosis. *Clin Microbiol Rev* 3: 247–268.

Peterson CM, Schuppert K, Kelly PC, Pappagianis D, 1993. Coccidioidomycosis and pregnancy. *Obstet Gynecol Surv* 48: 149–156.

Reiss E, Kaufman L, Kovacs JA, Lindsley MD, 2002. Clinical Immunomycology, pp. 559–583, *in*: Rose NR, Hamilton RG, Detrick B (eds.), *Manual of Clinical Laboratory Immunology*, 6th ed. American Society for Microbiology, Washington, DC.

Rosenstein NE, Emery KW, Werner SB, Kao A, Johnson R, Rogers D, Vugia D, Reingold A, Talbot R, Plikaytis BD, Perkins BA and Hajjeh RA, 2001. Risk factors for severe pulmonary and disseminated coccidioidomycosis: Kern County, California, 1995–1996. *Clin Infect Dis* 32: 708–715.

Smith CE, Saito MT, Simons SA, 1956. Pattern of 39,500 serologic tests in coccidioidomycosis. *J Am Med Assoc* 160: 546–552.

Stevens DA, Shatsky SA 2001 Intrathecal amphotericin in the management of coccidioidal meningitis. Seminars Respir Infect 16: 263–269.

Sunenshine RH, Anderson S, Erhart L, Vossbrink A, Kelly PC, Engelthaler D, Komatsu K, 2007. Public health surveillance for coccidioidomycosis in Arizona. *Ann NY Acad Sci* 1111: 96–102.

Sutton DA, 2007. Diagnosis of coccidioidomycosis by culture: Safety considerations, traditional methods, and susceptibility testing. *Ann NY Acad Sci* 1111: 315–325.

Symmers W St. C, 1965. Cases of Coccidioidomycosis seen in Britain, pp. 301–305, *in*: Ajello L (ed.), *Coccidioidomycosis*. University of Arizona Press, Tucson, AZ.

Talamantes J, Behseta SAM, Zender CS, 2007. Fluctuations in climate and incidence of coccidioidomycosis in Kern County, California. A review. *Ann NY Acad Sci* 1111: 73–82.

Tarcha EJ, Basrur V, Hung C-Y, Gardner MJ, Cole GT, 2006. A recombinant aspartyl protease of *Coccidioides posadasii* induces protection against pulmonary coccidioidomycosis in mice. *Infect Immun* 74: 516–527.

Tintelnot K, de Hoog GS, Antweiler E, Losert H, Seibold M, Brandt MA, van den Ende AH, Fisher MC, 2007. Taxonomic and diagnostic markers for identification of *Coccidioides immitis* and *Coccidioides posadasii*. *Med Mycol* 45: 385–393.

Untereiner WA, Scott JA, Naveau FA, Sigler L, Bachewich J, Angus A, 2004. The Ajellomycetaceae, a new family of vertebrate-associated Onygenales. *Mycologia* 96: 812–821.

Werner SB, Pappagianis D, Heindl I, Mickel A, 1972. A coccidioidomycosis outbreak of among archeology students in northern California. *New Engl J Med* 28: 507–512.

Xue J, Chen X, Selby D, Hung CY, Yu JJ, Cole GT, 2009. A genetically engineered live attenuated vaccine of *Coccidioides posadasii* protects BALB/c mice against coccidioidomycosis. *Infect Immun* 77: 3196–3208.

Zhu Y, Yang C, Magee DM, Cox RA, 1996. Molecular cloning and characterization of *Coccidioides immitis* antigen 2 cDNA. *Infect Immun* 64: 2695–2699.

WEBSITES CITED

The following presentation or publications are accessed at the listed URL.

Canine coccidioidomycosis http://www.vfce.arizona.edu/ValleyFeverInPets/Default.aspx.

Interstate shipment of etiologic agents June 7, 2006—Final Rule, Department of Transportation Pipeline and Hazardous Materials Safety Administration, Hazardous Materials: Infectious Substances; Harmonization with the United Nations Recommendations. A summary is provided at the URL http://www.asm.org/Policy/index.asp?bid=43417.

Select Agent listing. The regulation and additional information may be found at the URL http://www.cdc.gov/od/sap/.

QUESTIONS

The Answer Key to multiple choice questions may be found at the end of the book.

1. What structures in a wet mount of a sputum specimen would appear similar to the spherules of *Coccidioides* species?
   A. Budding cells of *Histoplasma capsulatum*
   B. *Candida albicans*
   C. Pollen grains
   D. Polymorphonuclear neutrophils

2. What precautions should you take in processing a mold culture to protect yourself and other laboratory personnel against possible exposure to *Coccidioides* species?
   A. Avoid preparing slide cultures in favor of tease mounts.
   B. Avoid the use of Petri plates in favor of screw cap culture tubes.
   C. Do not attempt to sniff cultures as is sometimes done in bacteriology.
   D. Work in a certified BSL 2 cabinet.
   E. All of the above.

3. What length of time may a diagnostic laboratory maintain a culture identified as *Coccidioides* species which was isolated from a patient clinical specimen?
   A. 14 days
   B. 7 days
   C. 90 days
   D. Use your best judgment, there is no set time limit

4. Which is true about the *Coccidioides* spherule?
   A. It is the sexual stage of the species.
   B. It is the infectious form of the species.
   C. It is both the sexual stage and the infectious form of the species.
   D. It is neither the sexual stage nor the infectious form of the species.

5. What is known about *C. immitis* compared with *C. posadasii*?
   A. *Coccidioides immitis* is the California endemic species.
   B. *Coccidioides posadasii* is endemic east and south of the Sierra Nevada range.
   C. The two species have the same morphology in tissue and in culture.
   D. The two species cause the same clinical forms of disease.
   E. All are correct.

6. Which of the following resembles the mold form of *Coccidioides* species?
   A. *Chrysosporium keratinophilum*
   B. *Malbranchea aurantiaca*
   C. *Rhinocladiella* species
   D. *Sepedonium* species

7. Describe the type of sporulation of the mold form of *Coccidioides* species. What is meant by rhexolytic cleavage and annular frills in connection with this type of sporulation?

8. The laboratory has reported back that the patient's VF titer is 1:32. What is the implication of this finding? How would you expect the titer to change after a favorable response to therapy?

9. What are the climate conditions of the lower Sonoran life zone? Why is *Coccidioides* found in association with Native American middens? What climatic disturbances have caused outbreaks of coccidioidomycosis outside of the known endemic areas?

10. Summarize the patient's status of acute pulmonary coccidioidomycosis that would influence the decision to start specific antifungal therapy. Discuss the relative merits of fluconazole versus amphotericin B for treatment of coccidioidal meningitis.

Chapter 6

Histoplasmosis

6.1 HISTOPLASMOSIS-AT-A-GLANCE

- *Introduction/Disease Definition.* Histoplasmosis is a community-acquired pulmonary infectious disease caused when an environmental disturbance aerosolizes microconidia of the soil-dwelling mold *Histoplasma capsulatum*. Conidia are then inhaled and convert to yeast forms in the lung, causing an infection that may be subclinical, influenza-like, or pneumonia.
- *Etiologic Agents. Histoplasma capsulatum*—North America and microfoci worldwide; *H.capsulatum* var. *duboisii*—Africa; *H. capsulatum* var. *farciminosum*—Africa, causes epizootic lymphangitis in equines.
- *Geographic Distribution.* The major endemic area is in the United States in states bordering the Mississippi and Ohio river valleys; from the St. Lawrence River in the north to the Rio Grande River in Texas; and smaller foci worldwide.
- *Ecologic Niche.* Soil mixed with bird droppings or bat guano; blackbird roosting sites, attics of old buildings, caves.
- *Epidemiology.* Exposure is common, ~20 million in the United States; 0.5 million new cases/year. Most exposed persons have a mild flu-like illness. Cases are sporadic or result from outbreaks. Outbreaks occur after disturbing the environment near bird roosts, construction, renovation, and demolition involving attics and belfries. It is a recreational risk to cave explorers.
- *Risk Groups/Factors.* Cleaning contaminated sites, for example, chicken coops or bird roosting areas; demolition and construction, installing heating/air conditioning; restoring old buildings; cave exploring. Extrapulmonary disease occurs in immunosuppressed persons including people living with AIDS.
- *Transmission.* The route of infection is via inhalation of dusts containing microconidia. Histoplasmosis is not transmissible from person to person.
- *Determinants of Pathogenicity.* Mold → yeast dimorphism; cell wall component α-(1 → 3) glucan; catalase; modulation of phagolysosome pH, calcium-binding protein-1, DRK1, a histidine kinase global regulator of dimorphism.
- *Clinical Forms.* Acute pulmonary, chronic pulmonary, extrapulmonary disseminated.
- *Therapy.* Itraconazole (ITC), 200 mg/day (oral) for 4–6 weeks, longer in immunosuppressed patients; amphotericin B is reserved for rapid disease progression or treatment failures.
- *Laboratory Detection, Recovery, and Identification.* Culture early morning sputum; identify cultures with DNA probe (AccuProbe®); serology—antibody tests; urine HPA antigen test for disseminated disease.

6.2 INTRODUCTION/DISEASE DEFINITION

Histoplasmosis is a primary pulmonary infectious disease caused by inhaling dusts containing microconidia of the soil-dwelling mold *Histoplasma capsulatum*. The causative fungus occurs in soil mixed with bird droppings or bat guano and can persist in such sites for months or years. Accumulations of bird droppings or bat guano also occur in attics and belfries of old buildings and can cause outbreaks during their renovation or demolition. Histoplasmosis has a well-defined endemic area in the

Fundamental Medical Mycology, First Edition. By Errol Reiss, H. Jean Shadomy and G. Marshall Lyon, III.
© 2012 Wiley-Blackwell. Published 2012 by John Wiley & Sons, Inc.

United States and smaller foci worldwide. The U.S. endemic area for histoplasmosisis is in the eastern and midwestern states, especially along the upper Mississippi, Ohio, and Missouri river valleys extending north along the St. Lawrence river valley and south along the Rio Grande River in Texas (Fig. 6.1).

When *H. capsulatum* microconidia are inhaled, they lodge in the lungs, germinate, convert to yeast forms, and cause lesions. Anyone can contract histoplasmosis, but especially persons whose occupation or recreational pursuits bring them in contact with dusts mixed with bird droppings or bat guano. Histoplasmosis, and all systemic mycoses caused by dimorphic fungal pathogens, are not contagious.

The disease exists in three major forms.

- *Acute or Primary Pulmonary Histoplasmosis.* This form most often causes influenza-like symptoms. Most persons who are infected recover without medical intervention. Residual pulmonary foci can persist, however, leaving open the possibility of later endogenous reactivation. (Please see Section 6.3, Case Presentation 1.)
- *Chronic Pulmonary Histoplasmosis.* This is most often seen in the setting of *COPD*.
- *Extrapulmonary-Disseminated Histoplasmosis.* This form affects multiple organs and can be fatal, especially to immunosuppressed persons, including those living with AIDS. (See Section 6.3, Case Presentation 2.)

6.3 CASE PRESENTATIONS

Case Presentation 1. Outbreak Among Members of a Wagon Train (Gustafson et al., 1981)

History

A migrant group of 85 teenagers and young adults were traveling by mule drawn Wagon Train from Tucson,

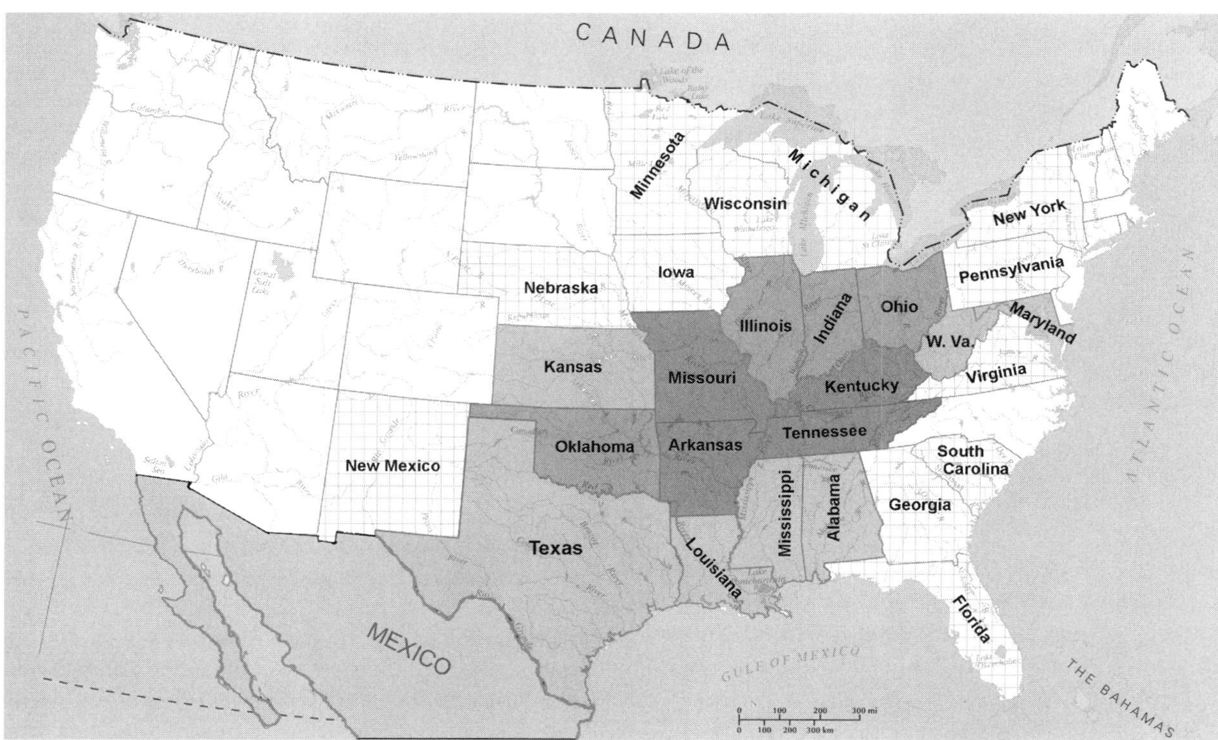

Figure 6.1 Map of the histoplasmosis endemic areas in the United States. The principal histoplasmosis endemic areas are the states of Missouri, Arkansas, Kentucky, Tennessee, Oklahoma, Illinois, Indiana, Ohio, Texas, Louisiana, Mississippi, Alabama, West Virginia (W. Va), and Maryland. States where histoplasmin skin test sensitivity levels meet or exceed 50% of the population in rural and urban areas are shaded 75% gray: Missouri, Kentucky, Tennessee, and Arkansas. States with histoplasmin skin test sensitivity of 50% or more in rural areas only are shaded 50% gray: Oklahoma, Illinois, Indiana, and Ohio. States with histoplasmin skin test sensitivity of 50% or more in one or more counties only are shaded 25% gray: Kansas, Texas, Louisiana, Mississippi, Alabama, West Virginia, and Maryland. States with peripheral endemic areas identified based on soil isolations of *Histoplasma capsulatum* are shown in light cross hatch shading: New Mexico, Nebraska, Minnesota, Iowa, Wisconsin, Michigan, New York, Pennsylvania, Virginia, South Carolina, Georgia and Florida.
Source: Data are from Ajello (1971).

Arizona, in March and arrived in Pennsylvania in November. It was an outdoor work and travel experience for troubled youth. Most came from states without endemic mycoses. They traveled 20 mi/day and slept in Indian-style wig-wams. Each group cared for its own wagon, horses, and mules.

On July 25th the Wagon Train camped in a field near the municipal building in Charleston, Tennessee. Next to that field was a wood lot that was a blackbird roost for many years. The field where they camped was part of the bird roost until it was cleared to make a park 5 years before the Wagon Train encampment.

Fifty-four of 85 members became ill during the first two weeks in August with a febrile disease, myalgias, and pulmonary symptoms. Thirty-three Wagon Train members were hospitalized in a period of one week with severe influenza-like symptoms (Table 6.1).

An epidemic and laboratory investigation was conducted by CDC. Soil samples were taken for analysis.

Laboratory and Clinical Findings

Chest radiographs showed pneumonitis, hilar adenopathy (16 of 32 cases). Only 5 showed a classic "snowstorm" appearance (*definition*: multiple nodules in both lung fields).

Acute and Convalescent Phase Serum Specimens Acute (hospital admission) and convalescent (3 weeks later) serum samples were analyzed for anti-fungal antibodies. Skin tests were applied 6 weeks after the outbreak. Of 30 patients sampled on admission, only 2 had detectable antibody titers. Twenty-eight of 30 patients showed a fourfold rise in complement fixing (CF) antibodies in the convalescent serum samples. No acute phase serum samples showed specific precipitins, but all 28 convalescent CF-positive patients had "m"-precipitins, and 30% had "h"-precipitins.

Table 6.1 Prominent Symptoms in the Wagon Train Outbreak of Acute Pulmonary Histoplasmosis

Symptom	Number of cases	Percentage of cases
Fever	47	87
Headache	46	85
Substernal chest pain	41	76
Pleuritic chest pain	38	70
Nonproductive cough	37	69
Stiff neck	31	57
Abdominal pain	30	56
Sore throat	28	52
Nasal congestion	25	48
Diffuse myalgias	25	46
Thigh pain	19	35
Diarrhea	10	19
Vomiting	8	15

Source: Gustafson et al. (1981). Used with permission from Elsevier Publishers, © 1981.

Skin Tests Sixty-seven of 85 of the Wagon Train members had positive skin tests to the fungal antigen, histoplasmin. Two patients with acute fungal infection had negative skin tests.

Cultures Bacterial, fungal, and viral cultures were not diagnostic. Serologic tests were performed for adenovirus, influenza, legionnaires' disease, mycoplasma, psittacosis, and respiratory syncytial virus.

Clinical Course

Patients recovered without specific therapy and returned to complete their trip.

Environmental Studies

Suspensions of 25 soil samples from the grassy field where the Wagon Train camped were passaged in mice for enrichment; 14 samples grew *H. capsulatum*.

Diagnosis

The diagnosis was acute, self-limited pulmonary histoplasmosis.

Comments

During an outbreak where exposure is limited to 1 or 2 days, the incubation time until first symptoms appear is 7–14 days. *Histoplasma capsulatum* persists in soil and, in this setting, the fungus persisted 5 years after the winter blackbird roost was removed and the 1 acre wood lot was cleared for a park. No construction or excavation occurred before or during the Wagon Train encampment.

Direct culture of soil suspensions is not very rewarding. Some State Departments of Health laboratories are able to conduct mouse passage of soil suspensions to determine the presence of *H. capsulatum* in soil but mouse passage is not a routine test. Mice filter out nonpathogenic soil microbiota, but *H. capsulatum* grows in the spleen and other organs.

> *Safety Note:* Soil testing carries the risk of exposure to a respiratory pathogen and should only be conducted by well-trained personnel under direct supervision. Soil samples should not be sent to testing laboratories without prior approval, and then only in compliance

with U.S. Department of Transportation (DOT) regulations regulating the shipment of hazardous biologic materials. For further information please contact: DOT's Hazardous Materials Information Center at 1-800-467-4922.

Some patients with acute histoplasmosis are skin test nonresponders. This is the anergy of infection and, when patients respond to therapy, the skin test will also turn to positive. *Note*: Skin test reagents are no longer available in the United States.

Case Presentation 2. Cough and Rash in a 38-yr-old Man Living with AIDS[1]

History

A 38-yr-old man with AIDS and a recent $CD4^+$ T-lymphocyte count of 6/μL presented with a 6 month history of nonproductive cough and a rash. He had just been released from prison after serving a 10 year sentence. He reported intermittent fevers to 40°C during the prior 6 months along with a 10 kg weight loss. He knew of no tuberculosis exposure. Medications included antiretroviral therapy, trimethaprim-sulfamethoxazole, and vitamins.

HIV infection was diagnosed 10 years earlier, without a history of opportunistic infections. Primary syphilis was treated 12 years previously with a single penicillin injection. He was a lifelong resident of Maryland.

Physical Exam

His temperature was 38.4°C, his pulse 138/ min, and the respiratory rate 24/ min. He had diffuse 0.5 cm diameter red-purple papular lesions over his face, arms, back, and trunk (Fig. 6.2). There was shotty lymphadenopathy present in the anterior and posterior cervical chains. Chest was clear to auscultation. The liver was slightly tender and percussed to 10 cm, but there was no splenomegaly.

Clinical and Laboratory Findings

Laboratory tests showed a hematocrit of 20%; a platelet count of 50,000/ mm^3. His white blood cell count was 3800 cells/mm^3, with a differential of 13% band forms, 80% PMNs, 7% lymphocytes. The chest radiograph showed no infiltrates.

Assessment and Differential Diagnosis This immunocompromised patient presented with a chronic rash, fevers, and cough. He also was pancytopenic. Skin involvement is associated with several fungal agents which cause disseminated disease in patients with AIDS. Histoplasmosis, endemic in Maryland, can present with similar skin lesions.

[1]Used with permission from Dr. Patricia Barditch-Crovo.

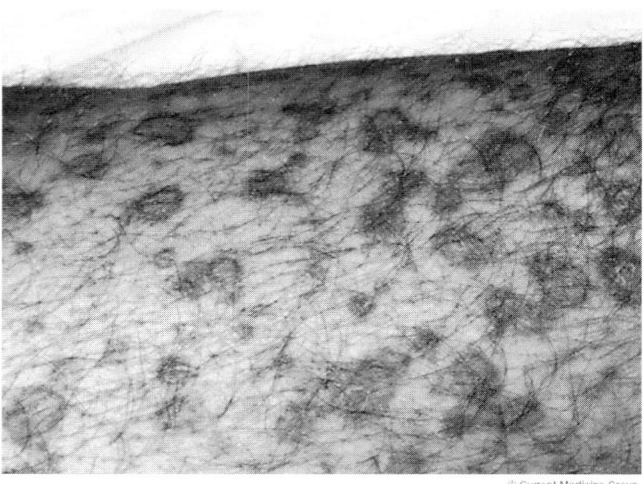

Figure 6.2 Diffuse plaques with fine scale on the leg of an AIDS patient with disseminated histoplasmosis. Similar lesions were present on the patient's back, chest, and arms.
Source: Figure 8.16 of the HIVpro Image Library and Mildvan, 2007. Used with permission from Dr. Donna Mildvan and with kind permission of Springer Science and Business Media.

Other infectious diseases with similar signs and symptoms compatible with the patient's history include bacillary angiomatosis, chronic meningococcemia, coccidioidomycosis, cryptococcemia, disseminated nontuberculous mycobacterial infection, fungemia due to *Candida* spp., nocardiosis, and secondary syphilis.

Diagnostic Procedure The patient underwent biopsy of a skin lesion and bone marrow biopsy. The GMS stain showed numerous yeast forms consistent with *H. capsulatum*. A DNA probe test on early colony growth was positive for *H. capsulatum. Histoplasma* polysaccharide antigen also was detected in the patient's urine in the HPA test.

Diagnosis

The diagnosis was progressive disseminated histoplasmosis, AIDS.

Clinical Course

The patient was treated with daily IV AmB (1 mg/kg/day) for a total of 500 mg; and was then switched to oral ITC. His fevers and skin lesions resolved and his thrombocytopenia gradually improved. He was discharged to home with continued ITC suppressive therapy.

Comment

Disseminated histoplasmosis can be acutely life-threatening. Progressive disseminated histoplasmosis may result from extension of primary infection, exogenous reinfection, or reactivation of an endogenous latent infection. Although it most commonly presents with respiratory tract symptoms, other organ systems such as bone marrow, gastrointestinal tract, brain, and pericardium often are involved. Most patients are diagnosed after 1–2 months of nonspecific symptoms such as fever and weight loss. Although entry of the fungus is through the respiratory tract, only about half of AIDS patients with disseminated disease demonstrate abnormalities on chest radiographs. Approximately 10% of patients have skin lesions. Enlargement of the liver and spleen along with bone marrow suppression are common findings. (Please see Section 6.9, Clinical Forms; Section 6.11, Therapy; and Section 6.12, Laboratory Detection, Recovery, and Identification). The *Histoplasma* polysaccharide antigen (HPA) test is available from MiraVista Laboratories, Indianapolis, Indiana.

6.4 ETIOLOGIC AGENTS

Histoplasma capsulatum, a soil-dwelling mold, grows in soil mixed with bird droppings or bat guano, for example, soil of chicken houses, blackbird roosts, and pigeon- or bat-occupied buildings and caves. Birds are not infected because of their high body temperatures but bats have a lower body temperature and can be infected. Infected bats can excrete the organism in their droppings. *Histoplasma capsulatum* demonstrates temperature-sensitive dimorphism, growing as a mold in the soil, and in the laboratory on artificial medium at ambient temperature. It converts to a yeast form during infection and at 37°C on enriched laboratory medium. (Please see Section 6.12, Laboratory Detection, Recovery, and Identification.)

Mycologic Data for *H. capsulatum*

Histoplasma capsulatum var. *capsulatum* Darling, *J Am Med Assoc* 46: 1285 (1906). Position in classification: family *Ajellomycetaceae*, order *Onygenales, Ascomycota*.

Genetic Relatedness to Other Fungi

Phylogenetic trees derived from sequence comparisons of the 28S rDNA and the ITS regions of rDNA indicate that *H. capsulatum, Blastomyces dermatitidis, Emmonsia crescens*, and *Paracoccidioides brasiliensis* are related members of the *Ajellomycetaceae* clade. Closely related varieties are *H. capsulatum* var. *duboisii* and *H. capsulatum* var. *farciminosum*.

Sexual Reproduction

Compatible (+) and (−) mating types of the *H. capsulatum* mold form can undergo sexual reproduction forming *cleistothecia* containing *asci* with 8 ascospores. The teleomorph of *H. capsulatum* is the ascomycete, *Ajellomyces capsulatus*. Mycologic studies to produce the teleomorph stage are not relevant for identification purposes in the clinical mycology laboratory.

Two varieties of *H. capsulatum* are var. *duboisii* and var. *farciminosum*. It is not clear yet that those classifications can stand scrutiny, as seen in phylogenetic analysis (please see below, "Phylogeography").

Histoplasma capsulatum var. *duboisii* has a colony and microscopic morphology similar to *H. capsulatum*. Yeast forms are larger, 8–15 μm in diameter, lemon-shaped with a narrow isthmus between mother and daughter cells. The mold forms of *H. capsulatum* and var. *duboisii* are indistinguishable. Variety *duboisii* causes a systemic mycosis endemic to sub-Saharan, western and central Africa. It is classically associated with cutaneous nodules and ulcers and osteolytic bone lesions, especially affecting the skull, ribs, and vertebrae (Loulergue et al., 2007). *Histoplasma capsulatum* var. *farciminosum* is the cause of epizootic lymphangitis in horses and also is endemic to Africa. (Please see Section 6.10, Veterinary Forms).

Phylogeography

Phylogeography reveals geographically isolated species. Phylogenetic relationships among the three varieties of *H. capsulatum* were evaluated using sequence variation in four protein-encoding genes sampling a collection of isolates from six continents (Kasuga et al., 2003). At least eight clades are identified: two from North America, two from South America, one each from Australia and The Netherlands (possibly from Indonesia, formerly a Dutch colony), one from Eurasia, and an African clade. All but one of these are genetically isolated groups, which may qualify them as phylogenetic species. The Eurasian clade originated from within a Latin American clade. *Histoplasma capsulatum* var. *capsulatum* isolates are present in all eight clades. The African clade included the *H. capsulatum* var. *duboisii* strains and isolates from the other two varieties. Variety *farciminosum* was distributed in three phylogenetic species! The three varieties of *Histoplasma* are not supported by phylogenetic analysis; instead there are several genetically distinct geographic populations or species. The radiation of *Histoplasma* started between 3 and 13 million years ago in South America.

6.5 GEOGRAPHIC DISTRIBUTION/ECOLOGIC NICHE

Histoplasma capsulatum grows in the soil mixed with bird droppings and bat guano, such as that found in chicken house litter, blackbird roosting sites, and blackbird-, pigeon-, or bat-occupied buildings and caves. Exposure occurs as a result of disturbing the environment in these microfoci. Persons also may become exposed as a result of the airborne dispersal of microconidia from a focus of disturbed soil. For example, office workers in a medical school building in Dallas, Texas, became symptomatic with acute histoplasmosis when the air handling system in their building did not exclude *H. capsulatum* microconidia, which were aerosolized during nearby construction (Luby et al., 2005). Soil in microfoci which were once bird roosting sites may remain contaminated for 5 years after the sites are cleared. (Please see Section 6.3, Case Presentation 1, the Wagon Train outbreak.)

United States

Histoplasma capsulatum is only found in certain geographic areas in the United States: the most heavily endemic areas are states bordering the Mississippi and Ohio river valleys of the central United States: Missouri, Illinois, Indiana, Kentucky, Tennessee, and Mississippi. The endemic area extends into the St. Lawrence River in the north and to the Rio Grande River on the Mexican border. Also note that the endemic area extends into Maryland. (Please see Section 6.3, Case Presentation 2) (Fig. 6.1).

Worldwide

Smaller endemic areas occur in Mexico, Puerto Rico, Jamaica, and other islands in the Caribbean, and Central America. In Southeast and South Asia cases are reported in Thailand and India. As shown in Fig. 6.3a, endemic areas are found in Brazil, western Colombia, northwest Ecuador, portions of Venezuela, Guyana, Surinam, and French Guiana. Endemic areas are located in the

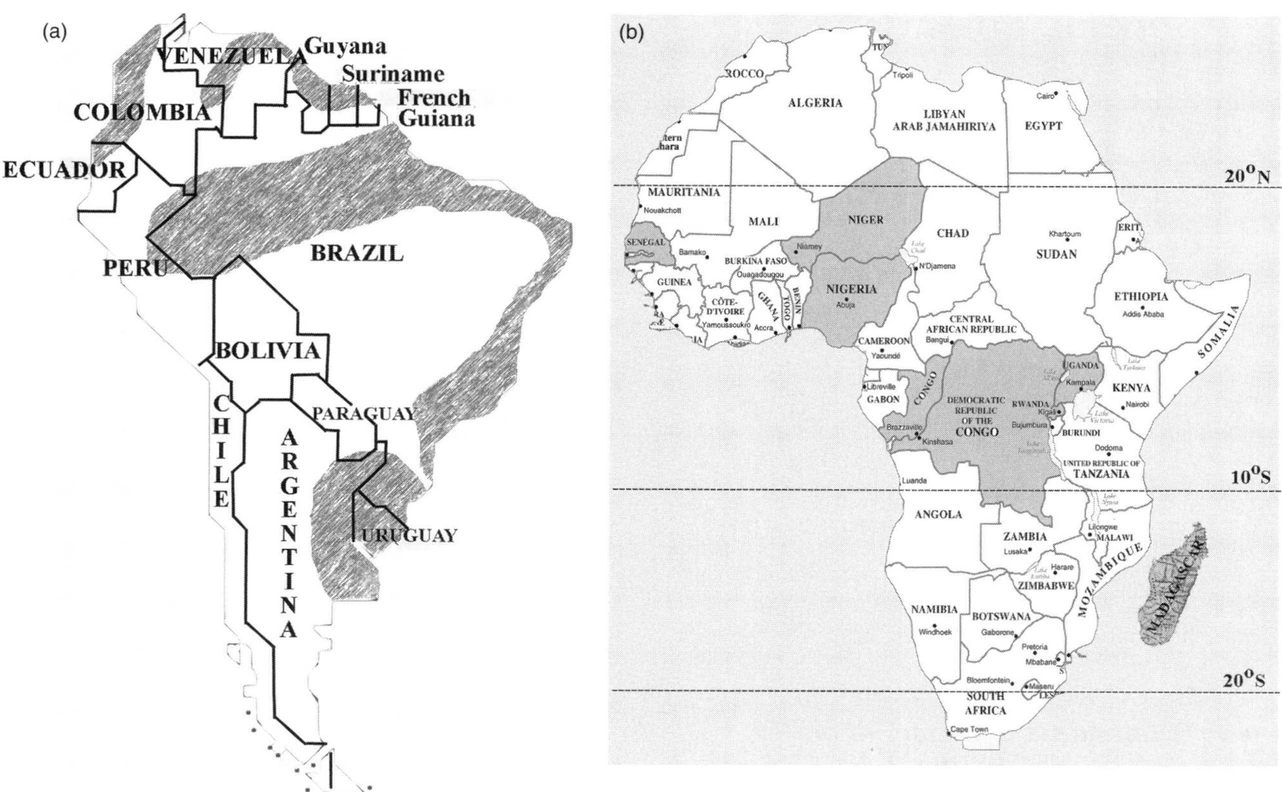

Figure 6.3 (a) Histoplasmosis in South America. The most prevalent areas for histoplasmosis in Latin America are in Venezuela, Ecuador, Brazil, Paraguay, Uruguay, and Argentina. In Brazil, endemic areas are located in the midwest and Southeast portions of the country. Data are from Guimarães et al. (2006). (b) African histoplasmosis (*Histoplasma capsulatum* var. *duboisii*) is endemic in Central and West Africa between the latitudes 15°N and 10°S and on the island of Madagascar (Gugnani and Muotoe–Okafor, 1997). Cases have been reported from Congo (Brazzaville), Democratic Republic of Congo, Niger, Nigeria, Senegal, and Uganda.

midwestern and southeastern portions of Brazil extending into southern Paraguay, Uruguay, and eastern Argentina (Guimarães et al., 2006). Reports of outbreak investigations in Brazil identified 161 cases of histoplasmosis (Zancopé-Oliveira et al., 2005). *H. capsulatum* isolates from Goias state were grouped by *RAPD* typing into a cluster with 100% identity. Isolates from Espirito Santo and Rio do Grande do Sul demonstrated 89% and 85% similarity, respectively (Zancopé-Oliveira et al., 2005). The high percentage of genetic similarity suggests that geographic isolation led to evolution into clades.

Africa

All three varieties of *H. capsulatum* are found in Africa (Fig. 6.3b): *H. capsulatum, H. capsulatum* var. *duboisii*, and also *H. capsulatum* var. *farciminosum*, the latter fungus causes equine epizootic lymphangitis. *Histoplasma capsulatum* var. *duboisii* is found in countries at latitudes bounded by the Tropic of Cancer and the Tropic of Capricorn, and also is present on the island of Madagascar.

6.6 EPIDEMIOLOGY

6.6.1 Incidence and Prevalence

A history of histoplasmin skin test surveys in endemic areas shows that high exposure is common in long-term residents. An estimated 20 million of the U.S. population is exposed. Ninety-five percent of those exposed had a mild flu-like illness. About 500,000 new cases of histoplasmosis occur annually in the United States. About 200,000 of these patients become clinically ill and 4000 require hospitalization (Rippon, 1988). The 2002 Nationwide Inpatient Hospital Sample found 3259 adults and 111 children were hospitalized with histoplasmosis for a total of 3370, or 15.8% lower than the earlier estimate of 4000 (Chu et al., 2006).

Histoplasmosis is a reportable disease in several states including Alabama, Arkansas, Delaware, Illinois, Indiana, Kentucky, Michigan, Minnesota, New York, Ohio, Pennsylvania, Rhode Island, and Wisconsin. Reporting is inadequate, probably because it does not require follow-up public health action. Regrettably, there does not seem to be a systematic effort to collate the data that are reported in those states.

To Do: Visit the websites of the Departments of Health of these states and determine the number of reported cases of histoplasmosis in the most recent 3 year period. Collate the results and estimate the burden of histoplasmosis.

In the heaviest endemic area, the central United States, there were 27 hospitalizations for histoplasmosis/million persons. In the United States, disseminated histoplasmosis was reported by the CDC HIV/AIDS surveillance to occur in 309 patients in 1995. Histoplasmosis accounted for between 5% and 20% of opportunistic infections in AIDS patients in the endemic areas of Indianapolis and Kansas City. In Houston in the 1980s histoplasmosis was the first opportunistic infection in 75% of patients with AIDS (reviewed by Cano and Hajjeh, 2001). The availability of combination antiretroviral therapy has significantly reduced AIDS-histoplasmosis in the United States.

Histoplasmosis has been reported from at least 60 countries and is globally distributed in small foci, whereas the United States contains the largest endemic areas. The rising epidemic of AIDS in Southeast Asia and Brazil contributes to additional cases of disseminated histoplasmosis.

6.6.2 Risk Groups/Factors

Persons at risk include those engaged in farming, gardening, roofing, cleaning chimneys, demolition and construction, installation or maintenance of heating and air conditioning units, restoration of old or abandoned buildings, and cave exploring. All of these activites in endemic areas carry the risk of exposure to the fungus near deposits of bird droppings or bat guano.

The very young and the elderly, especially those with preexisting lung disease (or adults who are heavy smokers) can develop more severe symptoms. Immunocompromised persons are highly likely to develop chronic or disseminated histoplasmosis. This risk group includes persons living with AIDS, leukemia patients, and those receiving systemic corticosteroids or cytotoxic chemotherapy for cancer or collagen-vascular diseases.

Importantly, histoplasmosis occurs both in isolated cases and in outbreaks. Isolated cases are sporadic and any person living in an endemic area is at risk. Outbreaks, such as in the Wagon Train (please see Section 6.3, Case Presentation 1), are related to specific sites that are foci of contaminated soil being disturbed for occupational or recreational purposes.

6.7 TRANSMISSION

The infectious propagules are microconidia stirred up and inhaled in dust near bird or bat habitats. Once in the lung, *Histoplasma* converts to the yeast form, that is, *H. capsulatum* demonstrates mold → yeast dimorphism.

Prevention

Preventive measures to avoid exposure to *H. capsulatum* are detailed in the document entitled

"Histoplasmosis–Protecting Workers at Risk" at the URL http://www.cdc.gov/niosh/docs/2005-109/.

In their review, Cano and Hajjeh (2001) provide the following guidelines to reduce exposure to *H. capsulatum*:

- Avoid contaminated sites: bird roosts, caves, chicken coops.
- Avoid activities such as cave exploring and the demolition of old buildings.
- Use personal protective equipment: respirators, disposable protective clothing, shoe covers, hoods, or helmets.
- Control aerosolized dusts by the use of water sprays.
- Decontaminate with disinfectants (e.g., formaldehyde solutions).

Disclaimer: There is no guarantee that following this guidance will prevent exposure to *H. capsulatum*. Formaldehyde solutions are toxic and should only be used by personnel trained in the handling of hazardous chemicals. Selection and use of respirators should be done after consultation with the National Institute of Occupational Safety and Health (NIOSH) (URL: http://www.cdc.gov/niosh/npptl/topics/respirators/default.html).

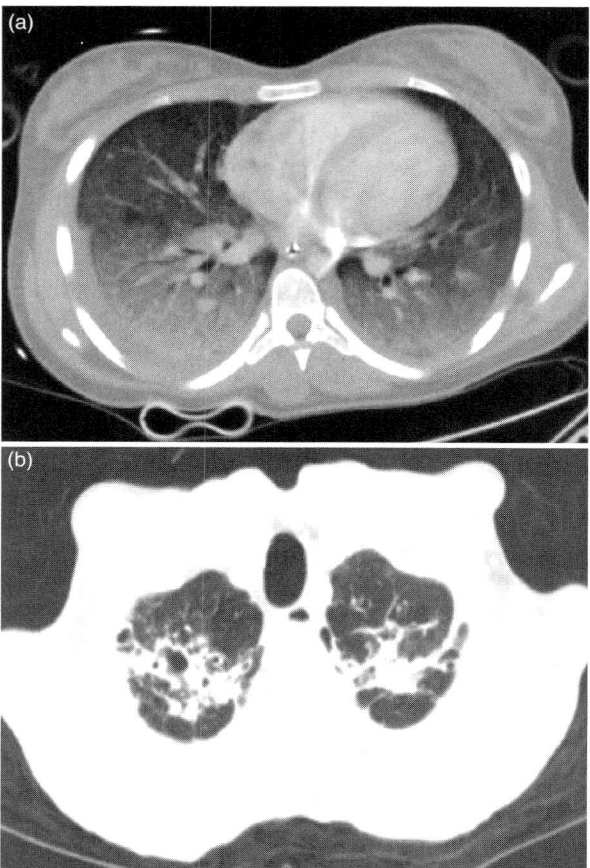

Figure 6.4 (a) CT chest scan, acute pulmonary histoplasmosis. (b) Chronic cavitary histoplasmosis.
Source: Used with permission from Dr. Chadi A. Hage, Pulmonary-Critical Care and Infectious Diseases, Roudebush VA Medical Center and Indiana University, Indianapolis, IN.

6.8 DETERMINANTS OF PATHOGENICITY

6.8.1 Host Factors

Histoplasmosis is similar to tuberculosis, in that it is an intracellular pathogen that evokes a *granulomatous* host response. Inhaled conidia germinate undergoing a dimorphic shift to the yeast form, which initially resist lysis by alveolar macrophages, neutrophils, and natural killer lymphocytes. Once phagocytosed, yeast forms divide into alveolar macrophages and are carried to lymph nodes draining the lungs, and then via the blood to tissues of the monocytic-phagocyte system: the spleen, liver, and bone marrow.

Effective T-cell-mediated immunity is required to control the disease. When adaptive T-cell immunity is induced 10–14 days after infection, macrophages become activated to kill yeast forms, limiting the infection. Some evidence indicates that natural killer lymphocytes are active against yeast forms. Recovery is accompanied by some degree of immunoprotection, but reinfection, as well as endogenous reactivation, can occur months or years later.

Hallmarks of acute histoplasmosis are the chest radiograph or CT scan with granulomatous foci in both lung fields (Fig. 6.4a) and blood and bone marrow monocytes engorged with intracellular yeast forms (Fig. 6.5).

Granulomas and Immunopathology

Histoplasma capsulatum evokes a granulomatous response in the lung. *Granulomas* persist for months or years in which the encircled *H. capsulatum* yeast forms are dormant but probably viable. Over time, pulmonary granulomas clear or form calcifications.

Delayed-Type Hypersensitivity

Associated with this pulmonary reaction is a classic type IV delayed-type hypersensitivity skin test reaction to histoplasmin, a culture filtrate of the mold form. Uncommonly, *Histoplasma* has the potential to induce granulomatous immunopathology; that is, mediastinal lymph nodes become fibrotic and can exert pressure on the bronchi, esophagus, or major blood vessels. Please see Section 6.9, Clinical Forms—Mediastinal Granuloma and Mediastinal Fibrosis.

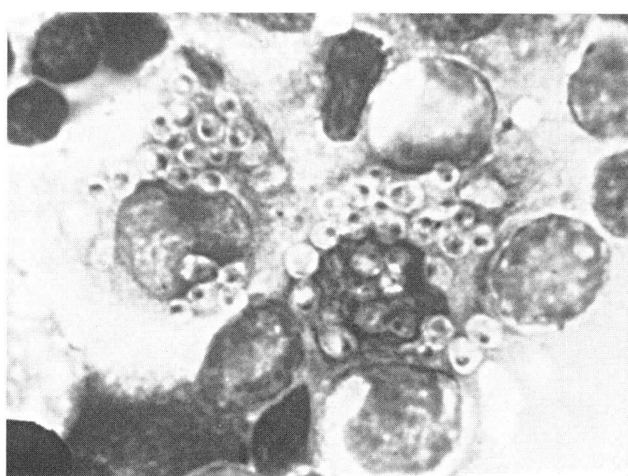

Figure 6.5 Intracellular *H. capsulatum* yeast forms within macrophages, bone marrow, Wright's stain. See insert for color representation of this figure.
Source: E. Reiss.

Protective T-Cell-Mediated Immunity

$CD4^+$ T lymphocytes activate macrophages for increased power to inhibit intracellular growth of engulfed *Histoplasma* yeast forms and to organize granulomas that wall-off and eventually kill them. Interferon-γ (IFN-γ), TNF-α, and IL-12 are important in mounting an effective Th1 immune response (Deepe et al., 2005). IL-12-dependent production of IFN-γ is crucial for protection against *H. capsulatum*. TNF-α, a product of activated macrophages, very likely plays an important role in host defense because acute life-threatening histoplasmosis has been reported in rheumatoid arthritis and Crohn's disease patients treated with TNF-α antagonists (Lee et al., 2002). TNF-α acts by supporting the production of NO (nitric oxide). That molecule can bind iron and thus deprive the fungus of that micronutrient. Granulocyte-macrophage-colony stimulating factor (GM-CSF) also is an essential component of defense in primary histoplasmosis.

The absence of intact T-cell-mediated immunity can result in progressive disseminated histoplasmosis.

Humoral immunity is not the critical limb of the immune response conferring protection in histoplasmosis. That said, an mAb produced against a 17 kDa histone-like surface exposed protein (H2b) was shown, in a murine model infection, to opsonize yeast forms for improved phagocytosis (Nosanchuk et al., 2003). Passive transfer in mice of mAbs to the histone H2b reduced fungal burden, decreased inflammation, and prolonged survival (Nosanchuk, 2005). The low concentrations of antibodies against H2b in immune serum suggest, however, that it is not an immunodominant antigen. Engineering an IgG class mAb against the H2b histone may provide adjunctive therapy in histoplasmosis. The IgG class would enable penetration to the alveoli.

Antibodies

Antibodies have great diagnostic value in histoplasmosis. (Please see Section 6.12, Laboratory Detection, Recovery, and Identification.)

6.8.2 Microbial Factors

α-(1 → 3)-D-Glucan

Its presence in the cell wall correlates with strain-specific virulence in a mouse model of infection. Furthermore, among the related dimorphic pathogens, *B. dermatitidis* and *P. brasiliensis*, the spontaneous loss of ability to produce α-(1 → 3)-D-glucan corresponds to a loss of virulence (Kugler et al., 2000). Two chemotypes of *H. capsulatum* yeast forms differ in their cell wall content of α-(1 → 3)-D-glucan. Chemotype I contains less of this polysaccharide, whereas chemotype II contains a major amount. The binding of dectin-1 present on the surface of macrophages to its ligand, β-(1 → 3)-D-glucan, in the *H. capsulatum* yeast form is masked by α-(1 → 3)-D-glucan, which occupies an outer cell wall layer (Rappleye et al., 2007). The result is evasion of phagocytosis.

Adherence to Monocytes

Heat shock protein 60 (Hsp60), usually performing as a chaperonin,[2] is located on the cell surface of yeast forms where it acts as an adhesin, binding to heterodimeric β2 or CD18 integrins on monocytes. Hsp60 also is characterized as a potent elicitor of T-cell-mediated immunity (Woods, 2003).

Calcium-Binding Protein 1 (Cbp1)

Cbp1 is secreted only by *H. capsulatum* yeast forms. Gene knock-out experiments showed that the *CBP1* gene is required for yeast forms to kill macrophages (Kugler et al., 2000). Its mechanism of action is either to scavenge calcium ions or, alternatively, to affect signaling processes of the fungus or of the host (Woods, 2003).

Catalase

The *H. capsulatum* major secreted protein antigen, "m"-antigen, is a catalase (Zancopé-Oliveira et al., 1999). Surface-located or secreted antioxidants function as a defensive tactic against reactive oxygen species produced by the oxidative burst of PMNs or macrophages. There are three unrelated catalases in *H. capsulatum*, but each

[2]Chaperonin. *Definition:* A category of protein that functions in protein folding and intercellular signaling. Chaperonins are also called "heat-shock proteins" due to their upregulated expression in microbes subjected to stress.

has a heme group in their structures (Guimarães et al., 2008). Growth and morphology of *H. capsulatum* were not affected by exposure to 1 mM H_2O_2, 20× higher than physiological concentration. MAbs reactive with *H. capsulatum* m-antigen localized the catalase at the surface of yeast forms.

DRK1*: A Histidine Kinase Gene Regulating Dimorphism*

A hybrid histidine kinase (*DRK1*) acts as a global regulator of dimorphism and virulence in both *H. capsulatum* and *B. dermatitidis* (Klein and Tebbits, 2007). Properties of Drk1 include:

- A sensor of environmental change in response to stress in the host affecting the fungus's ability to cause disease
- Required for morphogenesis from mold to yeast form, and also for the expression of genes related to pathogenicity
- Regulation of the expression of yeast-form-specific genes related to pathogenicity (e.g., *CBP1* and *YPS-3*).

Two-component phosphorelay systems are comprised of a histidine kinase (HK) component, which autophosphorylates in response to an environmental stimulus, and a response regulator (RR) component, which transmits the signal, resulting in an output such as activation of transcription, or of a mitogen-activated protein kinase (MAP kinase) cascade.

In response to the environmental stress, a sensor HK autophosphorylates a histidine residue and transfers the phosphate to an aspartate residue on a RR, leading to activation. The active RR controls transcription to elicit a cellular response from the HK. Two-component phosphorelay signaling systems in pathogenic fungi regulate cell wall biosynthesis, virulence factor expression, drug resistance, and dimorphism. For further reading and illustration of phosphorelay signaling please see Catlett et al. (2003).

Iron Acquisition

The ability of *H. capsulatum* yeast forms to acquire iron is essential for pathogenesis, and macrophage ability to withhold iron is an element of host defense. When iron availability is limited, low molecular mass hydroxamate siderophores bind ferric ions (Fe^{3+}). Several ferric reductase activities aid *H. capsulatum* yeast forms to gain access to ferrous ions: ferric reductase activity in the *H. capsulatum* cell membrane, secreted reduced glutathione-dependent enzymatic reductants, and secreted low molecular mass nonenzymatic reductants (Woods, 2003). They are all active under iron-limiting conditions.

pH Modulation of the Phagosome

Histoplasma capsulatum modulates the pH of the phagolysosome at ~pH 6. At that pH the acidic lysosomal hydrolases are not optimally active and transferrin is at a pH at which free ferrous iron is available to the yeast forms (Newman et al., 1994). At higher pH iron bound to transferrin is unavailable to *H. capsulatum*.

YPS3

Several yeast-form-specific genes are upregulated in the dimorphic transition from germinating conidia to yeast forms. The Yps3 protein, with a mass of 20 kDa, is bound to and secreted from cell-wall-exposed chitin (Bohse and Woods 2007a). The *YPS3* gene has a hypervariable region encoding a 5–6 amino acid tandem repeat element, which varies between 2 and 20 copies among strains and accounts for 7–43% of the gene's length. There is evidence that the copy number of the *YPS3* is in inverse proportion to virulence (Vincent el al., 1986; Bohse and Woods 2007a). Please see "RFLP Class and Pathogenicity" below). The intragenic variability occurs at the same point where Yps3 diverges from its homolog Bad1, the *Blastomyces dermatitidis* adhesin and virulence factor (please see Section 4.10, Determinants of Pathogenicity in Chapter 4). The corresponding region of *BAD1* contains a longer repetitive element, encoding a 24 amino acid repeat but composed of similar amino acids. Silencing the *YPS3* gene by means of RNA interference resulted in more rapid clearance of *H. capsulatum* yeast forms from mice after they were infected by either the intranasal or intraperitoneal routes (Bohse and Woods, 2007b). This result underlines the importance of the yeast-form-specific Yps3 for pathogenicity. Although it is surface located, there is no definite evidence of a role for Yps3 as an adhesin and its mechanism of action remains to be determined.

RFLP Class and Pathogenicity

Isolates of *H. capsulatum* can be classed into three general genotypes: class 1, 2, or 3, depending on mtDNA and whole genomic DNA cut with restriction endonucleases and then detected in Southern blots using radiolabeled rRNA as probes (Vincent et al., 1986). In order of decreasing pathogenicity they were class 2 > class3 > class1. Only RFLP class 2 strains express the Yps3 protein (Bohse and Woods, 2007a). Those strains show the smallest number of repeats, averaging two. The RFLP class 3 strains have 11–12 copies of the repeat. The RFLP class 1 strains have 18–20 copies of the repeat, whereas strains of *H. capsulatum* var. *duboisii* have 16–17 copies. The basis for their differential pathogenicity remains unknown.

6.9 CLINICAL FORMS

In the Wagon Train outbreak of acute pulmonary histoplasmosis (Section 6.3, Case Presentation 1), symptoms were nonspecific and histoplasmosis was one of the last diseases suspected. A prolonged febrile illness, with unusual radiographic findings, was the first clue that a virus was not involved. Five hospitalized Wagon Train members had "snowstorm" chest radiographs: multiple granulomatous foci. At the other end of the disease spectrum, progressive disseminated histoplasmosis was seen in the 38-yr-old AIDS patient who presented with skin lesions, despite the lack of chest radiograph abnormalities (Section 6.3, Case Presentation 2).

Acute Pulmonary Histoplasmosis

Light Inoculum

About 95% of persons infected with *H. capsulatum* experience inapparent, subclinical, or mild self-limited benign disease. When symptoms occur, they appear 3–17 days after exposure (average time is 10 days). Symptoms usually are mild and nonspecific, resembling a cold or influenza: fever, chills, dry cough, myalgia, headache, pleuritic chest pain, enlarged lymph glands, and fatigue (Table 6.1).

From 5% to 10% of patients with self-limited histoplasmosis may experience other symptoms:

- Arthritis and arthralgias.
- Mediastinal lymphadenopathy.
- Cutaneous symptoms including erythema nodosum and/or maculopapular rash.
- *Pericarditis*. Pericarditis in histoplasmosis is considered to be immune-mediated because pericardial fluid in the affected patients usually is sterile. Granulomatous inflammation in the mediastinal lymph nodes induces pericarditis in ~6% of patients with acute pulmonary histoplasmosis (Deepe, 2005; Kauffman 2003).
  - The singer/songwriter Bob Dylan was hospitalized on May 25, 1997, seriously ill with pericarditis due to *H. capsulatum*. While it is unknown exactly how Mr. Dylan contracted histoplasmosis, he recovered after 4–6 weeks of treatment and was released from a Los Angeles hospital over the weekend of May 31 to June 1. When asked about his plans for his recovery period, Mr. Dylan said, "I don't know what I'm going to do. I'm just glad to be feeling better. I really thought I'd be seeing Elvis soon."

Differential Diagnosis

In the Wagon Train outbreak other diseases to be ruled out included adenovirus, influenza, legionnaires' disease, *Mycoplasma*, psittacosis, and respiratory syncytial virus infection.

Pediatric Histoplasmosis

Children have a high rate of asymptomatic or mild infections. This is illustrated by an outbreak of histoplasmosis in a school in Ohio in 1970. In observation of Earth Day in Delaware, Ohio, an old bird roost in the junior high school courtyard was raked and cleared. The air intake of the school circulated the dust from the courtyard into the classrooms. The first cases appeared 7 days later and the outbreak peaked 15 days after the incident. Clinical illness occurred in 384 (40%) of the children and the faculty. The rate of skin test positivity in the school population was 84% (Brodsky et al., 1973). More than 3/4 of those affected were ill a week or less. There was a high rate of subclinical infection and, among those affected, a mild clinical course.

Immune-normal children with symptomatic acute histoplasmosis present with fever, malaise, mild cough, and nonpleuritic chest pain. These symptoms usually subside in 2–3 days (Kleinman 2003). Fewer than 5% of children's symptoms persist after 2 weeks, with loss of appetite, weight loss, and lethargy. Pericarditis may occur in ~10% of children with histoplasmosis and, as in adults, pericardial fluid usually is sterile (please see "*Pericarditis*" in bulleted list above).

Recovery in persons who develop symptoms usually occurs within a few weeks, even without antifungal therapy, but individuals may experience fatigue lasting for months.

Heavy Inoculum

Moderate-to-severe illness ensues. Hypoxemia may develop and require mechanical ventilation. Chest radiographs and CT scans show a bilateral reticulonodular pattern, sometimes referred to as the "snowstorm" pattern (Fig. 6.6). Pulmonary cavities form in 10–15% of these clinical cases. Without antifungal therapy there is a prolonged convalescence. Recommended therapy consists of AmB, with adjunctive corticosteroids. After recovery, persons have partial immunity to reinfection. Partial immunity can be overcome by a heavy exposure to conidia such as in a laboratory accident or in disturbing heavily contaminated soil.

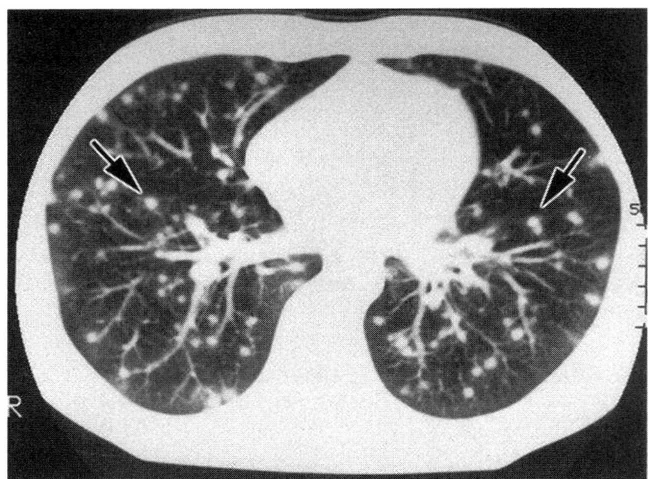

Figure 6.6 CT scan of multiple nodules in both lung fields in a Japanese tourist after visiting a cave inhabited by bats in Manaus, Brazil. Arrows point to lesions in both lung fields.
Source: Suzaki et al. (1995). Used with permission from Kansenshogaku Zasshi Publishers.

Reactivation Histoplasmosis

Granulomas may contain dormant but possibly viable yeast forms; however, the incidence of reactivation is very low, <0.1% (Vail et al., 2002).

Less Frequent Clinical Forms of Histoplasmosis

Chronic Pulmonary

Disease progresses over months or years in a small fraction of persons with histoplasmosis (1 in 2000, according to Goldman et al., 1999), affecting those with preexisting chronic obstructive pulmonary disease (COPD). A majority of these patients are cigarette smokers with emphysema. Symptoms include progressive lung infiltrates, fibrosis and cavitation, productive cough, dyspnea, chest pain, fevers, sweats, and fatigue, mimicking reactivation tuberculosis. Close to 80% of untreated chronic pulmonary disease improve without therapy, but the remainder will have progressive loss of respiratory function (Fig. 6.4b).

Progressive Disseminated Histoplasmosis

Persons living with AIDS or with other immune defects are at increased risk to develop extrapulmonary disseminated histoplasmosis (e.g., patients with lymphoma or those receiving immunosuppressive therapy following transplantation). Symptoms are those found in chronic histoplasmosis and, in addition, there may be weight loss, diarrhea, and/or open lesions in the mouth and nose. *Histoplasma capsulatum* yeast forms may invade the spleen, liver, bone marrow, adrenal glands, and/or gastrointestinal tract. Ten percent of AIDS-histoplasmosis patients present with a syndrome resembling septicemia (reviewed by Cano and Hajjeh, 2001).

Approximately 10–20% of patients living with AIDS who develop histoplasmosis present with septic shock, fever, hypotension, renal and hepatic failure, respiratory distress syndrome, and coagulopathy (de Francesco Daher et al., 2006). Even with treatment, the mortality rate for histoplasmosis in immunosuppressed patients is estimated at 33% whereas the rate in immune-normal patients is 17% (reviewed by Cano and Hajjeh, 2001). AIDS-histoplasmosis results from either fresh exposure or endogenous reactivation of dormant yeast forms in granulomatous foci in the lung. (Please see Section 6.3, Case Presentation 2).

Solitary Pulmonary Nodule

When due to *H. capsulatum*, such a lesion is a granuloma (or histoplasmoma) with inflammation at the periphery and fibrous tissue in the interior. It also is known as a "coin" lesion for its shape on chest radiographs (Fig. 6.7) (Scully et al., 1988). Eventually, the central portion calcifies. During the time when the lung nodule is uncalcified it is difficult to differentiate from a bronchogenic carcinoma. When surgically excised, it contains a granuloma with multinucleate giant cells and small round *Histoplasma* yeast forms (GMS stain) (Fig. 6.8).

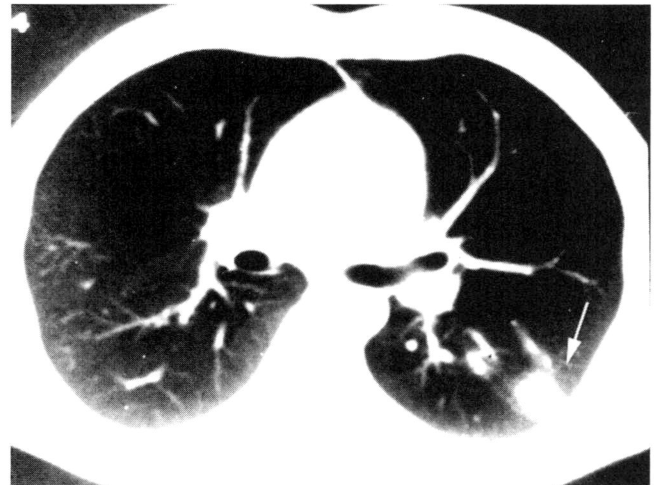

Figure 6.7 CT scan of solitary pulmonary nodule (*arrow*).
Source: Scully et al. (1988). Copyright © 1988 Massachusetts Medical Society, all rights reserved.

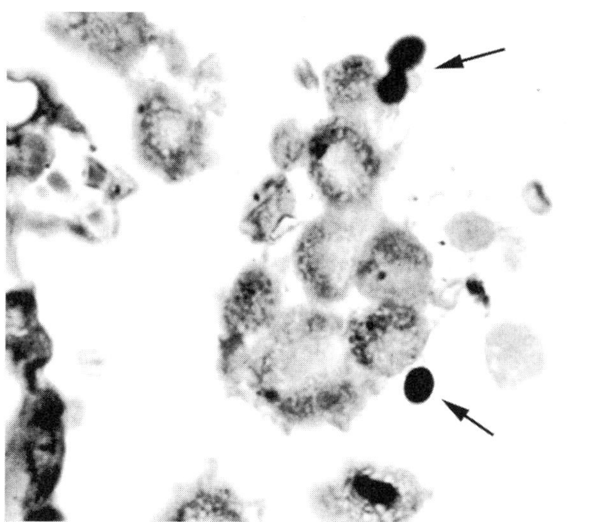

Figure 6.8 Solitary pulmonary nodule, histopathology, silver stain shows yeast forms (*arrows*).
Source: Scully et al. (1988). Copyright © 1988 Massachusetts Medical Society, all rights reserved.

Mediastinal Granuloma

This is an uncommon complication of healed pulmonary histoplasmosis in which the mediastinal[3] lymph nodes become enlarged and undergo caseation necrosis, with an outer fibrotic shell. The nodes may remain enlarged for months or longer and may exert pressure on adjacent structures: the bronchi, esophagus, or major blood vessels. The cause is thought to be an exaggerated inflammatory response to *H. capsulatum* antigens released into tissues surrounding lesions.

Mediastinal Fibrosis

This is a related and dire clinical form. An abnormal host response results in fibrotic envelopment of the structures of the mediastinum including the major airways, superior vena cava, and esophagus. Symptoms include hypoxemia, compression of the superior vena cava with resultant interference with blood flow, and difficulty in eating. Surgery often is unhelpful because it is difficult to identify the enveloped structures.

[3]Mediastinum. *Definition:* The mediastinum is in the center of the chest and contains the heart, thymus, and lymph nodes, along with portions of the aorta, vena cava, trachea, esophagus, and various nerves. It encompasses the area bordered by the breastbone (sternum) in front, the spinal column in back, the entrance to the chest cavity above, and the diaphragm below. Functionally, the mediastinum isolates the left and right lung from each other. *Source:* The Merck Manuals On-Line Medical Library. Porter RS et al. (eds.). Merck Research Laboratories, Whitehouse, NJ.

CNS Histoplasmosis

CNS histoplasmosis may present as chronic meningitis, intracranial mass lesions (histoplasmomas), or less commonly cerebritis consisting of unorganized regions of infection with a less organized tissue response (Deepe, 2005; Zalduondo et al., 1996). Symptoms may include headache, altered state of consciousness, and cranial nerve deficits. Seizures, ataxia, nuchal rigidity, and photophobia also may occur. Localized CNS infection is found in half of patients; the remainder have multisystem disseminated histoplasmosis with associated hepatosplenomegaly, lymphadenopathy, and mucocutaneous lesions. *Histoplasma* meningitis is accompanied by *pleocytosis* in the CSF in which lymphocytes predominate with cell counts of 10–100/μL. Reduced glucose and elevated protein are present in most patients. Histopathology reveals granulomatous inflammation. Histoplasmomas consist of dense fibrotic tissue surrounding a caseous center containing yeast forms. CT imaging may mistake this form for a malignancy or abscess. Histoplasmomas may or may not be associated with meningitis.

Presumed Ocular Histoplasmosis Syndrome (POHS)

This syndrome occurs mostly in patients coming from the Ohio, Missouri, and Mississippi river valley, a region endemic for histoplasmosis. Evidence of a fungal cause is based primarily on epidemiologic studies. Macular disease is associated with HLA-B7. A multifocal choroiditis develops, often with peripapillary scarring and occasionally with macular hemorrhage from choroidal neovascularization. Vitritis is absent. Patients may be largely asymptomatic unless the choroidal scars or neovascularization involve the optic nerve or fovea, producing decreased vision. Skin tests with histoplasmin are positive in 80% of U.S. cases. Chest radiographs often show characteristic old scars, frequently with calcification. Treatment is reserved primarily for neovascularization, in which case laser photocoagulation, often in conjunction with periocular corticosteroid injections, has been shown to be effective.

A scarcity of pathologic evidence exists linking *H. capsulatum* to POHS (reviewed by Prasad and van Gelder, 2005). *Histoplasma capsulatum* DNA was detected in the enucleated eye of a patient with bilateral POHS and a positive histoplasmin skin test. Using laser capture microdissection and PCR, *H. capsulatum* DNA was detected in macular and choroidal lesions in a pathology specimen with POHS. POHS is thought to be the result of a chronic reaction to antigenic localization of the fungus, acting as a nidus of inflammation. HLA subtypes (DRw2 and B7) are associated with POHS, suggesting a link between *H.*

capsulatum and genetic susceptibility. But it is unclear how such a link manifests as disease because there is no known crossreactivity between *H. capsulatum* antigens and ocular proteins. European patients with clinical POHS had no detectable exposure to *H. capsulatum*, or to other related infectious agents. In the absence of a mechanism to explain how exposure to *H. capsulatum* results in this ocular pathology, scientific skepticism exists, causing some investigators to suggest renaming the clinical syndrome "multifocal choroidopathy."

African Histoplasmosis (*H. capsulatum* var. *duboisii*)

Skin and skeleton are the most frequently affected organs, whereas pulmonary disease is uncommon (Gugnani and Muotoe-Okafor, 1997). Cutaneous lesions may be the only symptom. They usually are a cutaneous manifestation of widespread disseminated disease, involving bones and visceral organs. Skin lesions may be (i) superficial granulomas, either papules or nodules; (ii) subcutaneous granulomas that may rupture and heal spontaneously; and (iii) skin lesions arising from bone lesions and forming large granulomas, or draining sinuses. The skull, ribs, and vertebrae are most frequently involved and multiple bones may be infected. *Histoplasma capsulatum* var. *duboisii* produces granulomatous inflammation in the bone which may lead to sinus formation and cystic bone lesions. Osteolytic lesions may be seen in up to half of cases.

6.10 VETERINARY FORMS

Histoplasma capsulatum var. *farciminosum* has a colony and microscopic morphology similar to *H. capsulatum*. The var. *farciminosum* is the etiologic agent of epizootic lymphangitis, a systemic infection of equines affecting the subcutaneous lymph nodes and lymphatics of the neck. The disease is endemic in western, northern, and northeast Africa, the Middle East, India, and the Far East. Further information on this equine disease can be found in the publication *Foreign Animal Diseases-The Gray Book*, 7th ed. by the Committee on Foreign and Emerging Diseases of the United States Animal Health Association at the URL http://www.usaha.org/Publications.aspx.

The African variety of *H. capsulatum, H. capsulatum* var. *duboisii*, causes natural infections in humans and also in baboons. A well-documented case in a young adult female baboon *Papio* species was reported by Eggers (2001). Lesions are typically nodular, ulcerated, and exudative. Common sites of involvement include the face, ears, hands, feet, tail, scrotum, and buttocks. Regional draining lymph nodes often are enlarged. Skin lesions may extend to the adjacent bone, with resultant osteomyelitis. Granulomatous foci also may occur in liver and testis.

Histoplasmosis in Companion Animals

Histoplasmosis in companion animals (Brömel and Sykes, 2005; Edison et al., 2003 website), dogs and cats, is not contagious for other animals or humans and a diagnosis of histoplasmosis does not constitute grounds for euthanasia.

Canine Histoplasmosis

Respiratory signs of histoplasmosis in dogs include dyspnea, coughing, and abnormal lung sounds. Disseminated histoplasmosis may be accompanied by fever unresponsive to antibiotics, weight loss, depression, anorexia. Fungal invasion of the small intestine may result in the production of large amounts of watery diarrhea, and accompanying protein-losing enteropathy. Colonic involvement is associated with tenesmus, mucus, and fresh blood in the feces. Pale mucous membranes are common in dogs with gastrointestinal blood loss or bone marrow involvement. Other common findings are splenomegaly, hepatomegaly, visceral lymphadenomegaly, icterus, and ascites.

Feline Histoplasmosis

Cats with disseminated histoplasmosis often present with a fever unresponsive to antibiotics. Tachypnea, dyspnea, and abnormal lung sounds are common in cats, but coughing is rare. Respiratory signs can be the only signs if the infection is limited to the lungs. Weight loss, depression, anorexia, and pallor of mucous membranes also may be observed. Other common signs include peripheral lymphadenopathy, splenomegaly, and hepatomegaly. Ocular involvement is somewhat more common in cats than in dogs and may include conjunctivitis, granulomatous blepharitis, chorioretinitis, retinal detachment, and optic neuritis.

Therapy for Canine and Feline Histoplasmosis

Oral ITC is recommended for a moderate case of histoplasmosis (Brömel and Sykes, 2005), continued for at least 2 months beyond clinical resolution. Usually 4–6 months of therapy is required. Refractory cases with ocular and neurologic involvement may benefit from FLC, which gives better penetration of the eyes and CNS. Initial combination therapy with AmB and ITC may be helpful for severe pulmonary or progressive disseminated histoplasmosis. Ketoconazole is less effective, has more side effects than other azoles, and has more drug interactions, but can be used when cost is a limiting factor.

6.11 THERAPY

Mild forms of histoplasmosis do not generally require treatment (Goldman et al., 1999; Wheat et al., 2007). Moderately severe forms of histoplasmosis, such as acute pulmonary disease, or chronic pulmonary cavitary disease, are treated with oral ITC, 200–600 mg/day. Severe forms of histoplasmosis generally are treated initially with either AmB-deoxycholate at 1 mg/kg/day or a lipid preparation of AmB at 3.0–5.0 mg/kg/day. Other azole drugs including fluconazole (FLC), voriconazole (VRC), and posaconazole (PSC) have activity either as shown in case reports and case series, in vitro studies, or in animal models. Resistance to FLC may develop during therapy, and there are reports of VRC failing in patients who have previously failed FLC. Therefore, PSC may be a better alternative to ITC than either FLC or VRC. The echinocandin class of antifungal agents do not have activity against *H. capsulatum*.

Acute Pulmonary Histoplasmosis

- *Mild–Moderate.* In the Wagon Train outbreak, specific antifungal therapy was not necessary (Section 6.3, Case Presentation 1). Only 5% of infections require antifungal therapy. When symptoms persist for 1 month or more, therapy with oral ITC is advised: 200 mg 3×/day for 3 days, then 200 mg once or twice/day for 6–12 weeks.
- *Moderately Severe.* Treatment for these patients consists of lipid AmB (3.0–5.0 mg/kg/day, IV) for 2 weeks, then, after improvement, ITC, 200 mg 3×/day for 3 days, followed by ITC, 200 mg twice/day, for a total of 12 weeks. AmB-deoxycholate is recommended for patients at low risk for nephrotoxicity at 1 mg/kg/day, IV. Patients who develop respiratory distress or hypoxemia benefit from methylprednisolone (0.5–1.0 mg/kg/day, IV) for the first 1–2 weeks of therapy.

Complications of Acute Pulmonary Histoplasmosis

- *Pericarditis.* Nonsteroidal anti-inflammatory medication is recommended for mild symptoms. When symptoms do not subside after several days of therapy, steroids are indicated, for example, prednisone, 0.5–1.0 mg/kg/day up to a maximum of 80 mg/day, tapering the dosage in 1–2 weeks. Along with steroids, antifungal therapy is recommended—ITC 200 mg/3 × /day for 3 days then 2 × /day for 6–12 weeks. A similar regimen is recommended for patients with mild histoplasmosis who develop rheumatologic symptoms.
- *Mediastinal Granuloma.* A trial of AmB is indicated for these patients before considering surgery. Patients with mediastinal fibrosis may not benefit from antifungal therapy. Some patients with pulmonary vessel obstruction benefit from intravascular stents. If the differential diagnosis of mediastinal granuloma and mediastinal fibrosis is in doubt, treatment with ITC, 200 mg once or twice/day for 12 weeks is recommended.
- *Pulmonary Nodules.* Also known as histoplasmomas, these healing or healed lesions do not benefit from antifungal therapy.

Chronic Cavitary Histoplasmosis

A long course of azole therapy is necessary to treat these patients, consisting of 200 mg ITC 3×/day for 1 year or more. The reason to treat for more than 1 year is to reduce the risk of relapse. Patients who do not respond to oral azole therapy should receive AmB for a total dose of 35 mg/kg given over 12–16 weeks.

Progressive Disseminated Histoplasmosis

In the case of the AIDS patient with progressive disseminated histoplasmosis (Section 6.3, Case Presentation 2) induction therapy consisted of AmB (1 mg/kg/day) for a total of 500 mg, followed by maintenance therapy on oral ITC for 6 months. A similar treatment regimen, or beyond 6 months' duration, is recommended for patients who require prolonged high dose immunosuppressive therapy.

A more general recommendation for moderately severe to severe progressive disseminated histoplasmosis consists of liposomal AmB (3 mg/kg/day) for 1–2 weeks, followed by oral ITC, 200 mg 3 × /day for 3 days, followed by 200 mg 2 × /day for at least 12 months. AmB-deoxycholate is advised in patients at low risk for nephrotoxicity.

Patients who remain $CD4^+$ T-lymphocyte deplete or who require prolonged immunosuppressive therapy for a primary disease benefit from indefinite suppressive therapy with ITC at 200 mg/day. Persons living with AIDS who are compliant in taking combination ART respond better to progressive disseminated histoplasmosis compared with those who do not have access or are not compliant with ART. Two types of monitoring of these patients to gauge the effectiveness of therapy are blood concentrations of ITC and urine antigen HPA testing. Antigenuria subsiding over months to low but persistent concentrations in patients who have no other evidence of active histoplasmosis may indicate that suppressive therapy is no longer necessary. In persons living with AIDS, recovery of the

$CD4^+$ T-lymphocyte count to >150/μL is an indicator of immune system recovery such that ITC suppressive therapy may be discontinued.

Patients with a mild form of progressive disseminated histoplasmosis can be treated with 400 mg/day of oral ITC and, when improvement occurs, with a change in dose to 200 mg/day to complete a 6 month course.

Other Special Categories of Histoplasmosis Patients

Therapy recommendations for histoplasmosis in pregnancy or in children may be found in Wheat et al. (2007).

6.12 LABORATORY DETECTION, RECOVERY, AND IDENTIFICATION

Direct Examination

In acute pulmonary histoplasmosis, *H. capsulatum* may be seen microscopically in sputum or bronchial washings as very small ovoid yeast forms with dimensions of 2–4 μm diameter. When cells are budding, buds have a narrow base. Direct smears of sputum are stained with Wright's, GMS, or PAS stains. Yeast forms in sputum may be any yeast and further studies are indicated. In disseminated histoplasmosis, the *H. capsulatum* yeast forms may be seen within monocytes in thin films of peripheral blood, bone marrow, and also extracellularly in any body fluid.

Histopathology

The yeast form of *H. capsulatum* usually can be found in tissue removed from the diseased site, often within macrophages in granulomas. *Histoplasma capsulatum* does not stain well with hematoxylin and eosin (H&E). GMS or PAS stains are best for fixed, embedded tissues. The yeast forms stain brown to black with GMS stain (Fig. 6.9).

Bone Marrow Aspirate

Bone marrow aspirates or peripheral blood smears stained with Wright's stain display intracellular yeast forms (Fig. 6.5).

Differential Laboratory Diagnosis

Other yeast and a yeast-like fungi directly detected in clinical specimens which resemble *H. capsulatum* are *Candida glabrata* and the fission yeast form of *Penicillium marneffei*. Other microbes that can, at times, be confused

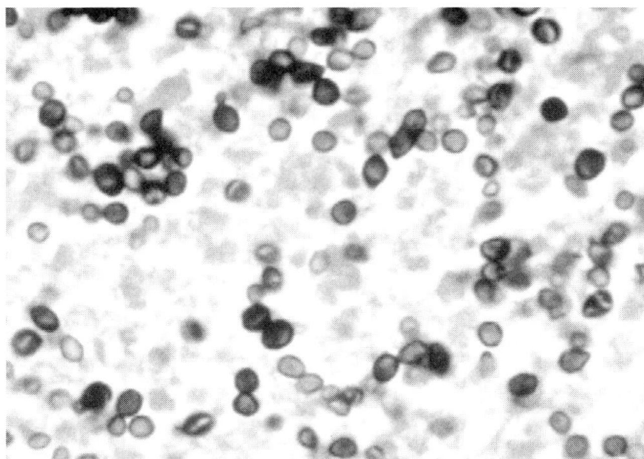

Figure 6.9 Histopathology of a tissue section showing yeast forms of *Histoplasma capsulatum* stained with GMS stain. Yeast forms are 2–4 μm in diameter with a narrow connection between mother and daughter cells (Chandler and Watts, 1987).
Source: Used with permissions from Dr. John C. Watts and from the American Society for Clinical Pathology, ©ASCP.

in tissues and exudates with *H. capsulatum* are *Pneumocystis jirovecii* and single endospores of *Coccidioides* species. Please see below "Microscopic Morphology" for look-alike molds that resemble the mold form of *H. capsulatum* cultures grown at 30°C.

Culture

Early Morning Sputum

The yield of sputum cultures is low in self-limited acute pulmonary histoplasmosis, but >50% positive in chronic pulmonary and ≥80% in progressive disseminated disease. Multiple sputum samples often are necessary to obtain a positive result. The *H. capsulatum* and *H. capsulatum* var. *duboisii* mycelial forms grow slowly when incubated at room temperature to 30°C on SDA or Mycosel agar. Colonies usually appear in 7–10 days, but may take longer, up to a month or more. Conidia appear after colonies have grown out and produced aerial hyphae. When a suspect culture appears on a Petri plate, transfer it to a screw cap test tube and proceed as follows: Young mycelial cultures are best identified by the nucleic acid probe test (AccuProbe®, Gen-Probe, Inc.). Further mycologic studies can be carried out to confirm the identity by microscopic morphology of the mold phase and, optionally, by attempting conversion to the yeast form on enriched medium.

Bone Marrow and Blood Cultures

These clinical specimens are useful in progressive disseminated histoplasmosis. Bone marrow cultures may be positive in >75% of samples, with positive blood cultures occurring in 50–70% of specimens. The better results are obtained with the lysis-centrifugation method. (Please see Chapter 2, Section 2.3.5.5 Types of Blood Culture Methods.) Growth out of blood cultures is slow and may take up to 4 weeks to become positive.

Safety Note: All manipulations should be conducted by trained and supervised personnel in a well-functioning, currently certified laminar flow biological safety cabinet. Personnel should wear adequate personal protective equipment. Suspected *H. capsulatum* cultures should be grown in agar slants (=slopes) in test tubes with cotton stoppers or screw cap closures. If colonies suspected of being *H. capsulatum* appear in Petri plate cultures, as soon as possible, transfer the growth to agar slant test tubes. The Petri plate culture should then be sealed with Parafilm® or Shrink Seals® (Scientific Device Laboratory, Des Plaines, IL). Shrink seals are colorless transparent cellulose bands made for this purpose. The Petri plate then is placed in a Ziploc® bag and stored at 4°C. When growth is established in the test tube, the Petri plate culture should be autoclaved, then discarded.

Nucleic Acid Probe Test

AccuProbe® is the *H. capsulatum* culture identification test (Gen-Probe, Inc.; http://www.gen-probe.com/prod_serv/culture.asp). This test format is ribosomal DNA–RNA hybridization. The principle is described in Chapter 2.

Colony Morphology

On SDA or other nonenriched medium, the colony usually is white and cottony, becoming tan with age. On enriched medium, such as blood agar, the colony is brownish, smooth, and sometimes heaped with little or no aerial hyphae. This type of colony usually produces few conidia (Fig. 6.10). Conversion to the yeast form in vitro, while a desirable confirmatory test, conversion on BHI + 5% sheep blood agar slants at 37°C is very slow and some strains are difficult to convert in vitro.

Shipment of Cultures to Reference Laboratories

Please consult the following sources:

- "Technical Instructions for the Safe Transport of Dangerous Goods by Air" issued by the International Civil Aviation Organization at the URL http://www.icao.int/icaonet/dcs/9284.html.
- "Packing Instruction 650" is issued by the International Air Transport Association (IATA) at the URL http://www.iata.org/NR/ContentConnector/CS2000/SiteInterface/sites/whatwedo/dangerousgoods/file/PI650.pdf.
- The U.S. Department of Transportation (DOT) regulates all modes of transportation within the United States. DOT has regulations for hazardous material shipment via ground transportation.
    - When questions arise about ground shipment of hazardous materials, including infectious agents, please refer to the DOT website URL: http://hazmat.dot.gov.
    - For more information, please contact the U.S. DOT's Hazardous Materials Information Center at 1-800-467-4922 from 9:00 A.M. to 5:00 P.M, Eastern time, Monday–Friday, except U.S. federal holidays.
- Please be aware that when shipping infectious agents via ground within the United States, the State DOT could possibly have additional regulations.

Microscopic Morphology

The *H. capsulatum* mold form grows as hyaline and septate, with branching hyphae that produce two types of

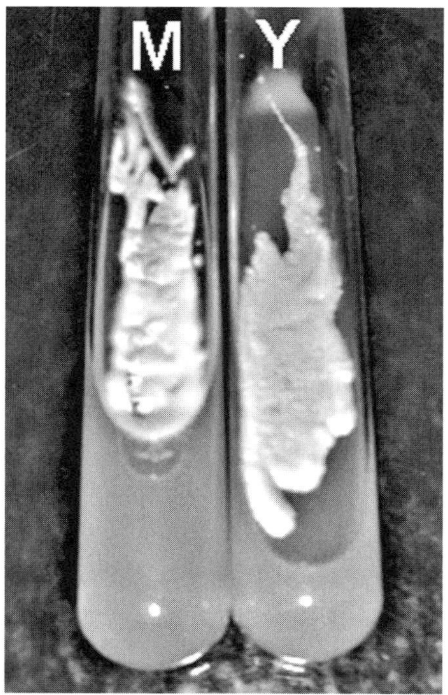

Figure 6.10 *Histoplasma capsulatum* colonies. "M," mycelial form, SDA; "Y," yeast form, BHI.
Source: Used with permission from Dr. Rosely M. Zancopé-Oliveira, Fundação Oswaldo Cruz, Rio de Janeiro, Brazil.

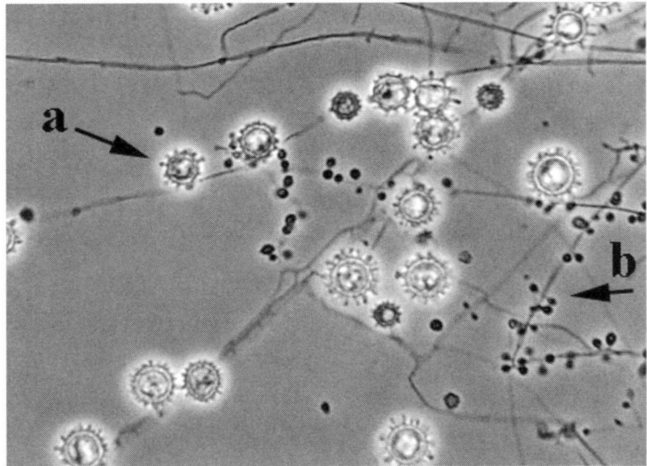

Figure 6.11 *Histoplasma capsulatum* microscopic morphology illustrates (a) tuberculate macroconidium and (b) microconidium.
Source: E. Reiss, CDC.

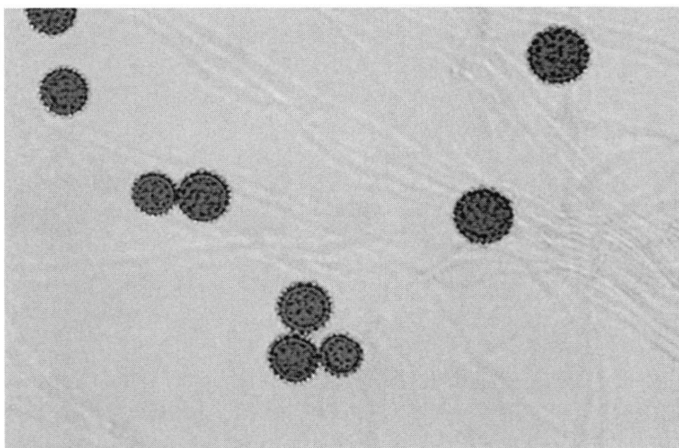

Figure 6.12 Tuberculate macroconidia are produced by look-alike *Chrysosporium* species (no microconidia are produced); 400×, lactofuchsin stain.
Source: E. Reiss, CDC.

conidia. (See Fig. 6.11.) Tease mounts are examined first; then if there is insufficient sporulation, a slide culture is prepared. Viewed microscopically, tease mounts and slide cultures illustrate *tuberculate* macroconidia and microconidia. These two types of conidia are (i) small, 4–6 μm diameter, round to pyriform, smooth-walled microconidia; and (ii) large, 8–18 μm diameter, round, thick-walled macroconidia surrounded by spiny projections termed tuberculations. Macroconidia are not believed to be the infectious propagules but have diagnostic importance in laboratory identification of the *H. capsulatum* mold form.

Tuberculate macroconidia also are produced by look-alike but monomorphic saprobes (e.g., *Sepedonium* and *Myceliophthora* species). The latter fungi do not cause disease in humans or animals (Fig. 6.12).

Dimorphism

When growing in the body during infection, *H. capsulatum* converts to a small budding yeast, 2–4 μm diameter, with a slender, hourglass-shaped constriction between mother and daughter cells. Morphogenesis to the yeast form can also be obtained when grown in vitro on enriched medium[4] at 37°C. Cysteine in the medium stimulates yeast form development.

Histoplasma capsulatum var. *duboisii*

Histopathology

Histoplasma capsulatum var. *duboisii* infection results in a pyogranulomatous inflammatory reaction in the dermis and subcutaneous tissues with numerous macrophages, multinucleate giant cells (both Langhans and foreign-body types), and small aggregates of neutrophils distributed throughout. In cases with ulceration, neutrophils will be more numerous. Within the cytoplasm of giant cells and histiocytes, there are numerous, round to ovoid, pale-staining organisms measuring from 8 to 15 μm in diameter, with thick walls, and single, narrow-based buds. A single, small, round to elongate, basophilic-staining body often can be observed near the center of the yeast. The yeast forms stain well with GMS or Gridley's fungal stain, and often are found in pairs and short chains. The identification of *H. capsulatum* var. *duboisii* generally can be made histologically by its unique morphologic appearance and resultant inflammation in tissue.

Differential diagnosis for *H. capsulatum* var. *duboisii* includes several other fungi. Yeast forms of *H. capsulatum* var. *capsulatum* measure only 2–4 μm in diameter. *Cryptococcus neoformans* yeast, slightly smaller than *H. capsulatum* var. *duboisii*, has a single bud, thin cell wall, and generally a wide, clear, unstained capsule that stains positive with mucicarmine stain. Immature or nonendosporulating spherules of *Coccidioides* species can be similar in shape but often are slightly larger (5–25 μm) than *H. capsulatum* var. *duboisii*. The presence of mature spherules containing endospores would clearly differentiate the two fungi. The *Blastomyces dermatitidis* yeast form, very similar in size and shape to that of *H. capsulatum* var. *duboisii*, may be differentiated by its broad-based budding and lack of chain formation. Ultrastructurally, *B. dermatitidis* has multiple nuclei, whereas *H. capsulatum* var. *duboisii* has a single nucleus.

[4]Enriched medium. *Definition*: Includes blood agar, BHI, SABHI, or BHI + 5% sheep blood agar. Further details of these agars are found in Chapter 2.

Serology

In the Wagon Train outbreak (Section 6.3, Case Presentation 1), admission serology and cultures were negative, but later, complement-fixing antibodies appeared and, finally, precipitins (immunodiffusion). Detectable antibody concentrations are produced within 10–21 days after the first symptoms appear and occur in 90% of patients (Goldman et al., 1999). If the disease is mild the antibody response is low.

Immunodiffusion

The double immunodiffusion test for histoplasmosis provides a specific diagnosis based on the demonstration of the diagnostic m- and h-immunoprecipitates. Kits for this test can be obtained from Immuno-Mycologics® Inc., Norman, Oklahoma. The m-precipitin is first to arise in infection, the h-precipitin less frequently, and more often in extrapulmonary dissemination. These reactions are very specific for histoplasmosis. Functionally, the h-antigen has glucosidase activity, whereas m-antigen has catalase activity. The immunodiffusion test will be positive 4–6 weeks after the first symptoms appear (Table 6.2).

Complement Fixation (CF)

Two antigens are used: histoplasmin and a whole formalin-killed yeast form antigen. CF antibody titers against the yeast form antigen of $\geq$1:32 or rising titers are significant. Crossreactions occur with sera from patients infected with *Blastomyces dermatitidis* or with *Coccidioides* species. When histoplasmin is used as a serologic antigen, CF titers usually are lower than those obtained with the yeast form antigen. When histoplasmin skin tests were available it was realized that a positive skin test will induce an increase in the CF antibody response to the histoplasmin antigen.

Urine Antigen (HPA) Test

Antigenuria precedes development of antibodies. A commercial test for urine antigen is most useful in disseminated histoplasmosis, especially persons living with AIDS, and sometimes early in acute pulmonary histoplasmosis (where there has been a heavy exposure to the fungus). This HPA test (*Histoplasma* polysaccharide antigen) is available from MiraVista Laboratories®, Indianapolis, Indiana (URL: http://www.miravistalabs.com/). The HPA test can also be useful in predicting relapse in patients receiving chronic suppressive therapy.

Histoplasmin Skin Test

At present there is no current source for histoplasmin, the skin test antigen.

This test has the following limitations:

- Positive reaction in healthy exposed persons.
- Anergy (negative skin test) in acute disseminated and chronic pulmonary histoplasmosis.
- Skin testing results in an elevation of complement-fixing antibodies. Draw blood ***before*** applying skin test.
- Crossreacts with *Blastomyces dermatitidis*.

The skin test is useful only in:

- Young children
- Known negative reactors with recent exposure [e.g., lab workers or spelunkers (cave explorers)]
- Residents outside an endemic area with recent travel in an endemic area

Summary

See Fig. 6.13. When a culture is available, the identification can be made with the AccuProbe DNA probe (Gen-Probe, Inc.) kit. Microscopic morphology of the mold form is effective but slow. When identification is based solely on the microscopic morphology, confirmation should be sought by attempting to convert the culture to the yeast form, even though this is a slow process.

When no culture is available, biopsy and histopathology and/or serology can be helpful. Precipitin tests in agar gel can turn positive approximately 4 weeks from the first clinical signs of illness. The complement fixation test is performed in some reference laboratories and

Table 6.2 Serologic Results in Acute Pulmonary Histoplasmosis[a]

Serologic test for histoplasmosis	Percent positive after the first appearance of clinical symptoms at		
	1 week	2 weeks	4 weeks
Complement fixation test with yeast antigen	7	66	77
Immunodiffusion test for h- and m-precipitins	0	0	50

[a]Data from Davies (1986).

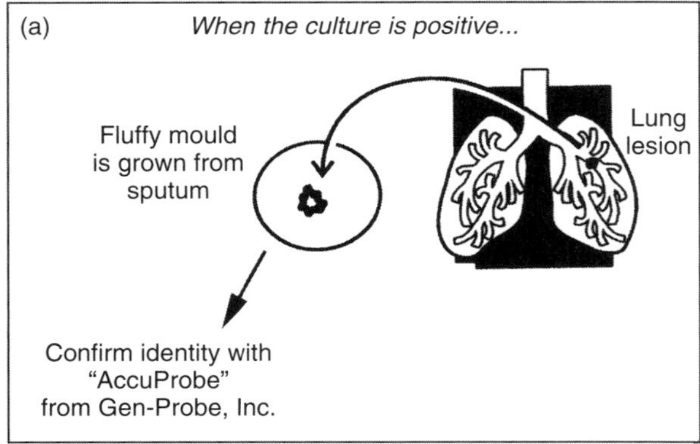

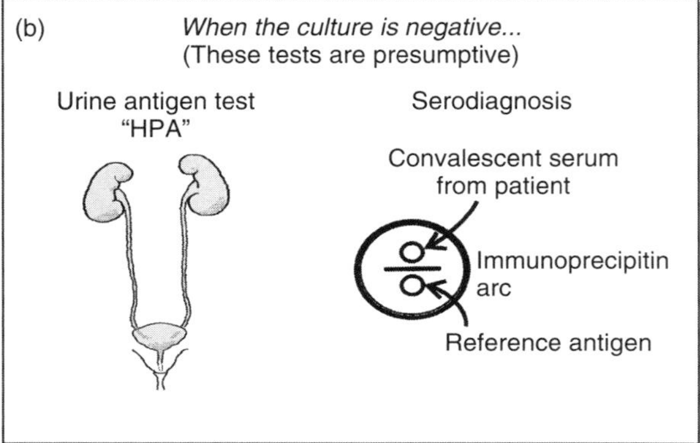

Figure 6.13 Diagnosis of histoplasmosis with or without a culture.
Source: E. Reiss, CDC.

may be superseded by EIA tests in the future. The HPA test for antigenuria is positive in the latent period before antibodies become evident and is indicated early in acute pulmonary histoplasmosis, where there has been exposure to a heavy inoculum, resulting in a moderate to severe infection, and to disseminated histoplasmosis in persons living with AIDS and other immunocompromising conditions. Skin test reagents are not available because there is no licensed manufacturer and in any case such tests have serious limitations.

SELECTED REFERENCES FOR HISTOPLASMOSIS

Ajello L, 1971. Distribution of *Histoplasma capsulatum* in the United States, Chapter 15, pp. 103–122, *in*: Ajello L, Chick EW, Furcolow ML (eds.), *Histoplasmosis—Proceedings of the Second National Conference*. CC Thomas Publishers, Springfield, IL.

Bohse ML, Woods JP, 2007a. Expression and interstrain variability of the *YPS3* gene of *Histoplasma capsulatum*. *Eukaryotic Cell* 6: 609–615.

Bohse ML, Woods JP, 2007b. RNA interference-mediated silencing of the *YPS3* gene of *Histoplasma capsulatum* reveals virulence defects. *Infect Immun* 75: 2811–2817.

Brodsky AL, Gregg MB, Lowenstein MW, Kaufman L, Mallison GF, 1973. Outbreak of histoplasmosis associated with the 1970 Earth Day activities. *Am J Med* 54: 333–342.

Brömel C, Sykes JE, 2005. Histoplasmosis in dogs and cats. *Clin Techniques Small Anim Pract* 20: 227–232.

Cano MV, Hajjeh RA, 2001. The epidemiology of histoplasmosis: A review. *Semin Respir Infect* 16: 109–118.

Catlett NL, Yoder OC, Turgeon BG, 2003. Whole-genome analysis of two-component signal transduction genes in fungal pathogens. *Eukaryotic Cell* 2: 1151–1161.

Chandler FW, Watts JC, 1987. *Pathologic Diagnosis of Fungal Infections*. ASCP Press, Chicago, Fig. 194 on p. 133–.

Chu J, Feudtner C, Heydon K, Walsh T, Zaoutis T, 2006. Hospitalizations for endemic mycoses: A population based national study. *Clin Infect Dis* 42: 822–825.

Davies SF, 1986. Serodiagnosis of Histoplasmosis, pp. 9–15 *in:* Sarosi GA (ed.), *Seminars in Respiratory Infections*, Vol. 1. Grune & Stratton, Orlando, FL.

Deepe GS, 2005. Histoplasmosis, Chapter 262, *in*: Mandell GL, Bennett JE, Dolin R. (eds.), *Principles and Practice of Infectious Disease*, 6th ed. Churchill-Livingstone, Edinburgh, UK.

Deepe GS, Wuthrich M, Klein BS, 2005. Progress in vaccination for histoplasmosis and blastomycosis: Coping with cellular immunity. *Med Mycol* 43: 381–389.

de Francesco Daher E, de Sousa Barros FA, da Silva Junior GB, Takeda CFV, Mota RMS, Ferreira MT, Martins JC, Oliveira SAJ, Gutierrez-Adrianzen OA, 2006. Risk factors for death in acquired immunodeficiency syndrome-associated disseminated histoplasmosis. *Am J Trop Med Hyg* 74: 600–603.

Eggers J, 2001. A case of histoplasmosis duboisii in a female baboon, *Papio* species. The Armed Forces Institute of Pathology, Department of Veterinary Pathology, Wednesday Slide Conference 2000–2001. Conference 23 Case I–N99–909 (AFIP 2749408). URL: http://www.afip.org/vetpath/WSC/wsc00/00wsc23.htm.

Goldman M, Johnson PC, Sarosi GA, 1999. Fungal pneumonias—the endemic mycoses. *Clin Chest Med* 20: 507–519.

Gugnani HC, Muotoe-Okafor F, 1997. African histoplasmosis: a review. *Rev Iberoamer Micol* 14: 155–159.

Guimarães AJ, Hamilton AJ, de M Guedes HL, Nosanchuk JD, Zancopé-Oliveira RM, 2008. Biological function and molecular mapping of M antigen in yeast phase of *Histoplasma capsulatum*. *PLoS One* 3: e3449.

Guimarães AJ, Nosanchuk J, Zancopé-Oliveira RM, 2006. Diagnosis of histoplasmosis. *Braz J Microbiol* 37: 1–13.

Gustafson TL, Kaufman L, Weeks R, Ajello L, Hutcheson RH Jr, Wiener SL, Lambe DW Jr, Sayvetz TA, Schaffner W, 1981. Outbreak of acute pulmonary histoplasmosis in members of a wagon train. *Am J Med* 71: 759–765.

Kasuga T, White TJ, Koenig G, McEwen J, Restrepo A, Castaneda E, Da Silva Lacaz C, Heins Vaccari EM, De Freitas RS, Zancope-Oliveira RM, Qin Z, Negroni R, Carter DA, Mikami Y, Tamura M, Taylor ML, Miller GF, Poonwan N, Taylor JW, 2003. Phylogeography of the fungal pathogen *Histoplasma capsulatum*. *Mol Ecol* 12: 3383–3401.

Kauffman CA, 2003. Histoplasmosis, pp. 285–298, *in*: Dismukes WE, Pappas PB, Sobel JD (eds.), *Clinical Mycology*. Oxford University Press, Oxford, UK.

Klein BS, Tebbets B, 2007. Dimorphism and virulence in fungi. *Curr Opin Microbiol* 10: 314–319.

Kleinman MB, 2003. *Histoplasma capsulatum* (Histoplasmosis), Chapter 264, *in*: Long SS, Pickering LK, Prober CG (eds.), *Principles and Practice of Pediatric Infectious Diseases*, 2nd ed. Churchill Livingstone (Elsevier), Philadelphia, PA.

KUGLER S, SEBGHATI TS, EISSENBERG LG, GOLDMAN WE, 2000. Phenotypic variation and intracellular parasitism by *Histoplasma capsulatum*. *Proc Natl Acad Sci USA* 97: 8794–8798.

LEE J-H, SLIFMAN NR, GERSHON SK, EDWARDS ET, SCHWIETERMAN WD, SIEGEL JN, WISE RP, BROWN SL, UDALL JN Jr, BRAUN MM, 2002. Life-threatening histoplasmosis complicating immunotherapy with tumor necrosis factor α antagonists infliximab and etanercept. *Arthritis Rheumatism* 46: 2565–2570.

LOULERGUE P, BASTIDES F, BAUDOUIN V, CHANDENIER J, MARIANI-KURKDJIAN P, DUPONT B, VIARD JP, DROMER F, LORTHOLARY O, 2007. Literature review and case histories of *Histoplasma capsulatum* var. *duboisii* infections in HIV-infected patients. *Emerging Infect Dis* 13: 1647–1652.

LUBY J, SOUTHERN JP, HALEY C, VAHLE K, MUNFORD R, HALEY R, 2005. Recurrent exposure to *Histoplasma capsulatum* in modern air-conditioned buildings. *Clin Infect Dis* 41: 170–176.

MILDVAN D, 2007. *International Atlas of AIDS*, 4th ed. Springer Science and Business Media, Philadelphia, PA.

NEWMAN SL, GOOTEE L, BRUNNER G, DEEPE GS Jr, 1994. Chloroquine induces human macrophage killing of *Histoplasma capsulatum* by limiting the availability of intracellular iron and is therapeutic in a murine model of histoplasmosis. *J Clin Invest* 93: 1422–1429.

NOSANCHUK JD, 2005. Protective antibodies and endemic dimorphic fungi. *Curr Mol Med* 5: 435–442.

NOSANCHUK JD, STEENBERGEN JN, SHI L, DEEPE GS Jr, CASADEVALL A, 2003. Antibodies to a cell surface histone-like protein protect against *Histoplasma capsulatum*. *J Clin Invest* 112: 1164–1175.

PRASAD AG, VAN GELDER RN, 2005. Presumed ocular histoplasmosis syndrome. *Curr Opin Ophthalmol* 16: 364–368.

RAPPLEYE CA, EISSENBERG LG, GOLDMAN WE, 2007. *Histoplasma capsulatum* alpha-(1,3)-glucan blocks innate immune recognition by the beta-glucan receptor. *Proc Natl Acad Sci USA* 104: 1366–1370.

RIPPON JW, 1988. Histoplasmosis, pp. 381–423, *in*: RIPPON JW (ed.), *Medical Mycology: Pathogenic Fungi and Pathogenic Actinomycetes*, 3rd ed. WB Saunders, Philadelphia, PA.

SUZAKI A, KIMURA M, KIMURA S, SHIMADA K, MIYAJI M, KAUFMAN L, 1995. An outbreak of acute pulmonary histoplasmosis among travelers to a bat-inhabited cave in Brazil. *Kansenshogaku Zasshi* 69: 444–449.

SCULLY RE, MARK EJ, MCNEELY WF, MCNEELY BU, 1988. Case records of the Massachusetts General Hospital (Case 49-1988): A 40-year-old man with a persistent nodular density in the left lower lobe. *New Engl J Med* 319: 1530–1537.

VAIL GM, YOUNG RS, WHEAT LJ, FILO RS, CORNETTA K, GOLDMAN M, 2002. Incidence of histoplasmosis following allogeneic bone marrow transplant or solid organ transplant in a hyperendemic area. *Transplant Infect Dis* 4: 148–151.

VINCENT RD, GOEWERT R, GOLDMAN WE, KOBAYASHI GS, LAMBOWITZ AM, MEDOFF G, 1986. Classification of *Histoplasma capsulatum* isolates by restriction fragment polymorphisms. *J Bacteriol* 165: 813–818.

WHEAT LJ, FREIFELD AG, KLEIMAN MB, BADDLEY JW, MCKINSEY DS, LOYD JE, KAUFFMAN CA, 2007. Clinical practice guidelines for the management of patients with histoplasmosis: 2007 update by the Infectious Diseases Society of America. *Clin Infect Dis* 45: 807–825.

WOODS JP, 2003. Knocking on the right door and making a comfortable home: *Histoplasma capsulatum* intracellular pathogenesis. *Curr Opin Microbiol* 6: 327–331.

ZALDUONDO FM, PROVENZALE JM, HULETTE C, GORECKI JP, 1996. Meningitis, vasculitis, and cerebritis caused by CNS histoplasmosis: Radiologic-pathologic correlation. *AJR Am J Roentgenol* 166: 194–196.

ZANCOPÉ-OLIVEIRA RM, MORAIS E SILVA TAVARES P, MUNIZ MM, 2005. Genetic diversity of *Histoplasma capsulatum* strains in Brazil. *FEMS Immunol Med Microbiol* 45: 443–449.

ZANCOPÉ-OLIVEIRA RM, REISS E, LOTT TJ, MAYER LW, DEEPE GS Jr, 1999. Molecular cloning, characterization, and expression of the M antigen of *Histoplasma capsulatum*. *Infect Immun* 67: 1947–1953.

WEBSITES CITED

EDISON L, LATIMER KS, BAIN PH, ROBERTS RE, 2003. Canine and feline histoplasmosis. http://www.vet.uga.edu/vpp/clerk/Edison/.

EGGERS J, 2001. A case of histoplasmosis duboisii in a female baboon, *Papio* species. The Armed Forces Institute of Pathology, Department of Veterinary Pathology, Wednesday Slide Conference 2000–2001. Conference 23 Case I—N99-909 (AFIP 2749408). http://www.afip.org/vetpath/WSC/wsc00/00wsc23.htm

Epizootic lymphangitis. Further information on this equine disease can be found Committee on Foreign and Emerging Diseases of the United States Animal Health Association, 2008. *Foreign Animal Diseases-The Gray Book*, 7th ed. USAHA: St. Joseph, MO at the URL: http://www.usaha.org/Publications.aspx.

Gen-Probe, Inc., at the URL http://www.gen-probe.com/prod_serv/culture.asp.

"Histoplasmosis—Protecting Workers at Risk" at the URL http://www.cdc.gov/niosh/docs/2005-109/.

Immuno-Mycologics Inc. at the URL http://www.immy.com/index.asp.

Mira Vista Laboratories, Inc. Source of the histoplasmosis polysaccharide antigen (HPA) urine antigen test for histoplasmosis. http://www.miravistalabs.com/.

"Packing Instruction 650" is issued by the International Air Transport Association (IATA): http://www.iata.org/NR/ContentConnector/CS2000/SiteInterface/sites/whatwedo/dangerousgoods/file/PI650.pdf.

Respirator selection criteria from the U.S. National Institute of Occupational Safety and Health (NIOSH), at the URL http://www.cdc.gov/niosh/npptl/topics/respirators/default.html.

"Technical Instructions for the Safe Transport of Dangerous Goods by Air" issued by the International Civil Aviation Organization: http://www.icao.int/icaonet/dcs/9284.html

U.S. Department of Transportation (DOT) rules governing shipment of hazardous materials, including infectious agents: http://hazmat.dot.gov

QUESTIONS

The Answer Key to multiple choice questions may be found at the end of the book.

1. A mold is isolated from a sputum specimen. The colony on SDA-Emmons at 30°C is slow growing and light brown in color. You make a tease mount. What would you expect to find if the mold is *Histoplasma capsulatum*?
   A. Arthroconidia
   B. Chlamydospores
   C. Hourglass-shaped yeast forms
   D. Tuberculate macroconidia
2. Which of the following produce conidia or macroconidia that may be mistaken for those of *Histoplasma capsulatum*?
   A. *Blastomyces dermatitidis*
   B. *Chrysosporium*
   C. *Myceliophthora*

D. *Sepedonium*

E. All of the above

3. Which of the following are hallmarks of acute pulmonary histoplasmosis?

1. "Halo sign" in chest radiograph
2. "Snowstorm" appearance of chest radiograph with multiple granulomas in both lungs
3. An inflammatory response with palisaded epithelioid cells and multinucleate giant cells
4. Peripheral blood smear: monocytes with phagocytosed yeast forms

Which are correct?

A. 1 + 3

B. 1 + 4

C. 2 + 3

D. 2 + 4

4. A solitary pulmonary nodule may result from

A. Bronchogenic carcinoma

B. Histoplasmosis

C. Tuberculosis

D. All of the above

E. None of the above

5. What proportion of persons living with AIDS who develop histoplasmosis present with normal appearing chest radiographs?

A. Less than 50%

B. About 50%

C. Greater than 50%

6. *Histoplasma capsulatum* and *Blastomyces dermatitidis* can be differentiated because the latter fungus:

A. Is more difficult to convert to the yeast form

B. Has a larger yeast form with a broad based bud

C. Produces tuberculate macroconidia

D. None of the above are true

E. All of the above are true

7. Which of the following antifungal agents is not indicated as primary therapy for histoplasmosis?

A. Amphotericin B

B. Echinocandins

C. Itraconazole

D. Lipid preparation of amphotericin B

8. Discuss serologic diagnosis of histoplasmosis with respect to the role of antigen and antibody detection. What antigens elicit antibodies during infection? Explain the order of appearance of the antibodies in pulmonary histoplasmosis. What is known about the nature of the antigen circulating in histoplasmosis and detected in the HPA test for urine antigen?

9. Explain the role of birds and bats in outbreaks of histoplasmosis.

10. The patient with central nervous system symptoms and an intracranial mass lesion failed to respond to antituberculosis therapy. The physician tested the CSF for cryptococci but none were found. The laboratory reported that *Sepedonium* grew from a sputum specimen. How could the laboratory finding prompt reconsideration of the fungal etiology?

Chapter 7

Paracoccidioidomycosis

7.1 PARACOCCIDIOIDOMYCOSIS-AT-A-GLANCE

- *Introduction/Disease Definition.* Paracoccidioidomycosis is a primary pulmonary mycosis with mucocutaneous dissemination, endemic to Central and South America. A long latent period, months or even years, may elapse between the time of infection and the development of clinical disease.
- *Etiologic Agent. Paracoccidioides brasiliensis* is a dimorphic fungal pathogen.
- *Geographic Distribution.* Central and South America, especially Southern Brazil.
- *Ecologic Niche. Paracoccidioides brasiliensis* is rarely isolated from the environment. It has been isolated from armadillos, a sentinel animal for an environmental microfocus of *P. brasiliensis*.
- *Epidemiology.* The annual estimated incidence is 1–3 /100,000 population in Brazil.
- *Risk Groups/Factors.* The male sex is at increased risk: for example, farmers, coffee growers, and lumbermen.
- *Transmission.* The route of infection is via inhalation of conidia from the environment.
- *Determinants of Pathogenicity.* Cell wall α-(1 → 3)-glucan resists lysis in the host; gp43 is an adhesin and a dominant antigen.
- *Clinical Forms.*
  - *Adults.* Mucocutaneous form: mouth ulcers, enlarged cervical nodes, adjacent skin lesions; also chronic pulmonary disease.
  - *Under 30 years old.* Infection of lymph nodes, spleen, liver, and bone marrow.
- *Therapy.* Ketoconazole, itraconazole (ITC), AmB, role for sulfa.
- *Laboratory Detection, Recovery, and Identification.*
  - Dimorphic: mold at 25°C; at 37°C converts to a yeast form with multipolar budding in a "pilot wheel" shape.
  - Direct examination: KOH prep from sputum, touch smears of mucocutaneous lesions; biopsy.
  - Culture from sputum and from mucocutaneous lesions.
  - Identify culture by rDNA gene sequence.
  - Serodiagnosis: antibodies against 43 kDa protein are diagnostic (EIA).
  - Genetic identification: gene targets are the ITS1, ITS2 regions of rDNA and exon 2 of *PbGP43*.

7.2 INTRODUCTION/DISEASE DEFINITION

Paracoccidioidomycosis is a chronic granulomatous disease beginning with a primary pulmonary focus, and disseminating with characteristic mucocutaneous-lymphatic involvement. Sites of dissemination are mucous membranes of the nose, mouth, and gastrointestinal tract, and the skin. *Paracoccidioides brasiliensis*, the etiologic agent, is a soil-dwelling mold. After inhalation, conidia germinate in the lung to develop into distinctive multipolar budding yeast forms. *Paracoccidioides brasiliensis* is capable of causing clinical disease in immune-normal as well as immunocompromised persons. This systemic, dimorphic fungal pathogen has an endemic area restricted to Mexico and Central and South America but is not uniformly distributed within the region. It is rarely isolated from the environment so that its ecologic niche remains

Fundamental Medical Mycology, First Edition. By Errol Reiss, H. Jean Shadomy and G. Marshall Lyon, III.
© 2012 Wiley-Blackwell. Published 2012 by John Wiley & Sons, Inc.

obscure. A characteristic of paracoccidioidomycosis is extreme latency measured in months or years from the time of infection to the appearance of clinical disease.

7.3 CASE PRESENTATION

7.3.1 Oral Lesions in a Legionnaire (Horré et al., 2002)

History

A 61-yr-old German man presented at a hospital in Southern Germany with a 10 year history of mouth swelling and ulcerations. From 1970 to 1980 he was employed as a legionnaire in Asia and Africa. During 1980–1990 he lived in the Mato Grosso state of Brazil. He suffered from chronic bronchitis for several years, which was diagnosed as a tropical pulmonary mycosis. He was alcoholic during his military service. In 1995 a Billroth II operation was performed because of relapsing gastric ulcers. Five years before the current admission he was treated for pyogenic ulcers and edematous erythema at the left angle of his mouth. *Staphylococcus aureus* was isolated. Histologic examination showed granulomatous lesions. He responded to antibiotic treatment and was discharged.

Clinical and Laboratory Findings

He attended clinic in January 2000 for persistent mouth lesions. A physical examination at that time revealed erythematous swollen lips and mucocutaneous pustules with ulcerations. Nodules surrounded by whitish plaques were observed on the hard palate and tongue. The lesions were painful and the patient was physically and psychologically impaired. His symptoms included night sweats.

Laboratory findings showed a microcytic anemia and an increased ESR. A biopsy of the oral lesions was submitted for culture and histology. Yeast forms were observed in the PAS-stained tissue section, and 2 weeks later a fungus grew in liquid *Leishmania* culture medium. Subcultures of the fungus on BHI agar were incubated at 30°C and 37°C. Growth at 37°C resulted in cream colored, yeast-like wrinkled colonies. Microscopically they were round yeast cells with multipolar buds. Daughter cells were connected to the mother cell by small bridges typical of *P. brasiliensis*. The mold form grew more slowly at 30°C and consisted of sterile, septate mycelia with terminal and intercalary *chlamydospores*. rDNA gene sequences were identical to the GenBank deposited sequence for *Paracoccidioides brasiliensis*.

Diagnosis

Diagnosis was chronic mucocutaneous paracoccidioidomycosis.

Therapy and Clinical Course

The patient was placed on ITC, 400 mg/day, and his oral lesions subsided over the next 4 weeks. Therapy was continued for another 5 months.

Comments

Although very rare in Europe, paracoccidioidomycosis has been reported there in patients with a history of residence in Brazil or Ecuador. Nodules with whitish plaques and pinpoint hemorrhages are characteristic of *P. brasiliensis* stomatitis. The primary infection with the fungus likely had occurred 10–20 years before the current admission, during the patient's residence in Brazil. *Such extreme latency is a feature of chronic mucocutaneous paracoccidioidomycosis*.

Differential Diagnosis for Paracoccidioidomycosis

Paracoccidioidomycosis must be differentiated from tuberculosis, which can coexist in 15–20% of cases (Restrepo and Tobón, 2005). Other diseases to be considered are:

- Actinomycosis
- Cancer (including lymphoma)
- Coccidioidomycosis
- Granuloma inguinale
- Histoplasmosis
- Leishmaniasis
- Leprosy
- Sarcoidosis
- Syphilis
- Wegener's granulomatosus

7.4 ETIOLOGIC AGENT

The etiologic agent is the soil-dwelling mold *Paracoccidioides brasiliensis*.

Paracoccidioides brasiliensis (Splend.) F.P. Almeida, *Spicilegium Pl. neglect*. 105: 316 (1930). Position in classification: anamorphic *Ajellomycetaceae*, *Onygenales*, *Ascomycota*. *Paracoccidioides brasiliensis* is the only species in the genus.

Taxonomically, based on sequence analysis of the LSU (28S) subunit and the ITS regions of rDNA, *P. brasiliensis* fits into a clade of the ascomycete order *Onygenales*, family *Ajellomycetaceae*, which includes the genus *Ajellomyces*. That genus contains the teleomorph forms of *Blastomyces dermatitidis, Histoplasma capsulatum*, and *Emmonsia* species (Guarro and Cano, 2002). Conidia of *P. brasiliensis* are *pyriform*, borne singly, attached directly to or on a short side branch of the hypha. They fit the definition of *micro-aleurioconidia*. Conidia are released by rupture of the supporting cell. Once inhaled, conidia germinate in the lung, converting to a distinctive multipolar budding yeast form.

Phylogenetic analysis based on *multilocus sequence typing* (MLST) discovered that *P. brasiliensis* is a species complex (Teixeira et al., 2009). Furthermore, a highly divergent clade including the standard tester strain "Pb01" was designated a separate species, *P. lutzii*. Members of *P. lutzii* are geographically localized to the central-western Brazilian states of Mato Grosso and Goiás. Morphologically, *P. lutzii* at 36°C produces large yeast forms with small multipolar buds.

7.5 GEOGRAPHIC DISTRIBUTION/ ECOLOGIC NICHE

Paracoccidioidomycosis is a rural disease endemic to Mexico and Central and South America, most often found in men aged 20–50 years, especially coffee growers of Colombia, Venezuela, and Brazil (Fig. 7.1). The disease is limited to areas between 20° N and 35° S, but is not distributed homogeneously. Most cases (~80%) are in Brazil, with at least 5000 reported cases (Laniado-Laborín, 2007). The greatest prevalence is in states in southern Brazil. Prevalence is estimated by reported death rates (Coutinho et al., 2002, Prado et al., 2009). The mortality rates and annual deaths for the Brazilian endemic area are shown in Table 7.1. Other countries where paracoccidioidomycosis is found are Colombia, Venezuela, northeast Argentina, and Southern Paraguay. Small highly endemic areas occur in Ecuador, Guatemala, Mexico, and Peru. In Mexico 70% of cases have been reported in the state of Acapulco (Laniado-Laborín, 2007). The climate and topography of endemic areas are cool, moist, subtropical forests, with a mean temperature of 18–23°C, an elevation of 500–1800 m, and 800–2000 mm of annual rainfall. The habitats support humid vegetation and are located near water sources (Bagagli et al., 2003). The ecologic niche of *P. brasiliensis* is uncertain, because it only rarely has been isolated from soil. *Paracoccidioides brasiliensis* was isolated in Brazil from wild nine-banded armadillos (*Dasypus novemcinctus*). Infected armadillos are likely sentinels for the presence of the fungus in the local environment.

7.6 EPIDEMIOLOGY

7.6.1 Incidence

Skin test surveys with paracoccidioidin[1] in the endemic areas have shown that up to 60% of the population is sensitized. A large number of infections may be asymptomatic and self-limiting. Annually, there are an estimated 225 serious cases of paracoccidioidomycosis, or 0.8 / 100,000 population. This figure is an underestimate because the disease is not reported through an organized system. The majority of cases occur in Brazil where incidence is estimated to be 1–3/100,000 population (Prado et al., 2009). A more accurate determination of the true incidence will require a national active surveillance program. The disease is not distributed homogeneously in Brazil; please see Fig. 7.1 and Section 7.5, Geographic Distribution/Ecologic Niche. Difficulty in identifying both habitat and patients is explained by scarcity of reports of finding the fungus in nature, lack of outbreaks involving clusters of patients, prolonged latency before obvious infection, and migration of individuals from the endemic area (please see Section 7.3, Case Presentation). Unlike other endemic mycotic agents, *P. brasiliensis* does not appear to be an AIDS-associated opportunistic pathogen.

A striking aspect of paracoccidioidomycosis is a long latent period from the time of primary exposure in Latin America and the occurrence of clinical symptoms which may not become overt for many years. That feature accounts for cases that arise in Europe, Japan, and the United States as, for example, in Section 7.3, Case Presentation. Possibly, the fungus remains dormant in residual lymph node lesions or lung granulomas until the host becomes debilitated or immunosuppressed.

7.6.2 Risk Groups/Factors

Although surveys of skin test hypersensitivity indicate equal exposure of the sexes, men are 10–100-fold more likely to contract paracoccidioidomycosis. Men are disproportionately affected because, in women, estrogen receptors on *P. brasiliensis* bind to estrogens and this binding prevents the conversion of mycelia to yeast forms, effectively limiting the infection (Loose et al., 1983). Estrogens, but not androgens, inhibit conversion of conidia of the mold form to the yeast form found in tissue. Prior to puberty the infection rate is equal between males and females. Young and middle-aged men who work as farmers, coffee growers, and lumbermen are at risk. Poor nutritional status, alcoholism, smoking, and

[1]Please see Section 7.12, Laboratory Detection, Recovery, and Identification, for a further description of paracoccidioidin.

Figure 7.1 Distribution of paracoccidioidomycosis in South America. The boundaries of the endemic areas are restricted by climate and topography (please also see Section 7.5, Geographic Distribution/Ecologic Niche). Most cases are reported from Brazil, followed by Colombia and Venezuela. A smaller endemic area ("1" on the map) consists of a portion of Paraguay and Uruguay. Area "2" on the map consists of a portion of Peru and Ecuador. Brazilian states of highest endemicity are as follows (reported mortality rate/1,000,000 population for the period 1996–2006): Rondônia (5.7); Mato Grosso (4.7); Acre (3.4); Tocantins (2.6); Parana (2.6); São Paulo state (1.2); Rio Grande do Sul (1.2); Minas Gerais, (1.2); Santa Catarina (1.0); Rio de Janeiro (0.6) (Prado et al., 2009). Brazilian states (Goiás, Para) shaded in a lighter color have reported mean mortality of 3–6 deaths/year. Amazonas state had no reported deaths from paracoccidioidomycosis/year. The other Brazilian states (unshaded) have $\leq$1 death reported/year. Colombia and Venezuela are notable for paracoccidioidomycosis endemic areas, as is the state of Acapulco in Mexico.

Source: Paracoccidioidomycosis endemic areas drawn from information and data in Brummer et al. (1993) and Prado et al. (2009).

Table 7.1 Paracoccidioidomycosis in South America[a,b]

Endemic area	State or geographic region	Mean deaths/yr (1996–2006)	Mortality rate/1,000,000 inhabitants
Brazil	São Paulo	46	1.2
	Paraná	25	2.6
	Mato Grosso	12	4.7
	Minas Gerais	21	1.2
	Rio Grandé do Sul	12	1.2
	Rio de Janeiro	9	0.6
	Rondônia	8	5.7
	Santa Catarina	5	1.0
	Goiás	5	<1
	Para	5	<1
	Mato Grosso do Sul	4	<1
	Espirito Santo	4	<1
	Tocantins	3	2.6
	Acre	2	3.4
Area 1 on Fig. 7.1	Southern Paraguay, northeast Argentina	Unknown	Unknown
Area 2 on Fig. 7.1	Parts of Ecuador and Peru	Unknown	Unknown
Colombia		Unknown	Unknown
Venezuela		Unknown	Unknown

[a]This table is to accompany the map in Fig. 7.1.
[b]Data from Brazil (Prado et al., 2009).

parasitic and mycobacterial diseases also increase the risk of *P. brasiliensis* infection.

7.7 TRANSMISSION

The route of infection is respiratory, following inhalation of conidia of the mold form. Paracoccidioidomycosis is not communicable between humans or from animals to humans.

7.8 DETERMINANTS OF PATHOGENICITY

7.8.1 Host Factors

Protective Th1 immunity is characterized by the reaction of lymphocytes from healthy sensitized[2] subjects or cured patients. When activated by *P. brasiliensis* antigens, a brisk production of IL-2 and IFN-γ ensues. The major effector cell in host defense is the macrophage. When there is IFN-γ production, the intracellular replication of yeast forms is halted. PMNs also exert a fungistatic effect against *P. brasiliensis* yeast forms, enhanced by IFN-γ and granulocyte-macrophage colony stimulating factor (GM-CSF).

The juvenile form of paracoccidioidomycosis, on the other hand, is on the Th2 pole of reactivity, with impaired responses to *P. brasiliensis* antigens: lower IFN-γ production, cutaneous anergy (negative delayed cutaneous hypersensitivity to paracoccidioidin), and a concomitant increase in Th2 responses—increased IL-4, IL-5, and IL-10. Corresponding to the depression in CMI there are poorly organized granulomas, more generalized systemic lesions with many yeast forms. There is a correlation between depressed CMI and the severity of paracoccidioidomycosis. Among parameters affected is a decreased $CD4^+/CD8^+$ T-lymphocyte ratio.

[2]They have a positive skin test (delayed cutaneous hypersensitivity) to paracoccidioidin.

7.8.2 Microbial Factors

Events leading to the morphogenesis from the mold form to the yeast form (found in vivo) are traced to genes controlling cell wall glucan and chitin synthesis. Other metabolic processes, for example, production of heat shock proteins and ornithine decarboxylase activity, also are involved in this morphogenesis (San-Blas and Nino-Vega, 2004).

α-(1 → 3)-d-Glucan

The higher the amount of α-(1 → 3)-glucan layer in the cell wall of the yeast form, the greater the pathogenic potential of the strain, according to numerous studies using experimental infection models. Macrophages are not able to digest the outer layer of α-(1 → 3)-glucan (San Blas, 1985).

gp43

The gp43 is a wall-associated and/or a secreted glycoprotein. The sequence of gp43 has homology to exo-ß-(1 → 3)-D-glucanase, but it lacks that enzymatic activity, which is presumed to be the result of the substitution in the enzyme active site from Asn-Glu-Pro (active) to Asn-Lys-Pro (reviewed by Borges-Walmsley et al., 2002). The gp43 is important in the immune response to *P. brasiliensis*.

Immunodominant Antigen Mediating Serologic Reactions and CMI Antibody titers to gp43 are diagnostic and prognostic (please see Section 7.12, Laboratory Detection, Recovery, and Identification—Serodiagnosis). Peripheral blood mononuclear cells of patients clinically cured after completing therapy for paracoccidioidomycosis were reconstituted in tissue cultures with gp43 peptides (Borges-Walmsley et al., 2002). These patients produced high concentrations of IFN-γ and TNF-α, both critical elements in protective T-cell-mediated immunity in paracoccidioidomycosis.

gp43 This is one of the immunodominant antigens in paracoccidioidin capable of eliciting delayed cutaneous hypersensitivity in skin tests (Rodrigues and Travassos, 1994; Travassos et al., 1995; Saraiva et al., 1996).

Mouse Protection Experiments Immunization with a synthetic oligopeptide based on the T-cell epitope of gp43 exerts a protective CMI response in mice challenged with live yeast forms.

Adhesin Acting as an adhesin, gp43 binds to laminin in the basal membrane of the extracellular matrix of mammalian tissues.

Melanin

Paracoccidioides brasiliensis yeast forms grown in the presence of the melanin precursor, L-DOPA, become melanized (Silva et al., 2009). In that state, they were more resistant to chemically generated NO, H_2O_2, and hypochlorite than nonmelanized yeast. Melanized yeast forms, phagocytosed by macrophage cell lines, were more resistant to killing than nonmelanized yeast forms.

Glyceraldehyde-3-phosphate dehydrogenase (GAPDH)

This enzyme was localized to the cell wall and cytoplasm of *P. brasiliensis* yeast forms (Barbosa et al., 2006). Recombinant GAPDH was capable of binding to extracellular matrix proteins—fibronectin, laminin, and type 1 collagen. As well, pneumocytes preincubated with recombinant GAPDH or anti-GAPDH antibodies reduced the adherence of *P. brasiliensis* yeast forms.

7.9 CLINICAL FORMS

Primary pulmonary infection probably occurs in the first and second decades of life with a benign course, often without symptoms. *Reactivation of quiescent lesions is considered the main mechanism of adult chronic disease*. The dormant period may be lengthy, measured in years, after the host becomes debilitated or immunosuppressed (please see Section 7.3, Case Presentation). Common presenting symptoms are mucosal ulcers in the upper respiratory and digestive tracts, mostly in the mouth and nose. The patient may describe difficulty in swallowing and vocal changes. Cutaneous lesions often are present on the face and limbs. Lymph nodes, especially cervical nodes, become enlarged.

Acute and Chronic Pulmonary Paracoccidioidomycosis

Acute pulmonary paracoccidioidomycosis is rare except in immunosuppressed patients. Typically, there is a chronic, progressive, pulmonary course. Symptoms of pulmonary paracoccidioidomycosis include productive cough, chest pain, shortness of breath, fever, and purulent or blood-tinged sputum. Accompanying these symptoms are weakness, malaise, fever, and weight loss (Restrepo and Tobón, 2005). After inhalation of *P. brasiliensis* conidia the initial reaction is alveolitis, with an influx of macrophages and a few giant cells. A pyogenic reaction with numerous PMNs then ensues, followed by granuloma formation. Granulomas may be poorly defined, and either solid or with centers of caseous necrosis. Granulomas contain epithelioid macrophages, multinucleate giant cells, lymphocytes, plasma cells, and fibroblasts. Numerous yeast forms are present inside giant cells or scattered in the granulomas.

Infiltration and nodules progress to confluent lesions, particularly in the lower lobes. Intra-alveolar lesions may be accompanied by interstitial fibrosis and small calcifications. Extensive fibrosis results in respiratory deficiency. Rarely, there are large cavitary masses termed paracoccidioidomas. Pulmonary radiographs reveal interstitial and alveolar infiltrates, pulmonary nodules, cavities, and fibrosis.

Acute or Subacute (Juvenile Form) Paracoccidioidomycosis

A series of 63 cases of paracoccidioidomycosis in children aged 2–15 years was reviewed by Pereira et al. (2004), collected during the period 1981–2001. The children presented to the hospital of the State University of Campinas, São Paulo, Brazil. Most children had disseminated disease, affecting more than one organ or more than two lymph node chains. Children and adolescents of both genders have involvement of the mononuclear phagocytic system *after dissemination from a primary pulmonary focus*. The ratio of males to females increased from parity to 5.5:1 in children aged 12 years or older. Obvious pulmonary symptoms were rare. Lymph node enlargement was the most frequent clinical manifestation. Hepatomegaly and splenomegaly were prevalent. Lymph node enlargement caused intestinal obstruction in three children. Verrucous skin lesions occurred in ~10% of cases. Bone lesions were found in approximately half of cases, predominantly of the multiple lytic type.

Diagnosis was made most often by lymph node biopsy and histopathology. Therapy was primarily with sulfa, requiring $\geq$18 months to achieve remissions. Mortality in patients who were followed was 9.5%. Seven of 63 patients were lost to follow-up.

Chronic Mucocutaneous Paracoccidioidomycosis

(Please see Section 7.3, Case Presentation.) This form (Almeida et al., 2003) afflicts mainly men, aged 29–50 years, in a ratio of men:women = 15:1. Chronic paracoccidioidomycosis presents with pulmonary involvement > oropharynx > lymph nodes. Oral lesions usually are secondary to lung involvement. Even when no lung lesions are evident, the fungus often can be isolated from sputum.

- Lesions on the oral mucosa are a common presenting symptom. Mouth ulcers develop with swelling that includes the gums. Oral lesions are granulomatous spreading slowly to cover large areas with erythematous finely granular hyperplasia, speckled with pinpoint hemorrhages, and a mulberry-like surface. The lips may become swollen. Lesions usually are multiple involving lip, gingival, buccal mucosa, tongue, and floor of the mouth. Tissue damage can destroy the epiglottis and uvula, perforating the hard palate and extending to the tongue, lip, and nose (Fig. 7.2).
- These granulomatous lesions drain into the cervical lymph nodes, which enlarge and may become massive, then suppurate and drain, forming sinus tracts. Numerous yeast forms are present in the drainage.
- In older patients there is a decline in obvious infection of the mononuclear phagocytic system and a more chronic, progressive course. Dual infection with *Mycobacterium tuberculosis* may complicate the clinical assessment. In aggressive infections the normal architecture of lymph nodes may be effaced, especially in T-dependent, paracortical areas, as in leprosy. Skin lesions develop as extensions from mucosal foci. They appear as ulcerative and crusted granulomas resembling pseudoepitheliomatous hyperplasia, so that skin cancer may be suspected. Later, new skin lesions may arise elsewhere. This predilection for the skin and mucosa has been explained as a preference of the fungus for slightly lower temperatures.
- Pulmonary fibrosis. Antifungal therapy effectively controls *P. brasiliensis* but prolonged post-therapy follow-up over a period of 3–5 years revealed a high percentage of patients with pulmonary fibrosis as a pathologic sequel (Tobón et al., 2003). Over half of patients had evidence of pulmonary fibrosis, either persisting or developing de novo during follow-up. Residual lesions not only persist but increase over time so that a sizable proportion of patients have permanent quality of life impairment as a result of dyspnea and cor pulmonale.

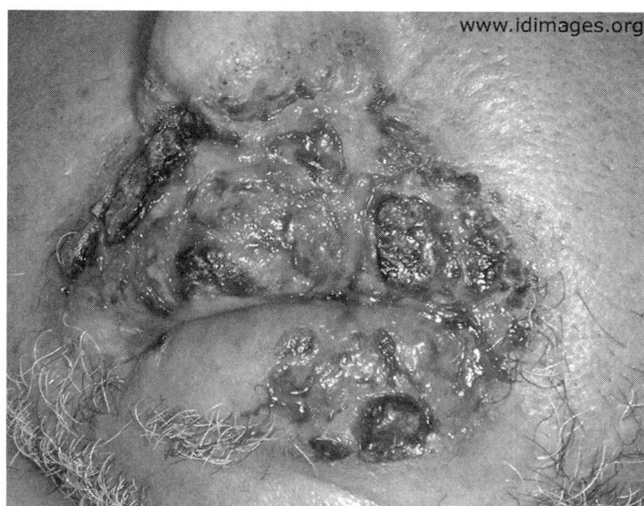

Figure 7.2 Facial lesions, paracoccidioidomycosis. The patient was born and lived in Brazil, and worked for many years as an agricultural laborer in a rural northeast section of the country. He had frequent contact with domesticated animals, livestock, and a freshwater source He smoked tobacco regularly until his throat began to bother him, and consumed several alcoholic beverages per night. The facial lesions are shown; the remainder of the examination was normal. He was admitted to the hospital for a course of AmB therapy and also received sulfamethoxazole. *Source:* Contributed by Rebeca M. Plank, M.D., originally published by the Partners in Infectious Disease Images at www.idimages.org, and used with permission from Rajesh T. Gandhi, M.D., Editor.

Disseminated Paracoccidioidomycosis

In comparison with other systemic mycoses, generalized dissemination is rare. When it occurs, hematogenous dissemination to the mesenteric lymph nodes, spleen, liver, adrenals, and bones, skin, and brain can lead to potentially life-threatening complications. Paracoccidioidomycosis is implicated as a major cause of adrenal insufficiency in South America.

7.10 VETERINARY FORMS

The nine-banded armadillo, *Dasypus novemcinctus*, harbors the disease and may be important as a sentinel animal for environmental microfoci of *P. brasiliensis*. Possibly, armadillos also may introduce the fungus into soil. Canine paracoccidioidomycosis is exceptional (Ricci et al., 2004).

7.11 THERAPY

Paracoccidioidomycosis is susceptible to azole antifungal therapy with ketoconazole or, more effectively, with ITC (Restrepo and Tobón, 2005; Yasuda, 2005). AmB is reserved for refractory cases. There is a history of some success using sulfa therapy, but azole therapy offers a more prompt response and less recurrence of disease.

Sulfa

Paracoccidioidomycosis is the only mycosis amenable to treatment with sulfa drugs (alone or combined with trimethoprim). Sulfadiazine is distributed by the Brazilian Health Ministry for this indication at no charge. It is used as oral therapy for non-life-threatening paracoccidioidomycosis, for a duration of 6–12 months. The dosage regimen consists of 100 mg/kg (maximum 6 g) every 6 h: two to three 500 mg tablets 4×/day.

Step-down maintenance therapy choices include sulfamethoxypyridazine, a slowly excreted sulfonamide, at a dose of 2 g every 12 h (children receive a half dose or a proportional dose) and sulfadoxine, a very slowly excreted sulfa given in single weekly doses of 1–2 g/week for adults and 25–50 mg/kg/week for children.

Duration of therapy for moderately severe and severe cases of paracoccidioidomycosis consists of 6–12 months of induction and 12–24 months of maintenance. Availability of more potent medications has reduced the use of sulfonamides. Treatment failure rates of < 25–30% have been reported in patients treated with sulfonamides, but in many cases failure is related to poor compliance, poor absorption, or premature step-down to maintenance treatment.

Amphotericin B

AmB has been in use since 1958 to treat the most severe cases of paracoccidioidomycosis and for patients unresponsive to other therapy. In such cases the daily dose should be increased as rapidly as possible to 1 mg/kg/day with an induction phase of 30–60 days or longer if necessary. In 47 patients with disseminated paracoccidioidomycosis a total dose of 2–3 g (30 mg/kg) effected clinical and serologic cure in 54% of patients (reviewed by Yasuda, 2005).

AmB is not curative by itself, and all patients treated with it should also receive maintenance sulfonamide or azole therapy. Sulfonamides alone, or AmB induction and sulfonamide maintenance therapy are not always successful, and the mortality rate is high (17–25%); improvement is obtained in 65–70% of the cases, and the remainder have relapses or fail to improve.

Azoles

Orally administered ketaconazole resulted in major improvement in 84–95% of the cases with only a 10% relapse rate after 5 years.

- Ketoconazole in doses of 200–400 mg/day are continued for a minimum of 6 months and for as long as 12–18 months, depending on the patient's response and the results of mycologic tests. Testing for hepatic dysfunction and gonadal alterations are necessary.
- ITC is superior to ketoconazole in that concentrations of ITC 10–100× lower than ketoconazole concentrations produce the same effects on fungal cells (reviewed by Yasuda, 2005). Findings from a group of approximately 300 patients showed that doses of ITC, 100–400 mg/day, are effective when administered for 6–24 months, with clinical improvement observed in ~91% of patients. Rates of treatment failure varied from 2.6% to 19% after 2–5 years of follow-up in patients who received 100–200 mg of ITC/day for 6–12 months. A dosage of 400 mg/day of primary therapy, for a duration of 12–18 months, is indicated for acute and disseminated chronic paracoccidioidomycosis (Yasuda, 2005).
- Voriconazole (VRC) is an effective azole to treat paracoccidioidomycosis, consistent with the high in vitro susceptibility of *P. brasiliensis* to VRC (Queiroz-Telles et al., 2007). Of 30 patients treated with VRC and 17 with ITC, the rate of complete or partial response was 88.6% for VRC and 94.4% for ITC therapy. The duration of therapy was ≥6 months. The dosage regimen for VRC was 200 mg 2×/day (oral), with a loading dose of 400 mg given twice on the first day. The dosage was reduced by 50% in patients weighing < 40 kg. The ITC dosage was 100 mg 2×/day (oral). There were no relapses, after 8 weeks of follow-up, among patients in both groups who had a complete therapeutic response.

CNS Paracoccidioidomycosis

AmB (administered either IV or intrathecally) or sulfadiazine produced comparable results, with a rate of treatment failure of 23–25% (Yasuda, 2005). A patient with cerebral abscess paracoccidioidomycosis responded well to VRC without the need for surgery (Queiroz-Telles et al., 2007). Although ITC has been used successfully to treat CNS paracoccidioidomycosis, its poor penetration into the CNS limits its use, whereas VRC penetrates well into the CNS.

7.12 LABORATORY DETECTION, RECOVERY, AND IDENTIFICATION

Paracoccidioides brasiliensis is a dimorphic, systemic, pathogenic fungus growing as a mold in soil and on laboratory medium at 25°C, and growing as a yeast form with multipolar buds at 37°C on enriched media.

Direct Examination

When available, sputum, exudates, and/or pus are examined in a KOH prep (preferably including Calcofluor white). Such specimens from most patients reveal *P. brasiliensis* yeast forms with thick cell walls. Repeat if results are negative and include digestion and concentration of sputum (Restrepo and Tobón, 2005). Identification is facilitated by the relatively large size of the yeast and multipolar budding. The mother cell may reach a maximum size of 42 μm diameter; with a mean size of 14–16 μm diameter with 4 or more buds/mother cell (Svidzinski et al., 1999). Budding cells can be distinguished from other fungal pathogens on the basis of narrow necks or pores, and buds that may completely surround the mother cell.

Histopathology

Histopathology is very important in view of the long period necessary to grow the fungus on mycologic media. The yeast form of *P. brasiliensis* in infected tissue appears as large (14–16 μm mean diameter) cells with *multiple* small (2–10 μm diameter) peripheral buds connected to the parent cell by narrow necks. They may be phagocytosed by multinucleate giant cells. This pattern is termed a "pilot wheel" formation and is highly characteristic of *P. brasiliensis*. They usually are visible in hematoxylin and eosin (H&E) stained smears (Fig. 7.3), but improved resolution of morphology requires special stains, such as GMS (Fig. 7.4). Morphologic criteria may not be sufficient when there are only single buds or no obvious buds. In that case, *Blastomyces dermatitidis* or *Coccidioides* species or small capsule cryptococci have to be ruled out.

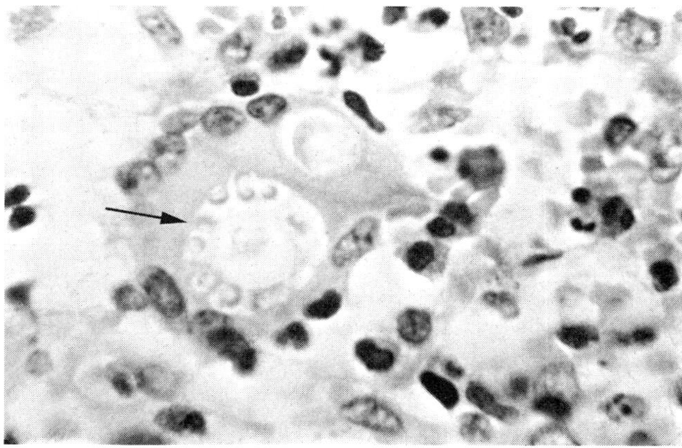

Figure 7.3 H&E stained tissue section shows *P. brasiliensis* tissue form: a large yeast with small multipolar buds (arrow).
Source: CDC.

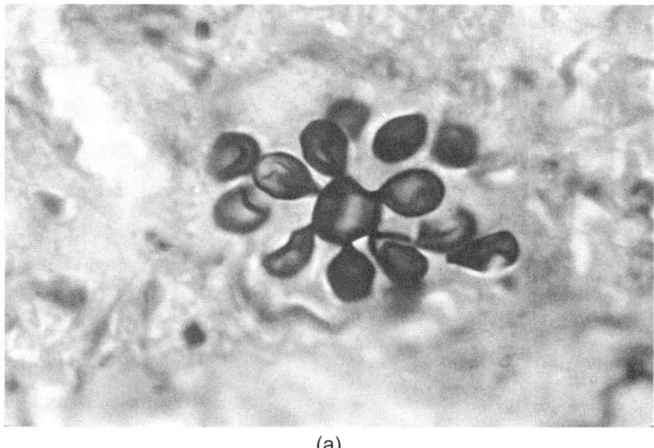

(a)

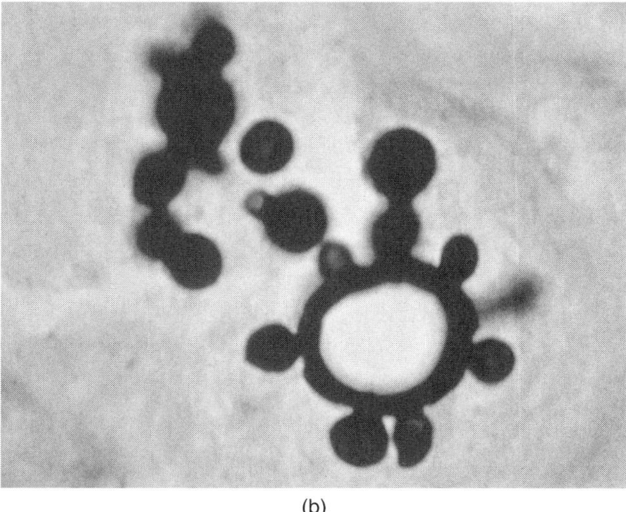

(b)

Figure 7.4 Histopathologic sections show typical "pilot-wheel" shape of *P. brasiliensis* yeast forms: (a) small mother cell with equal size buds (GMS stain) and (b) large mother cell with small buds (GMS stain).
Source: CDC.

Culture

Clinical specimens planted as primary cultures should be incubated at 25°C or 30°C on SDA with antibiotics, and at 32–37°C on BA or BHI agar + 5% sheep blood (referred to in following paragraphs as "enriched media").

Colony Morphology

Growth of *P. brasiliensis* at 25°C on SDA is *very slow*, sometimes taking up to 4 weeks to produce a white colony that is compact and folded. *Paracoccidioides brasiliensis* grows more rapidly on yeast extract agar with antibacterial antibiotics and cycloheximide. Culture tubes or plates should be held for up to 6 weeks to allow for the slow growth of this fungus. Mature colonies may be glabrous or floccose with a short nap of white aerial hyphae. The colonies turn brown with age.

Safety Note: Always handle cultures in a well-ventilated, periodically inspected, and certified biological safety cabinet. Transfer mycelial colonies to screw cap or cotton stoppered agar slants. Petri plates should be sealed with with Parafilm® or Shrink Seals® (Scientific Device Laboratory, Des Plaines, IL).

Microscopic Morphology

Viewed microscopically, when grown at 30°C, there are thin, hyaline, septate, branched hyphae that give rise to two conidial types.

- *Intercalary* and terminal bulging arthroconidia (Bustamante-Simon et al., 1985). These arthroconidia differentiate later into oval to pyriform small *aleurioconidia* (median length 3.6 μm).
- Single celled sessile conidia. Another conidial type also is found: these are single celled conidia, pear-shaped (3.9 μm × 3 μm) formed randomly along the fertile hyphae. They are not later developments of the arthroconidia.

Once the culture has grown from the clinical specimen on SDA or on enriched media, it is essential to transfer the culture to a nutrient-poor medium to encourage formation of conidia. A comparative growth study found that, of four *P. brasiliensis* mold isolates, *none* sporulated after 8 weeks incubation on SDA or CMA (Bustamante-Simon et al., 1985). Two media that stimulate sporulation are water agar and yeast extract agar.

Conversion to the Yeast Form In Vitro

Identification of *P. brasiliensis* is based on the conversion of the mold to the distinctive yeast form in vitro at 37°C, because of the difficulty in inducing sporulation of the mold form. At 37°C on yeast extract agar or enriched media, the growth is heaped, cream to tan, moist, soft, becoming waxy and yeast-like: large, round, thick-walled yeast (ranging in size from a mean 14–16 μm diameter up to 42 μm diameter) with single or multiple buds (1–10 μm diameter). Buds are attached to the mother cell by slender connections and may surround the mother cell presenting a classic "pilot's wheel" appearance. Two types of daughter cells occur: uniformly smaller than the mother cell, or variable in size, but larger than the mother cell. Yeast colonies also may contain hyphal elements.

Serodiagnosis: Double Agarose Immunodiffusion (ID) Test

This is the test most used in the serologic diagnosis of paracoccidioidomycosis (de Camargo and de Franco, 2000; de Camargo, 2008). The preferred ID test uses a standardized yeast form culture filtrate antigen, "Ag7," which contains abundant 43 kDa glycoprotein (gp43) as a major component (please see Section 7.8.2 Microbial Factors, for further information on this antigen). A multinational laboratory evaluation of the standardized Ag7 ID test showed it was 84.3% sensitive and 98.9% specific for paracoccidioidomycosis.

False-negative results occur in 2–3% of patients, usually in those with severe disease and pulmonary involvement. Two explanations for this lack of responsiveness (reviewed by de Camargo, 2008) are (i) the possibility of IgG with asymmetric antibodies directed against a mannose-oligosaccharide in only one Fab fragment, resulting in functionally univalent and non-precipitating antibodies, or (ii) the production of low-avidity IgG_2 antibodies also directed against carbohydrate epitopes. The precipitin reaction converts to positive after a satisfactory response to 1–2 months of antifungal therapy. Patients with severe disease and no serologic response may be diagnosed by direct examination of sputum specimens in a KOH prep to reveal the characteristic multipolar budding yeast forms.

Counterimmunoelectrophoresis (CIE)

This variation of precipitin analysis in gel has a sensitivity that is greater than or equal to that of the ID test, which, depending on the antigenic preparation used, varies between 77% and 97% sensitive and 95% specific. Some laboratories use CIE as a routine screen of serum specimens on grounds it may be faster than the ID test, but this is a marginal time difference.

Complement Fixation (CF)

The CF test for antibodies in paracoccidioidomycosis uses a yeast form antigen and is positive in 80–90% of patients with active disease. Crossreactions, however, may occur. There are only a few laboratories who continue to use this test.

EIA

A culture filtrate of *P. brasiliensis* adsorbed to wells of microtitration plates was evaluated for detecting antibodies by EIA in patients with paracoccidioidomycosis (Mendes-Giannini et al., 1984). They reported a sensitivity of 100% but specificity was 88% owing to crossreactions with sera from patients with histoplasmosis and lobomycosis. By using purified gp43 (then called "E_2") as antigen, crossreactions were eliminated provided that patient serum was first adsorbed with killed *H. capsulatum* yeast cells (Mendes-Giannini, 1990). Antibodies from paracoccidioidomycosis patients react primarily with peptide epitopes of the gp43; but some parcoccidioidomycosis patient sera react with carbohydrate epitopes, accounting for up to 45% of the total reactivity (Puccia and Travassos, 1991). Camargo et al. (1994) screened sera for antibodies against gp43 in an EIA format in which a monoclonal anti-gp43 was first adsorbed to polystyrene wells, followed by binding the gp43 antigen. In this way the geometry favored the capture of patient antibodies by protein epitopes of gp43. Analysis of the O.D. readings at 1:800–1:3200 patient serum dilutions found very large differences between homologous and heterologous serum specimens, aiding in the discrimination of antibody titers of paracoccidioidomycosis patients from titers of patients with histoplasmosis, lobomycosis, candidiasis, and aspergillosis.

Crossreactions with sera from histoplasmosis and lobomycosis patients are increased in the conventional EIA format where *P. brasiliensis* gp43 antigen is adsorbed directly to the plastic wells of a microtitration plate. This is because crossreactive carbohydrate epitopes of gp43 are more exposed by conformational changes when the antigen is bound directly to plastic. This format is still recommended for use because (i) crossreactions can be eliminated by deglycosylation of gp43, and/or (ii) using an appropriate O.D. cutoff for a positive test usually suffices to exclude crossreactions.

Reactivity in the EIA is influenced by periodate treatment of the antigen to inactivate the carbohydrate component. This has created a predicament because the carbohydrate component is crossreactive with sera from patients with other systemic mycoses, such as histoplasmosis or lobomycosis. However, when specificity is increased by removing the carbohydrate, the sera from some patients with adult form paracoccidioidomycosis no longer react in the EIA (Neves et al., 2003).

Antigenemia Tests

A species-specific murine monoclonal antibody, "17c," directed against gp43 was used in an inhibition EIA to detect the antigen in patient serum (Marques da Silva et al., 2004).

Application Note: Gp43 Antigenemia EIA. Serial dilutions of gp43 are added to normal human serum-buffer. The mAb "17c" is preincubated with each dilution overnight. Then they are added to wells of a microtitration plate to which a standard amount of gp43 was previously adsorbed. After incubation and washing goat–anti-mouse IgG–peroxidase is added to the plates and color is developed with a chromogenic substrate. The resulting inhibition curve serves as the standard curve against which the inhibition of patient serum is measured.

The sensitivity of inhibition-EIA ranges from 0.0053 to 30 μg of antigen/mL of serum. Antigen concentrations in patient serum higher than 1.35 μg/mL are considered a positive result. All 23 paracoccidioidomycosis patients studied had concentrations of serum–gp43 antigen above the cutoff point at the time of diagnosis (mean antigen concentration of 11.92 μg/mL). After 30 days of ITC therapy, the mean gp43 antigen concentration dropped to 7.23 μg/mL, decreasing further until the 12th month, when the mean antigen concentration was 0.57 μg/mL.

Most of the test reagents are "home-made" and there is no consistent supply of commercial kits for serology of this disease.

Paracoccidioidin

The paracoccidioidin skin test is in use as a diagnostic and epidemiologic tool in the endemic areas. The traditional paracoccidioidin is the Fava Netto Polysaccharide Antigen (FNPA) devised by Celeste Fava Netto and colleagues (Rodrigues and Travassos, 1994).

Application Note: FNPA Paracoccidioidin. The skin test antigen is produced by combining the biomass from five *P. brasiliensis* strains grown for 20 days at 36°C on solid medium composed of proteose peptone, 0.3%; peptone, 1%; meat extract, 0.5%; NaCl, 0.5%; yeast extract, 0.5%; glucose, 4%; and agar 1.8%, adjusted to pH 7.2–7.4. The combined washed cells are suspended in acetone, and then defatted by extraction 3× with diethyl ether. The defatted cells are then suspended in Veronal buffer, pH 7.5, and autoclaved for 25 min at 120°C. The cell-free supernatant is reserved in 5 mL aliquots as the skin test antigen. The carbohydrate:protein ratio of FNPA is 5:1. FNPA depleted of the gp43 antigen by immunoaffinity chromatography results in a marked reduction in its cutaneous reactivity, implicating gp43 as the major active principal of FNPA (Rodrigues and Travassos, 1994).

The gp43 was compared with the FNPA paracoccidioidin to elicit delayed cutaneous hypersensitivity in patients with paracoccidioidomycosis (Saraiva et al., 1996). Of 25 patients studied, 12 were anergic to both antigens. Of 13 paracoccidioidomycosis patients who demonstrated skin test responses to gp43 and/or to FNPA, 92.3% reacted against gp43 and 53.8% reacted against FNPA. The gp43 skin test diameters were significantly larger than those obtained with FNPA. Histology of the human skin test biopsies indicated a cell infiltrate typical of delayed-type hypersensitivity. The two preparations represent either a single molecule as elicitor of hypersensitivity (gp43) or a complex mixture of antigens (FNPA). Further research is needed to characterize the other proteins in FNPA that contribute to its reactivity and to determine what amount of reactivity in FNPA is the result of its gp43 content. The goal is to better understand the antigenic mosaic of *P. brasiliensis* so that a well-standardized antigenic complex is available for epidemiologic surveys.

Genetic Identification of P. brasiliensis

PCR directed to the *PbGP43* exon 2 and universal fungal primers ITS4 and ITS5 amplify a 634 bp PCR product including ITS1, 5.8S, and ITS2 (Hebeler-Barbosa et al., 2003). The resulting ITS sequences confirmed the similarity of *P. brasiliensis* isolated from 10 armadillos to that of 19 human clinical isolates. The *PbGP43* sequences were more polymorphic, but this gene segment is species specific and further indicates intraspecific variation in the immunodominant antigen gp43.

SELECTED REFERENCES FOR PARACOCCIDIOIDOMYCOSIS

Almeida OP, Jacks J Jr, Scully C, 2003. Paracoccidioidomycosis of the mouth: An emerging deep mycosis. *Crit Rev Oral Biol Med* 14: 377–383.

Bagagli E, Franco M, Bosco S de M, Hebeler-Barbosa F, Trinca LA, Montenegro MR, 2003. High frequency of *Paracoccidioides brasiliensis* infection in armadillos (*Dasypus novemcinctus*): An ecological study. *Med Mycol* 41: 217–223.

Barbosa MS, Báo SN, Andreotti PF, de Faria FP, Felipe MS, dos Santos Feitosa L, Mendes-Giannini MJ, Soares CM, 2006. Glyceraldehyde-3-phosphate dehydrogenase of *Paracoccidioides brasiliensis* is a cell surface protein involved in fungal adhesion to extracellular matrix proteins and interaction with cells. *Infect Immun* 74: 382–389.

Borges-Walmsley MI, Chen D, Shu X, Walmsley AR, 2002. The pathobiology of *Paracoccidioides brasiliensis*. *Trends Microbiol* 10: 80–87.

Brummer E, Castaneda E, Restrepo A, 1993. Paracoccidioidomycosis: An update. *Clin Microbiol Rev* 6: 89–117.

Bustamante-Simon B, McEwen JG, Tabares AM, Arango M,M, Restrepo-Moreno A, 1985. Characteristics of the conidia produced by the mycelial form of *Paracoccidioides brasiliensis*. *Sabouraudia* 23: 407–414.

Camargo ZP, Gesztesi J-L, Saraiva ECO, Taborda CP, Vicentini AP, Lopes JD, 1994. Monoclonal antibody capture enzyme immunoassay for detection of *Paracoccidioides* brasiliensis antibodies in paracoccidioidomycosis. *J Clin Microbiol* 32: 2377–2381.

Coutinho ZF, Silva D, Lazera M, Petri V, Oliveira RM, Sabroza PC, Wanke B, 2002. Paracoccidioidomycosis mortality in Brazil (1980–1995). *Cad Saude Publica* 18: 1441–1454.

de Camargo ZP, 2008. Serology of paracoccidioidomycosis. *Mycopathologia* 165: 289–302.

de Camargo ZP, de Franco MF, 2000. Current knowledge on pathogenesis and immunodiagnosis of paracoccidioidomycosis. *Rev Iberoam Micol* 17: 41–48.

Guarro J, Cano J, 2002. Phylogeny of Onygenalean fungi of medical interest. *Stud Mycol* 47: 1–4.

Hebeler-Barbosa F, Morais FV, Montenegro MR, Kuramae EE, Montes B, McEwen JG, Bagagli E, Puccia R, 2003. Comparison of the sequences of the internal transcribed spacer regions and *PbGP43* genes of *Paracoccidioides brasiliensis* from patients and armadillos (*Dasypus novemcinctus*). *J Clin Microbiol* 41: 5735–5737.

Horré R, Schumacher G, Alpers K, Sietz HM, Adler S, Lemmer K, de Hoog GS, Schaal KP, Tintelnot K, 2002. A case of imported paracoccidioidomycosis in a German legionnaire. *Med Mycol* 40: 213–216.

Laniado-Laborín R, 2007. Coccidioidomycosis and other endemic mycoses in Mexico. *Rev Iberoam Micol* 24: 249–258.

Loose DS, Stover EP, Restrepo A, Stevens DA, Feldman D, 1983. Estradiol binds to a receptor-like cytosol binding protein and initiates a biological response in *Paracoccidioides brasiliensis*. *Proc Natl Acad Sci USA* 80: 7659–7663.

Marques da Silva SH, Queiroz-Telles F, Colombo AL, Blotta MH, Lopes JD, Pires de Camargo Z, 2004. Monitoring gp43 antigenemia in paracoccidioidomycosis patients during therapy. *J Clin Microbiol* 42: 2419–2424.

Mendes-Giannini MJS, Bueno JP, Shikanai-Yassuda MA, Stolf AMS, Masuda A, Amato-Neto V, Ferreira AW, 1990. Antibody response to 43 kDa glycoprotein of *Paracoccidioides brasiliensis* as a marker for the evaluation of patients under treatment. *Am J Trop Med Hyg* 43: 200–206.

Mendes-Giannini MJS, Camargo ME, Lacaz CS, Ferreira AW, 1984. Immunoenzymatic absorption test for serodiagnosis of paracoccidioidomycosis. *J Clin Microbiol* 20: 103–108.

Neves AR, Mamoni RL, Rossi CL, de Camargo ZP, Blotta MH, 2003. Negative immunodiffusion test results obtained with sera of paracoccidioidomycosis patients may be related to low-avidity immunoglobulin G2 antibodies directed against carbohydrate epitopes. *Clin Diagn Lab Immunol* 10: 802–807.

Pereira RM, Bucaretchi F, Barison Ede M, Hessel G, Tresoldi AT, 2004. Paracoccidioidomycosis in children: Clinical presentation, follow-up and outcome. *Rev Inst Med Trop São Paulo* 46: 127–131.

Prado M, Barbosa da Silva M, Laurenti R, Travassos LR, Taborda CP, 2009. Mortality due to systemic mycoses as a primary cause of death or in association with AIDS in Brazil: A review from 1996–2006. *Mem Inst Oswaldo Cruz* 104: 513–521.

Puccia R, Travassos LR, 1991. 43-kilodalton glycoprotein from *Paracoccidioides brasiliensis*: Immunochemical reactions with sera from patients with paracoccidioidomycosis, histoplasmosis, and Jorge Lobo's disease. *J Clin Microbiol* 29: 1610–1615.

Queiroz-Telles F, Goldani LZ, Schlamm HT, Goodrich JM, Espinel-Ingroff A, Shikanai-Yasuda MA, 2007. An open-label comparative pilot study of oral voriconazole and itraconazole for long-term treatment of paracoccidioidomycosis. *Clin Infect Dis* 45: 1462–1469.

Restrepo A, Tobón AM, 2005. Chapter 266, *Paracoccidioides brasiliensis*, pp. 3062–3068, *in*: Mandell GL, Bennett JE, Dolin R (eds.), *Mandell, Douglas, and Bennett's Principles and Practice of Infectious Diseases*, 6th ed., Vol. 2. Elsevier Churchill Livingstone, Philadelphia, PA.

RICCI G, MOTA FT, WAKAMATSU A, SERAFIM RC, BORRA RC, FRANCO M, 2004. Canine paracoccidioidomycosis. *Med Mycol* 42: 379–383.

RODRIGUES EG, TRAVASSOS LR, 1994. Nature of the reactive epitopes in *Paracoccidioides brasiliensis* polysaccharide antigen. *J Med Vet Mycol* 32: 77–81.

SAN-BLAS G, 1985. *Paracoccidioides brasiliensis*: Cell Wall Glucans, Pathogenicity, and Dimorphism, pp. 235–254, *in*: MCGINNIS MR (ed.), *Current Topics in Medical Mycology*, Vol. 1. Springer-Verlag, New York.

SAN-BLAS G, NIÑO-VEGA G, 2004. Chapter 5, Morphogenesis of Agents of Endemic Mycoses, pp. 167–220, *in*: SAN-BLAS G, CALDERONE R (eds.), *Pathogenic Fungi: Structural Biology and Taxonomy*. Caister Academic Press, Wymondham, Norfolk, UK.

SARAIVA EC, ALTEMANI A, FRANCO MF, UNTERKIRCHER CS, CAMARGO ZP, 1996. *Paracoccidioides brasiliensis*-gp43 used as paracoccidioidin. *J Med Vet Mycol* 34: 155–161.

SILVA MB, THOMAZ L, MARQUES AF, SVIDZINSKI AE, NOSANCHUK JD, CASADEVALL A, TRAVASSOS LR, TABORDA CP, 2009. Resistance of melanized yeast cells of *Paracoccidioides brasiliensis* to antimicrobial oxidants and inhibition of phagocytosis using carbohydrates and monoclonal antibody to CD18. *Mem Inst Oswaldo Cruz* 104: 644–648.

SVIDZINSKI TIE, MIRANDA-NETO MH, SANTANA RG, FISCHMAN O, COLOMBO AL, 1999. *Paracoccidioides brasiliensis* isolates obtained from patients with acute and chronic disease exhibit morphological differences after animal passage. *Rev Inst Med Trop San Paulo* 41: 279–283.

TEIXEIRA MM, THEODORO RC, DE CARVALHO MJ, FERNANDES L, PAES HC, HAHN RC, MENDOZA L, BAGAGLI E, SAN-BLAS G, FELIPE MS, 2009. Phylogenetic analysis reveals a high level of speciation in the *Paracoccidioides* genus. *Mol Phylogenet Evol* 52: 273–283.

TOBÓN AM, AGUDELO CA, OSORIO ML, ALVAREZ DL, ARANGO M, CANO LE, RESTREPO A, 2003. Residual pulmonary abnormalities in adult patients with chronic paracoccidioidomycosis: Prolonged follow-up after itraconazole therapy. *Clin Infect Dis* 37: 898–904.

TRAVASSOS LR, PUCCIA R, CISALPINO P, TABORDA C, RODRIGUES EG, RODRIGUES M, SILVEIRA JF, ALMEIDA IC, 1995. Biochemistry and molecular biology of the main diagnostic antigen of *Paracoccidioides brasiliensis*. *Arch Med Res* 26: 297–304.

YASUDA MA, 2005. Pharmacological management of paracoccidioidomycosis. *Expert Opin Pharmacother* 6: 385–397.

QUESTIONS

The Answer Key to multiple choice questions may be found at the end of the book.

1. Which is the typical and most successful method to identify cultures of *Paracoccidioides brasiliensis*?
   A. Conversion to the yeast form
   B. Intercalary arthroconidia and sessile conidia are produced abundantly on SDA
   C. PCR sequencing

2. Temperature sensitive conversion to the yeast form (32–37°C) is facilitated by transferring the mold form to which agar?
   A. Malt extract
   B. Mycosel™
   C. SDA-Emmons
   D. Yeast extract

3. The yeast form of *Paracoccidioides brasiliensis* is characterized by
   A. A small yeast form with a slender hour glass-shaped neck between mother cell and a single bud
   B. Large yeast form with a wide neck between mother cell and a single bud
   C. Large or small yeast form with narrow neck connections to multipolar buds
   D. Pseudohyphae with whorls of yeast cells

4. Serodiagnosis of paracoccidioidomycosis is accomplished by a necessary test and one optional test. Next to each entry indicate if it is 1—necessary or 2—optional.
   A. Agar gel diffusion
   B. EIA using the gp43 antigen
   C. Complement fixation

5. Genetic identification of *Paraccoccidioides brasiliensis* is possible using PCR directed against
   A. 5.8S rDNA
   B. *BAD1*
   C. *EF*-1α
   D. *PbGP43* exon 2

6. If paracoccidioidomycosis is a New World disease, how can one explain cases that emerge in Europe years after travel or residence by the patient in Latin America?

7. What is known of the ecologic niche of *Paracoccidioides brasiliensis*? What animal may be a sentinel species to locate the environmental focus of this fungus?

8. Discuss the juvenile form of paracoccidioidomycosis. Is more than one organ system affected? What is the primary focus of the disease? How does the gender frequency change in children 12 years old or older? What is the most frequent clinical manifestation? How often are bone lesions found in this form of paracoccidioidomycosis?

9. Discuss the role of gender in chronic mucocutaneous paracoccidioidomycosis. What is the basis for gender preference? What is the primary site of infection? Describe the pathogenesis of oral lesions including the involvement of draining lymph nodes. What type of dual infection may complicate clinical assessment?

10. Follow-up of successfully treated paracoccidioidomycosis has revealed a pathologic sequel affecting quality of life of many patients. What is the nature of this problem?

11. In the case of severe paracoccidioidomycosis treated with amphotericin B, what is the outlook for relapses following completion of therapy? What is the role of maintenance therapy in preventing a relapse?

Chapter 8

Penicilliosis

8.1 PENICILLIOSIS-AT-A-GLANCE

- *Introduction/Disease Definition.* Penicilliosis is a pulmonary-disseminated mycosis of the host immunocompromised by HIV/AIDS or other immunosuppressive condition in individuals residing in or traveling through its Southeast Asia and Southwest China endemic regions.
- *Etiologic Agents. Penicillium marneffei* is a dimorphic environmental mold with a fission yeast tissue form. Other *Penicillium* species, not dimorphic, are exceptional causes of disease in humans and lower animals.
- *Geographic Distribution/Ecologic Niche. Penicillium marneffei* mycosis is endemic to Southeast Asia, especially in Thailand, Southern China, and Hong Kong. The disease is associated with bamboo rats and their burrows, but has not been isolated from other environmental sources.
- *Epidemiology.* Penicilliosis is an AIDS-associated opportunistic infectious disease and also may cause disease in other immunocompromised hosts.
- *Transmission.* The route of infection is via inhalation of airborne conidia.
- *Determinants of Pathogenicity.* These include dimorphism, production of superoxide dismutase, and secreted proteinase.
- *Clinical Forms.* The primary form is pulmonary, with hematogenous dissemination to multiple organ systems, including skin lesions.
- *Therapy.* Amphotericin B (AmB) is recommended for disseminated disease, with itraconazole (ITC) used for mild–moderate disease and for maintenance therapy.
- *Laboratory Detection, Recovery, and Identification.* Direct exam of touch smears of skin lesions and histopathology reveal a fission yeast. *Penicillium marneffei* grows on SDA at room temperature as a mold with brush-shaped conidiophores, and secretes red pigment into medium. Conversion to the fission yeast occurs in the host and at 37°C in vitro on enriched medium. Serum and urine antigen tests are prototypes but are not yet commercially available.

8.2 INTRODUCTION/DISEASE DEFINITION

The pathogen of interest in this chapter is *Penicillium marneffei*. Other *Penicillium* species are exceptional causes of disease in humans and lower animals. Pulmonary and pulmonary-disseminated penicilliosis is an AIDS-associated opportunistic, endemic mycosis geographically restricted to Southeast Asia and caused by *Penicillium marneffei*. Persons from other geographic areas living with AIDS or suffering from other immunosuppressive conditions have acquired penicilliosis during visits to the endemic area. A feature of this mycosis is frequent skin dissemination (Fig. 8.1).

Penicillium marneffei is a dimorphic fungus with an environmental form consisting of mold which produces conidia from brush-shaped conidiophores typical of *Penicillium* species. The mold form secretes a characteristic brick-red pigment into the growth medium. Conidia, once inhaled, convert to a fission yeast form in tissue. Disseminated penicilliosis is restricted to residents of or travelers to Southeast Asia and neighboring countries; as such, penicilliosis deserves consideration as a travel-medicine-related disease.

Immunosuppressed HIV-negative persons also are at risk who suffer from defects in T-cell-mediated

Fundamental Medical Mycology, First Edition. By Errol Reiss, H. Jean Shadomy and G. Marshall Lyon, III.
© 2012 Wiley-Blackwell. Published 2012 by John Wiley & Sons, Inc.

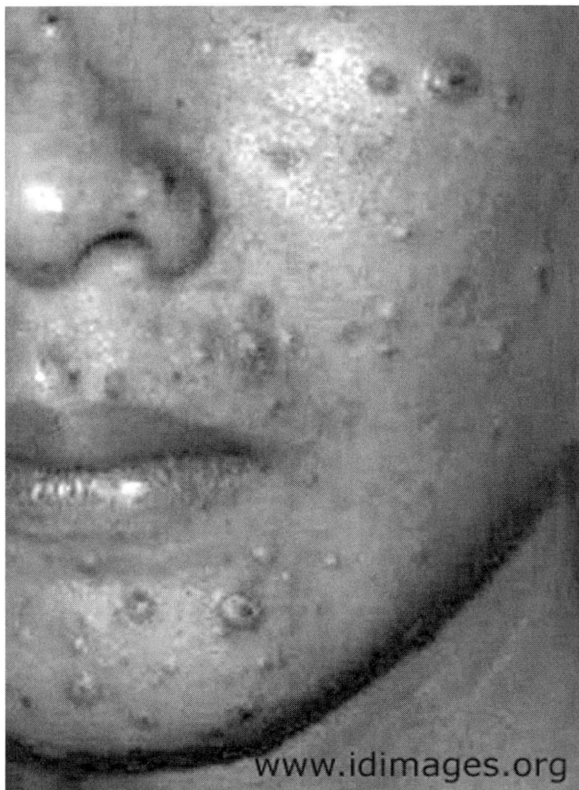

Figure 8.1 Umbilicated skin lesions on the face, a site of disseminated penicilliosis in a man with HIV/AIDS.
Source: Eric L. Krakauer, M.D., Ph.D., originally published by the Partners in Infectious Disease Images at www.idimages.org, and used with permission from Rajesh T. Gandhi, M.D., Editor.

immunity, are receiving cytotoxic cancer chemotherapy, or are receiving immunosuppressive therapy for management of stem cell, solid organ transplants, or for collagen vascular diseases. Diagnosis of penicilliosis must be considered for febrile immunocompromised patients with relevant clinical features, including umbilicated skin lesions, and a history of travel to Southeast Asia.

8.3 CASE PRESENTATION

A Businessman from Thailand with Cutaneous Lesions[1]

History

A 36-yr-old businessman from the Chiang Mai province of Thailand visiting the United States presented to the emergency department. Over the past 2 months, he developed a nonproductive cough followed by myalgias, fever, night sweats, and a 7 kg weight loss. In the past week, he noted several skin lesions on his extremities and face. He reported unprotected heterosexual activity with multiple partners.

Physical Exam

The patient is slender, appears chronically ill, but is in no acute distress. His vital signs include a temperature of 38.6°C; pulse, 94 beats/min; respiration, 22/ min (unlabored); blood pressure, 110/70; weight, 59 kg. The patient has scattered umbilicated 3–8 mm diameter cutaneous lesions on his forehead, upper extremities, and chest wall. The liver is enlarged 2 cm below the right costal margin. The spleen tip is palpable at the lower border of the L costal margin.

Radiographic Findings

His chest radiograph revealed multiple, small 3–5 mm diameter nodules in both lung fields.

Laboratory

Viewed microscopically, Wright–Giemsa stain and PAS-stained touch smears of a biopsied skin lesion revealed multiple 2–3 μm diameter fission yeast forms within macrophages.

Diagnosis

The clinical presentation and biopsy findings are characteristic of pulmonary-disseminated penicilliosis due to *Penicillium marneffei*.

Therapy and Clinical Course

The patient was hospitalized and was found to be HIV-positive, with a $CD4^+$ T-lymphocyte count of <200/μL. He was treated for the first week with AmB followed by outpatient therapy with ITC. This regimen led to a favorable response with resolution of cutaneous and pulmonary lesions. Combination antiretroviral therapy was initiated.

Comments

The chest radiograph is one of several types seen in penicilliosis. Please see Section 8.11, Clinical Forms—Radiographic Findings. AmB is used in more critically ill patients, those with multi-organ involvement, and numerous fission yeast forms on touch smears. ITC is used in patients with mild to moderate levels of infection who are compliant and ambulatory. Secondary prophylaxis with ITC is effective for preventing recurrent penicilliosis in persons living with AIDS. The probability of relapse in an AIDS patient not receiving such prophylaxis is high, particularly if the $CD4^+$ T-lymphocyte count is low.

[1]Used with permission from Dr. Thomas J. Walsh and Prof. Robert H. Rubin from Fungal Infections: Virtual Grand Rounds.

8.4 DIAGNOSIS

Touch smears of skin lesions reveal the pathogen, whose identity can be confirmed by growth at 25°C revealing a typical brush-shaped penicillus. The secretion of a brick-red pigment aids in the identification of *P. marneffei*. Sputum is another specimen for culture and identification. Blood smears may reveal intracellular fission yeast forms.

Differential Diagnosis

Cryptococcosis, histoplasmosis, and tuberculosis all occur in the endemic area for penicilliosis. *Histoplasma capsulatum* and *Penicillium marneffei* infections can occur as mixed infections or with *Cryptococcus neoformans*, so that cultures are required to confirm the diagnosis.

8.5 ETIOLOGIC AGENTS

Biverticillate penicillia are a homogeneous subgenus including *Penicillium marneffei*, the dimorphic endemic pathogen, other mycotoxigenic species contaminating food, and species used in biotechnology for enzyme production. The saprobic species are soil fungi and are cellulase producers with a role in degrading vegetation. This subgenus is noteworthy for the production of yellow, orange, or red pigments (Mapari et al., 2009).

Penicillium marneffei Segretain *Bull trimmest Soc Mycol Fr* 75: 416 (1960). Anamorphic member of the *Penicillium* species, subgenus *Biverticillium, Eurotiales, Ascomycota*.

Penicillium marneffei is classed in the *Penicillium* subgenus *Biverticillium* and among the sexual *Talaromyces* species with anamorphs in the *biverticillate* penicilli. Such classification is supported by sequence analysis of the mt rDNA and nuclear rDNA. *Penicillium marneffei* has no known sexual state.

Penicillium marneffei is a dimorphic fungus causing pulmonary-disseminated disease in persons with AIDS or other immunosuppressive conditions and for certain rat species indigenous to Southeast Asia. The mold grows rapidly secreting a brick-red pigment into the medium Microscopic morphology of the mold form shows a brush-shaped conidiophore typical of *Penicillium* species. At 37°C *P. marneffei* converts into a fission yeast.

Penicillium piceum, Raper & Fennell, *Mycologia* 40(5): 533 (1948).

Like *P. marneffei, P. piceum* is in the subgenus *Biverticillium*. They are phylogenetically close relatives, morphologically similar, and *P. piceum* may also produce red pigment in agar (Horré et al., 2001). *Penicillium piceum*, however, is a monomorphic mold and does not undergo a transition to a yeast form at 37°C. It has been isolated from blood cultures in humans and, in lower animals, from pig lung tissue, and mycotic abortion in cattle. Importantly, *P. piceum* grows better at 37°C than at 25°C. Genetic identification by sequencing the ITS region is definitive.

Penicillium purpurogenum Fleroff 1906. Anamorphic *Eupenicillium* in the subgenus *Biverticillium*.

Penicillium purpurogenum, a soil and plant contaminant able to grow at 37°C, is an exceptional opportunistic pathogen in that there are as few as two case reports extant of its isolation from human disease, both from pulmonary infections in patients with a primary hematologic disease. Disseminated penicilliosis due to *P. purpurogenum* was diagnosed in a German shepherd dog, including discospondylitis (Zanatta et al., 2006). Four different color pigments are known from this species.

A list of exceptional infections in humans in which other *Penicillium* species are implicated may be found in Cooper and Haycocks (2000).

8.6 GEOGRAPHIC DISTRIBUTION/ECOLOGIC NICHE

Penicilliosis is an endemic mycosis restricted to Southeast Asia, especially Thailand, and also is found in Southwest China including Hong Kong, parts of Malaysia, northeast India (Manipur State bordering Myanmar), Taiwan, and Vietnam (Fig. 8.2). *Penicillium marneffei* and opportunities for exposure are suspected in Indonesia, Myanmar, the Philippines, and Singapore, but case reports from these countries are scant. Most cases have been reported from northern Thailand, especially near Chiang Mai, and in Southwest China. Hong Kong is at the same latitude as Guangxi Province. The northern boundary of the endemic zone is believed to be the latititude of the Yangtse River in China extending south to the Equator. Increasingly, it is encountered out of the endemic area as a travel medicine problem. (Please see Section 8.3, Case Presentation.)

Seasonal variation of *P. marneffei* infections in AIDS patients in northern Thailand shows that *P. marneffei* but not *C. neoformans* infections are more frequent in the rainy than the dry season, providing a clue in locating the source of exposure to *P. marneffei* (Chariyalertsak et al., 1996). The ecologic niche of *P. marneffei* is uncertain apart from its association with certain rat species. *Penicillium marneffei* was isolated originally in 1956 from bamboo rats (*Rhizomys sinensis)* from the central highlands of Vietnam. Alternative hosts are *Rhizomys pruinosus* in Southern China and *Cannomys badius* in Vietnam and northeast India. One study found that *P. marneffei* could be isolated from 18 of 19 rats captured from different

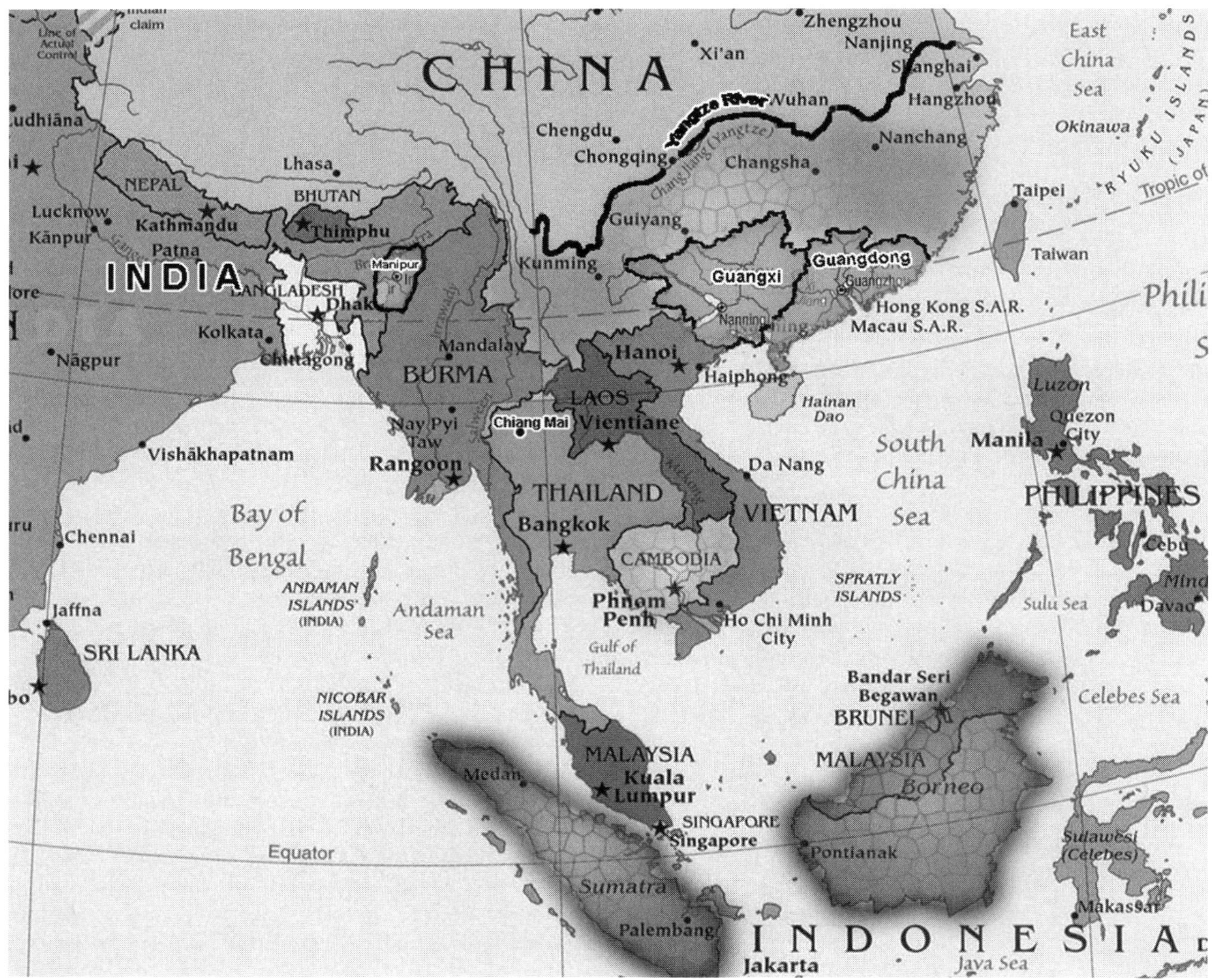

Figure 8.2 Endemic areas in Southeast Asia for penicilliosis. Known endemic areas are colored in light red. Those areas shown in a reticulated pattern are presumed to harbor *P. marneffei* The Thailand endemic area is centered on the northern city of Chiang Mai. Case series were reported in Hong Kong. Case reports from China include Guangxi and Guangdong provinces. Areas south of the Yangtse River are presumed to harbor *P. marneffei*. Other Southeast Asian countries reporting cases include Vietnam, Laos, Taiwan, and parts of Malaysia. Manipur State in northeastern India also has reported cases. Indonesia and Cambodia are suspected of harboring the fungus but the lack of case reports indicates either the rarity of autochthonous cases or lack of surveillance for the pathogen. See insert for color representation of this figure.

regions within the Guangxi Province of China. Curiously, the rats appeared well, and when they were euthanized and necropsied there were no gross pathologic changes in their lungs, liver, spleen, and mesenteric lymph nodes. Culture of these specimens, however, was positive for *P. marneffei* (Deng et al., 1986). *Penicillium marneffei* also was isolated from the soil of rat burrows and was experimentally inoculated to bamboo shoots and sugarcane.

There is no evidence that bamboo rats serve as a reservoir for human infection, but rather that the fungus is present in the soil and aerosolized conidia are the infectious propagules. No efforts to date have been successful in isolating *P. marneffei* from soil other than that associated with burrows of the aforementioned rodents. One reason proposed by Vanittanakom et al. (2006) is that *P. marneffei* does not survive well in nonsterile soil, possibly owing to its inability to compete well with other soil microbiota.

8.7 EPIDEMIOLOGY

8.7.1 Incidence

Penicilliosis is an AIDS-defining opportunistic disease in Southeast Asia such that the epidemic curve of penicilliosis follows the epidemic curve of AIDS. In 1995 the annual incidence in Thailand was 1300 cases. New cases of AIDS peaked in Thailand in the 1995–1996 period and then, with the advent of combination ART, new cases of AIDS and of penicilliosis began to subside. The Thai Ministry of Health estimated there were 6709 cases of disseminated penicilliosis during the decade 1984–1994. Penicilliosis is the third most frequent opportunistic disease of people living with AIDS in northern Thailand, after tuberculosis and cryptococcosis, and accounts for 15–20% of presenting AIDS illnesses in that region (Chaiwarith et al., 2007).

Penicilliosis is an emerging mycosis for persons with AIDS or other immunosuppressive conditions living in or traveling to the endemic area, and is dangerous to susceptible hosts regardless of the duration of their exposure or the time elapsed since their visit to one of the endemic areas.

8.7.2 Molecular Epidemiology

The origin of an isolate of *P. marneffei* may be traced to China or Thailand using multilocus sequence typing (MLST). The gene encoding the *P. marneffei* antigenic cell wall glycoprotein, MP1, is highly polymorphic with 12 different MP1 types (Lasker, 2006). Single-nucleotide polymorphisms were observed at 21 different locations in the *MP1* gene. This high degree of MP1 polymorphism suggests the sequence evolves to evade host immune responses. MLST using *MP1* and three housekeeping genes revealed clusters composed of isolates obtained only from China, only from Thailand, or from China and Thailand, the latter indicating mixing or common descent.

8.8 RISK GROUPS/FACTORS

Persons living with AIDS whose $CD4^+$ T-lymphocyte count is <200/μL are at risk: for example, the average $CD4^+$ T-lymphocyte count of persons presenting with penicilliosis at Chiang Mai University Hospital, Thailand, was 64/ μL (std dev 47/ μL) (Vanittanakom et al., 2006). HIV-negative immunosuppressed patients are at risk of penicilliosis if they have defects in T-cell-mediated immunity or are being treated with cytotoxic and/or immunosuppressive drugs for hematologic malignancies or for management of organ transplants or collagen-vascular disease. Penicilliosis must be considered for febrile immunocompromised patients with relevant clinical features, including umbilicated skin lesions and a history of travel to or residence in Southeast Asia.

8.9 TRANSMISSION

Transmission occurs via inhalation of airborne conidia. The ecologic niche of *P. marneffei* is unknown (aside from its association with bamboo rats and their burrows). Plausibly, as persons with penicilliosis travel outside the known endemic areas, the fungus will establish itself more widely in the environment.

8.10 DETERMINANTS OF PATHOGENICITY

8.10.1 Host Factors

Pulmonary alveolar and circulating monocytes have antifungal activity against *P. marneffei*. Circulating monocytes respond to *P. marneffei* conidia with an oxidative burst and superoxide production, which is significantly enhanced by the cytokine, macrophage-CSF (Roilides et al., 2003). Experiments with interferon-γ knock-out mice infected with *P. marneffei* demonstrated the importance of the Th1 cytokine, IFN-γ, in host defense. IFN-γ knock-out mice could not organize granulomas in their livers and spleens: instead there were disorganized masses of macrophages and fission yeast cells (Sisto et al., 2003).

8.10.2 Microbial Factors

Morphogenesis

The ability of *P. marneffei*, alone among the *Penicillium* species, to undergo temperature-sensitive dimorphism (Adrianopoulos, 2002) to the fission yeast tissue form is a pathogenic stratagem. The sequence of steps in the mold→ yeast dimorphism is:

1. At 37°C in vitro conidia germinate forming a germ tube.
2. Outgrowth of a hypha occurs, with branching.
3. Nuclear and cell divisions are coupled so that each hyphal cell contains a single nucleus.
4. A double septum separates each hyphal cell.
5. Cell wall material between the double septum is digested causing separation of individual cells. This process is arthroconidiation.
6. The resultant single cells continue to grow and divide as fission yeast.
7. Genes controlling the steps from arthroconidiation to the yeast form have not yet been identified but the gene *CDC42* homolog in *P. marneffei, cflA*, controls normal fission yeast cell division.

8. This is the *only known example of a mold first producing arthroconidia which then progress to a fission yeast form*.

The process of arthroconidiation, so well orchestrated in vitro, has not been detected when conidia are phagocytosed by macrophages.

Environmental Sensing

Penicillium marneffei, like other dimorphic fungal pathogens, has two p21-activated kinases that control important signaling and morphogenesis functions (Boyce et al., 2009). The *pakA* gene encodes a protein-activated kinase essential to germination of conidia and growth of fission yeast at 37°C. Conidia from Δ*pakA* strains fail to germinate in macrophages. At 25°C, PakA is not required for germination, for hyphal growth, or conidiogenesis. Some of those functions are served by a second protein activated kinase, PakB, which controls apical growth of hyphae, and conidiogenesis. Deletion of *pakB* leads to a failure of yeast cells to divide within macrophages. They are completely blocked in the link between septation and cell separation, instead growing as hyphae. At 25°C, Δ*pakB* strains do not produce conidia but instead produce a mix of hyphae and yeast cells.

Cell Wall: Mp1 Mannoprotein

The cloned *P. marneffei MP1* gene is unique without homologs in the sequence databases (Cao et al., 1998a). The protein it encodes is Mp1 containing 462 amino acid residues, with features in common with cell wall proteins of *Saccharomyces cerevisiae* and *Candida albicans*: there are two putative *N*-glycosylation sites, a serine- and threonine-rich region for *O*-glycosylation, and a putative *g*lycosyl*p*hosphatidyl*i*nositol attachment signal sequence.

Specific anti-Mp1 antibody probe of Western blots localized Mp1 to proteins with molecular masses of 58 and 90 kDa. Mp1 is a glycoprotein with affinity for concanavalin A, characteristic of a mannoprotein. Patients with penicilliosis caused by *P. marneffei* demonstrate antibodies against Mp1.

Enzymes: Defense Against Host Reactive Oxygen Species

- *Cu, Zn-Superoxide Dismutase*. As an intracellular pathogen of the mononuclear phagocytic system, *P. marneffei* shields itself from oxidative microbicidal pathways by producing Cu, Zn-superoxide dismutase and upregulating expression of this enzyme in the fission yeast form and after phagocytosis by macrophages (Thirach et al., 2007). The significance of this enzyme is underlined because human monocytes that have phagocytosed conidia undergo an oxidative burst, with resultant superoxide production (Roilides et al., 2003).
- *Catalase-Peroxidase. Penicillium marneffei* contains a bifunctional enzyme, Cpe A, capable of either reducing H_2O_2 with an external reductant (peroxidase activity) or converting it to H_2O and O_2 (catalase activity) (Pongpom et al., 2005). Expression of *cpeA* was higher in the yeast form, and the protein elicits an antibody response in patients infected with *P. marneffei*, suggesting a role in pathogenesis.

Eqolisin Proteinase

Penicillium marneffei secretes proteinases that were characterized as pepstatin-insensitive acid proteinases of the eqolisin family (Moon et al., 2006). Their role in pathogenesis awaits investigation.

8.11 CLINICAL FORMS

Pathophysiology

The portal of entry is the lung, with conidia of *P. marneffei* acting as the infectious *propagules*. The disease is characterized by systemic infection of the mononuclear-phagocytic system, resulting in a progressive febrile illness that is fatal if untreated. *Penicillium marneffei* has a predilection for skin dissemination affecting the face, upper trunk, and extremities. Penicilliosis also afflicts non-AIDS patients living with other immunocompromising conditions.

Pulmonary-Disseminated Penicilliosis

Penicilliosis is a primary pulmonary pathogen that disseminates hematogenously via the mononuclear-phagocytic system to the internal organs. Symptoms include fever, cough, weight loss, lymphadenopathy, and cutaneous, subcutaneous, and oropharyngeal lesions. The illness may progress for a month or more before patients seek medical care. Clinical signs include cough, fungemia, anemia, general lymphadenopathy, and hepatosplenomegaly. Signs and symptoms are similar to those seen in histoplasmosis, which also occurs naturally in the same geographic area. *Penicillium marneffei* is, like *Histoplasma capsulatum*, an intracellular pathogen of the mononuclear-phagocytic system.

Radiographic Findings

Pulmonary symptoms (cough, dyspnea) are present in about 50% of cases. Chest radiographs reveal diffuse reticulonodular infiltrates. Cavitary lesions occur and have been associated with hemoptysis, but the chest radiograph may be normal in half of cases. Chest radiography findings in a series of Thai patients with penicilliosis found a normal pattern in 78% of patients, with interstitial infiltrates associated with pleural effusions being the most common pattern in patients with abnormal chest radiographs. This pattern of more frequent interstitial infiltrates also is seen in patients with histoplasmosis treated in the same Thai hospital (Mootsikapun and Srikulbutr 2006) (Table 8.1). Similarities between the two endemic mycoses in Thai AIDS patients extended to both abnormal chest X-rays and the frequency of skin lesions.

Skin Lesions

Skin lesions, appearing in up to 85% of cases, are papules with necrotic centers or maculopapules, and are found most often on the forehead, trunk, upper extremities, and abdomen. They are commonly umbilicated papules with or without central necrosis, resembling *mulloscum contagiosum*. The lesions contain numerous yeast forms intra- and extracellularly. Skin lesions similar to those in penicilliosis (Fig. 8.1) also are found in cryptococcosis and histoplasmosis.

Osseous

Multiple lytic lesions may be present involving the ribs, long bones, skull, vertebrae, scapula, and/or the temporomandibular region.

Multi-organ Dissemination

This can include adrenals, bone, bone marrow, bowel, kidney, liver, lung, meninges, pericardium, skin, spleen, and tonsils.

Laboratory findings include fungemia and intracellular fission yeast in monocytes and macrophages. These signs and symptoms are accentuated in persons living with AIDS in whom the pathogenesis is acute, severe, and rapidly progressive. The sudden appearance of *molluscum contagiosum*-like lesions on the forehead, face, trunk, and hands are hallmarks in persons living with AIDS who develop penicilliosis.

Relapsing Penicilliosis

Recovery from primary penicilliosis requires maintenance antifungal therapy, for example, with ITC, to prevent recurrences which, absent maintenance therapy and/or recovery of $CD4^+$ T lymphocytes, can occur in half of the cases. The relapse rate for persons surviving invasive penicilliosis was 57% if they lacked access to combination ART, and if oral ITC was not available (Supparatpinyo et al., 1998). Where persons living with AIDS receive combination ART and their $CD4^+$ T-lymphocyte count is maintained at $\geq$100 cells/μL, the risk of relapse of penicilliosis is low (0.6 cases per 100 person-months) compared with 11.5 cases per 100 person-months before the advent of combination ART (Chaiwarith et al., 2007). Such patients may be spared the need for indefinite ITC maintenance therapy.

Laboratory infections

- Accidental exposure by means of a finger stick with a suspension of *P. marneffei* resulted in localized skin lesions and regional adenopathy in an immune-competent laboratory worker (Drouhet, 1993).
- An HIV-positive physician with no history of travel or residence in an endemic area contracted penicilliosis while attending a mycology training course at the Institut Pasteur, Paris. He never handled the culture,

Table 8.1 Chest Radiograph Findings of Patients with Histoplasmosis and Penicilliosis at Srinagarind Hospital, Thailand 1996–2002

Chest X-ray findings	Number of patients (%)	
	Histoplasmosis (n = 32)	Penicilliosis (n = 36)
Normal	19 (59.4)	25 (78.1)
Abnormal	13 (40.6)	11 (21.9)
Alveolar/patchy infiltration	2 (6.2)	2 (6.2)
Interstitial infiltration	7 (21.9)	5 (15.6)
Mixed alveolar/interstitial infiltration	0 (0.0)	1 (3.1)
Miliary infiltration	2 (6.2)	1 (3.1)
Pleural effusion	1 (3.1)	3 (9.4)
Cavitation	1 (3.1)	1 (3.1)

Source: Mootsikapun and Srikulbutr (2006). Used with permission from Elsevier Publishers, © 2006.

but live cultures of *P. marneffei* were distributed for study to other students (Drouhet, 1993).

Penicillium marneffei has to be considered a dangerous pathogen capable of infection via the respiratory route and by penetrating wounds. Consequently, *P. marneffei* should be handled by properly trained and supervised personnel in a separate procedure room under negative air pressure, reserved for respiratory pathogens, and **only** within a currently certified, well-functioning laminar flow biological safety cabinet (BSC). Personal protective equipment including well-fitting face masks or respirators as well as two pairs of gloves should be worn during manipulations of *P. marneffei*.

Handling living *P. marneffei* cultures by students in a teaching environment *cannot be recommended*. Research students are an exception provided they have received specific training for respiratory and blood-borne pathogens and are directly supervised by their research preceptor during their experiments with *P. marneffei*.

8.12 VETERINARY FORMS

Penicillium marneffei was first isolated in 1956 at the Pasteur Institute in Dalat, Vietnam, from the viscera of a bamboo rat (*Rhizomys sinensis*) in the highlands of central Vietnam. Surveys recovered *P. marneffei* from the internal organs of 10 (9.1%) of 110 bamboo rats examined from Manipur State, in northeast India bordering Myanmar, an area endemic for penicilliosis (Gugnani et al., 2004).

8.13 THERAPY

Without antifungal therapy penicilliosis is fatal in persons living with AIDS, and nearly so in HIV-negative immunocompromised persons. AmB monotherapy is effective, with ITC as maintenance therapy. AmB is advisable in cases of multi-organ dissemination. ITC, as primary therapy, is reserved for patients with mild or moderate disease, and for follow-up maintenance therapy and secondary prophylaxis. One regimen used is amphotericin B (0.6 mg/kg daily) for 2 weeks, followed by oral ITC (400 mg/day) for 10 weeks (Wong et al., 1999).

8.14 LABORATORY DETECTION, RECOVERY, AND IDENTIFICATION

Direct Examination

Rapid diagnosis is facilitated by microscopic examination including direct smears of scrapings and aspirates of skin lesions, sputum, bone marrow aspirates, peripheral blood smears, BAL fluid, CSF, urine, stool. Smear and culture of bone marrow aspirates and peripheral blood are fruitful specimens in approximately half of cases (Wong et al., 1999).

Histopathology

Histologic sections of skin lesions and fine needle aspirates of enlarged lymph nodes, kidney, and liver reveal small yeasts. The presence of a transverse septum, not budding, indicates a fission yeast and differentiates *P. marneffei* from *H. capsulatum* and *Cryptococcus neoformans. Penicillium marneffei* is well stained by PAS or GMS stains and does not stain with H&E, but monocytes engorged with fission yeast can be discerned in H&E stained sections. *Penicillium marneffei* is better seen with the Giemsa stain[2]. The appearance in tissue is of yeast-like septate cells—fission yeast that are round, oval, or elongated cells with each pair divided by a transverse septum. Each cell is 2–3 μm × 2–6.5 μm and is found intracellularly in macrophages (Fig. 8.3a). Extracellular fission yeast also are present in disseminated penicilliosis with occasional elongate or curved cells, 8–13 μm long, or short filaments 20 μm in length. Seen in histopathologic section, the fission yeasts of *P. marneffei* are *pleomorphic* in comparison with the budding yeast of *H. capsulatum* (Fig. 8.3b).

Culture

Penicillium marneffei, because it is dimorphic, can be recovered as a fission yeast from blood cultures incubated at 37°C. This mold is susceptible to and will not grow on medium containing cycloheximide. The fission yeast form will be propagated if transferred from blood cultures to SDA or BA and incubated at 37°C. Specimens that give a high yield of recovery in culture include blood cultures, bone marrow, and scrapings, aspirates, and biopsy of skin lesions. On SDA at ambient temperature, *P. marneffei* grows as a typical conidia-forming mold.

Colony Morphology: Mold Form

Penicillium marneffei grows rapidly on SDA *without cycloheximide* but with antibacterial agents at 25°C, producing a flat, downy, blue-gray-green mold colony with a white border and a brown-red reverse. During the

[2]Wright's stain, Giemsa stain, and Diff-Quik® are a group of related stains derived from the original Romanovsky stain, in that they use eosin Y and methylene blue or eosin and azure A or B to stain blood smears and malaria parasites. They are useful to visualize *P. marneffei* within macrophages and also have been used effectively to detect *Pneumocystis jirovecii*.

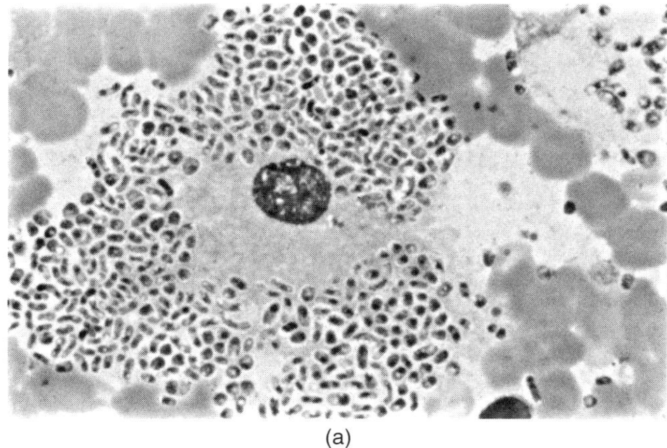

(a)

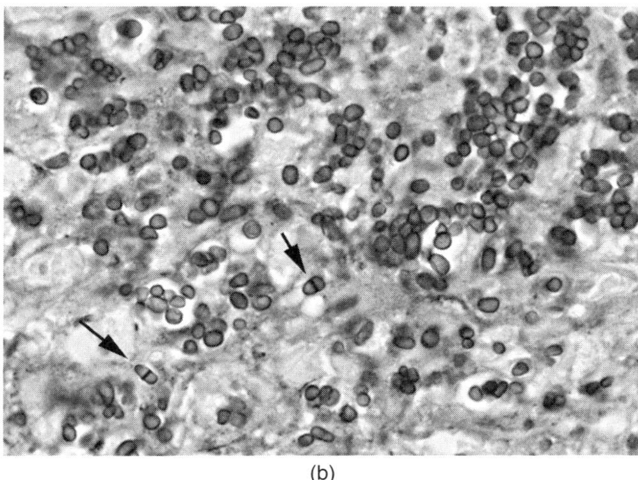

(b)

Figure 8.3 (a) The fission yeast morphology of *Penicillium marneffei*. Intracellular yeast within a macrophage. *Source:* Kenrad Nelson M.D., Johns Hopkins University School of Public Health, Baltimore, Maryland (Dismukes et al., 2003) and used with permission from Oxford University Press. (b) Fission yeast of *P. marneffei* as seen in histopathologic section, GMS stain. *Arrows* point to fission yeast division. Yeast cells are ovoid or elongate (3–8 μm length).
Source: Used with permission from Benjaporn Chaiwun, M.D., Department of Pathology, Faculty of Medicine, Chiang Mai University, Chiang Mai, Thailand.

first 24–48 h of growth, *P. marneffei* secretes a diffusible brick-red pigment into the surrounding agar (Fig. 8.4a). Differential diagnosis is based on colony and microscopic morphology.

- *Microsporum gallinae* produces a diffusible red pigment. It is an uncommon causative agent of *tinea corporis* and *tinea capitis* and more often isolated from chickens and other fowl. *Trichophyton rubrum* is well known for its port-wine red pigment which does not diffuse into the medium.

The following are pigment-producing *Penicillium* species:

- *Penicillium purpurogenum* produces an intense red or purple-red pigment and is typically biverticillate, symmetrical, with rough conidia.
- *Penicillium citrinum* produces pale yellow or pale brown to red-brown pigment. Its penicillus has defined columns of conidia.
- *Penicillium janthinellum* produces an orange-red or purple pigment, has strongly divergent conidial chains, does not form compact columns, and has phialides that taper abruptly with a long narrow neck.

None of these look-alike species are dimorphic.

Microscopic Morphology: Mold Form

First a word about nomenclature of the microscopic morphology. The conidial fruiting structure is the penicillus. Conidia are borne in chains on phialides. Phialides, in turn, are borne in *verticils* on a hyphal stalk, the stipe, either directly or on intervening supporting cells, the metulae. The penicillus is the entire structure supported by the conidiophore. In *P. marneffei*, the conidiophores are short, smooth, or finely roughened, 1.5–2 μm diameter, and up to 111 μm long. Metulae (4–5) arise from the tip of the conidiophore; one usually is much longer than the others and usually is septate. The penicillus is typically biverticillate, compact or irregular, either asymmetric or symmetric, and seldom branched (Fig. 8.4b).

Conidia arise from conspicuously pointed phialides. Conidial chains are not parallel, not formed in columns, but are distinct and tangled; conidia are smooth, oval to lemon shaped (2–3 μm × 2.3 μm). Other molds that have morphology similar to *P. marneffei* are *Paecilomyces* species (elongated phialides taper to a slender tube with conidia in long unbranched chains) and *Scopulariopsis* species (annellides singly or in groups with chains of thick-walled round conidia).

Colony Morphology: Fission Yeast Form

Cultures on blood agar at 37°C are smooth, beige colored, and *do not secrete red pigment* (Wong et al., 1999).

Microscopic Morphology: Fission Yeast

Morphogenesis to the fission yeast is accomplished at 37°C on blood agar or semisynthetic medium containing asparagine or casein hydrolyzate as the nitrogen source (Drouhet, 1993). Blood cultures transferred to SDA and incubated at 37°C will maintain the fission yeast forms. Yeast cells are ovoid or elongate (3–8 μm length) with

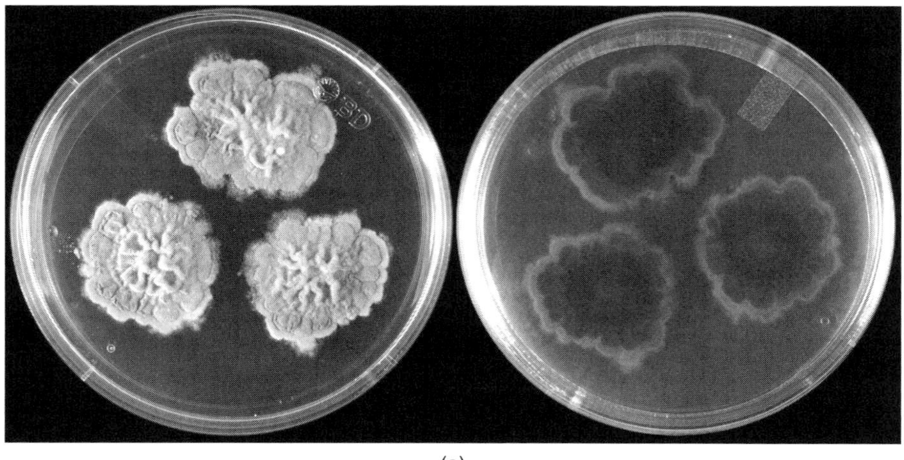

(a)

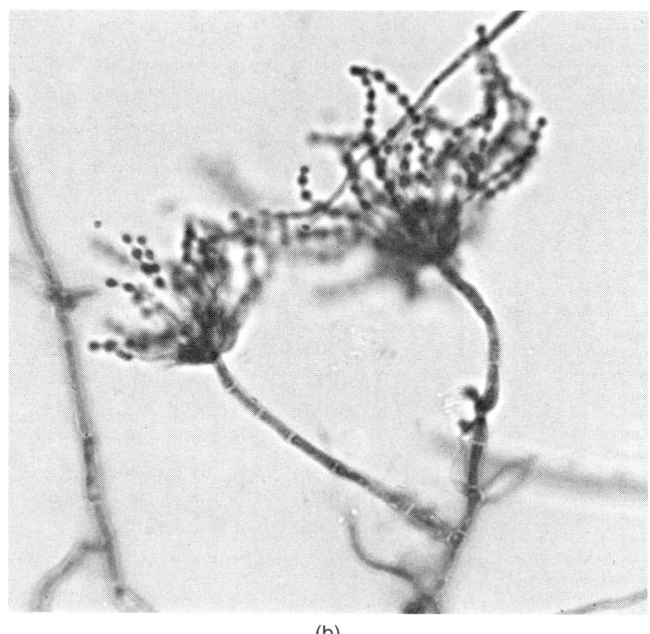

(b)

Figure 8.4 (a) *Penicillium marneffei* colony (mold form). Left: surface; Right: reverse showing diffusible red pigment. *Source:* Mr. Jim Gathany, Creative Arts Branch, CDC. See insert for color representation of this figure. (b) Microscopic morphology of the mold form of *Penicillium marneffei*. The brush-shaped conidiophore is typical of *Penicillium* species. Conidiophores are *biverticillate*. Penicilli are either asymmetric or symmetric, seldom branched. Conidia arise from pointed phialides. Conidial chains are short, often tangled. Conidia are smooth, ellipsoid (2.5–4 μm × 2–3 μm) (500×). *Source:* Used with permission from Dr. Arvind A. Padhye, CDC.

a transverse septum. Cell division is by *scissiparity* (*definition*: reproduction by fission) not budding. Cross-walls are formed before separation. The in vitro and in vivo appearances of the yeast form are similar. In vitro the yeast form colonies become pinkish-red but there is no diffusible pigment. Dangerous! Work in a well-ventilated biological safety cabinet. For further details of biosafety levels consult the CDC/NIH *Biosafety in Microbiological and Biomedical Laboratories* (BMBL) 5th ed., at the URL http://www.cdc.gov/od/ohs/biosfty/bmbl5/bmbl5toc.htm.

Serology

The choice of serologic methods is determined by the ability of the host to mount an antibody response. In AIDS, that ability is compromised so that antigen detection remains the most accurate surrogate marker for diagnosis. Investigators in the Southeast Asia endemic area developed antibody tests, tests for antigenemia, and antigenuria, summarized below.

Antibody Detection

An indirect EIA to detect antibodies in patients with penicilliosis caused by *P. marneffei* was designed using recombinant *P. marneffei* mannoprotein—Mp1 (Cao et al., 1998b). The EIA was positive in 12 of 17 cases of penicilliosis caused by *P. marneffei* in persons living with AIDS, and no false-positive reactions were detected in controls consisting of 90 local (Hong Kong) healthy blood donors, 20 patients with typhoid fever, and 55 patients with tuberculosis. Western blots constituted from antigenic complexes from *C. albicans, H. capsulatum*, and *Cryptococcus neoformans* and probed with anti-Mp1 antibodies failed to demonstrate any crossreactivity.

Antigenemia

Penicillium marneffei antigens circulate in the plasma and urine of penicilliosis patients. The antigens include carbohydrates, probably of cell wall origin, mycelial glycoproteins, and yeast form culture filtrate antigens (Chaiyaroj et al., 2003).

- A double antibody sandwich EIA was devised using rabbit polyclonal capture antibodies and, as indicator antibodies, two different mAbs derived from the mycelial culture filtrate and directed against carbohydrate epitopes (Trewatcharegon et al., 2000). Serum specimens included those from 53 Thai persons living with AIDS with culture-confirmed penicilliosis caused by *P. marneffei*, persons living with AIDS and diagnosed with candidiasis, cryptococcosis, or histoplasmosis, and sera from healthy adults from *P. marneffei* endemic areas.

  The sensitivity, specificity, of serum antigen detection when compared to culture-confirmed penicilliosis were 92.5% and 97.5%, respectively. The positive predictive value of the test was 89.1%, while the negative predictive value was 98.3%. Chaiyaroj et al. (2003) found no crossreactivity between their mAbs and antigens of *Aspergillus fumigatus*. The epitopes being detected were presumed to be carbohydrate but not crossreactive with the galactomannan of *A. fumigatus*. This observation leaves the characterization of the circulating carbohydrate antigen for further investigation.
- Another double antibody sandwich EIA was designed to detect antigenemia in penicilliosis (Panichakul et al., 2002). It used an mAb against 62 and 76 kDa *P. marneffei* proteins present in a yeast form culture supernatant. No crossreactions were found with antigens of *A. fumigatus, A. flavus, Candida albicans, Cryptococcus neoformans*, or *Histoplasma capsulatum*.

  The EIA was evaluated in serum from 18 patients with culture-proven penicilliosis, 23 patients with other fungal and bacterial infections, and healthy volunteer controls. The resultant sensitivity was 72%, with a specificity of 100%, a positive predictive value of 100%, and a negative predictive value of 97%.
- The *Aspergillus* EIA detecting galactomannan antigenemia, *GM-EIA* (Bio-Rad, Inc.), crossreacts with serum from patients living with AIDS and penicilliosis caused by *P. marneffei*. Although the antigenemia EIA reported by Chaiyaroj et al. (2003) did not apparently crossreact with antigens of *A. fumigatus*, others found that the GM-EIA produced positive results in 11 of 15 (73.3%) of culture-confirmed penicilliosis in persons living with AIDS in Taiwan (Huang et al., 2007). Patients with fungemia produced the highest antigenemia concentrations. No patients had aspergillosis nor were they taking the antibiotics piperacillin-tazobactam, or amoxicillin-clavulanate, known to give false-positive reactions in the GM-EIA. This antigen detection EIA could be used in its kit form in the endemic areas, with the caveat that it is not specific for *P. marneffei*.

Seemingly, more than one carbohydrate-containing antigen circulates in penicilliosis—one antigen is crossreactive with *A. fumigatus* galactomannan and at least one that is not. The comparison of antigenemia detection using the Bio-Rad GM-EIA and the alternative antigenemia detection using the Chaiyaroj et al. (2003) EIA would be useful in determining which gives better results.

Antigenuria

Antibodies raised in rabbits immunized with whole killed fission yeast forms of *P. marneffei* were used in an antigen-capture sandwich EIA to detect antigenuria (Desakorn et al., 1999). Thirty-seven HIV-positive patients from Northeast Thailand with culture-confirmed penicilliosis caused by *P. marneffei* were tested and 36 (97.3%) were positive for urine antigen, when the titer for a positive test was $\geq$1:40. False-positive antigenuria titers were few, occurring in 6 (2%) of 300 control urine samples including those from HIV-seropositive patients, those with other fungal infections (cryptococcosis, histoplasmosis, and candidiasis), and serum specimens from patients with bacterial infections. No crossreactivity was observed with culture supernatants of *Aspergillus fumigatus, A. terreus*, cryptococci, or *Histoplasma capsulatum*.

Genetic Identification

Molecular methods can identify dimorphic mold pathogens, including *P. marneffei* (Lindsley et al., 2001). Universal fungal primers ITS1 and ITS4 amplify the ITS2 region of rRNA genes. Sequencing that region provides a specific target for *P. marneffei*. The probe specific for *P. marneffei* that hybridizes with the target is, from 5′ to 3′, GGGTTGGTCACCACCATA. Either direct sequencing of the ITS2 region or use of the Luminex® hybridization method provides a convenient way to identify young cultures.

SELECTED REFERENCES FOR PENICILLIOSIS

Adrianopoulos A, 2002. Control of morphogenesis in the human fungal pathogen *Penicillium marneffei*. *Int J Med Microbiol* 292: 331–347.

Boyce KJ, Schreider L, Andrianopoulos A, 2009. In vivo yeast cell morphogenesis is regulated by a p21-activated kinase in the human pathogen *Penicillium marneffei*. *PLoS Pathog* 5(11): e1000678.

Cao L, Chan C-M, Lee C, Sai-Yin Wong S, Yuen K-Y, 1998a. MP1 encodes an abundant and highly antigenic cell wall mannoprotein in the pathogenic fungus *Penicillium marneffei*. *Infect Immun* 66: 966–973.

Cao L, Chen D-L, Lee C, Chan C-M, Chan K-M, Vanittanakom N, Tsang DNC, Yuen K-Y, 1998b. Detection of specific antibodies to an antigenic mannoprotein for diagnosis of *Penicillium marneffei* penicilliosis. *J Clin Microbiol* 36: 3028–3031.

Chaiwarith R, Charoenyos N, Sirisanthana T, Supparatpinyo K, 2007. Discontinuation of secondary prophylaxis against penicilliosis marneffei in AIDS patients after HAART. *AIDS* 212: 365–369.

Chaiyaroj SC, Chawengkirttikul R, Sirisinha S, Watkins P, Srinoulprasert Y, 2003. Antigen detection assay for identification of *Penicillium marneffei* infection. *J Clin Microbiol* 41: 432–434.

Chariyalertsak S, Sirisanthana T, Supparatpinyo K, Nelson KE, 1996. Seasonal variation of disseminated *Penicillium marneffei* infections in northern Thailand: A clue to the reservoir? *J Infect Dis* 173: 1490–1493.

Cooper CR Jr, Haycocks NG, 2000. *Penicillium marneffei*: An insurgent species among the penicillia. *J Eukaryot Microbiol* 47: 24–28.

Deng Z, Yun M, Ajello L, 1986. Human penicilliosis marneffei and its relation to the bamboo rat (*Rhizomys pruinosus*). *Med Mycol* 24: 383–389 .

Desakorn V, Smith MD, Walsh AL, Simpson AJH, Sahassananda D, Rajanuwong A, Wuthiekanun V, Howe P, Angus BJ, Suntharasamai P, White NJ, 1999. Diagnosis of *Penicillium marneffei* infection by quantitation of urinary antigen by using an enzyme immunoassay. *J Clin Microbiol* 37: 117–121.

Dismukes WE, Pappas PG, Sobel JD 2003, Figure 23.6, *Clinical Mycology*. Oxford University Press, Oxford, UK.

Drouhet E, 1993. Penicilliosis due to *Penicillium marneffei*: A new emerging systemic mycosis in AIDS patients traveling or living in Southeast Asia. *J Mycol Med* 4: 195–224.

Gugnani H, Fisher MC, Paliwal-Johsi A, Vanittanakom N, Singh I, Yadav PS, 2004. Role of *Cannomys badius* as a natural animal host of *Penicillium marneffei* in India. *J Clin Microbiol* 42: 5070–5075.

Horré R, Gilges S, Breig P, Kupfer B, de Hoog GS, Hoekstra E, Poonwan N, Schaal KP, 2001. Case report. Fungaemia due to *Penicillium piceum*, a member of the *Penicillium marneffei* complex. *Mycoses* 44: 502–504.

Huang Y-T, Hung C-C, Liao C-H, Sun H-Y, Chang S-C, Chen Y-C, 2007. Detection of circulating galactomannan in serum samples for diagnosis of *Penicillium marneffei* infection and cryptococcosis among patients infected with human immunodeficiency virus. *J Clin Microbiol* 45: 2858–2862.

Lasker BA, 2006. Nucleotide sequence-based analysis for determining the molecular epidemiology of *Penicillium marneffei*. *J Clin Microbiol* 44: 3145–3153.

Lindsley MD, Hurst SF, Iqbal NJ, Morrison CJ, 2001. Rapid identification of dimorphic and yeast-like fungal pathogens using specific DNA probes. *J Clin Microbiol* 39: 3505–3511.

LoBuglio KF, Taylor JW, 1995. Phylogeny and PCR identification of the human pathogenic fungus *Penicillium marneffei*. *J Clin Microbiol* 33: 85–89.

Mapari SA, Meyer AS, Thrane U, Frisvad JC, 2009. Identification of potentially safe promising fungal cell factories for the production of polyketide natural food colorants using chemotaxonomic rationale. *Microb Cell Fact* 8: 24.

Moon JL, Shaw LN, Mayo JA, Potempa J, Travis J, 2006. Isolation and properties of extracellular proteinases of *Penicillium marneffei*. *Biol Chem* 387: 985–993.

Mootsikapun P, Srikulbutr S, 2006. Histoplasmosis and penicilliosis: Comparison of clinical features, laboratory findings and outcome. *Int J Infect Dis* 10: 66–71.

Panichakul T, Chawengkirttikul R, Chaiyaroj SC, Sirisinha S, 2002. Development of a monoclonal antibody-based enzyme-linked immunosorbent assay for the diagnosis *of Penicillium marneffei* infection. *Am J Trop Med Hyg* 67: 443–447.

Pongpom P, Cooper CR, Vanittanakom N, 2005. Isolation and characterization of a catalase-peroxidase gene from the pathogenic fungus, *Penicillium marneffei*. *Med Mycol* 43: 403–411.

Roilides E, Lyman CA, Sein T, Petraitiene R, Walsh TJ, 2003. Macrophage colony-stimulating factor enhances phagocytosis and oxidative burst of mononuclear phagocytes against *Penicillium marneffei* conidia. *FEMS Immunol Med Microbiol* 36: 19–26.

Sisto F, Miluzio A, Leopardi O, Mirra M, Boelaert JR, Taramelli D, 2003. Differential cytokine pattern in the spleens and livers of BALB/c mice infected with *Penicillium marneffei*: Protective role of gamma interferon. *Infect Immun* 71: 465–473.

Supparatpinyo K, Perriens J, Nelson KE, Sirisanthana T, 1998. A controlled trial of itraconazole to prevent relapse of *Penicillium marneffei* infection in patients infected with the human immunodeficiency virus. *N Engl J Med* 339: 1739–1743.

Thirach S, Cooper CR, Vanittanakom P, Vanittanakom N, 2007. The copper, zinc superoxide dismutase gene of *Penicillium marneffei*: Cloning, characterization, and differential expression during phase transition and macrophage infection. *Med Mycol* 45: 409–417.

Trewatcharegon S, Chaiyaroj SC, Chongtrakool P, Sirisinha S, 2000. Production and characterization of monoclonal antibodies reactive with the mycelial and yeast phases of *Penicillium marneffei*. *Med Mycol* 38: 91–96.

Vanittanakom N, Cooper CR Jr, Fisher MC, Sirisanthana T, 2006. *Penicillium marneffei* infection and recent advances in the epidemiology and molecular biology aspects. *Clin Microbiol Rev* 19: 95–110.

Wong SS, Siau H, Yuen KY, 1999. *Penicilliosis marneffei*—West meets East. *J Med Microbiol* 48: 973–975.

Zanatta R, Miniscalco B, Guarro J, Gené J, Capucchio MT, Gallo MG, Mikulicich B, Peano A, 2006. A case of disseminated mycosis in a German shepherd dog due to *Penicillium purpurogenum*. *Med Mycol* 44: 93–97.

WEBSITES CITED

Biosafety in Microbiological and Biomedical Laboratories (BMBL), 5th ed. Available at the URL http://www.cdc.gov/od/ohs/biosfty/bmbl5/bmbl5toc.htm. Fungal Infections: Virtual Grand Rounds at the URL http://figrandrounds.org/case-studies/

QUESTIONS

The Answer Key to multiple choice questions may be found at the end of the book.

1. What is the most relevant description of the morphology of *Penicillium marneffei* seen in the Gram-stained smear from a blood culture bottle?

A. Fission yeast
B. Pleomorphic
C. Typical *Penicillium* mold
D. Yeast-like fungus

2. A specimen from a blood culture bottle that is suspected of being *Penicillium marneffei* is transferred to SDA and incubated at 37°C. Which of the following results would you expect?
A. Growth if the medium contains cycloheximide.
B. Medium containing blood is required for growth.
C. The mold form is the only form capable of growth on SDA.
D. The same form present in blood cultures would be maintained.

3. How does the morphology of *Penicillium marneffei* differ from saprophytic *Penicillium* species and similar molds?
A. Annellides appear singly or in groups with chains of thick-walled round conidia.
B. Conidiophores are biverticillate with short tangled conidial chains.
C. Elongated phialides taper to a slender tube with conidia in long unbranched chains.
D. Only *P. marneffei* produces a reddish pigment.

4. How should cultures suspected of being *Penicillium marneffei* be handled in the laboratory?
A. Only HIV-positive persons should take BSL 2 precautions with *P. marneffei*.
B. Since *Penicillium* species are common environmental molds, no special precautions are required.
C. Work in BSL 2 conditions, wear PPE, and treat the mold culture as a serious respiratory pathogen.
D. Yeast form cultures are dangerous whereas the mold form is not.

5. Which of the following areas have reported the most cases of penicilliosis marneffei?
A. Kyushu in Miyazaki Prefecture
B. Northern Thailand near Chiang Mai
C. Southwest China in Guangxi Province
D. Tamil Nadu State near Pondicherry

6. Discuss the geographic distribution and ecologic niche of penicilliosis. Include in your discussion the role of any sentinel animal species.

7. What primary diseases make a person susceptible to penicilliosis?

8. What is the route of infection with *Penicllium marneffei*? From that point, what is the most frequent site of dissemination?

9. What other diseases are in the differential diagnosis of penicilliosis?

10. Discuss the choices for primary therapy of penicilliosis, the role of maintenance therapy, and secondary prophylaxis.

Chapter 9

Sporotrichosis

9.1 SPOROTRICHOSIS-AT-A-GLANCE

- *Introduction/Disease Definition.* Following a scratch or prick from a splinter, thorn, or other vegetable matter containing conidia of *Sporothrix schenckii*, a chronic, subcutaneous nodule appears which may remain fixed or may form a chain of nodules tracing the path of lymphatic drainage from a primary lesion. Lesions are granulomatous or may suppurate, ulcerate, and drain. They may remain localized or spread to adjacent bones and joints.
- *Etiologic Agents. Sporothrix schenckii* is the major etiologic agent. Other genetic sister species, *S. brasiliensis, S. globosa*, and *S. mexicana*, have been reported.
- *Geographic Distribution.* Worldwide with some highly endemic areas: Brazil, India, Mexico, Japan, Peru, Uruguay, and South Africa.
- *Ecologic Niche.* Thorny plants, splinters, sphagnum moss, hay.
- *Epidemiology.* Most common subcutaneous mycosis worldwide with localized highly endemic areas.
- *Risk Groups/Factors.* Gardeners, plant nursery workers, agriculture workers, and children, in highly endemic areas; cat owners and veterinarians in zoonotic transmission regions.
- *Transmission.* The route of infection is via a penetrating injury or scratch with plant material or soil-contaminated object; rarely through inhalation of conidia; also from the scratch or bite from an infected domestic cat.
- *Determinants of Pathogenicity.* Melanized conidia, a 70 kDa surface antigen-adhesin; mold→yeast dimorphism; peptide-L-rhamno-D-mannan surface antigen.
- *Clinical Forms.* Fixed cutaneous, lymphocutaneous, or extracutaneous.
- *Human–Animal Interface.* Zoonotic sporotrichosis is transmitted by infected domestic cats to dogs and humans.
- *Therapy.* Cutaneous and lymphocutaneous: itraconazole (ITC) or alternative therapy with saturated solution of potassium iodide (SSKI) (oral) or terbinafine. Extracutaneous/systemic: AmB. Osteoarticular disease may be treated with ITC.
- *Laboratory Detection, Recovery, and Identification.* Often missed by direct examination, including histopathology, *S. schenckii* grows well as a mold on SDA with characteristic "daisy cluster" sporulation and melanized *sessile* conidia. Conversion to a yeast form occurs in vivo and in vitro at 35–37°C. Some yeasts are cigar shaped. Serologic tests are useful in extracutaneous sporotrichosis and when culture results are delayed.

9.2 INTRODUCTION/DISEASE DEFINITION

Sporotrichosis is an indolent, chronic, or subacute infection of cutaneous, subcutaneous tissues with spread along the draining lymphatics, and by direct extension to the skeletal system. Sporotrichosis is the most important subcutaneous mycosis in the United States. The disease is distributed worldwide, but regions of higher endemicity occur, for example, the Andean highlands of Peru, as a cat-associated zoonotic disease in Brazil, and on the east coast of Kyushu, Japan. The causative agent is *Sporothrix schenckii*, a dimorphic fungus occurring as a mold form in the environment where it is associated with decaying wood, hay, sphagnum moss, and thorny plants. *Sporothrix*

Fundamental Medical Mycology, First Edition. By Errol Reiss, H. Jean Shadomy and G. Marshall Lyon, III.
© 2012 Wiley-Blackwell. Published 2012 by John Wiley & Sons, Inc.

schenckii is introduced by a scratch or penetrating injury, usually by a thorn or splinter. Once implanted, *S. schenckii* displays temperature-sensitive dimorphism and converts to a yeast form in tissue.

Beginning in the cutaneous form as a nodule on an extremity, it becomes erythematous and may ulcerate. As the nodule grows, the lymphatics draining the area become cord-like and form a series of nodules tracing the path of lymphatic drainage from the primary lesion: the distinct "cording effect." The nodules may suppurate and drain. The inoculation site produces a lesion that closely resembles the primary stage of cutaneous syphilis and is termed "chancriform." Infection may be spread to adjacent joints and hematogenously to bones, skin, eyes, CNS, and the genitourinary tract.

The lymphocutaneous form occurs in the great majority of cases. Less commonly a fixed cutaneous lesion may develop. Rarely, *S. schenckii* produces a pulmonary-systemic infection, believed to be caused by inhalation of conidia of the fungus from nature. Dissemination can result in disease at weight-bearing joints (e.g., knee or hip), however, musculoskeletal and disseminated involvements are rare.

The clinical forms are:

- Fixed cutaneous: may heal spontaneously with scarring
- Lymphocutaneous: most common
- Extracutaneous: including osteoarticular, pulmonary, eyes, and meningeal forms.

Sporotrichosis also is known as Beurmann's disease, Schenck's disease, or rose gardener's disease.

9.3 CASE PRESENTATIONS

Case Presentation 1. Swollen Wrist in a 43-yr-old Steelworker: The Alcoholic Rose Gardener Syndrome (Kedes et al., 1964)

History

A 43-yr-old steelworker was admitted with a 6 month history of a painful swollen right wrist. He enjoyed both rose gardening and alcoholic beverages. During the summer before the current admission he suffered many pricks from rose thorns that sobriety might have avoided. In August he was treated with hot compresses and vitamin B_{12} injections. After 8 weeks and no improvement he entered another hospital and a surgeon removed a swelling from the wrist, but it was not examined. Six weeks later it returned and was excised again. A diagnosis of tuberculous synovitis was made and isoniazid and PABA were prescribed. He was admitted after no improvement was seen.

Clinical and Laboratory Findings

The ulnar side of the right wrist was red and swollen, the WBC count was 10,000/mm^3, and the differential was normal. The PPD test was positive.

Surgical Procedure

The tendon sheaths, synovium containing rice-like material, and scar tissue were removed.

Histopathology and Culture

Histologically there was extensive fibrosis, and the cellular infiltrate contained lymphocytes, macrophages, giant cells, and plasma cells. Ziehl–Neelsen, Gram, Grocott, and PAS stains were all negative. The ground tissue was plated on a variety of media and heavy growth of a mold grew on SDA at room temperature. A yeast form grew at 37°C on blood agar. Microscopic examination of the mold form demonstrated the floret arrangement of conidia characteristic of *Sporothrix schenckii*. (Please see Section 9.13, Laboratory Detection, Recovery, and Identification.)

Diagnosis

The diagnosis was articular sporotrichosis caused by *Sporothrix schenckii*.

Therapy and Clinical Course

The patient received oral drops of saturated solution of potassium iodide (SSKI), 4× daily, and griseofulvin 2 g/day for 6 months with no improvement. He then received 70–150 mg of AmB IV daily, for a total dose of 800 mg, until his blood urea nitrogen (BUN) rose to 46 mg/dL. Treatment was stopped for 12 days until the BUN returned to 11 mg/dL and then the patient received an additional course of 700 mg of AmB. The wrist was no longer swollen, red, or painful. The patient returned to work in the steel mill and has had no recurrence.

Comment

This was an early example of successful use of AmB in the treatment of articular sporotrichosis. SSKI is not an appropriate therapy for extracutaneous sporotrichosis and neither is griseofulvin. Note that direct examination including histopathology was negative, but that the fungus grew readily from biopsied tissue.

Case Presentation 2. A 51-yr-old Man with a Nonhealing Finger Wound and Regional Adenopathy (Sugar and Mattia, 1994)

History

The patient, a 51-yr-old man, worked in a rural area and had landscaped a sheep pen several months before the current admission. His hobbies included a rose garden and an aquarium. He was being treated for hypertension and used a beclomethasone inhaler for asthma. Ten years before admission his large bowel was resected to treat diverticulitis. Five years earlier he fractured his right hip.

A month before the current admission, he was chopping wood when a splinter scraped the dorsal surface of the left third finger. After 2 weeks, the wound ulcerated, and he was prescribed an unknown antibiotic for 7 days, without improvement. Nine days before the current admission he came to the emergency department with a superficial nonhealing ulcer on the left third finger. Cefadroxil was given but 2 days later the swelling worsened. Penicillin V was substituted with little improvement. The patient noticed red nodules on the forearm. He recalled no recent history of fever, chills, sweats, exposure to salt water, cat scratch, rat bite, rabbits, or travel outside of New England.

Clinical and Laboratory Findings

The patient appeared well. An erythematous lesion 2 cm × 3 cm with an oval ulcer was on the dorsal surface of the left third finger, with local edema extending over the dorsum of the hand. There was no purulence, exudate, loss of motion, or sensation. Two red indurated nodules occurred in linear order on the medial surface of the left forearm. Lymph nodes were slightly tender in the left epitrochlear[1] and axilla.

Diagnostic Procedures

Radiography showed soft tissue swelling but underlying bones were normal. Biopsy and culture were performed on the primary wound and nodular lesions. Histopathology showed acute and chronic inflammation in the dermis, focal necrosis, small granulomas, scattered giant cells, and fibrosis. Special stains for bacteria, mycobacteria, and fungi were negative. Three days later, fungal cultures at 25°C revealed branching septate hyphae, which developed single and floret clusters of pear-shaped conidia. Conversion occurred from the mycelial to the yeast form at 37°C.

[1]Epitrochlear. *Definition*: Above the trochlea—the notch within the hook-like proximal end of the ulna that slides in and out of the olecranon fossa of the humerus.

Diagnosis

The diagnosis was lymphocutaneous sporotrichosis caused by *Sporothrix schenckii*.

Therapy and Clinical Course

Therapy consisted of high dose ITC (dosage not specified). Two months later lesions were almost entirely resolved and lymphadenopathy had subsided.

Comment

If the initial splinter was not removed this could explain delayed healing but not the additional lesions. *Nocardia* species and *Mycobacterium* species are implicated in this clinical picture. Secondary lesions proximal to the primary lesion, the "sporotrichoid" pattern, is found in cutaneous nocardiosis. *Mycobacterium marinum* is in the differential diagnosis because the patient maintained a freshwater aquarium. The recent scratch might serve as a portal of entry for *M. marinum*. The sporotrichoid pattern of lesions also would apply. Cat scratch disease, caused by *Bartonella henselae*, also may result in nodular lymphangitis. If the patient had fever or systemic illness, then tularemia, anthrax, and melioidosis would have to be excluded. *Staphylococcus aureus* and *Streptococcus pyogenes* skin lesions would be accompanied by systemic signs and symptoms. Failure to observe yeast forms by direct microscopic examination of the exudates, or with special stains of histologic sections, is most often the case with sporotrichosis; however, the mold grows from such specimens on SDA. Conversion to the yeast form on BA or BHI is not difficult and provides important confirmation.

Case Presentation 3. Laboratory Acquired Sporotrichosis (Cooper et al., 1992b)

History

In late October 1989, a healthy 31-yr-old man was seeding flasks of media with conidial suspensions of *S. schenckii* from the 1988 multistate sporotrichosis epidemic in the United States. The procedures were conducted in a laminar flow BSC, but no hand protection was worn. Several days later the biomass was recovered, frozen, and ground in a mortar and pestle. No injury occurred during the procedure. Two weeks after working with *S. schenckii* in the BSC, a 3 mm diameter subcutaneous abscess developed on the tip of the middle finger of the microbiologist's right hand. It was lanced and over the next days it enlarged and ulcerated. Epitrochlear lymphadenopathy occurred and 4–6 satellite lesions appeared. A subcutaneous nodule appeared near the middle joint of the same finger.

Clinical and Laboratory Findings

Direct examination of smears of exudate of the primary lesion revealed numerous budding yeasts and cigar-shaped cells. Incubation of the specimen on PDA at 27°C resulted in a darkly pigmented mold colony; slide cultures revealed the conidia borne in a *sympodial* arrangement—the typical floret pattern—as well as melanized sleeves of sessile conidia along hyphae. (Please see Section 9.13, Laboratory Detection, Recovery, and Identification.)

Diagnosis

The diagnosis was lymphocutaneous sporotrichosis caused by *Sporothrix schenckii*.

Therapy and Clinical Course

The patient received oral potassium iodide for 2 months. The dosage was not specified. Recommended dosage regimens are discussed in Section 9.12, Therapy. The swelling subsided and treatment was discontinued. One month later swelling recurred near the middle joint of the same finger on which the original lesion was located. Oral potassium iodide treatment was reinitiated and continued for 2 months. All symptoms resolved and there were no further recurrences.

Comment

Radiograph of the right hand should have been performed to check for bone involvement. Infection occurred even in the absence of overt skin trauma, underscoring the need to take extreme caution in handling cultures of *S. schenckii*, including wearing of gloves and other personal protective equipment. (Before working with any pathogenic fungus please consult the CDC/NIH manual—*Biosafety in Microbiological and Biomedical Laboratories* (BMBL), 5th edition at the URL http://www.cdc.gov/od/ohs/biosfty/bmbl5/bmbl5toc.htm.

9.4 DIAGNOSIS

The diagnostic approach relies on direct examination, histopathology, and culture, with radiographic imaging studies to determine the involvement of extracutaneous sites. Specimens for identification include aspirated pus, scrapings, or biopsied tissue from cutaneous or subcutaneous lesions and, as needed, respiratory tract specimens and biopsy specimens or fluids from other infected sites (e.g., CSF, synovial fluid). Direct microscopic examination of the aforementioned specimens may be negative because, typically, there are few yeast cells in lesions. *Sporothrix schenckii* yeast forms are also difficult to locate in histopathologic sections. Cultural studies are thus of increased importance and *S. schenckii* mold form grows well on mycologic media, often within a few days. Confirmation of the fungus requires conversion to the yeast form on enriched medium at 37°C. When direct examination and cultures are negative, serologic tests may be useful. A commercial test is available and researchers have developed promising EIA tests. Please see Section 9.13, Laboratory Detection, Recovery, and Identification, for further information.

9.5 ETIOLOGIC AGENTS

The fungal species producing this disease is *Sporothrix schenckii*.

Sporothrix Schenckii. Hektoen & CF Perkins, *J Exp Med* 5: 80 (1900). Position in classification: *Ophiostomataceae, Ophiostomatales, Sordariomycetidae, Sordariomycetes, Ascomycota*.

Although it appears *hyaline* in tissue, it may be shown to have melanin in the hyphal cell walls with the Fontana–Masson stain. *Sporothrix schenckii* displays temperature-sensitive dimorphism, growing as a mold in soil and in the laboratory at 25–30°C, converting to a yeast form in infected tissue or when grown at 35–37°C on artificial media (e.g., BA). Mature cultures are *melanized* due to pigmentation in the sleeves of sessile conidia forming along hyphae.

Some strains of the fungus may grow at 35–37°C while others, isolated from the environment, grow only below that range. *Only those able to grow at the higher temperatures are believed to cause lymphocutaneous or extracutaneous disease*.

The population structure of *S. schenckii* was assessed in 60 isolates from different geographic regions by phylogenetic analysis of sequences from three protein coding loci (chitin synthase, ß-tubulin, and calmodulin) (Marimon et al., 2006). Three major phylogenetic species were discerned: one contained all of the European isolates, another the Brazilian isolates, and the third was associated with isolates from other South American countries and Africa. Three new species were described as *Sporothrix brasiliensis, S. globosa*, and *S. mexicana* (Marimon et al., 2007). Phenotypic features differentiating these species are the morphology of their *sessile pigmented conidia*, growth at 30°C, 35°C, and 37°C, and assimilation of sucrose, raffinose, and ribitol. Genetic analysis also found that *Sporothrix albicans, S. inflata*, and *S. schenckii* var. *luriei* are clearly different from *S. schenckii*.

9.6 GEOGRAPHIC DISTRIBUTION/ECOLOGIC NICHE

Sporothrix schenckii is found worldwide, in temperate, subtropical, and tropical regions, but evidence is accumulating that *S. schenckii* thrives in special ecologic subregions (reviewed by Lyon et al., 2003).

- *The Americas.* It is the single most common mycosis in the central highlands of Mexico. The southern Andean mountains are hyperendemic for lymphocutaneous sporotrichosis (Lyon et al., 2003). A zoonotic outbreak of sporotrichosis transmitted by domestic cats in Rio de Janeiro, Brazil, has been ongoing since 1998 (Schubach et al., 2004).
- *Asia and South Asia.* India and the eastern coast of Kyushu, the south island of Japan.

The ecologic niche of *S. schenckii* is as a saprobe on both living and dead vegetation, and usually is transferred via inoculation into the skin from soil, wood splinters, prairie or salt marsh hay, straw, thorns, and sphagnum moss.

9.7 EPIDEMIOLOGY

The numbers of cases worldwide have increased but this may be due primarily to an increased ability to diagnose the disease. Sporotrichosis is commonly recognized in the developing world. The extent of subclinical exposure in the community is unknown because the skin test reagent, sporotrichin, lacks specificity.

Australia

A cluster of sporotrichosis cases occurred in the Busselton–Margaret River region of Western Australia from 2000 to 2003 (Feeney et al., 2007). Hay distributed through a commercial supplier was the source of the outbreak.

India

A compilation of 205 cases of lymphocutaneous sporotrichosis from India found that 91 (44%) originated in West Bengal, 56 (28%) were from Himachal Pradesh, and 45 (22%) were from Assam (Randhawa et al., 2003).

Japan

Along the eastern coast of the southern island of Kyushu, sporotrichosis accounted for 1.5–1.7 visits per 1000 outpatient visits to a dermatology clinic, with the majority of cases (64%) among female patients from farming communities.

Mexico

National incidence and prevalence data are not available but, in one state, Jalisco, 822 sporotrichosis cases were compiled for the period 1960–1996 (Bustamante and Campos, 2001; Mayorga-Rodríguez et al., 1997).

South Africa

The largest known epidemic of lymphocutaneous sporotrichosis occurred in the Witwatersrand, South Africa, in the 1940s, when some 3300 gold miners were infected (reviewed by Quintal, 2000) after receiving scratches from contaminated mine timbers. Treatment of the timbers eliminated the problem, although several years later a smaller group of workers again contracted the disease. Of major interest, in no instance did the disease become disseminated.

South America

Population-based surveillance and a case–control study were conducted in the Andean mountain region of Peru in the vicinity of Abancay (altitude 2750 m) to estimate the burden of sporotrichosis and to determine risk factors for sporadic lymphocutaneous sporotrichosis (Lyon et al., 2003). Hospital records were reviewed for 1997 and 1998, and surveillance was conducted for a year beginning in September 1998. Children were disproportionately affected in the hyperendemic region for sporotrichosis in the vicinity of Abancay. The mean annual incidence was 156 cases/100,000 in the $\leq$15-yr-old population, highest among children aged 7–14 years, approaching 1 case per 1000 persons (Lyon et al., 2003). Children had an incidence three times higher than adults and were more likely to have lesions on the face and neck. The mean annual incidence in adults was 52 cases/100,000 population. In 1995–1997, 238 cases of sporotrichosis were recorded from the South Central highlands of Peru (Pappas et al., 2000).

United States

The largest multistate outbreak of sporotrichosis in the United States occurred between April and June 1988, when 84 cases of cutaneous sporotrichosis occurred in persons in 15 states who handled conifer seedlings packed in Pennsylvania with sphagnum moss that was harvested in Wisconsin (Centers for Disease Control, 1988; Dixon et al., 1991). The involved Pennsylvania nurseries ship seedlings and moss to 47 states. All clinical isolates had the same restriction fragment length polymorphism (RFLP) DNA pattern on agarose gel electrophoresis and matched the RFLP pattern from environmental isolates (sphagnum moss) from New York and Pennsylvania (Cooper et al., 1992a).

9.7.1 Risk Groups/Factors

Adult men or women engaged in outdoor occupations or avocations involving direct contact with soil, thorny plants, sphagnum moss, hay, and decaying wood are at increased risk of acquiring sporotrichosis, for example, floral, plant nursery, and forestry workers (thorny plants, sphagnum moss), agricultural laborers, gardeners or home gardeners (thorny plants), and feed and seed company employees (sphagnum moss, hay). Risk is increased when such individuals are compromised by alcoholism and/or diabetes.

Risk factors for sporotrichosis in children living in economically disadvantaged sections of the Andean highlands of Peru were determined (Lyon et al., 2003). They were playing in crop fields, owning a cat, or having a dirt floor in the home, all of which are related to increased chance of exposure to an environmental source. The rough living conditions included houses with exposed wooden ceiling beams, a potential source of *S. schenckii*. Infected cats have been implicated in several outbreaks or case reports of sporotrichosis in humans; the mode of transmission is through scratches, bites, or direct contact with lesions on the cat. (Please see Section 9.11, Human–Animal Interface).

Risk Factors for Extracutaneous Sporotrichosis

Men are at greater risk tied to alcoholism (please see Section 9.3, Case Presentation 1) and/or COPD or diabetes mellitus. A history of immunosuppression occasionally has been identified, and persons living with AIDS are at increased risk of sporotrichosis. A review of 19 patients with AIDS-sporotrichosis found the disease was likely to spread from cutaneous or subcutaneous sites to involve the joints, central nervous system, and multisystem disseminated disease (al-Tawfiq and Wools, 1998). Control of the disease with antifungal therapy was not optimal, only 5 patients had a complete response, 3 had a partial response, and the remainder were nonresponders. Those responding received either AmB or ITC or a combination of both; however, disease progressed despite receiving AmB in 4 patients.

9.8 TRANSMISSION

Traumatic implantation of the fungus from natural sites (e.g., plants, sphagnum moss, soil, splinters from decaying wood) is the most common route of infection. Several laboratory infections have been documented, including contact with experimentally infected animals and laboratory accidents.

Several reported cases involve transmission to humans from infected cats, insect stings (Miller and Keeling 2002), and bites from dogs, horses, rodents, parrots, and pecks from birds including hens. Cats develop serious and often fatal infection characterized by chronic ulcerative skin lesions (Schubach et al., 2008). Armadillo hunters have become infected; although the animals do not harbor the disease, the fungus may be found in the dry grass used in building their nests. Uncommonly, inhalation of conidia causes a respiratory infection. Usually this occurs in the setting of COPD and alcoholism or diabetes.

Occupational Sporotrichosis

A veterinarian acquired facial lesions due to *S. schenckii* after treating a facial abscess on a cat (Reed et al., 1993). There was no cat scratch or bite involved but instead contact with draining lesions seems to be all that was required for transmission. Veterinarians, veterinary technicians, and animal owners should wear gloves when handling cats with ulcerative lesions or open draining tracts. Reports of laboratory-associated infections with *S. schenckii* are rare, but the potential for such remains a possibility. Handling of *S. schenckii* cultures was associated with an infection (please see Section 9.3, Case Presentation 3). An eye infection caused by splash exposure to a suspension of *S. schenckii* was mentioned in Kauffman (1999).

9.9 DETERMINANTS OF PATHOGENICITY

9.9.1 Host Factors

Role of T-Cell-Mediated Immunity

Following a primary infection, mice are protected against a secondary subcutaneous challenge with *S. schenckii* (Tachibana et al., 1999). Subcutaneously immunized mice are protected from a lethal IV infection. Protection can be transferred to athymic nude mice with lymph node cells from immune donor mice, but not if they are depleted of $CD4^+$ T lymphocytes. Treatment of mice with carrageenan, a macrophage blocker, also abolishes protection. Acquired immunity against *S. schenckii* is mediated by macrophages activated by $CD4^+$ T lymphocytes. Immune lymph node cells expressed message for IFN-γ, TNF-α, and IL-10 after stimulation with killed *S. schenckii*.

Role of Innate Immunity

Nitric oxide (NO) is fungicidal against *S. schenckii* in vitro but when NO synthase deletion mutant mice

($iNOS^{-/-}$ mice) or mice treated with the NO synthase inhibitor *N*-ω-nitro-arginine were infected they displayed *increased* resistance (Fernandes et al., 2008). The mice had a reduced fungal load in tissues, restored T-cell activity, and increased IFN-γ production. Although NO may mediate in vitro killing of *S. schenckii* by macrophages, the activation of the NO system in vivo may contribute to immunosuppression.

Adherence to Extracellular Matrix Proteins

An important step in *S. schenckii* pathogenesis is when yeast forms adhere to the endothelial lining of blood vessels, migrate across the endothelium, and penetrate the subendothelial matrix. Initiating the process, yeast forms adhere to extracellular matrix proteins fibronectin, collagen, and laminin (Lima et al., 2001; Figueiredo et al., 2004). Adherence of *S. schenckii* yeasts to endothelial monolayers is modulated by cytokines. *Sporothrix schenckii* yeasts internalized by endothelial cells are found inside endocytic vacuoles without causing damage after 24 h of infection (Figueiredo et al., 2004). *S. schenckii* is able to adhere to and invade endothelial cells without significantly affecting cellular integrity.

Transforming growth factor, TGF-ß1, increases fungal migration across endothelial monolayers and inhibits fungus internalization (Figueiredo et al., 2007). TGF-ß1 increases exposure of fibronectin and laminin in the endothelial matrix, thus enhancing transendothelial migration. Antibodies to extracellular matrix proteins inhibit *S. schenckii* association with endothelial cells.

Activation of Dendritic Cells

Sporothrix schenckii strains of cutaneous origin are more potent activators of dendritic cells, inducing IFN-γ production as part of a strong Th1 response, whereas *S. schenckii* strains of visceral origin induce minimal dendritic cell activation (Uenotsuchi et al., 2006). Thus, differential activation of dendritic cells may influence the clinical form of disease.

9.9.2 Microbial Factors

Antigens

A 70 kDa glycoprotein extracted from *S. schenckii* yeast form cell walls is distributed uniformly on the cell surface and may have a role as an adhesin (Ruiz-Baca et al., 2009). *S. schenckii*-infected mice produce antibodies reactive with the 70 kDa antigen (Nascimento and Almeida, 2005). Monoclonal antibodies against the 70 kDa glycoprotein and fibronectin colocalized on the cell wall of *S. schenckii*, evidence that the 70 kDa glycoprotein is the ligand which binds to fibronectin in the extracellular matrix (Teixeira et al., 2009). The 70 kDa glycoprotein may be considered a virulence factor because it mediates adherence to extracellular matrix proteins of the host.

Cell Wall Polysaccharide Antigen

The outer cell wall of *S. schenckii* contains a thick, electron dense microfibrillar material. Peptido-L-rhamno-D-mannan is the major carbohydrate antigen of the outer cell wall and is shed into the culture medium. The epitope that is specific for *S. schenckii* within the peptido-L-rhamno-D-mannan and which does not crossreact with other fungi is a trisaccharide: α-L-Rha (1→4) [α-L-Rha 1→2]α-D-GlcA (GlcA = glucuronic acid). The GlcA residue is linked to an α-(1→2)-mannobiose. This in turn is attached by an alkali-labile *O*-glycosidic bond to serine or threonine of the peptide portion of the rhamnomannan (Penha and Bezerra, 2000).[2] This trisaccharide is proposed as the epitope that elicits specific antibodies against *S. schenckii* in patients with sporotrichosis.

Asteroid Bodies

Asteroid bodies[3], the star-shaped deposits of antigen–antibody complexes surrounding individual yeast cells in lesions of lymphocutaneous sporotrichosis, may protect the yeast forms from further attack by the immune system (da Rosa et al., 2008). IgG and IgM are localized in the spikes of the radiated crowns of the asteroid bodies. Yeast cells in the center of the bodies remain viable and can germinate in vitro, producing conidia after incubation at 22°C.

Factors Affecting Dimorphism

Dimorphism in *S. schenckii* responds to second messengers such as cAMP and calcium. A calcium/calmodulin-dependent signaling pathway may be involved in the regulation of dimorphism (Valle-Aviles et al., 2007). *SSCMK1*, a member of the calcium/calmodulin kinase family from *S. schenckii*, encodes a 45.6 kDa protein. Treatment of *S. schenckii* with inhibitors of calmodulin and calcium/calmodulin kinase inhibits budding by cells

[2] A diagram of this epitope is found in Lopes-Bezerra et al. (2006). Those authors also give a complete review of the discovery and progress with rhamnomannan.

[3] Asteroid bodies. *Definition*: Histopathologic sections of lesions in sporotrichosis may contain a central yeast surrounded by eosinophilic material arranged in a star shape. The rays, or "radiated crowns," of eosinophilic material contain immunoglobulins directed against the yeast forms. Asteroid bodies in sporotrichosis should not be confused with intracellular asteroid bodies seen in giant cells of granulomas, which are filamentous, and myelin figures that contain lipid.

when they are about to reenter the yeast cell cycle, favoring instead the yeast to mycelium conversion.

Sporothrix schenckii, like other fungi, uses signal transduction pathways to adapt to changes in its environment. The main intracellular receptors of environmental signals are heterotrimeric G proteins. A G-protein-α subunit of *S. schenckii* interacts with a catalytic phospholipase subunit (cPLA2) (Valentín-Berríos et al., 2009). This phospholipase acts on phospholipids to release lysophospholipids such as arachidonic acid. When arachidonic acid is added to *S. schenckii* yeast forms, budding is stimulated. Although this step does not affect the mycelium → yeast dimorphism, cPLA2 is needed for the further development of the yeast cell cycle, at the start of DNA synthesis.

Membrane Glycolipids

Ceramides[4] are sphingolipids present in the cell membranes of fungi. Glucosyl ceramides have been found in *Candida albicans, Cryptococcus neoformans, Aspergillus fumigatus, Histoplasma capsulatum, Paracoccidioides brasiliensis*, and *Sporothrix schenckii*. The mycelial form of *S. schenckii* expresses glucosylceramide, whereas the yeast form expresses both glucosyl- and galactosylceramides (Rhome et al., 2007; Toledo et al., 2000). Deletion of genes encoding glucosylceramide synthesis was shown in other fungi to result in defects in the cell cycle, formation of hyphae, and spore germination. Also, glucosylceramides are the target of defensins, cationic granule peptides isolated from insects and plants. Conceivably, defensins from PMN also may target these membrane glycolipids.

Melanin

Melanin in the cell walls of *S. schenckii* conidia and yeast cells may be an important factor in infection. *S. schenckii* produces brown sessile conidia, which are infectious propagules. *S. schenckii* synthesizes melanin via the 1,8-dihydroxynaphthalene pentaketide pathway. Two melanin-deficient mutants are an albino one that produces normal melanin on scytalone-amended medium, and a reddish-brown mutant that secretes melanin metabolites, including scytalone, into the medium (Romero-Martinez et al., 2000). Electron microscopy shows electron-dense granules in wild-type conidial walls, which are absent in the albino mutant, unless it is grown on scytalone-amended medium.

Melanized conidia of wild-type *S. schenckii* were less susceptible to killing by oxygen- and nitrogen-derived radicals and by UV light than were conidia of the mutants. Melanized conidia were more resistant to phagocytosis and killing by human monocytes and murine macrophages than were unmelanized mutant conidia. Melanin protects *S. schenckii* conidia against oxidative antimicrobial compounds and against attack by macrophages.

Treatment of both *S. schenckii* conidia and yeast cells with proteinases and concentrated hot acid yields dark particles, melanin ghosts, similar in shape to the original conidia or yeast cells (Morris-Jones et al., 2003). These melanin particles can serve as immunogens to produce anti-melanin mAbs. The mAbs bind to pigmented conidia and yeast cells grown in vitro and also to *S. schenckii* in infected tissues. Sera from sporotrichosis patients also contain antibodies that bind to melanin particles. These findings add to evidence that melanin may function as a pathogenic factor in *S. schenckii*, as has been shown in other melanized molds. Please see Part Five, Mycoses of Implantation—*Special Topic*: *Fungal Melanin*.

Thermotolerance

Sporothrix schenckii strains with the ability to grow at 37°C are capable of causing systemic disease. *Isolates from fixed cutaneous lesions multiply poorly above 35°C and appear unable to cause lymphatic or systemic involvement*.

Glucosamine-6-phosphate Synthase (GlcN-6-P Synthase): A Novel Target for Antifungal Drugs

GlcN-6-P synthase catalyzes the first step of hexosamine biosynthesis in fungi, converting fructose-6-phosphate into glucosamine-6-phosphate using glutamine as an amino group donor. Deletion of any of the four genes involved in UDP-glucosamine synthesis in yeast is lethal unless the yeast is supplemented with exogenous glucosamine or *N*-acetyl glucosamine. GlcN-6-P synthase inhibitors occur in two classes: oligopeptides containing analogs of L-glutamine—for example, N_3-(4-methoxyfumaroyl)-L-2,3-diaminopropanoic acid (FMDP)—and the second class, phosphorylated aminohexitols—for example, 2-amino-2-deoxy-D-mannitol-6-phosphate (ADMP) (Wojciechowski et al., 2005). Oligopeptides containing FMDP were poor inhibitors of *S. schenckii* growth, despite being strong inhibitors of GlcN-6-P synthase in vitro. ADMP and similar compounds, however, inhibited

[4]Ceramides. *Definition*: consist of a fatty acid chain attached through an amide linkage to sphingosine. Sphingosine (2-amino-4-octadecene-1,3-diol) is an 18-carbon amino alcohol with an unsaturated hydrocarbon chain, which is a primary part of sphingolipids, a class of cell membrane lipids. Glycosphingolipids are ceramides with glucose or galactose joined in a β-glycosidic bond at the 1-hydroxyl position of sphingosine. Glucosylceramide is a term for a glycosphingolipid containing glucose.

the growth of *S. schenckii* at concentrations of 500–625 μg/mL (González-Ibarra et al., 2010).

9.10 CLINICAL FORMS

By far, the most common form of sporotrichosis is lymphocutaneous, although other clinical forms occur such as fixed cutaneous, mucous membrane (mainly conjunctival and nasal), primary pulmonary, and other forms of extracutaneous disease. The usual portal of entry is percutaneous as a result of scrapes, scratches, or a penetrating injury which lodges conidia of *S. schenckii* from nature (thorns, splinters, etc.) into the subcutaneous tissue, usually of the back of the hand, arm, foot, or leg.

Fixed Cutaneous Sporotrichosis

Primary infection is generally restricted to the inoculation site. Lesions are variable in appearance and may be infiltrated and plaque-like, *verrucous,* scaly, macular or papular. They do not involve the local lymphatics, but remain "fixed" and rarely develop systemic disease. The lesions require treatment (please see Section 9.12, Therapy) (Fig. 9.1).

The differential diagnosis includes:

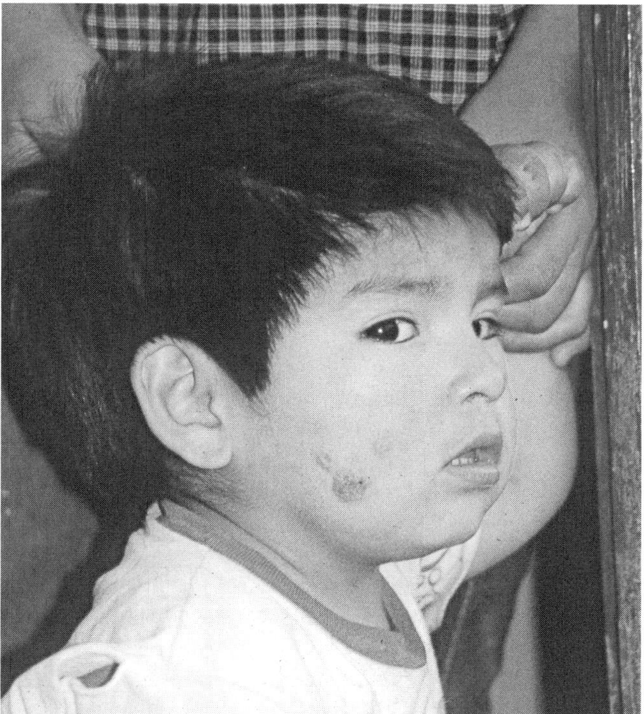

Figure 9.1 Lesion on the face of a child: fixed cutaneous sporotrichosis, Abancay, Peru.
Source: Dr. G. M. Lyon.

- Chromoblastomycosis
- Cutaneous anthrax
- Leishmaniasis
- Mycobacterial infections
- Nocardiosis
- Other systemic fungal infections (blastomycosis, paracoccidioidomycosis)
- Syphilis chancre

Lymphocutaneous Sporotrichosis

Up to 75% of cases are of this form of sporotrichosis. Traumatic implantation into the subcutaneous tissue results in the appearance of a small hard, nonattached nodule. After a period of time a subcutaneous nodule forms a *bubo*[5] that becomes attached to the overlying epidermis. The lesion becomes discolored, from pink through purple to black, and local or regional adenopathy develops. The bubo ulcerates and the skin becomes necrotic, forming the typical "sporotrichoid chancre." This lesion remains for up to months, then tends to heal with scarring as new buboes erupt, usually above the primary lesion, following the local lymphatic channels. The lymphatics become hardened and obvious to the observer, presenting the pathognomonic picture of the distinct cording effect of lymphocutaneous sporotrichosis (please see Fig. 1.2). Rarely, gummatous ulcerations may develop on wrists, forearms, or legs that form deep lesions, exposing underlying fascia, bones, and tendons.

Differential Diagnosis of Lymphocutaneous Sporotrichosis

Chronic ascending lymphangitis is a process characterized by a penetrating injury followed by an infection that gradually spreads up the lymphatic chain. The causative agents are usually found in soil, water, or thorny plants. The most common cause is *S. schenckii*, but the differential diagnosis includes other bacterial and fungal pathogens. The differential diagnosis includes:

- Cutaneous anthrax
- Leishmaniasis
- Mycobacterial infections
- Nocardiosis (*Streptomyces, Actinomadura, Nocardia* species)

[5]Bubo. *Definition*: (Greek—*boubôn*, "groin") (plural form, buboes) This is an inflammatory swelling of one or more lymph nodes, which usually suppurate and drain pus. The term usually is applied to a swelling in the groin, but applies also to the nodes draining the lesions of lymphocutaneous sporotrichosis.

- Nontuberculous mycobacterioses (*M. marinum, M. kansasii, M. chelonae*)
- Other systemic fungal infections: blastomycosis, chromoblastomycosis, paracoccidioidomycosis, and eumycetoma agents (e.g., *Pseudallescheria* and *Scedosporium* species)
- Pyoderma gangrenosum (Byrd et al., 2001)
- Syphilis chancre
- Tularemia

Extracutaneous and Disseminated Sporotrichosis

Extracutaneous sporotrichosis is uncommon, but is second only to the subcutaneous form, and usually is associated with underlying risk factors: alcoholism, diabetes, COPD, or HIV infection. The categories within this clinical form are the following:

- *Osteoarticular Sporotrichosis.* Osteoarticular structures are infected by direct inoculation, extension from cutaneous lesions, or hematogenously. A single joint, multiple joints, or bones may be involved. Tendosinovitis and bursitis also occur. The affected joint may be at the site of an overlying skin lesion; in bony structures remote from the initial lesion; or as an isolated finding. A risk factor is underlying alcoholism (please see Section 9.3, Case Presentation 1). The prognosis for restoration of joint function is poor especially if the diagnosis and treatment are delayed and also because of poor host response.
- *Pulmonary Sporotrichosis.* The route of infection is respiratory via aerosolized *S. schenckii* conidia when contaminated soil or plant material is disturbed. The risk factors are alcoholism, COPD, and AIDS. Chronic cavitary fibronodular disease may occur and the response to therapy is poor owing to a delayed diagnosis and underlying pulmonary disease. The differential diagnosis for pulmonary sporotrichosis includes:
  - Pulmonary tuberculosis
  - Other primary fungal diseases caused by dimorphic pathogens (blastomycosis, coccidioidomycosis, histoplasmosis, paracoccidioidomycosis)
  - Sarcoidosis

  Symptoms are similar to other fungal respiratory diseases and tuberculosis, with fever, night sweats, anorexia, weight loss, fatigue, and purulent sputum. Cavities may be seen by chest radiographs in the upper lobes of the lung. Without treatment, disease progresses slowly but is relentless.

Disseminated Sporotrichosis

The clinical presentation is one of multiple dispersed lymphocutaneous lesions and/or visceral dissemination. Although rare, patients living with AIDS are at increased risk, and the prognosis is poor, requiring induction therapy with AmB and suppressive ITC maintenance therapy. Other risk factors include diabetes, alcoholism, sarcoidosis, long-term corticosteroid therapy, and TNF-α antagonists.

TNF-α Antagonists Disseminated sporotrichosis was diagnosed in a 49-yr-old man treated with multiple immunosuppressants, including TNF-α antagonists etanercept and infliximab, for presumed inflammatory arthritis (Gottlieb et al., 2003). This case illustrates the potential for infectious complications related to the use of cytotoxic immunosuppressants and anticytokine agents.

Meningeal Sporotrichosis This most dire form of sporotrichosis may be an isolated event or, more likely, is the result of multisystem disseminated sporotrichosis. Meningeal sporotrichosis is a difficult disease to diagnose, where treatment options are limited and prognosis is poor.

Other forms of extracutaneous or disseminated sporotrichosis have been reported but are rare. These include the eye and adnexae (lacrimal and other appendages of the eye).

9.11 HUMAN-ANIMAL INTERFACE

Since 1998 an epidemic of sporotrichosis in domestic cats and zoonotic transmission has been ongoing in Rio de Janeiro (Galhardo et al., 2008). Before that time, between 1986 and 1997, only 13 sporotrichosis cases were seen at the Rio de Janeiro referral center for infectious diseases. From 1998 to 2004, 759 humans, 64 dogs, and 1503 cats were diagnosed with sporotrichosis (Schubach et al., 2008). Genotyping analysis showed that epidemic strains were genetically related, possibly originating from a common source. Most of the dogs (85%) and 83% of patients had contact with cats with sporotrichosis, and 56% of the patients received cat bites or scratches. *Sporothrix schenckii* was isolated from feline skin lesions and from their nasal and oral specimens. Canine sporotrichosis is a self-limited mycosis whereas feline sporotrichosis varied from subclinical to severe systemic disease with hematogenous dissemination. Dogs respond to treatment with ITC (Sykes et al., 2001). Sporotrichosis in cats preceded its occurrence in their owners and their canine contacts. Cat→human transmission implicates *S. schenckii* yeast forms as the infectious propagules, whereas with most other dimorphic pathogens, that role is served by conidia of the mold form.

The prevalence of sporotrichosis was four times higher among patients caring for animals (Barros et al., 2008a). Taking care of sick cats was the main factor associated with transmission of the disease to humans. In the Rio de Janeiro epidemic 81 children younger than 15 years were diagnosed with sporotrichosis between 1998 and 2004 (Barros et al., 2008b). A majority were 10–14-yr-old girls. The most frequent clinical form was lymphocutaneous, located on the upper limbs. ITC was the treatment of choice; 66 patients were cured, 9 were lost to follow-up, and 6 had spontaneous regression of the lesions. This is the largest reported series of childhood sporotrichosis with zoonotic transmission.

The potential role of cats in transmission of sporotrichosis in the endemic area of Abancay, Peru, was estimated by sampling oral and nasal swabs and nail clippings of one household cat in each of 85 neighborhoods (Kovarik et al., 2008). The prevalence of *S. schenckii* colonization was 2.38% in this cat population, suggesting a role for cats as a reservoir for sporotrichosis in Abancay and confirming earlier findings where cat ownership was a risk factor for sporotrichosis among children (Lyon et al., 2003).

9.12 THERAPY

Please also see Kauffman et al. (2007) for clinical practice guidelines of the Infectious Diseases Society of America.

Cutaneous and Lymphocutaneous Sporotrichosis

First-Line Therapy

First-line therapy consists of ITC at a dosage of 100–200 mg/day (oral) until 2–4 weeks after all lesions have resolved, usually for a total of 3–6 months. Patients who do not respond should receive 200 mg ITC 2×/day or terbinafine, at a dosage of 500 mg (oral) 2×/day (Chapman et al., 2004). For all conditions requiring ITC, determination of serum concentrations is recommended to ensure proper absorption and distribution. The $t_{1/2}$ of ITC is long and there is little variation over a 24 h period.

Oral Dosage Considerations ITC comes in two oral forms: a 100 mg capsule and a solution of 100 mg/10 mL. A loading dose of 200 mg 3×/day for 3 days is recommended to initiate therapy. After that, when the daily dose exceeds 200 mg it should be given in two divided doses. The powder form in capsules is taken with food to improve absorption. Normal stomach acidity improves absorption as does taking ITC with acidic liquids (e.g., orange juice, Coca-Cola, other soft drinks). ITC solution is given on an empty stomach and in this form increases blood levels but may not be well tolerated.

Second-Line Therapy

A saturated solution of potassium iodide (SSKI), taken orally, was the standard treatment for many years, even though the mechanism of action remains unknown. At present it is considered a second-line drug when ITC therapy fails or is not tolerated, or when socioeconomic conditions require the least expensive therapy option (Lyon et al., 2003). SSKI treatment is initiated with 5 drops 3×/day (with a standard eyedropper) usually added to water, fruit juice, or milk before drinking. The dosage is increased gradually until 40–60 drops 3×/day is reached (a dosage of 500–1500 mg/day for adults and 1/3 to 1/2 this dosage for children) (Sterling and Heymann, 2000). Treatment is continued for 6–10 weeks.

SSKI Clinical Response, Adverse Effects Clinical response in sporotrichosis to SSKI usually is achieved within 32 days. There is at least one report of lymphocutaneous sporotrichosis unresponsive to ITC but which responded well to SSKI therapy with 90% regression of lesions within 5 weeks of therapy (Sandhu and Gupta, 2003).

Side effects of SSKI therapy usually are mild but occasionally can be serious and can result in poor compliance: mild nausea, urticarial rash, metallic taste, and salivary gland enlargement. Common acute side effects are diarrhea, nausea, vomiting, and stomach pain. With prolonged use, symptoms of iodism or potassium toxicity may ensue. Iodism is associated with burning mouth, increased watering of the mouth, metallic taste, soreness in teeth and gums, and headache. Potassium toxicity is associated with confusion, arrhythmia, hand numbness, or general weakness. Other side effects are listed in Sterling and Heymann (2000).

Alternative Therapy for Fixed-Cutaneous and Lymphocutaneous Sporotrichosis

Terbinafine Terbinafine is receiving increased attention as an alternative to ITC to treat fixed-cutaneous and lymphocutaneous sporotrichosis. Fifty patients for whom ITC was contraindicated or resulted in adverse pharmacologic interactions were treated with terbinafine (250 mg/day) (Francesconi et al., 2009). The patients' disease was traced to the cat-transmitted zoonotic sporotrichosis epidemic in Rio de Janeiro. Thirty-nine patients (78%) had lymphocutaneous and 11 (22%) had fixed-cutaneous sporotrichosis. Almost all the patients presented comorbidities and used a medication interacting with ITC. Sporotrichosis was cured in 86% of patients receiving terbinafine alone, and 96% when adjunctive therapy was needed. Five patients received cryotherapy, curettage, or both. Most patients were cured within a mean period of 14 weeks. There were no recurrences of the mycosis after ~37 weeks of post-therapy follow-up.

Fluconazole Fluconazole is a less effective therapeutic agent for sporotrichosis and is reserved for situations in which patients are intolerant of ITC. A dosage of 400–800 mg FLC/day for 6 months is suggested.

Local Hyperthermia When neither ITC nor SSKI can be given (e.g., in pregnancy or for nursing mothers), local hyperthermia is indicated for fixed-cutaneous or lymphocutaneous lesions consisting of weeks of daily heat generated by a pocket warmer or infrared source for 20–30 min 3 times daily to warm the tissue to ~42–43°C.

AmB AmB is not indicated for lymphocutaneous sporotrichosis because the disease is localized and non-life-threatening in the patient with normal immune and endocrine functions.

Therapy for Systemic and Disseminated Sporotrichosis

Some antifungal agents that are successful in treating fixed cutaneous and lymphocutaneous sporotrichosis are not effective for systemic or disseminated disease including FLC and SSKI. There are no published data, as yet, regarding extended spectrum azoles, voriconazole (VRC) and posaconazole (PSC). VRC is not active in vitro against *S. schenckii* but PSC does have in vitro activity.

Pulmonary Sporotrichosis

Treatment options include ITC or AmB, depending on the severity of symptoms. When feasible, surgical resection is combined with antifungal therapy.

For severe pulmonary sporotrichosis primary therapy consists of AmB as a lipid formulation at 3–5 mg/kg/day. AmB-deoxycholate (0.7–1.0 mg/kg/day) also may be used. Pending a favorable clinical response, therapy can be stepped-down to ITC at 200 mg 2×/day (oral) for a total of at least 12 months. Maintenance therapy is appropriate in the clinical setting of alcoholism, diabetes, and/or COPD. For more moderate disease, ITC may be used at 200 mg 2×/day (oral) for 12 months or more. Pulmonary sporotrichosis in an AIDS patient with concurrent *Pneumocystis* pneumonia improved with cotrimoxazole and ITC therapy (Losman and Cavanaugh 2004).

Osteoarticular Sporotrichosis

Primary therapy consists of ITC (200 mg 2×/day) for 12 months or more. Alternatively, AmB may be given as a lipid formulation at 3–5 mg/kg/day (IV), or AmB-deoxycholate, 0.7–1.0 mg/kg/day (IV) for a total dose of 1–2 g. Pending a favorable response, therapy may be switched to ITC [200 mg 2×/day (oral)] for at least 12 months.

Meningeal Sporotrichosis

Primary therapy consists of AmB lipid formulation at 5 mg/kg/day for 4–6 weeks followed by step-down ITC therapy (200 mg 2×/day) for at least 12 months. Suppressive therapy with ITC (200 mg/day) is recommended to prevent a relapse for patients with AIDS or another immunosuppressive condition.

Disseminated Sporotrichosis

Primary therapy is with AmB lipid formulation (3–5 mg/kg/day, IV). AmB-deoxycholate (0.7–1.0 mg/kg/day) could also be used but was not preferred by Kauffman et al. (2007). ITC (200 mg 2×/day) suppressive therapy is provided after the patient responds to AmB. A total period of at least 12 months of therapy is recommended, or indefinitely, for patients with AIDS or another immunosuppressive condition. Based on experience with other deep-seated mycoses, it is possible to discontinue therapy for sporotrichosis patients with AIDS who were treated with ITC for at least 1 year and whose $CD4^+$ T-lymphocyte counts remain >200 cells/μL for 1 year.

Sporotrichosis During Pregnancy

Lymphocutaneous sporotrichosis can be treated with local hyperthermia. AmB is reserved for extracutaneous infection. When extracutaneous sporotrichosis must be treated during pregnancy, AmB lipid formulation is administered (3–5 mg/kg/day, IV) or AmB-deoxycholate (0.7–1 mg/kg/day). Azoles should be avoided during pregnancy.

Therapy for Children with Fixed-Cutaneous or Lymphocutaneous Sporotrichosis

Primary therapy consists of ITC for children up to 20 kg at a dosage of 5 mg/kg/day. SSKI may be used as second-line therapy starting with 1 drop/day and increasing up to 10 drops, 3 times daily. Disseminated sporotrichosis in children is treated with AmB (0.7 mg/kg/day) and, pending a clinical response, therapy is stepped-down to ITC (5 mg/kg/day).

9.13 LABORATORY DETECTION, RECOVERY, AND IDENTIFICATION

Sporothrix schenckii is a dimorphic fungus producing yeast cells in tissue and on enriched media at 37°C. Specimens most commonly collected for identification are aspirated pus, scrapings, or biopsied tissue from cutaneous or subcutaneous lesions. Respiratory tract specimens include sputum, mucous membrane specimens, particularly from the nasal sinuses, and material from the lower respiratory tract. When appropriate, biopsy specimens are taken from other infected sites: CSF, synovial fluid, and multiple systemic sites during disseminated infection.

Direct Examination

Direct microscopic examination of tissues or exudates is most often negative, as the fungus does not typically produce large numbers of yeast cells. (Please see Section 9.3, Case Presentations.) Persistence in studying a number of slides may reveal some of the yeast forms. Swabs taken from open lesions and placed in sterile normal saline overnight at 37°C often exhibit large numbers of characteristic yeast cells. When seen, the yeast cells are elongated, "cigar" shaped to the oval. Less often seen are round cells, or yeast surrounded by PAS-positive material (please see below: "Asteroid Bodies"). Since it is difficult to observe yeast cells in direct examination of exudates or biopsy material, it is most useful to perform cultural studies.

Histopathology

Sporothrix schenckii are difficult to locate in tissue, and more often only granulomatous changes may be found. When present, yeast forms can be stained with PAS or GMS and are viewed as small spindle- to cigar-shaped, or round to oval cells, 1–3 μm × 3–10 μm (Fig. 9.2).

Asteroid Bodies

The asteroid body is a rare but characteristic structure that aids in the diagnosis of lymphocutaneous sporotrichosis (da Rosa et al., 2008). The H&E staining reaction reveals these eosinophilic asteroid bodies. The eosinophilic material, formed by antibodies or antigen–antibody complexes, is an example of the *Splendore–Hoeppli* reaction, which is suggestive but not pathognomonic for sporotrichosis.[6]

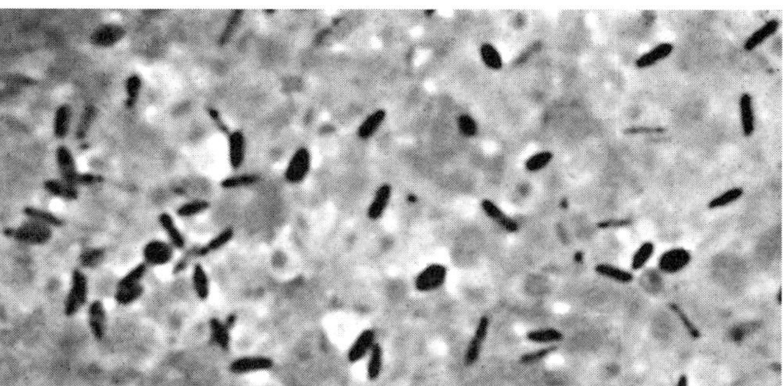

Figure 9.2 *Sporothrix schenckii* in an exudate, PAS stain. The yeast forms have a cigar shape or fusiform (spindle) shape.
Source: H. J. Shadomy.

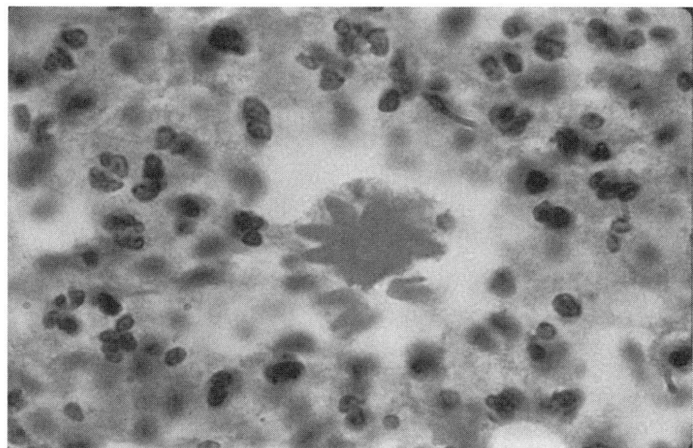

Figure 9.3 Asteroid body in sporotrichosis. Stellate projections of eosinophilic material surround a *S. schenckii* yeast form cell, histopathology, H&E stain. See insert for color representation of this figure.
Source: H. J. Shadomy.

Visualization of the ray-like projections can lead to detecting asteroid bodies in adjacent sections, a clue in the diagnosis of sporotrichosis (Fig. 9.3).

Culture

Direct microscopy is often negative but *S. schenckii* grows well on mycologic media. Appropriate material (e.g., ground up tissue biopsy material or drainage from skin lesions or, preferably, aspirated pus) should be cultured on SDA-Emmons (with antibiotics but without cycloheximide) and held at 25–30°C for up to 4 weeks before discarding. Growth, however, usually is seen within 3–5 days. *Note*: The conidia of *S. schenckii*

[6]Splendore–Hoeppli phenonomenon. *Definition:* The Splendore–Hoeppli phenomenon describes the eosinophilic, pseudomycotic structures composed of necrotic debris and immunoglobulin. In the United States, it is most frequently found in association with botryomycotic infections caused by *Staphylococcus aureus* and *Pseudomonas aeruginosa*. In the tropics, it can be found surrounding schistosome eggs, microfilariae, and a variety of fungi, including *S. schenckii*.

are dangerous to inhale. Always work in a properly functioning BSC, not on the open bench.

Confirmation of the fungus requires conversion to the yeast form on BA or BHI–sheep blood agar at 37°C. Growth in the range of 35–37°C is an important criterion for pathogenicity. Isolates incapable of growth at 37°C are considered to have much reduced or no pathogenic potential. Yeast colonies may appear wrinkled and an off-white, cream color. Antibacterial antibiotics should be added to the media to reduce contaminating bacteria. Viewed microscopically, yeast forms of *S. schenckii* in vitro are similar to those seen in histopathologic sections (Fig. 9.4).

Colony Morphology

Colonies (Fig. 9.5) begin as moist, yeast-like, membranous, and white or cream colored, later becoming dark brown to black within 1–2 weeks. The surface becomes wrinkled and felt-like. Colonies may appear that have both the light and dark segments. Intensity of colony color is directly related to the numbers and pigmentation of the conidia. Colony color is influenced by the growth medium. Clinical isolates that grew as darkly pigmented on cornmeal, PDA, Mycosel, and malt agar were gray or taupe on SDA (Dixon et al., 1991).

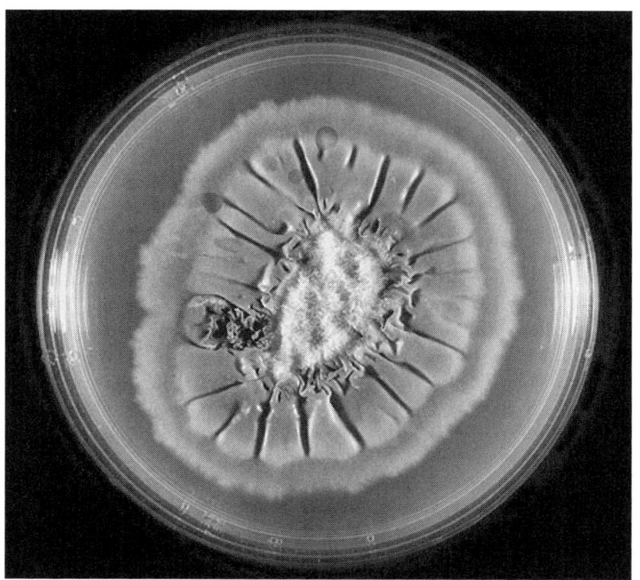

Figure 9.5 Colony morphology of *S. schenckii* mold form. Dark color areas of the colony are due to the presence of melanized *sessile* conidia arising directly from hyphae (holoblastic conidiogenesis).
Source: Used with permission from Indiana Pathology Images™, © 2004.

Microscopic Morphology

Oval conidia are formed on small, cylindrical *denticles* in a *sympodial* arrangement on apically swollen conidiophores arising at a right angle to the hyphae, resulting in a flower-like (floret or daisy cluster) arrangement (Fig. 9.6). The small oval conidia are thin walled and measure a mean length to width of 2.0–2.4 μm × 1.4 μm. Environmental isolates can have longer conidia, 3.5–4.3 μm length (Dixon et al., 1991). A second type of conidia is sessile: brown, thick-walled, ovoid, and sometimes triangular conidia are produced directly from the main hyphal axis, in abundance, so as to cover the hyphae in a cylindrical sleeve (Fig. 9.7).

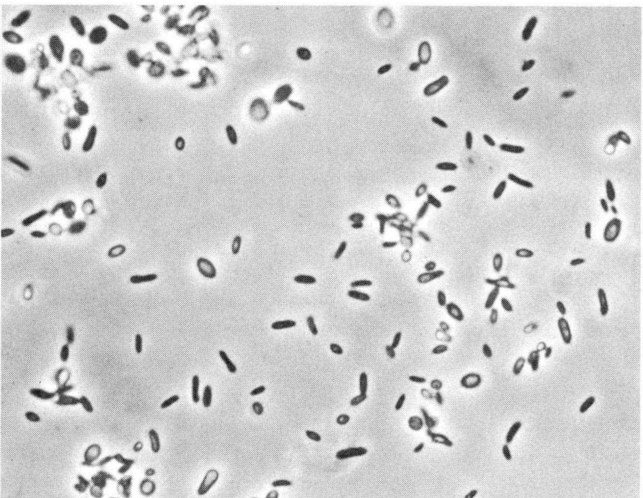

Figure 9.4 *Sporothrix schenckii* yeast form; microscopic morphology, in vitro, phase contrast.
Source: Used with permission from Dr. Glenn S. Bulmer, Tatay City, the Philippines.

Differential Diagnosis *Ophiostoma stenoceras* produces conidia in a floret pattern similar to *S. schenckii*, but it does not grow at 37°C, cannot be converted to the yeast form, and is nonpathogenic. *O. stenoceras* grows as white to cream-colored colonies that do not turn brown to black. *O. stenoceras* is homothallic, producing crescent-shaped ascospores within spherical black perithecia with necks up to 1000 μm long. These perithecia were observed in PDA slide cultures or on CMA plates (Dixon et al., 1991). Other fungi that may be differentiated from *S. schenckii* are *Aureobasidium pullulans, Trichosporon* species, and *Sporothrix* species, which do not grow at 37°C and are not dimorphic.

Sporothrix schenckii var. *luriei*, a rare variety of *S. schenckii*, was first described from a patient in Africa in 1969. It has been identified in at least one other case, from India in 1992. The fungus differs from *S. schenckii* in tissue: yeast forms that multiply by budding and the more striking oval, elongated, hyaline thick-walled fungal

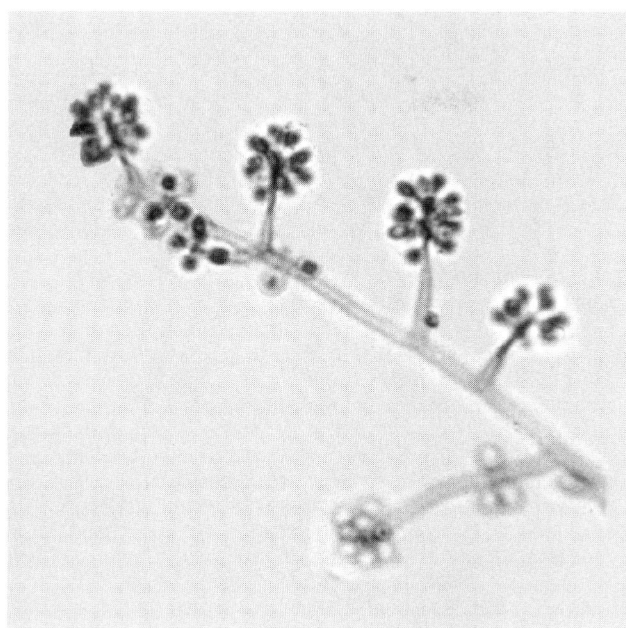

Figure 9.6 *Sporothrix schenckii* conidiophores with florets of conidia borne sympodially on tapering denticles (holoblastic conidiogenesis). Conidia are ovoid to club shaped, ranging in size: 2.5–5.5 μm × 1.5–2.5 μm, 400×.
Source: E. Reiss, CDC.

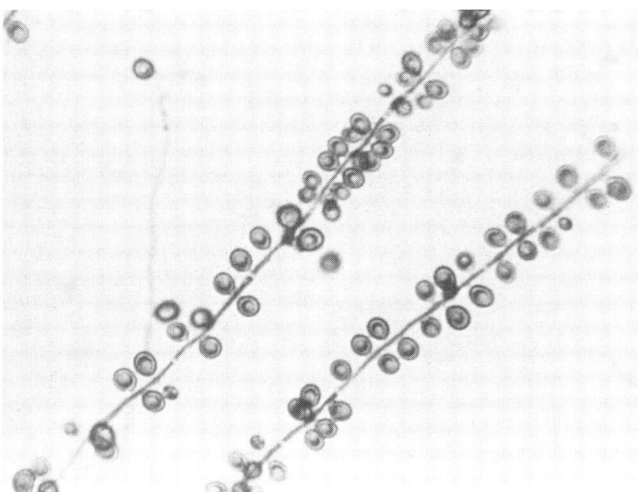

Figure 9.7 Sessile conidia, melanized, borne directly as a sleeve along hyphae of *S. schenckii*.
Source: H. J. Shadomy.

cells 10–30 μm in diameter, many dividing by equatorial septations into two daughter cells. Cultures begin as tan colored, but become white upon maintenance in the laboratory. Two types of conidia are formed. The first is identical to *S. schenckii*, while the second is longer (up to 10 μm) and usually thick walled. Molecular identification studies have found this variety is distinct from *S. schenckii* and is most likely a separate species, based on phylogenetic analysis based on the calmodulin gene (Marimon et al., 2007).

Serodiagnosis

A specific and quantitative measure of the serologic response in sporotrichosis is useful in both establishing a diagnosis and monitoring the response to therapy. Serologic tests are useful in sporotrichosis when the yeast forms cannot be seen by direct microscopy or histopathology, there is a delay in growth of the mold form in culture, and in cases of extracutaneous sporotrichosis. A commercial test, the latex agglutination test for sporotrichosis, is available in the United States (LA-Sporo Antibody System, Immunomycologics, Inc.).

Sporotrichoidal meningitis is an uncommon form of fungal meningitis and is a diagnostic dilemma. In their series of 7 such cases, Scott et al. (1989) reported that all had *S. schenckii* antibodies in their CSF and serum. The titers were revealed by commercial latex agglutination test and a "homebrew" EIA. They recommend antibody assays for sporotrichosis in patients with chronic meningitis for which no etiologic diagnosis is established by routine tests.

Knowledge gained from immunochemistry of the peptido-L-rhamno-D-mannan cell wall polysaccharide of *S. schenckii* was applied in an EIA to detect antibodies in patient sera and evaluated in 92 Brazilian patients (Bernardes-Engemann et al., 2005). The Con A-binding fraction of rhamnomannan, termed SsCBF (*S. schenckii* Con-A binding fraction) contains an epitope that is specific for *S. schenckii*, containing rhamnose, mannose, and glucuronic acid (please see Section 9.9, Determinants of Pathogenicity, for details). The test specificity was 80% with crossreactions observed at low O.D. values with sera from some patients with aspergillosis, chromoblastomycosis, cryptococcosis, histoplasmosis, and paracoccidioidomycosis. Possibly these crossreactions could be excluded by increasing the cutoff for a positive test from 2 standard deviations above the mean of the normal amount of antibodies in healthy volunteers to 3 standard deviations or by using a kinetic EIA. That, however, might affect the sensitivity of the test, which was reported as 90%. Titers rose in proportion to disease severity and declined with a favorable response to therapy.

Another EIA was developed for serodiagnosis of sporotrichosis by affixing mycelial-phase *S. schenckii* exoantigens to the microtitration plate (Almeida-Paes et al., 2007a). Antibodies produced during sporotrichosis and after ITC therapy were compared using this exoantigen EIA (Almeida-Paes et al., 2007b). Overall, 78% of patients had detectable levels at diagnosis, declining to 62.9% in patients during therapy.

Genetic Identification

A nested PCR assay for specific for *S. schenckii* used a target sequence in the 18S rDNA (Hu et al., 2003). Tissues of experimentally infected mice and clinical biopsy specimens of 12 confirmed sporotrichosis patients were evaluated with the nested PCR. All 5 infected mice and 11 of the 12 clinical specimens were positive reactors. The future availability of such tests may depend on a commercial panel of tests for endemic mycoses, including sporotrichosis, or the extension of the AccuProbe® (Gen-Probe Inc., San Diego, CA) approach, which has proved useful with other systemic dimorphic, fungal pathogens.

SELECTED REFERENCES FOR SPOROTRICHOSIS

al-Tawfiq JA, Wools KK, 1998. Disseminated sporotrichosis and *Sporothrix schenckii* fungemia as the initial presentation of human immunodeficiency virus infection. *Clin Infect Dis* 26: 1403–1406.

Almeida-Paes R, Pimenta MA, Pizzini CV, Monteiro PC, Peralta JM, Nosanchuk JD, Zancopé-Oliveira RM, 2007a. Use of mycelial-phase *Sporothrix schenckii* exoantigens in an enzyme-linked immunosorbent assay for diagnosis of sporotrichosis by antibody detection. *Clin Vaccine Immunol* 14: 244–249.

Almeida-Paes R, Pimenta MA, Monteiro PC, Nosanchuk JD, Zancopé-Oliveira RM, 2007b. Immunoglobulins G, M, and A against *Sporothrix schenckii* exoantigens in patients with sporotrichosis before and during treatment with itraconazole. *Clin Vaccine Immunol* 14: 1149–1157.

Barros MB, Schubach AO, Schubach TM, Wanke B, Lambert-Passos SR, 2008a. An epidemic of sporotrichosis in Rio de Janeiro, Brazil: Epidemiological aspects of a series of cases. *Epidemiol Infect* 136: 1192–1196.

Barros MB, Costa DL, Schubach TM, do Valle AC, Lorenzi NP, Teixeira JL, Schubach Ade O, 2008b. Endemic of zoonotic sporotrichosis: profile of cases in children. *Pediatr Infect Dis J* 27: 246–250.

Bernardes-Engemann AR, Costa RC, Miguens BR, Penha CV, Neves E, Pereira BA, Dias CM, Mattos M, Gutierrez MC, Schubach A, Oliveira Neto MP, Lazéra M, Lopes-Bezerra LM, 2005. Development of an enzyme-linked immunosorbent assay for the serodiagnosis of several clinical forms of sporotrichosis. *Med Mycol* 43: 487–493.

Bustamante B, Campos PE, 2001. Endemic sporotrichosis. *Curr Opin Infect Dis* 14: 145–149.

Byrd DR, El-Azhary RA, LE Gibson, Roberts GD, 2001. Sporotrichosis masquerading as pyoderma gangrenosum: Case report and review of 19 cases of sporotrichosis. *J Eur Acad Dermatol Venereol* 15: 581–584.

Centers for Disease Control (CDC), 1988. Multistate outbreak of sporotrichosis in seedling handlers. *MMWR Morb Mortal Wkly Rep* 37: 652–653.

Chapman SW, Pappas P, Kauffmann C, Smith EB, Dietze R, Tiraboschi-Foss N, Restrepo A, Bustamante AB, Opper C, Emady-Azar S, Bakshi R, 2004. Comparative evaluation of the efficacy and safety of two doses of terbinafine (500 and 1000 mg/day) in the treatment of cutaneous or lymphocutaneous sporotrichosis. *Mycoses* 47: 62–68.

Cooper CR Jr, Breslin BJ, Dixon DM, Salkin IF, 1992a. DNA typing of isolates associated with the 1988 sporotrichosis epidemic. *J Clin Microbiol* 30: 1631–1635.

Cooper CR, Dixon DM, Salkin IF, 1992b. Laboratory acquired sporotrichosis. *J Med Vet Mycol* 30: 169–171.

da Rosa WD, Gezuele E, Calegari L, Goñi F, 2008. Asteroid body in sporotrichosis. Yeast viability and biological significance within the host immune response. *Med Mycol* 46: 443–448.

Dixon DM, Salkin IF, Duncan RA, Hurd NJ, Haines JH, Kemna ME, Coles FB, 1991. Isolation and characterization of *Sporothrix schenckii* from clinical and environmental sources associated with the largest U.S. epidemic of sporotrichosis. *J Clin Microbiol* 29: 1106–1113.

Feeney KT, Arthur IH, Whittle AJ, Altman SA, Speers DJ, 2007. Outbreak of sporotrichosis, Western Australia. *Emerging Infect Dis* 13: 1228–1231.

Fernandes KS, Neto EH, Brito MM, Silva JS, Cunha FQ, Barja-Fidalgo C, 2008. Detrimental role of endogenous nitric oxide in host defence against *Sporothrix schenckii*. *Immunology* 123: 469–479.

Figueiredo CC, Deccache PM, Lopes-Bezerra LM, Morandi V, 2007. TGF-beta1 induces transendothelial migration of the pathogenic fungus *Sporothrix schenckii* by a paracellular route involving extracellular matrix proteins. *Microbiology* 153: 2910–2921.

Figueiredo CC, De Lima OC, De Carvalho L, Lopes-Bezerra LM, Morandi V, 2004. The in vitro interaction of *Sporothrix schenckii* with human endothelial cells is modulated by cytokines and involves endothelial surface molecules. *Microb Pathog* 36: 177–188.

Francesconi G, Valle AC, Passos S, Reis R, Galhardo MC, 2009. Terbinafine (250 mg/day): An effective and safe treatment of cutaneous sporotrichosis. *J Eur Acad Dermatol Venereol* 23: 1273–1276.

Galhardo MC, De Oliveira RM, Valle AC, Paes Rde A, Silvatavares PM, Monzon A, Mellado E, Rodriguez-Tudela JL, Cuenca-Estrella M, 2008. Molecular epidemiology and antifungal susceptibility patterns of *Sporothrix schenckii* isolates from a cat-transmitted epidemic of sporotrichosis in Rio de Janeiro, Brazil. *Med Mycol* 46: 141–151.

González-Ibarra J, Milewski S, Villagómez-Castro JC, Cano-Canchola C, López-Romero E, 2010. *Sporothrix schenckii*: Purification and partial biochemical characterization of glucosamine-6-phosphate synthase, a potential antifungal target. *Med Mycol* 48: 110–121.

Gottlieb GS, Lesser CF, Holmes KK, Wald A, 2003. Disseminated sporotrichosis associated with treatment with immunosuppressants and tumor necrosis factor-alpha antagonists. *Clin Infect Dis* 37: 838–840.

Hu S, Chung WH, Hung SI, Ho HC, Wang ZW, Chen CH, Lu SC, Kuo TT, Hong HS, 2003. Detection of *Sporothrix schenckii* in clinical samples by a nested PCR assay. *J Clin Microbiol* 41: 1414–1418.

Kauffman CA, 1999. Sporotrichosis. *Clin Infect Dis* 29: 231–237.

Kauffman CA, Bustamante B, Chapman SW, Pappas PG, 2007. Clinical practice guidelines for the management of sporotrichosis: 2007 Update by the Infectious Diseases Society of America. *Clin Infect Dis* 45: 1255–1265.

Kedes LH, Siemienski J, Braude AI, 1964. The syndrome of the alcoholic rose gardener. Sporotrichosis of the radial tendon sheath. Report of a case cured with amphotericin B. *Ann Intern Med* 61: 1139–1141.

Kovarik CL, Neyra E, Bustamante B, 2008. Evaluation of cats as the source of endemic sporotrichosis in Peru. *Med Mycol* 46: 53–56.

Lima OC, Figueiredo CC, Previato JO, Mendonça-Previato L, Morandi V, Lopes-Bezerra LM, 2001. Involvement of fungal cell wall components in adhesion of *Sporothrix schenckii* to human fibronectin. *Infect Immun* 69: 6874–6880.

Lopes-Bezerra LM, Schubach A, Costa RO, 2006. *Sporothrix schenckii* and sporotrichosis. *An Acad Bras Cienc* 78: 293–308

Losman JA, Cavanaugh K, 2004. Cases from the Osler Medical Service at Johns Hopkins University. Diagnosis: *P. carinii* pneumonia and primary pulmonary sporotrichosis. *Am J Med* 117: 353–356.

Lyon GM, Zurita S, Casquero J, Holgado W, Guevara J, Brandt ME, Douglas S, Shutt K, et al, 2003. Population-based surveillance and a case–control study of risk factors for endemic lymphocutaneous sporotrichosis in Peru. *Clin Infect Dis* 36: 34–39.

Marimon R, Cano J, Gené J, Sutton DA, Kawasaki M, Guarro J, 2007. *Sporothrix brasiliensis, S. globosa*, and *S. mexicana*, three new *Sporothrix* species of clinical interest. *J Clin Microbiol* 45: 3198–3206.

Marimon R, Gené J, Cano J, Trilles L, Dos Santos Lazéra M, Guarro J, 2006. Molecular phylogeny of *Sporothrix schenckii*. *J Clin Microbiol* 44: 3251–3256.

Mayorga-Rodríguez JA, Barba-Rubio J, Muñoz-Estrada VF, Rangel-Cortés A, García-Vargas A, Magaña-Camarena I, 1997. Esporotricosis en el estado de Jalisco, estudio clínico-epidemiológico (1960–1996). *Dermatología Rev Mex* 41: 105–108.

Miller SD, Keeling JH, 2002. Ant sting sporotrichosis. *Cutis* 69: 439–442.

Morris-Jones R, Youngchim S, Gomez BL, Aisen P, Hay RJ, Nosanchuk JD, Casadevall A, Hamilton AJ, 2003. Synthesis of melanin-like pigments by *Sporothrix schenckii* in vitro and during mammalian infection. *Infect Immun* 71: 4026–4033.

Nascimento RC, Almeida SR, 2005. Humoral immune response against soluble and fractionate antigens in experimental sporotrichosis. *FEMS Immunol Med Microbiol* 43: 241–247.

Pappas PG, Tellez I, Deep AE, Nolasco D, Holgado W, Bustamante B, 2000. Sporotrichosis in Peru: Description of an area of hyperendemicity. *Clin Infect Dis* 30: 65–70.

Penha CV, Bezerra LM, 2000. Concanavalin A-binding cell wall antigens of *Sporothrix schenckii*: A serological study. *Med Mycol* 38: 1–7.

Quintal D, 2000. Sporotrichosis infection on mines of the Witwatersrand. *J Cutan Med Surg* 4: 51–54.

Randhawa HS, Chand R, Mussa AY, Khan ZU, Kowshik T, 2003. Sporotrichosis in India: First case in a Delhi resident and an update. *Indian J Med Microbiol* 21: 12–16.

Reed KD, Moore FM, Geiger GE, Stemper ME, 1993. Zoonotic transmission of sporotrichosis: Case report and review. *Clin Infect Dis* 16: 384–387.

Rhome R, McQuiston T, Kechichian T, Bielawska A, Hennig M, Drago M, Morace G, Luberto C, Del Poeta M, 2007. Biosynthesis and immunogenicity of glucosylceramide in *Cryptococcus neoformans* and other human pathogens. *Eukaryot Cell* 10: 1715–1726.

Romero-Martinez R, Wheeler M, Guerrero-Plata A, Rico G, Torres-Guerrero H, 2000. Biosynthesis and functions of melanin in *Sporothrix schenckii*. *Infect Immun* 68: 3696–3703.

Ruiz-Baca E, Toriello C, Perez-Torres A, Sabanero-Lopez M, Villagomez-Castro JC, Lopez-Romero E, 2009. Isolation and some properties of a glycoprotein of 70kDa (Gp70) from the cell wall of *Sporothrix schenckii* involved in fungal adherence to dermal extracellular matrix. *Med Mycol* 47: 185–196.

Sandhu K, Gupta S, 2003. Potassium iodide remains the most effective therapy for cutaneous sporotrichosis. *Dermatol Treat* 14: 200–202.

Schubach A, Barros MB, Wanke B, 2008. Epidemic sporotrichosis. *Curr Opin Infect Dis* 21: 129–133.

Schubach TM, Schubach A, Okamoto T, Barros MB, Figueiredo FB, Cuzzi T, Fialho-Monteiro PC, Reis RS, Perez MA, Wanke B, 2004. Evaluation of an epidemic of sporotrichosis in cats: 347 cases (1998–2001). *J Am Vet Med Assoc* 224: 1623–1629.

Scott EN, Kaufman L, Brown AC, Muchmore HG, 1987. Serologic studies in the diagnosis and management of meningitis due to *Sporothrix schenckii*. *N Engl J Med* 317: 935–940.

Sterling JB, Heymann WR, 2000. Potassium iodide in dermatology: A 19th century drug for the 21st century—uses, pharmacology, adverse effects, and contraindications. *J Am Acad Dermatol* 43: 691–697.

Sugar AM, Mattia AR, 1994. Case records of the Massachusetts General Hospital. Weekly Clinicopathological Exercises. July 21, 1994 Case 28-1994—A 51-year-old man with a nonhealing finger wound and regional lymphadenopathy. *N Engl J Med* 331: 181–187.

Sykes JE, Torres SM, Armstrong PJ, Lindeman CJ, 2001. Itraconazole for treatment of sporotrichosis in a dog residing on a Christmas tree farm. *J Am Vet Med Assoc* 218: 1440–1443, 1421.

Tachibana T, Matsuyama T, Mitsuyama M, 1999. Involvement of $CD4^+$ T cells and macrophages in acquired protection against infection with *Sporothrix schenckii* in mice. *Med Mycol* 37: 397–404.

Teixeira PA, de Castro RA, Nascimento RC, Tronchin G, Torres AP, Lazéra M, de Almeida SR, Bouchara JP, Loureiro y Penha CV, Lopes-Bezerra LM, 2009. Cell surface expression of adhesins for fibronectin correlates with virulence in *Sporothrix schenckii*. *Microbiology* 155(Pt 11): 3730–3738.

Toledo MS, Levery SB, Straus AH, Takahashi HK, 2000. Dimorphic expression of cerebrosides in the mycopathogen *Sporothrix schenckii*. *J Lipid Res* 41: 797–806.

Uenotsuchi T, Takeuchi S, Matsuda T, Urabe K, Koga T, Uchi H, Nakahara T, Fukagawa S, Kawasaki M, Kajiwara H, Yoshida S, Moroi Y, Furue M, 2006. Differential induction of Th1-prone immunity by human dendritic cells activated with *Sporothrix schenckii* of cutaneous and visceral origins to determine their different virulence. *Int Immunol* 18: 1637–1646.

Valentín-Berríos S, González-Velázquez W, Pérez-Sánchez L, González-Méndez R, Rodríguez-Del Valle N, 2009. Cytosolic phospholipase A2: A member of the signalling pathway of a new G protein alpha subunit in *Sporothrix schenckii*. *BMC Microbiol* 9: 100.

Valle-Aviles L, Valentin-Berrios S, Gonzalez-Mendez RR, Rodriguez-Del Valle N, 2007. Functional, genetic and bioinformatic characterization of a calcium/calmodulin kinase gene in *Sporothrix schenckii*. *BMC Microbiol* 7: 107.

Wojciechowski M, Milewski S, Mazerski J, Borowski E, 2005. Glucosamine-6-phosphate synthase, a novel target for antifungal agents. Molecular modelling studies in drug design. *Acta Biochim Pol* 52: 647–653.

WEBSITE CITED

Biosafety in Microbiological and Biomedical Laboratories (BMBL), 5th edition. Available at the URL http://www.cdc.gov/od/ohs/biosfty/bmbl5/bmbl5toc.htm.

QUESTIONS

The Answer Key to multiple choice questions may be found at the end of the book.

1. Which laboratory approach is most likely to lead to a diagnosis of sporotrichosis?
   A. Direct examination of pus from skin lesions
   B. Culture of pus from skin lesions
   C. Serology
   D. Histopathology of biopsied skin lesion
2. Which classic sign of lymphocutaneous sporotrichosis can lead to a presumptive diagnosis?
   A. Circular lesions with active borders, scaling, and pruritus
   B. Chain of subcutaneous nodules trace lymphatic drainage from a lesion on an extremity

C. Tumefaction, with sinus tracts draining granules

D. Umbilicated papules resembling molluscum contagiosum

3. Colonies of *Sporothrix schenckii* begin as moist, yeast-like, white or cream colored, later becoming dark brown to black within 1–2 wks. What causes the dark brown to black color to develop?

A. Black conidia are formed on small denticles in a floret arrangement on apically swollen conidiophores.

B. Brown, sessile, ovoid conidia are produced directly on hyphae, covering the hyphae in a cylindrical sleeve.

C. *Sporothrix schenckii* is a melanized mold.

D. While not a melanized mold, *S. schenckii* secretes a black pigment.

4. Which of the following is/are histopathologic clues to the identity of *Sporothrix schenckii*?

1. Cigar-shaped yeast forms
2. Multipolar budding yeast forms
3. Round to pleomorphic thick-walled cells with internal septations.
4. Splendore–Hoeppli reaction

A. 1 + 2

B. 1 + 3

C. 1 + 4

D. 2 + 3

5. A culture is identified as *Sporothrix* based on the microscopic morphology of the pattern of asexual sporulation. However, there is no growth at 37°C and no dimorphism at 35°C. What could this fungus be?

A. *Cladophialophora carrionii*

B. *Fonsecaea pedrosoi*

C. Nonpathogenic *Sporothrix* species

D. *Rhinocladiella aquaspersa*

6. Can sporotrichosis be considered a zoonosis? Indicate in your discussion any large outbreak in which a lower animal was implicated as the source of infection.

7. Considering that sporotrichosis is worldwide in distribution, how can it be considered to be an endemic mycosis?

8. Discuss the clinical experience using oral saturated solutions of potassium iodide as therapy for sporotrichosis. What is the current recommended antifungal therapy for fixed-cutaneous and lymphocutaneous sporotrichosis?

9. List and describe three extracutaneous forms of sporotrichosis, giving risk factors for each and the recommended therapeutic options.

10. How can serology aid in the diagnosis of sporotrichosis? What tests for fungal antibodies are recommended in a patient with chronic meningitis for whom no etiologic diagnosis is available from routine tests?

Chapter 10A

Less Frequent Mycoses Caused by Dimorphic Environmental Molds: Adiaspiromycosis

Two diseases, adiaspiromycosis and lobomycosis (see Chapter 10B), and their causative agents, *Emmonsia* species and *Lacazia loboi*, respectively, differ from the definition of dimorphism as it is understood in medical mycology: thermally induced morphogenesis in which a hyphal form exists in the environment and another distinct form usually, but not necessarily, a yeast form, is induced by a shift to 37°C, such as occurs in the host during infection. The hyphal form of *Emmonsia* species is well known in the environment, but after inhalation of conidia, in tissue the conidium enlarges by *isotropic* growth but does not replicate. Furthermore, adiaspiromycosis is distributed worldwide and does not have a well-demarcated endemic area. The etiologic agent of lobomycosis, *Lacazia loboi*, has a yeast-like tissue form and a defined endemic area but an environmental mold form, if it exists, has not been discovered. Both of these etiologic agents are classed among the *Ajellomycetaceae*, a family that includes primary, dimorphic environmental molds: *Blastomyces dermatitidis, Histoplasma capsulatum*, and *Paracoccidioides brasiliensis*. Close taxonomic relatedness with other dimorphic endemic pathogens is a criterion for placing adiaspiromycosis and lobomycosis in Part Two of this text.

10A.1 ADIASPIROMYCOSIS AT-A-GLANCE

- *Introduction/Disease Definition.* Adiaspiromycosis is a pulmonary disease mostly affecting burrowing rodents. Inhalation of conidia by humans results in the formation of nonreplicating but enlarging adiaspores, which may reach 700 μm diameter in size. They evoke a granulomatous response in the host.
- *Etiologic Agents. Emmonsia crescens* and *Emmonsia parva* are dimorphic environmental molds. Genetic studies place them in the *Ajellomycetaceae*, the same family as *Blastomyces dermatitidis, Histoplasma capsulatum*, and *Paracoccidioides brasiliensis*.
- *Geographic Distribution/Ecologic Niche. Emmonsia parva* is isolated from a few animal species in small geographic ranges in North America, Asia, Australia, and eastern Europe. *Emmonsia crescens* is known from many animal species and from soil worldwide.
- *Epidemiology/Risk Groups/Factors. Emmonsia crescens* is most often isolated from humans. *Emmonsia parva* is uncommon even in lower animals. Adiaspiromycosis is rarely diagnosed in humans, with fewer than 60 reported cases. Farmers, plant nursery workers, and rural workers in renovation and demolition of old buildings are at increased risk of exposure. Immunocompromised persons also are at risk.
- *Transmission.* The route of transmission is via inhalation of conidia from soil.
- *Clinical Forms.* Human pulmonary adiaspiromycosis presents as a solitary granuloma, localized granulomas, or diffuse, pulmonary-disseminated granulomatous disease, depending on the quantity of conidia that are inhaled and the host response.

Fundamental Medical Mycology, First Edition. By Errol Reiss, H. Jean Shadomy and G. Marshall Lyon, III.
© 2012 Wiley-Blackwell. Published 2012 by John Wiley & Sons, Inc.

- *Veterinary Forms.* Burrowing mammals are the major species affected.
- *Therapy.* Human pulmonary adiaspiromycosisis usually is self-limited, and antifungal therapy usually is unnecessary. Ketoconazole, fluconazole (FLC), and AmB have been successful in severe or progressive disease.
- *Laboratory Detection, Recovery, and Identification. Emmonsia crescens* has a maximum growth temperature of 37°C and has larger adiaspores, often ≥100 μm diameter. *Emmonsia parva* grows up to 40°C. Adiaspores are viewed microscopically in pulmonary biopsy specimens and smears of BAL fluid. Culture on PDA at 25°C reveals the mold form: hyaline hyphae, short conidiophores with terminal aleurioconidia. At 37°C adiaspores are produced by *E. crescens* on phytone yeast extract agar.

10A.2 INTRODUCTION/DISEASE DEFINITION

Adiaspiromycosis is a pulmonary disease of burrowing rodents, other mammals, and rarely of humans, caused by the dimorphic molds—*Emmonsia crescens* and *Emmonsia parva* (Sigler, 2006). The disease usually is asymptomatic and self-limited. Human pulmonary disease presents as either (i) a solitary granuloma, (ii) localized granulomas, or (iii) diffuse, bilateral disseminated granulomatous disease. Extrapulmonary dissemination is exceptional. Severity of disease depends on the infectious dose of conidia. Following inhalation from the environment, conidia enlarge to produce the tissue form: a large round or oval thick-walled nonreplicating adiaspore. These uni- or multinucleate cells are found in enlarged cyst-like formations in the lungs of animals and humans. The term "adiaspore" derives from the Greek, indicating enlargement without multiplication. Importantly, adiaspores lack internal spores and are not analogous to the spherules of *Coccidioides* species. They are viable resting cells which, when transferred to temperatures <37°C, are capable of germinating into the mold form.

10A.3 CASE SUMMARIES

Case 1

Severe respiratory distress with diffuse bilateral pulmonary granulomatous disease occurred in a man, more than 1 month after he worked in a poorly ventilated shed in São Paulo, Brazil (Barbas Filho et al., 1990). The shed contents included rat and bat droppings. The patient presented with cough, dyspnea, fever, and night sweats. Chest radiographs showed bilateral interstitial infiltrates in a reticular pattern. CT scan indicated disseminated pulmonary nodules, 1–3 mm diameter, resembling miliary tuberculosis. An open lung biopsy showed granulomas with an unidentified organism at the center, initially thought to be a helminth. A collapsed adiaspore can appear as a roundworm. The patient was prescribed thiabendazole and had a positive clinical response. Laboratory studies on culture collection isolates indicated that *E. crescens* was resistant to thiabendazole, but that *E. parvum* had a MIC of 1 μg/mL. Reexamination of the histologic sections resulted in the identification of adiaspores establishing the diagnosis as adiaspiromycosis. He became asymptomatic and 1 year later the chest radiograph was normal.

Case 2

A 30-yr-old immune-normal man related several weeks of weakness, cough, fever, and weight loss of 10 kg and was admitted to a hospital in France (Dot et al., 2009). The chest radiograph revealed diffuse bilateral interstitial pneumonia in a micronodular pattern. High resolution CT scan of the chest showed disseminated pulmonary nodules. Bronchoalveolar lavage (BAL) fluid contained 13,330 leukocytes/μL. Direct examination and culture of BAL fluid were negative. Panfungal PCR was conducted on the BAL fluid using His3 and His4 primers spanning a 613 bp region from 18S to 28S rDNA. BLAST analysis identified the fungus as *Emmonsia crescens*. A transbronchial biopsy sample processed for histopathology revealed PAS positive adiaspores (50–100 μm diameter) within granulomas. He was treated with oral itraconazole (ITC) (200 mg/day). At the 1 month follow-up clinic visit after hospital discharge, his respiratory function was improved and the chest radiograph showed the lesions were stable.

Case 3

A 40-yr-old man living with AIDS had recovered from *Pneumocystis* pneumonia but presented again with cough and dyspnea (Turner et al., 1999). A chest radiograph revealed consolidation in the upper lobe of the right lung, along with small diffuse patches in both lungs. Cultures of bronchoalveolar lavage fluid on SDA at room temperature grew a white-buff mold with small *aleurioconidia*. When incubated at 37°C, the conidia swelled and the hyphae became irregularly swollen and distorted. Some conidia grew in size and became thick-walled, uninucleate adiaspores measuring 12–15 μm. These characteristics were consistent with *Emmonsia parva*. The patient was treated with AmB and recovered gradually.

10A.4 DIAGNOSIS

Diagnosis is made by histologic examination of biopsy specimens revealing adiaspores surrounded by a combined suppurative-granulomatous infiltrate. The fungus does not have fastidious nutritional requirements (Sigler, 1996) but recovery from clinical specimens may be difficult. The fungus grows from clinical specimens as hyphae at room temperature. When the mold form sporulates, conidia can be transferred to higher temperatures to undergo the dimorphic switch: they swell and convert to the adiaspore form at 37°C or 40°C. As shown in Case 2 above, PCR-sequencing can be a successful alternative diagnostic test when there is no growth from clinical specimens.

Differential Diagnosis

Based on the histopathologic appearance the differential includes helminthic parasites, aspirated starch granules of lentil pneumonia, or large form pathogens: *Coccidioides* species or *Rhinosporidium seeberi* (Watts and Chandler, 1990). Based on the radiographic findings the differential includes tuberculosis, fungal pneumonia, bacterial pneumonia, and lung tumor. A 56-yr-old man living in England was diagnosed with right upper lobe lung adenocarcinoma with multiple subpleural nodules in the upper, middle, and lower lobes and extensive lymphadenopathy (Denson et al., 2009). An adiaspore of *E. crescens* was identified in a subpleural granuloma. If this second pathology was not correctly identified it might have been misinterpreted as widespread aggressive cancer with lymph node metastases.

10A.5 ETIOLOGIC AGENTS

Two species of *Emmonsia* are described: *E. parva* and *E. crescens*. Sequences of the 18S (SSU) rDNA support grouping *E. crescens* with *Blastomyces dermatitidis* and *Histoplasma capsulatum* in the family *Ajellomycetaceae, Onygenales, Ascomycota* (Sigler, 1996).

Emmonsia parva (C.W. Emmons & Ashburn) Cif. & A.M. Corte, *Mycopathologia* 10: 314 (1959). There is no known teleomorph stage.

Emmonsia crescens C.W. Emmons & Jellison, *Ann NY Acad Sci* 89: 98 (1960). The teleomorph is heterothallic: *Ajellomyces crescens* Sigler, *J Med Vet Mycol* 34: 303 (1996). This teleomorph of *Emmonsia crescens* is based on sexual compatibility among 12 of 22 strains, which produced stellate *gymnothecia* with helically coiled appendages and small, globose ascospores (Sigler, 1996).

Phylogenetic trees produced from a combined 5.8S and 28S rDNA data set indicate that all isolates of *Emmonsia* spp., *Ajellomyces capsulatus* (*H. capsulatum*), *P. brasiliensis*, and *A. dermatitidis* (*B. dermatitidis*), are members of a single clade (Peterson and Sigler, 1998). The *E. crescens* strains fall into a strongly supported clade with closely related Eurasian and North American subgroups. The strongest affinity for *E. crescens* seems to be with *B. dermatitidis*. The colony morphology subgroups are similar, but not identical, to the phylogenetic subgroups (please Section 10A.13, Laboratory Detection, Recovery, and Identification). Isolates previously identified as *E. parva* are phylogenetically more diverse, probably including additional separate genetic species.

10A.6 GEOGRAPHIC DISTRIBUTION/ECOLOGIC NICHE

Emmonsia parva is isolated from relatively few animal species in narrow geographic ranges in North America, Asia, Australia, and Eastern Europe. *Emmonsia crescens* is known from over 96 animal species and from soil worldwide (please see Section 10A.11, Veterinary Forms). Reports of human adiaspiromycosis cases originate from South and Central America, Israel, Europe, and the United States. The usual clinical isolate is *E. crescens*, as *E. parva* rarely infects humans.

10A.7 EPIDEMIOLOGY/RISK GROUPS/FACTORS

Emmonsia crescens is the species most often isolated from humans. *E. parva* is relatively uncommon even in lower animals. Despite its worldwide occurrence in rodents and small wild animals, adiaspiromycosis is rarely diagnosed in humans, with fewer than 60 cases in the literature, but it is probably underdiagnosed because of failure to grow the fungus from clinical material. Now that PCR-sequencing is increasingly being used, more cases are likely to be reported. Farmers, workers in soil, and people working in plant nurseries are at increased risk of exposure to the fungus. Please also see Section 10A.3, Case Summaries. Workers in construction, demolition, and renovation of old buildings in rural areas are at risk. Based on reports of two cases, there is a suspicion that patients living with AIDS are at increased risk to develop adiaspiromycosis (reviewed by Pfaller and Diekema, 2005).

10A.8 TRANSMISSION

Inhalation of conidia aerosolized from soil is believed to be the main route of transmission.

10A.9 DETERMINANTS OF PATHOGENICITY

Unknown.

10A.10 CLINICAL FORMS

Three forms of human pulmonary adiaspiromycosis are solitary granuloma, localized granulomatous disease, and diffuse, pulmonary-disseminated granulomatous disease (Pfaller and Diekema, 2005). Severity depends on host factors and the number of conidia inhaled, because fungal replication does not occur in vivo. Most adiaspiromycosis cases are asymptomatic and localized. Pulmonary nodules may be detected by chest radiographs, or are an incidental finding in surgical lung specimens or of autopsy material. Patients with disseminated granulomatous, pulmonary adiaspiromycosis may develop fever, cough, and dyspnea resulting from compression and displacement of distal airways and lung tissue by the expanding granulomas. Other organs rarely may be involved (e.g., the skin, peritoneum, and bone).

Cutaneous Adiaspiromycosis

Three cases of cutaneous adiaspiromycosis in the absence of pulmonary disease are known (Stebbins et al., 2004). The tissue biopsies demonstrated adiaspores identical to those seen in typical adiaspiromycosis.

10A.11 VETERINARY FORMS

A useful list of animal clinical isolates of *Emmonsia* species may be found in Sigler (1996). We should not be surprised that infection of wild animals with *Emmonsia* species may be widespread. Although there are relatively few field surveys this subject has interested investigators for a long time. Jellison (1950) summarized the findings of a field survey in Montana he conducted with Dr. Chester W. Emmons. They reported histopathologic and, in several instances, cultural evidence of *Emmonsia* species in the lungs of the following animal species: beaver, cottontail rabbit, mink, muskrat, pine marten, pine squirrel, rock rabbit, skunk, weasel, white-tail mouse, and wood rat. Pulmonary adiaspiromycosis was diagnosed in 7 of 25 striped skunks trapped in Alberta, Canada (Albassam et al., 1986). Infection with *E. crescens* was surveyed among wildlife in Southwest England (Borman et al., 2009). Almost one-third of 90 animals, most of which had been killed by road traffic or by predators, had evidence of infection determined by microscopic evaluation of lung specimens in histopathologic sections and KOH preps. Attempts to culture *E. crescens* from infected lungs were unsuccessful, but it was identified by PCR and sequencing of rDNA from adiaspores dissected from animal lung tissue. Lesions were evident in the lungs of 19 of 55 otters, and adiaspores were also found in specimens from weasels, stoats, a red fox, pine marten, mole, mice, and a rat.

10A.12 THERAPY

Human pulmonary adiaspiromycosis is self-limited because the adiaspores do not replicate. Accordingly, antifungal therapy usually is unnecessary. Mortality, however, can be high in disseminated forms of the disease: 3 fatal cases occurred in a series of 9 reviewed in Barbas Filho et al. (1990). Therapy with ketoconazole, FLC, and AmB was successful in severe or progressive infection in immunocompromised patients. A 2-yr-old girl with bilateral pulmonary adiaspiromycosis was successfully treated with AmB (Nuorva et al., 1997). When disease is relentless in the face of antifungal therapy partial surgical resection has been applied (Wellinghausen et al., 2003).

10A.13 LABORATORY DETECTION, RECOVERY, AND IDENTIFICATION

The two species, *E. crescens* and *E. parva*, are separated by their maximum growth temperatures and the different sizes of their adiaspores. Most human cases are caused by *E. crescens*, which has a maximum growth temperature of 37°C and larger adiaspores, often ≥100 μm diameter. *Emmonsia parva* cultures switched to 40°C produce adiaspores at that temperature.

Direct Examination

Adiaspores may be recovered from bronchoalveolar lavage fluid and dissected from biopsy specimens, usually from lung.

Histopathology

In tissue sections only adiaspores are seen, similar to but larger than adiaspores produced in vitro at 37–40°C (Fig. 10A.1). Adiaspores are visualized by staining with PAS, H&E, or GMS stains. Other histochemical stains occasionally have been used: mucicarmine, picrosirius, or Congo red, including polarized light microscopy (dos Santos et al., 2000).

In human lung tissue, adiaspores may reach 700 μm diameter in size. Their interior usually is empty but may

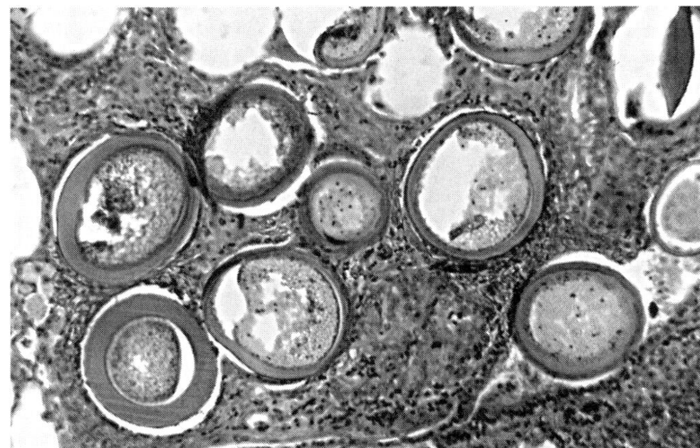

Figure 10A.1 Histopathology of adiaspiromycosis. Mouse lung showing thick-walled and empty adiaspores. The etiologic agent in humans is *Emmonsia crescens*.
Source: Used with permission from Prof. Lynne Sigler, University of Alberta, Edmonton, Canada. This image appeared in Sigler (2006, Fig 38.4a) and is used with permission from Wiley-Blackwell Publishers.

contain small eosinophilic globules along the inner surface of the walls. The empty nature of adiaspores seen in histopathologic sections is probably an artifact of fixation, embedment, and sectioning. In fact, adiaspores of *E. crescens* are multinucleate, resting structures that are capable of germinating to produce hyphal growth at temperatures lower than 37°C.

The adiaspore wall is divided into three layers. In contrast to spherules of *C. immitis*, the adiaspores of *E. crescens* are much larger, have a thicker wall, appear empty, and do not contain endospores. No other fungal pathogen is known to contain walls as thick as adiaspores of *E. crescens* (Pfaller and Diekema, 2005). Each adiaspore is surrounded by epithelioid and giant cell granulomas, which are themselves enveloped in a dense, fibrous capsule. All the granulomas are at a similar stage of development, because after exposure and inhalation, there is no in vivo replication. The differential in the histologic diagnosis of adiaspiromycosis is: spherule of *Coccidioides* species; myospherulosis (please see definition in Chapter 5); collapsed roundworm cysts (as indicated in the Section 10A.3, Case Summaries); and aspiration lentil pneumonia including starch granules.

Culture

Adiaspore-containing biopsy material (Sigler, 1996) may be minced and plated to PDA or phytone yeast extract agar, and will produce the septate mold form at temperatures below 37°C.

Colony Morphology

Two groups are known for *E. Crescens*: *group I* —faster-growing yellowish-white to orange-white, with a dense woolly center, broad smooth margin, and a gray-brown reverse; and *group II*: growth is coarse, powdery with pale orange to gray-orange aerial mycelium over a reddish-gray surface with an irregular margin and a reddish brown reverse. *Emmonsia parva* isolates resemble *group I* of *E. crescens* but grow more slowly. A group of granular *E. parva* strains resemble *group II*. Colony diam, after 21 days, ranges from 48 to 82 mm for *E. crescens* and from 36 to 85 mm for *E. parva*.

Microscopic Morphology

The mold form (Fig. 10A.2) of the two species grown at room temperature consists of indistinguishable, hyaline, septate hyphae that develop short conidiophores, each bearing a terminal conidium (aleurioconidium). The swollen tip of the conidiophore may bear one to three spine-like pegs, each forming a solitary conidium. Conidia are subglobose, ovoid, or pyriform measuring 2.5 μm × 3.5 μm (Sigler, 1996). The wall is smooth or finely roughened.

Adiaspore Form At 37°C adiaspores are produced by *E. crescens* and may grow to 70 μm in vitro and to 700 μm in vivo. *Emmonsia parva* grows in the hyphal form at 37°C and only produces adiaspores at 40°C that are ≤25 μm in vitro and 40 μm in vivo. Adiaspores are produced on phytone–yeast extract agar. Blood supplementation does not enhance adiaspore production.

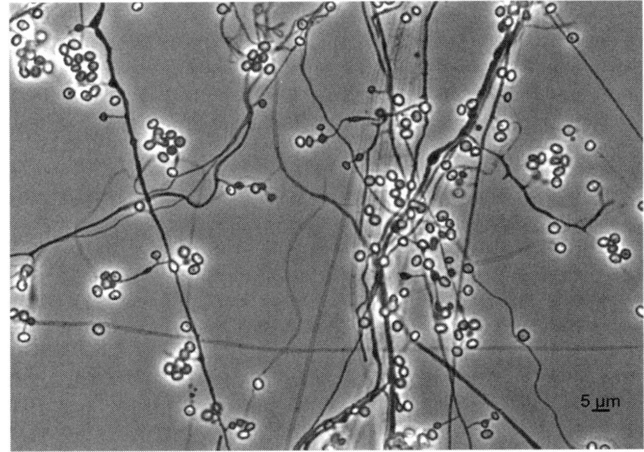

Figure 10A.2 *Emmonsia crescens* mold form, microscopic morphology.
Source: Used with permission from Prof. Lynne Sigler, University of Alberta, Edmonton, Canada. Identifier: UAMH 125 sc 14d 25C.

Genetic Identification

PCR-sequencing can be a successful alternative diagnostic modality when cultures fail to grow (please see Section 10A.3, Case 2). Even when there is growth, identification of this unusual fungus is not simple. PCR-sequencing was used in a case of atypical bilateral granulomatous pneumonia, where a cytocentrifuge smear of lung wash fluid viewed microscopically revealed macrophage-ingested, scarcely budding, fungal elements (Wellinghausen et al., 2003). A mold grew from a transbronchial brush biopsy specimen and was identified as *E. crescens* by PCR using primers directed at a 491 bp segment of 18S rDNA. The sequence obtained was 100% homologous to that from *Ajellomyces crescens* in GenBank.

SELECTED REFERENCES FOR ADIASPIROMYCOSIS

ALBASSAM MA, BHATNAGAR R, LILLIE LE, ROY L, 1986. Adiaspiromycosis in striped skunks in Alberta, Canada. *J Wildlife Dis* 22: 13–18.

BARBAS FILHO JV, AMATO MB, DEHEINZEILIN D, SALDIVA PH, DE CARVALHO CR, 1990. Respiratory failure caused by adiaspiromycosis. *Chest* 97: 1171–1175.

BORMAN AM, SIMPSON VR, PALMER MD, LINTON CJ, JOHNSON EM, 2009. Adiaspiromycosis due to *Emmonsia crescens* is widespread in native British mammals. *Mycopathologia* 168: 153–163.

DENSON JL, KEEN CE, FROESCHLE PO, TOY EW, BORMAN AM, 2009. Adiaspiromycosis mimicking widespread malignancy in a patient with pulmonary adenocarcinoma. *J Clin Pathol* 62: 837–839.

DOS SANTOS VM, DOS REIS MA, ADAS SJ, SALDANNA JC, TEIXEIRA VP, 2000. Contribution to the morphologic diagnosis of lung adiaspiromycosis. *Rev Soc Bras Med Trop* 33: 493–497.

DOT JM, DEBOURGOGNE A, CHAMPIGNEULLE J, SALLES Y, BRIZION M, PUYHARDY JM, COLLOMB J, PLÉNAT F, MACHOUART M, 2009. Molecular diagnosis of disseminated adiaspiromycosis due to *Emmonsia crescens*. *J Clin Microbiol* 47: 1269–1273.

JELLISON WL, 1950. Haplomycosis in Montana rabbits, rodents, and carnivores. *Public Health Rep* 65: 1057–1063.

NUORVA K, PITKÄNEN R, ISSAKAINEN J, HUTTUNEN NP, JUHOLA M, 1997. Pulmonary adiaspiromycosis in a two year old girl. *Clin Pathol* 50: 82–85.

PETERSON SW, SIGLER L, 1998. Molecular genetic variation in *Emmonsia crescens* and *E. parva*, etiologic agents of adiaspiromycosis, and their phylogenetic relationship to *Blastomyces dermatitidis* (*Ajellomyces dermatitidis*) and other systemic fungal pathogens. *J Clin Microbiol* 36: 2918–2925.

PFALLER AA, DIEKEMA DJ, 2005. Unusual fungal and pseudofungal infections of humans. *J Clin Microbiol* 43: 1495–1504.

SIGLER L, 2006. Adiaspiromycosis and other infections caused by *Emmonsia* species, pp. 809–824, *in*: MERZ WG, HAY RJ (eds.), *Topley and Wilson's Microbiology and Microbial Infections*, 10th ed., *Volume* 5, *Medical Mycology*. Wiley-Blackwell Publishers, London.

SIGLER L, 1996. *Ajellomyces crescens* sp. nov., taxonomy of *Emmonsia* species, and relatedness with *Blastomyces dermatitidis* (teleomorph *Ajellomyces dermatitidis*). *J Med Vet Mycol* 34: 303–314.

STEBBINS WG, KRISHTUL A, BOTTONE EJ, COHEN S, 2004. Cutaneous adiaspiromycosis: A distinct dermatologic entity associated with *Chrysosporium* species. *Am Acad Dermatol* 51(Suppl 5): S185–S189.

TURNER D, BURKE M, BASHE E, BLINDER S, YUST I, 1999. Pulmonary adiaspiromycosis in a patient with acquired immunodeficiency syndrome. *Eur J Clin Microbiol Infect Dis* 18: 893–895.

WATTS JC, CHANDLER FW, 1990. Adiaspiromycosis. An uncommon disease caused by an unusual pathogen. *Chest* 97: 1030–1031.

WELLINGHAUSEN N, KERN WV, HAASE G, ROZDZINSKI E, KERN P, MARRE R, ESSIG A, HETZEL J, HETZEL M, 2003. Chronic granulomatous lung infection caused by the dimorphic fungus *Emmonsia* sp. *Int J Med Microbiol* 293: 441–445.

QUESTIONS

The Answer Key to multiple choice questions may be found at the end of the book.

1. What type of mycosis is adiaspiromycosis ?
   A. A disease of burrowing rodents and rarely of humans
   B. A systemic mycosis caused by a dimorphic environmental mold
   C. The tissue form demonstrates isotropic growth
   D. Usually self-limited, not requiring antifungal therapy
   E. All of the above
2. What definition best fits the description of an adiaspore?
   A. Conidium that enlarges but does not replicate
   B. Spherule
   C. Sporangium
   D. The sexual stage of *Emmonsia crescens*
3. What is known about culturing the causative agent of adiaspiromycosis?
   A. Growth occurs on PDA at 30°C, producing conidia.
   B. Growth occurs on PDA but without sporulation.
   C. Growth occurs on SDA with production of ballistoconidia.
   D. The causative agent cannot be cultured in the laboratory.
4. Which of the following is the closest relative to *Emmonsia crescens*?
   A. *Blastomyces dermatitidis*
   B. *Coccidioides* species
   C. *Pneumocystis jirovecii*
   D. *Rhinosporidium seeberi*
5. The number of reported cases of adiaspiromycosis in the world literature are
   A. Fewer than 100
   B. 100–1000
   C. 1000–10,000
   D. More than 10,000
6. Which of the following best describes the histologic appearance of adiaspores?
   A. Bounded by a very thin cell wall
   B. Phagocytosed by multinucleate giant cells
   C. Surrounded by a thick fibrous granuloma
   D. Variation in sizes is due to their different stages of cell division

7. Describe the geographic distribution and risk factors for adiaspiromycosis. Explain the pathogenesis of the disease and whether timely therapy is an important consideration.

8. Given that cases may arise when abnormal chest radiographs are found in asymptomatic patients, what other pathogens and nonliving foreign substances may masquerade as adiaspores?

9. How are animal and human cases apportioned between the two species *Emmonsia parva* and *Emmonsia crescens*?

10. Given that the interior of adiaspores appears empty, is it possible to use PCR-sequencing to establish their identity?

Chapter 10B

Less Frequent Mycoses Caused by Dimorphic Environmental Molds (Endemic Mycoses): Lobomycosis (Jorge Lôbo's Disease)

10B.1 LOBOMYCOSIS-AT-A-GLANCE

- *Introduction/Disease Definition.* Lobomycosis is a chronic cutaneous mycosis with an endemic region restricted to persons living in the Amazon rain forest ecosystem.
- *Etiologic Agent.* The disease is caused by *Lacazia loboi*, a noncultivatable fungus known only to grow in the yeast-like form. Phylogenetic analysis places it in the *Ajellomycetaceae*, in a clade with *Paracoccidioides brasiliensis. Lacazia loboi* also is related to other dimorphic, endemic, environmental molds.
- *Geographic Distribution.* The countries in the Amazon rain forest ecosystem.
- *Ecologic Niche. Lacazia loboi* has not been isolated from nature, but its occurrence as a pathogen of marine mammals is suggestive of an aquatic habitat.
- *Epidemiology.* The highest number of cases occur in inhabitants of the Brazilian Amazon River basin in persons who live near the border of rivers, creeks, or other aquatic environments.
- *Risk Groups/Factors.* Agricultural workers who are exposed to both soil and vegetation in the tropical rain forest environment are at risk. Exceptionally, aquarium workers exposed to *Lacazia*-infected dolphins also are at risk.
- *Transmission.* Introduction of the fungus during a traumatic lesion such as snake bite, insect bite, ray poisoning, or a wound from splinters.
- *Determinants of Pathogenicity.* Abnormal immunoregulation in the host may be a risk factor or a consequence of this mycosis.
- *Clinical Forms.* After a scrape, cut, or superficial wound, the lesion begins as an indurated papule on the ear lobes, buttocks, lumbosacral and scapular areas, or extremities. Very slowly, over months or years, keloidal, gummatous, verrucous, or ulcerative lesions evolve. The net result may be extreme disfigurement.
- *Veterinary Forms.* Cutaneous lesions on the Atlantic bottlenose dolphin. Lobomycosis is an emerging epizootic disease in some dolphin populations.
- *Therapy.* Surgical excision of small lesions. Reports of successful antifungal therapy are exceptional.
- *Laboratory Detection, Recovery, and Identification Lacazia loboi* is noncultivatable, so diagnosis is based on histopathology of lesional biopsy specimens. Viewed with GMS stain, there are clusters of yeasts in small chains linked by thin connecting tubules.

Fundamental Medical Mycology, First Edition. By Errol Reiss, H. Jean Shadomy and G. Marshall Lyon, III.
© 2012 Wiley-Blackwell. Published 2012 by John Wiley & Sons, Inc.

10B.2 INTRODUCTION/DISEASE DEFINITION

Lobomycosis is a chronic cutaneous mycosis caused by the noncultivatable fungus *Lacazia loboi*. Persons at risk are those living in the major endemic area, the Amazon rain forest, near river banks, or other aquatic environments. Beginning with a scrape or a puncture, a wart-like lesion evolves slowly over months or years. The most frequently affected anatomic sites are on the ear lobes, buttocks, elbows, or legs. The lesions are granulomatous and keloid-like, nodular, or, at times, crusted plaques. They are restricted to the epidermis and dermis and are asymptomatic. Older lesions become verrucous and ulcerative. Satellite lesions may be spread by autoinoculation, rarely by draining lymphatics.

Penetration to internal organs or mucous membranes is exceptional, and the disease is not fatal. Over years, or even decades, in persons without access to primary care, large areas of disfiguring skin tumors may develop. Surgical removal is the only certain cure and applies when lesions are small. Lesions covering large areas may not be amenable to surgery and, thus far, reports of successful response to antifungal therapy are exceptional. Lobomycosis also is a disease of dolphins, underlining the importance of the aquatic habitat as a source of the fungus.

Other names for lobomycosis are Lacaziosis, Jorge Lôbo's disease, and keloidal blastomycosis.

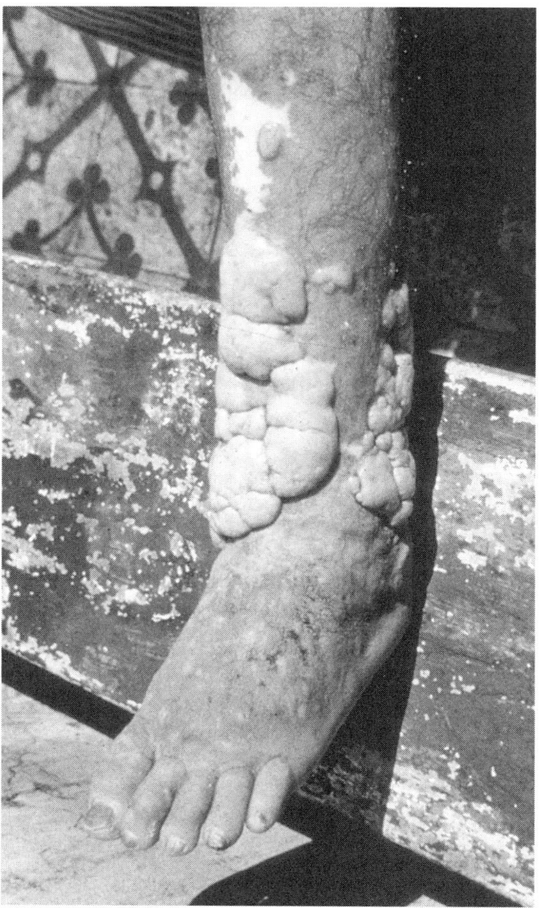

Figure 10B.1 B.1 Lobomycosis: keloid-like lesions on leg. *Source:* Used with permission from Dr. Arvind A. Padhye, CDC.

10B.3 CASE PRESENTATIONS

Case Presentation 1. Multiple Lesions on the Right Leg of a 93-yr-old Rubber Collector from the Brazilian Amazon (Talhari et al., 2009)

History

A 93-yr-old man attended the Dermatology Clinic in Manaus, Brazil, and presented with multiple cutaneous lesions on his right lower leg, which evolved insidiously over a period of 45 years. The patient related that he worked as a rubber collector for many decades in the Brazilian Amazon region.

Physical Examination

Multiple, firm, keloid-like lesions with a confluent tendency were present on the lower right leg. Ulcerated lesions were also present. No lymphadenopathy was detected (Fig.10B.1).

Laboratory

The surface of a keloid-like lesion was gently scraped with a sterile scalpel blade. The scrapings were placed on a microscope slide and coverslipped. No KOH was added. High power (400×) microscopic examination revealed round cells singly and in chains with thick, refractile walls. Small budding yeast were seen with narrow tubules between the mother and daughter cells. A biopsied specimen revealed numerous, diffuse inflammatory granulomas in the dermis containing histiocytes and giant cells with intracellular yeast. The GMS stain confirmed the presence of yeast cells consistent with *L. loboi* (Fig. 10B.2).

Diagnosis

The occupation, place of residence of the patient, clinical features, cytology, and histopathology were consistent with lobomycosis caused by *L. loboi*.

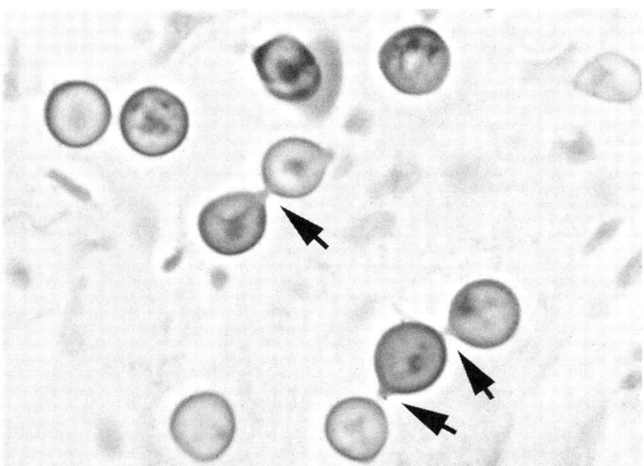

Figure 10B.2 B.2 Histopathology of lobomycosis, skin biopsy, GMS stain. Thin tubules connect yeast cells (arrows).
Source: H. J. Shadomy.

Therapy

As the lesions were too extensive for surgical excision, the patient received suppressive therapy with ITC, 300 mg/day.

Comment

The slow but inexorable progress of the lesions over many decades is instructive that the disease does not invade beyond the dermis. Exfoliative cytology may be useful in the diagnosis of lobomycosis. In this case it was possible to make a presumptive diagnosis without stains or adding KOH, but confirmatory histopathology remains the gold standard. Antifungal chemotherapy for lobomycosis has, with rare exception, been unsuccessful (please see Section 10B.13, Therapy).

Case Presentation 2. Nodules in the Arm of a 66-yr-old Rubber Worker in the Amazon Region (Chrusciak-Talhari et al., 2007)

History

A healthy 66-yr-old Brazilian man presented with a 10 year history of an asymptomatic mass of nodules affecting his left arm. The lesion first appeared as a papule after the patient recalled pulling a tick from the site. For many years he had been a rubber worker in the Amazon region.

Physical Examination

Four ill-defined, smooth, shiny, elastic nodules were present on the medial aspect of the right arm.

Laboratory Findings

A skin biopsy specimen was obtained and stained with H&E and GMS (Fig.10B.2). The lesional biopsy specimen revealed numerous diffuse inflammatory granulomas in the dermis containing histiocytes and giant cells with numerous intracellular thick-walled cells. An asteroid body also was seen, as it is in other mycoses. The GMS stain showed the typical yeasts of lobomycosis in chains of uniform round to oval cells.

Diagnosis

The diagnosis was lobomycosis caused by *Lacazia loboi*.

Therapy and Clinical Course

The nodules were completely excised, and the patient received oral ITC, 200 mg/day, for 6 months in an attempt to prevent a relapse.

Comment

The extremely slow evolution of lesions and the lack of spread to internal organs are characteristics of lobomycosis. The patient's residence in the Amazonian region underlines that river basin as the major endemic area for lobomycosis. Asteroid bodies also may be seen in lobomycosis. Optimal treatment for localized lesions is wide surgical excision, with sufficient margins free of infection to avoid recurrence. There are only isolated reports of clinical response to antifungal chemotherapy (please see Section 10B.13, Therapy).

Case Summary. Pustule on the Chest Wall of a 42-yr-old Man Living in the United States (Burns et al., 2000)

Seven years before seeking a surgical consult in the State of Georgia in the United States, the patient related he had traveled to Venezuela. On that occasion he recalled walking under the Canaima Angel Falls, several times exposing himself to high water pressures. A small pustule developed on the skin of his right chest wall surrounded by an erythematous area, and having a keloidal appearance. He requested removal of the lesion for cosmetic reasons. There was no recurrence. Histopathologic review of the lesional biopsy demonstrated histiocytes, multinucleated giant cells, and typical globose cells connected by narrow tubes. When viewed using GMS stain their appearance was characteristic of *Lacazia loboi*.

10B.4 DIAGNOSIS

Diagnosis is made by visualization of the fungus in lesional skin biopsy specimens using fungal stains (i.e., GMS stain). The differential diagnosis includes anergic cutaneous leishmaniasis, chromoblastomycosis, dermatofibrosarcoma protuberans, fibroma, Kaposi's sarcoma, keloids, leprosy, neurofibromas, paracoccidioidomycosis, and metastatic lesions.

10B.5 ETIOLOGIC AGENT

Lacazia loboi Taborda P, Taborda V, McGinnis MR, *J Clin Microbiol* 37: 2031 (1999). The position in classification of this noncultivatable fungus was determined when biopsied tissues containing *L. loboi* from human lobomycosis patients were used as a template for PCR amplification of the 18S rDNA and the *CHS2* gene (Herr et al., 2001). *Lacazia loboi* sequences form a sister clade with the dimorphic pathogen *Paracoccidioides brasiliensis* within the *Ajellomycetaceae* in the order *Onygenales*. This order also includes *Ajellomyces dermatitidis* (= *Blastomyces dermatitidis*), *Ajellomyces capsulatus* (= *Histoplasma capsulatum*), and *Ajellomyces crescens* (= *Emmonsia crescens*). Based on this relationship, *L. loboi* is probably dimorphic, presenting a yeast form in tissue, with an as yet undiscovered mold form in nature (Mendoza et al., 2001).

10B.6 GEOGRAPHIC DISTRIBUTION/ECOLOGIC NICHE

Lobomycosis has an endemic area in the Amazon rain forest in South America. Cases occur in Brazil, Bolivia, Colombia, Ecuador, French Guiana, Guyana, Honduras, Peru, Surinam, and Venezuela. The disease is encountered in tropical, humid, or subtropical forested areas at elevations above 200 meters and with >200 cm of annual rainfall.

The occurrence of the disease in the Atlantic bottlenose dolphin, *Tursiops truncatus*, widens the endemic area. Cases have occurred in Florida, on the Texas coast, the Bay of Biscay off the Spanish–French Atlantic coast, and the south Brazilian coast. The ecologic niche of *Lacazia loboi* is unknown because it is noncultivatable. The various hypotheses include that it is an aquatic species, or that it is an obligate intracellular pathogen of an unidentified lower animal.

10B.7 EPIDEMIOLOGY

The endemic region for human lobomycosis ranges from Central America south to Brazil, with the highest number of cases in inhabitants of the Brazilian Amazon River basin. Most infections occur in persons who live near the border of rivers, creeks, or other aquatic environments. Lobomycosis is uncommon but there are localities of high endemicity. It is the most common subcutaneous mycosis in Manaus, population 1 million, midway from the coast along the Amazon River (Talhari et al., 1988).

Another local focus of lobomycosis is among Amer-Indians living in the Orinoco and Amazon basins on the Colombian–Venezuelan border. The prevalence of lobomycosis is 8.5% among the Amoruas tribe from the Casanare region near the Orinoco River (Rodríguez-Toro and Tellez, 1992). Lobomycosis also is diagnosed in the United States and in Europe, in patients with a history of travel to the endemic countries.

10B.8 RISK GROUPS/FACTORS

There is a clear preponderance of cases in men (90% of reported cases). Gender susceptibility, however, is directly related to occupation. The Cayabi people in the Brazilian Amazon have the highest prevalence of lobomycosis. Their adult women are affected because they engage in agricultural work (reviewed by Paniz-Mondolfi et al., 2007). Patients are typically residents of rural areas, often agricultural workers who are exposed to both soil and vegetation in the tropical rain forest environment.

10B.9 TRANSMISSION

The mode of transmission is likely through trauma. Patients report suffering a previous traumatic lesion such as snake bite, insect bite, or wounds from splinters. An aquatic habitat for *L. loboi* is another reasonable speculation on grounds that it is a pathogen of dolphins and because humans relate a history of visiting aquatic habitats. At least one case of dolphin-to-human transmission has occurred: a European aquarium worker in France became infected after handling a dolphin with *Lacazia* skin lesions (Symmers, 1983).

Autoexperiment

A laboratory technician volunteered to be inoculated in the knee with dermal extracts of a human case of lobomycosis

(Borelli, 1961). Progress of the lesion was monitored. It was 2 mm in diameter after 1 month. After 15 months the lesion was 1 cm in diameter. After 4 years the lesion measured 33 mm in diameter with a 4 mm satellite lesion.

10B.10 DETERMINANTS OF PATHOGENICITY

10B.10.1 Host Factors

TGF-β[1] is a regulatory cytokine which, when overexpressed, can promote fibrosis and also may exert a suppressive effect on macrophage function. TGF-β1 was localized in formalin-fixed skin biopsies from patients with lobomycosis using immunohistochemical staining for TGF-β1 and for the macrophage marker CD68 (Xavier et al., 2008). Intense immunolabeling for TGF-β was observed in cases of lobomycosis and may explain in part the weak reactivity to CD68, as TGF-β is able to inhibit macrophage activity, reducing microbicidal efficacy. The increased collagen observed in lesions from lobomycosis may relate to the ability of TGF-β to induce fibrosis, contributing to the keloid appearance of the lesions.

10B.10.2 Microbial Factors

These are unknown.

10B.11 CLINICAL FORMS

The clinical appearance varies among patients but usually begins as an indurated papule, following a scrape, cut, or penetrating but superficial wound (please see Section 10B.9, Transmission). Anatomic sites affected are ear lobes, buttocks, lumbosacral and scapular areas, elbows, and legs. Variably sized hard cutaneous plaques and multiple or solitary nodules evolve very slowly. As the papule matures new lesions may develop either adjacent to, or distant from, the initial papule. Depending on the appearance of the lesions different forms are described:

- Keloidal (Fig. 10B.1)
- Initial papules or nodules
- Gummatous
- Verrucous (seen only in very chronic lesions)
- Ulcerative (uncommon)

Over many years the skin tumors may reach a large size and a hard consistency (= keloidal). They may become widespread either as a result of autoinoculation or, according to one theory, via hematogenous spread. The larger lobulated lesions have well-defined borders. *Lacazia loboi* is always localized in the cutaneous and subcutaneous tissues with no attachment to underlying tissue or bone. The tendency to be restricted to the epidermis and dermis has been attributed to the idea that the fungus may not grow well or at all at 37°C. No systemic symptoms occur. Spread to internal organs or mucosa is exceptional. There have been no deaths reported from lobomycosis in humans.

10B.12 VETERINARY FORMS

Lobomycosis occurs in the Atlantic bottlenose dolphin, *Tursiops truncatus*, in the western and eastern Gulf of Mexico, the Surinam River, the Brazilian coast, and the Bay of Biscay in Europe. This underlines the importance of an aquatic reservoir for the fungus. Some have suggested that *L. loboi* is an obligate intracellular pathogen of some lower animal species, as yet undiscovered (Lupi et al., 2005).

Between 2003 and 2004, Atlantic bottlenose dolphins were captured, 75 in the Indian River Lagoon, Florida, and 71 in estuarine waters near Charleston, South Carolina (Reif et al., 2006). Each animal was photographed, lesions were biopsied, and the dolphins then were released. Nine of 30 (30%) dolphins captured in the southern portion of Florida's Indian River Lagoon had lesions which, when biopsied and stained with GMS, were diagnosed as lobomycosis. This disease may be occurring in epidemic proportions among dolphins in the Indian River Lagoon. The southern portion of the lagoon is characterized by freshwater intrusion and lower salinity, and may be an environmental stressor, but specific factors are unknown. A *Lacazia*-like disease was observed in a pod of dolphins living in the tropical lagoon of Mayotte between Mozambique and Madagascar (Kiszka et al., 2009). *Dolphins are a sentinel species for this uncommon but possibly emerging mycosis*.

[1]TGF-ß. *Definition*: This regulatory cytokine can act as a switch to regulate inflammation (Wahl,1994). Released locally from platelets early in an inflammatory response, TGF-ß recruits and activates leukocytes. Later, it also is generated by inflammatory cells and helps reduce inflammation and promote tissue repair through fibroblast recruitment and matrix synthesis. Excess TGF-ß within a lesion, however, is associated with unresolved inflammation and promotion of fibrosis. Moreover, TGF-ß inhibits macrophage ability to generate nitric oxide (NO), a pivotal mechanism by which macrophages inhibit a variety of pathogens.

10B.13 THERAPY

Surgical excision of infected tissue is the only proven means of achieving clinical and mycologic cure, and then only when the lesions remain localized and are amenable to wide surgical excision. Relapse is not uncommon. A single case report of a 45-yr-old man with keloid lesions of lobomycosis on his face, arms, and legs, of 32 years duration, was successfully treated with combination therapy consisting of clofazimine (100 mg/day) and ITC (100 mg/day) (Fischer et al., 2002). Complete clinical and histopathologic remission of the disease was achieved persisting into 3 years of follow-up exams. Clofazimine (Lamprene®, Novartis) is used to treat leprosy and nontuberculous mycobacterial infections. The drug has serious abdominal adverse effects in some patients as described in the prescribing information available from Novartis. Currently, clofazimine is only available from the U.S. FDA as an IND (Investigational New Drug).

10B.14 LABORATORY DETECTION, RECOVERY, AND IDENTIFICATION

Direct Examination

Scrapings from a lesion may be examined in a KOH prep as described in Chapter 2.

Vinyl Adhesive Tape Mount

This type of mount can be used advantageously because a feature of lobomycosis, like some other mycoses, is transepidermal elimination of infectious organisms.

Procedure (Miranda and Silva, 2005): Wearing latex gloves, apply a 19 mm width strip of clear vinyl adhesive tape to the lesion. Exert gentle pressure to ensure close contact of scales to the adhesive. Remove the tape from the lesion and place it, glue face down, on a glass slide containing a drop of a 10% KOH in 40% DMSO/water, vol/vol. After 5 min, when the scales have cleared, gently press the slide, tape facing down, against blotting paper to eliminate excess solution. View the mount in the microscope at 100× to 400× magnification for the presence of yeast forms, singly, some budding, with short tubules connecting mother and daughter cells. Small chains of yeast cells also may be seen.

Histopathology

Lesional biopsy material is stained with PAS, H&E, or GMS. H&E stains may not demonstrate cell walls sufficiently, the PAS stain is excellent for the cell and cell wall, and GMS is best for demonstrating the intercellular connecting tubules. Viewed microscopically in stained lesional biopsy sections, *L. loboi* consists of masses of thick-walled spherical or lemon-shaped yeast cells. They occur singly and in branched or unbranched chains of three or more cells connected by short tubules, or necks, similar to beads on a string. The yeast are 5–12 μm in diameter and are readily detected with most fungal stains. The differential diagnosis includes *P. brasiliensis*. It is noteworthy that the multipolar budding cells of *P. brasiliensis* also are joined by short tubules to the mother cell.

Tissue Response

Lesions in the dermis display a grenz zone,[2] dilated and newly formed vessels, fibrosis, and a dense granulomatous infiltrate consisting of lymphocytes, foamy histiocytes rich in intracellular yeasts, epithelioid cells, numerous Langhans giant cells, some eosinophils, and plasma cells (Xavier et al., 2008). The affected epidermis often has the appearance of pseudoepitheliomatous hyperplasia.

Culture

All attempts to isolate this pathogen in culture have failed. *Lacazia loboi* can be propagated by passaging the yeast cells obtained from a confirmed case of human lobomycosis in the footpads of BALB/c mice (Vilani-Moreno et al., 2003).

SELECTED REFERENCES FOR LOBOMYCOSIS

Borelli D, 1961. Lobomicosis experimental. *Dermatol Venez* 3: 72–82.

Burns RA, Roy JS, Woods C, Padhye AA, Warnock DW, 2000. Report of the first human case of lobomycosis in the United States. *J Clin Microbiol* 38: 1283–1285.

Chrusciak-Talhari A, Talhari C, de Souza Santos MN, Ferreira LC, Talhari S, 2007. Nodular lesions on the arm. *Arch Dermatol* 143: 1323–1328.

Fischer M, Chrusciak Talhari A, Reinel D, Talhari S, 2002. Successful treatment with clofazimine and itraconazole in a 46 year old patient after 32 years duration of disease. *Hautarzt* 53: 677–681.

Herr RA, Tarcha EJ, Taborda PR, Taylor JW, Ajello L, Mendoza L, 2001. Phylogenetic analysis of *Lacazia loboi* places this previously uncharacterized pathogen within the dimorphic Onygenales. *J Clin Microbiol* 39: 309–314.

Kiszka J, Van Bressem MF, Pusineri C, 2009. Lobomycosis-like disease and other skin conditions in Indo-Pacific bottlenose dolphins *Tursiops aduncus* from the Indian Ocean. *Dis Aquat Organ* 84: 151–157.

[2]Grenz zone. *Definition*: Narrow area of uninvolved dermis between the epidermis and a dermal inflammatory or neoplastic infiltrate. *Grenz* is German for "border."

LUPI O, TYRING SK, MCGINNIS MR, 2005. Tropical dermatology: Fungal tropical diseases. *J Am Acad Dermatol* 53: 931–995 (quiz on pp. 952–954).

MENDOZA L, AJELLO L, TAYLOR JW, 2001. The taxonomic status of *Lacazia loboi* and *Rhinosporidium seeberi* has been finally resolved with the use of molecular tools. *Rev Iberoam Micol* 18: 95–98.

MIRANDA MF, SILVA AJ, 2005. Vinyl adhesive tape also effective for direct microscopy diagnosis of chromomycosis, lobomycosis, and paracoccidioidomycosis. *Diagn Microbiol Infect Dis* 52: 39–43.

PANIZ-MONDOLFI AE, REYES JAIMES O, DÁVILA JONES L, 2007. Lobomycosis in Venezuela. *Int J Dermatol* 46: 180–185.

REIF JS, MAZZOIL MS, MCCULLOCH SD, VARELA RA, GOLDSTEIN JD, FAIR PA, BOSSART GD, 2006. Lobomycosis in Atlantic bottlenose dolphins from the Indian River Lagoon, Florida. *J Am Vet Med Assoc* 228: 104–108.

RODRÍGUEZ-TORO G, TELLEZ N, 1992. Lobomycosis in Colombian Amer Indian patients. *Mycopathologia* 120: 5–9.

SYMMERS WS, 1983. A possible case of Lôbo's disease acquired in Europe from a bottle-nosed dolphin (*Tursiops truncatus*). *Bull Soc Pathol Exot Filiales* 76: 777–784.

TALHARI C, CHRUSCIAK-TALHARI A, DE SOUZA JV, ARAÚJO JR, TALHARI S, 2009. Exfoliative cytology as a rapid diagnostic tool for lobomycosis. *Mycoses* 52: 187–189.

TALHARI S, CUNHA MG, SCHETTINI AP, TALHARI AC, 1988. Deep mycoses in Amazon region. *Int J Dermatol* 27: 481–484.

VILANI-MORENO FR, BELONE ADE F, ROSA PS, MADEIRA S, OPROMOLLA DV 2003 Evaluation of the vital staining method for *Lacazia loboi* through the experimental inoculation of BALB/c mice. Med Mycol 41: 211–216.

WAHL SM 1994 Transforming growth factor beta: the good, the bad, and the ugly. J Exp Med 180: 1587–1590.

XAVIER MB, LIBONATI RM, UNGER D, OLIVEIRA C, CORBETT CE, DE BRITO AC, QUARESMA JA 2008 Macrophage and TGF-beta immunohistochemical expression in Jorge Lobo's disease. Hum Pathol 39: 269–274.

QUESTIONS

The Answer Key to multiple choice questions may be found at the end of the book.

1. What is the definition of lobomycosis?
   A. Chronic cutaneous mycosis endemic to persons living in the South American rain forest ecosystem.
   B. Following a puncture wound with a thorn or splinter, a chain of nodules develops along the path of lymphatic drainage from the primary lesion.
   C. Inhalation of conidia results in a pulmonary disease, mostly affecting burrowing rodents and, rarely, humans.
   D. Slow, relentless subcutaneous mycosis, most often affecting the foot and leg accompanied by an indurated swelling, draining sinus tracts, discharging grains.

2. The lesions of lobomycosis are
   A. Circular lesions on the smooth skin with active borders, scaling, and pruritus.
   B. Granulomatous, keloid-like, nodular or crusted plaques in the epidermis and dermis. Older lesions become verrucous and ulcerative.
   C. Initially a small subcutaneous, painless swelling; later, fibrosis occurs, with multiple nodules connected by sinus tracts lined by red granulation tissue.
   D. Primary pulmonary infection often without symptoms for months or years, then mucosal ulcers appear in the upper respiratory and digestive tracts, mostly in the mouth and nose.

3. Histopathologic examination of biopsied lesion tissue from a patient with lobomycosis will reveal
   A. Large, up to 100 μm diameter, thick-walled nonbudding cells surrounded by a fibrous capsule.
   B. Monopolar budding yeast in a "footprint shape" with a distinctive scar and irregular fragments of hyphae.
   C. Multipolar budding yeast.
   D. Thick-walled round or lemon-shaped yeast cells, singly and in branched or unbranched chains of three or more cells connected by short tubules.

4. What role does a vinyl adhesive tape mount have in the diagnosis of lobomycosis?
   A. No role because only a lesional biopsy can reveal the fungus in tissue.
   B. Appropriate, because in this mycosis there is transepidermal elimination of infectious organisms.
   C. Can be done if the fibrous capsule of the lesion is first scraped with a scalpel.
   D. Possibly if the local area is first anesthetized because the lesions are painful.

5. Biopsied lesion tissue from a patient with lobomycosis, if minced and planted on medium, will
   A. Grow as a yeast form at 37°C, only if 5% defibrinated sheep blood is added to BHI medium.
   B. Grow as the mold form on SDA with cycloheximide.
   C. Grow in biphasic blood culture bottles.
   D. Not yield growth because *Lacazia* cannot be cultivated.

6. The endemic regions for lobomycosis include
   1. Brazilian Amazon River basin
   2. Casanare region near the Orinoco River
   3. Eastern coast of Kyushu, Japan
   4. Northeast Thailand near Chiang Mai

   Which are correct?
   A. 1 + 2
   B. 1 + 3
   C. 1 + 4
   D. 2 + 3
   E. 2 + 4

7. Show on a map where outbreaks of lobomycosis have occurred in marine mammals. What species is most affected? What can the marine mammal disease tell us about the habitat and range of *Lacazia loboi*? Is it possible that *Lacazia* is a zoonosis?

8. Explain the approach to therapy of lobomycosis, describing the role for surgery and the record of success using systemic antifungal agents.

9. Describe the five different lesional types found in lobomycosis.

10. Discuss the position in classification of *Lacazia loboi*, indicating the nearest neighboring species.

Part Three

Systemic Mycoses Caused by Opportunistic Yeasts and *Pneumocystis*

Fundamental Medical Mycology, First Edition. By Errol Reiss, H. Jean Shadomy and G. Marshall Lyon, III.
© 2012 Wiley-Blackwell. Published 2012 by John Wiley & Sons, Inc.

Chapter 11

Candidiasis and Less Common Yeast Genera

11.1 CANDIDIASIS-AT-A-GLANCE

- *Introduction/Disease Definition.* Cutaneous, mucosal, systemic, or disseminated disease caused by yeasts of the genus *Candida*.
- *Etiologic Agents.* The major agents are *C. albicans, C. glabrata, C. tropicalis, C. parapsilosis*, and *C. krusei*.
- *Ecologic Niche.* Mucosae and skin are the ecologic niches of these *endogenous commensal* members of the normal human microbiota; some species survive in the environment.
- *Epidemiology. Candida* species are opportunistic pathogens and a leading cause of hospital bloodstream infections. They also cause oral-esophageal disease in AIDS patients.
- *Risk Factors. Candida* converts from a commensal to a pathogen in patients who are compromised by immunosuppression, cancer chemotherapy, intravascular catheters, cardiothoracic or gastrointestinal surgery, or HIV infection.
- *Transmission.* The route of transmission is via endogenous activation, or exogenously from fomites, or from person-to-person contact.
- *Determinants of Pathogenicity.* These include morphogenesis to the hyphal form; formation of biofilms, phenotype switching, expression of adhesins, and secretion of proteinases and/or phospholipases. Antifungal drug resistance can contribute to pathogenicity.
- *Clinical Forms.*
    - Cutaneous: skin and nails.
    - Mucosal: oral, vulvovaginal, gastrointestinal.
    - Invasive: candidemia, endocarditis, hepatosplenic, and multi-organ disseminated candidiasis.
- *Therapy.* Fluconazole (FLC), itraconazole (ITC), voriconazole (VRC), micafungin (MCF), caspofungin (CASF), or amphotericin B (AmB).
- *Laboratory Detection, Recovery, and Identification.* Culture, preferably from a sterile site; identification using morphologic, biochemical, or molecular genetic tests.

11.2 INTRODUCTION/DISEASE DEFINITION

Candidiasis includes a spectrum of disease from cutaneous, mucosal, systemic, or multisystem disseminated disease caused by the yeast *Candida albicans* or other *Candida* species. Disease may be acute or chronic. *Candida* species are *opportunistic* pathogens, causing disease primarily in debilitated or immunocompromised patients. The most common etiologic agent is *Candida albicans*, an endogenous commensal member of the mucosal microbiota of humans which may be isolated from 30–50% of healthy human beings.

Other species may be endogenous or exogenous in nature. The significance of any *Candida* species isolated ultimately rests with the physician, who will consider (i) the specimen type—Is it from a normally sterile site? and (ii) the number of specimens recovered from the same site having the same organism. *Candida* species are significant contributors to hospital-acquired bloodstream infections which, if not controlled, may

Fundamental Medical Mycology, First Edition. By Errol Reiss, H. Jean Shadomy and G. Marshall Lyon, III.
© 2012 Wiley-Blackwell. Published 2012 by John Wiley & Sons, Inc.

cause systemic or disseminated candidiasis. Mucosal candidiasis, while not life-threatening, is significant for (i) oropharyngeal and esophageal disease in AIDS patients and (ii) vulvovaginitis, which may be recurrent in a subset of women.

There are five *Candida* species of major medical importance: *C. albicans*, C. *glabrata, C. tropicalis, C. parapsilosis*, and *C. krusei. Candida albicans* remains the most clinically important. The relative ranking of the remaining four major species depends on the hospital and the geographic region. (Please see Section 11.6, Epidemiology.) There are at least 13 less commonly encountered *Candida* species (Vazquez and Sobel, 2003). Please also see Section 11.12, Laboratory Detection, Recovery, and Identification. For candidiasis hypersensitivity syndrome, please see Section 11.9, Clinical Forms.

11.3 CASE PRESENTATIONS

Case Presentation 1. Fever in a 16-yr-old Boy with T-Cell Lymphoma (McNeill et al., 1992)

See Fig. 11.1

History

The patient was diagnosed with diffuse T-cell lymphoma and responded well to chemotherapy. Three months after the diagnosis, his inguinal lymph nodes became enlarged but responded to chemo- and radiotherapy of his groin. Two years later, he underwent an elective autologous bone marrow transplant. Pretransplant therapy consisted of cyclophosphamide and total body irradiation. Pancytopenia, low-grade fever, mouth ulcers, and diarrhea ensued, and his fever was unresponsive to antibacterial agents. Three days after receiving the transplant he developed a generalized maculopapular rash.

Diagnostic Procedures

Cultures were taken 3 days after transplantation and, after incubation, a yeast was isolated from blood, urine, and tracheal aspirate specimens. The next day he developed tachycardia and hypotension. His chest radiograph showed bilateral interstitial infiltrates, and the patient required a ventilator.

Diagnosis

The diagnosis was *Candida tropicalis* sepsis with possible disseminated candidiasis.

Therapy and Clinical Course

Six days post-transplant, IV amphotericin B (AmB) therapy was begun. Subsequent blood cultures were negative, but mannan antigenemia[1] measured by *sandwich EIA* continued to be positive. Severe thrombocytopenia and bilateral lung nodular opacities persisted. Later, he developed right-sided pneumothorax and, despite IV vancomycin therapy, bacteremia due to coagulase-negative *Staphylococcus*. A varicella-zoster skin eruption also occurred. Gradual improvement in his white blood cell count indicated engraftment had occurred (Fig. 11.1, middle graph). However, despite aggressive support, respiratory failure occurred, and the patient died a month after the diagnosis of candidemia.

Postmortem findings showed no evidence of lymphoma, but numerous abscesses were seen in liver, kidneys, spleen, heart, and bone marrow. Histologic examination of autopsy specimens showed numerous *pseudohyphae*. The reason for progression of disease despite continued AmB is unclear.

Comment

Engraftment had started, blood cultures became negative, but infected foci persisted in the liver and spleen and could not be eradicated.

- *What are early warning signs of disseminated candidiasis?* Skin rash as seen in this patient, which, if scraped and mounted, might have indicated yeast forms or pseudohyphae.
- The mannan EIA, although a research prototype in this case, is sold in Europe as Platelia® *Candida* Ag EIA kit (Bio-Rad, France). (Please see Section 11.12, Laboratory Detection, Recovery, and Identification—Serodiagnosis–Antigen Detection.) This test is not FDA-approved at this time.
- A simple noninvasive test for disseminated disease is a fundoscopic exam with an ophthalmoscope looking for cottony growths on the retina (please see Section 11.9, Clinical Forms—Endophthalmitis). However, <5% of patients with bloodstream infections will have evidence of retinal lesions. Even after engraftment, lesions may persist in the liver and spleen, where they are difficult to eradicate and may cause chronic disseminated or hepatosplenic candidiasis. *Candida tropicalis* is as pathogenic as *C. albicans* and is a frequent clinical isolate in cancer patients.

[1]Mannan is the readily soluble antigenic outer layer of the *Candida* cell wall. Please see Section 11.10.2 Microbial Factors.

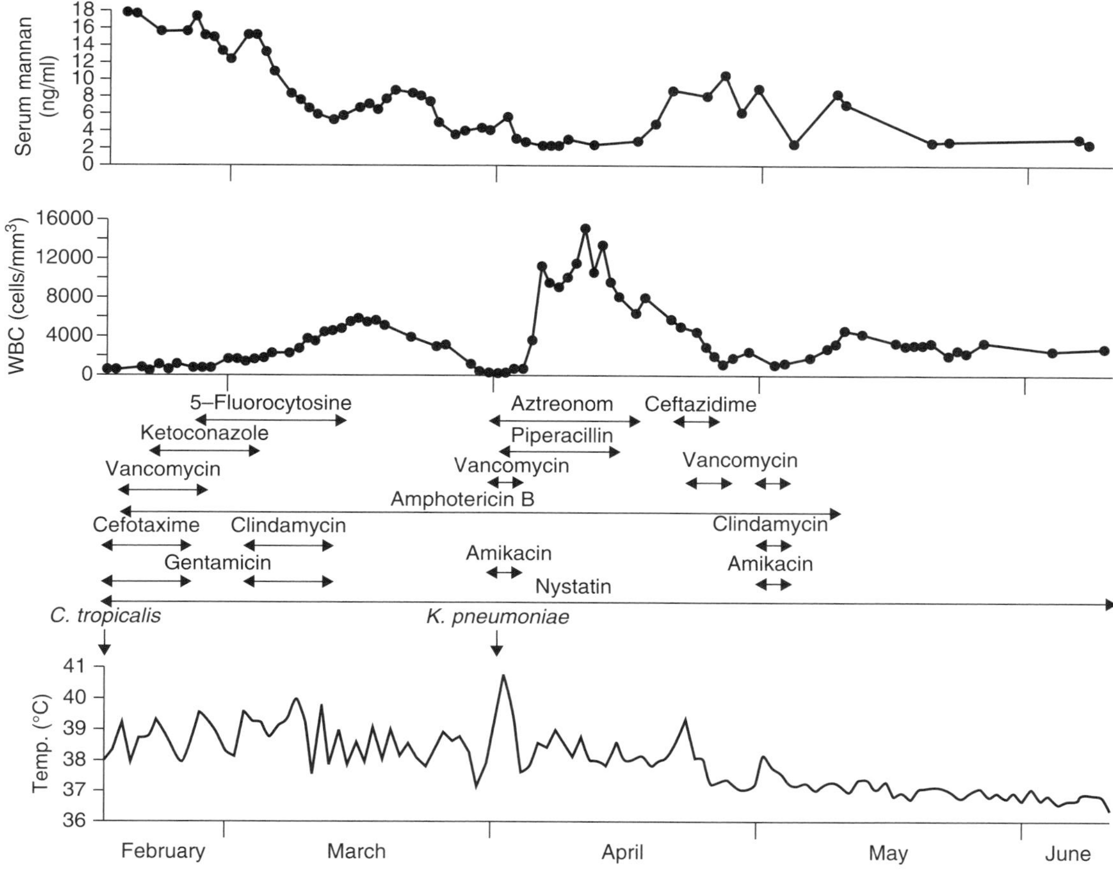

Figure 11.1 Clinical course of candidiasis in a 16-yr-old boy with T-cell lymphoma.
Source: McNeill et al. (1992). Used with permission from Wolters Kluwer Health.

Case Presentation 2. Skin Rash in a 32-yr-old HIV-Positive, Diabetic Man

History

A 32-yr-old HIV-positive male attended the AIDS clinic for recurrent *thrush*, which was treated with oral clotrimazole troches. He responded well and was asymptomatic until he became ketoacidotic following an alcoholic binge. He was hospitalized for dehydration and received IV fluids for 72 h. During that time he developed a fever of 38.2°C and a diffuse rash of nodules and papules on his trunk.

Diagnostic Procedure

Upon removal of the IV catheter a small gelatinous mass was seen at the tip, from which *C. albicans* was cultured. A blood culture also was positive for *C. albicans*.

Diagnosis

The diagnosis was *Candida albicans* sepsis with probable disseminated candidiasis.

Therapy

Treatment was begun with IV AmB for acute disseminated candidiasis, but the patient was unable to tolerate the drug. The regimen was changed to fluconazole (FLC).

Clinical Course

Blood cultures became negative. His skin lesions diminished in size and began to heal. The patient was discharged on oral FLC therapy. He continued to attend the AIDS clinic periodically. His skin cleared and he remained free of other signs or symptoms of candidiasis.

Comment

The coincidence of HIV infection, diabetes, and alcoholism is uncommon, but this combination of factors increases the risk of deep-seated fungal infection. In that setting either candidiasis or, less commonly, mucormycosis can occur. Whereas cutaneous candidiasis can occur as the primary site of infection, in this patient it is more likely to have been a cutaneous manifestation of deep-seated candidiasis.

Case Presentation 3. Disseminated Candidiasis in a 66-yr-old Woman After 20 weeks in the Intensive Care Unit (Krogh-Madsen et al., 2006)

History

The patient underwent surgery for removal of her gallbladder in a local hospital but during surgery damage was done to the liver artery, gall ducts, and spleen. The patient was transferred to the tertiary care center for emergency surgery for removal of her spleen and suture of the liver artery.

Diagnostic Procedures

Specimens from catheters and body fluids were tested for microbial growth. A single episode of *Enterococcus faecalis* bacteremia occurred. The patient was colonized with *Candida tropicalis* in the airways and suffered two episodes of *C. tropicalis* candidemia on hospital days 8 and 43. Of more significance was colonization and infection with *Candida glabrata*, which grew from numerous specimens obtained shortly after admission and afterwards. The sites and frequency of isolation of *C. glabrata* were as follows: candidemia occurred on day 81. After that, the yeast was found in almost every blood culture until the patient died on hospital day 144. Cultures of urine and wound drainage frequently were positive during the period in the ICU. Colonization of airways and feces remained positive throughout the hospital stay. Sterile site tissue samples were positive on four occasions.

Diagnosis

The diagnosis was disseminated candidiasis due to *Candida glabrata*.

Therapy

Oral AmB therapy, 50 mg, 3x/day, was administered in an effort to reduce gastrointestinal colonization, along with antibiotics. Voriconazole (VRC), 250 mg 2x/day, IV was started on day 42 for 12 days, and given orally thereafter until day 109. The patient received caspofungin (CASF) on days 9 to 144 (70 mg IV loading dose on the first day, and 30 mg/day thereafter).

Antifungal Susceptibility Tests Tests were conducted on the *C. glabrata* isolates recovered from the patient. The E-test revealed that MICs for AmB were >2 μg/mL (often 4–8 μg/mL, and sometimes 24–32 μg/mL). Four isolates of *C. glabrata* had reduced susceptibility to CASF (MIC >2 μg/mL). One isolate was resistant to all azoles.

Clinical Course

After emergency surgery upon admission, numerous reoperations were necessary for bleeding and organ dysfunction. The patient required mechanical ventilation and hemodialysis. Twelve weeks after admission she required a liver transplant but died of multi-organ failure and shock after 20 weeks in the intensive care unit.

Comment

This case, despite aggressive management, illustrates the persistence of *C. glabrata* fungemia in the face of timely combination antifungal therapy with extended spectrum azoles (voriconazole) and ß-glucan synthase inhibitors (caspofungin). *Candida glabrata* is notable for its rapid acquisition of reduced susceptibility to azole antifungal agents, but isolates with reduced susceptibility to AmB are rare. The E-test is among the more reliable methods for measuring AmB susceptibility. The lack of success of antifungal therapy is probably attributable to reduced in vitro susceptibility to azoles and CASF, but host factors are a major influence on clinical outcome: in this case the poor clinical condition of the patient after numerous abdominal surgeries. The "90/60 rule" is a useful yardstick to predict clinical outcome (Rex and Pfaller, 2002). If the isolate is susceptible to a given drug one could predict a 90% cure rate (10% of patients may be too sick for a susceptibility result to impact outcome), whereas infections due to resistant isolates respond approximately 60% of the time.

Case Presentation 4. Abdominal Pain and Fever in a 19-yr-old Man with Acute Myelogenous Leukemia (AML) After an Autologous Peripheral Blood Stem Cell Transplant (SCT) (Sorà et al., 2002).

History

This 19-yr-old man with AML in first remission underwent an autologous peripheral blood SCT after graft failure of a matched unrelated donor transplant.

Post-transplant Clinical Course

A few days after the SCT, his PMN count was <100 cells/μL, and he developed fever (= febrile neutropenia) and abdominal pain. He was started on therapy with imipenem, vancomycin, and amikacin. After 72 h of persistent fever, liposomal AmB was started. Blood cultures yielded *Pseudomonas* resistant to imipenem, aminoglycosides, and quinolones, and he was diagnosed with disseminated *P. aeruginosa* infection.

On day 10 after the peripheral blood SCT, laparoscopy was performed because of worsening abdominal pain. Culture of peritoneal fluid was positive for *P. aeruginosa*. He improved with antibiotic therapy and his fever resolved on day 19 after the SCT. During treatment with *granulocyte-CSF,* neutropenia resolved, and a bone marrow aspirate revealed complete remission of leukemia.

On day 30 after the SCT, the patient was still receiving liposomal AmB, vancomycin, and imipenem. *Candida albicans* was isolated from multiple sites: stool and a central venous catheter. A total body CT scan revealed multiple and diffuse low-density areas in his liver and spleen (microabscesses) and fluid collection near the ascending colon.

Ten days later (on day 40 after the SCT), the patient underwent a hemicolectomy because of increasing abdominal pain. Intestinal and hepatic biopsies and culture revealed *C. albicans* and *Pseudomonas aeruginosa*. These organisms also were isolated from the peritoneal fluid. Ketoconazole was added to antifungal therapy with liposomal AmB.

Diagnosis

The diagnosis was hepatitis and peritonitis due to polymicrobial infection with *P. aeruginosa* and *C. albicans* (hepatosplenic candidiasis).

On day 100 after SCT, fever persisted, and a repeat CT scan revealed no improvement of hepatosplenic involvement. Antibiotic and antifungal therapy was stopped; caspofungin (CASF) was started at 50 mg/day IV After 30 days, the fever resolved and his condition improved.

CASF was continued for 60 days until he was discharged. A CT scan performed 30 days after the end of this therapy showed that lesions in the liver and spleen persisted in the absence of symptoms. Six months later, an MRI revealed hepatosplenic areas with low signal intensity without perilesional ring, that was interpreted as chronic healed lesions.

Comment

This patient had chronic disseminated candidiasis, also referred to as hepatosplenic candidiasis. Leukemia patients are prone to develop this clinical form of invasive candidiasis, presenting with fever that returns when antifungal therapy is discontinued or, in this case, persists when the PMN count returns to normal. The response rate for AmB in treating hepatosplenic candidiasis ranges from 40% to 80% (reviewed by Masood and Sallah, 2005). When it became apparent that combination therapy with an older generation azole, ketoconazole, and lipid AmB were not effective, therapy was changed to CASF with a good outcome and no side effects during prolonged treatment Please see Section 11.9, Clinical Forms—Chronic Disseminated (Hepatosplenic) Candidiasis for further discussion of this clinical form including guidelines for therapy.

11.4 DIAGNOSIS

As noted in Section 11.3, Case Presentations, a variety of diagnostic procedures apply in suspected invasive candidiasis. Fundoscopic examination with an ophthalmoscope, biopsy of skin lesions, culture of catheter tips, blood cultures, and serum mannan EIA are some examples. Imaging methods, such as CT scan and ultrasonography, are able to detect microabscesses in the liver, spleen, and kidneys. The differential diagnosis includes bacterial sepsis or endocarditis, bacterial meningitis, and invasive mycoses caused by other yeast, mold, or dimorphic species. Morphologic, biochemical, and genetic methods to identify the major *Candida* species are found in Section 11.12, Laboratory Detection, Recovery, and Identification, and in Hazen and Howell (2007).

11.5 ETIOLOGIC AGENTS AND THEIR ECOLOGIC NICHES

This section introduces the major and less common *Candida* species. Their classification, general characteristics, and ecologic niches are described.

Candida albicans is the species most commonly isolated in disease, probably because it is a *commensal* in humans. The other *Candida* species may be endogenous and/or exogenous, surviving on the skin and on environmental surfaces. Often they may be found in natural environments including soil, water, and plants. The four other most frequent species found as pathogens are *C. glabrata*, *C. tropicalis*, *C. parapsilosis*, and *C. krusei. Candida krusei* is included although its prevalence in invasive candidiasis is low (<3%) (Pfaller and Diekema, 2007). The proportion of each species causing disease varies depending on the hospital and geographic area. Please see Section 11.6, Epidemiology.

11.5.1 Classification of *Candida* Species

Medically important yeasts of the genus *Candida* are classed in the family *Saccharomycetaceae*, order *Saccharomycetales*, class *Saccharomycetes* (syn: *Hemiascomycetes*), subphylum *Saccharomycotina* in the phylum *Ascomycota*. The phylogeny was determined by multigene analysis (Diezmann et al., 2004; Taylor, 2006). Three clades are recognized:

- *Clade 1.* The *Candida albicans* clade includes *C. tropicalis* and *C. parapsilosis*.
- *Clade 2.* The *Clavispora-Metschnikowia* clade contains *C. guilliermondii, Clavispora* (*Candida*) *lusitaniae*, and *Debaryomyces hansenii*.
- *Clade 3.* The *Saccharomycetaceae* clade includes *Saccharomyces cerevisiae* and the pathogens *C. glabrata* and *C. krusei* (teleomorph of the latter is *Issatchenkia orientalis*).

The presence of pathogens in all three clades infers that pathogenicity evolved independently on different occasions.

Candida albicans

Candida albicans (C.P. Robin) Berkhout *De Schimmelgesl. Monilia, Oidium, Oospora en Torula, Disset. Utrecht*: 44 (1923).

The *Candida albicans* genome database is located at http://www.candidagenome.org/.

Candida albicans is a diploid yeast with two pairs of 8 chromosomes. Its genome size is about 16 Mb (haploid), about 30% greater than the genome of *Saccharomyces cerevisiae*. Mating types and conditions for conjugation and a *parasexual* cycle are known (Soll and Daniels, 2007), but true sexual reproduction has either been lost or the right set of conditions has so far eluded researchers. From a morphologic viewpoint, *C. albicans* yeast[2] cells multiply by budding and undergo morphogenesis; that is, yeast cells germinate producing characteristic germ tubes that then grow out as either pseudohyphae or true hyphae. On cornmeal agar (CMA) *C. albicans* produces characteristic terminal round resting cells called chlamydospores. A closely related, phenotypically very similar genetic species, *C. dubliniensis*, is a minor pathogen (please see Section 11.5.2, Less Common *Candida* Species of Clinical Importance).

Ecologic Niche *Candida albicans* is a yeast that leads an inconspicuous existence primarily on the mucosae of humans. The mouth, lower intestinal tract, and female genital tract are the most frequent habitats for *C. albicans*. Unlike most pathogenic fungi, which are soil dwellers with humans as an accidental host, *C. albicans* is an *endogenous commensal* microbe and does not survive well in the environment. Although not part of the normal microbiota of healthy *glabrous* skin, it may be isolated from intertrigenous areas such as toe clefts and fingers.

[2] "Blastoconidium" is the formal mycologic term for a yeast. For brevity, however, the term "yeast" will also be used in the text.

Candida glabrata

Candida glabrata. (H.W. Anderson) S.A. Meyer & Yarrow (1978).

The *C. glabrata* genome map may be found at http://www.ncbi.nlm.nih.gov/genomeprj/12362.

Candida glabrata is a small, haploid, monomorphic yeast with 13 chromosomes and a genome size of 12.3 Mb. There is a rising prevalence of *C. glabrata* as a cause of candidemia and invasive candidiasis in hospitals worldwide; please See Section 11.6, Epidemiology. Two minor genetic species are recognized:

- *Candida nivariensis.* Atypical *C. glabrata*-like yeast were isolated from three patients in Spain (Alcoba-Florez et al., 2005). Based on the sequences for the D1/D2 and ITS regions of rDNA, these yeasts were 94–96% identical to *C. glabrata* and 99% similar to a *Candida* species on *Hibiscus* species flowers in Canada. The new species, *C. nivariensis*, is non-pigmented on CHROMagar® *Candida* and ferments trehalose.
- *Candida bracarensis.* Sequencing of the D1/D2 rDNA region showed a 94.8% similarity with *C. glabrata*. Three (2.2%) of 137 clinical isolates from the United States that were identified as *C. glabrata* were positive with a *C. bracarensis* probe (Bishop et al., 2008).

Ecologic Niche *C. glabrata* is rare on the smooth skin of healthy individuals and is a minor, but obligatory mucosal commensal of warm-blooded animals.

Candida krusei

Candida krusei (Castell.) Berkhout (1923).

Candida krusei is diploid, with 3–5 chromosomes. It is capable of growing both as a yeast and as pseudohyphae. On cornmeal agar, CMA, *C. krusei* grows as extremely long, rarely branched pseudohyphae. Its teleomorph is *Issatchenkia orientalis*. The clinical importance of *C. krusei* may be underestimated. In one clinical center *C. krusei* accounted for 24% of candidemias in patients with hematologic malignancies (Hachem et al., 2008).

Ecologic Niche The ecologic niches of *C. krusei* include soil, water, plants, and other natural sources.

Candida parapsilosis

Candida parapsilosis (Ashford) Langeron & Talice (1932).

Candida parapsilosis is diploid or aneuploid with 14 chromosomes and a genome size of 16 Mb. Both yeast and pseudohyphae are produced. On CMA, *C. parapsilosis* grows as elongated, curved pseudohyphae with blastoconidia at the septa. This species is particularly associated with bloodstream infections in neonates and also with catheter-associated candidemia and total parenteral nutrition.

Candida parapsilosis isolates are divisible into three groups (I, II, and III) by random amplified polymorphic DNA (RAPD) analysis, multilocus enzyme electrophoresis, sequence analysis of the D1/D2 and ITS1 regions of rDNA, and classical DNA–DNA homology. Group I is the major group, *C. parapsilosis* (Tavanti et al., 2005). Sequences of several housekeeping genes from the three groups gave products that were characteristic for separate species. Group II was designated *C. orthopsilosis* and the remaining group, Group III, was designated *C. metapsilosis*.

Ecologic Niche *Candida parapsilosis* is found on healthy human skin but also survives well in the environment. The ecologic niche of *C. parapsilosis* is broad including humans and domestic animals. Interestingly, it was recovered from marine environments off the coast of Florida, the Bahamas, and the Indian Ocean (Fell and Meyer, 1967).

Candida tropicalis

Candida tropicalis. (Castell.) Berkhout (1923).

Candida tropicalis is diploid, with 10–12 chromosomes and a haploid genome size of 15 Mb. Yeast and pseudohyphal forms are produced. *C. tropicalis* is most frequently encountered in cancer patients and is a common cause of fungemia in the United States, Europe, India, and perhaps worldwide. It is suspected of displaying a trend toward reduced susceptibility to FLC and AmB in cancer centers (Krcmery and Barnes, 2002). Animal model infections indicate *C. tropicalis* is at least as pathogenic as *C. albicans*.

Ecologic Niche *Candida tropicalis* is found on the skin and in the digestive tracts of healthy humans. Natural sources of *C. tropicalis* include organically enriched soil and aquatic environments. *C. tropicalis* is used in industry for the preparation of polyester, polyamide, and perfume, and in the formation of xylitol, a sugar alcohol that can replace sucrose.

11.5.2 Less Common *Candida* Species of Clinical Importance

Candida dubliniensis

Candida dubliniensis is phenotypically similar to *C. albicans* and is also known as a "genetic species," in that it is primarily identified only by molecular genetic analysis. It is difficult to identify in the clinical laboratory using gross phenotypic characteristics because, like *C. albicans*, it produces germ tubes, pseudohyphae, and chlamydospores. Differences in D-xylose and α-methyl-D-glucose assimilation have been used to differentiate these species biochemically, although molecular methods are more definitive (Ellepola et al., 2003). *Candida dubliniensis* was discovered in Ireland in 1995 by dental investigators who found that about 5% of *C. albicans* strains isolated from AIDS patients with OPC had characteristics on the genetic level that constituted a new species: for example, electrokaryotype, rDNA sequence, length of the Group 1 intron in the 25S rDNA (McCullough et al., 1999), and absence of the repetitive element, Ca3. *C. dubliniensis* can be differentiated from *C. albicans* by PCR amplification of the IS 1 element followed by agarose gel electrophoresis, or by the use of species-specific DNA probes directed to the ITS2 rDNA region (Elie et al., 1998).

C. dubliniensis is less pathogenic than *C. albicans* (Sullivan et al., 2005). It is a minor constituent of the oral microbiota of normal individuals and causes OPC in persons living with AIDS. Relatively high prevalence rates of *C. dubliniensis* have been found in patients suffering from diabetes. *C. dubliniensis* has been identified in only 1–2% of yeast isolated from blood cultures. *Candida dubliniensis* has a reduced capacity to produce hyphae, which may contribute to a lower pathogenic potential. Nearly all *C. dubliniensis* isolates recovered from blood are susceptible to commonly used antifungal agents and it is unlikely that identification of either *C. albicans* or *C. dubliniensis* in a patient would result in any differences in treatment.

C. dubliniensis may, however, display a higher tendency to develop azole resistance than *C. albicans*. Replacement of *C. albicans* with *C. dubliniensis* having decreased in vitro susceptibility to FLC was described in 6 of 30 cases of OPC in HIV-positive patients who received prolonged FLC therapy (Martinez et al., 2002).

Candida guilliermondii

Candida guilliermondii (teleomorph—*Pichia guilliermondii*) is reported to have a genome size of 12 Mb. Chains of *C. guilliermondii* blastoconidia form sparse pseudohyphae on CMA. This species causes a variety of deep-seated infections in cancer patients, in the ICU, among surgical patients, and, less frequently, in IV drug users. *Candida guilliermondii* infections are most common in oncology patients. *Candida guilliermondii* is subdivided into two species: *Pichia caribbica* (*Candida fermentati*) and *Candida carpophila*. Most strains of *C. guilliermondii* are morphologically and biochemically indistinguishable from *Candida famata* (teleomorph: *Debaromyces hansenii*), and DNA- based methods are required to differentiate the two species.

Ecologic Niche *Candida guilliermondii* has been isolated from environmental surfaces and from the skin and nails of healthcare workers (Medeiros et al., 2007). In nature, *C. guilliermondii* and its teleomorph *Pichia guilliermondii* have been recovered from municipal wastes, from sewage sludge, and in marine environments, even in deep-sea hydrothermal systems of the Mid-Atlantic Rift (Gadanho and Sampaio, 2005). *Candida guilliermondii* represented 1.4% of the *Candida* species clinical isolates in a large collection from many international centers (Pfaller et al., 2006). Sixty-eight percent to 77% of the *C. guilliermondii* isolates were susceptible to FLC.

Candida lusitaniae

Candida lusitaniae (teleomorph—*Clavispora lusitaniae*) has a genome size of 16 Mb. Yeast and pseudohyphae are produced including distinctly curved pseudohyphae on CMA. Most reports of human infections with this species have been in patients with hematologic malignancies or in stem cell transplant recipients. *Candida lusitaniae* can develop secondary resistance to AmB.

Candida orthopsilosis *and* Candida metapsilosis

Clinical microbiologists are seeking to understand the clinical importance of these species which are related to *Candida parapsilosis. Candida orthopsilosis* has been recovered from blood, nails, skin, lungs, urine, and indwelling catheters. Compared with *C. parapsilosis*, the sister species account for a small fraction of yeast isolated from candidemia patients. For example, in a 2002–2003 surveillance at hospitals in Barcelona, Spain, *C. parapsilosis* accounted for 19% of candidemias, whereas the prevalence for *C. metapsilosis* and *C. orthopsilosis* was 1.7% and 1.5%, respectively (Gomez-Lopez et al., 2008).

Candida pelliculosa

Candida pelliculosa (teleomorph—*Pichia anomala*) is an uncommon cause of neonatal septicemia. As a contaminant in beer bottling plants, *C. pelliculosa* is implicated in beer-spoiling biofilms. *Pichia anomala* is a free-living ascomycete yeast found in plants, fruits, and soil and is associated with food and feed products, either as a production organism or as a spoilage yeast. The ability to grow in preserved food and feed environments is due to its capacity to grow under low pH, high osmotic pressure, and low oxygen tension. The sexual stage includes the production of hat-shaped ascospores (Passoth et al., 2006).

Candida rugosa

An animal pathogen causing mastitis in cattle, *C. rugosa* has emerged as a rare human pathogen in burn patients and those being treated with polyene antifungal agents (e.g., AmB or oral nystatin prophylaxis). Most *C. rugosa* strains are polyene-resistant. *Candida rugosa* is of major industrial importance because of its lipases, one of the enzymes most frequently used in biotransformations (Domínguez de María P et al., 2006).

Candida famata, Candida kefyr, Candida lipolytica, *and* Candida norvegensis

These are exceptionally rare human pathogens. Their medical importance is described in Krcmery and Barnes (2002).

11.6 EPIDEMIOLOGY

The source of candidiasis is primarily *endogenous*, found within or on the skin or mucosae of the human body. Direct contact between individuals or via fomites are known routes of transmission of strains in the hospital and community. Various *Candida* species are isolated from a range of anatomic sites, including the skin, throat, and vagina. Oropharyngeal and esophageal candidiasis are opportunistic infections in persons living with AIDS. *Candida* species frequently can be isolated from stool, indicating the gastrointestinal tract is an important endogenous source. The predominant *Candida* species in a given instance often depends on the type of disease and varies according to the reporting hospital and geographic region.

11.6.1 Major Types of Candidiasis

Mucosal Candidiasis

Oropharyngeal Candidiasis (OPC) The appearance of OPC in HIV-positive patients heralds the onset of AIDS, corresponding to the decrease in the $CD4^+$ T-lymphocyte count below 200/μL and plasma viral loads of more than 100,000 copies/mL (Nokta, 2008). In countries where HIV-seropositive persons have access to combination antiretroviral therapy (ART), the incidence of OPC has significantly subsided since 1997, when combination ART was introduced. OPC in patients receiving ART suggests a failure of the antiretroviral therapy. Even in the era of ART, patients with a late HIV diagnosis (less than 6 months before an AIDS diagnosis) and thus not in care, were 3.5 times as likely to have an opportunistic infection than those in care (Hanna et al., 2007). Within this group of AIDS patients in New York City, the rate of oroesophageal and bronchial candidiasis was 3.3/100 diagnoses.

Vulvovaginal Candidiasis (VVC) About 50–75% of women in good health with normal immune systems will have an episode of VVC in their lifetime related to antibiotic use, high estrogen oral contraceptive use, hormone replacement therapy, or pregnancy. Chronic or recurrent VVC afflicts 5–10% of women (Fidel, 2005). (Please see Section 11.9, Clinical Forms—Vulvovaginal Candidiasis.)

Cutaneous Candidiasis Surveillance for cutaneous candidiasis is lacking at present. The groups at risk to develop cutaneous candidiasis are diverse, including infants with diaper dermatitis, incontinence-associated dermatitis in the elderly, intertrigo in obese persons, and diabetic foot ulcers unresponsive to antibiotics (Heald et al., 2001).

Invasive Candidiasis

Systemic or Disseminated Candidiasis Depending on the severity of disease, invasive candidiasis includes systemic candidiasis (one organ system involved) and disseminated candidiasis (multisystem disease). Hepatosplenic candidiasis is a form of chronic disseminated candidiasis that occurs in patients who have recovered from neutropenia. Persistent fever is indicative of *Candida* slowly multiplying in the liver, spleen, kidney, and lung.

Candidemia The most useful information on the burden of invasive candidiasis comes from surveillance for *candidemia* (*definition*: a bloodstream infection (BSI) with a *Candida* species). The National Healthcare Safety Network , URL http://www.cdc.gov/nhsn/, is a reporting system of small, medium, and large hospitals. During 1980–1990, the system reported that *Candida* species were the fifth leading cause of BSIs hospital-wide and the fourth in the ICUs (Beck-Sague and Jarvis 1993). These estimates are supported by data from other surveillance systems (Rangel-Frausto et al., 1999; Trick et al., 2002). Between 1992 and 1998 a total of 1579 BSIs were reported from >50 U.S. medical centers (Pfaller et al., 1999). More recent data estimate that *Candida* species are the fourth leading cause of hospital-acquired BSIs reaching to 8–10% of all BSIs acquired in the hospital (Pfaller and Diekema, 2007). Candidemia is shifting from a concentration in the ICU to the general hospital population, such that between 1998 and 2000, 36% of *Candida* species BSIs were in the ICU whereas 28% of cases had a community onset (Hajjeh et al., 2004).

Secular Trends in Candidemia (1979–2000) Between 1979 and 2000 there was a 207% increase in the number of cases of fungal *sepsis* linked to:

- An increasing number of patients receiving chemotherapy for cancer or immunosuppressive therapy for maintenance of stem cell or solid organ transplants, resulting in either neutropenia and/or reduced T-cell-mediated immunity.
- Patients with a disruption of normal anatomic barriers, either mucosal or cutaneous, caused by injury or surgery (e.g., cardiothoracic or gastrointestinal surgery).
- Those with metabolic endocrine disorders (e.g., diabetes mellitus).
- Treatment of seriously ill patients with long-term, multiple broad-spectrum antibiotics.
- Increases in low birthweight infants and another demographic shift toward a larger elderly population.

Population-Based Active Surveillance for Candidemia in the United States (1992–1993 and 1998–2000) Prospective surveillance was conducted in 1992–1993 in hospitals in Atlanta, Georgia, and San Francisco, California (Kao et al., 1999) and in 1998–2000 in Baltimore, Maryland, and Connecticut (Hajjeh et al., 2004). In the latter period, 1143 cases of candidemia were reported (Table 11.1). Eleven percent of candidemia cases occurred in neutropenic patients, 38% occurred in patients in ICUs, and 8% were in HIV-positive persons. Seventy-eight percent of the patients with candidemia had intravenous catheters in place at the time of candidemia diagnosis, and 26% had received abdominal surgery.

Over the two study periods (1992–1993 and 1998–2000), the proportion of candidemia caused by *C. albicans* remained constant at 45–46%. *Candida glabrata* surpassed *C. parapsilosis* as the second most frequent yeast species isolated from blood cultures (Fig. 11.2). The incidence of candidemia varied according to the geographic region: 3/100,000 population in Connecticut, rising to 24/100,000 population in Baltimore, Maryland (Hajjeh et al., 2004). This incidence was significantly greater in African-Americans and in certain demographics, for example, infants and the elderly. In the African-American population the incidence among infants <1 yr old was 158/100,000 and among the elderly >65 yr old, it was 100/100,000. The incidence rate of invasive candidiasis rose during the period 1996–2003. Data from the U.S. National Hospital Discharge Survey indicated an incidence rate in 1996 of 20 cases/10,000 hospital discharges, which increased to 24 cases/10,000 discharges in 2003 (Pfaller and Diekema, 2007).

Incidence of *Candida* BSIs in Neonatal Intensive Care Units (NICUs) Septicemia is the most common severe infection encountered in the NICU. Depending on the institution, *Candida* species rank second to fourth as

Table 11.1 Candidemia in Baltimore, Maryland, and Connecticut hospitals, 1998–2000, Shown According to Primary Disease, Contributing Factors, and Mortality at 30 days After an Episode of Candidemia

Primary disease, contributing factor(s), mortality	Number (%) of isolates					
	C. albicans	Non-*C. albicans*	*C. glabrata*	*C. tropicalis*	*C. parapsilosis*	All species[a]
Malignancy	115 (23)	151 (25)	69 (25)	38 (29)	29 (20)	266 (24)
Neutropenia	53 (11)	67 (10)	17 (6)	21 (16)	19 (13)	120 (11)
Transplant recipient	20 (4)	41 (7)	19 (7)	9 (6)	8 (5)	61 (5)
Diabetes	132 (26)	190 (31)	102 (37)	38 (28)	36 (24)	322 (29)
Liver disease	122 (25)	113 (19)	45 (17)	31 (23)	25 (17)	235 (21)
Dialysis[b]	70 (14)	100 (16)	39 (14)	28 (21)	22 (15)	170 (15)
Renal failure	171 (34)	224 (37)	103 (38)	54 (40)	45 (30)	395 (35)
Immunosuppressive therapy	212 (44)	261 (44)	120 (45)	66 (51)	53 (37)	473 (44)
Surgery in last 3 months	273 (55)	275 (46)	127 (47)	60 (45)	64 (44)	548 (50)
Outpatient acquired[c]	117 (23)	208 (33)	83 (30)	48 (34)	58 (38)	325 (28)
ICU at time of diagnosis	204 (40)	205 (33)	98 (36)	43 (31)	49 (32)	409 (36)
Central venous catheter in place at time of diagnosis	388 (83)	464 (84)	197 (83)	109 (85)	116 (81)	852 (78)
Died within 30 days of culturing	205 (40)	204 (33)	100 (36)	57 (40)	28 (18)	409 (36)
Total	516 (45)	626 (55)	275 (24)	141 (12)	153 (13)	1143

[a] Data from Table 1, Hajjeh et al. (2004).

[b] Any patient who received any form of dialysis within 3 months prior to this candidemia episode.

[c] Positive blood culture either prior to or on the day of hospital admission.

Source: Data from Table 1, Hajjeh et al. (2004). Used with permission from the American Society for Microbiology.

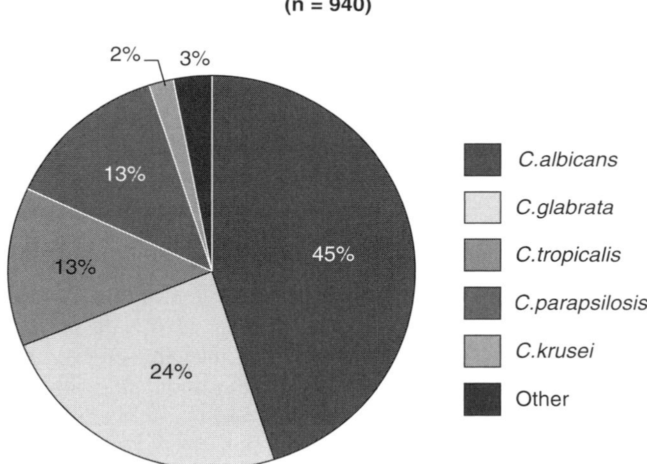

Figure 11.2 Candidemia rate in Baltimore, Maryland, and Connecticut hospitals, 1998–2000 (data from Hajjeh et al., 2004).

the most frequent cause of late-onset sepsis in extremely low birthweight (ELBW, <1000 g) and low birthweight (LBW, <1500 g) infants. Occurring ≥72 h after birth, late-onset sepsis is referred to as nosocomial sepsis to emphasize the possibility of horizontal transmission in the NICU. *Candida parapsilosis* has emerged as the predominant fungal pathogen causing BSIs among LBW infants in some NICUs and, in others, ranks second to *C. albicans*. The other leading causes of late-onset sepsis are, in order of frequency, coagulase-negative *Staphylococcus* > *S. aureus* > Gram-negative bacteria. The incidence of systemic candidiasis in hospitalized neonates within the United States was estimated using the 2003 Kids' Inpatient Database from the Healthcare Cost and Utilization Project (Zaoutis et al., 2007). The overall incidence of invasive candidiasis in neonates is 13–16 cases/10,000 neonatal admissions.

Mortality Associated with Candidemia Of 669 patients with candidemia in the 1998–2000 surveillance study, 271 (40%) died (Hajjeh et al., 2004). Of these 15% died within 48 h of the first blood culture and 50% died within 7 days. *Candida* species associated with these fatal cases were, in order of decreasing percent crude mortality,[3] *C. tropicalis* (40%) = *C. albicans* (40%)> *C. glabrata* (36%)>*C. parapsilosis* (18%).

Internationally, data from 106 hospitals in seven European countries collected between September 1997 and December 1999 (Tortorano et al., 2004, 2006) found

[3] Crude mortality. *Definition*: Death from all causes in patients with candidemia.

that the mean percent crude mortality 30 days after diagnosis of candidemia ranged, in descending order, as follows: *C. krusei* (55.3%) > *C. glabrata* (45%) > *C. tropicalis* (41.4%) > *C. albicans* (38.5%) > *C. parapsilosis* (25.9%). Within each species, however, there was a wide range among different hospitals owing to the patient population and the standard of care.

Pappas et al. (2003) found several factors associated with mortality in a prospective study. Among adults, factors associated with 30 day mortality included an APACHE II score of greater than 18, a cancer diagnosis, the presence of a urinary catheter, male gender, receipt of corticosteroids, and an arterial catheter in place. Infection with *C. parapsilosis* was associated with lower mortality (Table 11.2).

It is noteworthy that *C. albicans*, regarded as the most pathogenic *Candida* species, has the highest attack rate of candidemia, but in terms of clinical outcome is not the most lethal. Outcome is multifactorial, based on host factors and prompt diagnosis. Although *C. albicans* retains a high level of susceptibility to FLC, this can be compromised if that species is embedded in a biofilm on an intravenous catheter.

Delays in treating candidemia or treatment with an agent against which the isolate is resistant are independent predictors of mortality. On the other hand, lower mortality of candidemia patients is associated with removal of intravenous catheters, and a timely and adequate (5 days or longer) period of appropriate antifungal therapy (Pfaller and Diekema, 2007).

The incidence of candidiasis has increased in NICUs, and invasive candidiasis is associated with significant morbidity and mortality in NICUs (Zaoutis et al., 2007). ELBW neonates with invasive candidiasis were twice as likely to die as ELBW neonates without candidiasis. The mortality rate attributable to candidiasis among ELBW neonates was 11.9%. Candidiasis in ELBW infants was not associated with an increased length of hospital stay but was associated with a mean increase in total charges of $39,045 (U.S.). Among infants with a birthweight ≥1000 g, those who had candidiasis did not experience a significant increase in mortality, but their length of stay and charges attributable to candidiasis were 16 days and $122,302 (U.S.), respectively. Crude mortality caused by *Candida* species infections in the NICU is estimated to be 25–30% with an attributable mortality of 13%. Premature infants are susceptible to candidiasis because of an immature immune system and immature skin that is not an efficient barrier to *Candida*.

Trends in Species Distribution in Candidemia

National Studies ***Candida glabrata*** *incidence increases*. Beginning in 1995, *C. parapsilosis* was replaced by *C. glabrata* as the most common non-*C. albicans* yeast species isolated from blood cultures in a U.S. population-based surveillance for candidemia (Hajjeh et al., 2004). Recovery of *C. glabrata* from blood cultures over an 9 year period (1992–2000) continued to increase in the United States from 12% to 24% of all candidemias (Kao et al., 1999; Hajjeh et al., 2004). Such increases may be due, in part, from better detection of *C. glabrata* by certain blood culture methods and media, including the BacT/Alert system and BACTEC Mycosis IC/F broth medium (Pfaller and Diekema, 2007) (please see Chapter 2, Section 2.3.5.5, Types of Blood Culture Methods). Other factors influencing the rise in *C. glabrata* BSIs include previous azole exposure, patient age, and underlying disease. Older patients (>60 yr old) are increasingly colonized by *C. glabrata*. Treatment with piperacillin-tazobactam and vancomycin tends to result in a replacement microbiota, including *C. glabrata*. Because of reduced susceptibility of *C. glabrata* to FLC, prior treatment with that drug, once thought to be the major or only selection factor for this species, is now regarded as one of several factors underlying the

Table 11.2 Factors Associated with Mortality Among Adult Patients with Candidemia

Factor	Mortality in associated[a] patients (%)	Mortality in unassociated patients (%)	P value
APACHE II score >18	62.1	26.1	<0.001
Corticosteroids	56.3	37.7	<0.001
Arterial catheter	55.1	36.5	<0.001
Cancer	50.1	40.2	0.002
Urinary catheter	49.2	31.0	0.004
Male sex	45.6	39.8	0.004
C. parapsilosis	23.8	45.6	<0.001

[a]"Associated" defined as patients who are positive for the factor in column 1.

Source: Pappas et al. (2003). Copyright © 2003. Used with permission of the University of Chicago Press.

increased percentage of *C. glabrata* BSIs in the United States.

***Candida krusei** emerges as a cause of candidemia.* The emergence of this uncommon *Candida* species in stem cell transplant patients receiving FLC prophylaxis is related to the species' innate resistance to that drug. As is the case for rising incidence of *C. glabrata* fungemia, the use of piperacillin-tazobactam and vancomycin also may be important as a cause of a resistant replacement microbiota, including *C. krusei*.

International Studies International laboratory-based surveillance in 127 centers in 39 countries during an 8 year period (1997–2003) revealed trends in the distribution of *Candida* species associated with candidemia, other sterile site isolations, candiduria, and genital isolations. *Candida albicans* isolation declined from 73.8% to 62.3% (Pfaller and Diekema, 2007). The rank order of the species did not change but there were increases in the percentage of non-*albicans Candida* species isolated: *C. glabrata* from 11% to 12%; *C. tropicalis* from 4.6% to 7.5%; *C. parapsilosis* from 4.2% to 7.3%; and *C. krusei* from 1.7% to 2.7%. The relative percentage of *C. albicans* isolates recovered varied among the countries, from a low in Latin America of 37% of isolates, to a high of 70% in Norway. *C. albicans* remains the dominant bloodstream *Candida* species isolated from hospitals in seven European countries in patients from surgical wards, the ICU, and NICU, with solid tumors, and those with HIV infection (Fig. 11.3) (Tortorano et al., 2006). An exception is that *C. albicans* accounts for less than 40% of *Candida* bloodstream isolates in patients with hematologic malignancy. *Candida tropicalis* is more often isolated from leukemia patients than in other patient groups. *Candida parapsilosis* is second to *C. albicans* as the cause of sepsis in the NICU.

Trends in Susceptibility to Antifungal Agents In vitro susceptibilities of *Candida* species to FLC and VRC were determined in the *disk diffusion* test for isolates from 124 hospitals in the Global Antifungal Surveillance Study, 1997–2005: an 8.5 year analysis of *Candida* species susceptibility (Pfaller et al., 2007). A summary of the susceptibility of six different *Candida* species to all classes of antifungal agents is shown in Table 11.3 (Pappas et al., 2009).

Candida krusei All isolates of *C. krusei* are innately resistant to FLC (but not to VRC). Resistance is related to diminished sensitivity of the target enzyme in the ergosterol biosynthetic pathway, lanosterol-14α-demethylase, to the effects of FLC. In *C. krusei*, VRC binds more efficiently to this enzyme than does FLC, resulting in better clinical success using VRC for treatment of *C. krusei* infections. *Candida krusei* is innately resistant to flucytosine (5FC).

Candida albicans *C. albicans* isolates remain more susceptible to FLC than do other *Candida* species, with fewer than 2% of clinical isolates demonstrating in vitro resistance. In the 1998–2000 surveillance study (Hajjeh et al., 2004), in vitro resistance of *C. albicans* bloodstream isolates to FLC remained low ($\leq$2%); only 5 of 423 *C. albicans* isolates tested were resistant in vitro to FLC and 3 were *susceptible (dose-dependent)*.

Candida glabrata *Candida glabrata* is both prevalent and demonstrates in vitro resistance to both FLC and VRC, requiring a different clinical treatment decision:

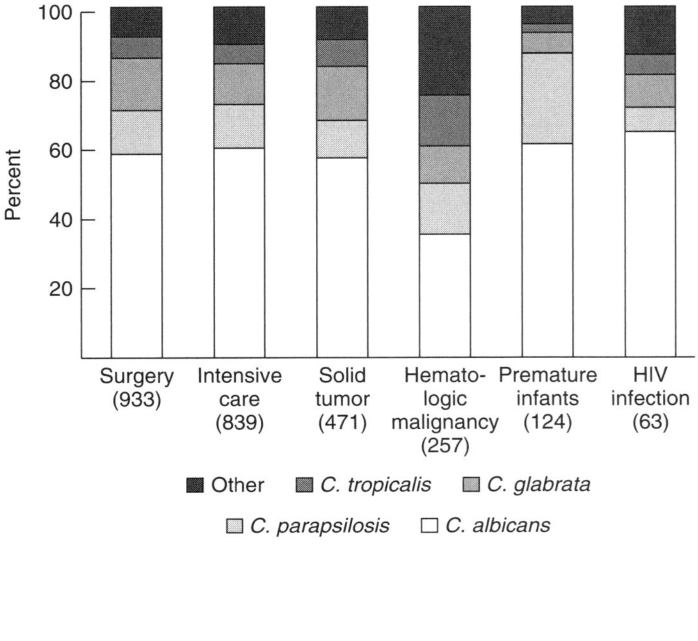

Figure 11.3 European Confederation of Medical Mycology (ECMM) survey: distribution of the four most frequent bloodstream isolates of *Candida* species according to underlying disease and level of hospital care. *Candida glabrata* was the most frequent non-*albicans* isolate in surgical (16%) and solid tumor (16%) patients, whereas *C. parapsilosis* was isolated with the highest frequency in premature neonates (29%). In the hematologic malignancy group, the large "other" category (24% of BSIs), *C. krusei* accounted for half of the non-*albicans Candida* species.
Source: Tortorano et al. (2006). Used with permission from Elsevier Publishers, Ltd., © 2006.

Table 11.3 General Patterns of Drug Susceptibility[a] of Major *Candida* Species

Species	Fluconazole	Itraconazole	Voriconazole	Posaconazole	Flucytosine	Amphotericin B	Echinocandins
C. albicans	S	S	S	S	S	S	S
C. tropicalis	S	S	S	S	S	S	S
C. parapsilosis	S	S	S	S	S	S	S to R[b]
C. glabrata	S-DD to R	S-DD to R	S-DD to R	S-DD to R	S	S to I	S
C. krusei	R	S-DD to R	S	S	I to R	S to I	S
C. lusitaniae	S	S	S	S	S	S to R	S

[a]I, intermediately susceptible; R, resistant; S, susceptible; S-DD, susceptible dose-dependent.
[b]Echinocandin resistance among *C. parapsilosis* isolates is uncommon.
Source: Pappas et al. (2009). Copyright © 2009, used with permission of the University of Chicago Press.

15% of *C. glabrata* in the disk diffusion study were resistant in vitro to FLC, with 9% VRC resistant. Of concern, patients with *C. glabrata* fungemia treated previously with FLC failed to respond clinically to VRC (reviewed by Pfaller and Diekema, 2007). In *C. glabrata* the main mechanism of resistance is to upregulate CDR *efflux pumps*, which efficiently pump out both triazoles. For these reasons, primary therapy for *C. glabrata* infections is an echinocandin, for example, caspofungin (CASF) or micafungin (MCF).

Candida guilliermondii The in vitro FLC-resistance of 10% of *C. guilliermondii* clinical isolates also influences therapeutic choices for that uncommon cause of invasive candidiasis. VRC has a place in treating *C. guilliermondii* fungemia in isolates showing susceptibility to that azole, either as primary therapy or as step-down therapy after an initial response to liposomal AmB or an echinocandin.

Candida parapsilosis Echinocandins [CASF, MCF, and anidulafungin (ANF)] which target ß- (1→3)-D-glucan synthesis are potent anti-*Candida* agents. An echinocandin is the recommended primary therapy for patients with moderately severe to severe illness or patients who have had recent azole exposure (Pappas et al., 2009). An exception is that resistance of *C. parapsilosis* fungemia treated with echinocandins occurs, but this is uncommon. First-line therapy for *C. parapsilosis* fungemia in clinically stable patients is FLC, but patients who are treated with echinocandins and who respond to that therapy should continue with the echinocandin (Pappas et al., 2009).

Candida tropicalis *C. tropicalis* BSIs collected over a 4 year period at hospitals in Paris, France, revealed a persistent clone with reduced susceptibility to 5FC comprising 35% of the candidemia isolates (Desnos-Ollivier et al., 2008). The clone was recognized by DNA subtyping of two microsatellites and the *URA3* gene sequence. 5FC is rarely used as monotherapy, but is usually given in combination with AmB for patients with invasive disease, such as *Candida* endocarditis or meningitis. Occasionally, it is used to treat urinary tract candidiasis (Pappas et al., 2009).

Susceptibility of Rarely Isolated *Candida* Species

Resistance to both AmB and FLC is observed in strains of *C. krusei, C. lusitaniae, C. guilliermondii, C. inconspicua*, and *C. sake* (Pfaller et al., 2003). *Candida lipolytica* has reduced susceptibility to FLC and is resistant to AmB. Decreased susceptibility to FLC has been reported for *C. dubliniensis, C. inconspicua*, and *C. lambica*, and decreased susceptibility to AmB and FLC has been reported for *C. rugosa*. Compared with more common *Candida* species, these rare *Candida* species are less susceptible to older systemic antifungal agents. As such they pose a continued threat for immunocompromised patients. Isolates of of some species, *C. guilliermondii, C. lipolytica*, and *C. inconspicua*, show cross-resistance to all azoles. These rare species may possess multidrug efflux pumps or the ability to alter the target enzyme (lanosterol-14α-demethylase), resulting in azole resistance.

Scale and Trend: The Relative Burden of Invasive Candidiasis Compared with Invasive Aspergillosis

Over the period 1991–2003, the incidence of invasive candidiasis remained steady but invasive aspergillosis declined (Pfaller and Diekema, 2007). Earlier (1980–1996), there was an alarming 357% increase reported for fatal cases of invasive aspergillosis (McNeil et al., 2001); but from 1997 deaths from invasive aspergillosis were 0.42/100,000 population, declining to 0.25 deaths/100,000 population in 2003. Between 1996 and 2003 the number of cases of invasive aspergillosis was tenfold lower than invasive candidiasis cases. Invasive aspergillosis was declining in incidence from 4.1 cases/100,000 population in 2000 to 2.2/100,000 in 2003. The estimated burden of invasive aspergillosis is 8000 cases per year compared with an estimated 63,000 cases of invasive candidiasis.

Two data sets compiled by the U.S. Transplant-Associated Infection Surveillance Network help clarify the cumulative incidence of candidiasis and aspergillosis in stem cell transplant (SCT) and solid organ transplant recipients. A 5 year surveillance for invasive fungal infections among 15 U.S. centers for organ transplantation found that most common invasive fungal infections (IFIs) were invasive candidiasis (53%), invasive aspergillosis (19%), cryptococcosis (8%), non-*Aspergillus* molds (8%), endemic fungi (5%), and mucormycosis (2%) (Pappas et al., 2010). The 1 year incidence was highest for invasive candidiasis (1.95%) and aspergillosis (0.65%). A parallel 5 year surveillance for invasive fungal infections in SCT recipients enrolled at 23 U.S. transplantation centers found the cumulative incidence (6 months post-transplant) of invasive aspergillosis was 1.25% versus 1% for invasive candidiasis (Kontoyiannis et al., 2010).

These trends were influenced by expanded, more effective, and less toxic antifungal therapy as well as improved immunosuppressive regimens for cancer therapy and maintenance of solid organ and stem cell transplants. Thus, the mortality from invasive aspergillosis has declined but that from invasive candidiasis has remained constant.

11.7 RISK GROUPS/FACTORS

11.7.1 Invasive Candidiasis

The incidence of invasive candidiasis has risen in recent years because of its relation to a combination of one or more *iatrogenic* factors and as the number of risk factors increase, the probability of invasive candidiasis increases.

Intensive Care Unit Patients

These patients can experience anatomic disruption because of

- Burns
- Cardiothoracic surgery (open heart surgery especially valves)
- Gastrointestinal surgery that may lead to increased translocation across the gut
- Other penetrating injury
- Prematurity and low birthweight (<1500 g)/(risk factors in the NICU).
- Solid organ transplants

Immunosuppressive/Cytotoxic Therapy

Invasive and acute disseminated candidiasis occurs in the setting of serious immune compromise, that is, the lack of a functioning immune system. Serious immune compromise is a consequence of the use of drugs and/or radiation to treat a primary disease such as hematologic malignancy, solid tumors, autoimmune disease, or for the maintenance of a stem cell or solid organ transplant. Such therapy reduces the white blood cell (WBC) count to concentrations below 1000/μL, usually described as neutropenia. Prolonged neutropenia occurs when the PMN count falls below 1000/μL for 1 week or more, whereas profound neutropenia corresponds to less than 100 PMNs/μL. In summary the risk factors are:

- Cytotoxic and/or immunosuppressive therapy
- Prolonged or profound neutropenia
- Mucosal erosion in the gastrointestinal tract because of cytotoxic chemotherapy
- Systemic steroid therapy

Risk Factors Common to Both Groups

- Intravascular catheters for nutritional support: total parenteral nutrition (including persons being treated at home with IV infusion therapy). *Candidemia is almost always accompanied by the presence of intravascular catheters, especially central venous catheters*. Catheters bypass the anatomic barrier of the skin, providing a plastic surface for adherence of *Candida* species and biofilm formation. The nutritional support of total parenteral nutrition (TPN) provides a rich growth medium for yeast. Additionally, TPN weakens the enteric mucosae, increasing the risk of translocation of *Candida* from the gut.
- Mechanical ventilation.
- Prolonged broad-spectrum antibacterial antibiotic therapy. The number of drugs used and the duration of therapy are important factors.
- Prolonged hospital stays.

Other Risk Factors

These include sharing of needles and "works" by injection drug users, and accompanying debilitation; endocrine disorders (e.g., diabetes mellitus); hemodialysis; and HIV disease. Invasive candidiasis is rare in HIV-seropositive patients as PMN function is intact until late in AIDS.

11.7.2 Mucosal Candidiasis

Oropharyngeal Candidiasis (OPC)

OPC and oroesophageal candidiasis, without other symptoms, may occur in HIV-seropositive persons when the $CD4^+$ T-lymphocyte count falls below 200/μL. Oral candidiasis also is seen in both normal and diabetic individuals who wear dentures (please see: Section 11.9, Clinical Forms—Denture Stomatitis.).

Vulvovaginal Candidiasis (VVC)

An episode of VVC is triggered by the use of antibiotics, high-estrogen oral contraceptives, hormone replacement therapy, or pregnancy (Fidel, 2005). Diabetes mellitus also is a predisposing factor. The cause of most cases of recurrent VVC is idiopathic. (Please see Section 11.9, Clinical Forms—Vulvovaginal Candidiasis.) Sexual transmission of *C. albicans* occurs, but its significance in disease is not known.

11.7.3 Cutaneous Candidiasis

Risk factors include alcoholism, diabetes mellitus, obesity, and the need to wear diapers (infants, incontinence). *C. albicans* may cause skin and nail infections (onychomycosis) in occupations requiring frequent immersion of hands in water or hand-washing or wearing of rubber gloves (e.g., bar waiters, cooks, dish washers, healthcare workers).

11.8 TRANSMISSION

Community-Onset Candidiasis

The increased use of home infusion therapy involving intravenous catheters for nutritional support means that a rising proportion of *Candida* sepsis is community acquired. Although a minor contributor to sepsis in the community, injection drug users may develop *Candida* sepsis after injection of brown heroin diluted in lemon juice. Lemon juice provides a medium in which *Candida* species can survive.

Hospital-Acquired or Nosocomial Candidiasis

Invasive candidiasis is primarily a hospital-acquired disease. The source of the infection may be endogenous or exogenous.

- *Endogenous Source.* Intravenous catheters may be seeded with *Candida* from the skin or through the bloodstream from the gastrointestinal tract.
- *Exogenous Source.* Horizontal transfer of *Candida* species can result from healthcare workers' hands.

In hospitals and institutions, *Candida* species are transmitted to patients from healthcare workers' hands and from contaminated invasive devices such as IV catheters for nutritional support, transducers for monitoring blood pressure, or ventriculoperitoneal shunts. Other exogenous sources are improper disposal of urine collection bags, infusion pumps, propofol anesthetic, and topical skin emollients. Prosthetic heart valves also may develop growths of *Candida* and must be replaced.

Transmission in the neonatal intensive care unit (NICU) is associated with healthcare workers' (HCWs) hands in installation and maintenance of indwelling catheters and from a variety of other sources (e.g., contaminated glycerol suppositories, stethoscopes). Such a common source can result in horizontal transmission leading to clusters of candidemia caused by the same *Candida* strain or strains circulating in the NICU (Reiss et al., 2008). Colonization of the hands of HCWs varies according to the hospital; in one study 47% of the nurses and 44% of the physicians in the NICU were colonized with *C. parapsilosis*, the predominant *Candida* species present (Verduyn Lunel et al., 1999). Gastrointestinal colonization with *Candida* species in the NICU was estimated to be 45.5% in infants who later developed candidemia, compared with 21.9% colonization in noncase infants. Crowding in the neonate nursery also may contribute to transmission of *C. albicans* and related *Candida* species.

Endogenous or Familial Sources

Many or most infections with *C. albicans* are caused by the patient's own *endogenous* mucosal strain, which converts to become a pathogen as a result of decreased host resistance resulting from chemotherapy for cancer and/or immunosuppressive therapy; immune or endocrine disorders; or breakdown in the integrity of the skin and/or mucosae resulting from injury or surgery.

Similar or identical genotypes of *C. albicans* oral and gastrointestinal isolates were shared among family members, as detected by multilocus sequence typing (MLST) (Bougnoux et al., 2006). The prevalence of colonization was 47% among members of 45 families. *C. albicans* accounted for 88% of the isolates. *C. glabrata* was identified in 9% of carriers, whereas for *C. parapsilosis* the extent of colonization was 4%. This study indicates there is a high level of oral and gastrointestinal carriage of *C. albicans* and that *C. glabrata* also is involved in colonization. Furthermore, there is a strong tendency for strains to be shared among family members. Vertical transmission of *C. albicans* from mothers to their newborns can occur at birth as a result of passage through the vagina but it is not the primary means of *Candida* acquisition for neonates.

The less frequently encountered *Candida* species occur as minor constituents of the host microbiota, but are more apt to be exogenously acquired from the environment. *Candida* species survive on hard, inanimate surfaces, glass and stainless steel, as well as on fabrics and can be transmitted from those sources to human hands (Kumamoto and Vinces, 2005).

Other Transmission Sources

Opportunities for transmission of *C. albicans* arise due to casual or intimate contact between individuals. Sexual transmission of *C. albicans* occurs, but males are most often asymptomatic.

Prevention of Invasive Candidiasis

Three low-tech approaches are effective and should be used before resorting to prophylactic, *preemptive*, or *empiric* antifungal drug therapy. They are educational programs to increase vigilance regarding hand hygiene; adherence to recommended care of intravenous catheters; and prudent use of antibacterial antibiotics. If infection control procedures are in place and candidemia persists, then consideration is given to prophylactic antifungal therapy. Fluconazole, 400 mg/day (in adults) during neutropenia, is effective in decreasing *C. albicans, C. tropicalis*, and *C. parapsilosis* bloodstream infections (BSIs).

11.9 CLINICAL FORMS

See Table 11.4. (Please also see Section 11.7, Risk Groups/Factors, for risk factors associated with clinical forms of candidiasis.) *Candida* species vary in their response to antifungal agents and in their pathogenesis, so it is important to identify the species involved when isolates are obtained from specimens other than skin, respiratory, and urinary tracts (unless requested) in order to guide clinical treatment decisions. Also, the rationale for antifungal susceptibility testing is strengthened because of the broader range of antifungal agents with potent activity against *Candida* species, and the increased percentage of *Candida* species other than *C. albicans* involved in candidiasis, some of which have decreased susceptibility to FLC or to other antifungal agents.

Table 11.4 Summary of Clinical Forms of Candidiasis

Superficial cutaneous
Diaper rash
Intertrigo
Onychomycosis
Onychia
Paronychia
Superficial mucosal
Denture stomatitis
Oropharyngeal (thrush)
Hyperplastic (*Candida* leukoplakia)
Pseudomembranous
Angular cheilitis (perleche)
Chronic mucocutaneous candidiasis
Mucosal
Esophageal
Gastrointestinal
Urinary tract
Vulvovaginal
Invasive (systemic)
Candidemia
Central nervous system
Endocarditis
Endophthalmitis
Peritonitis
Renal
Respiratory tract
Invasive (multisystem, disseminated)
Acute disseminated
Multisystem
Skin lesions
Chronic disseminated
Hepatosplenic

Superficial Candidiasis

Superficial candidiasis includes common conditions, for example, cutaneous candidiasis, chronic mucocutaneous candidiasis, oropharyngeal candidiasis, paronychia and onychomycosis, and vulvovaginal candidiasis. Most individuals with superficial candidiasis have some underlying predisposition. *Candida albicans* is the *Candida* species most often associated with superficial candidiasis (Hay, 1999). Other species—*C. tropicalis, C. parapsilosis, C. glabrata*, or *C. guilliermondii*—also may be involved. *C. albicans* is a common commensal on the mucosae of the mouth, gastrointestinal tract, and vagina. *C. albicans* may not survive for long periods in the environment but can be isolated when there is frequent human contact, on sinks, cups, and so on. Oral colonization may originate at birth or in the first year of life. The occurrence of *Candida* on the skin and mucosae is increased by various factors, for example, hospitalization, bottle feeding, antibiotic therapy, and immunosuppression.

Cutaneous

On *glabrous* skin cutaneous candidiasis normally starts with erythematous fine papules, which then develop into white pustules.

Diaper Rash Diaper rash is a benign dermatitis seen in infants and may occur concomitantly with oropharyngeal candidiasis. It may begin in the perianal area as erythematous confluent plaques and may progress forward where there is diaper contact. It may extend to the upper back and abdomen. The hallmark finding that differentiates cutaneous candidiasis from bacterial infection is the presence of satellite lesions. A similar clinical form may occur in the incontinent elderly and/or those confined to a hospital or nursing home for prolonged periods (Fig. 11.4).

Intertrigo Intertrigo is found where two skin surfaces are in close proximity, causing a warm, moist area in which the *Candida* yeast can multiply. Areas favorable for the development of intertrigo are the axilla, submammary, umbilical, perineum, and under rings on hands (Fig. 11.5). Intertrigo also is seen in the elderly and obese individuals, particularly those with diabetes mellitus.

Nail Disease (Onychomycosis) *Candida* species cause between 1% and 32% of toenail onychomycosis, and 51–70% of fingernail infections (Gupta et al., 2003) either as the primary pathogen or combined with dermatophytes or other molds. *Candida* onychomycosis is seen in chronic mucocutaneous candidiasis patients and in other immunocompromised patients, including HIV-positive persons. When there is an underlying immunodeficit, distal-lateral-subungual-onychomycosis may be the initial symptom but may progress to total dystrophic disease involving the entire nail plate. Involvement of the surrounding soft tissues may occur. (Please see Chapter 21 for a description of the various types of onychomycosis.)

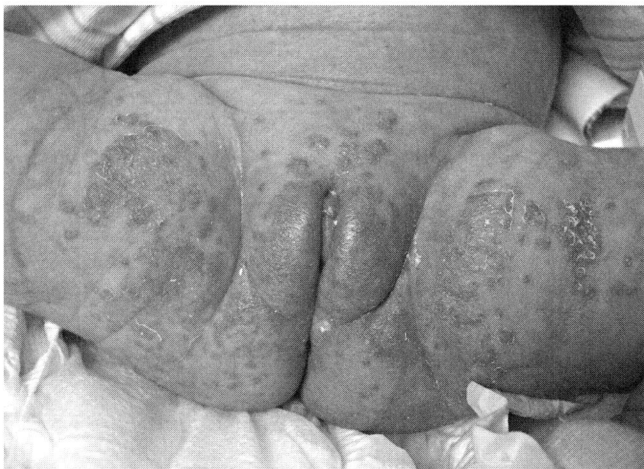

Figure 11.4 Diaper rash in a 3-wk-old infant. The diaper area was inflamed with confluent bright red papules and plaques with scattered pustules, overlying scale, and satellite lesions at the periphery. A KOH prep of skin scrapings revealed abundant hyphae. The infant also had thrush (oral candidiasis).
Source: Copyright Douglas Hoffman M.D. Used with permission from Dermatlas at www.dermatlas.org.

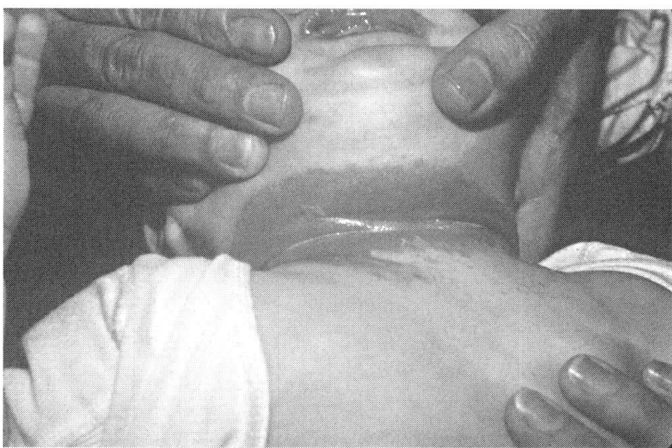

Figure 11.5 *Candida* intertrigo on an infant's neck.
Source: H. J. Shadomy.

Onychia consists of a separation of the nail from the underlying bed (onycholysis). Paronychia is disease of the tissue surrounding the nail (eponychium) and may be acute, with redness and painful swelling, or chronic, with little erythema and little pain or swelling. It is common to isolate bacteria or other fungi from the same site, and diagnosis is made primarily by the predominance of *Candida* in KOH preps and culture. (Fig. 11.6).

Persons with normal immune and endocrine systems whose nails are constantly wet or have been damaged may develop *Candida* onycholysis. Onycholysis may be either distal or lateral, with or without *paronychia*.

Cutaneous Manifestation of Disseminated Candidiasis As seen in Section 11.3, Case Presentations, candidemia and invasive candidiasis often are accompanied by a generalized maculopapular rash which, if scraped and observed microscopically, yield *Candida* species blastoconidia and/or hyphae (please see Section 11.12, Laboratory Detection, Recovery, and Identification—Direct Examination).

Mucosal Candidiasis

Denture stomatitis, oropharyngeal candidiasis (OPC), thrush, and VVC are the most common forms of mucosal candidiasis, followed by esophageal disease. *Candida*

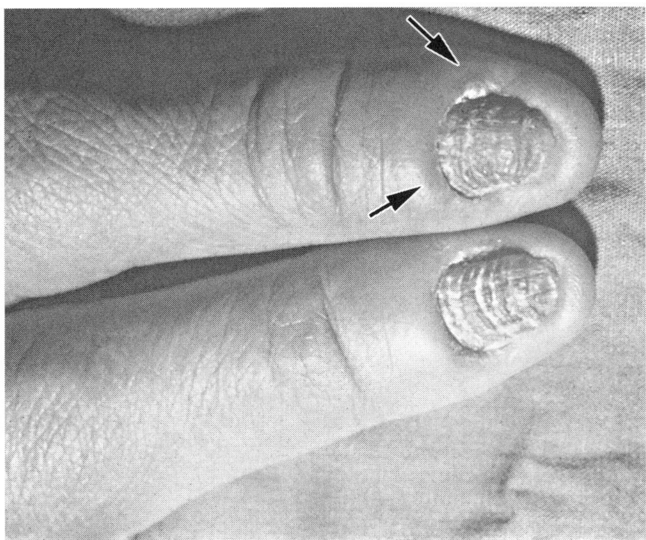

Figure 11.6 *Candida* onychia with paronychia, with digital folds. *Arrows* point to swelling.
Source: H. J. Shadomy.

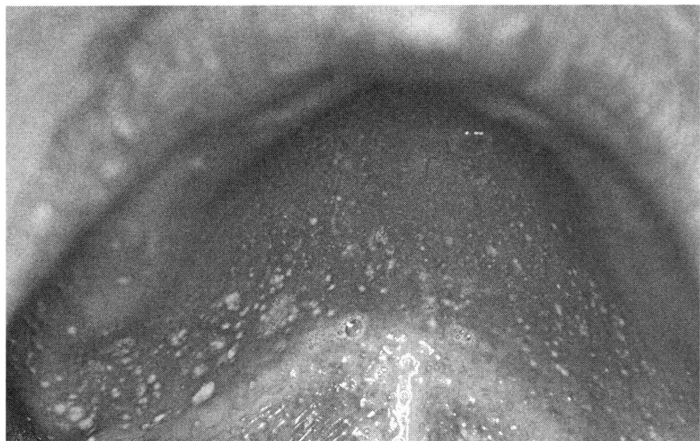

Figure 11.7 *Candida* denture stomatitis (Scully et al., 2004). See insert for color representation of this figure.
Source: Used with permissions from Prof. Crispian Scully, Eastman Dental Institute for Oral Health, University of London, and from Taylor and Francis Books, UK. Copyright © 2004.

albicans is most frequently encountered and, less commonly, *C. glabrata*.

Denture Stomatitis Denture stomatitis (Fig. 11.7) is chronic atrophic candidiasis or "denture sore mouth." It is common beneath complete upper dentures, especially if worn during sleep. In denture stomatitis, candidiasis usually occurs only as chronic erythema, without an adherent white membrane.

Oropharyngeal Candidiasis (OPC) and Thrush Oropharyngeal candidiasis, pseudomembranous type (Fig. 11.8) usually is symptomatic, causing pain, a burning sensation, and a loss of appetite and, if untreated, can lead to wasting. It occurs on the buccal mucosae and tongue (*thrush*), as well as the gums, palate, tonsillar area, and uvula. Thrush is characterized by white curd-like patches on the tongue and other oral mucosal surfaces. Patches on the tongue may become confluent if not treated, and when removed, red, bleeding or "weeping" areas may be seen. Thrush also may be seen in cancer patients and some patients receiving inhalant steroids for treatment of asthma.

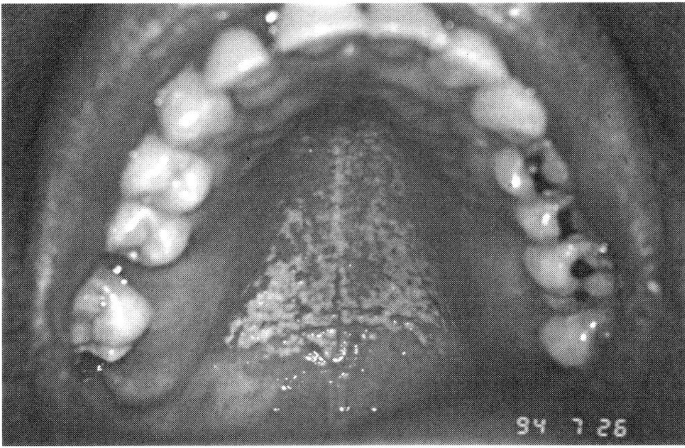

Figure 11.8 Oropharyngeal candidiasis, pseudomembranous type.
Source: Used with permission of HIVdent.org © Dr. David Reznik, Director, Oral Health Center, Infectious Disease Program of Grady Health Systems, Atlanta, Georgia.

- *Pseudomembranous OPC.* Frequently the infected area is covered with a loosely adherent white membrane or curd-like plaques which, when removed, display a moist red base. This form of OPC is a common early indication of the transition of HIV infection to AIDS (Reznik, 2005).
- *Erythematous OPC.* This form is characterized by red, flat, subtle lesion(s) on the dorsal surface of the tongue or on the hard or soft palate. The patient reports oral burning when eating salty or spicy foods. Scraping affected tissues and examination of KOH preps reveal *Candida* hyphae or, more likely, yeast forms.
- *Hyperplastic (*Candida *Leukoplakia) OPC.* Hyperplastic candidiasis is characterized by erythematous redness of palate, tongue, and gingival and oral mucosae in general. In AIDS, white hyperplastic lesions may occur on the palate, buccal mucosae, oropharynx, and lateral borders of the tongue. These lesions also may be scraped to demonstrate *Candida* organisms, but the plaques are not easily removed as in the pseudomembranous type.

- *Angular Cheilitis (Perleche).* This disease is characterized by painful white and red cracks at the corners of the lips.

Chronic Mucocutaneous Candidiasis (CMC)

This is a rare, disfiguring, and, in some patients, severe granulomatous form of candidiasis. Beginning early in childhood, generally starting in the mouth, the disease spreads over time to skin, hair, and nails. Persistent and recurrent infections of the mucous membranes, skin, and nails last for years in untreated patients. Lesions in the mouth and oropharynx can extend into the esophagus. Vulvovaginal candidiasis may also occur. Mucous membrane disease is similar in symptoms to that of other patients with OPC or VVC. Invasion of the deep tissues is rare. The skin lesions are red, raised, and hyperkeratotic, but are not considered painful. The nails can be severely affected and disfigured. The overall clinical picture is one of disfigurement and accompanied by the social rejection patients feel when people avoid touching them, possibly fearing contagion.

Classification systems for CMC have proposed seven distinct groups, underlining that this disease is comprised of several different syndromes. A pattern of familial inheritance is noted in approximately 25% of patients.

Clinical Syndromes of Mucocutaneous Candidiasis (Kirkpatrick, 2001)

- Chronic oral candidiasis
  - Iron deficiency
  - HIV infection
  - Denture stomatitis
  - Inhaled corticosteroid use
- Familial chronic mucocutaneous candidiasis
- Autoimmune-polyendocrinopathy-candidiasis-ectodermal dystrophy (please see text)
- Chronic localized candidiasis
- Chronic mucocutaneous candidiasis with thymoma
- Candidiasis with chronic keratitis
- Candidiasis with hyper-IgE syndrome

CMC Associated with Endocrinopathy Recognizing the role played by endocrine dysfunction in CMC, workers in the field proposed the term "autoimmune-polyendocrinopathy-candidiasis-ectodermal dystrophy (APECED) (Meriluoto et al., 2001). APECED is an autosomal recessive disease associated with mutations in *AIRE* (for "autoimmune regulator"). This gene encodes a proline-rich 58 kDa protein with two zinc finger motifs characteristic of transcription regulators, and a nuclear targeting signal indicating its cellular location. Expression level is highest in the thymus, lymph nodes, spleen, and peripheral blood cells and also is present in some nonlymphoid tissues. The exact function of AIRE is unknown, but it appears to interfere with T-cell self-tolerance. The underlying deficit results in autoimmune disease manifested as circulating tissue-specific autoantibodies which attack the endocrine organs—adrenals, gonads, pancreas, parathyroids, and thyroid, with resultant organ dysfunction. APECED occurs worldwide and at increased frequency in isolated populations such as Finns, Iranian Jews, and Sardinians (Meriluoto et al., 2001).

Other Clinical Forms of CMC Localized CMC involving both hands with hyperkeratosis and cutaneous horn formation is not accompanied by endocrine dysfunction. Another group with diffuse CMC includes persons with the most widespread and severe disease, with no known familial history.

Humoral and complement-mediated immunity is unaffected in CMC but CMI defects are commonly present in the form of anergy to *Candida*-specific antigens and to common recall antigens. Patients vary in their pattern of reactivity to these antigens. Therapy consists of long-term oral FLC.

Mucosal Candidiasis

Esophageal

Patients with esophageal candidiasis complain about difficult or painful swallowing. Retrosternal pain and burning may indicate ulcerative esophageal disease.

Gastrointestinal

Colonization of the gastrointestintal tract with *Candida* species is common in healthy persons and increases markedly after systemic broad spectrum antibiotic therapy. Isolation of *Candida* species from stool specimens in immune-normal individuals is not indicative of disease in and of itself. Disease is rare and is associated with *Candida* penetration of superficial erosions resulting in nausea, pain, and diarrhea.

Urinary Tract

Candiduria is rare in otherwise healthy individuals but is well known in hospitalized patients with indwelling drainage devices. The catheters are the portal of entry of *Candida* into the urinary drainage. Women are at greater risk because of vaginal colonization with *Candida* species.

Infection can occur when *Candida* migrate into the bladder but symptomatic *Candida* cystitis is rare. The simple isolation of a *Candida* species from urine is, in and of itself, not indicative of disease.

Urinary tract candidiasis is caused by several *Candida* species, including *C. albicans, C. tropicalis, C. parapsilosis, C. glabrata*, and occasionally other species. *Candida* spp. are isolated both in urinary tract infection and in pyelonephritis. Colonization of the urinary tract by *Candida* often is asymptomatic but can lead to invasion of the bladder wall. Instrumentation of the renal pelvis by catheterization may lead to *Candida* pyelitis. Hematogenous candidiasis may lead to renal abscesses. Urethral candidiasis in men usually results from sexual contact with women with *Candida* vaginitis, while that seen in women generally is an extension of *Candida* vaginitis.

Vulvovaginal Candidiasis (VVC)

Symptoms include burning, itching, soreness, abnormal discharge, and/or dyspareunia (*definition*: pain in the vagina or pelvis experienced during sexual intercourse). VVC is most frequently associated with antibiotic use or high estrogen oral contraceptive use, hormone replacement therapy, or the third trimester of pregnancy. Patients receiving long-term systemic antibiotic treatment develop mucosal candidiasis when an antibiotic-resistant replacement microbiota grows, including yeasts. The same is observed in cancer patients undergoing chemotherapy. Women with *tinea cruris* may have candidiasis rather than dermatophytosis. A single oral dose of 150 mg of FLC is sufficient to treat an episode of VVC but *C. albicans* colonization persists because FLC is fungistatic and not fungicidal.

Candidiasis Hypersensitivity Syndrome

The chronic *Candida* syndrome or candidiasis hypersensitivity syndrome is not covered here. For complete discussion see American Academy of Allergy, Asthma and Immunology Physician Reference Materials and their Position Statement 14, "Candidiasis Hypersensitivity Syndrome" a portion of which is excerpted below. Position paper #14 was superseded in 1999 by position paper #35 which can be read at the URL http://www.aaaai.org/members/academy_statements/position_statements/ps35.asp.

> *The symptoms are described as wide-ranging, involving multiple systems, and include fatigue, lethargy, depression, inability to concentrate, hyperactivity, headaches, skin problems, including urticaria, gastrointestinal symptoms such as constipation, abdominal pain, diarrhea, gas, and bloating, respiratory tract symptoms, and symptoms involving urinary tract and reproductive organs.*
>
> Candida *germs live in every person's body especially in mucous membranes. Accordingly, vaginal and other smears and culture for* Candida *don't help. Therefore diagnosis is suspected from the patient's history and confirmed by his response to treatment.*
>
> *Critique: The Practice Standards Committee finds multiple problems with the Candidiasis Hypersensitivity Syndrome.*

1. *The basic elements of the syndrome would apply to almost all sick patients at some time. The complaints are essentially universal; the broad treatment program would produce remission in most illnesses regardless of cause.*
2. *Elements of the proposed treatment program are potentially dangerous.*

For further reading please see Blonz (1986).

Invasive Candidiasis (Systemic)

Candidemia

Candidemia, a BSI, usually is acquired in the hospital setting and may follow such procedures as introduction of IV or urinary catheters, corticosteroid or antibacterial antibiotic therapy, and neutropenia. Mucositis[4] in the gastrointestinal tract from cytotoxic chemotherapy, bowel surgery, or other abdominal surgery can allow *Candida* species to invade the bloodstream. Any of the major *Candida* species may be isolated from patients with deep-seated candidiasis. The *Candida* species distribution in candidemia is discussed in Section 11.6, Epidemiology. Early onset candidemia occurs within 48 h of hospital admission and is classed as community onset. Late onset candidemia, occurring $\geq$72 h after admission, is classed as hospital associated or nosocomial candidemia.

Central Nervous System (CNS) Candidiasis

CNS candidiasis is seen in low birthweight neonates with candidemia and in patients with hematologic malignancies, complicated neurosurgery, and intracerebral prosthetic devices, such as ventriculoperitoneal shunts. Aside from rare complications of neurosurgery or accidental injury, CNS candidiasis is caused by hematogenous spread and is a frequent site of disseminated disease. *Most often CNS findings are multiple microabscesses and granulomas that are not evident on imaging*. Brain tissue and/or the meninges may be involved. The signs and symptoms of acute *Candida* meningitis often are indistinguishable from bacterial meningitis.

[4]A side effect of chemotherapy that causes inflammation of the mucosal tissues that line the mouth, throat, stomach, intestines, and rectum.

Endocarditis

Endocarditis often is preceded by candidemia in patients with abnormal heart valves or those who have undergone heart valve replacement. Two additional contributing factors are IV catheters for nutritional support and the habits of injection drug users (i.e., using contaminated needles and syringes). Positive blood cultures can provide a clue to this disease, supplemented by echocardiography. Prosthetic valve endocarditis is the most common form of *Candida* endocarditis with *C. parapsilosis* notable as a causative agent. Therapy consists of surgery combined with antifungal therapy.

Endophthalmitis

The eyes (Pick et al., 1988; Sallam et al., 2006; Spellberg and Edwards, 2002), as well as the heart and kidneys, are common sites of involvement in hematogenously disseminated candidiasis. There has been a decline in *Candida* chorioretinitis and endophthalmitis to an estimated prevalence of 2.5% in patients with hematogenously disseminated candidiasis (reviewed by Sallam et al., 2006). The reasons are due to prompt early treatment when *Candida* is detected in blood cultures and because FLC, used as prophylaxis and primary therapy, has superior ocular penetration compared to therapy with AmB. The consequences of unrecognized endophthalmitis are severe, requiring a formal ophthalmologic assessment after detection of candidemia, to prevent potential loss of vision. If the intial review is negative, another review should follow in 2 weeks.

The PanOptic™ ophthalmoscope is advantageous because it (i) can be used in ambient lighting; (ii) can look through small, undilated pupils; (iii) provides a 5× larger view of the fundus than a standard ophthalmoscope in an undilated eye; (iv) enables a 25° field-of-view versus the standard 5°; and (v) increases magnification by 26% over a standard scope. The hallmark of *Candida* chorioretinitis is a fluffy, creamy-white lesion at the level of the retina and choroid that is usually associated with the vitreous. Lesions are commonly multiple and can be bilateral. The inflammation progresses into the vitreous and sometimes includes intravitreal puff ball-like lesions that are vitreous abscesses (Fig. 11.9).

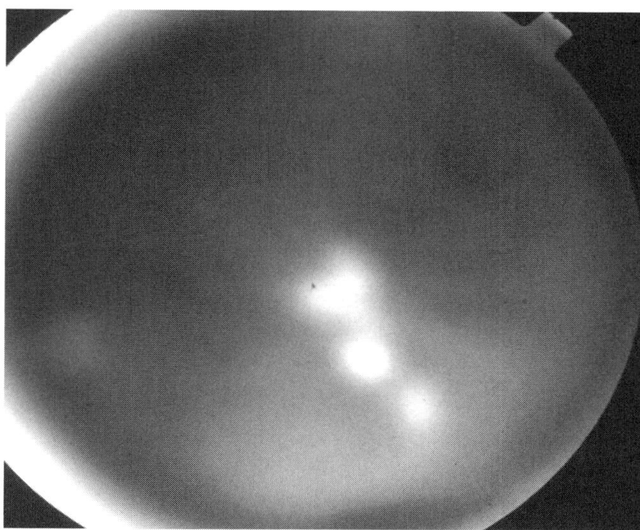

Figure 11.9 Fundoscopic image of the classic appearance of *Candida* endophthalmitis with fluffy white lesions extending into the vitreous.
Source: Used with permission from Dr. Susan L. Lightman M.D., Professor and University Chair of Clinical Ophthalmology, Moorfields Eye Hospital, London, England, UK.

Peritonitis

Candida peritonitis is rare and may be caused by gastrointestinal perforation through injury or surgery, producing a mixed infection including both bacteria and *Candida*. Pancreas transplantation is associated with *Candida* infection of the pancreas and intraabdominal abscess. Peritonitis in chronic ambulatory peritoneal dialysis patients also may occur due to growth of the fungus in the fibrinous debris of biofilm in the catheter tip. The organism may be cultured but usually is not seen in smears.

Renal Disease

Candida species have a tropism for the kidneys and it is a frequent site in disseminated candidiasis.

Respiratory Tract

Invasion of the lung parenchyma or pneumonia caused by *Candida* species is rare. When it does occur, *Candida* causes either a local or diffuse bronchopneumonia from endobronchial inoculation of the lung, or a hematogenously seeded, finely nodular diffuse infiltrate. Because it is not uncommon to find yeasts, including *Candida*, in the bronchi, the simple finding of yeasts is not diagnostic of disease. The only definitive method for diagnosis is biopsy with histopathology.

Invasive Candidiasis (Multisystem/Disseminated)

Disseminated candidiasis may be either acute or chronic. (Please see Section 11.7, Risk Groups/Factors—Invasive Candidiasis.)

Acute Disseminated Candidiasis

This begins with abrupt onset of fever, fungemia, occasionally with skin rash, and sometimes shock. Prompt therapy is essential. Sites of dissemination include the eyes, kidneys, liver, lung, heart, skin, and central nervous system.

Symptoms and signs of acute disseminated candidiasis (whether or not patients are immunosuppressed) include:

- Candidemia
- Candiduria
- Fever unresponsive to antibiotics
- Focal microabscesses in the brain
- Hematuria
- Muscle pain
- New or changed heart murmurs
- Skin rash (diffuse cutaneous nodules or papules)
- White fluffy retinal plaques

Chronic Disseminated (Hepatosplenic) Candidiasis

This can occur in patients who have recovered from neutropenia. Typically this initially presents with a fever that occurs as the neutropenia is resolving, when antifungal therapy has been discontinued, or persists when the PMN count returns to normal. Persistent fever is indicative of *Candida* lesions growing slowly over a period of weeks or months in the liver, spleen, kidney, and lung. White-yellow nodules between 1 and 3 cm in diameter are microabscesses containing PMNs and yeast forms. After antifungal therapy, there is an infiltrate of mononuclear cells and a fibrovascular capsule forms with necrotic fungi at the center. This is followed by fibrosis and scar tissue. Candidemia is less likely to be present (please see Section 11.3, Case Presentation 4) (Fig. 11.10) (Halkic and Ksontini, 2007).

Diagnostic Aids for Hepatosplenic Candidiasis
At present, diagnostic aids for this form of candidiasis consist of diagnostic imaging [magnetic resonance imaging (MRI) and high resolution computed tomography (CT)], increased serum alkaline phosphatase, and liver biopsy. Using MRI and histopathology, the disease was detected in 23 (4.5%) of 514 patients with acute leukemia (Sallah et al., 1999).

*I*mmune *r*econstitution *i*nflammatory *s*yndrome (IRIS) can explain the enigmatic entity of chronic disseminated (hepatosplenic) candidiasis in approximately 4.5% of leukemia patients after their recovery from neutropenia. Please see Chapter 12, Section 12.10, Therapy, for a discussion of IRIS.

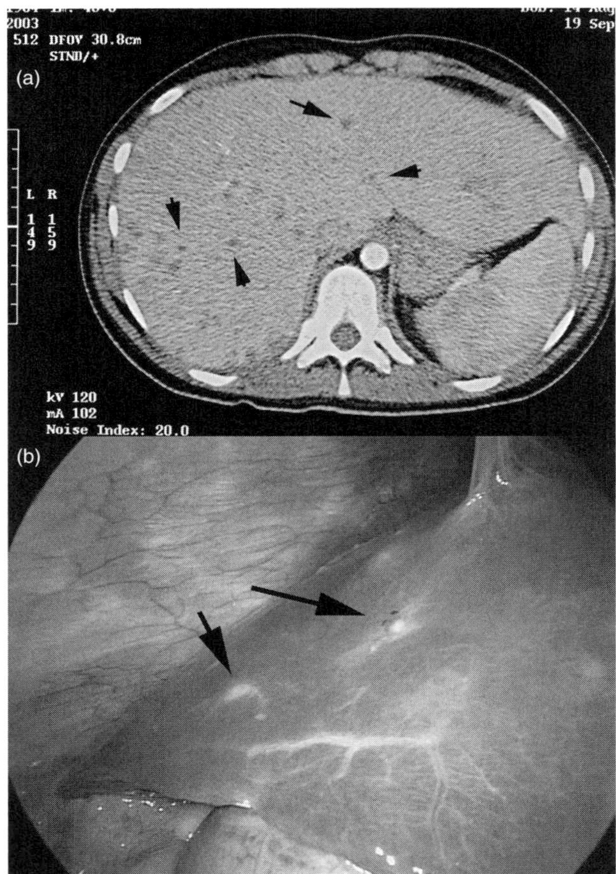

Figure 11.10 Chronic disseminated (hepatosplenic) candidiasis: *Candida albicans* microabscesses shown (*arrows*) in (a) CT scan, abdomen, and (b) on the liver of a 19-yr-old woman with acute myelomonocytic leukemia. She underwent five cycles of chemotherapy and during the last cycle developed neutropenia lasting 22 days. A fever peaking at 39.5°C ensued along with an increased serum alkaline phosphatase, 335 U/L. The abdominal CT scan showed multiple hypodense lesions in the liver and spleen (*panel a, arrows*)—consistent with hepatosplenic microabscesses. Diagnostic laparoscopy was performed to obtain biopsy specimens of the hepatic lesions (*panel b, arrows*) which, after PCR-sequencing, identified the agent as *C. albicans*. The patient received FLC for 1.5 months. Her body temperature returned to normal and alkaline phosphatase decreased to 185 U/L. Follow-up CT scan showed a decrease in the size and number of lesions.
Source: Halkic and Ksontini (2007). Used with permissions from Dr. Nermin Halkic, University of Lausanne Medical School, and from the Massachusetts Medical Society. Copyright © 2007, all rights reserved.

Recommended Therapy for Hepatosplenic Candidiasis Fluconazole is recommended for clinically stable patients (Pappas et al., 2009). Acutely ill patients or those with refractory disease can benefit from lipid AmB. The response rate for AmB in treating hepatosplenic candidiasis ranges from 40% to 80%

(reviewed by Masood and Sallah, 2005). Echinocandins are recommended as alternative therapy (e.g., caspofungin) for a minimum of 2 weeks. When appropriate, therapy should switch to FLC and should continue for weeks or months until calcifications occur or lesions resolve. Patients continuing to receive chemotherapy, or patients with stem cell transplants, should continue to receive therapy to prevent relapse. Lesions may recur if therapy is stopped prematurely.

11.10 DETERMINANTS OF PATHOGENICITY

Pathogenicity of *C. albicans* and of other *Candida* species is multifactorial. In *C. albicans*, morphogenesis to the pseudohyphal and hyphal forms occurs during invasion of host tissues, and all three forms—yeast, pseudohyphal, and hyphal forms—are commonly present in infected tissue. Pseudohyphae and hyphae generally are too large for *endophagocytosis* and adhere better to host tissues than do yeast cells.

11.10.1 Host Factors

(Please also see Section 11.7, Risk Groups/Factors.)

Role of Host Defenses in Candidiasis

Normal intact anatomic barriers, along with the innate and T-cell-mediated limbs of the immune response are important for a complete defense against candidiasis. Defense against mucosal candidiasis is linked to CMI, whereas innate phagocytic immunity is key to defense against invasive, disseminated candidiasis. Vulvovaginal candidiasis (VVC) is a special case that defies these generalities [please see "Immune Response in Vulvovaginal Candidiasis (VVC)" below].

Innate Immunity

Early recognition of *C. albicans* and *Candida* species, as they attempt to breach host defenses, is mediated by pattern recognition receptors (PRRs) in the innate immune system, which recognize several pathogen associated molecular patterns (PAMPs). The PRRs consist of integrins,[5] lectins, the macrophage mannose receptor,[6] a ß-mannose receptor, ß-glucan receptor (dectin), and Toll-like receptors (TLRs) (Table 11.5). (Vonk et al., 2006). The ß-mannose receptor, galectin 3, present on mature macrophages, is responsible for adherence and endocytosis of *Candida*. Galectin 3 binds to ß-(1→2)-linked mannose residues which are present in *C. albicans* but not in *Saccharomyces* (Jouault et al., 2006). One result of these interactions is to rapidly filter out *Candida* from the bloodstream by their adherence to Kupffer cells in the liver and to macrophages in the marginal zone of the spleen.

Chief among effector cells of innate immunity are the "professional phagocytes," mainly polymorphonuclear neutrophilic granulocytes (PMNs) and monocytes. Pathogen recognition by innate immune reactions (Table 11.5) induces a systemic inflammatory response including production of TNF-α, IFN-γ, and granulocyte colony stimulating factor (G-CSF). The result is to augment the killing of *Candida* cells by PMNs. Yeast cells are first phagocytosed and then killed by oxidative microbicidal pathways. PMN killing of *Candida* yeast and hyphae is potentiated by nonoxidative means: defensins, calprotectin, and lactoferrin.

Oxidative Microbicidal Pathways that Kill Candida Yeast

The oxidative pathway functions principally in PMNs, with macrophages occupying a secondary role:

- *Myeloperoxidase (MPO)–Peroxide–Halide Pathway*. Hydrogen peroxide combines with chloride ion to form hypochlorous acid (the same compound as household bleach), which damages the target yeast cell.
- *MPO-Independent Oxidative Mechanism*. Secondary in importance to the MPO-dependent pathway, mature macrophages are capable of utilizing the reactive oxygen species superoxide, O_2^-, to produce hydroxyl radical: $HO^\bullet$. *Candida* yeast and pseudohyphae phagocytosed by macrophages are subject to this reactive oxygen species-mediated killing.
- *Reactive Nitrogen Species*. Activated PMNs produce nitric oxide (NO), a soluble, highly labile free-radical gas. Activated macrophages also produce NO. Within the phagocytic vacuole, in the presence of superoxide, NO is converted to peroxynitrite, $ONOO^-$. Studies in experimentally infected animals argue for a protective role for reactive nitrogen species, but at this time their role in defense against fungi in human mycoses is not well established.

These oxidative pathways, collectively, result in increased oxygen consumption by PMNs, the respiratory

[5]Integrin. *Definition*: Cell surface receptors for extracellular matrix molecules.

[6]The mannose receptor family, including the macrophage mannose receptor, contains three types of extracellular domains: an *N*-terminal cysteine-rich domain that binds to sulfated galactose and sulfated galactosamine; a fibronectin type II domain involved in collagen binding; and eight tandemly arranged C-type lectin domains responsible for Ca^{2+} dependent binding to terminal D-mannose, L-fucose, or *N*-acetyl glucosamine (Gazi and Martinez-Pomares, 2009).

Table 11.5 Pathogen Associated Molecular Patterns (PAMPs) of *Candida* Species that Are Recognized by Pattern Recognition Receptors (PRRs) of the Innate Immune System

PAMP	PRR		
	Lectin	Toll-like receptor (TLR)	Integrin
Mannan or mannose	Mannose receptor	TLR-4	
ß-(1→2) mannose oligosaccharides	S-lectin, galectin-3		
ß- (1→3)glucan	Dectin-1	TLR-2	
Cell wall	Dendritic cell-specific intercellular adhesion molecule (ICAM)		
Opsonized yeast			CD11b/CD18, CR3

burst, occurring immediately after phagocytosis and persisting for up to 3 hours.

Host Defense Peptides Are Candida*-cidal Effectors*

A variety of antimicrobial proteins and peptides are expressed by PMNs, by eosinophils, and by mast cells and make up a nonoxidative microbicidal pathway. Categorically they are cationic granule proteins (a comprehensive review is given by Levy, 2004). Host defense peptides are also secreted by epithelial cells in the oral cavity and in the airways. They are active in saliva and in airway fluid (Diamond et al., 2008). Some of these factors, which have attracted attention because of their potent anti-*Candida* activity, are summarized:

- *Calprotectin.* This calcium and zinc binding peptide (8–14 kDa) is released upon death from PMNs at inflammatory sites, a process called "holocrine secretion."
- *Defensins.* These peptides are approximately 30 amino acids long, positively charged, and proteinase resistant with three internal disulfide bonds. Defensins are classed as either α or ß according to the arrangement of the disulfides. Both classes have a similar three-stranded folded structure. They cause direct lysis of the target cell by forming ion channels in the cell membrane causing leakage of cell contents.
- *Lactoferrin.* By chelating iron, lactoferrin deprives microbes of that essential micronutrient. It also may have direct anti-*Candida* activity.
- *Histatin.* A component of saliva from humans, histatins are a group of α-helical host defense peptides (24 amino acids long and histidine-rich) with potent activity against yeast and hyphal forms of *C. albicans* (Edgerton and Koshlukova, 2000). Histatin 5 is the most potent member of this family. It binds to a *Candida* surface receptor which then translocates into the cytoplasm, where histatins interfere with cellular processes leading to leakage of ATP, with resultant death of the target cell. They are a unique antifungal defense in the oral cavity. The *Candida* surface receptor for histatin was identified as a heat shock protein, ssa2.

Killing of Candida *Pseudohyphal and Hyphal Forms*

Oxidative and nonoxidative candidacidal mechanisms function against these forms which may be too large for endophagocytosis. PMNs adhere to the target and release granule contents onto its surface, termed extracellular degranulation.

Adaptive Immunity in Candidiasis

Adaptive immunity[7] is not an absolute requirement for protection against invasive candidiasis. Even in the complete absence of T and B lymphocytes, phagocytes remain capable of defense against *Candida*. Children with pure B-lymphocyte deficiency are not more susceptible to mucosal or invasive candidiasis (Fidel, 2002). Investigators have been surprised by these findings and have sought to better understand how adaptive immunity might function both positively and negatively in mucosal and invasive candidiasis.

[7]Adaptive immunity. A period of several days are required for clonal expansion and differentiation of naïve lymphocytes into effector T lymphocytes and antibody-secreting B lymphocytes that target a pathogen for elimination. During this latent period, specific immunological memory is established. This ensures a rapid reinduction of antigen-specific antibody and armed effector T cells on subsequent encounters with the same pathogen, providing long-lasting protection against reinfection (Janeway et al., 2001).

Humoral Immunity

Antibody-mediated (humoral) immunity is not required to defend against candidiasis, but can increase the clearance efficiency of phagocytes. The role of humoral immunity is in opsonizing *Candida* for phagocytosis. Few reports show that antibodies, with or without complement, can kill *Candida* in the absence of phagocytosis. An exception is that the passive transfer of recombinant Fc-free human antibodies against *C. albicans* Hsp90 (please see Section 11.10.2, Microbial Factors—Hsp90) increases survival of infected mice (reviewed by Vonk et al., 2006). Opsonizing antibodies also are important in the phagocytosis of *Candida* by antigen-presenting dendritic cells.

T-Cell-Mediated Immunity (CMI)

T lymphocytes protect mucosal tissues from infection. Oropharyngeal candidiasis in AIDS patients occurs when the $CD4^+$ T-lymphocyte count falls below 200/μL. Children with chronic mucocutaneous candidiasis have *Candida*-specific T-lymphocyte unresponsiveness (Fidel, 2002).

T lymphocytes that function along the Th1 pathway secrete IL-2, IFN-γ, and lymphotoxin-α (LTα).[8] These cytokines enhance innate immunity by stimulating phagocytosis, intracellular killing, and production of opsonic IgG_1, and IgG_3, whose Fc portions bind well to phagocyte Fc receptors (Spellberg and Edwards, 2002). Th1 immunity is protective against both mucosal and disseminated candidiasis.

Th17 cells are important in their role in protection against extracellular mucosal pathogens. T-lymphocyte precursors of the CD4 phenotype differentiate under the influence of inflammatory cytokines (IL-1, IL-6, and transforming growth factor β) into Th17 lymphocytes, so named because they produce IL-17. Once they are mature, they are stabilized by another cytokine, IL-23 (Peck and Mellins, 2010; van de Veerdonk et al., 2009). In the maturation of the immune response, Th17 cells are a bridge between innate and adaptive immunity. Many cell types have IL-17 receptors which, when engaged, elicit a range of antimicrobial peptides from mucosal epithelial cells (e.g., ß-defensins). Growth factors and chemoattractants secreted in response to IL-17 are neutrophil chemotactic factor, macrophage inflammatory proteins (MIP1 and MIP2), and granulocyte colony stimulating factor (G-CSF). An innate immune stimulus to the development of Th17 lymphocytes is the binding of ß-(1→3)-D-glucan to dectin-1 on antigen-presenting cells. *C. albicans* mannan also induces Th17 cells via mannan binding to the mannose receptor on monocytes.

Patients with chronic mucocutaneous candidiasis are defective in *Candida*-specific Th17 responses, which helps explain why they are very susceptible to cutaneous and mucosal candidiasis. Genetic defects in dectin-1 or CARD9 are associated with chronic onychomycosis and mucosal candidiasis. As mentioned above, dectin-1 is a key PRR for fungal ß-(1→3)-D-glucan. CARD9, caspase-associated recruitment domain, is a protein interaction domain that activates or suppresses members of the caspase proteinases which are essential for apoptosis and inflammation. CARD9 mediates signals from ligand-engaged PRRs (dectin-1 and macrophage mannose receptor) to downstream signaling pathways which activate proinflammatory cytokines that are potent inducers of Th17 responses. The autosomal recessive form of susceptibility to chronic mucocutaneous candidiasis is associated with mutations in CARD9.

The susceptibility of HIV patients to oropharyngeal candidiasis may also be explained by the depletion of Th17 cells known to be active in mucosal immunity. Dissecting the immune response to other fungal pathogens will further delineate the role of Th17 cells in immunity and inflammation.

Candida albicans yeast phase cells are easily phagocytosed and induce secretion of IL-12 by phagocytes, stimulating Th1 responses. Hyphal forms of *Candida*, however, inhibit IL-12 and induce IL-4, which stimulates futile Th2 responses (Spellberg and Edwards, 2002). This is because hyphal forms often are too large for endophagocytosis, causing phagocytes to degranulate, showering the hyphal wall with hydrolases and defensins. This is accompanied by upregulation of IL-10, stimulating a Th2 response. Damage to the host endothelium by *Candida* hyphal forms releases prostaglandins, which are known to induce Th2 immunity. Thus *C. albicans*, by its shape-shifting ability, can influence the host immune response along a futile pathway. Th2 cells secrete IL-4, IL-10, and IL-13, which suppress phagocytosis and intracellular killing. So Th2 immunity weakens the host response.

The first lines of defense in candidemia are Kupffer cells in the liver and circulating PMNs. Tissue fixed macrophages provide local defense, which can be increased by opsonizing *Candida* with complement or IgG_1 and IgG_3. The Th1 response boosts phagocytosis and killing. Conversion to hyphal forms complicates killing by their tendency to escape from the phagolysosome. Further readings on the broad topic of clinical and experimental immunology of candidiasis can be found in Fidel and Huffnagle (2005).

[8]LTα is similar to TNF-α (produced by macrophages) in that it activates PMNs and increases adherence to endothelial cells and promotes extravasation of leukocytes into tissues.

Immune Response in Vulvovaginal Candidiasis (VVC)

Studies of local or systemic immune responses in women with recurrent VVC have shown little or no deficiency in T-cell-mediated or antibody-mediated immunity (reviewed by Fidel, 2005). HIV-positive women who experience OPC are at no more increased risk of VVC than HIV-negative women. Systemic T-cell responses to *Candida* antigens were similar in women with or without symptomatic VVC.

The consensus after numerous studies is that no protection is provided from Th1 responses or antibodies in VVC, leading to the conclusion that no part of adaptive immunity is helpful. $\gamma\delta$ T cells are present; these cell types are associated with downregulation of immune responses. This lack of responsiveness is designed to prevent inflammation against a commensal yeast. Absent a protective role for adaptive immunity, innate immunity has moved to the center stage, especially the role of vaginal epithelial cells. On their own, the anti-*Candida* activity of vaginal epithelial cells requires cell–cell contact, a membrane event, and is fungistatic not fungicidal. The result of cell contact between vaginal epithelial cells and *Candida* is downregulation, a noninflammatory effect allowing the presence of commensal yeast without evoking an inflammatory PMN infiltrate which would result in symptomatic VVC.

Women with recurrent VVC have the lowest threshold of tolerance and their epithelial cells produce signals that attract PMNs. Women with a history of infrequent VVC have a higher threshold of immunoregulation (i.e., downregulation) than those with recurrent VVC. Women with no history of VVC have the highest level of tolerance of *C. albicans* and keep it in a commensal state. A high level of tolerance is an innate response related to the threshold of *Candida* colonization necessary to trigger vaginal epithelial cells to produce signals that attract PMNs. The nature of these signals awaits further investigation. Therefore, VVC results from an aggressive innate response, with infiltration of PMNs, leading to symptoms.

Experimental Intravaginal Challenge Studies

When a live intravaginal challenge, in the form of a bolus of *C. albicans* yeast, was administered to a group of young women who had never had VVC, only a few percent developed any symptoms. Those who did develop symptoms had a cellular infiltrate consisting of PMNs. PMNs are the main effector cells in defense against invasive candidiasis operating through oxidative and nonoxidative fungicidal mechanisms at other body sites, but are not effective at killing *Candida* in the vaginal environment (Fidel, 2005).

Another experimental group consisted of women who had a brief history of VVC. Live intravaginal challenge of these women with *C. albicans* yeast cells resulted in an increased percentage with symptoms compared to the first experimental group and was accompanied by an exclusively PMN infiltrate. These women have a *low tolerance* for the commensal yeast and their vaginal epithelia produce signals that attract PMNs. PMN infiltration correlates with susceptibility and symptoms. Thus, women who become symptomatic have a low inhibitory vaginal epithelial response related to immunoregulation (or downregulation, or tolerance).

11.10.2 Microbial Factors

Morphogenesis

Candida albicans grown in vitro is a yeast at 25°C and below pH 5; growth in the hyphal form is favored at 37°C and pH >5.5. A mix of yeast forms and hyphal forms is usually present in lesions, indicating a mechanism whereby *C. albicans* and other dimorphic *Candida* species can utilize signaling pathways to modulate form (Whiteway and Bachewich, 2006). Yeast forms are well suited to circulate in the bloodstream and invade deep tissues. Hyphal forms are better equipped for invading across tissue planes. Not only are hyphal forms too large for endophagocytosis, but when morphogenesis occurs within the phagolysosome the hyphal forms can rupture it and burst the phagocyte.

When a phagocyte tries to engulf a target cell larger than itself it reorganizes its cytoskeleton in a reaction known as "frustrated phagocytosis." The result is degranulation, releasing hydrolytic enzymes and defensins, which damage the target cell (Spellberg and Edwards, 2002).

The pathogenic process is initiated when host defenses are compromised. Depending on the type of compromise (please see Section 11.7, Risk Groups/Factors), *C. albicans* will cause mucosal or invasive disease. *Candida* generation time is slow in mucosal niches. However, when there is a breach in anatomic barriers or a reduction in immune surveillance, *C. albicans* yeast cells may produce germ tubes and insinuate itself into the tissues, aided by secreted aspartyl proteinases and phospholipase.

When host defenses are compromised, *C. albicans* crosses the mucosal borders and invades the bloodstream as a prelude to causing disease in the internal organs especially the kidney, spleen, liver, eyes, brain, and heart. Alone among the *Candida* species, *C. albicans* produces a germ tube and then begins to grow in the hyphal form. In diseased tissue one finds a mix of yeast forms, pseudohyphae, and true hyphae. In pseudomembranous OPC, hyphae are numerous and extend into the *spinous cell layer* (de Repentigny et al., 2004). (Please see Section 11.9, Clinical Forms, for other forms of mucosal candidiasis.)

Phenotype Switching

High-frequency phenotype switching is detected by changes in colony morphology. The prototypical switch was the "white" to "opaque" switch described by Soll and colleagues (reviewed by Soll, 2006). The opaque phenotype was later found to have important properties related to mating in *C. albicans*.

Special Topic: Mating Type and Parasexuality

The mating type locus in *C. albicans*, a diploid yeast, is on chromosome 5. Only homozygous isolates express a mating type, "a" or "α." The first step in mating is the loss of heterozygosity at chromosome 5 resulting in the expression of mating type (please see "Genomic Instability" below). When phenotype switching occurs from the white to the opaque form, *C. albicans* is orders of magnitude more capable of undergoing mating. Having satisfied the loss of heterozygosity requirement, and converted to the opaque phenotype, the next requirement is that both a and α mating types are in proximity. Next, mediated by pheromones, conjugation tubes are expressed by the yeast cells leading to karyogamy and a short-lived tetraploid stage. Meiosis has not been detected in *C. albicans*; instead, fusants shed chromosomes to return to a diploid state. A subset of these cells also undergoes multiple gene conversion events bearing a resemblance to meiotic recombination, and most remain trisomic for one to several chromosomes. This process defines a "parasexual" type of genetic recombination. A definitive review of the types of genetic diversity found in *C. albicans* is that of Selmecki et al. (2010).

Further restrictions appear to be that mating does not occur at 37°C, but was observed to occur efficiently on newborn mouse skin, which has a lower surface temperature. The implications for pathogenicity of this odd system of mating, appear to be that recombination can occur in the global population of *C. albicans*, albeit at a low level. The large predominance of the heterozygous a/α diploid among clinical *C. albicans* isolates is related to their superior colonizing ability and pathogenicity.

Phenotype switching is a pathogenic factor (Vargas et al., 2000). Phenotype switch variants of *C. albicans* were isolated from oral swabs of HIV-positive patients at a higher frequency than among oral isolates from immune-normal individuals. Switch variants (e.g., "Irregular wrinkle," "Star," "Ring," "Myceliated") from HIV-positive patients had higher average proteinase (SAP1, SAP3) activity than those from healthy individuals. *C. albicans* strains colonizing HIV-positive individuals who had not yet required antifungal drug therapy also displayed decreased susceptibility to antifungal drugs, including VRC, FLC, and AmB.

Genome Instability

A diploid yeast with eight chromosomes, *C. albicans* is clonal, with limited ability to recombine genes between strains. Even without a complete sexual cycle, *genomic instability* is common in *C. albicans* and may function in selecting mutants exposed to FLC therapy and other environmental stresses. Aneuploidy and isochromosomes serve as a mechanism of azole resistance in *C. albicans*. Examples of aneuploidy are chromosome 5 (C5) monosomy and C4 trisomy, as revealed by electrokaryotyping, a technique that separates entire chromosomes by pulsed field gel electrophoresis. C5 contains genes that regulate azole susceptibility: *ERG 11* and *TAC1* (a transcription factor). Clinical *C. albicans* isolates were found which possess two extra copies of the left arm of C5, sometimes in the form of an isochromosome: a centromere flanked by two left arms of C5. Twenty-one percent of FLC-resistant strains appear to have this isochromosome. This important discovery was revealed by microarray technology (Selmecki et al., 2006).

Biofilm Formation

Candida albicans and *C. parapsilosis* can become embedded in biofilms formed on biomedical implant devices [e.g., catheters (intravenous and urinary), endotracheal tubes, implants (lens, breast, denture), pacemakers, prostheses (voice, heart valve, knee), stents].

Structure A biofilm is a structured microbial community embedded in extracellular matrix material (EMM), secreted by the microbe(s). Within that community *C. albicans* develops a dense network of yeasts, pseudohyphae, and hyphae. The innermost layer in contact with the biomaterial is a monolayer of yeast cells, overlaid by a network of microcolonies and hyphal forms. Secretion of the EMM begins with the basal layer and continues through the rest of the biofilm. Biofilms may harbor "joint tenants" of *Candida* and bacteria.

Function Once embedded, *Candida* species are much less susceptible to killing by antifungal agents than free-floating or planktonic cells. The explanation for this decreased susceptibility is that the biofilm matrix is an impediment to penetration by drugs, and because efflux pumps, of both of the ABC transporter and major facilitator types, are upregulated by biofilm-embedded *C. albicans*. (Please see below, Antifungal Drug Resistance Mechanisms.")

Adhesins

Adherence is the first task for a microbe to initiate pathogenesis. Not surprisingly, *C. albicans* has evolved more than one receptor–ligand interaction to facilitate adherence to epithelial and endothelial cells.

The ALS Gene Family

Candida albicans has surface adhesins that help initiate pathogenesis. The Als adhesin protein is present on the *C. albicans* cell surface. There are 8 *ALS* genes which contain tandem repeats of 108 bp in a variable number of copies according to the *C. albicans* strain (Hoyer et al., 2008).

Structure Als3 adhesin is anchored to ß-(1→6)-D-glucan in the cell wall. There is structural homology with the *Saccharomyces cerevisiae* α-agglutinin protein which mediates contact between a and α mating types. The double knockout strain, *als* 3Δ/*als* Δ, has reduced adherence to epithelial and endothelial cells. Als3 is localized on germ tubes and hyphae. When germ tubes are cultured in RPMI medium, Als3 coats the entire germ tube surface and into the hyphae, as detected with Als3-specific monoclonal antibodies.

Function Binding of *C. albicans* to endothelial cells is followed by endocytosis, leading to invasion of tissues adjacent to the lumen of blood vessels. Receptor–ligand interactions govern endocytosis such that Als3 binds to N-cadherin on the endothelial cell surface inducing endocytosis. Furthermore, when grown as a biofilm on silicone elastamer material, the *als* 3Δ/*als* Δ knockout strain had half the biomass of the wild-type strain, indicating that Als3 plays a role in biofilm formation as well as in mediating adherence to epithelial and endothelial cells (Zhao et al., 2006).

Hsp90

Cells are equipped with several classes of molecular chaperones which facilitate proper folding of proteins upon synthesis, and refolding under conditions of denaturing stress (Young et al., 2001). Heat shock protein 90 kDa (Hsp90) is essential to maintain the activity of numerous signaling proteins and plays a key role in cellular signal transduction networks. Hsp90 chaperones calcineurin, a normal regulator of cell signaling, aiding in circuits that shape adaptive phenotypes.

Structure Hsp90 is a homodimer with its main inter-subunit contacts within the COOH-terminal 190 residues (Young et al., 2001).

Function Mechanisms that allow *Candida albicans* to cope with drug-induced stress are dependent upon Hsp90 function (Cowen, 2009). Under azole stress, loss of function of Erg3 helps *Candida* to survive by blocking a toxic sterol that would otherwise accumulate when the azole inhibits Erg11, encoding lanosterol demethylase. Hsp90 stabilizes calcineurin, enabling calcineurin-dependent stress responses that allow *Candida* to survive exposure to fungistatic azoles.

Recombinant Antibody Against Hsp90 Human recombinant antibodies were genetically engineered from cDNA of patients who recovered from invasive candidiasis (Matthews et al., 2003). Recombinant Mycograb® (NeuTec Pharma) is directed against an immunodominant epitope located in the 47 kDa carboxy terminal fragment of *C. albicans* Hsp90. It consists of covalently linked variable regions of the heavy and light immunoglobulin chains. Synergy was determined between AmB and Mycograb in murine candidiasis, prompting a clinical trial of lipid AmB and Mycograb in 119 patients with confirmed invasive candidiasis (Pachl et al., 2006). Clinical, mycologic, and culture-confirmed clearance of the infection was significantly greater in the combination therapy group compared to the group receiving lipid-AmB alone and, as well, attributable mortality was significantly reduced in the combination therapy group.

Hwp1

This adhesin is the product of the *HWP* (hyphal wall protein 1) gene. As the name implies, Hwp1 is expressed only in the hyphal phase of *C. albicans*.

Structure and Function This adhesin is anchored into the cell wall ß-(1→3)-D-glucan. The means by which Hwp1 mediates adherence is remarkable because it contains a proline-rich region near the N terminus which serves as a substrate for transaminases of buccal epithelial cells. The result is covalent coupling of *C. albicans* to the surface of epithelia (Staab et al., 1999).

Secreted Aspartyl Proteinases (SAPs)

Aiding *C. albicans* invasion of tissues are the secreted acid proteinases (SAPs) (Schaller et al., 2004). SAPs are capable of degrading cell membranes, eroding host surface molecules, and attacking immunoglobulins. The *SAP* gene family encodes 10 aspartyl proteinases located on 5 different chromosomes.

Structure Synthesized as preproenzymes, they are processed by cleavage of the N-terminal signal peptide in the endoplasmic reticulum, translocated to the Golgi where further cleavage of the propeptide occurs before secretion. The functional molecular mass range of the SAPs is 35–48 kDa.

Function *C. albicans* SAPs have broad specificity against a range of host surface proteins including collagen, fibronectin, human stratum corneum, laminin, and mucin. Other host proteins susceptible to degradation by SAPs are α_2-macroglobulin and different classes of immunoglobulin, including secretory IgA. Several SAPs are secreted into the extracellular milieu.

Attention focuses on SAP2 as the most abundant aspartyl proteinase. SAP2 expression is regulated by positive feedback from the peptide proteolytic products of its activity. SAPs are differentially expressed during different types and, perhaps, different stages of candidiasis. Individual SAPs have distinct pH optima, so that taken as a group, proteolytic activity can occur over the range from pH 2 to pH 7. This broad range is important because mucosal pH varies from neutral (oral mucosae) to acidic (vaginal mucosae).

Cell Wall Mannan

The readily soluble outer layer of the *Candida* cell wall is important as a major antigenic determinant and a PAMP. As such, mannan is a ligand serving as an adhesin and as a source of epitopes. Small quantities of mannan in the low ng/mL range can be detected in the serum of immunosuppressed patients with invasive candidiasis. (Please see Section 11.12, Laboratory Detection, Recovery, and Identification—Serodiagnosis–Antigen Detection.)

Structure The structural features of mannan are illustrated in Fig. 11.11 (Shibata et al., 2007). Mannan is a peptidophosphomannan with the polysaccharide portion *N*-linked via a glycopeptide bond to the peptide moiety. The polysaccharide portion consists of a linear α (1→6) linked backbone from which the antigenic oligomannoside epitopes are suspended. The epitopes are 4 to 6 mannose units in length connected by α (1→2)-mannose and α (1→3) linkages. In some *C. albicans* strains, two and probably three additional β (1→2) linkages extend from the α (1→2)-linked side chains to determine serotype. A specificity (Shibata et al., 2003, 2007; Hobson et al., 2004). A separate subdomain contains *O*-phosphono-linked ß (1→2) oligomannosides.

The antigenic factors associated with mannose epitopes were defined by serologic methods (reviewed by Suzuki, 2002). Factor antisera defined 10 antigenic factors of which Factor 5 and Factor 6 are described in the legend for Fig. 11.11.

Functions

Recognition by the Innate Immune System (Poulain and Jouault, 2004)

- *Macrophage Mannose Receptor (MMR).* Now called simply the mannose receptor, it is the main receptor required for efficient uptake, phagocytosis, and antigen presentation of mannan resulting in production by macrophages and monocytes of IL−1β, IL-6, and GM-CSF (granulocyte-macrophage colony-stimulating factor), the latter secreted by macrophages, endothelial cells, and T lymphocytes.
- *S-lectin.* Known as galectin-3, this receptor is essential to phagocytosis. It is specific for the cell wall β-(1→2) oligomannosides of *C. albicans* mannan.
- *TLR2 and TLR4.* These are the major members of the IL-1R/TLR (Toll-like receptor) superfamily involved in *C. albicans* recognition via mannan ligands. Apart from mannan per se, a small molecule, peptidolipomannan (PLM), also is involved in TLR recognition (Poulain and Jouault, 2004). PLM is a glycolipid present in and released from the *C. albicans* cell wall. PLM binds to the macrophage membrane and induces tumor necrosis factor-alpha (TNF-α), which requires TLR2 expression on the cells.

Role of Mannan in Adaptive Immunity Experimental studies have shown that mAbs can react with ß-(1→2) oligomannosides present over the entire surface of the *C. albicans* yeast cell (Mochon and Cutler, 2005). These mAbs exert host protection by acting as opsonins which fix complement, priming the target cell for phagocytosis and killing by late complement components. Once the protective property of natural oligomannosides was realized, a synthetic ß-(1→2)-mannotriose was produced and conjugated to synthetic *Candida* peptides for use as a potential vaccine (Xin et al., 2008).

Serotypes Although serotypes A and B in *C. albicans* have been characterized, thus far there has been no clear association of serotype with pathogenicity. Serotype B strains are more likely to be resistant to 5FC.

Phospholipase

Phospholipase B (Plb1) is the major phospholipase secreted by *C. albicans* hyphal and yeast forms. It acts upon the substrate, glycerol-phospholipids, by cleaving the ***sn1*** and ***sn2*** phosphodiester linkages to release free fatty acids (Ghannoum, 2000).

Structure Plb1 is a glycoprotein with a molecular mass of 84 kDa. The specific activities of the enzyme are, firstly, a hydrolase activity—fatty acid release. The second activity is the production of phospholipids by means

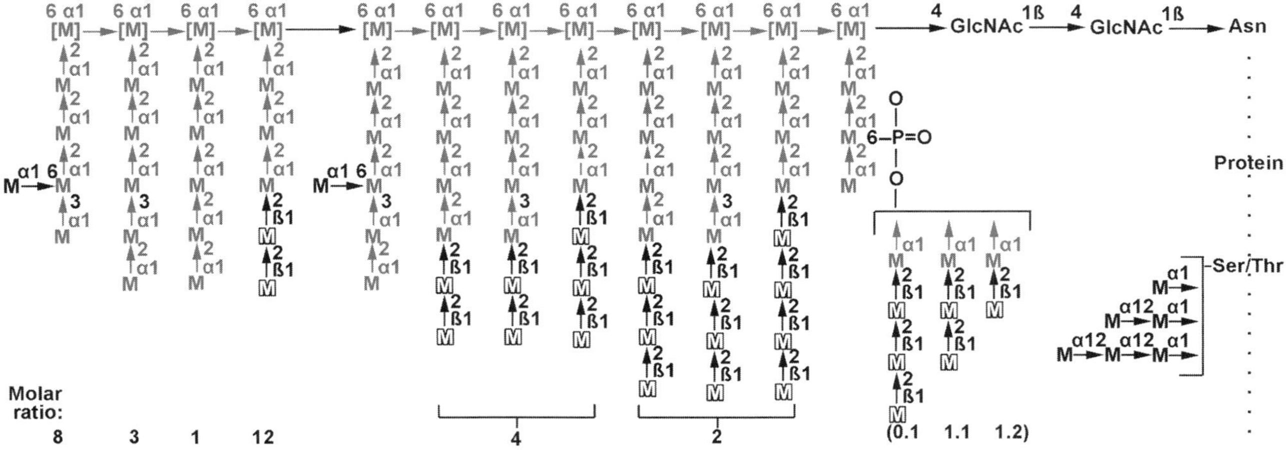

Figure 11.11 A model for *Candida albicans* serotype A mannan (Shibata et al., 2007). The polysaccharide portion of mannan (M, mannose) is linked to protein via a glycopeptide bond between *N*-acetyl-D-glucosamine and asparagine of the peptide moiety. Mannan is built upon a linear α-(1→6) mannan backbone with mannose oligosaccharide side chains carrying antigenic determinants disposed along the backbone. A second domain of ß-(1→2) oligomannose chains is linked indirectly to the mannan backbone via phosphodiester bonds. A third domain of short α-(1→2) linked oligomannosides is joined by alkali labile *O*-glycosidic bonds to serine or threonine of the peptide moiety. Antigenic *Factor 5* is present in *C. albicans* serotypes A and B and in *Candida tropicalis*. This epitope consists of ß-(1→2) -linked mannose oligosaccharides in both the neutral and in the phosphodiester-linked mannan domains. *Factor 6* is present only in *C. albicans* serotype A and its epitope is a mannohexaose, containing α-(1→2)- and ß-(1→2) mannose units linked directly to the mannan backbone, and not through a phospohodiester bond. Another novel epitope consists of a single α-(1→6)-mannose side chain in the α-linked linear mannotetraose and mannopentaose. *Source:* Shibata et al. (2007). Used with permission from the Biochemical Society (www.biochemj.org). Copyright © 2007.

of transferring a free fatty acid to a lysophospholipid—the transacylase activity. Thus, in *C. albicans* both hydrolase and transacylase activities reside in the same enzyme, Plb1.

Function Using targeted gene disruption *C. albicans* null mutants were created which failed to secrete Plb1. The parent phospholipase-producing strain causes fatal infections in mice, whereas the null mutant is avirulent. Deletion of *PLB1* does not produce any detectable effects on adherence of *Candida* to human endothelial or epithelial cells, but the ability of the null mutant to penetrate host cells is greatly reduced. This is powerful evidence that *PLB1* is necessary for virulence; the mechanism appears to be the direct effect of Plb1 on penetrating cell membranes of *Candida*. During experimental infections in mice, Plb1 was localized by immunofluorescence to the invading hyphal tips of the fungus. Also, human patients with systemic candidiasis produce antibodies that react with purified Plb1. The inference is that Plb1 is an invasin, helping *C. albicans* to invade across mucosal borders and tissue planes.

Antifungal Drug Resistance Mechanisms

Ergosterol Biosynthesis Inhibitors (Azoles) Most fungi contain ergosterol in the cell membrane (syn: plasmalemma) as an essential component, analogous to cholesterol in the mammalian cell membrane. Ergosterol is a major lipid constituent of the fungal plasmalemma contributing to membrane stability, fluidity, and proper functioning of membrane-bound enzymes (e.g., nutrient transport proteins and chitin synthase). Within the yeast plasmalemma there are lipid rafts enriched in sphingolipid/ergosterol microdomains where plasmalemma proteins are located at the cell surface. Ergosterol also is important:

- As a permeability barrier against chemical stresses.
- In normal receptor-mediated endocytosis. This is a function in which a ligand binds specifically to its receptor at the plasmalemma, leading to its internalization into small endocytic vesicles. This applies to uptake of extracellular nutrients and other functions including recycling of receptors and transporters.
- As a major component of secretory vesicles, with an important role in mitochondrial respiration. Loss of ergosterol results in membrane destabilization, loss of selective permeability, leakage of cell contents, and altered drug susceptibilities.

Any disruption of ergosterol synthesis has multiple harmful effects on the above cell functions, leading to cell stasis.

Ergosterol synthesis has been the most promising target for the development of antifungal agents, mainly of the azole class. (Please see Chapter 3A for description of these drugs.) Of these, FLC has been the most effective agent for treatment of mucosal candidiasis, with important roles in prophylaxis, and as primary therapy for deep-seated candidiasis in non-neutropenic patients.

Azoles are regarded as fungistatic, depending in part on host defenses to achieve a clinical cure. Long-term therapy with azoles can lead to the selection of resistant strains or variants. This has been shown in the historical treatment of CMC with ketoconazole and the treatment of recurrent OPC with FLC in persons living with AIDS (Lasker et al., 2001; Makarova et al., 2003). The mechanism of action of FLC and other triazole antifungal agents is by inhibiting cytochrome P450-dependent 14 α-lanosterol demethylase, a key enzyme in the pathway of ergosterol biosynthesis.

Azoles form a one-to-one complex with P450 lanosterol demethylase, characteristic of tight-binding inhibitors. The triazole moiety binds to the sixth coordination position of heme, while substituent groups interact with the apoprotein of the enzyme. Besides inhibiting binding of lanosterol, this mechanism inhibits the binding of O_2, thereby preventing the demethylation of lanosterol (Hitchcock, 1993).

Fluconazole has much greater affinity for the fungal enzyme, 14 α-lanosterol demethylase, than the analogous mammalian enzyme in the cholesterol biosynthetic pathway. FLC is 10,000-fold more selective for fungal P450-demethylase than the corresponding mammalian enzyme, contributing to its efficacy and relatively low toxicity in humans.

Fortunately, since 1990 when FLC was first introduced, resistance to the drug has remained below 2% of bloodstream isolates of the major pathogen, *C. albicans*. (Please see Section 11.6, Epidemiology, for trends in susceptibility of *Candida* species to azoles.) Nonetheless, in vitro and clinical resistance has emerged and complicates clinical management.

Resistance is defined as (i) clinical resistance that is dependent on many host factors (i.e., lack of a functioning immune system) and (ii) microbial in vitro resistance—the innate or acquired resistance to an individual drug or class of drugs. Acquired resistance in fungi is related to exposure to drug pressure, because *Candida* species are unable to exchange genetic material by transduction (via phages) or by conjugation via plasmids (extrachromosomal DNA).

Candida species are tested for FLC susceptibility according to standard microtitration plate methods recommended by the CLSI, which also has established minimum inhibitory concentration (MIC) breakpoints for *C. albicans* susceptible strains: ≤ 2 μg/mL; susceptible dose-dependent strains: 4 μg/mL; and resistant strains: ≥ 8 μg/mL. Clinical breakpoints for other *Candida* species may be found in Table 3B.1.

Molecular Mechanisms of Azole Resistance

Efflux Pumps These are normal mechanisms for pumping drugs and other chemical stresses out of the cell. The ATP-dependent efflux pump is upregulated in resistant strains of various *Candida* species. Genes encoding the *A*TP *b*inding *c*assette: ABC transporters in *Candida albicans* are Ca*CDR1* (*Candida* drug resistance-1) and Ca*CDR2* genes. Analogous genes occur in other *Candida* species. ABC transporters pump out not only FLC and other azoles, but also a wide range of compounds including antifungal agents and metabolic inhibitors. Examples are azoles and cycloheximide in *C. albicans* and azoles, terbinafine, 5FC, and cycloheximide in *C. glabrata* (Cannon et al., 2009). Although multiple mechanisms contribute to clinical *C. albicans* FLC resistance, high-level resistance in clinical isolates often correlates with overexpression of mRNA for Ca*CDR1* and Ca*CDR2*.

A secondary class of efflux pumps includes those that depend on the proton-motive force across the plasma membrane, belonging to a family of major facilitators (MFS). The corresponding gene in *C. albicans* is *CaMDR1* and its expression has been detected in azole-resistant clinical isolates. Despite the involvement of CaMdr1 in the azole resistance of certain clinical *C. albicans* isolates and a strong association between expression of the *C. dubliniensis* MFS transporter, Mdr1, and FLC resistance, there is much stronger association between azole resistance and the expression of ABC efflux pumps (Cannon et al., 2009).

Alterations Affecting the Target Enzyme: Lanosterol Demethylase The cytochrome P450-dependent enzyme Erg11 demethylates lanosterol. Resistance is mediated by:

- Overexpression of *ERG11* (35% of resistant strains)
- Mutation in the enzyme so that drug does not bind with avidity
- Conversion of the two alleles of *ERG11* to the more resistant allele by "loss of heterozygosity"

Alterations at Other Points in the Ergosterol Biosynthetic Pathway An example is the loss of function mutation of *ERG3*, which encodes sterol $\Delta^{5,6}$-desaturase.

Genome Instability *Candida albicans* is not genetically stable and can, as a stress response, produce genotypic and phenotype variants. These switch variants have a range of altered morphology and other characteristics including reduced susceptibility to azole antifungals, *even without previous exposure to these agents*. (Please see Section 11.10.2, Microbial Factors—Genome Instability.

Role for ß-(1→3)-D-glucan Resistance to Fluconazole in Biofilms Biofilms on catheter tips containing embedded

C. albicans are difficult to cure without removal of the IV device (Nett et al., 2007). Marked FLC and AmB resistance occurs in biofilm-embedded *C. albicans* both in vitro and in vivo. The ß-(1→3)-D-glucan content in *C. albicans* cell walls in biofilms is increased compared with *planktonic* yeast, which explains the reduced susceptibility of biofilm-embedded *C. albicans* to the actions of most azole drugs. Regulation of expression of the ß-glucan synthase gene, *FKS1*, affects the antifungal drug susceptibility of biofilm-embedded *C. albicans* (Nett et al., 2010). Reduced expression increased susceptibility to AmB, ANF, and 5FC, whereas overexpression reduced susceptibility to ANF and AmB. Thus, sequestering of antifungal agents by glucan acts as a multidrug resistance mechanism.

11.11 THERAPY

General Considerations of Differences in In Vitro Antifungal Susceptibility Among *Candida* Species

Antifungal drug resistance in *C. albicans, C. tropicalis*, and *C. parapsilosis* remains low. A current assessment of in vitro resistance to FLC among the clinically important *Candida* species found the following: *C. albicans*, 1.2%; *C. glabrata*, 5.9%; and less than 1% each for *C. parapsilosis* and *C. tropicalis* (Lyon et al., 2010). Resistance to FLC is predictive of resistance to voriconazole (VRC), however VRC binds more tightly to lanosterol demethylase than FLC does. Echinocandins have an entirely different mode of action than azoles. *Candida krusei* is innately FLC resistant, but is exceptional in that *C. krusei* isolates remain susceptible to VRC. Thus, of the major clinically important *Candida* species, only *C. glabrata* has shown significant in vitro resistance to FLC.

Global surveillance for *Candida* species susceptibility among 22 sentinel sites during 2001–2007 found decreased susceptibility of *C. glabrata* to FLC in all geographic regions, ranging from 62.8% susceptibile in Africa and the Middle East to 76.7% susceptible in Latin America (Pfaller et al., 2010). VRC was more active than FLC (range, 82–84% susceptible) against *C. glabrata* but, as the susceptibility to FLC decreases, so does susceptibility to VRC. Fortunately, both *C. glabrata* and *C. krusei* remain susceptible to echinocandins. Laboratories are advised to perform antifungal susceptibility testing for resistance to FLC for *C. glabrata* isolates from blood and sterile sites and for other *Candida* species from patients who have failed to respond to antifungal therapy or when azole resistance is suspected. *C. glabrata* isolates found to be susceptible in vitro to FLC may be treated with that drug as primary therapy, or as step-down therapy after primary therapy with an echinocandin. Patients at high risk of *C. glabrata* infections include elderly patients, patients with cancer, or those with diabetes.

In general, in vitro resistance to echinocandins remains low; however, *C. parapsilosis* strains have a reduced susceptibility to echinocandins (MIC_{50} of 0.5 μg/mL for CASF) compared with *C. albicans* (MIC_{50} of 0.03 μg/mL), (Lyon et al., 2010). For this reason FLC is recommended as primary therapy for *C. parapsilosis* candidemia.

Therapy for Superficial Candidiasis (Cutaneous)

Therapy for nails (onychia and paronychia), intertrigo and diaper rash are listed in Table 11.6.

Therapy for Mucosal Candidiasis

Therapy for oropharyngeal, denture stomatitis, and esophageal and gastrointestinal candidiasis are discussed in Table 11.7. Chronic mucocutaneous candidiasis requires continuous systemic therapy with ITC, FLC, VRC, or PSC.

Therapy for Vulvovaginal Candidiasis (VVC)

Please see Table 11.8.

Therapy for Urinary Tract Candidiasis

Recommended treatment for asymptomatic and symptomatic urinary tract candidiasis is described in Table 11.9. For patients with asymptomatic urinary tract candidiasis undergoing urologic procedures, FLC, 200–400 mg (3–6 mg/kg/day) or AmB-deoxycholate (0.3–0.6 mg/kg/day) is recommended for several days before and after surgery.

Therapy for patients with symptomatic cystitis is FLC [200 mg (3 mg/kg)/day] for 2 weeks. Alternative therapy is recommended for patients with FLC-resistant yeasts and consists of AmB-deoxycholate (0.3–0.6 mg/kg for 1–7 days) or 5FC, 25 mg/kg, 4×/day for 7–10 days. Patients with pyelonephritis as a site for disseminated candidiasis are treated in the manner described for acute systemic candidiasis (Table 11.10).

Table 11.6 Therapy for Superficial Candidiasis (Cutaneous)

Agent	Comment
Nails (onychia and paronychia)	
ITC (oral) FLC	Oral ITC is the drug of choice and may be given as pulse therapy. Alternatively, oral FLC can be used but requires continuous and prolonged therapy.
Intertrigo/diaper rash	
Topical antifungal agents: Azoles Clotrimazole Econazole Miconazole Oxiconazole Sulconazole Ciclopiroxolamine Haloprogin Polyenes Nystatin AmB	Initial therapy for diaper rash should promote local dryness, avoid occlusion, and provide good local hygiene. Topical therapy is the same as for intertrigo. Systemic therapy with FLC or ITC is rarely appropriate but is considered for widespread disease or complicated situations (i.e., uncontrolled diabetes mellitus).

Table 11.7 Therapy for Mucosal Candidiasis

Agent	Comment
Oral candidiasis	
Oropharyngeal candidiasis (OPC, thrush)/denture stomatitis	
CLT (troche) NYT (suspension, pastilles) FLC (oral), ITC (oral, capsules, or solution) AmB oral suspension ANF (IV) CASF (IV) MCF (IV) AmB (IV)	For mild infections of the oropharynx; topical agents are administered as "swish and swallow" or oral troche. In moderate to severe disease, or in AIDS patients, or for esophageal involvement, systemic therapy is indicated. For refractory disease, ITC, VRC, PSC, or AmB suspension is recommended. Bioavailability of ITC is unpredictable.
Esophageal/gastrointestinal	
FLC (oral) AmB (IV) ANF (IV) CASF (IV) MCF (IV) AmB-deoxycholate (IV) ITC (oral, capsules, or solution), PSC, or VRC.	Patients with cancer usually experience mucocutaneous candidiasis only, whereas the severely immunosuppressed or patients with advanced AIDS ($CD4^+$ count $<100/\mu L$) may have more permanent immunosuppression leading to relapses of OPC and esophageal disease. Agents (left column) are efficacious in these patients. Patients with AIDS often experience a relapse of the infection, and some require frequent treatment and/or chronic suppressive therapy. FLC is most often used in this setting. All the echinocandins have a higher associated relapse rate than FLC.
Peritonitis	
Treatment is discussed in systemic peritonitis. Please see Table 11.10.	

Abbreviations. Amphotericin B, AmB; clotrimazole, CLT; flucytosine, 5FC; fluconazole, FLC; itraconazole, ITC; ketoconazole, KTC; nystatin, NYT; voriconazole, VRC; anidulafungin, ANF; caspofungin, CASF; micafungin, MCF; posaconazole, PSC.

Table 11.8 Therapy[a] for Vulvovaginal Candidiasis (VVC)

Local azoles (topical) *Butoconazole 2% cream *CLT, 1% cream or vaginal tablets (100 mg, 500 mg) *Miconazole 2% cream or vaginal suppository (100, 200, or 1200 mg) Econazole 150 mg vaginal tablet Fenticonazole 2% cream *Tioconazole 2% and 6.5% cream Terconazole (0.4% and 0.8% cream), 80 mg vaginal suppository NYT, 100,000 U vaginal tablet Oral FLC, single dose of 150 mg	*Candida glabrata* often fails to respond to azoles but there have been encouraging results using 600 mg boric acid capsules intravaginally or with topical 5FC. Maintenance therapy for recurrent VVC may be oral FLC (150 mg/week for 6 months). FLC orally also is used in regimens of variable duration including single tablet/dose therapy for single episodes or recurrent VVC.

[a]Asterisk (*) indicates availability in at least one form without prescription.

Table 11.9 Therapy for Urinary Tract Candidiasis

Urinary tract (asymptomatic)

Candiduria in catheterized patients usually represents catheter or superficial bladder colonization only. It is common, invariably asymptomatic, of little clinical significance (rarely complicated by either ascending or invasive candidiasis) and should not be treated.

Asymptomatic candiduria should only be treated in neutropenic patients, in patients undergoing elective urinary instrumentation, and following renal transplantation. *Candida albicans* is less often responsible for candiduria episodes whereas *C. glabrata* may be responsible for 20–30%.

Urinary tract (symptomatic)

Oral FLC Oral VRC IV AmB (topical AmB bladder irrigation also may be effective)	Symptomatic lower urinary tract *Candida* infections are rare and should be treated, especially in noncatheterized patients. Oral 5FC is effective but rarely indicated because of emergence of resistance. Ascending pyelonephritis is uncommon, serious, and may be complicated by candidemia and disseminated infection. Therapy consists of relieving any urinary obstruction and treatment with FLC or systemic AmB.

Candidemia (Non-neutropenic Patients)

Recommended therapy (Table 11.10) for most adult patients is either FLC [loading dose of 800 mg (12 mg/kg), then 400 mg (6 mg/kg) daily] or an echinocandin (i.e., CASF: loading dose of 70 mg, then 50 mg daily; MCF: 100 mg/day). All echinocandins are considered interchangeable in their efficacy against *Candida* species. Echinocandins are recommended for moderately severe to severe illness. FLC is recommended for patients who are less critically ill with no recent azole exposure. Clinically stable patients receiving an echinocandin may be switched to FLC if the isolate is susceptible. BSI due to *C. glabrata* should be treated with an echinocandin, pending in vitro susceptibility test to determine if the isolate is susceptible to FLC. FLC is preferred for *C. parapsilosis* BSI.

Second-line alternatives are AmB-deoxycholate (0.5–1 mg/kg/day) or lipid-AmB[9] (3–5 mg/kg). Clinically stable patients may be switched to FLC, if their isolates are susceptible. VRC has no obvious advantages compared with FLC, except that it supplies additional mold and *C. krusei* coverage. Therapy for candidemia with no detectable hematogenously disseminated foci should continue for 2 weeks after blood cultures turn negative and symptoms have subsided. Intravenous catheter removal is recommended.

[9]Please see Chapter 3A for a discussion of lipid derivatives of AmB.

Table 11.10 Therapy for Deep Candidiasis (Systemic)

Agent	Comment
Candidemia	
FLC Echinocandins AmB	*Non-neutropenic patient*—Antifungal treatment is always indicated in these patients. Candidemia is associated with an IV catheter in up to 80% of patients. Removal of the catheter is always indicated to assist in clearing the bloodstream; but, in patients with surgically implanted catheters, this may not always be practical. For clinically stable patients, start treatment with either an echinocandin or FLC. For *C. albicans*, continue FLC. Otherwise, standard therapy with an echinocandin or high-dose FLC should be used until identification and susceptibility testing of minimum inhibitory concentration (MIC) are complete. There is no clear correlation between MIC and clinical outcome or response to treatment of candidemia. AmB can be used if either echinocandins or FLC are contraindicated or have failed. AmB is second-line therapy because of its increased toxicity. Combinations of FLC or AmB with 5FC may be useful in some patients.
AmB Echinocandins VRC (IV) FLC 5FC	*Neutropenic patient*—AmB or an echinocandin is used until recovery from leukopenia, which is critical for successful outcome. Removal of IV catheters may be important, in particular, for *C. parapsilosis* fungemia in cancer patients. Infections with *C. krusei* or *C. glabrata* typically require larger doses of AmB. FLC therapy alone may be helpful in some patients. 5FC may be used as adjunctive therapy but this drug has a potential for bone marrow suppression and there is no readily available IV formulation.
Lipid-based AmB	Use in patients with renal impairment.
Chronic disseminated (hepatosplenic) candidiasis	
FLC VRC Echinocandin AmB	FLC is used for clinically stable patients. Unstable patients should receive lipid-AmB, until they become stable, then therapy should be changed to FLC until lesions resolve (2–14 months). Alternatively, an echinocandin may be used for several weeks, then therapy can be changed to FLC. Maintain patients who will receive chemotherapy and transplantation on anti-*Candida* therapy until immune functions return.
Endocarditis	
FLC AmB 5FC	Most cases of *Candida* endocarditis are caused by *C. albicans* or *C. parapsilosis*, two species that are generally susceptible to FLC. Chronic suppressive (indefinite) therapy with FLC is successful in selected patients with both native and prosthetic valve endocarditis. A course of AmB followed by long-term suppressive FLC therapy is indicated in patients where cardiac surgery is unusually risky.
Endophthalmitis	
FLC VRC AmB 5FC Echinocandin	Combined AmB and 5FC, or FLC monotherapy are recommended primary systemic therapy choices for intraocular candidiasis. Alternative systemic therapy choices are lipid-AmB (3–5 mg/kg/day); VRC (6 mg/kg/12 h for 2 doses then 3–4 mg/kg/12 h); or an echinocandin. VRC 1 week of IV treatment results in intraocular concentrations exceeding 18× the MIC_{90} of *C. albicans*.

(*continued*)

Table 11.10 (*Continued*)

Agent	Comment
	AmB is the most common antifungal agent given by an intravitreal route at a dose of 5–10 μg in 50–100 μL.
Peritonitis	
AmB (IV or intraperitoneal) FLC 5FC (although its role is unclear)	Systemic peritonitis is common following spontaneous or surgical bowel perforation. Peritoneal cultures often are polymicrobial, and the significance of *Candida* isolation is not always obvious. In patients undergoing peritoneal dialysis, removal of the dialysis catheter is usually required for successful therapy.

Abbreviations: Amphotericin B, AMB; flucytosine, 5FC; fluconazole, FLC; itraconazole, ITC; voriconazole, VRC.

Empiric therapy for *suspected* invasive candidiasis in non-neutropenic patients is similar to that for confirmed candidiasis. Empiric therapy is advised for critically ill patients with risk factors for invasive candidiasis and fever of unknown origin after evaluating culture data from non-sterile sites. Use of an echinocandin is preferred where there has been recent azole therapy, in severe illness, or for patients at risk of infection with *C. glabrata* or *C. krusei*. AmB-deoxycholate and lipid-AmB are second-line alternatives.

Therapy for Acute and Chronic Disseminated Candidiasis

Therapy for patients with acute disseminated candidiasis should follow the treatment recommendations for patients with candidemia. Empiric therapy may be started in neutropenic patients with persistent fever despite several days of broad-spectrum antibiotic treatments. Chronic disseminated (i.e., hepatosplenic) candidiasis in clinically stable patients may be treated with FLC, 400 mg (6 mg/kg/day). Severely ill patients should receive lipid-AmB (3–5 mg/kg/day) or AmB-deoxycholate (0.5–0.7 mg/kg/day). When patients become stable they may be switched to FLC. Alternative therapy consists of an echinocandin for several weeks, followed by FLC. Patients who are clinically stable after several weeks of primary therapy may be switched to FLC. Therapy is continued until lesions resolve, which may take months. Patients who are going to receive chemotherapy and transplantation should be maintained on therapy until immune functions are restored.

Therapy guidelines for other serious forms of systemic candidiasis such as infected pacemaker, endophthalmitis, suppurative thrombophlebitis, osteomyelitis, CNS candidiasis, septic arthritis, or pericarditis may be found in Pappas et al. (2009). Although the current guideline for treatment of *Candida* endophthalmitis calls for AmB with 5FC as preferred primary therapy Sallam et al. (2006) state that although inflammation improves AmB penetration into the eye, systemic AmB (even lipid-AmB), as monotherapy, does not achieve therapeutic concentrations in the eye and is not effective in managing *Candida* retinitis extending into the vitreous. Oral 5FC achieves therapeutic concentrations in the vitreous and aqueous humor and can be used in combination with AmB.

11.12 LABORATORY DETECTION, RECOVERY, AND IDENTIFICATION

Direct Examination

Mounts of smears (including peripheral blood), scrapings, and biopsy specimens are suitable for direct examination of yeasts.

Unstained Specimens

Specimens include wet mounts of sediment from centrifuged urine or cerebrospinal fluid (CSF), aspirates, or scrapings. To process, add 1 drop of saline or 10% KOH to aspirates or scrapings, add a coverslip, and observe microscopically. In scrapings, epithelial cells also may be seen when mounted in saline (Fig. 11.12).

Stained Specimens

Although not essential, staining (e.g., with lactophenol-cotton blue or lactofuchsin) is helpful for direct mounts. Specimens stained with the following stains also are used in the clinical mycology laboratory:

- *Gram Stain.* Young *Candida* organisms are Gram-positive, as are all fungi (Fig. 11.13).
- Wright's or Giemsa stain for peripheral blood smears.

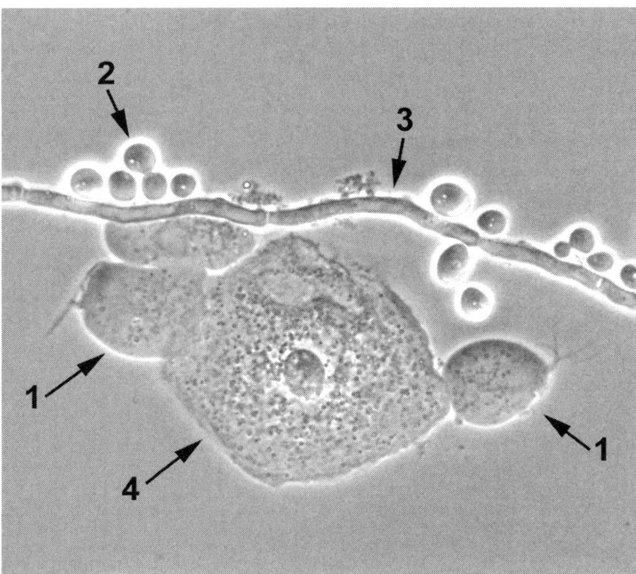

Figure 11.12 Wet mount of *Candida* in vaginal swab showing *Trichomonas vaginalis* (1), *Candida* species yeast cells (2), and a pseudohypha (3), also showing an epithelial cell (4), phase contrast.
Source: Photomicrograph courtesy of Dr. Jeffrey P. Massey and Dr. Sandip Shah, Michigan Department of Community Health, Lansing, Michigan.

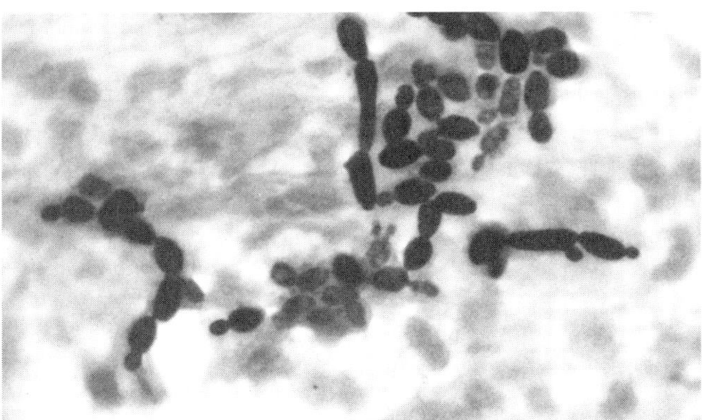

Figure 11.13 Gram-stained smear from a patient with oropharyngeal candidiasis showing yeasts with pseudohyphae. 1000×.
Source: Used with permission from Dr. Uma M. Tendolkar, M.D., LTM Medical College, Mumbai.

- Heat-fixed smears stained with Loeffler methylene blue or Calcofluor white. Calcofluor White also may be combined with a KOH prep. This method increases visibility but requires a fluorescence microscope. (Please see "Presumptive Identification of Some Major *Candida* Species by Staining Reactions with Calcofluor White," later in this chapter.)

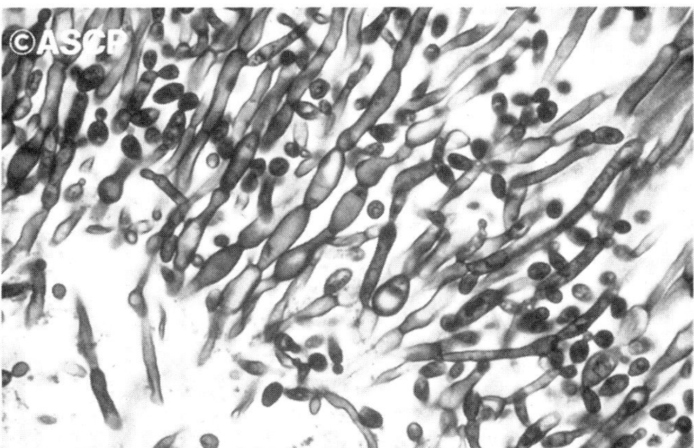

Figure 11.14 *Candida* species in tissue, illustrating yeast and pseudohyphae. GMS stain, 250 ×.
Source: Chandler and Watts (1987). Used with permission from Dr. John C. Watts and from the American Society for Clinical Pathology, © ASCP.

Stained specimens will show some of the following:

- Budding yeast cells. Please see Chapter 1, Fig. 1.3.
- Pseudohyphae. Please see a description of pseudohyphae in Chapter 1, Fig. 1.4, and further examples in Figs. 11.13 and 11.14
- Short elements of true hyphae.

Histopathology

Specimens for histopathologic analysis consist of touch smears of scrapings from cutaneous or mucocutaneous lesions, pus, sediment of centrifuged urine or CSF, or paraffin-embedded tissue sections. The best stains for fungi in tissue are PAS and GMS (Fig. 11.14). Hematoxylin and eosin (H&E)-stained smears demonstrate *Candida* elements poorly. That stain is recommended primarily for detecting *Histoplasma capsulatum* in tissue, although it is one of the most frequently used stains in the pathology laboratory. Papanicolaou-stained smears occasionally are sent to the microbiology laboratory for confirmation of the presence of a fungus.

Culture

Primary isolation from clinical specimens is accomplished as follows.

Sputum, Urine, Biopsy, CSF, Aspirates, Tissue, Skin, and Other Clinical Specimens

Fresh specimens must be used to eliminate the overgrowth of *Candida* species by bacteria. Prolonged standing also may result in multiplication of *Candida* giving a false impression of the original amount of yeast present. Isolation from nonsterile sites may be indicative of *Candida* infection, or the yeast may be a contaminant or commensal. Finding *Candida* in fresh urine specimens might be indicative of systemic or disseminated disease (please see Section 11.9, Clinical Forms—Urinary Tract). Although most species grow rapidly on laboratory media, *C. glabrata* and *C. guilliermondii* may require up to 2–3 weeks for colonies to appear.

Blood

Positive blood cultures should not be ignored, but transient candidemia due to seeding of blood from an intravascular catheter must be ruled out. Better blood culture methods include the lysis-centrifugation method, for example, the Wampole Isolator® tube (Inverness Medical Professional Diagnostics, Princeton NJ), and the BacT/Alert® blood culture system. Blood culture methods are discussed in Chapter 2. Time is of the essence in determining whether a yeast isolate is *C. albicans* or another *Candida* species.

Empiric antifungal therapy is started once a blood culture is positive for yeast. Typically, a subculture requires 24 hours of incubation before a germ tube test can be done. The germ tube test takes ≤3 hours. A further 24 hours of incubation often is needed for the definitive identification of *Candida* species. Short-cut methods are described below for more rapid identifiction of *C. glabrata* and, presumptively, for *C. krusei* (please also see below, "Genetic Identification of *Candida* Species").

Tests for candidemia may be negative during disseminated candidiasis. Among patients with leukemia, blood cultures have been positive for between 23% and 82% of patients (depending on the series) who developed disseminated candidiasis. Results depend on the type of patient, culture method used, and frequency of sampling.

Why are blood cultures often negative? Candida albicans is a "sticky" organism and is filtered out of venous circulation even in immunosuppressed patients, because of antibody-independent receptor–ligand interactions between *Candida* and host phagocytes. Rapid removal of yeasts from the bloodstream in the liver and spleen is mediated by tissue-fixed macrophages: Kupffer cells in the liver and macrophages in the splenic marginal zone. (Please see Section 11. 10.1, Host Factors—Innate Immunity, for details of the PAMPs and PRRs).

Media for Recovery of Candida *Species*

Sabouraud (SDA) or SDA-Emmons agar + chloramphenicol or penicillin and streptomycin (to inhibit bacteria) are used in most mycology laboratories. *Candida* species grow readily on blood agar and brain–heart infusion (BHI) agar. Media containing cycloheximide prohibit growth of many yeasts and should not be used if yeast is suspected; however, *Candida albicans* is able to grow on media containing 0.04% cycloheximide (Mycosel™, formulation is given in Table 2.1). Special media for individual species also are used.

CHROMagar*Candida*® is suitable for the primary isolation of yeasts and the presumptive identification of some *Candida* species. Distinctive colors are formed by colonies of certain (but not all) species. CHROMagar is also very helpful for identifying multiple yeast species which occurs in 3–10% of blood cultures. The medium is selective because it contains chloramphenicol, and the colors are formed as a result of a mixture of chromogenic substrates for enzymes that are specific for individual species. Incubation of isolates at either 30°C or at 37°C produces colonies of the same color for a given species. Colors appear more intense at 37°C and intensify daily at both temperatures, peaking at 72 hours of incubation (Hospenthal et al., 2006). Typical characteristics of *Candida* species are:

C. albicans or *C. dubliniensis*	*Green*. Medium sized, smooth, matte colony, with slight green halo in surrounding medium.
C. krusei	*Pink-rose*. Large, spreading, rough, with pale pink to white edge.
C. tropicalis	*Blue to blue-gray*. Paler pink edge; may have dark brown to purple halo that diffuses into agar.
C. glabrata	*Variable color and small colonies*. Therefore CHROMagar is unsuitable for definitive identification.
C. parapsilosis	*Ivory, pink, lavender*. Most variable in color and morphology, therefore not suitable for identification with CHROMagar *Candida* medium.

Morphologic and Biochemical Tests to Identify *Candida* Species from Clinical Specimens

See Fig. 11.15. A general reference is Hazen and Howell (2007).

Assimilation and Fermentation Tests

Further tests often are essential to identify the yeast species. A widely used test is the API 20C AUX II (bioMérieux, Marcy l'Etiole, France). Traditional tests such as the Wickerham broth assimilation and fermentation tests are perhaps the most reliable, but rarely are used in the clinical laboratory because of the complexity and time required to obtain results.

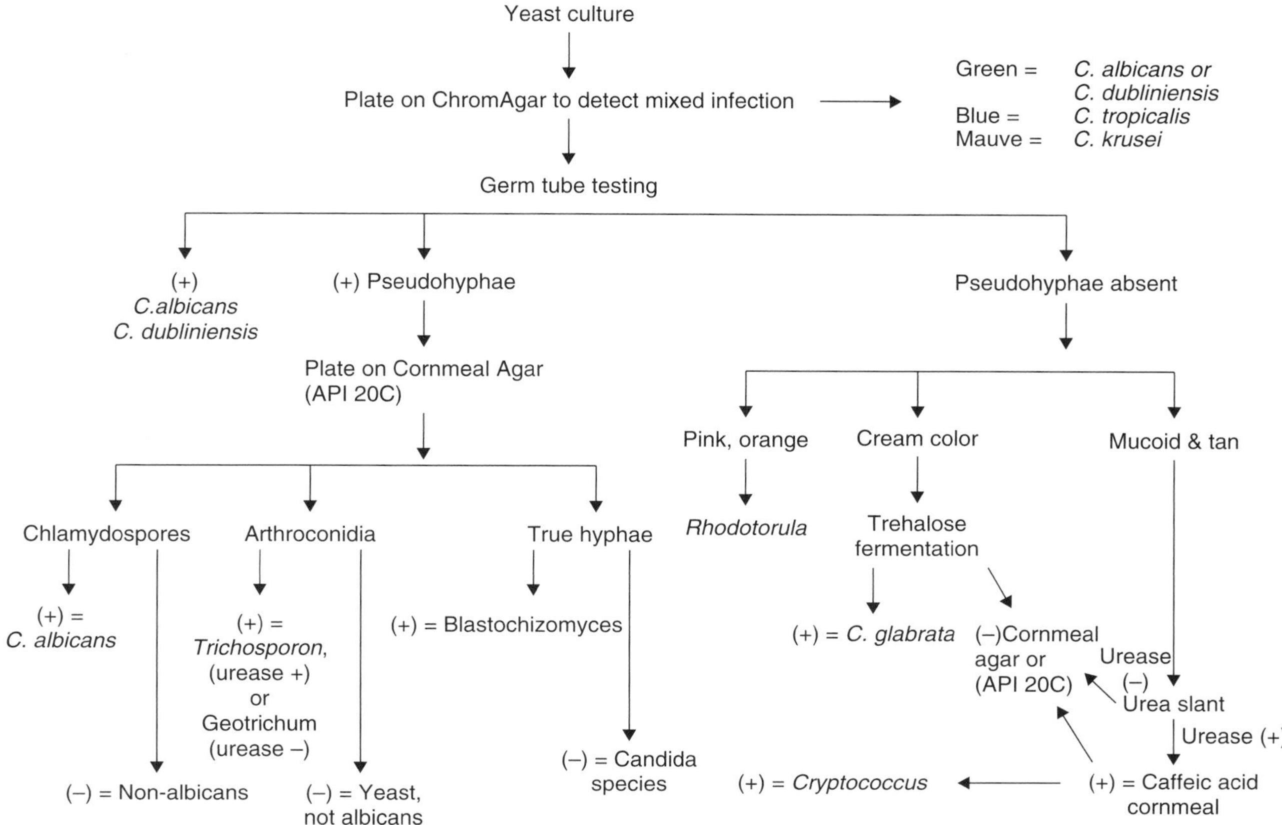

Figure 11.15 Flow chart for yeast identification. API 20c® (bioMérieux, Inc., Durham, NC) is a panel of 19 carbon assimilation substrates. Caffeic acid–Cornmeal agar, CHROMagar *Candida* are explained in Chapter 2. *Blastoschizomyces capitatus* is now known as *Geotrichum capitatum*.
Source: H. J. Shadomy, Nancy Warren Ph.D., G. Marshall Lyon, M.D.

Alternative Tests

Many alternative, rapid tests have been developed and, where available, may be employed after receiving a thorough evaluation (e.g., rapid assimilation of trehalose by *Candida glabrata*). A variety of these tests are described in Chapter 2, Section 2.3.7.2, Rapid Tests for Yeast Identification.

Presumptive Identification of Some Major Candida *Species by Staining Reactions with Calcofluor White*

Application Note: Rapid (2 hours or less) presumptive identification of *C. albicans, C. glabrata*, and *C. krusei* can be accomplished using Calcofluor white-containing medium (Harrington and Williams, 2007). The medium is 1% proteose peptone and 0.1% glucose in deionized water. For a solid medium, add agar to a final concentration of 1.5%. Supplement the medium with Calcofluor white from a 1% stock solution to give a final concentration of 0.0025% before autoclaving. (Calcofluor white: Fluorescent Brightener 29, Sigma Chemical Co., St. Louis, MO, cat # F-6259.)

Log phase *Candida* cells grown on primary isolation medium, tripticase soy–blood agar, chocolate agar, or SDA are suspended in sterile saline. A drop of yeast cell suspension is placed on the Calcofluor white-containing agar medium, coverslipped, and incubated for 2 hours at 37°C. The cells are visualized with a fluorescence microscope using epi-illumination, a UV excitation filter (330–380 nm), and an oil immersion lens. *C. albicans* is detected after 2 hours of incubation by germ tube formation (Fig. 11.16), *C. glabrata* is detected without incubation by weak fluorescence at the poles of cells, and *C. krusei* by weak fluorescence of the entire cell wall but brighter fluorescence at the poles of cells.

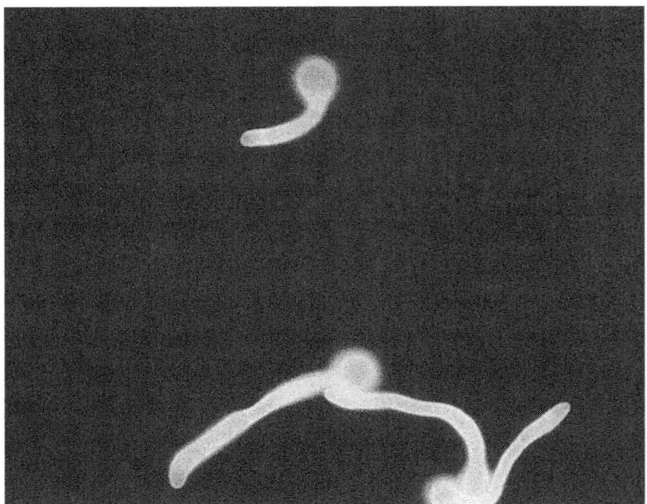

Figure 11.16 Germ tube formation by *Candida albicans*, Calcofluor White stain.
Source: Used with permission from Dr. Brian J. Harrington, University of Toledo Health Sciences, Toledo, Ohio.

Identification of the Most Prevalent *Candida* Species

Candida albicans

Germ Tube Test *Candida albicans* produces germ tubes when incubated in serum or specialized medium at 35°C. Under these conditions *C. albicans* is alone among the medically important yeasts to produce germ tubes.[10] Pooled human serum is a good germ tube formation medium, and other media may be used, for example, sheep, fetal bovine, or newborn calf serum, or bovine serum albumin.

If human serum is used, it must be pooled to dilute *Candida* antibodies and tested so that there is no interference with germ tube formation. (*Note*: When handling human serum, observe CDC precautions against bloodborne pathogens, for example, hepatitis viruses or HIV.) The germ tube test is performed from a previously isolated colony and must be read within 4 h. Beyond that time the germ tubes may become so entwined that proper microscopic examination is not possible. Care must be taken to differentiate budding yeasts and pseudohyphae from germ tubes (Fig. 11.16).

In addition to genotype differences, *C. albicans* and the sister species, *C. dubliniensis*, can be separated by their colony and microscopic morphology on birdseed agar or Pal's (sunflower seed) agar. On these media *C. dubliniensis* colonies have a rough texture due to abundant pseudohyphae and chlamydospores, whereas *C. albicans* colonies are smooth and yeast-like. Birdseed agar is described in Chapter 2.

[10]Please see Section 11.5, Etiologic Agents, for information about the genetic sister species *C. dubliniensis*.

Cornmeal–Tween 80 or Rice–Tween Agar Incubation of yeast cultures at room temperature (25°C) is used to demonstrate hyphae and pseudohyphae with chlamydospores (chlamydoconidia by some authors). Please see Fig. 2.1 in Chapter 2. Many ellipsoid blastoconidia in whorls (= verticils) at the junction between pseudohyphal cells help confirm *C. albicans*. Round, thick-walled chlamydospores borne terminally or laterally on hyphae or pseudohyphae are a distinctive feature of *C. albicans* (and the rarer *C. dubliniensis*). Chlamydospore formation usually requires 48 hours of incubation after primary isolation of the yeast.

The use of the *Dalmau* technique is advised for optimum chlamydospore production. (Please see the Chapter 2, Section 2.37, Media and Tests for Yeast Identification.)

Candida glabrata

Candida glabrata is the only *Candida* species that does not form pseudohyphae or hyphae, but instead remains in the small yeast form both in culture and in tissue. It also contains a trehalase enzyme that rapidly splits trehalose into two glucose molecules. This characteristic is the principle behind the *Glabrata* RTT, a rapid trehalase test (Fumouze Diagnostics, Levallois-Perret Cedex, France).

> *Application Note: Glabrata RTT.* To perform the RTT, add 25 μL of the yeast suspension in distilled water to each of three wells containing trehalose, maltose, and sugar-free basic medium, respectively, the latter serves as a control. After 10 min at ambient temperature, 25 μL of detection reagent is added to the wells (glucose oxidase, peroxidase, and a chromogenic substrate). Observe the results after incubation for 5–10 min at room temperature. When an orange color develops only in the well containing trehalose, the yeast is identified as *C. glabrata*. If other wells also give positive reactions, the result is interpreted as a yeast, not *C. glabrata*.

Candida krusei

Grown on cornmeal–Tween 80 agar at 25°C for 72 hours, *C. krusei* produces pseudohyphae in a cross-matchstick formation and blastoconidia that are elongated (Fig. 11.17). This is the only *Candida* species that can grow on medium without vitamin supplementation. The species has innate reduced susceptibility to FLC contributing to its appearance among the "top five" clinically important *Candida* species.

Figure 11.17 *Candida krusei*. CMA, elongate pseudohyphae in a "cross-matchstick" pattern.
Source: CDC.

Candida parapsilosis

Grown on CMA + Tween 80 agar at 25°C for 72 hours, blastoconidia are observed along the pseudohyphae either singly or in small clusters. The pseudohyphae tend to be curved and the presence of large hyphae ("giant cells") may be observed.

Candida tropicalis

Grown on CMA + Tween 80 agar at 25°C for 72 hours, long pseudohyphae are formed producing single or small groups of blastoconidia. True hyphae may be produced, as well as occasional thin-walled, tear-drop shaped "chlamydospores," not to be confused with the thick-walled, round chlamydospores or *C. albicans* of *C. dubliniensis*.

Serodiagnosis–Antigen Detection

The Platelia® *Candida* antigen sandwich EIA (Bio-Rad Laboratories, Marnes-la-Coquette, France) uses mAbs to detect *Candida* species mannan which circulates in the low ng/mL range during invasive candidiasis. The EIA is considered to be positive when ≥0.5 ng of mannan per milliliter is detected in two consecutive blood specimens. Based on clinical signs of sepsis and positive blood cultures, invasive candidiasis was observed in 12 (6.5%) of 184 infants admitted to an Italian NICU in a 2 year period (2004–2006) (Oliveri et al., 2008). The Platelia mannan test was positive in 11 of those infants. The remaining infant was infected with *C. parapsilosis*, a species, along with *C. krusei*, which is not detected by the current version of the mannan antigen detection assay. Further evaluation of the test is expected, and additional mAbs with broader specificity may help to broaden the range of species detected.

Genetic Identification of *Candida* Species

PNA-FISH® Test Kit [*Definition*: Peptide-nucleic acid-fluorescent in situ hybridization (AdvanDx Inc., Woburn, MA).] The *C. albicans*/*C. glabrata* PNA FISH is a multicolor, qualitative nucleic acid hybridization assay for *Candida albicans* and *C. glabrata* isolated from blood cultures. The test is performed after a blood culture turns positive and a Gram stain of cells from the blood culture reveals yeast. The 2.5 h PNA-FISH assay uses fluorescently labeled peptide nucleic acid (PNA) probes that target the species-specific ribosomal RNA (rRNA) of *C. albicans* and *C. glabrata*. Results are visualized by fluorescence microscopy. Green fluorescing cells identify *C. albicans* whereas red fluorescing cells identify *C. glabrata*. The absence of fluorescence indicates another species is present in the positive blood culture.

PNA probes for *C. krusei, C. parapsilosis*, and *C. tropicalis* have also been evaluated for the ability to identify these species from colonies resuspended from agar plates or slants. The results indicated that the PNA-FISH method is highly sensitive, rapid (2.5 h), and accurate for the identification of *Candida* to the species level (Reller et al., 2007). A further development is the "Yeast Traffic Light PNA-FISH," which combines these probes such that, in a blood smear, *C. albicans* and *C. parapsilosis* display green fluorescence, *C. tropicalis* fluoresces in yellow, and *C. glabrata* and *C. krusei* display red fluorescence. A negative result indicates the absence of these five organisms.

LightCycler® Septi*Fast* Test (Roche Diagnostics) This test is a system-based approach to detect and identify 25 bloodstream pathogens including five major *Candida* species (*C. albicans, C. tropicalis, C. glabrata, C. parapsilosis, C. krusei*) and *Aspergillus fumigatus* from 3 mL of patient blood in 6 hours. No incubation or culture steps are required. The system includes a MagNA Lyser instrument, consumables for lysis of the yeast, and a separate kit for sample preparation. The internal transcribed spacer (ITS) region of rDNA is the target for real-time PCR conducted in the LightCycler 2.0 instrument. Amplification is combined with specific melting point analysis of the probe–amplicon complex to deliver rapid species results. Data analysis is facilitated with a software tool to consolidate data points into a patient report. However, sensitivity and specificity have not been determined for this test in clinical practice.

Research using DNA probes directed to the ITS2 region of rDNA and an enzyme immunoassay detection format (Elie et al., 1998), as well as TaqMan® exonuclease assays (Shin et al., 1997, 1999), a multianalyte profiling system (Luminex®) (Das et al., 2006), and ITS2 rDNA sequencing (Ellepola and Morrison, 2005; Ellepola et al., 2003) demonstrate the utility of this region for the discrimination of closely related *Candida* species.

Applied Biosystems Inc. (ABI) MicroSeq D2 LSU rDNA Sequencing Kit This system is based on PCR amplification of the D1/D2 region of 26S (LSU) rDNA, cycle sequencing to produce fluorescently labeled partial sequences, which are then analyzed by capillary electrophoresis on an ABI genetic analyzer. The sequence obtained is assembled, edited with MicroSeq® ID Analysis Software v2.0, and then compared to a proprietary database of 289 yeast species (MicroSeq D2 fungal library version 1.4.2 and version 2.0) This system is used to identify a positive culture from a clinical specimen.

Analysis of the D1/D2 region of the 26S rDNA is sufficient to identify most ascomycetous yeast species, including *Candida* species, as well as basidiomycetous yeasts. In an evaluation of the MicroSeq D2 LSU method compared to conventional identification of clinical yeasts with the API 20C AUX assimilation test, 100 isolates (representing 19 species of *Candida*) were sequenced, and 98% (97/99) gave results concordant with the identifications made by the API 20C AUX method (Hall et al., 2003). Overall, nucleic acid sequencing identified 93.9% of the clinical isolates of yeast species to the correct genus and species. However, when the sequences for discrepant isolates were matched with those in GenBank, 99.2% of identifications agreed with the identification provided by the MicroSeq database and only 94.7% of the phenotypic identifications were concordant. This finding underlines that genotypic identification can be more accurate than phenotype.

11.13 LESS COMMON OPPORTUNISTIC YEAST GENERA

11.13.1 *Geotrichum capitatum* (*Blastoschizomyces capitatus*)

The taxonomic status of this yeast is vexed by controversy because of its unusual morphology (reviewed in Polacheck et al., 1992). Initially the yeast was named *Trichosporon capitatum. Trichosporon* is a basidiomycetous yeast genus. The pattern of conidiogenesis was described as consisting of annelloconidia that undergo schizolytic division and then form blastoconidia. On that basis, Salkin et al. (1985) established the genus *Blastoschizomyces*. De Hoog et al. (1986) stated that the fungus should be referred to as *Geotrichum capitatum* because, although there was a high incidence of annellidic cells, sympodial rachides[11] were the main characteristic of this yeast. This difference of opinion left the yeast in an ambivalent taxonomic status, referred to as either *Blastoschizomyces captitatus* or as *Geotrichum capitatum*.

Etiologic Agent

Blastoschizomyces capitatus (Diddens & Lodder) Salkin, MA Gordon, Sams. & Rieder, *Mycotaxon* 22: 378 (1985)

Geotrichum capitatum (Diddens & Lodder) Arx, in von Arx et al., *Stud Mycol* 14: 32 (1977) (anamorph)

Dipodascus capitatus de Hoog et al., 1986, is the teleomorph of *Geotrichum capitatum*

Position in classification: *Dipodascaceae, Saccharomycetales, Saccharomycete*s, *Ascomycota*

The presence of micropores in the septa points to membership in the class *Hemiascomycetes* in the order *Saccharomycetales*. The characteristic of *Geotrichum*, as observed in *G. candidum*, is up to 50 micropores (each approximately 9 nm diameter) in the septum between cells, referred to as multiperforate septa. Confirmation of this classification was the discovery of a *Dipodascus* teleomorph of *G. capitatum* (de Hoog et al., 1986).

Sequencing the D1/D2 region of 26S rDNA and query of the GenBank database resulted in the identification of three clinical isolates from the same hospital in Japan as *Dipodascus capitatus* (teleomorph of *Geotrichum capitatum*), (Ersoz et al., 2004). The utility of molecular diagnosis should alleviate the taxonomic uncertainty for this fungus.

Further molecular taxonomic studies of this yeast resulted in other nomenclature changes (de Hoog and Smith, 2004). The classification of *G. capitatum* was revised by erecting another teleomorph and anamorph name for this yeast: *Magnusiomyces capitatus* (teleomorph) and *Saprochaete capitata* (anamorph), based on sequence homology and because it has smaller, four-spored asci, without subsequent mitoses.

Ecologic Niche

Natural environmental sources include wood pulp, milk, and curd in cheese-making. Few articles have studied the ecology of this yeast and because of its changing nomenclature and similar morphology to other yeast species (please see "Laboratory" below) further ecologic studies are needed coupled with modern molecular identification methods.

[11]Rachis. *Definition*: The central axis of hyphal growth from which branches radiate (think of a feather shape). As the branches form one higher than the next the form is called sympodial. The plural of rachis is rachides.

Epidemiology

Geotrichum capitatum infections are reported more frequently in Europe (85% of all cases) than in the United States (10%). The majority of European cases occur in Italy, Spain, and France, suggesting that climate may influence disease ecology. These Mediterranean cases represent approximately three-fourths of all cases worldwide (Girmenia et al., 2005). *Trichosporon* species infections, on the other hand, seem to be uniformly distributed on all continents.

Thirty-five cases of invasive *G. capitatum* mycosis occurred in Italy over a 20 year period. Acute myeloid leukemia was the primary disease in 74% of the cases. The incidence of *G. capitatum* infections in acute leukemia patients was 0.5%. Overall, 74.3% of cases had positive blood cultures; 57% of cases were fatal. An estimated 99 cases were accumulated in the world literature up to 2004; 91.7% of these occurred in patients with hematologic disease. The crude mortality rate of patients with invasive *G. capitatum* was 55.7%.

Surveillance for *G. capitatum* in Italian hospitals over a 6 year period found the yeast was isolated from superficial sites (sputum, oral swab, stool, urine) in 26 patients (Girmenia et al., 2005). For eight (31%) of these colonization was transient, but 13 (50%) patients developed invasive infections. Five other patients had diffuse alveolar infiltrates, with sputum yielding *G. capitatum*. The isolation of *G. capitatum* from nonsterile sites in neutropenic patients should increase suspicion that it may be the cause of invasive disease.

Surveillance in seven Spanish tertiary-care hospitals between 1992 and 2002 revealed 26 patients who developed invasive *G. capitatum* disease (Martino et al., 2004). Most patients (92%) had acute leukemia and developed infection during the period of neutropenia. Fungemia (20 cases) was a common finding, with frequent visceral dissemination. The 30 day crude mortality was 57% among patients with systemic infection. Twenty-one patients were treated with AmB or lipid-AmB. Other treatments consisted of FLC, ITC, VRC, terbinafine, CASF, or flucytosine (5FC). Twelve patients received two or more antifungal drugs. Thirteen patients died a median of 6 days after the diagnosis of *G. capitatum* mycosis, and it was the contributing cause of death, confirmed by postmortem examination of disseminated multi-organ infection. The remaining 10 patients responded to therapy, with successful outcome linked to removal of the central venous catheter within 5 days after the onset of infection, stable clinical status before the onset, and receipt of prior prophylactic or empiric antifungal therapy.

Clinical Forms

Both *G. capitatum* and *Trichosporon* species cause infections that are clinically similar to invasive candidiasis, but the former are associated with higher bloodstream recovery rates, more frequent deep organ involvement, and a worse prognosis. *G. capitatum* is a very uncommon cause of severe systemic infection in immunocompromised patients occurring, for the most part, in patients with hematologic malignancies. Infection with *G. capitatum* occurs mainly in neutropenic patients, with fungemia and multi-organ (including brain) dissemination, and a mortality rate of 60–80%. Blood cultures usually are positive. As with *Trichosporon*, a chronic disseminated form of *G. capitatum* infection, similar to chronic disseminated candidiasis, may be observed upon recovery from neutropenia.

Susceptibility to Antifungal Agents

In vitro susceptibility tests indicate that *G. capitatum* has high levels of susceptibility to AmB and to VRC. Using the CLSI M27A2 broth microdilution method, 65% of isolates tested were susceptible to ITC, the remainder were susceptible-dose-dependent (Girmenia et al., 2003). On the basis of in vitro susceptibility, therefore, FLC (91.3% susceptible) and not ITC would be recommended for therapy. Most isolates are susceptible to 5FC, predicting it would be useful in combination therapy. Recent clinical experience, however, in leukemia patients with invasive *G. capitatum* disease found that when AmB was combined with 5FC or FLC there was no apparent additional therapeutic benefit (Pfaller and Diekema, 2004). The excellent in vitro activity of VRC suggests that it may be useful for the therapy of *G. capitatum* infections. Removal of central venous catheters, treatment with VRC, or high-dose FLC plus AmB are recommended for treatment of this rare but potentially fatal infection.

Laboratory

The high rate of isolation of *G. capitatum* from the bloodstream (~74%) during a deep infection is in contrast to the yield in blood cultures of *Candida* species (<50%).

Colony Morphology

Rapid growth is yeast-like, white to cream color, smooth to wrinkled, with radiating edges, later with short aerial hyphae. This yeast is resistant to cycloheximide.

Microscopic Morphology

Geotrichum capitatum produces true hyphae and pseudohyphae. It has sympodial rachides on slender conidiogenous cells (Fig. 11.18). Annelloconidia are produced which are short, cylindrical, to slightly club-shaped, with a flat basal scar and broadly rounded tips. Their shape differentiates them from arthroconidia, which also are produced, and which are truncate or rounded at both ends.

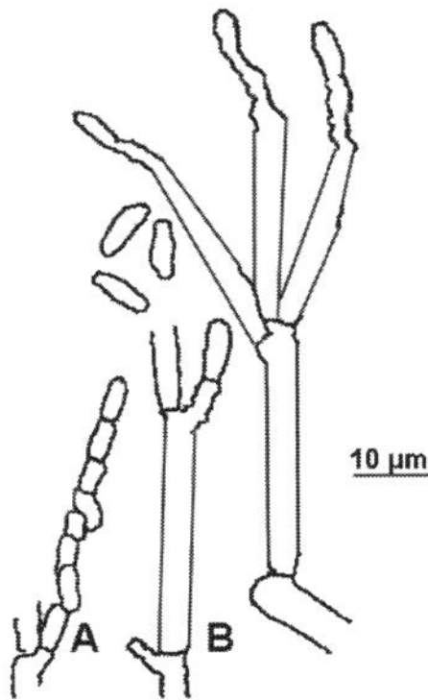

Figure 11.18 Morphology of *Geotrichum capitatum*: (A) *arthroconidia* and (B) *sympodial rachides* with annelloconidia. *Source:* Guého et al. (1987). Adapted with permission from the American Society for Microbiology.

Differentiation of G. capitatum *from Morphologically Similar Yeasts*

Geotrichum capitatum and *Trichosporon* species are difficult to differentiate; however, most *Trichosporon* species are urease positive. *Trichosporon* species form arthroconidia as well as blastoconidia. *G. capitatum* produces arthroconidia and annelloconidia, which may be misidentified as arthroconidia or blastoconidia. The two species usually are differentiated by their carbohydrate assimilation patterns. *G. capitatum* is nonfermentative and assimilates only glucose, galactose, glycerol, D,L-lactate, and succinate. Additionally, *G. capitatum* can grow at temperatures up to 45°C.

G. capitatum is sometimes misidentified as *Candida krusei*; however, *G. capitatum* grows in the presence of cycloheximide and assimilates galactose, whereas *C. krusei* does not.

G. capitatum can easily be distinguished from *G. candidum*; the latter yeast has rapidly growing colonies and hyphae which often are dichotomously branched at the colony margin, is unable to grow at 40°C, and utilizes D-sorbitol. It is further distinguished from *G. capitatum* by its capacity to assimilate D-xylose.

11.13.2 Rhodotorula Species

Rhodotorula species are so named because of the color of the colonies, which varies from pink to coral, or orange to red. The colony color is due to the production of carotenoid pigments.

Etiologic Agents

Rhodotorula species of medical importance include *R. glutinis, R. mucilaginosa* (formerly *R. rubra*), and *R. minuta*.

Rhodotorula mucilaginosa (A Jörg.) FC Harrison, *Proc Trans R Soc Canada, Ser* 3, 21: 349 (1928) (anamorph)

Rhodotorula glutinis (Fresen.) FC Harrison [as "*glutinus*"], *Proc Trans R Soc Canada, Ser* 3, 21: 349 (1928) (anamorph); *Rhodosporidium* (teleomorph)

Rhodotorula minuta (Saito) FC Harrison (1928) (anamorph)

Other *Rhodotorula* species are omitted here because they are of doubtful medical importance.

Position in classification: *Sporidiobolaceae, Sporidiales, Urediniomycetes, Basidiomycota*.

Ecologic Niche

R. mucilaginosa is regarded as a ubiquitous yeast, present in various natural and artificial environs. The ecologic niche of *Rhodotorula* species includes air, soil, and the phyllosphere (leaf surfaces). *Rhodotorula* species are found in milk and cheese products. Air sampling devices may trap *Rhodotorula. R. glutinis* is a common yeast in leaf litter (Sampaio et al., 2007).

Rhodotorula species are found in lakes and ocean water. The most frequently encountered yeasts from beach sand in Southern Florida were *Candida tropicalis* and *R. mucilaginosa*. They were more abundant in the beaches highly frequented by humans (Vogel et al., 2007). *Candida parapsilosis, Rhodotorula* species, and *Trichosporon* species were isolated from water samples and marine invertebrates from the French Mediterranean (Gautret et al., 1999). Several *Rhodotorula* species are found in ice and melt water from Alpine glaciers, but these are different species than those associated with human colonization and infection. So the label of psychrophile may be too broad when applied to the entire genus *Rhodotorula*.

R. rubra (now classed as *R. mucilaginosa*) was isolated from air samples in an ICU (Gniadek and Macura, 2007). Gels used for therapeutic ultrasound units were found to be contaminated with *R. mucilaginosa* (Schabrun et al., 2006). In the home, *Rhodotorula* species are found growing on shower curtains, bathtub grouts, and toothbrushes.

Isolations from Nonsterile Site Specimens

Hand cultures of medical students indicated *Rhodotorula mucilaginosa* and *Candida parapsilosis* were the yeasts most frequently isolated, with the percentage rising to 30% of medical students in their clinical year. A control group of engineering students had a hand culture prevalence of 9%, considered to be the baseline for asymptomatic carriage by nonmedical personnel (Silva et al., 2003). *Rhodotorula* species are commensals on the skin, nails, and mucous membranes, including oral cavity, and may be isolated from stool and rectal swabs.

Evidence for Pathogenicity in Humans

Rhodotorula species are viewed as contaminants when isolated from nonsterile site clinical specimens. *Rhodotorula* is seen as an opportunistic pathogen in the immunocompromised host, especially in patients receiving cancer chemotherapy. Catheters may provide the portal of entry for fungemia, with hematogenous spread to cause endocarditis, peritonitis, meningitis, and endophthalmitis. Search of the world literature uncovered 128 cases of *Rhodotorula* mycosis, of which 79% were fungemia, 7% endophthalmitis, and 6 cases (5%) of peritonitis associated with chronic ambulatory peritoneal dialysis (Tuon and Costa, 2008). Overall mortality was 12.6%. *R. mucilaginosa* was the most commonly isolated species (74% of cases) followed by *R. glutinis*. AmB was the drug of choice in treating *Rhodotorula* fungemia. Ocular disease was treated with topical AmB but loss of vision ensued. A retrospective survey covering 1996–2004 found 25 cases of *Rhodotorula* species fungemia in a large São Paulo, Brazil, tertiary-care hospital (de Almeida et al., 2008). All were caused by *R. mucilaginosa*. Forty percent of these fungemic patients were bone marrow transplant recipients. *Rhodotorula* sepsis was considered the likely cause of death in two patients.

Susceptibility to Antifungal Agents

Clinical isolates of *Rhodotorula* species are highly susceptible in vitro to AmB, at dosages of ≤ 1 μg/mL, and 5FC at ≤ 4 μg/mL (Pfaller and Diekema, 2004). The MICs of FLC, CASF, and MCF are all classed as resistant and these antifungal agents should not be used to treat infections with *Rhodotorula* species. ITC is moderately active ($MIC_{90} = 1$ μg/mL). AmB has excellent activity against *Rhodotorula* and, combined with catheter removal, is an optimal approach to managing *Rhodotorula* infections.

Laboratory

When isolated from a nonsterile body site, the likelihood is that *Rhodotorula* is a commensal or contaminant but, possibly, its presence indicates a disease process in a debilitated and/or immunosuppressed host. *Rhodotorula* species share physiologic and morphologic characteristics with *Cryptococcus* species: both are round to oval, budding, encapsulated yeasts which produce urease and are nonfermentative. *Rhodotorula* species differ from cryptococci, being pigmented, and in their inability to assimilate inositol. When a capsule is present it is typically small, unlike that of *Cryptococcus neoformans*.

Colony Morphology

On SDA, cultures of *Rhodotorula* species grow rapidly at 30°C, are coral pink, smooth, and moist to mucoid. Colony color varies on a spectrum of pink to orange to red. These yeast are inhibited by cycloheximide and growth at 37°C is variable.

Microscopic Morphology

Rhodotorula mucilaginosa is an ellipsoid, monopolar budding yeast, producing a limited amount of pseudohyphae. *R. glutinis* grows as ellipsoid, multipolar budding yeast, producing rudimentary pseudohyphae, and undergoes sexual reproduction with mycelial clamp connections and teliospores (de Hoog et al., 2000). *R. minuta* grows as an ellipsoid, monopolar budding yeast and no pseudohyphae are produced. These are nonfermentative yeasts that produce urease. Carbohydrate assimilation by the most frequent clinical isolate, *R. mucilaginosa*, is positive for glucose, sucrose, xylose, raffinose, and trehalose.

11.13.3 *Saccharomyces cerevisiae*

The yeast of commerce, *Saccharomyces cerevisiae* (Baker's yeast), has been a friend to humankind for millennia. In modern medicine, however, reports of *Saccharomyces* fungemia have made clear that *Saccharomyces* can be pathogenic in an immunocompromised or debilitated human host. *Saccharomyces* on the patient's skin or on a healthcare worker's hands can enter the bloodstream from the site of insertion of a central venous catheter. *Saccharomyces* bloodstream infections also may originate from gastrointestinal translocation. The gastrointestinal source is problematic in view of the use of *Saccharomyces* as probiotic therapy (please see next paragraph). *Saccharomyces* fungemia carries the risk of hematogenous dissemination, endocarditis, and, in some cases, a fatal outcome.

In a review of 60 cases of *Saccharomyces* fungemia in the literature extant to 2004, Muñoz et al. (2005) found that 60% of the affected patients were in the ICU, 71% received enteral or parenteral nutritional support. Probiotic use of *Saccharomyces* was documented in 26 patients (45%). The iatrogenic aspect of this development is related to the approval of *Saccharomyces boulardii* (Ultralevura®, Bristol-Myers-Squibb, Inc.) as a probiotic to treat antibiotic-induced *Clostridium difficile* diarrhea.

Etiologic Agent

Saccharomyces cerevisiae Meyen ex EC. Hansen (1883)

Position in classification: *Saccharomycetaceae, Saccharomycetales, Saccharomycetes* (syn: Hemi-ascomycetes), subphylum *Saccharomycotina, Ascomycota*

Ecologic Niche

Yeast of commerce. *Saccharomyces cerevisiae* is a common colonizer of mucosal surfaces and may be present in the microbiota of the human gastrointestinal tract, the respiratory tract, and vagina.

Transmission

Portals of entry include translocation of ingested yeast from the gastrointestinal or oral mucosae, or contamination during insertion or manipulation of IV catheters.

Risk Factors

The probiotic Ultralevura (*S. boulardii*) should be considered a risk factor for nosocomial fungemia in patients with predisposing underlying conditions. Those conditions include a central venous catheter, being treated in the ICU, receiving broad-spectrum antibiotics, and other risk factors associated with candidemia (please see Section 11.7, Risk Groups/Factors). Severe immunosuppression is not a requisite. Loss of integrity of the gastrointestinal tract is a risk factor underlined by the observation that the antibody titer against *S. cerevisiae* can function as a diagnostic test for Crohn's disease (Bernstein et al., 2001).

Clinical Forms

The most frequent, serious clinical form is fungemia, which, in a survey of reported cases, resulted in endocarditis in six patients (Muñoz et al., 2005). The crude mortality rate for patients with *S. cerevisiae* fungemia was 29.5% (18 patients) (Muñoz et al., 2005). *S. cerevisiae* can cause various clinical forms, for example, cellulitis, empyema, esophagitis, fever of unknown origin, liver abscess, peritonitis, pneumonia, urinary tract infection, and vaginitis. Septic shock in patients with *Saccharomyces* fungemia is exceptional (Hennequin et al., 2000; Lolis et al., 2008)

Therapy

The important interventions are to stop probiotic *Saccharomyces*, initiate antifungal therapy, and, if possible, remove the central venous catheter. *S. cerevisiae* is highly susceptible to AmB and to 5FC. Azole susceptibility is lower; the MIC_{90} for FLC ranges from 8 to 128 μg/mL, and to ITC, the MIC is 1–16 μg/mL (reviewed by Muñoz et al., 2005). The antifungal agent of choice is AmB, but FLC treatment has been successful because many isolates have MICs in the susceptible range. CASF also has been successful as monotherapy (Lolis et al., 2008).

Laboratory

Saccharomyces boulardii is a particular strain of *S. cerevisiae*, and not a separate species.

Colony Morphology

Saccharomyces cerevisiae grows rapidly on SDA or CMA at 30–37°C as smooth, moist, white to cream yeast colonies.

Microscopic Morphology

Saccharomyces cerevisiae is an ellipsoid, multipolar budding yeast, occasionally with short pseudohyphae.

Differentiation from Other Yeasts

S. cerevisiae ferments glucose, maltose, and sucrose. *S. cerevisiae* grows at 37°C but does not grow on medium containing cycloheximide. A specific assimilation biocode identifies this organism in the API 20C AUX yeast identification system.

Sexual Reproduction

Sexual reproduction can be observed when *S. cerevisiae* is grown on V8 juice agar or acetate ascospore agar. The ascospores are round and located in asci. Each ascus contains 1–4 ascospores. Asci do not rupture at maturity.

Ascospore Agar (Remel Div., Fisher Scientific, Inc.) Acetate ascospore agar contains (per liter) potassium acetate, 10 g; yeast extract, 2.5 g; glucose, 1 g; agar, 30 g; final pH = 6.5.

Ascospore Stains

- *Gram Stain.* Ascospores are gram negative while vegetative cells are Gram positive.
- *Acid-Fast Stain.* The Kinyoun method is used (as for mycobacteria). Decolorize with acid-ethanol (Larone, 2002) (conc. hydrochloric acid: ethanol = 3 mL: 97 mL, vol/vol).

Safety Note: Use utmost care when handling concentrated hydrochloric acid, **protect eyes and skin**. Work in a chemical fume hood. Wear eye protection, gloves, and a lab coat. Add acid slowly to alcohol.

Ascospores stain red, other cells are counterstained green.

- *Ascospore Stain.* This is a malachite green stain with a safranin counterstain (Larone, 2002). The ascospores stain green, yeast cells are red colored.

SELECTED REFERENCES FOR CANDIDIASIS

Alcoba-Florez J, Mendez-Alvarez S, Cano J, Guarro J, Perez-Roth E, del Pilar Arevalo M, 2005. Phenotypic and molecular characterization of *Candida nivariensis* sp. nov., a possible new opportunistic fungus. *J Clin Microbiol* 43: 4107–4111.

American Academy of Allergy, Asthma and Immunology Position Paper on Idiopathic Environmental Intolerances, 1999. (Includes Candidiasis Hypersensitivity Syndrome.) http://www.aaaai.org/media/resources/academy_statements/position_statements/ps35.asp.

Beck-Sague C, Jarvis WR, 1993. Secular trends in the epidemiology of nosocomial fungal infections in the United States, 1980–1990. National Nosocomial Infections Surveillance System. *J Infect Dis* 167: 1247–1251.

Bernstein CN, Orr K, Blanchard JF, Sargent M, Workman D, 2001. Development of an assay for antibodies to *Saccharomyces cerevisiae:* easy, cheap and specific for Crohn's disease. *Can J Gastroenterol* 15: 499–504.

Bishop JA, Chase N, Magill SS, Kurtzman CP, Fiandaca MJ, Merz WG, 2008. *Candida bracarensis* detected among isolates of *Candida glabrata* by peptide nucleic acid fluorescence in situ hybridization: Susceptibility data and documentation of presumed infection. *J Clin Microbiol* 46: 443–446.

Blonz ER, 1986. Is there an epidemic of chronic candidiasis in our midst ? *JAMA* 256: 3138–3139.

Bougnoux ME, Diogo D, François N, Sendid B, Veirmeire S, Colombel JF, Bouchier C, Van Kruiningen H, d'Enfert C, Poulain D, 2006. Multilocus sequence typing reveals intrafamilial transmission and microevolutions of *Candida albicans* isolates from the human digestive tract. *J Clin Microbiol* 44: 1810–1820.

Cannon RD, Lamping E, Holmes AR, Niimi K, Baret PV, Keniya MV, Tanabe K, Niimi M, Goffeau A, Monk BC, 2009. Efflux-mediated antifungal drug resistance. *Clin Microbiol Rev* 22: 291–321.

Cowen LE, 2009. Hsp90 orchestrates stress response signaling governing fungal drug resistance. *PLoS Pathog* Aug 5(8):e1000471.

Das S, Brown TM, Kellar KL, Holloway BP, Morrison CJ, 2006. DNA probes for the rapid identification of medically important *Candida* species using a multianalyte profiling system. *FEMS Immunol Med Microbiol* 46: 244–250.

de Almeida GM, Costa SF, Melhem M, Motta AL, Szeszs MW, Miyashita F, Pierrotti LC, Rossi F, Burattini MN, 2008. *Rhodotorula* spp. isolated from blood cultures: Clinical and microbiological aspects. *Med Mycol* 46: 547–556.

de Hoog GS, Smith MT, 2004. Ribosomal gene phylogeny and species delimitation in *Geotrichum* and its teleomorphs. *Stud Mycol* 50: 489–515.

de Hoog GS, Guarro J, Gené J, Figueras MJ, 2000. *Atlas of Clinical Fungi*, 2nd ed. Centralbureau voor Schimmelcultures; Baarn, The Netherlands (CD version 2004.11).

de Hoog GS, Smith MT, Guého E, 1986. A revision of the genus *Geotrichum* in its teleomorphs. *Stud Mycol* 29: 1–131.

de Repentigny L, Lewandowski D, Jolicoeur P, 2004. Immunopathogenesis of oropharyngeal candidiasis in human immunodeficiency virus infection. *Clin Microbiol Rev* 17: 729–759.

Desnos-Ollivier M, Bretagne S, Bernède C, Robert V, Raoux D, Chachaty E, Forget E, Lacroix C, Dromer F, 2008. Clonal population of flucytosine-resistant *Candida tropicalis* from blood cultures, Paris, France. *Emerging Infect Dis* 14: 557–565.

Diamond G, Beckloff N, Ryan LK, 2008. Host defense peptides in the oral cavity and the lung: Similarities and differences. *J Dent Res* 87: 915–927.

Diezmann S, Cox CJ, Schönian G, Vilgalys RJ, Mitchell TG, 2004. Phylogeny and evolution of medical species of *Candida* and related taxa: A multigenic analysis. *J Clin Microbiol* 42: 5624–5635.

Domínguez de María P, Sánchez-Montero JM, Sinisterra JV, Alcántara AR, 2006. Understanding *Candida rugosa* lipases: An overview. *Biotechnol Advances* 24: 180–196.

Edgerton M, Koshlukova SE, 2000. Salivary histatin 5 and its similarities to the other antimicrobial proteins in human saliva. *Adv Dent Res* 14: 16–21.

Elie CM, Lott TJ, Reiss E, Morrison CJ, 1998. Rapid identification of *Candida* species with species-specific DNA probes. *J Clin Microbiol* 36: 3260–3265.

Ellepola AN, Morrison CJ, 2005. Laboratory diagnosis of invasive candidiasis. *J Microbiol* 43: 65–84.

Ellepola AN, Hurst SF, Elie CM, Morrison CJ, 2003. Rapid and unequivocal differentiation of *Candida dubliniensis* from other *Candida* species using species-specific DNA probes: Comparison with phenotypic identification methods. *Oral Microbiol Immunol* 18: 379–388.

Ersoz G, Otag F, Erturan Z, Aslan G, Kaya A, Emekdas G, Sugita T, 2004. An outbreak of *Dipodascus capitatus* infection in the ICU: Three case reports and review of the literature. *Japan J Infect Dis* 57: 248–252.

Fell J, Meyer SA, 1967. Systematics of yeast species in the *Candida parapsilosis* group. *Mycopathol Mycol Appl* 32: 177–193.

Fidel PL Jr, 2005. Immunity in vaginal candidiasis. *Curr Opin Infect Dis* 18: 107–111.

Fidel PL, 2002. Immunity to *Candida*. *Oral Dis* 8: 69–75.

Fidel PL Jr, Huffnagle GB, 2005. *Fungal Immunology—From an Organ Perspective*. Springer Science, New York.

Gadanho M, Sampaio JP, 2005. Occurrence and diversity of yeasts in the mid-Atlantic ridge hydrothermal fields near the Azores Archipelago. *Microbial Ecol* 50: 408–417.

Gautret P, Cosson J, Kauffmann-Lacroix C, Rodie MH, Charron M, Jacquemin JL, 1999. Isolation and molecular typing of yeasts in Mediterranean waters and marine invertebrates from southern France. *J Mycol Med* 9: 162–165.

Gazi U, Martinez-Pomares L, 2009. Influence of the mannose receptor in host immune responses. *Immunobiology* 214: 554–561.

Ghannoum MA, 2000. Potential role of phospholipases in virulence and fungal pathogenesis. *Clin Microbiol Rev* 13: 122–143.

Girmenia C, Pagano L, Martino B, D'Antonio D, Fanci R, Specchia G, Melillo L, Buelli M, Pizzarelli G, Venditti M, Martino P, 2005. Invasive infections caused by *Trichosporon* species and *Geotrichum capitatum* in patients with hematological malignancies: A retrospective multicenter study from Italy and review of the literature. *J Clin Microbiol* 43: 1818–1828.

Girmenia C, Pizzarelli G, D'Antonio D, Cristini F, Martino P, 2003. In vitro susceptibility testing of *Geotrichum capitatum*: Comparison of the E-test, disk diffusion, and Sensititre colorimetric methods with the NCCLS M27-A2 broth microdilution reference method. *Antimicrob Agents Chemother* 47: 3985–3988.

Gniadek A, Macura AB, 2007. Intensive care unit environment contamination with fungi. *Adv Med Sci* 52: 283–287.

Gomez-Lopez A, Alastruey-Izquierdo A, Rodriguez D, Almirante B, Pahissa A, Rodriguez-Tudela JL, Cuenca-Estrella M, 2008. Prevalence and susceptibility profile of *Candida metapsilosis* and *Candida orthopsilosis*: Results from population-based surveillance of candidemia in Spain. *Antimicrob Agents Chemother* 52: 1506–1509.

Guého E, de Hoog GS, Smith MT, Meyer SA, 1987. DNA relatedness, taxonomy, and medical significance of *Geotrichum capitatum*. *J Clin Microbiol* 25: 1191–1194.

Gupta AK, Ryder JE, Baran R, Summerbell RC, 2003. Nondermatophyte onychomycosis. *Dermatol Clin* 21: 257–268.

Hachem R, Hanna H, Kontoyiannis D, Jiang Y, Raad I, 2008. The changing epidemiology of invasive candidiasis. *Cancer* 112: 2493–2499.

Hajjeh RA, Sofair AN, Harrison LH, Lyon GM, Arthington-Skaggs BA, Mirza SA, Phelan M, Morgan J, Lee-Yang W, Ciblak MA, Benjamin LE, Sanza LT, Huie S, Yeo SF, Brandt ME, Warnock DW, 2004. Incidence of bloodstream infections due to *Candida* species and in vitro susceptibilities of isolates collected from 1998 to 2000 in a population-based active surveillance program. *J Clin Microbiol* 42: 1519–1527.

Halkic N, Ksontini R, 2007. Images in clinical medicine—hepatosplenic candidiasis. *N Engl J Med* 356:c4 (January 25).

Hall L, Wohlfiel S, Roberts GD, 2003. Experience with the MicroSeq D2 large-subunit ribosomal DNA sequencing kit for identification of commonly encountered, clinically important yeast species. *J Clin Microbiol* 41: 5099–5102.

Hanna DB, Gupta LS, Jones LE, Thompson DM, Kellerman SM, Sackoff JE, 2007. AIDS-defining opportunistic illnesses in the HAART era in New York City. *AIDS Care* 19: 264–272.

Harrington BJ, Williams DL, 2007. Rapid, presumptive identification of *Torulopsis* (*Candida*) *glabrata* and *Candida krusei* using Calcofluor white. *Lab Med* 38: 227–231.

Hay RJ, 1999. The management of superficial candidiasis. *J Am Acad Dermatol* 40(No. 6Pt 2): S35–S42.

Hazen KC, Howell SA, 2007. *Candida, Cryptococcus*, and Other Yeasts of Medical Importance, pp. 1762–1788, *in*: Murray PR, Baron EJ, Jorgenson JH, Landry ML, Pfaller MA (eds.), *Manual of Clinical Microbiology*, 9th ed. ASM Press, Washington, DC.

Heald AH, O'Halloran DJ, Richards K, Webb F, Jenkins S, Hollis S, Denning DW, Young RJ, 2001. Fungal infection of the diabetic foot: Two distinct syndromes. *Diabet Med* 18: 567–572.

Hennequin C, Kauffmann-Lacroix C, Jobert A, Viard JP, Ricour C, Jacquemin JL, Berche P, 2000. Possible role of catheters in *Saccharomyces boulardii* fungemia. *Eur J Clin Microbiol Infect Dis* 19: 16–20.

Hitchcock CA, 1993. Resistance of *Candida albicans* to azole antifungal drugs. *Biochem Soc Trans* 21: 1039–1047.

Hobson RP, Munro CA, Bates S, MacCallum DM, Cutler JE, Heinsbroek SE, Brown GD, Odds FC, Gow NA, 2004. Loss of cell wall mannosylphosphate in *Candida albicans* does not influence macrophage recognition. *J Biol Chem* 279: 39628–39635.

Hospenthal DR, Beckius ML, Floyd KL, Horvath LL, Murray CK, 2006. Presumptive identification of *Candida* species other than *C. albicans, C. krusei*, and *C. tropicalis* with the chromogenic medium CHROMagar *Candida*. *Ann Clin Microbiol Antimicrob* 5: 1–5.

Hoyer LL, Green CB, Oh SH, Zhao X, 2008. Discovering the secrets of the *Candida albicans* agglutinin-like sequence (ALS) gene family—a sticky pursuit. *Med Mycol* 46: 1–15

Janeway CA Jr, Travers P, Walport M, Shlomchik MJ, 2001. *Immunobiology: The Immune System in Health and Disease*, 5th ed. Garland Science, New York. (Available online at the NCBI Bookshelf, http://www.ncbi.nlm.nih.gov/bookshelf/br.fcgi?book = imm.

Jouault T, El Abed-El Behi M, Martínez-Esparza M, Breuilh L, Trinel PA, Chamaillard M, Trottein F, Poulain D, 2006. Specific recognition of *Candida albicans* by macrophages requires galectin-3 to discriminate *Saccharomyces cerevisiae* and needs association with TLR2 for signaling. *J Immunol* 177: 4679–4687.

Kao AS, Brandt ME, Pruitt WR, Conn LA, Perkins BA, Stephens DS, Baughman WS, Reingold AL, Rothrock GA, Pfaller MA, Pinner RW, Hajjeh RA, 1999. The epidemiology of candidemia in two United States cities: Results of a population-based active surveillance. *Clin Infect Dis* 29: 1164–1170.

Kirkpatrick CH, 2001. Chronic mucocutaneous candidiasis (immunology for the pediatrician). *Ped Infect Dis J* 20: 197–206.

Kontoyiannis DP, Marr KA, Park BJ, Alexander BD, Anaissie EJ, Walsh TJ, Ito J, Andes DR, Baddley JW, Brown JM, Brumble LM, Freifeld AG, Hadley S, Herwaldt LA, Kauffman CA, Knapp K, Lyon GM, Morrison VA, Papanicolaou G, Patterson TF, Perl TM, Schuster MG, Walker R, Wannemuehler KA, Wingard JR, Chiller TM, Pappas PG, 2010. Prospective surveillance for invasive fungal infections in hematopoietic stem cell transplant recipients, 2001–2006: Overview of the Transplant-Associated Infection Surveillance Network (TRANSNET) Database. *Clin Infect Dis* 50: 1091–1100.

Krcmery V, Barnes AJ, 2002. Non-albicans *Candida* spp. causing fungaemia: Pathogenicity and antifungal resistance. *J Hosp Infect* 50: 243–260.

Krogh-Madsen M, Cavling Arendrup M, Helset L, Knudsen JD, 2006. Amphotericin B and caspofungin resistance in *Candida glabrata* isolates recovered from a critically ill patient. *Clin Infect Dis* 42: 938–944.

Kumamoto CA, Vinces MD, 2005. Alternative *Candida albicans* lifestyles: Growth on surfaces. *Annu Rev Microbiol* 59: 113–133.

Larone DH, 2002. *Medically Important Fungi, A Guide to Identification*, 4th ed. ASM Press, Washington, DC.

Lasker BA, Elie CM, Lott TJ, Espinel-Ingroff A, Gallagher L, Kuykendall RJ, Kellum ME, Pruitt WR, Warnock DW, Rimland D, McNeil MM, Reiss E, 2001. Molecular epidemiology of *Candida albicans* strains isolated from the oropharynx of HIV-positive patients at successive clinic visits. *Med Mycol* 39: 341–352.

Lolis N, Veldekis D, Moraitou H, Kanavaki S, Velegraki A, Triandafyllidis C, Tasioudis C, Pefanis A, Pneumatikos I, 2008. *Saccharomyces boulardii* fungaemia in an intensive care unit patient treated with caspofungin. *Crit Care* 12:414.

Levy O, 2004. Antimicrobial proteins and peptides: Anti-infective molecules of mammalian leukocytes. *J Leukoc Biol* 76: 909–925.

Lyon GM, Karatela S, Sunay S, Adiri Y, Candida Surveillance Study Investigators, 2010. Antifungal susceptibility testing of *Candida* isolates from the *Candida* surveillance study. *J Clin Microbiol* 48: 1270–1275.

Makarova NU, Pokrowsky VV, Kravchenko AV, Serebrovskaya LV, James MJ, McNeil MM, Lasker BA, Warnock DW, Reiss E, 2003. Persistence of oropharyngeal *Candida albicans* strains with reduced susceptibilities to fluconazole among human immunodeficiency virus-seropositive children and adults in a long-term care facility. *J Clin Microbiol* 41: 1833–1837.

Martinez M, López-Ribot JL, Kirkpatrick WR, Coco BJ, Bachmann SP, Patterson TF, 2002. Replacement of *Candida albicans* with *C. dubliniensis* in human immunodeficiency virus-infected patients with oropharyngeal candidiasis treated with fluconazole. *J Clin Microbiol* 40: 3135–3139.

Martino R, Salavert M, Parody R, Tomás JF, de la Cámara R, Vázquez L, Jarque I, Prieto E, Sastre JL, Gadea I, Pemán J, Sierra J, 2004. *Blastoschizomyces capitatus* infection in patients with leukemia: Report of 26 cases. *Clin Infect Dis* 38: 335–341.

MASOOD A, SALLAH S, 2005. Chronic disseminated candidiasis in patients with acute leukemia: Emphasis on diagnostic definition and treatment. *Leuk Res* 29: 493–501.

MATTHEWS RC, RIGG G, HODGETTS S, CARTER T, CHAPMAN C, GREGORY C, ILLIDGE C, BURNIE J, 2003. Preclinical assessment of the efficacy of mycograb, a human recombinant antibody against fungal HSP90. *Antimicrob Agents Chemother* 47: 2208–2216.

MCCULLOUGH MJ, CLEMONS KV, STEVENS DA, 1999. Molecular and phenotypic characterization of genotypic *Candida albicans* subgroups and comparison with *Candida dubliniensis* and *Candida stellatoidea*. *J Clin Microbiol* 37: 417–421.

MCNEIL MM, NASH SL, HAJJEH RA, PHELAN MA, CONN LA, PLIKAYTIS BD, WARNOCK DW, 2001. Trends in mortality due to invasive mycotic diseases in the United States, 1980–1997. *Clin Infect Dis* 33: 641–647.

MCNEILL MM, GERBER AR, MCLAUGHLIN DW, VEGA RA, WINN K, KAUFMAN L, KEYSERLING HL, JARVIS WR, 1992. Mannan antigenemia during invasive candidiasis caused by *Candida tropicalis*. *Ped Infect Dis J* 11: 493–496.

MEDEIROS EAS, LOTT TJ, COLOMBO AL, GODOY P, COUTINHO AP, BRAGA MS, NUCCI M, BRANDT ME, 2007. Evidence for a pseudo-outbreak of *Candida guilliermondii* fungemia in a university hospital in Brazil. *J Clin Microbiol* 45: 942–947.

MERILUOTO T, HALONEN M, PELTO-HUIKKO M, KANGAS H, KORHONEN J, KOLMER M, ULMANEN I, ESKELIN P, 2001. The autoimmune regulator: A key toward understanding the molecular pathogenesis of autoimmune polyendocrinopathy–candidiasis–ectodermal dystrophy. *Keio J Med* 50: 225–239.

MOCHON BA, CUTLER JE, 2005. Prospects of vaccines for medically important fungi. *Med Mycol* 43: 97–115.

MUÑOZ P, BOUZA E, CUENCA-ESTRELLA M, EIROS JM, PÉREZ MJ, SÁNCHEZ-SOMOLINOS M, RINCÓN C, HORTAL J, PELÁEZ T, 2005. *Saccharomyces cerevisiae* fungemia: An emerging infectious disease. *Clin Infect Dis* 40: 1625–1634.

NETT JE, CRAWFORD K, MARCHILLO K, ANDES DR, 2010. Role of Fks1p and matrix glucan in *Candida albicans* biofilm resistance to an echinocandin, pyrimidine, and polyene. *Antimicrob Agents Chemother* 54: 3505–3508.

NETT J, LINCOLN L, MARCHILLO K, MASSEY R, HOLOYDA K, HOFF B, VAN HANDEL M, ANDES D, 2007. Putative role of beta-1,3 glucans in *Candida albicans* biofilm resistance. *Antimicrob Agents Chemother* 51: 510–520.

NOKTA M, 2008. Oral manifestations associated with HIV infection. *Curr HIV/AIDS Rep* 5: 5–12.

OLIVERI S, TROVATO L, BETTA P, ROMEO MG, NICOLETTI G, 2008. Experience with the Platelia *Candida* ELISA for the diagnosis of invasive candidosis in neonatal patients. *Clin Microbiol Infect* 14: 391–393.

PACHL J, SVOBODA P, JACOBS F, VANDEWOUDE K, VAN DER HOVEN B, SPRONK P, MASTERSON G, MALBRAIN M, AOUN M, GARBINO J, TAKALA J, DRGONA L, BURNIE J, MATTHEWS R, 2006. A randomized, blinded, multicenter trial of lipid-associated amphotericin B alone versus in combination with an antibody-based inhibitor of heat shock protein 90 in patients with invasive candidiasis. *Clin Infect Dis* 42: 1404–1413.

PAPPAS PG, ALEXANDER BD, ANDES DR, HADLEY S, KAUFFMAN CA, FREIFELD A, ANAISSIE EJ, BRUMBLE LM, HERWALDT L, ITO J, KONTOYIANNIS DP, LYON GM, MARR KA, MORRISON VA, PARK BJ, PATTERSON TF, PERL TM, OSTER RA, SCHUSTER MG, WALKER R, WALSH TJ, WANNEMUEHLER KA, CHILLER TM, 2010. Invasive fungal infections among organ transplant recipients: Results of the Transplant-Associated Infection Surveillance Network (TRANSNET). *Clin Infect Dis* 50: 1101–1111.

PAPPAS PG, KAUFFMAN CA, ANDES D, BENJAMIN DK Jr, CALANDRA TF, EDWARDS JE Jr, FILLER SG, FISHER JF, KULLBERG BJ, OSTROSKY-ZEICHNER L, REBOLI AC, REX JH, WALSH TJ, SOBEL JD, 2009. Clinical practice guidelines for the management of candidiasis: 2009 Update by the Infectious Diseases Society of America. *Clin Infect Dis* 48: 503–535.

PAPPAS PG, REX JH, LEE J, HAMILL RJ, LARSEN RA, POWDERLY W, KAUFFMAN CA, HYSLOP N, MANGINO JE, CHAPMAN S, HOROWITZ HW, EDWARDS JE, DISMUKES WE, NIAID Mycoses Study Group, 2003. A prospective observational study of candidemia: Epidemiology, therapy, and influences on mortality in hospitalized adult and pediatric patients. *Clin Infect Dis* 37: 634–643.

PASSOTH V, FREDLUND E, DRUVEFORS UA, SCHNÜRER J, 2006. Biotechnology, physiology and genetics of the yeast *Pichia anomala*. *FEMS Yeast Res* 6: 3–13.

PECK A, MELLINS ED, 2010. Precarious balance: Th17 cells in host defense. *Infect Immun* 78: 32–38.

PFALLER MA, DIEKEMA DJ, 2007. Epidemiology of invasive candidiasis: A persistent public health problem. *Rev Infect Dis* 20: 133–163.

PFALLER MA, DIEKEMA DJ, 2004. Rare and emerging opportunistic fungal pathogens: Concern for resistance beyond *Candida albicans* and *Aspergillus fumigatus*. *J Clin Microbiol* 42: 4419–4431.

PFALLER MA, DIEKEMA DJ, GIBBS DL, NEWELL VA, BARTON R, BIJIE H, BILLE J, CHANG SC, DA LUZ MARTINS M, DUSE A, DZIERZANOWSKA D, ELLIS D, FINQUELIEVICH J, GOULD I, GUR D, HOOSEN A, LEE K, MALLATOVA N, MALLIE M, PENG NG, PETRIKOS G, SANTIAGO A, TRUPL J, VANDEN ABEELE AM, WADULA J, ZAIDI M, 2010. Geographic variation in the frequency of isolation and fluconazole and voriconazole susceptibilities of *Candida glabrata*: An assessment from the ARTEMIS DISK Global Antifungal Surveillance Program. *Diagn Microbiol Infect Dis* 67: 162–171.

PFALLER MA, DIEKEMA DJ, GIBBS DL, NEWELL VA, MEIS JF, GOULD IM, FU W, COLOMBO AL, RODRIGUEZ-NORIEGA E, 2007. Results from the ARTEMIS DISK Global Antifungal Surveillance study, 1997 to 2005: An 8.5-year analysis of susceptibilities of *Candida* species and other yeast species to fluconazole and voriconazole determined by CLSI standardized disk diffusion testing. *J Clin Microbiol* 45: 1735–1745.

PFALLER MA, DIEKEMA DJ, MENDEZ M, KIBBLER C, ERZSEBET P, CHANG SC, GIBBS DL, NEWELL VA, 2006. *Candida guilliermondii*, an opportunistic fungal pathogen with decreased susceptibility to fluconazole: Geographic and temporal trends from the ARTEMIS DISK antifungal surveillance program. *J Clin Microbiol* 44: 3551–3556.

PFALLER MA, DIEKEMA DJ, MESSER SA, BOYKEN L, HOLLIS RJ, JONES RN, 2003. In vitro activities of voriconazole, posaconazole, and four licensed systemic antifungal agents against *Candida* species infrequently isolated from blood. *J Clin Microbiol* 41: 78–83.

PFALLER MA, MESSER SA, HOLLIS RJ, JONES RN, DOERN GV, BRANDT ME, HAJJEH RA, 1999. Trends in species distribution and susceptibility to fluconazole among blood stream isolates of *Candida* species in the United States. *Diagn Microbiol Infect Dis* 33: 217–222.

PICK JJ, KNOT EAR, SCHOONEVELD MJ, RIETRA PJGM, 1988. Case reports—candidemia, look at the eyes. *Intensive Care Med* 14: 173–175.

POLACHECK I, SALKIN IF, KITZES-COHEN R, RAZ R, 1992. Endocarditis caused by *Blastoschizomyces capitatus* and taxonomic review of the genus. *J Clin Microbiol* 30: 2318–2322.

POULAIN D, JOUAULT T, 2004. *Candida albicans* cell wall glycans, host receptors and responses: Elements for a decisive crosstalk. *Curr Opin Microbiol* 7: 342–349.

RANGEL-FRAUSTO MS, WIBLIN T, BLUMBERG HM, SALMAN L, PATTERSON J, RINALDI M, PFALLER M, EDWARDS JE Jr, JARVIS W, DAWSON J, WENZEL RP, 1999. National epidemiology of mycoses survey (NEMIS): Variations in rates of bloodstream infections due to *Candida* species in seven surgical intensive care units and six neonatal intensive care units. *Clin Infect Dis* 29: 2253–2258.

Reiss E, Lasker BA, Iqbal N, James MJ, Arthington-Skaggs BA, 2008. Molecular epidemiology of *Candida parapsilosis* sepsis in neonatal intensive care units. *Infect Genetics Evol* 8: 103–109.

Reller ME, Mallonee AB, Kwiatkowski NP, Merz WG, 2007. Use of peptide nucleic acid–fluorescence in situ hybridization for definitive, rapid identification of five common *Candida* species. *J Clin Microbiol* 45: 3802–3803.

Rex JH, Pfaller MA, 2002. Has antifungal susceptibility testing come of age ? *Clin Infect Dis* 35: 982–989.

Reznik DA, 2005. Perspective—oral manifestations of HIV disease. *Top HIV Med* 13: 143–148.

Salkin IF, Gordon MA, Samsonoff WA, Rieder CL, 1985. *Blastoschizomyces captitatus*, a new combination. *Mycotaxon* 22: 373–380.

Sallah S, Semelka RC, Wehbie R, Sallah W, Nguyen NP, Vos P, 1999. Hepatosplenic candidiasis in patients with acute leukaemia. *Br J Haematol* 106: 697–701.

Sallam A, Lynn W, McCluskey P, Lightman S, 2006. Endogenous *Candida* endophthalmitis. *Expert Rev Anti Infect Ther* 4: 675–685.

Sampaio A, Sampaio JP, Leao C, 2007. Dynamics of yeast populations recovered from decaying leaves in a nonpolluted stream: A 2-year study on the effects of leaf litter type and decomposition time. *FEMS Yeast Res* 7: 595–603.

Schabrun S, Chipcase L, Rickard H, 2006. Are therapeutic ultrasound units a potential vector for nosocomial infection ? *Physiother Res Int* 11: 61–71.

Schaller M, Korting HC, Hube B, 2004. Virulence Factors that Promote Invasion of *Candida albicans*, pp. 97–127, *in*: San-Blas G, Calderone RA (eds.), *Pathogenic Fungi–Host Interactions and Emerging Strategies for Control*. Caister Academic Press, Norfolk, England, UK.

Scully C, Flint S, Porter SR, Moos K, 2004. *Oral and Maxillofacial Diseases*, 3rd ed. Taylor and Francis, London, Fig. 4.22 on p.118.

Selmecki A, Forche A, Berman J, 2010. Genomic plasticity of the human fungal pathogen *Candida albicans*. *Eukaryot Cell* 9: 991–1008.

Selmecki A, Forche A, Berman J, 2006. Aneuploidy and isochromosome formation in drug-resistant *Candida albicans*. *Science* 313: 367–370.

Shibata, N, Suzuki A, Kobayashi H, Okawa Y, 2007. Chemical structure of the cell wall mannan of serotype A and its difference in yeast and hyphal forms. *Biochem J* 404: 365–372.

Shibata N, Kobayashi H, Okawa Y, Suzuki S, 2003. Existence of novel ß-1,2 linkage-containing side chain in the mannan of *Candida lusitaniae*, antigenically related to *Candida albicans* serotype A. *Eur J Biochem* 270: 2565–2575.

Shin JH, Nolte FS, Holloway BP, Morrison CJ, 1999. Rapid identification of up to three *Candida* species in a single reaction tube by a 5′ exonuclease assay using fluorescent DNA probes. *J Clin Microbiol* 37: 165–170.

Shin JH, Nolte FS, Morrison CJ, 1997. Rapid identification of *Candida* species in blood cultures by a clinically useful PCR method. *J Clin Microbiol* 35: 1454–1459.

Silva V, Zepeda G, Rybak ME, Febre N, 2003. Yeast carriage on the hands of medicine students. *Rev Iberoam Micol* 20: 41–45.

Soll DR, 2006. The Mating Type Locus and Mating of *Candida albicans* and *Candida glabrata*, pp. 89–112, *in*: Heitman J, Filler SG, Edwards JE Jr (eds.), *Molecular Principles of Fungal Pathogenesis*. ASM Press, Washington, DC.

Soll DR, Daniels KJ, 2007. *MAT*, Mating, Switching, and Pathogenesis in *Candida albicans, Candida dubliniensis* and *Candida glabrata*, pp. 225–245, *in*: Heitman J, Kronstad JW, Taylor JW, Casselton LA (eds.), *Sex in Fungi*. ASM Press, Washington, DC.

Sorà F, Chiusolo P, Piccirillo N, Pagano L, Laurenti L, Farina G, Sica S, Leone G, 2002. Successful treatment with caspofungin of hepato-splenic candidiasis resistant to liposomal amphotericin B. *Clin Infect Dis* 35: 1135–1136.

Spellberg B, Edwards JE Jr, 2002. The pathophysiology and treatment of *Candida* sepsis. *Curr Infect Dis Rep* 4: 387–399.

Staab JF, Bradway SD, Fidel PL, Sundstrom P, 1999. Adhesive and mammalian transglutaminase substrate properties of *Candida albicans* Hwp1. *Science* 283: 1535–1538.

Sullivan DJ, Moran GP, Coleman DC, 2005. *Candida dubliniensis*: Ten years on. *FEMS Microbiol Lett* 253: 9–17.

Suzuki S, 2002. Chapter 3, Serological Differences Among Pathogenic *Candida* spp, pp. 29–36, *in*: Calderone RA (ed.), *Candida and Candidiasis*. ASM Press, Washington, DC.

Tavanti A, Davidson AD, Gow NAR, Maiden MCJ, Odds FC, 2005. *Candida orthopsilosis* and *Candida metapsilosis* spp. nov. to replace *Candida parapsilosis* Groups II and III. *J Clin Microbiol* 43: 284–292.

Taylor JW, 2006. Evolution of Human-Pathogenic Fungi: Phylogenies and Species, pp. 113–132, *in:* Heitman J, Filler SG, Edwards JE Jr (eds.), *Molecular Principles of Fungal Pathogenesis*. ASM Press, Washington, DC.

Tortorano AM, Kibbler C, Peman J, Bernhardt H, Klingspor L, Grillot R, 2006. Candidaemia in Europe: Epidemiology and resistance. *Int J Antimicrob Agents* 27: 359–366.

Tortorano AM, Peman J, Bernhardt H, Klingspor L, Kibbler CC, Faure O, Biraghi E, Canton E, Zimmermann K, Seaton S, Grillot R, 2004. Epidemiology of candidaemia in Europe: Results of 28-month European Confederation of Medical Mycology (ECMM) hospital-based surveillance study; ECMM Working Group on Candidaemia. *Eur J Clin Microbiol Infect Dis* 23: 317–322.

Trick WE, Fridkin SK, Edwards JR, Hajjeh RA, Gaynes RP, National Nosocomial Infections Surveillance System Hospitals, 2002. Secular trend of hospital-acquired candidemia among intensive care unit patients in the United States during 1989–1999. *Clin Infect Dis* 35: 627–630.

Tuon FF, Costa SF, 2008. *Rhodotorula* infection. A systematic review of 128 cases from literature. *Rev Iberoam Micol* 25: 135–140.

van de Veerdonk FL, Gresnigt MS, Kullberg BJ, van der Meer JW, Joosten LA, Netea MG, 2009. Th17 responses and host defense against microorganisms: An overview. *BMB Rep* 42: 776–787.

Vargas K, Messer SA, Pfaller M, Lockhart SR, Stapleton JT, Hellstein J, Soll DR, 2000. Elevated phenotypic switching and drug resistance of *Candida albicans* from human immunodeficiency virus-positive individuals prior to first thrush episode. *J Clin Microbiol* 38: 3595–3607.

Vazquez JA, Sobel JD, 2003. Candidiasis, pp. 143–187, *in*: Dismukes WE, Pappas PG, Sobel JD (eds.), *Clinical Mycology*. Oxford University Press, Oxford, UK.

Verduyn Lunel FM, Meis JF, Voss A, 1999. Nosocomial fungal infections: Candidemia. *Diagn Microbiol Infect Dis* 34: 213–220.

Vonk AG, Netea MG, van der Meer JWM, Kullberg BJ, 2006. Host defence against disseminated *Candida albicans* infection and implications for antifungal therapy. *Expert Opin Biol Ther* 6: 891–903.

Vogel C, Rogerson A, Schatz S, Laubach H, Tallman A, Fell J, 2007. Prevalence of yeasts in beach sand at three bathing beaches in South Florida. *Water Res* 41: 1915–1920.

Whiteway M, Bachewich C, 2006. Signal Transduction in the Interactions of Fungal Pathogens and Mammalian Hosts, pp. 143–162, *in*: Heitman J, Filler SG, Edwards JE Jr, Mitchell A (eds.), *Molecular Principles of Fungal Pathogenesis*. ASM Press, Washington, DC.

Xin H, Dziadek S, Bundle DR, Cutler JE, 2008. Synthetic glycopeptide vaccines combining beta-mannan and peptide epitopes induce protection against candidiasis. *Proc Natl Acad Sci USA* 105: 13526–13531.

Young JC, Moarefi I, Hartl FU, 2001. Hsp90: A specialized but essential protein-folding tool. *J Cell Biol* 154: 267–273.

Zaoutis TE, Heydon K, Localio R, Walsh TJ, Feudtner C, 2007. Outcomes attributable to neonatal candidiasis. *Clin Infect Dis* 44: 1187–1193.

Zhao X, Daniels KJ, Oh SH, Green CB, Yeater KM, Soll DR, Hoyer LL, 2006. *Candida albicans* Als3p is required for wild-type biofilm formation on silicone elastomer surfaces. *Microbiology* 152: 2287–2299.

QUESTIONS

The Answer Key to multiple choice questions may be found at the end of the book.

1. Where is the ecologic niche of *Candida albicans*?
   A. It is a member of the aerosopora, and conidia in the environment are inhaled.
   B. It is one of the few foodborne fungal pathogens.
   C. On the floor of shower rooms and in fitness centers.
   D. Oral, intestinal, and vaginal mucosae of humans.

2. Which is not a risk factor for invasive candidiasis?
   A. Broad-spectrum antibiotic therapy
   B. HIV-positive individuals
   C. IV catheters for nutritional support
   D. Prolonged hospital stays

3. A Gram-stained smear of a mucosal swab from a patient suspected of mucosal candidiasis will reveal
   A. Budding yeast with or without pseudohyphae.
   B. *Candida* doesn't react with Gram stain and requires special fungal stains.
   C. Gram-negative staining reaction if yeast is present.
   D. Pseudohyphae will always be present if the culture is a *Candida* species.

4. A yeast is seen in a smear from a blood culture. Arrange the following in the order of most rapid to slowest test to result in a specific diagnosis.
   A. CHROMagar*Candida* culture
   B. PNA-FISH test
   C. SDA-Emmons and API20c panel
   D. SDA-Emmons culture and germ tube test

5. Which of the following noninvasive diagnostic methods is useful in hepatosplenic candidiasis?
   A. Blood cultures
   B. HRCT
   C. Liver biopsy
   D. Serum galactomannan

6. What is the estimated annual burden of invasive candidiasis in the United States?
   A. 600 cases/yr
   B. 6000
   C. 60000
   D. 600000

7. What is meant by an intravascular device pathogen? How do yeasts colonize these devices? What happens to their susceptibility to antifungal agents? What role do healthcare workers play in transmission of *Candida*?

8. What risk factors for invasive candidiasis are common to intensive care unit patients and those receiving immunosuppressive/cytotoxic therapy?

9. For each of these antifungal agents indicated for suspected invasive candidiasis, explain how the causative *Candida* species and status of the patient would influence your clinical treatment decision for primary therapy: AmB (or lipid-AmB), fluconazole, voriconazole, micafungin, and caspofungin.

10. What is chronic disseminated candidiasis seen in some leukemia patients recovering from neutropenia? How often is this form of disseminated candidiasis encountered? How is this clinical form diagnosed and what is the recommended antifungal therapy? What is the presumed role of immune reconstitution inflammatory syndrome in this disease?

Chapter 12

Cryptococcosis

12.1 CRYPTOCOCCOSIS-AT-A-GLANCE

- *Introduction/Disease Definition.* Chronic meningitis and meningoencephalitis caused by encapsulated yeasts.
- *Etiologic Agents. Cryptococcus neoformans* is the major species. *Cryptococcus gattii* causes fewer than 5% of cases worldwide. Both species are basidiomycetes with a predominant yeast form and a less obvious hyphal form, the latter producing infectious basidiospores.
- *Geographic Distribution. C. neoformans* is found in urban and rural areas worldwide; *C. gattii* is endemic to parts of Australia, South Africa, Southern California, and Vancouver Island, British Columbia, but also is found in parts of Asia, India, and Europe.
- *Ecologic Niche. C. neoformans* lives in soil mixed with pigeon droppings; *C. gattii* is associated with decayed hollows of *Eucalyptus* and other trees.
- *Epidemiology.* The incidence of cryptococcosis follows the epidemic curve of AIDS. In non-AIDS patients it may occur in those with Hodgkin's disease, after immunosuppression for solid organ transplantation or for collagen-vascular diseases, and also in persons with no known underlying immunosuppressive condition.
- *Transmission.* Respiratory exposure is common; but disease is uncommon, linked to host CMI, in most cases.
- *Determinants of Pathogenicity.* Polysaccharide capsule, melanin, mannitol.
- *Clinical Forms.* Meningitis, meningoencephalitis, pulmonary (in descending order of prevalence). AIDS-cryptococcosis includes cryptococcemia and skin dissemination.
- *Therapy.* AmB + flucytosine (5FC) with fluconazole (FLC) maintenance therapy. FLC is used in resource-poor countries.
- *Laboratory Detection, Recovery, and Identification.* CSF specimens in an India ink mount reveal an encapsulated yeast. Culture: grows as an encapsulated yeast that is urease (+); produces the brown colony effect on birdseed agar. Cryptococcal latex agglutination (LA) is a test for antigen in CSF or serum. The antigen titers are diagnostic and prognostic.

12.2 INTRODUCTION/DISEASE DEFINITION

Cryptococcosis is an opportunistic deep-seated mycosis found worldwide, caused by encapsulated yeasts of two *Cryptococcus* species: *C. neoformans* and *C. gattii*. The life cycle of the two species includes budding yeast, typical sexual reproduction, and monokaryotic fruiting; both of the latter result in a hyphal phase in the environment with the production of infectious basidiospores. Infection occurs via inhalation of desiccated yeast and/or basidiospores in dusts in the vicinity of pigeon droppings (*C. neoformans*) or in proximity to *Eucalyptus*, other trees, and rotting wood (*C. gattii*).

Exposure is common but disease is uncommon and linked to CMI and other host factors. The fate of the inhaled cryptococci is determined by the host's susceptibility. In the immune-normal host exposure is sub clinical

Fundamental Medical Mycology, First Edition. By Errol Reiss, H. Jean Shadomy and G. Marshall Lyon, III.
© 2012 Wiley-Blackwell. Published 2012 by John Wiley & Sons, Inc.

and results in either killing of the cryptococci or long-term latency. In the immunocompromised host the inhaled cryptococci germinate in the lung forming budding yeast which produce a high molecular weight, viscous, acidic polysaccharide capsule: glucuronoxylomannan. A subacute or chronic disease ensues. The pulmonary phase often is transient, possibly leading to long-term latency; then the fungus tends to disseminate to the central nervous system producing chronic meningitis or meningoencephalitis that may prove fatal if untreated. The incidence of cryptococcosis follows the epidemic curve of AIDS and has subsided in countries where antiretroviral therapy is generally available. The global burden of cryptococcal meningitis is estimated to be 957, 900 cases each year, resulting in 624,700 deaths in persons living with AIDS (please see Section 12.6, Epidemiology).

Excluding AIDS-related cryptococcosis, patients at risk are those receiving chemotherapy, for example, for Hodgkin's lymphoma, for solid tumors or immunosuppressive therapy for maintenance of solid organ transplants, or for treating collagen-vascular diseases. Of the non-AIDS cryptococcosis, 30–50% of cases have no known underlying cause (Perfect, 2005).

The major etiologic agent is *Cryptococcus neoformans* var. *grubii*, less frequently *C. neoformans* var. *neoformans* (northern Europe). *C. gattii* accounts for a small minority of cases, is geographically restricted, and has increased potential to infect immune-normal persons. The extent of *C. gattii* microfoci is not completely understood. The species is associated with *Eucalyptus camaldulensis* (red river gum), other subtropical trees, and in the vicinity of Vancouver, British Columbia, Canada. Cases have been found in areas where *Eucalyptus* trees occur naturally and near tree plantations in Australia, Mexico, South Africa, and Zaire.

12.3 CASE PRESENTATIONS

Case Presentation 1. Headache, Fever and Altered Mental Status After a Liver Transplant (Fishman et al., 2008)

History

A 45-yr-old man was admitted for an orthotopic liver transplant. At 18 years old he developed hepatitis C, the source of which was instruments used for tattoos. He suffered bouts of jaundice becoming more frequent 6 years before this admission. He developed alcoholic cirrhosis 2 years before admission. The HCV viral load was 900,000 RNA copies/mL but he was not treated with antivirals because of alcohol use. Four months before this admission he had multiple paracenteses at another hospital, removing 26 liters of ascites.

Physical Examination

His temperature and blood pressure were normal. The skin was jaundiced and covered with tattoos. The chest radiograph was normal.

Liver transplant surgery was without incident and his liver function tests were stable. Conditioning therapy consisted of azathioprine, tacrolimus, methyl prednisolone, trimethoprim-sulfa, and clotrimazole.

Clinical and Laboratory Findings

Between the 13th and 16th hospital day he developed headaches. His chest radiograph showed clear lungs and a small pleural effusion on the right side. His condition worsened with confusion, slurred speech, difficulty walking, and incontinence. Chest radiographs showed no change. On the 19th day his temperature rose to 39°C with chills. His medication consisted of cyclosporine, mycophenolate mofetil, prednisone, propanolol, trimethoprim-sulfa, omeprazole, vitamins, ITC, and valgancyclovir.

A MRI of the brain showed abnormal hyperintensity of the basal ganglia, thought to be due to chronic liver disease. On the 20th hospital day his temperature was 39.3°C. The patient's speech became unintelligible. Examination revealed photophobia and stiffness of the neck. An EEG showed generalized disorganization and slowing of background activity.

A lumbar puncture was performed on the 20th hospital day. The CSF glucose concentration was 48 mg/dL (normal 50–75 mg/dL); protein, 184 mg/dL (normal 5–55). An India ink prep of the CSF revealed encapsulated yeast cells. Budding yeast were noted on the manual complete blood count. *Cryptococcus neoformans* was isolated from the CSF and from blood cultures. A serum specimen obtained before the transplant indicated a cryptococcal latex agglutination titer of 1:4.

Diagnosis

The diagnosis was cryptococcal meningitis and cryptococcemia reactivated by immunosuppression for liver transplantation.

Therapy and Clinical Course

Liposomal AmB and 5FC was started on the day of diagnosis. His neurologic condition deteriorated despite therapy and his cryptococcal antigen titer in the serum and CSF remained elevated for months. Worsening renal failure required hemodialysis. Episodes of Gram-negative sepsis followed. The family decided to withdraw aggressive therapy and dialysis and the patient died on the 179th hospital day.

Comments

The finding of a low latex agglutination titer in the pretransplant serum specimen suggested this was a case of low-level infection reactivated by immunosuppressive therapy. A high mortality rate in cryptococcosis is associated with abnormal mental status, renal failure, and cryptococcemia. Liver failure is a risk factor for poor prognosis. Brain imaging studies were not revealing. The earliest finding of meningitis is enhancement of the leptomeningeal structures on contrast MRI which was not present in this case.

Case Presentation 2. A 22-yr-old Man with Hodgkin's Disease and Headache (Korfel et al., 1998)

History

This HIV-seronegative man related a 6 month history of fever, sweating, and weight loss.

Physical and Laboratory Findings

He was diagnosed with Hodgkin's disease. The right lung hilus, left lung, and bone marrow findings were indicative of *Ann Arbor stage IVB*. A bone marrow preparation indicated Hodgkin's and Reed–Sternberg cells. He was started on chemotherapy but after 1 week he developed continuing fever, nausea, and a frontal headache. His body temperature was 39°C, pulse 118, and his blood pressure was 150/80. A CT scan of his brain showed no abnormalities. A CSF specimen contained 576 cells/mL (85% neutrophils), 75 mg/100 mL protein, and 39.8 mg/100 mL glucose. India ink stain of the CSF sediment revealed many encapsulated budding yeast. The yeast was cultured from CSF. The cryptococcal antigen titer (latex agglutination test) in both CSF and serum was 1:640.

Diagnosis

The diagnosis was Hodgkin's disease and cryptococcal meningitis.

Therapy and Clinical Course

Immediately after the diagnosis of cryptococcal meningitis he was started on IV AmB-deoxycholate (1 mg/kg/day) and 5FC (150 mg/kg/day). A week later the lumbar puncture was repeated and culture of his CSF remained positive for *C. neoformans*. His clinical condition deteriorated and he became comatose. The CSF concentration of AmB after administration of 60 mg on day 6 showed 0.068 μg/mL in CSF. The patient was then switched to liposomal AmB at 2 mg/kg/day and 5FC was continued. This resulted in an increased CSF concentration of AmB of 0.11 μg/mL. Over the next few days he regained consciousness and alertness. Liposomal AmB was continued for 4 weeks and then therapy was switched to AmB-deoxycholate for an additional 4 weeks. Therapy was then stepped down to 100 mg/day of fluconazole (FLC) maintenance, but he developed a recurrence of cryptococcal meningitis. Another 6 weeks of AmB-deoxycholate was administered. The patient then underwent several courses of cancer chemotherapy resulting in remission of Hodgkin's disease. He was placed on 300 mg/day of ITC maintenance therapy. Follow-up after 5 years found that he remained in complete remission of Hodgkin's with no recurrence of cryptococcosis. He remained on anticonvulsion medication because of recurrent seizures.

Comments

His first symptoms appeared only 1 week after starting chemotherapy, suggesting the possibility of endogenous reactivation of a quiescent focus of infection. The successful management of both Hodgkin's disease and of cryptococcosis is noteworthy. The patient's CSF had increased protein, reduced glucose, and greatly increased WBCs. These changes in protein and glucose concentrations are typical in cryptococcal meningitis in non-AIDS patients.

Normal CSF values are:

- Opening pressure: 50–180 mm H_2O
- Appearance: clear, colorless
- CSF total protein: 15–45 mg/dL
- Gamma globulin: 3–12% of the total protein
- CSF glucose: 50–80 mg/dL (or approximately 2/3 of serum glucose level)
- CSF cell count: 0–5 WBCs, no RBCs.
- Chloride: 110–125 mEq/L

This patient was unable to sustain sufficient AmB concentration to be fungicidal in his CSF, prompting an increased loading dose made tolerable by replacing AmB with liposomal AmB. Later, on FLC maintenance, he suffered a relapse. Although FLC has excellent activity against *C. neoformans* and excellent penetration into the CSF, in the absence of a fully functioning immune system the dose of FLC probably was insufficient to inhibit regrowth of cryptococci. The use of ITC as maintenance therapy was because it also is a mold-active agent. ITC is not recommended to treat meningitis or for prophylaxis to prevent cryptococcal meningitis because it has poor penetration into the CSF. Finally, lingering neurologic deficits sometimes accompany recovery from cryptococcosis.

Case Presentation 3. Cavitary Lung Lesions and Meningitis in a 39-yr-old Man Living with AIDS (Schubert et al., 1997)

History

A 39-yr-old man, living with AIDS, was admitted to the hospital for a routine check-up. He related a mild cough with headache for the past 8 days. Eight months previously he had *Pneumocystis* pneumonia which resolved without pulmonary residuals, and also esophageal candidiasis, 6 months previously.

Physical and Laboratory Findings

His chest radiograph showed three solitary pulmonary nodules of 0.5–2.5 cm diameter in the middle lobe of the right lung which were revealed by CT scan to be air-filled cavities. His hemoglobin level was 10.5 g/dL, with a WBC count of 3100/ mm^3. The $CD4^+$ lymphocyte count was 89/μL, and his HIV-viral load was 142,000 copies/mL. Bronchial washings were collected and a bronchial brush biopsy was performed. Cytologic preparation of the bronchial washings stained with Giemsa, PAS, and mucicarmine showed a mild inflammatory reaction and partially encapsulated yeast. Histologic examination of the bronchial biopsy specimen showed a granulomatous reaction with epithelioid giant cells and variously sized intra- and extracellular yeasts. These stained brightly with mucicarmine, a characteristic of *C. neoformans*.

Culture of the bronchoscopy specimens was positive for yeast after 5 days and was confirmed as *C. neoformans* by microscopic morphology and culture methods. (Please see Section 12.13, Laboratory Detection, Recovery, and Identification.) Examination of the CSF revealed 43 leukocytes/mL and a cryptococcal latex agglutination antigen titer of 1:128. There was mild elevation of CSF protein, 1.0 g/L. Histologic examination of the CSF sediment showed an increase in mononuclear cells with single capsule-deficient yeasts. Cultures of CSF were positive for *C. neoformans* after 4 days.

Diagnosis

The diagnosis was pulmonary cavities and disseminated cryptococcosis in a person living with AIDS.

Therapy and Clinical Course

During the hospital stay he developed fever (38.3°C) and a worsening headache. He received 0.7 mg/kg/day IV of AmB-deoxycholate, and 400 mg/day of oral FLC. The patient's meningeal symptoms subsided rapidly, and both CSF and blood cultures became negative after 3 weeks. At that time he was switched to maintenance therapy of FLC, 100 mg/day. He remained free of symptoms 6 months later and his CSF cryptococcal antigen titer declined to 1:16. The pulmonary cavities almost completely closed with minimal scarring.

Comments

AIDS-associated cryptococcosis usually occurs when the $CD4^+$ T-lymphocyte counts fall below 100/μL. Abnormalities in the CSF often are observed in cryptococcosis but may be normal in AIDS patients except for the presence of yeast. This case underscores the role of the lungs as the portal of entry for *C. neoformans* and that pulmonary cryptococcosis can be symptomatic. Even so, pulmonary cavities are rare in AIDS-cryptococcosis.

His cryptococcal antigen titer was lower than that seen in many cases of AIDS-cryptococcosis, possibly owing to this isolate being a small capsule form, or the disease having been diagnosed at an early stage. The recommended therapy for AIDS-cryptococcosis where the clinical form includes meningoencephalitis is induction therapy with AmB-deoxycholate (0.7–1 mg/kg/day) plus 5FC (100 mg/kg/day for 2 weeks, in 4 divided doses) followed by FLC (400 mg/day) for a minimum of 8 weeks (Perfect et al., 2010).

Maintenance therapy with FLC is necessary in persons living with AIDS to prevent recurrences. If patients have access to combination ART and are compliant so that their $CD4^+$ T- lymphocyte count recovers to >100/μL it may be possible to discontinue maintenance therapy.

Case Presentation 4. Skin Lesions in a 32-yr-old Man Living with AIDS (Baker and Reboli, 1997)

History

This man presented with multiple umbilicated papules on his face and, to a lesser extent, on his trunk, buttocks, arms, and legs. Curettage of the gelatinous core of a papule was examined with Wright's stain and revealed many encapsulated yeast. Later, cultures of the CSF, blood, and skin grew *C. neoformans*.

Diagnosis

The diagnosis was cutaneous and disseminated cryptococcosis in a man living with AIDS.

Therapy and Clinical Course

The patient received a total dose of 1.5 g of AmB-deoxycholate, IV, and the lesions gradually resolved. He was then switched to daily oral FLC.

Comments

Dissemination to the skin occurs frequently in AIDS-cryptococcosis and may be its first presenting symptom. (Please see Section 12.10, Clinical Forms—Cutaneous Cryptococcosis.)

Differential Diagnosis for Cryptococcosis

Other diseases to be considered in the differential diagnosis of cryptococcoma are brain tumor, lymphoma, toxoplasmosis, tuberculosis, and, of the fungi, aspergilloma, *Cladophialophora*, and, in IV drug users, cerebral mucormycosis (Dubey et al., 2005). The differential diagnosis of chronic meningitis includes brucellosis, multi-infarct dementia, neurosyphilis, toxoplasmosis, tuberculosis, viral meningoencephalitis, and chronic meningitis due to other bacterial or fungal agents (e.g., coccidioidomycosis, histoplasmosis).

12.4 ETIOLOGIC AGENTS

Pathogenic cryptococci occur as two species: *Cryptococcus neoformans* is the major pathogen, whereas *C. gattii* accounts for <5% of cases of cryptococcosis (Table 12.1). One of the first clues that cryptococci are basidiomycetes was when Shadomy and Utz (1966) reported clamp connections in hyphae produced by the Coward strain of *C. neoformans*, followed by histopathologic observations on hyphal formation in that strain (Lurie and Shadomy, 1971).

Cryptococcus neoformans (San Felice) Vuill., *Rev Gén Sci Pures Appl* 12: 747–750 (1901). Position in classification: anamorph—*C. neoformans, Tremellaceae, Tremellales, Basidiomycota*; teleomorph—*Filobasidiella neoformans* Kwon-Chung, *Mycologia* 67: 1197 (1976).

Two Varieties of *Cryptococcus neoformans*

Cryptococcus neoformans var. *grubii* (serotype A) is by far the most prevalent, is the major cause of AIDS-cryptococcosis, and occurs worldwide. *Cryptococcus neoformans* var. *neoformans* (serotype D) is implicated in a minor number of cases and is found most often in northern Europe.

The four serotypes—A and D associated with *C. neoformans*, and B and C associated with *C, gattii*—can be delineated with the Iatron agglutinin kit (Mitsubishi Kagaku Iatron, Inc., Tokyo, Japan[1]) but that is reserved for epidemiologic studies. As shown in Table 12.1, *C. neoformans* and *C. gattii* can readily be differentiated by the ability of the latter to grow on canavanine–glycine–bromothymol blue (CGB) medium. *C. gattii* has a restricted geographic distribution and ecologic niche as discussed below (please see Section 12.5, Geographic Distribution/Ecologic Niche), and generally is seen in immunocompetent individuals where it tends to produce a more protracted infection, requiring longer therapy and a greater incidence of sequelae. *C. gattii* is not commonly found in AIDS patients.

Cryptococcus gattii. (Vanbreus. & Takashio) Kwon-Chung & Boekhout, in Kwon-Chung, Boekhout, Fell & Diaz, *Taxon* 51: 806 (2002).

Gatti and Eeckels (1970) reported an unusual strain of *C. neoformans*, isolated in 1966, from a Congolese Bantu child who developed convulsive seizures as his first symptom. Three months later the child entered a pediatric ward with signs and symptoms of cryptococcal meningitis, that is, stiffness of neck, headache, nausea, vomiting, fever, and positive Kernig's and Brudzinski's[2] signs. The name *C. neoformans* var. *gattii* was proposed for this strain (Vanbreuseghem and Takashio, 1970). The *gattii* isolates produce oval to lemon-shaped yeast cells.

In 1975 Kwon-Chung reported the teleomorph state of *C. neoformans* as *Filobasidiella neoformans*. This was facilitated when crosses were made between isolates of *C. neoformans* serotypes A and D with the opposite mating type of the type cultures (*C. neoformans* serotype D). These matings produced fertile basidiospores externally on a basidium (Kwon-Chung, 1975). (Please see Fig. 1.13 depicting sexual reproduction in pathogenic cryptococci: *Filobasidiella neoformans* and *F. bacillispora*.) Soon thereafter it was found that isolates of the *gattii* variety (serotypes B and C) would not produce fertile basidiospores when crossed with *F. neoformans* (Kwon-Chung, 1976). Serotypes B and C, however, crossed among themselves producing bacilliform basidiospores. *Cryptococcus gattii* then was designated a separate species by Kwon-Chung with the teleomorph *Filobasidiella bacillispora*, the sexual state of serotypes B and C (Kwon-Chung, 1976).

[1]This product is not currently available from this source.

[2]These are two signs associated with meningitis. Kernig's sign is the inability to extend the patient's knees beyond 135 degrees without causing pain. For Brudzinski's sign, with the patient supine, the physician places one hand behind the patient's head and places the other hand on the patient's chest. The physician then raises the patient's head (with the hand behind the head) while the hand on the chest restrains the patient and prevents the patient from rising. Flexion of the patient's lower extremities (hips and knees) constitutes a positive sign. Brudzinski's neck sign has more sensitivity than Kernig's sign.

Table 12.1 Two Varieties of *Cryptococcus neoformans* and the Second Pathogenic species, *C. gattii*

Characteristic	*C. neoformans* var. *grubii*	*C. neoformans* var. *neoformans*	*C. gattii*
Serotype(s)	A	D	B, C
Main geographic distribution	Worldwide (major agent in AIDS patients)	Northern Europe	Tropical and subtropical; also in British Columbia
Mating type(s) isolate	a and α	a and α	a and α
Perfect state	*F. neoformans* var. *grubii*	*F. neoformans* var. *neoformans*	*F. bacillispora*
CGB medium[a]	No growth	No growth	Growth
CDBT medium[b] (thymine assimilation)	No growth	Growth (orange color change)	Growth (blue-green color change)

[a]Canavanine–glycine–bromothymol blue medium.

[b]Creatinine–dextrose–bromothymol blue–thymine medium.

The primary morphology of *C. neoformans* is as a yeast, which occurs in tissues and in the environment. Basidiospores, however, are formed *only* by the hyphal form. Although *C. neoformans* and *C. gattii* can exist as yeast and hyphal forms, it is not considered a dimorphic pathogen. That term, in medical mycology, refers to thermally induced morphogenesis, where one form is not found in the environment but is induced by a shift to 37°C, such as occurs in the host during infection.

Speculation about whether sexual reproduction resulting in basidiospores in *Cryptococcus* species is a laboratory phenomenon has been replaced with more certainty that basidiospore formation occurs in the natural environment, and that basidiospores are infectious propagules (Ellis and Pfeiffer, 1990; Lin and Heitman, 2006). Many basidiomycetes release basidiospores which are borne on air currents. Bird guano filtrate agar is suitable for mating of *C. neoformans*. Genotype analysis of *Cryptococcus* species reveals meiotic recombination, proof that mating occurs naturally, resulting in the airborne dispersion of basidiospores that are sized appropriately (diameter of $\leq 2\ \mu m$) for inhalation and deposition in the alveoli. Basidiospores are also formed by an atypical method of sexual reproduction: monokaryotic fruiting, which is discussed below in "Life Cycle of *Cryptococcus* Species".

Yeast Systematics Within the Basidiomycetous Yeasts

Medically important yeasts that are basidiomycetes include *Malassezia* species, *Trichosporon* species, *C. neoformans*, and *C. gattii*. The teleomorphs of the latter two species, *Filobasidiella neoformans* and *F. bacillispora*, are classified based on the D1/D2 regions of the 28S (LSU) rDNA (Fell et al., 2000). They are related to the *Tremellales*, jelly fungi that grow as frilly masses on rotting wood. The clade containing the two pathogenic *Filobasidiella* species is set apart from the jelly fungi.

Life Cycle of *Cryptococcus* Species

See Fig. 12.1. Sporulation in *C. neoformans* results in the production of basidiospores borne externally on a basidium. This is accomplished by sexual reproduction and also by same-sex mating, termed monokaryotic fruiting, which is an atypical sexual process. During sexual reproduction the teleomorph state is formed.

Sexual Reproduction

Typical

Cryptococcus neoformans is heterothallic with a bipolar mating system that consists of two mating types, α and a (reviewed by Wickes, 2002). Under the influence of pheromones, two haploid cells of opposite mating type make contact and undergo plasmogamy to form a diploid. Diploid cells begin to differentiate into dikaryotic hyphae with clamp connections, typical of basidiomycetes. Hyphae contain paired nuclei, one from each parent, which divide synchronously. As the hypha grows, the nuclei partition into new hyphal cells assisted by the clamp connections. Next, the tips of hyphae differentiate to form a basidium, where karyogamy and meiosis occur. Additional mitotic divisions produce four genetically random chains of basidiospores on the basidium surface. They are dispersed in air currents and are infectious propagules. Furthermore, the progeny of fruiting display quantitative differences in the amount and length of hyphae produced (Lin et al., 2005; Lin and Heitman 2006). The haploid basidiospores germinate into yeast cells which divide by mitosis. *Cryptococcus gattii* has a similar life cycle.

Atypical

Monokaryotic fruiting offers an explanation for why the preponderance of wild-type *C. neoformans* is of mating type α (Lin et al., 2005; Lin and Heitman, 2006). By

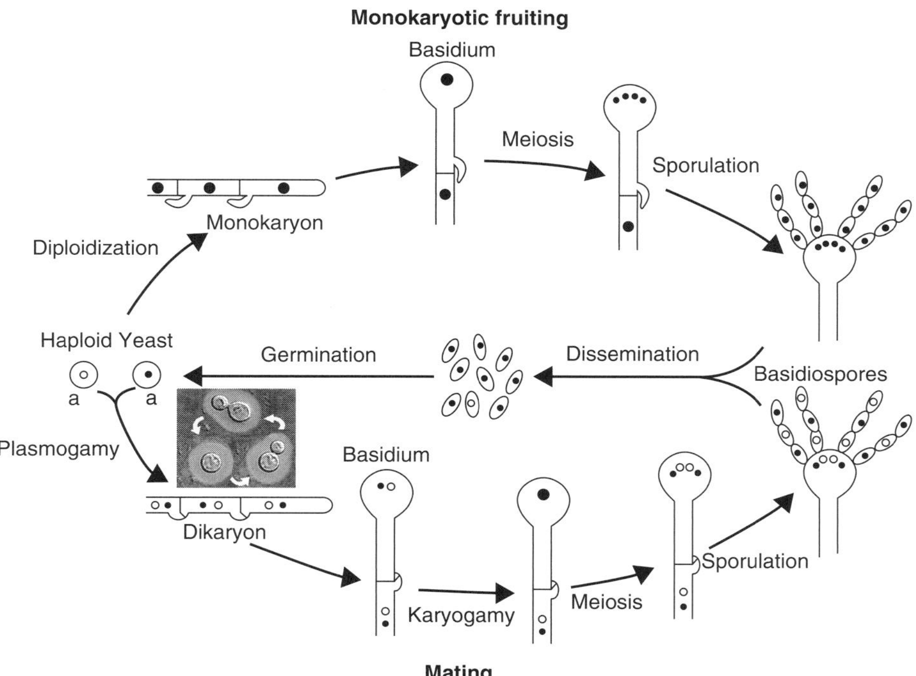

Figure 12.1 The *C. neoformans* life cycle includes asexual growth (mitosis), sexual reproduction, and monokaryotic fruiting (Lin and Heitman, 2006). Monokaryotic fruiting (*upper panel*) Yeast cells of one mating type (i.e., two genetically distinct α mating type yeasts) become diploid α/α cells. The diploid cells develop into hyphae with rudimentary clamp connections. A basidium develops, meiosis occurs, and haploid basidiospores are produced in four chains. Mitosis results in blastoconidia and chlamydospore formation during both fruiting and mating. Sexual reproduction (*lower panel*): a and α mating type yeast produce pheromones which attract them. Plasmogamy then occurs resulting in a stable dikaryon phase. The two nuclei divide synchronously and migrate together during hyphal growth with clamp connections. Later, a terminal basidium develops where karyogamy and meiosis take place producing four haploid cells. These haploid basidiospores divide by mitosis forming chains that extrude through the surface of the basidium. The teleomorph is *Filobasidiella neoformans*. The inset indicates that asexual reproduction of yeast occurs via budding.
Source: Used with permission from the *Annual Review of Microbiology*.

this means, two α mating type yeast cells undergo nuclear fusion to form a diploid. This cell then shifts its shape and develops a hyphal form. A terminal basidium is formed into which the diploid cell migrates. Then four chains of haploid basidiospores are produced from the basidium. This type of fruiting during α–α same-sex mating was originally thought to be strictly mitotic and asexual but all the hallmarks of mating including meiosis have been identified, as indicated below.

Evidence for Meiotic Recombination During Monokaryotic Fruiting

This evidence (Lin et al., 2005) was obtained in α–α diploid *C. neoformans* yeast produced from two α haploid parents each carrying a dominant resistance marker. Restriction-fragment length polymorphism (RFLP) analysis of 20 markers from 6 linkage groups on 5 chromosomes found recombination in the majority of chromosomes at a rate similar to that of conventional a–α mating and sporulation. All progeny showed unique genotypes, none identical to the parents, revealing a high frequency of recombination throughout the genome during fruiting.

Key Points

- Hyphal cells produced during fruiting are mononucleate, with unfused clamp connections, whereas mating hyphal cells are dikaryotic, linked by fused clamps.
- Because of the preponderance of α strains, the *Cryptococcus* population is largely unisexual and α–a mating may not usually occur in nature.
- Fruiting and mating are both stimulated by nitrogen starvation, desiccation, darkness, and pheromones.
- Monokaryotic fruiting could occur through self-diploidization, or cell–cell fusion of genetically distinct partners of like mating type.

What evidence is there that monokaryotic fruiting (same-sex mating) occurs in nature? (i) Natural diploid αA–Dα hybrids arose by fusion between two α cells of different serotypes (A and D). (ii) Environmental isolates of *C. neoformans* were found that are diploids of serotype A with two copies of the *MAT*α allele. Several were intravarietal allodiploid hybrids produced by fusion of two genetically distinct α cells through same-sex mating (Lin et al., 2009).

What are the advantages of same-sex mating in Cryptococcus? Meiosis, which accompanies same-sex mating, increases genetic diversity. Basidiospores produced during fruiting are dispersed in the environment where they can act as infectious propagules.

Gene Genealogy Within *Cryptococcus neoformans* and *C. gattii*

A consensus multilocus typing (MLST) scheme was developed for genotyping and phylogenetic analysis of *C. neoformans* and *C. gattii* by the Cryptococcal Working Group of the International Society for Human and Animal Mycology (ISHAM) (Meyer et al., 2009). The MLST scheme uses seven unlinked genetic loci including *CAP59, GPD1, LAC1, PLB1, SOD1, URA5*, and the IGS1 region. Three of the selected genes code for cryptococcal virulence factors: the polysaccharide capsule (*CAP59*), melanin synthesis (*LAC1*), and phospholipase B (*PLB1*). Allele and sequence type information can be accessed at the URL www.mlst.net. A consensus genotype nomenclature is shown in Table 12.2, depicting seven major haploid molecular types based on the Working Group report and that of Ngamskulrungroj et al. (2009).

Table 12.2 Genotype Nomenclature for *C. neoformans* and *C. gattii*[a]

Species/variety/hybrid	Serotype	Haploid molecular type
C. neoformans var. *grubii*	A	VNI/VNB
	A	VNII
AD hybrid	AD	VNIII[b]
C. neoformans var. *neoformans*	D	VNIV[c]
C. gattii	B	VGI[d]
	B	VGII[e,f]
	B/C	VGIII
	C	VGIV

[a] Meyer et al. (2009); Ngamskulrungroj et al. (2009).
[b] Most AD hybrids are diploid or aneuploid.
[c] The basal lineage of *C. neoformans* var. *neoformans*.
[d] Most prevalent (excluding United States).
[e] The basal lineage of *C. gattii*.
[f] Prevalent in the United States and British Columbia.

As shown in Table 12.2, three major clades of *C. neoformans* are VNI/VNB, VNII, and VNIV. There are four major clades of *C. gattii*: VGI, VGII, VGIII, and VGIV. The VNIV and VGII clades are basal to the *C. neoformans* and the *C. gattii* clades, respectively. The best practice genotyping method is MLST but the framework of molecular types is based on earlier work on *C. gattii* using various typing methods (Sorrell et al., 1996; Fraser et al., 2005), on *C. neoformans* var. *grubii* (Litvintseva et al., 2006) and *C. neoformans* var. *neoformans* (Meyer et al., 1999).

Rounding out the genotypes are molecular type VNIII consisting of *C. neoformans* serotype AD hybrids, and a unique clade, VNB of *C. neoformans* var. *grubii*, occurring in Botswana (reviewed by Ngamskulrungroj et al., 2009).

Cryptococcus neoformans AD Hybrids

Cryptococcus neoformans var. *grubii* (serotype A) and *C. neoformans* var. *neoformans* (serotype D) are varieties that are monophyletic lineages that, according to the phylogenetic species concept, may reflect cryptic species. More than 90% of clinical isolates from patients with cryptococcosis are strains of serotype A. Strains of serotype D are also found globally, but they are more prevalent in Europe. Serotype AD strains are hybrids of these two varieties and are commonly isolated from clinical and environmental samples.

Origin of *Cryptococcus neoformans* AD Hybrids with the *MAT*a Allele

Half of AD hybrid strains isolated from Asia, Europe, and the United States possess a rare serotype A *MAT*a allele (Litvintseva et al., 2007). *Cryptococcus neoformans* AD strains are diploid or aneuploid and contain two sets of chromosomes and two mating type alleles: *MAT*a and *MAT*α (one from each serotype). Serotype A strains with the *MAT*a allele, however, are extremely rare, with less than 0.1% prevalence. A notable exception is a unique population of serotype A in Botswana (the VNB molecular type, in Table 12.2), in which 25% of the isolates possessed the *MAT*a allele. AD strains with the genotype aA–Dα cluster with isolates from Botswana, whereas AD hybrids that do not possess the *MAT*a allele cluster with globally widespread isolates of serotype A. It appears that the AD hybrids of the aA–Dα genotype originated in sub-Saharan Africa from a cross between strains of serotypes A and D and that this hybid has increased fitness, which has enabled the rare serotype A *MAT*a genome to migrate from Botswana and become globally distributed.

Evolutionary Divergence and Geographic Spread

Evolutionary divergence was studied using MLST data and maximum-likelihood estimations. Comparing the two species, the *C. gattii* lineages (12.5 million years ago) evolved later than the major *C. neoformans* lineages, *C. neoformans* var. *grubii* and *C. neoformans* var. *neoformans* (24.5 million years ago). The most recent speciation event took place around 4.7 million years ago splitting the two *monophyletic* lineages within *C. neoformans* var. *grubii*, VNI and VNII. Surveys of South American isolates revealed extensive genetic diversity among strains of *C. gattii*, along with both mating types in nature and evidence of recombination, suggesting the evolutionary origin of *C. gattii* is in South America.

Overall, there is a lack of geographic concordance with the phylogenetic analysis, arguing for the recent global dispersal of the *Cryptococcus* species complex. The bias of *MAT*α over *MAT*a limits the ability of the yeast to undergo sexual reproduction and recombination. As a result, clonal populations occur in several cryptococcal habitats, including Australia, Thailand, and Canada where the *MAT*a is scarce.

The genetic diversity among the seven haploid major molecular lineages supports the phylogenetic species concept, by which strains that form the same monophyletic groups are considered as independent species. By this standard the molecular types in *C. gattii* should be considered as varieties.

Asexual reproduction is the main reproductive mode in the *Cryptococcus* species complex, but sexual reproduction appears to occur in some natural populations, as well as same-sex mating (please see above, "Sexual Reproduction—Atypical"). There is evidence of sexual reproduction among *C. gattii* VGII both globally and in local populations of Australia. Frequent recombination in the population of VGII could explain how this genotype was able to expand its range to include Vancouver Island (please see Section 12.6, Epidemiology). The other major molecular types, however, show a low degree or absence of recombination.

Other *Cryptococcus* Species of Uncertain Pathogenicity

Species other than *C. neoformans* usually are considered nonpathogenic. *Cryptococcus albidus, C. curvatus*, and *C. laurentii* rarely are seen, even in severely immunocompromised patients, and their clinical significance is not well understood.

12.5 GEOGRAPHIC DISTRIBUTION/ECOLOGIC NICHE

The ecologic niches of *C. neoformans* and *C. gattii* are distinctly different. *C. neoformans* is globally distributed and is associated with soil containing droppings of the common pigeon. *C. gattii* is found in association with the decayed hollows of mature trees including eucalypts and 54 other tree species in many parts of the world (Springer and Chaturvedi, 2010).

Geographic Distribution and Ecologic Niche of *Cryptococcus neoformans*

Cryptococcus neoformans var. *grubii* (capsule serotype A) occurs worldwide as a saprobe in soil mixed with excreta of the common pigeon, *Columba livia*. The pigeon originated in Southern Europe and North Africa and was introduced worldwide from those areas. Pigeons themselves do not become infected. *C. neoformans* var. *grubii* is the variety most often isolated from AIDS patients worldwide. *C. neoformans* var. *neoformans* (capsule serotype D), also found in pigeon droppings, is reported to be common in northern Europe. A survey of cryptococcosis in Europe from 1997 to 2001 found that up to 30% of *C. neoformans* isolated from patients in Europe were AD hybrids (Litvintseva et al., 2007). Approximately 7.5% of *C. neoformans* in North America isolated from the environment are AD hybrids.

C. neoformans is found near pigeon roosting sites, areas around statues, and so on. The stimulatory effect of pigeon droppings in enhancing the growth of cryptococci is related to the presence of creatinine, a normal metabolic waste product that the yeast can utilize as a nitrogen source. *C. neoformans* also has been isolated from cow's milk and plants, as well as skin and feces of normal humans.

Both *C. neoformans* and *C. gattii* are capable of growth on pigeon guano as the sole nutrient source (Nielsen et al., 2007). *C. neoformans* var. *neoformans* and var. *grubii* exhibit prolific mating on pigeon guano medium, exceeding that obtained on V8 juice mating medium. Basidiospore analysis indicates the occurrence of sexual reproduction. *C. gattii* strains, however, show significantly reduced mating on pigeon guano medium, because that is not their natural niche.

Other ecologic niches for *C. neoformans* have been identified. One of the more curious ones was the recovery of *C. neoformans* var. *grubii* (serotype A) from a tree trunk and decaying wood decorating the cage of striped grass mice in the Nocturnal Department of the Antwerp Zoo (Bauwens et al., 2004). Several of the mice were infected and one died of cryptococcosis. Possibly the wood or cage contents was exposed to bird droppings at one time and the fungus persisted in the decaying wood.

Geographic Distribution of *Cryptococcus gattii*

Initially reported from tropical and subtropical regions including Africa, Australia, Europe, the United States, and South America, the ever-expanding list of countries reporting *C. gattii* now includes Austria, Canada, China, India, The Netherlands, and South Korea.

Ecologic Niche of *Cryptococcus gattii*

The decayed hollows of mature trees provide a specialized niche and refuge well suited to *C. gattii* serotype B. The ecologic niche of *C. gattii* serotype C is not known. Like eucalypts, other angiosperm tree species acting as hosts for *C. gattii* have been extensively exported from their native areas, including *Ficus* spp. and *Terminalia spp*. (almond) trees. Dispersal of *C. gattii* from its specialized niche in trees is facilitated by export of wood products, air and water currents, as well as by birds, animals, and insects. *Cryptococcus gattii* can survive in fresh and salt water for long periods, providing opportunities for exposure to marine mammals, such as porpoises (Springer and Chaturvedi, 2010).

C. gattii (capsule serotype B) has a close association with the Red River gum tree, *Eucalyptus camaldulensis*. The fungus has been isolated in the vicinity of eucalypts in Australia by Dr. David H. Ellis. *E. camaldulensis* is a common and widespread tree along watercourses over much of mainland Australia. *E. camaldulensis* is generally dominant in the community, forming pure forests or woodlands. On lower levels of the floodplain, it is usually the only tree species present. In higher areas, it may be associated with black box (*Eucalyptus largiflorens*) in the South or coolibah (*E. microtheca*) in the North.

Eucalypts also are found in southern California, southern United States, Mexico, southern Europe, South Africa, and Zaire, which may explain the occurrence of *C. gattii* disease in those areas. Cryptococci were recovered from the hollows of living trees in northern Brazil, including the pink shower tree (*Tabelbuia rosea*), fig tree, and pottery tree. *C. neoformans* and *C. gattii* were recovered either together or separately in 6 of 32 of these trees. A recent study also found both species associated with several different genera of trees in Rio de Janeiro, Brazil and from almond trees (*Terminalia catappa*) in Colombia (Callejas et al., 1998). *C. gattii* also was isolated from Douglas fir, cedar, and alder trees on Vancouver Island, British Columbia, Canada (please see Section 12.6, Epidemiology). *C. gattii* is occasionally found in fermenting fruit juice, drinking water, rotting wood, and air. Because of its restricted geographic distribution *C. gattii* accounts for <5% of all cryptococcosis cases and is uncommonly isolated from AIDS patients.

12.6 EPIDEMIOLOGY

Exposure

Exposure to *C. neoformans* is thought to be widespread or universal. The route of infection is via inhalation of dusts containing aerosolized desiccated yeast, basidiospores, and pigeon droppings, or soil dusts mixed with bird droppings.

Latency

Serologic surveys in children in New York City found that, by 2 years of age, a majority of those studied had antibodies demonstrable by *Western blot* to various *C. neoformans* protein antigens (Goldman et al., 2001). Specificity controls were included and showed lack of correlation between reactivity against *C. neoformans* and that against *Candida albicans*, the common commensal yeast. Antibodies against capsular glucuronoxylomannan were tested and are ubiquitous in HIV-negative adults, suggesting that infection with this yeast is subclinical but persistent, or is the result of repeated exposures (reviewed in Casadevall and Pirofski, 2005). Further evidence of latency is that DNA subtypes of *C. neoformans* var. *grubii* recovered from African expatriates in Europe were highly similar to African strains. Nine patients who lived in Africa a median of 110 months before moving to France showed significant clustering with African DNA subtypes, compared to European *C. neoformans* strains. Patients had acquired their *C. neoformans* strains long before their clinical diagnoses were made, consistent with reactivation disease (Garcia-Hermoso et al., 1999).

There are few documented outbreaks of cryptococcosis in humans from a particular natural source (except please see below, "Cryptococcosis Caused by *Cryptococcus gattii* on Vancouver Island, British Columbia").

Studies of normal individuals and pigeon breeders living in the same geographic area showed that significantly greater numbers of pigeon breeders had antibodies to *C. neoformans* than did individuals not associated with pigeons. The incidence of diagnosed cryptococcosis among pigeon breeders is the same as the incidence in the general population. Actual numbers of subclinically exposed persons in the general population, however, are not known.

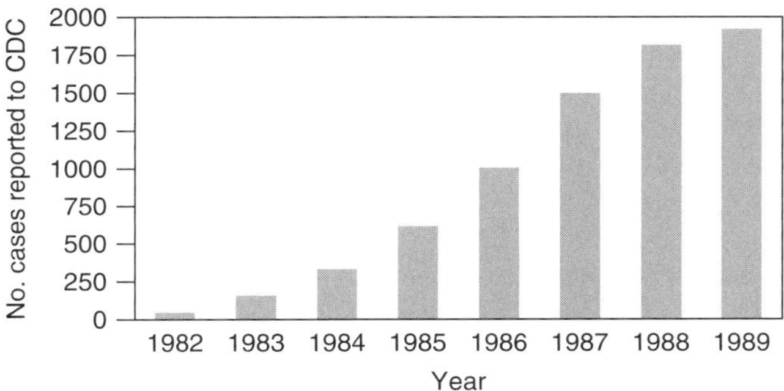

Figure 12.2 Cryptococcosis cases associated with AIDS patients in the United States reported to the CDC 1982–1989. The rise in cases follows the epidemic curve of AIDS.

Cryptococcosis and the Epidemic Curve of AIDS

Cryptococcosis incidence closely follows the epidemic curve of AIDS (Fig. 12.2). In the United States cryptococcosis reached a plateau in 1989–1991 and began to decline in 1992 because of the advent of antiretroviral therapy. This downward trend continued because of the availability of improved and combination antiviral therapy (ART). A total of 1491 cases of cryptococcosis were identified by population-based surveillance in Atlanta and Houston over the period 1992–2000 (Fig.12.3) (Mirza et al., 2003). The majority of cases (1322, 89%) occurred in persons known to be HIV infected; 124 cases (8%) occurred in persons with negative HIV antibody test results. Among AIDS patients in the United States, before the advent of combination ART, cryptococcosis was the defining illness in 5% of AIDS patients, with an overall 5–10% prevalence. The prevalence is even higher in central Africa, and cryptococcosis was the initial defining illness in 88% of AIDS patients in Harare, Zimbabwe (Perfect, 2005).

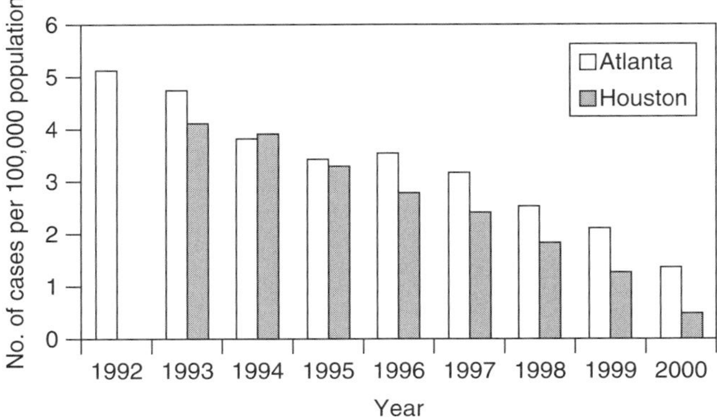

Figure 12.3 Incidence of cryptococcosis in the Atlanta, Georgia, and Houston, Texas, metropolitan areas, 1992–2000 (Mirza et al., 2003).

The global burden of cryptococcal meningitis among persons living with AIDS was estimated by geographic region according to published reports and expert opinion (Park et al., 2009). Published incidence ranges from 0.04% to 12% per year among persons with HIV. Sub-Saharan Africa had the highest yearly burden estimate with a median incidence of 3.2% or 720,000 cases. Median incidence was lowest in Western and Central Europe and Oceania ($\leq$0.1% each). The 3 month mortality rate is estimated to be 9% in high-income regions, with a 55% rate in low- and middle-income regions, and a 70% rate in sub-Saharan Africa. Globally, an estimated 957, 900 cases of cryptococcal meningitis occur each year among persons living with AIDS, resulting in 624,700 deaths.

Cryptococcosis Caused by *Cryptococcus gattii* on Vancouver Island, British Columbia

A new endemic area for *C. gattii* was identified in the coastal Douglas fir biogeoclimatic zone on the southeast coast of Vancouver Island. British Columbia–CDC issued a Health Alert on September 2, 2002 "Backgrounder—Cryptococcal Disease on Vancouver Island."

Prior to 1999, British Columbia reported 2 or 3 cases/year of cryptococcosis. Since 1999, more than 50 human cases, including one fatality, were reported in residents of, or visitors to, Vancouver Island. Cases have occurred in various places on the island's east coast. The etiologic agent was identified as *Cryptococcus gattii*.

The British Columbia Centre for Disease Control identified the ecologic niche of *C. gattii* on the central east coast. Environmental sampling found *C. gattii* in the soil and air and growing on Douglas fir, alder trees, and a cedar tree. Since 2000, more than 35 cases of cryptococcosis were reported on Vancouver Island among dogs, cats, llamas, a ferret, and wild porpoises (Kidd et al., 2004). *C. gattii* is reputed to survive for months in salt water.

Seventy-eight feline cases and 51 canine cases of proven or probable *C. gattii* infection were identified between 1999 and 2003. The pets had a history of living on or traveling to Vancouver Island. These companion animals had a high level of morbidity and mortality despite specific antifungal therapy (Duncan et al., 2006).

How did this new strain of C. gattii evolve?

Evidence and Speculation About the Genealogy of *Cryptococcus gattii* VGIIa Genotype

The Vancouver Island (Fraser et al., 2005) major genotype VGIIa strain is identical across 30 different genetic loci to a strain from Seattle isolated from human sputum in the 1970s and another strain collected in San Francisco in 1992 from a *Eucalyptus* tree. Thus, this major genotype was present in the Pacific Northwest for at least 30 years. A minor genotype, VGIIb, from Vancouver Island was identical across 30 loci with a fertile Australian isolate indicating the minor outbreak genotype may have originated in Australia. The proportion of 14 identical:16 divergent loci shared between the major and minor genotypes from Vancouver Island could be explained as Mendelian segregation of unlinked loci following meiosis. The minor outbreak strain genotype occurs in multiple locations in Australia, but the major outbreak genotype occurs in the Pacific Northwest.

An explanation for the origin of *C. gattii* VGIIa is that it is a meiotic recombinant resulting from same-sex mating ($\alpha-\alpha$) between the minor outbreak strain (source Australia) with a local as yet undiscovered strain (possibly also in Australia) giving rise to the major outbreak strain as progeny. In general the absence of inter-VG hybrid strains indicate that meiosis in *C. gattii* tends to occur within rather than between VG molecular types. Further proof of this is that analysis of the *MAT* α locus of the two Vancouver Island genotypes indicates they differ by a widespread low level polymorphism, indicating they bear different but related *MAT* α alleles. Experiments in mice showed that the major outbreak VGIIa strain was hypervirulent compared with the minor outbreak isolate.

All 31 cryptococcal isolates collected between 1987 through 1998 from Vancouver Island were identified as *C. neoformans*, so that the increased reports of *C. gattii* represent true emergence of the species in the region (MMWR, 2010). An explanation for the emergence of cryptococcosis caused by *C. gattii* may be because of climate change, changes in land use, or to a more susceptible human population. The average annual incidence of *C. gattii* on Vancouver Island of 25.1/ million population is one of the highest in the world. Australia reported an incidence of cryptococcal infection of 140/ million population among aboriginals in Arnhemland, Northern Territory, in 1976–1992; 77.8% of these cases were caused by *C. gattii*.

Spread of Reports of *Cryptococcus gattii* Cases to Mainland Canada and Northwest United States

Widescale environmental sampling of air, water, soil, swabs of trees, and so on in mainland British Columbia, islands in the Strait of Georgia, and Washington, found 3% of non-Vancouver Island samples positive for *C. gattii* serotype B, most were VGIIa, a few were of the VGI genotype. Air sampling on Vancouver Island detected particles small enough to be either desiccated *Cryptococcus* yeast cells or basidiospores. Between 1999 and 2007, 218 cases of *C. gattii* cryptococcosis were reported in British Columbia, Canada, an incidence of 5.8/ million population (Galanis and Macdougall, 2010). Cases plateaued on Vancouver Island in 2002 but increased on the mainland since 2005. Of 176 cases, 76.6% presented with pulmonary symptoms, 7.8% with a CNS syndrome, and 10% with both. The VGIIa strain was responsible for 106 or 86.3% of confirmed cases; 8 cases were due to VGI and 9 to VGIIb. Older patients ($\geq$50 years old) were more likely to be infected with VGIIa or VGIIb.

During 1999–2006, 322 persons in British Columbia were hospitalized with cryptococcosis (*C. neoformans* and *C. gattii*). Of these 191 or 59.3% were HIV-negative, an average rate that rose from 10 in 1999 to 38 in 2006. At 5.8 hopsitalizations/million population, the average annual hospitalization rate was higher for cryptococcosis in HIV-negative patients, than those with AIDS. Persons without HIV were likely to be older, with the highest incidence in the 70–79 year age group, and to have pulmonary cryptococcosis. *Cryptococcus gattii* is notable for affecting immune-normal persons, causing pneumonia, and, when invading the CNS, to cause brain cryptococcomas.

Closer analysis of the patients from British Columbia indicates that many had some underlying condition affecting immune function. The case/fatality ratio of patients who died with *C. gattii* infection was 8.7%. Of the case patients who died, fully 73.7% had an underlying medical condition that may have contributed to a fatal outcome including cancer, COPD, asthma, liver disease, diabetes, HIV infection, and lung transplant. Nine or 47% were immunocompromised.

In 2006, a human case of cryptococcosis due to the Vancouver Island outbreak strain *C. gattii* VGIIa was diagnosed on Orcas Island, Washington and between 2006 and 2008, 9 cases caused by the same strain were diagnosed in residents of four additional counties in Washington. *Cryptococcus gattii* infection is considered by Washington State to be a notifiable rare disease of public health significance (Datta et al., 2009). Between 2004 and 2008, 19 cases of *C. gattii* disease were diagnosed in Oregon but were caused by a different VGII genotype that the Vancouver outbreak strain. Some of the Oregon patients were infected with a new previously unknown genotype, VGIIc. By July 2010, 60 human cases of *C. gattii* disease were reported from California, Idaho, Oregon, and Washington (MMWR, 2010).

The U.S. cases of *C. gattii* occurred mostly among HIV-negative persons, only 8% were patients with HIV-AIDS, but overall 81% had some predisposing condition, for example, solid organ transplant recipients, people on oral steroids, and patients with a non-HIV-immunosuppressive condition, lung or kidney disease, or cancer. Nine percent had no underlying condition.

Fifty-seven percent had pneumonia and 43% presented with meningitis. Fifteen (33%) of 45 of these cases with known outcomes died.

Wild and domestic animals are sentinels for surveillance of *C. gattii* cryptococcosis and they were diagnosed before human cases were found in British Columbia, Washington, and Oregon, California, and Hawaii (MMWR, 2010). Disease in animals is characterized by upper respiratory symptoms, CNS involvement, ocular disease, lymphadenopathy, pneumonia, and subcutaneous nodules. Of domestic animals, cats are prone to develop this disease but cryptococcosis occurs in several other animal species (e.g., alpacas, dogs, ferrets, and porpoises).

Cryptococcosis in the Gauteng Province of South Africa

Gauteng is a land-locked province in the northeast section of the country. It is the economic hub of South Africa where financial, manufacturing, transport, technology, and telecommunications sectors are concentrated. Johannesburg is at its geographic center. The province is highly urban with a population of nearly 9 million. Most of Gauteng is on a high-altitude grassland. Between Johannesburg and northeast to Pretoria there are low ridges and hills. The northern part of Gauteng is more subtropical, due to its lower altitude and is mostly dry savanna habitat. Even though the province is at a subtropical latitude, the climate is comparatively cooler, especially in Johannesburg, at 1700 m above sea level. The Red River gum, *Eucalyptus camaldulensis*, is an invasive species in Gauteng, occurring in medium dense stands.

The overall incidence of cryptococcosis in Gauteng, determined for the period 2002–2004, was 15.6 cases/100,000 in the general population with an associated mortality of 27% (McCarthy et al., 2006). Among HIV-infected persons, the rate was 95/100,000, and among persons living with AIDS the rate was 14/1000. Of 2753 incident cases, 47 cases or 2.6% involved *C. gattii* (Morgan et al., 2006). Twenty-four of these cases were related to persons living with AIDS. The clinical form, HIV status, intercurrent disease, or history of other opportunistic infections or outcome was indistinguishable between patients with *C. gattii* versus those infected with *C. neoformans*.

Cryptococcosis in China and India

In contrast to the experience in Europe and North America, clinical isolates from China collected during 1980–2006 found a large majority (71%) of the 129 cryptococcosis patients had no underlying risk factor; only 8.5% were from AIDS patients (Chen et al., 2008). More than half of 39 cryptococcosis patients from India treated during a 1-year period had no known impairment of their immune system (Jain et al., 2005). Nine percent of the isolates were identified as *C. gattii*, the remainder were *C. neoformans*.

12.7 RISK GROUPS/FACTORS

Besides persons living with AIDS, individuals with other immunosuppressive conditions are at risk of cryptococcosis, including patients being treated for Hodgkin's lymphoma (please see Section 12.3, Case Presentation 2) and patients receiving immunosuppressive therapy for solid organ transplants or systemic corticosteroid therapy for collagen-vascular diseases, sarcoidosis, and, rarely, diabetes mellitus. Patients on antibody therapies that neutralize TNF-α also are at increased risk for cryptococcosis. Aside from AIDS, 30–50% of patients have no discernible underlying condition (percentage varies according to the study). In a large case–control study, specific factors found to be associated with increased risk for cryptococcosis include smoking and working in outdoor construction or landscaping. Receipt of FLC in the previous 3 months was associated with a decreased risk (Hajjeh et al. 1999).

Solid Organ Transplant Patients

Cryptococcosis is a significant opportunistic infection in solid organ transplant recipients, with a prevalence rate ranging from 0.26% to 5% and an overall mortality of 42% (Silveira and Husain, 2007). When cryptococcosis occurs it is usually >6 months after transplantation, acquired as a pulmonary infection and presenting most often as meningitis. Antirejection steroid therapy is likely the primary risk factor. Newer lymphocyte depletion therapies also are implicated: rabbit antithymocyte globulin or alemtuzumab used for induction therapy or preconditioning and to treat acute organ rejection. Two doses of either antibody preparation are associated with an increased risk of cryptococcosis (Silveira et al., 2007).

Sources of infection during the immediate post-transplant period are:

- Colonized or infected donor organ tissues
- Low-level infection brought about by debilitation and organ failure before the transplant period
- Reactivation of an asymptomatic focus in the lung

Cryptococcal pneumonia in solid organ transplant patients can take the form of nodules, bilateral pulmonary infiltrates, and pleural effusion, and lead to acute respiratory failure requiring mechanical ventilation.

Certain immunosuppressive agents—tacrolimus, cyclosporine, and sirolimus—demonstrate in vitro activity against *Cryptococcus* and those receiving tacrolimus may be significantly less likely to have CNS involvement.

An immune reconstitution inflammatory syndrome (IRIS) has been identified in 4.8% of solid organ transplant recipients being treated for cryptococcosis (please see Section 12.12, Therapy). Reduction in immunosuppression and response to antifungal therapy can sometimes be accompanied by worsening of the clinical symptoms. Cultures may be negative and the symptoms may be caused by a Th1 proinflammatory response. This immune reconstitution syndrome may result in renal allograft rejection.

12.8 TRANSMISSION

Cryptococcus neoformans is not communicable but is transmitted to humans and animals via inhalation of the fungus into the lungs. Encapsulated yeast cells with cell diameter of 4–20 μm, once inhaled, may be too large to penetrate the defenses of the upper respiratory tract. Desiccated yeast cells (diameter of less than 2 μm), however, may act as infectious propagules. Basidiospores of the sexual stage of the yeast with diameter of ≤3 μm are capable of reaching the alveoli, where they act as infectious propagules (Ellis and Pfeiffer, 1990). Studies, both intentional (laboratory animals) and accidental (laboratory accidents), have shown that the fungus does not disseminate from dermal inoculation, but instead develops a localized lesion in immune-normal individuals.

12.9 DETERMINANTS OF PATHOGENICITY

12.9.1 Host Factors

Normal anatomic barriers and the mucociliary elevator prevent most inhaled desiccated yeast and basidiospores from reaching the alveoli. Those that manage to arrive there are controlled by the immune-normal host. *Cryptococcus neoformans* is eliminated or else it is confined to the lung within granulomas, persisting in a dormant but possibly viable state. Immunocompromised individuals have difficulty killing the inhaled yeast and basidiospores, resulting in pneumonitis and increasing the probability of extrapulmonary dissemination. By evading opsonophagocytosis, encapsulated cryptococci enter the bloodstream, bind to microvascular endothelial cells of the blood–brain barrier, and traverse that barrier into the CNS.

Clinical and experimental evidence underlines the critical importance of CMI in defense against cryptococci. Most cases of disseminated cryptococcosis are associated with the $CD4^+$ T-lymphocyte depletion that is the hallmark of AIDS, or cytoreductive therapy for cancer or immunosuppressive therapy for a stem cell or solid organ transplant. Protective immunity depends on Th1 responses: $CD4^+$ T lymphocytes with a lymphokine profile including IFN-γ and IL-12 are required to initiate protective immune functions (Shoham and Levitz, 2005).

Early Events in the Encounter of Cryptococci *with Cells of the Immune System*

Early in infection there is an influx of PMNs into the infected tissues. Phagocytosis is triggered by recognition of cryptococci by complement or antibodies. Phagocytosis cannot occur if the cryptococci are not opsonized. PMNs kill cryptococci via oxidative microbicidal pathways, cationic granule peptides–defensins, and cytosolic calprotectins. Calprotectin, abundant in the cytosol of PMNs, also is present on the membrane of monocytes. There is a direct relationship between calprotectin expression and the ability of alveolar macrophages to increase production of TNF-α. Soluble calprotectin exerts an antimicrobial effect by Zn chelation.

PMNs PMNs modulate the immune response by producing proinflammatory cytokines, including TNF-α, IL-1β, IL-6, and IL-8. A major role for TNF-α is to prime cells for increased chemokine secretion, with resultant leukocyte migration into infected tissue.

Macrophages and Dendritic Cells Key effector cells in the early stages of the immune response to cryptococci are macrophages and dendritic cells. Both have the capacity to phagocytose cryptococci, process and present antigens, and activate adaptive immunity. *Cryptococcus neoformans* and its capsular polysaccharide are recognized by receptors on these cell types including mannose receptors (MRs), CD14, TLR-2, TLR-4, and CD18 (Shoham and Levitz, 2005). *C. neoformans* capsular polysaccharide, glucuronoxylomannan (GXM), acts as a PAMP and can bind to TLR-4 on macrophages.

Dendritic cells respond to *C. neoformans* by utilizing the MRs to bind mannoproteins on the cryptococcal cell wall and then to phagocytose cryptococci. They secrete cytokines and migrate to lymphoid tissues to present processed antigens to T lymphocytes. The induction of T-cell responses by dendritic cells is more efficient than that induced by alveolar or peritoneal macrophages.

Macrophages respond to *C. neoformans* with the release of proinflammatory signaling molecules, such as chemokines and IL-1. Secretion of IL-1 regulates proliferation and activation of T lymphocytes, which are important in mediating pulmonary clearance. Fc γ receptors on macrophages occur in two types: FcγRIIa bind to cryptococci opsonized with antibodies and activate an inflammatory response. The second receptor type, FcγRIIb, binds directly to cryptococcal capsular GXM, resulting in immunosuppressive signaling (Monari et al., 2006). It is thought that the two types of receptors modulate the immune response, for example, to prevent bacterial lipopolysaccharide endotoxin (LPS) from inducing endotoxin shock.

C. neoformans continues to replicate in the macrophage phagosome and to secrete or slough off capsular GXM, which accumulates as vesicles (Alvarez and Casadevall, 2006). The yeast is able to exit the macrophage through an extrusion in the phagosome and, by that means, to spread the infection. One theory asserts that macrophages containing phagocytosed cryptococci may traverse the blood–brain barrier, acting as a Trojan horse, to infect the CNS (reviewed by Voelz and May, 2010).

NK cells bind to cryptococci releasing cytolytic compounds, such as perforin, which lead to direct killing. NK cells produce perforin, a protein stored in granules, which they release after contact with a target cell, such as cryptococci. NK cells may further contribute to host defense by secreting IFN-γ, thereby stimulating macrophages to kill the yeast.

Special Topic: Chemokines

Chemokines are a large superfamily of small cytokines that have two major functions. (i) They guide leukocytes via chemotactic gradients in tissue. (ii) A subset of chemokines increases the binding of leukocytes via their integrins to ligands at the endothelial cell surface, facilitating adhesion and extravasation of leukocytes in tissue (Oppenheim and Horuk, 2001).

Many chemokines from the CC, CXC, and CX3C subfamilies have proinflammatory activity. They attract particular leukocytes into sites of injury or infection. Nearly all members of this category are secreted by activated macrophages, and by other cell types. Stimuli that induce their release include IL-1, TNF-α, and IFN-γ.

Most known CXC chemokines are proinflammatory; some attract PMNs and have a prominent role in acute inflammation. When PMNs encounter chemokines of this type, they adhere to endothelial cells and migrate (extravasate) into the underlying tissue, following the chemokine gradient.

CXC receptors are expressed on activated T cells (particularly Th1 cells), B cells, monocytes, and NK cells. By chemoattracting these cell types, chemokines (e.g., IP-10 or I-TAC) promote chronic inflammation, particularly the development of Th1 acquired immune responses.

A majority of CC chemokines are proinflammatory, and they produce their effects through receptors on activated T cells, immature dendritic cells, and other mononuclear cell types. The proinflammatory CC chemokines (e.g., RANTES, MIP-1α) generally promote chronic inflammation and acquired immune responses. MIP-1α has been implicated in chronic pulmonary immune responses.

Inbred SJL/J mice have a higher natural resistance to pulmonary challenge with cryptococci than C57BL/6J mice, operated as a control strain (Guillot et al., 2008). Resistant SJL/J mice display an increased expression of a range of chemokines of the CX and CC classes, including MIP-1α. This enhanced innate immune response elicits a higher neutrophil influx and increased concentration of TNF-α, IFN-γ, and IL-12, which are the components of a strong protective Th1 response.

Adaptive Immunity

Both CD4$^+$ and CD8$^+$ T lymphocytes combine to facilitate cellular infiltration into infected tissue. In vitro, CD4$^+$ and CD8$^+$ T lymphocytes directly inhibit fungal growth by attaching to cryptococci. CD4$^+$ and CD8$^+$ T lymphocytes exert a direct antimicrobial effect against *C. neoformans* acting via granulysin, a secreted protein, to induce membrane permealization and lysis.

Th1-associated cytokines are essential for protective immunity against cryptococci, whereas Th2 immunity is not protective. Upregulation of Th1 cytokines, TNF-α and IFN-γ, results in improved control of the infection. Indirectly, CD4$^+$ T lymphocytes activate macrophages, marshalling them to organize granulomas. Encapsulated yeast can be encircled in tissues by the formation of granulomas, aided by the fusion of macrophages into multinucleate giant cells. Killing of phagocytosed cryptococci by activated macrophages is mediated by phagosome–lysosome fusion, phagosome acidification, iron sequestration, and lysosomal hydrolase degradation of fungal proteins. Extracellular killing is mediated by release of antifungal peptides and nitric oxide (Shoham and Levitz, 2005).

Human macrophages activated in vitro with the Th1 cytokine, TNF-α, or the Th17 cytokine, IL-17, exhibit a much lower yeast intracellular proliferation rate than that observed in cells treated with the Th2 cytokines IL-4 or IL-13 (Voelz et al., 2009). The risk of cryptococcosis increases during the course of HIV infection corresponding to a loss of the Th1 response. Maintaining a balance between Th1-Th2-Th17 effector functions appears to be essential to a successful host defense against cryptococci.

Humoral Immunity

Considering the glycocalyx of *C. neoformans* and *C. gattii* contains the high molecular mass, acidic, capsular GXM polysaccharide, it is not surprising that humoral immunity plays a role in cryptococcosis. Nonopsonized cryptococci are not well recognized by phagocytes. Human serum and saliva inhibit *C. neoformans* growth. Alternative complement activation is thought to be the dominant pathway for opsonization of cryptococci. The alternative pathway deposits C3b on the capsular surface but its conversion to the inactive form, C3bi, is rapid and this is the major form bound. As such, C3bi cannot participate in the activation of C5, thus reducing development of the membrane attack complex. Importantly, the binding of C3bi to complement receptor 3 (CR3) is sufficient to stimulate phagocytosis. Complement activation also generates C3a and C5a, which assist in leukocyte trafficking. *C. neoformans* resists complement-mediated pore formation and lysis by the late complement components.

Antibody-opsonized yeast cells are recognized by Fc receptors on the surfaces of macrophages, neutrophils, and dendritic cells but opsonic capsule-binding antibody is not consistently present during infection. MAbs of the correct idiotype and isotype enhance internalization and killing of *C. neoformans* by macrophages and PMNs. By promoting phagocytosis, anticapsular antibody enhances antigen presentation to T lymphocytes. "Designer" mAbs are being considered as adjunctive therapy in cryptococcosis (Dadachova et al., 2007).

12.9.2 Microbial Factors

Capsular Polysaccharide

The capsular polysaccharide of *C. neoformans* and *C. gattii* is a glucuronoxylomannan (GXM), the major virulence factor of these two species. The high molecular weight acidic polysaccharide is analogous to similar capsules of the bacterial agents of meningitis, all of which are T-independent antigens: *Neisseria meningitidis, Hemophilus influenzae*, and *Streptococcus pneumoniae*. They elicit poor protective antibody responses unless the carbohydrate is conjugated to a protein carrier (Lagos et al., 2009).

Function *Why is the capsule considered a virulence factor?* The GXM capsule has the following immunomodulatory properties (Zaragoza et al., 2009):

- *Phagocytosis.* Encapsulated cryptococci are less efficiently phagocytosed than acapsular mutants. Once phagocytosed by macrophages, GXM accumulates in small vesicles, resulting in alteration of macrophage function.
- *Antibody Production.* GXM induces a state of dose-dependent immunologic tolerance in mice measured by the inhibition of splenic plaque forming cell responses (which equates to the inhibition of antibody production).
- *Th1 Cytokines.* Infection with a large capsule strain causes failure in mice to upregulate Th1 cytokines IL-2 and IFN-γ.
- *Complement.* GXM induces a drastic decrease in C5a receptors in PMNs, interfering with complement mediated inflammation.
- *Proinflammatory Cytokines.* GXM induces monocytes to produce IL-10, a potent downregulator of proinflammatory cytokines TNF-α and IL-1ß.
- *TNF-α.* In PMNs, GXM induces L-selectin shedding with consequent reduced TNF-α receptor expression and reduced responsiveness to TNF-α.
- *Apoptosis.* A high proportion of apoptotic cells are found in rat lungs and spleen during *C. neoformans* infection. GXM binds to TLR4, which in turn induces expression of the Fas ligand (FasL) on antigen presenting cells. Binding of FasL to Fas on T cells results in a caspase cascade and T cell apoptosis.

Upon inhalation into the lungs *C. neoformans* rapidly develops a capsule that inhibits phagocytosis (Fig. 12.4.). The yeast then invades tissue and, if not contained, may disseminate via the bloodstream into the CNS.

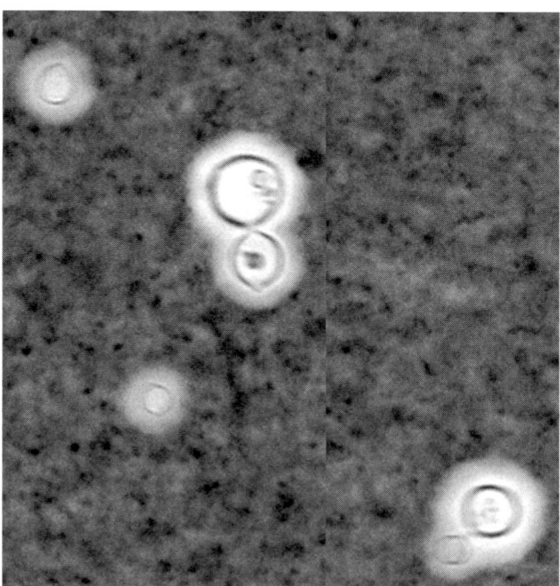

Figure 12.4 *Cryptococcus neoformans*, India ink prep shows the polysaccharide capsule as a clear zone surrounding the yeast cells. Note the characteristic narrow base budding between mother and daughter cell.
Source: Used with permission of Steven J. O'Connor, M.D., Houston, Texas.

In summary, GXM is antiphagocytic and poorly immunogenic, and acapsular strains have diminished virulence. GXM inhibits leukocyte migration, enhances HIV infection in human lymphocytes, and promotes L-selectin shedding from neutrophils (Cherniak et al., 1998). GXM downregulates protective cytokines including TNF-α and upregulates production of the Th2 cytokine, IL-10. GXM, through binding to TLR4, increases apoptosis of T lymphocytes.

Structure The capsule is a high molecular weight, acidic, viscous polysaccharide composed of glucuronic acid/mannan/xylose/*O*-acetyl (GXM). The capsule represents approximately 88% of the readily soluble cell envelope carbohydrate, the remainder consisting of antigenic galactoxylomannan and mannoprotein (~12%).

The typical GXM consists of a linear α- (1 → 3)-D-mannan with β-D-xylosyl (Xyl*p*), β-D-glucopyranosyluronic acid (Glc*p*A), and 6-*O*-acetyl substituents (Fig. 12.5). The *O*-acetyl substituents are major antigenic determinants from all serotypes (A, B, C, D). Type-specific serologic activity is lost after de-*O*-acetylation. A simple structural relationship exists between GXMs of reference isolates for the four serotypes. They are all comprised of a core repeat unit to which (1 → 2)-linked and (1 → 4)-linked β-D-Xyl*p* units are added in increments of one to four residues. In this way, molar ratios of Xyl/Man/GlcA in serotypes D, A, B, and C have been assigned as 1:3:1, 2:3:1, 3:3:1, and 4:3:1, respectively (Cherniak et al., 1998).

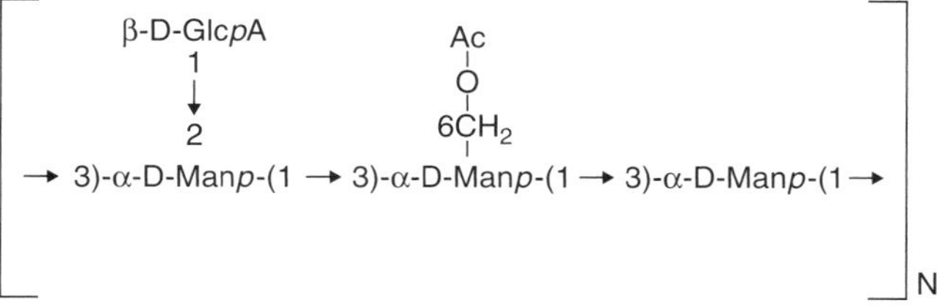

Figure 12.5 The core repeat unit structure of the cryptococcal capsular polysaccharide, glucuronoxylomannan, GXM (Cherniak et al., 1998). A simple relationship exists between GXMs of the four serotypes (serotypes A, B, C, D) They are comprised of a core repeat unit, shown here, to which (1 → 2)-linked and (1 → 4)-linked β-D-Xyl *p* units are added in increments of one to 4 units. In this way, molar ratios of Xyl/Man/GlcA in serotypes D, A, B, and C are 1:3:1, 2:3:1, 3:3:1, and 4:3:1, respectively. GXM from serotypes A and D are mainly substituted with Xyl*p* at O-2, whereas GXM from serotypes B and C are substituted at O-2 and at O-4. GlcA, glucuronic acid; Man, mannose, Xyl, xylose. ("*p*" denotes the pyranose form.) In addition to the Xyl *p* substituents, the GXM is *O*-acetylated at the C-6 position of mannosyl units (Ac). The molecular weight of GXM is very high, in the range of 1700–7000 kDa.
Source: Adapted with permission from the American Society for Microbiology.

Melanin

Cryptococcus neoformans expresses phenoloxidase (syn: laccase), an enzyme in the pathway of melanin biosynthesis. The deposition of small quantities of melanin in the cell wall is a pathogenic determinant for *Cryptococcus*. Yeast cells do not appear pigmented in tissue, however, they react with Fontana–Masson stain which binds to melanin. Colonies of the yeast on SDA are white. Special media containing melanin precursors demonstrate the in vitro ("brown colony") effect. Please see Section 12.13, Laboratory Detection, Recovery, and Identification, and Chapter 2.

Many human fungal opportunistic pathogens are melanized fungi, and melanin is a common virulence factor in fungal plant pathogens. Disruption of *CNLAC1* results in a melanin (−) mutant of *C. neoformans* which is less virulent in mice, and reverse mutants regain their ability to produce disease (Salas et al., 1996). The gene for laccase, *CNLAC1*, was cloned and characterized as capable of oxidizing DOPA or dopamine to dopaminequinone. *C. neoformans* substrates for laccase possibly include other neurotransmitters found in the brain. Laccase is associated with the production of immunomodulatory catecholamines from brain (Panepinto and Williamson, 2006). This may help explain the neurotropism of the fungus, as the brain provides substrates for melanin production, which may be a factor contributing to meningitis.

Evidence for the production of melanin during cryptococcal infection was obtained when reverse transcriptase–PCR detected the *CNLAC1* transcript in the CSF of infected rabbits, indicating laccase was produced in the infected host (Salas et al., 1996), and melanin ghosts retaining the ellipsoid yeast shape were chemically isolated from tissues of experimentally infected rodents (Rosas et al., 2000). Melanized cryptococci (i) resist ingestion and killing by macrophages; (ii) neutralize the free radical effects of oxidants produced by macrophages (e.g., hypochlorite) (Buchanan and Murphy, 1998); and (iii) are less susceptible to killing by microbicidal peptides. Melanin appears to enhance virulence by protecting the yeast against the immune system (reviewed by Casadevall et al., 2000). Melanin impairs antibody formation, decreases lymphoproliferation, and downregulates TNF-α production (Shoham and Levitz, 2005).

Regulation and Functions of Laccase

C. neoformans laccase, a copper-dependent enzyme, is located in the cell wall and is encoded by *CNLAC1* and

CNLAC2. Deletion of *CNLAC1* alone results in significant virulence attenuation. Laccase displays broad activity to oxidize polyphenolic compounds and iron (Zhu and Williamson, 2004). Laccase stands at the nexus of two functions important for pathogenesis: melanin synthesis and iron acquisition. The activity and expression of laccase is influenced by iron. Melanin participates in ferric iron reduction, and ferrous ion binds to melanin. This function increases redox buffering, which may protect fungal cells from oxidative killing such as that mediated by hypochlorite produced by macrophages and neutrophils (Jacobson, 2000; Jacobson and Hong, 1997). Laccase may also act during infection as an iron scavenger.

Laccase, itself, may have an antioxidant ferrous iron oxidase activity, distinct from melanin synthesis. This function would protect cryptococci from the generation of hydroxyl radicals by phagocytes. It does this by oxidizing Fe(II) to Fe(III) before the reactive electron of the Fe(II) can reduce H_2O_2 to hydroxyl radicals (Jacobson, 2000). This production of Fe(III) may contribute to iron transport (Jung and Kronstad, 2008).

Immunocytochemistry established that laccase is highly expressed early in pulmonary infection, where it may contribute to virulence by catalyzing the formation of melanin precursors and by protecting yeast against alveolar macrophages by acting as an iron scavenger (Garcia-Rivera et al., 2005).

Regulation of the enzyme in response to the environmental signals of nutrient starvation and multivalent cations is mediated through signal transduction pathways. A mutant defective in laccase activity was localized to a disruption of *CLC1* encoding voltage-gated chloride channels which activate copper transporting enzymes. Addition of copper restored laccase activity and melanin production to the Δ*clc1* mutant (reviewed by Panepinto and Williamson, 2006). Exogenous copper induces transcription of laccase via the action of a copper-dependent transcription factor.

Laccase is repressed by the glucose analog, 2-deoxyglucose, which can be phosphorylated but not further metabolized. Glucose, or a phosphorylated derivative, is a potential second messenger signaling glucose repression of *CNLAC1* transcription in *C. neoformans*. Regulation of *CNLAC1* expression by carbon source favors an environmental role of laccase in *Cryptococcus* degradation of lignin found in the bark of trees for use as a source of carbon.

D-Mannitol

Cryptococcus neoformans produces large amounts of D-mannitol in vitro (Wong et al., 1990). Studies in experimentally infected rabbits measured quantities of serum mannitol which were proportional to the fungal burden. By increasing osmolarity of the surrounding fluid, cerebral edema may be caused by D-mannitol accumulation, leading to neurologic damage. Moreover, D-mannitol may exert a protective effect for the yeast, similar to that of melanin, by scavenging oxidative products of macrophages.

Does survival in soil-dwelling amebae explain evolution of pathogenicity in C. neoformans? While studying the mouse gastrointestinal tract as a model for the portal of entry of *C. neoformans*, Neilson et al. (1978) observed *Acanthamoeba polyphaga* growing on *C. neoformans* cultured from mouse feces. The amebae were engorged with yeast cells. Later they discovered that *Acanthamoeba palestinensis* isolated from pigeon droppings ingested and killed 99.9% of *C. neoformans* cells after 7 days of incubation (Ruiz et al., 1982). More recently, Steenbergen et al. (2001) incubated *C. neoformans* with *Acanthamoeba castellanii* and showed that the yeast were phagocytosed and then they replicated in the amebae's phagocytic vacuole. Phagocytosed cryptococci produce capsular polysaccharide, which accumulates in vesicles, resulting in killing of amebae and replication of the yeast.

There is a parallel between this sequence of events and those observed in *C. neoformans*-infected macrophages because polysaccharide-containing vesicles are deleterious to macrophages. Furthermore, Steenbergen et al. (2001) observed that melanized cryptococci, even acapsular forms, were more resistant to killing by amebae. This research suggests that *C. neoformans* has evolved as an effective pathogen by surviving within soil-dwelling amebae. The intracellular existence underscores a natural role for capsule and melanin, even where cryptococci in the environment may have very small capsules.

Iron Regulation

Pathogenic fungi, like other microbes, must compete with the host for iron. Ferrous ion, Fe(II), is needed by fungi for a variety of host processes (e.g., cytochrome P450 mediated electron transfer reactions). The *CIR1* gene in *C. neoformans* ("*Cryptococcus* iron regulator") was identified in the genome sequence (Jung and Kronstad, 2008). Δ*cir1* mutants (deletion mutants) show poor growth at 37°C, reduced capsular polysaccharide, and increased melanin production. Microarray experiments showed that Cir1 acts as the major regulator of *C. neoformans* iron responses, both as repressor and activator. Cir1 activates a high affinity iron permease, encoded by the *CFT1* gene. Cft1 is required for the reductive iron uptake of both inorganic ferric iron and iron from transferrin. Given that transferrin is a major iron carrier in the mammalian host, *CFT1* may play a key role for iron acquisition during infection (Jung et al., 2008).

12.10 CLINICAL FORMS

Cryptococcosis is influenced by the ability of the host to mount an immune response. Pathogenesis also is influenced by the *Cryptococcus* species involved (please see below, "Observations of Clinical Forms of Infection with *Cryptococcus gattii*").

Effect of Infectious Dose in the Immune-Normal Individual

Inhalation of a few yeast is thought to result in minimal subclinical infection that is rapidly resolved but may lead to cryptococci persistent in the lung in a dormant but possibly viable state. A larger inoculum may produce infection in the lungs that may or may not be contained, depending on the host's immune status. In some instances the infection will result in extrapulmonary dissemination including meningitis.

Cryptococcosis in the Immunocompromised Host

Patients receiving immunosuppressive therapy to maintain solid organ transplants or to treat Hodgkin's lymphoma or collagen-vascular diseases become vulnerable to cryptococcosis. $CD4^+$ T-lymphocyte-depleted HIV-positive persons are at increased risk. Meningitis or meningoencephalitis may, in these patients, disseminate hematogenously, resulting in cryptococcemia and seeding of other organs including the appearance of skin lesions (please see Section 12.7, Risk Groups/Factors). When cryptococcosis occurs in the immediate post-transplant period, it may be caused by preexisting low-level infection acquired before transplantation because of comorbidities associated with organ failure (Fishman et al., 2008). Otherwise, the transplanted organ may harbor dormant but viable cryptococci.

Colonization, Transient

Immune-normal healthy individuals may demonstrate transient asymptomatic carriage of *C. neoformans* from the respiratory tract after exposure to the fungus in the environment. The fungus, at times, may be isolated as a transient member of sputum and the skin or other nonsterile sites. It is not, however, part of the normal endogenous microbiota. Infection occurs in a susceptible host following inhalation of desiccated yeast forms and basidiospores. The usual result from such environmental exposure is subclinical or self-limited disease. Evidence of some infections may be found only on chest radiographs. Invasive cryptococcosis may occur in some patients, however, producing pneumonia with a productive cough, chest pain, fever, and weight loss.

Colonization, Persistent

Long-term asymptomatic colonization of the bronchi may occur. *C. neoformans* may be isolated from the sputum repeatedly over months and years in patients with preexisting chronic lung disease but who are otherwise apparently immune-normal. They show no evidence of active invasion of the lung parenchyma, have negative cryptococcal serum antigen titers, and their urine and CSF cultures are negative (Perfect, 2005).

Mouse Model of Allergic Bronchopulmonary Cryptococcosis

Low-dose infection of C57BL/6 mice with *C. neoformans* produces a chronic allergic bronchopulmonary mycosis (Hernandez et al., 2005). Chronic fungal infections can develop when the balance of cellular immunity is shifted from Th1 to Th2 immune responses. C57BL/6 mice develop a chronic pulmonary fungal infection, accompanied by upregulation of Th2 immunity. Susceptibility of C57BL/6 mice correlates with high levels of IL-5 secretion and low levels of IFN-γ and IL-2. Elevated IL-5 in the lungs promotes the development of pulmonary eosinophilia and tissue damage.

Pulmonary Cryptococcosis

In immune-normal hosts, chest radiographs commonly show well-defined, noncalcified single or multiple nodules. An initial presentation may be that of radiographic lesion(s) for which a lung malignancy must be ruled out. Other radiographic findings may include indistinct mass infiltrates, hilar lymphadenopathy, lobar infiltrates, pleural effusions, and lung cavities (Perfect, 2005). If there is a positive test for serum cryptococcal antigen a workup for CNS disease is warranted.

Pulmonary cryptococcosis in persons living with AIDS presents with fever, cough, dyspnea, and pleuritic chest pain. Radiographic findings include lymphadenopathy or pleural effusions. Often there are diffuse mixed interstitial and intraalveolar infiltrates resembling those of patients with *Pneumocystis jirovecii* pneumonia (Perfect, 2005). In Section 12.3, Case Presentation 2, the patient's pulmonary cryptococcosis extended to three air-filled cavities.

Central Nervous System (CNS) Cryptococcosis

Cryptococcal meningitis has an insidious onset and is the most common clinical form of cryptococcosis. The symptoms are headache, memory loss noted by family members and friends, dizziness and irritability, and visual disturbances. Intermittent headaches unresponsive to analgesics and mental confusion often are the critical symptoms that prompt seeking of medical attention. The primary pulmonary infection often has resolved by the time this form of the disease is identified.

Cryptococci cross the blood–brain barrier, are present in the CSF, and infect the subarachnoid space. In a subset of patients cryptococci invade the brain, in which case the disease is properly termed meningoencephalitis, wherein one may find invasion of the cerebral cortex, brain stem, cerebellum, and meninges. Diminishing visual acuity and coma may occur in the later stages of disease.

Detection of CNS cryptococcosis is via lumbar puncture. Typical findings are increased pressure due to fluid accumulation, clear fluid, somewhat elevated cell count, rarely >800 WBCs/mm^3 with lymphatic *pleocytosis* (may be normal early in disease), increased protein concentration, decreased sugar concentration (10–40 mg/dL), yeast cells (sparse to numerous), and a positive cryptococcal latex agglutination test. The CSF parameters in cryptococcosis in the immunocompromised host, particularly in persons living with AIDS, may be normal for protein and glucose concentration and without pleocytosis.

Cryptococcoma (*definition*: a large focal collection of cryptococci usually found in the brain). Most infections are diffuse, but brain invasion can occur, producing localized solid tumor-like masses. Symptoms may mimic the radiologic appearance of a growing neoplasm (Fig.12.6.) The inflammatory response varies from strong to minimal:

- Strong: granulomatous (a fibrous reaction appearing as a tumor)
- Minimal: gelatinous (mucoid cysts with little or no cellular reaction)

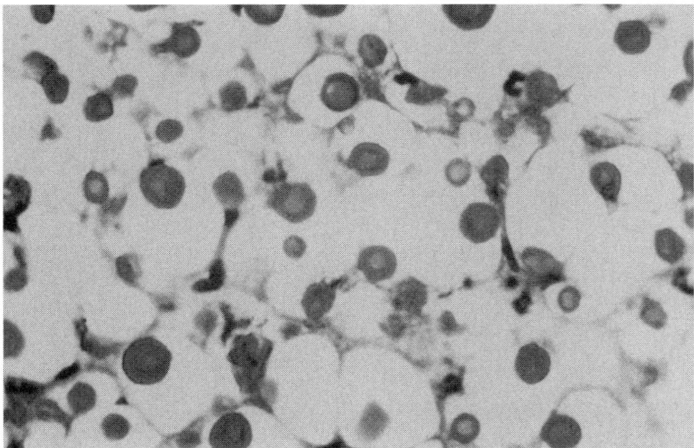

Figure 12.6 Histopathology of cryptococcosis, brain, mucicarmine stain. The wall and polysaccharide capsule are stained bright red. The capsule is retracted from surrounding tissue during fixation and embedment, leaving an empty space. See insert for color representation of this figure.
Source: H. J. Shadomy.

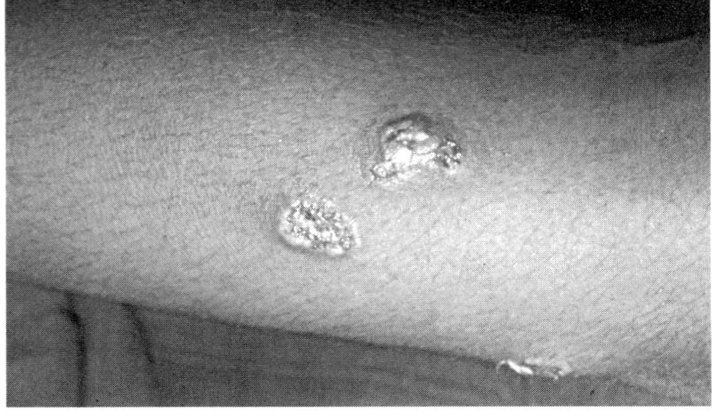

Figure 12.7 Molluscum-like lesions evolving into nodulo-ulcerative forms on the forearm in a case of disseminated cryptococcosis.
Source: Used with permission from Dr. Uma M. Tendolkar, LTM Medical College, Mumbai.

Cutaneous Cryptococcosis

Skin lesions occur via hematogenous spread of the fungus from a pulmonary site and are found most frequently in immunosuppressed patients (Fig. 12.7). Please see Section 12.3, Case Presentation 4. They are seen in 5–15% of patients and often indicate a poor prognosis. Although many lesions may be seen, most frequently only a solitary one is found. Lesions appear as either cellulitis, small papules, pustules, subcutaneous masses, or ulcers. *Cryptococcus neoformans* may be identified in specimens taken from the open skin lesions. (Please see Section 12.13. "Laboratory Detection, Recovery, and Identification"). In persons living with AIDS, skin lesions are the *second most common sites* of disseminated cryptococcosis. Lesions frequently may be seen on the head and neck. Umbilicated papules in AIDS patients may resemble *molluscum contagiosum* or Kaposi's sarcoma.

Primary Cutaneous Cryptococcosis

Cryptococcosis cases associated with skin lesions reported in the French National Registry amounted to 28 patients with primary cutaneous cryptococcosis (Neuville et al., 2003). These patients were older and half of them had no underlying immunocompromising condition. There was no evidence of dissemination and usually only a solitary skin lesion on an exposed area. Other associated factors were a history of a skin injury, outdoor activities, and/or contact with birds or bird droppings. Seventy-one percent of the isolates were *C. neoformans* var. *neoformans* (serotype D). The shift towards serotype D is thought to be related to differences in dermotropism or temperature tolerance between serotypes A and D.

Ocular Cryptococcosis

Ocular cryptococcosis is uncommon but cryptococcal meningitis patients may develop (i) visual loss by direct invasion of the optic nerve resulting in rapid visual loss or (ii) slow loss of vision over weeks or months related to increased intracranial pressure (Rex et al., 1993). Contributing factors are papilledema, elevated CSF opening pressure, and positive CSF India ink preparations. Reducing intracranial pressure is effective if initiated early.

Osseous Cryptococcosis

Bone lesions are seen in 10–15% of patients but are very difficult to diagnose. Radiographic findings show one or more areas of osteolysis. They may involve the pelvis, bony prominences, cranial bones, ribs, and vertebrae. Over time, a sanguinopurulent exudate with numerous cryptococci accumulates in the adjacent soft tissue.

Visceral Cryptococcosis

Any organ or tissue may become infected via hematogenous spread. Although rare, this is an extremely serious finding, often found only at autopsy.

Prostate

As with some other fungal diseases (e.g., blastomycosis), the prostate can be a persistent nidus of infection in some patients after antifungal treatment, and is the focus of relapse in these individuals. A case of cryptococcal prostatitis in a heart transplant recipient is described by Sax and Mattia (1994).

Observations of Clinical Forms of Infection with *Cryptococcus gattii*

Pulmonary cryptococcosis and meningoencephalitis are seen in *C. gattii* infections (Perfect, 2005; Bromilow and Corcoran, 2007) which occur in apparently immune-normal persons. Because of their immune status they mount a greater immune response and, generally, have a better prognosis with significantly less mortality. Findings associated with *C. gattii* are cerebral mass lesions and/or hydrocephalus, papilledema, and high cryptococcal antigen titers. A subgroup of patients with *C. gattii* meningoencephalitis has multiple enhancing lesions observed on CT scans, complications of hydrocephalus, intracranial hypertension, cranial neuropathies, and a worse prognosis.

12.11 VETERINARY FORMS

Cryptococcosis was first identified in animals in 1902 from the pulmonary lesion of a horse. It is a subacute or chronic granulomatous infection in a number of animal species. The fungus is primarily a respiratory or CNS disease in dogs, cats, horses, and dolphins.

Cats

Feline cryptococcosis usually presents as nasal cavity involvement with sneezing, nasal discharge, fleshy masses protruding from the nostrils, or subcutaneous swelling over the bridge of the nose. The infection may spread from the ethmoid sinuses through the cribriform plate causing meningoencephalitis, optic neuritis, and retinitis. Cutaneous lesions adjacent to the nose are common but widespread cutaneous and subcutaneous lesions are rare, resulting from hematogenous dissemination (Acha and Szyfres, 2001; O'Brien et al., 2004). Feline immunodeficiency virus (FIV) infection is a risk factor but is not obligatory. FIV-positive cats have a more rapid and disseminated clinical course. The major pathogen isolated is *C. neoformans* but *C. gattii* also is reported. Primary infection of the sinonasal region likely results from the inhalation of cryptococci or its basidiospores, with the infection penetrating overlying bones to invade the subcutaneous space. Invasive nasal malignancy is in the differential diagnosis.

Cats without neurologic involvement respond well to monotherapy with FLC, effecting cure after a mean of 4 months of therapy. Cats with CNS involvement, or those refractory to FLC, respond well to AmB therapy (O'Brien et al., 2006). The overall success rate for both clinical forms was 76%. Where there is eye involvement as part of systemic disease, the infection may be controlled systemically but the retina may have become severely damaged. After cessation of therapy a relapse rate of 17% was observed, even though the latex agglutination titer had decreased to zero. Possibly, mycologic cure had not been achieved, and maintenance therapy with FLC may need to be prolonged.

Dogs

Canine cryptococcosis disproportionally affects large breed dogs <4 years old; older dogs are rarely affected. Certain breeds may be more often infected, for example, American Cocker Spaniels, Doberman Pinschers, and German Shepherds. Dogs may have obvious signs of nasal disease but often present with neurologic or ocular signs. About 60% of dogs in one series had nasal and localized disease, whereas of the remainder, the major category was CNS and disseminated cryptococcosis

(O'Brien et al., 2004). The success rate for treating canine cryptococcosis in one series of 11 dogs was 55%. In both dogs and cats large cumulative doses of AmB could be administered subcutaneously with minimal nephrotoxicity (O'Brien et al., 2006).

Koalas

A high prevalence of nasal colonization and subclinical infection, also including clinical cryptococcosis, occurs in captive and free-living koala populations (Krockenberger et al., 2003). The coincidence of the ecologic niche of *C. gattii* in the decayed hollows of trees and the preferred food source of *Eucalyptus* leaves by koalas sets the stage for a unique host–parasite relationship among koalas, eucalypts, and *C. gattii*. Although cryptococcosis is not a leading cause of death of koalas, it is probably more common in koalas than in domestic animals or people in Australia. A series of 43 koalas with *C. gattii* cryptococcosis was reported from captive populations in Sydney and captive and wild populations from the north coast of New South Wales, Australia (Krockenberger et al., 2003). Typically, koalas with cryptococcosis presented with chronic upper and/or lower respiratory tract disease (13 koalas), with neurologic signs (13 koalas), or as unexplained deaths (6 koalas). Pulmonary disease was evident as severe dyspnea. Seizures were the leading neurologic signs. Sixteen cases (37%) had meningitis and of these 10 had encephalitis. Two koalas had localized cutaneous lesions, most likely arising from penetrating injuries. The histopathologic appearance of lesions ranged from large aggregations of cryptococci with minimal host response, to pyogranulomatous and granulomatous inflammation. Widespread dispersal of capsular polysaccharide was seen in tissues from many cases.

Other Animals

In dairy cattle, mastitis is the usual finding. Other animals, for example, foxes, ferrets, cheetahs, monkeys, goats, and sheep, have been reported to have cryptococcosis (Acha and Szyfres, 2001). A captive Atlantic bottlenose dolphin (*Turciops truncatus*) maintained at the U.S. Navy Marine Mammal facility in San Diego, California, developed bronchopneumonia with pleuritis which was treated unsuccessfully with ITC (Miller et al., 2002). Necropsy revealed extensive coalescing granulomatous lesions in both lungs which histopathologic examination revealed as large capsule cryptococci. The species was identified as *C. gattii*, which occurs in southern California associated with *Eucalyptus calmodulensis*.

Cryptococcosis is not transmitted between humans and animals. Although pigeon and other bird droppings enhance growth of *C. neoformans*, birds are not subject to this disease due to their higher body temperature.

12.12 THERAPY

Factors for successful treatment include prompt diagnosis, control of the underlying disease, and antifungal therapy.

Meningitis Therapy in HIV-Negative Patients

(See Perfect et al., 2010.)

i. Recommended therapy consists of an initial dose of AmB-deoxycholate (0.7–1.0 mg/kg/day), plus 5FC, 100 mg/kg/day, in four divided oral doses daily for 6 weeks or longer, if needed, until the patient is clinically stable. Therapy is generally continued until weekly cultures are negative for 2 weeks.

Meningitis Therapy in People Living with AIDS

(See Perfect et al., 2010.) The majority of patients who develop cryptococcosis have $CD4^{+}$T-lymphocyte counts <50 cells/μL.

i. Recommended therapy consists of an initial dose of AmB-deoxycholate (0.7–1.0 mg/kg/day) combined with 5FC, 100 mg/kg/day in four divided doses for >2 weeks for patients with normal renal function. The 5FC dose is adjusted for those with renal impairment. The addition of 5FC to AmB during acute treatment results in a more rapid sterilization of CSF. AmB combined with FLC at 400 mg/day is inferior to AmB and 5FC but better than AmB alone.

ii. Patients who develop renal dysfunction during therapy benefit from lipid formulations of AmB, at 4–6 mg/kg/day.

iii. After successful induction therapy, defined as clinical improvement and negative CSF culture, follow-up therapy should consist of FLC, 400 mg/day for 8 weeks, followed by maintenance therapy for at least 1 year.

iv. The risk of recurrence is low when patients successfully complete initial therapy, remain asymptomatic, and have a sustained increase (>6 months) in their $CD4^{+}$ T-lymphocyte count of >200 cells/μL, after ART. Maintenance therapy should be resumed if the $CD4^{+}$ count falls≤200 cells/μL.

v. Although relapse during antifungal maintenance therapy is uncommon, cryptococci may persist in the prostate, serving as the reservoir for a recurrence of infection. After prostatic massage, urine specimens may be directly examined and cultured (Perfect, 2005).

Immune Reconstitution Inflammatory Syndrome (IRIS)

Immunologic recovery (Singh and Perfect 2007) can be detrimental and contribute to worsening disease. Immune restoration in HIV-positive persons may stimulate an exuberant inflammatory response. The clinical picture is one of local and systemic inflammation with both beneficial and harmful effects on a quiescent deep-seated mycosis. The basis for IRIS is thus an inflammatory response triggered by resolving immunosuppression.

Initiation of Combination ART

After beginning combination ART in HIV-positive persons, the viral count subsides and a repopulation of memory T cells occurs. Four to six weeks later there is an increase in naive $CD4^+$ T lymphocytes which express Th1 functionality. These events coincide with the clinical development of IRIS, in the period 2 months after initiation of combination ART.

Immunoregulation of T Lymphocytes Activation of precursor $CD4^+$ T lymphocytes causes them to differentiate into Th1 cells, which produce IFN-γ and elicit proinflammatory responses. Th2 cells produce anti-inflammatory and immunosuppressive cytokines (e.g., IL-10). Another class of T lymphocytes, Th3 cells, secrete transforming growth factor ß. Th3 cells in concert with Th2 cells inhibit the function of Th1 cells. This interplay maintains homeostasis of the immune system.

Cytokines Involved in IRIS Certain cytokines and growth factors are elevated at the time of IRIS including TNF-α, granulocyte colony stimulating factor (G-CSF), and vascular endothelial growth factor (VEGF-A) (Boulware et al., 2010). VEGF-A increases vascular permeability and stimulates chemotaxis of macrophages, $CD4^+$ memory T cells, and is costimulatory for IFN-γ-secreting Th1 cells. G-CSF aids in clearance of cryptococci from the CSF by increasing the activity of neutrophils and macrophages. Increased CSF inflammation results from the upregulation of these cytokines, increased WBCs, and protein, setting the stage for the development of IRIS.

Transplant Recipients

Cytokines responsible for graft rejection are major targets for immunosuppressive therapy, including the calcineurin inhibitors tacrolimus and mycophenolate mofetil. The mAb alemtuzumab suppresses $CD4^+$ T lymphocytes, and corticsteroids suppress inflammation. Deliberate immunosuppression in transplant recipients has a dominant suppressive effect on T-cell responses. When they are withdrawn, a shift toward a proinflammatory response ensues, especially if a pathogen is present in host tissues during the immune suppression period. The stage is set for a pathological production of Th1 cytokines, particularly IFN-γ.

Candidiasis

IRIS can explain the enigmatic entity of chronic disseminated (hepatosplenic) candidiasis in $\sim$4.5% of leukemia patients after recovery from neutropenia.

Cryptococcosis

IRIS has been described for *Histoplasma capsulatum, Pneumocystis jirovecii*, and aspergillosis, but it is best known in *C. neoformans* infection. Cryptococcal IRIS presents after immunity is restored (i) triggering a relapse of a partially treated infection or to persistent cryptococcal antigens, or (ii) unmasking a subclinical infection present at the time when ART is initiated. As many as 30–35% of HIV-infected patients develop active cryptococcosis after initiation of combination ART. There is a higher risk of IRIS if a patient was treated for a fungal infection within 30–60 days of initiation of combination ART.

IRIS-related cryptococcosis appears with lymphadenitis, enhancing CNS lesions, and increased intracranial pressure. Granulomatous changes evoked by Th1 cells are present. This response is the same as that which normally contains and resolves an infection. But in the setting of IRIS the exuberant response results in necrosis in the lymph nodes and in suppurative lesions. Renal transplant recipients who develop IRIS related cryptococcosis also may lose their transplanted kidney because of the upregulation of Th1 immunity.

The rate of cryptococcosis in AIDS patients in Uganda is exponentially higher than it is in North America in the era of combination ART (Kambugu et al., 2008). Mortality among patients with AIDS and cryptococcosis remains high in Uganda due, in part, to IRIS related complications after combination ART is initiated. Of 170 patients with cryptococcal meningitis, half survived to initiate ART 5 weeks after a cryptococcosis diagnosis. A median of 8 weeks after beginning ART, 33 (39%) developed IRIS with CNS manifestations (Boulware et al., 2010). The clinical IRIS events were headache, photophobia, vomiting, meningismus, papilledema, and seizures. Lumbar puncture revealed elevated CSF opening pressure, a median of 305 mm. The patients' CSF was culture-negative for cryptococci. Because cryptococcal antigen persists for months, in the setting of immune restoration, an exaggerated inflammatory response is directed at the persisting antigen burden.

Pneumocystosis

An estimated 5% of patients treated for *Pneumocystis* pneumonia develop recurrent disease 5–17 days after initiation of ART or discontinuation of adjunctive corticosteroid therapy.

A surrogate marker for IRIS is hypercalcemia, a recognized response to developing granulomas brought about by overproduction of vitamin D by activated macrophages.

Diagnosis of IRIS-Cryptococcosis

The diagnosis is made by clinical or radiologic imaging evidence of inflammation, contrast-enhancing lesions in CT or MRI scans, *pleocytosis* in the CSF, >5 WBCs/μL, increased intracranial pressure, unexplained hypercalcemia, coupled with negative cultures.

Approach to Clinical Management of IRIS

This phenomenon reflects a poorly regulated inflammatory response rather than failure of treatment to kill the fungal pathogen. An approach to consider is to delay the initiation of ART 4–10 weeks until antifungal therapy has controlled the infection by clearing the CSF of cryptococci. In the case of transplant recipients, withdrawal of immunosuppressive therapy during treatment for invasive fungal infection may result in loss of the allograft. If acute IRIS is suspected, whether in AIDS or solid organ transplant recipients, patients should be treated with steroids to decrease the inflammatory response.

12.13 LABORATORY DETECTION, RECOVERY, AND IDENTIFICATION

(“How-to” instructions for certain tests in this section may be found in the Chapter 2.)

Cryptococcus neoformans is not part of the normal endogenous microbiota, and demonstration of the fungus in culture from a sterile site clinical specimen is evidence of infection. If the clinical specimen is from a nonsterile site (e.g, skin or sputum), further studies, at times including biopsies, are performed to determine whether disease is present and tissue has been invaded.

Direct Examination

Material sent for laboratory studies may include CSF, tissue biopsy and exudates, sputum, bronchial washings, blood, urine, pus, and skin scrapings. Cryptococci may be demonstrated in wet preparations, appearing as a round budding yeast (4–8 μm diameter). When freshly isolated from the host cryptococci have large capsules. The entire cell diameter including capsule may be 15–20 μm with the cell diameter of 7 μm. Yeast cells vary considerably in both size and shape. On Gram stain, *Cryptococcus* takes up crystal violet heterogeneously unlike other yeast which stain Gram-positive. Additionally, the capsule can be visualized as a faint, red halo around the yeast.

India Ink Preparation

An India ink preparation of the sediment from centrifuged CSF is useful because, when positive, it affords a rapid, presumptive diagnosis facilitating timely treatment decisions.[3] A caveat is this test’s relatively low sensitivity compared with the latex agglutination (LA) test (please see below, “Immunodiagnosis of Cryptococcosis”). Observed microscopically with India ink, the fungus demonstrates the polysaccharide capsule and budding typical of the genus *Cryptococcus*. The capsule and small pore, or isthmus, separating the mother and daughter cell, as well as the release of the daughter cell while still small, differentiate *C. neoformans* and *C. gattii* from other fungi such as *Blastomyces dermatitidis* yeast forms. A narrow to wide capsule may be demonstrated, and the fungal size may extend to 20 μm diameter (please see Fig 12.5). Occasionally specimens may contain acapsular *C. neoformans* yeast cells.

Histopathology

Tissue specimens are best stained with Mayer’s mucicarmine, which stains the polysaccharide capsule red. The staining procedure also condenses the capsule around the cell, often giving the impression of deeply stained cells with a clear area surrounding them. This clear area was filled by capsule prior to fixation and embedment (Fig 12.7). PAS also is a useful stain for this fungus. H&E, the most commonly used stain in histopathology, may be used but demonstrates the fungi less well. With this stain, a clear zone surrounding the cryptococci is the result of fixation whereby the capsule has collapsed.

Culture

When cultured on SDA, SDA-Emmons, blood agar, or similar laboratory media, *C. neoformans* and *C. gattii* grow rapidly, usually maturing within 3 days, although some strains grow more slowly, and culture media should be kept for up to 2–4 weeks at room temperature. When

[3]A standardized permanent, preservative-fixed control for use with India ink is sold by Scientific Device Laboratory, Des Plaines, Illinois. URL: www.scientificdevice.com. This suspension of *Cryptococcus* is grown to maximize capsule production. Each vial contains 1 mL of fluid containing many organisms. India ink negative control is a 1 mL suspension of *Candida albicans*.

growth occurs, colonies formed are flat or occasionally heaped, shiny, moist, and often mucoid to the point of resembling *Klebsiella*. On SDA at 25°C, *C. neoformans* grows as round budding yeast varying in size from 2 to 10 4 μm in diameter. *C. gattii*, in addition to round cells, produces oval-to-lemon shaped cells, 3–5 × 3–7 μm in diameter with more cells becoming elongated at 37°C (Kwon-Chung and Bennett, 1992). The fungus grows well at both 25°C and 37°C and is cream to pale tan in color. Specimens suspected of containing yeasts should **not** be grown on media containing cycloheximide. Antibiotics to inhibit bacterial growth should be used. Continued subculture of the fungus in the laboratory usually leads to loss of the capsule.

Selective Media

Carbohydrate Assimilation *Cryptococcus neoformans* and *C. gattii* (as well as nonpathogenic *Cryptococcus* species) are nonfermentative yeasts. The assimiliation profile of *C. neoformans* in the API 20C kit (BioMérieux Vitek, Durham, NC) panel is shown in Table 12.3. *C. gattii* displays the same profile. Occasionally, nonpathogenic *Cryptococcus* species will grow in clinical specimens from nonsterile sites and their identification is helpful to rule out *C. neoformans* or *C. gattii*: these are *C. albidus, C. laurentii, C. luteolus, C. terreus*, and *C. uniguttulatus*. None of these species produce the brown colony effect on birdseed agar. Some important assimilation or other tests to differentiate *Cryptococcus* species are not in the API 20C panel; for example, *C. neoformans* assimilates galactitol (dulcitol) but *C. uniguttulatus* does not. *Cryptococcus uniguttulatus* liquefies gelatin but *C. neoformans* does not. In addition, *C. uniguttulatus* does not grow at 37°C. Charts depicting tests to differentiate various *Cryptococcus* species may be found in de Hoog et al. (2000), Hazen and Howell 2007, and Larone (2002).

Misidentification of *C. neoformans* with the API 20C can result when phenoloxidase-positive *Trichosporon* species are scored as *C. neoformans* (Freydiere et al., 2001). This anomaly can be clarified by the microscopic morphology on cornmeal agar or rice–Tween agar, because *Trichosporon* species produce true mycelia that disarticulate to form arthroconidia.

Urease Production Identification of the genus *Cryptococcus* may be performed by placing a loopful of a colony on Christensen's urea agar. Pathogenic members of this genus produce urease, split urea, releasing NH_3, increasing the pH, and causing a pH-related color change in the agar from pale yellow agar to pink/magenta, at times within hours. The rapid urease test for *Cryptococcus neoformans* (time required: 15 min) is detailed in Chapter 2, Section 2.3.7.2, Rapid Tests for Yeast Identification—Urease Test for *Cryptococcus neoformans*.

Table 12.3 Assimilation Test Results[a] Obtained with the API 20C® Test Panel and *Cryptococcus neoformans*

Substrate	Strains giving a positive reaction (%)
Adonitol (ribitol)	71
Arabinose	14
Cellobiose	10
Galactose	93
Glucose	100
Glycine	0
Inositol	97
Lactose	0
Maltose	99
MDG (α-methyl-D-glucoside)	99
Melitizose	97
N-acetylglucosamine	88
Raffinose	88
Sorbitol	100
Sucrose	99
Trehalose	92
Xylitol	1
Xylose	91
α-Ketoglutarate	100

[a]This table is found in the API 20C package insert, BioMérieux Vitek, Durham, NC.

Melanin Production *C. neoformans* may be identified by the "brown colony" effect. A loopful of the colony is placed on "birdseed" agar. The birdseed, *Guizotia abyssinica* (also called Niger seed or thistle seed), extract may be simply made or obtained commercially [please see Chapter 2, Section 2.3.7.2, Rapid Tests for Yeast Identification—Birdseed (Niger Seed) Agar Test for *Cryptococcus neoformans*] also containing chloramphenicol. On this medium, colonies of *C. neoformans* become deep tan to brown in color after 2–3 days. Other species remain cream to pale tan in color (Fig. 12.8). Care must be taken not to exceed 0.1% glucose in the medium, as this sugar represses phenoloxidase (laccase), the enzyme causing browning of the colony. Glucose–salts medium containing dihydroxyphenylalanine (DOPA) can substitute for birdseed agar. The formula for this medium is found in Chapter 2.

Caffeic Acid Test Disks After incubation at 37°C for between 30 min and 4 h, the disk will turn brown in the case of a positive reaction. Please see Chapter 2 for details of this test.

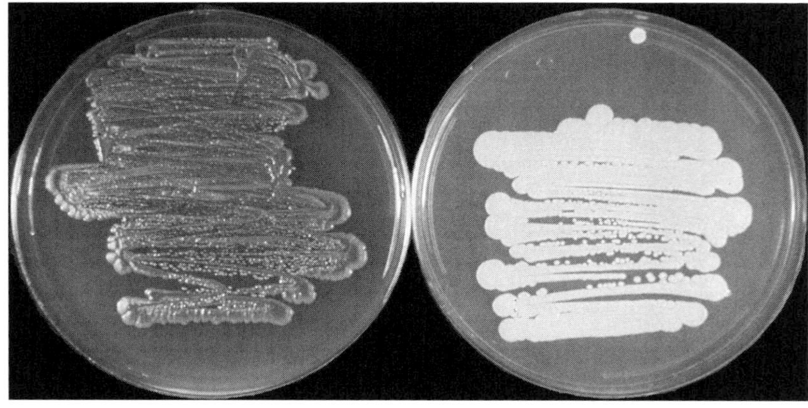

Figure 12.8 Birdseed agar medium demonstrates the brown colony effect of *Cryptococcus neoformans* (left) and growth of the same culture on SDA, (right) no pigment is produced. See insert for color representation of this figure.
Source: H. J. Shadomy.

"CGB" Medium Medium containing canavanine–glycine–bromothymol blue (CGB) may be used to differentiate the two species (Kwon-Chung et al., 1982). *Cryptococcus neoformans* maintains a yellow color in the medium, whereas *C. gattii* turns the medium blue. Most *C. gattii* isolates assimilate glycine whereas this amino acid is assimilated by only ~20% of *C. neoformans* serotype A and not by serotype D.

Principle: L-Canavanine is an arginine analog. All *C. gattii* isolates tested are resistant to L-canavanine, whereas one-third of *C. neoformans* are resistant and these isolates do not assimilate glycine. In summary, *C. gattii* is resistant to L-canavanine and assimilates glycine resulting in a rise in pH of the medium from pH 5.8 to above neutrality and turning bromothymol blue ($pK_a = 7.10$) a blue color. *C. neoformans* are susceptible to L-canavanine, or if not, they do not assimilate glycine, and the medium remains yellow. This medium is not generally used in the clinical laboratory, unless involved in an environmental study. Please see Chapter 2 for the formula.

Cryptococcus neoformans var. *grubii* cannot be differentiated from var. *neoformans* in the clinical laboratory, only by serotyping or DNA strain subtyping (Meyer et al., 2003).

Optimization of Capsule Size in Cryptococcus *Species*

Culture conditions favoring capsule production are the following:

- *Diluted Sabouraud Medium.* Capsule induction is accomplished simply by incubating cryptococci in a 1/10 dilution of Sabouraud broth in H_2O. These results are improved further by using PBS, MOPS, or HEPES buffer as the diluent at 0.05 M and pH 7.3. These formulations may be used in broth or agar form and incubated either at 30°C or 37°C.
- *Serum.* Incubating cryptococci in agar medium consisting of 10% fetal bovine serum diluted in PBS will induce capsule formation at 37°C.
- *CO_2.* Incubation of the yeast in RPMI medium in a routine laboratory tissue culture CO_2 incubator.
- *Limited Iron Medium.* Depleting iron in a chemically defined medium provides a stimulus for inducing capsule formation. Details of these formulations are given in Zaragoza and Casadevall (2004).

Immunodiagnosis of Cryptococcosis

Cryptococcal Latex Agglutination (LA) Test

This rapid test is used to detect capsular polysaccharide in the serum, CSF, or urine of patients and is considered the "gold standard" for the diagnosis of cryptococcosis. It is a highly sensitive and specific, diagnostic, and prognostic test. The kit contains rabbit anti–cryptococcal antibodies adsorbed to polystyrene beads. These are the "latex particles." When mixed with the dilutions of patient specimen, an agglutination titer of 1:4 is suggestive of infection, while a 1:8 or higher titer is indicative of active infection. Titers decline proportional to a favorable response to therapy except in patients living with AIDS. Their elevated titers may decline slowly even with clinical improvement.

Rare false-positive results have been reported from patients infected with *Trichosporon asahii*. This problem may be eliminated if morphologic studies are performed on the fungal colony. False-positive results also may be seen in patients with severe rheumatoid arthritis. Nonspecific agglutination has been seen in CSF and sera from patients without rheumatoid arthritis, probably by interfering proteins. Interference can be reduced or eliminated by proteinase and/or heat treatment (with EDTA) of the specimen (Stockman and Roberts, 1983).

The LA test may be positive when an India ink preparation is negative. This is due to low numbers of yeasts, poor quality specimens, or unknown reasons. Some reports in AIDS patients have demonstrated budding yeasts in 82% of specimens whereas cryptococcal antigen in the CSF was positive in 100% of these cases.

Further details about the cryptococcal LA test and commercial suppliers are given in Chapter 2.

GENERAL REFERENCES FOR CRYPTOCOCCOSIS

Calderone RA, Cihlar R, 2002. *Fungal Pathogenesis: Principles and Clinical Applications*. Marcel Dekker, New York. Chapters including: Genetic Basis of Pathogenicity in *Cryptococcus neoformans*, Brian L. Wickes and K. J. Kwon-Chung; Humoral Immunity and *Cryptococcus neoformans*, Arturo Casadevall; *Cryptococcus neoformans* and Macrophages, Thomas S. Harrison and Stuart M. Levitz; and Strain Variation and Clonality in *Candida* spp. and *Cryptococcus neoformans*, Jianping Xu and Thomas G. Mitchell.

Casadevall A, Perfect JR, 1998. *Cryptococcus neoformans*. ASM Press, Washington, DC.

Hazen KC, Howell SA, 2007. *Candida, Cryptococcus*, and Other Yeasts of Medical Importance, pp. 1762–1788, *in*: Pfaller MA (ed.), *Manual of Clinical Microbiology* 9th ed., Vol. 2, Section VIII, Mycology, Warnock DW (ed.) ASM Press, Washington, DC.

SELECTED REFERENCES FOR CRYPTOCOCCOSIS

Acha PN, Szyfres B, 2001. Mycoses, pp. 303–356, *in*: *Zoonoses and Communicable Diseases Common to Animals and Humans*, 3rd ed., Vol. 1. Scientific and Technical Publication No. 580. Pan American Health Organization, Washington, DC.

Alvarez M, Casadevall A, 2006. Phagosome extrusion and host–cell survival after *Cryptococcus neoformans* phagocytosis by macrophages. *Curr Biol* 16: 2161–2165.

Baker DJ, Reboli AC, 1997. Images in clinical medicine: Cutaneous cryptococcosis. *N Engl J Med* 336: 998.

Bauwens L, Vercammen F, Wuytack C, van Looveren K, Swinne D, 2004. Isolation of *Cryptococcus neoformans* in Antwerp Zoo's nocturnal house. *Mycoses* 47: 292–296.

Boulware DR, Bonham SC, Meya DB, Wiesner DL, Park GS, Kambugu A, Janoff EN, Bohjanen PR, 2010. Paucity of initial cerebrospinal fluid inflammation in cryptococcal meningitis is associated with subsequent immune reconstitution inflammatory syndrome. *J Infect Dis* 202: 962–970.

Bromilow J, Corcoran T, 2007. *Cryptococcus gattii* infection causing fulminant intracranial hypertension. *Br J Anaesth* 99: 528–531.

Buchanan KL, Murphy JW, 1998. What makes *Cryptococcus neoformans* a pathogen? *Emerging Infect Dis* 4: 71–83.

Callejas A, Ordoñez N, Rodriguez MC, Castañeda E, 1998. First isolation of *Cryptococcus neoformans* var. *gattii*, serotype C, from the environment in Colombia. *Med Mycol* 36: 341–344.

Casadevall A, Pirofski L, 2005. Prospects of vaccines for medically important fungi. Feasibility and prospects for a vaccine to prevent cryptococcosis. *Med Mycol* 43: 667–680.

Casadevall A, Rosas AL, Nosanchuk JD, 2000. Melanin and virulence in *Cryptococcus neoformans*. *Curr Opin Microbiol* 3: 354–358.

Chen J, Varma A, Diaz MR, Litvintseva AP, Wollenberg KK, Kwon-Chung KJ, 2008. *Cryptococcus neoformans* strains and infection in apparently immunocompetent patients, China. *Emerging Infect Dis* 14: 755–762.

Cherniak R, Valafar H, Morris LC, Valafar F, 1998. *Cryptococcus neoformans* chemotyping by quantitative analysis of ^{1}H nuclear magnetic resonance spectra of glucuronoxylomannans with a computer-simulated artificial neural network. *Clin Diagn Lab Immun* 5: 146–159.

Dadachova E, Bryan RA, Huang X, Ortiz G, Moadel T, Casadevall A, 2007. Comparative evaluation of capsular polysaccharide-specific IgM and IgG antibodies and F(ab')2 and Fab fragments as delivery vehicles for radioimmunotherapy of fungal infection. *Clin Cancer Res* 13(18 Pt 2): 5629s–5635s.

Datta K, Bartlett KH, Baer R, Byrnes E, Galanis E, Heitman J, Hoang L, Leslie MJ, MacDougall L, Magill SS, Morshed MG, Marr KA, 2009. Spread of *Cryptococcus gattii* into Pacific Northwest region of the United States. *Emerging Infect Dis* 15: 1185–1191.

de Hoog GS, Guarro J, Gené J, Figueras MJ, 2000. *Atlas of Clinical Fungi*, 2nd ed. Centralbureau voor Schimmelcultures, Baarn, The Netherlands (CD version 2004.11).

Dubey A, Patwardhan RV, Sampth S, Santosh V, Kolluri S, Nanda A, 2005. Intracranial fungal granuloma: Analysis of 40 patients and review of the literature. *Surg Neurol* 63: 254–260.

Duncan C, Stephen C, Campbell J, 2006. Clinical characteristics and predictors of mortality for *Cryptococcus gattii* infection in dogs and cats of southwestern British Columbia. *Can Vet J* 47: 993–998.

Ellis DH, Pfeiffer TJ, 1990. Ecology, life cycle, and infectious propagule of *Cryptococcus neoformans*. *Lancet* 336: 923–925.

Fell JW, Boekhout T, Fonseca A, Scorzetti G, Statzell-Tallman A, 2000. Biodiversity and systematics of basidiomycetous yeasts as determined by large-subunit rDNA D1/D2 domain sequence analysis. *Int J System Evol Microbiol* 50: 1351–1371.

Fishman JA, Gonzalez RG, Pranada JA, 2008. Case Records of the Massachusetts General Hospital. Case 11–2008. A 45-year-old man with changes in mental status after liver transplantation. *N Engl J Med* 358: 1604–1613.

Fraser JA, Giles SS, Wenink EC, Geunes-Boyer SG, Wright JR, Diezmann S, Allen A, Stajich JE, Dietrich FS, Perfect JR, Heitman J, 2005. Same-sex mating and the origin of the Vancouver Island *Cryptococcus gattii* outbreak. *Nature* 437: 1360–1364.

Freydiere A-M, Guinet R, Boiron P, 2001. Yeast identification in the clinical microbiology laboratory: Phenotypical methods. *Med Mycol* 39: 9–33.

Galanis E, Macdougall L, 2010. Epidemiology of *Cryptococcus gattii*, British Columbia, Canada, 1999–2007. *Emerging Infect Dis* 16: 251–257.

Garcia-Hermoso D, Janbon G, Dromer F, 1999. Epidemiological evidence for dormant *Cryptococcus neoformans* infection. *J Clin Microbiol* 37: 3204–3209.

Garcia-Rivera J, Tucker SC, Feldmesser M, Williamson PR, Casadevall A, 2005. Laccase expression in murine pulmonary *Cryptococcus neoformans* infection. *Infect Immun* 73: 3124–3127.

Gatti F, Eeckels R, 1970. An atypical strain of *Cryptococcus neoformans* (San Felice) Vuillemin 1894. I. Description of the disease and of the strain. *Ann Soc Belges Med Trop Parasitol Mycol* 50: 689–693.

Goldman DL, Khine H, Abadi J, Lindenberg DJ, Pirofski L, Niang R, Casadevall A, 2001. Serologic evidence for *Cryptococcus neoformans* infection in early childhood. *Pediatrics* 107: e66.

Guillot L, Carroll SF, Homer R, Qureshi ST, 2008. Enhanced innate immune responsiveness to pulmonary *Cryptococcus neoformans* infection is associated with resistance to progressive infection. *Infect Immun* 76: 4745–4756.

Hajjeh RA, Conn LA, Stephens DS, Buaghman W, Hamill R, Graviss E, Pappas PG, Thomas C, Reingold A, Rothrock G, Hutwagner LC, Schuchat A, Brandt ME, Pinner RW, 1999. Cryptococcosis: Population-based multistate active surveillance and risk factors in human immunodeficiency virus-infected persons. *Clin Infect Dis* 179: 449–454.

Hazen KC, Howell SA, 2007. *Candida, Cryptococcus*, and Other Yeasts of Medical Importance, pp. 1762–1788, *in*: Murray PR, Baron EJ, Jorgenson JH, Landry ML, Pfaller MA (eds.), *Manual of Clinical Microbiology*, 9th ed. ASM Press, Washington, DC.

Hernandez Y, Arora S, Erb-Downward JR, McDonald RA, Toews GB, Huffnagle GB, 2005. Distinct roles for IL-4 and IL-10 in regulating T2 immunity during allergic bronchopulmonary mycosis. *J Immunol* 174: 1027–1036.

Jacobson ES, 2000. Pathogenic roles for fungal melanins. *Clin Microbiol Rev* 13: 708–717.

Jacobson ES, Hong JD, 1997. Redox buffering by melanin and Fe(II) in *Cryptococcus neoformans*. *J Bacteriol* 179: 5340–5346.

Jain N, Wickes BL, Keller SM, Fu J, Casadevall A, Jain P, Ragan MA, Banerjee U, Fries BC, 2005. Molecular epidemiology of clinical *Cryptococcus neoformans* strains from India. *J Clin Microbiol* 43: 5733–5742.

Jung WH, Kronstad JW, 2008. Iron and fungal pathogenesis: A case study with *Cryptococcus neoformans*. *Cell Microbiol* 10: 277–284.

Jung WH, Sham A, Lian T, Singh A, Kosman DJ, Kronstad JW, 2008. Iron source preference and regulation of iron uptake in *Cryptococcus neoformans*. *PLoS Pathog* 4: e45.

Kambugu A, Meya DB, Rhein J, O'Brien M, Janoff EN, Ronald AR, Kamya MR, Mayanja-Kizza H, Sande MA, Bohjanen PR, Boulware DR, 2008. Outcomes of cryptococcal meningitis in Uganda before and after the availability of highly active antiretroviral therapy. *Clin Infect Dis* 46: 1694–1701.

Kidd SE, Hagen F, Tscharke RL, Huynh M, Bartlett KH, Fyfe M, MacDougall L, Boekhout T, Kwon-Chung KJ, Meyer W, 2004. A rare genotype of *Cryptococcus gattii* caused the cryptococcosis outbreak on Vancouver Island (British Columbia, Canada). *Proc Natl Acad Sci USA* 101: 17258–17263.

Korfel A, Menssen HD, Schwartz S, Thiel E, 1998. Cryptococcosis in Hodgkin's disease: Description of two cases and review of the literature. *Ann Hematol* 76: 283–286.

Krockenberger MB, Canfield PJ, Malik R, 2003. *Cryptococcus neoformans* var. *gattii* in the koala (*Phascolarctos cinereus*): A review of 43 cases of cryptococcosis. *Med Mycol* 41: 225–234.

Kwon-Chung KJ, 1976. A new species of *Filobasidiella*, the sexual state of *Cryptococcus neoformans* B and C serotypes. *Mycologia* 68: 943–946.

Kwon-Chung KJ, 1975. A new genus, *Filobasidiella*, the perfect state of *Cryptococcus neoformans*. *Mycologia* 67: 1197–2000.

Kwon-Chung KJ, Bennett JE, 1992. *Medical Mycology*. Lea and Febiger, Philadelphia, pp. 428, 435.

Kwon-Chung KJ, Polacheck I, Bennett JE, 1982. Improved diagnostic medium for separation of *Cryptococcus neoformans* var. *neoformans* (serotypes A and D) and *Cryptococcus neoformans* var. *gattii* (serotypes B and C). *J Clin Microbiol* 15: 535–537.

Lagos R, Munoz A, Levine MM, Watson W, Chang I, Paradiso P, 2009. Immunology of combining CRM (197) conjugates for *Streptococcus pneumoniae, Neisseria meningitis* and *Haemophilus influenzae* in Chilean infants. *Vaccine* 27: 2299–2305.

Larone DH, 2002. *Medically Important Fungi, A Guide to Identification*, 4th ed. ASM Press, Washington, DC.

Lin X, Heitman J, 2006. The biology of the *Cryptococcus neoformans* species complex. *Annu Rev Microbiol* 60: 69–105.

Lin X, Patel S, Litvintseva AP, Floyd A, Mitchell TG, Heitman J, 2009. Diploids in the *Cryptococcus neoformans* serotype A population homozygous for the alpha mating type originate via unisexual mating. *PLoS Pathog* 5: e1000283.

Lin X, Huang JC, Mitchell TG, Heitman J. 2006. Virulence attributes and hyphal growth of *C. neoformans* are quantitative traits and the *MAT*α allele enhances filamentation. *PLoS Genetics* 2: e187.

Lin X, Hull CM, Heitman J, 2005. Sexual reproduction between partners of the same mating type in *Cryptococcus neoformans*. *Nature* 434: 1017–1021.

Litvintseva AP, Lin X, Templeton I, Heitman J, Mitchell TG, 2007. Many globally isolated AD hybrid strains of *Cryptococcus neoformans* originated in Africa. *PLoS Pathog* 3(8): e114.

Litvintseva AP, Thakur R, Vilgalys R, Mitchell TG, 2006. Multilocus sequence typing reveals three genetic subpopulations of *Cryptococcus neoformans* var *grubii* (serotype A), including a unique population in Botswana. *Genetics* 172: 2223–2238.

Lurie HI, Shadomy HJ, 1971. Morphological variations of a hypha-forming strain of *Cryptococcus neoformans* (Coward strain) in tissues of mice. *Sabouraudia* 9: 10–14.

McCarthy KM, Morgan J, Wannemuehler KA, Mirza SA, Gould SM, Mhlongo N, Moeng P, Maloba BR, Crewe-Brown HH, Brandt ME, Hajjeh RA, 2006. Population-based surveillance for cryptococcosis in an antiretroviral-naive South African province with a high HIV seroprevalence. *AIDS* 20: 2199–2006.

Meyer W, Aanensen DM, Boekhout T, Cogliati M, Diaz MR, Esposto MC, Fisher M, Gilgado F, Hagen F, Kaocharoen S, Litvintseva AP, Mitchell TG, Simwami SP, Trilles L, Viviani MA, Kwon-Chung J, 2009. Consensus multi-locus sequence typing scheme for *Cryptococcus neoformans* and *Cryptococcus gattii*. *Med Mycol* 47: 561–570.

Meyer W, Castañeda A, Jackson S, Huynh M, Castañeda E, IberoAmerican Cryptococcal Study Group, 2003. Molecular typing of IberoAmerican *Cryptococcus neoformans* isolates. *Emerging Infect Dis* 9: 189–195.

Meyer W, Marszewska K, Amirmostofian M, Igreja RP, Hardtke C, Methling K, Viviani MA, Chindamporn A, Sukroongreung S, John MA, Ellis DH, Sorrell TC, 1999. Molecular typing of global isolates of *Cryptococcus neoformans* var. *neoformans* by polymerase chain reaction fingerprinting and randomly amplified polymorphic DNA—a pilot study to standardize techniques on which to base a detailed epidemiological survey. *Electrophoresis* 20: 1790–1799.

Miller WG, Padhye AA, van Bonn W, Jensen E, Brandt ME, Ridgway SH, 2002. Cryptococcosis in a bottlenose dolphin (*Tursiops truncatus*) caused by *Cryptococcus neoformans* var. *gattii*. *J. Clin Microbiol* 40: 721–724.

Mirza SA, Phelan M, Rimland D, Graviss E, Hamill R, Brandt ME, Gardner T, Sattah M, de Leon GP, Baughman W, Hajjeh RA, 2003. The changing epidemiology of cryptococcosis: An update from population-based active surveillance in 2 large metropolitan areas, 1992–2000. *Clin Infect Dis* 36: 789–794.

MMWR, 2010. Emergence of *Cryptococcus gattii*—Pacific Northwest, 2004–2010. *MMWR Morb Mortal Wkly Rep* Jul 23; 59: 865–868.

Monari C, Kozel TR, Paganelli F, Pericolini E, Perito S, Bistoni F, Casadevall A, Vecchiarelli A, 2006. Microbial immune suppression mediated by direct engagement of inhibitory Fc receptor. *J Immunol* 177: 6842–6851.

Morgan J, McCarthy KM, Gould S, Fan K, Arthington-Skaggs B, Iqbal N, Stamey K, Hajjeh RA, Brandt ME, 2006. *Cryptococcus gattii* infection: Characteristics and epidemiology of cases identified in a South African province with high HIV seroprevalence. *Clin Infect Dis* 43: 1077–1080.

Neuville S, Dromer F, Morin O, Dupont B, Ronin O, Lortholary O, 2003. Primary cutaneous cryptococcosis: A distinct clinical entity. *Clin Infect Dis* 36: 337–347.

Ngamskulrungroj P, Gilgado F, Faganello J, Litvintseva AP, Leal AL, Tsui KM, Mitchell TG, Vainstein MH, Meyer W, 2009. Genetic diversity of the *Cryptococcus* species complex suggests that *Cryptococcus gattii* deserves to have varieties. *PLoS One* 4(6): e5862.

Neilson JB, Ivey MH, Bulmer GS, 1978. *Cryptococcus neoformans*: pseudohyphal forms surviving culture with *Acanthamoeba polyphaga*. *Infect Immun* 20: 262–266.

Nielsen K, de Obaldia AL, Heitman J, 2007. *Cryptococcus neoformans* mates on pigeon guano: Implications for the realized ecological niche and globalization. *Eukaryot Cell* 6: 949–959.

O'Brien CR, Krockenberger MB, Martin P, Wigney DI, Malik R, 2006. Long-term outcome of therapy for 59 cats and 11 dogs with cryptococcosis. *Aust Vet J* 84: 384–392.

O'Brien CR, Krockenberger MB, Wigney DI, Martin P, Malik R, 2004. Retrospective study of feline and canine cryptococcosis in Australia from 1981 to 2001: 195 cases. *Med Mycol* 42: 449–460.

Oppenheim JJ, Horuk R, 2001. Chapter 11, Chemokines, *in:* Parslow TG, Stites DP, Terr AI, Imboden JB (eds.), *Medical Immunology*, 10th ed. Lange Medical Books/McGraw-Hill Medical Publishing Division, New York.

Panepinto JC, Williamson PR, 2006. Intersection of fungal fitness and virulence in *Cryptococcus neoformans*. *FEMS Yeast Res* 6: 489–498.

Park BJ, Wannemuehler KA, Marston BJ, Govender N, Pappas PG, Chiller TM, 2009. Estimation of the current global burden of cryptococcal meningitis among persons living with HIV/AIDS. *AIDS* 23: 525–530.

Perfect JR, 2005. Chapter 261, *Cryptococcus neoformans*, *in:* Mandell GL, Bennett JE, Dolin R. (eds.), *Mandell, Douglas, and Bennett's Principles and Practice of Infectious Diseases*, 6th ed., Vol 2. Elsevier-Churchill Livingstone, London.

Perfect JR, Dismukes W, Dromer F, Goldman DL, Graybill JR, Hamill RJ, Harrison TS, Larsen RA, Lortholary O, Nguten MH, Pappas PG, Powderly WG, Singh N, Sobel JD, Sorrell TC, 2010. Clinical practice guidelines for the management of cryptococcal disease: 2010 Update by the Infectious Diseases Society of America. *Clin Infect Dis* 50: 291–322.

Rex JH, Larsen RA, Dismukes WE, Cloud GA, Bennett JE, 1993. Catastrophic visual loss due to *Cryptococcus neoformans* meningitis. *Medicine (Baltimore)* 72: 207–214.

Rosas AL, Nosanchuk JD, Feldmesser M, Cox GM, McDade HC, Casadevall A, 2000. Synthesis of polymerized melanin by *Cryptococcus neoformans* in infected rodents. *Infect Immun* 68: 2845–2853.

Ruiz AA, Neilson JB, Bulmer GS, 1982. Control of *Cryptococcus neoformans* in nature by biotic factors. *Sabouraudia* 20: 21–29.

Salas SD, Bennett JE, Kwon-Chung KJ, Perfect JR, Williamson PR, 1996. Effect of the laccase gene CNLAC1, on virulence of *Cryptococcus neoformans*. *J Exp Med* 184: 377–386.

Sax PE, Mattia AR, 1994. Case records of the Massachusetts General Hospital. Weekly clinicopathological exercises. Case 7–1994. A 55-year-old heart transplant recipient with a tender, enlarged prostate gland. *N Engl J Med* 330: 490–496.

Schubert S, Lebeau A, Zell R, Goebel F, 1997. Cavitary cryptococcoma of the lungs and meningitis by *Cryptococcus neoformans* in a patient with AIDS. *Eur J Med Res* 2: 173–176.

Shadomy HJ, Utz JP, 1966. Preliminary studies on a hypha forming mutant of *Cryptococcus neoformans*. *Mycologia* 58: 383–390.

Shoham S, Levitz SM, 2005. The immune response to fungal infections. *Br J Haematol* 129: 569–582.

Silveira FP, Husain S, 2007. Fungal infections in solid organ transplantation. *Med Mycol* 45: 305–320.

Silveira FP, Husain S, Kwak EJ, Linden PK, Marcos A, Shapiro R, Fontes P, Marsh JW, de Vera M, Tom K, Thai N, Tan HP, Basu A, Soltys K, Paterson DL, 2007. Cryptococcosis in liver and kidney transplant recipients receiving anti-thymocyte globulin or alemtuzumab. *Transplant Infect Dis* 9: 22–27.

Singh N, Perfect JR, 2007. Immune reconstitution syndrome associated with opportunistic mycoses. *Lancet Infect Dis* 7: 395–401.

Sorrell TC, Chen SC, Ruma P, Meyer W, Pfeiffer TJ, Ellis DH, Brownlee AG, 1996. Concordance of clinical and environmental isolates of *Cryptococcus neoformans* var. *gattii* by random amplification of polymorphic DNA analysis and PCR fingerprinting. *J Clin Microbiol* 34: 1253–1260.

Springer DJ, Chaturvedi V, 2010. Projecting global occurrence of *Cryptococcus gattii*. *Emerging Infect Dis* 16: 14–20.

Steenbergen JN, Shuman HA, Casadevall A, 2001. *Cryptococcus neoformans* interactions with amoebae suggest an explanation for its virulence and intracellular pathogenic strategy in macrophages. *Proc Natl Acad Sci USA* 98: 15245–15250.

Stockman L, Roberts GD, 1983. Corrected version specificity of the latex test for cryptococcal antigen: A rapid, simple method for eliminating interference factors. *J Clin Microbiol* 17: 945–947.

Vanbreuseghem R, Takashio M, 1970. An atypical strain of *Cryptococcus neoformans* (San Felice) Vuillemin 1894. II. *Cryptococcus neoformans* var. *gattii* var. nov. *Ann Soc Belge Med Trop* 50: 695–702.

Voelz K, May RC, 2010. Cryptococcal interactions with the host immune system. *Eukaryot Cell* 9: 835–846.

Voelz K, Lammas DA, May RC, 2009. Cytokine signaling regulates the outcome of intracellular macrophage parasitism by *Cryptococcus neoformans*. *Infect Immun* 77: 3450–3457.

Wickes BL, 2002. The role of mating type and morphology in Cryptococcus neoformans pathogenesis. *Int J Med Microbiol* 292: 313–329.

Wong B, Perfect JR, Beggs S, Wright KA, 1990. Production of the hexitol D-mannitol by *Cryptococcus neoformans* in vitro and in rabbits with experimental meningitis. *Infect Immun* 58: 1664–1670.

Zaragoza O, Casadevall A, 2004. Experimental modulation of capsule size in *Cryptococcus neoformans*. *Biol Proced Online* 6: 10–15.

Zaragoza O, Rodrigues ML, de Jesus M, Frases S, Dadachova E, Casadevall A, 2009. The capsule of fungal pathogen *Cryptococcus neoformans*. *Adv Appl Microbiol* 68: 133–216.

Zhu X, Williamson PR, 2004. Role of laccase in the biology and virulence of *Cryptococcus neoformans*. *FEMS Yeast Res* 5: 1–10.

WEBSITES OF INTEREST OR CITED

Stedman's On-Line Medical Dictionary with free access to Stedman's Medical Dictionary, 27th Edition at the URL http://www.stedmans.com/section.cfm/45.

QUESTIONS

The Answer Key to multiple choice questions may be found at the end of the book.

1. A CSF specimen comes to the laboratory for workup in mycology. Which of the following tests would be relevant for this specimen?
   A. India ink prep of centrifuged pellet
   B. *Cryptococcus* latex agglutination test for antigen on supernatant after centrifugation
   C. Culture of centrifuged pellet on SDA-Emmons
   D. Culture of centrifuged pellet on BHI agar
   E. All of the above

2. Rank the following tests in descending order of their speed as presumptive tests for *Cryptococcus neoformans* and *C. gattii*.
   A. Birdseed agar

B. Caffeic acid disc test for phenoloxidase

C. India ink test on CSF

D. Rapid urease test

3. Which is true about both *Cryptococcus neoformans* and *C. gattii?*

A. Prefer certain trees for their habitat

B. Monomorphic yeasts

C. Yeast and hyphal forms in life cycle

D. Evenly distributed, worldwide

E. All of the above

4. *Cryptococcus neoformans* is inhibited by canavanine, whereas *C. gattii* is not and can use glycine as a carbon source. This is the principle of what differential medium?

A. CGB

B. Czapek–Dox

C. Lowenstein–Jensen

D. Modified Leonian's agar

5. Why is the cryptococcal capsule considered to be a virulence factor?

A. It is antiphagocytic, tolerogenic, and interferes with macrophage function.

B. It is the only fungal pathogen of the central nervous system.

C. The polysaccharide circulates during infection.

D. There is no clinically useful test for antibodies against the capsular polysaccharide.

E. All of the above.

6. Which best describes the ecologic niche of *Cryptococcus neoformans?*

A. Despite the worldwide occurrence of cryptococcosis, the causative agent cannot be isolated from the environment.

B. Normal inhabitants of the skin and mucosae of warm-blooded animals and humans.

C. Worldwide in soil mixed with bat droppings.

D. Worldwide in soil mixed with pigeon droppings

7. Where in the central nervous system is *C. neoformans* found? What is the difference in clinical presentation between *C. neoformans* and *C. gattii* in cryptococcosis?

8. What is the evidence that basidiospores are infectious propagules of *Cryptococcus* species?

9. What events can explain how *C. neoformans* and *C. gattii* have expanded their ecologic niches?

10. A campaign to treat AIDS-cryptococcosis in Africa with fluconazole is under way. In this effort what consideration should be given to IRIS? How should the initiation of antiretroviral therapy and fluconazole treatment of cryptococcosis be staged to minimize the effect of IRIS?

Chapter 13

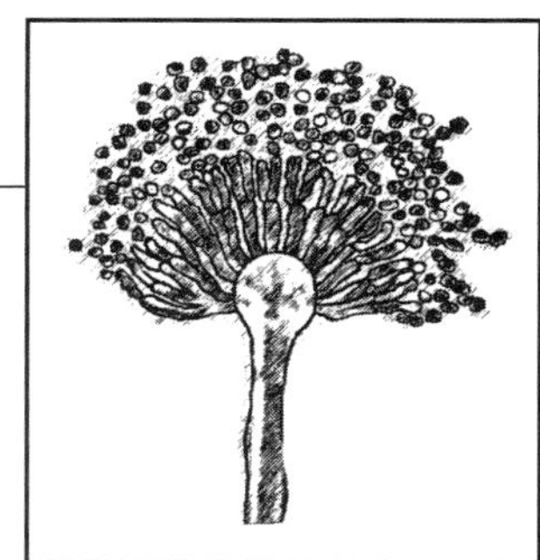

Pneumocystosis

13.1 PNEUMOCYSTOSIS-AT-A-GLANCE

- *Introduction/Disease Definition.* *Pneumocystis jirovecii* pneumonia (PCP) affects immunocompromised adults and children especially persons living with AIDS. Symptoms are shortness of breath, fever, and nonproductive cough. In HIV-positive persons symptoms can last weeks to months. The acute phase clinical picture includes dyspnea, tachypnea, and hypoxemia.
- *Etiologic Agent.* *Pneumocystis jirovecii* is an atypical ascomycete fungus.
- *Ecologic Niche.* Worldwide, *P. jirovecii* is an endogenous opportunist restricted to humans; it cannot be continuously cultured in vitro; and there is no known environmental reservoir.
- *Epidemiology.* PCP is an AIDS-associated opportunistic infection that mainly afflicts HIV-positive persons when their $CD4^{+}$ T-lymphocyte cell count falls <200 μL.
- *Risk Groups/Factors.* Other persons at risk are those receiving cancer chemotherapy or immunosuppressive therapy for autoimmune collagen-vascular diseases, for stem cell and solid organ transplants, and in malnourished children.
- *Transmission.* Person–person via inhalation of droplet aerosols.
- *Determinants of Pathogenicity.* *P. jirovecii* adheres to type 1 alveolar cells and provokes the host inflammatory response. Antigenic variation in the major surface glycoprotein evades immune defenses
- *Clinical Forms.* Pneumonia (PCP). The fungus proliferates extracellularly, filling alveolar spaces with a foamy exudate, resulting in reduced oxygenation, a thickened *interstitium*[1] leading to fibrosis, and respiratory failure. The diffuse lung damage resembles adult respiratory distress syndrome.
- *Therapy.* Trimethoprim-sulfamethoxazole (TMP-SMX) is the drug of choice for both prophylaxis and primary therapy.
- *Laboratory Detection, Recovery, and Identification.* *Pneumocystis jirovecii* is unable to be continuously cultivated in vitro; diagnosis is made via microscopy of stained smears of induced sputum, BAL fluid, or biopsy; PCR diagnostics.

13.2 INTRODUCTION/DISEASE DEFINITION

Pneumocystis jirovecii is an atypical fungus that cannot be continuously cultivated in vitro. It colonizes the lower respiratory tract of humans and, in the setting of serious immune compromise, causes *Pneumocystis* pneumonia (PCP). The symptoms of PCP are shortness of breath, fever, and a nonproductive cough. The symptoms in HIV-positive patients are subtle, lasting weeks to months. Acutely ill patients have dyspnea, tachypnea, and hypoxemia. *P. jirovecii* adheres to type 1 alveolar cells, and when conditions in the host are permissive, a mix of large numbers of adherent trophic forms and cysts (= asci) multiply. The resulting infection evokes an inflammatory response, filling the alveoli with a foamy exudate. The result is decreased oxygenation, a thickened interstitium, and, eventually, respiratory failure. This diffuse lung damage resembles adult respiratory distress syndrome.

[1] Interstitium. *Definition*: The matrix or supporting tissue of an organ.

Fundamental Medical Mycology, First Edition. By Errol Reiss, H. Jean Shadomy and G. Marshall Lyon, III.
© 2012 Wiley-Blackwell. Published 2012 by John Wiley & Sons, Inc.

Pneumocystis jirovecii, in PCP, does not invade the lung parenchyma. The host inflammatory response varies according to the primary disease: inconspicuous in persons living with AIDS and, in less compromised patients, the PMN inflammatory response fails to protect but instead increases alveolar damage and respiratory impairment. Mortality in the first *P. jirovecii* infection of persons living with AIDS is 10–20%, and 30–60% in cancer patients or transplant recipients. Extrapulmonary dissemination is rare, seen only in overwhelming infection or severe immunosuppression, and is associated with aerosolized pentamidine prophylaxis.

There are several important unknowns about *Pneumocystis* species including:

- *Why they cannot be continuously cultivated outside the mammalian lung*
- *Location of the ecologic niche in nature, if any*
- *The basis for narrow host specificity* (each *Pneumocystis* species is host-specific)

PCP remains the most common opportunistic infection in persons with AIDS. Other groups at risk for PCP are those being treated with cytotoxic cancer chemotherapy or immunosuppressive therapy for maintenance of stem cell or organ transplants or for collagen-vascular diseases.

Humans are the reservoir of infection and the mode and route of infection is human–human via inhalation of droplet aerosols. There is no known environmental reservoir. Asymptomatic carriage of *P. jirovecii* is not universal but transient among healthy normal adults and children. Young children, elderly adults, persons with chronic lung disease, and HIV-positive persons are the likely reservoirs and are typically asymptomatic. When, in HIV-positive persons, the $CD4^+$ T-lymphocyte count falls below 200/μL, *P. jirovecii* can multiply causing PCP.

Seropositivity is nearly universal with primary infection occurring in the first year of life. *Pneumocystis* DNA can be detected in 20% of otherwise healthy adults (Peterson and Cushion, 2005). As population data accumulate, *P. jirovecii* may be considered a member of the normal microbiota, but carriage of *P. jirovecii* in healthy adults is transient so that disease occurs only when host resistance and the immune system are compromised.

Azole and polyene antifungal agents are not active against *P. jirovecii*. Lacking ergosterol in its cell membranes, *P. jirovecii* has another sterol, pneumocysterol, which is methylated at position Δ14 so that lanosterol demethylase inhibitors such as the azole antifungal drugs are not active against *P. jirovecii*. Primary therapy consists of trimethoprim-sulfamethoxazole (TMP-SMX).

13.3 CASE PRESENTATION

Fever and Shortness of Breath in a 49-yr-old Man Living with AIDS
(Kovacs et al., 2001)

History

A 49-yr-old Hispanic man was diagnosed as HIV-positive in 1995 and received antiretroviral therapy but continued to experience poor viral control. In 2001 he was diagnosed with PCP and was treated with TMP-SMX until he developed a rash. In April 2001 the patient presented in the emergency department with fever and shortness of breath of 2 weeks duration. He was taking combination antiretroviral therapy (ART), as well as ITC (antifungal but not anti-PCP), atovaquone (anti-PCP), and aerosolized pentamidine (anti-PCP). Compliance with therapy was judged questionable. The shortness of breath was treated with antibiotics but 2 weeks later he reported no improvement.

Clinical and Laboratory Findings

The chest radiograph showed diffuse interstitial infiltrates and a mediastinal mass. The $CD4^+$ T-lymphocyte count was 4 cells/μL, with a viral load of 250,000 copies/μL. Biopsy of the mediastinal mass revealed a large B-cell lymphoma. A BAL smear failed to show any pathogen. The patient developed respiratory failure requiring a ventilator and was transferred to the NIH Clinical Center for further treatment.

Therapy and Clinical Course

A chest radiograph showed diffuse infiltrates, both interstitial and alveolar. During the next 4 days he received therapy for the B-cell lymphoma, but his respiratory status worsened. BAL was performed again. Many *P. jirovecii* trophic forms were seen by DFA (direct fluorescent monoclonal antibody). PCR showed 18,460 copies of the *P. jirovecii* major surface glycoprotein/20 μL of BAL fluid. He received IV pentamidine but did not improve and was switched to IV TMP-SMX. The DFA and PCR remained positive but slow improvement occurred over the next 9 days. The patient then developed a rash and was switched back to IV pentamidine for 2 weeks. After 5 weeks of therapy PCR found only 6 copies/20 μL of the *P. jirovecii* gene in BAL fluid. The patient was weaned after 6 weeks of ventilator support and his therapy for lymphoma was continued.

Comments

The patient was receiving secondary prophylaxis with atovaquone and aerosolized pentamidine, which are not as effective as TMP-SMX. Independent risk factors for decreased 2 month survival are recurrent PCP, the need for ventilator support, and therapy other than TMP-SMX. While most clinicians will initiate empiric anti-*Pneumocystis* therapy in a patient such as this, it is important to confirm *Pneumocystis* as quickly as possible because many other infections can present like pneumocystosis. This is especially important in a patient who became allergic to the preferred therapy.

Diagnosis

Diagnosis was made based on microbiologic examination of induced sputum and BAL fluid smears, or by histopathology of transbronchial or open lung biopsy specimens.

Differential Diagnosis Many diseases have a similar presentation, including mycobacterial, fungal, viral, or bacterial pneumonias, heart failure, Kaposi's sarcoma, and pulmonary emboli.

13.4 ETIOLOGIC AGENT

The etiologic agent of human pneumocystosis is *Pneumocystis jirovecii*.

Pneumocystis jirovecii Frenkel [as "*jiroveci*;"], *Nat Cancer Inst Monograph* 43: 16 (1976).

Position in classification: *Pneumocystis* is one of the most enigmatic microbes known. The history of its discovery and classification is found in Redhead et al. (2006). It is a basal ascomycete in the *Ascomycota*, subphylum *Taphrinamycotina* in the order *Pneumocystidales*. The molecular biology of *Pneumocystis* is summarized by Beck and Cushion (2009).

Proof that Pneumocystis jirovecii *Is a Fungus*

The phylogenetic position of *Pneumocystis* species is based on sequence analysis of 18S rDNA, other nuclear genes (e.g., translation elongation factor 3), and mitochondrial genes. This analysis places *Pneumocystis* species in an early divergent line in the kingdom Fungi near the base dividing basidiomycete and ascomycete lineages. Phylogenetic trees of 18S rDNA show neighboring branches contain the fission yeast *Schizosaccharomyces pombe* and the dimorphic plant pathogen *Taphrina deformans*.

Pneumocystis jirovecii is an atypical fungus because:

- It is unable to be continuously cultivated in vitro, unlike most pathogenic fungi.
- *Pneumocystis* species are host specific and do not cross-infect other species.
- No environmental niche has been found.
- PCP is a *communicable disease* via the respiratory route.
- It is not susceptible to amphotericin B or to azole antifungal drugs.

Pneumocystis *Life Cycle*

Microscopic observations of *P. jirovecii* reveal two life-cycle forms—trophic forms and cysts. Both are found in the human lung (cysts will be called asci in keeping with fungal nomenclature). No environmental niche is known. During PCP trophic forms outnumber asci by a ratio of 10:1. As well, trophic forms are believed to be the infectious propagules, spread by droplet aerosols. They are haploid, mononuclear, 1–4 μm in diameter, and contain a single mitochondrion (Thomas and Limper, 2007). Their shape is ameboid with cytoplasmic projections known as filopodia by which they adhere to type 1 pneumocytes (Aliouat-Denis et al., 2009). It is hypothesized, but there is doubt, that trophic forms are capable of binary fission in the alveolar space. Otherwise reproduction may occur only through mating and ascus formation, resulting in a thick-walled ascus (8–10 μm diameter).

Morphologic (ultrastructural) and molecular genetic evidence exists for sexual reproduction, which occurs in the alveolar space. The *Pneumocystis* genome has genes encoding receptors for a and α mating type pheromones, supporting the existence of mating types and that conjugation of *Pneumocystis* trophic forms occurs in vivo (reviewed by Aliouat-Denis et al., 2009). The sequence of events in the sexual cycle is illustrated in Fig. 13.1. Haploid trophic forms of opposite mating types conjugate (plasmogamy), forming a diploid zygote (karyogamy), which is a round, thin-walled cell. Meiosis produces 4 haploid nuclei—the pre-ascus form. Mitosis results in 8 haploid ascospores. The ascus matures in three stages corresponding to the above events: early, intermediate, and late, containing 2, 4, and 8 ascospores. The ascus wall thickens, as it matures, reaching a size of 8–10 μm in diameter. Eight individual ascospores are clearly delineated. Each contains a single nucleus, a rounded mitochondrion, rough endoplasmic reticulum, and numerous ribosomes. Ascospores may be spherical, banana-shaped, or irregular. Once mature, ascospores are extruded from the ascus (cyst) probably through a pore. Expulsion of ascospores may be facilitated by residual fluid from the ascus, leaving behind a collapsed, empty

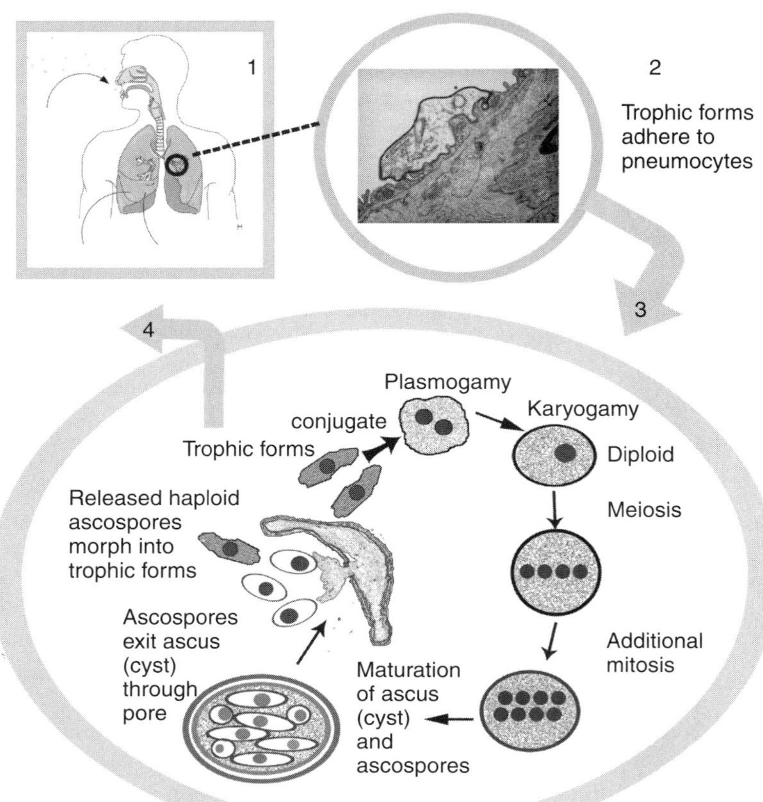

Figure 13.1 The *Pneumocystis* life cycle. (Please see the text for information about the intracellular organelles, size, and other properties of trophic and ascus forms.) 1, Droplet aerosols spread the fungus from person to person. 2, Trophic forms occur in the alveolar space, but to cause disease they must bind to type 1 pneumocytes, the outer layer of the lung epithelium. 3, Mating occurs in the alveolar space by the conjugation of trophic forms of opposite mating types. After plasmogamy and karyogamy, the zygote, a round, thin-walled cell, undergoes meiosis. Mitosis follows, resulting in 8 nuclei which develop into ascospores. The ascus (cyst) wall thickens as it matures with 8 ascospores clearly delineated. They may appear round or elongate depending on the plane of section. Once mature, ascospores are extruded from the ascus (cyst) probably through a pore. Expulsion of ascospores leaves a collapsed, empty ascus (cyst) remnant. The ascospores differentiate into 8 haploid trophic forms. 4, Trophic forms divide by binary fission in the alveolar space. *Inset*: A trophic form adheres to a type 1 pneumocyte.
Source: Limper et al. (1997). Used with permissions from Dr. Andrew A. Limper and Elsevier Publishers, Ltd. Copyright © 1997.

ascus remnant. The ascospores then differentiate into 8 haploid trophic forms (Itatani, 1994; Aliouat-Denis et al., 2009). They escape into the extracellular alveolar space and may resume vegetative growth and reproduce by fission as trophic forms. Trophic forms must attach to the type 1 pneumocyte of the alveolar epithelium, an essential step for reproduction and initiation of pathogenesis. Type 1 pneumocytes form the structure of the alveolar wall; but *P. jirovecii* remains extracellular.

Strict Host Specificity

Pneumocystis species display strict host specificity and there is no known cross-infectivity among mammals; for example, human *Pneumocystis* cannot infect the rat and vice versa. Phylogeny of *Pneumocystis* from 18 primate species confirms this host specificity. Phylogenetic analysis of mtLSU rDNA and dihydropteroate synthase (DHPS) gene showed that no nonhuman primate harbored *P. jirovecii*. Genetic divergence in primate-derived *Pneumocystis* organisms varied according to the phylogenetic divergence existing among their host species. *P. carinii* is reserved for the rat *Pneumocystis* species.

13.5 GEOGRAPHIC DISTRIBUTION/ECOLOGIC NICHE

Pneumocystis jirovecii is worldwide in distribution and, so far, the mammalian lung appears to be its sole environmental niche: there is no known free-living form. Some reports of detecting *Pneumocystis* DNA in air surrounding apple orchards or on the surface of pond water have not been substantiated, and it is now becoming apparent that there may be no ecologic niche outside the human host. The reservoir for *P. jirovecii* is an unsettled question. Healthy human adults may be transiently colonized. Silent reservoirs may occur in infants who acquire a primary *Pneumocystis* infection very early in life, in patients with chronic respiratory disease, and in elderly adults. *Pneumocystis* airway colonization is prevalent in patients with chronic obstructive pulmonary disease (COPD) reaching 37–55% (reviewed by Catherinot et al., 2010).

Lungs of people in Santiago, Chile, who died of violent causes and of diseases causing a rapid demise in the street were tested by nested PCR and *P. jirovecii* was identified in 50 (64.9%) of 77 individuals (Ponce et al., 2010). The PCR results were supported by immunofluorescent microscopic analysis. These findings allow for the hypothesis that a mild *P. jirovecii* pulmonary infection is prevalent in more than half of the general adult population and that immune-normal adults develop frequent self-limited reinfections throughout life. Pending confirmation by other similar investigations, the general population would have to be regarded as the reservoir for circulation of *P. jirovecii* to susceptible individuals.

13.6 EPIDEMIOLOGY

Humans show near universal seropositivity, acquired at an early age. Transient colonization occurs even among healthy adults, and *P. jirovecii* may be considered part of normal human microbiota. The occurrence of different *P. jirovecii* genotypes in recurrent infections and clusters of PCP cases in urban communities indicates that airborne transmission results in temporary asymptomatic carriage or, if the host is immunocompromised, the initiation of PCP. Colonization is detectable by PCR even when no *P. jirovecii* is seen on respiratory specimens. It is infrequent in healthy adults and increased where there is an underlying chronic illness or one of various respiratory disorders (e.g., COPD, smokers).

Children, infants especially, appear to be an important reservoir of *P. jirovecii*, based on observations that *P. jirovecii* is common in normal infants with and without respiratory symptoms. Primary exposure to *P. jirovecii* is widespread, occurring early in life as measured by seroprevalence of 85% of children by the age of 20 months (Peterson and Cushion, 2005). Colonization, detected in nasopharyngeal aspirates, is positive in ~15% of children with respiratory symptoms (Huang et al., 2006).

HIV-Associated PCP

In the 1980s PCP was the AIDS-defining disease in two-thirds of adults and adolescents with AIDS in the United States and 75% of HIV-infected persons were predicted to develop PCP during their lifetime. The advent of PCP prophylaxis in 1989 and combination ART in 1996 witnessed a decline in PCP of 3.4%/year from 1992 to 1995, and then 21.5%/year during 1996–1998 (Huang et al., 2006). The incidence in Western Europe and the United States declined to 2–3 cases/100 person-years (Fig. 13.2) (Morris et al., 2004). New cases occur in individuals who are not aware of their HIV-positive status, or those not receiving care for HIV infection.

Even in the era of effective combination ART, PCP remains the most frequent serious opportunistic infection in persons living with AIDS in the developed world. Mortality in persons living with AIDS without combination ART and who contracted PCP was 63%, compared to 25% in those taking combination ART (Steele et al., 2005). Whereas most healthy persons do not have detectable *Pneumocystis* in respiratory specimens, among HIV-positive adults, 10–69% are colonized by *P. jirovecii*, depending on the study technique and population studied.

The CDC Adult and Adolescent Spectrum of HIV Disease Project analyzed access to and compliance with medical care of 1073 HIV-positive persons who developed PCP during the period 1999–2001 (Morris et al.,

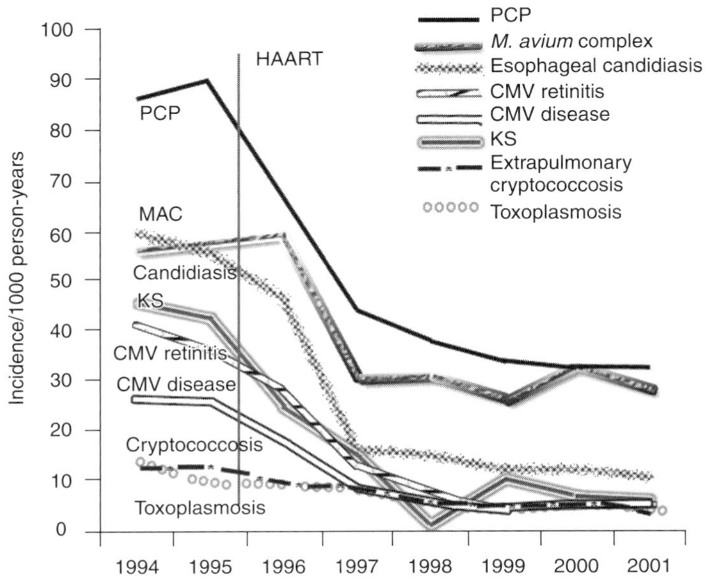

Figure 13.2 Yearly opportunistic infection rates per 1000 person-years. CDC Adult and Adolescent Spectrum of Disease Project, 1994–2001. CMV, cytomegalovirus; HAART, highly active antiretroviral therapy; KS, Kaposi's sarcoma; MAC, *Mycobacterium avium* complex; PCP, *Pneumocystis* pneumonia.
Source: Adapted from Morris et al. (2004).

2004). Of these, almost 44% of PCP cases occurred in patients not receiving medical care, who probably did not know they were HIV-positive. A further 41% of patients who were prescribed prophylaxis did not adhere to treatment, or PCP developed despite their taking medications. The latter group of "breakthrough" PCP might be understood as having been infected with a *P. jirovecii* strain which had acquired resistance to TMP-SMX (please see Section 13.11, Therapy). Finally, 9.4% of patients were under medical care but had not been prescribed prophylaxis when their CD4$^+$ T-lymphocyte count fell below 200/μL.

PCP in Developing Countries

Asia

The number of PCP cases in Thailand increased each year from 1992 to 2000 and peaked in 2000 at 6255 cases (Sritangratanakul et al., 2004). PCP was diagnosed in 47% of 291 Cambodian and Vietnamese HIV/AIDS patients with acid-fast-negative sputum smears (Vray et al., 2008). During the period 1985–2004 three series of HIV/AIDS patients in Hong Kong reported that PCP was the most frequent AIDS-related opportunistic infection (Wong et al., 2006).

Africa

Africa is a special case because PCP was thought to be rare as cases are not diagnosed and reported in an organized system because of limitations in laboratory infrastructure for diagnosis, costs of bronchoscopy, and the need for specialized equipment and training. Because of the lack of infrastructure for diagnosis, pneumonia is treated empirically, resulting in low estimates of the burden of PCP. Deaths from bacterial pneumonia and tuberculosis appear to occur at higher CD4$^+$ T-lymphocyte counts, so that AIDS patients may not live long enough with their disease to develop PCP. Africans may have increased resistance to PCP, as seen when comparing PCP rates among AIDS patients in the United States of African-American and Caucasian backgrounds.

- *Ethiopia.* PCP was diagnosed by immunofluorescence in 30% of HIV-infected patients with respiratory symptoms and sputa smear-negative for acid fast bacilli (reviewed by Davis et al., 2008)
- *South Africa.* Autopsies found histologic evidence of PCP in the lungs of 328 of 8421 (3.9%) deceased South African miners (reviewed by Davis et al., 2008). Coinfection was present 33% of the time, most often with *Cryptococcus neoformans* or bacterial pneumonia. The rate of PCP infection increased from 9.1 cases per 1000 population in 1996 to 66 cases per 1000 population in 2000, yet the disease was not suspected prior to death in 89% of patients
- *Uganda.* Thirty-eight percent of 83 HIV-positive patients admitted to hospital with pneumonia, who were negative for tuberculosis, were diagnosed with PCP after bronchoscopy with BAL. High rates of PCP have been reported in African children, most commonly in the age group <6 months old, ranging from 14% to 49% according to the country and particular study (Morris et al., 2004).

Latin America

Pneumocystosis was diagnosed in 19.9% of 168 people living with HIV/AIDS who were immigrants to Houston, Texas, from Central America during 1994–1998 (Bouckenooghe and Shandera, 2002).

Brazil

PCP was the second most common opportunistic infection seen annually in a surveillance conducted from 1993 to 2002 in Rio de Janiero (Soares et al., 2006). Of persons living with AIDS, 24–29% were diagnosed with PCP (Morris et al., 2004).

PCP in HIV-Negative Persons

Incidence and prevalence of PCP in collagen-vascular diseases being treated with immunosuppressive therapy are indicated in Table 13.1 (Ward and Donald, 1999). The prevalence of PCP according to the type of immunosuppressive condition was reported in a 12 year study period (non-AIDS) of discharge data from acute care hospitals in California (Ward and Donald, 1999). Crude mortality, during their hospitalization, for patients with collagen-vascular diseases and PCP in this series was 45.7%.

An estimate of the incidence of PCP in collagen-vascular diseases treated with immunosuppressive therapy as shown in Table 13.1 was derived by determining the number of cases of PCP in a particular acquired immunodeficiency per 1000 hospitalizations with the said acquired immunodeficiency per year (Sowden and Carmichael, 2004). A risk factor for PCP in these autoimmune diseases is that most patients have taken prednisone in excess of 15 mg/day or equivalent doses of another corticosteroid. Steroids deplete CD4$^+$ T-lymphocytes, with resultant immune dysfunction. Also involved as a risk factor are cyclophosphamide, azathioprine, methotrexate, and cyclosporine. Cyclophosphamide has brought marked improvement in patients with Wegener's granulomatosis, but increases the risk of PCP.

Table 13.1 Incidence of PCP According to the Type of Immunosuppressive Condition (non-AIDS) in a 12 year Period in California Hospitals (Ward and Donald, 1999)

Primary disease	Primary immunosuppressive condition underlying PCP[a] in a 12 year period	Incidence of PCP per 1000 cases of the primary disease
Systemic lupus erythematosus (SLE)	94	1.2
Rheumatoid arthritis (RA)	39	0.2
Wegener's granulomatosis	31	8.9
Inflammatory myopathy	26	2.7
Polyarteritis nodosa	21	6.5
Scleroderma	12	0.8
Total	*223*	

Role of DNA Genotyping

Strain typing is necessary to track transmission of *P. jirovecii*. Two DNA targets are the (i) large subunit of mitochondrial rDNA genes, which is not very informative because of a limited number of genotypes, and (ii) ITS typing based on SNPs at several points in ITS1 and ITS2 subregions of rDNA (Helweg-Larsen, 2004). Genotyping has revealed that persons living with AIDS were sometimes infected with more than one, and up to as many as seven genotypes. Some genotypes were more common, five types accounting for more than half of all subtypes.

Future of PCP

The decline in AIDS-associated PCP in developed countries and elsewhere with improved access to medical care and combination ART may be short-lived because of the advent of multidrug resistant strains of HIV-1. Additionally, DHPS mutants of *P. jirovecii* continue to accumulate under drug pressure from TMP-SMX and dapsone.

13.7 RISK GROUPS/FACTORS

$CD4^+$ T-lymphocyte deficiency is a primary risk factor for developing PCP. The risk increases exponentially the lower the $CD4^+$ T-lymphocyte cell count falls below 200 cells/μL. Risk factors also include a $CD4^+$ T-lymphocyte percentage <14%, a previous diagnosis of PCP, oropharyngeal candidiasis, bacterial pneumonia, weight loss, and higher plasma HIV RNA.

Children

Serologic evidence indicates that primary infection with *P. jirovecii* occurs early in infancy with 85% seroconversion by the age of 20 months. Approximately one-third of normal children younger than 2 years old may have *P. jirovecii* DNA in nasopharyngeal aspirates detected by PCR, but most are healthy. *P. jirovecii* DNA was detected in 24 of 74 (32%) infants with mild respiratory infections, and molecular typing of the *DHPS* gene and ITS region of rDNA indicated the strain carried by infants matched those of adults in the same community (Peterson and Cushion, 2005). Infants and children with the primary infection may serve as a reservoir for *P. jirovecii* in the community, including the preservation of *DHPS* mutant strains. Several studies have reported low levels of *P. jirovecii* detected histologically in children with SIDS, higher than that of control groups of children, but correlation does not mean causation. Children clear *P. jirovecii* after they reach 6 months old. Clusters or outbreaks of PCP have occurred in immunosuppressed children with malignancies or transplant recipients. For children born with HIV, the incidence of PCP is highest between 3 and 6 months of age.

Adults

HIV patients are at greatest risk of PCP, followed by HIV-negative patients receiving immunosuppressive therapy for solid organ or stem cell transplants, in those with cancer receiving chemotherapy, and other persons receiving chronic immunosuppressive therapy. Among HIV-positive individuals, smoking increases the risk of *Pneumocystis* colonization and infection (reviewed by Catherinot et al., 2010). Persons living with AIDS who recovered from an episode of PCP persistently or intermittently harbor *P. jirovecii* DNA in their BAL fluid for periods of 0.5–15 months (Peterson and Cushion, 2005). Healthcare workers exposed to AIDS patients may have higher *P. jirovecii* antibody titers than the general population.

PCP in HIV-negative persons accounts for an estimated one-third of cases. Several underlying conditions are associated with increased rates of carriage of *P. jirovecii*, including asthma, other chronic lung diseases, COPD, cystic fibrosis, Epstein–Barr virus disease, lupus erythematosus, recipients of high-dose systemic corticosteroids for brain neoplasms, inflammatory and collagen-vascular disease (rheumatoid arthritis), thyroiditis, transplant recipients, and ulcerative colitis.

Without prophylaxis, rates of PCP are 5–25% in transplant patients, 2–6% in patients with collagen-vascular disease, and 1–25% in patients with cancer.

Rat models depict with clarity the role of $CD4^+$ T-lymphocyte immunosuppression on the susceptibility to PCP, which increases when corticosteroids deplete the $CD4^+$ T-lymphocyte population, and resolves when drug is withdrawn and $CD4^+$ T-lymphocytes recover.

13.8 TRANSMISSION

Person–person transmission through inhalation of droplet aerosols is the most likely mode of transmission. Air sampling in hospital rooms of patients with PCP led to the detection by PCR of *Pneumocystis* DNA. Serology of normal healthy children has long indicated that infection of young children is common and *Pneumocystis* DNA has been detected in nasopharyngeal secretions of infants with mild respiratory infections. These findings in young children suggest they are one reservoir for infectious *Pneumocystis* in the community. Healthcare workers who are immune-normal may also become colonized through contact with patients with PCP. Deep nasal swab specimens from a sick child and two contact healthcare workers were PCR-positive for *P. jirovecii*, but similar specimens from noncontact healthcare workers were negative (Vargas et al., 2000).

13.9 DETERMINANTS OF PATHOGENICITY

13.9.1 Host Factors

Inflammation is needed to eliminate *P. jirovecii* proliferation in the alveoli, yet the very same host response, when it becomes exuberant, causes injury. *P. jirovecii* cell components evoke production of inflammatory mediators during pathogenesis, resulting in lung injury and impaired oxygenation.

Role of Phagocytes and Lymphocytes in Protection, Inflammation, and Lung Damage

Alveolar macrophage The first line of defense in the lung is phagocytosis by alveolar macrophages which adhere to *P. jirovecii* through multiple pattern recognition receptors (PRRs), *e.g.*: mannose receptor; dectin-1; Toll-like receptor-2 (TLR-2). Once phagocytosed, *P. jirovecii* are degraded after phagosome-lysosome fusion. As a side-effect, a variety of inflammatory mediators are produced which aid in eradicating *P. jirovecii*, but also harm host cells (Kelly and Shellito, 2010). Alveolar macrophages are potent anti- *P. jirovecii* cells, leading to death and elimination of *P. jirovecii*. Macrophages and lung epithelial cells release pro-inflammatory cytokines and chemokines, *e.g.*: tumor necrosis factor α (TNF-α), and IL-8, reactive oxidant species, and eicosanoids. The stimuli for their production are PAMPs, the mannose portion of the *P. jirovecii* major surface glycoprotein (MSG), and cell wall ß-(1 → 3)-glucan.

1. *TLR2*. TLR2 is the major PRR in the inflammatory response to *Pneumocystis* infection and inflammation plays a major role in lung damage in PCP. $TLR2^{-/-}$ knock-out mice with PCP have decreased inflammation in the lungs but show more severe symptoms and increased organism burden (Wang et al., 2008). In comparison, wild-type mice with PCP have more severe inflammation, but a lower organism burden, and a higher survival rate, implying that inflammation is only one of the host defense mechanisms against *Pneumocystis*.

Activated TLR2 transduces signals through a myeloid differentiation factor 88 (MyD88) pathway, activating transcription factor NF-κB, which upregulates production of proinflammatory cytokines such as TNF-α and chemokines (e.g., IL-8 and matrix metalloproteinase-9). The result is that $TLR2^{-/-}$ mice with PCP have less TNF-α than wild-type mice. Additionally, TLR2 collaborates with macrophage dectin-1 and mannose receptors to mediate inflammatory responses to *Pneumocystis*. The numbers of neutrophils and macrophages are decreased in $TLR2^{-/-}$ mice with PCP, underlining that they are affected by TLR2-activation (Wang et al., 2008).

2. *Dectin 1*. Dectin-1 is a PRR on alveolar macrophages that recognize the cell wall β-glucan of fungi, including *Pneumocystis*. β-glucan binds ECM proteins, which, in turn, coordinate binding to dectin-1 on alveolar macrophages. Once bound to dectin-1, *P. jirovecii* is subject to nonopsonic phagocytosis and killing via generation of reactive oxygen species. The reaction of β-glucan with dectin-1 also upregulates TNF-α and IL-8, through NF-kB translocation.

P. jirovecii ß-glucan also binds to glycosphingolipid lactosylceramide[2] on lung epithelial cells, initiating the release of inflammatory cytokine, IL-8. The concentration gradient of IL-8 attracts PMNs. The PMN concentration

[2]Lactosylceramide. *Definiton*: Non-acidic di- and oligoglycosphingolipids, that is, with two or more carbohydrate moieties attached to a ceramide unit, are vital components of mammalian cellular membranes. They are less abundant than other lipids, except in epithelial and neuronal cells, but are extremely important for cell function. The most important and abundant of the diosylceramides is β-D-galactosyl-(1 → 4)-β-D-glucosyl-(1 → 1′)-ceramide, termed lactosylceramide (LacCer). *Source:* http://www.lipidlibrary.co.uk/index.html.

in BAL fluid correlates more closely with impaired oxygenation than the burden of *P. jirovecii* (Huang et al., 2006).

3. *Scavenger Receptor A* (*SRA*). SRA is a PRR on macrophages that regulates the inflammatory response to *Pneumocystis*. SRA-deficient mice produced significantly more TNF-α, IL-12, and IL-18 in response to *Pneumocystis* infection than wild-type mice.

4. Intracellular Killing of *Pneumocystis* by Alveolar Macrophages. Once phagocytosed, hydrogen peroxide is primarily responsible for the killing of *Pneumocystis* (Kelly and Shellito, 2010). Since alveolar macrophages require lymphocyte-derived products, such as IFN-γ, for maximum effect, immunosuppression has severe consequences during infection. Moreover, numbers of alveolar macrophages are influenced by the degree of immunosuppression and the infectious burden of pneumocystis organisms. Immunosuppressed *Pneumocystis*-infected rats had significant decreased numbers of alveolar macrophages compared with immunosuppressed non infected controls.

During HIV infection macrophage mannose receptors are reduced in numbers, leaving macrophages less able to phagocytose *P. jirovecii*. The loss of alveolar macrophages during PCP infection is partly due to apoptosis. Patients living with AIDS or those being treated for hematologic malignancy or for solid tumors have impaired macrophage responses, which suppress clearance of *P. jirovecii*.

CD4$^+$ T-Lymphocytes CD4$^+$ T-lymphocytes are critical to resolve PCP by recruiting and activating macrophages and other effector cells, which are responsible for elimination of *Pneumocystis*. PCP is observed when the CD4$^+$ T-lymphocyte count is <200 cells/μL. Research (reviewed by Kelly and Shellito, 2010) has focused on T-cell activation through costimulatory molecules. Loss of the costimulatory receptors CD2 and CD28 result in lethal PCP associated with an accumulation of CD8$^+$ T-lymphocytes in the lungs, reduced antibody titers, and faulty regulation of cytokines IL-10 and IL-15, affecting immune downregulation.

PMN and CD8$^+$ T-Lymphocytes Severe PCP is characterized by increased CD8$^+$ T-lymphocytes and PMN lung inflammation, leading to diffuse alveolar damage, impaired oxygenation, and respiratory failure. PMNs release proteinase and oxidant species that injure alveolar epithelial and capillary epithelial cells. In the absence of inflammation, *P. jirovecii* have little direct effect on lung function. Moreover, it was shown in PCP that elevated concentrations in BAL fluid of IL-6 and IL-8 correlated with the proportion of PMNs and inversely with the oxygenation index, thus underlining the correlation between influx of PMNs and loss of respiratory capacity (Tasaka et al., 2010).

A controversial series of experiments in murine pneumocystosis cast doubt on PMNs as the *cause* of damage to lung tissue. Using gene knock-outs affecting the production of reactive oxygen species (ROS), nitric oxide synthase, and chemokine receptor CXCR2 controlling accumulation of intraalveolar PMNs, Swain et al. (2004) found that knock-out mice and wild-type mice had the same level of pulmonary damage. Moreover, wild-type and knock-out mice had the same burden of pneumocystis organisms. Whereas PMNs are a valid marker of lung damage during pneumocystosis these experiments in mice found that neither neutrophils nor ROS appear to cause tissue damage.

There is controversy as to whether CD8$^+$ T lymphocytes are protective or detrimental in PCP. Simian immunodeficiency virus-infected macaques develop a prolonged *Pneumocystis* colonization state and increased concentrations of proinflammatory mediators and Th2-type cytokines in their BAL fluid. This chronic inflammation produces a progressive pulmonary decline similar to that seen in emphysema. As the burden of *Pneumocystis* increases, a CD8$^+$ T-lymphocyte- and neutrophil-dominant response may develop and amplify inflammation-mediated pulmonary damage (Shipley et al., 2010).

CD8$^+$ T-cells can be protective against *Pneumocystis* depending on their T-cytotoxic (Tc)1-like phenotype, which upregulate concentrations of endogenous IFN-γ. IFN-γ delivery via overexpression using gene transfer resulted in clearance of *Pneumocystis* in the absence of CD4$^+$ T-lymphocytes and was associated with increased recruitment of IFN-γ-producing CD8$^+$ T-lymphocytes (reviewed by Kelly and Shellito, 2010).

B-Lymphocytes B-Lymphocytes provide activities essential for host defense against *P. jirovecii*. Opsonization by antibodies increases phagocytosis and killing by alveolar macrophages. Concentrations of IgA and IgG against *Pneumocystis* in BAL fluid were lower in HIV-positive patients than in controls, raising the possibility that decreased local antibody production may contribute to PCP (Jalil et al., 2000). Other evidence for a possible protective role of antibodies in PCP comes from experiments in mice immunized intranasally with soluble *Pneumocystis* antigens, then depleted of CD4$^+$ T-lymphocytes, and infected with *Pneumocystis* (Pascale et al., 1999). Immunized mice, but not controls, cleared their infection and produced IgA in their BAL fluid reactive with a 55–60 kDa *Pneumocystis* antigen.

Pneumocystis jirovecii *Alteration of Pulmonary Surfactants*

Pneumocystis jirovecii are embedded in alveolar exudates consisting of extracellular matrix (ECM) proteins: fibronectin, vitronectin, and collectin[3] surfactant proteins A and D (SP-A and SP-D). Pulmonary surfactant is a complex of lipids and proteins secreted by alveolar type II cells (Kelly and Shellito, 2010). The four major surfactant proteins are divided into two groups: the hydrophobic proteins SP-B and SP-C control surface tension, whereas the hydrophilic proteins SP-A and SP-D are important in innate lung immunity. Increased amounts of hydrophilic SP-A and SP-D are found in PCP with the former corresponding to increased resistance to *P. jirovecii* and the latter promoting exuberant infection and inflammation. *Pneumocystis* major surface glycoprotein, MSG, binds to SP-A and SP-D, which enhances *P. jirovecii* binding to alveolar macrophages (Huang et al., 2006). *Pneumocystis* infection results in downregulation of SP-B and SP-C, resulting in significant increases in surface tension, contributing to hypoxemic respiratory failure in PCP patients (Kelly and Shellito, 2010).

Alveolar Epithelial Cells Contrary to earlier beliefs that these cells are only involved in gas exchange, evidence has accumulated that they are a rich source of inflammatory mediators. Airway epithelial cells recognize infectious agents and initiate signal transduction pathways resulting in production of cytokines and chemokines which recruit inflammatory cells such as PMNs to the site of infection. The binding of *Pneumocystis* trophic forms to type 1 alveolar cells is essential in establishing infection. *Pneumocystis* asci (cysts) are found in expectorated sputum so that it is reasonable for *Pneumocystis* cell wall ß-(1 → 3)-D-glucan to interact with epithelia in the lower respiratory tract. Airway epithelial cells lack dectin-1 receptors present in macrophages. Instead, the glycosphingolipid ß-lactosylceramide in the cell membrane is a receptor for β-glucan and is essential to the activation of cytokine signaling pathways. Following binding of ß-glucan, airway epithelial cells induce changes in cytosolic calcium influx which initiates activation of two major MAPK[4] pathways resulting in the release of IL-8 (Carmona et al., 2010). This mechanism of chemokine generation helps us to understand how inflammatory cells are recruited to the lung during PCP, which, if not regulated, results in the exuberant inflammation that causes damage.

[3]Collectins. *Definition:* Soluble pattern recognition receptors (PRRs) belonging to the superfamily of collagen-containing C-type lectins, for example, mannan-binding lectin (MBL), surfactant protein A, (SP-A) and surfactant protein D (SP-D). They are major modulators playing a key role in innate defense against invading microbes. Functionally, collectins usually are trimers. Each monomer has four structural domains: a cysteine-rich N terminus, a collagen domain, and a coiled-coil neck domain, a C-type lectin domain that is also called a carbohydrate recognition domain (CRD). Collectins selectively bind to specific complex carbohydrates of microbes using their CRDs.

Immune Reconstitution Inflammatory Syndrome (IRIS)

IRIS is prevalent in AIDS patients when there is a rapid recovery of $CD4^+$ T-lymphocytes after antiretroviral therapy, and in patients whose immune recovery accompanies cessation of systemic steroids or cancer treatment. In PCP-related IRIS, patients do not have severe PCP, so that severity is proportional to the extent of immune recovery. Murine models of IRIS indicate that the pathology involves parenchymal lung inflammation, influx of macrophages and lymphocyte recruitment into BAL fluid, and impaired surfactant function (reviewed by Kelly and Shellito, 2010).

13.9.2 Microbial Factors

Major Surface Glycoprotein (MSG)

This molecule is multifunctional and essential in the pathogenesis of *P. jirovecii*.

- Serial expression of single isoforms of MSG helps *P. jirovecii* evade the host immune response.
- MSG binds to extracellular matrix (ECM) proteins of the alveoli, facilitating firm binding of *P. jirovecii* trophic forms to type 1 pneumocytes.
- Mannose residues in the MSG bind to mannose receptors on alveolar macrophages, leading to phagocytosis and inflammation.

Structure

Serial Expression of MSG Isoforms MSGs have a molecular mass of 95–120 kDa and are a family of proteins encoded by ~80 highly polymorphic genes repeated and distributed among all *P. jirovecii* chromosomes, clustered in the subtelomeric region. The MSG gene array may comprise ~10% of the genome (Thomas and Limper, 2007).

Function

- Only a single isoform is expressed and antigenic variation occurs by switching expression of MSG genes

[4]MAPK. *Definition*: Mitogen-activated protein kinase. This is a cascade that leads directly to the phosphorylation and activation of transcription factors in the nucleus, which induce new gene expression. For a further discussion see Janeway et al. (2001).

to evade the host immune response. Antibodies are evoked by MSG within a 3 week period after initiation of disease but by then the switch to another MSG isoform evades this antibody response.

- There is tight binding of *P. jirovecii* to type 1 pneumocytes. After inhalation, *P. jirovecii* does not invade host cells, but attaches tightly to type 1 pneumocytes in the lower respiratory tract. *P. jirovecii* trophic forms adhere tightly to this cell type. The host cells appear vacuolated and eroded but preserve their barrier function. This essential step in pathogenesis occurs when the MSG expressed by *P. jirovecii* binds to host proteins (fibronectin, vitronectin) of the alveolar ECM. By coating the surface of *P. jirovecii* they mediate firm attachment to integrin receptors on type 1 pneumocytes.

  Attachment of *P. jirovecii* to the alveolar epithelium is essential for trophic growth and transition to sexual reproduction. After trophic forms bind to type 1 alveolar pneumocytes desquamation and a foamy eosinophilic exudate (visible on H&E stain) is produced. Damage to the alveolus is indirect, mediated by the influx of PMNs and $CD8^+$ T-lymphocytes with the resultant inflammatory response (Fig. 13.3).
- Evading host immune attack, *P. jirovecii* sheds MSG, which blocks phagocytosis by competing for the mannose receptor on alveolar macrophages (Pop et al., 2006).

In sum, the MSG and its mannosyl residues play a role in attachment to lung epithelia and in evasion of phagocytosis by alveolar macrophages.

Signal Transduction Pathways

In fungi these pathways convert extracellular signals to intracellular events by a sequence of phosphorylations of protein kinases: these are the *m*itogen-*a*ctivated *p*rotein *k*inases, or MAPK pathways. Kinase genes control life cycle stages including sexual reproduction, cell wall integrity, and the invasive growth mode (Thomas and Limper, 2007).

- The pheromone-induced MAPK pathway in *P. jirovecii* is activated in trophic forms and drives mating, which results in ascus formation.
- Another MAPK pathway is induced by binding of trophic forms to type 1 pneumocytes via host ECM proteins (collagen, fibronectin, and vitronectin). This binding is essential to the proliferation of trophic forms in the host.

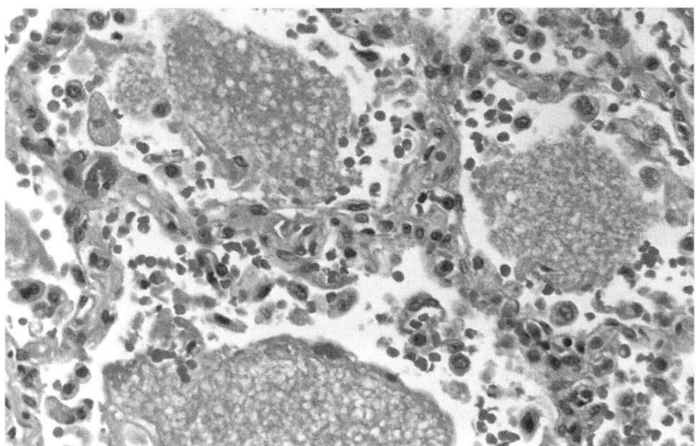

Figure 13.3 *Pneumocystis jirovecii* pneumonia (PCP). Alveoli fill with a foamy eosinophilic exudate, producing a honeycomb appearance. Cysts (asci) cannot be detected with the H&E stain. *Source:* Dr. Edwin P. Ewing, Jr., CDC PHIL.

Ascus (Cyst) Wall

The *P. jirovecii* cell wall is a complex carbohydrate structure. The major ascus wall structural component is ß-(1 → 3)-glucan, also containing side chains of ß-(1 → 4) and ß-(1 → 6) glucose residues. Glucan serves a dual role in providing structural support and in evoking a host inflammatory response. The result is immune-mediated damage to alveolar epithelia (Thomas and Limper, 2007).

The ß-(1 → 3)-D-glucan of *P. jirovecii* binds to alveolar macrophages, inducing a proportional release of TNF-α. ß-(1 → 3)-D-glucan also binds directly to alveolar epithelia, leading to inflammation (please see Section 9.1 Host Factors). Echinocandin inhibitors of ß-(1 → 3)-glucan have been capable of clearing ascus forms of *P. carinii* from the lungs of infected rats. It is of great interest to see if echinocandins (e.g., CASF, MCF, and ANF) can be effective in treating PCP.

Genomics

The genome of rat *Pneumocystis, P. carinii*, reveals a genome of 8 Mb of DNA divided into 15 linear chromosomes, ranging from 100 to 700 kb. The majority of genes are studded with introns, up to 9 per gene, more so than in yeast (Thomas and Limper, 2007). Only a single copy of the rDNA gene locus is present in *P. jirovecii*, in contrast to other simple eukaryotes, and this finding remains an enigma. Cushion (2004) speculated that this finding may explain why *P. jirovecii* seems to be slow to reach a point of fulminant infection.

13.10 CLINICAL FORMS

PCP

Clinical signs and symptoms of PCP common to persons living with AIDS and in HIV-negative immunosuppressed persons are cough, weakness, loss of appetite, expectoration, sweats, weight loss, bloody sputum, chest pain, and dyspnea on exertion. Clinical signs include fever, rales, tachycardia, and tachypnea. Chest radiographs show bilateral abnormalities and interstitial and alveolar opacities, but also may be normal in the early stages.

No combination of signs or symptoms is pathognomonic and, because therapy is potentially toxic, microbiologic confirmation of PCP is necessary. Evidence in stained respiratory specimens consists of demonstrating cysts (asci) and/or trophic forms. Specimens for diagnostic tests include induced sputum and bronchoscopy with BAL (please see Section 13.12, Laboratory Detection, Recovery, and Identification).

The symptoms in HIV-infected persons develop slowly and progress with dyspnea, fever, nonproductive cough, and chest pain that worsens over days or weeks. The pulmonary examination of mild cases in HIV-positive patients may be normal at rest, but with exertion there is tachypnea, tachycardia, and diffuse dry ("cellophane") rales. The fulminant pneumonia seen in non-HIV patients is less common. Oropharyngeal candidiasis is commonly present in HIV-positive patients. Extrapulmonary disease is rare but can occur in any organ and is associated with aerosolized pentamidine prophylaxis.

Hypoxemia and interstitial infiltrates in an HIV-positive person whose $CD4^+$ T-lymphocyte count falls <200 cells/μL prompts the need for empiric therapy for PCP. HIV-positive patients have higher arterial oxygen tension and their BAL specimens contain more *P. jirovecii* organisms and fewer PMNs than in HIV-negative patients with PCP.

Hypoxemia ranges from mild (room air arterial oxygen pO_2 of ≥ 70 mm Hg or alveolar–arterial O_2 difference [A-a] $DO_2 < 35$ mm Hg) to moderate ([A-a] $DO_2 > 35$ and <45 mm Hg), to severe ([A-a] $DO_2 > 45$ mm Hg). Chest radiographs early in disease may be normal but later findings include diffuse, bilateral, symmetric, interstitial infiltrates from the hila in a butterfly pattern (Kaplan et al., 2009). Atypical presentations include nodules, blebs, cysts, asymmetric disease, upper lobe localization, and pneumothorax (collapsed lung). Pneumothorax in a patient with HIV raises the suspicion of PCP.

Concurrent Pulmonary Infections

Cavitation, intrathoracic adenopathy, and pleural effusion are uncommon and, if present, may indicate an alternative diagnosis. Between 13% and 18% of patients with PCP have another complicating pulmonary infection: tuberculosis, Kaposi's sarcoma, or bacterial pneumonia (Castro et al., 2007). Concurrent pulmonary infections in immunocompromised patients (HIV-negative) are common, up to 50%, including cytomegalovirus, *Candida*, and tuberculosis (Sowden and Carmichael, 2004).

If untreated, PCP has a fatal outcome in persons living with AIDS. Improved therapy including the use of adjuvant corticosteroids has reduced mortality to 10–20%. From 1989 to 2003, 1 month mortality in the United States in AIDS from PCP was 15%. Independent risk factors for decreased 2 month survival are low PaO_2, therapy with other than TMP-SMX, and positive BAL for CMV culture (Helweg-Larsen, 2004).

Statistics on the mortality of PCP in AIDS patients are elusive, because diagnosis based on AIDS-defining opportunistic infections are no longer cataloged. For that reason one must rely on regional or local studies to gain an impression of trends in mortality from PCP.

In a cohort of 197 AIDS patients admitted to San Francisco General Hospital between 1997 and 2002 with confirmed PCP due to DHPS mutants with decreased susceptibility to TMP-SMX, 14.3% died within 60 days (please see Section 13.11, Therapy). Of these, 9.7% of the deaths were attributable to PCP. In the same hospital, of 40 AIDS patients infected with wild-type *P. jirovecii*, all-cause death claimed 7.5%, with 5% attributable to PCP (Crothers et al., 2005). This result underlines that better management has reduced mortality from PCP, but raises the specter of drug resistance.

Recurrent PCP

There is still a high rate of recurrent PCP in persons living with AIDS. About half of patients who have episodes of PCP retain the same genotype. A patient may be simultaneously infected with more than one genotype, complicating interpretation of whether the source for recurrence is exogenous or endogenous. PCR genotyping of the ITS region of rDNA shows that infections often are not clonal and that coinfections with multiple genotypes are common (Helweg-Larsen, 2004).

PCP in Immunosuppressed Patients (Non-AIDS)

PCP patients who are HIV-negative have a worse prognosis including ICU admission in 31–60%, mechanical ventilation in 14–64%, and overall mortality of 19–47%, rising to 50–71% in intensive care (Sowden and Carmichael, 2004). Poor prognosis is associated with tachypnea, tachycardia, elevated CRP, and/or mechanical ventilation. Mortality from PCP depends on the underlying pathology—63% in Wegener's granulomatosis, 58% in inflammatory myopathy, 48% in polyarteritis nodosa, 31% in rheumatoid arthritis (RA), and 17% in systemic sclerosis (Sowden and Carmichael, 2004). Whereas there is a high rate of relapse in persons living with AIDS, in the HIV-negative immunosuppressed group relapse is uncommon because tapering of corticosteroids or other immunosuppressive agents restores protective $CD4^+$ T-lymphocytes. The prerequisite for PCP in the immunosuppressed (non-AIDS) patient is lymphocytopenia (<1000 cells/μL). PCP patients who were immunosuppressed (non-AIDS) had $CD4^+$ T-lymphocyte counts of <300/μL.

Example

In one study reporting seven HIV-negative patients who developed PCP after immunosuppressive therapy, the mean $CD4^+$ T-lymphocyte count was 90.6/μL (range 20–182/μL) (Glück et al., 2000). The underlying diseases were non-Hodgkin's lymphoma, systemic lupus erythematosus, rheumatoid arthritis, Wegener's granulomatosis, and mixed connective tissue disease with vasculitis and dermatomyositis. All of these patients survived.

Radiography

The typical chest radiographic presentation of PCP is bilateral perihilar or diffuse symmetric interstitial pattern, which may be finely granular, reticular, or ground-glass in appearance (Fig. 13.4) (Aviram et al., 2003). HRCT is very helpful in detecting PCP in symptomatic patients with equivocal radiographic findings. The classic HRCT finding in PCP is extensive ground-glass attenuation. It often is distributed in a patchy or geographic fashion, with a predilection for the central, perihilar regions of the lungs. Cystic lung disease is observed in up to one-third of cases and may be complicated by pneumothorax (Aviram et al., 2003). Gallium scans show increased pulmonary uptake. Both findings should prompt a diagnostic bronchoscopy. Radiographic findings were described in a cohort of AIDS patients with PCP by high-resolution CT of the lung (Fujii et al., 2007):

> *Ground-glass opacities sparing the lung periphery (41% of episodes) or displaying a mosaic pattern (29%); or being nearly homogeneous (24%); ground-glass opacities associated with air-space consolidation (21%); or with cystic formation (21%); or with linear-reticular opacities (18%); patchily and irregularly distributed (15%); associated with solitary or multiple nodules (9%); and associated with parenchymal cavity lesions (6%).*

Pathology

Inflammation in the lung causes more damage than the direct effects of *P. jirovecii*. When macrophages phagocytose *P. jirovecii* they release IL-8, a chemokine, which signals PMN cells to enter the site of inflammation. PMNs mediate lung injury via secretion of proteinases, oxidants, and cationic proteins (Thomas and Limper, 2007). This influx is responsible for impaired oxygenation more than the burden of *P. jirovecii*. Moreover, $CD8^+$ T-lymphocytes inflict direct injury on lung tissue contributing to increased inflammation.

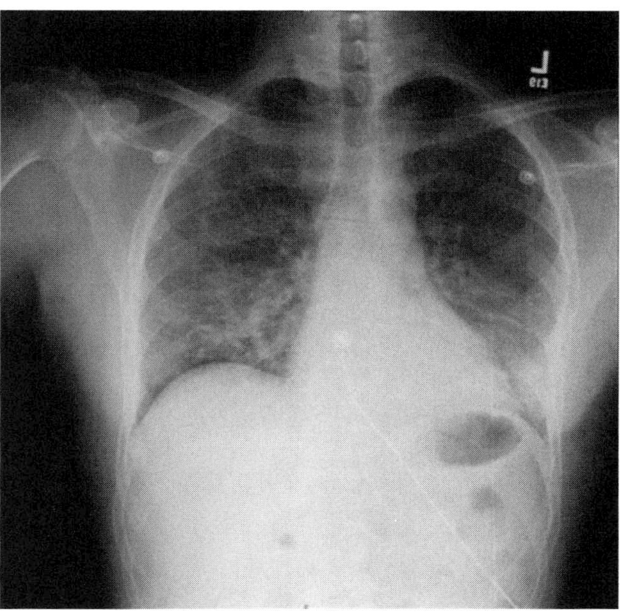

Figure 13.4 Chest radiograph of PCP in a person living with AIDS. This 46-yr-old male with AIDS, $CD4^+$ lymphocyte count of 25 cells/μL, presented with shortness of breath and nonproductive cough. The radiograph shows bilateral patchy infiltrates with a primarily interstitial pattern, right lung worse than left. A smear of bronchoalveolar lavage fluid revealed *Pneumocystis jirovecii*. *Source:* G. Marshall Lyon, M.D.

As the $CD4^+$ T-lymphocyte count declines to <200/μL, *P. jirovecii* multiply and fill the alveolar lumens. At this stage the histopathologic finding is a foamy eosinophilic alveolar exudate (Sowden and Carmichael, 2004). This intraalveolar exudate consists of coalesced macrophages and remnant *P. jirovecii* making a honeycomb pattern. By scanning and transmission EM the foamy exudate is a collection of vacuolated and degenerated trophic forms and asci.

With increasing disease severity there also may be hyaline membrane formation and interstitial fibrosis with edema. Impaired oxygenation occurs with an elevated alveolar–arterial (PAO_2-PaO_2)[5] oxygen gradient, impaired diffusing capacity, and alterations in lung compliance, total lung capacity, and vital capacity.

[5]PAO_2-PaO_2. *Definition*: PAO_2 is the alveolar partial pressure of O_2 (measured in mm Hg): PaO_2 is the arterial partial pressure of O_2. The term alveolar–arterial PO_2 is called the alveolar–arterial PO_2 difference or "A-a gradient" for short. The A-a gradient answers the important question: Are the lungs transferring oxygen properly from the atmosphere to the pulmonary circulation? If the A-a gradient is elevated, the answer is **no**. If the A-a gradient is normal the answer is **yes**. (*Source:* Martin, 1999.)

- This picture of diffuse lung damage resembles that of adult respiratory distress syndrome (ARDS).
- Production of surfactant phospholipids, which normally function to reduce alveolar surface tension, are decreased during PCP, so that the lungs are stiffer and ventilation becomes more difficult.
- Pneumocystosis, in the end stage, may follow a pathway common to other restrictive lung diseases, resulting in extensive interstitial fibrosis in which contiguous cyst-like airspaces impart a "honeycomb" or sponge-like appearance to the lung.

Extrapulmonary Pneumocystosis

Before the era of AIDS this clinical form was considered extremely uncommon. Now some 90 cases have been reported in AIDS patients with PCP. Extrapulmonary pneumocystosis occurs in patients with advanced HIV infection who are receiving no prophylaxis or only aerosolized pentamidine.

The main sites of involvement are lymph nodes, spleen, liver, bone marrow, gastrointestinal tract, eyes, thyroid, adrenal glands, and kidneys. The clinical picture varies from incidental findings at autopsy to rapid multisystem dissemination. Histopathology reveals necrotic areas filled with foamy material. Fungal stains (GMS) or immunofluorescence reveal numerous *P. jirovecii* (Walzer and Smulian, 2005).

13.11 THERAPY

Primary prophylaxis is considered first and then therapy for PCP.

Pneumocystis jirovecii is not susceptible to polyene or azole antifungal agents. The finding that neither AmB or a range of azole antifungal are ineffective against *P. jirovecii*, perhaps more than other findings, seemed to underline, erroneously as it turned out, that this microbe was not a fungus. Sterols analyzed from lung tissue of a human case of PCP retain the methyl group at the Δ14 position: there is no Δ14 demethylation (Giner et al., 2004). That is critical to understanding why azole antifungal agents are not active against *P. jirovecii*, because lanosterol Δ14 demethylase is the target of azole antifungal agents. Sterols retaining the Δ14 methylation are functional in *P. jirovecii*. AmB, which binds to ergosterol, does not bind or binds inefficiently to pneumocysterol and, like azoles, is ineffective in therapy.

Prophylaxis

Chemoprophylaxis of PCP has been cited as one of the "major successes in the field of infectious diseases in the past 25 years" (Kovacs et al., 2001). The standard practice of PCP prophylaxis among HIV-infected patients, bone marrow transplant recipients, cancer patients, and other immunosuppressed individuals is a regimen that has saved innumerable lives, particularly in resource-poor nations.

The World Bank has cited PCP prophylaxis as a simple intervention costing less than $20/ year that has had a significant effect in delaying the onset of PCP, up to that time the most common initial AIDS-defining event, and that PCP prophylaxis positively influenced survival (World Bank, 2006).

- HIV-infected adults and adolescents, including pregnant women and those on ART, should receive prophylaxis against PCP if their $CD4^+$ T-lymphocyte count is <200 cells/μL, or if they have a history of oropharyngeal candidiasis (Kaplan et al., 2009). Persons with a $CD4^+$ T-lymphocyte percentage of <14% or a history of an AIDS defining illness should be considered for prophylaxis. Consideration also should be given to starting prophylaxis when a patient's $CD4^+$ T-lymphocyte count is between 200 and 250 cells/μL.
- TMP-SMX is recommended, at a dosage of one double-strength tablet/day. One single-strength tablet also is effective and is better tolerated. One double-strength tablet 3×/week also is effective. TMP-SMX, at a dose of one double-strength tablet/day, confers cross-protection against toxoplasmosis and against some respiratory bacterial pathogens. Lower doses of TMP-SMX also provide some protection.
- Patients with adverse reactions that are not life-threatening should continue to receive TMP-SMX, if possible. Resuming therapy after an adverse reaction has resolved is advisable. A gradual increase in dose or a reduced frequency may be beneficial in patients who discontinued therapy because of fever and a rash. Up to 70% of patients can tolerate reintroduced TMP-SMX.
- Patients intolerant of TMP-SMX can be given alternative prophylaxis with dapsone, dapsone plus pyrimethamine plus leucovorin, aerosolized pentamidine administered by the Respiragard II nebulizer (Marquest, Englewood, CO), and atovaquone (Kaplan et al., 2009). Atovaquone is as effective as aerosolized pentamidine or dapsone but more expensive. Patients seropositive for *Toxoplasma gondii* who cannot tolerate TMP-SMX should receive dapsone plus pyrimethamine plus leucovorin. One drawback to aerosolized pentamidine is it does not confer protection against extrapulmonary pneumocystosis.

Immunosuppressed HIV-negative patients receiving immunosuppressive therapy for 1 month should have $CD4^+$ T-lymphocyte counts determined when the steroid dose is $\geq$15 mg prednisolone equivalent/day; >3 months of corticosteroid therapy; or when the total lymphocyte count is $\leq$300/μL (Thomas and Limper, 2004).

The major cause of prophylaxis failure is noncompliance. Adverse effects occur with TMP-SMX prophylaxis. After 13 months of prophylaxis half of patients experience at least one adverse effect: usually a skin rash. A switch to a non-TMP-SMX drug is associated with treatment failure even when there is compliance; for example, in bone marrow transplant patients the relative risk of developing PCP on alternative therapy was 18.8 compared with TMP-SMX (Helweg-Larsen, 2004). A summary of causes of failure of prophylaxis includes noncompliance, insufficient drug levels—possibly linked to development of resistance, or a switch to alternative therapy.

When patients living with AIDS have responded to ART with an increase in $CD4^+$ T-lymphocyte counts of >200 cells/μL for >3 months, prophylaxis may be discontinued. Discontinuation reduces pill burden, potential for toxicity, drug interactions, and selection of resistant variants.

Therapy

Trimethoprim-sulfamethoxazole (TMP-SMX; syn: cotrimoxazole, Bactrim) is the first line therapy for PCP and has both excellent tissue penetration and high efficacy without the frequency of adverse reactions when primary therapy consists of aerosolized pentamidine. TMP-SMX is as effective as parenteral pentamidine and more effective than other regimens. Patients who develop PCP despite TMP-SMX prophylaxis are effectively treated with standard doses of TMP-SMX.

Primary therapy consists of TMP-SMX (Jacobs and Guglielmo, 2007) at doses of 15–20 mg/kg/day of trimethoprim and 75–100 mg/kg/day of sulfamethoxazole in 3 or 4 divided doses administered intravenously or orally, depending on the severity of disease, for 3 weeks. The dose must be adjusted for abnormal renal function. Patients who have allergy to TMP-SMX should be considered for desensitization, and then treated with TMP-SMX.

Patients with confirmed or suspected PCP and moderate to severe disease (room air pO_2 of <70 mm Hg or arterial–alveolar O_2 gradient >35 mm Hg) should receive adjunctive corticosteroids within 72 h after starting PCP therapy. Prednisone[6] should be given at a dose of 40 mg 2× daily for 5 days, then 40 mg/day through days 6 to 11, then 20 mg/day on days 12 to 21. For PCP patients who are immunosuppressed but HIV-negative, a dose of 60 mg or more of prednisone/day has a better outcome than lower doses (Thomas and Limper, 2004). Methyl prednisolone at 75% of the respective prednisone dose can be used if parenteral administration is necessary.

A substantial number of cases fail to improve under optimal therapy, and mortality for PCP requiring ventilator support has not changed in the 1994–2004 period. Overall prognosis for patients whose degree of hypoxemia requires ICU or mechanical ventilation remains poor. Survival in up to 50% of patients requiring ventilator support has been reported in recent years.

Adverse Effects

Failure on TMP-SMX is related to the need to cease treatment because of toxicity. Up to 60% of patients develop rash, neutropenia, gastrointestinal upset, and/or liver enzyme disturbances.

Alternative Therapy

The alternatives are pentamidine, dapsone, clindamycin-primaquine, or atovaquone. Parenteral pentamidine results in more severe side effects: hypoglycemia, neutropenia, thrombocytopenia, and orthostatic hypotension. Refer to the guideline for alternative therapeutic regimens (Kaplan et al., 2009).

Mechanism of Action: TMP-SMX

Sulfamethoxazole (SMX) inhibits dihydropteroate synthase (DHPS). This enzyme catalyzes dihydrofolate formation from *p*-aminobenzoic acid. SMX is a structural analog of *p*-aminobenzoic acid. Trimethoprim inhibits dihydrofolate reductase (DHFR), which catalyzes formation of tretrahydrofolate from dihydrofolate. The combination of TMP-SMX is thought to be synergistic, but the inhibition of DHFR enzyme has little role in the antifungal action and SMX is mostly responsible for inhibition of *P. jirovecii*.

The therapeutic effect of TMP-SMX in humans *is almost completely due to the effect of SMX as an inhibitor of DHPS* (Helweg-Larsen, 2004). Thus, DHPS mutants are of major clinical relevance.

Dihydropteroate Synthase (DHPS) Mutants

Resistance to TMP-SMX and sulfones (dapsone) is related to mutations in *DHPS*. Most all of the DHPS resistant mutants have nonsynonymous amino acid substitutions:

[6]Prednisone is a prodrug that is converted in the liver into prednisolone. Both are synthetic glucocorticoid hormones. They are intermediate acting, broad anti-inflammatory medications. They have less activity than dexamethasone or betamethasone, but greater activity than hydrocortisone.

alanine for threonine substitution at position 55 and a serine for proline at position 57 (Thomas and Limper, 2004, 2007). Drug pressure selects for these mutants which are considered to confer some clinical resistance to therapy. DHPS mutations are an independent predictor of all-cause mortality at 3 months. DHPS mutants increase the risk of PCP treatment failure with TMP-SMX or dapsone.

A switch to a mutant strain from wild type can occur between the first and second episode of PCP, indicating the DHPS mutants are selected in vivo under drug pressure. Disease resulting from a DHPS mutant is likely to have reduced 3 month survival after a diagnosis of PCP but is not invariably associated with treatment failure (Helweg-Larsen, 2004). A majority of patients whose *P. jirovecii* strain contained DHPS mutations respond and survive after treatment with TMP-SMX (Thomas and Limper, 2004). A trend to require mechanical ventilation was reported for patients whose PCP is caused by DHPS mutants (Crothers et al., 2005).

Detection of DHPS Mutants

Direct sequencing of PCR-amplified *DHPS* from clinical samples reveals nucleotide changes at codon 55, 57, and 60. Mutant DHPS strains are being detected in patients seemingly unexposed to TMP-SMX in the United States, indicating their accumulation in the human reservoir (Helweg-Larsen, 2004) The full effect of such mutants on clinical resistance to high-dose TMP-SMX is as yet unclear. Widespread use of TMP-SMX in Africa, because it is inexpensive to treat a variety of childhood infections and for prophylaxis in AIDS, increases the risk of development of DHPS mutant strains and clinical sulfa resistance.

13.12 LABORATORY DETECTION, RECOVERY, AND IDENTIFICATION

Specific diagnosis is necessary because other diseases have similar signs and symptoms. The clinical examination, blood tests, and chest radiographs are not definitive for PCP. Moreover, *P. jirovecii* is unable to be continuously cultivated in vitro. Therefore, examination of stained smears of induced sputum or BAL fluid is necessary to confirm the diagnosis. Histopathology of lung biopsy tissue is an alternative approach. PCR diagnosis is an option.

Direct Examination

Spontaneously produced sputum has low sensitivity and is not a suitable specimen. Induced sputum is preferred for patients with AIDS. This is accomplished when the patient inhales hypertonic (3% saline) from a high-flow nebulizer for 15–30 min and is then asked to cough. Patients who are immunosuppressed because of transplantation or treatment of autoimmune disorders should have samples obtained by bronchoscopy. Induced sputum specimens have a low sensitivity in these patients.

Microscopy

Only few *P. jirovecii* may be present in upper respiratory tract secretions. The current approach is to examine induced sputum and if negative to proceed to BAL. Sputum induced by nebulized saline is less invasive and has good sensitivity. Hypertonic saline-induced sputum has a lower and variable sensitivity compared with bronchoscopy with BAL and depends on the experience of personnel.

Microscopic evidence of *P. jirovecii* is effectively obtained by bronchoscopy with BAL. Fiberoptic bronchoscopy with BAL is the most commonly used diagnostic procedure. Sampling of the lower respiratory tract by bronchoscopy with BAL is usually required to isolate *P. jirovecii* and make a definite microbiologic diagnosis. When combined with stains or immunofluorescence, cytocentrifuge smears of BAL fluid have a sensitivity of 81–90% and specificity of 90–100%.

There is a high burden of *P. jirovecii* in persons living with AIDS, but lower numbers in other categories of immunosuppressed patients. Open lung biopsy is definitive but is the method of last resort. Lung biopsies have been of declining importance because of the success of these other two methods (Kovacs et al., 2001).

Tinctorial Methods

Smears stained with Giemsa, Diff-Quik®, or Wright's stain reveal both cysts (asci) and trophic forms but do not stain the ascus (cyst) wall. Commercially available immunofluorescent stains also are effective. Gomori methenamine silver (GMS) and toluidine blue O are very specific for ascus walls and allow good contrast between asci and background staining of BAL fluid (Fig. 13.5).

Toluidine blue O stain is reported to have low sensitivity, 49% in one study (Tiley et al., 1994). The ascus wall also is stained with Gram–Weigert and cresyl violet.

Wright's stain, Giemsa stain, and Diff-Quik are a group of related stains derived from the original Romanovsky stain, in that they use eosin Y and methylene blue or eosin and azure A or B to stain blood smears and malaria parasites. One may expect they will have similar properties with respect to staining *P. jirovecii*.

To evaluate the efficacy of staining methods for diagnosis of PCP, four hospital laboratories evaluated Diff Quik, Calcofluor white, GMS, and immunofluorescent

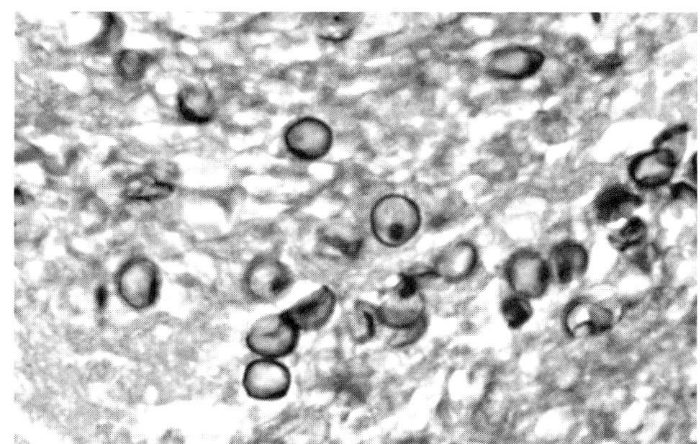

Figure 13.5 Histopathology of PCP showing characteristic cysts (asci) with cup forms and dot-like cyst wall thickenings (GMS stain).
Source: Dr. Edwin P. Ewing, Jr. (Image No. 960), CDC PHIL.

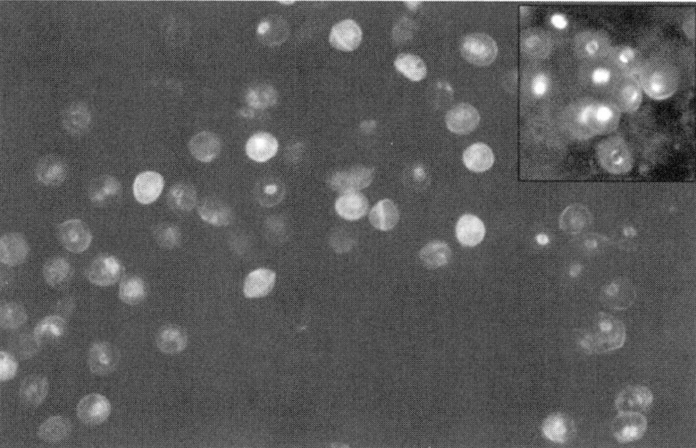

Figure 13.6 *Pneumocystis* pneumonia. A cytocentrifuge smear of BAL fluid, Calcofluor white stain. *Inset*: The double parenthesis-shaped, or comma-shaped, areas in the asci (cysts) have brighter fluorescence (Harrington, 2008).
Source: Used with permissions from Dr. Brian J. Harrington and the American Society for Clinical Pathology, © ASCP.

stains on 313 replicate clinical respiratory specimens, most of which were cytocentrifuge smears of BAL fluid (Procop et al., 2004) (Table 13.2).

More true positive specimens were detected by the Merifluor® *Pneumocystis* immunofluorescence but this number was not significantly different from those detected by Calcofluor white or GMS. The number of true positives detected by GMS and Calcofluor white were not significantly different from each other. Turning to Diff-Quik stain, true positives were significantly lower than results with the other three stains. Merifluor *Pneumocystis* stain had significantly more false positives than GMS, Calcofluor white, and Diff-Quik. Diff-Quik is not an effective means of screening for *P. jirovecii* as a primary staining method because of low sensitivity. Although the Merifluor method was most sensitive the higher number of false positives reduced the positive predictive value to 81.9%. If it is used as a primary screening method positive results should be confirmed with either Calcofluor white or GMS. Both of these latter stains are highly specific for *P. jirovecii*, with acceptable positive and negative predictive values.

The direct fluorescent antibody method (DFA) (Merifluor *Pneumocystis*, Meridian Bioscience, Cincinnati, OH) utilizes monoclonal antibody staining and has certain advantages compared with colorimetric stains:

- It excludes *Candida* which may be morphologically confused with *P. jirovecii*.
- The Meridian Merifluor® *Pneumocystis* test detects both trophic forms and asci (cysts).

Calcofluor White This fluorescent brightener (Fig. 13.6) stains the ascus (cyst) wall against a dark background, and highlights kidney bean-shaped (double parenthesis-shaped, comma-shaped, or crushed ping-pong balls) areas that have brighter fluorescence (Harrington, 2008). The brighter stained structures are the thickened portion of the ascus wall that preferentially takes up the stain. The distinctive appearance of the asci (cysts) makes it easy to differentiate them from other fungi of similar size (5–6 μm diameter) such as yeasts. Smears stained with fluorescent antibodies can be overstained with Calcofluor white solution. Calcofluor can be applied to formalin fixed paraffin embedded tissues. Commercial

Table 13.2 Comparison of Four Stains to Detect *Pneumocystis jirovecii* in Respiratory Clinical Specimens[a]

Stain	Per cent			
	Sensitivity	Specificity	Positive predictive value	Negative predictive value
Calcofluor white	73.8	99.6	98	93.4
MeriFluor®, immunofluorescence Meridian Biosciences)	90.8	94.7	81.9	97.5
Diff-Quik (modified Wright's stain)	48.4	99.6	96.9	88.0
GMS	76.9	99.2	96.2	94.2

[a]Data from Procop et al. (2004).

sources for Calcofluor and application notes are found in Chapter 2.

In summary, the diagnostic sensitivity of stained smears of respiratory tract specimens depends on specimen quality and the observer's experience. Sensitivity for induced sputum ranges from <50% to >90%; bronchoscopy with BAL fluid, 90–99%; transbronchial biopsy, 95–100%; and open lung biopsy, 95–100%. Therapy can be started before the diagnosis is confirmed.

Histopathology

Invasive lung biopsy procedures are used less often but have a high level of sensitivity because of access to more tissue. When there are coinfections, lung biopsy procedures can aid in their diagnosis and that of Kaposi's sarcoma. Stains used for routine tissue sections (H&E) or the Diff-Quik modification of Wright's stain are applicable and reveal the characteristics of PCP: foamy eosinophilic exudate (Fig. 13.3). (Please also see Section 13.9.1, Host Factors.) The GMS stain is applicable to tissue sections and stains the ascus (cyst) wall (Fig. 13.5).

Fungitell® Test for ß-(1 → 3)-d-glucan

Measurement of ß-(1 → 3)-D-glucan levels in the serum, plasma, or BAL fluid may be useful in the diagnosis of PCP (Marty et al., 2007). The "Fungitell" test (Associates of Cape Cod, Inc., East Falmouth, MA.) does not distinguish between glucan from various fungi. Sixteen patients had serum ß-glucan measurements near the time of diagnosis of PCP. ß-Glucan testing was positive (≥80 pg/mL) in 15 of 16 (93.8%) patients with a median ß-glucan value greater than 500 pg/mL (range, 141 to >500 pg/mL), regardless of the relative quantity of cysts observed by immunofluorescence in BAL fluid. Because it is noninvasive, the Fungitell test may be useful as a rapid presumptive screen, before confirmatory microbiologic tests. The Fungitell test is performed as a service by Beacon Diagnostics Laboratory, East Falmouth, Massachusetts.

PCR Diagnosis

At first PCR methods were applied to BAL specimens but, as the method's sensitivity increased, *P. jirovecii* DNA was detected in nasopharyngeal aspirates, oral rinses, and oropharyngeal swabs. Deep nasal swab specimens are also suitable specimens for PCR. They are obtained using a sterile, saline-moistened, cotton-tipped plastic swab, inserted approximately 4–5 cm into one nostril at an oblique angle toward the turbinates (avoiding the nasopharynx), rotated when resistance is found, extracted, and placed into a tube containing 0.5 mL of sterile saline.

PCR detection of *P. jirovecii* DNA in oral wash samples from HIV-positive patients has high sensitivity and specificity (sensitivity 78%, specificity 100%) but *interpretation of PCR data is complicated because PCR cannot distinguish between colonization, carriage, asymptomatic infection, and subclinical infection*.

Illustrative Example *Pneumocystis jirovecii* DNA was detected by nested PCR in healthy older adults (>65 years old) in 14 (12.8%) of 109 oropharyngeal wash specimens, in 7 (10.6%) of 66 nasal swab specimens, and in 14 (21.5%) of 65 paired oropharyngeal wash and nasal swab specimens (Vargas et al., 2010).

PCR results are influenced by the patient population. Comorbidity, such as COPD, increases carriage of *P. jirovecii*. The use of TMP-SMX prophylaxis will reduce numbers of *P. jirovecii*. A negative PCR test reasonably rules out active PCP with the exception when inhibitors of PCR in the specimen cause false-negative results. The following examples illustrate progress in PCR diagnosis of pneumocystosis.

Nested PCR targeting the mt LSU rDNA was compared to single cycle PCR directed against that gene target and also against other gene targets including *DHPS, MSG*, and 18S rDNA genes (Robberts et al., 2007). In addition, real-time TaqMan PCR was directed against the 5S rDNA gene target. Various patient specimens were tested including sputum, tracheal aspirates, BAL, fresh biopsy tissue, and formalin-fixed paraffin embedded tissue. Of these various tests the nested PCR of the mt LSU rDNA was superior having a sensitivity of 78% and a specificity of 100%. Real-time PCR had a high number of false-negative results in this study. Indirect immunofluorescence tested on the same panel of clinical specimens had high sensitivity, 97.5%, but low specificity, 50.0%, with evidence of high false-positive results.

Real-time PCR with primers complementary to *P. jirovecii* heat shock protein 70 (*HSP70*), and a 5′-FAM-dye labeled fluorescent probe, was performed with the Rotor-Gene® (Qiagen, Inc.) using as specimens BAL fluid from 60 patients with cytochemically confirmed PCP. The copy number of *HSP70* ranged from 13 to 18,608 copies/patient specimen. Among 71 patients with alternative diagnoses, 6 episodes (8%) were positive for *P. jirovecii HSP70* with copy numbers ranging from 6 to 590 per specimen. The sensitivity and specificity were calculated to be 98% and 96%, respectively (Huggett et al., 2008).

Detection of mRNA by reverse transcriptase PCR provides information about the viability of *P. jirovecii*, since mRNA is short-lived and rapidly degraded. The target transcript was the *Phsb1* gene, and when the assay was applied to respiratory specimens it demonstrated a sensitivity of 100% and a specificity of 86% on BAL fluid specimens, and a sensitivity of 65% with a specificity of 85% on induced sputum (reviewed by Huang et al., 2006).

The optimization of sample preparation for PCR and further refinements in real-time quantitative PCR and reverse transcriptase PCR can be expected to improve the diagnostic efficacy of these tests.

SELECTED REFERENCES FOR PNEUMOCYSTOSIS

Aliouat-Denis CM, Martinez A, Aliouat el M, Pottier M, Gantois N, Dei-Cas E, 2009. The *Pneumocystis* life cycle. *Mem Inst Oswaldo Cruz* 104: 419–426.

Aviram G, Fishman JE, Boiselle PM, 2003. Thoracic manifestations of AIDS. *Appl Radiol* 32: 11–21.

Beck JM, Cushion MT, 2009. *Pneumocystis* Workshop: 10th Anniversary Summary. *Eukaryot Cell* 8: 446–460.

Bouckenooghe AR, Shandera WX, 2002. The epidemiology of HIV and AIDS among Central American, South American, and Caribbean immigrants to Houston, Texas. *J Immigr Health* 4: 81–86.

Carmona EM, Lamont JD, Xue A, Wylam M, Limper AH, 2010. *Pneumocystis* cell wall beta-glucan stimulates calcium-dependent signaling of IL-8 secretion by human airway epithelial cells. *Respir Res* 11: 95.

Castro JG, Manzi G, Espinoza L, Campos M, Boulanger C, 2007. Concurrent PCP and TB pneumonia in HIV infected patients. *Scand J Infect Dis* 39: 1054–1058.

Catherinot E, Lanternier F, Bougnoux ME, Lecuit M, Couderc LJ, Lortholary O, 2010. *Pneumocystis jirovecii* pneumonia. *Infect Dis Clin North Am* 24: 107–138.

Crothers K, Beard CB, Turner J, Groner G, Fox M, Morris A, Eiser S, Huang L, 2005. Severity and outcome of HIV-associated *Pneumocystis* pneumonia containing *Pneumocystis jirovecii* dihydropteroate synthase gene mutations. *AIDS* 19: 801–805.

Cushion MT, 1998. Taxonomy, genetic organization, and life cycle of *Pneumocystis carinii*. *Semin Respir Infect* 13: 304–312.

Cushion MT, 2004. Comparative genomics of *Pneumocystis carinii* with other protists: Implications for life style. *J Eukaryot Microbiol* 51: 30–37.

Davis JL, Fei M, Huang L, 2008. Respiratory infection complicating HIV infection. *Curr Opin Infect Dis* 21: 184–190.

Fujii T, Nakamura T, Iwamoto A, 2007. *Pneumocystis* pneumonia in patients with HIV infection: Clinical manifestations, laboratory findings, and radiological features. *J Infect Chemother* 13: 1–7.

Giner JL, Zhao H, Amit Z, Kaneshiro ES, 2004. Sterol composition of *Pneumocystis jirovecii* with blocked 14 alpha-demethylase activity. *J Eukaryot Microbiol* 51: 634–643.

Glück T, Geerdes-Fenge HF, Straub RH, Raffengerg M, Lang B, Lode H, Schölmerich J, 2000. *Pneumocystis carinii* pneumonia as a complication of immunosuppressive therapy. *Infection* 28: 227–230.

Harrington B, 2008. Staining of cysts of *Pneumocystis jiroveci* (*P. carinii*) with the fluorescent brighteners Calcofluor white and Uvitex 2B: A review. *Lab Med*. 39: 731–735.

Heckman DS, Geiser DM, Eidell BR, Stauffer RL, Kardos NL, Hedges SB, 2001. Molecular evidence for the early colonization of land by fungi and plants. *Science* 293: 1129–1133.

Helweg-Larsen J, 2004. *Pneumocystis jiroveci*. Applied molecular microbiology, epidemiology and diagnosis. *Danish Med Bull* 51: 251–273.

Huang L, Morris A, Limper AH, Beck JM, and Participants ATSPW, 2006. An Official ATS Workshop summary: Recent advances and future directions in *Pneumocystis* pneumonia (PCP). *Proc Am Thoracic Soc* 3: 655–664.

Huggett JF, Taylor MS, Kocjan G, Evans HE, Morris-Jones S, Gant V, Novak T, Costello AM, Zumla A, Miller RF, 2008. Development and evaluation of a real-time PCR assay for detection of *Pneumocystis jirovecii* DNA in bronchoalveolar lavage fluid of HIV-infected patients. *Thorax* 63: 154–159.

Itatani CA, 1994. Ultrastructural demonstration of a pore in the cyst wall of *Pneumocystis carinii*. *J Protozool* 80: 644–648.

Jacobs RA, Guglielmo BJ, 2007. Anti-Infective Chemotherapeutic & Antibiotic Agents, Chapter 37, Sulfanilamides and Antifolate Drugs, pp. 1606–1607, *in*: McPhee SA, Papadakis MA, Tierney, KM Jr (eds.), *Lange Current Medical Diagnosis & Treatment 2007*, 46th ed. McGraw-Hill/Medical, New York.

Jalil A, Moja P, Lambert C, Perol M, Cotte L, Livrozet JM, Boibieux A, Vergnon JM, Lucht F, Tran R, Contini C, Genin C, 2000. Decreased production of local immunoglobulin A to *Pneumocystis carinii* in bronchoalveolar lavage fluid from human immunodeficiency virus-positive patients. *Infect Immun* 68: 1054–1060.

Janeway CA Jr, Travers P, Walport M, Shlomchik MJ, 2001. *Immunobiology* (5th ed.), *The Immune System in Health and Disease*. Garland Science, New York. (Available at the NCBI Bookshelf, http://www.ncbi.nlm.nih.gov/bookshelf/br.fcgi?book = imm.)

Kamboj M, Weinstock D, Sepkowitz KA, 2006. Progression of *Pneumocystis jirovecii* pneumonia in patients receiving echinocandin therapy. *Clin Infect Dis* 43: e92–e94.

Kaplan JE, Benson C, Holmes KK, Brooks JT, Pau A, Masur H, 2009. Guidelines for prevention and treatment of opportunistic infections in HIV-infected adults and adolescents. *MMWR Morb Mortal Wkly Rep* 58: 1–198.

Kelly MN, Shellito JE, 2010. Current understanding of *Pneumocystis* immunology. *Future Microbiol* 5: 43–65.

Kovacs JA, Gill VJ, Meshnick S, Masur H, 2001. New insights into transmission, diagnosis, and drug treatment of *Pneumocystis carinii* pneumonia. *JAMA* 286: 2450–2460.

Limper AH, Thomas CF Jr, Anders RA, Leof EB, 1997. Interactions of parasite and host epithelial cell cycle regulation during *Pneumocystis carinii* pneumonia. *J Lab Clin Med* 130: 132–138.

Martin L, 1999. Chapter 5, PaO_2, SaO_2 and Oxygen Content, *in*: *All You Really Need to Know to Interpret Arterial Blood Gases*, 2nd ed. Lippincott Williams & Wilkins, Philadelphia.

Marty FM, Koo S, Bryar J, Baden LR, 2007. (1 → 3)-ß-D-glucan assay positivity in patients with *Pneumocystis (carinii) jirovecii* pneumonia. *Ann Intern Med* 147: 70–72.

Morris A, Lundgren JD, Masur H, Walzer PD, Hanson DL, Frederick T, Huang L, Beard CB, Kaplan JE, 2004. Current epidemiology of *Pneumocystis* pneumonia. *Emerging Infect Dis* 10: 1713–1720.

Pascale JM, Shaw MM, Durant PJ, Amador AA, Bartlett MS, Smith JW, Gregory RL, McLaughlin GL, 1999. Intranasal immunization confers protection against murine *Pneumocystis carinii* lung infection. *Infect Immun* 67: 805–809.

Peterson JC, Cushion MT, 2005. *Pneumocystis*: Not just pneumonia. *Curr Opin Microbiol* 8: 393–398.

Ponce CA, Gallo M, Bustamante R, Vargas SL, 2010. *Pneumocystis* colonization is highly prevalent in the autopsied lungs of the general population. *Clin Infect Dis* 50: 347–353.

Pop SM, Kolls JK, Steele C, 2006. *Pneumocystis*: Immune recognition and evasion. *Int. J Biochem Cell Biol* 38: 17–22.

Procop GW, Haddad S, Quinn J, Wilson ML, Henshaw NG, Reller LB, Artymyshyn RL, Katanik MT, Weinstein MP, 2004. Detection of *Pneumocystis jirovecii* in respiratory specimens by four staining methods. *J Clin Microbiol* 42: 3333–3335.

Redhead SA, Cushion MT, Frenkel JK, Stringer JR, 2006. *Pneumocystis* and *Trypanosoma cruzi*: Nomenclature and typifications. *J Eukaryot Microbiol* 53: 2–11.

ROBBERTS FJL, LIEBOWITZ LD, CHALKLEY LJ, 2007. Polymerase chain reaction detection of *Pneumocystis jirovecii*: Evaluation of 9 assays. *Diagn Microbiol Infect Dis* 58: 385–392.

SHIPLEY TW, KLING HM, MORRIS A, PATIL S, KRISTOFF J, GUYACH SE, MURPHY JE, SHAO X, SCIURBA FC, ROGERS RM, RICHARDS T, THOMPSON P, MONTELARO RC, COXSON HO, HOGG JC, NORRIS KA, 2010. Persistent pneumocystis colonization leads to the development of chronic obstructive pulmonary disease in a nonhuman primate model of AIDS. *J Infect Dis* 202: 302–312.

SOARES EC, SARACENI V, LAURIA L de M, PACHECO AG, DUROVNI B, CAVALCANTE SC, 2006. Tuberculosis as a disease defining acquired immunodeficiency syndrome: Ten years of surveillance in Rio de Janeiro, Brazil. *J Bras Pneumol* 32: 444–448.

SOWDEN E, CARMICHAEL AJ, 2004. Autoimmune inflammatory disorders, systemic corticosteroids and *Pneumocystis* pneumonia: A strategy for prevention. *BMC Infect Dis* 4: 42.

SRITANGRATANAKUL S, NUCHPRAYOON S, NUCHPRAYOON I, 2004. *Pneumocystis* pneumonia: An update. *J Med Assoc Thai* 87(Suppl 2): S309–S317.

STEELE C, SHELLITO JE, KOLLS JK, 2005. Immunity against the opportunistic fungal pathogen *Pneumocystis*. *Med Mycol* 43: 1–19.

SWAIN SD, WRIGHT TW, DEGEL PM, GIGLIOTTI F, HARMSEN AG, 2004. Neither neutrophils nor reactive oxygen species contribute to tissue damage during *Pneumocystis* pneumonia in mice. *Infect Immun* 72: 5722–5732.

TASAKA S, KOBAYASHI S, KAMATA H, KIMIZUKA Y, FUJIWARA H, FUNATSU Y, MIZOGUCHI K, ISHII M, TAKEUCHI T, HASEGAWA N, 2010. Cytokine profiles of bronchoalveolar lavage fluid in patients with pneumocystis pneumonia. *Microbiol Immunol* 54: 425–433.

THOMAS CF Jr, LIMPER AH, 2004. Medical progress—*Pneumocystis* pneumonia. *N Engl J Med* 350: 2487–2498.

THOMAS CF Jr, LIMPER AH, 2007. Current insights into the biology and pathogenesis of *Pneumocystis* pneumonia. *Nature Rev Microbiol* 5: 298–308.

TILEY SM, MARRIOTT DJE, HARKNESS JL, 1994. An evaluation of four methods for the detection of *Pneumocystis carinii* in clinical specimens. *Pathology* 26: 325–328.

VARGAS SL, PONCE CA, GIGLIOTTI F, ULLOA AV, PRIETO S, MUÑOZ MP, HUGHES WT, 2000. Transmission of *Pneumocystis carinii* DNA from a patient with *P. carinii* pneumonia to immunocompetent contact health care workers. *J Clin Microbiol* 38: 1536–1538.

VARGAS SL, PIZARRO P, LÓPEZ-VIEYRA M, NEIRA-AVILÉS P, BUSTAMANTE R, PONCE CA, 2010. *Pneumocystis* colonization in older adults and diagnostic yield of single versus paired noninvasive respiratory sampling. *Clin Infect Dis* 50: e19–e21.

VRAY M, GERMANI Y, CHAN S, DUC NH, SAR B, SARR FD, BERCION R, RAHALISON L, MAYNARD M, L'HER P, CHARTIER L, MAYAUD C, 2008. Clinical features and etiology of pneumonia in acid-fast bacillus sputum smear-negative HIV-infected patients hospitalized in Asia and Africa. *AIDS* 22: 1323–1332.

WALZER PD, SMULIAN AG, 2005. Chapter 269, *Pneumocystis* species, *in*: MANDELL GL, BENNETT JE, DOLIN R (eds.), *Mandell, Douglas and Bennett's Principles and Practice of Infectious Diseases*, 6th ed., Part 2. Elsevier Churchill-Livingstone, Philadelphia.

WANG SH, ZHANG C, LASBURY ME, LIAO CP, DURANT PJ, TSCHANG D, LEE CH, 2008. Decreased inflammatory response in Toll-like receptor 2 knockout mice is associated with exacerbated *Pneumocystis* pneumonia. *Microbes Infect* 10: 334–341.

WARD MM, DONALD F, 1999. *Pneumocystis carinii* pneumonia in patients with connective tissue diseases: The role of hospital experience in diagnosis and mortality. *Arthritis Rheum* 42: 780–789.

WONG KH, LEE SS, CHAN KC, 2006. Twenty years of clinical human immunodeficiency virus (HIV) and acquired immunodeficiency syndrome (AIDS) in Hong Kong. *Hong Kong Med J* 12: 133–140.

World Bank, 2006. *Disease Control and Priorities Project—Infectious Diseases*. The International Bank for Reconstruction and Development/The World Bank, Washington, DC.

QUESTIONS

The Answer Key to multiple choice questions may be found at the end of the book.

1. Match the following staining methods with the results of direct examination of respiratory specimens for *Pneumocystis jirovecii*.

A. Calcofluor white (Fungi-Fluor®)

B. Diff-Quik

C. GMS

D. MeriFluor® *Pneumocystis* fluorescent antibodies

1. Most sensitive

2. Highlights kidney bean-shaped areas of ascus

3. Not a good primary screen because of low sensitivity

4. Ascus wall stained

2. Which of the following methods of obtaining a respiratory specimen has the highest sensitivity to detect *P. jirovecii*?

A. All are equivalent

B. Bronchoalveolar lavage

C. Induced sputum

D. Transbronchial biopsy

3. Which component of trimethoprim-sulfamethoxazole has the most potent activity to treat PCP?

A. Both are needed

B. Sulfamethoxazole

C. Trimethoprim

4. Which one or more of the following is/are the most important reason(s) why *Pneumocystis* was regarded as a protozoan, and not a fungus?

1. It could not be cultured in vitro.

2. PCP does not respond to AmB or to azole antifungal agents.

3. The environmental source is unknown.

4. There are neither budding yeasts nor hyphae seen in lung tissue.

Which are correct?

A. 1 + 2

B. 2 + 3

C. 2 + 4

D. 3 + 4

5. Which of the following has caused scientists to reclassify *P. jirovecii* as a fungus?

A. Air sampling has captured *P. jirovecii* DNA, suggesting an environmental source.

B. Microscopic evidence that the cyst is really an ascus with ascospores.

C. *Pneumocystis jirovecii* can be cultured in vitro for a few generations.

D. Phylogenetic analysis of 18S rDNA, translation elongation factor 3, and mitochondrial genes.

6. What groups in the population are likely to be the reservoir for *P. jirovecii*?

7. What drug classes are effective against PCP? Are there any drugs developed as antifungal agents that have activity against *P. jirovecii*?
8. What is the current chemoprophylaxis for PCP in HIV-positive patients? What impact has it had on the incidence of PCP? Has the use of chemoprophylaxis resulted in resistant strains of *P. jirovecii*? What is the major gene target for mutations affecting drug susceptibility?
9. If *P. jirovecii* does not invade the lung parenchyma, how does it cause potentially life-threatening pneumonia? How does the patient's immune status affect the pathogenesis of disease.
10. Compare the expected mortality after the first episode of PCP in HIV-positive patients with cancer patients. What other groups are at risk to develop PCP?

Part Four

Systemic Mycoses Caused by Opportunistic Hyaline Molds

Molds are omnipresent in the environment, worldwide, producing abundant, asexual spores that are easily dispersed in the *aerospora* and are routinely inhaled, causing no infections[1] in persons with intact immune and endocrine systems. Their asexual spores are *conidia,* except for the *Mucoromycotina*, where they are produced in sporangia, and are called sporangiospores.

Several of these molds have no known sexual stage and are called "hyphomycetes", an older classification referring to mitosporic fungi that produce their conidia on simple conidiophores arising directly from vegetative hyphae. We will not use that term. The term "hyaline" refers to the lack of cell wall pigment in vivo or in vitro. Exposure to these molds is universal, but disease is uncommon and linked to host factors:

- Broad-spectrum antibiotic therapy
- Chronic granulomatous disease of childhood
- Chronic lung disease (e.g., COPD, cystic fibrosis)
- Diabetes mellitus
- Immunosuppresive therapy
  - Systemic corticosteroid therapy
  - Cytotoxic drugs, X-irradiation used to treat patients for management of hematologic malignancy, organ and stem cell transplantation, and solid tumors

What are the major genera of opportunistic hyaline molds causing human disease? They are *Aspergillus, Fusarium, Pseudallescheria/Scedosporium*, and *Rhizopus*.[2] The morphologic features of opportunistic hyaline molds as they invade tissue are shown in Table 14.1. When causing systemic infections, these molds tend to invade blood vessels; that is, they are *angioinvasive*, causing hemorrhage, thrombosis, and infarcts. In the correct setting they can cause fulminant, rapidly progressing, and devastating disease.

Invasive fungal infections caused by hyaline molds are a major complication of treating persons with hematologic malignancies, those receiving stem cell transplants, and, to a lesser degree, those receiving solid organ transplants. The therapy for their primary disease with immunosuppressive agents creates the clinical setting in which these normally benign molds can cause rapidly progressing invasive sinopulmonary and extrapulmonary disease, which is fatal if not diagnosed promptly and treated with the appropriate antifungal agent or combination of agents.

The frequency of hyaline mold mycosis in hematopoietic stem cell transplant (SCT) recipients is related to changes in transplant procedures:

- Delayed immune reconstitution in patients receiving transplants with T-cell depletion
- Increased risk and severity of graft versus host disease (GVHD) in allogeneic SCTs

[1]Fungal conidia are allergenic. Please see Chapter 1, Section 1.1, Topics Not Covered, or Receiving Secondary Emphasis.

[2]Fungi containing melanin in their cell walls, are discussed in Chapters 18–20.

Fundamental Medical Mycology, First Edition. By Errol Reiss, H. Jean Shadomy and G. Marshall Lyon, III.
© 2012 Wiley-Blackwell. Published 2012 by John Wiley & Sons, Inc.

- Increased upper age limit for transplants with HLA-mismatched or unrelated donors
- Nonmyeloablative transplants
- Peripheral blood instead of bone marrow as a source of stem cells

GVHD is associated with invasive mold mycosis because it prolongs the immunodeficient state and because of the need for high-dose corticosteroids and other powerful immunosuppressants. For example, use of infliximab (anti-TNF-α) is associated with aspergillosis and mucormycosis. Recipients of nonmyelobalative transplant have a higher frequency of hyaline mold mycosis associated with severe acute GVHD, chronic extensive GVHD, and cytomegalovirus disease. The periods of greatest risk are when there is prolonged neutropenia or in non-neutropenic patients with GVHD. Solid organ transplant recipients including liver, lung, and heart transplants are at risk for invasive mold mycosis.

Aspergillus species are by far the most frequent mold mycoses in transplant recipients, but infections with *Fusarium* species, *Scedosporium* species, and *Mucorales* are becoming more common. Changes in transplant procedures involving more severe immunosuppression is playing a role. The mycoses with non-*Aspergillus* species molds tend to be disseminated with a poor prognosis because of their greater resistance to antifungal agents.

The clinical forms of infectious disease caused by opportunistic hyaline molds include:

- Acute invasive sinusitis and rhinocerebral dissemination
- Cutaneous infections both primary and superinfection of burns
- Invasive pulmonary disease with or without extrapulmonary dissemination

Chapter 14

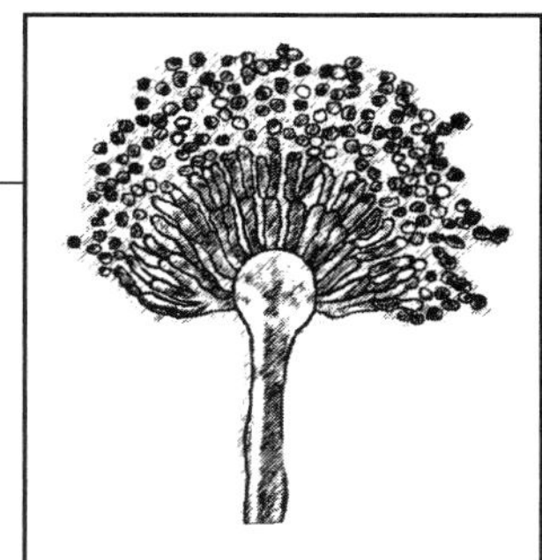

Aspergillosis

14.1 ASPERGILLOSIS AT-A-GLANCE

- *Introduction/Disease Definition.* Aspergillosis includes a broad range of clinical forms: the serious form is invasive pulmonary disease in the immunocompromised host caused by *Aspergillus fumigatus* and related species.
- *Etiologic Agents.* In order of prevalence they are *A.fumigatus* > *A. flavus* > *A. terreus* > *A. niger*. *A. fumigatus*, the major pathogen, is thermotolerant, growing up to 48°C, and is found in composts at even higher temperatures.
- *Geographic Dsitribution/Ecologic Niche.* Worldwide, aspergilli are among the most common environmental molds. *Aspergillus fumigatus* is a cellulose decomposer in composts. Conidia are present in the air, including hospital air. *A. flavus* is found naturally in postharvest infestation of grains and nuts.
- *Epidemiology.* Invasive aspergillosis is associated with the immunocompromised host, especially patients with hematologic malignancy, stem cell, or lung transplants.
    - *Incidence*. The sharpest increase was in the early 1990s, followed by a gradual decline with the advent of extended spectrum triazole therapy (e.g., voriconazole), improved diagnostic imaging, and an antigenemia test.
- *Risk Groups/Factors.* They include *CGD* in children, corticosteroid therapy, cytomegalovirus, cystic fibrosis, cytotoxic drugs, diabetes, GVHD, hematologic malignancy, organ transplants, prolonged neutropenia, radiation therapy, solid tumors, stem cell transplants, and surgical wound infections.
- *Transmission.* Infection is via the respiratory route—inhalation of airborne conidia. Less commonly infection may occur by conidia alighting on injured skin, by penetrating injuries, or via fomites such as contaminated or nonsterile tape used to fasten bandages or catheters.
- *Determinants of Pathogenicity.* Angioinvasive; pathogenic factors—catalase, superoxide dismutase, proteinase(s), phospholipase, hemolysin, and gliotoxin.
- *Clinical Forms.*
    - Invasive pulmonary aspergillosis in the immunocompromised host—tracheobronchitis and extrapulmonary dissemination
    - Allergic bronchopulmonary aspergillosis
    - Colonizing aspergilloma
    - Intoxication: ingestion of peanuts, corn contaminated with *A. flavus* toxin (aflatoxin)
    - Keratitis (post-traumatic)
    - Otomycosis
    - Sinusitis
- *Therapy.* Primary therapy with voriconazole (VRC) is superior to amphotericin B (AmB), to AmB-lipid derivatives, or to itraconazole (ITC). Choices for salvage therapy include monotherapy with AmB, caspofungin (CASF), or posaconazole (PSC), or combination therapy.
- *Laboratory Detection, Recovery, and Identification.* Sterile site isolation is preferred; growth from a nonsterile site may be a contaminant, except in patients with neutropenia or GVHD. Identification is via direct microscopic examination of respiratory secretions or histopathologic sections; antigen detection—galactomannan EIA, PCR from clinical specimens, genetic identification for uncommon species, or nonsporulating isolates.

Fundamental Medical Mycology, First Edition. By Errol Reiss, H. Jean Shadomy and G. Marshall Lyon, III.
© 2012 Wiley-Blackwell. Published 2012 by John Wiley & Sons, Inc.

14.2 INTRODUCTION/DISEASE DEFINITION

Aspergillosis is an opportunistic invasive mycosis caused by hyaline molds (Table 14.1)—selected members of the genus *A.* (*A.fumigatus* > *A.flavus* > *A.terreus* > *A.niger*) and *rarely* by other *Aspergillus* species. Aspergillosis encompasses a broad range of clinical forms including:

- Invasive aspergillosis:
  - Pulmonary aspergillosis (the most important clinical form); tracheobronchitis
  - Extrapulmonary dissemination; bones and joints, cardiac, cerebral abscess, cutaneous, endocarditis, esophageal and gastrointestinal, eyes, renal
- Colonizing aspergilloma
- Allergic bronchopulmonary aspergillosis
- Other distinct clinical forms including endocarditis following implantation of prosthetic heart valves, intoxication from mycotoxins, otomycosis, and sinusitis, including allergic and invasive forms

Aspergillus species are among the most common fungi in the human environment. The mean conidia/m^3 in outdoor air is variable; a mean of 45 conidia/m^3 was measured over a 10 year period in the vicinity of a U.S. hospital (Falvey and Streifel, 2007) (Table 14.2). The density of *Aspergillus* conidia in hospital air is usually much lower than outside air and is influenced by nearby construction. *A. fumigatus*, the major etiologic agent, is thermotolerant and can thrive in the elevated temperatures of composts where it functions as a cellulose decomposer. Conidia counts near composting facilities are higher than in ambient air (please see Section 14.5, Geographic Distribution/Ecologic Niche).

Table 14.2 Air Sampling for *Aspergillus* Species at a Minnesota Hospital over a 10 year Period

Hospital area	*Aspergillus* colony forming units/m^3 (37°C incubation temperature)	Percentage of air samples positive for *Aspergillus*
Lobby	6.7	66
Pediatric BMT unit with HEPA filtration	1.2	27[a]
General patient care areas	2.3	54
Outdoor site	45	94

[a] *Aspergillus* present despite HEPA filtration.

Source: Falvey and Streifel (2007). Used with permission from Elsevier Publishers, © 2007.

14.3 CASE PRESENTATIONS

Case Presentation 1. Cough in a 24-yr-old Man After Renal Transplantation (Burton et al., 1972)

History

Ten weeks after receiving a renal allograft from his sister the patient, a 24-yr-old man, was discharged from the hospital with normal creatinine clearance. His daily regimen included 30 mg prednisone and 150 mg azathioprine. He cleaned his pet pigeon's coop, and 2 days later developed a cough and night sweats.

The patient was readmitted 5 days after his hospital discharge. His serum creatinine doubled and acute rejection was considered. The prednisone dose was increased to 200 mg/day. Cough, chest pain, and a 38.9°C fever

Table 14.1 Morphologic Features of Fungi that Occur as Hyaline Hyphae in Tissue

Property of hyphae	*Aspergillus* species	*Rhizopus* and other *Mucorales*	*Pseudallescheria* and *Scedosporium*	*Fusarium* species
Width (μm)	3–6	5–20	2–5	3–8
Contours	Parallel	Irregular	Parallel	Parallel
Branching pattern	Dichotomous	Haphazard	Dichotomous but more haphazard than *Aspergillus*	Right-angle but acute angle also occurs
Orientation of branches	Parallel	Random	Random	Random
Frequency of septations	Frequent	Rare, or not seen	Frequent	Frequent
Angioinvasion	Yes	Yes	Yes	Yes

Source: Adapted from Chandler and Watts (1987) Table 10, p. 87. Used with permission from the American Society for Clinical Pathology. Copyright © 1987, ASCP.

ensued. A sputum specimen grew *S. pneumoniae* and penicillin G was prescribed. A chest radiograph revealed a small infiltrate in his right upper lobe. Five days after readmission, his cough worsened, he developed shaking chills and a spiking 41°C fever. Eighteen days after readmission his chest radiograph showed a progressing infiltrate and cavity; within 7 days he became moribund.

Open lung biopsy and fresh touch smear of his lung showed branching septate hyphae with conidiophores (*stipes* with *vesicles*). Culture on SDA agar grew a blue-green mold. Histology of the lung section (with PAS stain) confirmed a fungus.

Antibiotics and azathioprine were stopped, prednisone was reduced to 15 mg/day, and AmB-deoxycholate was started and advanced to 40 mg/day.

Diagnosis

The diagnosis was invasive pulmonary aspergillosis, secondary to immunosuppressive therapy for maintenance of renal allograft.

Clinical Course

His fever continued and his pulmonary lesions became cavities. Ten days after initiation of AmB, he developed acute endophthalmitis. The diagnosis was revised to include invasive pulmonary aspergillosis with CNS dissemination. A global aspirate yielded a fungus. Enucleation of his eye was performed.

More neurologic signs developed. His brain scan was consistent with abscesses. AmB was maintained at 1.2 mg/kg/day and he slowly improved over the next 10 weeks. The patient insisted that the AmB dose be widened to every 14 days. The total dose over 12 months of therapy was 2.5 g. His renal allograft was accepted, and there was no recurrence of aspergillosis.

Comment

Diagnosis was facilitated by the presence of hyphae with a conidiophore and conidia in the fresh touch preparation of lung. Even though diagnosis by biopsy was made only 9 days after first minimal symptoms, it was almost too late. Questions relative to this case included the following:

1. *What is the case/fatality rate for invasive aspergillosis in kidney transplant recipients who receive standard antifungal therapy?* According to Abbott et al. (2001), who surveyed 33,420 renal transplant recipients in the U.S. Renal Data System from 1994 to 1997, the mortality of those hospitalized for aspergillosis was 12% after 1 year, 38% after 2 years, and 60% after 3 years post-transplant. With the advent of extended spectrum triazoles, such as VRC, in the late 1990s survival rates are expected to increase (please see Section 14.11, Therapy).

2. *When will Aspergillus be seen in tissue sections with conidiophores and sporing heads?* It will be seen when the lung lesion is adjacent to an air space (a rare occurrence and fortuitous for the diagnosis of this case).

3. *What is the differential diagnosis for invasive aspergillosis?* It is infection with other hyaline molds including *Fusarium* species, members of the *Mucorales*, or *Pseudallescheria/Scedosporium* species.

4. *What preventative measures can be taken in the post-transplant period?* With a solid organ transplant, immunosuppression is maintained for life, so that the period of risk for opportunistic infections is extended. In this patient exposure to dusts in the pigeon coop likely resulted in inhalation of large numbers of *Aspergillus* conidia. Solid organ transplant recipients should be advised to avoid occupational or recreational exposure to sources of *Aspergillus* conidia, such as gardening, caring for pet birds, or poultry raising.

Case Presentation 2. A 29-yr-old Man with Fever Unresponsive to Antibiotics Following Conditioning Therapy for an Allogeneic Stem Cell Transplant (Klont et al., 2006)

History

A 29-yr-old man in remission of myelodysplastic syndrome[1] was referred for a T-cell-depleted hematopoietic stem cell transplant (SCT) from an unrelated donor.

Pre-transplant Conditioning and Clinical Course

Conditioning therapy consisted of cyclophosphamide, antithymocyte globulin, and total body irradiation. Twelve days before transplantation he developed fever and received meropenem. Blood cultures yielded *Staphylococcus epidermidis*, and teicoplanin was added.

Laboratory Findings

Plasma samples were tested 2×/week for galactomannan (GM, *Aspergillus* EIA; Bio-Rad). The GM index increased from 0.6 to 2.6 one day before transplantation.

[1]Myelodysplastic syndrome. *Definition*: A diverse collection of hematologic conditions united by ineffective production of blood cells and varying risks of transformation to acute myelogenous leukemia. Myelodysplastic syndromes are bone marrow stem cell disorders resulting in ineffective hematopoiesis with irreversible quantitative and qualitative defects in hematopoietic cells. The course of disease is often chronic with gradually worsening cytopenia due to progressive bone marrow failure. Approximately one-third of patients with MDS progress to AML within months to a few years.

Radiography

High-resolution CT (HRCT) imaging revealed diffuse infiltrates in both lungs, with a *halo sign* in the upper left lobe. There were no pulmonary symptoms, but there was dull percussion of the left upper lung. These findings suggested probable invasive aspergillosis. The protocol prescribed VRC.

Diagnosis

The diagnosis was invasive pulmonary aspergillosis.

Antifungal Therapy and Clinical Course

Because the patient did not tolerate voriconazole for possible aspergillosis at another hospital, IV caspofungin (CASF) was started the day of SCT with a loading dose of 70 mg on the first day, and then 50 mg/day. Cyclosporin A was added to prevent GVHD. On the third and sixth day of CASF treatment, the GM index increased to 17.4 and 32.6, and then declined. There was no evidence of clinical deterioration, and blood cultures yielded only *S. epidermidis*. A follow-up HRCT scan 1 week after the start of CASF showed slight progression of the infiltrate in the left upper lobe. After 2 weeks of CASF therapy, kidney and liver function remained normal, but a second HRCT scan revealed further progression of the lesion in the left lung.

He developed respiratory insufficiency and multi-organ failure on the 24th day of CASF treatment, coinciding with granulocyte recovery, and was transferred to the ICU. Therapy was changed to liposomal amphotericin B (200 mg/day IV), but his condition deteriorated. He died 6 days later—30 days after the allogeneic SCT. The GM index at death was 5.7.

Tissue of the upper lobe of the left lung obtained at autopsy was positive for septate hyphae, and cultures yielded *Aspergillus fumigatus*. The minimal effective concentration (MEC) of CASF was 0.125 μg/mL (susceptible), and the MIC of VRC was 0.5 μg/mL (susceptible).

Comments

The patient was previously treated with VRC at another hospital for possible pulmonary aspergillosis. Very likely he had a low-level infection or colonization before the current hospitalization. Colonization is a risk factor for development of invasive pulmonary aspergillosis. The appearance of the halo sign during the conditioning period before SCT indicated that the infection had already progressed. Salvage therapy with liposomal AmB was probably too late to reverse the progress of invasive aspergillosis in this case. The choice of primary therapy is important or decisive and salvage therapy has limited potential to achieve a cure of invasive pulmonary aspergillosis.

Hyphae invade tissue by attacking blood vessels. They are angioinvasive with a predilection for the elastic lamina of blood vessels, causing hemorrhage, mycotic emboli, and infarction of the tissue served. The concentration of circulating GM corresponds with clinical outcome in patients treated with AmB or mold-active azoles, decreasing in patients who respond to therapy, and stable or increasing in patients for whom therapy failed. This correlation is less clear for patients treated with the echinocandin, CASF. In this case, the level of circulating GM increased to a peak GM index of 32.6 and did not correspond to a clinical response to CASF therapy. This GM index is higher than that observed in many patients with invasive aspergillosis treated with drugs other than CASF (range 0.3–15.5).

The initial increase in circulating GM did not correspond to clinical deterioration, nor did it coincide with important changes in kidney or liver function, leaving treatment with CASF as the most likely explanation for the initial increase in circulating GM. Cyclosporin A therapy could increase the release of GM. Cyclosporin has antifungal activity against *A. fumigatus* and is synergistic in vitro when combined with CASF (Klont et al., 2006).

Increased circulating GM in patients treated with echinocandins is due to the mechanism of action of the drug, which fragments *Aspergillus* hyphae and increases branching at the hyphal tips. GM is released at the hyphal tips, which combined with loss of cell wall integrity at the growth points contributes to a total increase in serum GM.

14.3.1 Diagnosis

The most common symptoms of invasive pulmonary aspergillosis are nonspecific: dyspnea, chest pain, and fever, thus presenting a diagnostic problem. Prompt diagnosis is important because extrapulmonary dissemination is common in allogeneic transplant recipients with the CNS involved in a significant number of patients. Histopathology and culture remain the means of confirming the diagnosis. Radiology is an important noninvasive diagnostic aid in invasive pulmonary aspergillosis. The halo sign is an early phenomenon in more than 80% of patients, with the *air crescent* sign appearing in more than half of patients in 1–2 weeks, although these signs are less common in solid organ transplant recipients. The halo sign is regarded as indicating a better response to therapy and to improved survival. Bronchoalveolar lavage (BAL) is indicated when there are radiologic abnormalities in allogeneic SCT recipients with respiratory symptoms. Microscopy and culture of BAL fluid may be a low yield process, because many patients receive mold-active prophylactic or empirical therapy. Increased sensitivity is gained by use of nonculture methods including GM-EIA and Genetic Identification. Please see Section 14.12, Laboratory Detection, Recovery, and Identification.

14.4 ETIOLOGIC AGENTS

The genus *Aspergillus* is large, comprised of at least 150 species. Monographs are devoted to their description and classification (Klich and Pitt, 1988). Here we narrow the focus to the major clinically important species: *A. fumigatus, A. flavus, A. terreus*, and *A. niger*. Of these, *A. fumigatus* and, to a lesser extent, *A. flavus* are the most frequent clinically encountered species in invasive aspergillosis. *A. terreus* is an uncommon cause of invasive disease, except in some hospitals (please see Section 14.6, Epidemiology and Risk Groups/Factors) but has decreased susceptibility to AmB and, as such, is more difficult to treat with that drug. *A. nidulans*, while a rare cause of infection in persons with cancer, is more frequent among children with chronic granulomatous disease (CGD).

Not all *Aspergillus* molds are a threat to human health. The koji mold (*A. oryzae*) produces kojic acid, a metabolite with antibacterial and antifungal properties. *A. oryzae* is very important in Japan for the production of *sake* (rice wine), miso (soy bean paste), *shoyu* (soy sauce), *amasake* (sweet rice milk), and *shouchu* (distilled rice, brown sugar, or sweet potato liquor, which is similar to brandy).

Characteristics of the Genus *Aspergillus*

Aspergilli are *hyaline* molds with distinctive asexual or vegetative reproductive structures. Their conidiophores and conidiogenous cells consist of a stalk or "stipe" terminating in a conidial head in the shape of a vesicle covered with rows of phialides. *Aspergillus* is an *anamorphic* genus. The genus name derives from the similarity of the sporing head to an aspergillum,[2] the priest's instrument used in the Catholic mass: a perforated globe used to sprinkle holy water. For further characteristics please see Section 14.12, Laboratory Detection, Recovery, and Identification.

Mycologic Data

The *Aspergillus* species described are classed as *Trichocomaceae, Eurotiales, Eurotiomycetidae, Ascomycetes, Ascomycota*.

[2]*Definition*: From the Latin, *aspergere*, to sprinkle.

Aspergillus fumigatus Fresenius *Beitr Mykol*: 81 (1863) (anamorph).

Neosartorya fumigata O'Gorman. Fuller & Dyer (2009) (teleomorph).

Position in classification: *Trichocomaceae, Eurotiales, Ascomycota*.

A. fumigatus is heterothallic and compatible mating types, *MAT-1* and *MAT1-2*, are required for sexual reproduction (O'Gorman et al., 2009). These genes encode transcription factors which regulate sexual reproduction. Other genes in the high mobility group (HMG) domain encode sex pheromones and their receptors. In nature there is a 1:1 ratio of *MAT-1* and *MAT-1-2* isolates. Mating was achieved after 6 months incubation of compatible strains seeded on oatmeal agar plates and incubated at 30°C in the dark. Light-yellow cleistothecia (150–500 μm diameter) were formed along the junction where parental hyphae came into contact. Squash preps of cleistothecia released greenish-white lenticular (shaped like a biconvex lens) ascospores 4–5 μm in diameter with ridged ornamentation similar to ascospores of *Neosartorya assulata*. Single ascospore cultures were genotyped by RAPD, resulting in 53–67% of the progeny with unique genotypes, different from both parental strains, indicating meiotic recombination.

Aspergillus fumigatus is haploid, has 8 chromosomes, and a genome size of 28–29 Mb (two genomes have been sequenced) (Fedorova et al., 2008). As with other fungal pathogens, genomic analysis has uncovered genetic sister species of *A. fumigatus*. Molecular identification was used to reexamine a large collection of *A. fumigatus* clinical isolates from human and animal sources deposited at the Center for Pathogenic Fungi at Chiba University, Japan. Sequences of portions of the β-tubulin, hydrophobin, and calmodulin genes were subjected to phylogenetic analysis (reviewed by Balajee et al., 2007). Species within the section *Fumigati* could be divided into five clades, three of which contain clinical isolates. They are clade I, typical strains of *A. fumigatus* and *A. arvii*; clade II, species including *A. lentulus* and *A. fumisynnematus*; and clade IV, atypical strains of *A. fumigatus* including *A. viridinutans*. The implication of this work is that when *A. fumigatus* is identified on morphologic grounds alone it should be reported as "*A. fumigatus* complex." Please see Section 14.12, Laboratory Detection, Recovery, and Identification, for the genetic identification of species within the *A. fumigatus* complex and for morphologic identification of the following clinically important *Aspergillus* species:

Aspergillus flavus Link, *Magazin Ges naturf Freunde, Berlin* 3(1–2): 16 (1809) (anamorph).

Aspergillus terreus Thom and Church, *Am J Bot* 5: 85–86 (1918) (anamorph).

Aspergillus niger Tiegh, *Ann Sci Nat Bot Ser* 5, 8: 240 (1867) (anamorph).

14.5 GEOGRAPHIC DISTRIBUTION/ECOLOGIC NICHE

Aspergillus species are found worldwide as saprobes where they are active decomposers of organic matter. Aspergilli grow well on many substrates including hay, grain, and decaying vegetation. They can decompose cloth, leather, textiles, cheeses, and cured meats. *Aspergillus fumigatus* grows well at 45°C, making it a common microbe in compost piles, playing an important role in cellulose decomposition. *A. fumigatus* also grows and sporulates at colder than ambient temperatures as a blue-green mold on bread and fruit in the refrigerator. The colony color of *A. fumigatus* is also described as gray-turquoise. *Penicillium* species also are blue-green and grow in the refrigerator, and are differentiated from *Aspergillus* species by morphology of their asexual reproductive conidiophores and conidiogenous cells. Illustrations of the colony and microscopic morphology of *Aspergillus* species and, for comparison purposes, a *Penicillium* species are found in Section 14.12, Laboratory Detection, Recovery, and Identification.

A. flavus is a human, animal, and plant pathogen. This species is responsible for the postharvest infestation of corn, peanuts, rice, cottonseed, and tree nuts. About half of the *A. flavus* strains produce a potent toxin and carcinogen, *aflatoxin*, and its contamination of food and feed is of major economic and public health importance (Bennett and Klich, 2003). *A. flavus* invades seeds in the field, but it also may develop in seeds and grains of other crop species when they are improperly stored (Diener et al., 1987).

Aspergillus terreus is worldwide in distribution found in soils, more abundantly in tropical and subtropical regions, including the southern and southwestern United States. As a causative agent, *A. terreus* is notable for its innate resistance to AmB. *A. terreus* may be present in potted plants in the hospital and showerheads in hospital rooms and hospital water storage tanks (Lass-Flörl et al., 2000; Steinbach et al., 2004).

Mold in the Refrigerator

Moldy refrigerated foods and swabs of refrigerator surfaces were obtained from 66 Wisconsin households (Torrey and Marth, 1977a,b). They found that 49% of the food isolates (68 total isolates) were penicillia and 38% were aspergilli. Molds isolated from the refrigerator surfaces (11 isolates) were 36% *Aspergillus*, 27% *Penicillium*, 19% *Cladosporium*, and 9% *Rhizopus* species.

Aspergillus in Outdoor Air

Disposal of organic waste is an increasing problem and dumping in water and incineration are no longer environmentally justified practices. Composting reduces organic waste to (essentially) carbon dioxide and water. Sewage waste is mixed with wood chips and stored in piles that are periodically turned to provide oxygen to the microbes using the mixture for food. *A. fumigatus* has an important role in the high-temperature phase of composts producing conidia that become airborne in enormous numbers when the piles are moved (Burge and Rogers, 2000). Quantitative studies found that *A. fumigatus* is the dominant fungus in composted wood chips (Dillon et al., 2007).

Airborne conidial concentrations near composted wood chips (for all fungi) were reported to be up to 1.4×10^6 CFU/m^3 (Millner et al., 1994). *A. fumigatus* conidia concentration in the air ranged between 5000 and 7×10^4 CFU/m^3 at yard waste composting sites in four New Jersey communities during periods of high activity (Millner et al., 1994), and a mean *Aspergillus* conidia count of 1525 CFU/m^3 (range 0–59,227) at a suburban yard waste composting facility in northern Illinois (Hryhorczuk et al., 2001).

Molds in Certain Occupational Environments

Conidial concentration in poultry houses can vary from <100 CFU/m^3 to 2060 CFU/m^3 (in spring). This finding has implications for Case Presentation 1, describing the patient who cleaned his pet pigeon's coop after being discharged from the hospital with a renal transplant.

Aspergillus in the Hospital Environment

Hospital air often contains conidia of *A. fumigatus, A. flavus, A. terreus*, and/or *A. niger*. The density of fungal spores/m^3 is variable, usually much lower than outside air and the spore count is influenced by construction near patients at risk. In a 10 year period of air sampling at a Minnesota hospital, the mean *Aspergillus* conidia/m^3 (% of samples positive for *Aspergillus*) of air was in the hospital lobby, 6.7 (66%); the pediatric SCT ward (with HEPA filtration, 27%) 1.2; in general patient care areas, 47 (54%) (see Table 14.2) (Falvey and Streifel, 2007). An earlier edition of the Centers for Disease Control (CDC) Healthcare Infection Control Practices Advisory Committee "Guidelines for Preventing Health-care-Associated Pneumonia" (Tablan et al., 2004) advised that the spore (conidia) count should not exceed 500–700/m^3. However, there is no official standard and this statistic was omitted from the 2003 guideline. The CDC recommends steps to minimize accumulation of fungal spores via a protective environment for housing neutropenic patients including the following:

- *Directed Room Airflow*
  - HEPA filtration of incoming air.
  - High ($\geq$12) air changes per hour.
  - Laminar airflow (LAF) is not recommended routinely in the protective environment.
  - Positive air pressure in patient rooms in relation to the corridor.
  - Well-sealed room.
- *Solid Surfaces*. Sources of mold include:
  - Condensation in HVAC systems, during the "off cycle," stimulates fungi on filters or insulation in the duct-work.
  - Conidia circulate through ventilation systems unless removed by HEPA filters.
  - Fungi grow under vapor barriers in buildings.
  - Moisture (leaks, vapor barriers) increase mold growth.
  - Wet cellulose acoustic ceiling tiles can be a source.
- *Other Sources of Potential Infection*. Showerheads, shower stalls, and cold water holding tanks.

Prevention of Infection in Immunocompromised Hosts

Ventilation

Allogeneic SCT recipients should be placed in rooms with >12 air changes/h and point-of-use HEPA filters capable of removing particles > 0.3 μm diameter. This requirement is optional in patients with autologous SCT transplants, but should be considered in those who have prolonged neutropenia, a major risk factor for invasive aspergillosis. HEPA filtration prevents particles including *Aspergillus* conidia from the external air supply from entering the room.

Laminar airflow (LAF) ensures that conidia brought into the room by healthcare workers or on fomites is rapidly removed. LAF rooms are those where HEPA-filtered air enters via one wall and moves in even, parallel, unidirectional flow to exit via the opposite wall.

Hospital Construction

When construction is in progress, units housing transplant or immunocompromised patients should be protected by placing appropriate barriers, moving high risk patients, using HEPA filtration, and proper exhausting of construction dust.

14.6 EPIDEMIOLOGY AND RISK GROUPS/FACTORS

Patients with invasive pulmonary aspergillosis fall into four general categories, with the common theme of immunosuppressive therapy. Approximately 30% are those with stem cell transplants, 30% are those with hematologic malignancies, and 10% are patients who have received solid organ transplants. The remaining 10% are patients with another underlying pulmonary disease (Maschsmeyer et al., 2007). The incidence of invasive pulmonary aspergillosis can be as high as 11–13% in the high-risk patient populations. The patient group with the greatest prevalence of invasive pulmonary aspergillosis is comprised of neutropenic cancer patients. Neutropenic patients (n = 240) not receiving prophylaxis with an agent active against *Aspergillus* species had a 6.25% *incidence* of invasive aspergillosis in one multicenter randomized clinical trial of prophylaxis (reviewed by Maschmeyer et al., 2007).

Over the period 1980–1996 there was an alarming 357% increase reported for fatal cases of invasive aspergillosis (McNeil et al., 2001). The incidence of invasive aspergillosis increased in the early 1990s attributed to more patients receiving solid organ and stem cell transplants, accompanied by intensified chemotherapy. Persons living with the late stage of AIDS also are susceptible to invasive aspergillosis. In the 1990s individual tertiary care centers reported an incidence of invasive aspergillosis of 10% or higher (reviewed by Jantunen et al., 2008). The incidence of invasive aspergillosis among 395 patients who received peripheral blood grafts was 8% (Martino et al., 2002).

A survey of 50 studies reviewing 1941 patients with invasive aspergillosis from 1995 to 1999 (Lin et al., 2001) indicated the mortality rate was very high, reaching between 80% and 90% in the patients at highest risk including allogeneic SCT recipients (Fig. 14.1). The mortality rate was ~60% when disease was confined to the lung, approaching 100% when dissemination outside the lung occurred before antifungal therapy was started.

From 1997 on, deaths from invasive aspergillosis were 0.42/100,000 population, declining to 0.25 deaths/100,000 population in 2003. Invasive aspergillosis was declining in incidence from 4.1 cases/100,000 population in 2000 to 2.2/100,000 in 2003. Between 2000 and 2005 there was a decline in invasive aspergillosis, which became an uncommon opportunistic infection in patients with acute leukemia or myelodysplastic syndrome who received SCTs. The incidence of invasive aspergillosis in that group declined to 2% among those patients who did not relapse in their primary disease after transplant. A multicenter U.S. study in the period 2000–2002 found an incidence of invasive pulmonary aspergillosis of 2.3%

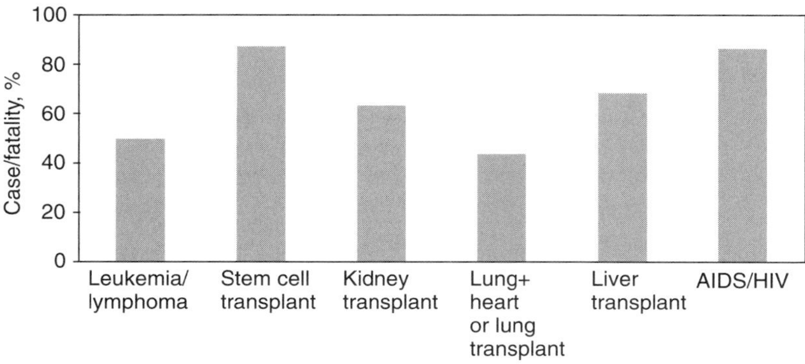

Figure 14.1 Case-fatality rates for patients with aspergillosis, according to underlying diseases or conditions determined from patient data from a review of the literature after 1995.
Source: Lin et al. (2001). Used with permission from the University of Chicago Press.

in HLA-matched sibling transplants and 3.9% in matched unrelated donor transplants (Ascioglu et al., 2004).

Factors Influencing the Incidence of Invasive Aspergillosis

The estimated burden of invasive aspergillosis in the United States is 8000 cases/year (Pfaller and Diekema, 2007). This trend is influenced by expanded, more effective, and less toxic antifungal therapy as well as improved immunosuppressive regimens for cancer therapy and maintenance of solid organ and stem cell transplants. Factors contributing to better outcomes in transplant patients are improved tissue typing techniques allowing better donor–recipient matches and preemptive treatment of CMV antigenemia. Other changes in transplantation practices also may *add* to the incidence of invasive aspergillosis: more common use of peripheral blood stem cells instead of marrow grafts, more patients receiving matched unrelated donor grafts, an increasing age of patients, and reduced intensity conditioning.

Advances in Antifungal Therapy

Beginning in the mid-1990s lipid formations of AmB, because of their reduced nephrotoxicity, became preferred antifungal agents for treating invasive aspergillosis, leading to improved survival rates. This downward trend in incidence continues and is related to better management of the patient's primary disease and because of advances in the diagnosis of and therapy for invasive aspergillosis. Notably, the use of therapy with the extended spectrum triazole VRC has led to measurable improvement in outcome. In the randomized, unblinded clinical trial reported by Herbrecht et al. (2002) patient survival after 12 weeks was 59% in the AmB treated group versus 75% in patients receiving primary therapy with VRC (Fig. 14.2). VRC is now recommended as primary therapy for invasive pulmonary and extrapulmonary aspergillosis in most patients (Walsh et al., 2008).

Benefits have also accrued because of increased choices for alternative and salvage therapy. Recommendations for alternative therapy include liposomal AmB, CASF, MCF, PSC, or ITC (Walsh et al., 2008). ß-(1 → 3)-D-Glucan synthase inhibitors (e.g., CASF) have had some success as salvage therapy in treating aspergillosis refractory to AmB or azoles (Maertens et al., 2004).

Better methods contributing to earlier clinical treatment decisions include diagnostic imaging and detection of antigenemia. CT scan evidence of the halo sign and air crescent sign is helpful. Nonculture methods are in use or

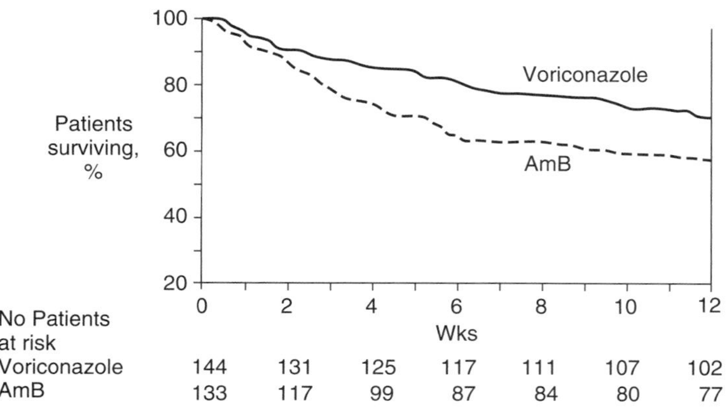

Figure 14.2 Survival curves for the population with proven or probable aspergillosis according to treatment with amphotericin B or voriconazole.
Source: Herbrecht et al. (2002). Used with permission of the Massachusetts Medical Society. Copyright © 2002, all rights reserved.

being validated to measure (i) galactomannan antigenemia and (ii) the nonantigenic metabolite ß-(1 → 3)-D-glucan with the Fungi-Tell® kit. PCR detection of *Aspergillus* DNA directly in clinical specimens is nearer to clinical utility because of attempts to achieve a consensus method for real-time PCR from blood specimens. Other factors influencing the decline in invasive pulmonary aspergillosis are the use of strict isolation and HEPA-filtered air in the immediate post-transplant period.

Even with these improvements the overall case-fatality rate is still high, estimated at 58%, even approaching 90% in allogeneic SCT recipients with GVHD, and in patients with extrapulmonary dissemination and cerebral abscess. *Aspergillus fumigatus* accounts for 50–60% of cases with *A. flavus, A. terreus*, and *A. niger* accounting for ~10–15% each. *A. nidulans, A. ustus*, and other rarely encountered species account for <2% of cases. Two periods of risk for invasive aspergillosis are neutropenia induced by the immunosuppressive regimen prior to transplantation and, later, GVHD. Severely immunocompromised patients, especially those with prolonged neutropenia are at highest risk for invasive aspergillosis. Clinical settings in which prolonged neutropenia apply include deliberate immunosuppressive therapy:

- Following stem cell transplantation
- Immunosuppression as a consequence of late-stage AIDS
- Management of solid organ transplants
- Remission-induction in hematologic malignancy

What is prolonged neutropenia? When the PMN count is <1000/ μL for a week or more. Profound neutropenia ≤100 PMN/μL.

After engraftment takes place, late onset invasive aspergillosis (41–180 days post-transplant) may occur if/when there is GVHD, corticosteroid therapy, secondary neutropenia, and cytomegalovirus (CMV) disease. Very late invasive aspergillosis (>6 months after transplantation) is associated with chronic GVHD and CMV disease.

Figure 14.3 summarizes the response to standard therapy according to the underlying disease. Patients receiving allogeneic SCTs have the poorest prognosis, whereas those with solid organ transplants have the best clinical response.

Invasive Aspergillosis According to the Primary Disease or Condition

Major groups at risk for invasive aspergillosis are patients with:

- Hematologic malignancy
- Acute leukemia
- Solid tumors
- Recipients of transplants—solid organs (liver, lung, heart, and kidney) and hematopoietic stem cell transplants (SCTs)
- Defective neutrophil function

These patient categories are considered in more detail below.

Invasive Aspergillosis in Hematologic Malignancy

Leukemia patients represent the largest group of patients at high risk of developing invasive aspergillosis. In a review of 1941 cases of invasive aspergillosis from 50 studies during 1995–1999, leukemia and lymphoma were the underlying disease in 42.6% of cases; greater than SCT recipients (25.8%) and solid organ transplant recipients (13%) (Lin et al., 2001). The mortality attributable to invasive aspergillosis in leukemia patients is 30–40%.

Invasive Aspergillosis in Patients with Acute Leukemia

A survey of 20 European hospitals over a 1 year period revealed 123 cases of invasive aspergillosis (Denning et al., 1998). Of these 49% were from acute myelogenous leukemia patients. Survival 3 months after diagnosis was 36%. Since 1997 there has been a declining trend in the annual number of cases of invasive aspergillosis related to (i) improvements in the chemotherapy regimen for patients with leukemia that result in a shorter period of neutropenia, and (ii) the empiric use of VRC (Herbrecht et al., 2005; Lin et al., 2001). Improved outcome in leukemia patients with invasive aspergillosis, compared with other immunosuppressed patients, is explained by their ability to recover from neutropenia, whereas stem cell or organ transplant recipients have a relatively slower rate of recovery of their immune status (Fig. 14.1).

The comparison of AmB and VRC as primary therapy was evaluated in 277 patients with proven or probable invasive aspergillosis (Herbrecht et al., 2002) (Fig. 14.2). Patients were enrolled from 95 centers in 19 countries for the clinical trial, which took place between 1997 and 2001. Of these, 118 were acute leukemia patients. They were divided into two treatment arms: 58 patients received AmB and 60 received VRC. Sixty-three percent of leukemia patients with invasive aspergillosis who received VRC had a successful outcome compared to 38.1% of leukemia patients receiving AmB. Since the late 1990s there has been an improvement in survival linked to changes in clinical management of leukemia and improved antifungal therapy.

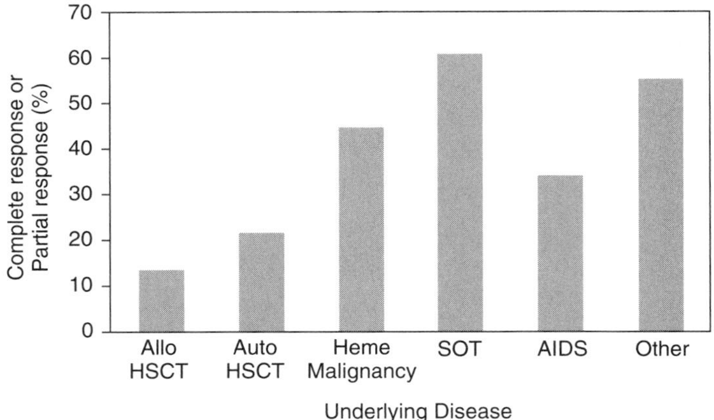

Figure 14.3 Invasive aspergillosis: response to therapy by underlying disease. Complete or partial responses (y-axis) to standard therapy of aspergillosis in each of the patient populations listed.
Source: Data are from Patterson et al. (2000). Used with permission from Elsevier Publishers, Ltd. Copyright © 2000.

Invasive Aspergillosis in Patients with Solid Tumors

Neutropenia is not as common in solid tumor patients as it is in patients with hematologic malignancy, unless the former group become neutropenic after receiving stem cell transplantation (Fig. 14.3).

Risk Factors for Invasive Aspergillosis in Solid Organ Transplant Recipients

The contributing factors are long-term immunosuppressive therapy (e.g., the use of high-dose steroids, antilymphocyte therapy) and viral infections, especially CMV infection. The incidence of invasive aspergillosis varies according to the transplanted organ. Kidney/pancreas transplants have moderate risk—with incidence ranging from 1% to 8%. Heart recipients' risk for invasive fungal infection is estimated at between 1% and 14%. The highest incidence of aspergillosis is in lung transplant recipients with incidence ranging from 6% to 16%. The cumulative incidence of aspergillosis at 12 months post-transplant was: for lung, 2.4%; for heart, 0.8%; for liver, 0.3%; and for kidney, 0.1% in a multicenter surveillance program including 4110 patients in 19 U.S. centers (reviewed by Silveira and Husain, 2007). Mortality at 3 months following the diagnosis of aspergillosis ranged from 10% (lung transplantation) to 66.7% for heart and kidney transplant recipients.

Aspergillosis in Recipients of Solid Organ Transplants

Liver Transplant Recipients Invasive aspergillosis occurs later in the post-transplant period, is less likely to be associated with CNS infection, and is associated with lower mortality, compared with invasive aspergillosis in the early 1990s. Patients receiving liver transplants who developed invasive aspergillosis in the 1990–1995 period ($n = 26$) were compared to a similar size group of patients receiving liver transplants and who developed invasive aspergillosis during the 1998–2001 period ($n = 23$) (Singh et al., 2003). Twenty-three percent of the *Aspergillus* infections in the earlier cohort occurred ≥90 days post-transplant, compared with 55% of such infections in the later cohort. Patients transplanted in the 1990–1995 period were significantly more likely to have disseminated aspergillosis and CNS involvement. The mortality rate was higher for the earlier cohort (92%) than in the more recent period (60%).

The incidence of invasive aspergillosis among liver transplant recipients in early 1990s and the late 1998–2001 periods was unchanged: 1.06% in the earlier cohort and 1.18% in the later period. Surgical advances and better management have reduced post-transplant liver failure. CMV infection in organ transplant recipients is occurring later because of ganciclovir prophylaxis. The important decrease in CNS involvement in the cohort from the 1998–2001 period may be credited to an increased use of high-resolution CT scans of the chest leading to earlier diagnosis. A functioning hepatic allograft is a major defense against the dissemination of *Aspergillus* because the mononuclear phagocytic system in the liver (Kupffer cells) is protective against hematogenous spread. Less hepatic dysfunction in the late 1990s to early 2000s may explain a lower risk of disseminated aspergillosis with CNS involvement.

The mortality rate was lower among patients who had received lipid-AmB instead of AmB-deoxycholate but better surgical and clinical management, including improved imaging studies, in recent years may be more relevant than the direct impact of the therapeutic regimen. The currently used immunosuppressive agents, calcineurin inhibitors (e.g., cyclosporine, tacrolimus) and rapamycin (= sirolimus), have in vitro activity against *Aspergillus* and this also may contribute to a decrease in disseminated aspergillosis. Advances in prophylaxis and combination antifungal therapy have had further impact.

A survey of 67 North American active liver transplant programs found that 72% used prophylaxis targeted toward high-risk patients (Singh et al., 2008). Indications for targeted prophylaxis induced retransplantation, requirement for dialysis, reexploration, fulminant hepatic failure, *Candida* colonization, prolonged ICU stay, or ventilation requirement. Leading choices for mold-active prophylaxis were echinocandins, voriconazole, and lipid-AmB. The timing of prophylaxis was for the initial hospital stay in about half of patients, for 1 month post-transplant in 19%, and for 3 months in 8.5%.

Lung Transplant Recipients Most often the infection occurs in the allograft (Posfay-Barbe et al., 2004). Patients receiving heart–lung or lung transplants are at high risk for infection because of the exposure of the transplanted lung to the environment. Infectious causes account for up to half of deaths in these patients. Lung transplant recipients have impaired lung immunity, decreased cough and mucociliary clearance, and abnormal lymphatic drainage. The severity of illness before the transplant is a factor including the need for intensive care and/or poor nutritional status.

Risk factors for infection after thoracic transplantation are related to the organ(s) transplanted (heart–lung versus double-lung versus single lung) and to the patient's pretransplant condition and underlying disease. The timing of aspergillosis is similar to that observed after transplantation of other solid organs: early (<1 month), intermediate (1–6 months), and late (>6 months). Infection may be difficult to differentiate from chronic rejection of the lung.

Post-transplant risk factors for infection during the early period involve the surgical site, central venous catheters, or inhalation of conidia in hospital air. Patients with *Aspergillus* airway colonization in the first 6 months post-lung-transplant are 11 times more likely to develop invasive aspergillosis than those without colonization (Silveira and Husain, 2007). Colonization of the donor's respiratory tract also poses a risk. *Aspergillus* colonization develops in up to 46% of lung recipients.

Cystic fibrosis patients with pretransplant airway colonization have a higher risk of post-transplant tracheobronchitis (Silveira and Husain, 2007). Tracheobronchitis is found almost exclusively in lung transplant recipients at the anastomotic site. The airways and bronchi are involved without extension to the lungs. Tracheobronchitis usually occurs in the first 3 months post-transplant. Routine surveillance bronchoscopy is essential for early diagnosis. If not treated early, the infection may extend to the lung parenchyma and potentially disseminate. Pathologic features include necrosis, ulceration, and pseudomembrane formation. Tracheobronchitis is a form of invasive aspergillosis and is its most common presentation following lung transplantation.

Fungal infections occur frequently in lung/heart–lung transplant recipients and may be newly acquired or may reactivate from latent infection in lung transplant recipients. *Aspergillus* species may colonize airways (25% of the transplant recipients) or progress to invasive disease (~ 5%). Colonization at the time of transplantation increases risk but the presence of fungi preoperatively is not a contraindication to transplantation.

More than half of all diagnoses are made in the first 6 months after transplantation. The incidence of progression from airway colonization to invasive aspergillosis is less than 5%. Most patients with isolated tracheobronchitis respond to antifungal therapy and/or surgical debridement. The survival rate of invasive aspergillosis in lung transplant recipients, however, is less than 50%.

Heart Transplant Recipients *Aspergillus* species are the opportunistic pathogens with the highest attributable mortality among cardiac transplant patients, with fungal pneumonia being the most frequently encountered clinical form of aspergillosis (Montoya et al., 2003). Invasive aspergillosis should be suspected in heart transplant recipients who develop fever and respiratory symptoms during the first 3 months after transplantation, who have a positive culture for *Aspergillus* species, and abnormal radiologic findings (in particular, nodules) even in the absence of neutropenia.

Before the introduction of cyclosporine therapy the overall incidence and mortality of invasive aspergillosis following cardiac transplantation was reportedly as high as 27% and 60%, respectively. Changes in post-cardiac-transplant therapy since 1980 have been, in approximate chronological order, cyclosporine, OKT3, mycophenolate mofetil-tacrolimus, prophylactic gancyclovir, and inhaled AmB. The net result of these advances has reduced the overall incidence and mortality to 8% and 36%, respectively.

Kidney Transplant Recipients Fungal pneumonia is associated with bolus steroid therapy and renal allograft rejection crisis. Corticosteroid therapy increases the risk of invasive aspergillosis because it interferes with phagocyte function. A total of 33,420 renal transplant recipients in the U.S. Renal Data System from July 1994 to June 1997 were analyzed and fungal infections were most associated with diagnoses of pneumonia (57 patients, 19.8% of pneumonia), meningitis (23 patients, 7.6%), and urinary tract infection (29 patients, 10.3%) (Abbott et al., 2001). Opportunistic mycoses were led by candidiasis, aspergillosis, cryptococcosis, and mucormycosis. Mucormycosis and aspergillosis were independently associated with increased mortality and length of hospital stay. Most fungal infections in renal transplant recipients occur later in the post-transplant period (66% occurred by 6 months post-transplant, but only 22% by 2 months), and resulted in decreased patient survival. In addition to the long-term systemic corticosteroids for antirejection therapy, recipients with diabetes, prolonged pretransplant dialysis, chronic GVHD, and tacrolimus immunosuppression are at increased risk for fungal infections.

Invasive Aspergillosis in Hematopoietic Stem Cell Transplant Recipients (SCT) SCT recipients are those for whom donor cells are harvested from peripheral blood, umbilical cord blood, or bone marrow. The type of transplant is important: autologous SCT patients have a longer period of neutropenia before engraftment occurs, whereas those who receive matched unrelated SCTs are at greater risk after engraftment occurs because of GVHD. In the 1990s changes in transplant practices and antifungal therapy influenced the epidemiology of opportunistic fungal infections. The resulting control of candidiasis through use of prophylactic FLC increased the incidence of invasive aspergillosis against which FLC is ineffective. Incidence is higher in allogeneic SCT recipients, both early in the post-transplant period and later if/when they develop chronic GVHD and require intensive immunosuppression. The incidence of invasive pulmonary aspergillosis in allogeneic SCT recipients in the 1990s was 10% within the first year post-transplant, with an associated mortality of 60%.

Improvements in transplant practices, better control of CMV disease, and improvements in antifungal therapy have resulted in marked improvements in prevention and control of invasive aspergillosis in SCT patients. A multicenter survey in the United States in the period 2001–2002 found an incidence for invasive aspergillosis of 2.3% in HLA-matched sibling transplants and 3.9% in unrelated transplants (Morgan et al., 2005). A significantly decreased mortality in SCT recipients with invasive aspergillosis occurred beginning in 2001–2002 because transplant practices changed to the use of non-myeloablative conditioning, and the choice of peripheral blood stem cells instead of bone marrow derived stem cells. Two other factors driving improved survival are the more prompt diagnosis of invasive aspergillosis and extensive use of VRC.

Experience in a major cancer center in Seattle, Washington, is illustrative of these trends (Upton et al., 2007). Lipid derivatives of AmB were replaced by VRC by 2002, and CASF was added to VRC as combination primary therapy. At present, there is no clear consensus whether this combination offers improved therapeutic response compared with VRC monotherapy. VRC also replaced ITC as maintenance therapy. After these changes took place, survival at 90 days after a diagnosis of invasive aspergillosis was 45% for patients diagnosed in 2002–2004 compared with 22% survival in earlier years. The probability of survival 1 year after diagnosis of invasive aspergillosis in the period 2002–2004 was 30%, compared to 15% in the 1999–2001 period. The probability of death attributable to invasive aspergillosis was 18% for patients diagnosed in 2002–2004 compared with 25% for those diagnosed in 1999–2001.

Sixteen medical centers from North America reported 250 proven or propbable invasive fungal infections among SCT recipients during 2004–2007 (Neofytos et al., 2009). Invasive aspergillosis was the most frequent such infection, accounting for 59.2% of cases; invasive candidiasis, 24.8%; and mucormycosis, 7.2%. Mortality in patients with invasive aspergillosis after 6 weeks was 21.5%, rising to 35.5% after 12 weeks. That latter mortality rate was lower than that for candidiasis in this cohort of SCT patients, 48.9%. Outcomes of invasive aspergillosis have improved corresponding to improved diagnosis and therapy.

Risk factors related to all-cause mortality in SCT recipients were several, including impaired pulmonary function pretransplant, elevated bilirubin and creatinine clearance, corticosteroid therapy of >2 mg/kg/day, and invasive aspergillosis disseminating outside the lungs or occurring >40 days post-transplant. An intervention is needed to decrease the burden of late community-acquired invasive aspergillosis in allogeneic SCT recipients. The risk factor of chronic GVHD is important for late and very late onset invasive aspergillosis.

A review of risk factors for invasive aspergillosis at a major cancer treatment center in Washington State identified 187 patients who developed invasive aspergillosis out of 1495 who received allogeneic SCTs (Marr et al., 2002a). The rate of developing invasive aspergillosis was 10.5% per year among recipients during the study period. Survival 1 year after aspergillosis was approximately 20%. Factors related to significantly greater risk for developing invasive aspergillosis were age older than 40 years and if the underlying disease was aplastic anemia, multiple myeloma, or myelodysplastic syndrome.

Risk factors for invasive aspergillosis varied according to the period post-transplant, early (<40 days), late (41–180 days), and very late (> 6 months) post-transplant.

Early. The risk of invasive aspergillosis in the early period was linked to:

- Underlying disease. First remission of hematologic malignancy, aplastic anemia, or myelodysplastic syndrome.
- Source of stem cells. Cord blood as the source increased risk. Matched related peripheral blood stem cells were protective relative to bone marrow derived.
- Concurrent CMV disease.

Late. Invasive aspergillosis in the late period was related to acute GVHD, CMV disease, and corticosteroid treatment of GVHD. Other risk factors included the underlying disease (multiple myeloma) and the receipt of CD34-selected stem cells.

Very Late. In the very late period, > 6 months after transplant, the risk of invasive aspergillosis was related to extensive chronic GVHD and secondary neutropenia. Associated risks were receipt of unrelated or HLA-mismatched or unrelated donor stem cells and CMV disease.

T-cell deficiencies can also predispose to invasive aspergillosis because normally T cells activate alveolar macrophages which can kill aspergilli. T-cell deficiencies can occur as a result of hematopoietic stem cell transplantation and also may result from prolonged high-dose corticosteroid therapy.

Invasive Aspergillosis Where the Risk Factor Is Defective Neutrophil Function — Chronic Granulomatous Disease of Childhood (CGD)

(National Primary Immunodeficiency Resource Center, New York. http://npi.jmfworld.org/.) CGD is an inherited defect in which polymorphonuclear neutrophilic granulocytes (PMNs) are unable to kill certain microbes including aspergilli. In these patients, PMNs migrate normally and phagocytose the microbe. Killing of the phagocytosed cell is prevented by a specific defect in the oxidative microbicidal pathway: there is no oxidative burst because the PMNs are unable to produce hydrogen peroxide and other oxygen metabolites.

Some microbes (e.g., pneumococci and streptococci) can produce hydrogen peroxide themselves and can assist in their own destruction. But catalase-positive microbes neutralize hydrogen peroxide and the defective PMNs are left without resources to kill the invader. Catalase-positive microbes include *Aspergillus* species, *Escherichia coli, Pseudomonas, Serratia*, and staphylococci.

Consequently, children with CGD have an increased susceptibility to recurrent, serious infections by catalase-positive microbes. Such infections usually involve the skin, soft tissues, respiratory tract, lymph nodes, liver, spleen, or bones and require weeks to months of antimicrobial therapy because the normal mechanism of pathogen removal is absent.

Summary of Risk Factors for Invasive Aspergillosis

Examples of patients at risk are those receiving cancer chemotherapy and/or stem cell transplantation for hematologic malignancy, other stem cell transplant recipients, patients with solid tumors, or those receiving therapy for maintenance of solid organ transplants where two of three criteria are satisfied:

- <1000 PMNs/μL of peripheral blood for 1 week or more
- Cytotoxic drug therapy (e.g., cyclophosphamide, methotrexate)
- High-dose prolonged corticosteroid therapy (e.g., prednisone) or prolonged therapy with other immunosuppressive agents (e.g., azathioprine and its product 6-mercaptopurine, cyclosporine, and mycophenolate mofetil, tacrolimus).

Summarizing, risk factors for invasive aspergillosis are related to the following:

- Stem Cell Transplant Recipients.
  - Acute GVHD in the post-transplant period
  - Extensive chronic GVHD
  - GVHD associated with donor lymphocyte infusions
  - Higher age of recipient
  - HLA mismatch
  - Neutropenia during the conditioning phase prior to transplantation
  - Unrelated donor
  - Patients receiving cord blood stem cell allografts after reduced intensity conditioning had a 10% incidence of IPA (reviewed by Jantunen et al., 2008)
- *Net State of Immunosuppression.* Corticosteroid dose, lymphopenia, alemtuzumab use (an mAb directed against CD52, used to treat chronic B-lymphocytic leukemia). Genetic factors such as IL-10 promoter gene polymorphisms.
- *Concurrent Viral Infections and Related Therapy.* CMV infection, respiratory virus infection, ganciclovir use.
- *Other Factors.* Defective neutrophil function (e.g., chronic granulomatous disease, alcohol use, IV drug use, marijuana use—*Aspergillus* species can contaminate marijuana), pentamidine-induced diabetes, prior fungal infection, prolonged antibiotic therapy, and underlying lung disease including tuberculosis.

14.7 TRANSMISSION

Transmission occurs from the environment to the host with the inhalation of conidia as the main route of infection. *Aspergillus* species conidia are small enough (2.5–3 μm) to reach the alveoli or paranasal sinuses. These conidia can circulate in hospital air and enter chest wounds following surgery. For example, after an outbreak of sternal surgical-site *A. flavus* infections in a Belgian hospital following cardiac surgery, a mycologic survey was conducted of air and surfaces in the surgical ward and other patient care areas. Contamination of more than 100 CFU per contact plate was frequently observed in some areas of the surgical ward (Heinemann et al., 2004).

In the operating room, conidia circulating in the ventilation system can enter chest wounds and cause endocarditis (reviewed by Pasqualotto and Denning, 2006). Aspergillosis is not observed in patients housed in LAF rooms. Immunocompromised patients should be shielded from dusty renovation conditions. For further details please see Section 14.5, Geographic Distribution/Ecologic Niche—*Aspergillus* in the Hospital Environment. Another means of transmission is via nonsterile tape used to fasten catheters to the skin of premature infants (James et al., 2000).

14.8 DETERMINANTS OF PATHOGENICITY

14.8.1 Host Factors

The sequence of innate and adaptive immune responses in the lung to inhaled *Aspergillus* conidia is summarized in Table 14.3 and expanded in this section. Most of these functions are downregulated or absent in the immunocompromised host.

Pathogenesis

Aspergillus fumigatus conidia are 2.5–3 μm in diameter, small enough, once inhaled, to reach all levels of the respiratory tract. The initial contact of *A. fumigatus* conidia with the mucociliary elevator of the respiratory epithelia is met by exclusion of the conidia, but *A. fumigatus* toxins can inhibit ciliary activity, and its proteinases can damage epithelial cells. Some of the conidia are deposited in the distal airway lumen and the alveoli. There they can bind to lung surfactant molecules, to complement components, and fibrinogen. Conidia that evade phagocytosis can germinate, form hyphae, and release proteinases which erode the lung epithelia, exposing the underlying matrix. Next, in the immunocompromised neutropenic host, the hyphae can invade the lung parenchyma.

In the encounter with alveolar macrophages, conidia are able, through receptor–ligand interactions, to bind via the mannose receptor, a ß-glucan (dectin-1) receptor, and a receptor that is inhibited by chito-oligosaccharides. Phagocytosed and sequestered in the phagosome, the conidia swell but at that stage are not seemingly inhibited by the macrophage. Killing starts 6–8 h after phagocytosis at a time when the germ tube is formed. Killing occurs at a slow rate within macrophages, with only a 10% reduction in viability after 6 h (Latgé, 2001). Chitinase in the lung and the plasma increase during aspergillosis, implying that glycosyl hydrolases may be responsible for digesting the inner conidial wall.

When germinated conidia manage to escape from the macrophage, or germinate without having been phagocytosed, an influx of PMNs is required as the second line of defense. Macrophages recognize pathogen-associated

Table 14.3 Sequential Activity of Host Factors in the Response to *Aspergillus* Infection[a]

Host factor	Function
Mucociliary elevator	Clears *Aspergillus* conidia from the airways.
Alveolar macrophages	Conidia penetrating to the alveoli are phagocytosed. (Macrophages are not well activated before T cells stimulate them). Macrophages produce TNF-α, which increases vascular permeability and endothelial adhesiveness for leukocytes and platelets, allowing an influx of fluid, cells, and proteins that aid in host defense. TLR4 and dectin-1 on the macrophage surface bind to PAMPs on *Aspergillus*. This activates transcription factors, which induce inflammatory cytokines and activate T lymphocytes.
Opsonization	Complement activation and IgG opsonize conidia before and after germination, aiding in their phagocytosis.
PMNs	Conidia that escape macrophage phagocytosis or germinate and burst through macrophages encounter PMNs, active killers of *Aspergillus* using reactive oxygen species (ROS) and nonoxidative microbicidal peptides.
Plasminogen	Local blood clots serve to confine the infection but must be lysed to prevent infarcts. This is accomplished by activation of plasminogen.
Adaptive immunity, $CD4^+$ T-lymphocytes	T cells provide two signals that activate macrophages: IFN-γ and the CD40 ligand, which sensitizes macrophages via contact with CD40 on the macrophage. T cells produce IL-18, IFN-γ, and IL-12. IL-18 is first to be detected in the Th1 response and stimulates production of IFN-γ.
Activated macrophages	They fuse their lysosomes efficiently to phagosomes, so that lysosomal hydrolases act on ingested *Aspergillus* conidia and hyphal elements. Activated macrophages also make oxygen radicals and nitric oxide (NO) with potent antimicrobial activity.

[a]Most of the above functions are downregulated or absent in the immunocompromised host.

molecular patterns (PAMPs) on *Aspergillus* hyphae and produce cytokines that trigger inflammation and attract PMNs. If successful, PMNs will extravasate into the tissues and rapidly degranulate in the presence of the young hyphae. The effector molecules are oxidants and, as yet unknown, nonoxidative granule enzymes. Killing by PMNs occurs swiftly within 2 h of incubation with hyphae.

Numerical and Functional Neutropenia

During the period of prolonged and profound neutropenia due to conditioning immunosuppression, or before engraftment, numbers of PMNs may not be sufficient to prevent tissue invasion by *Aspergillus* hyphae. Even though SCT patients may have normal numbers of PMNs after engraftment there *may be no influx* of them into lung tissue because of defects in their recruitment and function. This situation defines *functional neutropenia*. Numeric recovery of PMNs does not translate into functional recovery. A reconstitution period for innate and adaptive immunity follows SCT.

The slow recovery of adaptive immune cells, including T-lymphocytes, results in low or absent production of signaling and activating lymphokines (Stergiopoulou et al., 2007). There is a necessary period required for the recovery of innate and adaptive immune responses following stem cell infusion, during which there is impaired recruitment of PMNs. The period for recovery varies from a few weeks to almost a year depending on the immune system cell type. Thus, in SCT patients, even after engraftment, *Aspergillus* species can grow and invade blood vessels. Neutrophil chemotaxis is induced by a combination of chemokines including IL-8 and MIP-1α (macrophage inflammatory protein-1α). MIP-1α has a chemotactic effect on IFN-γ-activated PMNs. If $CD4^+$ T-lymphocytes are not recovered after SCT, IFN-γ secretion will be suboptimal and PMN chemotaxis may be weakened.

Corticosteroids reduce the effectiveness of killing of conidia phagocytosed by alveolar macrophages and blunt the production of proinflammatory cytokines IL-1α, TNF-α, and the chemokine MIP-1α, all of which are important for recruiting PMNs and monocytes. Corticosteroids also steer the adaptive T-cell response along a Th2 path which does not provide protection against fungi.

If either the alveolar macrophages or PMNs are impaired in their numbers or activation, the stage is set for invasive pulmonary aspergillosis. *Aspergillus* species grow unchecked in neutropenic patients and SCT recipients (during the period of functional neutropenia) across tissue planes, invading blood vessels, and causing intraalveolar hemorrhage. The lung tissues are invaded by numerous *Aspergillus* hyphae growing in a radial pattern. The radiographic halo sign around a pulmonary nodule denotes hemorrhage or edema and the nodule is tissue infarcted by *Aspergillus*. This pattern of tissue injury in neutropenic patients underlines the importance of PMNs in defense against *Aspergillus*.

There are several types of tissue damage in invasive pulmonary aspergillosis:

(i) Hemorrhagic infarction—erythrocytes in a region of infarcted lung tissue
(ii) Granulomas—organized aggregates of mononuclear cells, macrophages, lymphocytes, and neutrophils with or without caseous necrosis
(iii) Inflammatory necrosis—a mixed PMN and monocytic infiltrate and necrosis of the pulmonary architecture

Non-neutropenic patients mount an inflammatory response with an influx of PMNs resulting in inflammatory necrosis; angioinvasion *is seldom observed* (Stergiopoulou et al., 2007). Where the immune response is relatively intact, a fibrosing, granulomatous response is seen, with numerous giant cells and the *Aspergillus* hyphae forced into distorted shapes.

Bronchial colonization by aspergilli can occur. *Aspergillus* can exist for months in ectatic (dilated, inelastic) bronchi in cysts or lung cavities, or poorly draining paranasal sinuses, *without causing disease*. Pulmonary colonization can stimulate IgE and IgG.

Aspergilli can also grow in preexisting pulmonary cavities as an aspergilloma (fungus ball). (Please see aspergilloma, in Section 14.9, Clinical Forms.)

Adaptive Immunity

There is a strong link between innate and adaptive immunity, wherein the activation of PMNs by IFN-γ is important in their effective mobilization at the site of *Aspergillus* hyphae attempting to invade the lung parenchyma. Granulocyte-stimulating factors and IFN-γ stimulate phagocytosis. $CD4^+$ T-lymphocytes assert protective immunity by the production of the Th1 cytokines, IFN-γ, IL-18, and IL-12. IL-18 is first to be detected in the Th1 response. IL-18 secretion stimulates production of IFN-γ and activates NK cells. Invasive aspergillosis is enhanced by a Th2 response which increases IL-4 and IL-10 and decreases IFN-γ.

The normal activation of T-cell-mediated immunity is weak or absent in the immunocompromised host, leading to disease progression. GVHD itself inhibits T-cell recovery and function because of immunosuppressive cytokine production—IL-10 and transforming growth factor-ß. IL-10 inhibits the secretion of proinflammatory cytokines, notably TNF-α. TNF-α induces chemotactic chemokines and adhesion molecules.

Genetic Factors Influence Susceptibility to Aspergillosis

Plasminogen The fibrinolytic system controlled by plasminogen is linked to the host response to infectious microbes. Genetic variation in the plasminogen gene influences the outcome of experimental murine aspergillosis (Zaas et al., 2008). Haplotype-based genetic analysis of survival data identified the plasminogen gene, *PLG*, as important in host susceptibility to invasive aspergillosis. Inbred mouse strains differed in survival curves after experimental *A. fumigatus* infection. Sequences of the *PLG* gene in susceptible and resistant mouse strains found a single SNP that altered an amino acid in a domain that regulates binding of plasminogen to fibrin and to plasmin. The result is a physiologic effect on lysis of clots and protelolytic activity against extracellular matrix proteins.

A SNP causing a single amino acid substitution in human *PLG* increases the risk of developing invasive aspergillosis in SCT recipients (Zaas et al., 2008). The *PLG* gene was sequenced from 20 donor and recipient pairs. One SNP encoded a nonconservative substitution from a neutral (asparagine) to an acidic amino acid (aspartic acid) in a loop region connecting the 4th to 5th kringle[3] domain of plasminogen, capable of altering the alignment of these domains and affecting ligand binding.

The genotype Asp472Asn was determined for the SCT cohort of 83 patients. Those who were homozygous or heterozygous for Asn472 were *at increased risk of developing invasive aspergillosis*. There was a constant increase in hazard ratio over the 40–365 days post-engraftment period for recipients who carried at least one Asn allele. A gene dosage effect was found such that homozygous AsnAsn patients were at a 5.6-fold increased risk compared to 3.0-fold risk for heterozygous AsnAsp patients.

Immunofluorescence microscopy showed how plasminogen might interact with the pathogen when murine plasminogen was seen to bind to swollen *A. fumigatus* conidia and hyphae. Polymorphisms in plasminogen might increase its binding to the pathogen and, once there, its activation could facilitate tissue invasion. The fibrinolytic pathway also may mediate hemorrhage and tissue damage.

Toll-like Receptor 4 (TLR4) Polymorphisms in TLR4 haplotypes of donors of allogeneic SCT are associated with an increased risk of invasive aspergillosis in recipients (12% vs. 1%) (Bochud et al., 2008). Increased risk covered a 36 month post-transplant period. Alterations in TLR influence the ability to recognize ligands on the *A. fumigatus* cell surface. Recall that TLRs are transmembrane proteins on the surface of immune cells that detect conserved PAMPs from various microbes. They interact with adapter proteins to activate transcription factors which induce inflammatory cytokines and activate adaptive immunity. *TLR4* haplotype S4 in SCT donors (haplotype frequency, 6%) is associated with an increased risk of invasive aspergillosis. This haplotype is present in carriers of 2 SNPs in strong linkage disequilibrium that influence TLR4 function.

[3]Kringle domain. *Definition*: Autonomous protein domains that fold into large loops stabilized by three disulfide bridges. These are important in protein–protein interactions with blood coagulation factors. The name "kringle" comes from a Scandinavian pastry that these structures resemble.

14.8.2 Microbial Factors

Aspergillus fumigatus is highly prevalent in the environment as a carbon and nitrogen recycler in its role in the high temperature phase of composting, where temperatures reach 48–66°C. In addition to thermotolerance, *A. fumigatus* possesses nutritional versatility: that is, it can grow on various substrates and has no special nutritional requirements. From its ecologic niche, where it grows rapidly and sporulates profusely, its conidia are dispersed in air currents and are among the most frequent fungal conidia in the *aerospora*.

Rapid Germination, Size of Conidia, Growth Rate, and Thermotolerance

Conidia of *A. fumigatus* germinate more rapidly and have a higher percentage of germination than those of *A. flavus* and *A. niger* (Rhodes, 2006). Germ tubes appear 1 h earlier in *A. fumigatus* compared to the other two species. These properties of a higher germination rate, an increased percentage germination, and smaller size conidia (2–3 μm diameter) favor penetration to the lower respiratory tract. The rapid growth rate at or above 37°C is well suited to its being a competitive opportunistic pathogen.

Frequent and Chronic Environmental Exposure

As a result of the frequent presence of *Aspergillus* conidia in outdoor air, invasive aspergillosis may be seen as a *chronic disease caused by continuous exposure to small quantities of conidia*. In the immunocompromised host, conidia may overcome the innate immune responses, germinate, and grow into a branched, septate mycelium while invading the lung tissue.

The immune-normal host's innate immune system effectively kills *A. fumigatus*. The main difficulty in studying the virulence of *A. fumigatus* is that *disease in the immune-normal host is exceptional*. The lack of a single indispensable virulence factor has led to the view that all *A. fumigatus* requires to cause disease in the immunocompromised host is its ability to grow rapidly at ≥37°C, produce large numbers of conidia small enough to distribute themselves throughout the lower respiratory tract, and the apparent lack of any special nutritional requirement (Latgé, 2001).

Potential virulence factors have been studied but no single factor is known to be indispensable in the pathogenesis of experimental infections in mice. Various candidate virulence factors are described below. They include ribotoxin, hemolysin, secondary metabolites such as gliotoxin, proteinases (including serine proteinase, aspartyl proteinase, metalloproteinase, dipeptidyl peptidases), catalase, superoxide dismutases, and phospholipases (Askew, 2008; Hohl and Feldmesser, 2007; Latgé, 2001; Taylor et al., 2009). Many of these candidate virulence factors are listed in Table 14.4. *A. fumigatus* toxins inhibit ciliary activity and proteinases can damage host tissue. *A. fumigatus* can adhere to host proteins through different cell wall receptors. But no single factor, in a null mutant, has been shown to be critical in the pathogenesis in experimentally infected animals. While some or all of these factors may have incremental activity, virulence is multifactorial.

Initiation of Infection

Adhesins Conidia of *A. fumigatus* contain sialic acid residues and other surface proteins which adhere to fibrinogen, to laminin the basement membrane glycoprotein, and to fibronectin the extracellular matrix component. *A. fumigatus* conidia and hyphae also adhere to lung surfactant A and D, mannose-binding lectin, complement components, and immunoglobulins. Epithelial and endothelial cells are known to internalize conidia, possibly serving as a focus of infection.

Conidial Wall Molecules Conidial cell wall molecules (Latgé, 2001) resist phagocytosis. The outer

Table 14.4 *Aspergillus fumigatus* Factors Implicated in Pathogenesis

Stage of infectious process	Nature of the pathogenic factor	Effect on host
Conidia, the infectious propagule	Small size, rapid germination	2–3 μm diameter, penetrates into lower respiratory tract
	Green pigment, a type of melanin	Polyketide synthase mutants have low virulence
	Rodlet proteins	Scavenge reactive oxygen species
Growth, temperature tolerance	Rapid growth of hyphae	Faster than other *Aspergillus* species
	Grows and sporulates up to 48°C	
Adherence to host surfaces	Surface adhesins	Bind to basement membrane laminin and to extracellular matrix fibronectin.
	Sialic acid residues on conidial surface	Bind to fibrinogen
	ß-(1 → 3)-D-Glucan in hyphal wall	Binds to dectin 1 on macrophages, neutrophils
Secretion of damaging metabolites	Gliotoxin	Depresses epithelial ciliary motion; induces apoptosis of macrophages, T cells
	Verruculogen	Tremorogenic; reduces resistance of epithelia to invasion
Signaling and stress response	Protein kinase (PKA)	Imbalance disrupts fungus's ability to sense and respond to host-specific stressors
	Calcineurin	Deletion of *cnaA* stunts hyphal growth
Interaction with neutrophils	Catalase oxidative stress responses	Mutations in catalase genes increase fungus's susceptibility to oxidative stress
Metabolism	Siderophores	Enable growth in vivo, an iron-poor environment
	Methylcitrate synthase	Reduced virulence of a Δ*mcsA* mutant; detoxifies proprionyl CoA, a product of protein catabolism
Mammalian tissue degradation	Elastase, serine, metallo-, and aspartyl proteinases	Increase lung matrix colonization and degrade host proteins (e.g., elastin)
	Asp-hemolysin	Hemolytic and cytotoxic to macrophages, endothelium
	Phospholipase C	Possible role in degrading lung surfactant
	Restrictocin	Ribonucleotoxin, interferes with PMN damage to hyphae

Sources: Askew (2008), Hohl and Feldmesser (2007), Latgé (2001), and Taylor et al. (2009).

layer of *A. fumigatus* conidia is composed of hydrophobic proteins—a rodlet layer. Deletion mutants, Δ *rodA*, elicit a weak inflammatory response in infected mice. Adjacent to the rodlet layer is a green pigment layer synthesized by the dihydroxynaphthalene (DHN) melanin pathway. Melanin production in *A. fumigatus* conidia encoded by *ALB1* is developmentally regulated, and the 7 kb transcript is detected only during conidiogenesis (Tsai et al., 1998). Δ*alb1* deletion mutants have reduced complement activation on conidial surfaces, and the white conidia of the *alb1* disruptant are damaged more efficiently by phagocytes than dark green conidia. The reduction of mouse virulence in the white conidia mutant is 20–50%, compared with the wild-type but the albino mutants can still cause disease in mice.

Secretion of Damaging Products

Gliotoxin Gliotoxin (Fig 14.4), a metabolite of *A. fumigatus*, exerts immunosuppressive activity on cells of the immune system. It is a lipid-soluble, thioreactive epipolythiodixopiperazine molecule. Gliotoxin, in laboratory experiments, depresses ciliary movement by epithelial cells and induces apoptosis of macrophages and T lymphocytes.

Verruculogen Verruculogen is a tremorogenic metabolite of many *A. fumigatus* strains; it reduces transepithelial resistance of nasal epithelial cells.

Signaling and Stress Response

Protein Kinase (PKA) The cAMP-dependent protein kinase (PKA) is key to sensing carbon sources and environmental stress. Faulty regulation of the PKA pathway, either by disrupting its activity (Δ*gpaB*, or Δ*pkaC*1) or allowing unrestrained PKA activity (Δ*pkaR*), attenuates virulence in mice. An imbalance in *A. fumigatus* PKA signaling in either direction affects the pathogenesis of aspergillosis, by disrupting the fungus's ability to sense and respond to host-specific stressors.

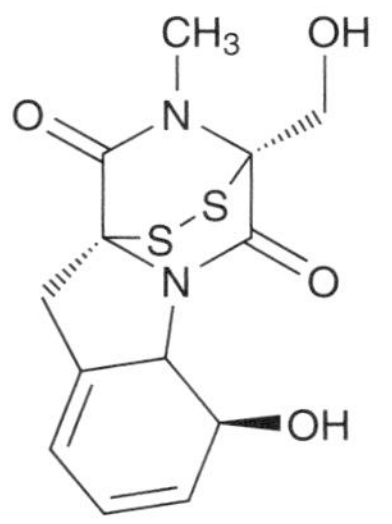

Figure 14.4 Gliotoxin.
Source: PubChem http://pubchem.ncbi.nlm.nih.gov, a service of the National Center for Biotechnology Information and the National Library of Medicine, U.S.A.

Calcineurin Calcineurin is a Ca^{2+}-calmodulin-activated protein phosphatase and an important mediator of calcium signaling and stress responses in eukaryotes. Deletion of *cnaA*, encoding the catalytic A subunit, stunts the growth of *A. fumigatus* hyphae, resulting in loss of virulence. PKA and calcineurin are essential to viability of aspergilli so they are not, strictly speaking, virulence factors. They are, however, potential targets for novel antifungal pharmaceuticals.

Interaction with Neutrophils: Oxidative Stress Response

A. fumigatus produces mannitol, catalase, and superoxide dismutase, all of which neutralize the oxidative microbicidal pathway [*e.g.*: myeloperoxidase (MPO)-peroxide-halide], a major mechanism of fungicidal activity by PMNs. *A. fumigatus* detoxifies oxidative threats via glutathione synthesis and oxidoreductase activity. The protein products of four catalase and four superoxide dismutase genes degrade host H_2O_2 and superoxide radicals. Catalase genes (*catA, cat1, cat2*) of *A. fumigatus* modify the sensitivity of the organism to oxidative stress. Neutrophils from patients with CGD and MPO deficiency are unable to kill *A. fumigatus* hyphae, in contrast to those from healthy patients. The antihyphal activity could be reconstituted with H_2O_2.

Metabolism

Siderophores *A. fumigatus* uses siderophores (negatively charged molecules that bind ferric iron) for iron acquisition and intracellular storage. Disrupting siderophore biosynthesis shows that *A. fumigatus* relies heavily on these low-molecular-weight chelators for growth in the host's iron-limited environment.

Methylcitrate Synthase Amino acid metabolism results in toxic accumulation of propionyl-CoA, which is metabolized by methylcitrate synthase, McsA, of the methyl citrate cycle. Greatly reduced virulence of a Δ*mcsA* mutant implies that *A. fumigatus* relies on protein degradation for food in vivo, making the fungus vulnerable to propionyl-CoA accumulation.

Mammalian Tissue Degradation

Proteinases *A. fumigatus* produces a variety of secreted proteinases involved in promoting lung matrix colonization and/or degrading soluble host proteins. Elastase is produced by virulent *A. fumigatus* strains and attacks elastin, a major structural component of lung tissue. One or more proteinases with elastinolytic activity are produced during infection including a serine proteinase and a metalloproteinase. Aspartyl proteinase produced by growing hyphae can attack elastin, collagen, and fibronectin.

Hemolysin Asp-hemolysin encoded by *asp-HS* is a cytolysin secreted by *A. fumigatus* which interacts with specific receptors on the cell membrane, probably with human low density lipoprotein (Ebina et al., 1994). This low molecular mass 14 kDa protein has hemolytic activity on sheep and rabbit erythrocytes and cytotoxic effects on macrophages and endothelial cells in vitro (Rementeria et al., 2005).

RNAse, Restrictocin The ribonucleotoxin, restrictocin, is an 18 kDa RNase acting on large subunit rRNA. Culture filtrates from restrictocin deletion mutants were unable to suppress neutrophil-mediated hyphal damage, indicating a role for restrictocin in resistance to PMNs.

Phospholipase Clinical isolates produce more phospholipase C than environmental isolates (Birch et al., 2004). This is significant because ~80% of lung surfactant is composed of phospholipid.

14.9 CLINICAL FORMS

Aspergillosis encompasses the *broadest range* of mycotic disease.

Intoxication

A potent mycotoxin, aflatoxin, is a product of *A. flavus* growing on moldy grain or peanuts. Please see the Appendix at the end of this chapter for an overview. This topic is outside the scope of present material, which is restricted to infections.

Allergic Bronchopulmonary Aspergillosis (ABPA)

Please see the Appendix for an overview. This topic is outside the scope of present material, which is restricted to infections.

Colonizing Aspergilloma

Chronic pulmonary disorders such as healed tuberculosis, sarcoidosis, and emphysema may result in colonization of preexisting pulmonary cavities with *Aspergillus* fungus balls. Aspergilloma is defined as a fungus ball or mycetoma in the lung caused by an *Aspergillus* species; a mass of fungal hyphae embedded in a matrix of cell debris and fibrin, contained in a preexisting pulmonary cavity. The cavity wall may become thickened and gradually the fungus ball may move freely in the cavity. Chest radiographs show a thick-walled, cavitated lesion. If the fungus ball begins to fill the entire cavity the *air crescent* sign may be evident. This radiographic appearance is accompanied by multiple serum precipitins.

Clinical findings include cough and hemoptysis. Most patients tolerate the fungal colonization and there is little change on the annual chest film. CT scan can demonstrate the abnormalities better than chest radiographs. In the event of severe hemorrhage, surgical resection or bronchial artery embolization is necessary. Postsurgical complications may ensue.

Occasionally fungus balls can occur in a previously normal lung in the setting of mild to moderate immunosuppression (i.e., diabetes, emphysema) or after the resolution of neutropenia. This clinical form is termed *chronic necrotizing aspergillosis* and is a progressive infection that requires antifungal therapy (Marr et al., 2002b).

Keratitis

Aspergillus species are an occasional etiologic agent of this locally invasive infection of the cornea following a penetrating injury or corneal surgery. *A. fumigatus* contaminating a donor cornea also has been reported (Brasnu et al., 2007). *Aspergillus* keratitis is occupationally associated with agricultural workers. This infection is characterized by ocular pain and the potential for rapid vision loss. It is an ophthalmologic emergency that requires slit lamp examination, assessment of the depth of infection, and initiation of topical antifungal therapy (Walsh et al., 2008). A corneal transplant may be necessary if the infection progresses despite antifungal therapy, or if there is the risk of corneal perforation. *Aspergillus* species are an infrequent cause of fungal keratitis in contact lens wearers who are not compliant with proper antiseptic cleaning procedures. Please also see Section 14.11, Therapy—For Keratitis and Section 14.12, Laboratory Detection, Recovery, and Identification.

Otomycosis

Otomycosis is a benign condition of *Aspergillus* growing on debris in the external auditory canal (especially *A. niger*).

Sinusitis

Allergic Fungal Sinusitis

Allergic sinusitis typically occurs in persons who are *atopic*. The symptoms are allergic rhinitis, nasal polyps, recurrent sinusitis, and sometimes including asthma (Marr et al., 2002b). *Aspergillus* species are among the fungi implicated in allergic sinusitis, but melanized molds are believed to more often be involved. Please see Chapter 18 for further material on fungal sinusitis.

Invasive Sinusitis

Most often occurring in the host immunocompromised by SCT and neutropenia, invasive sinusitis is a distinct disease requiring aggressive diagnostic work-up and antifungal therapy (Marr et al., 2002b). Hyphae can invade mucosa and bone, resulting in hemorrhage, infarcts, and spread into the orbit of the eye and brain. Biopsy, culture, and antifungal susceptibility testing are important because some *Aspergillus* species involved in invasive sinusitis may have reduced susceptibility to AmB.

Tracheobronchitis

Localized in the trachea and bronchi, this clinical form of aspergillosis results in mucoid casts containing *Aspergillus* hyphae, which cause airway obstruction and appear clinically as bilateral lower lobe infiltrates. Obstructing bronchial aspergillosis is found in persons living with AIDS and in lung transplant recipients. Diagnosis requires bronchoscopy which also may be therapeutic if mucus plugs can be removed. Systemic antifungal therapy is appropriate and case reports have observed a clinical response to aerosolized AmB (Marr et al., 2002b).

Invasive Pulmonary Aspergillosis in the Immunocompromised Host

Invasive pulmonary aspergillosis is accompanied by fever, pleuritic chest pain, and acute pneumonia, but the clinical presentation can be variable (Marr et al., 2002b). Pneumonia may be preceded by invasive fungal sinusitis. Hemoptysis can accompany the pneumonia because of pulmonary infarction owing to the *angioinvasive* tendency of *Aspergillus*. On CT scan there may be pulmonary infiltrates or nodules. Certain radiographic findings are helpful when present: the halo sign caused by hemorrhage surrounding a nodule is not, however, specific for aspergillosis. The follow-up air crescent sign is caused by necrosis around a nodule, with air filling the space between the devitalized tissue and the surrounding lung parenchyma (Fig. 14.5). The halo sign can be observed early in *Aspergillus* pneumonia in more than 80% of patients, followed by the air crescent sign within 1–2 weeks.

Criteria for proven invasive aspergillosis follow the definition of an invasive fungal disease defined by the European Organization for Research and Treatment of Cancer and the NIAID Mycoses Study Group (de Pauw et al., 2008):

- *Microscopic Analysis of Sterile Material*. Histopathologic, cytopathologic, or direct microscopic examination of a specimen obtained by needle aspiration or biopsy in which hyphae are seen accompanied by associated tissue damage. Septate hyphae with acute angle branches are a clue, as well as the presence of sporing heads if the lesion is near an air space.
- *Culture of Sterile Material.* Recovery of *Aspergillus* species by culture of a specimen obtained by a sterile procedure from a normally sterile and clinically or radiologically abnormal site consistent with an infectious process, excluding BAL fluid, a cranial sinus cavity specimen, or urine. The characteristic colony and microscopic morphology are described in Section 14.12, Laboratory Detection, Recovery, and Identification.

Without prompt treatment multifocal pulmonary infiltrates expand and consolidate. Hematogenous dissemination can occur to the brain, heart, and liver, also producing necrotizing skin ulcers. Unless diagnosed and treated promptly, a fatal course can occur in 1–3 weeks. Invasive aspergillosis is notable for a high case/fatality ratio (please see Section 14.6, Epidemiology and Risk Groups/Factors). The most useful diagnostic procedures for invasive pulmonary aspergillosis are:

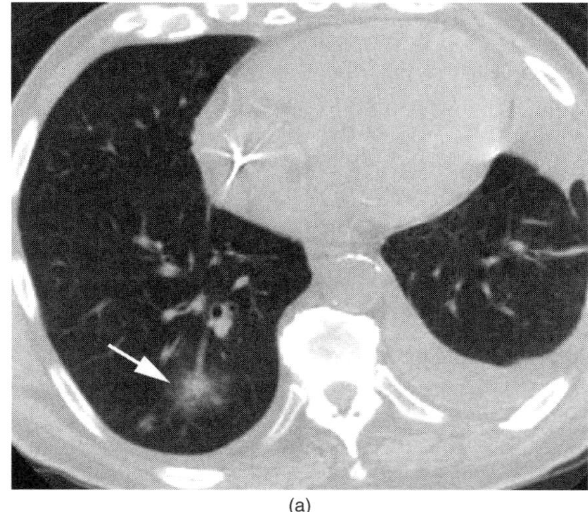

(a)

Figure 14.5 (a) CT scan of lung showing "halo" sign in AML patient with acute pulmonary aspergillosis, represented by ground-glass attenuation (*arrow*), surrounding nodular opacities. (b) CT scan of lung showing the "air crescent" sign (*arrows*): 1, transverse; 2, frontal; and 3, sagittal planes of section. This sign marks a later stage in invasive pulmonary aspergillosis and indicates a response to therapy. A crescent of air is seen above the *Aspergillus* fungus ball. The halo and air crescent signs are not specific for aspergillosis.
Source: Used with permission from Dr. Eugene Berkowitz, Department of Radiology, Emory University School of Medicine, Atlanta, Georgia.

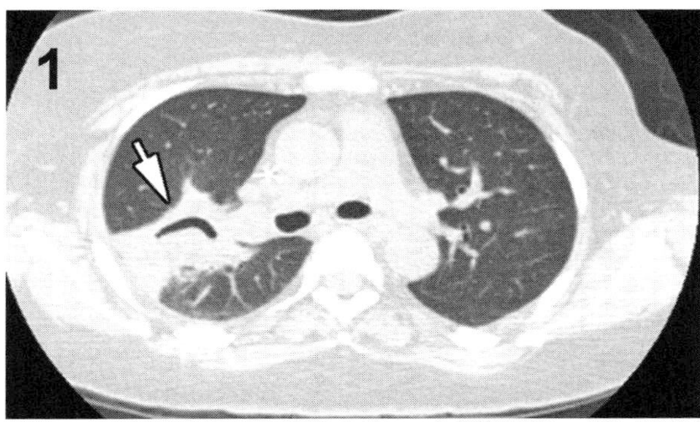

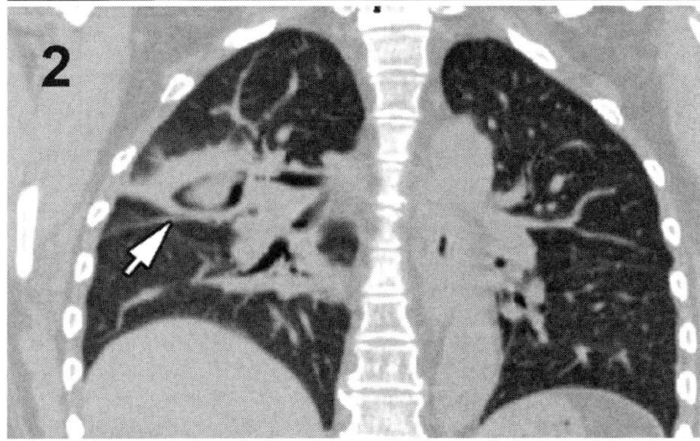

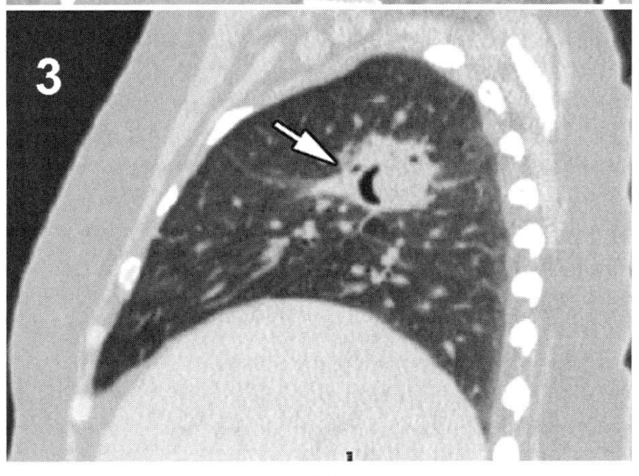

(b)

Figure 14.5 (*Continued*. Please see Figure legend on the previous page.)

- Microscopy and culture. Cultures of respiratory tract, including sputum, are important in high-risk patients; nasal sinus swab when culture-negative for bacteria but positive for *Aspergillus* species.
- Lung biopsy.
- Chest radiography.
- Computed tomography of chest (i.e., screening CT scans in febrile neutropenic patients). The halo and air crescent signs implicate aspergillosis.
- Bronchoalveolar lavage (BAL).
- Percutaneous needle biopsy of lung or other lesions.
- Antigen detection (cell wall galactomannan) by EIA (*Aspergillus* Platelia® EIA, Bio-Rad, Inc.).

The differential diagnosis for invasive pulmonary aspergillosis includes infections with *Fusarium* species, *Mucorales*, or *Pseudallescheria/Scedosporium* species.

Cerebral Invasive Aspergillosis

Cerebral invasive aspergillosis, a dire clinical form, most often is the result of hematogenous spread from a pulmonary focus, rather than arising as a direct extension of sinonasal disease. The risk factors for cerebral infection are the same as for pulmonary infection.

- Immunosuppression (the primary risk factor).
- Systemic corticosteroid therapy or chemotherapy.
- Severe hematologic disease (e.g., leukemia or lymphoma) or stem cell transplants involve much higher risk than other sources of immunosuppression.
- Solid organ transplant
- Late-stage AIDS
- Diabetes mellitus

The pathophysiology of *Aspergillus* sets it apart from other infectious organisms. Because it is angioinvasive, cerebral infection with aspergilli will quickly lead to acute infarcts with or without hemorrhage. If the patient survives long enough, these areas of infarct can progress to infectious cerebritis or abscess. Cerebral invasive aspergillosis can develop as single or multiple lesions. Further information on cerebral invasive aspergillosis is found in Mackenzie et al. (2002).

Other Forms of Invasive Aspergillosis

These may occur as single site infections or via hematogenous dissemination of pulmonary aspergillosis.

Cardiac

Cardiac invasion by *Aspergillus* species may present as pericarditis, endocarditis, or myocarditis (Walsh et al., 2008). *Aspergillus* endocarditis takes the form of valvular or endocardial infection. Valvular vegetations may develop on prosthetic valves but also occur on normal valves (i.e., in injection drug users). Valve infection carries a high risk of emboli and their lodgment particularly in the CNS. *Aspergillus* mural endocarditis may result from extrapulmonary dissemination or involvement of the mitral valve. *Aspergillus* myocarditis usually results from extrapulmonary dissemination and presents as myocardial infarction or cardiac arrhythmia. *Aspergillus* pericarditis results from a direct extension of invasive pulmonary disease, extension from a myocardial lesion, or intraoperative contamination. Cardiac aspergillosis carries a poor prognosis. Management requires both antifungal therapy and surgical resection of lesions.

Cutaneous

Cutaneous aspergillosis occurs secondary to hematogenous dissemination in immunocompromised patients, less commonly as a primary infection resulting from the application of contaminated dressings on burn patients, percutaneous infections in neonates, or by *Aspergillus* conidia alighting on vascular sites in the operating room. ITC concentrates in the skin and has the potential to treat cutaneous aspergillosis (Walsh et al., 2008).

Endophthalamitis

Aspergillus endophthalmitis is a dire infection, one that may result in loss of vision and destruction of the eye. Mechanisms of infection are hematogenous dissemination, penetrating injury, or as the result of a contaminated surgical procedure. Diagnosis requires ophthalmoscopic examination and culture of specimens of the vitreous or aqueous humor. Therapy requires both antifungal therapy (intravitreal and systemic AmB, or VRC) and vitrectomy.

Esophageal and Gastrointestinal Aspergillosis

Aspergillosis of the esophagus and gastrointestinal tract is relatively common in advanced cases of hematogenously disseminated invasive aspergillosis and is associated with high morbidity and mortality.

Osteomyelitis and Septic Arthritis

Osteomyelitis is the fourth most common site of infection for aspergillosis after pulmonary, sinus, and cerebral infections (Kirby et al., 2006). *Aspergillus* osteomyelitis may be subdivided into vertebral, head, rib, and long bone infections; however, therapy recommendations have been similar. *Aspergillus* osteomyelitis occurs primarily in immunocompromised patients due to hematogenous dissemination but may also be acquired by immunocompetent patients as a result of surgery, IV drug use, trauma, and previous aspergillosis (Stratov et al., 2003). *Aspergillus* osteomyelitis can be difficult to treat, often requiring prolonged antifungal therapy and surgical intervention. The treatment plan includes surgical resection of devitalized bone and cartilage combined with systemic therapy with VRC preferred (Walsh et al., 2008). Sixty-three cases of osteomyelitis and septic arthritis due to *Aspergillus* species were found in the world literature for the period 1997–2003 (Kirby et al., 2006). Of these 40 cases had a successful outcome and in 23 cases therapy was unsuccessful.

Peritonitis

Aspergillus peritonitis is a complication of chronic ambulatory peritoneal dialysis, but is less frequent than *Candida* species as a cause of fungal peritonitis. Removal of the dialysis catheter combined with intraperitoneal and intravenous AmB has led to successful outcomes (Walsh et al., 2008).

Renal

Renal aspergillosis is usually the result of hematogenous dissemination or due to a contaminated surgical procedure. Single or multiple parenchymal abscesses may cause hematuria, ureteral obstruction, perinephric abscess, or passing of fungal balls or fungal elements in the urine. Urinary tract infection by *Aspergillus* spp. may evolve through three patterns: ascending infection from the lower urinary tract, bilateral renal parenchymal involvement secondary to hematogenous dissemination, and a localized obstructive uropathy related to *Aspergillus* casts in the renal pelvis. Acute renal colic[4] occurred in a diabetic man who also was an injection drug user (Pérez-Arellano et al., 2001). During hospitalization he eliminated, per urethra, multiple fungus balls identified as due to *A. flavus*. No other body sites were affected. Therapy with ITC led to complete recovery. Several patients with renoureteric aspergillosis also had the double risk factor of IV drug use and diabetes mellitus (reviewed by Pérez-Arellano et al., 2001). Inadvertent infusion of aspergilli in contaminated IV "works" combined with poorly controlled diabetes can predispose to colonization of the urinary tract. This is because of the functional neutropenia of diabetic PMNs, papillary necrosis (which is associated with diabetes), and glycosuria favoring the growth of *Aspergillus*.

14.10 VETERINARY FORMS

Animals affected by aspergillosis include poultry, birds, penguins, marine mammals, dogs, and cats. Aspergillosis is a common avian disease, particularly in captive waterfowl, wading birds, penguins, raptors, pheasants, and passerines. Among pet birds aspergillosis is seen in Amazon parrots and African grey parrots. The causative species is usually *A. fumigatus*, although *A. flavus* and *A. niger* infections also occur.

[4]Flank pain caused by obstruction to the flow of urine, often caused by kidney or ureteral stones.

Clinical Forms of Avian Aspergillosis

Usually occurring in the respiratory tract, avian aspergillosis sometimes disseminates to the CNS. Three common forms are:

1. Diffuse lower respiratory tract disease
2. Syringeal granuloma
3. Focal CNS granuloma (brain)

Diffuse Lower Respiratory Tract Disease

The bird presents as dyspneic (open-beak breathing, increased respiratory rate or effort at rest, prolonged tachypnea following manual restraint, wide-based stance on perch, tail bob, choanal exudate) and also may have the "sick bird" appearance: fluffed and less active.

Syringeal Granuloma

Birds with syringeal granuloma usually have only a single lesion but also may have lower respiratory disease. The classic sign is a change in the voice, found most commonly among psittacines, passerines, and waterfowl. If large enough, the syringeal granuloma will obstruct the airway, resulting in signs of dyspnea.

Cerebral Granulomas

Focal granulomas in the brain can occur in any avian species but are most often seen in waterfowl. CNS signs such as ataxia or torticollis may be present and the lower respiratory tract also may be involved.

Diagnosis and Laboratory Tests

Epidemiology, therapy, prevention, and control of avian aspergillosis are covered in Guillot et al. (2006) and Kearns and Loudis (2003).

Aspergillosis in Companion Animals

Sinonasal aspergillosis is a frequent cause of nasal discharge in otherwise healthy, young to middle-aged dogs (Peeters and Clercx, 2007). A local immune dysfunction is suspected in affected animals. Diagnosis is made by direct visualization of fungal plaques during endoscopy or histopathologic observation of fungal elements. Effective treatment of nasal aspergillosis in dogs is difficult (Clercx, 2000). Therapeutic approaches have included surgery as well as systemic and topical antifungal agents. Systemic antifungal therapy requires prolonged regimens because of poor efficacy, with clinical cures obtained in only about half of patients. Topical therapy has greater success and has improved management of this previously intractable condition. A noninvasive technique using nonsurgically placed catheters to infuse the topical drug (e.g., enilconazole emulsion or clotrimazole) into the nasal cavities and frontal sinuses is done under general anesthesia. The prognosis is good for dogs that are mycologically cured at the end of antifungal therapy.

Primary infection of the sinonasal region in cats results from the inhalation of fungi, with the infection penetrating overlying bones to invade the subcutaneous space. These lesions are typically the result of cryptococcosis or aspergillosis and must be distinguished from invasive nasal malignancies.

14.11 THERAPY

The recommendations for therapy of various clinical forms of aspergillosis are contained in the clinical practice guidelines of the Infectious Diseases Society of America (Walsh et al., 2008). Of the species encountered in invasive aspergillosis, *A. terreus* as well as some *A. flavus* strains are clinically resistant to AmB. Rarely, *A. fumigatus* isolates with reduced azole susceptibility have been described (Mortensen et al., 2010). Uncommon species such as *A. glaucus, A. lentulus, A. nidulans, A. ustus*, and *A. versicolor* also may demonstrate decreased susceptibility to AmB.

Therapy for Invasive Pulmonary Aspergillosis

A wider variety of mold-active anti-*Aspergillus* compounds is available with extended spectrum triazoles. In 2001, voriconazole (VRC) supplanted polyene antifungal agents as primary therapy for most patients. The choice of primary therapy is very important because salvage therapy has limited potential to achieve a cure. The duration of therapy should be several weeks or months. Measuring the serum galactomannan concentration by EIA can serve as a surrogate marker of response to therapy.

Amphotericin B

AmB-deoxycholate is no longer the drug of choice for primary therapy of invasive pulmonary aspergillosis, superseded by less nephrotoxic lipid formulations (lipid complexes of AmB, liposomal AmB) (reviewed by Groll and Walsh, 2002). Liposomal AmB is recommended as alternative primary therapy, and for salvage therapy in patients whose disease is refractory to voriconazole. *A. terreus* has innately reduced susceptibility to AmB. The mean MIC against AmB was measured at 3.37 μg/mL, and the minimum fungicidal concentrations (MFCs) are elevated beyond the AmB concentration achievable in plasma: a 24 h MFC of 7.03 μg/mL and a mean 48 h MFC of 13.4 μg/mL (reviewed by Steinbach et al., 2004).

Voriconazole (VRC)

VRC is licensed as primary therapy for invasive aspergillosis and is recommended for most patients. VRC therapy led to better responses, improved survival, with fewer severe side effects than the previous standard primary therapy with AmB. Herbrecht et al. (2002) studied a total of 391 patients with definite or probable aspergillosis recruited at 95 centers in 19 countries between July 1997 and October 2000. Figure 14.2 illustrates survival curves for the population with proven or probable aspergillosis according to treatment with AmB or VRC. VRC has excellent penetration into the CNS and may be superior to AmB in patients infected with *A. terreus*.

Formulation VRC is available as tablets or as a sulfobutyl-ether cyclodextrin solution for IV administration. There is concern about the accumulation in plasma of the cyclodextrin ether vehicle in patients with renal insufficiency. That concern does not apply to the oral formulation, which does not contain a cyclodextrin solubilizer. VRC is widely distributed in tissues; however, the CNS concentration is approximately 50% of plasma levels.

Dosage The recommended daily dose of VRC for adults is 6 mg/kg IV twice daily for the first day of therapy, followed by 4 mg/kg, 2×/day. Oral therapy has not been separately validated in clinical trials. The oral dose can be approximated to the standard IV dosage by using 4 mg/kg/dose rounded up to the nearest pill size and given 2×/day. Dosages for pediatric patients may be higher, 7 mg/kg, 2x/day, because of increased metabolic clearance.

Adverse Effects VRC use can be associated with transient visual disturbances including photopsia (flashes of light) and visual hallucinations. Hepatotoxicity may be dose-limiting.

Posaconazole (PSC) (Noxafil®, Schering-Plough, Inc.)

PSC is structurally related to ITC.

Indications PSC is indicated for prophylaxis in preventing invasive aspergillosis in neutropenic patients with acute myelogenous leukemia and in SCT recipients with GVHD. It is also effective as salvage therapy for refractory invasive aspergillosis.

Dosages The oral formulation is administered as 200 mg, 3×/day and for salvage therapy, 800 mg in two or four divided doses. The measurement of plasma drug levels is recommended to optimize drug dosages for efficacy.

Formulations and Bioavailability Only the oral formulation has been evaluated for aspergillosis. PSC has saturable kinetics so that loading doses cannot be used. Steady-state plasma concentrations may take up to a week to be achieved. Absorption from the gastrointestinal tract is enhanced if the drug is administered with fatty food.

Adverse Effects Greater toxicity has been observed in patients taking PSC who have acute leukemia or myelodysplasia.

Itraconazole (ITC) (Sporanox®, Janssen Pharmaceutica)

Indications ITC is no longer recommended as primary therapy for invasive pulmonary aspergillosis. It is recommended as an alternative empiric therapy for patients who are persistently neutropenic and who have prolonged fever unresponsive to antibiotics. As salvage therapy, it is indicated in patients who are refractory to or intolerant of AmB lipid formulations. ITC is not recommended as salvage therapy in patients unresponsive to VRC because of the same mechanism of action and erratic bioavailability. ITC is recommended as a second choice for prophylaxis in patients at high risk for invasive pulmonary aspergillosis: SCT recipients with GVHD, or in patients with acute myelogenous leukemia or myelodysplastic syndrome. Other forms of aspergillosis for which ITC has shown therapeutic benefit are aspergilloma, ABPA, and allergic *Aspergillus* sinusitis.

Formulation and Bioavailability ITC is lipophilic and is supplied as a liquid solubilized in hydroxypropyl-ß-cyclodextrin for both oral and IV administration. Absorption from oral capsules is favored by low gastric pH and dietary lipids, so its effectiveness is improved by taking it with an acidic cola beverage and food. Monitoring the plasma concentration of ITC by bioassay or HPLC is recommended to document its absorption because of its variable bioavailability.

Dosages For adults, the recommended dosage for capsules of ITC is 400 mg/day. For the IV liquid formulation, the recommended adult dosage is 200 mg, 2×/day for 2 days, followed by 200 mg/day for 12 days.

Echinocandin Derivatives

Echinocandins are a second-line choice for therapy of invasive pulmonary aspergillosis, when patients are intolerant of or unresponsive to primary therapy. Caspofungin (CASF) (Cancidas®, Merck) or micafungin (Micamine®,

Astellas) are promising as salvage monotherapy or combined with other agents (i.e., VRC). Success in salvage therapy with micafungin alone or in combination therapy occurred in between 24% and 26% of 98 SCT patients with invasive aspergillosis, with an additional 12 patients achieving stable disease (Kontoyiannis et al., 2009).

These antifungal agents inhibit synthesis of ß-(1 → 3)-D-glucan, an essential fibrillar cell wall component: CASF, micafungin, and anidulafungin (Eraxis®, Pfizer). The potential for human toxicity is low because the drug target is absent in the mammalian host.

Indications CASF is approved in the United States and the European Union for invasive aspergillosis in patients who are refractory to or intolerant of other approved therapy. Eighty-three patients with invasive aspergillosis who failed therapy with AmB or lipid-AmB were treated with CASF by IV infusion of a 70 mg loading dose followed by 50 mg/day. A favorable clinical response occurred in 37 (45%) patients, including 32 of 64 (50%) with pulmonary aspergillosis and 3 (23%) of 13 with extrapulmonary disseminated aspergillosis (Maertens et al., 2004). Reversal of immunosuppression is critical in the prognosis of patients with invasive aspergillosis. The favorable clinical response to CASF was lower in allogeneic SCT recipients who were receiving ongoing immunosuppression, and in patients who were neutropenic at the initiation of CASF therapy. Despite neutropenia, clinical improvement occurred before neutrophil recovery in some patients. CASF is also recommended for empiric antifungal therapy for high-risk patients neutropenic for 10 or more days, with fever unresponsive to antibiotics, and for presumptive therapy for invasive aspergillosis.

Combination Therapy for Invasive Aspergillosis

Monotherapy for invasive aspergillosis is still associated with treatment failures, so that interest has increased in the potential of combination therapy. Echinocandins offer an opportunity for combination therapy without the threat of antagonism, which theoretically exists between azoles and polyenes. A prospective trial of CASF and another antifungal agent was conducted in 53 patients who were refractory to or intolerant of primary therapy with an antifungal agent other than an echinocandin derivative (Maertens et al., 2006). The patients were very ill with a high index of poor prognostic factors. Most were refractory to previous therapy and over half were neutropenic. Combination therapy was successful in 55% of patients by the end of treatment and in 49% by the 84th day following initiation of therapy. The proportion of patients with a favorable clinical response was greater in patients receiving regimens of CASF and VRC (54%) and CASF and a polyene (50%) than those receiving CASF and ITC (29%).

A multicenter study in solid organ transplant recipients compared outcomes in 40 patients who received VRC plus CASF as primary therapy for invasive aspergillosis with those in 47 patients who received a lipid-AmB monotherapy (reviewed by Singh and Pursell, 2008). Overall survival at 90 days was 67.5% in the group receiving combination therapy versus 51% in the monotherapy group. Added benefit from combination therapy was seen as a lower mortality rate in a subgroup of patients with renal failure. Additionally, a study examined VRC + CASF versus VRC for salvage therapy of invasive aspergillosis (Marr et al., 2004). This study showed a 3 month survival of 70% in the combination patients versus 35% in the monotherapy group.

Micafungin and anidulafungin have activity against *Aspergillus* species, but as yet are not approved for aspergillosis. An ongoing prospective clinical research study evaluating VRC versus VRC plus anidulafungin will help determine whether combination therapy is superior to monotherapy for invasive aspergillosis.

Formulation and Bioavailability All echinocandin derivatives are only available for IV infusion. Their $t_{1/2}$ is 10–15 h, allowing for a once/day dosing regimen. Echinocandins distribute into all major organs, except that concentrations in the CNS are low. CASF and micafungin are degraded in the liver and excreted in the urine and feces, whereas anidulafungin is degraded in the plasma.

Dosages The recommended dosage for CASF in adults is a 70 mg loading dose on the first day, followed by 50 mg/day. The drug is administered IV slowly over 1 h.

Adverse Effects All echinocandins are well tolerated, with very few patients discontinuing therapy because of side effects. Increased liver enzymes, gastrointestinal upsets, and headaches have been reported. CASF increased histamine concentrations in whole blood and human mast cells suggesting that infusion-related reactions associated with CASF may be mediated by histamine release secondary to CASF (Cleary et al., 2003).

Adjunctive Therapy with Cytokines and Anticytokines

Successful therapy depends on reversing underlying immunosuppression and prompt restoration of granulocytes. Neutropenic patients who are not receiving granulocyte colony stimulating factor (G-CSF) and granulocyte-macrophage colony stimulating factor (GM-CSF) as part of their cancer chemotherapy may shorten the period of neutropenia by adding these factors to the therapeutic regimen (Walsh et al., 2008). Moreover, these factors and IFN-γ increase chemotaxis, phagocytosis, and oxidative microbicidal processes of granulocytes, monocytes, and macrophages. Cytokines and anticytokines (IFN-γ, IL-12, and anti-IL-4) stimulate the Th1 pathway of adaptive immunity and may exert a cooperative effect with antifungal therapy (Groll and Walsh, 2002).

Granulocyte transfusions may be beneficial in patients with proven or probable invasive aspergillosis who require this as a temporary measure until they recover from neutropenia. Such transfusions should only be given from cytomegalovirus-negative donors.

Prophylaxis

Mold-active prophylaxis using ITC is associated with a lower frequency of invasive aspergillosis but has significant adverse effects and drug interactions. Oral PSC prophylaxis is recommended for SCT recipients with GVHD and in neutropenic cancer patients who are at high risk for invasive aspergillosis, with ITC recommended as an alternative (Walsh et al., 2008).

Current Therapy Practice in Solid Organ Transplant Patients — Liver Transplants

A survey conducted during a 4 month period in 2006 of 67 North American liver transplant programs found that slightly more than half (53%) of sites used a single antifungal agent as primary therapy: with VRC, 83%; AmB-lipid complex, 33.3%, CASF, 19.4%; liposomal AmB, 11.8%; and other echinocandins, 14.3% of sites (Singh et al., 2008). This underlines the broader range of therapeutic choices now available. Combination antifungal therapy consisted of VRC plus an echinocandin in 73% of sites; echinocandin plus lipid-AmB in 33% of sites; and VRC plus lipid-AmB in 18%. Salvage therapy for invasive aspergillosis was usually combination therapy used at 80% of sites; 58% of centers added a new mold-active agent to an existing drug, and 22% selected a new combination altogether.

Therapy for Aspergilloma

Aspergillomas in the lung are usually managed conservatively as they remain asymptomatic in many patients. Surgical resection is reserved for patients carefully evaluated for risks, when there is danger of life-threatening hemoptysis from an *Aspergillus* fungus ball adjacent to a blood vessel. Therapy with ITC or VRC has some potential for therapeutic benefit with minimal risk. Long-term therapy with ITC or VRC is required for chronic cavitary pulmonary aspergillosis.

Therapy for Endocarditis

Surgical removal and replacement is indicated for a heart valve infected with *Aspergillus* species. However, indefinite suppressive therapy with VRC or PSC should be considered because of the potential for recurrent infections.

Therapy for Tracheobronchitis

Prompt treatment is needed for ulcerative lesions and to prevent disruption of the anastomotic site and loss of a transplanted lung. Bronchoscopic examination and CT scanning are important for diagnosis and to evaluate spread of the infection. VRC is the choice for primary therapy. Liposomal AmB is the choice of polyenes for this indication. Aerosolized AmB or lipid-AmB complex (ABLC) can deliver high drug concentrations to the anastomotic site but this approach is still not standardized. Tracheobronchitis rarely occurs in patients with AIDS. VRC is the treatment of choice; however, even with therapy these patients often die from the infection.

Therapy for Keratitis

Therapy consists of topical treatment with AmB or natamycin drops administered hourly, although studies demonstrating efficacy of either agent are limited. Favorable reports using oral ITC for mold keratitis suggest that it, or oral VRC, may be a useful addition to topical therapy. A review comparing therapy for fungal keratitis in 370 patients, with clinical cure as the endpoint, found that of all treatments silver sulphadiazine ointment had the lowest proportion of treatment failures followed by ITC, miconazole, chlorhexidine, or econazole. The drug with the most treatment failures was natamycin (Florcruz and Peczon, 2008). The power of the individual patient groups was too small to determine significance of the results.

Penetration of topical agents into the corneal stroma is usually poor, and fungal infiltration deep within the stroma is frequently unresponsive to therapy. Progressive disease, unresponsive to therapy, may require a corneal transplant to prevent blindness. Clearly, further clinical trials are warranted to find a better way to treat this serious condition. As in many mycoses, fungi are among the last microbes to be considered, only after the failure of antibiotic therapy for bacterial infection or antiviral therapy. Thus, the diagnosis often is delayed, further diminishing chances for a successful outcome. Recently, in vivo confocal microscopy with the Heidelberg Retina Tomograph II has been evaluated for noninvasive, early diagnosis of fungal keratitis (Brasnu et al., 2007). This improved diagnostic modality may facilitate earlier diagnosis and a more successful treatment plan.

Drug Resistant Species

Azole-Resistant Aspergillus fumigatus

Soil samples from the vicinity of hospitals in Copenhagen yielded four *A. fumigatus* isolates displaying elevated azole MICs: ITC ≥ 8 μg/mL; VRC = 4 μg/mL (Mortensen et al., 2010). All contained the same mutation in the gene encoding lanosterol 14-α-demethylase, the key enzyme in ergosterol biosynthesis. A single resistance mechanism, substitution of leucine to histidine at codon 98 of the *cyp*51A gene, combined with a tandem repeat of 34 bp in the promoter region (TR/L98H), was found in 94% of the resistant isolates, including those obtained from azole-naive human patients. The means by which resistance was acquired in soil isolates was linked to the use of azole agricultural fungicides. *A. fumigatus* isolates with the TR/L98H resistance genotype were also recovered from patients in Spain, France, Norway, the United Kingdom, The Netherlands, and Belgium (Mortensen et al., 2010).

Aspergillus terreus *— An Emerging Pathogen*

Cancer treatment centers in certain localities have a high percentage of *A. terreus* as a cause of invasive aspergillosis. In a major cancer treatment center in Houston, Texas, 300 patients developed confirmed invasive aspergillosis in the years 1995–2001. Of these, *A. fumigatus* was isolated from 90 patients (30%), *A. terreus* from 70 patients (23%), *A. flavus* from 65 patients (22%), and *A. niger* from 30 patients (10%). Other *Aspergillus* species were isolated from the remaining 45 patients (15%) (Hachem et al., 2004). *A. terreus* was hospital-acquired more frequently than *A. fumigatus*, defined as the first signs and symptoms occurring ≥ 14 days after hospital admission. The environmental niche in the hospital, if one existed, could not be determined. Of 32 patients with *A. terreus* disease treated with AmB-deoxycholate or lipid derivatives of AmB, a clinical response occurred in 5 (14.6%), whereas the response to polyenes in patients with *A. fumigatus* disease was 8 of 33 or 24%. *A. terreus* isolates display reduced susceptibility to polyenes in vitro and in vivo. Salvage therapy with PSC was promising in both *A. terreus* and *A. fumigatus* disease.

Another locality where *A. terreus* has become a troublesome pathogen is Innsbruck in the Austrian Tyrol (Lass-Flörl et al, 2005). The Medical University's hematology ward, until 1999, was operated as an "open window" hospital. During that time *A. terreus* accounted for 30% of the invasive aspergillosis cases. The hospital environment was monitored and *A. terreus* was cultured from potted plants near the patient care areas (Lass-Flörl et al., 2000). Based on RAPD genotyping, the *A. terreus* genotypes from four patients matched those of the potted plant isolates. Five additional patients were infected by genotypes not found in the environment. Starting in 2000 patients with hematologic malignancies at the Innsbruck hospital were housed in LAF rooms. Despite that level of air filtration, 11 patients in 2000–2004 developed invasive *A. terreus* disease, and 37 patients developed invasive aspergillosis caused by other *Aspergillus* species. Possibly the patients were colonized by *A. terreus* before entering the hospital.

Aspergillus lentulus *— A Sibling Species of* A. fumigatus

Poorly sporulating variants were identified by *rodlet A* and *β-tubulin* sequences as a new species, *A. lentulus* (Balajee et al., 2005). Hyphal growth rates of the *A. lentulus* isolates are equivalent to those of *A. fumigatus*, but unlike *A. fumigatus, A. lentulus* grows poorly at 45°C and not at all at 48°C. The vesicles of their conidial heads are smaller and phialides are shorter than those of *A. fumigatus*. Conidium characteristics (color, size, shape, and ornamentation) of the variants are similar to those of *A. fumigatus. A. lentulus* was isolated from invasive aspergillosis in SCT recipients (Balajee et al., 2005). These isolates had reduced in vitro susceptibility to multiple antifungal agents and patients from whom isolates were recovered all had poor clinical outcomes and died with invasive aspergillosis despite antifungal therapy. Twelve *A. lentulus* isolates, most from sputum specimens, showed high MICs with respect to AmB [geometric mean (GM) MICs, 4.5–16 μg/mL], ITC (GM MICs, 6–10.25 μg/mL), and VRC (GM MICs, 3–7.5 μg/mL) (Alcazar-Fuoli et al., 2008).

14.12 LABORATORY DETECTION, RECOVERY, AND IDENTIFICATION

Direct Examination

Specimens for direct examination include sputum, bronchoalveolar lavage fluid, fine needle aspirates, thorascopic[5] biopsy, and other biopsy materials. In the

[5]Thoracoscopic chest biopsy. *Definition*: Involves inserting a lighted tube (thoracoscope) through a 1.25–2.5 cm incision in the chest wall to directly visualize the lung. The biopsy is then taken with other

case of fungal keratitis, specimens are obtained using surgical blades, blunt platinum spatulas, or calcium alginate swabs dipped in trypticase soy broth.

The morphology of hyphal elements may be seen on microscopic examination of KOH preps of fresh clinical material. *Aspergillus* hyphae in sputum or nasal scrapings are hyaline, closely septate, and *dichotomously* branched at an acute angle. Addition of Calcofluor white M2R (Fungi-Fluor®, Polysciences, Inc.) enhances visibility when viewed in a fluorescence microscope. The presence of septate hyphae prompts further steps to culture the fungus to distinguish it from other molds. Bloody or mucoid specimens should be first treated with Mucolyse®—dithiothreitol in phosphate buffer (Pro-Lab Diagnostics, Richmond Hill, Canada).

Histopathology

H&E stains hyphal elements but with low contrast between hyphae and tissue cells. *Aspergillus* hyphae are *easily missed in hematoxylin and eosin stained tissue*. The fungal-specific stains GMS and PAS are preferred. Of these, PAS stain reveals tissue structure and the inflammatory response but GMS stain may be more sensitive. Calcofluor also may be applied to tissue sections. With PAS or GMS stain the hyphae are 3–6 μm wide, closely septate, and dichotomously branched at acute angles (Fig. 14.6). *A. terreus*, like *Fusarium* species, is able to sporulate in tissue, termed "adventitious sporulation," resulting from its ability to produce accessory conidia on short sterigmata directly from hyphae.

Aspergillus species invade blood vessels causing mycotic thrombi which may be seen in tissue. When there is little host inflammatory response, such as in invasive aspergillosis of the immunocompromised host, the fungus proliferates in tissue in parallel or radial arrays. When there is a host inflammatory response, the hyphae may be distorted and forced into bizarre "varicose" shapes, resembling those of the *Mucorales*. Aspergilli rarely produce conidial heads in host tissue, but these can be seen where the infection is growing near an air space.

Hyphae of other hyaline molds (e.g., *Fusarium* species and *Pseudallescheria/Scedosporium* species) may resemble the aspergilli, both in being angioinvasive and in forming dichotomous, acute angle branching. Hyphae of the *Mucorales* are broader, twisted, and infrequently septate.

Further discussion of the histopathology of invasive pulmonary aspergillosis by Dr. Kazutoshi Shibuya, MD, may be be found at the aspergillosis website, URL http://www.aspergillus.org.uk/indexhome.htm?education/index.php~main.

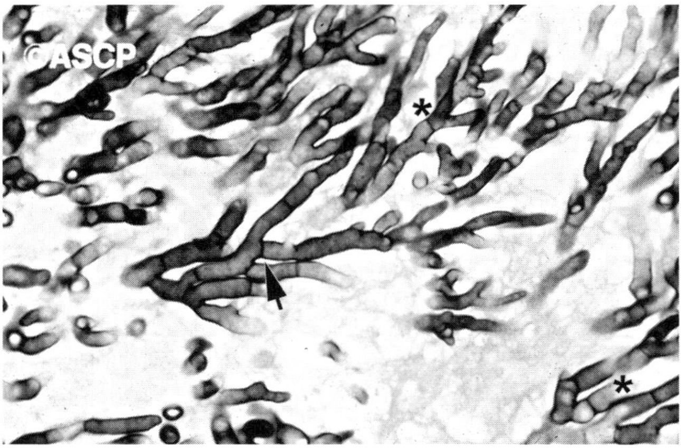

Figure 14.6 Histopathology—aspergillosis, lung. *Arrow*: acute angle branch; asterisks indicate septa; GMS stain, 250×.
Source: Chandler and Watts (1987). Used with permission from the American Society for Clinical Pathology. Copyright © 1987.

Culture

Cautious interpretation of cultures is advised because *A. fumigatus* and other *Aspergillus* species growing from specimens obtained from nonsterile sites may result from contamination, or the presence of conidia in lung or sinus may be colonizing and not invading tissue. The significance of isolations of *Aspergillus* species from body fluids or secretions is as follows:

- Sputum, urine, stool, or wound, cultures per se have little diagnostic significance because of the common environmental presence of aspergilli. Repeated isolation from these sites, however, should be reported.
- Bronchoalveolar lavage fluid or transbronchial brush biopsy cultures are more significant.
- Normally sterile sites by biopsy: these cultures are highly significant and should be promptly reported.
- The specificity of an isolate of *Aspergillus* from the respiratory tract of a neutropenic patient or SCT recipient is very high.
- Blood cultures for *Aspergillus*. The clinical significance of *Aspergillus* fungemia in invasive aspergillosis is not clearly established. One series of patients treated in the hematology–oncology service over a 17 year period found fungemia in 9 of 89 (10.1%) patients with culture-confirmed invasive pulmonary aspergillosis (Girmenia et al., 2001). Fungemia was detected a median of 5 days from the onset of clinical signs of infection. A comparison between invasive aspergillosis patients with or without fungemia showed that fungemic

instruments inserted through 1–2 similar size incisions. This procedure is conducted under general anesthesia and requires the insertion of a chest tube.

patients were similar to those without positive blood cultures regarding clinical course and outcome. The diagnostic role of blood cultures in invasive aspergillosis is limited because blood cultures are frequently negative and when they are positive this usually occurs after a clinical or microbiologic diagnosis has been made. *Aspergillus* fungemia does not appear to correlate with a poorer prognosis.

Media for Growth of Aspergilli

SDA-Emmons Agar This and the original SDA are the media of choice in most clinical microbiology laboratories for mold isolation. Antibiotics, chloramphenicol and gentamicin, are incorporated in the medium when the specimen is from a nonsterile site. Aspergilli do not grow on media containing cycloheximide as, for example, "Mycosel" and "Mycobiotic" medium. On SDA-Emmons agar sporulation occurs in 48–72 h at 30–37°C. Tease mounts are the first choice or, if they are not revealing, *slide cultures* are advantageous for observing asexual reproductive structures: the sporing heads. Table 14.5 shows important *Aspergillus* species and their color on SDA. The colony color of an *Aspergillus* species is influenced by the culture medium and by strain–strain variation; that is, *A. fumigatus* can vary from blue-green to gray-turquoise.

- Czapek–Dox agar is well-accepted for speciation of aspergilli but not generally used in the clinical laboratory.
- Malt extract agar often is used in air sampling devices such as the Andersen Sampler. Both Czapek–Dox and malt extract agars are the media of choice of *Aspergillus* specialists to demonstrate sporulation and colony color (Klich and Pitt, 1988).

Culture at 25°C on SDA-Emmons agar results in a rapidly growing mold (blue-green to gray-turquoise in the case of *A. fumigatus*) with a typical asexual reproductive structure consisting of *foot cell*, *stipe*, club-shaped *vesicle*, and *phialoconidia* (Fig.14.7) (Klich and Pitt, 1988). Note the difference between the conidiogenous cell of *Aspergillus* (vesicle) and *Penicillium* (no vesicle) (Fig. 14.8).

Table 14.5 Colony Color on SDA Agar in Selected *Aspergillus* Species of Medical Importance[a]

Species	Surface	Reverse
A. flavus	Yellow-green	Pale to light yellow
A. fumigatus	Blue-green to gray-green, to olive	Pale to light yellow
A. nidulans	Dark green to dark olive-buff	Purple to olive
A. niger	White to yellow becoming black to deep brown	White to pale yellow
A. terreus	Cinnamon to brown	Pale yellow to brown

[a]Data obtained from Table 2, St. Germain and Summerbell (1996).

Methods

- *Morphology*. Molds are identified by their macroscopic and microscopic characteristics. Microscopic morphology is dependent on a tease mount or the slide culture method.
- *Genetic ID*. When the culture is poorly sporulating or nonsporulating, the sequence of the D1/D2 of the 28S rDNA or ITS region of rDNA is PCR-amplified and sequenced and the sequence submitted to BLAST to compare with those deposited in Gen-Bank.

Morphology of Some Major Aspergilli of Medical Importance

General Characteristics

The asexual reproductive structure of the aspergilli is a swollen apical vesicle attached to a long stipe, which is connected to the vegetative hypha by a basal portion called a foot cell (Fig. 14.7). The foot cell is part of the same cell as the stipe. The simplest sporing head for *Aspergillus* consists of a swollen vesicle, partially or fully covered with a layer of elongate phialides from which conidia are extruded. Some species, including the major pathogen, *A. fumigatus*, contain a single layer of phialides; that is, they are *uniseriate*; whereas in others, the phialides are connected to the vesicle via a single layer of intermediary cells, the *metulae*, and are termed *biseriate*. Numerous phialoconidia are produced in long chains, which, according to the species, may radiate or occur in compact columns. The phialoconidia are readily airborne. Some species, in addition to phialoconidia, may produce either *hülle cells* or *sclerotia* (*A. flavus*). Hülle cells are thick-walled sterile cells produced as intercalary or terminal cells. Some *Aspergillus* species (e.g., *A. nidulans*) are homothallic and may produce cleistothecia on ordinary isolation medium (SDA). Hülle cells are found surrounding cleistothecia in that species.

Aspergillus fumigatus

A. fumigatus is thermotolerant and grows and sporulates up to 48°C and also below 20°C.

Colony Morphology For colonies on SDA-Emmons agar, typical isolates are fast growing, blue-green

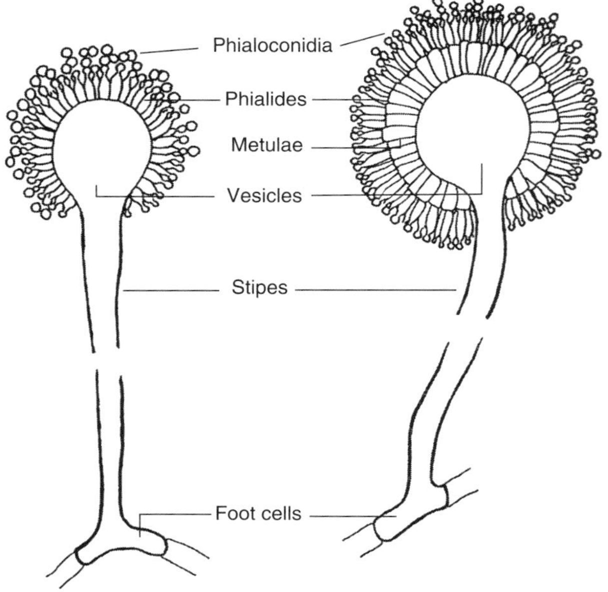

Figure 14.7 Diagram of *Aspergillus* species asexual reproduction, showing a uniseriate and a biseriate species. In *A. fumigatus*, a uniseriate species, chains of phialoconidia are short and are easily dispersed. The foot cell is integral to base of the stipe. Of medically important species, *A. terreus* is biseriate.
Source: Klich and Pitt (1988). Used with permission from Dr. Ailsa D. Hocking, CSIRO Food and Nutritional Sciences, North Ryde NSW, Australia.

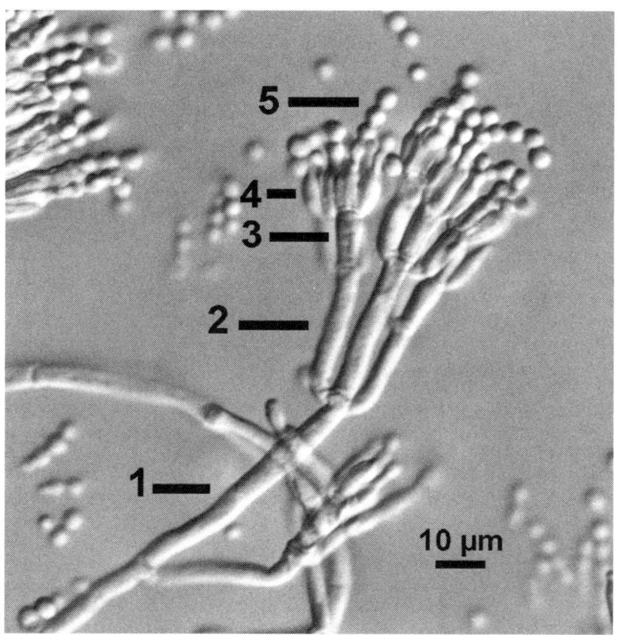

Figure 14.8 *Penicillium* conidiophore morphology (DIC microscopy). Species of *Penicillium* are recognized by their brush-like conidiophores, which may be simple or branched, terminating in clusters of phialides producing chains of phialoconidia There is no vesicle. A "complex" isolate with a biverticillate shaped penicillus is shown: 1, stalk or stipe; 2, ramus (pl: rami) a supporting cell; 3, metula, a second type of supporting cell; 4, phialide; 5, phialoconidia. (Numbers 2–4 make up the penicillus.)
Source: Used with permission from Robert Simmons, Ph.D., Director, Biological Imaging Core Facility, Department of Biology, Georgia State University, Atlanta, Georgia.

to gray-turquoise, with the reverse uncolored to cream. The color results from the pigmentation of conidia. The colony texture is velvety and wooly. The outer edge of the colony is white, which is the area of active mycelial growth (Fig. 14.9).

Microscopic Morphology See Fig. 14.10. The stipes are 225–350 μm long, smooth walled, and hyaline. Vesicles are 15–25 μm wide, hyaline to slightly green in color, and pyriform, *spathulate*, or *subclavate* in shape. Conidial heads are uniseriate (no *metulae* in contrast to *A. flavus*) with phialides concentrated on the upper surface of the vesicle. Conidia are readily dispersed but occasionally are in long chains. Phialoconidia are 2.4–3.0 μm long, spherical to broadly ellipsoidal, with the surface finely roughened to rough.

Aspergillus flavus

The second most frequent clinically encountered species, *A. flavus*, is an important plant pathogen. Several species are morphologically similar to *A. flavus*. Care should be taken in separating them (Klich and Pitt, 1988).

Colony Morphology On SDA-Emmons agar at 25°C, the colony color is olive green to yellow (parrot) green owing to the colored conidia (Fig. 14.9). The colony reverse is uncolored to pale buff. Black sclerotia may be present.

Microscopic Morphology See Fig. 14.11. Stipes are 450–1000 μm long, hyaline, with walls finely rough to

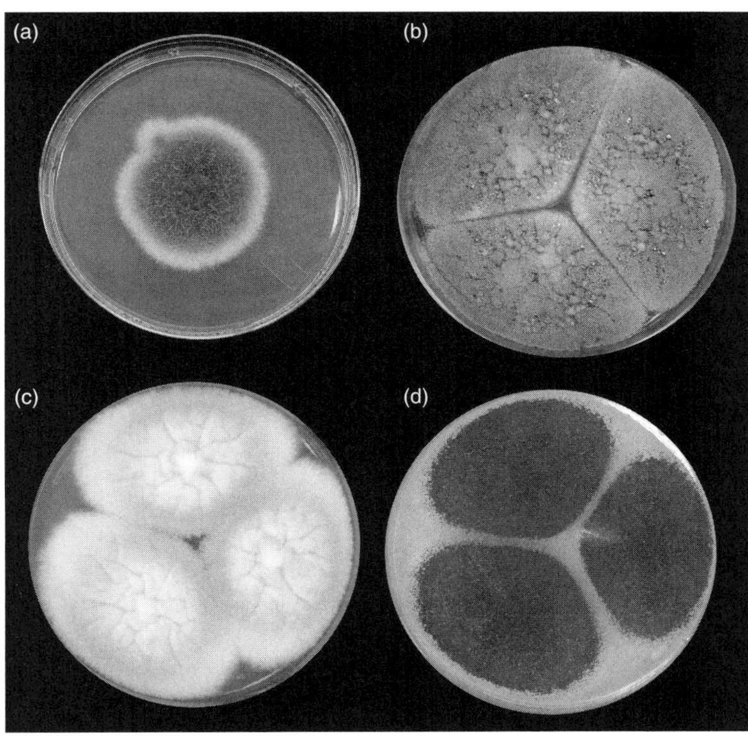

Figure 14.9 Colony morphology of four *Aspergillus* species. (a) *Aspergillus fumigatus*. The surface is velvety or wooly ("velutinous" or "floccose," respectively). *Source:* G. Marshall Lyon, M.D., Emory University Hospital. (b) *Aspergillus flavus*, 7 da, CzapekYeast Extract agar (CYA). (c) *Aspergillus terreus*, 7 da, CYA. (d) *Aspergillus niger*, 7 da, CYA. *Source:* (*b–d*) Used with permission from Stephen W. Peterson, Ph.D., Microbial Genomics and Bioprocessing Unit, National Center for Agricultural Utilization Research, U.S. Department of Agriculture, Peoria, Illinois. See insert for color representation of this figure.

rough. Vesicles are 30–35 μm wide, hyaline, globose, subglobose to pyriform in shape. Aspergilli are uniseriate or biseriate. Phialoconidia are 3.8–5 μm in diameter, globose to subglobose, with a surface texture finely rough to rough at high power.

Aspergillus terreus

Colony Morphology See Fig. 14.9. On SDA-Emmons colonies are cinnamon-brown, rarely bright orange-brown. The colony reverse is pale yellow.

Microscopic Morphology Conidiophore stipes are smooth walled and hyaline (Fig. 14.12). Vesicles are subspherical, 10–20 μm in diameter. The phialides are biseriate and are present in the upper part of the vesicle. Metulae are as long as the phialides. Conidial heads are long, fan-shaped, and compact, ranging from 30 to 50 μm in diameter and 150 to 500 μm in length. Conidia are smooth walled, spherical to broadly ellipsoidal, 1.5–2.5 μm in diameter, and brown. A second type of "accessory conidia" (aleuroconidia) occur singly and are attached by short pedicels directly to the hyphae submerged in the agar.

Aspergillus niger

Colony Morphology See Fig. 14.9. For the colony color, white mycelium becomes black due to conidia.

Microscopic Morphology See Fig. 14.13. Conidial heads radiate. Conidiophore stipes are smooth walled, hyaline, or pigmented. Vesicles are subspherical, 50–100 μm in diameter. Conidiogenous cells are biseriate. Metulae are twice as long as the phialides. Conidia are brown, covering the entire vesicle. Conidia are ornamented with warts and ridges and are subspherical, 3.5–5.0 μm in diameter.

Serology

Antibody tests are useful diagnostic aids in the allergic bronchopulmonary and aspergilloma forms of disease

EIA to Detect Serum Galactomannan (GM) in Invasive Aspergillosis.

GM production is proportional to disease activity and the GM concentration may be prognostic. A persistent GM concentration despite antifungal therapy is predictive of an unfavorable outcome. The cut off level for a positive test originally set at O.D. (optical density) of 1.5 has been revised down to 0.5. As a guide, serum samples are tested twice a week. GM also may be evaluated in BAL fluid.

The Platelia® *Aspergillus* EIA (Bio-Rad Laboratories, Inc.) is an approved antigenemia test to detect serum-GM in invasive aspergillosis. The Platelia test is a double antibody sandwich EIA utilizing a rat monoclonal antibody to detect ß-(1 → 5)-D-galactofuranose determinants

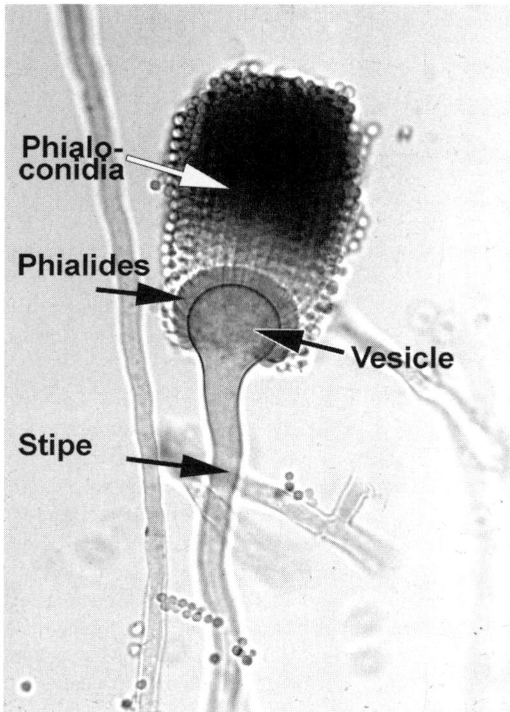

Figure 14.10 *Aspergillus fumigatus* microscopic morphology. The prominent conidiogenous cell is a vesicle with phialides. This uniseriate species produces blue-green conidia that are easily dispersed in air.
Source: Used with permission from Dr. George Barron, Department of Environmental Biology, Ontario Agricultural College, University of Guelph, Ontario, Canada.

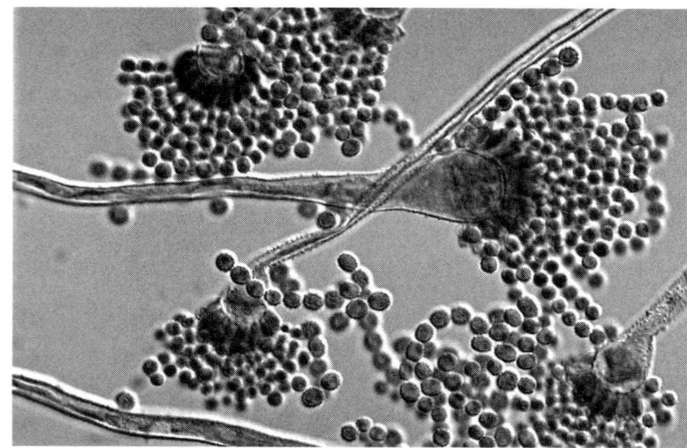

Figure 14.11 *Aspergillus flavus* microscopic morphology. The conidia are larger than those of *A. fumigatus*. The stipe of *A. flavus* is roughened. The arrangement of the conidial heads is uniseriate and biseriate.
Source: E Reiss, CDC.

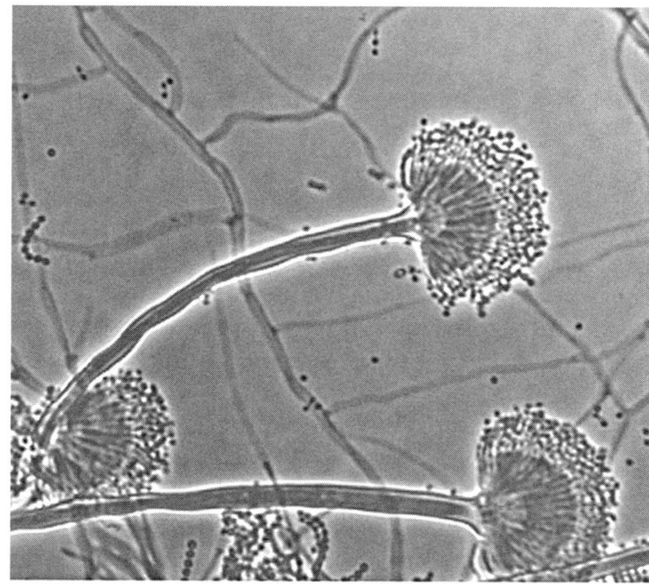

Figure 14.12 *Aspergillus terreus* microscopic morphology, 400× phase contrast, lactofuchsin stain.
Source: E. Reiss, CDC.

of GM, the readily soluble outer cell wall layer present in clinically important *Aspergillus* species. GM is heat stable and occurs as soluble immune complexes which are dissociated by boiling for 3 min in EDTA solution (provided in the kit). The GM is released into the fluid phase and recovered after centrifugation. The kit provides a negative serum control, a threshold control for a borderline reaction, and a 1 ng/mL GM positive serum control.

Results are expressed as an index determined by the ratio of the $\text{O.D.}_{\text{sample}}/\text{mean O.D.}_{\text{with threshold serum}}$. Threshold serum is normal human serum spiked with a borderline amount of GM: "Cutoff Control" serum. An index of ≥ 0.5 is interpreted as a positive result. A ratio of < 0.5 is considered negative. For quality control the ratio obtained with the positive control serum (1 ng GM/mL) should be > 2, and the ratio obtained with the negative control serum should be < 0.5. (The test is not useful with human urine specimens.) Patient serum specimens with a ratio of ≥ 0.5 should be reconfirmed by retesting the same sample.

A prospective evaluation with serial serum specimens from hematology–oncology patients found 27 proven cases of invasive aspergillosis (Maertens et al., 1999). Serum-GM was positive a mean of 5.4 days before clinical suspicion in 18 patients with proven invasive aspergillosis. A group of 25 patients were tested at intervals and the day the test became positive with respect to the first indication of clinical suspicion varied from -23 to $+4$ days. A false-positive rate of 7.4% was determined.

Lowering the cutoff for positivity from O.D. of 1.5 to 0.5 increased sensitivity from 76% to 97% but decreased

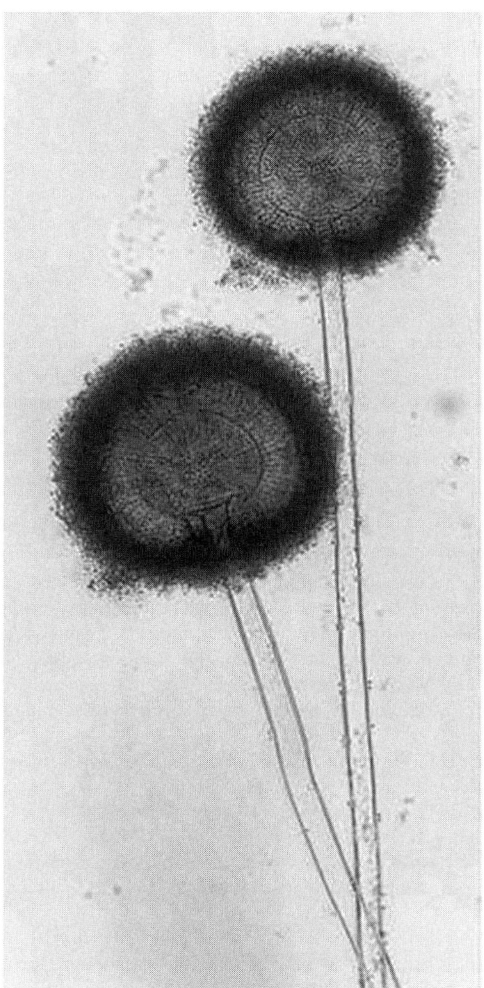

Figure 14.13 *Aspergillus niger*, microscopic morphology, black conidia cover the entire vesicle, 400×.
Source: E. Reiss, CDC.

specificity by 7% from 98% to 91% (Maertens et al., 2007). Requiring two consecutive samples with an O.D. index of $\geq$0.5 resulted in high test accuracy and improved the positive predictive value. At an O.D. cutoff of 0.5, the antigen test result was positive during the week before conventional diagnosis in 65% of cases, and during the week of diagnosis in 80% of cases.

False-positive reactions are considered due to subclinical infection, intestinal colonization, crossreaction with cyclophosphamide, and/or adsorption of dietary GM from a damaged gut. At achievable concentrations in serum, some lots of the antibiotic combination piperacillin–tazobactam yield a positive GM test. A strict definition of invasive aspergillosis should be established to exclude false-positive serologic results (i.e., demonstration of the fungus by culture or histology from a sterile site). It is prudent to report a positive test result if a retest of the same sample confirms a ratio of $\geq$0.5.

Invasive aspergillosis may be diagnosed by testing BAL fluid using the GM test. In these studies, GM testing of BAL fluid yielded a sensitivity of 75–100% and a specificity of 87–100%, compared to serum GM testing which had a sensitivity of 25–47% and a specificity of 93–97% (Becker et al., 2003; Clancy et al., 2007; Meersseman et al., 2008; Musher et al., 2004). Whereas false-positive results have been associated with airway colonization with either *Aspergillus* spp. or *Penicillium* spp., the increased sensitivity of BAL testing has led many centers to adopt this as as adjunct to traditional means of diagnosis.

Genetic Identification of Aspergilli

PCR for mycoses is hindered by a lack of guidelines and standardization. Nonetheless, PCR is promising as an early indicator of invasive aspergillosis and may precede standard histopathology and culture methods (Halliday et al., 2006). The common target for panfungal primers is the rDNA gene cluster; followed by post-PCR analysis of species-specific polymorphisms. The ITS2 region of rDNA contains sufficient sequence diversity to facilitate identification of seven medically important *Aspergillus* species: *A. fumigatus, A. flavus, A. terreus, A. niger, A. nidulans, A. ustus*, and *A. versicolor* (Hinrikson et al., 2005). Postamplification analysis has utilized real-time analysis, that is, LightCycler® or the Luminex® assay; the latter is able to identify six *Aspergillus species* (Etienne et al., 2009).

Progress in Direct Detection of Aspergillus *DNA in Blood Specimens*

The optimal blood fraction for detecting *Aspergillus* DNA is not established among the various choices: serum, plasma, white cell pellet, and whole blood. Serum is advantageous for simplicity of sample preparation. DNA extraction methods have settled on high-speed cell disruption using lysing matrices and chaotropic agents to give an efficient yield of DNA. The methods are, at present, nonstandardized, with each laboratory using different kits. The approach taken by Halliday et al. (2006) uses whole blood and processes it with a red cell lysis buffer. Next, the pellet is placed in sorbitol buffer to stabilize spheroplasts and is digested with lyticase, a ß $(1 \rightarrow 3)$-D-glucanase, to produce spheroplasts. Spheroplasts are centrifuged and resuspended in spheroplast lysis buffer with proteinase K. After digestion and heat treatment, a spin filter column is used to capture and elute the DNA. The DNA is then PCR-amplified in a two-stage nested PCR and detected by agarose gel electrophoresis.

The nested PCR assay evaluated by Halliday et al. (2006) tested blood specimens from 95 febrile neutropenic episodes in patients with leukemia and SCT recipients. Two positive results (intermittent or consecutive) defined a positive episode, of which there were a total of 13 episodes. The sensitivity, specificity, positive predictive value, and negative predictive value of the PCR test for proven or probable invasive aspergillosis were 100%, 75.4%, 46.4%, and 100%, respectively. PCR positivity preceded standard diagnosis by a mean of 14 days. Two *consecutive* positive PCR results occurred in 61.5% of 13 episodes. If clinical treatment decisions were based on two positive PCR tests, use of empiric treatment could have been reduced by up to 37%.

Genetic Identification of *Aspergillus fumigatus* Species Complex (Section *Fumigati*)

The section *Fumigati* (Alcazar-Fuoli et al., 2008) now contains 25 different species and molecular methods are needed for their correct identification. Clinical isolates in the section *Fumigati* were identified based on partial sequences of the β-tubulin and *rodlet A* genes: *Neosartorya hiratsukae, Neosartorya pseudofischeri, Aspergillus viridinutans, Aspergillus lentulus, Aspergillus fumigatiaffinis*, and *Aspergillus fumisynnematus*. High MICs of AmB for *A. lentulus* and *A. fumigatiaffinis* and high ITC, VRC, and ravuconazole MICs for *N. pseudofischeri* and *A. viridinutans* have potential clinical significance and justify the further application of molecular identification to the section *Fumigati*.

SELECTED REFERENCES FOR ASPERGILLOSIS

Abbott KC, Hypolite I, Poropatich RK, Hshieh P, Cruess D, Hawkes CA, Agodoa LY, Keller RA, 2001. Hospitalizations for fungal infections after renal transplantation in the United States. *Transplant Infect Dis* 3: 203–211.

Alcazar-Fuoli L, Mellado E, Alastruey-Izquierdo A, Cuenca-Estrella M, Rodriguez-Tudela JL, 2008. *Aspergillus* section *Fumigati*: Antifungal susceptibility patterns and sequence-based identification. *Antimicrob Agents Chemother* 52: 1244–1251.

Ascioglu S, Rex JH, de Pauw B, Bennett JE, Bille J, Crokaert F, Denning DW, Donnelly JP, Edwards JE, Erjavec Z, Fiere D, Lortholary O, Maertens J, Meis JF, Patterson TF, Ritter J, Selleslag D, Shah PM, Stevens DA, Walsh TJ, 2004. Defining opportunistic invasive fungal infections in immunocompromised patients with cancer and hematopoietic stem cell transplants: An international consensus. *Clin Infect Dis* 34: 7–14.

Askew DS, 2008. *Aspergillus fumigatus*: Virulence genes in a street-smart mold. *Curr Opin Microbiol* 11: 331–337.

Balajee SA, Houbraken J, Verweij PE, Hong SB, Yaghuchi T, Varga J, Samson RA, 2007. *Aspergillus* species identification in the clinical setting. *Stud Mycol* 59: 39–46.

Balajee SA, Gribskov JL, Hanley E, Nickle D, Marr KA, 2005. *Aspergillus lentulus* sp. nov., a new sibling species of *A. fumigatus*. *Eukaryot Cell* 4: 625–632.

Becker MJ, Lugtenburg EJ, Cornelissen JJ, van der Schee C, Hoogsteden HC, de Marie S, 2003. Galactomannan detection in computerized tomography-based broncho-alveolar lavage fluid and serum in haematological patients at risk for invasive pulmonary aspergillosis. *Br J Haematol* 121: 448–457.

Bennett JW, Klich M, 2003. Mycotoxins. *Clin Microbiol Rev* 497–516.

Birch M, Denning DW, Robson GD, 2004. Comparison of extracellular phospholipase activities in clinical and environmental *Aspergillus fumigatus* isolates. *Med Mycol* 42: 81–86.

Bochud PY, Chien JW, Marr KA, Leisenring WM, Upton A, Janer M, Rodrigues SD, Li S, Hansen JA, Zhao LP, Aderem A, Boeckh M, 2008. Toll-like receptor 4 polymorphisms and aspergillosis in stem-cell transplantation. *N Engl J Med* 359: 1766–1777.

Brasnu E, Bourcier T, Dupas B, Degorge S, Rodallec T, Laroche L, Borderie V, Baudouin C, 2007. In vivo confocal microscopy in fungal keratitis. *Br J Ophthalmol* 91: 588–591.

Burge HA, Rogers CA, 2000. Outdoor allergens. *Environ Health Perspect* 108 (Suppl 4): 653–659.

Burton JR, Zachery JB, Bessin R, Rathbun HK, Greenough WB 3rd, Sterioff S, Wright JR, Slavin RE, Williams GM, 1972. Aspergillosis in four renal transplant recipients. Diagnosis and effective treatment with amphotericin B. *Ann Intern Med* 77: 383–388.

CAST, 2003. Mycotoxins: Risks in plant, animal, and human systems, report 139. Council for Agricultural Science and Technology, Ames, Iowa.

Chandler FW, Watts JC, 1987. *Pathologic Diagnosis of Fungal Infections*, Table 10, p. 87. ASCP Press, Chicago.

Clancy CJ, Jaber RA, Leather HL, Wingard JR, Staley B, Wheat LJ, Cline CL, Rand KH, Schain D, Baz M, Nguyen MH, 2007. Bronchoalveolar lavage galactomannan in diagnosis of invasive pulmonary aspergillosis among solid-organ transplant recipients. *J Clin Microbiol* 45: 1759–1765.

Cleary JD, Schwartz M, Rogers PD, de Mestral J, Chapman SW, 2003. Effects of amphotericin B and caspofungin on histamine expression. *Pharmacotherapy* 23: 966–973.

Clercx C, 2000. Treatment of canine nasal aspergillosis. Please see websites cited.

de Pauw B, Walsh TJ, Donnelly JP, Stevens DA, Edwards JE, Calandra T, Pappas PG, Maertens J, Lortholary O, Kauffman CA, Denning DW, Patterson TF, Maschmeyer G, Bille J, Dismukes WE, Herbrecht R, Hope WW, Kibbler CC, Kullberg BJ, Marr KA, Mu noz P, Odds FC, Perfect JR, Restrepo A, Ruhnke M, Segal BH, Sobel JD, Sorrell TC, Viscoli C, Wingard JR, Zaoutis T, Bennett JE, 2008. Revised definitions of invasive fungal disease from the European Organization for Research and Treatment of Cancer/Invasive Fungal Infections Cooperative Group and the National Institute of Allergy and Infectious Diseases Mycoses Study Group (EORTC/MSG) Consensus Group. *Clin Infect Dis* 46: 1813–1821.

Denning DW, Marinus A, Cohen J, Spence D, Herbrecht R, Pagano L, Kibbler C, Kermery V, Offner F, Cordonnier C, Jehn U, Ellis M, Collette L, Sylvester R, 1998. An EORTC multicentre prospective survey of invasive aspergillosis in haematological patients: Diagnosis and therapeutic outcome. *J Infect* 37: 173–180.

Diener UL, Cole RJ, Sanders TH, Payne GA, Lee LS, Klich MA, 1987. Epidemiology of aflatoxin formation by *Aspergillus flavus*. *Ann Rev Phytopathol* 25: 249–270.

Dillon HK, Boling DK, Miller JD, 2007. Comparison of detection methods for *Aspergillus fumigatus* in environmental air samples in an occupational environment. *J Occup Environ Hyg* 4: 509–513.

EBINA K, SAKAGAMI H, YOKOTA K, KONDO H, 1994. Cloning and nucleotide sequence of cDNA encoding Asp-hemolysin from *Aspergillus fumigatus*. *Biochim Biophys Acta* 1219: 148–150.

ETIENNE KA, KANO R, BALAJEE SA, 2009. Development and validation of a microsphere-based Luminex assay for rapid identification of clinically relevant aspergilli. *J Clin Microbiol* 47: 1096–1100.

FALVEY DG, STREIFEL AJ, 2007. Ten-year air sample analysis of *Aspergillus* prevalence in a university hospital. *J Hosp Infect* 67: 35–41.

FEDOROVA ND, KHALDI N, JOARDAR VS, MAITI R, AMEDEO P, ANDERSON MJ, CRABTREE J, SILVA JC, BADGER JH, ALBARRAQ A, ANGIUOLI S, BUSSEY H, BOWYER P, COTTY PJ, DYER PS, EGAN A, GALENS K, FRASER-LIGGETT CM, HAAS BJ, INMAN JM, KENT R, LEMIEUX S, MALAVAZI I, ORVIS J, ROEMER T, RONNING CM, SUNDARAM JP, SUTTON G, TURNER G, VENTER JC, WHITE OR, WHITTY BR, YOUNGMAN P, WOLFE KH, GOLDMAN GH, WORTMAN JR, JIANG B, DENNING DW, NIERMAN WC, 2008. Genomic islands in the pathogenic filamentous fungus *Aspergillus fumigatus*. *PLoS Genet* 4: e1000046.

FLORCRUZ NV, PECZON I Jr, 2008. Medical interventions for fungal keratitis. *Cochrane Database Syst Rev* 23: CD004241.

GIRMENIA C, NUCCI M, MARTINO P, 2001. Clinical significance of *Aspergillus* fungaemia in patients with haematological malignancies and invasive aspergillosis. *Br J Haematol* 114: 93–98.

GROLL AH, WALSH TJ, 2002. Invasive fungal infections in the neutropenic cancer patient: Current approaches and future strategies. *Infect Med* 19: 324–334.

GUILLOT J, LE LOC'H G, ARNÉ P, FÉMÉNIA F, CHERMETTE R, 2006. Avian Aspergillosis. Please see websites cited.

HACHEM RY, KONTOYIANNIS DP, BOKTOUR MR, AFIF C, COOKSLEY C, BODEY GP, CHATZINIKOLAOU I, PEREGO C, KANTARJIAN HM, RAAD II, 2004. Aspergillus terreus: An emerging amphotericin B-resistant opportunistic mold in patients with hematologic malignancies. *Cancer* 101: 1594–1600.

HALLIDAY C, HOILE R, SORRELL T, JAMES G, YADAV S, SHAW P, BLEAKLEY M, BRADSTOCK K, CHEN S, 2006. Role of prospective screening of blood for invasive aspergillosis by polymerase chain reaction in febrile neutropenic recipients of haematopoietic stem cell transplants and patients with acute leukaemia. *Br J Haematol* 132: 478–486.

HEINEMANN S, SYMOENS F, GORDTS B, JANNES H, NOLARD N, 2004. Environmental investigations and molecular typing of *Aspergillus flavus* during an outbreak of postoperative infections. *J Hosp Infect* 57: 149–155.

HERBRECHT R, MOGHADDAM A, MAHMAL L, NATARAJAN-AME S, FORNECKER L-M, LETSCHJER-BRU V, 2005. Invasive aspergillosis in the hematologic and immunologic patient: New findings and key questions in leukemia. *Med Mycol* 43 (Suppl 1): s239–s242.

HERBRECHT R, DENNING DW, PATTERSON TF, BENNETT JE, GREENE RE, OESTMANN JW, KERN WV, MARR KA, RIBAUD P, LORTHOLARY O, SYLVESTER R, RUBIN RH, WINGARD JR, STARK P, DURAND C, CAILLOT D, THIEL E, CHANDRASEKAR PH, HODGES MR, SCHLAMM HT, TROKE PF, DE PAUW B, 2002. Voriconazole versus amphotericin B for primary therapy of invasive aspergillosis. *N Engl J Med* 347: 408–415.

HINRIKSON HP, HURST SF, DE AGUIRRE L, MORRISON CJ, 2005. Molecular methods for the identification of *Aspergillus* species. *Med Mycol* 43 (Suppl 1): s129–s137.

HOHL TM, FELDMESSER M, 2007. *Aspergillus fumigatus*: Principles of pathogenesis and host defense. *Eukaryot Cell* 6: 1953–1963.

HRYHORCZUK D, CURTIS L, SCHEFF P, CHUNG J, RIZZO M, LEWIS C, KEYS N, MOOMEY M, 2001. Bioaerosol emissions from a suburban yard waste composting facility. *Ann Agric Environ Med (Poland)* 8: 177–185.

JAMES MJ, LASKER BA, MCNEIL MM, SHELTON M, WARNOCK DW, REISS E, 2000. Use of a repetitive DNA probe to type clinical and environmental isolates of *Aspergillus flavus* from a cluster of cutaneous infections in a neonatal intensive care unit. *J Clin Microbiol* 38: 3612–3618.

JANTUNEN E, NIHTINEN A, ANTTILA VJ, 2008. Changing landscape of invasive aspergillosis in allogeneic stem cell transplant recipients. *Transpl Infect Dis* 10: 156–161.

KEARNS KS, LOUDIS B (eds.), 2003. Avian Aspergillosis, *in: Recent Advances in Avian Infectious Diseases*, an electronic publication at the website of the International Veterinary Information Service, Ithaca, NY (www.ivis.org) A1902.0903 (Access restricted to veterinary professionals).

KIRBY A, HASSAN I, BURNIE J, 2006. Recommendations for managing *Aspergillus* osteomyelitis and joint infections based on a review of the literature. *J Infect* 52: 405–414.

KLICH MA, PITT JI, 1988. *A Laboratory Guide to Common Aspergillus Species and Their Teleomorphs*. Commonwealth Scientific and Industrial Research Organization, North Ryde, NSW, Australia.

KLONT RR, MENNINK-KERSTEN MA, RUEGEBRINK D, RIJS AJ, BLIJLEVENS NM, DONNELLY JP, VERWEIJ PE, 2006. Paradoxical increase in circulating *Aspergillus* antigen during treatment with caspofungin in a patient with pulmonary aspergillosis. *Clin Infect Dis* 43: e23–e25.

KONTOYIANNIS DP, RATANATHARATHORN V, YOUNG JA, RAYMOND J, LAVERDIÈRE M, DENNING DW, PATTERSON TF, FACKLAM D, KOVANDA L, ARNOLD L, LAU W, BUELL D, MARR KA, 2009. Micafungin alone or in combination with other systemic antifungal therapies in hematopoietic stem cell transplant recipients with invasive aspergillosis. *Transpl Infect Dis* 11: 89–93.

LASS-FLÖRL C, GRIFF K, MAYR A, PETZER A, GASTL G, BONATTI H, FREUND M, KROPSHOFER G, DIERICH MP, NACHBAUR D, 2005. Epidemiology and outcome of infections due to *Aspergillus terreus*: 10-year single centre experience. *Br J Haematol* 131: 201–207.

LASS-FLÖRL C, RATH PM, NIEDERWIESER D, KOFLER G, WÜRZNER R, KREZY A, DIERICH MP, 2000. *Aspergillus terreus* infections in haematological malignancies: Molecular epidemiology suggests association with in-hospital plants. *J Hosp Infect* 46: 31–35.

LATGÉ J-P, 2001. The pathobiology of *Aspergillus fumigatus*. *Trends Microbiol* 9: 382–389.

LIN SJ, SCHRANZ J, TEUTSCH SM, 2001. Aspergillosis case-fatality rate: Systematic review of the literature. *Clin Infect Dis* 32: 358–366.

MACKENZIE JD, SODICKSON AD, OH K, SCHWARTZ RB, SELTZER SE, 2002. Please see websites cited.

MAERTENS JA, KLONT R, MASSON C, THEUNISSEN K, MEERSSEMAN W, LAGROU K, HEINEN C, CRÉPIN B, VAN ELDERE J, TABOURET M, DONNELLY JP, VERWEIJ PE, 2007. Optimization of the cutoff value for the *Aspergillus* double-sandwich enzyme immunoassay. *Clin Infect Dis* 44: 1329–1336.

MAERTENS J, GLASMACHER A, HERBRECHT R, THIEBAUT A, CORDONNIER C, SEGAL BH, KILLAR J, TAYLOR A, KARTSONIS N, PATTERSON TF, AOUN M, CAILLOT D, SABLE C, 2006. Multicenter, noncomparative study of caspofungin in combination with other antifungals as salvage therapy in adults with invasive aspergillosis. *Cancer* 107: 2888–2897.

MAERTENS J, RAAD I, PETRIKKOS G, BOOGAERTS M, SELLESLAG D, PETERSEN FB, SABLE CA, KARTSONIS NA, NGAI A, TAYLOR A, PATTERSON TF, DENNING DW, WALSH TJ, 2004. Efficacy and safety of caspofungin for treatment of invasive aspergillosis in patients refractory to or intolerant of conventional antifungal therapy. *Clin Infect Dis* 39: 1563–1571.

MAERTENS J, VERHAEGEN J, DEMUYNCK H, BROCK P, VERHOEF G, VANDENBERGHE P, VAN ELDERE J, VERBIST L, BOOGAERTS M, 1999. Autopsy-controlled prospective evaluation of serial screening for circulating galactomannan by a sandwich enzyme-linked immunosorbent assay for hematological patients at risk for invasive aspergillosis. *J Clin Microbiol* 37: 3223–3228.

Marr KA, Boeckh M, Carter RA, Kim HW, Corey L, 2004. Combination antifungal therapy for invasive aspergillosis. *Clin Infect Dis* 39: 797–802.

Marr KA, Carter RA, Boeckh M, Martin P, Corey L, 2002a. Invasive aspergillosis in allogeneic stem cell transplant recipients: changes in epidemiology and risk factors. *Blood* 100: 4358–4366.

Marr KA, Patterson T, Denning D, 2002b. Aspergillosis: Pathogenesis, clinical manifestations, and therapy. *Infect Dis Clin North Am* 16: 875–894.

Martino R, Subirá M, Rovira M, Solano C, Vázquez L, Sanz GF, Urbano-Ispizua A, Brunet S, de la Cámara R, 2002. Invasive fungal infections after allogeneic peripheral blood stem cell transplantation: Incidence and risk factors in 395 patients. *Br J Haematol* 116: 475–482.

Maschmeyer G, Haas A, Cornely OA, 2007. Invasive aspergillosis: Epidemiology, diagnosis and management in immunocompromised patients. *Drugs* 67: 1567–1601.

McNeil MM, Nash SL, Hajjeh R A, Phelan MA, Conn L A, Plikaytis BD, Warnock DW, 2001. Trends in mortality due to invasive mycotic diseases in the United States, 1980–1997. *Clin Infect Dis* 33: 641–647.

Meersseman W, Lagrou K, Maertens J, Wilmer A, Hermans G, Vanderschueren S, Spriet I, Verbeken E, van Wijngaerden E, 2008. Galactomannan in bronchoalveolar lavage fluid: A tool for diagnosing aspergillosis in intensive care unit patients. *Am J Respir Crit Care Med* 177: 27–34.

Millner PD, Olenchock SA, Epstein E, Rylander R, Haines J, Walker J, Ooi BL, Horne E, Maritato M, 1994. Bioaerosols associated with composting facilities. *Compost Sci Utilization* 2 (no 4): 6–57.

Montoya JG, Chapparo SV, Celis D, Cortes JA, Leung AN, Robbins RC, Stevens DA, 2003. Invasive aspergillosis in the setting of cardiac transplantation. *Clin Infect Dis* 37(Suppl 3): s281–s292.

Morgan J, Wannemuehler KA, Marr KA, Hadley S, Kontoyiannis DP, Walsh TJ, Fridkin SK, Pappas PG, Warnock DW, 2005. Incidence of invasive aspergillosis following hematopoietic stem cell and solid organ transplantation: Interim results of a prospective multicenter surveillance program. *Med Mycol* 43(Suppl 1): s49–s58.

Mortensen KL, Mellado E, Lass-Flörl C, Rodriguez-Tudela JL, Johansen HK, Arendrup MC, 2010. Environmental study of azole-resistant *Aspergillus fumigatus* and other aspergilli in Austria, Denmark, and Spain. *Antimicrob Agents Chemother* 54: 4545–4549.

Musher B, Fredricks D, Leisenring W, Balajee SA, Smith C, Marr KA, 2004. *Aspergillus* galactomannan enzyme immunoassay and quantitative PCR for diagnosis of invasive aspergillosis with bronchoalveolar lavage fluid. *J Clin Microbiol* 42: 5517–5522.

Neofytos D, Horn D, Anaissie E, Steinbach W, Olyaei A, Fishman J, Pfaller M, Chang C, Webster K, Marr K, 2009. Epidemiology and outcome of invasive fungal infection in adult hematopoietic stem cell transplant recipients: Analysis of Multicenter Prospective Antifungal Therapy (PATH) Alliance registry. *Clin Infect Dis* 48: 265–273.

O'Gorman CM, Fuller HT, Dyer PS, 2009. Discovery of a sexual cycle in the opportunistic fungal pathogen *Aspergillus fumigatus*. *Nature* 457(7228): 471–474.

Pasqualotto AC, Denning DW, 2006. Post-operative aspergillosis. *Clin Microbiol Infect* 12: 1060–1076.

Patterson TF, Kirkpatrick WR, White M, Hiemenz JW, Wingard JR, Dupont B, Rinaldi MG, Stevens DA, Graybill JR, 2000. Invasive aspergillosis: Disease spectrum, treatment practices, and outcomes. I3 *Aspergillus* Study Group. *Medicine (Baltimore)* 79: 250–260.

Peeters D, Clercx C, 2007. Update on canine sinonasal aspergillosis. *Vet Clin North Am Small Anim Pract* 37: 901–916,vi.

Pérez-Arellano JL, Angel-Moreno A, Belón E, Francès A, Santana OE, Martín-Sánchez AM, 2001. Isolated renoureteric aspergilloma due to *Aspergillus flavus*: Case report and review of the literature. *J Infect* 42: 163–165.

Pfaller MA, Diekema DJ, 2007. Epidemiology of invasive candidiasis: A persistent public health problem. *Clin Microbiol Rev* 20: 133–163.

Posfay-Barbe KM, Green MDL, Michaels MG, 2004. Chapter 103, Lung and Heart–Lung Transplant Patients, pp. 1109–1110, *in*: *Cohen & Powderly*: *Infectious Diseases*, 2nd ed. Elsevier, Philadelphia.

Rementeria A, López-Molina N, Ludwig A, Vivanco AB, Bikandi J, Pontón J, Garaizar J, 2005. Genes and molecules involved in *Aspergillus fumigatus* virulence. *Rev Iberoam Micol* 22: 1–23.

Rhodes JC, 2006. *Aspergillus fumigatus*: Growth and virulence. *Med Mycol* 44: 577–581.

Silveira FP, Husain S, 2007. Fungal infections in solid organ transplantation. *Med Mycol* 45: 305–320.

Singh N, Pursell KJ, 2008. Combination therapeutic approaches for the management of invasive aspergillosis in organ transplant recipients. *Mycoses* 51: 99–108.

Singh N, Wagener MM, Cacciarelli TV, Levitsky J, 2008. Antifungal management practices in liver transplant recipients. *Am J Transpl* 8: 426–431.

Singh N, Limaye AP, Forrest G, Safdar N, Muñoz P, Pursell K, Houston S, Rosso F, Montoya JG, Patton P, Del Busto R, Aguado JM, Fisher RA, Klintmalm GB, Miller R, Wagener MM, Lewis RE, Kontoyiannis DP, Husain S, 2006. Combination of voriconazole and caspofungin as primary therapy for invasive aspergillosis in solid organ transplant recipients: A prospective, multicenter, observational study. *Transplantation* 81: 320–326.

Singh N, Avery RK, Munoz P, Pruett TL, Alexander B, Jacobs R, Tollemar JG, Dominguez EA, Yu CM, Paterson DL, Husain S, Kusne S, Linden P, 2003. Trends in risk profiles for and mortality associated with invasive aspergillosis among liver transplant recipients. *Clin Infect Dis* 36: 46–52.

Slavin RG, Hutcheson PS, Chauhan B, Bellone CJ, 2004. An overview of allergic bronchopulmonary aspergillosis with some new insights. *Allergy Asthma Proc* 25: 395–399.

St. Germain G, Summerbell S, 1996. *Identifying Filamentous Fungi*. Star Publishing Co., Belmont, CA.

Steinbach WJ, Perfect JR, Schell WA, Walsh TJ, Benjamin DK Jr, 2004. In vitro analyses, animal models, and 60 clinical cases of invasive *Aspergillus terreus* infection. *Antimicrob Agents Chemother* 48: 3217–3225.

Stergiopoulou T, Meletiadis J, Roilides E, Kleiner DE, Schaufele R, Roden M, Harrington S, Dad L, Segal B, Walsh TJ, 2007. Host-dependent patterns of tissue injury in invasive pulmonary spergillosis. *Am J Clin Pathol* 127: 349–355.

Stratov I, Korman TM, Johnson PD, 2003. Management of *Aspergillus* osteomyelitis: Report of failure of liposomal amphotericin B and response to voriconazole in an immunocompetent host and literature review. *Eur J Clin Microbiol Infect Dis* 22: 277–283.

Tablan OC, Anderson LJ, Besser R, Bridges C, Hajjeh R, 2004. Guidelines for preventing health-care-associated pneumonia: Recommendations of CDC and the Healthcare Infection Control Practices Advisory Committee (HICPAC) 2003. *MMWR Morb Mortal Wkly Rep* 53(RR03): 1–36.

Taylor R, Dagenais T, Keller NP, 2009. Pathogenesis of *Aspergillus fumigatus* in invasive aspergillosis. *Clin Microbiol Rev* 22: 447–465.

Torrey GS, Marth EH, 1977a. Isolation and toxicity of molds from foods stored in homes. *J Food Protection* 40: 187–190.

Torrey GS, Marth EH, 1977b. Temperatures in home refrigerators and mold growth at refrigeration temperatures. *J Food Protection* 40: 393–397.

Tsai H-F, Chang YC, Washburn RG, Wheeler MH, Kwon-Chung KJ, 1998. The developmentally regulated *alb1* gene of *Aspergillus fumigatus*: Its role in modulation of conidial morphology and virulence. *J Bacteriol* 180: 3031–3038.

Upton A, Kirby KA, Carpenter P, Boeckh M, Marr KA, 2007. Invasive aspergillosis following hematopoietic cell transplantation: Outcomes and prognostic factors associated with mortality. *Clin Infect Dis* 44: 531–540.

Walsh TJ, Anaissie EJ, Denning DW, Herbrecht R, Kontoyannis DP, Marr KA, Morrison V, Segal B, Steinbach W, Stevens D, Burik J-A, Wingard J, Patterson T, 2008. Treatment of aspergillosis: Clinical practice guidelines of the Infectious Diseases Society of America. *Clin Infect Dis* 46: 327–360.

Williamson EC, Millar MR, Steward CG, Cornish JM, Foot AB, Oakhill A, Pamphilon DH, Reeves B, Caul EO, Warnock DW, Marks DI, 1999. Infections in adults undergoing unrelated donor bone marrow transplantation. *Br J Haematol* 104: 560–568.

Zaas AK, Liao G, Chien JW, Weinberg C, Shore D, Giles SS, Marr KA, Usuka J, Burch LH, Perera L, Perfect JR, Peltz G, Schwartz DA, 2008. Plasminogen alleles influence susceptibility to invasive aspergillosis. *PLoS Genet* 4: e1000101.

WEBSITES CITED

Aspergillus website: http://www.aspergillus.org.uk/

Bioaerosols Associated with Composting Facilities. U.S. Composting Council, 1993 report. p. 5, 18–29. http://compostingcouncil.org/index.cfm.

Childhood chronic granulomatous disease (CGD), National Primary Immunodeficiency Resource Center, New York. http://npi.jmfworld.org/.

Clercx C, 2000. Treatment of canine nasal aspergillosis. http://www.aspergillus.org.uk/indexhome.htm?secure/veterinary/canine.html~main.

Guillot J, Le Loc'h G, Arné P, Féménia F, Chermette R. Avian Aspergillosis. A PowerPoint at the *Aspergillus* website, URL www.aspergillus.org.uk/education/GuillotsympISHAM06.ppt.

Kearns KS, Loudis B (eds.), 2003. *Recent Advances in Avian Infectious Diseases*. It can be viewed by veterinary professionals and paraprofessionals at the website of the International Veterinary Information Service, Ithaca NY: www.ivis.org, A1902.0903.

Mackenzie JD, Sodickson AD, Oh K, Schwartz RB, Seltzer SE (April 8, 2002). Cerebral invasive aspergillosis. http://brighamrad.harvard.edu/education/online/tcd/bwh-query-keyword.html.

Merck & Co. Open label trial of caspofungin salvage therapy for invasive aspergillosis. http://www.cancidas.com/cancidas/shared/documents/english/pi.pdf).

National Primary Immunodeficiency Resource Center, New York. http://npi.jmfworld.org/.

Samson S. The International Commission on *Penicillium* and *Aspergillus*. The website of the Commission resides at the Centraalbureau voor Schimmelcultures in Baarn, The Netherlands. There is ample information about the mycology of the aspergilli. http://www.cbs.knaw.nl/icpa/index.htm.

Shibuya K. Histopathology of invasive pulmonary aspergillosis. An electronic article at the aspergillosis website, URL http://www.aspergillus.org.uk/indexhome.htm?education/index.php~main.

QUESTIONS

The Answer Key to multiple choice questions may be found at the end of the book.

1. For each of the following specimens indicate whether the isolation of *Aspergillus* has little (1), moderate (2), or great (3) diagnostic significance:
   A. Bronchoalveolar lavage fluid or transbronchial brush biopsy
   B. Normally sterile sites by biopsy
   C. Respiratory tract of a neutropenic patient or SCT recipient
   D. Sputum, urine, stool, or wound cultures

2. A blue-green mold is isolated from a clinical specimen. Microscopically its asexual reproductive structure is a conidiophore, brush-like, simple or branched, with clusters of phialides producing chains of phialoconidia (no vesicle). Which of the following is likely to be its identity?
   A. *Aspergillus flavus*
   B. *Aspergillus* species because they are all blue-green
   C. *Aspergillus terreus*
   D. *Penicillium* species

3. What percentage of blood cultures from patients with invasive pulmonary aspergillosis are apt to be positive?
   A. <5%
   B. 25%
   C. 50%
   D. >50%

4. Which of the following is a preferred medium for isolation of *Aspergillus* species from a nonsterile site specimen?
   A. BHI agar
   B. Malt extract agar
   C. Mycosel or Mycobiotic agar
   D. SDA + chloramphenicol

5. Which of the following is true about the Platelia® *Aspergillus* EIA to detect galactomannan (GM) antigenemia?
   1. Serum specimens with a ratio of $\geq$0.5 should be reconfirmed by retesting the same sample.
   2. The cellular origin of GM is the cell membrane.
   3. The cellular origin of GM is the cell wall.
   4. The threshold for a positive result is the ratio obtained with the positive control serum (1 ng GM/mL).
      A. 1 + 2
      B. 1 + 3
      C. 2 + 4
      D. 3 + 4

6. Next to each statement indicate if it refers to the radiographic (1) halo sign or (2) the air crescent sign in invasive pulmonary aspergillosis (IPA).
   A. Air fills the space between devitalized and living lung tissue.
   B. Caused by necrosis around a nodule.
   C. Follows the halo sign within 1–2 weeks in more than half of patients with IPA.

D. It is a pulmonary nodule with hemorrhage or edema.

E. Occurs in approximately 80% of patients with IPA.

F. Occurs within the first 2 weeks of clinical disease.

7. The recent downward trend in invasive pulmonary aspergillosis in transplant centers corresponds to the approval of which antifungal drug for treatment?

A. Amphotericin B

B. Fluconazole

C. Griseofulvin

D. Voriconazole

8. Which statement is true about salvage therapy for invasive aspergillosis?

A. Caspofungin is approved for patients who are refractory to or intolerant of other approved therapy.

B. Reversal of immunosuppression is critical in the prognosis of patients with invasive aspergillosis.

C. The choice of primary therapy is critical because salvage therapy is successful in only a minority of cases.

D. All of the above.

E. None of the above.

9. What is the ecologic niche of *Aspergillus fumigatus*?

A. Plant pathogen

B. Cannot grow above 37°C

C. Common in air in the environment

D. Occurrence in hospital air is exceptional

10. Which of the following precautions is not routinely recommended to protect neutropenic patients from *Aspergillus*?

A. Directed room airflow

B. HEPA filtration of incoming air

C. High ($\geq$12) air changes/hour

D. Laminar airflow

E. Positive air pressure in patient rooms in relation to the corridor

11. Even with improvements in diagnosis and therapy, the overall case-fatality rate of invasive pulmonary aspergillosis is still high. What is the overall case-fatality ratio? How is the case-fatality ratio affected in patients in allogeneic stem cell transplant recipients with GVHD and in those with extrapulmonary dissemination and cerebral abscess?

12. Referring to stem cell transplant (SCT) recipients, define and describe the three periods of increased risk of developing invasive pulmonary aspergillosis.

13. Discuss the role for lipid-AmB, itraconazole, posaconazole, voriconazole, and echinocandins for the following indications in aspergillosis: mold-active prophylaxis in neutropenic cancer patients and in SCT recipients with GVHD, primary therapy for invasive pulmonary aspergillosis, alternative primary therapy, second-line primary therapy, and salvage therapy. Are there any data to support a role for combination therapy in invasive aspergillosis?

APPENDIX

Intoxication

Aflatoxins occur in crops before and after harvest. Postharvest contamination may occur if crop drying is delayed and during storage if excess moisture allows mold growth. Aflatoxins are detected occasionally in milk, cheese, corn, peanuts, cottonseed, nuts, almonds, figs, spices, and various other foods and feeds. Milk, eggs, and meat are sometimes contaminated if the producing animal consumes aflatoxin-contaminated feed. The commodities with the highest risk of aflatoxin contamination are corn, peanuts, and cottonseed.

Humans are exposed to aflatoxins by consuming foods contaminated by infestation with strains of *Aspergillus flavus* or *A. parasiticus*. There are four major aflatoxins of significance as contaminants of food and feed: B1, B2, G1, and G2, and two additional metabolites, M1 and M2. Aflatoxicosis is a hepatic disease. Of the different aflatoxins and metabolites only aflatoxin B1 is designated a cancer causing agent by the International Agency for Research on Cancer (IARC). Contaminated food and feed are not permitted in the marketplace in developed countries, but there is concern for the possible adverse effects of long-term exposure to low levels of aflatoxins in the food supply. Acute aflatoxicosis in humans occurs in developing countries, for example, Uganda and India. The syndrome is characterized by vomiting, abdominal pain, pulmonary edema, convulsions, coma, and death with cerebral edema and fatty involvment of the liver, kidneys, and heart.

Conditions increasing the risk of acute aflatoxicosis in humans include limited availability of food, environmental conditions that favor postharvest fungal infestation of crops and commodities, and lack of regulatory systems for aflatoxin monitoring and control. Epidemiologic studies in Asia and Africa found a positive association between dietary aflatoxins and liver cell cancer.

Turkey X-Disease

In 1960 more than 100,000 young turkeys on poultry farms in England died within a few months from an apparently new disease termed "Turkey X disease." A survey of the early outbreaks showed that they were associated with Brazilian peanut meal. The suspect peanut meal was highly toxic to poultry and ducklings with symptoms typical of Turkey X disease. The toxin-producing fungus was identified as *Aspergillus flavus* and the toxin was given the name "aflatoxin" by virtue of its origin (*A. flavus*). Aflatoxins also are produced by strains of *A. parasiticus*.

Further information can be found at the website of the Cornell University Poisonous Plants database at the URL http://www.ansci.cornell.edu/plants/toxicagents/aflatoxin/aflatoxin.html. For further general information on various mycotoxins, please consult CAST (2003).

Allergic Bronchopulmonary Aspergillosis (ABPA)

Because ABPA is an allergic disease and not an infectious process, it is not in the scope of the present text. A subset of patients with asthma produce sputum plugs containing *A. fumigatus*. This allergic disease is also found in patients with cystic fibrosis, but its laboratory parameters and clinical findings in cystic fibrosis are not the same as in non-cystic fibrosis patients (Slavin et al., 2004). A similar form of allergic aspergillosis is extrinsic allergic alveolitis. Reduced respiratory capacity is the result of immune complex-mediated damage to bronchi and bronchioles. If untreated, episodes lead to decreased respiratory capacity, fixed pulmonary shadows, and bronchiectasis.[6]

The seven primary diagnostic criteria for ABPA are:

- Episodic bronchial obstruction (asthma)
- Peripheral blood eosinophilia
- Positive scratch or prick tests to *Aspergillus* antigens
- Positive serum precipitins
- High total serum IgE
- History of pulmonary infiltrates (transient or fixed)
- Central bronchiectasis

Secondary diagnostic criteria include *Aspergillus* species in sputum samples using stain and/or culture, a history of expectoration of brown plugs or flecks, elevated *Aspergillus*-specific IgE, and Arthus reaction (Gell and Coombs Type III skin reactivity) to *Aspergillus* antigen.

ABPA may progress through clinical stages of acute corticosteroid-responsive asthma to corticosteroid-dependent asthma, to fibrotic end-stage lung disease, with honeycomb lung.

Prevention is to avoid opportunities for exposure, for example, farmers exposed to moldy silage. If that form of avoidance is not sufficient, a change in occupation may be considered. Now it is known that ABPA occurs as commonly in rural and urban dwellers and that exposure is necessary but not sufficient to cause disease in atopic or cystic fibrosis patients. There appears to be a HLA-DR and DQ gene linked disease susceptibility (Slavin et al., 2004).

Therapy consists of corticosteroids and ITC (Walsh et al., 2008). ITC (200 mg twice/day, oral, for 16 weeks) treatment allows an increased interval between

[6]Bronchiectasis. *Definition*: An acquired disorder of the large bronchi (airways), which become dilated after destructive infections of the lungs. Rarely, it may be congenital.

corticosteroid courses, reduces eosinophilia and IgE concentration, and improves exercise tolerance and pulmonary function. A similar benefit is seen in cystic fibrosis patients with ABPA. Corticosteroids block antigen–antibody immune complex mediated damage and block the arachidonic acid pathway, minimizing late phase reactions resistant to antihistaminics (prostaglandins and leukotrienes).

Chapter 15

Fusarium Mycosis

15.1 *FUSARIUM* MYCOSIS-AT-A-GLANCE

- *Introduction/Disease Definition.* Certain *Fusarium* species from among a large genus of plant pathogens cause locally invasive disease in immune-normal persons—onychomycosis, keratitis, and infections in persons with serious burns or in those receiving peritoneal dialysis. Immuncompromised patients may develop paronychia, invasive sinusitis, and pulmonary and extrapulmonary hematogenously disseminated disease with skin lesions and fungemia. Immunocompromised patients at risk are those with neutropenia, with severe T-cell immunodeficiency, and SCT recipients with severe GVHD.
- *Etiologic Agents.* The most frequent species isolated from clinical specimens are *Fusarium solani* > *F. oxysporum* > *F. verticillioides* (formerly *moniliforme*) > *F. proliferatum*.
- *Geographic Distribution/Ecologic Niche. Fusarium* species are found worldwide in all climate zones in soil, in plant debris, and on plants. Dispersal of conidia occurs via the air and in water, including hospital water, systems (especially *F. oxysporum*).
- *Epidemiology.*
  - *Fusarium* species are the most frequent mold causing keratitis (especially *F. solani*). It is second to *Aspergillus* as a cause of invasive mold disease in neutropenic patients with hematologic malignancy and in stem cell transplant (SCT) patients. Rates vary widely among institutions.
  - *Risk Groups/Factors.* For onychomycosis: walking barefoot. For keratitis: agricultural labor, other workers exposed to potential eye injury, and in soft contact lens wearers. For abdominal sepsis: peritoneal dialysis. For sinopulmonary and disseminated *Fusarium* mycosis: prolonged and profound neutropenia, GVHD, and corticosteroid therapy.
- *Transmission. Fusarium* mycosis is noncommunicable. Conidia are inhaled, ingested, or acquired through a penetrating injury or burn. Aerosols from hospital water supply or other water source also may be involved in transmission. Conidia are introduced by walking barefoot and by local abrasions, which cause nail infections and, rarely, tinea pedis.
- *Determinants of Pathogenicity.* Some *Fusarium* species produce mycotoxins but their role in infection is unknown. *Fusarium* species are angioinvasive, resulting in hemorrhage, thrombosis, and infarcts.
- *Clinical Forms.* In immune-normal persons: nail infections, keratitis, and secondary infections in burn patients. In immunocompromised hosts: hematogenously disseminated disease with skin lesions, invasive sinusitis, nail infections with paronychia, and pulmonary infection.
- *Veterinary Forms.* Penetrating injuries cause keratitis in dogs and horses. *Fusarium* toxin (fumonisin B_1) causes leukoencephalomalacia (hole-in-the-head syndrome) in horses and rabbits; pulmonary edema and hydrothorax in swine.
- *Therapy.* Deep-seated *Fusarium* species mycosis is refractory to treatment, particularly during neutropenia, with a high case/fatality ratio. AmB-lipid derivatives and/or voriconazole (VRC) have some success as therapy for disseminated disease. Natamycin drops are used for keratitis; endophthalmitis is treated with topical, intracameral, and systemic VRC, or systemic AmB-lipid complex. VRC or nystatin cream is applied to skin lesions.

Fundamental Medical Mycology, First Edition. By Errol Reiss, H. Jean Shadomy and G. Marshall Lyon, III.
© 2012 Wiley-Blackwell. Published 2012 by John Wiley & Sons, Inc.

- *Laboratory Detection, Recovery, and Identification.* Skin lesions and blood cultures are the main sources of specimens for diagnosis: biopsy of skin lesions for histopathology and culture; KOH preparations for nail clippings, skin scrapings, and corneal specimens. Species identification is difficult using microscopic morphology; it is better to use DNA-based identification with respect to the database, FUSARIUM-ID.

15.2 INTRODUCTION/DISEASE DEFINITION

Fusarium species are common fungi found worldwide in soil, in plant debris, and as important plant pathogens. These fungi produce potent mycotoxins (fumonisins, trichothecenes) but their role in human infections is unknown. *Fusarium* species cause eye disease (keratitis), onychomycosis, and, less commonly, infections in burn patients and in those receiving continuous ambulatory peritoneal dialysis. Invasive sinusitis and pulmonary and hematogenously disseminated *Fusarium* mycosis can occur in the neutropenic host immunocompromised by therapy for hematologic malignancy or for maintenance of stem cell transplants (SCTs). Solid organ transplant patients have a lower incidence and more localized disease, with a better outcome. Mortality associated with *Fusarium* mycosis in SCT patients is high—63–90%. The incidence is much lower than invasive aspergillosis but is rising because of:

- A larger number of persons with prolonged neutropenia
- Successful fluconazole prophylaxis of yeast infections
- Changes in transplant procedures—peripheral blood stem cells, increasing age of persons receiving SCT, more HLA-mismatched or unrelated donors, and other developments
- Improved control of cytomegalovirus infections in immunocompromised hosts

Possible routes for disseminated infections include inhalation, ingestion, and entry through skin trauma. *Fusarium* mycosis is community-acquired but there is also a connection between human disease and *Fusarium* in hospital water sources. Invasive *Fusarium* mycosis resembles invasive aspergillosis but the former is characterized by a high incidence of positive blood cultures and skin lesions. *Fusarium* species are less susceptible than aspergilli to AmB, but there is a promise of better control with VRC and, possibly, with posaconazole (PSC). Azoles are used as monotherapy or in combination with lipid-AmB. Success depends on recovery from neutropenia.

15.3 CASE PRESENTATION

Skin Lesions in a 32-yr-old Woman with Leukemia (Durand-Joly et al., 2003)

History

This 32-yr-old woman was hospitalized for induction chemotherapy of relapsing ALL on June 14th. She was placed on broad spectrum antibiotics and G-CSF. On July 1st she was aplastic, left the hospital against medical advice, but returned on day 6 with fever and severe neutropenia lasting 16 days. Her antibiotic regimen was changed and a central venous catheter was removed and cultured. The catheter and a blood culture failed to grow a pathogen. On hospital day 10, 4 days after the onset of fever, vesicular and necrotic lesions appeared on both calves and the next day extended to all leg surfaces.

Laboratory Findings

Skin biopsy and blood cultures were performed. Fungal hyphae were present in the skin biopsy. AmB-lipid complex, 5 mg/kg body weight per day, was added on day 12 and maintained until day 30. Biopsy and blood cultures grew a mold, identified as *Fusarium oxysporum*. The culture grew rapidly on SDA at 37°C, producing colonies with white to purple aerial mycelium. The culture produced short phialides with fusiform, septate (3–5 septa) macroconidia, and both terminal and *intercalary* chlamydospores.

Therapy and Clinical Course

VRC was added on day 18 after the rash first appeared. The dose was 4 mg/kg/day, IV. Fever persisted and skin lesions covered her entire body. After 9 days on the combined AmB-lipid complex and VRC the skin lesions improved and the patient defervesced. On day 28 the patient recovered from neutropenia (>1000 PMNs/μL). The patient left the hospital before being discharged and continued to take 400 mg of oral VRC. There was no recurrence when she attended clinic 9 months later.

Diagnosis

The diagnosis was disseminated *Fusarium* mycosis.

Comments

Skin lesions and fungemia were the only clinical manifestations, and remain the major basis for a diagnosis (Torres et al., 2003). The case illustrates an improved clinical response with the combination of lipid-AmB and VRC.

15.3.1 Diagnosis

Disseminated *Fusarium* mycosis should be considered in a patient with febrile neutropenia and hematologic malignancy when there is digital paronychia, especially with onychomycosis, digital ulcer or eschar, and/or suggestive skin lesions. The evaluation should include blood cultures, radiologic examination of the lungs and sinuses, skin biopsy and culture (Bodey et al., 2002). The differential diagnosis includes invasive aspergillosis and, because of similar microscopic morphology, *Acremonium* species infections.

15.4 ETIOLOGIC AGENTS

Fusarium solani is the most common species recovered from clinical specimens followed by *F. oxysporum, F. verticillioides* (formerly *moniliforme*), and *F. proliferatum*. The *F. solani* species complex (FSSC) contains approximately 18 medically relevant phylogenetically distinct species. On the other hand, *Fusarium oxysporum* species complex (FOSC) 3-a is a single widespread clonal lineage (O'Donnell et al., 2007).

Fusarium solani (Mart.) Sacc., *Michelia* 2 (no.7): 296 (1881) (anamorph). The teleomorph is *Nectria haematococca*. Position in classification: *Nectriaceae, Hypocreales, Hypocreomycetidae, Sordariomycetes, Ascomycota*.

Fusarium oxysporum E.F. Sm. & Swingle (anamorph). The teleomorph genus is *Gibberella*.

Fusarium verticillioides (Sacc.) Nirenberg (1976), (anamorph). The teleomorph is *Gibberella fujikuroi*.

Fusarium proliferatum (Matsush.) Nirenberg ex Gerlach & Nirenberg (1976) (anamorph). The teleomorph genus is *Gibberella*.

Members of the genus *Fusarium* are difficult to speciate because specialized media may be needed to induce conidia, and their interpretation requires expertise not ordinarily available in the clinical laboratory. Specific identification is made accurate and precise by PCR-sequencing with respect to the database FUSARIUM-ID (O'Donnell et al., 2010). PCR sequencing has revealed far more species diversity than was estimated by morphologic identification. (Please see Section 15.12, Laboratory Detection Recovery and Identification).

15.5 GEOGRAPHIC DISTRIBUTION/ECOLOGIC NICHE

Fusarium species are found worldwide in all climate zones as common fungi in soil, in plant debris, and as important plant pathogens causing crown rot, head blight, scab on cereal grains, vascular wilts, root rots, cankers, and other diseases of important crops such as sugarcane and rice. In addition to damaging crops, they secrete powerful mycotoxins that contaminate food and feed, consumption of which causes illness and even death (Bennett and Klich 2003; Bhatnagar et al., 2002).

Fusarium species have an efficient means of airborne spore dispersal but are not usually isolated from air sampling in urban areas. Some *Fusarium* species are associated with an aquatic habitat. Hospital water supply systems are implicated as a potential source of *Fusarium. Fusarium* species were recovered from 162 (57%) of 283 water system samples in a Houston, Texas, cancer center. *Fusarium* was prevalent in sink drains, less commonly isolated from sink faucet aerators, and in shower heads. *F. solani* was isolated from the hospital water tank. Aerosolization of *Fusarium* species was documented after running the showers (Anaissie et al., 2001).

15.6 EPIDEMIOLOGY

15.6.1 Incidence and Prevalence

Fusarium species, after aspergillosis, are the second most frequent cause of invasive mold mycosis in hematologic malignancy and SCT patients (Dignani and Anaissie, 2004). The occurrence of *Fusarium* mycosis is highly localized to some institutions and not to others. A search of the world literature from January 2000 to April 2007 found 63 cases of pulmonary or disseminated *Fusarium* mycoses (Stanzani et al., 2007).

Hematologic Malignancy

The most frequent clinical setting for *Fusarium* mycosis is the neutropenic patient being treated for a hematologic malignancy. Of 84 such patients who developed *Fusarium* mycosis, slightly more than half had acute leukemia (56%), and most (83%) were neutropenic at diagnosis (reviewed by Nucci and Anaissie, 2007). The attack rate among leukemia patients was estimated at 0.06%. The following example gives a good impression of the scale of *Fusarium* mycosis compared to other invasive mold infections in patients with hematologic malignancy.

A study of patients in 18 hematology wards of Italian hospitals during the period 1999–2003 revealed 310 cases of invasive aspergillosis, 14 cases of mucormycosis, and 14 cases of invasive *Fusarium* mycosis. Of 11,802 hematology patients the incidence of *Fusarium* mycosis was 0.1% compared with 2.5% for invasive aspergillosis (Pagano et al., 2006). Although the incidence may be low, the invasive fungal infection associated mortality for *Fusarium* mycosis was 53% compared with 42% for invasive aspergillosis.

Stem Cell Transplant (SCT) Patients

The incidence of *Fusarium* mycosis is estimated at 6 cases per 1000 SCTs, lowest (1.5 to 2/1000) among autologous recipients, and highest (20/1000) among recipients of mismatched related donor allogeneic SCTs (Nucci and Anaissie, 2007). In a review of 259 cases of *Fusarium* mycosis reported up to 2001 there were 61 cases in SCT patients (reviewed by Nucci, 2003). Of these, 47 (77%) had disseminated disease, with positive blood cultures in almost half of those affected. Skin lesions were present in most patients. Persistently neutropenic patients with disseminated *Fusarium* mycosis have a very poor prognosis with mortality approaching 100% (Nucci and Anassie, 2007).

Solid Organ Transplant Recipients

Fusarium mycosis is rare in this group, and when it occurs it is usually late in the post-transplant period. Disease may be locally invasive and responds better than that found in neutropenic cancer patients. Of 6 cases in the English literature up to 2001, 2 occurred on the foot, 2 were other localized cutaneous disease, and the remaining 2 persons had infectious endocarditis and pulmonary nodules, respectively (Sampathkumar and Paya, 2001). Of local cutaneous *Fusarium* mycosis in this series, the time after transplantation varied from 21 weeks to 5 years.

Keratitis

Fusarium species are the molds most frequently isolated from mycotic keratitis. For the causative agents of keratitis in the 259 cases of *Fusarium* mycosis updated with 35 additional cases published between 2001 and 2005, *Fusarium solani* was the most frequent etiologic agent (50%), followed by *F. oxysporum* (20%) and *F. verticillioides* and *F. moniliforme*[1] (10% each) (Nucci and Anaissie, 2007). Fungal keratitis is more often found in tropical or subtropical climates; in Southern Florida fungal keratitis comprises up to 35% of microbial keratitis cases, compared with 1% in New York (Chang et al., 2006).

The most common fungi isolated from contact lens cases are *Fusarium* (41%) and *Candida* species (14%), followed by *Curvularia* (12%) and *Aspergillus* (12%) (reviewed by Imamura et al., 2008). A multistate outbreak of *Fusarium* keratitis occurred in the United States in 2005–2006 (Chang et al., 2006). Case patients steadily increased beginning in June 2005, peaked in April 2006, and then declined in May and June 2006. In February 2006, several clusters of patients with *Fusarium* keratitis were also reported in Singapore and in Hong Kong. By June 2006 there were 164 cases in 33 states. One-hundred-forty-six patients were confirmed as contact lens wearers, with 115 users of "Renu with Moisture Loc" cleaning and storage solution. Corneal transplants were needed in 55 patients (34%). The epidemiologic investigation revealed that *Fusarium* was not recovered in the factory, warehouse, solution filtrate, or unopened "Renu with Moisture Loc" solution bottles.

Genotyping was determined by multilocus sequence typing (MLST) augmented with sequence from the second largest RNA polymerase subunit (Zhang et al., 2006). The U.S. clinical isolates represented 19 unique genotypes including 12 genotypes (30 isolates) within the *Fusarium solani* species complex (FSSC) and 5 genotypes (7 isolates) within the *Fusarium oxysporum* species complex (FOSC). Thus, most keratitis cases were caused by *F. solani*. Genotyping of the keratitis outbreak isolates indicated they clustered in *F. solani* species complex Groups 1 and 2, known from previous work to occur in sink and shower drains. This finding and the diversity of genotypes in the outbreak led to the conclusion that extrinsic contamination of solution bottles or lens cases occurred with the source of *Fusarium* being water in patients' homes.

[1] *Fusarium moniliforme* is synonymous with *F. verticillioides*.

15.6.2 Risk Factors

Hematogenously Disseminated Fusarium Mycosis

Persons experiencing prolonged (up to 3 weeks duration) and profound neutropenia in acute leukemia, mismatched or unrelated donor SCT patients, those with GVHD, and patients with multiple myeloma are at risk to develop *Fusarium* mycosis. Three periods of risk for SCT patients are a median of 16 days post-transplant during neutropenia; between day 61 and 80 for acute GVHD; and >360 days post-transplant owing to chronic GVHD (Nucci, 2003). It is prudent for neutropenic in-patients to avoid showering to minimize exposure to mold in hospital water systems.

Fusarium *Keratitis*

Eye injuries involving plant material are specifically implicated in fungal keratitis. In addition to a penetrating injury, soft contact lens wearers are at increased risk who do not comply with proper disinfection and maintenance of the lenses. Another risk factor is the use of topical corticosteroids, which slows the response to antifungal therapy.

Continuous ambulatory peritoneal dialysis (CAPD) is a risk factor for fungal disease, of which *Fusarium* is a minor contributor, but one that is difficult to diagnose and treat.

15.7 TRANSMISSION

Three sources of *Fusarium* species exposure are known.

1. *Inhalation of Conidia from the Aerospora in the Community*. A review of 70 patients at a cancer treatment center in Texas between 1987 and 1997 found that most developed invasive *Fusarium* mycosis within 3 days of admission. Patient isolates were diverse and could not be related by RAPD genotyping to environmental isolates obtained by air sampling from within and outside the hospital (Raad et al., 2002). During the rainy summer, outdoor air concentrations of *Fusarium* conidia were highest, coinciding with the peak incidence of *Fusarium* mycoses. The source of infection in patients with febrile neutropenia who develop invasive *Fusarium* disease is likely community-acquired via the respiratory route and may correspond to the seasonal pattern of air dispersal of conidia.

2. *Aerosols Containing Conidia from the Hospital Water System*. *Fusarium* species conidia are present in air and also in aquatic environments. Hospital water supply systems, showers, and sink drains are implicated as potential sources of *Fusarium* (Anaissie et al., 2001; Raad et al., 2002). DNA strain genotyping discovered that a geographically widespread clonal lineage was responsible for >70% of all clinical isolates of *F. oxysporum* (O'Donnell et al., 2004). Similar genotypes were isolated from water systems of three U.S. hospitals, further underlining the importance of that source of environmental exposure.

3. *Spread of* Fusarium *from Nail Infections*. An important source may be endogenous in patients with onychomycosis (Bodey et al., 2002). During the period of neutropenia invasive disease may evolve from a preexisting, even superficial cutaneous lesion.

15.8 DETERMINANTS OF PATHOGENICITY

15.8.1 Host Factors

Human PMNs and monocyte-derived macrophages were compared for the relative ability to damage hyphae of *F. solani* and *F. oxysporum* (Winn et al., 2003). *Fusarium solani* was less susceptible to damage by monocytes.

Role of the Immune System in Fusarium *Keratitis*

Mouse models of *Fusarium* keratitis are characterized by a profound neutrophil infiltrate. Systemic steroid treatment causes unregulated fungal growth and subsequent destruction of the cornea (reviewed in Tarabishy et al., 2008), underlining the essential role of the immune response in controlling fungal growth and halting the development of keratitis. Mouse Δ*MyD88* knock-out mutants are unable to clear an experimentally induced *F. oxysporum* keratitis. MyD88 is a common adapter protein relaying signals from Toll-like receptors (TLRs) on the surface of epithelial cells through a classic signaling pathway to stimulate production of proinflammatory cytokines. TLRs are also present on PMNs, dendritic cells, and macrophages.

In normal mice, PMNs infiltrate the corneal lesion and the anterior chamber of the eye and gradually, over 48 h, kill the fungal hyphae. TLR4, although not involved in PMN recruitment, participates in killing fungi when expressed on PMNs, possibly by stimulating the production of reactive oxygen species. The cytokine receptor IL-1R1, by stimulating production of *CXC chemokines,* is also important in recruiting PMNs. An explanation for the host defense, and also the damage brought about by fungal activity and immunopathology, was provided by Tarabishy et al., (2008). Events leading to keratitis in immune-normal animals involves the sequence:

1. Recognition of *Fusarium* by resident cells in the corneal stroma and production of IL-1.
2. IL-1R1/MyD88-dependent CXC chemokine production and recruitment of PMN from peripheral vessels to the central cornea.
3. TLR4-dependent antifungal killing by PMNs.
4. Tissue damage and corneal opacification due to fungal enzymes and mycotoxins or by products of PMN degranulation.
5. Eventual corneal scarring and resolution of inflammation.

15.8.2 Microbial Factors

The *Fusarium solani* species complex (FSSC) contains approximately 18 medically relevant phylogenetically distinct species. On the other hand, *F. oxysporum* species complex (FOSC) 3-a is a single widespread clonal lineage. Because of this diversity, experimental studies employing *Fusarium* species should be based on the multilocus haplotype nomenclature (O'Donnell et al., 2007). *Fusarium* species are capable of rapid growth, including microcycle conidiation.[2] They produce dormant resistant cells and attach to and penetrate soft contact lenses.

[2]Microcycle conidiation. *Definition*: Conidia germinate and germlings spontaneously produce more conidia with no intervening mycelial phase.

Role of Signaling Cascades

The *Fusarium oxysporum FGB1* gene encodes a heterotrimeric G protein β subunit that regulates hyphal development by signaling through a cAMP-dependent pathway (Prados-Rosales et al., 2006). The Δ*fgb1* mutant of *F. oxysporum* has reduced extracellular proteinase activity. A mitogen-activated protein kinase (MAPK) cascade (*P*athogenicity MAP *K*inase or PMK1 cascade) signaling pathway controls fungal pathogenicity and is remarkably conserved among fungi. The *Fusarium* ortholog of the pathogenicity MAPK cascade (Fmk1) is a signaling module essential for virulence in all fungal plant pathogens. Fmk1 positively regulates adherence to fibronectin. Fibronectin is the blood plasma and matrix glycoprotein involved in binding of fungal pathogens to host cells. The double mutant Δ*fmk1*/Δ*fgb1* of *F. oxysporum* is avirulent for mice. Thus, signaling pathways that control proteinase secretion and adherence to host tissues are critical for pathogenicity. The dual function of adherence and proteinase secretion may help explain the angioinvasive property of *Fusarium* during pathogenesis.

Biofilm[3] formation by Fusarium solani *and* F. oxysporum

The 2005–2006 multistate U.S. and international outbreak of keratitis in soft contact lens wearers stimulated interest in the biofilm-forming ability of *Fusarium* species (Chang et al., 2006). Of the most common fungi isolated from contact lens cases *Fusarium* > *Candida* > *Curvularia* and *Aspergillus, Fusarium* species display the greatest ability to invade the cornea (reviewed by Imamura et al., 2008). Three human keratitis-associated *Fusarium* isolates were compared for the ability to form biofilms on Lotrafilcon A® soft contact lenses. *F. solani* FSSC 1-b (MRL8609) and *F. oxysporum* FOSC 3-a (MRL8996) were recent clinical isolates from the 2006 keratitis outbreak associated with " Renu with Moisture Loc" lens cleaning and disinfectant solution. Both isolates formed robust biofilms composed of a homogeneous layered mesh of hyphal elements. Attachment to the lens surface ranged from a loose association of conidia and hyphae to firmly attached hyphae that were difficult or impossible to remove. Suboptimal activity of "Renu with MoistureLoc" against *Fusarium* infections may be due to the ability of these pathogens to form resistant biofilms.

[3]Biofilms. *Definition*: Microbial communities that are embedded in an extracellular matrix of polysaccharide.

Fusarium *Toxins*

Fusarium mycotoxins (Bennett and Klich 2003; Bhatnagar et al., 2002) and their effects are well studied when they are ingested in contaminated food and feed, but their involvement during infection has not been well studied.

Fumonisins The most abundant toxin in the fumonisin group is fumonisin B_1. Fumonisins are produced by a number of *Fusarium* species, notably *F. verticillioides* (formerly *F. moniliforme*, teleomorph is *Gibberella fujikuroi*). Please see Section 15.10, Veterinary Forms.

Trichothecenes The trichothecenes comprise a family of more than 60 sesquiterpenoid metabolites produced by a number of fungal genera, including *Fusarium* species. The trichothecenes are very potent inhibitors of eukaryotic protein synthesis. Different trichothecenes interfere with initiation, elongation, and termination stages or protein synthesis. All trichothecenes inhibit peptidyl transferase by binding to the same ribosome-binding site. Diacetoxyscirpenol, deoxynivalenol, and T-2 are the best studied of the trichothecenes produced by *Fusarium* species.

Effect of Fusarium *Mycotoxins on T, B, and NK Cell Functions*

Fusarium mycotoxins were tested on human peripheral blood mononuclear cells from different blood donors for their effects on T and B lymphocytes in a proliferation assay, antibody-dependent cellular cytotoxicity (ADCC), and natural killer (NK) cell activity (Berek et al., 2001). The concentrations used were similar to those found in normal human peripheral blood (0.2–1800 ng/mL). Among the eight mycotoxins tested, T-2 toxin, fusarenon X, nivalenol, and deoxynivalenol exerted the highest immunosuppressive effects. T-2 toxin and fusarenon X inhibited each of the in vitro reactions in a dose-dependent manner. Nivalenol and deoxynivalenol had the greatest inhibition of mitogen-induced blast transformation and the NK reaction.

15.9 CLINICAL FORMS

Invasive *Fusarium* Mycosis in Immune-Normal Hosts

The clinical forms seen in immune-normal persons are:

- Keratitis
- Onychomycosis

- Peritonitis in patients with chronic renal failure being maintained on continuous ambulatory peritoneal dialysis (CAPD)
- Severe burns (rarely)

Keratitis

Penetrating injury, introducing bits of fungus-containing plant material, remains the major route of infection leading to fungal keratitis. The widespread use of soft contact lenses has seen a surge in fungal keratitis, associated with improper care and disinfection, or the presence of preexisting ocular defects.

Fungal keratitis is a medical emergency and rapid initiation of aggressive therapy is needed to halt the disease process and limit the extent of corneal scarring and loss of vision (Thomas and Geraldine, 2007). Early signs and symptoms of fungal keratitis include redness, tearing, pain, sensitivity to light, discharge, decreased vision, and a white corneal infiltrate. Serrated margins of the lesions with raised slough, and color other than yellow are associated with fungal keratitis. Among the agents of mycotic keratitis, infection with *Fusarium* species is a risk factor for perforation of the cornea, through Descemet's membrane, into the anterior chamber (Vemuganti et al., 2002). Corneal lesions are suppurative and ulcerative. *Fusarium solani* infection may completely destroy an eye in a few weeks. The infection is usually severe and corneal ulcers develop with perforation, deep extension, and malignant glaucoma (Thomas, 2003).

In the early stages of *Fusarium* species keratitis, however, hyphae grow horizontally in the cornea, in contrast to a vertical growth pattern observed in cases of *Aspergillus* species keratitis (Xie et al., 2008). This observation led investigators to recommend the use of lamellar keratoplasty[4] in early stages of *Fusarium* keratitis instead of the more invasive and problematic penetrating keratoplasty.

Prevention of Keratitis in Soft Contact Lens Wearers A suboptimal hygiene practice contributing to the 2006 outbreak of *Fusarium* keratitis was *reusing contact lens solution already in the lens case* (Chang et al., 2006). A decline of the disinfecting ability of the solution in the contact lens case was related to drying, dilution, and sorption of the antimicrobial component alexidine to the soft contact lenses. The epidemic curve returned to baseline following the withdrawal of "Renu with Moisture Loc" from the market. The following advice is intended to reduce the chance of *Fusarium* keratitis and, for that matter, keratitis of any etiology (Ahearn et al., 2008):

- Avoid "topping off" disinfectant in the lens case.
- Avoid diluting the disinfectant with water.
- Discard used disinfection solution.
- Disinfect lenses for the recommended time.
- Empty, rinse, and store the lens case in fresh disinfectant, avoiding evaporation between uses.
- Remove sticky residue in the contact lens case and outside of the container bottle.
- Replace or sterilize the contact lens case weekly, by boiling or with H_2O_2 solution.
- Rub lenses manually as part of the disinfectant procedure.
- Wash hands before handling lenses and accessories.

Onychomycosis

Fusarium is a cause of superficial white onychomycosis and proximal, distal, or lateral subungual onychomycosis (Hay, 2007). In the extreme there is total dystrophic onychomycosis. Dense white coloration is common. *F. solani* is also associated with melanonychia. A total of 71 isolates of *Fusarium* species from patients with onychomycosis were identified by PCR of the 311 bp fragment of the 28S rDNA. *F. oxysporum* was most frequently identified (54%); *F. proliferatum* (10%) and *F. solani* accounted for only 8% (Ninet et al., 2005). *Paronychia* in an immunocompromised patient may become the route of invasion into the deep tissues.

Tinea Pedis

Fusarium species cause macerated interdigital infection, most often in tropical environments, and can be confused with dermatophytosis especially with *Trichophyton interdigitale*.

Peritonitis in Patients with Chronic Renal Failure and Chronic Ambulatory Peritoneal Dialysis (CAPD)

Peritonitis is a threat to patients with end-stage renal disease who receive CAPD. A permanent indwelling catheter provides microbes with access to nutrient-rich medium and an opportunity to adhere to and invade the peritoneum. Fungal peritonitis was estimated to account for 7.8% of all episodes of peritonitis in patients receiving CAPD (Eisenberg et al., 1986). Of 88 patients with fungal peritonitis surveyed between 1966 and 1984, six cases were identified as caused by *Fusarium* species. It was the most

[4]Keratoplasty. *Definition*: Operation to restore vision by replacing a diseased portion of a person's cornea with healthy cornea from a donor. The operation may involve the full thickness of the cornea (penetrating keratoplasty) or only a superficial layer (lamellar keratoplasty).

frequently isolated mold pathogen in that series. There are many etiologic agents of fungal peritonitis (Shin et al., 1998). Most cases are caused by *Candida* species and are treated with fluconazole. Peritoneal catheter removal and systemic antifungal therapy may be required to treat fungal peritonitis. The goal is to avoid operative procedures and the necessity to change to hemodialysis.

Superinfection of Burns and in Diabetic Ulcers

Fusarium species can invade diabetic ulcers and severe burns in immune-normal hosts. The diagnosis can be confused with aspergillosis. Spread to normal skin and invasion of the deep tissues may develop.

Invasive *Fusarium* Mycosis in Immunocompromised Hosts

Acute Paronychia

Fusarium species are involved in acute paronychia. Paryonchia or digital ulcer may precede development of disseminated *Fusarium* mycosis. Patients at risk because of their immunosuppressed status should be examined for onychomycosis, cultured to determine the etiologic agent, and treated appropriately.

Skin Lesions from Hematogenous Dissemination

Painful skin lesions are present in up to 85% of patients being treated for hematologic malignancy and who develop a positive blood culture for *Fusarium* species. This frequency is much higher than skin lesions in aspergillosis, which are few and large (2–5 cm diameter) (Bodey et al., 2002). The setting includes persistent neutropenia, which often is profound (<100 PMNs/μL). The skin rash suggestive of *Fusarium* species hematogenous dissemination often presents at an early stage of the disease with very typical "target" lesions—large red or gray annular lesions (1–2 cm diameter), and is widely scattered on the trunk and limbs. The lesions develop central ulceration that progresses to a thick black eschar with a rim of erythema (Bodey et al., 2002; Hay, 2007). They can involve any site but are more often found on the extremities. Three types are found, including bullae or target lesions, multiple subcutaneous nodules, and ecthyma (*Definition*: shallow lesions with crusts or scabs, followed by discoloration and scarring). Thrombosis of the dermal blood vessels by *Fusarium* hyphae leads to extravasation of erythrocytes, dermal necrosis, and epidermal ulceration.

Invasive *Fusarium* Mycosis in Solid Organ Transplant Recipients

Fusarium species are second to aspergillosis in causing invasive disease in patients with hematologic malignancies. Solid organ transplant recipients, however, only rarely develop *Fusarium* mycosis and, when it occurs, it more often takes the form of localized soft tissue disease involving the lower extremities. These infections respond better to therapy than those in neutropenic cancer patients (Sampathkumar and Paya, 2001).

Disseminated *Fusarium* Mycosis

The portal of entry for disseminated infection (Dignani and Anaissie, 2004) is unknown but possible routes include inhalation of conidia, ingestion, entry through skin trauma, or with chronic nail infection as the source. The pathogenesis of disease unfolds with one or more of these signs and symptoms:

- Fever unresponsive to antibiotics and prolonged and profound neutropenia, especially in patients with leukemia, or those receiving SCTs.
- Invasive sinusitis.
- Paronychia surrounding infected toenails, skin lesions with cellulitis at sites of skin breakdown.
- Pleuritic chest pain, cough, hemoptysis. These symptoms are indistinguishable from invasive pulmonary aspergillosis.
- Extrapulmonary dissemination. Organs most frequently affected are skin (70–85%) and lungs and sinuses (70–80%). The central nervous system may also become involved.

Positive blood cultures and skin lesions are features that distinguish disseminated *Fusarium* mycosis from invasive aspergillosis.

Species Involved in Disseminated Fusarium *Mycosis*

Fusarium solani, F. oxysporum, and *F. verticillioides* are the species most often involved. They may be isolated from nail infections, skin lesions, foot and leg ulcers, brain, kidney, lung, and blood, as well as from peritoneal dialyzate, BAL fluid, and bronchial washings.

15.10 VETERINARY FORMS

Fumonisin toxins produced by *Fusarium* species, which can taint animal feed, affect animals in different ways by interfering with sphingolipid metabolism. They cause leukoencephalomalacia (hole in the head syndrome) in equines and rabbits; pulmonary edema and hydrothorax in swine; and hepatotoxic and carcinogenic effects and apoptosis in rat liver. In humans, there is a possible link with esophageal cancer.

Aspergillus species and *Fusarium* species are the most frequent molds isolated from fungal keratitis in horses caused by penetrating injury with plant material.

15.11 THERAPY

Table 15.1 summarizes the clinical forms of *Fusarium* mycosis and their recommended therapy choices (Young, 2007). *Fusarium* species demonstrate decreased in vitro susceptibility measured against AmB and the extended spectrum triazoles—posaconazole (PSC) and voriconazole (VRC)—compared with the aspergilli, other hyaline molds, or endemic dimorphic fungi (Sabatelli et al., 2006). In addition, *Fusarium* species have high MICs against echinocandin derivatives. This profile of decreased susceptibilities complicates clinical management.

Invasive Pulmonary and Extrapulmonary Disseminated *Fusarium* Mycosis

Antifungal agents appropriate to treat *Fusarium* mycosis are AmB-lipid formulations, VRC, and PSC. Echinocandins have no activity against this genus. If neutropenia does not resolve, the mortality approaches 100% regardless of the antifungal agent used (Malani and Kauffman, 2007).

Fusarium mycosis is seen as "breakthrough" infection in neutropenic patients during empiric or prophylactic therapy with AmB, FLC, or ITC. Proper recognition of the causative agent and prompt treatment are keys to success. *Fusarium* species are less susceptible to AmB. The optimal therapy for invasive fusariosis remains unclear because there have been no controlled drug trials. The lipid formulation of AmB with or without an azole antifungal is commonly used. VRC is FDA-approved in the United States as salvage therapy for *Fusarium* mycosis resistant to AmB monotherapy.

Voriconazole (VRC)

VRC (Vfend®, Pfizer) received U.S. FDA approval in 2002 for first-line treatment of invasive aspergillosis and as salvage therapy for infections with the mold pathogens *Scedosporium apiospermum* and *Fusarium* species. Since then, two large series of *Fusarium* mycosis treated with VRC were reported. Case records were reviewed for 73 patients with invasive *Fusarium* mycosis who received VRC primary therapy or, in most cases, as salvage therapy. The patients were from two sources: a Pfizer database for the period 1996–2002, and the French National Reference Center at the Institut Pasteur, Paris, for the period 2002–2009 (Lortholary et al., 2010). Isolates of the *F. solani* complex were the most frequent causative species and differences in response rates for all *Fusarium* species were not significantly different. Most patients (73%, 53/73) had disseminated disease; pulmonary/sinus infections affected 15%. Blood cultures were positive in approximately half of patients with disseminated *Fusarium* mycosis. More than half of the patients' primary disease was leukemia and another 18% were SCT recipients. Neutropenia was present or recent in more than half of the patients. Complete or partial responses to VRC therapy were achieved in 47% (34/73) of patients.

Slightly less than 10% of patients also received caspofungin (CASF), liposomal AmB, terbinafine, PSC, or white blood cell transfusion, concurrently or after VRC therapy, but there was no significant difference in responses for patients receiving VRC monotherapy or combination therapy. Patients remaining neutropenic at the end of VRC therapy had only a 5% positive clinical response compared to a 63% positive response in non-neutropenic patients. A total of 43 (59%) patients died, 51% with progressive invasive fungal infection. In the setting of persistent neutropenia no antifungal regimen appears to be effective. Pending a formal clinical trial for *Fusarium* mycosis, combination therapy with lipid-AmB and VRC or another azole is considered the best choice.

A series of 44 patients with invasive *Fusarium* mycosis during 1998–2009 and treated at a major cancer treatment center in Texas were reviewed (Campo et al., 2010). Over 80% of these patients were neutropenic at diagnosis. Most patients (84%) received combination therapy with lipid formulation AmB and a triazole. The mortality at 12 weeks attributable to *Fusarium* mycosis was 50%. The conclusion reached from these two large series is that *Fusarium* mycosis in patients with leukemia and/or SCT has an extremely poor prognosis where there is fungemia or persistent neutropenia.

Posaconazole (PSC)

PSC (Noxafil®, Schering-Plough, Inc.) has limited in vitro activity against *F. oxysporum*, MIC_{50} of 2–4 μg/mL; *F. solani*, MIC_{50} of >8–32 μg/mL; and *F. verticillioides* MIC_{50} of >8 μg/mL (reviewed by Schiller and Fung, 2007) Although this in vitro activity of PSC does not appear excellent, clinical success was obtained in treating invasive *Fusarium* mycosis in almost half of 21 patients,

Table 15.1 Therapy for *Fusarium* Species Mycosis[a]

Infection site	Drug	Dosage	Relative efficacy	Comments
Invasive *Fusarium* mycosis	AmB	1.0–1.5 mg/kg IV daily for 2–12 weeks	First-choice agent	Surgery may be needed. Use lipid formulation for patients with nephrotoxicity or infusion toxicity.
	Voriconazole	6 mg/kg every 12 h for first 24 h; then 4 mg/kg every 12 h. When oral medication is tolerated, 200 mg is taken every 12 h; continue treatment for several months, or for a minimum of 2 weeks after recovery from neutropenia. Resume if patient becomes neutropenic again after an infection.	Alternative agent	Oral form is taken 1 h before or after meal.
Keratitis/corneal ulcer	Natamycin 5% suspension	1 drop every 1–2 h for for the first 2 days; then decrease gradually over 3–6 weeks	First-choice agent	May require surgery.
Skin lesions (as part of invasive disease)	Treat primarily as invasive disease; continue topical therapy until lesions are dry scabs			
	Nystatin	Cream twice daily to affected areas	Adjunct to IV AmB	—
	AmB	Topical compound twice daily to affected areas	Adjunct to IV AmB	Compound locally

[a]Data are from Young (2007). Used with permission from BC Decker Publishers. Copyright © 2007.

most with hematologic malignancy, who became intolerant of or failed to respond to AmB-lipid formulations (Raad et al., 2006). PSC was continued for a median of 59 days with success in 48% (10 of 21 patients). When *Fusarium* infection remained confined to the lung, 3 of 4 (75%) patients responded to PSC. Leukemia patients who remained neutropenic and those who received SCTs were poor responders; recovery from myelosuppression being key to a positive response to antifungal agents.

Adjunctive immunotherapy with G-CSF and GM-CSF is also helpful in neutropenic patients with invasive *Fusarium* mycosis. Survival is associated with recovery from neutropenia and lack of GVHD. Corticosteroid use impairs the response to therapy. Removal of intravenous catheters is advisable in all cases of fungemia.

Combination Therapy for Disseminated Fusarium *Mycosis*

Patients with leukemia and/or SCT who are diagnosed with *Fusarium* mycosis often fail to respond to lipid-AmB monotherapy. The failure rate was estimated at about 70% (Langner et al., 2008). Some additional success has accompanied salvage therapy when another antifungal agent was added to the AmB regimen. These have included VRC, PSC, echinocandins, and terbinafine as well as cytokines—IFN-γ and GM-CSF. The results with AmB combined with VRC or PSC appear to be promising, provided that the PMN function of the patient can be restored, as discussed above. Echinocandins might act in synergy with AmB. Disseminated fusariosis in a child with leukemia was diagnosed by chest CT scan and positive blood cultures (Vagace et al., 2007). Primary monotherapy with liposomal AmB resulted in no improvement but, after addition of CASF, there was progressive resolution of pulmonary lesions.

Keratitis

Patients with *Fusarium* keratitis respond slowly to antifungal therapy over a period of weeks. Topical natamycin (5%) or AmB (0.15%), applied hourly, for several days (mean duration 20 days) (Thomas, 2003) is the usual first-line therapy for superficial fungal keratitis. When lesions are more deep-seated, systemic antifungal therapy is added to the topical treatment consisting of oral ITC, 200 mg/day, or VRC (please see below). Clinical improvement after therapy for *Fusarium* species keratitis in a series of patients treated between 1984 and 1994 was for: ketoconazole, 60%; ITC, 54%; AmB, 56%; and 44% of those treated with natamycin were responders (Thomas, 2003). Corneal transplant surgery is indicated if corneal infection continues despite appropriate antifungal therapy.

Patients may initially be treated for bacterial or herpetic keratitis, and when their conditions worsen, they may be referred to an eye center with more experience treating fungal keratitis (Jeng et al., 2007). Four such patients were placed on standard therapy of topical natamycin drops every hour. When corneal scraping cultures revealed *Fusarium* species, topical AmB drops were started on an hourly basis combined with oral VRC, unless the patient was a nursing mother for whom the latter drug is contraindicated. Despite this therapy two patients required therapeutic penetrating keratoplasties for progressive keratomycoses, whereas the remaining two resolved their infection in 1–2 months. The MICs of the various antifungal agents tested against the patient isolates indicated reduced susceptibility.

Success in treating *Fusarium* keratitis resistant to standard polyene therapy and unresponsive to VRC was achieved in three patients with PSC (Tu et al., 2007). The patients failed to improve after antifungal therapy and penetrating keratoplasty. Two patients developed endophthalmitis. PSC was administered orally in these patients for periods from 6 weeks to 6 months, at doses of 200 mg 4×/day and in one patient via local eye drops several times/day. In this group of three patients, the infection and inflammation resolved. No other systemic or intracameral medications were administered after the addition of PSC.

Natamycin, 0.5%, is the only available ophthalmic topical antifungal agent approved for use in the United States. It is the initial drug of choice for *Fusarium* keratitis but there has been a high failure rate. *F. solani* is the most consistently resistant species in vitro. Surveys of *Fusarium* isolates show the best in vitro responses to AmB, but there have been clinical failures. Some in vitro resistance was also observed with VRC, PSC, and ravuconazole. The MICs (all 8 μg/mL or less) more closely approach levels achievable in the eye and cornea.

Recalcitrant cases of *Fusarium* endophthalmitis have responded to topical, intracameral, and systemic VRC, and to treatment with systemic AmB-lipid complex. In some cases of *Fusarium* keratitis, however, VRC is either ineffective or poorly tolerated. PSC exhibits good tissue penetration and shows clinical efficacy in recalcitrant cases of *Fusarium* keratitis resistant to or intolerant of VRC. Pending its approval for this indication, PSC may prove useful as first-line therapy for *Fusarium* keratitis.

15.12 LABORATORY DETECTION, RECOVERY, AND IDENTIFICATION

The major means to diagnose disseminated *Fusarium* species mycosis are recovery of the fungus from the blood and biopsy and culture of skin lesions.

Direct Examination

For direct examination of skin, hair, nails, and corneal scrapings please refer to Chapter 2, Section 2.3.3.1, Wet Mount—KOH Prep and Fluorescent Brighteners Used with or without KOH.

Nails

Samples of skin, nail fragments, and material from under the infected nail are removed for microscopy. A sample may be taken by lightly scraping the nail near the infected area, or by shaving off a piece of nail using a small blade. The specimen is placed on a slide with 10% KOH solution, coverslipped, and gently heated. The skin and nail cells are dissolved, leaving the hyphae, which are observed microscopically. Imaging is enhanced by adding Calcofluor white to the preparation and observation in a fluorescence microscope.

Skin Lesions

Skin scrapings taken with a sterile scalpel are digested in 10% KOH. Punch biopsies are used for deeper skin lesions. The resulting material is divided between a KOH prep, culture, and fixation and embedment for histopathology.

Keratitis

Scrapings are made with a sterile blade or spatula from the base and edges of the corneal ulcer, by corneal biopsy, or at the time of performing a penetrating keratoplasty. Direct microscopy of corneal scrapings is accomplished with a KOH prep with Calcofluor white for enhanced visibility of fungal elements, Gram stain, or use of PAS or GMS stains. A detailed table describing the relative advantages and sensitivity of direct microscopy using various tinctorial methods is found in Thomas (2003).

Histopathology

Histopathology can lead to a tentative identification in less than one day and is very helpful in the diagnosis of hyaline mold mycoses (Liu et al., 1998). Biopsy of skin lesions and KOH preps of nail clippings indicate septate hyaline hyphae with an irregular diameter and both 45° and 90° branching. While helpful in establishing a diagnosis of invasive mold infection, it is difficult to determine the infecting fungal genus based only on histopathologic examination. *Adventitious sporulation* (arising in an unusual location, in this case, in tissue) consists of phialides, phialoconidia, and, rarely, multicellular conidia. The size and shape of conidia and phialides formed in tissue overlap between *Fusarium* and *Paecilomyces* species. Preparations of minced tissue, as a KOH–Calcofluor prep or as a GMS-stained smear, can increase the chance to demonstrate adventitious sporulation because, in contrast to tissue sections, the three-dimensional profile is maintained.

Culture

Specimens suspected of containing a hyaline mold such as *Fusarium* species are planted on SDA *without cycloheximide* and incubated at 25–30°C. PDA is a useful adjunct to improve chances of inducing sporulation. Corneal specimens are seeded to agar using "C" shaped streaks, so as to differentiate the infecting isolate from inadvertent contaminants. *Fusarium* cultures can be difficult to identify by morphologic means because sporulation is variable on ordinary mycologic media such as SDA or PDA.

Why is species identification important? There may be a cluster of cases from a common source in the hospital and, secondly, to exclude confusion with morphologically similar molds, that is, *Acremonium* species. *How can clinical laboratories identify the genus so that further molecular identification can be facilitated?* When a hyaline mold grows from a clinical specimen clues to the presence of a *Fusarium* species are the following:

- Specimen source: if the culture is from an eye infection, nail infection, a blood culture, a skin lesion or a burn patient.
- The initial colony color is pink or purple.
- Lack of obvious features that point to an *Aspergillus* species or to one of the *Mucorales*. Those features include the typical stipe, vesicle, metulae, and phialoconidia of the aspergilli or sporangia with sporangiospores of one of the *Mucorales*.

Use a dichotomous key to aid in the differential diagnosis, that is, "non-dematiaceous fungi producing solitary *aleurioconidia* and typically isolated from deep body sites" (St. Germain and Summerbell, 1996).

Colony Morphology

Fusarium species grow rapidly and develop a cottony to wooly appearance. Surface colors range from white, yellow, pink, purple, and tan. The purple color is an indication that the culture may be *Fusarium*. The reverse is pale, red, violet, brown, or blue.

Microscopic Morphology

Morphologic criteria used for identification are presence and shape of *microconidia*, presence and location of chlamydospores, shape and number of septations within *macroconidia*, and shape of their *foot cells*. Other characteristics include presence of *sporodochia* (*definition* macroconidia produced on phialides in a "superficial cushion mass" sometimes encased in slime) and *sclerotia*.

Specialists (Nelson et al., 1983) recommend carnation leaf agar to study the microscopic morphology of *Fusarium* species. This medium is ideal for the production of macroconidia on sporodochia arising directly from the leaf pieces but is found only in a specialized reference laboratory. *Fusarium* species sporulate poorly on media known to medical mycologists (e.g., SDA or PDA) and these media are not optimal for morphologic studies on *Fusarium* species. Growth on SDA or PDA is, however, sufficient for molecular identification purposes. Mycologists interested in pursuing morphologic identification of *Fusarium* species should consult the Dr. Nelson's manual (Nelson et al., 1983).

Sickle-shaped, "fusoid" (spindle-shaped) multicellular macroconidia are borne on phialides, which may be long or short, with barely visible collarettes (Fig. 15.1). Macroconidia have foot cells with some type of heel. This is accepted as the most definitive characteristic of *Fusarium*.

Fusarium solani

Colony Morphology Growth is rapid with a cottony white-cream surface and a cream reverse.

Microscopic Morphology See Fig. 15.1. Curved (sickle shaped) macroconidia are thick walled with 3–5 septa borne on short phialides. Macroconidia cluster into sporodochia, which develop a blue-green coloration. Carnation leaf agar may be necessary to demonstrate sporodochia. One to three-celled (primarily one-celled) microconidia are produced on long phialides. Chlamydospores may be present and abundant. Distinguishing features are a cream color colony and long phialides bearing microconidia.

Fusarium oxysporum

Colony Morphology Growth is rapid with a cottony surface, aerial mycelia tinged with purple, and a colorless to dark purple reverse.

Microscopic Morphology Phialides are short in contrast to *F. solani*. Macroconidia have a slight sickle shape and are thin-walled with a foot-shaped basal cell. Sporodochia are tan or orange. Chlamydospores are in pairs. When few or no macroconidia are produced *Fusarium* has to be distinguished from *Acremonium* species. Microscopic morphology of *Acremonium* reveals slender phialides producing clusters of oblong to ovoid unicellular conidia (St. Germain and Summerbell, 1996).

Blood Cultures An increased frequency of positive blood cultures for *Fusarium* species, with recovery rates up to 50%, compares favorably with *Aspergillus* species ($\leq$5% recovery). A possible explanation for this is that *Fusarium* species are capable of adventitious sporulation, resulting in conidia circulating in the bloodstream.

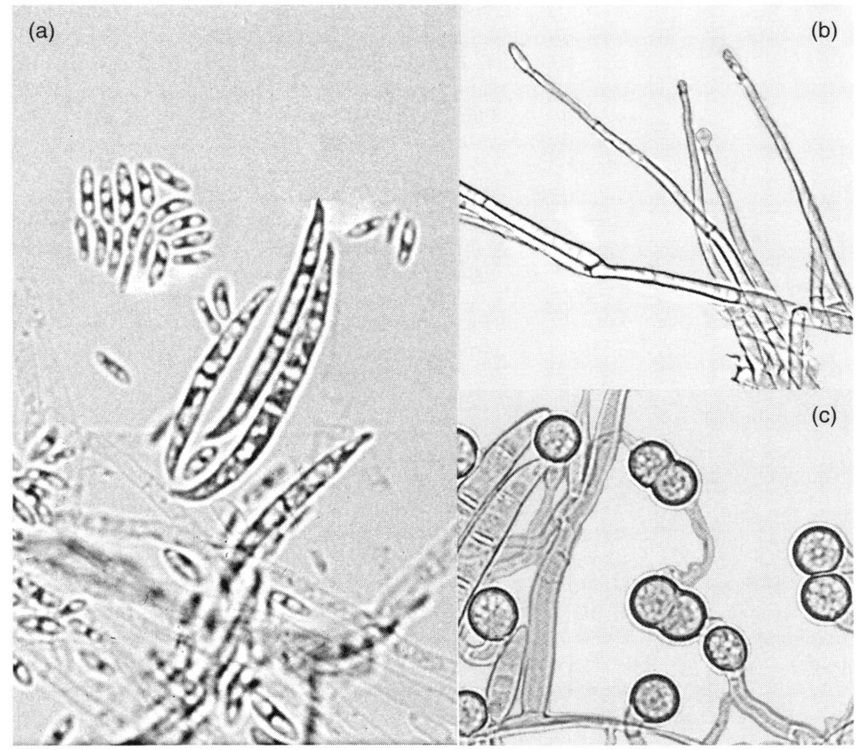

Figure 15.1 *Fusarium solani* microscopic morphology. Species of *Fusarium* typically produce both macro- and microconidia. Conidiophores are unbranched or branched phialides The phialides *of F. solani* bearing microconidia are long, compared to those of *F. oxysporum*. (*Panel* a). Macro- and microconidia. Macroconidia are curved and thick-walled. Macroconidia of *F. solani* are abundant, thick-walled, 3- to 5-septate (usually 3-septate). Microconidia vary from sparse to abundant, generally are single-celled, and oval to kidney-shaped.
Source: Used with permission from Dr. Yoshistugu Sugiura. (*Panel* b). Phialides which bear microconidia. (*Panel* c) Chlamydospores are formed singly and in pairs. They are abundant in most isolates. (Panels (b) and (c)) were adapted from Nelson et al. (1983), Fig. 63, with permission from Penn State Press.

Molecular Identification of Fusarium *Species*

The requirement for morphologic identification to species level in *Fusarium* has been relieved by advances in molecular identification.

***FUSARIUM* ID** This is a three-locus DNA sequence database for molecular identification of 69 clinically important *Fusarium* species (O'Donnell et al., 2010). It is comprised of partial sequences from translation elongation factor 1α (*EF*−1α), the largest subunit of RNA polymerase (*RPB1*), and the second largest subunit of RNA polymerase (*RPB2*) Phylogenetic analysis of the data set places clinically relevant *Fusarium* spp. into eight species complexes, which are monophyletic lineages. Clinical isolates can be identified by submitting a BLAST query to the database, "FUSARIUM-ID" at the URL http://isolate.fusariumdb.org/welcome.php?a=intro

Many sequences have been placed on the BLAST server. A useful resource contains the technical specifications for DNA extraction, PCR and sequencing primers, clean-up of the amplicon for sequencing, and identification with respect to the database (O'Donnell et al., 2010).

The following six species complexes constitute the *Gibberella* clade: *F. incarnatum-equiseti* (FIESC), *F. sambucinum* (FSAMSC), *F. chlamydosporum* (FCSC), *F. tricinctum* (FTSC), *Gibberella fujikuroi* (GFSC), and *F. oxysporum* (FOSC). Clinically important *F. verticilloides* and *F. proliferatum* are placed in the *Gibberella fujikuroi* species complex containing at least 34 known and 20 undescribed species. The *F. incarnatum-equiseti* species complex (FIESC) also contains marked cryptic speciation. Of the foregoing six complexes *F. tricinctum* and *F. sambucinum* complexes are very rarely isolated from clinical specimens.

The two remaining complexes are *F. solani* (FSSC) and *F. dimerum* (FDSC). Multilocus sequence analysis found that the *F. solani* species complex (FSSC) contains several new phylogenetic species that await taxonomic treatment.

An example of the importance of sequence based analysis is that it has been able to identify the widespread *F. oxysporum* clonal lineage (FOSC 3-a = sequence type 33) (O'Donnell et al., 2004). Molecular studies discovered this single widespread clonal lineage accounting for 72% (n = 63) of the clinical isolates of this mold, and 17 of 18 *F. oxysporum* cultures isolated from hospital water systems in three different U.S. cities. Very low genetic diversity of the species was observed. *Fusarium solani* FSSC 1-a and 2-d are also common in water systems in hospitals where they pose a risk for nosocomial transmission. Clinical and hospital water system isolates were highly related suggesting that water sources inside or outside a hospital may be a reservoir for *Fusarium* species (O'Donnell et al., 2004).

SELECTED REFERENCES FOR FUSARIUM MYCOSIS

AHEARN DG, ZHANG S, STULTING RD, SCHWAM BL, SIMMONS RB, WARD MA, PIERCE GE, CROW SA Jr, 2008. *Fusarium* keratitis and contact lens wear: Facts and speculations. *Med Mycol* 46: 397–410.

ANAISSIE EJ, KUCHAR RT, REX JH, FRANCESCONI A, KASAI M, MÜLLER FM, LOZANO-CHIU M, SUMMERBELL RC, DIGNANI MC, CHANOCK SJ, WALSH TJ, 2001. Fusariosis associated with pathogenic *Fusarium* species colonization of a hospital water system: A new paradigm for the epidemiology of opportunistic mold infections. *Clin Infect Dis* 33: 1871–1878.

BEREK L, PETRI IB, MESTERHÁZY, TÉREN J, MOLNÁR J, 2001. Effects of mycotoxins on human immune functions in vitro. *Toxicology in Vitro* 15: 25–30.

BENNETT JW, KLICH M, 2003. Mycotoxins. *Clin Microbiol Rev* 16: 497–516.

BHATNAGAR D, YU J, EHRLICH KC, 2002. Toxins of Filamentous Fungi, pp. 167–206, *in*: BREITENBACH M, CRAMERI R, LEHRER SB (eds.), *Fungal Allergy and Pathogenicity, Volume 81, Chemical Immunology*. Karger Press, Basel.

BODEY GP, BOKTOUR M, MAYS S, DUVIC M, KONTOYIANNIS D, HACHEM R, RAAD I, 2002. Skin lesions associated with *Fusarium* infection. *J Am Acad Dermatol* 47: 659–666.

CAMPO M, LEWIS RE, KONTOYIANNIS DP, 2010. Invasive fusariosis in patients with hematologic malignancies at a cancer center. *J Infect* 60: 331–337.

CHANG DC, GRANT GB, O'DONNELL K, WANNEMUEHLER KA, NOBLE-WANG J, RAO CY, JACOBSON LM, CROWELL CS, SNEED RS, LEWIS FMT, SCHAFFZIN JK, KAINER MA, GENESE CA, ALFONSO EC, JONES DB, SRINIVASAN A, FRIDKIN SK, PARK BJ, for the Fusarium Keratitis Investigation Team, 2006. Multistate outbreak of *Fusarium* keratitis associated with use of a contact lens solution. *JAMA* 296: 953–963.

DIGNANI MC, ANAISSIE E, 2004. Human fusariosis. *Clin Microbiol Infect* 10(Suppl 1): 67–75.

DURAND-JOLY I, ALFANDARI I, BENCHIKH Z, RODRIGUE M, ESPINEL-INGROFF A, CATTEAU B, CORDEVANT C, CAMUS D, DEI-CAS E, BAUTERS F, DELHAES L, DE BOTTON S, 2003. Successful outcome of disseminated *Fusarium* infection with skin localization treated with voriconazole and AmB-lipid complex in a patient with acute leukemia. *J Clin Microbiol* 41: 4898–4900.

EISENBERG ES, LEVITON I, SOEIRO R, 1986. Fungal peritonitis in patients receiving peritoneal dialysis. Experience with 11 patients and review of the literature. *Rev Infect Dis* 8: 309–321.

GEISER DM, JIMÉNEZ-GASCO MDM, KANG S, MAKALOWSKA I, VEERARAGHAVAN N, WARD TJ, ZHANG N, KULDAU GA, O'DONNELL K, 2004. FUSARIUM-ID v. 1.0: A DNA sequence database for identifying *Fusarium*. *Eur J Plant Pathol* 110: 473–479.

HAY RJ, 2007. *Fusarium* infections of the skin. *Curr Opin Infect Dis* 20: 115–117.

IMAMURA Y, CHANDRA J, MUKHERJEE PK, LATTIF AA, SZCZOTKA-FLYNN LB, PEARLMAN E, LASS JH, O'DONNELL K, GHANNOUM MA, 2008. *Fusarium* and *Candida albicans* biofilms on soft contact lenses: Model development, influence of lens type, and susceptibility to lens care solutions. *Antimicrob Agents Chemother* 52: 171–182.

JENG BH, HALL GS, SCHOENFIELD L, MEISLER DM, 2007. The *Fusarium* keratitis outbreak: Not done yet ? *Arch Ophthalmol* 125: 981–983.

LANGNER S, STABER PB, NEUMEISTER P, 2008. Posaconazole in the management of refractory invasive fungal infections. *Ther Clin Risk Manag* 4: 747–758.

LIU K, HOWELL DN, PERFECT JR, SCHELL WA, 1998. Morphologic criteria for the preliminary identification of *Fusarium, Paecilomyces*, and *Acremonium* species by histopathology. *Am J Clin Pathol* 109: 45–54.

LORTHOLARY O, OBENGA G, BISWAS P, CAILLOT D, CHACHATY E, BIENVENU AL, CORNET M, GREENE J, HERBRECHT R, LACROIX C, GRENOUILLET F, RAAD I, SITBON K, TROKE P, 2010. International retrospective analysis of 73 cases of invasive fusariosis treated with voriconazole. *Antimicrob Agents Chemother* 54: 4446–4450.

MALANI AN, KAUFFMAN CA, 2007. Changing epidemiology of rare mould infections. *Drugs* 67: 1803–1812.

NELSON PE, TOUSSOUN TA, MARASAS WFO, 1983. *Fusarium Species—An Illustrated Manual for Identification*. Pennsylvania State University Press, University Park, PA. (It is out of print but is in the collections of many university libraries.)

NINET B, JAN I, BONTEMS O, LÉCHENNE B, JOUSSON O, LEW D, SCHRENZEL J, PANIZZON RG, MONOD M, 2005. Molecular identification of *Fusarium* species in onychomycoses. *Dermatology* 210: 21–25.

NUCCI M, 2003. Emerging moulds: *Fusarium, Scedosporium*, and zygomycetes in transplant recipients. *Curr Opin Infect Dis* 16: 607–612.

NUCCI M, ANAISSIE E, 2007. *Fusarium* infections in immunocompromised patients. *Clin Microbiol Rev* 20: 695–704.

O'DONNELL K, SUTTON DA, RINALDI MG, SARVER BA, BALAJEE SA, SCHROERS HJ, SUMMERBELL RC, ROBERT VA, CROUS PW, ZHANG N, AOKI T, JUNG K, PARK J, LEE YH, KANG S, PARK B, GEISER DM, 2010. An internet-accessible DNA sequence database for identifying *Fusaria* from human and animal infections. *J Clin Microbiol* 48: 3708–3718.

O'DONNELL K, SARVER BAJ, BRANDT M, CHANG DC, NOBLE-WANG J, PARK BJ, SUTTON DA, BENJAMIN L, LINDSLEY M, PADHYE A, GEISER DM, WARD TJ, 2007. Phylogenetic diversity and microsphere array-based genotyping of human pathogenic fusaria, including isolates from the multistate contact lens-associated U.S.A. keratitis outbreaks of 2005 and 2006. *J Clin Microbiol* 45: 2235–2248.

O'DONNELL K, SUTTON DA, RINALDI MG, MAGNON KC, COX PA, REVANKAR SG, SANCHE S, GEISER DM, JUBA JH, VAN BURIK JH, PADHYE A, ROBINSON JS, 2004. Genetic diversity of human pathogenic members of the *Fusarium oxysporum* complex inferred from gene genealogies and Aflp analyses: Evidence for the recent dispersion of a geographically widespread clonal lineage and nosocomial origin. *J Clin Microbiol* 42: 5109–5120.

PAGANO L, CAIRA M, CANDONI A, OFFIDANI M, FIANCHI L, MARTINO B, PASTORE D, PICARDI M, BONINI A, CHIERICHINI A, FANCI R, CARAMATTI C, INVERNIZZI R, MATTEI D, MITRA ME, MELILLO L, AVERSA F, VAN LINT MT, FALCUCCI P, VALENTINI CG, GIRMENIA C, NOSARI A, 2006. The epidemiology of fungal infections in patients with hematologic malignancies: The SEIFEM-2004 study. *Haematologica* 91: 1068–1075.

PRADOS-ROSALES RC, SERENA C, DELGADO-JARANA J, GUARRO J, DI PIETRO A, 2006. Distinct signalling pathways coordinately contribute to virulence of *Fusarium oxysporum* on mammalian hosts. *Microbes Infect* 8: 2825–2831.

RAAD I, TARRAND J, HANNA H, ALBITAR M, JANSSEN E, BOKTOUR M, BODEY G, MARDANI M, HACHEM R, KONTOYIANNIS D, WHIMBEY E, ROLSTON K, 2002. Epidemiology, molecular mycology, and environmental sources of *Fusarium* infection in patients with cancer. *Infect Control Hosp Epidemiol* 23: 532–537.

RAAD II, HACHEM RY, HERBRECHT R, GRAYBILL JR, HARE R, CORCORAN G, KONTOYIANNIS DP, 2006. Posaconazole as salvage treatment for invasive fusariosis in patients with underlying hematologic malignancy and other conditions. *Clin Infect Dis* 42: 1398–1403.

SABATELLI F, PATEL R, MANN PA, MENDRICK CA, NORRIS CC, HARE R, LOEBENBERG D, BLACK TA, MCNICHOLAS PM, 2006. In vitro activities of posaconazole, fluconazole, itraconazole, voriconazole, and amphotericin B against a large collection of clinically important molds and yeasts. *Antimicrob Agents Chemother* 50: 2009–2015.

Sampathkumar P, Paya CV, 2001. *Fusarium* infection after solid-organ transplantation. *Clin Infect Dis* 32: 1237–1240.

Schiller DS, Fung HB, 2007. Posaconazole: An extended-spectrum triazole antifungal agent. *Clin Ther* 29: 1862–1886.

Shin JH, Lee SK, Suh SP, Ryang DW, Kim NH, Rinaldi MG, Sutton DA, 1998. Fatal *Hormonema demtioides* peritonitis in a patient on continuous ambulatory peritoneal dialysis: Criteria for organism identification and review of other known fungal etiologic agents. *J Clin Microbiol* 36: 2157–2163.

Stanzani M, Tumietto F, Vianelli N, Baccarani M, 2007. Update on the treatment of disseminated fusariosis: Focus on voriconazole. *Ther Clin Risk Manag* 3: 1165–1173.

St Germain G, Summerbell R, 1996. *Identifying Filamentous Fungi—A Clinical Laboratory Handbook*. Star Publishing Co., Belmont, CA.

Tarabishy AB, Aldabagh B, Sun Y, Imamura Y, Mukherjee PK, Lass JH, Ghannoum MA, Pearlman E, 2008. MyD88 regulation of *Fusarium* keratitis is dependent on TLR4 and IL-1R1 but not TLR2. *J Immunol* 181: 593–600.

Thomas PA, 2003. Fungal infections of the cornea. *Eye* 17: 852–862.

Thomas PA, Geraldine P, 2007. Infectious keratitis. *Curr Opin Infect Dis* 20: 129–141.

Torres HA, Raad II, Kontoyiannis DP, 2003. Infections caused by *Fusarium* species. *J Chemother* 15 (Suppl 2): 28–35.

Tu EY, McCartney DL, Beatty RF, Springer KL, Levy J, Edward D, 2007. Successful treatment of resistant ocular fusariosis with posaconazole (SCH-56592). *Am J Ophthalmol* 143: 222–227.

Vagace JM, Sanz-Rodriguez C, Casado MS, Alonso N, Garcia-Dominguez M, de la Llana FG, Zarallo L, Fajardo M, Bajo R, 2007. Resolution of disseminated fusariosis in a child with acute leukemia treated withcombined antifungal therapy: A case report. *BMC Infect Dis* 7: 40.

Vemuganti GK, Garg P, Gopinathan U, Naduvilath TJ, John RK, Buddi R, Rao GN, 2002. Evaluation of agent and host factors in progression of mycotic keratitis: A histologic and microbiologic study of 167 corneal buttons. *Ophthalmology* 109: 1538–1546.

Winn RM, Gil-Lamaignere C, Maloukou A, Roilides E, EUROFUNG Network, 2003. Interactions of human phagocytes with moulds *Fusarium* spp. and *Verticillium nigrescens* possessing different pathogenicity. *Med Mycol* 41: 503–509.

Xie L, Zhai H, Shi W, Zhao J, Sun S, Zang X, 2008. Hyphal growth patterns and recurrence of fungal keratitis after lamellar keratoplasty. *Ophthalmology* 115: 983–987.

Young J-A, 2007. Section 7, XXXVIII Mycotic Infections in the Compromised Host, Table 7. Treatment of Infections Caused by *Fusarium* Species, *in*: Dale DC, Federman D (eds.), *ACP Medicine*, 3rd ed, BC Decker Publisher, Philadelphia.

Zhang N, O'Donnell K, Sutton DA, Nalim FA, Summerbell RC, Padhye AA, Geiser DM, 2006. Members of the *Fusarium solani* species complex that cause infections in both humans and plants are common in the environment. *J Clin Microbiol* 44: 2186–2190.

WEBSITE CITED

http://isolate.fusariumdb.org/welcome.php?a=intro.

QUESTIONS

The Answer Key to multiple choice questions may be found at the end of the book.

1. What morphologic marker is useful in identifying a mold as a *Fusarium* species?

A. Acute angle branching hyphae

B. Microconidia formed on small denticles in a floret arrangement on apically swollen conidiophores

C. Right angle branching hyphae

D. Sickle-shaped multicelled macroconidia

2. If a culture suspected of being a *Fusarium* species produces only microconidia, which other genus of hyaline molds should be considered in the differential diagnosis?

A. *Acremonium* species

B. *Chrysosporium* species

C. *Penicillium* species

D. *Trichoderma* species

3. Which *Fusarium* species is most frequently isolated from keratitis associated with contact lens use?

A. *F. oxysporum*

B. *F. proliferatum*

C. *F. solani*

D. *F.verticillioides*

4. Referring to fresh clinical isolates, what morphologic characteristics of *F. oxysporum* differentiate it from *F. solani*?

1. Colony reverse is colorless to dark purple.
2. Colony reverse is cream.
3. Microconidia are borne on short phialides.
4. Microconidia are produced from long simple or branched phialides.

Which are correct?

A. 1+3

B. 1+4

C. 2+3

D. 2+4

5. Which of the following is true about anatomic site isolates of *Fusarium* species from an immunocompromised patient?

1. Skin: If isolated from a lesion may indicate disseminated disease.
2. Skin: *Fusarium* species are common environmental molds so it is not significant.
3. Nail infections: *Fusarium* species are not dermatophytes so it is not significant.
4. Paronychia: Spread from the nail bed to the deep tissues may occur.

Which are correct?

A. 1+3

B. 1+4

C. 2+3

D. 2+4

6. Evidence from genotyping discovered that a geographically widespread clonal lineage was responsible for >70% of all clinical isolates of:

A. *F. oxysporum*

B. *F. proliferatum*

C. *F. solani*

D. *F. verticillioides*

7. What specimens would be useful in detection and recovery of *Fusarium* species in a patient with febrile neutropenia unresponsive to antibiotics? If an isolate produces only microconidia, what other fungal genus is in the differential diagnosis? In that case, how could *Fusarium* be identified to the species level?

8. The patient has febrile neutropenia and hematologic malignancy. What signs of *Fusarium* mycosis would tend to differentiate it from aspergillosis?

9. Recovery from neutropenia is key to success in treating disseminated *Fusarium* mycosis. What is the success in treating this mycosis with AmB monotherapy? What role is there for alternative primary therapy and for AmB combination therapy? Do any combinations raise the question of possible drug antagonisms?

10. The 2006 multistate outbreak of *Fusarium* keratitis associated with a defective contact lens disinfecting storage solution brought to the forefront information about the etiology, ecologic niche, and preventive steps to be taken. Of the two major species isolated in the outbreak, what is known about the prevalent genotypes of *F. solani* versus *F. oxysporum*? What is known about the ecologic niche of *Fusarium* species in urban indoor environments? Review the nine steps to take to prevent *Fusarium* keratitis in contact lens users.

11. What group is at greatest risk of developing disseminated *Fusarium* mycosis? In that group, what is the relative incidence of aspergillosis, *Fusarium* mycosis, and mucormycosis?

Chapter 16

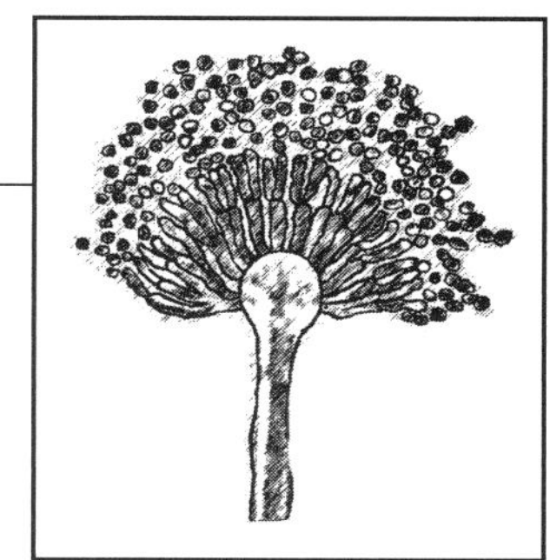

Pseudallescheria/Scedosporium Mycosis

16.1 *PSEUDALLESCHERIA/ SCEDOSPORIUM* MYCOSIS-AT-A-GLANCE

- *Introduction/Disease Definition.* Subcutaneous (including eumycotic mycetoma), sinopulmonary, and pulmonary-disseminated infections caused by the molds in the *Pseudallescheria boydii* species complex and other *Scedosporium* species.
- *Etiologic Agents.* Environmental molds: *Pseudallescheria boydii, Scedosporium apiospermum, S. prolificans*, and minor species.
- *Geographic Distribution/Ecologic Niche.* Worldwide agents of sinopulmonary and pulmonary-disseminated mycosis in immunocompromised hosts. Causes of eumycetoma in North America. These species are isolated from agricultural soil, sewage, and polluted water receiving runoff from livestock raising areas.
- *Epidemiology.* An emerging mycosis in immunocompromised hosts, third in prevalence behind aspergillosis and *Fusarium* mycosis. Approximately 1–27% (depending on study design) of opportunistic mold mycoses are caused by non-*Aspergillus* species molds.
- *Risk Groups/Factors.* Immune-normal host: occupational or recreational exposure to penetrating injuries, contact lens wearers, near-drowning accident victims, and cystic fibrosis patients. Immunocompromised host: patients with hematologic malignancy and stem cell and organ transplant recipients.
- *Transmission.* Traumatic implantation or inhalation of conidia, aspiration of conidia in water (near-drowning accidents).
- *Determinants of Pathogenicity.*
  - *Adventitious Sporulation.* Conidia produced in tissue by *S. apiospermum* and *S. prolificans* contribute to hematogenous spread.
  - *Cell Wall α-D-Glucan.* The cell wall α-D-glucan of conidia and hyphae is recognized by Toll-like receptor 2 (TLR2) leading to phagocytosis or endocytosis. *S. prolificans* glycoproteins on conidia bind to epithelial cells.
  - *Enzymes.* Wall-associated alkaline and acid phosphatases and a secreted metallopeptidase.
  - *Melanin. S. prolificans* is melanized, which protects the pathogen.
- *Clinical Forms.* Eumycetoma, fungal sinusitis, pneumonia, fungus ball in a preexisting pulmonary cavity, extrapulmonary dissemination, cerebral abscess. May be neurotropic with the CNS as a site of dissemination.
- *Veterinary Forms.* Dogs: keratitis, nasal granulomas, abdominal eumycetoma, multisystem disseminated disease. Equines: keratitis, nasal granulomas, hoof disease.
- *Therapy. Pseudallescheria/Scedosporium* species are less susceptible to AmB than *Aspergillus fumigatus*; voriconazole (VRC) is the preferred therapy; *S. prolificans* is refractory to most monotherapy but may be responsive to combination therapy.
- *Laboratory Detection, Recovery, and Identification.* Specimens for direct examination and culture: grains, skin scrapings, exudates, sputum, sinus drainage, invasive biopsy including drainage from closed cerebral abscess, and blood cultures. Growth is rapid on SDA. May demonstrate sexual reproduction:

Fundamental Medical Mycology, First Edition. By Errol Reiss, H. Jean Shadomy and G. Marshall Lyon, III.
© 2012 Wiley-Blackwell. Published 2012 by John Wiley & Sons, Inc.

cleistothecia with ascospores after 2–3 weeks incubation, or after transfer to cornmeal agar. Asexual reproduction is *enteroblastic* with single large *obovoid annelloconidia*. Sometimes conidiophores are in small clusters: synnemata. Histopathology: difficult to distinguish from *Aspergillus* species. Sometimes chlamydospores and conidia are produced in vivo; *S. prolificans* gives a positive Fontana-Masson staining reaction in tissue for melanin.

16.2 INTRODUCTION/DISEASE DEFINITION

The *Pseudallescheria boydii* species complex consists of the major pathogens *P. boydii* and *Scedosporium apiospermum*, which cause subcutaneous and opportunistic sinopulmonary mold mycosis. The related species, *S. prolificans*, a melanized mold, is also responsible for pulmonary and systemic infections. Clinical forms of disease include the following:

1. Subcutaneous infections. are caused by *Pseudallescheria* and *Scedosporium* species in immune-normal and immunocompromised persons worldwide in developed and developing countries where occupation, recreation, or accidents expose individuals to abrasions or penetrating injuries with thorns or splinters containing conidia. Eumycetoma is a special category of subcutaneous mycosis and is discussed separately in Chapter 19.

2. Respiratory forms. of *P. boydii* and *Scedosporium* mycosis occur in individuals who are not receiving immunosuppressive therapy:

- Allergic bronchopulmonary mycosis
- Cerebral abscess in near-drowning victims, following aspiration of a conidia–water suspension
- Chronic fungal sinusitis
- Colonization in the lungs of cystic fibrosis patients and of children with chronic suppurative pulmonary disease
- Fungus ball colonization of preexisting pulmonary cavities

In the immunocompromised patient:

- Sinopulmonary and extrapulmonary disseminated mycosis occur primarily in neutropenic or otherwise immunosuppressed patients with hematologic malignancies or solid organ or hematopoietic stem cell transplants (SCTs). Dissemination to the CNS is particularly refractory to treatment.

3. Other less common clinical forms. are arthritis, endocarditis, keratitis, and osteomyelitis affecting immune-normal and immunocompromised persons. The major clinical and mycologic problems of *P. boydii* species complex mycosis are as follows:

- They are emerging mycoses resulting from increased use of immunosuppressive regimens to treat cancer and to maintain stem cell and solid organ transplants.
- The causative fungi often are not recognized, thus failing to obtain an early diagnosis.
- New species of medical importance have come to light as a result of multilocus sequencing and *cladistic analysis* (e.g., *Scedosporium aurantiacum*).
- Knowledge of the species is important in guiding clinical treatment decisions because this genus is more refractory to antifungal agents than other hyaline molds. Because of this they may cause "breakthrough" infections in patients receiving prophylactic antimold therapy.
- *S. prolificans* is resistant to monotherapy with various antifungal agents.
- Knowledge is lacking of the pathobiology of *Pseudallescheria* and *Scedosporium* species because of insufficient research with these pathogens.

16.3 CASE PRESENTATIONS

Case Presentation 1. Subcutaneous Nodules on the Forearm of a 61-yr-old Man (Stur-Hofmann et al., 2011)

History

The patient, whose hobby is gardening, had a multiyear history of systemic corticosteroid therapy for chronic obstructive pulmonary disease. He developed diabetes mellitus and osteoporosis, the latter requiring spinal surgeries. One month before admission two slowly growing subcutaneous nodules appeared on his right forearm. He reported no pruritis, pain, or fever.

Physical and Laboratory Examination

Two brownish tumors, 30 mm in diameter, were present on the the right forearm. Biopsy and a PAS-stained section of one of these showed granulomas with multinucleate giant cells containing septate hyphae. Culture on SDA produced a white-gray mold which, on tease mount, showed branching hyphae with short terminal and lateral conidiophores bearing single ovoid conidia, indicative of *Scedosporium apiospermum*. The isolate was susceptible to VRC.

Therapy and Clinical Course

The right nodule was excised and the left nodule kept as an indicator of response to therapy. The patient received oral VRC with a loading dose of two 400 mg tablets on the first day, followed by 200 mg, 2×/day. Systemic corticosteroids were stopped and replaced by inhaler bronchodilator monotherapy. The patient experienced some nausea and photosensitivity, which subsided upon completion of therapy. After 6 weeks of VRC the lesion size was reduced by 90% and was undetectable clinically and by sonography after 3 months. Therapy was continued for another 3 months. No new lesions were observed at a follow-up clinic visit 5 months after cessation of therapy.

Comments

Penetrating injury from thorny plants or splinters during the patient's gardening hobby is a reasonable explanation for the initiation of this mycosis. Likely risk factors are his immunosuppressed status resulting from chronic systemic corticosteroid therapy and diabetes mellitus.This case illustrates that persons in developed countries can develop *Pseudallescheria/Scedosporium* mycosis and that it is not restricted to rural environments in developing countries.

Case Presentation 2. Invasive Pulmonary Infection in a 51-yr-old Woman with Breast Cancer (Strickland et al., 1998)

History

Three years after undergoing a left modified-radical mastectomy and four courses of doxorubicin and cyclophosphamide, this 51-yr-old woman developed a chest wall mass diagnosed as recurrent breast cancer. Radiation and chemotherapy did not prevent a recurrence 6 months later. The patient was conditioned for an autologous SCT but 5 days before the transplant she developed *C. difficile* diarrhea and was treated with metronidazole. Fever continued with neutropenia and further IV antibiotics were administered.

Laboratory Procedure

Blood cultures were positive for bacteremia.

Therapy and Clinical Course

She was placed on *TPN* and the SCT was conducted. Fever persisted and the patient was placed on IV liposomal AmB. Seven days post-transplant a chest radiograph revealed bilateral pleural effusions and a pericardial effusion. Faced with persistent fever, a thoracentesis was done but the fluid was negative for cytology and cultures. On the 14th post-transplant day she developed severe hypotension and respiratory failure requiring intubation. Catheter port and blood cultures grew a mold and despite an increase in AmB to 64 mg/day, she died on the 16th post-transplant day.

An autopsy revealed metastatic cancer. Numerous infarcts were present in the spleen, kidneys, and brain, which, when prepared for histology, showed septate hyphae. Blood, lung, and liver cultures grew a mold identified as *Scedosporium prolificans*, the same mold that was cultured during life on the 15th post-transplant day.

Diagnosis

The diagnosis was metastatic cancer and disseminated *Scedosporium prolificans* mycosis.

Comments

In the absence of cultures, empiric therapy with AmB seemed appropriate but was unsuccessful for treating this infection, probably because this mold has *decreased susceptibility* to the drug. VRC, alone or combined with terbinafine, may be more promising. (Please see Section 16.12, Therapy.)

Case Presentation 3. Mycotic Cerebral Aneurysm[1] in a Near-Drowned Child (Messori et al., 2002)

History

A previously healthy 3-yr-old girl was admitted to the pediatric ICU in coma after near-drowning in polluted water. She developed fever and was treated with broad-spectrum antibiotics.

Radiologic and Laboratory Procedures

Four weeks later neurologic examination revealed right hemiparesis and a CT scan showed right frontal and left frontotemporal lesions, consistent with abscesses. The left-sided lesion was drained and excised yielding, on culture, *Scedosporium apiospermum*.

[1]Cerebral aneurysm. *Definition*: a balloon-like outpouching or sac that arises from an artery around the brain. The aneurysm grows from a weakened wall of an artery. If the aneurysm bursts and bleeds into the space around the brain, this is called a subarachnoid hemorrhage. A subarachnoid hemorrhage is a catastrophic event that requires emergency medical treatment at a specialized medical center.

Diagnosis

The diagnosis was cerebral abscess due to *Scedosporium apiospermum*.

Therapy and Clinical Course

Intravenous ITC was started but despite treatment the CT scan showed progressing infection and new abscesses. On the 8th postoperative day, her neurologic condition deteriorated, associated with extensive cerebral hemorrhage and hydrocephalus. A contrast enhanced CT scan confirmed aneurysms of the basilar and right posterior cerebral arteries. Rapid deterioration and clinical brain death ensued and the patient expired. No autopsy was performed.

Comments

The case underlines the tendency of *Scedosporium apiospermum*, after aspiration of polluted water, to display a neurotropism and, once inside the CNS, to develop as a cerebral abscess. Knowledge of mycotic intracranial aneurysms is mostly derived from experience with aspergillosis where massive destruction of the elastic lamina of the arterial wall has been documented. *Scedosporium* also has an affinity for blood vessels, but few cases resulting in cerebral aneurysms have been described.

16.4 DIAGNOSIS

The clinical presentation is similar to that of other pulmonary or systemic mycoses caused by *hyaline* molds producing septate hyphae. *Pseudallescheria* and *Scedosporium* are distinctive in that when they cause eumycetoma they form "grains," compact masses of white fungal hyphae, which are discharged from sinus tracts. When causing pulmonary or other systemic infection, growth in tissue is characterized by branching septate hyphae. *P. boydii* is identified morphologically in culture when it is shown to form sexual fruiting bodies, cleistothecia. *S. apiospermum* may grow in vitro in both a typical mold with single obovoid conidia produced from annellides, or as bundles of hyphae, the synnemata. *S. prolificans*, a related species, is melanized. Several genetic species of the *Pseudallescheria* species complex are differentiated by genetic identification, also aided by biochemical tests. These are discussed in Section 16.4, Etiologic Agents, and Section 16.13, Laboratory Detection, Recovery, and Identification. The differential diagnosis includes other mycoses caused by hyaline molds (e.g., aspergillosis, *Fusarium* mycosis).

16.5 ETIOLOGIC AGENTS

Mycologic Data

Pseudallescheria boydii (Shear) McGinnis, AA Padhye & L Ajello, *Mycotaxon* 14: 97 (1982) *P. boydii* (teleomorph) is a single species genus. The sexual form including cleistothecia can be observed when clinical isolates are allowed to grow on SDA or PDA for 2–3 weeks.

Scedosporium apiospermum Sacc. ex Castell. & Chalm., *Manual of Tropical Medicine* (London): 1122 (1919).

Scedosporium prolificans (Hennebert & B.G. Desai) E. Guého & de Hoog *Journal de Mycologie Médicale* 1: 8 (1991) *Scedosporium inflatum* (synonym of *S. prolificans*).

Position in classification: *Ascomycota, Euascomycetes, Microascales, Microascaceae, Pseudallescheria* (genus), *Scedosporium* (genus). *Pseudallescheria boydii* and *Scedosporium* species are classed in the *Ascomycota* in the order *Microascales*—characterized by *cleistothecia*-producing molds. Within that order they occur in a single family, the *Microascaceae. P. boydii* is a *homothallic* fungus.

Scedosporium apiospermum is heterothallic and compatible mating types produce fertile ascomata typical of the genus *Pseudallescheria*. The teleomorph *is Pseudallescheria apiosperma* (Gilgado et al., 2009).

Scedosporium prolificans is mitosporic; no sexual stage is known at this time.

Pseudallescheria and *Scedosporium* species are agents of eumycetoma and opportunistic mycoses in the immunocompromised host, as well as other clinical forms (please see Section 16.10, Clinical Forms). *P. boydii, S. apiospermum*, and the related species, *S. prolificans*, are the major etiologic agents. *P. boydii* and *S. apiospermum*, formerly considered as the teleomorph and anamorph, respectively, of *P. boydii* are now regarded as separate species. The evidence for this is DNA sequence analysis of four loci, ß tubulin (two loci), calmodulin gene (one locus), and the ITS region of rDNA (Gilgado et al., 2008) (Fig. 16.1).

How could the laboratory determine if an isolate is S. apiospermum *or* P. boydii *in the absence of formation of the ascoma, which may take 2–3 weeks*? *P. boydii* assimilates ribose, whereas *S. apiospermum* does not, as indicated in Fig. 16.2. Reference laboratories with access to multilocus sequence typing can make this determination, but the timeliness of such a procedure is, at present, not relevant for clinical treatment decisions.

Closely related to *P. boydii* are the minor species *P. angusta, P. ellipsoidea*, and *P. fusoidea*. These genetic

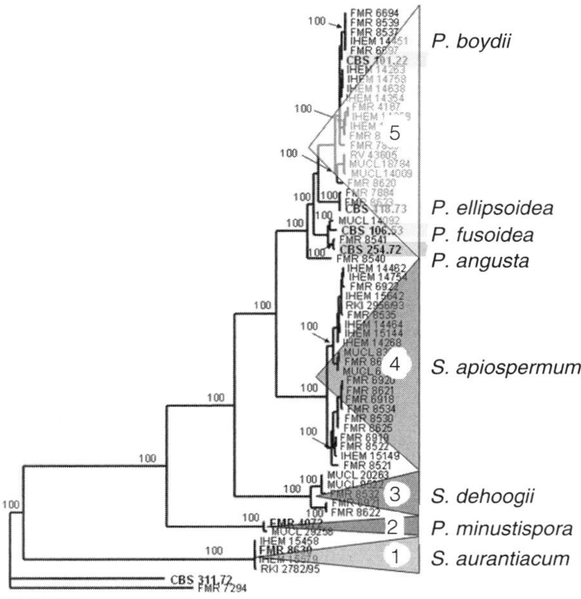

Figure 16.1 Dendrogram of relationships in the *Pseudallescheria* species complex, determined using *multilocus sequence typing (MLST)*. Clades are numbered 1–5, according to the type of sexual or asexual reproduction *Clade 1* (*S. aurantiacum*) is a clonal population of asexual reproducing strains with near identical MLST genotypes. *Clade 2* (*P. minutispora*) is minor meiosporic species. *Clade 3* (*S. dehoogii*) is a small cluster of clonal genotypes, asexually reproducing, with little genetic variation. *Clade 4* (*S. apiospermum*) is a large diverse group with genetic variation suggestive of a sexually recombining population. *Clade* 5, the major *Pseudallescheria* clade, is genetically distinct from *S. apiospermum*. It contains four described species, of which *P. boydii* is the major and most clinically important; *P. ellipsoidea, P. fusoidea*, and *P. angusta* are uncommon. Each of these four species is an inbreeding population with little intraspecific genetic variation. *Source:* Used with permission from Dr. Josep Guarro, Facultat de Medicina i Ciències de la Salut, Universitat Rovira i Virgili, Tarragona, Spain (Guarro and Kano, 2007).

species are distinguishable *only* by sequence analysis, for example, of the *TUB* region of the tubulin gene. Other species now recognized are *S. aurantiacum, S. dehoogii*, and *P. minutispora. S. aurantiacum* may be separated from *P. boydii* and *S. apiospermum* because it is thermotolerant at 45°C, and because the colony reverse is in shades of yellow due to a diffusible pigment.

P. boydii and *S. apiospermum* cultures take on a brown color in vitro on PDA (dark gray to brownish gray on SDA) because of their pigmented conidia *but are not pigmented when infecting the human host*.

The neighboring species, *S. prolificans*, is noted for its decreased susceptibility to antifungal agents. It is morphologically distinct from *S. apiospermum* in that it has an inflated base in the conidiophore and is a melanized fungus giving a positive histopathologic reaction for melanin with the Fontana–Masson stain. *S. prolificans* also may be differentiated because it does not assimilate ribitol, L-arabinitol, and sucrose, all of which are assimilated by *S. apiospermum* and *P. boydii. S. prolificans*, in the immunocompromised host, causes rapidly progressing disseminated disease with a high case/fatality ratio. This mold has in unusual circumstances been responsible for airborne transmission in a hematology–oncology ward (Guerrero et al., 2001). (Please see Section 16.7, Epidemiology.)

What is the clinical relevance of the minor species in the Pseudallescheria *species complex?* Their true clinical importance will emerge as more data accumulate on their prevalence in clinical specimens. Significantly, *S. aurantiacum* has been isolated from cystic fibrosis patients and from a large collection of clinical isolates of *Scedosporium* species from Australia (Delhaes et al., 2008). Compared with *P. boydii* and *S. apiospermum* isolates, *S. aurantiacum* is more pathogenic for mice, and less susceptible to AmB, ITC, and VRC (Alastruey-Izquierdo et al., 2007; Gilgado et al., 2009).

Scedosporium isolates in an Australian collection were identified to the species level (Delhaes et al., 2008). *S. prolificans* accounted for 75 patient episodes (83 of 146 isolates; 56.9%), *S. apiospermum* for 25 (33 isolates; 22.6%) episodes, and *S. aurantiacum* for 23 episodes (30 isolates; 20.6%). ITS sequencing confirmed the identity of *S. aurantiacum* because there is a 5–10% sequence variation in the ITS region between *S. aurantiacum* and *S. apiospermum. S. aurantiacum* is prevalent in this collection and may have more general significance as DNA-based identification is applied to other collections. As well, there are morphologic and physiologic characters that may aid in identification (Table 16.1).

Clinical forms and susceptibility profiles of *S. aurantiacum* mycosis are similar to those of *S. apiospermum. S. aurantiacum* caused osteomyelitis, invasive sinusitis, keratitis, and pneumonia; but no deaths resulted. Invasive disease with *S. aurantiacum* occurred in six patients: two were immunosuppressed, the remainder had chronic lung disease. Low VRC and PSC MICs against *S. aurantiacum* suggest these agents may be effective therapies (Heath et al., 2009. About half of *S. apiospermum* and *S. aurantiacum* isolates were from the respiratory tract, including the lungs, compared with 20% for *S. prolificans*. Conversely, *S. prolificans* accounted for *all* blood isolates; of the remaining ones 57.2% of isolates were from skin or soft tissue and 66.7% were from the eye.

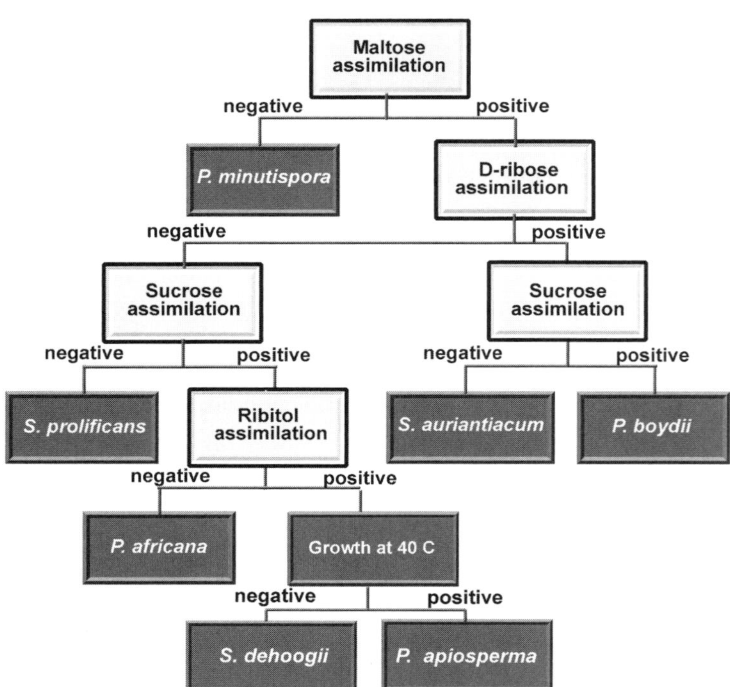

Figure 16.2 Physiologic separation of *Pseudallescheria* species complex members (*P. boydii, Scedosporium apiospermum, S. aurantiacum*) and *S. prolificans. P. boydii* may be separated from *S. apiospermum* by the former species' positive assimilation of D-ribose. (The teleomorph of *S. apiospermum* is *P. apiosperma*.) *Source:* Used with permission from Dr. Josep Guarro Facultat de Medicina i Ciències de la Salut, Universitat Rovira i Virgili, Tarragona, Spain (Guarro and Kano, 2007).

Table 16.1 Morphologic and Physiologic Characteristics of Species Identified by Molecular Phylogenetic Analysis in the *Pseudallescheria/Scedosporium* Group

Species	Diffusible pigment on PDA at 25°C	Maximum growth temperature (°C)	Teleomorph development of ascomata	Syanamorph	
				Shape of conidiogenous cells	Most common shape of conidia borne on vegetative hyphae
P. boydii	–	40	±	Cylindrical	Globose
P. ellipsoidea	–	40	+	Cylindrical	Ellipsoidal
S. aurantiacum	+ (yellow)	45	–	Cylindrical or slightly flaskshaped	Obovoid
S. prolificans	–	40	–	Flask shaped	Globose

Source: Gilgado et al. (2005). Adapted with permission from American Society for Microbiology.

16.6 GEOGRAPHIC DISTRIBUTION/ECOLOGIC NICHE

The *P. boydii* complex is found in temperate climates with an ecologic niche in moist, poorly aerated soil—sediments of polluted ditches, ponds, manure, and sewage. These fungi are salt-tolerant and have been recovered from brackish or salt water, sewage, soil, swamps, coastal tidelands, chicken manure, bird guano, and cattle and bat feces. Rare in unpolluted environments, *P. boydii* and *Scedosporium* species are normally isolated from sewage, ponds, and lakes with abundant waterfowl, and runoff containing poultry or cattle manure. Recovery from the environment is related to organic pollution, including that from human sources. Their production of slimy conidia and ascospores infers poor airborne dispersal in favor of water or adherence to arthropods. Increased water pollution and higher temperatures create environmental conditions favorable for the emergence of these species (de Hoog et al., 1994; Ranier and de Hoog, 2006).

S. apiospermum infections occur worldwide and cases are evenly geographically distributed, whereas *S. prolificans* mycoses appear to be more geographically prevalent in Australia, northern Spain, and the southern United States than *S. apiospermum* mycoses. *S. prolificans* is isolated from soil and animals, including cats and horses, and seems more restricted than the *P. boydii* complex to soils and potted plants.

Prevalence of *Scedosporium* infections is relatively low in most geographic areas. Unlike *A. fumigatus*, which is frequently isolated from ambient air, including indoor air, *P. boydii* and *Scedosporium* species are typically not found in such samples (O'Bryan, 2005). *Scedosporium* spp. have rarely been isolated from hospital air or from indoor or outdoor surface samples, raising questions of the mode of transmission (please see Section 16.7, Epidemiology). There are, however, reports of hospital-based clusters in hematology–oncology wards (Guerrero et al., 2001) during periods of building construction or renovation (Alvarez et al., 1995). *Aspergillus* conidia are in greater numbers and grow faster, overgrowing *Scedosporium* species such that selective media with antifungal agents are needed to recover *Scedosporium* from the environment.

16.7 EPIDEMIOLOGY

Invasive *Pseudallescheria* and *S. apiospermum* infections in the immunocompromised host are rare, sporadic, and associated with a high mortality.

S. prolificans mycosis is even more rare than that caused by *S. apiospermum*. Among leukemia patients, disseminated *S. prolificans* infection during induction therapy for AML was reported in Australia and in Spain. Cases in the world literature involving *P. boydii* and *Scedosporium* species revealed 435 cases in the period 1940–2007 (Cortez et al., 2008).

The originating locality was determined for 370 *Scedosporium* species identified by the Fungus Testing Laboratory, San Antonio, Texas, during the period 2000–2007. Most cases originated from California and Texas (Cortez et al., 2008). Moderate numbers of isolates originated in the Midwestern states of Wisconsin and Ohio. Few cases were from New England, the mountain West, and desert Southwest. The Southeastern United States had few cases except for Florida. The anatomic sites for these 370 *Scedosporium* isolates included thorax/lungs, 222; sinuses, 31; bones/joints, 31; eyes, 25; blood, 12; and CNS, 11.

The annual incidence of invasive scedosporiosis in an Australian study was 1 per million population (Heath et al., 2009). Underlying malignancy (40%) and organ transplantation (29%) were common in patients with invasive infections, but 11.3% had no comorbidity. Independent predictors of invasive disease were stem cell transplantation, leukemia, neutropenia, corticosteroid use, and diabetes. Twenty-four percent of patients had disseminated disease. Nine of 25 (36%) lung transplant patients had confirmed *Scedosporium* mycosis, 4 with dissemination. Crude 90 day mortality and attributable mortality in patients with invasive disease were both 30.6%. Independent predictors of death were *S. prolificans* and leukemia.

Immunocompromised Patients

Hematology–Oncology Patients and SCT Recipients

Pseudallescheria and *Scedosporium* mycoses occur in neutropenic cancer patients and in heavily immunosuppressed SCT recipients. The most common anatomic sites involved are the lungs and soft tissues, with occasional extension to bone. At a Seattle, Washington, cancer center, among SCT recipients treated in a span of 15 years from 1985 to 1999, 10 SCT patients developed *Scedosporium* mycosis (Panackal and Marr, 2004). Most cases occurred during the first 30 days post-transplant in the pre-engraftment period, with a fatal outcome in all 10 patients within 1 month following the diagnosis.

More recently of 22 SCT recipients with *Scedosporium* species infections gleaned from the world literature between 1985 and 2003, the overall mortality rate was 68% (15 of 22) (61.5% for patients with *S. apiospermum* infection, and 77.8% for patients with *S. prolificans* infection) (Husain et al., 2005). Mortality rates differed significantly between the patients treated with AmB, ITC, or VRC. Considering AmB as the comparison treatment, receipt of VRC was associated with a strong trend toward better survival. The outcome of ITC therapy, on the other hand, was not significantly different from AmB.

Oportunistic invasive mold infections in allogeneic SCT recipients, including *Scedosporium* mycosis, now occur later in the post-transplant period and are associated with potent immunosuppressive therapy for chronic GVHD.

A review of 25 *Scedosporium* infections from 1989 to 2006 at a major cancer center in Houston, Texas, found all occurred in patients with a hematologic malignancy (Lamaris et al., 2006). *S. apiospermum* and *S. prolificans* were the causative agents in 21 and 4 patients, respectively. The incidence of *Scedosporium* infection increased from 0.82 cases per 100,000 patient-inpatient days (in 1993–1998) to 1.33 cases per 100,000 patient-inpatient days (in 1999–2005). *S. prolificans* infection occurred only after 2000. Dissemination occurred in 16 patients (64%). The 12 week mortality rates were 70% and 100% for *S. apiospermum* and *S. prolificans* infection, respectively. Risk factors in the immunosuppressed population were lymphopenia, neutropenia, steroid treatment, and breakthrough infection with 74% of patients receiving AmB treatment. All *S. prolificans* infections occurred in the setting of neutropenia compared with 43% of *S. apiospermum* infection. Pneumonia preceded dissemination in most patients. Colonization with *Scedosporium* species was also identified in 26 patients.

Of all fungal blood cultures reported during 1996–2002 at a major U.S. cancer treatment center, molds other than *Aspergillus* or *Fusarium* accounted for 1% of the total or 44 positive cultures (Lionakis et al., 2004). Five of these cultures were identified as *Scedosporium* species; four were from proven invasive infections in patients with hematologic malignancy and/or who received allogeneic SCTs. *Scedosporium* species mycoses are rare, even in the most susceptible hosts; they can give positive blood cultures (providing good evidence of invasive disease) and, because they are less susceptible to all classes of antifungal agents than other hyaline molds, they have a high associated mortality.

Solid Organ Transplant Recipients

A transplant center in Pittsburgh, Pennsylvania, recorded 7 cases of *Scedosporium* mycosis among solid organ transplant recipients in the 1987–1999 period with 16 additional cases gleaned from the English literature during 1976–1999 (Castiglioni et al., 2002). The total of 23 patients included liver (4), kidney (8), heart (8), lung (2), and heart–lung (1) transplant recipients. *Scedosporium* mycosis was diagnosed in a range of 0.4–156 months after transplant. The clinical presentation included disseminated disease (8), skin lesions (3), lung disease (5), endophthalmitis (1), meningitis (1), cerebral abscess (3), mycotic aneurysm (1), and sinusitis (1). Almost half of the patients had CNS involvement. Of 22 patients with known outcome, 16 (72.7%) died including all with disseminated *Scedosporium* mycosis. Five of six who survived had infections localized to the skin, fungus ball in the nasal sinus, or a solitary pulmonary nodule. A patient with a brain abscess was successfully treated with VRC and surgical drainage. Three patients had airway colonization only, and on ITC prophylaxis, did not progress to invasive disease.

Of 330 lung and heart–lung transplants conducted at an Australian center between 1986 and 1999, 7 patients with abnormal airways developed pulmonary *Scedosporium* mycosis (Tamm et al., 2001). Of these, 5 of 7 were receiving ITC suppressive therapy because of *Aspergillus* respiratory cultures. Three of these 7 patients survived for more than 3 years post-transplantation, during which time *Scedosporium* could be isolated from their respiratory secretions; that is, they were persistently infected.

A multicenter U.S. study found 13 cases of *Scedosporium* mycosis following heart, liver, kidney, kidney/pancreas, or small bowel transplants during the period 1999–2003 (Husain et al., 2005). Non-*Aspergillus* species molds accounted for 27% of these infections including 3 patients who were infected with *S. apiospermum* and one with *S. prolificans*. An additional 44 patients who were solid organ transplant recipients and who developed invasive *Scedosporium* mycoses were identified from the English literature during 1985–2003. *Scedosporium* mycoses accounted for ~25% of the non-*Aspergillus* mold mycoses in solid organ transplant recipients, occurring later in the post-transplant period as a complication of immunosuppressive therapy for GVHD. The mortality rate for *Scedosporium* mycoses in solid organ transplant patients was estimated at 58%, with *S. prolificans* mycosis more likely to have a fatal outcome.

VRC is used as suppressive therapy in patients colonized with *S. apiospermum* before undergoing lung transplantation. Colonization of the recipient may be regarded as an indication for double-lung transplantation, because *S. apiospermum* in the native lung can disseminate in the post-transplant period (Sahi et al., 2007).

Pseudallescheria or *Scedosporium* Mycosis in Persons Who Are Not Deliberately Immunosuppressed

- *Atopy.* This category of patient may develop chronic fungal sinusitis or allergic bronchopulmonary mycosis. The *Pseudallescheria* species complex is involved in a small minority of patients with this clinical form.
- *Cystic Fibrosis.* Local pulmonary colonization with *Pseudallescheria* species complex occurs in an estimated 8.6% of cystic fibrosis patients, second in incidence to *Aspergillus* species among the molds. *S. apiospermum* species colonizes the respiratory tract of up to 10% of patients with cystic fibrosis and children with chronic suppurative lung disease.[2]
- *Fungus Ball.* A fungus ball consisting of *Pseudallescheria* or *Scedosporium* hyphae may develop in a preexisting pulmonary cavity or cyst (Hoshino et al. 2007). Members of the *Pseudallescheria* species complex are a rare cause of fungal sinusitis. Approximately 25 such cases were reported in the world literature (Bates and Mims, 2006).
- *Eumycetoma. P. boydii* white grain eumycetomas occur mostly in temperate zones and are responsible for 10% of reported cases. (Please see Chapter 19).
- *Near-Drowning.* Persons rescued from near-drowning in polluted waters may have aspirated *Pseudallescheria* or *Scedosporium* conidia, which, after hematogenous dissemination, can manifest as cerebral abscess or other form of CNS involvement. Within the two decades 1997–2007, 21 cases were

[2] *Definition*: Usually occurring in children, chronic suppurative lung disease is an excessively prolonged moist cough without high resolution CT scan of the chest evidence of bronchiectasis (Chang et al., 2008).

reported (reviewed by Cortez et al., 2008). Survival in these patients has been poor because of delayed diagnosis and, until the advent of extended spectrum azoles, lack of efficacious drugs with good CNS penetration. Significantly, *S. prolificans* is not associated with *Scedosporium* mycosis in near-drowning victims, indicating it does not occupy the same ecologic niche in polluted waters as *S. apiospermum*.

- *Subcutaneous Cyst (Nonmycetoma).* Please see Section 16.3 Case Presentation 1. This clinical form does not demonstrate the typical triad of tumefaction and draining sinus tracts with grains that is associated with eumycetoma.

16.7.1 Risk Groups/Factors

Both immunocompetent and immunocompromised patients are at risk. Immunocompetent persons have contracted *P. boydii* infections after traumatic implantation (subcutaneous cyst, mycetoma, or keratitis) and after aspiration of polluted water in near-drowning accidents. Immunocompromised persons at risk include cancer patients, recipients of solid organ or SCT transplants whose induction therapy with systemic corticosteroids and cytotoxic drugs induces prolonged neutropenia, or who, in the month after engraftment, must be immunosuppressed to treat GVHD.

16.8 TRANSMISSION

Two major routes are traumatic implantation and inhalation. In both instances conidia are the infectious propagules. A minor route of infection is aspiration of conidia in water, the result of near-drowning accidents.

16.9 DETERMINANTS OF PATHOGENICITY

16.9.1 Host Factors

Pulmonary Pseudallescheria *Species Mycosis*

Conidia alight on the nasal mucosa or are inhaled into the lungs. If not cleared by the mucociliary elevator, or phagocytosed and killed by alveolar macrophages, the conidia germinate, triggering an influx of PMNs which, using oxidative and nonoxidative microbicidal processes, phagocytose and kill the *germlings* and residual conidia. Individuals with impaired host responses resulting from immunosuppressive therapy, cystic fibrosis, chronic suppurative lung disease, or preexisting pulmonary cavities may be unable to eradicate the fungus. A state of colonization of the bronchial tree or fungus ball formation may ensue.

Where there is prolonged neutropenia, or intensive immunosuppression to control chronic GVHD, hyphae may invade the lung parenchyma, demonstrating an angioinvasive tendency to cause hemorrhage, thrombosis, and infarction. If unchecked at this stage by timely therapeutic intervention, hematogenous spread can result in disseminated disease. Cerebral abscess is a dire form that may arise via hematogenous dissemination, or related to aspiration of conidia owing to a near-drowning accident, or by direct extension in the setting of invasive fungal sinusitis (reviewed in Cortez et al., 2008).

Host factors that augment PMN responses in *Pseudallescheria* species mycosis are IFN-γ, GM-CSF (Gil-Lamaignere et al., 2005), and interleukin-15 (IL-15). IL-15 enhanced PMN-induced hyphal damage of both *Fusarium* spp. and *S. prolificans*, but not *S. apiospermum*. IL-15 also enhanced the PMN oxidative respiratory burst in response to *S. prolificans* (Winn et al., 2005).

16.9.2 Microbial Factors

Cell Wall Constituents

The α-D-glucan of *P. boydii* is located on the surface of conidia and the hyphal cell wall. It is a branched polysaccharide consisting of linear α (1→4)-D-glucopyranosyl residues substituted at position 6 with α-D-glucopyranosyl branches. The α-D-glucan acts as a PAMP (pathogen-associated molecular pattern) that is recognized by Toll-like receptor 2 (TLR2) leading to the phagocytosis of conidia (Bittencourt et al., 2006).

α-D-Glucan also induced cytokine secretion (TNF-α) by murine macrophages and both TNF-α and IL-12 secretion by dendritic cells also involving TLR2[3] and routing the immune response along the protective Th1 limb. Enzymatic removal of α-D-glucan from *P. boydii* conidia showed that it is essential to conidial phagocytosis by macrophages.

S. prolificans mycelial glycoproteins contain a hexasaccharide consisting of α-linked rhamnotriose units linked α-(1→3) to a mannobiose (Barreto-Bergter et al., 2008). The reducing mannosyl residue was substituted at position 6 with ß-D-galactopyranosyl-α-D-glucose. The rhamnomannan on *P. boydii* conidia acts as a ligand capable of binding to a receptor on the surface of an epithelial cell line (HEp-2 cells) leading to the endocytosis of the conidia (Pinto et al., 2004). Thus, it has the property of an invasin.

[3]The reader will recall that TLR activation initiates a signaling cascade through a pathway shared by IL-1R via adaptor protein MyD88, leading to NFκB transcription factor activation and the induction of proinflammatory cytokines.

Adventitious Sporulation

Development of conidiophores and conidia in tissue is considered to be a factor responsible for the high yield of *S. prolificans* from blood cultures (reviewed in Cortez et al., 2008). This type of in vivo sporulation is also observed in *S. apiospermum*-infected tissues and bronchial washings (Walts, 2000).

Enzymes

Cell-wall-associated alkaline and acid phosphatases of *P. boydii* are characterized but their role in pathogenesis is unknown (Kiffer-Moreira et al., 2007). *P. boydii* secretes a 28 kDa metallopeptidase (Silva et al., 2006).

Melanin

Without doubt, *P. boydii* and *S. apiospermum* are hyaline molds. Their conidia, however, are light brown. *S. prolificans*, in contrast, has olive-gray-to-black mycelia and is likely to be melanized. Melanin confers protection to the pathogen.

Certain aspects of the pathogenesis of the *Pseudallescheria* species complex have been elucidated. Among these are the structure–function of conidia and hyphal cell surface carbohydrates, which appear to act as invasins, the secretion of a metallopeptidase, and the presence of surface-exposed phosphatases. The importance of adventitious sporulation in the *P. boydii* species complex and the involvement of melanin in *S. prolificans* infections require more validation with microbiologic and histopathologic methods.

16.10 CLINICAL FORMS

Eumycetoma

Pseudallescheria boydii and *Scedosporium* species are the major causative agents of eumycetoma in North America ("white" and "yellow" grain mycetoma). Please see Chapter 19.

Mycotic Keratitis

Scedosporium species keratitis is rare but may be the most common site of infection by this species in immune-normal persons. *Scedosporium* species may cause post-traumatic mycotic keratitis, also including infection following laser in situ keratomileusis (LASIK). A search of the world literature assembled a panel of 28 such cases (Wu et al., 2002). The majority were the result of foreign body injury. Corneal ulcers can develop rapidly over a period of less than 1 week after an accidental eye injury. The clinical response to ophthalmic surgery and antifungal therapy can be slow. Treatment consists of topical natamycin, in combination with topical miconazole and local or systemic AmB. Although *P. boydii* and *Scedosporium* are innately less susceptible to AmB in vitro, it has been used with clinical success (Wu et al., 2002). Systemic ITC therapy also has been used with success to treat deep keratitis (D'Hondt et al., 2000).

Respiratory Tract Involvement

Airway Colonization

Preformed cavities (e.g., those occurring in tuberculosis, sarcoidosis, cystic fibrosis, poorly draining nasal sinuses) may provide a niche for persistent colonization or fungus ball formation by the *Pseudallescheria* species complex. These species may uncommonly be involved as agents of allergic bronchopulmonary mycosis. Fungal respiratory colonization is encountered in cystic fibrosis patients. *Pseudallescheria/Scedosporium* was isolated from respiratory secretions in 11 out of 128 patients (8.6%), second only to *Aspergillus* species as a cause of mold colonization in cystic fibrosis patients (Cimon et al., 2000).

Fungus Ball in the Lung

A lung transplant recipient with native lung mycetoma developed pneumonia and mycetoma in both lungs (Castiglioni et al., 2002). Two of 13 solid organ transplant recipients who developed *Pseudallescheria* or *Scedosporium* infections had evidence of fungus ball in the lung (reviewed by Panackal and Marr, 2004).

Fungus Ball in the Nasal or Paranasal Sinus

Invasive fungal sinusitis was observed in a liver transplant recipient, requiring surgical debridement and *S. apiospermum* was recovered in culture of the sinus aspirate (Castiglioni et al., 2002). Fungal sinusitis may be chronic, in the immune-normal host, or invasive in immunocompromised patients. In the former, nasal sinus surgery may be curative, but in the latter a surgical approach combined with antifungal therapy is necessary.

Sinopulmonary Pseudallescheria *Complex Mycosis*

This occurs almost exclusively in the immunocompromised host.[4] For example, patients treated for

[4]Except please see below: "Pulmonary-Disseminated *Pseudallescheria* Complex Mycosis in Immune-normal individuals: Victims of Near-Drowning Accidents."

hematologic malignancy or to maintain stem cell or solid organ transplants are at risk during periods of febrile neutropenia unresponsive to antibiotics, and also later in the post-transplant period concomitant with the use of strong immunosuppressants to treat chronic GVHD. The symptoms include fever, dyspnea, cough, sputum production including hemoptysis, pleuritic chest pain, diffuse pulmonary infiltrates, nodules, and bronchopneumonia. In the absence of prompt and effective therapy, the disease may run a fatal course in 3–4 weeks after the first appearance of symptoms.

Achieving successful therapy is difficult in the hematology–oncology patient who is persistently neutropenic. This is especially so in *S. prolificans* mycosis. The empiric use of AmB may not be appropriate for *Pseudallescheria* or *Scedosporium* species because of their innately reduced susceptibility to the drug. Surgical debridement and therapy with VRC is the current standard of care. Recovery from neutropenia is an important factor. The use of GM-CSF or granulocyte infusions may be useful adjunctive therapies.

Extrapulmonary Dissemination

Skin lesions appear as a sign of hematogenous dissemination and become necrotic. Extrapulmonary dissemination may result in cerebral abscess, meningitis, endocarditis, ophthalmitis, or *cellulitis*. *Pseudallescheria/Scedosporium* species are neurotropic opportunists with a predilection for the CNS as a site of dissemination (Rainer and de Hoog, 2006). CNS signs of headache and neurologic deficits indicate the possibility of cerebral abscess. There is a more frequent occurrence of cerebral abscess as a site of dissemination of *P. boydii* and *Scedosporium* species, in comparison with other molds (O'Bryan, 2005). This site of dissemination is common in the immunocompromised patient and in the survivor of a near-drowning accident. Fungemia is detected by blood culture in approximately 40% of patients with hematogenously disseminated *S. prolificans* species mycosis, but in fewer than 5% of patients with *S. apiospermum* mycosis (reviewed in Cortez et al., 2008) *S. prolificans* is the most common cause of disseminated mycosis caused by a melanized mold (phaeohyphomycosis) (Revankar et al., 2002). Immunocompromise is the major risk factor but cases also are reported in immune-normal patients. The outcome of antifungal therapy remains poor, with an overall mortality rate of 79%.

Pseudallescheria and *Scedosporium* species are *angioinvasive*, resulting in tissue necrosis. Postmortem examinations have found histopathologic evidence of mycotic emboli and necrosis in the lung and other internal organs.

Pulmonary-Disseminated Pseudallescheria *Complex Mycosis in Immune-Normal Individuals: Victims of Near-Drowning Accidents*

Immune-normal individuals who are victims of near-drowning accidents can develop pulmonary and pulmonary-disseminated *Pseudallescheria* and *Scedosporium* mycosis after aspiration of polluted water containing the organism: for example, polluted ponds or swine waste retention ponds. The neurotropic tendency of *Pseudallescheria* and *Scedosporium* is underlined by the association of aspiration of polluted water and the subsequent development of cerebral abscess. Once seeded into the bloodstream these fungi tend to develop in the CNS. A table summarizing cases of cerebral abscess with these fungi following near-drowning accidents is found in Wilichowski et al. (1996). The differential diagnosis for infections following near-drowning in polluted waters includes *Aspergillus* species, *Pseudallescheria, Scedosporium*, or the amebae *Acanthamoeba*, and *Naegleria*.

16.11 VETERINARY FORMS

Canine

Pseudallescheria/Scedosporium spp. cause keratitis, abdominal eumycetoma, and multisystem disseminated disease in dogs. Canine nasal granuloma due to *S. apiospermum* presented as chronic, bilateral nasal discharge unresponsive to antibiotics (Cabañes et al., 1998). Biopsy specimens contained granules with numerous hyaline to pale brown hyphae. The dog responded to systemic ketoconazole. German shepherd dogs are prone to develop invasive fungal infections including those due to *Pseudallescheria/Scedosporium* species. *P. boydii* caused disease that disseminated from the abdominal cavity to the bones, joints, eye, and kidneys (Elad et al., 2010). Another German shepherd dog developed discospondylitis and osteomyelitis caused by *S. apiospermum* (Hugnet et al., 2009).

Equine

Pseudallescheria/Scedosporium species cause equine keratitis (Friedman et al., 1989) and are an uncommon fungal cause of nasal granuloma (Brearley et al., 1986). White line disease in horses was linked to *Pseudallescheria/Scedosporium* species. This is a destructive disease in which the firm horn tissue of the hoof becomes crumbly, "cottage-cheese-like," and deteriorates. Ten thoroughbred horses were diagnosed with onychomycosis and seven of these showed severe inner hoof wall fissures

(Kuwano et al., 1998). *S. apiospermum* or *P. boydii* was isolated from the fissure cavities, the terminal horn of the white line, and the terminal horn-like laminae of the metaplastic white line-like tissue.

Other Animal Species

A stranded male northern elephant seal pup undergoing rehabilitation was found severely ill (Haulena et al., 2002). The animal was euthanized and *S. apiospermum* was cultured from brain, kidney, and subcutaneous tissue.

16.12 THERAPY

Pseudallescheria boydii, Scedosporium apiospermum, and *S. prolificans* have reduced in vitro susceptibility to AmB, and a suspicion of these fungi is grounds to alter therapy. Surgical debridement and administration of effective antifungal agents (e.g., VRC) has provided better treatment, even including control of disseminated infections. *S. prolificans* infections are less susceptible to treatment with VRC and remain challenging.

In Vitro Susceptibilities

Scedosporium apiospermum

VRC and other mold-active triazoles[5] exert good in vitro activity; the MIC_{50} of VRC is 0.25 μg/mL. Little or no in vitro activity against *S. apiospermum* was observed with the allylamine terbinafine or with flucytosine. The MIC_{50} observed with AmB of 4 μg/mL suggests reduced susceptibility. Caspofungin (CASF) and anidulafungin, among the echinocandins, have good in vitro activity against *S. apiospermum*. Combinations of antifungal agents tested against *S. apiospermum* show a low degree of in vitro synergism, and more additive effects for combinations of VRC and micafungin, and AmB and micafungin (reviewed in Cortez et al., 2008).

Scedosporium prolificans

Regardless of the method used, in vitro susceptibility of *S. prolificans* to various classes of antifungal agents including polyenes, triazoles, allylamines, echinocandins, and the pyrimidine flucytosine demonstrate uniformly decreased susceptibility with MIC_{50} values >8 μg/mL. In vitro results showed no antagonism or synergistic effects for *S. prolificans* exposed to terbinafine and ITC, terbinafine and VRC, VRC with CASF (synergy), ravuconazole with CASF (synergy), or AmB and CASF (synergy) (Cuenca-Estrella et al., 2008).

[5]Triazoles. The term is defined in Chapter 3A.

Clinical Experience with Antifungal Therapy of *Pseudallescheria* Species Complex Mycosis

VRC has had more success in treating *Pseudallescheria/Scedosporium* mycosis than other azoles and AmB. VRC monotherapy was evaluated in 107 patients with *Scedosporium* species mycosis (Troke et al., 2008). Seventy (65%) patients had *S. apiospermum* infections, and 35 (33%) had *S. prolificans* mycosis. This species differentiation is important because *S. prolificans* is innately more resistant to all classes of antifungal drugs. About one-quarter of these patients received VRC as primary therapy, the remainder received it as salvage or compassionate therapy. The categories of primary diseases were solid organ transplant recipients (22%), those with hematologic malignancy (21%), and 15% were a miscellaneous group of near-drowning victims, diabetics, and a patient with idiopathic pulmonary fibrosis. Clinical responses according to the site of infection are shown in Table 16.2 (Troke et al., 2008). Of patients receiving primary therapy, 61% had a successful response; and nearly as many who received the drug as salvage therapy were also successes. Best responses were seen in localized cutaneous or subcutaneous mycosis (91%), or in those with bone infections (79%). The lowest response rate occurred in CNS infections (43%). Survival for SCT patients was 40%; hematologic malignancy, 45%; and solid organ transplant recipients, 63%. Median survival time for therapy successes was 252 days; and for failures, 21 days. Forty-three (40%) patients died, 73% of the deaths were attributed to *Scedosporium* mycosis. Successful responses to therapy were observed for 45/70 (64%) of *S. apiospermum*-infected and 16/36 (44%) of *S. prolificans*-infected patients. Thus, even when VRC monotherapy was used to treat the previously intractable *S. prolificans* infections, the success rate was promising.

Seven immunocompromised children with *Scedosporium* species infections who failed conventional therapy were treated with VRC, on compassionate use (Walsh et al., 2002). Two children with *S. prolificans* failed to respond. Of the remaining five children with *S. apiospermum* infections of the lung, CNS, or disseminated disease, three had a complete response; the remaining two children with CNS disease failed therapy.

Suppressive therapy with VRC may not be sufficient to prevent breakthrough infection with *S. prolificans* (Tong et al., 2007). A 61-yr-old man with AML who received a matched allogeneic peripheral blood stem cell transplant from a sibling was receiving immunosuppressants for a grade 2 GVHD. He was placed on VRC suppressive therapy (200 mg, 2×/day). The patient's AML relapsed, he received another peripheral blood

Table 16.2 *Scedosporium* Mycosis Treated with Voriconazole: Clinical Response and Site of Infection in 107 Patients

Site(s) of infection, number (%)	Number with successful outcome/total number (%)	Number of patients alive/total number (number of deaths from *Scedosporium* mycosis)
CNS, 21 (20)	9/21 (43)	10/21 (8)
Disseminated, 23 (21)	11/23 (48)	10/23 (12)
Bone, 19 (18)	15/19 (79)	18/19 (1)
Lung, sinus,[a] 26 (24)	14/26 (54)	14/26 (9)
Other body sites,[b] 7 (7)	2/7 (29)	5/7 (2)
Skin/subcutaneous, 11 (10)	10/11 (91)	8/11 (0)
Total, 107	61/107 (57)	65/107 (32 died of invasive fungal infection)

[a]Lungs (25), sinus (1).

[b]Other body sites: eye (3), ear (2), and vocal cord (1). *Source:* Troke et al. (2008). Adapted with permission from the American Society for Microbiology.

stem cell infusion, and was maintained on VRC for 11 months when he developed fever and blood cultures positive for a mold, identified as *S. prolificans.* In vitro susceptibility indicated the isolate was resistant to all antifungal agents but synergy was demonstrated with VRC and terbinafine on checkerboard testing. VRC and terbinafine were continued for another 4 months until the leukemia relapsed. The patient died but had no evidence of *Scedosporium* infection at autopsy. This case illustrates that in vitro susceptibility testing of combinations of antifungal agents may help guide such therapy in cases unresponsive to monotherapy.

The median time from lung transplantation in three patients to the onset of disseminated *Scedosporium* mycosis was 14 months (Sahi et al., 2007). One cystic fibrosis patient with a history of *S. apiospermum* respiratory colonization received a double-lung transplant. Suppressive therapy with VRC was continuous during the post-transplant period; the patient developed fungal endophthalmitis. Antifungal therapy was extended to include CASF and terbinafine, but the endophthalmitis required enucleation and the patient developed multi-system disseminated *S. apiospermum* mycosis. Salvage therapy was begun with oral PSC (200 mg, 4×/day) also including GM-CSF as an immunoadjuvant. The patient initially responded but later pulmonary *S. apiospermum* mycosis developed and despite increasing the dose of PSC and adding liposomal AmB the patient expired. This case is a commentary on the difficulty of treating this refractory mold mycosis, even with combination therapy.

A second cystic fibrosis patient, who also was previously colonized with *Scedosporium* species, received a double-lung transplant while on VRC suppressive therapy. She had a routine clinical course and continued indefinite maintenance therapy with VRC. A third patient, who also had *Scedosporium* colonization, pretransplant, had a routine clinical course while on VRC suppressive therapy and as of 4 months post-transplant did not develop an invasive mold mycosis.

Posaconazole (PSC)

Clinical experience with PSC in treatment of *Pseudallescheria* species complex mycosis is limited to individual case reports. A 41-yr-old diabetic man with acute lymphoblastic leukemia developed sinusitis during a period of neutropenia (Mellinghoff et al., 2002). Cultures of sinus drainage grew *Aspergillus niger, A. flavus, Candida albicans*, and *Alternaria* species. After 3 weeks of oral ITC therapy, he developed worsening headaches and epistaxis. A black eschar in both nasal passages, when biopsied, showed septate hyphae. The patient was started on AmB and immunoadjuvant G-CSF. His condition worsened and a CT scan of the head revealed two brain abscesses; one of which was surgically excised. His neutropenia resolved but there was no other improvement. New abscesses were visualized and *S. apiospermum* grew from a specimen from the first abscess. The MIC for PSC was determined to be 0.5 μg/mL, and the patient was started on oral PSC monotherapy. His condition improved within 1 week and after that he continued to receive PSC for 12 months. The cerebral lesions resolved and the patient functions independently and remains free of leukemia. A factor contributing to success in this case was recovery from neutropenia. Based on in vitro testing, PSC may be useful for the treatment of *Scedosporium* and *Pseudallescheria* species infections. However, not all isolates appear sensitive; thus, PSC cannot be recommended for all infections.

Echinocandins

Several determinations of in vitro MICs against *Scedosporium* species indicate these molds are susceptible to echinocandins. Combinations of azoles and echinocandins are synergistic when tested in vitro against *S. apiospermum*, especially the combination if ITC and CASF (Cuenca-Estrella et al., 2008). Turning to *S. prolificans*, the most active combination was ravuconazole plus CASF, which was synergistic against 42% of *S. prolificans* isolates tested.

16.13 LABORATORY DETECTION, RECOVERY, AND IDENTIFICATION

Eumycetoma

Please see Chapter 19 for specimen processing for this clinical form involving *Pseudallescheria* and other agents of eumycetoma.

Direct Examination

Specimens including biopsy material, BAL fluid, bronchial washings, CSF, sputum, and tracheal aspirates are examined in 10% KOH preps. Detection is improved with fluorescent brighteners (e.g., Calcofluor). Hyphae, branching at acute angles, will be seen but do not provide a differential diagnosis with respect to other hyaline molds. Gram-stained smears show branched, septate hyaline hyphae.

Histopathology

P. boydii and *S. apiospermum* in tissue sections are observed with GMS or PAS stains. Hyphae are septate, 2–5 μm in diameter, branching at acute angles, and are nonpigmented. Hyphae in tissue *may show an irregular branching pattern* in contrast to *dichotomous* branching of *Aspergillus* species. Conidia production in tissue is another point of differentiation from aspergilli. Two types of conidia may be found in histopathologic sections: terminal or *intercalary* chlamydosopores (up to 20 μm diameter) and, secondly, hyphae in tissue may produce ovoid to pyriform brown pigmented conidia, (5 × 10 μm). Conidia (annelloconidia) are situated terminally on hyphae or are attached by short, thin *annellophores*. Sporulation of *S. prolificans* may occur in vivo, for example, in bronchial washings where *annellides* with conidia may be observed, thus expediting the diagnosis (Perfect and Schell, 1996). *S. prolificans* is pigmented in tissue when revealed by Fontana–Masson stain.

Culture

Culture is necessary for a specific identification of the causative agent because of the scarcity of distinguishing features on direct examination or histopathology. These molds are uncommonly encountered at the mycology bench of the clinical microbiology laboratory. Background information as to the source and primary disease of the patient are useful in the overall assessment of the significance of cultures, especially from sputum or other nonsterile site. Nonetheless, isolation of *Pseudallescheria* or *Scedosporium* species is potentially highly significant, requiring a prompt report to the attending physician.

Growth Characteristics of the Pseudallescheria *Species Complex*

SDA with cycloheximide and chloramphenicol is a suitable growth medium, as is blood agar. When the specimen is from a nonsterile site, steps may be taken to improve the yield of *Pseudallescheria* species complex because of the potential for the rapid overgrowth of *Aspergillus* or *Candida* species. They include addition of benomyl (10 μg/mL) to the agar, in addition to cycloheximide. Growth in the presence of cycloheximide may be slow, and *S. prolificans* is sensitive to that protein synthesis inhibitor. Medium containing low-dose AmB may be helpful if *Pseudallescheria* or *Scedosporium* species are suspected.

Pseudallescheria boydii

For *P. boydii* and *S. apiospermum* growth is rapid at 25°C on SDA, with the upper limit of 42°C. Growth is permissive in media containing ≤8 μg/mL of cycloheximide. The only evidence for melanization by *P. boydii* or *S. apiospermum* is a melanin-like pigment diffusing from their pigmented conidia. *P. boydii* is an exception to the general rule that the *teleomorph* form is not seen in the clinical laboratory. *Pseudallescheria boydii* is *homothallic* and *cleistothecia* containing sexually derived ascospores are observed in patient isolates, when primary cultures are allowed to incubate for 2–3 weeks. Often it is necessary to passage cultures on media such as PDA to induce fruiting.

Carbohydrate assimilation patterns assist in identifying members of the *Pseudallescheria* species complex (Fig. 16.2).

Colony Morphology The *P. boydii* colony surface is white, becoming dark gray to brownish gray (on SDA) with a pale reverse also containing brown to black zones. The hyphae are hyaline, and the dark color is due to brown conidia. Fontana–Masson stain is negative. In histologic sections hyphae are hyaline. Grains produced in mycetoma are white.

Microscopic Morphology This teleomorph is homothallic. After 2–3 weeks of incubation brown cleistothecia 100–300 μm in diameter may be produced on nutritionally deficient media (e.g., PDA, CMA). In nature, the cleistothecia deliquesce, releasing a mass of slimy ascospores that are then spread by insects and arthropods. Ascospores are ovoid to ellipsoid, pale yellow-brown to copper colored, and measuring 4 × 5 to 7 × 9 μm.

Scedosporium apiospermum

Colony Morphology Rapid growing, the surface is grayish-white to olive-gray, becoming darker with a wooly or cottony texture on SDA. The colony reverse is pale with brownish-black zones. *S. apiospermum* shows weak growth in the presence of cycloheximide, an ingredient of Mycosel or Mycobiotic medium.

Microscopic Morphology Conidiophores of *S. apiospermum* occur singly or in bundles, the synnemata.

- ***Scedosporium Type.*** This is the major type of sporulation consisting of septate hyaline hyphae (2.5–5 μm diameter), which give rise to short or elongate conidiophores produced singly or in small clusters, each with a single conidium (Fig. 16.3). The conidia are produced in an enteroblastic manner from annellides. The conidia are unicellular (4–9 × 6–10 μm), obovoid, pale brown, and sticky. Annellides are difficult to see unless high power and Calcofluor white are used. Phase contrast or differential interference micrsocopy[6] also are helpful to reveal fine structural details like annellides.
- ***"Graphium" Synanamorph*** At the edge of a colony, later in the growth phase, a formation may be seen of erect stiff bundles of olive brown hyphae terminating in a brush of conidiogenous cells with a cluster of conidia. This type of asexual reproductive structure is a *synnema* (*definition*: a cylindrical compact bundle of conidiophores).

Scedosporium prolificans

S. prolificans is noteworthy for producing positive blood cultures. When first isolated from clinical materials growth is moderate to rapid on SDA at 25°C in 5 days on medium *without* cycloheximide. *Scedosporium prolificans* is inhibited by cycloheximide. The upper limit of growth is 45°C.

[6]Learn about differential interference microscopy at the URL http://olympusmicro.com/primer/techniques/dic/dicintro.html.

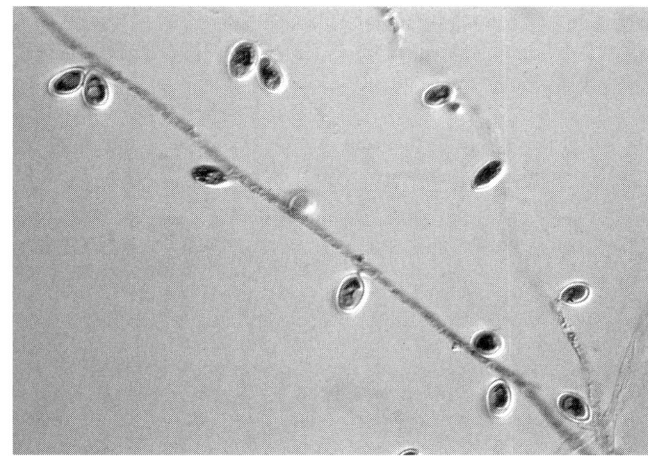

Figure 16.3 *Scedosporium apiospermum* microscopic morphology. Conidia are unicellular, pale brown, sticky, 4–9 by 6–10 μm, *obovoid*. Conidiophores are produced singly or in small clusters, each with a single conidium. *Source:* E. Reiss, CDC.

Colony Morphology *Scedosporium prolificans* colonies are flat, spreading, with a suede-like moist surface that is white turning to brown-olive-gray to black. The colony reverse turns a pale dark brown.

Microscopic Morphology The conidiophore is flask shaped with a swollen base and conidia that form a small cluster of single-celled conidia, 2–5 × 3–13 μm, obovoid to pyriform, hyaline to pale brown (Fig. 16.4). *S. prolificans* does not produce synnemata.

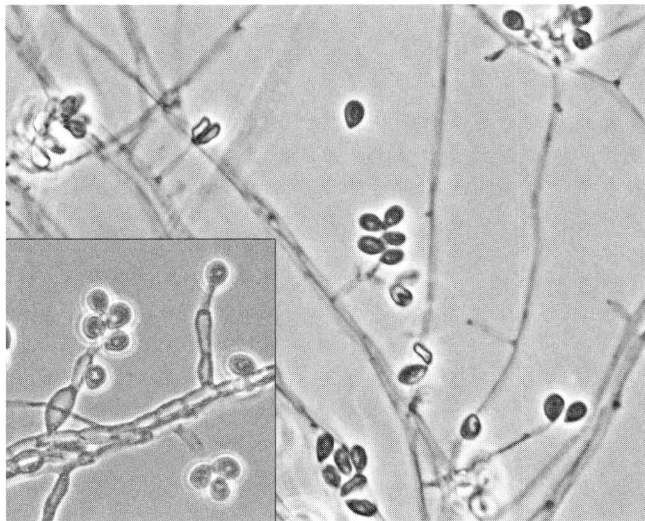

Figure 16.4 *Scedosporium prolificans* microscopic morphology. Conidia are borne in small groups on basally swollen, flask-shaped annellides (*inset*), singly or in clusters. Conidia are single-celled, pale-brown, ovoid to pyriform, 2–5 × 3–13 μm. Phase contrast, lactofuchsin stain. *Source:* E. Reiss, CDC.

SELECTED REFERENCES FOR *PSEUDALLESCHERIA/SCEDOSPORIUM* MYCOSIS

Alastruey-Izquierdo A, Cuenca-Estrella M, Monzon A, Rodriguez-Tudela JL, 2007. Prevalence and susceptibility testing of new species of *Pseudallescheria* and *Scedosporium* in a collection of clinical mold isolates. *Antimicrob Agents Chemother* 51: 748–751.

Alvarez M, Ponga BL, Rayon C, Garcia-Gala J, Roson-Porto MC, Gonzalez M, Martinez-Suarez JV, Rodriguez-Tudela JL, 1995. Nosocomial outbreak caused by *Scedosporium prolificans (inflatum)*: Four fatal cases in leukemic patients. *J Clin Microbiol* 33: 3290–3295.

Barreto-Bergter E, Sassaki GL, Wagner R, Souza LM, Souza MVAR, Pinto MR, da Silva MID, Gorin PAJ, 2008. The opportunistic fungal pathogen *Scedosporium prolificans*: Carbohydrate epitopes of its glycoproteins. *Int J Biol Macromol* 42: 93–102.

Bates DD, Mims JW, 2006. Invasive fungal sinusitis caused by *Pseudallescheria boydii*: Case report and literature review. *Ear Nose Throat J* 85: 729–737.

Bittencourt VCB, Figueiredo RT, da Silva RB, Mourao-Sa DS, Fernandez PL, Sassaki GL, Mulloy B, Bozza MT, Barreto-Bergter E, 2006. An {alpha}-glucan of *Pseudallescheria boydii* is involved in fungal phagocytosis and toll-like receptor activation. *J Biol Chem* 281: 22614–22623.

Brearley JC, McCandlish IA, Sullivan M, Dawson CO, 1986. Nasal granuloma caused by *Pseudallescheria boydii*. Equine Vet J 18: 151–153.

Cabañes FJ, Roura X, García F, Domingo M, Abarca ML, Pastor J, 1998. Nasal granuloma caused by *Scedosporium apiospermum* in a dog. *J Clin Microbiol* 36: 2755–2758.

Castiglioni B, Sutton DA, Rinaldi MG, Fung J, Kusne S, 2002. *Pseudallescheria boydii* (synanamorph *Scedosporium apiospermum*) infection in solid organ transplant recipients in a tertiary medical center and review of the literature. *Medicine* 81: 333–348.

Chang AB, Redding GJ, Everard ML, 2008. Chronic wet cough: Protracted bronchitis, chronic suppurative lung disease and bronchiectasis. *Pediatr Pulmonol* 43: 519–531.

Cimon B, Carrere J, Vinatier JF, Chazalette JP, Chabasse D, Bouchara JP, 2000. Clinical significance of *Scedosporium apiospermum* in patients with cystic fibrosis. *Eur J Microbiol Infect Dis* 19: 53–56.

Cortez KJ, Roilides E, Quiroz-Telles F, Meletiadis J, Antachopoulos C, Knudsen T, Buchanan W, Milanovich J, Sutton DA, Fothergill A, Rinaldi MG, Shea YR, Zaoutis T, Kottilil S, Walsh TJ, 2008. Infections caused by *Scedosporium* spp. *Clin Microbiol Rev* 21: 157–197.

Cuenca-Estrella M, Alastruey-Izquierdo A, Alcazar-Fuoli L, Bernal-Martinez L, Gomez-Lopez A, Buitrago MJ, Mellado E, Rodriguez-Tudela JL, 2008. In vitro activities of 35 double combinations of antifungal agents against *Scedosporium apiospermum* and *Scedosporium prolificans*. *Antimicrob Agents Chemother* 52: 1136–1139.

de Hoog GS, Marvin-Sikkema FD, Lahpoor GA, Gottshcall JC, Prins RA, Gueho E, 1994. Ecology and physiology of the emerging opportunistic fungi *Pseudallescheria boydii* and *Scedosporium prolificans*. *Mycoses* 37: 71–78.

D'Hondt K, Parys-Van Ginderdeuren R, Foets B, 2000. Fungal keratitis caused by *Pseudallescheria boydii* (*Scedosporium apiospermum*). *Bull Soc Belge Ophtalmol* 277: 53–56.

Delhaes L, Harun A, Chen SC, Nguyen Q, Slavin M, Heath CH, Maszewska K, Halliday C, Robert V, Sorrell TC, Auscedo Study Group, Meyer W, 2008. Molecular typing of Australian *Scedosporium* isolates showing genetic variability and numerous *S. aurantiacum*. *Emerging Infect Dis* 14: 282–290.

Elad D, Perl S, Yamin G, Blum S, David D, 2010. Disseminated pseudallescheriosis in a dog. *Med Mycol* 48: 635–638.

Friedman DS, Schoster JV, Pickett JP, Dubielzig RR, Czuprynski C, Knoll JS, Wolfgram LJ, 1989. *Pseudallescheria boydii* keratomycosis in a horse. *J Am Vet Med Assoc* 195: 616–618.

Gil-Lamaignere C, Winn RM, Simitsopoulou M, Maloukou A, Walsh TJ, Roilides E, 2005. Interferon gamma and granulocyte-macrophage colony-stimulating factor augment the antifungal activity of human polymorphonuclear leukocytes against *Scedosporium* spp.: Comparison with *Aspergillus* spp. *Med Mycol* 43: 253–260.

Gilgado F, Gene J, Cano J, Guarro J, 2010. Heterothallism in *Scedosporium apiospermum* and description of its teleomorph *Pseudallescheria apiosperma* sp. nov. *Med Mycol* 48: 122–128.

Gilgado F, Cano J, Gene J, Serena C, Guarro J, 2009. Different virulence of the species of the *Pseudallescheria boydii* complex. *Med Mycol* 47: 371–374.

Gilgado F, Cano J, Gene J, Sutton DA, Guarro J, 2008. Molecular and phenotypic data supporting distinct species statuses for *Scedosporium apiospermum* and *Pseudallescheria boydii* and the proposed new species *Scedosporium dehoogii*. *J Clin Microbiol* 46: 766–771.

Gilgado F, Cano J, Gené J, Guarro J, 2005. Molecular phylogeny of the *Pseudallescheria boydii* species complex: Proposal of two new species. *J Clin Microbiol* 43: 4930–4942.

Guarro J, Kano J, 2007. Report of the ECMM/ISHAM Working Group *Pseudallescheria/Scedosporium* infections (only the most recent meeting proceedings are on line). http://www.scedosporium-ecmm.com.

Guerrero A, Torres P, Duran MT, Ruiz-Díez B, Rosales M, Rodriguez-Tudela JL, 2001. Airborne outbreak of nosocomial *Scedosporium prolificans* infection. *Lancet* 357: 1267–1268.

Haulena M, Buckles E, Gulland FM, Lawrence JA, Wong A, Jang S, Christopher MM, Lowenstine LJ, 2002. Systemic mycosis caused by *Scedosporium apiospermum* in a stranded northern elephant seal (*Mirounga angustirostris*) undergoing rehabilitation. *J Zoo Wildl Med* 33: 166–171.

Heath CH, Slavin MA, Sorrell TC, Handke R, Harun A, Phillips M, Nguyen Q, Delhaes L, Ellis D, Meyer W, Chen SC, 2009. Population-based surveillance for scedosporiosis in Australia: Epidemiology, disease manifestations and emergence of *Scedosporium aurantiacum* infection. *Clin Microbiol Infect* 15: 689–693.

Hoshino S, Tachibana I, Kijima T, Yoshida M, Kumagai T, Osaki T, Kamei K, Kawase I, 2007. A 60-year-old woman with cough, fever, and upper-lobe cavitary consolidation. *Chest* 132: 708–710.

Hugnet C, Marrou B, Dally C, Guillot J, 2009. Osteomyelitis and discospondylitis due to *Scedosporium apiospermum* in a dog. *J Vet Diagn Invest* 21: 120–123.

Husain S, Muñoz P, Forrest G, Alexander BD, Somani J, Brennan K, Wagener MM, Singh N, 2005. Infections due to *Scedosporium apiospermum* and *Scedosporium prolificans* in transplant recipients: Clinical characteristics and impact of antifungal agent therapy on outcome. *Clin Infect Dis* 40: 89–99.

Kiffer-Moreira T, Pinheiro A, Pinto M, Esteves F, Souto-Padrón T, Barreto-Bergter E, Meyer-Fernandes J, 2007. Mycelial forms of *Pseudallescheria boydii* present ectophosphatase activities. *Arch Microbiol* 188: 159–166.

Kuwano A, Yoshihara T, Takatori K, Kosuge J, 1998. Onychomycosis in white line disease in horses: Pathology, mycology and clinical features. *Equine Vet J Suppl* (26): 27–35.

Lamaris GA, Chamilos G, Lewis RE, Safdar A, Raad II, Kontoyiannis DP, 2006. *Scedosporium* infection in a tertiary care Cancer Center: A review of 25 cases from 1989–2006. *Clin Infect Dis* 43: 1580–1584.

Lionakis MS, Bodey GP, Tarrand JJ, Raad II, Kontoyiannis DP, 2004. The significance of blood cultures positive for emerging saprobic moulds in cancer patients. *Clin Microbiol Infect* 10: 922–925.

MELLINGHOFF IK, WINSTON DJ, MUKWAYA G, SCHILLER GJ, 2002. Treatment of *Scedosporium apiospermum* brain abscesses with posaconazole. *Clin Infect Dis* 34: 1648–1650.

MESSORI A, LANZA C, DE NICOLA M, MENICHELLI F, CAPRIOTTI T, MORABITO L, SALVOLINI U, 2002. Mycotic aneurysms as lethal complication of brain pseudallescheriasis in a near-drowned child: A CT demonstration. *AJNR Am J Neuroradiol* 23: 1697–1699.

O'BRYAN TA, 2005. Pseudallescheriasis in the 21st century. *Expert Rev Anti Infect Ther* 3: 765–773.

PANACKAL AA, MARR KA, 2004. *Scedosporium/Pseudallescheria* infections. *Semin Respir Crit Care Med* 25: 171–181.

PERFECT JR, SCHELL WA, 1996. The new fungal opportunists are coming. *Clin Infect Dis* 22 (Suppl 2): S112–S118.

PINTO MR, DE SÁ ACM, LIMONGI CL, ROZENTAL S, SANTOS ALS, BARRETO-BERGTER E, 2004. Involvement of peptidorhamnomannan in the interaction of *Pseudallescheria boydii* and HEp2 cells. *Microbes and Infection* 6: 1259–1267.

RANIER J, DE HOOG GS, 2006. Molecular taxonomy and ecology of *Pseudallescheria, Petriella* and *Scedosporium prolificans*(Microascaceae) containing opportunistic agents on humans. *Mycol Res* 110: 151–160.

REVANKAR SG, PATTERSON JE, SUTTON DA, PULLEN R, RINALDI MG, 2002. Disseminated phaeohyphomycosis: Review of an emerging mycosis. *Clin Infect Dis* 34: 467–476.

SAHI H, AVERY RK, MINAI OA, HALL G, MEHTA AC, RAINA P, BUDEV M, 2007. *Scedosporium apiospermum* (*Pseudoallescheria boydii)* infection in lung transplant recipients. *J Heart Lung Transplant* 26: 350–356.

SILVA B, SANTOS A, BARRETO-BERGTER E, PINTO M, 2006. Extracellular peptidase in the fungal pathogen *Pseudallescheria boydii*. *Curr Microbiol* 53: 18–22.

STRICKLAND LB, SANDIN RL, GREENE JN, AHMAD N, 1998. A breast cancer patient with disseminated *Scedosporium prolificans* infection. *Infect Med* 15: 849–852.

STUR-HOFMANN K, STOS S, SAXA-ENENKEL M, RAPPERSBERGER K, 2011. Primary cutaneous infection with *Scedosporium apiospermum* successfully treated with voriconazole. *Mycoses* 54: e201–e204.

TAMM M, MALOUF M, GLANVILLE A, 2001. Pulmonary *Scedosporium* infection following lung transplantation. *Transpl Infect Dis* 3: 189–194.

TONG SY, PELEG AY, YOONG J, HANDKE R, SZER J, SLAVIN M, 2007. Breakthrough *Scedosporium prolificans* infection while receiving voriconazole prophylaxis in an allogeneic stem cell transplant recipient. *Transp Infect Dis* 9: 241–243.

TROKE P, AGUIRREBENGOA K, ARTEAGA C, ELLIS D, HEATH CH, LUTSAR I, ROVIRA M, NGUYEN Q, SLAVIN M, CHEN SCA, 2008. Treatment of scedosporiosis with voriconazole: Clinical experience with 107 patients. *Antimicrob Agents Chemother* 52: 1743–1750.

WALSH TJ, LUTSAR I, DRISCOLL T, DUPONT B, RODEN M, GHAHRAMANI P, HODGES M, GROLL AH, PERFECT JR, 2002. Voriconazole in the treatment of aspergillosis, scedosporiosis and other invasive fungal infections in children. *Pediatr Infect Dis J* 21: 240–248.

WALTS AE, 2000. *Pseudallescheria*: An underdiagnosed fungus? *Diagn Cytopathol* 25: 153–202.

WILICHOWSKI E, CHRISTEN H-J, SCHIFFMANN H, SCHULZ-SCHAEFFER W, BEHRENS-BAUMANN W, 1996. Fatal *Pseudallescheria boydii* panencephalitis in a child after near-drowning. *Pediatr Infect Dis J* 15: 365–370.

WINN RM, GIL-LAMAIGNERE C, ROILIDES E, SIMITSOPOULOU M, LYMAN CA, MALOUKOU A, WALSH TJ, 2005. Effects of interleukin-15 on antifungal responses of human polymorphonuclear leukocytes against *Fusarium* spp. and *Scedosporium* spp. *Cytokine* 31: 1–8.

WU Z, YING H, YIU S, IRVINE J, SMITH R, 2002. Fungal keratitis caused by *Scedosporium apiospermum*. *Cornea* 21: 519–522.

WEBSITE CITED

Differential interference microscopy at the URL http://olympusmicro.com/primer/techniques/

QUESTIONS

The Answer Key to multiple choice questions may be found at the end of the book.

1. What microscopic findings in a mold culture will place *Pseudallescheria/Scedosporium* into the differential diagnosis?
   A. Either haphazard or acute angle branching
   B. Growth in medium containing cycloheximide
   C. Growth in medium containing Benomyl
   D. Production of single, pale brown, obovoid annelloconidia
   E. All of the above
2. Sometimes the conidiophores of *Scedosporium* species are clustered in bundles called
   A. Pycnidia
   B. Sclerotia
   C. Sporodochia
   D. Synnemata
3. The appearance in tissue sections of hyphae with short thin annellophores bearing an obovoid, light brown conidium is a diagnostic for
   A. *Aspergillus terreus*
   B. *Fusarium solani*
   C. *Scedosporium* species
   D. *Sporothrix schenckii*
4. The relationship of *Pseudallescheria boydii* and *Scedosporium apiospermum* is
   1. *P. boydii* assimilates ribose, whereas *S. apiospermum* does not.
   2. *P. boydii* is the causative agent of eumycetoma whereas *S. apiospermum* is not.
   3. *P. boydii* is the teleomorph form and *S. apiospermum* is its anamorph form.
   4. These two species are related members of the *Microascaceae*.

   Which are correct?
   A. 1 + 2
   B. 1 + 3
   C. 1 + 4
   D. 2 + 3
   E. 2 + 4
   F. 3 + 4
5. How does *S. prolificans* differ from *S. apiospermum* in the following categories: microscopic morphology, pigmentation, assimilation pattern, susceptibility to antifungal agents,

frequency as a cause of disease in humans, geographic distribution, and ecologic niche?

6. Discuss the appearance of the teleomorph form, *P. boydii*, in cultures from clinical specimens. Is ascoma development a reliable characteristic, and is it timely enough to influence treatment decisions? What media stimulate *P. boydii*?
7. Describe the clinical forms of *Pseudallescheria/Scedosporium* mycosis that occur in cystic fibrosis patients and victims of near-drowning accidents. Why should the diagnostic work-up of extrapulmonary *Pseudallescheria/Scedosporium* mycosis include the central nervous system?
8. What is the preferred primary therapy for pulmonary-disseminated *Pseudallescheria/Scedosporium* mycosis? What is the success rate and how is that success influenced if the species is *S. prolificans*? What do the in vitro MIC data reveal about drug combinations that may be promising therapy options?
9. What is the clinical relevance of genetic species within the *Pseudallescheria/Scedosporium* group?
10. What is the ecologic niche of *Pseudallescheria/Scedosporium* species? What properties of the conidia make them either more or less capable of airborne dispersion?

Chapter 17A

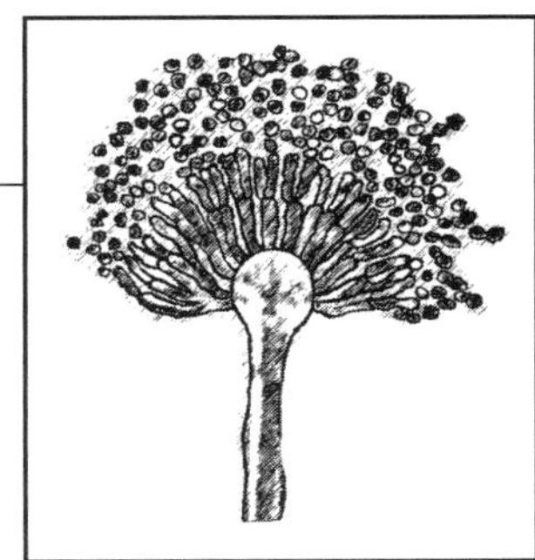

Mucormycosis

Mucormycosis and Entomophthoramycosis

The hyaline molds which are the agents of mucormycosis are classed in the order *Mucorales* in the *Mucoromycotina*, a subphylum of lower fungi (Hibbett et al., 2007). Formerly, the disease caused by these molds was referred to as zygomycosis but, in view of the revised higher level classification of fungi, it is appropriate to refer to the disease as mucormycosis, a name well known in clinical medicine. The former phylum *Zygomycota* is no longer taxonomically valid (Hibbett et al., 2007).

Entomophthoramycosis is caused by members of the subphylum *Entomophthoromycotina—Conidiobolus* species and *Basidiobolus* species—and is discussed in Chapters 17B and 17C.

Biology of the *Mucorales*

Before dwelling on the individual disease agents, let us address the general properties of this unique group of hyaline molds. The *Mucorales* are rapid growers, thermotolerant, and known in lay terms as "pin molds". Their asexual reproductive structures, sporangia borne on sporangiophores, are large enough to be seen with the unaided eye (Fig. 17A.1).

Mucorales Vegetative Growth, Morphology, and Asexual Reproduction

The *Mucorales* grow rapidly and branch as *aseptate* (coenocytic) or sparsely septate hyphae. The sparse septa facilitates cytoplasmic streaming and intrahyphal transport, resulting in a faster growth rate than that of the ascomycetes; for example, *Rhizopus oryzae* was observed to grow at the rate of 3 mm/h at a temperature of 36°C (Dijksterhuis and Samson, 2006). They rapidly overgrow the substrate at a rate of 20 mm/day, compared with much lower growth rates of higher fungi. The genome sequence of *R. oryzae* revealed a whole genome duplication; doubling of metabolism genes could contribute to this growth rate (Ma et al., 2009).

Some genera produce root-like rhizoids connected by hyphae called stolons. Asexual reproductive structures are important characters for identification purposes (Fig. 17A.1). Spores are produced in globose sporangia, which develop at the tips of a stalk, the sporangiophore. Most sporangiophores with sporangia of the *Mucorales* are between 500 μm and 2 mm in length, large enough to be seen with the unaided eye as "pin molds." In some genera [e.g., *Lichtheimia* (formerly *Absidia)* and *Apophysomyces*] there is a swelling just below the sporangium called the apophysis. The sporangium has two compartments. The first compartment is the columella, a swollen extension of the sporangiophore that protrudes into the sporangium. The columella is separated from the second or spore-containing compartment by a membrane. During maturation the cytoplasm of the sporangium divides into hundreds of sporangiospores. When the sporangium is mature, the sporangium wall deliquesces (dissolves), releasing the sporangiospores.

Variety in the sporangia is an important character as well. Some genera produce smaller sporangioles, containing few sporangiospores, whereas in other genera (e.g., *Cunninghamella*) only a single spore is formed from a series of sporangioles attached by short denticles to a globose vesicle (Dijksterhuis and Samson, 2006).

Definitions of the structural features of *Mucorales*:

- Apophysis. *Definition*: A swelling of the sporangiophore just below the sporangium

Fundamental Medical Mycology, First Edition. By Errol Reiss, H. Jean Shadomy and G. Marshall Lyon, III.
© 2012 Wiley-Blackwell. Published 2012 by John Wiley & Sons, Inc.

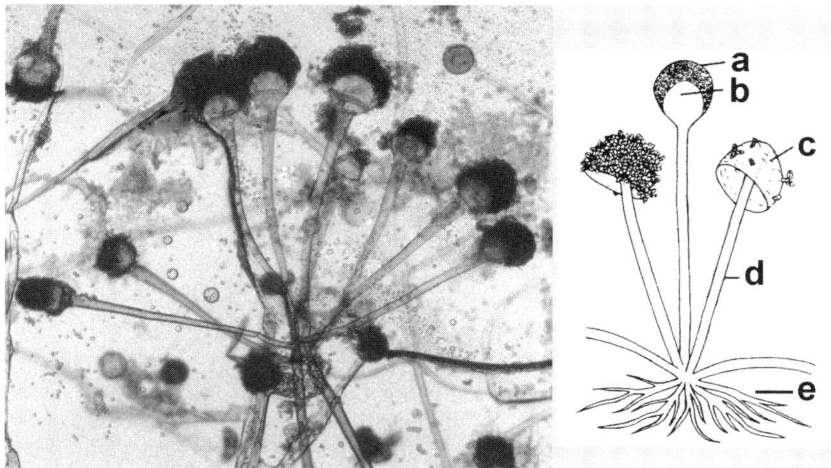

Figure 17A.1 Structural features of *Rhizopus* species. A cluster of *Rhizopus oryzae* illustrates the "pin mold morphology." The sporangiophores and sporangia may be up to 1.5 mm in length so as to be visible to the unaided eye. The sporangiophores, shown here in a cluster of 8, may be simple or branched, rising from stolons opposite rhizoids. Columellae and apophysis (the latter scarcely visible) together are round or oval.

Source: Used with permission from Dr. Robert Simmons, Director, Biological Imaging Core Facility, Department of Biology, Georgia State University, Atlanta, Georgia. *Inset*: (a) Sporangium, (b) columella, (c) collapsed columella (after release of sporangiospores), (d) sporangiophore, (e) rhizoids. *Source for inset:* Kendrick (1992). Used with permissions from Dr. Bryce Kendrick and Focus Publishing R. Pullins Co.

- Columella. *Definition*: A vesicle or central part inside the sporangium, continuous with the stalk or sporangiophore. Sometimes the columella has one or more projections.
- Rhizoids. *Definition*: Root-like structures that adhere stolons to the substrate, such as an agar surface.
- Sporangia are the asexual spore-forming structures of the *Mucorales*; the differences among them aid in identifying the species. Sporangia are borne on the sporangiophores.
- Sporangiole. *Definition*: A small, usually spherical sporangium that contains only one or two sporangiospores.
- Sporangiophore. *Definition*: A stalk that arises from the vegetative hypha and terminates in the columella supporting the sporangium. The sporangiophore may be single or branched. Branching may be monopodial[1] or sympodial[2]
- Sporangiospores are formed inside sporangia by cleavage of a multinucleate cytoplasm.
- Stolon. *Definition*: A term borrowed from botany; the stolon is a lateral hypha (a "runner") that grows horizontally along the substrate.

[1]Monopodial. *Definition*: Branching with a single, main, dominant and central axis and reduced lateral branches.

[2]Sympodial. *Definition*: Branching without a main axis but with several more-or-less equal lateral branches.

Sexual Reproduction in the *Mucoromycotina*

The zygospore is the hallmark of *Mucoromycotina* (Figs. 17A.2, 17A.3). Under the influence of pheromones (e.g., trisporic acid), hyphae of opposite mating types grow toward each other and differentiate into specialized branches (gametangia), which then fuse (plasmogamy). A multinucleate dikaryotic zygosporangium forms, surrounded by a thick cell wall ornamented with ridges, and often held by two suspensor cells. A single zygospore develops within the zygosporangium. The zygospore usually *persists in a dormant state for weeks or months before germinating*. When conditions are favorable, nuclear fusion (karyogamy) occurs within the zygospore, producing a diploid nucleus, the zygote. Meiosis occurs *only* during germination into a haploid form, which quickly produces a short germosporangium and then releases recombined[3] haploid spores. The haploid spores germinate into a vegetative mycelium and quickly resume production of sporangiophores and haploid sporangiospores. Sometimes *homothallic* species will undergo plasmogamy and develop zygospores on common mycologic media used in the clinical laboratory. The warty zygospore with suspensor cells is shown in Fig. 17A.3.

[3]Recombination. *Definition*: The reassortment of genes produced as a result of crossing-over during meiosis.

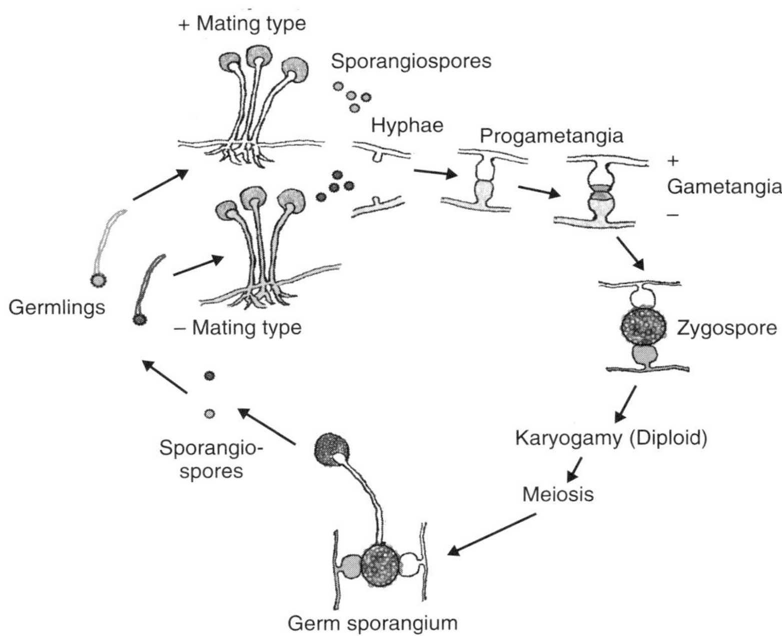

Figure 17A.2 Sexual and asexual reproduction in *Rhizopus* species. Please see text for an explanation.

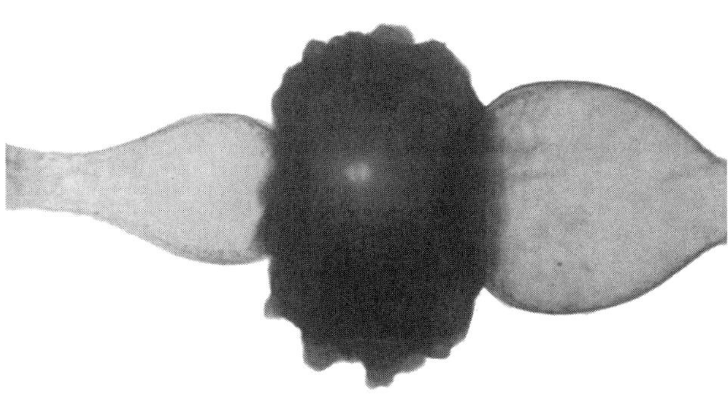

Figure 17A.3 Zygospore with suspensor cells, *Mucorales*. *Source:* CDC.

17A.1 MUCORMYCOSIS-AT-A-GLANCE

- *Introduction/Disease Definition.* Diseases caused by hyaline molds classed in the *Mucorales* have distinct clinical forms listed with the *major* risk groups:
  - Cutaneous disease in persons with penetrating injuries, injection drug users, or burn patients
  - Gastrointestinal disease in premature infants, malnourished children
  - Multi-organ disseminated disease spread from a single organ focus, in immunocompromised patients or those receiving deferoxamine iron chelation therapy
  - Pulmonary disease in the immunocompromised host
  - Rhinocerebral disease (a major clinical form) occurring in poorly controlled diabetics with ketoacidosis and also in immunosuppressed cancer patients and organ transplant recipients
- *Etiologic Agents.* The pathogenic *Mucorales* are, in approximate order of clinical prevalence, *Rhizopus, Mucor, Lichtheimia* (formerly *Absidia*), *Rhizomucor, Cunninghamella*, and *Apophysomyces*.
- *Geographic Distribution Ecologic Niche. Mucorales* occur worldwide as saprobes in the soil; they cause postharvest rot of fruit. They are plant pathogens on fruit and vegetables.
- *Epidemiology.* There are an estimated 500 mucormycosis cases/year in the United States. An increasing secular trend may result from an aging population, obesity, more immunosuppressive conditions, and an increase in type 2 diabetes. Breakthrough mucormycosis is reported in some patients receiving voriconazole (VRC) prophylaxis to prevent other mold infections.
- *Risk Groups/Factors.* Diabetic ketoacidosis, burns or other wounds, cancer chemotherapy, stem cell transplant (SCT), or solid organ transplant recipients; in infants (prematurity) and children (malnutrition); and deferoxamine iron chelation therapy.
- *Transmission.* Airborne sporangiospores are inhaled or alight on open wounds or burns.
- *Determinants of Pathogenicity.* These molds can grow and then invade tissues and blood vessels when ferric iron is released from transferrin by acidosis, or as a result of deferoxamine therapy. They are rapid growers and thermotolerant, producing lipase and proteinases.

- *Clinical Forms.* Invasive fungal sinusitis, rhinocerebral, pulmonary, gastrointestinal, cutaneous, necrotizing fasciitis, and multi-organ dissemination, especially to the brain.
- *Veterinary Forms.* Mucormycosis occurs in livestock, zoo animals, and pet dogs. Clinical forms include ulcers in the rumen omasum and abomasum in cattle and sheep; and in bovines: respiratory infection, abortion, and cranial infections. Gastric ulceritis and disseminated disease is known in dogs and various birds.
- *Therapy.* Success depends on early diagnosis, surgical debridement, restoration of metabolic balance, rapidly advancing doses of lipid-AmB; potential benefit from posaconazole (PSC).
- *Laboratory Detection, Recovery, and Identification.* Direct examination (KOH prep) for aseptate or sparsely septate hyphae in scrapings from skin, nasal sinus, or hard palate; or histologic stain (GMS) of same specimens, or frozen sections of sinonasal or other biopsy material.

17A.2 INTRODUCTION/DISEASE DEFINITION

Mucormycosis

Caused by hyaline molds in the order *Mucorales*, disease is a rapidly progressing tissue-destroying infection in a susceptible host. Persons at risk include poorly controlled diabetics with ketoacidosis and patients receiving chemotherapy for cancer or immunosuppressive therapy for maintenance of stem cell or solid organ transplants. A tender red patch on the cheek of a diabetic patient is a danger sign of mucormycosis. The major clinical form is invasive fungal sinusitis, which, without prompt diagnosis by biopsy and treatment, can lead to rhinocerebral mucormycosis. The disease begins in a susceptible host when spores germinate in the nasal sinuses and grow rapidly as *aseptate* or sparsely septate hyphae, invading blood vessels, growing through tissue, cartilage, and bone to the paranasal sinuses, through the lamina papyracea to the orbit of the eye, or to the ethmoid sinuses to the meninges and brain. The hyphae have a predilection for the elastic lamina of blood vessels, causing hemorrhage, thrombosis, and infarcts.

Other clinical forms include (i) necrotizing cellulitis in burn patients or patients with other wounds; and (ii) pulmonary disease and/or invasive nasal sinusitis secondary to immunosuppressive therapy for leukemia, solid tumors, or for SCTs. Control of mucormycosis depends on prompt diagnosis including direct examination of biopsied tissues, surgical debridement, restoration of metabolic balance, and prompt initiation of therapy with lipid-AmB, possibly in combination with PSC.

These molds pose little or no threat to persons with normal immune and endocrine systems, with certain exceptions: accident victims with penetrating injuries or burns, injection drug users, patients receiving iron chelation therapy with deferoxamine, and those with surgical wounds.

The causative agents are common, primitive, environmental, hyaline molds in the order *Mucorales: Rhizopus, Rhizomucor, Mucor, Cunninghamella, Apophysomyces, Lichtheimia*, and other rare species. The *Mucorales* are thermotolerant with some species growing up to 54°C. They are monomorphic and have aseptate or sparsely septate hyphae. *Mucorales*, often referred to as "pin molds" or "bread molds," are universal in the environment in soil worldwide, and in nature are economically important agents of spoilage of fruits and vegetables. In the laboratory they are known as "lid lifters" because their rapid aggressive growth presses up against the lid of the Petri plate.

During infection these molds grow as very broad (8–15 μm) aseptate or sparsely septate hyphae, branching at right angles, at times appearing twisted and distorted. They grow in various directions through tissues so that the plane of section will show hyphae in cross section, oblique and longitudinal sections.

17A.3 CASE PRESENTATIONS

Case Presentation 1. Facial Swelling in a 66-yr-old Woman (Scully et al., 1982)

History

This 66-yr-old woman previously had a prefrontal lobotomy 28 years ago. She had diabetes mellitus for 6 years, and injected insulin daily. No urine tests were done, but up to the present, there were no complications. Nine months prior to admission she suffered from otitis media. Three weeks previously she had toothache and two teeth were extracted. The day before the current admission she developed a left-sided facial droop, hallucinations, and lethargy.

Clinical and Laboratory Findings

Her temperature was 38.2°C, blood pressure was 200/100. She was overweight and dehydrated. The left maxillary area was red, swollen, and warm; with left upper eyelid droop, left facial droop, and left facial nerve palsy. Her left pupil was dilated and fixed. Her urine tested 3+ sugar (hyperglycemic), 3+ ketones (ketotic). The WBC count

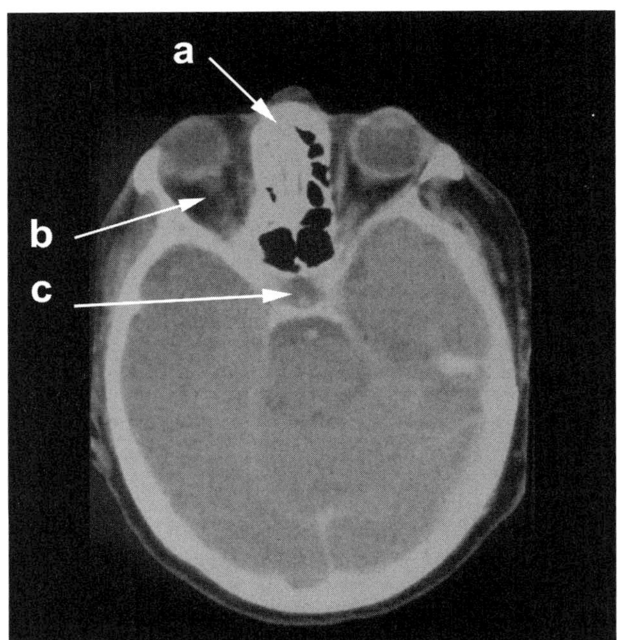

Figure 17A.4 Mucormycosis. CT scan, head showing (*arrows*) (a) opacity in left ethmoid air sacs; (b) engorged blood vessels in the retrobulbar space; and (c) opacity in hypophyseal stalk surrounded by CSF (Scully et al.,1982).
Source: Used with permission from the Massachusetts Medical Society. Copyright © 1982, all rights reserved.

was 23,300/ mm^3, with PMNs = 87%. A CT scan of her head suggested a paranasal sinus process with retroorbital extension. There was opacity in the left ethmoid air sacs and engorged blood vessels in the retrobulbar space indicating an acute infection (Fig. 17A.4). In the emergency ward, following the CT scan, her nose was anesthetized, and 2 mL of thick bloody matter was extracted by needle. Histopathologic examination with fungal stain (GMS) demonstrated branching, nonseptate hyphae (Fig. 17A.5). The presence of large, nonseptate hyphae in the nasal discharge indicated a mucoraceous mold, but no cultures were taken so the identity of the etiologic agent is unknown.

Diagnosis

The diagnosis was rhinocerebral mucormycosis secondary to diabetic ketoacidosis.

Therapy and Clinical Course

An insulin drip was started to restore metabolic balance. Surgical debridement of the left maxillary and ethmoid sinus was performed and she received AmB-deoxycholate (2 g), 1 mg/kg/day in the first week. Afterwards, the patient developed complete facial paralysis, then hemiparesis, and died 2 days later.

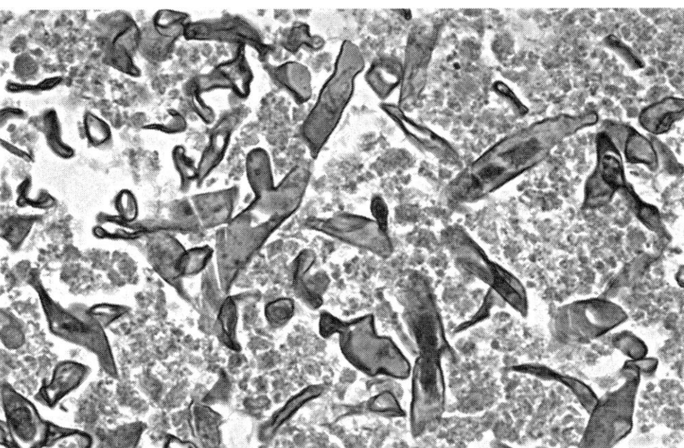

Figure 17A.5 Histopathology of mucormycosis due to *Mucor circinelloides*. The broad, twisted, nonseptate (*coenocytic*) hyphae are the major clue that this is not aspergillosis (GMS stain).
Source: Used with permission from Dr. Kazutoshi Shibuya, Department of Pathology, Toho University School of Medicine, Tokyo.

Comments

Although appropriate surgical, medical, and antifungal therapy were undertaken in a timely manner, the rapid infectious process could not be halted. The presence of protopsis, with a dilated and fixed pupil, indicates that the optic nerve and muscles controlling the eye were invaded. Although not available when this person was treated in 1982, lipid derivatives of AmB are preferred as primary therapy in order to increase the circulating blood level of the drug (please see "Case Presentation 2").

Case Presentation 2. Persistent Upper Respiratory Infection in a 57-yr-old Woman (Paul et al., 2006)

History

A 57-yr-old woman presented with a 2–3 week history of an upper respiratory infection. A complete blood count and differential showed she was pancytopenic with circulating blasts. She was diagnosed with Philadelphia chromosome-positive acute lymphoblastic leukemia (Ph+ ALL) and underwent induction chemotherapy with daunorubicin, vincristine, L-asparaginase, and prednisone. She initially did well but was readmitted a month later with fatigue and chills for the previous 5 days.

Clinical and Laboratory Findings

The patient was febrile to 38.0°C and neutropenic (300 leukocytes/μL) and was started empirically on vancomycin, cefepime, and liposomal AmB (5 mg/kg/day). A chest radiograph showed a right upper lobe (RUL) consolidation. CT scan of the chest confirmed a RUL pleural based mass consolidation and small pulmonary nodules, mostly in the left lower lobe. She underwent fine-needle aspiration of the RUL lesion, which revealed broad nonseptate hyphae consistent with a mucoraceous mold. Vancomycin and cefepime were discontinued as neutropenia resolved, and she was maintained on liposomal AmB. A chest CT scan 1 week later showed an increased lesion size with new cavitation. A short course of VRC resulted in further progression of the RUL mass. Liposomal AmB was resumed at a higher dose (7.5 mg/kg/day). The patient was maintained on imatinib (Gleevec®, Novartis) for Ph+ ALL therapy.

Her condition deteriorated, and she was referred for further treatment of mucormycosis and for evaluation as a candidate for allogeneic SCT for Ph+ ALL. The patient received PSC, an azole active against *Mucoromycotina*. However, the RUL mass increased in size.

Surgery

Because of its large size and pulmonary parenchymal necrosis with cavitation, surgery was required in addition to antifungal therapy. The patient underwent a RUL apical segmentectomy with a partial pleurectomy through a right thoracotomy for complete resection of the large fungal mass. The resected lung tissue grew *Cunninghamella bertholletiae*.

Diagnosis

The diagnosis was pulmonary mycosis due to *Cunninghamella bertholletiae* secondary to acute lymphocytic leukemia.

Clinical Course

The patient recovered rapidly and was maintained on PSC and imatinib. She underwent a successful myeloablative allogeneic SCT 4 months after surgical resection of the RUL fungal mass. She was conditioned for the SCT with cyclophosphamide and total body irradiation. The patient received both tacrolimus and sirolimus for GVHD prophylaxis. A chest CT scan before transplantation showed no signs of recurrent mycosis. She was discharged on an immunosuppressive regimen of tacrolimus and sirolimus and was maintained on PSC with no recurrent infection despite her immunosuppressed state.

Comment

Mucoromycotina are not generally susceptible to VRC. The clinical response to PSC was initially poor but after surgical resection, suppressive therapy with PSC was continued to good effect. This patient responded well to surgery and PSC, making it possible to proceed with an allogeneic SCT without a recurrence of *Cunninghamella* infection. *Cunninghamella* infections are uncommon and, compared with other *Mucorales*, have reduced susceptibility to antifungal agents.

17A.4 DIAGNOSIS

Mucormycosis is usually superimposed on an underlying condition and occurs when the immune or endocrine system has been disrupted. (Please see Section 17A.7, Epidemiology and Risk Groups/Factors). Regardless of the causal mold in this group, the clinical syndromes and pathologic findings are similar. Growth of mucoraceous fungi in tissue is rapid and can be fatal without prompt intervention. Diagnosis of the *Mucorales* must be rapid, relying on direct microscopic examination of scrapings or exudates, histopathologic examination of sinonasal or other biopsy material, and diagnostic imaging. Large, sparsely septate (*coenocytic*) hyphae, some with right angle branching, often folded, twisted, or distorted, may be seen in scrapings and around or within blood vessels and in nearby tissue, because these fungi are angioinvasive. Cultures are not always positive, even when typical hyphal elements are seen microscopically.

Differential Diagnosis

For rhinocerebral mucormycosis, other infectious diseases in the differential diagnosis are bacterial orbital cellulitis, cavernous sinus thrombosis, and central nervous system aspergillosis. Pulmonary mucormycosis must be differentiated from other fungal pneumonias—especially aspergillosis and *Pseudallescheria/Scedosporium* mycosis.

17A.5 ETIOLOGIC AGENTS

Rhizopus species are the frequent cause of mucormycosis, accounting for approximately half of the cases; with *Rhizopus oryzae* (syn: *R. arrhizus*) the most common species, followed by *R. microsporus* var. *rhizopodiformis*, the second most frequent clinically encountered species. *Mucor* species account for approximately 18% of reported cases (Roden et al., 2005), followed by *Cunninghamella bertholletiae* (7%), *Apophysomyces elegans* (6%), and *Lichtheimia* (formerly *Absidia*) species, 5%. Other less common agents are shown in Table 17A.1. *Rhizomucor pusillus*, formerly *Mucor pusillus*, is included as it is believed to account for a significant number of cases previously identified by its former name.

The predominant human pathogen of the *Mucorales, Rhizopus oryzae*, often is referred to as *Rhizopus arrhizus*, a synonym. *R. arrhizus* and *Rhizopus oryzae*

Table 17A.1 Classification of Pathogenic *Mucoromycotina* and *Entomophthoromycotina*

Mucoromycotina *Mucorales*	*Entomophthoromycotina*
Mucoraceae	*Entomophthorales*
Lichtheimia (formerly *Absidia*) *corymbifera*	*Conidiobolus*
	C. coronatus
Apophysomyces	*C. incongruus*
Mucor	*Basidiobolus*
M. circinellinoides	*B. ranarum*
M. hiemalis	
M. racemosus	
M. rouxianus	
Rhizomucor	
R. pusillus	
R. miehei (animal pathogen)	
Rhizopus	
R. oryzae	
R. azygosporus	
R. microsporus	
R. stolonifer	
Cunninghamellaceae	
C. bertholletiae	

remained separate species until they were found to have 95% DNA–DNA homology. The correct designation for *R. arrhizus/oryzae* is still in dispute. The lack of type material and the lack of a definitive description and illustrations have led some individuals to conclude that *R. oryzae* is most appropriate, with *R. arrhizus* being reduced to synonymy. The literature contains references to this mold under both designations, so it is necessary to understand they are the same fungus. A discussion of taxonomic changes resulting from phylogenetic analysis may be found in Boekhout et al. (2009).

Mucor species once were believed to be the second most common causes of mucormycosis. This misperception was because many case reports did not provide culture confirmation, identifying "mucormycosis" only on the appearance of hyphae in tissue sections. The nomenclature of the mucoraceous fungi has changed over time. Cases of *Mucor pusillus* infection, although relatively few, were reallocated when the thermophilic *Mucorales* were assigned to the *Rhizomucor* genus. *Lichtheimia corymbifera* was formerly called *Absidia corymbifera*.

Here the major species involved in mucormycosis are listed in alphabetic order, not in order of clinical importance. They are identified by the microscopic morphology of their sporangophores. Photomicrographs of the most important species are found in Section 17A.13, Laboratory Detection, Recovery, and Identification.

Apophysomyces elegans PC Misra, KJ Srivast. & Lata, *Mycotaxon* 8: 377 (1979).

Cunninghamella bertholletiae Stadel, *Über neuen Pilz, Cunn. bertholletiae* (Diss., Kiel): 1–35 (1911). Formerly this mold was known as *Mucor pusillus*.

Lichtheimia corymbifera (briefly termed *Mycocladus corymbifer*, and before that *Absidia corymbifera*) Hoffmann K, Walther G, Voigt K, *Mycol Res* 542: 3 (2009). Isolated from soil, decaying vegetation, grain. Phylogenetic analysis has prompted the name change to *Lichtheimia corymbifera* (Boekhout et al., 2009). *Absidia*-like fungi are divided into three lineages: *Absidia sensu stricto* grows at temperatures below 40°C (mesophilic); *Leptomyces* (mesophilic); and *Lichtheimia*, which are thermotolerant species with good growth at body temperature.

Mucor species. There are 60 or more species in the genus *Mucor*. A medically important species is *Mucor circinelloides* Tiegh., *Ann Sci Nat*, *Bot*, *Ser*. 6, 1: 94 (1875). *Mucor* species are common in soil or dung, decaying vegetation, stored grains, and *Mucor* rot of apples.

Rhizomucor pusillus (Lindt) Schipper, *Stud Mycol* 17: 54 (1978).

Rhizopus oryzae (syn: *R. arrhizus*) A Fisch *Rabenhorst's Kryptogamen-Flora von Deutschland, Oesterreichs und der Schweiz*, Edn 2 (Leipzig) 1(4): 233 (1892).

17A.6 GEOGRAPHIC DISTRIBUTION/ECOLOGIC NICHE

The *Mucorales* are found worldwide on fruit, on bread, and in habitats such as soil and decaying organic matter. The *Mucorales* grow on leaf litter, dung, and ripe fruit, which contain soluble nutrients. They depend on the appearance of suitable substrates, which are soon exhausted. Extracellular proteinases and lipases as well as amylases often are secreted but only a few species are celluloytic or chitinolytic. These fungi tend to be thermotolerant growing at >37°C (Table 17A.2). Mucoraceous fungi causing disease grow rapidly on any carbohydrate substrate, producing asexual sporangiospores that become airborne and may alight on exposed tissue or be inhaled into the nasal turbinates or lungs.

Knowledge of the activities of *Mucorales* in the environment aids in understanding how they function as opportunistic pathogens (Table 17A.3) (Dijksterhuis and Samson, 2006). Many *Mucorales* are regarded as primary and secondary "sugar fungi." They are fast-growing and utilize simple compounds. Garrett (1951) proposed a theory of succession during decomposition in which

Table 17A.2 Morphologic Characteristics of *Mucoromycotina* of Medical Importance

Morphologic *characteristics*	*Lichtheimia (Absidia) corymbifera*	*Apophysomyces elegans*	*Cunninghamella bertholletiae*	*Mucor circinelloides*	Mucor *ramosissimus*	*Rhizomucor pusillus*	*R. microsporus var. rhizopodiformis*	*Rhizopus oryzae*
Presence of rhizoids	May be present	Present	May be present	Absent	Absent	Present, usually poorly developed	Present	Present
Rhizoid placement	On stolon between sporangiophores	Opposite sporangiophores	Opposite sporangiophores	Absent	Absent	On stolon between sporangiophores	Opposite sporiangiophores	Opposite sporangiophores
Sporangiophore branching	Freely branched in whorls of 2–5; swollen where merged with columella	Single and unbranched, arise from "foot cell"	Whorl of short lateral branches	Branched	Sympodial branching, no swelling in apex where it merges with sporangium	Single, *freely branched*, not swollen where merged with columella	Singly or in tufts of up to 4	Unbranched, singly or in tufts of up to 4
Length of sporangiophore	Up to 450 μm	Up to 300 μm	Varied	Not determined	Up to 2000 μm	Up to 1000 μm	200–1000 μm	500–3500 μm
Apophysis	Long, funnel shaped	Distinct, bell shaped	Absent	Absent	Absent	Absent or very small	Distinct, angular	Inconspicuous
Columella	Dome shaped to short ovoidal	Dome shaped to short ovoidal	Absent	Globose	Globose	Subglobose to pyriform	Pyriform	Ellipsoid
Sporangium (diameter)	Pyriform, 100–150 μm	Pyriform 20–58 μm	Swollen vesicles, 30–65 μm with denticles supporting round to ovoid sporangiola, (7–11 μm)	Globose 25–80 μm; wall dissolves and may leave basal collarettes	Globose, 80 μm	Globose up 60–100 μm	~100 μm	Globose up to 200 μm
Shape of sporangiospores	Smooth, round to ellipsoid	Smooth, subglobose, may be cylindrical	Round to oval finely echinulate	Ellipsoid	Subglobose to broadly ellipsoid	Globose, smaller than *R. oryzae*	Globose to ovoid, smooth	Varied, mostly lemon shaped
Maximum growth temperature (°C)	48–52°C, most other *Absidia* and *Lichtheimia* species are not as thermotolerant	Up to 43°C	45°C	Poor growth at 37°C, not above 39°C	Up to 36°C	54–58°C	50–52°C	40–45°C

References: de Hoog et al. (2000); Larone (2002); O'Donnell (1979).

Table 17A.3 Industrial Bioconversions and Food Spoilage by *Mucoromycotina* of Medical Importance

Mucoromycotina species	Industrial bioconversion	Spoilage
Lichtheimia corymbifera (formerly Absidia)	Used in process to produce fexofenadine (Allegra®), a drug used to relieve common cold symptoms.	Sunflower seeds, nuts; pathogenic for growing peaches
Cunninghamella elegans	Polyunsaturated acids by bioconversion of orange peels.	Brazil nuts, kola nuts
Mucor circinelloides	Grows in 15% NaCl. Used in starter cultures to prepare *su-fu*, a Chinese fermented soybean cake, then ripened in brine.	Cheese, yams, mangoes, meat cereals, soybeans
Rhizomucor pusillus	Acid proteinases used for meat and fish processing.	Meat, nuts, spices, sunflower seeds, mung beans
Rhizopus oryzae	Produces high levels of lactate from glucose, which is used as a food additive and to produce biodegradable plastics.	Soft rot of avocados, cassava, jackfruit, sweet potato, yams; spoils peanuts
	Produces lactic acid from recycled office paper.	
	Pectinases are used in winemaking and preserves.	

fungi act in sequence: first are sugar fungi, followed by cellulolytic, ligninolytic, and then secondary sugar fungi. Primary sugar fungi are first colonizers of sugar-containing plant material, growing and sporulating rapidly to avoid competition by competing soil microbiota; for example, *Rhizopus stolonifer* acts on over-ripe fruit.

Secondary sugar fungi also are confined to soluble substrates by growing near other fungi that release them. *Mucor hiemalis* may be found on rotting wood, deriving nutrients from the activities of wood-rotting ligninolytic basidiomycetes. While this theory has been criticized for its rigidity, it provides a framework for understanding an environmental role for *Mucorales*.

The ability of *Mucorales* to break down starch, fats, and proteins, and their capacity for limited fermentative metabolism, is used in Oriental food fermentations. Rapid growth and production of lipases and proteinases are activities that help explain their rapid progress in the infectious process. For example, Tempeh, the Indonesian food, is made from dehulled, partially cooked soybeans. Vinegar is added, lowering the pH to favor the growth of tempeh mold over competitors. A starter culture of *Rhizopus oligosporus* is added. The beans ferment for 24–36 h at 30°C, after which they are knit together by a mat of white mycelia. Gray or black patches of sporulation are normal on mature tempeh.

Apophysomyces elegans is distributed in tropical and subtropical climates. Wounds contaminated with soil are the most common type of infection. Most patients who develop cutaneous–subcutaneous mucormycosis from this fungus are immune-normal. Other clinical forms involving *A. elegans*, such as rhinocerebral or pulmonary disease, are associated with the typical risk factors (please see Section 17A.7, Epidemiology and Risk Groups/Factors). Approximately 6% of all cases of mucormycosis are caused by *A. elegans*.

Cunninghamella species are well-known environmental molds, isolated from soil, peat, sewage, water, air, flowers, and other vegetation worldwide. They also have been cultured from various seeds and nuts including Brazil nuts, peas, and the seeds of spinach, parsley, flax, peppers, pumpkins, lettuce, carrots, and musk melons (Ribes et al., 2000). *C. bertholletiae*, the only human pathogen of this genus, is widespread and is thus often considered to be a clinical contaminant. *Cunninghamella* infections reported in the literature worldwide represented 7% of all cases of mucormycosis (Roden et al., 2005). Most cases involved pulmonary or pulmonary-disseminated disease.

Lichtheimia corymbifera (formerly *Absidia*) is a saprobe distributed worldwide in soil and decaying vegetation and has been isolated from grains, nuts, seeds, and cotton. It also has been cultured from house dust. Growing well under wet, humid conditions, it was cultured from waterlogged grasslands, swamps, mangrove mud, human sewage, animal dung, and bird and bat guano (Ribes et al., 2000). *Lichtheimia corymbifera* was recovered from dried grass in the right boot of the Tyrolean Ice Man, with an age estimated at 5300 years. *L. corymbifera* is the etiologic agent of 5% of mucormycosis cases. *Absidia*-like molds fall into three groups based on temperature tolerance: *Absidia* (*sensu stricto*) does not grow above 40°C. *Lichtheimia* consists of thermotolerant species that show good growth at body temperature (Boekhout et al., 2009).

Mucor species are globally distributed soil-dwelling saprobes. Spores have been found in air samples or dust from both homes and hospitals. *M. circinelloides* was isolated from cereals, nuts, flour, and oranges associated with an outbreak of onychomycosis (Sutherland-Campbell and Plunkett, 1934). *Mucor* species are relatively avirulent and are not very thermotolerant. Case findings in the English language literature over the period 1885–2004 reported 85 cases (18%) out of a total of 465 patients for whom microbiologic evidence of a *Mucor* species as the etiologic agent of mucormycosis was obtained (Roden et al., 2005).

There are 14 or more species in the genus *Rhizopus*. *Rhizopus* species cause soft rot of sweet potatoes and fruit, for example, cherries, figs, peaches, pears, grapes, and strawberries. *R. oryzae* growing on jackfruit, *Artiocarpus heterophyllus*, causes economic damage to this edible tropical fruit. *Rhizopus* spores have been found in house dust and from *settle plates* placed in kitchen sites, consistent with their role of "bread mold." Spores from *R. oryzae* and *R. microsporus* were recovered from air samples or dust from HVAC systems. *Rhizopus* spp. also have been found as part of the microbiota of dog hair, peaking in occurrence during the summer months. *Rhizopus* enzymes are used in industrial processes for making alcohol, fumaric acid, and lactic acid. The combined *Rhizopus* species are isolated from at least half of all mucormycosis cases according to the largest series (Roden et al., 2005).

Rhizomucor species are found worldwide in the air, soil, and organic matter. They are isolated from garden composts and municipal wastes, cultivated mushroom beds, manure, guano, leaf mold, and grass. *R. pusillus* also has been recovered from various foods including grains, seeds, nuts, and beans. Spores from *Rhizomucor* are small and have been isolated from air samples collected outdoors and in hospital air (Ribes et al., 2000). Despite its widespread occurrence in nature, *R. pusillus* is a rare cause of disease in humans accounting for 4% of all mucormycosis cases. Pathogenicity in *R. pusillus* is linked to its thermotolerance, allowing it to grow in febrile patients at temperatures above 39–40°C. The temperature range of growth for *R. pusillus* is from 20°C to 58°C.

Summary of Relative Clinical Importance

A large group of clinical isolates from different regions of the United States were identified by morphologic and molecular methods (ITS region of rDNA) (Alvarez et al., 2009). *Rhizopus oryzae* accounted for approximately half of the collection. The remainder were *Rhizopus microsporus* (22.1%), *Mucor circinelloides* (9.5%), *Lichtheimia* (formerly *Absidia*) *corymbifera* (5.3%), *Rhizomucor pusillus* (3.7%), *Cunninghamella bertholletiae* (3.2%), *Mucor indicus* (2.6%), *Cunninghamella echinulata* (1%), and *Apophysomyces elegans* (0.5%).

17A.7 EPIDEMIOLOGY AND RISK GROUPS/FACTORS

Mucormycosis is rare and a population-based surveillance indicated an incidence of 1.7 cases/million population, or ~500 cases/year in the United States (Spellberg et al., 2005). Autopsy series found 1–5 cases/10,000 autopsies, or about 10–50-fold less common than candidiasis or aspergillosis. A review of the English literature from 1885 to the 1990s revealed 929 confirmed mucormycosis cases (Roden et al., 2005). (Please also see Section 17A.6, Geographic Distribution/EcologicNiche for the species distribution.)

- Risk factors for the 929 patients with proven mucormycosis were sorted according to the presence or absence of a known underlying disease. The majority of mucormycosis patients had underlying conditions. The leading conditions were diabetes mellitus, 36%; malignancy, 17%; solid organ transplants, 7%; deferoxamine therapy, 6%; and injection drug use, 5%.
- Patients with no known underlying condition comprised 176 or 19% of the total cases. Of these, penetrating injuries accounted for 25%; surgery, 18%; burns, 6%; and motor vehicle accidents, 3%.
- Approximately 65% of mucormycosis cases occur in males but there is no explanation for this preference (Roden et al., 2005).

Five clinical forms of mucormycosis and their *major* associated risk factors are (in alphabetic order, not in order of incidence):

- Cutaneous—Disrupted barriers due to penetrating injury with soil contamination, burns, maceration by moist surface, or direct access via IV catheters or subcutaneous injections, or surgical wounds.
- Disseminated–Deferoxamine therapy.
- Gastrointestinal—Malnutrition, prematurity.
- Pulmonary–Neutropenic patients including those with hematologic malignancy and stem cell transplantation.
- Rhinocerebral—This category usually includes invasive sinusitis, sino-orbital disease, and true rhinocerebral disease. The major risk factor is diabetic ketoacidosis and also immunosuppression.

Trend

The rising incidence of diabetes and cancer in the increasingly obese and elderly U.S. population is linked to an increase over the past two decades of mucormycosis. The most recent survey in 2005 (CDC National Center for Health Statistics, 2005, website) estimated that 7% of the U.S. population or 21 million persons have diabetes and more than 6 million do not know they have the disease. No study has determined the number of mucormycosis cases/100,000 patients with diabetes, so the trend in incidence of mucormycosis among patients with diabetes is not known. Mucormycosis is, surprisingly, the diabetes-defining illness in 16% of newly diagnosed patients with diabetes.

The potential influence of statin drugs on reducing the incidence of mucormycosis in patients with diabetes was proposed as a hypothesis by Kontoyiannis (2007). Statins induce an apoptosis-like cell death in *Mucor racemosus*. Lovastatin has in vitro activity against a range of *Mucoromycotina* and has in vivo activity in a model of mucormycosis.

Over the decades of the 1980s to 1990s there was a rising proportion of mucormycosis cases in immunocompromised patients: those with hematologic malignancy, SCT recipients, and patients with solid organ transplants (Roden et al., 2005). Illustrative of this trend is the experience in a tertiary care center in Taiwan where, in the period 1986–2003, there were 39 cases of proven or probable mucormycosis; 74% were in immunocompromised patients whereas the remainder had diabetes mellitus. Mortality of mucormycosis in immunocompromised patients was 65.5%, with a mean time from diagnosis to death of 10 days. The incidence of mucormycosis increased from 1.86/100,000 discharges in 1986–1991 to 4.13/100,000 discharges in 1998–2003 (Wang et al., 2006).

Forty-one cases of rhinoorbital or rhino-orbital–cerebral mucormycosis were treated at the UCLA Medical Centers during the period 1994–2006 (Reed et al., 2008). Eighty-three percent were diabetics, of whom 16 patients (47%) received no medication for diabetes at the onset of mucormycosis. Less than half of the diabetic patients (41%) were acidotic when admitted. One-third of this patient population had cancer, two were solid organ transplant recipients, and two received SCTs. Of the culture-positive specimens, all were identified as *Rhizopus* species except for one *Rhizomucor* species. In addition to surgical debridement, outcome was linked to the use of AmB monotherapy or combination therapy of AmB with caspofungin (CASF) (Reed et al., 2008). Forty-five percent of patients receiving AmB monotherapy survived at 30 days after hospital discharge, whereas all six of those receiving combination therapy survived. In the subgroup with CNS dissemination, 25% of those on AmB monotherapy survived, whereas all four such patients receiving combination therapy survived. Although echinocandins have poor activity in vitro against *Mucorales*, clinical experience suggests they can be effective in combination therapy (please see Section 17A.12, Therapy).

The use of VRC as prophylaxis and treatment for molds since 2002 correlates with at least 23 cases of breakthrough mucormycosis. The most common clinical form was pulmonary mucormycosis, followed in prevalence by disseminated disease. The associated mortality was approximately 83%. Some patients received VRC prophylaxis for 49–210 days before developing mucormycosis (reviewed by Chayakulkeeree et al., 2006). In a major cancer treatment center in Texas, for a series of 27 patients with mucormycosis, VRC prophylaxis was an independent risk factor with an odds ratio of 10.4 (Kontoyiannis et al., 2005). Considering other risk factors, the possibility of mucormycosis is greater in immunosuppressed patients who develop sinusitis while receiving VRC prophylaxis, especially those with diabetes and malnutrition.

Hematologic Malignancies and SCTs

Major cancer treatment centers in Seattle, Washington, and Houston, Texas, tracked an increase in mucormycosis cases during the periods 1985–1999 and 1995–1999, respectively (Marr et al., 2002; Kontoyiannis et al., 2005). Mucormycosis in SCT recipients, in more than half of cases, occurred >90 days post-transplant. Even so, the relative proportion of mucormycosis versus aspergillosis in SCT patients weighs heavily toward aspergillosis with 294 cases of aspergillosis diagnosed during 1985–1999 versus 29 cases of mucormycosis in the Seattle cancer center (Marr et al., 2002). Risk factors for mucormycosis are treatment of GVHD with systemic steroids, prolonged neutropenia, and/or diabetes mellitus. In a 5 year surveillance program, mucormycosis occurred in 8% of SCT recipients with fungal infections, compared with aspergillosis in 44%, and candidiasis in 28%. The incidence of mucormycosis post-SCT was found to be 0.25% in the first 6 months (Kontoyiannis et al., 2010).

Solid Organ Transplant Recipients

Prospective surveillance in 11 medical centers in the United States and Europe identified 53 liver and heart transplant recipients with invasive mold mycoses over a 3 1/2 year period between 1998 and 2002 (Husain et al., 2003). Invasive mold infections were due to *Aspergillus* species in 37 (69.8%) patients. Three patients (5.7%)

had mucormycosis, all with a fatal outcome. However, a more recent study showed that mucormycosis was found in 2.3% of solid organ transplant recipients with fungal infections. The incidence was estimated to be 0.08% of all transplants conducted over the 5 year period (Pappas et al., 2010). In this study, all organ transplant types seemed to be equally affected.

Nine cases of mucormycosis were diagnosed and treated at a transplant center in Austria, out of a series of 2878 solid organ transplants (kidney, pancreas, islets, liver, heart, lung, and bowel) performed between 1995 and 2006 (Stelzmueller et al., 2008). An increase in non-*Aspergillus* invasive mold infections followed the introduction of CASF and VRC therapy. These agents are not generally active against *Mucoromycotina*.

Mucormycosis in Kidney Transplant Recipients Six renal transplant recipients (Forrest and Mankes, 2007) with biopsy-proven invasive mucormycosis received amphotericin B lipid complex (ABLC) in doses >5 mg/kg between 2000 and 2004 at a medical center in Maryland. All six patients had diabetes mellitus, were receiving immunosuppressive agents, and underwent surgery in addition to ABLC therapy. Three patients who survived had significant surgical debridement, reduction of immunosuppression to minimal prednisone, and received a prolonged course of ABLC at 10 mg/kg/day. All survivors lost graft function during the course of their therapy. In renal transplant recipients, as in all cases of invasive mold infection, the early diagnosis of invasive mucormycosis is urgent along with early therapy with surgical debridement, reduced immunosuppression, and the use of high doses of lipid-AmB formulations.

Mucormycosis in Lung Transplant Recipients Lung transplant recipients have an incidence of mucormycosis of 13.7–14/1000 patients (Silveira and Husain, 2008).

Mucormycosis in Patients Requiring Chelation Therapy

Changes in the dialysis process have eliminated aluminum excess and, combined with the routine use of erythropoietin, have eliminated the need for chelation therapy, thereby reducing the risk of mucormycosis. Patients with myelodysplastic syndrome and those with other hematologic disease who develop iron overload and require deferoxamine are still at risk for mucormycosis.

Overall mortality from mucormycosis was 84% in the 1950s and improved starting in the 1960s when AmB came into regular use. Overall mortality in the 1990s was still no less than 40%.

17A.8 TRANSMISSION

Sporangiospores of the *Mucorales* are carried on air currents and inhaled from the environment or alight on open wounds or burns. An unusual circumstance is when spores are accidently injected with insulin injections, via manipulation of indwelling catheters, or by injection drug users.

17A.9 DETERMINANTS OF PATHOGENICITY

17A.9.1 Host Factors

PMN Activity

Chemotactic activity of neutrophils from diabetic patients is significantly reduced compared with PMNs from healthy controls, and this trend is accentuated when the glucose concentration is >12 mmol/L. Transmigration of PMNs through endothelial cells is impaired in parallel with increased levels of circulating advanced glycosylation end (AGE) products (Peleg et al., 2007). Excess glucose combines with free amino acids on circulating or tissue proteins, a nonenzymatic process. AGE-conjugated to plasma proteins inhibits the PMN functions of transendothelial migration and production of reactive oxygen metabolites. PMNs have a receptor for AGE products—"RAGE." The binding of AGE to its receptor reduces PMN nitric acid production.

Ketone bodies (acetone, acetoacetate) inhibit the chemiluminescence of PMNs in response to zymosan (Saeed and Castle, 1998). That assay is a surrogate for the PMN oxidative burst. In summary, diabetes is accompanied by consistent defects in PMN chemotaxis, phagocytosis, and oxidative microbicidal activities. PMNs of diabetics are impaired in the ability to adhere to endothelial cells and to migrate to the site of inflammation. Ketosis further diminishes the ability of PMNs to produce reactive oxygen species. The net result is functional neutropenia.

17A.9.2 Microbial Factors

17A.9.2.1 Iron Utilization at Acidic pH

Patients in diabetic ketoacidosis, acidosis from other causes, *or* being treated with the iron chelator deferoxamine have an increased incidence of mucormycosis. *Mucorales* species utilize deferoxamine as a siderophore to supply iron to the fungus. Patients in systemic acidosis have elevated serum iron, released from iron binding proteins during acidosis. *Rhizopus oryzae* secretes a high-affinity iron permease during growth on low free iron conditions but its role in pathogenesis is not known.

Proteinases

Rhizopus oryzae secretes an aspartyl proteinase (Sap) and contains in its genome a suite of *SAP1-SAP4* genes (Farley and Sullivan, 1998). In vitro, secretion of the enzyme occurs at pH 4.5; but ribosomal gene transcripts are detectable at pH 7. At present there is no certainty that *R. oryzae* Sap is secreted during infection. An endoprotease Arp (alkaline *Rhizopus* protease) was purified from the *Rhizopus microsporus* var. *rhizopodiformis* recovered from a nonfatal case of rhino-orbital mucormycosis. The proteinase is active at ≥pH 6.0.

Lipase

Lipase of *R. oryzae* and related *Rhizopus* species are being investigated for use in biotransformations to produce biodiesel fuel from plant oils (Kaieda et al., 1999). The direct involvement of *R. oryzae* lipase in pathogenesis has not been investigated. A case of surgical wound mucormycosis caused by *R. microsporus* var. *rhizopodiformis* in a renal transplant recipient implicated lipase as a factor in the disease (Nordén et al., 1991). Histologic examination of debrided material from the surgical wound revealed necrosis of the fat tissue and massive hyphal invasion of the perirenal fat. Further more, in vitro experiments showed the clinical isolate was capable of high extracellular lipase production.

Bacterial Endosymbionts Produce Toxins

Endosymbiotic bacteria of *Rhizopus* species produce toxins that enhance fungal pathogenicity. Rhizonin is a hepatotoxic cyclopeptide isolated from *Rhizopus microsporus*. Rhizonin is produced by bacteria belonging to the genus *Burkholderia*, which were localized within the fungal cytosol (Partida-Martinez et al., 2007). Rhizonin causes degeneration and necrosis to renal tubular epithelium and liver tissue in rats. The growing rates of antimicrobial resistance may have selected for multidrug-resistant endosymbiotic bacteria of *Rhizopus* species, which could partially explain the emergence of mucormycosis. This hypothesis, if it can be proved, is the first example of potentiation of pathogenicity of opportunistic fungi by a bacterial endosymbiont (Chamilos et al., 2007).

Damage to Endothelial Cells

R. oryzae germlings[4] adhere to and damage human umbilical vein endothelial cells in vitro. This endothelial cell damage requires direct contact and phagocytosis of the fungus (Ibrahim et al., 2005).

[4]Germlings. *Definition*: Germinated conidia or, in this case, sporangiospores.

Cell Wall Composition

R. oryzae has an *FKS* gene (glucan synthase gene) and CASF inhibits ß-glucan synthase activity in crude *R. oryzae* membrane preparations (Lamaris et al., 2008). In vitro synthesis of ß-(1→3)-D-glucan was achieved in *R. oryzae* extracts, which was inhibited by CASF. The ß-glucan synthase of *R. oryzae* required a 1000-fold higher concentration of CASF to achieve 50% inhibition compared with the ß-glucan synthase of susceptible fungi, *Candida albicans* and *Aspergillus fumigatus*. Thus, it appears that one explanation for the in vitro resistance of *Mucoromycotina* to echinocandins is low affinity for glucan synthase.

There is a dearth of information about the ß-(1→3)-D-glucan content of *Mucorales* cell walls. Cell wall composition analysis of *Mucorales* is based on old studies which found that in place of ß-glucan they have de-*O*-acetylated chitin—chitosan. In *Mucor rouxii*, a model organism, the fibrillar components of the wall are chitin and chitosan. These fibrils are embedded in matrix materials, composed mostly of polysaccharides rich in glucuronic acid. Two such polyuronides in *M. rouxii* are mucoran (a heteropolymer of D-glucuronic acid, D-mannose, D-galactose, and L-fucose) and mucoric acid (a homopolymer of D-glucuronic acid) (Dow et al., 1983). As the cell walls of fungi have become an antifungal drug target, updated structure analysis of these *Mucorales* wall polymers is essential, parallel to development of inhibitors of their biosynthesis.

Antiserum raised against an extracellular polysaccharide (EPS) fraction obtained from *Mucor hiemalis* was found to react in EIA against EPS fractions from *L. corymbifera, M. circinelloides, M. racemosus, R. oryzae*, and *R. pusillus* among other *Mucorales*, but was not crossreactive with *Aspergillus fumigatus, Fusarium*, or *Candida* species (de Ruiter et al., 1992). While the EPS from *M. hiemalis* was not characterized further, it is tempting to speculate that it may be the mucoran heteropolymer described earlier (Dow et al., 1983). Antiserum specific for *Mucorales* is an important reagent for tissue diagnosis. Moreover, *Mucorales* EPS contribute to damage produced when *R. oryzae* or other *Mucorales* are endocytosed by endothelial cells.

At the present state of knowledge only a tentative explanation can be offered of how pathogenic factors from *Mucoromycotina* function in the pathogenesis of mucormycosis. Fragmentary knowledge of the microbial factors warrants more research on their characterization. *Mucorales* species are thermotolerant and rapid growing (because of their *aseptate* or pauci-septate method of hyphal extension) and are able to use simple carbon sources, especially glucose, for growth. If they can find a source of ferric ion, and the host is neutropenic, or the neutrophil response is blunted, they are capable of binding to and damaging endothelial cells. Their aggressive growth leads to deep extension of the infection through perivascular channels, invading muscle, fat, cartilage, and eroding or infecting bone (Chan et al., 2000).

17A.10 CLINICAL FORMS

Rhinocerebral

Rhinocerebral is the most frequent clinical form of mucormycosis accounting for up to half of all cases and found most often in the setting of diabetic ketoacidosis or acidosis from other causes. A more precise term is rhino-orbito–cerebral mucormycosis. This form is also found in patients with prolonged neutropenia, undergoing SCTs, or during steroid therapy for GVHD, much less frequently in solid organ transplant recipients.

Inhaled spores are deposited in the nose or paranasal sinuses where they germinate and hyphae invade rapidly. Sinusitis or periorbital cellulitis are the first symptoms of disease followed by eye or facial pain and numbness. *A red patch on the cheek of a diabetic is a warning sign of rhinocerebral infection*. Blurred vision and soft tissue swelling then occur. The infected tissue becomes red and swollen, then turns purple, progressing finally to a black eschar due to thrombosis of local blood vessels and infarction of the surrounding tissues. Extension through the hard palate can result in painful necrotic ulcers which, when scraped, can reveal the hyphal elements.

Fever is variable and may be absent in up to half of cases. If untreated, spread is from the ethmoid sinus eroding the bone of the sinus wall and spreading into the orbit. Proptosis is caused by invasion of the muscles controlling the eye. Disease progression produces new signs and symptoms—periorbital swelling, orbital cellulitis, blurred vision or loss of vision, external ophthalmoplegia, diplopia, ptosis, proptosis, chemosis, eyelid gangrene, retinal detachment, and endophthalmitis (Chayakulkeeree et al., 2006). Rapid extension occurs across bone, cartilage, and muscle to adjacent tissues. Bilateral proptosis is an ominous sign of cavernous sinus thrombosis. A clot in the cavernous sinuses (one in each hemisphere) causes paralysis of the cranial nerves which traverse through the cavernous sinus. This infection is life-threatening and requires immediate treatment with an appropriate antimicrobial agent and surgical drainage of the primary source of the infection, for example, the sphenoid sinus or facial abscess. Sometimes the internal carotid arteries are invaded by the fungal hyphae.

Posterior spread from the orbit or sinuses to the brain indicates a grave prognosis. Vision loss can result from invasion of the optic nerve or arteriolar invasion resulting in infarction. Facial numbness and pupillary dilation can occur if cranial nerves 5 and 7 are involved. Bloody nasal discharge indicates that invasion of the CNS has occurred through the ethmoid sinus and cribriform plate to the meninges and brain or through the nasal turbinates, the sphenoidal sinus into the cavernous sinus causing thrombosis, and extensive cerebral infarcts. Overall mortality is ~62% in patients with rhinocerebral mucormycosis, whereas sinus disease without cerebral involvement has a much lower mortality rate, 16% (Chayakulkeeree et al., 2006).

Diagnosis of Rhinocerebral Mucormycosis

Up to half of cases are diagnosed postmortem. Surgical exploration and biopsy of the suspected area of infection are necessary in high-risk patients. CT and MRI may appear normal early in infection. Later, they are important methods to define the extent of disease. Histopathologic evidence is essential to establish the diagnosis because culture results are equivocal: these molds are common contaminants and, on the other hand, may fail to grow from biopsy tissue. (Please see Section 17A.13, Laboratory Detection, Recovery, and Identification.) At the same time empiric therapy with AmB or, preferably, lipid formulations of AmB should be started because time is of the essence.

Surgical Intervention in Rhinocerebral Disease

Preoperative imaging defines the extent of disease. Contrast-enhanced CT scans show edematous mucosa, fluid filling the ethmoid sinuses, destruction of periorbital tissues, and bony margins. MRI can identify the intradural and intracranial extent of disease, cavernous sinus thrombosis, or thrombosis of the cavernous portion of the internal carotid artery (Mohindra et al., 2007). Surgical debridement is necessary in sinusoidal disease. Surgical findings include polyps in the paranasal sinuses, necrotic tissue, serosanguinous discharge, and/or black eschar carpeting the turbinates and hard palate. Orbital pathology is treated by removal of the lamina papyracea or enucleation of the eye. Surgery for intracranial disease can take the form of burr hole and aspiration of cerebral abscess, or craniofacial resection to debulk a frontal lobe mass (Mohindra et al., 2007).

Pulmonary

Patients at greatest risk are neutropenic cancer patients and those receiving SCTs or lung transplants. The route of transmission may be via inhalation of sporangiospores or by hematogenous spread to the lungs. Patient symptoms include dyspnea, cough, and chest pain. The clinical presentation is less specific than rhinocerebral disease and diagnosis of pulmonary mucormycosis is more difficult. The hyphae are angioinvasive, causing necrosis as they invade across tissue planes, resulting in cavities or hemoptysis. Radiographic findings include nodules, cavities, and lobar consolidation toward the great vessels of the mediastinum. Invasion of pulmonary vessels can lead to wedge-shaped infarcts. Without treatment patients may die of disseminated mucormycosis within 2–3 weeks. The overall mortality rate in pulmonary mucormycosis is 76% (Chayakulkeeree et al., 2006).

Cutaneous

Disruption of the normal anatomic barrier of the skin provides the portal of entry. This may include penetrating injuries involving soil, burns, or crushing wounds such as in auto accidents. Diabetics and neutropenic cancer patients are especially vulnerable. The consequences of cutaneous inoculation with *Mucorales* may be underestimated in that these lesions can extend deeply to the tendons, muscle, and bone and disseminate hematogenously with a high associated mortality (Roden et al., 2005). Infection may occur at the injection site for insulin or catheter insertion sites.

Necrotizing Cellulitis and Fasciitis

Associated with serious burns or other wounds, *R. oryzae* or other *Mucorales* can rapidly grow through fat, muscle, fascia, and bone. This clinical form occasionally has been associated with contaminated surgical dressings and splints. *Rhizopus microsporus* var. *rhizopodiformis* was isolated from a surgical wound covered by an adhesive ostomy bag. In that instance the epidemiologic investigation revealed that karaya, a natural gum adhesive used to hold the ostomy bag in place, was the likely source of *Rhizopus* (LeMaile-Williams et al., 2006).

An outbreak of neonatal mucormycosis in four infants caused by *R. microsporus* was traced to a box of contaminated tongue depressors used to splint limbs of the neonates in the intensive care unit (Mitchell et al., 1996). All infants were treated with liposomal AmB. Two infants died, one with autopsy confirmed cutaneous-disseminated mucormycosis. A third infant required amputation of the forearm to prevent further spread of the fungal infection. The fourth infant's cutaneous mucormycosis was successfully treated, but death occurred from other causes related to prematurity.

Apophysomyces elegans is implicated in invasion of cutaneous and subcutaneous tissues resulting from a penetrating injury with soil contamination. This fungus is more prevalent in tropical and subtropical areas. The clinical form presents as necrotizing cellulitis. The combination of tropical locale and cutaneous infection was underlined by case reports following the December 26, 2004 tsunami in Southeast Asia. Injuries suffered by one man in that disaster presented as multiple wounds, which worsened on antibiotic therapy, developing fever and an elevated PMN count. Histologic examination of debrided tissue showed invasion of necrotic vessel walls by aseptate or sparsely septate broad fungal hyphae, associated with arterial thrombosis and necrosis of muscle and fat. Necrotizing fasciitis of the wounds required surgical resection. *A. elegans* grew from the debrided tissue. This patient responded to the surgery and liposomal AmB therapy (Andresen et al., 2005).

Despite treatment with liposomal AmB, cutaneous mucormycosis may progress with painful swelling and redness or blackening of the involved tissues. Thrombosis with extensive necrosis of the involved tissues is common. Lesions may appear black and hemorrhagic or may show white fungal overgrowth on the infected tissues. Acute and chronic inflammatory exudates often appear with necrotic debris and thick "anchovy paste-like" pus. Exudates also may be serous, with few inflammatory cells and little necrotic debris (Ribes et al., 2000). If unchecked, disease may progress to osteomyelitis and multi-organ disseminated mucormycosis.

Gastrointestinal

These rare infections occur in premature neonates and in malnourished infants and children. The fungus may be acquired by ingestion; the sites involved are the stomach, colon, and ileum. The infection is acute and rapidly fatal, usually diagnosed at autopsy.

Disseminated

Pulmonary mucormycosis in neutropenic cancer patients is the site most often leading to extrapulmonary dissemination, especially to the brain causing cerebral abscess, but also to the heart, spleen, and other organs. Dissemination to the brain presents as sudden neurologic deficits or coma, with mortality approaching 100%. Deferoxamine therapy is a definite factor predisposing to disseminated mucormycosis. Primary therapy or prophylaxis with VRC has been associated in several case reports with development of disseminated mucormycosis; VRC has little or no activity against the *Mucorales*.

17A.11 VETERINARY FORMS

Mucormycosis is seen in livestock including horses, cattle, particularly feedlot cattle, deer, and hogs. Zoo animals also are affected including llamas, primates, birds, and marine mammals. Companion animal veterinary practice encounters mucormycosis in dogs. In cattle and sheep, ulcers in the rumen omasum and abomasum are seen. In feedlot cattle the disease often is fatal. Bovine respiratory infection, abortion, and cranial infections also occur, often mimicking tuberculosis with caseocalcareous lesions in bronchial and mediastinal lymph nodes. Gastric ulceritis and fatal disseminated mucormycosis have been reported in dogs and various birds.

Lichtheimia corymbifera (formerly *Absidia*) is the most frequent agent of mucormycosis among animals and was isolated in cases of mycotic bovine abortion, bovine mastitis, gastric disease in swine, and avian mucormycosis. *Lichtheimia* spp. are the predominant fungi cultured from skin scrapings of normal farm animals and also cause mucormycosis in guinea pigs, horses, and sheep.

Rhizomucor species are significant animal pathogens and are a common cause of bovine infertility, mycotic abortion, and mastitis in cattle, as well as disease in the newborn calf. *Rhizomucor* species cause pulmonary mycoses (also affecting other organs) in pigs, horses, cows, dogs, rabbits, ferrets, ducks, and seals. *Mucor amphibiorum* is the causative agent of ulcerative dermatitis and pneumonitis in the platypus, *Ornithorynchus anatinus* (Munday et al., 1998).

17A.12 THERAPY

Early therapeutic intervention is essential in mucormycosis. Extensive debridement of all necrotic tissue is necessary concomitant with antifungal therapy. Successful treatment requires control of the underlying disease whenever possible, including restoring metabolic balance in diabetics and cessation of treatment with deferoxamine. Diabetic ketoacidosis is controlled with insulin, correction of acidosis with sodium bicarbonate, and rehydration.

AmB-deoxycholate

AmB-deoxycholate has been the mainstay of antifungal therapy for mucormycosis. AmB-deoxycholate in high doses, 1–1.5 mg/kg/day (for a course of 2.5–3 g total), is required following surgery, as the fungi are relatively resistant. Liposomal AmB is preferable because its reduced nephrotoxicity permits a higher dosing regimen (10–15 mg/kg/day). A series of patients with hematologic malignancy treated with liposomal AmB had an associated survival rate of 67%, compared with 39% for AmB-deoxycholate therapy (Spellberg et al., 2005). The consensus is that liposomal AmB should be the preferred primary therapy for mucormycosis. AmB lipid complex, an alternative lipid formulation of AmB, has five-fold lower penetration into the CNS than AmB-deoxycholate and a lower response rate than AmB-deoxycholate for treating mucormycosis (Malani and Kauffman, 2007).

Azoles, in general, demonstrate low activity against the *Mucorales*. Mold-active prophylaxis or treatment of invasive mold infections with VRC has led to breakthrough mucormycosis because *Mucoromycotina* are not susceptible to that drug (Marty et al., 2004). (Please also see Section 17A.7, Epidemiology and Risk Groups/Factors.)

Posaconazole (PSC)

PSC (Noxafil®) is the first azole to have activity against *Mucorales* with efficacy comparable to liposomal AmB. Salvage therapy with PSC has been successful in mucormycosis cases and as primary therapy for pulmonary-disseminated *R. microsporus* disease (reviewed by Peel et al., 2008). A series of 91 patients with mucormycosis were reviewed in which salvage therapy consisted of PSC. Complete or partial responses occurred in 60% (reviewed by Malani and Kauffman, 2007). PSC is available only in oral formulation with a daily dose of up to 800 mg.

In vitro susceptibilities of some representative *Mucorales* species to AmB and azole antifungal agents are shown in Table 17A.4 (Singh et al., 2005). Almyroudis et al. (2007) found that the percentage of *R. oryzae* isolates tested that were susceptible in vitro to various antifungal agents was AmB, 100%; ITC, 50%; and PSC, 64%. Of nine *Lichtheimia corymbifera* isolates tested, all were susceptible to the three aforementioned antifungal agents. Of 13 *Cunninghamella* spp. isolates tested, 63%, 29%, and 75% were susceptible to AmB, ITC, and PSC, respectively. *Mucor circinelloides* isolates were all susceptible to AmB, but none were susceptible to ITC or PSC. In their series, and in general, none of the 11 *Mucorales* species tested were susceptible in vitro to CASF, representative of the echinocandin class of drugs. MICs of VRC were invariably higher than for PSC. In summary, AmB is still the most effective drug against the *Mucorales*, but if intolerance to it or to lipid-AmB derivatives occurs, then azoles are second-line drugs, pending the results of antifungal susceptibility testing, with a distinct preference for PSC.

Echinocandins

The echinocandin class of ß-(1→3)-D-glucan synthase inhibitors demonstrates minimal in vitro activity against *Mucorales*. Caspofungin (CASF) treatment of systemic fungal infections has led in a few instances to breakthrough mucormycosis. Echinocandins have no activity against the fungi that cause mucormycosis in standard in vitro susceptibility tests. However, *R. oryzae* expresses

Table 17A.4 In Vitro Susceptibility of *Mucorales* Species

Mucorales spp.	MIC (μg/mL)[a,b]						MIC (μg/mL)[b]		
	AmB		Itraconazole		Voriconazole		Posaconazole[c]		
	Geometric mean	Range	Geometric mean	Range	Geometric mean	Range	Range	MIC_{50}	MIC_{90}
Rhizopus oryzae	0.04	0.03–0.06	1.41	0.25–4	5.66	4 to >8	0.03–1	0.25	1
Mucor circinelloides	0.03	0.03	0.56	0.25–1	5.66	4 to >8	1–2	Not done	n.d.
Lichtheimia corymbifera	0.05	0.03–0.25	0.62	0.125–2	8	8 to >8	0.06–0.25	n.d.	n.d.
Rhizomucor pusillus	0.16	0.125–0.25	0.07	0.03–0.125	4	2–8	0.06–1[d]	n.d.	n.d.
Cunninghamella bertholletiae	0.35	0.25–0.5	2	1–4	>8	>8	0.06–1	0.5	n.d.

[a] CLSI M38 A broth microdilution method (Singh et al., 2005).

[b] The MICs were determined by visual inspection for the first well with complete growth inhibition; MIC_{50}, concentration required for 50% growth inhibition; MIC_{90}, concentration required for 90% growth inhibition

[c] M38 A broth microdilution method (Almyroudis et al., 2007).

[d] *Rhizomucor* species.

the target enzyme for echinocandins. Patients with rhino-orbital or rhinocerebral mucormycosis treated with the combination of AmB and CASF had significantly improved outcomes 30 days after hospital discharge and improved long-term survival, compared with patients treated with AmB monotherapy (Reed et al., 2008). These clinical data are supported by mouse models of mucormycosis, which have repeatedly demonstrated improved survival of mice treated with combinations of echinocandins and AmB formulations (Ibrahim et al., 2008).

Deferasirox (Exjade®, Novartis) Iron Chelation Therapy

Knowledge of the harmful effects of the siderophore deferoxamine, and the dependence by *Mucorales* on a source of free ferric ion, prompted investigators to examine other iron chelators that bind free ferric ions but cannot be used by *R. oryzae* as a siderophore. The objective is to deny *Mucorales* the ability to scavenge free iron released under acidotic conditions. Deferasirox chelates iron away from *R. oryzae* and exerts fungicidal activity in vitro against *R. oryzae, R. microsporus*, and *Mucor ramosissimus* (Ibrahim et al., 2007). Mice with diabetic ketoacidosis injected with *R. oryzae* and treated with deferasirox have improved survival and a decreased tissue fungal burden. The mechanism of action is via induction of an iron-starvation response. Deferiserox is also synergistic in mice receiving liposomal AmB. Kidneys of deferasirox-treated mice had no hyphae whereas those of mice treated with placebo had extensive aseptate hyphae. Deferasirox treatment also enhanced the penetration of PMNs into the tissues as part of the host inflammatory response to infection with *R. oryzae*.

Deferasirox is approved by the U.S. Food and Drug Administration as salvage therapy for rhinocerebral mucormycosis. A 40-yr-old man with diabetic ketoacidosis and advanced brainstem and cavernous sinus mucormycosis continued to deteriorate after months of liposomal AmB treatment, but had a successful outcome after salvage deferasirox was added (Reed et al., 2006). A clinical research study in humans is planned comparing lipid AmB with or without deferasirox.

17A.13 LABORATORY DETECTION, RECOVERY, AND IDENTIFICATION

Direct Examination

Because time is of the essence in intervening in rapidly progressing mucormycosis, reliance is placed on presumptive identification of the *Mucorales* by direct microscopic examination and histopathology of scrapings or exudates from the nares, hard palatal lesions, burns or other wounds, sputum or BAL fluid, and by culture of this material. KOH preps are appropriate. In such specimens broad, twisted, nonseptate *(coenocytic)* hyphae are the major clue that this is not aspergillosis. Fluorescent brighteners including Calcofluor (Fungi-Fluor®, Polysciences, Inc.) or Blankophor® (Lanxess Corp., Leverkusen, Germany) can be helpful in direct examination of the above specimens, in combination with *KOH preps* (Rüchel and Schaffrinski, 1999). Please see Chapter 2, Section 2.3.3.1, Wet Mount—Fluorescent Brighteners Used with or Without KOH, for further explanation.

Histopathology

The most useful stain for these fungi is GMS, as it readily demonstrates the hyphae. In tissue, large (5–15 μm diameter), hyaline, nonseptate hyphae are seen with the *Mucorales*. Usually short hyphal elements are seen near or within blood vessels. Recall that having no or rare septa, the hyphae may invade tissue in many directions rather than straight lines and are bizarre ("varicose") in shape, wrinkled or collapsed. Cross sections of the hyphae may be mistaken for yeast cells. Branching is haphazard and *is typically at right angles in contrast to acute angle branching in the aspergilli* (Fig. 17A.5). The only septa in the *Mucoromycotina* are isolated reproductive structures and walled-off vacuolated regions of the mycelium.

Preparation of Biopsied Tissue for Culture

Tissue grinds risk disruption of delicate *Mucorales* hyphae with loss of viability. Mincing or use of the Seward Stomacher® is recommended, as described in Chapter 2, Section 2.3.2, Specimen Collection.

Mycotic thrombi may be seen within blood vessels; however, blood cultures rarely are positive. Hyphae also may be seen invading the perineurium, the loose facial sheath surrounding nerve tissue (Frater et al., 2001). Fluorescent brighteners (Fungi-Fluor, Blankophor) are alternative stains for tissue sections, which, in combination with indirect fluorescent antibody staining, can aid in detecting mixed infections, for example, *Aspergillus* species and a *Mucorales* species (Rüchel and Schaffrinski, 1999).

Culture

Materials for culture include nonsterile and sterile site specimens. Isolation from nonsterile sites such as sputum, BAL fluid, scrapings or aspirates from the nasal sinuses, hard palate lesions, and burns or other wounds should be interpreted cautiously, in the light of clinical signs and symptoms, because of the frequent occurrence of these environmental molds in the *aerospora*. Isolations from these sites in diabetic and/or immunosuppressed patients should be reported without delay. Growth of *Mucorales* from sterile site biopsy specimens is likely to be linked to invasive disease and requires immediate notification of the physician.

Modified SDA, with antibiotics, is the medium of choice. The *Mucorales* are rapid growing on media with antibiotics but without antifungal agents such as cycloheximide, to which they are generally susceptible. However, most *Mucorales* species grow in media containing the fungicide Benomyl, 10 μg/mL. Some *Mucorales* may grow at elevated temperatures (Table 17A.2). Therefore, to aid in species differentiation, cultures are best cultured at 25°C, 35°C, 40°C, and 45°C. Cultures grow rapidly, filling a Petri plate or tube with a cottony colony within a few days. The mycelium in some isolates reaches the lid of the Petri plate, earning the name "lid lifters" for the *Mucorales*. Sporangiophores with sporing heads may be seen with the unaided eye as "pin molds."

Failure to Grow

Mucorales are notorious for failing to grow from histopathologically confirmed biopsy specimens, to the extent of 76.9% in one series of patients (Wang et al., 2006). Ominously, cultures positive for *Mucorales* have been a preterminal finding in acute fatal cases. Processing may influence poor recovery of *Mucorales* from both autopsy and premortem specimens because tissue grinding may kill their sparsely septate fragile hyphae (cytoplasm leaks out). Mincing the tissue is the simplest alternative. Another less destructive method for homogenizing tissue samples to recover infectious agents is the Seward Stomacher®, which uses paddles to crush samples retained in a plastic bag. This method could lead to higher yields. Recovery also may improve if minced tissue slices are placed directly on culture medium or baited with bread, to stimulate fungal growth. Despite invasion of blood vessels by the *Mucorales*, blood cultures rarely are positive. Even so, these molds can be recovered from blood cultures and that modality should not be overlooked (Chan-Tack et al., 2005).

Failure to Sporulate

Most members of the *Mucorales* readily sporulate on standard media such as SDA, CMA, or PDA but there are exceptions. *Apophysomyces elegans* and *Saksenaea vasiformis* may only produce sterile hyphae under routine culture conditions. Sporulation may be stimulated by first growing cultures on cornmeal–sucrose–yeast extract agar, then transferring them to a 1% agar block immersed in water (Wieden et al., 1985).

Microscopic morphology

(Please see "*Mucorales* Vegetative Growth, Morphology, and Asexual Reproduction," in the Introduction to this chapter.) *Mucorales* have broad hyphae (6–15 μm wide) and sparse to no septa (nonseptate/*coenocytic*). *When septa are seen, they usually are on sporangiophores just below the columella*. Septa sometimes may be seen microscopically in old cultures.

Tease Mount

The teasing process may damage the characteristic sporangia. For that reason the slide culture method is desirable to see the delicate and sometimes subtle morphologic characteristics. The various *Mucorales* genera and species are differentiated primarily by the presence of arched stolons, location of *rhizoids*, branched or unbranched *sporangiophores*, appearance of the *apophysis*, if present, shape of the *columella*, size and shape of the *sporangium*, and shape of *sporangiospores*. Table 17A.2 shows the main morphologic differences in the most common pathogens of this group.

Rhizopus oryzae *(*syn*:* R. arrhizus*)*

Colony Morphology Rapid growth of a white mycelium occurs on SDA-Emmons, turning shades of brown to black-gray according to the number of sporangiospores produced. The entire sporangiophore with sporangium are large enough for the entire colony to be a "lid lifter" and even to grow out of the Petri plate. It grows up to 45°C. Growth is inhibited by cycloheximide.

Microscopic Morphology *Rhizopus oryzae* (Fig. 17A.6) has wide stolons and brown rhizoids with 4–8 branches. Sporangiophores arise singly or in groups of up to 4 from the stolon opposite rhizoid tufts. The sporangium is globose to subglobose with a flattened base, above a columella. Sporangiophores are up to 3500 μm long by 18 μm wide, bearing gray-black sporangia up to 200 μm in diameter. These are large enough to be seen by the unaided eye as "pin molds" or "black bread molds." The columella is ellipsoid to subglobose and gray in color and collapses at maturity. The apophysis is inconspicuous. Sporangia are fragile and easily disrupted, releasing many lemon-shaped sporangiospores up to 8 μm in diameter (Ellis and Hermanis, 2003; Richardson and Koukila-Kähkölä, 2007). All *Rhizopus* species produce striated or grooved sporangiospores. This character is useful in differentiating *Rhizopus* species from *Lichtheimia* (formerly *Absidia*) and *Mucor* species, which produce smooth sporangiospores.

Rhizopus microsporus

Colony Morphology The *R. microsporus* group contains small-sized members of the genus. *Rhizopus microsporus* var. *rhizopodiformis* produces floccose aerial mycelium that is dark gray, gray-brown, or black, often in a speckled pattern. It is thermophilic, growing up to 50–52°C.

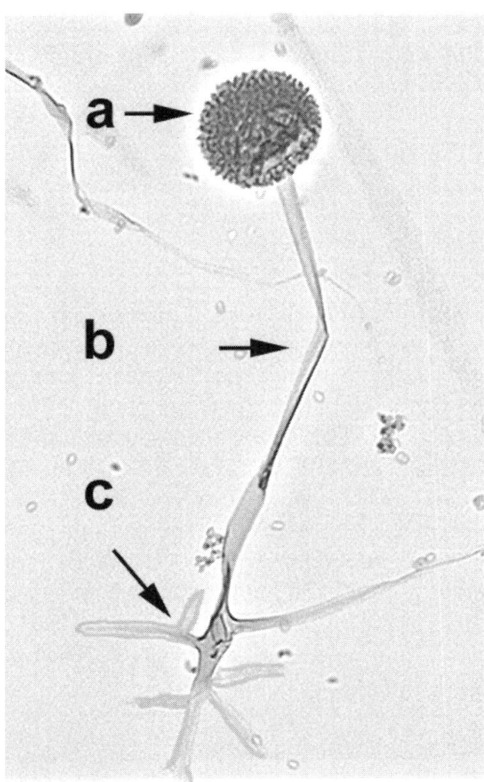

Figure 17A.6 *Rhizopus oryzae* microscopic morphology, showing (a) sporangium with sporangiospores, (b) sporangiophore, and (c) rhizoids.
Source: E. Reiss.

Microscopic Morphology *Rhizopus microsporus* var. *rhizopodiformis* sporangiophores do not exceed 800–1000 μm in length with small sporangia, 100 μm in diameter (Ribes et al., 2000). Rhizoids and sporangiophores originate from stolons opposing each other. Rhizoids vary from very branched to simple, measuring 80–100 μm in length. Sporangiophores are brown, in clusters from one to four, each terminating in a single round sporangium 50–60 μm in diameter. Columellae are elongate to pyriform. Apophyses are well defined and angular. Sporangiospores, 4–6 μm in diameter, have surface striations and spines.

Rhizomucor pusillus

(Formerly this mold was known as *Mucor pusillus*.) *Rhizomucor* is a thermophilic genus of the family *Mucoraceae*, containing three known species: *R. pusillus, R. miehei*, and *R. tauricus*. Of these, *R. pusillus* is the only opportunistic human pathogen. The diagnostic differential is generally between *Rhizomucor* and the other *Mucorales*, which produce globose sporangia—*Rhizopus* and *Mucor* species. *Rhizomucor pusillus* differs markedly from *Rhizopus oryzae* by its branched sporangiophores and short rhizoids. The presence of internodal rhizoids in *R. pusillus* and its high thermotolerance differentiate it from *Mucor* species.

Colony Morphology This rapid growing white mold turns dark gray to brown as sporulation occurs. It grows up to 54–58°C. Growth is inhibited by cycloheximide in the medium.

Microscopic Morphology Similar to *Rhizopus* species, *Rhizomucor pusillus* (Fig. 17A.7) produces rhizoids, but they are short and occur between, and not opposite, the base of the sporangiophores. *Rhizomucor pusillus* has *sympodially* branched sporangiophores with oval to pear-shaped columellae, and a septum at the base of the spherical, brown, or gray sporangia. A collarette is visible at the base of the sporangium after spore dispersal. This collarette is a remnant of the columella. The sporangiospores (3–5 μm diameter) are smaller than those of *R. oryzae*.

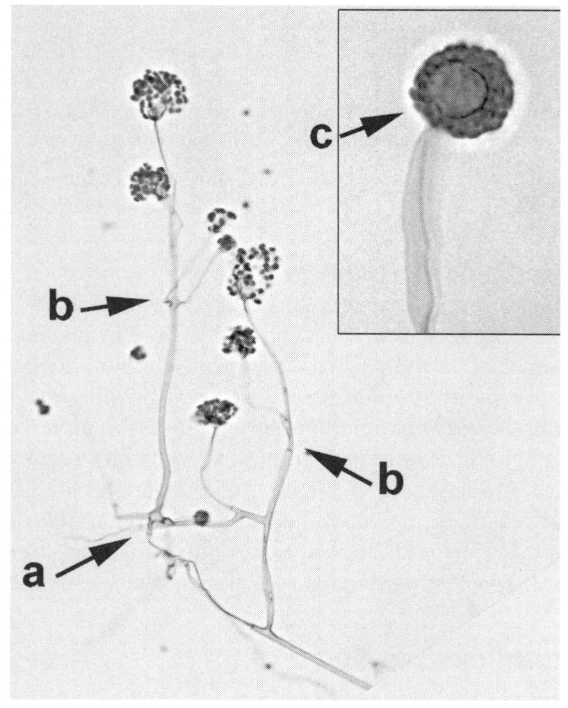

Figure 17A.7 *Rhizomucor pusillus* microscopic morphology. *Sympodially* branched sporangiophores with oval to pear-shaped columellae, and a septum at the base of the spherical, brown, or gray sporangia are characteristics of this species. A collarette is visible at the base of the sporangium after spore dispersal. This collarette is a remnant of the columella. The sporangiospores (3–5 μm diameter) are smaller than those of *R. oryzae*. (a) Short rhizoids; (b) branched sporangiophore; (c) (*Inset*) prominent columella.
Source: E Reiss, CDC.

Mucor *Species*

There are 60 or more species in the genus *Mucor*. Sporangia are globose and columellate, borne terminally on simple or branched sporangiophores. Rhizoids are lacking. *Mucor circinelloides* and *M. ramosissimus* are rare causative agents of mucormycosis. They are common in soil or dung, decaying vegetation, and stored grains.

Colony Morphology *Mucor circinelloides* grows slowly at 37°C and not above 39°C, resulting in pale gray-yellowish colonies, turning brown at 37°C.

Microscopic Morphology *Mucor circinelloides* sporangiophores are sympodially branched and often *circinate*. Sporangiophores bear spherical sporangia, 25–80 μm in diameter, and are deliquescent (they liquefy). Columellae are spherical but may vary in shape. Sporangiospores are smooth walled, 4.5–7 μm long, ellipsoid, and hyaline (Ribes et al., 2000).

Cunninghamella bertholletiae

Colony Morphology *Cunninghamella bertholletiae* produces a tall, aerial mycelium, 500–2000 μm in length. Colonies vary in color, with white, yellow, and light olive noted. Colonies become dark gray and powdery as sporulation occurs. Growth is up to 45°C and, as with some *Mucorales*, the mold is sensitive to cycloheximide.

Microscopic Morphology Wide ribbon-like hyphae give rise to simple sporangiophores (Fig. 17A.8). Swollen vesicles develop as lateral branches from the main sporangiophore trunk or are terminal. Vesicles give rise to globose (7–11 μm diameter) to ovoid sporangiola, which develop on short swollen denticles. Each sporangiole bears a single echinulate sporangiospore.

Apophysomyces elegans

Colony Morphology This rapid growing white mold turns buff with age. The aggressive growth, filling a Petri plate in 2 days, places it with other *Mucorales* as a "lid lifter." *Apophysomyces elegans* grows up to 43°C and is not sensitive to cycloheximide.

Microscopic Morphology *A. elegans* (Fig. 17A.9) does not readily sporulate on routine mycologic media, instead producing sterile *aseptate* or *pauci-septate* hyphae. Special nutrient-deficient growth conditions, high temperatures of incubation, and, if needed, prolonged incubation can induce sporulation. An inoculated 1% agar block in water stimulates sporulation in 3 weeks when incubated at 25°C, or in less time when the temperature

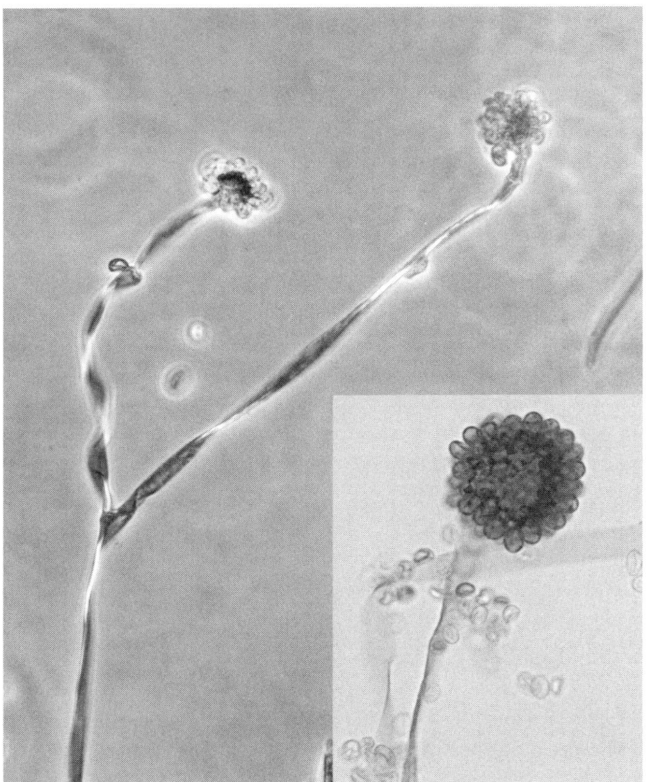

Figure 17A.8 *Cunninghamella bertholletiae*, microscopic morphology. *C. bertholettiae* has wide ribbon-like hyphae giving rise to simple sporangiophores. Swollen vesicles develop as lateral branches from the main sporangiophore trunk or are terminal. Vesicles give rise to globose (7–11 μm diameter) to ovoid sporangiola which develop on short swollen denticles (*inset*). Each sporangiole bears a single *echinulate* sporangiospore.
Source: E. Reiss, CDC.

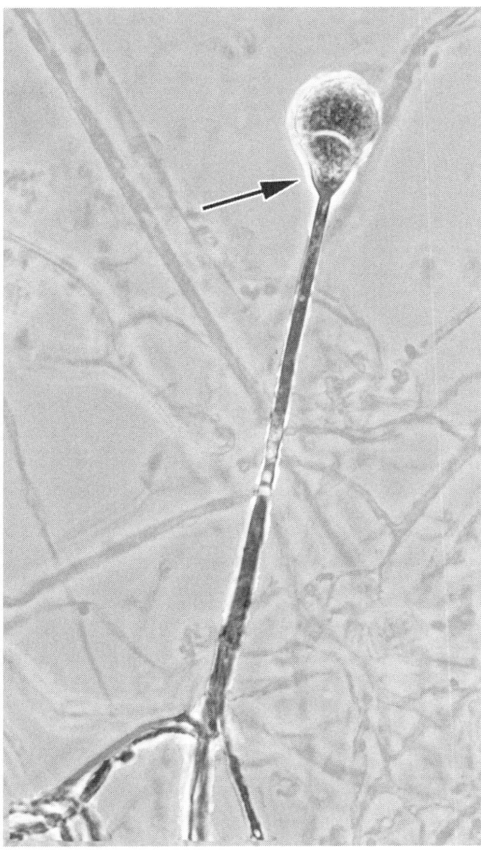

Figure 17A.9 *Apophysomyces elegans* microscopic morphology. *A. elegans* produces single unbranched sporangiophores arising from the ends of stolon-like hyphae. *Apophysomyces* resembles *Lichtheimia (Absidia)* species, but has a more prominent apophysis (*arrow*), funnel or bell shaped, and a hyphal segment analogous to the foot cell of *Aspergillus*.
Source: E. Reiss, CDC.

of incubation is 30°C. Supplementation of 1% agar with a small amount of yeast extract may assist in sporulation, which, under these conditions, is optimal at 37°C and 10–12 days of incubation (Wieden et al., 1985). A cornmeal glucose–sucrose–yeast extract agar block in water produces adequate sporulation in 2 weeks when incubated at 26°C or only 5–7 days at 32°C.

When sporulating, *A. elegans* produces single unbranched sporangiophores arising from the ends of stolon-like hyphae[5] (Wieden et al., 1985). The sporangiophores below the prominent apophyses are grayish-brown, thick walled, up to 300 μm in length. The apophyses are dark, bell shaped, or champagne glass shaped. Sporangiophores may arise from a foot cell also containing a tuft of rhizoids. The sporangia are *pyriform*, borne at the tips of sporangiophores arising above a dome-shaped columella. They are white turning to yellow-brown and 20–50 μm in diameter. The sporangial wall dissolves with age, releasing multiple oval, to oblong, smooth, light brown sporangiospores, 5–8 × 4–5 μm. In general, *Apophysomyces* resembles *Lichtheimia* (formerly *Absidia*) species but has a more prominent apophysis, funnel or bell shaped, and a hyphal segment reminiscent of the foot cell of *Aspergillus*.

[5]Stolon-like hyphae. *Definition*: Hyphae that rise into the air and then touch the substratum at various points.

Lichtheimia corymbifera *(Formerly Absidia)*

Colony Morphology *L. corymbifera* is the major *Lichtheimia* species known to be a human pathogen. It grows readily on routine mycologic media—SDA-Emmons agar with cloramphenicol and gentamicin is appropriate for this fungus. Its growth is stimulated by thiamine supplementation. *L. corymbifera* is capable of growth up to 48–52°C. At 37°C, *L. corymbifera* produces

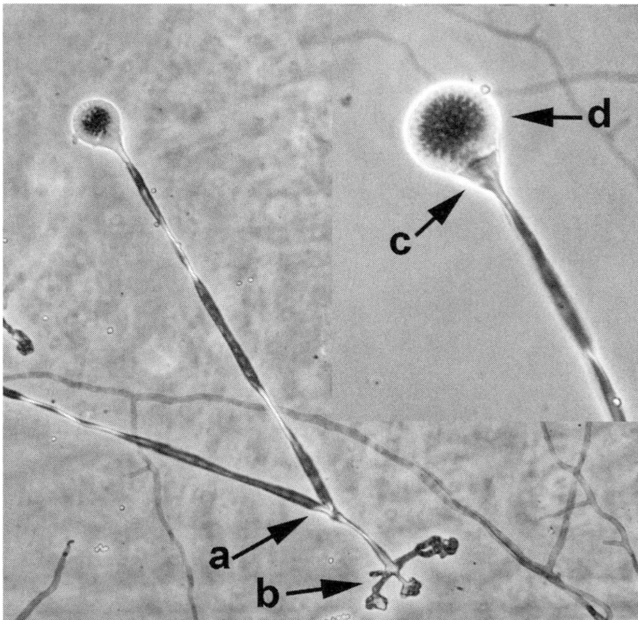

Figure 17A.10 *Lichtheimia (Absidia) corymbifera* microscopic morphology. The sporangium is relatively small, borne terminally on branched sporangiophores. Sporangiospores are round to elliptical and measure 2–3 × 3–4 μm. (a) Branched sporangiophores; (b) rhizoids; *inset*, (c) funnel shaped *apophysis*; (d) sporangium with sporangiospores. Lactofuchsin stain, 200×phase, *Inset*: Sporangium at 400×phase. *Source:* E. Reiss, CDC.

a woolly colony that may cover the entire Petri plate in 24 h. The colony is first white, then light gray, turning to gray-brown to green-tinted-beige with a colorless reverse. Growth is inhibited by cycloheximide.

Microscopic Morphology Hyphae are broad, often ≥10 μm in diameter, sparsely septate, with stolons and internodal rhizoids (Fig. 17A.10). Sporangiophores arise along the stolon (between rhizoids) and are hyaline to light gray in color. They are branched simply or arranged in whorls or clusters called corymbs,[6] accounting for the species name. The sporangium is pyriform (100–150 μm diameter) becoming gray-brown to green-beige with age, with a characteristic conical or funnel-shaped apophysis (Richardson and Koukila-Kähkölä, 2007). The columella is hemispheric to short-ovoidal in shape. The sporangium is relatively small, borne terminally on branched sporangiophores. Asexual sporangiospores are round to ellipsoid and measure 2–3 by 3–4 μm.

[6]Corymbiform. *Definition*: (From botany) a flat-topped or convex inflorescence in which the individual flower stalks grow upward from various points on the main stem to approximately the same height.

Genetic Identification of *Mucorales*

The notorious tendency of these molds to fail to grow from biopsy tissue and the often confusing histopathologic appearance of hyaline molds create a powerful rationale for the application of PCR-sequencing approach.

PCR-sequencing, histopathology, and culture evidence were compared in biopsy specimens from the lungs, nasal sinuses, or bronchi of 57 immunocompromised patients suspected of having a mold mycosis (Rickerts et al., 2007). Mold hyphae were detected using histopathology in about half of the patient specimens. *Aspergillus* was isolated in specimens from 14 patients. Cultures recovered *R. oryzae* and *Rhizomucor pusillus* from 2 of 6 specimens. PCR was superior to culture in detecting molds in 26 of 27 histopathology-positive biopsy specimens versus culture-positive results in 17 of 27 specimens. The gene target for PCR was the 18S rDNA. The *Mucorales* PCR had positive results in five histopathology-positive biopsy specimens and in a further specimen with no histopathologic evidence of a mold. PCR-sequencing identified *R. pusillus* in four biopsy specimens; and one each of *Lichtheimia corymbifera* (formerly *Absidia*) and *Rhizopus microsporus*. PCR-sequencing of one biopsy tissue revealed a double infection with *A. fumigatus* and *R. microsporus*. PCR-sequencing of fresh biopsy specimens is justified when direct examination or histopathology reveals fungal hyphae, in order to guide the choice of therapy.

SELECTED REFERENCES FOR MUCORMYCOSIS

Almyroudis NG, Sutton DA, Fothergill AW, Rinaldi MG, Kusne S, 2007. In vitro susceptibilities of 217 clinical isolates of zygomycetes to conventional and new antifungal agents. *Antimicrob Agents Chemother* 51:2587–2590.

Alvarez E, Sutton DA, Cano J, Fothergill AW, Stchigel A, Rinaldi MG, Guarro J, 2009. Spectrum of zygomycete species identified from clinically significant specimens in the United States. *J Clin Microbiol* 47:1650–1656.

Andresen D, Donaldson A, Choo L, Knox A, Klaassen M, Ursic C, Vonthethoff L, Krilis S, Konecny P, 2005. Multifocal cutaneous mucormycosis complicating polymicrobial wound infections in a tsunami survivor from Sri Lanka. *Lancet* 365:876–878.

Boekhout T, Gueidan C, de Hoog S, Samson R, Varga J, Walther G, 2009. Fungal taxonomy: New developments in medically important fungi. *Curr Fungal Infect Rep* 3:170–178.

Chamilos G, Lewis RE, Kontoyiannis DP, 2007. Multidrug-resistant endosymbiotic bacteria account for the emergence of zygomycosis: A hypothesis. *Fungal Genet Biol* 44:88–92.

Chan LL, Singh S, Jones D, Diaz EM Jr, Ginsberg LE, 2000. Imaging of mucormycosis skull base osteomyelitis. *Am J Neuroradiol* 21:828–831.

Chan-Tack KM, Nemoy LL, Perencevich EN, 2005. Central venous catheter-associated fungemia secondary to mucormycosis. *Scand J Infect Dis* 37:925–927.

Chayakulkeeree M, Ghannoum MA, Perfect JR, 2006. Zygomycosis: The re-emerging fungal infection. *Eur J Clin Microbiol Infect Dis* 25: 215–229.

de Hoog GS, Guarro J, Gené J, Figueras MJ, 2000. *Atlas of Clinical Fungi*, 2nd ed. Centraalbureau voor Schimmelcultures, Baarn, The Netherlands (CD version 2004.11).

de Ruiter GA, van Bruggen-van der Lugt AW, Nout MJ, Middelhoven WJ, Soentoro PS, Notermans SH, Rombouts FM, 1992. Formation of antigenic extracellular polysaccharides by selected strains of *Mucor* spp., *Rhizopus* spp., *Rhizomucor* spp., *Absidia corymbifera* and *Syncephalastrum racemosum*. *Antonie van Leeuwenhoek* 62:189–199.

Dijksterhuis J, Samson RA, 2006. Zygomycetes, pp. 415–436, *in*: Blackburn C de W (ed.), *Food Spoilage Microorganisms*. CRC Press, Boca Raton, FL.

Dow JM, Darnal DW, Villa VD, 1983. Two distinct classes of polyuronide from the cell walls of a dimorphic fungus, *Mucor rouxii*. *J Bacteriol* 155:1088–1093.

Ellis DH, Hermanis R, 2003. Kaminski's digital image library of medical mycology. Please see "Websites Cited" at the end of this chapter.

Farley PC, Sullivan PA, 1998. The *Rhizopus oryzae* secreted aspartyl proteinase gene family: An analysis of gene expression. *Microbiology* 144:2355–2366.

Forrest GN, Mankes K, 2007. Outcomes of invasive zygomycosis infections in renal transplant recipients. *Transplant Infect Dis* 9:161–164.

Frater JL, Hall GS, Procop GW, 2001. Histologic features of zygomycosis. Emphasis on perineural invasion and fungal morphology. *Arch Pathol Lab Med* 125:375–378.

Garrett SD, 1951. *Soil Fungi and Soil Fertility*. Pergamon Press, Oxford, UK.

Hibbett DS, Binder M, Bischoff JF, Blackwell M, Cannon PF, Eriksson OE, Huhndorf S, James T, Kirk PM, Lücking R, et al., 2007. A higher-level phylogenetic classification of the fungi. *Mycol Res* 111(Pt 5):509–547.

Husain S, Alexander BD, Munoz P, Avery RK, Houston S, Pruett T, Jacobs R, Dominguez EA, Tollemar JG, Baumgarten K, Yu CM, Wagener MM, Linden P, Kusne S, Singh N, 2003. Opportunistic mycelial fungal infections in organ transplant recipients: Emerging importance of non-*Aspergillus* mycelial fungi. *Clin Infect Dis* 37:221–229.

Ibrahim AS, Gebremariam T, Fu Y, Edwards JE Jr, Spellberg B, 2008. Combination echinocandin–polyene treatment of murine mucormycosis. *Antimicrob Agents Chemother* 52:1556–1558.

Ibrahim AS, Gebermariam T, Fu Y, Lin L, Husseiny MI, French SW, Schwartz J, Skory CD, Edwards JE Jr, Spellberg BJ, 2007. The iron chelator deferasirox protects mice from mucormycosis through iron starvation. *J Clin Invest* 117:2649–2657.

Ibrahim AS, Spellberg B, Avanessian V, Fu Y, Edwards JE Jr, 2005. *Rhizopus oryzae* adheres to, is phagocytosed by, and damages endothelial cells in vitro. *Infect Immun* 73:778–783.

Kaieda M, Samukawa T, Matsumoto T, Ban K, Kondo A, Shimada Y, Noda H, Nomoto F, Ohtsuka K, Izumoto E, Fukuda H, 1999. Biodiesel fuel production from plant oil catalyzed by *Rhizopus oryzae* lipase in a water-containing system without an organic solvent. *J Biosci Bioeng* 88:627–631.

Kendrick B, 1992. *The Fifth Kingdom*, 3rd ed., p. 32, Fig. 3.4. Focus Publishing, R. Pullins Co., Newburyport, MA.

Kontoyiannis DP, Marr KA, Park BJ, Alexander BD, Anaissie EJ, Walsh TJ, Ito J, Andes DR, Baddley JW, Brown JM, Brumble LM, Freifeld AG, Hadley S, Herwaldt LA, Kauffman CA, Knapp K, Lyon GM, Morrison VA, Papanicolaou G, Patterson TF, Perl TM, Schuster MG, Walker R, Wannemuehler KA, Wingard JR, Chiller TM, Pappas PG, 2010. Prospective surveillance for invasive fungal infections in hematopoietic stem cell transplant recipients, 2001–2006: Overview of the Transplant-Associated Infection Surveillance Network (TRANSNET) Database. *Clin Infect Dis* 50:1091–1100.

Kontoyiannis DP, 2007. Decrease in the number of reported cases of zygomycosis among patients with diabetes mellitus: A hypothesis. *Clin Infect Dis* 44:1089–1090.

Kontoyiannis DP, Lionakis MS, Lewis RE, Chamilos G, Healy M, Perego C, Safdar A, Kantarjian H, Champlin R, Walsh TJ, Raad II, 2005. Zygomycosis in a tertiary-care cancer center in the era of *Aspergillus*-active antifungal therapy: A case–control observational study of 27 recent cases. *J Infect Dis* 191:1350–1360.

Lamaris GA, Lewis RE, Chamilos G, May GS, Safdar A, Walsh TJ, Raad II, Kontoyiannis DP, 2008. Caspofungin-mediated beta-glucan unmasking and enhancement of human polymorphonuclear neutrophil activity against *Aspergillus* and non-*Aspergillus* hyphae. *J Infect Dis* 198:186–192.

Larone D, 2002. *Medically Important Fungi A guide to identification*, 4th ed., ASM Press, Washington, DC.

LeMaile-Williams M, Burwell LA, Salisbury D, Noble-Wang J, Arduino M, Lott T, Brandt ME, Iiames S, Srinivasan A, Fridkin SK, 2006. Outbreak of cutaneous *Rhizopus arrhizus* infection associated with karaya ostomy bags. *Clin Infect Dis* 43:e83–e88.

Ma LJ, Ibrahim AS, Skory C, Grabherr MG, Burger G, Butler M, Elias M, Idnurm A, Lang BF, Sone T, Abe A, Calvo SE, Corrochano LM, Engels R, Fu J, Hansberg W, Kim JM, Kodira CD, Koehrsen MJ, Liu B, Miranda-Saavedra D, O'Leary S, Ortiz-Castellanos L, Poulter R, Rodriguez-Romero J, Ruiz-Herrera J, Shen YQ, Zeng Q, Galagan J, Birren BW, Cuomo CA, Wickes BL, 2009. Genomic analysis of the basal lineage fungus *Rhizopus oryzae* reveals a whole-genome duplication. *PLoS Genet* 5(7):e1000549.

Malani AN, Kauffman CA, 2007. Changing epidemiology of rare mould infections. *Drugs* 67:1803–1812.

Marr KA, Carter RA, Crippa F, Wald A, Corey L, 2002. Epidemiology and outcome of mould infections in hematopoietic stem cell transplant recipients. *Clin Infect Dis* 34:909–917.

Marty FM, Cosimi LA, Baden LR, 2004. Breakthrough zygomycosis after voriconazole treatment in recipients of hematopoietic stem-cell transplants. *N Engl J Med* 350:950–952.

Mitchell SJ, Gray J, Morgan MEI, Hocking MD, Durbin GM, 1996. Nosocomial infection with *Rhizopus microsporus* in preterm infants: Association with wooden tongue depressors. *Lancet* 348:441–443.

Mohindra S, Monhindra S, Gupta R, Bakshi J, Gupta SK, 2007. Rhinocerebral mucormycosis: The disease spectrum in 27 patients. *Mycoses* 50:290–296.

Munday BL, Whittington RJ, Stewart NJ, 1998. Disease conditions and subclinical infections of the platypus (*Ornithorhynchus anatinus*). *Philos Trans R Soc Lond B Biol Sci* 353:1093–1099.

Nordén G, Björck S, Persson H, Svalander C, Li XG, Edebo L, 1991. Cure of zygomycosis caused by a lipase-producing *Rhizopus rhizopodiformis* strain in a renal transplant patient. *Scand J Infect Dis* 23:377–382.

O'Donnell KL, 1979. *Zygomycetes in Culture*. Department of Botany, University of Georgia, Athens GA.

Pappas PG, Alexander BD, Andes DR, Hadley S, Kauffman CA, Freifeld A, Anaissie EJ, Brumble LM, Herwaldt L, Ito J, Kontoyiannis DP, Lyon GM, Marr KA, Morrison VA, Park BJ, Patterson TF, Perl TM, Oster RA, Schuster MG, Walker R, Walsh TJ, Wannemuehler KA, Chiller TM, 2010. Invasive fungal infections among organ transplant recipients: Results of the Transplant-Associated Infection Surveillance Network (TRANSNET). *Clin Infect Dis* 50:1101–1011.

Partida-Martinez LP, de Looss CF, Ishida K, Ishida M, Roth M, Buder K, Hertweck C, 2007. Rhizonin, the first mycotoxins isolated from the zygomycota, is not a fungal metabolite but is produced by bacterial endosymbionts. *Appl Environ Microbiol* 73:793–797.

Paul S, Marty FM, Colson YL, 2006. Treatment of cavitary pulmonary zygomycosis with surgical resection and posaconazole. *Ann Thorac Surg* 82:338–340.

Peel T, Daffy J, Thursky K, Stanley P, Buising K, 2008. Posaconazole as first line treatment for disseminated zygomycosis. *Mycoses* 51:542–545.

Peleg AY, Weerarathna, T, McCarthy JS, Davis TME, 2007. Common infections in diabetes: Pathogenesis, management and relationship to glycaemic control. *Diabetes Metab Res Rev* 23:3–13.

Reed C, Bryant R, Ibrahim AS, Edwards J Jr, Filler SG, Goldberg R, Spellberg B, 2008. Combination polyene–caspofungin treatment of rhino-orbital–cerebral mucormycosis. *Clin Infect Dis* 47:364–371.

Reed C, Ibrahim A, Edwards JE Jr, Walot I, Spellberg B, 2006. Deferasirox, an iron-chelating agent, as salvage therapy for rhinocerebral mucormycosis. *Antimicrob Agents Chemother* 50:3968–3969.

Ribes JA, Vanover-Sams CL, Baker DJ, 2000. Zygomycetes in human disease. *Clin Microbiol Rev* 13:236–301.

Richardson MD, Koukila-Kähkölä P, 2007. Chapter 122, *Rhizopus, Rhizomucor, Absidia*, and Other Agents of Systemic and Subcutaneous Zygomycoses, pp. 1839–1856, *in:* Murray PR, Baron EJ, Jorgensen JH, Landry ML, Pfaller MA (eds.), *Manual of Clinical Microbiology*, 9th ed., Vol. 2. ASM Press, Washington, DC.

Rickerts V, Mousset S, Lambrecht E, Tintelnot K, Schwerdtfeger R, Presterl E, Jacobi V, Just-Nubling G, Bialek R, 2007. Comparison of histopathological analysis, culture, and polymerase chain reaction assays to detect invasive mold infections from biopsy specimens. *Clin Infect Dis* 44:1078–1083.

Roden MM, Zaoutis TE, Buchanan WL, Knudsen TA, Sarkisova TA, Schaufele RL, Sein M, Sein T, Chiou CC, Chu JH, Kontoyiannis DP, Walsh TJ, 2005. Epidemiology and outcome of zygomycosis: A review of 929 reported cases. *Clin Infect Dis* 41:634–653.

Rüchel R, Schaffrinski M, 1999. Versatile fluorescent staining of fungi in clinical specimens by using the optical brightener Blankophor. *J Clin Microbiol* 37:2694–2696.

Saeed FA, Castle GE, 1998. Neutrophil chemiluminescence during phagocytosis is inhibited by abnormally elevated levels of acetoacetate: Implications for diabetic susceptibility to infections. *Clin Diagn Lab Immunol* 5:740–743.

Scully RE, Mark EJ, NcNeely BU, 1982. Case records of the Massachusetts General Hospital (Case 38–1982) Weekly clinicopathological exercises. A 66-year-old diabetic woman with sinusitis and cranial-nerve abnormalities. *N Engl J Med* 307:806–813.

Silveira FP, Husain S, 2008. Fungal infections in lung transplant recipients. *Curr Opin Pulm Med* 14:211–218.

Singh J, Rimek D, Kappe R, 2005. In vitro susceptibility of 15 strains of zygomycetes to nine antifungal agents as determined by the NCCLS M38-A microdilution method. *Mycoses* 48:246–50.

Spellberg B, Edwards E Jr, Ibrahim A, 2005. Novel perspectives on mucormycosis: Pathophysiology, presentation, and management. *Clin Microbiol Rev* 18:556–569.

Stelzmueller I, Lass-Floerl C, Geltner C, Graziadei I, Schneeberger S, Antretter H, Mueller L, Zelger B, Singh N, Pruett TL, Margreiter R, Bonatti H, 2008. Zygomycosis and other rare filamentous fungal infections in solid organ transplant recipients. *Transplant Int* 21:534–546.

Sutherland-Campbell H, Plunkett OA, 1934. *Mucor* paronychia. *Arch Dermatol Syphilol* 30:651–658.

Wang J-L, Hsiao C-H, Chang S-C, Hsueh P-R, 2006. Diagnostic challenge of zygomycosis in compromised hosts. *Med Mycol* 44:19–24.

Wieden MA, Steinbronn KK, Padhye AA, Ajello L, Chandler FW, 1985. Zygomycosis caused by *Apophysomyces elegans*. *J Clin Microbiol* 22:522–526.

WEBSITES CITED

CDC National Center for Health Statistics, 2005. CDC National Diabetes Fact Sheet. www.cdc.gov/diabetes.

Ellis D, Hermanis R, 2003. Kaminski's Digital Image Library of Mycology. Women's and Children's Hospital of North Adelaide, South Australia. Published on line at the URL http://www.mycology.adelaide.edu.au/gallery/kaminski/

QUESTIONS

The Answer Key to the multiple choice questions may be found at the end of the book.

1. Which statement best describes the growth of *Mucorales* species on laboratory media?
   A. Cannot grow above 37°C
   B. Lid lifter
   C. Requires enriched medium
   D. Slow grower
2. "Sporangiophores that are unbranched with rhizoids opposite the sporangiophores" describes which *Mucorales* species?
   A. *Cunninghamella bertholletiae*
   B. *Mucor circinelloides*
   C. *Lichtheimea corymbifera*
   D. *Rhizopus oryzae*
3. What role do temperature tolerance tests play in identification of *Mucorales* species?
   A. Can differentiate *Mucor* species from *Rhizopus oryzae*.
   B. Most *Mucorales* species have the same maximum growth temperature.
   C. *Mucorales* species can grow in the refrigerator.
   D. Have no role.
4. What is the definition of a columella? Is it a . . .
   A. Branched hypha, root-like in appearance, which grows into the substrate
   B. Dome-like expansion at the apex of a sporangiophore extending into a sporangium
   C. Funnel shaped swelling at the tip of a sporangiophore, below the sporangium
   D. Suspensor cell supporting the zygospore
5. The most common species isolated from mucormycosis is
   A. *Cunninghamella bertholletiae*
   B. *Mucor* species
   C. *Rhizomucor pusillus*
   D. *Rhizopus oryzae*
6. Hyphae of the *Mucorales* species are considered to be aseptate or sparsely septate. What is the function of septa in these molds?

A. Isolate reproductive structures and wall-off vacuolated regions of the mycelium.
B. Old hyphae lose their septa.
C. Prevent nuclei from migrating.
D. There are no septa, only folded hyphae.

7. What effect does Benomyl (10 μg/mL) in the medium have on most *Mucorales* species?
A. Does not inhibit
B. Growth inhibitor
C. Growth stimulant
D. Inhibits bacteria, not fungi

8. A tender red patch on the cheek of a diabetic patient is a sign of what fungal disease?
A. Infection with any hyaline mold
B. Mucormycosis
C. Rosacea
D. Tinea corporis

9. What clinical specimens are relevant to obtain from a patient suspected of rhinocerebral mucormycosis? Is it important to rely on direct examination, or is it sufficient to await the results of culture studies? Explain.

10. Why is there a low yield for growth of *Mucorales* from biopsy tissue specimens? How can this effect be minimized?

11. What three measures are essential to take to control rhinocerebral mucormycosis in a patient with diabetic ketoacidosis? How does the ketotic state affect the availability of iron needed as a micronutrient by *Mucorales*? How has knowledge of the increased risk of mucormycosis in patients receiving deferoxamine chelation treatment led to the development of a novel antifungal agent active in mucormycosis?

12. What is the outcome of AmB monotherapy in patients with rhino-orbital mucormycosis? How is the outcome influenced if there is CNS dissemination? What effect is seen on survival in mucormycosis patients receiving combination AmB and caspofungin therapy?

13. Why is the phylum *Zygomycota* no longer a valid taxonomic classification?

Chapter 17B

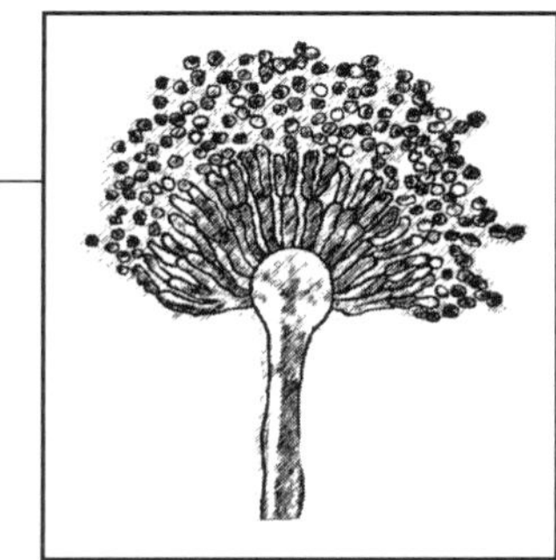

Entomophthoramycosis Caused by *Basidiobolus ranarum*

The term entomophthoramycosis is used to distinguish this disease from clinical forms caused by the *Mucorales*, especially to distinguish it from rhino-orbito–cerebral mucormycosis. In that respect we use the term "rhinoentomophthoramycosis." The *Entomophthoromycotina* cause two different mycoses—(i) entomophthoramycosis caused by *Basidiobolus ranarum* and (ii) nasofacial or rhinoentomophthoramycosis, caused primarily by *Conidiobolus coronatus*. These diseases occur mainly in tropical and subtropical Asia, Africa, and Latin America. The major causative agents, *Basidiobolus ranarum* and *Conidiobolus coronatus*, are environmental fungi found worldwide, including in the United States. Unlike the *Mucorales*, they do not have an affinity for blood vessels, but cause disease in the subcutaneous tissues, nasal mucosae, and surrounding integument. The sporadic nature of their occurrence among people who may have limited access to appropriate medical care has obscured the epidemiology and most other aspects of these diseases.

17B.1 ENTOMOPHOTHORAMYCOSIS CAUSED BY *BASIDIOBOLUS* RANARUM-AT-A-GLANCE

- *Introduction/Disease Definition.* Subcutaneous nodules become enlarged swellings on the trunk, arm, and shoulder, caused by a single species, *Basidiobolus ranarum*.
- *Etiologic Agent. Basidiobolus ranarum*, a *homothallic* member of the *Entomophthoromycotina*.
- *Geographic Distribution/Ecologic Niche.* This mycosis is found in tropical Asia, Africa, and South America, especially in Brazil. *Basidiobolus ranarum* is found in soil, plant detritus, and feces from insectivorous reptiles (frogs, toads, lizards) and bats.
- *Epidemiology.* Attacks primarily children living in subtropical and tropical areas.
- *Risk Groups/Factors.* Children <15 years old, male sex.
- *Transmission.* Cuts, abrasions, possibly via insect bites.
- *Determinants of Pathogenicity.* Secreted proteinase, lipase, and DNase but their role in disease is unknown.
- *Clinical Forms.* Subcutaneous nodules become enlarged swellings on trunk, upper arm, shoulder, axilla, buttocks, and, in temperate zones, intestinal masses.
- *Veterinary Forms.* Cutaneous lesions in horses; gastrointestinal and disseminated disease in dogs; cutaneous disease in frogs.
- *Therapy.* Itraconazole (ITC), ketoconazole; variably effective—SSKI, AmB, and flucytosine.
- *Laboratory Detection, Recovery, and Identification.* Direct examination of biopsy material in a KOH prep shows sparsely septate broad hyphae. Histopathology shows a granulomatous infiltrate and broad, sparsely septate hyphae with a sheath of eosinophilic Splendore–Hoeppli material. Culture on SDA, BHI, or Czapek agar produces sparsely septate hyphae and a variety of asexual mitospores: ballistoconidia with papillae (alone or with a corona of secondary conidia), and passively released adhesive *capilloconidia*, chlamydospores. This

Fundamental Medical Mycology, First Edition. By Errol Reiss, H. Jean Shadomy and G. Marshall Lyon, III.
© 2012 Wiley-Blackwell. Published 2012 by John Wiley & Sons, Inc.

species is homothallic, producing zygospores with characteristic beaks.

17B.2 INTRODUCTION/DISEASE DEFINITION

Basidiobolus ranarum causes a chronic subcutaneous mycosis affecting the trunk, upper arm, shoulder, and axilla. The disease especially affects children living in subtropical and tropical areas. No underlying immunodeficit is required for susceptibility to *B. ranarum*. Beginning as a subcutaneous nodule, the disease evolves into a woody, hard swelling. Patients present with a single, painless, sharply circumscribed, hard, subcutaneous mass. If left untreated the mass grows larger and can extend around the affected limb. Spread of lesions to other body sites is exceptional. The causative agent is a single species within the *Entomophthoromycotina*—*B. ranarum*—which is found in the environment worldwide in leaf litter and in feces of insectivorous amphibians and bats.

Basidiobolus ranarum is believed to play a role in the environment as a recycler of stressed insects and arthropods. The mode of transmission is through minor cuts, abrasions, scratches, and, possibly, by insect bites. One-half to three-quarters of cases occur in children 10 years old or younger. Males predominate at a ratio of 5:3. *Basidiobolus ranarum* is now recognized as an uncommon cause of intestinal mass lesions in patients, both adults and children, living in temperate zones. (Please see Section 17B.3, Case Presentation 2.)

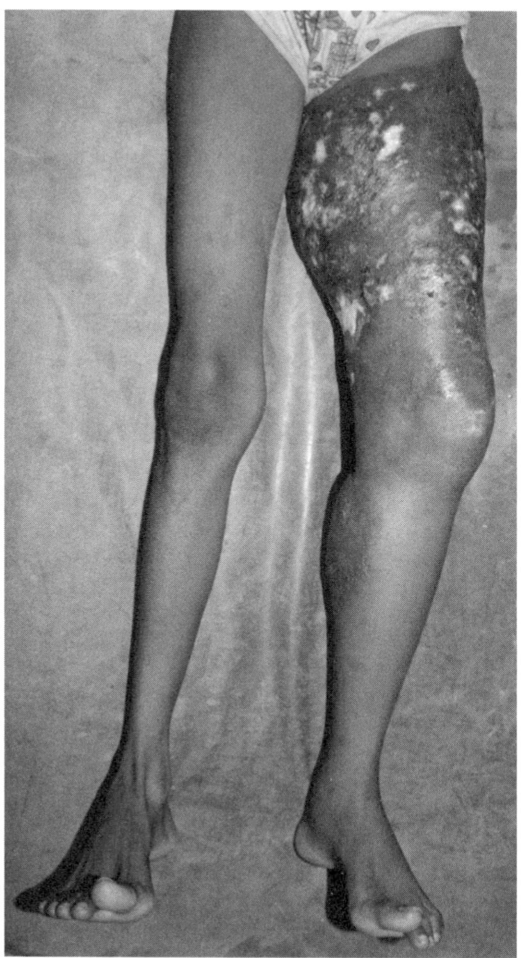

Figure 17B.1 Extensive swelling of the left thigh and leg of 8-yr-old girl with *Basidiobolus ranarum* infection.
Source: Reproduced from Mathew et al. (2005), with permissions from Dr. S. Kumaravel, Pondicherry Institute of Medical Sciences, Pondicherry, India, and Wiley–Blackwell Publishers.

17B.3 CASE PRESENTATIONS

Case Presentation 1. Lesions with Swelling in the Left Thigh of an 8-yr-old Girl (Mathew et al., 2005)

History

An 8-yr-old girl presented at the clinic in Puducherry, southern India, with deforming plaque-like lesions on the left thigh and leg with multiple ulcerations of 1 year duration (Fig. 17B.1). Two weeks before admission she developed fever. The lesion started in the previous year as a painless nodule, which increased in size and ulcerated. She received native and allopathic drugs with no clinical response. There was no history of local injury, tuberculosis, or other systemic illness.

Physical Examination

She appeared pale but with no other systemic abnormality. A large hyperpigmented and rubbery plaque extended over the entire thigh and upper leg. There were multiple ulcers, 0.5–1 cm in diameter, with purulent discharge.

Laboratory Findings

The total leukocyte count was 13,600/ mm^3 with a differential count of PMNs, 48; lymphocytes, 32; and eosinophils 20. Blood glucose was 100 mg %. Peripheral blood smear showed microcytic hypochromic anemia.

Direct Examination

A KOH prep of the biopsy tissue showed branching, sparsely septate hyphae.

Culture

Biopsy tissue cultured on SDA grew folded, creamy, gray, glabrous colonies after incubation at 37°C for 6–7 days. Microscopy indicated broad, *aseptate*, non-pigmented, branching hyphae and zygospores with beak-like structures and single-celled conidia (please see Section 17B.13, Laboratory Detection, Recovery, and Identification). The fungus was identified as *Basidiobolus ranarum*. Culture for bacteria yielded *Staphylococcus aureus* and *Pseudomonas aeruginosa*.

Histopathology

Examination of the biopsy showed ulceration of the skin and necrosis. A granulomatous reaction with dense eosinophilic infiltration was observed in the lower dermis and subcutis. Within the necrotic area, PAS stain indicated broad aseptate hyphae surrounded by eosinophilic Splendore–Hoeppli material (Fig. 17B.2).

Diagnosis

The diagnosis was subcutaneous entomophthoramycosis due to *Basidiobolus ranarum* with secondary bacterial infection.

Therapy and Clinical Course

Antibiotics were started to treat the bacterial infections. The patient was treated with SSKI for 6 weeks with a poor response and new ulcerations. Oral itraconazole (ITC, 5 mg/kg body weight) was added and the lesions started to regress 1 week later. Progressive thyromegaly was noted and SSKI was stopped after 16 weeks. ITC therapy resulted in a complete resolution after 15 weeks of therapy and was continued for another 4 weeks. The patient was disease-free at the 1 year follow-up clinic visit (Fig. 17B.3).

Comments

This case describes the classic clinical presentation of subcutaneous basidiobolomycosis. Although typical in the anatomic site of the lesions and the occurrence in a child, girls are less frequently afflicted than boys. Emphasis is placed on the importance of the combination of laboratory direct examination, histopathology, and culture. The problems associated with the use of SSKI to treat this disease are indicated, with better clinical response of longstanding, grossly deforming lesions achieved with oral ITC. Surgical debridement may not be necessary.

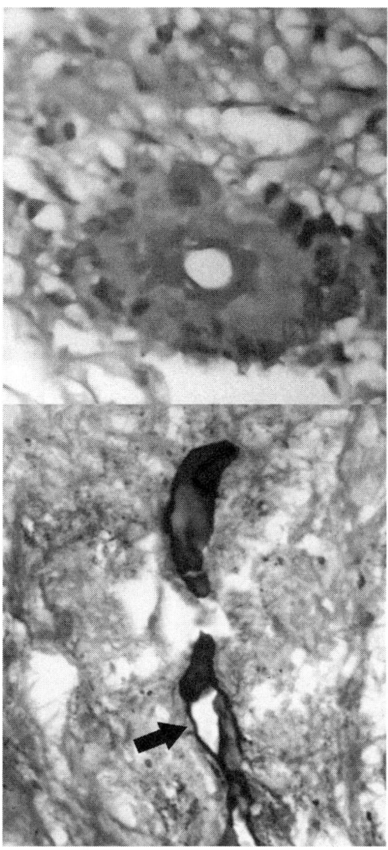

Figure 17B.2 Histopathology of basidiobolomycosis in the leg of an 8-yr-old girl. (*Upper panel*) Splendore–Hoeppli phenomenon on H&E staining and (*lower panel*) broad aseptate fungal hyphae (*arrow*), GMS stain (400×). See insert for color representation of this figure.
Source: Reproduced from Mathew et al. (2005), with permissions from Dr. S. Kumaravel and Wiley–Blackwell Publishers.

Case Presentation 2. Gastrointestinal Mass in a 52-yr-old Man in Arizona (Lyon et al., 2001)

History

A 52-yr-old man with controlled diabetes mellitus saw his physician because of a 10 day duration of abdominal pain. A gastroenterologist was consulted who diagnosed diverticulitis and prescribed ciprofloxacin. However, symptoms continued.

Clinical and Laboratory Findings

A CT scan demonstrated sigmoid colon diverticula with a 6 cm × 10 cm inflammatory mass. Laboratory studies included WBC = 12,800 cells/mm^3 with eosinophils = 14.3%. The patient underwent resection of the descending and sigmoid colon with partial excision of the mass.

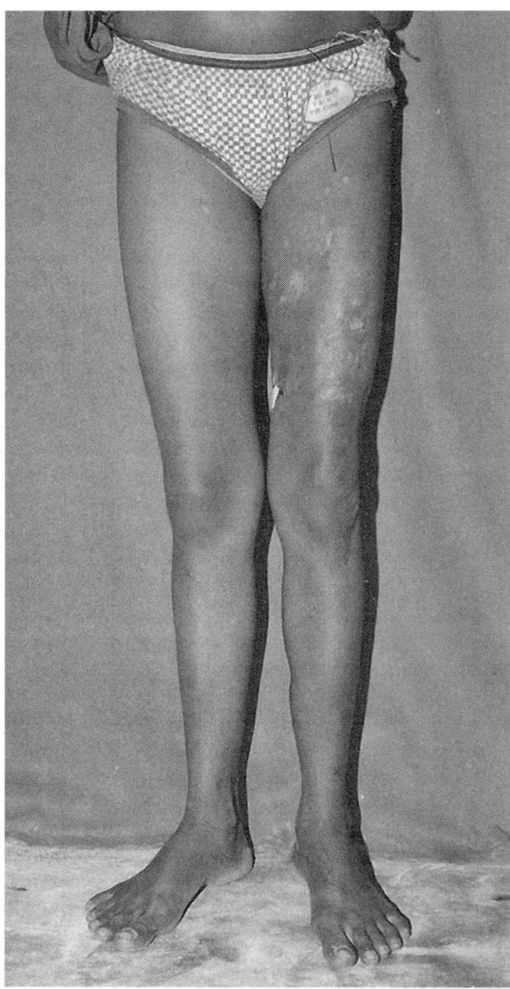

Figure 17B.3 Clinical photograph showing resolution of ulcers and swelling with return of limb circumference to normal after itraconazole therapy
Source: Reproduced from Mathew et al. (2005), with permissions from Dr. S. Kumaravel and Wiley–Blackwell Publishers.

Histopathology indicated bowel-wall thickening and dense eosinophilic granulomatous infiltrates containing many pieces of broad hyphae. Culture of surgical specimens was negative but serum antibodies were positive for *B. ranarum*.

Diagnosis

The diagnosis was gastrointestinal mass caused by *Basidiobolus ranarum*, an emerging clinical form of entomophthoramycosis.

Therapy and Clinical Course

The patient was started on ITC at 400 mg/day/6 months. At the 6 month clinic visit, at the time of reanastamosis, a 12 cm segment of diseased large bowel was seen and resected. Histopathology of the resected bowel showed necrotizing granulomas with septate hyphae. Treatment with ITC was continued for 400 mg/day for 4 months (total of 10 months). He remained asymptomatic at the follow-up visit 4 months later.

Comment

This is one of seven cases of gastrointestinal *B. ranarum* disease seen in Arizona from 1994 to 1999. Worldwide, only 24 cases of *B. ranarum* gastrointestinal mass have been reported (please see Section 17B.6, Epidemiology). The majority of all cases of *B. ranarum* infection are seen in South Asia, East and West Africa, and South America. Serology is a useful adjunct when cultures are negative. An effective immunodiffusion test that distinguishes basidiobolomycosis from conidiobolomycosis was used in this case, but is not commercially available (Kaufman et al., 1990).

Differential Diagnosis of Basidiobolomycosis

Cancer, lymphatic filariasis, mycetoma, pyogenic abscess, and tuberculosis also are possible diagnoses.

17B.4 ETIOLOGIC AGENT

Basidiobolus ranarum—Eidam 1886 *Beitr Biol Pfl* 4: 194 (1886).

Basidiobolus ranarum was listed in the subphylum *Entomophthoromycotina* in the *Entomophthorales* but is temporarily classed in the order *Basidiobolus* until its close relationship to the flagellated genus *Olpidium* is resolved (Boekhout et al., 2009; Hibbett et al., 2007). *B. ranarum* has an earth-like odor. (*Caution*: Sniffing fungi growing in tubes or plates is dangerous and prohibited!)

Basidiobolus ranarum is a prolific producer of various conidia types and zygospores. This homothallic species produces zygospores with characteristic "beaks." Zygospores germinate into hyphae or directly produce conidiophores containing a subapical apparatus that forcibly discharges round ballistoconidia. This is the major type of conidium formed. Alternatively, germination results in *capilloconidia* borne on the end of long narrow conidiophores. These conidia lack the propelling apparatus and are passively released. While lacking a means of forcible discharge, capilloconidia produce an apical sticky adhesive that aids in their dispersal via passing insects and arthropods.

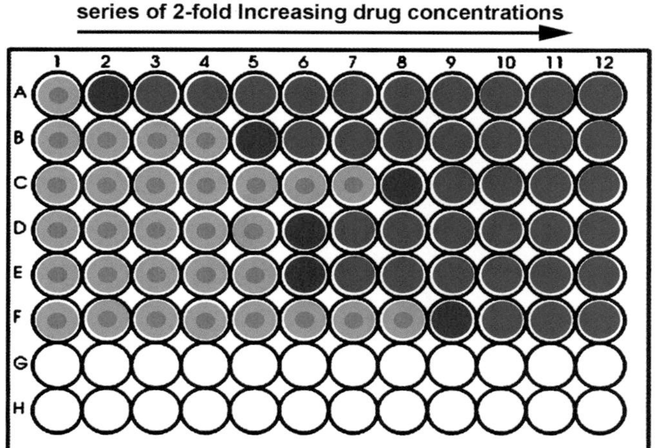

Figure 3B.2 YeastOne Sensititre plate. See page 113 for text discussion.

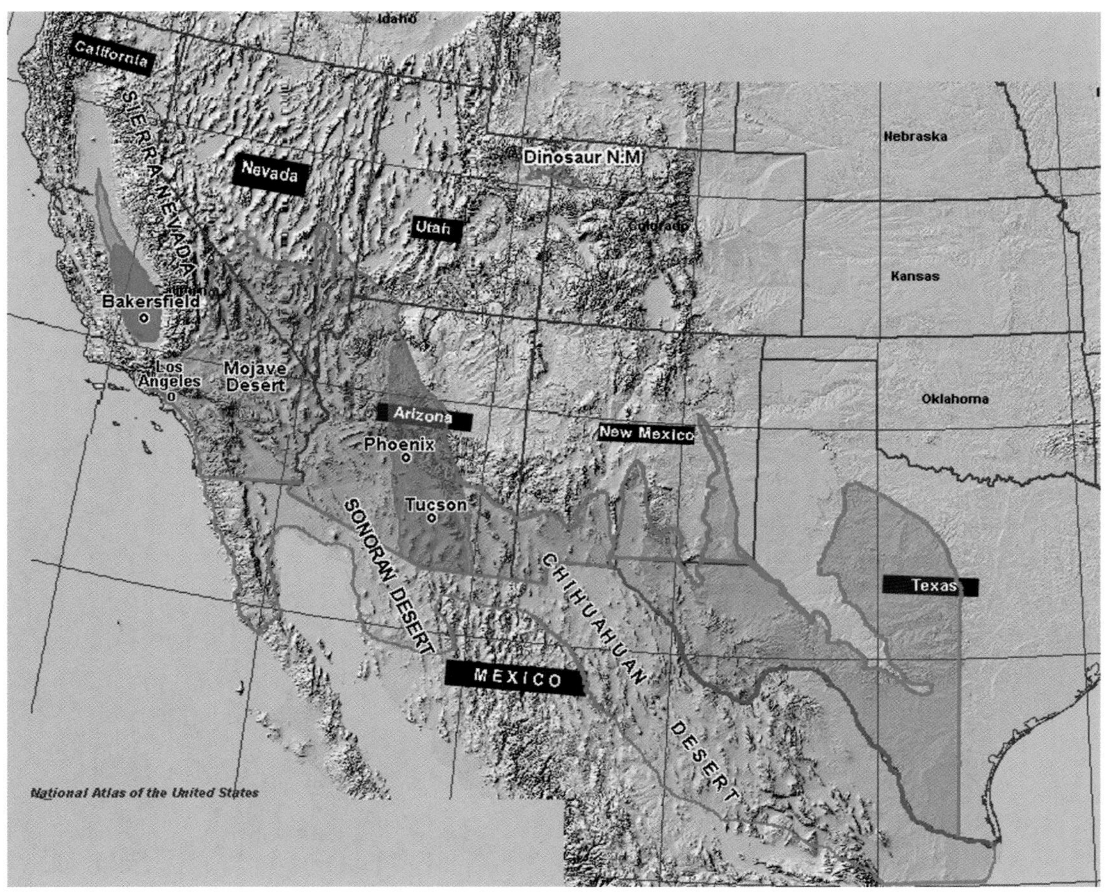

Figure 5.2 Map depicts the coccidioidomycosis endemic areas. See page 146 for text discussion.

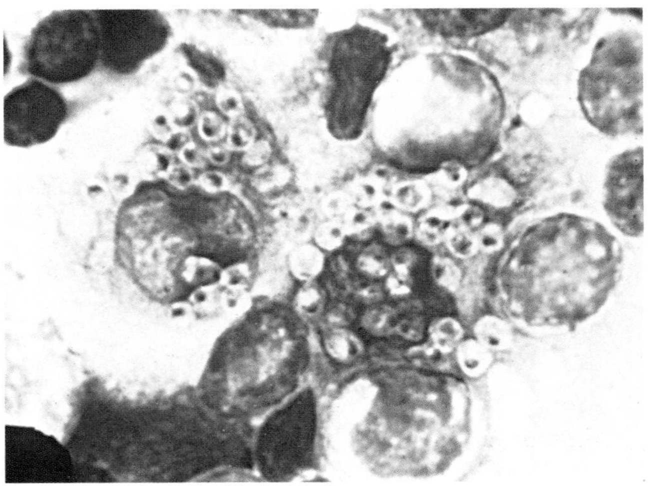

Figure 6.5 Intracellular *H. capsulatum* yeast forms within macrophages, bone marrow. See page 173 for text discussion.

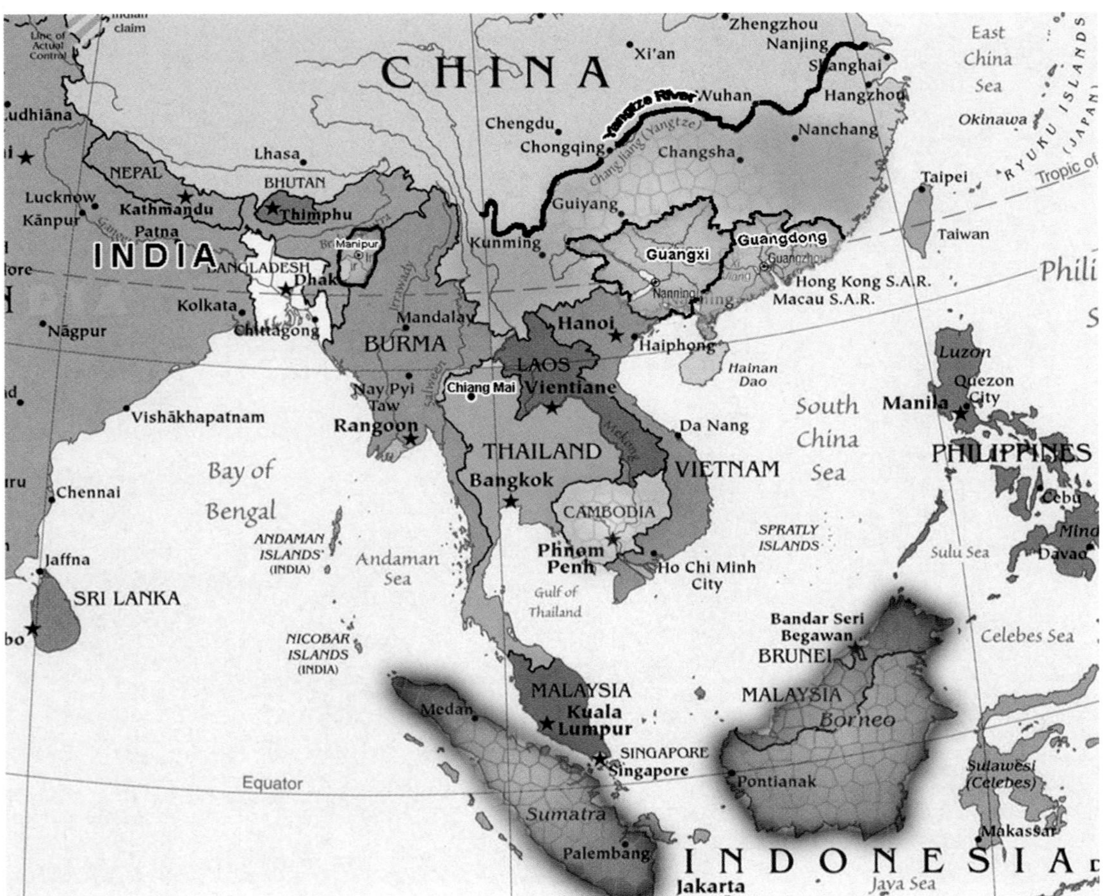

Figure 8.2 Geographic distribution of penicilliosis. See page 204 for text discussion.

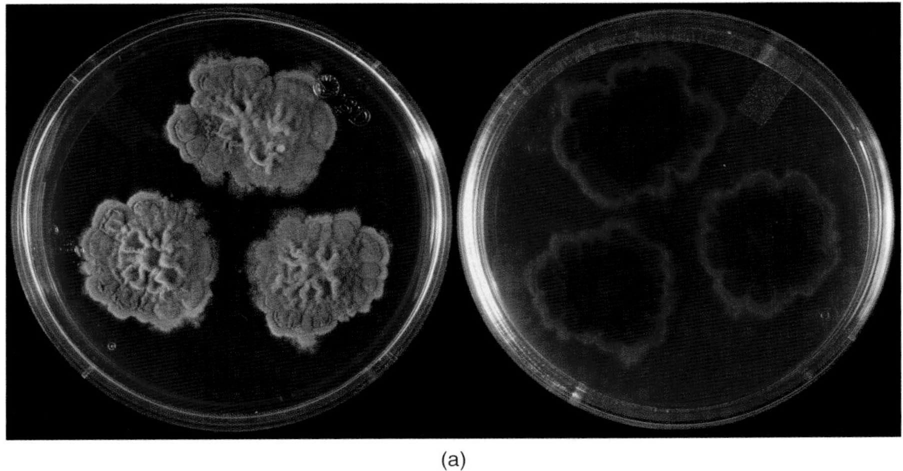

(a)

Figure 8.4 *Penicillium marneffei*, mold form, surface and reverse. See page 210 for text discussion.

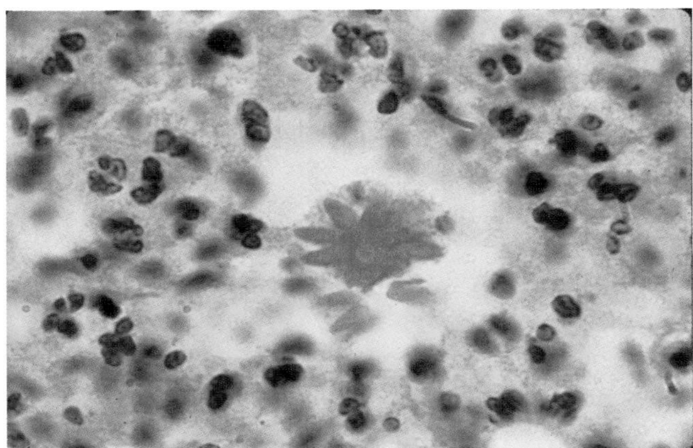

Figure 9.3 Asteroid body in sporotrichosis. See page 227 for text discussion.

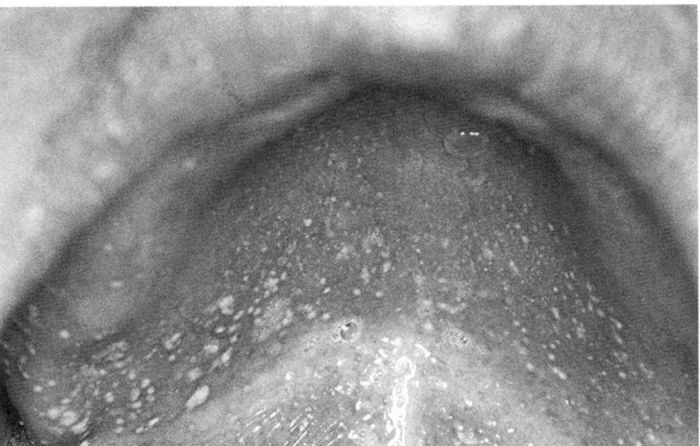

Figure 11.7 *Candida* denture stomatitis. See page 268 for text discussion.

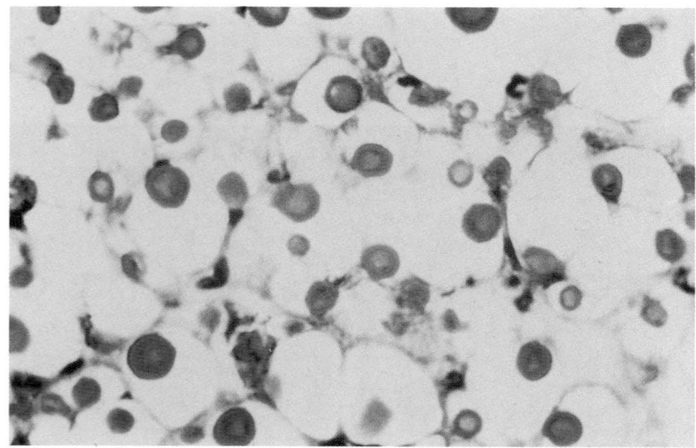

Figure 12.6 Histopathology of cryptococcosis, brain, mucicarmine stain. See page 322 for text discussion.

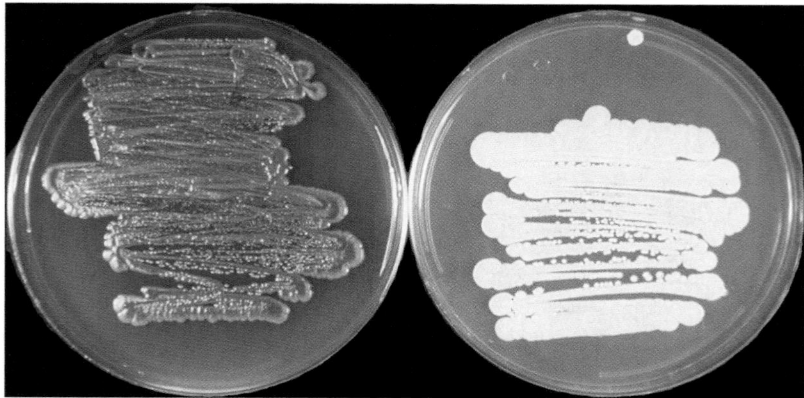

Figure 12.8 Brown colony effect of *Cryptococcus neoformans* on birdseed agar. See page 328 for text discussion.

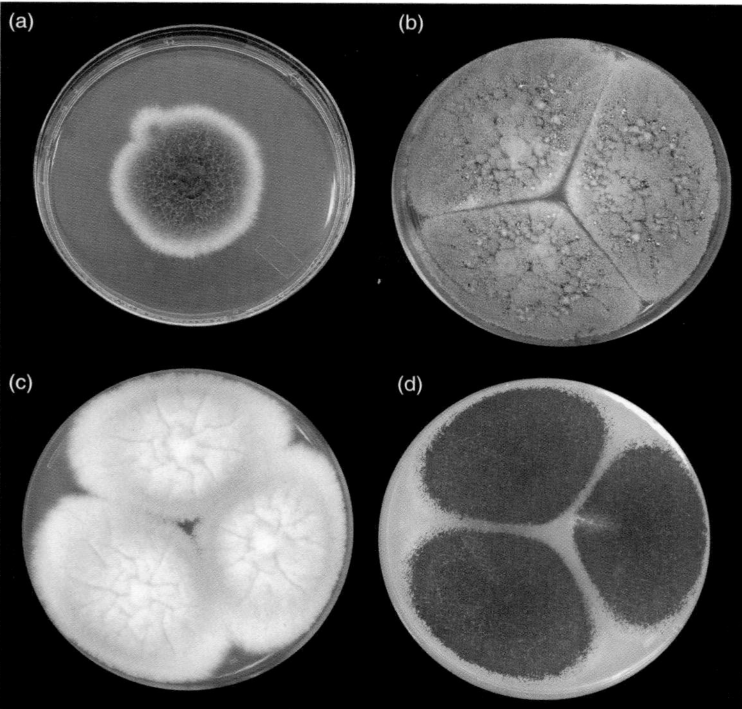

Figure 14.9 Four *Aspergillus* species colony morphology. See page 387 for text discussion.

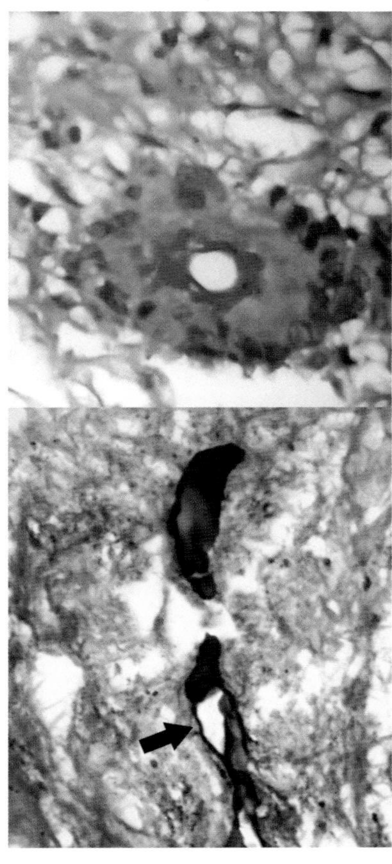

Figure 17B.2 Histopathology of basidiobolomycosis, showing (*upper panel*) Splendore–Hoeppli phenomenon on H&E staining and (*lower panel*) broad aseptate fungal hyphae on GMS stain. See page 459 for text discussion.

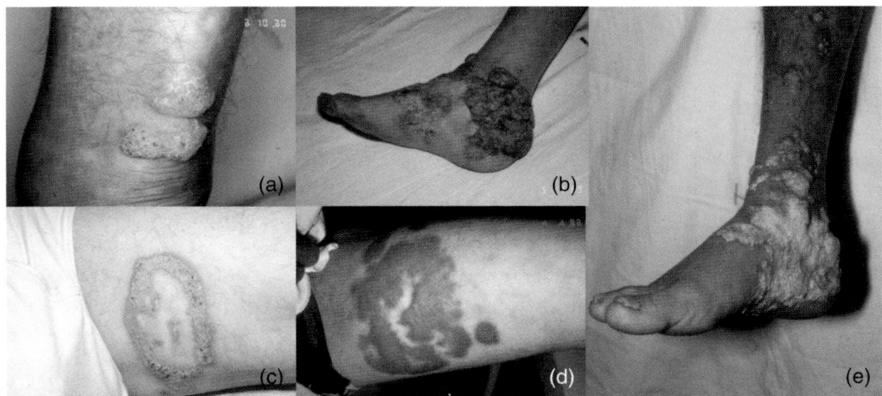

Figure 18.2 Five lesion types of chromoblastomycosis. See page 485 for text discussion.

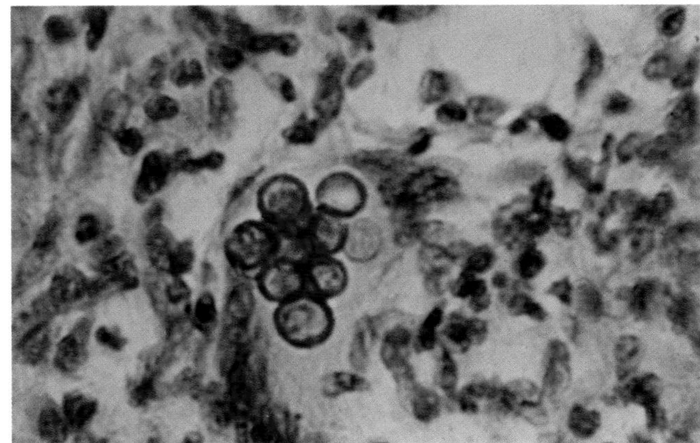

Figure 18.4 Histopathology of muriform cells. See page 488 for text discussion.

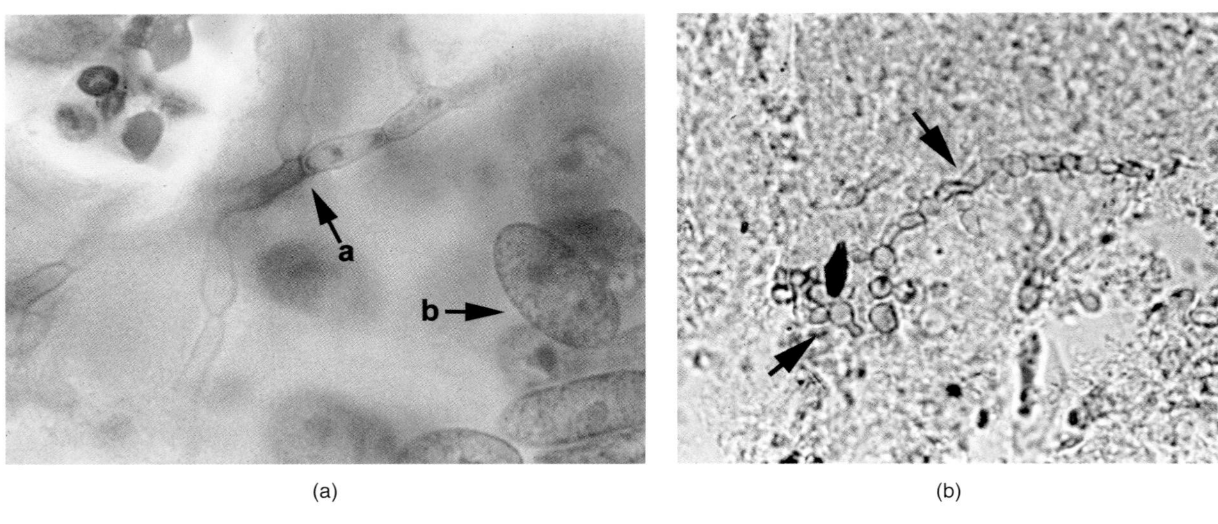

(a) (b)

(c)

Figure 19.3 (*Panel A*) Histopathology of a fine needle aspirate (FNA) from the phaeohyphomycotic cyst. (*Panel B*) *Wangiella (Exophiala) dermatitidis* from nail scraping, (*Panel C*) *Wangiella (Exophiala) dermatitidis* colony. See page 496 for text discussion.

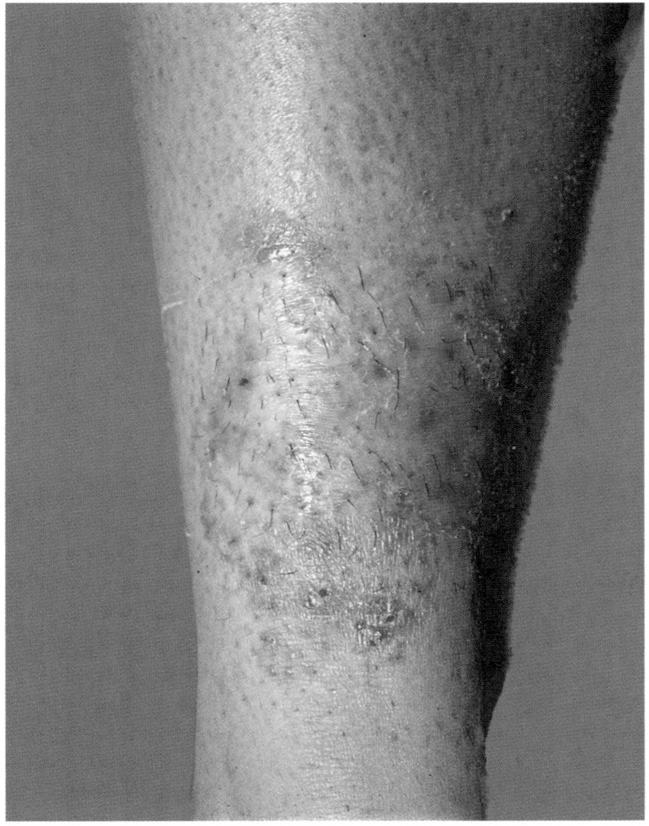

Figure 21.9 Majocchi's granuloma, shin area, color. See page 546 for text discussion.

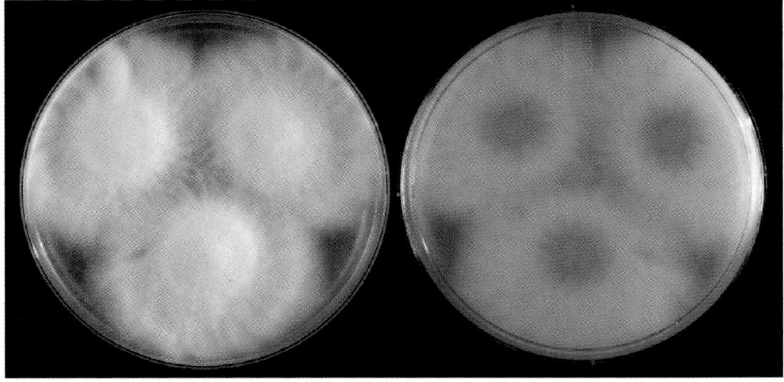

Figure 21.17 *Microsporum canis* colony: (*left*) surface and (*right*) reverse. See page 557 for text discussion.

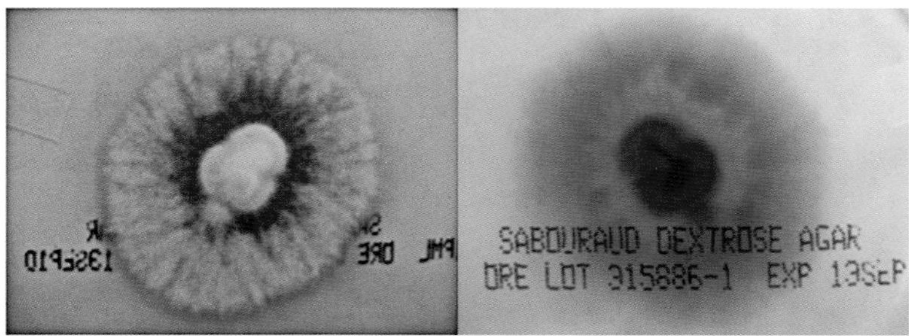

Figure 21.18 *Trichophyton interdigitale* (cottony phenotype): surface and reverse. See page 558 for text discussion.

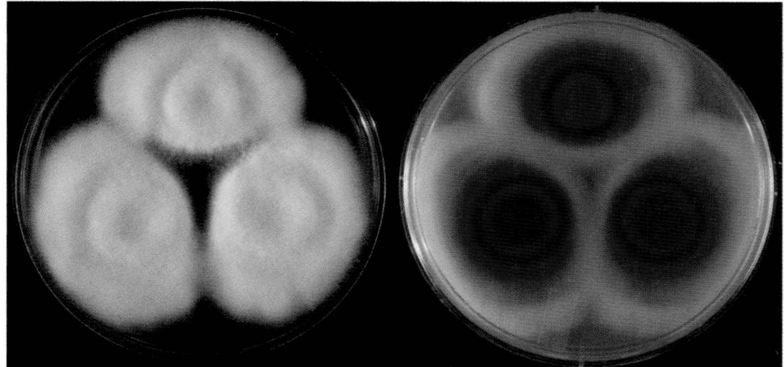

Figure 21.21 *Trichophyton rubrum* colony: surface and reverse. See page 560 for text discussion.

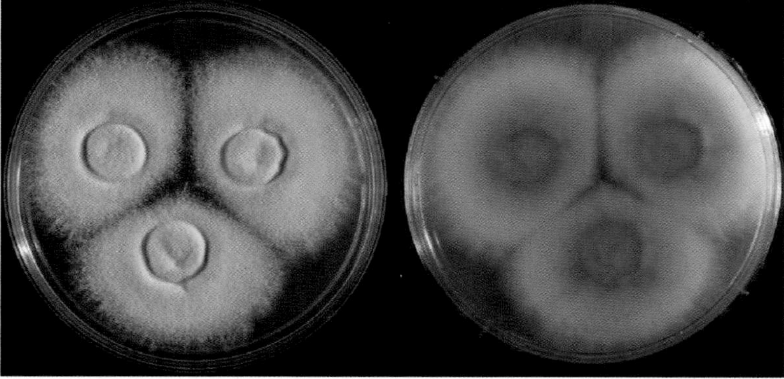

Figure 21.23 *Trichophyton tonsurans* colony: (*left*) surface and (*right*) reverse. See page 561 for text discussion.

In a further display of versatility, the conidia (ballistoconidia, capilloconidia) can, in a repetitive fashion, germinate to directly produce further conidiophores. Within the ballistoconidia and capilloconidia, internal cleavage can occur resulting in multiple uninucleate endospores which are released when the original conidial wall deliquesces (Dykstra and Bradley-Kerr, 1994).

17B.5 GEOGRAPHIC DISTRIBUTION/ECOLOGIC NICHE

Basidiobolus ranarum occurs worldwide in soil, in decaying vegetation, and in the gastrointestinal tract of reptiles, fish, amphibians, and bats. *B. ranarum* is most easily isolated from the intestinal tract and fecal pellets of insectivorous reptiles and amphibians. Despite its ease in isolation from these sources, the role of *B. ranarum* in the intestinal tract of its amphibian and reptile hosts is unknown. There is no doubt about the widespread colonization of amphibian gastrointestinal tract. Examination of fecal pellets from 169 animals representing 8 species and 4 genera of frogs and toads in west-central Florida found that *Basidiobolus* was isolated in 69% of the animals examined (Nelson et al., 2002).

Probably, the role of *B. ranarum* in nature is to recycle insect and arthropod cadavers. The capilloconidia and, potentially, ballistoconidia are carried on the exoskeletons of insects, mites, and other arthropods into their amphibian and reptile hosts. The capilloconidia produce an apical sticky droplet of adhesive that aids in dispersal by arthropods (e.g., mites and termites). Zygospores may be seen within mite cadavers (Manning et al., 2007). The widespread distribution of the fungus in nature and its uncommon cause of disease in humans suggests it is a weak pathogen, or that pathogenic strains are a fraction of the total.

17B.6 EPIDEMIOLOGY

The *Entomophthorales* cause disease in tropical and subtropical Africa and South Asia and Latin America. Only rarely, patients who contract the disease have not visited or resided in countries known for entomophthoramycosis. The disease most often affects children, less commonly adolescents, and rarely adults. The majority of the cases in Africa have been reported from Nigeria in West Africa (male:female = 3:1) and Uganda in East Africa (male:female = 3:2) where the most frequent sites of lesions are the perineum, buttocks, trunk, and thighs (Gugnani, 1999).

Basidiobolomycosis is endemic in Southern India. Of 8 cases of *B. ranarum* disease seen in Tamil Nadu State in a 4 year period, the patients ranged in age from 15 months to 30 years old; most were children under 6 years old and 6 were males (Krishnan et al., 1998). The thigh and/or gluteal region was affected in 5 cases. Case finding in Thailand over the period 1985–2001 revealed 5 female patients with culture-proven subcutaneous basidiobolomycosis (Chiewchanvit et al., 2002).

In recent years *B. ranarum* has been isolated in intestinal mass disease in the United States (please see Section 17B.3, Case Presentation 2) and reports of that clinical form have originated from other countries as well. Gastrointestinal basidiobolomycosis is very rare; in their review, Nemenqani et al. (2009) reported 3 Saudi Arabian cases and found an additional 15 cases in the world literature: 2 from Nigeria, 4 from Brazil, 1 from Kuwait, and 8 from the United States. The U.S series included 7 Arizona cases, 4 of which were a cluster in the period December 1998 to April 1999 (Lyon et al., 2001). Pediatric gastrointestinal basidiobolomycosis is even more exceptional, the largest series of cases involved 6 children (Al Jarie et al., 2003).

17B.7 RISK GROUPS/FACTORS

Depending on the causal fungus, age and sex of the patient are risk factors for infection with *Entomophthorales* species. *Basidiobolus ranarum* disease is seen primarily in male children.

17B.8 TRANSMISSION

Transmission is via minor cuts, abrasions, scratches, and possibly by insect bites. Gastrointestinal disease may occur after ingestion of spores of the *B. ranarum* from an environmental source.

17B.9 DETERMINANTS OF PATHOGENICITY

17B.9.1 Host Factors

No underlying immunodeficit is required for susceptibility to *B. ranarum*. The basis for a predilection for male children has not been investigated.

17B.9.2 Microbial Factors

Basidiobolus haptosporus (syn: *ranarum*) (strain 89-3-24) produces high levels of a single-strand specific extracellular DNase when grown on yeast extract–peptone broth. It has an optimum pH of 8.5 (Neelam and Desai 2000).

No connection has been made between this enzyme and the pathogenesis of disease.

Basidiobolus (strain N.C.L. 97.1.1) secretes an alkaline proteinase with an optimum pH of 10.0. Production of the enzyme in broth culture occurred during morphogenesis involving enlargement and internal division of hyphal segments (Ingale et al., 2002).

A survey of *B. ranarum* isolates for lipase activity indicated zones of precipitation in Tween-containing agar in all isolates tested (Okafor et al., 1986). Further experiments are needed to extend this observation and link it to disease pathogenesis.

Although these pilot experiments to characterize secreted enzymes of *B. ranarum* achieved success, there has been no plan to investigate their role in pathogenesis of disease.

17B.10 CLINICAL FORMS

Subcutaneous

B. ranarum causes a chronic subcutaneous mycosis affecting the trunk, upper arm, shoulder, and axilla. The disease especially affects children living in subtropical and tropical areas. Males are disproportionately affected. One-half to three-quarters of cases occur in children 10 years old or younger. Beginning as a subcutaneous nodule, the disease evolves into a woody hard swelling. Patients present most often with a single, painless, sharply circumscribed, hard, subcutaneous mass. The affected limb may swell to 2–3 times its normal size. In young children, the most common sites are the buttocks or thighs, the "bathing trunks" area, with spread to contiguous subcutaneous tissues. Smaller satellite lesions can occur in the same circumscribed area. Small lesions may be dispersed over the underlying muscle. Extensive lesions may be painful. There is generally no bone involvement but regional lymph nodes may become enlarged. Only rarely is there a history of trauma and underlying illness.

Gastrointestinal Mass

The range and etiology of *B. ranarum* disease has expanded to the temperate zone including another clinical form, gastrointestinal mass, affecting both adults and children (Lyon et al., 2001; Al Jarie et al., 2003) (Please see Section 17B.3, Case Presentation 2.) This form presents as abdominal pain with constipation. Risk factors have been identified including smoking, rantidine medication, and/or diabetes mellitus. Therapy consists of surgical resection and either AmB or azoles (Nemenqani et al., 2009). There is a useful but not commercially available immunodiffusion test for antibodies against *B. ranarum* developed at the CDC (Kaufman et al., 1990).

Gastrointestinal basidiobolomycosis was diagnosed in 6 patients <14 years old in Saudi Arabia between 2000 and 2002 (Al Jarie et al., 2003). All patients had abdominal pain and fever as presenting symptoms and were misdiagnosed initially: 2 as appendicitis with appendicular mass; 2 as abdominal tuberculosis; and 2 as lymphoma. They had leukocytosis with significant eosinophilia. Abdominal CT scan findings revealed gastrointestinal tract masses that in most cases involved adjacent organs. Three patients underwent partial surgical resection of fungal masses, followed by long-term (>1 year) ITC therapy. They had continuous radiologic and clinical improvement. One child improved on ITC therapy without surgery. Two other children had widespread basidiobolomycosis involving more than one organ. They did not respond to antifungal therapy and died. Fungal cultures identified as *B. ranarum* were obtained from 4 of these pediatric patients.

Other

Nasofacial *B. ranarum* disease has been reported, which should be differentiated on mycologic grounds from *C. coronatus* disease (Ghorpade et al., 2006).

17B.11 VETERINARY FORMS

Veterinary forms include cutaneous lesions in horses and gastrointestinal and disseminated disease in dogs. An epizootic of cutaneous entomophthoramycosis caused by *B. ranarum* occurred in a breeding facility for African clawed frogs (*Hymenochirus curtipes*) in central California (Groff et al., 1991). The disease was highly lethal and the target organ was the integument, which was infected on the surface and intraepidermally. No internal disease could be found. The tissue form consisted almost entirely of globose cells suggestive of conidia, 10–15 μm in diameter. When plated to SDA, pure cultures of vegetative hyphae grew in 2–3 days at 25°C. The identity of *B. ranarum* was based on the production of sporangiospores and smooth, thick-walled, beaked zygospores.

17B.12 THERAPY

Because it is effective and inexpensive, SSKI (oral) is used for *B. ranarum* disease consisting of 30 mg/kg/day for 3–6 months taken in a single daily dose or divided into 3 daily doses for 6–12 months. Patients not responding to SSKI have responded to ketoconazole at 400 mg/day for 3–6 months or, more recently, to ITC (please see Section 17B.3, Case Presentation 1). Surgical excision is an important part of therapy coupled with an antifungal regimen. However, spontaneous resolution is said to occur.

In Case Presentation 2 in Section 17B.3, surgery was performed, and without the addition of antifungal agents, the patient was cured of gastrointestinal *B. ranarum* disease. Three adults were suspected of colon cancer, but were diagnosed with gastrointestinal basidiobolomycosis (Nemenqani et al., 2009). One patient had a hemicolectomy and received ITC. The second patient had a ultrasound-guided needle biopsy which revealed fungal hyphae. She was treated with lipid-AmB without surgery. The mass regressed and she received maintenance therapy with ketoconazole. The third case was treated with surgery and voriconazole.

17B.13 LABORATORY DETECTION, RECOVERY, AND IDENTIFICATION

Direct Examination

Broad, septate, infrequently branching hyphae are seen in tissue macerated in a KOH prep.

Histopathology

Deep tissue is useful in locating the fungal hyphae. In tissue these fungi *are more regularly septate* than the *Mucorales*, and the hyphae may be less tortuous. Branch points are rare. When stained with H&E, a characteristic eosinophilic sleeve of Splendore–Hoeppli material may surround the hyphae. This material is stained bright red in PAS-stained sections. GMS counterstained with H&E better outlines the cell wall of *B. ranarum* (van den Berk et al., 2006). The lesions are granulomatous, including histiocytic granulomas, microabscesses with eosinophils, and epithelioid giant cell granulomas containing fungal elements.

Culture

Minced tissue specimens should be plated directly, without prolonged refrigeration, which reduces the yield (Gugnani, 1999). The *Entomophthorales* grow and sporulate well on SDA (within 5 days) and best on Czapek–Dox or water agar with penicillin and streptomycin but *without cycloheximide*. Brain–heart infusion agar also has been a successful growth medium. *Basidiobolus ranarum* grows rapidly at 30°C and is capable of growth at 37°C. The culture is phototrophic.

Colony Morphology

Unlike the *Mucorales, B. ranarum* grows as thin, flat, yellow or pale gray colonies with radial folds. After several days the colony is sparsely covered with white aerial hyphae. Soon spores are forcibly expelled, covering the sides of the culture tube or lid of the Petri plate. At maturity colonies of *Basidiobolus* are buff to grayish brown, with white aerial hyphae and a white reverse.

Microscopic Morphology

Microscopically, *B. ranarum* demonstrates wide hyphae with occasional septa. Upon sporulation, many more septa are produced and several spore types are formed (Fig. 17B.4):

1. Single-celled ballistoconidia are formed on short conidiophores that enlarge apically forming swollen areas, which contain the propelling apparatus. At maturity, the uninucleate ballistoconidia are blown out at the top of the vesicle below the sporangiole. The spores are forcibly discharged toward a light source. Each discharged spore contains a basal papilla.
2. Discharged ballistoconidia may develop a number of short extensions that give rise to a corona of secondary spores.
3. Ballistoconidia can themselves divide internally, giving rise to approximately 12 mononuclear spores. In that sense the ballistoconidia can develop into sporangia (Dykstra, 1994).
4. Other nonswollen sporoangiophores produce club-shaped capilloconidia with knob-like tips containing "sticky" adhesive material. They are passively released, perhaps attaching themselves to a passing mite or insect.
5. Chlamydospores also are formed.
6. Following the appearance of the asexual spores, this homothallic species undergoes sexual reproduction. Many round, intercalary, usually smooth, zygospores (20–50 μm diameter) are formed. They have thick walls and prominent beak-like appendages on one side.

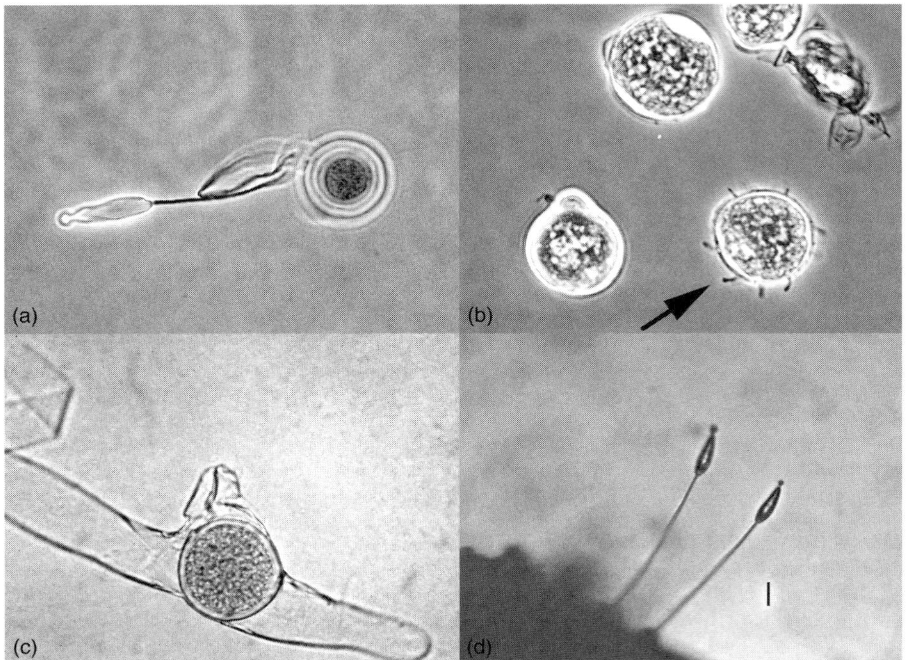

Figure 17B.4 Composite photomicrograph of asexual and sexual reproduction in the *Entomophthorales* pathogenic for humans. (a) Development of forcibly discharged conidium of *Basidiobolus ranarum*. *Source:* E. Reiss, CDC. (b) One spored *villous* (*arrow*) sporangioles of *Conidiobolus coronatus* function as ejectable ballistoconidia. *Source:* E. Reiss, CDC. (c) Zygospores of *B. ranarum* (20– 50 μm diameter) are thick walled and have prominent beak-like appendages on one side. *Source:* Used with permission from Dr. Arvind A. Padhye, CDC. (d) Capilloconidia of *B. ranarum* growing on leaf litter, bar 50 μm (unstained) (Manning et al., 2007). *Source:* Used with permissions from Dr. Arthur A. Callaghan, Staffordshire University, Stoke-on-Kent, United Kingdom, and from Elsevier Publishers, Ltd. Copyright © 2007.

SELECTED REFERENCES FOR ENTOMOPHTHORAMYCOSIS CAUSED BY *BASIDIOBOLUS RANARUM*

Al Jarie A, Al-Mohsen I, Al Jumaah S, Al Hazmi M, Al Zamil F, Al Zahrani M, Al Modovar E, Al Dayel F, Al Arishii H, Shehrani D, Martins J, Al Mehaidib A, Rossi L, Olaiyan I, Le Quesne G, Al-Mazrou A, 2003. Pediatric gastrointestinal basidiobolomycosis. *Pediatr Infect Dis J* 22: 1007–1014.

Boekhout T, Gueidan C, de Hoog S, Samson R, Varga J, Walther G, 2009. Fungal taxonomy: New developments in medically important fungi. *Curr Fungal Infection Rep* 3: 170–178.

Chiewchanvit S, Khamwan C, Pruksachatkunakorn C, Thamprasert K, Vanittanakom N, Mahanupab P, Lertprasertsuk N, Fooanant S, 2002. Entomophthoromycosis in Maharaj Nakorn Chiang Mai Hospital. *J Med Assoc Thai* 85: 1089–1094.

Dykstra MJ, 1994. Ballistosporic conidia in *Basidiobolus ranarum*: The influence of light and nutrition on the production of conidia and endospores (sporangiospores). *Mycologia* 86: 494–501.

Dykstra MJ, Bradley-Kerr B, 1994. The adhesive droplet of capilloconidia of *Basidiobolus ranarum* exhibits unique ultrastructural features. *Mycologia* 86: 336–342.

Ghorpade A, Sarma PSA, Iqbal SM, 2006. Elephantine nose due to rhinoentomophthoromycosis. *Eur J Dermatol* 16: 87–89.

Groff JM, Mughannam A, McDowell TS, Wong A, Dykstra MJ, Frye FL, Hedrick RP, 1991. An epizootic of cutaneous zygomycosis in cultured dwarf African clawed frogs (*Hymenochirus curtipes*) due to *Basidiobolus ranarum*. *J Med Vet Mycol* 29: 215–223.

Gugnani HC, 1999. A review of zygomycosis caused by *Basidiobolus ranarum*. *Eur J Epidemiol* 15: 923–929.

Hibbett DS, Binder M, Bischoff JF, Blackwell M, Cannon PF, Eriksson OE, Huhndorf S, James T, Kirk PM, Lücking R, et al, 2007. A higher-level phylogenetic classification of the Fungi. *Mycol Res* 111(Pt 5): 509–547.

Ingale SS, Rele MV, Srinivasan MC, 2002. Alkaline protease production by *Basidiobolus* (NCL 97.1.1): Effect of "darmform" morphogenesis and cultural conditions on enzyme production and preliminary enzyme characterization. *World J Microbiol Biotechnol* 18: 403–408.

Kaufman L, Mendoza L, Standard PG, 1990. Immunodiffusion test for serodiagnosing subcutaneous zygomycosis. *J Clin Microbiol* 28: 1887–1890.

Krishnan SG, Sentamilselvi G, Kamalam A, Das KA, Janaki C, 1998. Entomophthoromycosis in India—a 4-year study. *Mycoses* 41: 55–58.

Lyon GM, Smilack JD, Komatsu KK, Pasha TM, Leighton JA, Guarner J, Colby TV, Lindsley MD, Phelan M, Warnock DW, Hajjeh RA, 2001. Gastrointestinal basidiobolomycosis in Arizona: Clinical and epidemiological characteristics and review of the literature. *Clin Infect Dis* 32: 1448–1455.

Manning RJ, Waters ST, Callaghan AA, 2007. Saprotrophy of *Conidiobolus* and *Basidiobolus* in leaf litter. *Mycological Res* 111: 1437–1449.

Mathew R, Kuruvilla S, Kuruvilla S, Varghese RG, Shashikala, Srinivasan S, Mani MZ, 2005. Successful treatment of extensive basidiobolomycosis with oral itraconazole in a child. *Int J Dermatol* 44: 572–575.

NEELAM A, DESAI VS, 2000. Purification and characterization of the single-strand-specific and guanylic-acid-preferential deoxyribonuclease activity of the extracellular nuclease from *Basidiobolus haptosporus*. *Eur J Biochem* 267: 5123–5135.

NELSON RT, COCHRANE BJ, DELIS PR, TESTRAKE D, 2002. Basidiobolasis in Anurans in Florida. *J Wildlife Dis* 38: 463–467.

NEMENQANI D, YAQOOB N, KHOJA H, AL SAIF O, AMRA NK, AMR SS, 2009. Gastrointestinal basidiobolomycosis: An unusual fungal infection mimicking colon cancer. *Arch Pathol Lab Med* 133: 1938–1942.

OKAFOR JI, GUGNANI HC, TESTRAKE D, YANGOO BG, 1986. Extracellular enzyme activites of *Basidiobolus* and *Conidiobolus* isolates on solid media. *Mykosen* 30: 404–407.

VAN DEN BERK GE, NOORDUYN LA, VAN KETEL RJ, VAN LEEUWEN J, BEMELMAN WA, PRINS JM, 2006. A fatal pseudo-tumour: Disseminated basidiobolomycosis. *BMC Infect Dis* 6: 140.

QUESTIONS

These are combined for *Entomophthorales* and appear at the end of Chapter 17C.

Chapter 17C

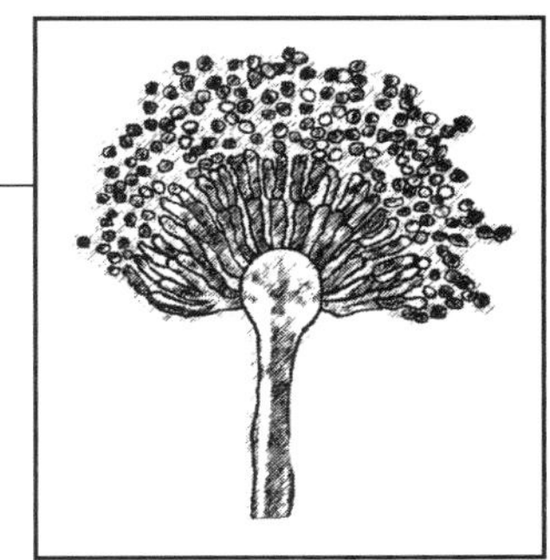

Entomophthoramycosis Caused by *Conidiobolus* Species

17C.1 ENTOMOPHTHORAMYCOSIS CAUSED BY *CONIDIOBOLUS* SPECIES-AT-A-GLANCE

- *Introduction/Disease Definition.* Entomophthoramycosis caused by *Conidiobolus* species is a chronic granulomatous, cutaneous–subcutaneous mycosis localized to the nasofacial region.
- *Etiologic Agents. Conidiobolus coronatus, C. incongruus*.
- *Geographic Distribution/Ecologic Niche.* The disease is found in many countries in a belt between the Tropic of Cancer and Tropic of Capricorn. *Conidiobolus* species live on leaf litter and can parasitize a variety of insects and also derive nutrients from arthropod cadavers.
- *Epidemiology.* Sporadic disease occurs in developing countries in the tropics and subtropics and as a travel medicine related disease in Europe and the United States.
- *Risk Groups/Factors.* Specific risk factors are unknown; disease occurs in immune-normal persons.
- *Transmission.* Inhalation or direct inoculation of spores into nose.
- *Determinants of Pathogenicity. Conidiobolus coronatus* is a potent producer of an alkaline proteinase.
- *Clinical Forms.* In nasofacial disease, fungal spores are inhaled or inoculated into the nose. They germinate, grow slowly, and spread from the nasal turbinates into the paranasal sinuses. Chronic sinusitis may occur. Over many months tumefactions develop involving skin of the nose, upper lip, eyelids, and face resulting in disfigurement. The disease can spread to the mouth, pharynx, and larynx. Multisystem dissemination is exceptional and may be associated with immunosuppression.
- *Therapy.* Various antifungal agents are used including AmB, itraconazole (ITC), ketoconazole, fluconazole (FLC), and, in developing countries, SSKI. There is no standard treatment protocol. Combination therapy with an azole and SSKI has been used.
- *Laboratory Detection, Recovery, Identification.* Direct examination of biopsy tissue can be done. *Conidiobolus* species grow at 30–35°C on SDA with antibiotics, without cycloheximide. Characteristic ballistoconidia on short conidiophores are expelled and will coat the lid of a Petri plate.

17C.2 INTRODUCTION/DISEASE DEFINITION

Entomophthoramycosis caused by *Conidiobolus* species is a chronic, granulomatous, subcutaneous mycosis localized to the nasofacial region caused by *Conidiobolus coronatus* or *C. incongruus*. These fungi are members of a genus found worldwide living on leaf litter and parasitizing a variety of insects and arthropods. The disease is rare, affecting primarily adults in rural, economically marginal areas, and is geographically distributed in subtropical and tropical countries. Disease is initiated when spores of the fungus are inhaled into or are implanted on the nasal mucosa. The spores germinate and the fungus grows slowly, spreading from the nasal turbinates through interosseus sutures and foramina to involve internally the nasopharynx, oropharynx, larynx, palate,

Fundamental Medical Mycology, First Edition. By Errol Reiss, H. Jean Shadomy and G. Marshall Lyon, III.
© 2012 Wiley-Blackwell. Published 2012 by John Wiley & Sons, Inc.

and the paranasal sinuses (Martinson, 1972). This leads to chronic sinusitis and, if untreated, the lesion(s) grow slowly externally over several months into tumefactions that may involve the entire nose, upper lip, eyelids, and face, resulting in severe disfigurement. Lesions do not spread below the level or the angle of the mouth. Tumefactions are covered by erythematous, hard skin. The first culture-confirmed cases in humans were described in 1965 in the Grand Cayman island of the Caribbean and in Zaire, Africa. Disseminated disease is extremely rare.

Diagnosis is made by clinical, histopathologic, radiologic, and mycologic examinations. The disease is amenable to treatment with various antifungal agents including AmB, ITC, ketoconazole, FLC and, in developing countries, SSKI. While these treatments may result in clinical improvement, clinical and mycologic cure may prove elusive. A promising clinical response has been achieved with VRC. Unlike mucormycosis caused by the *Mucorales*, which is an acute, life-threatening infectious disease, entomophthoramycosis is chronic and "benign" in the sense it is not life-threatening. The disfigurement and impairments of nasal blockage and dysphagia can hardly be considered benign.

17C.3 CASE PRESENTATION

Facial Swelling with Tumefaction in a 36-year-old Male Rice Farmer (Yang et al., 2010)

History

The patient, a 36-yr-old male rice farmer from Jiang Xi Province, China, presented with diffuse nasofacial swelling and involvement of the right paranasal sinuses (Yang et al., 2010) (Fig. 17C.1). He experienced nasal stuffiness and dizziness over the previous 10 months and underwent deviated septum surgery and antibiotic therapy with no benefit. The lesion spread to involve the whole nose, forehead, both cheeks, and lip. In addition to the nasal obstruction the patient reported severe itching. He attended the clinic at a hospital in Beijing.

Physical Examination

Swelling extended bilaterally over his cheeks and lower part of the upper lip. The medial canthus of both eyes was covered by swelling. The skin over the swelling was smooth and shiny, but without ulceration or increased local temperature. The swelling felt rubbery on local palpation. Anterior rhinoscopy showed hypertrophy and congestion in the right inferior turbinate. The left inferior and middle turbinates indicated medium hypertrophy and congestion.

Imaging Studies

A CT scan of the paranasal sinuses showed soft tissue swelling of the nose on the right side. The bone structure of the sinus wall remained intact.

Laboratory Tests

Specimens of cartilage-like tissue were obtained from the right nasal cavity by endoscopy. PAS-stained sections revealed eosinophilic granulomas with thin-walled broad hyphae surrounded by an eosinophilic sheath (Splendore–Hoeppli reaction). Culture of biopsy specimens on SDA at 25°C resulted in rapid growth of cream colored colonies. After 72 h the walls of the culture tube were covered with conidia (ballistoconidia). Microscopic observation revealed conidiophores and terminal spherical conidia, multiple secondary conidia, and villous conidia (See Fig. 17B.4). The ITS1 region of rDNA was PCR-amplified and sequenced. A BLAST search revealed the identity as *Conidiobolus coronatus*.

Diagnosis

The diagnosis was nasofacial entomophthoramycosis caused by *Conidiobolus coronatus*.

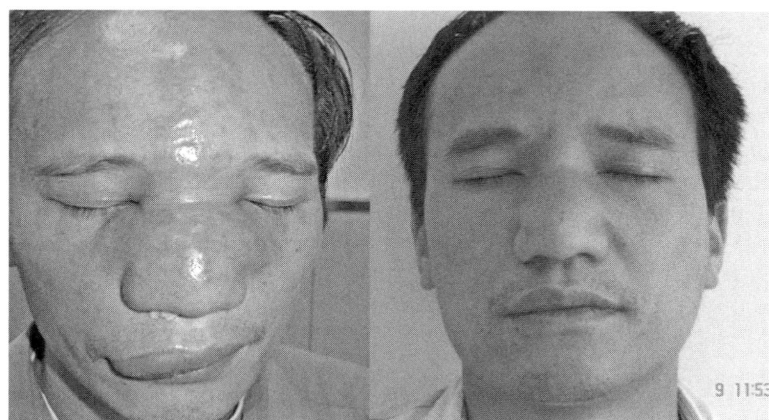

Figure 17C.1 Rhinofacial conidiobolomycosis caused by *Conidiobolus coronatus* (Yang et al., 2010). (*Left panel*) Before treatment; (*right panel*) after combination therapy with itraconazole and terbinafine. *Source:* Reproduced with permissions from Dr. Masataro Hiruma, M.D., Juntendo University, Tokyo, and from John Wiley and Sons, Publishers.

Therapy and Clinical Course

The patient was treated with an oral saturated solution of potassium iodide (SSKI) and 2 tablets of Septrin® (sulfamethoxazole, 200 mg; sulfadiazine, 200 mg; and trimethoprim, 80 mg) 2×/day. After 2 weeks the lesion became softer and reduced in size and itching subsided. There was no further improvement and the patient reported nausea, headache, and sore throat. Because of his indigent status he received donated ITC, 200 mg 2×/day (oral). Improvement was seen after 2 weeks but after 2 months there was no further benefit. Oral terbinafine was added to the regimen at 150 mg/day. The combination therapy led to improvement but terbinafine was not available without charge after 2 months. The patient was admitted to University Hospital and received IV ITC at 400 mg/day for 2 days then 200 mg/day for a total of 3 days. Thereafter the patient was maintained on oral ITC for 6 months. The lesion gradually disappeared and the patient has been asymptomatic at the 1 year follow-up clinic visit.

Differential Diagnosis

Angioneurotic edema, dermal leishmaniasis, dermal malignancy, lepromatous leprosy, rhinoscleroma, and systemic lupus erythematosus.

17C.4 ETIOLOGIC AGENTS

Conidiobolus coronatus (Costantin) A. Batko, *Entomophaga, Mémoires hors série* 2: 129 (1964).

Conidiobolus incongruus Drechsler, *Am J Bot* 47: 370 (1960).

Conidiobolus species are classed in the subphylum *Entomophthoromycotina*, order *Entomophthorales*.

Conidiobolus coronatus is known primarily as a recycler living on leaf litter. It is a member of a genus of insect pathogens and may play a role in recycling insect and arthropod cadavers. *C. coronatus* has been isolated from soil and rotting plant matter, in temperate, subtropical, and tropical areas, even though the disease in humans seems to be geographically concentrated in the latter two climate zones. *C. incongruus* is a less common cause of entomophthoramycosis.

17C.5 GEOGRAPHIC DISTRIBUTION

Entomophthoramycosis has been reported between the Tropic of Cancer and the Tropic of Capricorn in Africa, the Americas, Asia, Australasia, the Caribbean, India, and the Middle East.

17C.6 ECOLOGIC NICHE

Most species of *Conidiobolus* and *Basidiobolus* were initially described from woodland leaf litter or from amphibian excreta (reviewed by Manning et al., 2007). Although sometimes thought to be geographically restricted to the subtropical and tropical climate zones, both *C. coronatus* and *B. ranarum* were isolated from larch litter in a field in Staffordshire, England. *Conidiobolus* species are a group of fungi occurring in the upper layer of forest litter, plant detritus, decomposing in soil, dead fruit bodies of mushrooms, and decaying wood. *C. coronatus* was isolated from soil, rotten wood, keratin, grass leaves, and living ferns (Batko, 1974). As an insect pathogen, *C. coronatus* shows little or no evidence of host specificity and has been isolated from various homopteran species but also from *Coleoptera, Lepidoptera*, or *Diptera*. *C. coronatus* can probably attack any stressed insect (Freimoser et al., 2003).

Experiments in which *Basidiobolus* and *Conidiobolus* species isolates were grown on carbohydrates or on insect and arthropod cadavers indicated these species were *specialized to grow on arthropod cadavers as main substrates* (Manning et al., 2007). Both species could utilize chitin as a carbon source. *C. coronatus* produces proteinases and chitinases in response to insect cuticles. *B. ranarum* produced robust ballistoconidia on plant fragments. Although often linked to frogs or other amphibians, interaction of *B. ranarum* with those species is probably not a necessity.

Isolation of *Conidiobolus* and *Basidiobolus* from Leaf Litter

Fresh leaf litter (Manning et al., 2007) collected to a depth of 0–10 cm was homogenized in 0.05% aqueous Tween 20 and passaged through a 1 mm diameter mesh sieve. A filter paper disk was then charged with the filtrate and was held as a canopy over a Petri plate of malt extract agar. Ballistoconidia ejected from this source germinated on the agar layer.

17C.7 EPIDEMIOLOGY

The sporadic occurrence of *Conidiobolus* species mycosis in an underserved population in the developing world has complicated efforts to gain an understanding of its incidence and prevalence. Entomophthoramycosis is the second most common deep mycosis seen in Southern India, next to eumycetoma. Experience in Tamil Nadu State at the Madras General Hospital found 10 culture-confirmed cases over a 4 year period, 1993–1996

(Krishnan et al., 1998). Of these, 8 were due to *Basidiobolus*. The 2 patients with nasofacial disease caused by *Conidiobolus coronatus* were 45 and 50 years old. Their disease was of 3 months duration. All cases were treated successfully with oral SSKI.

Conidiobolus mycosis can appear outside of what is thought of as the endemic areas, as an example of travel medicine. A Hispanic man from Veracruz, Mexico, was diagnosed and treated in South Carolina where he worked as a farm laborer (de Ment et al., 2005).

17C.8 RISK GROUPS/FACTORS

Entomophthoramycosis usually occurs in the setting of poverty and malnutrition, in economically marginal countries. Persons afflicted usually have no known underlying immune defect.

17C.9 TRANSMISSION

Infection begins with the inhalation of *C. coronatus* or *C. incongruus* spores, or their direct inoculation into the nares with or without injury. Insect bites also are suspected of transmitting the fungus.

17C.10 DETERMINANTS OF PATHOGENICITY

The angioinvasive, tissue destroying pattern of the *Mucorales* is quite different from the granulomatous reaction in *C. coronatus* disease. Interest in microbial determinants of pathogenicity of *C. coronatus* has come from entomologists, who conduct gene discovery on *C. coronatus* using expressed tagged sequences (Freimoser et al., 2003). To date this approach has discovered a trypsin-like proteinase. Compared with *Basidiobolus ranarum* and to other *Mucorales, C. coronatus* is a potent producer of extracellular proteinases, especially serine proteinase activity, induced in vitro on collagen and peptone (Freimoser et al., 2003) (please see Special Topic: Serine Proteinases, below). The alkaline serine protease of *C. coronatus* is involved in its conidial discharge (Phadatare et al., 1989). Production of proteinases, whether related to digestion of insects or arthropods, or in the explosive discharge of ballistoconidia, also may function in the pathogenesis of human disease.

Special Topic: Serine Proteinases. Serine proteinases are a diverse group of bacterial, fungal, and animal enzymes which have the common property of an active site composed of serine, histidine, and aspartic acid, which come together in a folded structure. This amino acid triad is the essential element of the active site, but in other respects, the primary amino acid sequence and three-dimensional structure of the *C. coronatus* proteinase, subtilisin, and the pancreatic serine proteinases (e.g., trypsin) are completely different and likely evolved independently.

The secreted alkaline serine proteinase of *C. coronatus* has sufficiently high activity and thermostability that make it suitable for commercialization as a detergent additive (Bhosale et al., 1995). Stable up to 40°C, the thermal stability is increased by the presence of Ca^{2+}. Occurring in two isoforms, proteinase I contains a 7 amino acid peptide which, when cleaved, yields the second isoform, proteinase II (Phadtare et al., 1996). The molecular mass of proteinase I is 23 kDa, with a pI of 9.9 and an optimum pH of 9.7–10.

17C.11 CLINICAL FORMS

Entomophthoramycosis of the Midline Face: Rhinoentomophthoramycosis

The nasal mucosa and submucosa become infected by germinating ballistoconidia of *C. coronatus* or *C. incongruus* (Valle et al., 2001; Sharma et al., 2003). Lesions spread slowly internally from the nasal turbinates into the paranasal sinuses and externally bilaterally (usually) to the skin of the nose, glabella, cheeks, and upper lip. Symptoms include nasal obstruction, nasal discharge, or chronic sinusitis. A progressive facial swelling may involve the upper lip and face.

Without access to primary care and referral for diagnosis and treatment, the tumefactions that develop over months (or over years in persons without access to proper medical care) can result in extreme disfigurement. The skin overlying the lesions consists of erythematous, woody, hard, uniform, smooth swellings. Intraoral involvement, with swelling of the hard and soft palate, may progress to the pharynx and larynx with resultant dysphagia or laryngeal obstruction.

Disseminated *C. coronatus* Disease

A case of fatal rhino-orbito–cerebral entomophthoramycosis occurred in a immune-normal woman in Pretoria, South Africa (Hoogendijk et al., 2006a). The status of the patient was determined to be immune-normal based on a negative HIV test and her CD4/CD8 T-lymphocyte ratio. *Conidiobolus coronatus* was diagnosed based on the appearance of sparsely septate hyphae in a mass removed from the maxillary sinus, antemortem demonstration of hyphae and spores in the peripheral blood, and real-time PCR. *C. coronatus* was not, however, recovered in culture. The patient was unsuccessfully treated with AmB-deoxycholate and, when renal intolerance developed, with FLC. Smears made from blood culture bottles revealed sparsely septate hyphae and spores.

Scanning EM of these specimens revealed, in addition to the hyphae seen by conventional microscopy, *villous* sporangiospores at different stages of development (Hoogendijk et al., 2006b). These micrographs are impressive in their details of sporangiospore development and of the villous projections. This is the first report of the recovery of *C. coronatus* from blood cultures.

Pulmonary-disseminated entomophthoramycosis caused by *C. incongruus* occurred in a patient was hospitalized for 2 months in Maryland, U.S.A., for treatment of lymphocytic leukemia (Walsh et al., 1994). This case illustrated that disease caused by *Conidiobolus* species is not geographically restricted and that, in the immunocompromised host, disease may be more invasive than what is considered to be its usual pattern.

17C.12 VETERINARY FORMS

Veterinary forms of rhinoentomophthoramycosis due to *C. coronatus* or *C. lamprauges* occur in horses (Humber et al., 1989) causing ulcerative, granulomatous nodules—"kunkers"—in the nasofacial area. Most often this equine disease occurs in warmer climates including the southern United States, Australia, Brazil, Costa Rica, Colombia, and India. Apart from surgical removal of the accretions in the nasopharynx, there have been no reports of successful antifungal treatment of this equine disease.

17C.13 THERAPY

Saturated Solution of KI (SSKI)

Because of its low cost, SSKI has been used, taken orally for 4 weeks to 1 year or more. The oral dose of SSKI is 2–8 g/day and a clinical response usually occurs in 1–4 weeks: for example, 5 drops/day in milk, water, or fruit juice, increased gradually to a maximum 40–50 drops/day. Determining the endpoint of therapy is not obvious because of the different clinical response rates among patients. Recurrences are due to noncompliance with the daily routine, intolerance to KI, or premature cessation of therapy.

Azoles

Clinical responses have been obtained with oral ITC (200–400 mg/day), ketoconazole (200–600 mg/day) for 3–4 months, FLC (100–200 mg/day), and AmB.

In vitro MICs of six *C. coronatus* isolates were determined using a microtitration plate format with RPMI 1640 as the nutritional support. The inoculum was a suspension of small hyphae. Under these conditions all isolates showed decreased susceptibility, bordering on frank resistance to AmB, FLC, flucytosine, ITC, and ketoconazole (Guarro et al., 1999). Additional tests are needed with germinated ballistoconidia using the E-test for AmB and CLSI standard test conditions for azole antifungals.

AmB and Flucytosine

Conidiobolus incongruus isolated from a patient with disseminated entomophthoramycosis was tested in a broth macrodilution method. The MICs indicated resistance to AmB, >18.5 μg/mL, and to flucytosine, >322 μg/mL (Walsh et al., 1994).

Combination Therapy of Oral SSKI with Oral Azoles

This combination has produced lasting effects when there is an insufficient response to SSKI alone (Thomas et al., 2006). Relapses occur, so that therapy should be continued for a month or more after remission of symptoms. Surgical resection is not by itself curative, but has a place after a clinical response to antifungal therapy is evident (i) to debulk mass lesions and (ii) in reconstructive surgery after clinical and mycologic cure.

One advanced case that responded only partially to ITC, responded to a 2 month course of VRC which produced clinical and mycologic cure with no recurrence at the 2 year follow-up (Moore et al., 2007).

17C.14 LABORATORY DETECTION, RECOVERY, AND IDENTIFICATION

Direct Examination

Laboratory diagnosis is made by histopathologic and/or mycologic examination of 10% KOH preps of minced biopsy specimens. The growing edge of the lesion is more likely to contain viable fungus. The usual approach is intranasally or through the buccal sulcus to avoid a facial scar.

Histopathology

The hyphae in biopsy specimens stain poorly with H&E and with special fungal stains, PAS and GMS. The H&E stain is preferred because it reveals the brightly staining Splendore–Hoeppli[1] material outlining the

[1] Hyphae are surrounded by a band of eosinophilic material, the Splendore–Hoeppli phenomenon.

fungal elements, which may be large, thin-walled hyphae. Branching is rare, and there are occasional septa.

An inflammatory granulomatous reaction is seen which penetrates the dermis and subcutaneous tissue. Microabscesses contain PMNs and eosinophils, surrounded by palisading histiocytes[2] and occasional fibroblasts. The central portion shows broad, thin-walled, irregularly shaped hyphae with occasional septa and infrequent branching at nearly right angles. The predominant mononuclear infiltrate consists of lymphocytes, macrophages, and multinucleate giant cells, some containing fungal elements. Also present are plasma cells and numerous eosinophils. The histopathologic appearance is similar to that of subcutaneous entomophthoramycosis caused by *B. ranarum*.

Culture

Colony Morphology

Culture of the causative fungus is confirmatory, but it may fail to grow from biopsy material. *Conidiobolus* species display rapid growth, 3–4 days at 30°–35°C, on SDA into flat, cream-colored, *glabrous* colonies, with radial folds. Mature colonies become tan to brown with a white reverse. Numerous, round, sticky ballistoconidia are ejected covering the sides of the culture tube or the cover of the Petri plate. The conidiophores are said to be phototrophic and eject their spores in the direction of a light source; however, there has been no recent investigation of functional phototropism in *C. coronatus* or *C. incongruus*.

Microscopic Morphology

Hyphae are 6–15 μm in diameter and may show septa when mature. Sporangiophores (60–90 μm tall) are scarcely differentiated from vegetative hyphae. One-spored sporangioles function as ejectable ballistoconidia. They are round to pyriform with a prominent papilla. The sporangiole surface is smooth or *villous* (*definition*: covered with small fine hairs) (See Fig. 17B.4). *Conidiobolus coronatus* produces villous sporangioles but is not homothallic and *does not produce zygospores*. Some sporangioles germinate producing numerous secondary sporangioles which form a "crown" for which the species was named. *C. incongruus*, a homothallic species, produces zygospores without conjugation beaks but does not produce villous sporangioles.

[2]Palisaded granulomas of the skin. Local recruitment of histiocytes promotes lysis of collagen bundles resulting in a mucinous matrix referred to as a palisaded granuloma. Histiocytes are arranged in palisades (lined up as in a wall) around the granuloma.

Serology

There is an immunodiffusion test to detection antibodies specific for *C. coronatus* and *B. ranarum* but it is not commercially available (Kaufman et al., 1990).

Genetic Identification

Identification as *C. coronatus* was confirmed in one case by sequence analysis of the LSU D2 region of rDNA (de Ment et al., 2005).

SELECTED REFERENCES FOR ENTOMOPHTHORAMYCOSIS CAUSED BY *CONIDIOBOLUS* SPECIES

Batko A, 1974. Phylogenesis and Taxonomic Structure of the Entomophthoraceae, pp. 209–305, *in*: Nowinski C (ed.), *Ewolucja Biologiczna: Szkice teoretyczne i metodologiczne*. Polska Akademia Nauk, Instytyt Filozofii i Socjologii. Warsaw: Ossolineum. (USDA Technical Translation TT-77-54076 Translated from Polish by I. Jampoler.)

Bhosale SH, Rao MB Deshpande VV, Srinivasa MC, 1995. Thermal stability of high-activity alkaline protease from *Conidiobolus coronatus* (NCL 86.8.20). *Enzyme Microb Technol* 17: 136–139.

de Ment SH, Davis MS, Morris GL, Langom MG, Scates KW, 2005. Nasofacial zygomycosis: A case report of *Conidiobolus coronatus* in a Hispanic immigrant in South Carolina. *J SC Med Assoc* 101: 93–96.

Freimoser FM, Screen S, Hu G, St Leger R, 2003. EST analysis of genes expressed by the zygomycete pathogen *Conidiobolus coronatus* during growth on insect cuticle. *Microbiology* 149: 1893–1900.

Guarro J, Aguilar C, Pujol I, 1999. In-vitro antifungal susceptibilities of *Basidiobolus* and *Conidiobolus* spp. strains. *J Antimicrob Chemother* 44: 557–560.

Hoogendijk CF, van Heerden WEP, Pretorius E, Vismer HF, Jacobs JF, 2006a. Rhino-orbital entomophthoramycosis. *Int J Oral Maxillofac Surg* 35: 277–280.

Hoogendijk CF, Pretorius E, Marx J, van Heerden WEP, Imhof A, Schneemann M, 2006b. Detection of villous conidia of *Conidiobolus coronatus* in a blood sample by scanning electron microscopy investigation. *Ultrastruct Pathol* 30: 53–58.

Humber RA, Brown CC, Kornegay RW, 1989. Equine zygomycosis caused by *Conidiobolus lamprauges*. *J Clin Microbiol* 27: 573–576.

Kaufman L, Mendoza L, Standard PG, 1990. Immunodiffusion test for serodiagnosing subcutaneous zygomycosis. *J Clin Microbiol* 28: 1887–1890.

Krishnan SG, Sentamilselvi G, Kamalam A, Das KA, Janaki C, 1998. Entomophthoromycosis in India–a 4–year study. *Mycoses* 41: 55–58.

Manning RJ, Waters ST, Callaghan AA, 2007. Saprotrophy of *Conidiobolus* and *Basidiobolus* in leaf litter. *Mycological Res* 111: 1437–1449.

Martinson FD, 1972. Clinical epidemiological and therapeutic aspects of entomophthoromycosis. *Ann Soc Belge Méd Trop* 52: 329–342.

Moore MK, Murphy D, James C, 2007. Nasofacial phycomycosis: Persistence pays. *J Craniofacial Surg* 18: 448–450.

Phadatare S, Srinivasan MC, Deshpande M, 1989. Evidence for the involvement of serine protease in the conidial discharge of *Conidiobolus coronatus*. *Arch Microbiol* 153: 47–49.

Phadtare S, Rao M, Deshpande V, 1996. A serine alkaline protease from the fungus *Conidiobolus coronatus* with a distinctly different structure than the serine protease subtilisin. *Carlsberg Arch Microbiol* 166: 414–417.

Sharma NL, Mahajan VK, Singh P, 2003. Orofacial conidiobolomycosis due to *Conidiobolus incongruus*. *Mycoses* 46: 137–140.

Thomas MM, Bai SM, Jayaprakash C, Jose P, Ebenezer R, 2006. Rhinoentomophthoromycosis. *Indian J Dermatol Venereol Leprol* 72: 296–299.

Valle AC, Wanke B, Lazéra MS, Monteiro PC, Viegas ML, 2001. Entomophthoramycosis by *Conidiobolus coronatus*. Report of a case successfully treated with the combination of itraconazole and fluconazole. *Rev Inst Med Trop São Paulo* 43: 233–236.

Yang X, Li Y, Zhou X, Wang Y, Geng S, Liu H, Yang Q, Lu X, Hiruma M, Sugita T, Ikeda S, Ogawa H, 2010. Rhinofacial conidiobolomycosis caused by *Conidiobolus coronatus* in a Chinese rice farmer. *Mycoses* 53: 369–373.

Walsh TJ, Renshaw G, Andrews J, Kwon-Chung J, Cunnion RC, Pass HI, Taubenberger J, Wilson W, Pizzo PA, 1994. Invasive zygomycosis due to *Conidiobolus incongruus*. *Clin Infect Dis* 19: 423–430.

QUESTIONS

The Answer Key to multiple choice questions may be found at the end of the book.

1. Match each selection to the mycosis that it best describes.
   A. A nodule evolves into a woody, hard subcutaneous swelling affecting the trunk, upper arm, shoulder, and axilla.
   B. Fibrotic swelling is usually seen on the foot, with fistulas that form to the surface and drain pus containing visible grains, which are masses of fungal hyphae.
   C. Lesion(s) in the nasofacial region grow externally into tumefactions involving the nose upper lip, eyelids, and face.
   D. Plaque-like, or nodular, verrucous, crusted, ulcerated, and dense dermal fibrotic lesions occur on any part of the body, but usually on the arms, legs, face, or trunk.
   1. *Basidiobolus* entomophthoramycosis
   2. Chromoblastomycosis
   3. *Conidiobolus* entomophthoramycosis
   4. Eumycetoma

2. A tissue biopsy specimen comes to the mycology bench from a subcutaneous swelling on the bathing trunks area of a boy living in Nigeria. Which of the following steps would you take to detect and recover the causative fungus?
   A. Grind the tissue, make a KOH prep, and plate it on Mycosel agar.
   B. Grind the tissue, make a KOH prep, and plate it on SDA.
   C. Mince the tissue, make a KOH prep, and plate it on Mycosel agar.
   D. Mince the tissue, make a KOH prep, and plate it on SDA.

3. What type of sporulation would you expect to observe in a culture of *Basidiobolus ranarum*?
   A. Ballistoconidia, forcibly ejected onto the cover of the Petri plate.
   B. Club-shaped capilloconidia with knob-like tips containing "sticky" adhesive material.
   C. Round, intercalary, smooth, thick-walled zygospores with beak-like appendages on one side.
   D. First A and B, then C will occur.

4. What are the growth characteristics of *Basidiobolus* on laboratory media?
   A. Very slow growth
   B. Rapid growth
   C. Rapid and phototropic
   D. Requires enriched medium in order to sporulate

5. What is the recommended therapy for subcutaneous entomophthoramycosis caused by *Basidiobolus ranarum*?
   A. Topical therapy with SSKI
   B. Oral therapy with SSKI
   C. Oral SSKI followed by surgery
   D. Amphotericin B

6. Histopathology of *Basidiobolus* mycosis will demonstrate
   A. Hyphae and Splendore–Hoeppli material (PAS stain)
   B. Hyphae that are better outlined with GMS counterstained with H&E
   C. Large, weakly stained distorted hyphae surrounded by eosinophilic material (H&E stain)
   D. All of the above

7. Which of the following is true about the mycosis caused by *Conidiobolus*?
   A. A woody, hard subcutaneous swelling appears in the trunk, upper arm, shoulder, and axilla.
   B. Beginning as chronic sinusitis, the lesion enlarges involving the nose, upper lip, eyelids, and face but does not spread below the mouth.
   C. Conidia are inhaled and grow isotropically in the lung but do not divide.
   D. This chronic cutaneous mycosis is caused by a noncultivatable fungus.

8. What is the geographic distribution of entomophthoramycosis caused by *Conidiobolus*?
   A. Amazon Rain Forest ecosystem
   B. Between the Tropic of Cancer and Tropic of Capricorn
   C. The lower Sonoran life zone
   D. Worldwide

9. Which of the following is a risk factor for entomophthoramycosis caused by *Conidiobolus*?
   A. Age below 18 years
   B. Immunosuppression
   C. Living in poverty and malnourishment, in economically marginal countries
   D. Living near aquatic environments

10. The growth characteristics of *Conidiobolus coronatus* include
   A. Grows readily from biopsy material.
   B. One-spored villous sporangioles eject ballistoconidia.
   C. Smooth, thick-walled zygospores have beak-like appendages on one side.

D. Tall sporangiophores are well-differentiated from the vegetative hyphae.

11. Diagnosis of nasofacial entomophthoramycosis is made by

A. Clinical appearance and biopsy showing large, thin-walled hyphae, rarely branched and with few septa.

B. Culture of the fungus is confirmatory but is not essential to the diagnosis.

C. Paracoccidioidomycosis is in the rule-out category.

D. Radiology is important.

E. All are correct.

12. Discuss the pros and cons of SSKI therapy for nasofacial entomophthoramycosis. What is the clinical experience with azole therapy? Is there a place for combination SSKI and azole therapy? Include in your discussion of antifungal agents the problem of recurrences. What is the role for surgery in this mycosis?

Part Five

Mycoses of Implantation

Introduction to Mycoses of Implantation

The fungal etiologic agents presented in Chapters 18 and 19 discuss subcutaneous and other infections caused by melanized molds and their yeast-like relatives. Chapter 20 contains agents of a distinct mycosis caused by both melanized and hyaline molds. The presence of swelling, usually on an extremity, with sinuses draining pus containing compact fungal masses or "grains" typifies eumycetoma. While it is clear that sporotrichosis is a mycosis of implantation, it shares a common theme of mold-to-yeast dimorphism, and there are areas of high endemicity. For those reasons sporotrichosis is included with the other endemic, dimorphic mold pathogens in Chapter 9.

Melanized Molds

Melanized molds are a heterogeneous group of over 100 species in 60 genera of environmental molds with the common theme of melanin pigment in their cell walls. The term "melanized" is preferred to "dematiaceous."[1] Fungal melanin is a pathogenic factor protecting these fungi from the immune system. (Please see below, Special Topic: Fungal Melanin.) Even the term "mold" is applied cautiously to these etiologic agents because some of them grow as both yeast and hyphal forms, for example, members of the *Exophiala* genus.

Chromoblastomycosis and phaeohyphomycosis are mycoses caused by melanized fungi:

- Chromoblastomycosis is a cutaneous and subcutaneous mycosis caused by a phylogenetically related group of melanized molds with a distinctive tissue form, "muriform cells," consisting of round cells that enlarge and divide by forming internal septations, not by budding.
- Phaeohyphomycosis includes cutaneous–subcutaneous cysts, cerebral abscess, other deep-seated sites of infection, even fungemia, and a third category—fungal sinusitis. Subcutaneous phaeohyphomycosis is a mycosis of implantation which displays lesions of cystic, nodular, verrucous, ulcerated, or plaque types.
- *Why does phaeohyphomycosis include such a diverse group of clinical forms?* Because it is a disease definition based on the appearance of fungi in tissue (a *histopathologic* definition): to fit melanized molds and their yeast-like relatives that grow in tissue, or in the nasal sinuses, as melanized hyphae, or as a mix of hyphae and yeast forms.
- Eumyotic mycetoma (eumycetoma) is a subcutaneous mycosis caused by melanized *or* hyaline fungi and is discussed in the Chapter 20.

The characteristic clinical form of infection with dark-pigmented fungi is the chronic, indolent subcutaneous lesion developing into more extensive subcutaneous lesions which may remain localized or spread to adjacent underlying tissues.

Lesions in chromoblastomycosis are a mix of suppurative and granulomatous types. Subcutaneous lesions in phaeohyphomycosis are more likely to be cysts, which may disseminate from a cutaneous focus. Other clinical forms resulting from infection with black molds are allergic sinusitis, keratitis, deep-seated mycoses including cerebral abscess, and fungemia.

[1]Dematiaceous. *Definition*: Dark conidia and/or hyphae, usually brown or black. Dematiaceous is an older description of mitosporic molds within the *Fungi Imperfecti*, in the form-family *Dematiaceae*. The use of DNA-based phylogenetic analysis has rendered the term *Fungi Imperfecti* obsolete because taxonomic relationships among fungi can be interpreted from *cladistic* analysis even when the sexual state of the fungus has not been described.

Fundamental Medical Mycology, First Edition. By Errol Reiss, H. Jean Shadomy and G. Marshall Lyon, III.
© 2012 Wiley-Blackwell. Published 2012 by John Wiley & Sons, Inc.

Special Topic: Fungal Melanin. Definition: Melanins are dark, generally black, macromolecules composed of phenolic or indolic monomers, complexed with protein and, sometimes, with carbohydrates. The most frequent type of fungal melanin is dihydroxynaphthalene ("DHN") melanin (Butler and Day, 1998). Melanin is usually located in the fungal cell wall including conidial walls. Chemical properties of melanin are its insolubility in cold or boiling water, organic solvents, and hot or cold concentrated acids. Melanins can be bleached by hydrogen peroxide and may be solubilized and degraded by hot alkali solutions.[2]

Fungi synthesize *DHN melanin* via a pentaketide pathway (Butler and Day, 1998). DHN melanin does not contain nitrogen whereas mammalian melanin, based on a dihydroxyphenylalanine (DOPA) precursor, does contain nitrogen. An exception to the DHN formulation of fungal melanin is the DOPA melanin synthesized by *Cryptococcus neoformans* and *C. gattii* and deposited in cell walls of those yeasts.

Fungal DHN melanin forms a barrier layer consisting of granules or fibrils in the cell wall which resists digestion by host enzymes. Melanin protects fungi against UV irradiation and radioactivity. For example, a change toward melanized fungi is reported to have occurred in radioactive soil in the vicinity of the 1986 disaster at the Chernobyl nuclear reactor in the Ukraine (cited by Butler and Day, 1998).

- *Melanin Protection Against Oxidants*. Melanins are effective scavengers of free radicals (reviewed by Nosanchuk and Casadevall, 2006). Their antioxidant function may be particularly important at 37°C where the activity of superoxide dismutase is reduced. Melanized fungi are less susceptible than nonmelanized cells to killing by hypochlorite and to oxygen- and nitrogen-derived radicals. Melanized cells are thus protected against the neutrophil oxidative burst. Please see Section 18.9.2, Microbial Factors, for evidence of inhibition of reactive oxygen species and nitric oxide by *Fonsecaea pedrosoi* melanin.
- *Melanin Binds to Certain Antifungal Drugs*. Fungal melanin granules in the cell wall act in two ways to reduce the effectiveness of certain antifungal agents—AmB and caspofungin: (i) actual chemical binding and (ii) the closely spaced melanin granules reduce cell wall porosity, thereby preventing these large molecules from reaching their target sites (Nosanchuk and Casadevall, 2006). AmB has good activity against most clinically important melanized fungi in vitro, but clinical resistance is common. *Scedosporium prolificans, Curvularia* spp., and *Exophiala* spp. display in vitro resistance to AmB. The lack of clinical efficacy of echinocandins against these fungi, and their relative resistance to AmB, could be the result of the dense deposits of melanin in their cell walls. Azoles, in contrast, display broadest in vitro activity against melanized fungi. This drug class does not bind to melanin.

Infections with melanized molds can occur in persons with no known underlying immune or endocrine deficit but immunocompromised persons are at increased risk including recipients of solid organ transplants or patients receiving systemic corticosteroid therapy. Another group of melanized molds are implicated in "sick building syndrome" but they are not discussed here (please see Chapter 1, Section 1.1, Topics Not Covered, or Receiving Secondary Emphasis).

Morphology of Melanized Molds

Melanized molds may be classed as follows:

- *Molds that Are Monomorphic*. That is, they grow in the mold form in the environment and in tissues. For example, *Alternaria alternata, Bipolaris* species, *Curvularia lunata*, and *Exserohilum* species are all members of the *Pleosporaceae* characterized by their production of large multicelled *poroconidia*.
- *Black Yeasts*. Some melanized etiologic agents occur as a mix of hyphae and yeast forms in tissue, for example, *Wangiella* (*Exophiala*) *dermatitidis, Exophiala jeanselmei*, and *Exophiala spinifera*. Form development in these black yeasts is not a temperature sensitive switch to the yeast form at 37°C. Instead, a mix of yeast, hyphae forms and *moniliform* hyphae coexist in culture at room temperature and in tissue.
- *Dimorphic Species*. Other melanized species demonstrate a true dimorphic switch from the normal mold form at room temperature to a distinct tissue form. Agents of chromoblastomycosis are molds in the environment and, during infection, develop as round cells known as *muriform* cells or *sclerotic bodies*, which divide by laying down septa in internal cleavage planes. Factors influencing the muriform cell form are described in "Section 18.9.2, Microbial Factors".

[2]*Safety Note*: There is no indication to use concentrated acids or alkali in the clinical laboratory to identify melanin. Do not heat concentrated acids or alkali solutions. Always wear PPE when working with chemicals including proper eye protection.

Genetic Relationships of Melanized Molds and Their Relatives

Phylogenetic analysis of a large group of clinically relevant melanized fungi has shown they are members of the *Ascomycota*, in the order *Chaetothyriales*, family *Herpotrichiellaceae* (de Hoog et al., 2000). Gene genealogies among the black yeast-like fungi and agents of chromoblastomycosis were determined by sequence comparisons of the 18S rDNA and the ITS2 subregion of rDNA (Caligiorne et al., 2005). Members of the *Fonsecaea* and *Phialophora* genera as well as *Cladophialophora carrionii* and *Cladiophialophora bantiana* are united in a single clade composed of molds of increased pathogenicity. *Wangiella* (*Exophiala*) *dermatitidis* is a related core species of thermophilic and neurotropic black yeast-like fungi.

Chapter 18

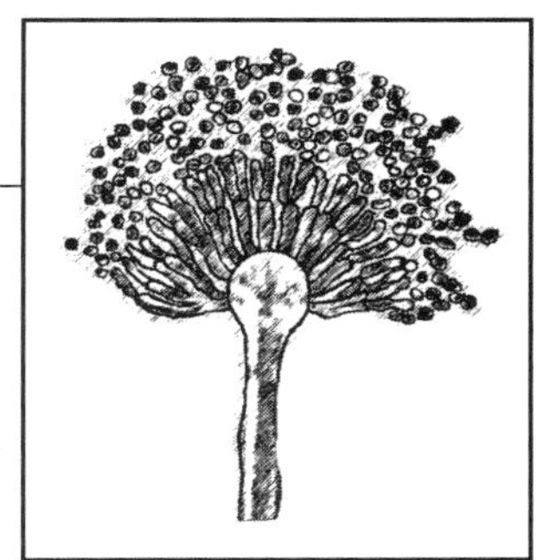

Chromoblastomycosis

18.1 CHROMOBLASTOMYCOSIS-AT-A-GLANCE

- *Introduction/Disease Definition.* Chromoblastomycosis is a cutaneous and subcutaneous mycosis characterized by chronic, indolent granulomas, especially on the feet and legs.
- *Etiologic Agents.* Dimorphic black molds: The major pathogens are *Fonsecaea pedrosoi* and *Cladophialophora carrionii* and, less commonly, *Botryomyces caespitosus, Exophiala jeanselmei, Phialophora verrucosa, and Rhinocladiella aquaspersa*, among other species.
- *Geographic Distribution.* Worldwide, including the United States; more common in Mexico, the Caribbean, Central and South America, Africa, notably in Madagascar, also in China and India. Sporadic cases occur in temperate climates, even among urban dwellers.
- *Etiologic Niche.* Decaying wood, plants, and soil.
- *Epidemiology.* Chromoblastomycosis occurs worldwide; most patients are men >30 years old.
- *Risk Group/Factors.* Adult males are at increased risk, especially where agricultural laborers work barefoot.
- *Transmission.* Via transcutaneous trauma with vegetable matter: thorns or splinters.
- *Determinants of Pathogenicity.* Melanin in the cell wall, secreted proteinase, wall associated phosphatase; conversion to muriform cells that are resistant to lysis. Calcium ions and the micronutrient Mn^{2+} are implicated in form development of muriform cells.
- *Clinical Form.* Chronic, indolent, cutaneous and subcutaneous granulomas, especially on feet and legs. Over many months, without treatment, disruption of the lymphatic channels can cause the affected limb to swell.
- *Therapy.* Small lesions—surgery and antifungal therapy to prevent recurrence; larger lesions are difficult to treat—itraconazole (ITC) and/or terbinafine (TRB), for periods of 6–12 months or more. Posaconazole (PSC) shows success in a small series of cases.
- *Laboratory Detection, Recovery, and Identification.* The chromoblastomycosis agents are dimorphic. The mold forms grow as a gray "mouse fur" colony with a black reverse. Microscopically each species has different asexual reproductive structures, all containing brown pigment. For the histopathology, the tissue form of all causative species consists of golden-brown round cells: *muriform* cells, also known as sclerotic bodies, fumagoid bodies, or "copper pennies." These divide by internal cleavage planes not by budding.

18.2 INTRODUCTION/DISEASE DEFINITION

Chromoblastomycosis is a cutaneous and subcutaneous mycosis of implantation producing either plaque-like, or nodular, *verrucous*, crusted, ulcerated, and dense dermal fibrotic lesions (may be on pedicles and resembling cauliflowers). The lesions occur on any part of the body, but usually on the arms, legs, face, or trunk. If untreated, lesions slowly spread over large areas. This disease progresses very slowly, remains in the subcutaneous tissues, and is not life-threatening. It is considered the "most superficial" deep mycosis when compared with phaeohyphomycosis, sporotrichosis, or eumycetoma. The

Fundamental Medical Mycology, First Edition. By Errol Reiss, H. Jean Shadomy and G. Marshall Lyon, III.
© 2012 Wiley-Blackwell. Published 2012 by John Wiley & Sons, Inc.

major causative agents of chromoblastomycosis are the black molds *Fonsecaea pedrosoi* and *Cladophialophora carrionii*. Seventy to 90% of cases are caused by *Fonsecaea pedrosoi*. Sporadic cases have been caused by other melanized molds (please see Section 18.5, Etiologic Agents). All are dimorphic with a tissue form consisting of single to multiple golden to dark brown, round, and septate single to multiple *muriform* cells which, on histologic examination, are the hallmark of the disease.

18.3 CASE PRESENTATION

Swelling in the Left Leg of a 67-yr-old Man (Brown and Pasvol, 2005)

History

A 67-yr-old man was admitted to the hospital with a 12 year history of swelling of his left leg. He had emigrated to the United Kingdom from Jamaica at the age of 31 years. When the patient was 41 years old, a golf-ball–sized nodule was excised from the dorsum of his left foot, and he was told he had a fungal infection, but did not receive further treatment. At the age of 55 years, the patient noticed a nodule on his left shin. He did not seek medical attention until the lesion had spread and become so extensive that he was unable to walk (Fig. 18.1).

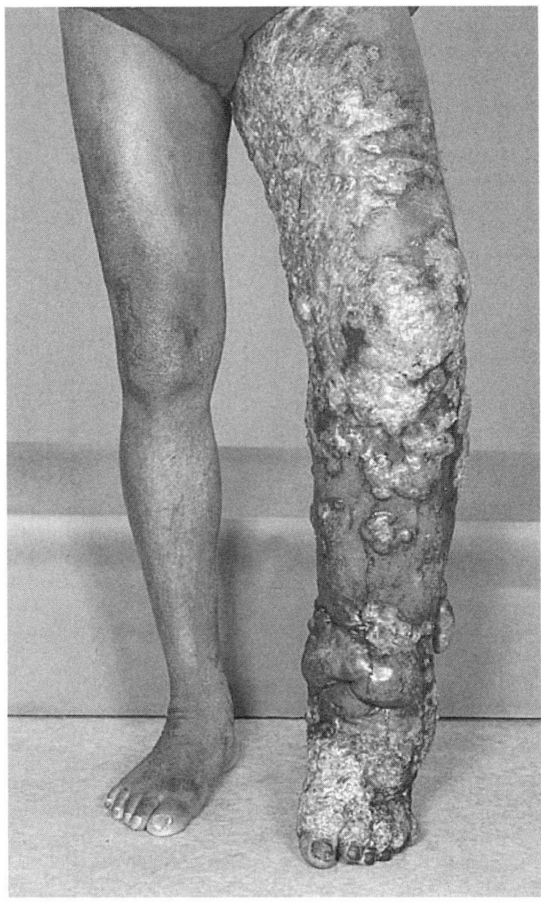

Figure 18.1 Swelling in the left leg of a 67-yr-old man. *Source:* Brown and Pasvol (2005). Used with permission from the Massachusetts Medical Society, Copyright © 2005, all rights reserved.

Diagnostic Procedure

A skin biopsy revealed a suppurative and granulomatous infiltrate with clusters of brown nonbudding round fungal cells (muriform cells), diagnostic of chromoblastomycosis. A *Rhinocladiella* species was cultured from skin scrapings. (Please see Section 18.12, Laboratory Detection, Recovery, and Identification, for an illustration.)

Diagnosis

The diagnosis was chromoblastomycosis due to *Rhinocladiella* species.

Therapy and Clinical Course

The patient was treated for 24 months with oral ITC and TRB resulting in improved mobility and substantial drying of the lesions, but warty changes, hyperpigmentation, and lymphedema persisted.

Comment

The disease may have been contracted many years prior to his immigration to the United Kingdom, when he was residing in Jamaica. Although less common, *Rhinocladiella* species are recognized causative agents of chromoblastomycosis. (Please see Section 18.5, Etiologic Agents.) When monotherapy fails to achieve clinical and mycologic cure, combination therapy with ITC and TRB may be more effective. Cryosurgery and adjunctive thermal therapy may improve the healing process. (Please see Section 18.11, Therapy)

18.4 DIAGNOSIS

Diagnosis is made by clinical presentation, visualization of muriform cells in tissue, culture of a melanized fungus, and its laboratory identification. Chromoblastomycosis is identified by the histopathologic presence of golden brown muriform cells in subcutaneous tissue. These cells display *isotropic*[1] growth. That is, the cells enlarge in all

[1]Isotropic. *Definition*: Having the same properties in all directions.

dimensions and divide internally by laying down septa in different planes. Occasionally small amounts of hyphae may be seen in epithelial tissue, at times confusing identification of the disease. The etiologic agent, one of several species of melanized molds, is identified by culture.

Differential Diagnosis

The following diseases are included in the rule-out for the diagnosis of chromoblastomycosis: cutaneous sarcoidosis, cutaneous paracoccidioidomycosis, keratocanthoma, leishmaniasis, leprosy, lobomycosis, lupus erythematosus, protothecosis, sporotrichosis, squamous cell carcinoma, verrucous tuberculosis, and, in superficial cases, Bowen's disease, psoriasis, tinea corporis, and verruga vulgaris.

18.5 ETIOLOGIC AGENTS

The most commonly identified agents are *Fonsecaea pedrosoi* and *Cladophialophora carrionii*. Sporadic cases have been caused by *Phialophora verrucosa, Rhinocladiella aquaspersa, Botryomyces caespitosus, Exophiala jeanselmei, E. spinifera, Aureobasidium pullulans*, and *Wangiella (Exophiala) dermatitidis*. Other genera and species occasionally have been identified (Queiros-Telles et al., 2009).

The position in classification of the following four agents of chromoblastomycosis is *Ascomycota, Chaetothyrales, Herpotrichiellaceae*.

Cladophialophora carrionii. (Trejos) de Hoog, Kwon-Chung & McGinnis, in de Hoog, Guého, Masclaux, Gerrits van den Ende, Kwon-Chung et al., *J Med Vet Mycol* 33(5): 345 (1995).

Fonsecaea pedrosoi. (Brumpt) Negroni, *Revista Inst Bacteriol 'Dr. Carlos G. Malbrán'*: 424 (1936). *F. pedrosoi*, the major etiologic agent of chromoblastomycosis, was reinvestigated and a sister species, *F. monophora*, was discovered (de Hoog et al., 2004). rDNA encoding the ITS (intergenic transcribed spacer) region was sequenced. *F. pedrosoi* differs from *F. monophora* by seven mutations in the ITS1 subregion and six in ITS2. Dendrograms produced by sequence alignments and neighbor-joining algorithms with ITS sequence data showed that the majority of the strains analyzed were clustered in a well-delimited clade, conforming to the *F. pedrosoi* type species (de Hoog et al., 2004). *Fonsecaea compacta*, formerly accorded species status, was determined to be a *dysgonic* mutant of *F. pedrosoi*.

Morphologically, the genus *Fonsecaea* is defined by the presence of indistinct melanized conidiophores with blunt, scattered denticles bearing conidia singly or in short chains, which eventually become branched. The chains contain a maximum of three conidia each. *F. monophora* cannot be distinguished by physiologic or biochemical tests from *F. pedrosoi* because they are also morphologically very similar. *F. pedrosoi* strains are isolated from warm-blooded hosts, and while it occurs in the environment, its recovery may require passage through a mammal. It appears that *F. monophora* is a human pathogen, as well, but is more easily isolated from the environment (de Hoog et al., 2004). Whether the distinctions between the sister species, based on few isolates of *F. monophora*, rise to the level of clinical relevance remains unknown.

Phialophora verrucosa (Medlar) *Mycologia* 7(4): 203 (1915). *P. verrucosa* is the major pathogen in the genus *Phialophora*. A neighbor joining tree of the ITS region of rDNA found that *Phialophora* species occur in a subgroup containing agents of subcutaneous mycoses (de Hoog et al., 2000). It is a *monophyletic* group with phialides and collarettes in all strains.

P. verrucosa and *P. americana* are related but distinct clades based on morphology and sequence analysis of the ITS subregion of rDNA (Yan, 1995). *P. americana* produces phialides with deep, vase-shaped collarettes, whereas the collarettes of *P. verrucosa* are funnel shaped. *P. americana* forms a strongly supported clade that does not include *P. verrucosa* in a phylogeny based on a combined data set of ß-tubulin-ITS-18S rDNA sequences (Untereiner et al., 2008). *Phialophora parasitica* is genetically variable, whereas *P. richardsiae* is genetically homogeneous, and thus the four species identities established on morphologic grounds also were separate species based on molecular analysis.

Rhinocladiella aquaspersa. (Borelli) Schell, McGinnis & Borelli, *Mycotaxon* 17: 343 (1983).

Phylogenetic analysis of the black yeasts and their close relatives supports their ecologic niche and role in pathogenesis. Phylogenetic analysis of the SSU rDNA from black yeasts and their close relatives in the *Herpotrichiellaceae* showed the species clustered in five clades (Caligiorne et al., 2005). These clades agree with earlier analysis of the ITS region of rDNA. The dendrogram clusters *Fonsecaea, Phialophora*, and *Cladophialophora* species into a single clade (clade 2), showing the phylogenetic similarity of family members with a similar clinical form of disease.

In addition to clade 2, containing the chromoblastomycosis agents, the core species of the five clades are

Wangiella (*Exophiala*) *dermatitidis, Cladophialophora bantiana, E. spinifera, E. nigra*, and *Coniosporium perforans*. The phylogeny is consistent with the ecologic niches of these black yeasts and their close relatives. This grouping is ecologically meaningful. Clade 1 is a thermophilic yeast group centered around the neurotropic *W. (Exophiala) dermatitidis*. Clade 2, as mentioned, is the group causing chromoblastomycosis and brain disease. Clade 3 is the *E. spinifera/E. jeanselmei* complex. Clade 4 is a meso- to psychrophilic group of *Exophiala* species found in showers and ocean waters. Clade 5 contains meristematic species inhabiting rocks.

18.6 GEOGRAPHIC DISTRIBUTION/ ECOLOGIC NICHE

Chromoblastomycosis is seen primarily in tropical and subtropical countries of Central and South America, the Caribbean, and Africa (López Martínez and Méndez Tovar, 2007). Countries and areas with the highest prevalence are:

- *Western Hemisphere.* Brazil (especially in the Amazon region), Mexico (in Oaxaca and Veracruz), Venezuela, the Dominican Republic, Cuba, and Colombia.
- *Africa.* Madagascar, the Natal state of South Africa, Botswana, Cameroon, Nigeria, Rwanda, and Zambia (Esterre et al., 1997).
- *Asia.* Japan and China.
- *Oceania.* Malaysia and Australia where, in the latter, there are a small number of cases due to *C. carrionii*.

Cases have, however, been reported worldwide: in Europe—from Russia, the Czech Republic, Romania, and the former East Germany. Interestingly, the first reported case was in a man from Boston, Massachusetts. The predominant fungus isolated from chromoblastomycotic lesions varies with geographic location. In Brazil, Costa Rica, Colombia, and other Latin American countries, the prevailing fungus isolated is *Fonsecaea pedrosoi*. In Japan and other Asian countries *F. pedrosoi* also is the major cause, followed by *Phialophora verrucosa*. The majority of cases in the United States are caused by *P. verrucosa*. Dry endemic regions harbor *C. carrionii*, such as southern Madagascar, in Australia, and parts of Venezuela.

Agents of chromoblastomycosis including *Cladophialophora, Fonsecaea, Phialophora*, and *Rhinocladiella* isolates are found in the environment in soil, wood, vegetation and decaying matter, and often on plants with barbs or thorns. The Pará state in the northeastern Amazon region of Brazil has a high prevalence of chromoblastomycosis (Salgado et al., 2004). *F. pedrosoi* was isolated from thorns of *Mimosa pudica* at the site where a woman being treated for chromoblastomycosis reported receiving a puncture with such a thorn. *Mimosa pudica* is a common low-growing roadside plant with an average height of 0.5 m. The fungus was observed by scanning electron microscopy on thorns removed from the plant, suggesting a fungus–plant interaction.

Cladophialophora carrionii is the major causative agent of chromoblastomycosis in semiarid northwestern Venezuela, where rural people often are goatherders. In the course of their work they can be pricked by cactus thorns, known to harbor *C. carrionii* (de Hoog et al., 2000). The high incidence of chromoblastomycosis in rural Venezuela is explained by a homogeneous, genetically susceptible population and opportunities for exposure to infection.

18.7 EPIDEMIOLOGY AND RISK GROUPS/FACTORS

Chromoblastomycosis occurs worldwide, although endemic areas are concentrated in tropical or subtropical areas. Seventy to 90% of patients are males over the age of 30 years. Children are rarely affected. Interestingly, in Japan the ratio of males to females contracting chromoblastomycosis is approximately equal. Diagnoses are most often made in males living and working in rural areas. A lack of appropriate protective clothing is a risk factor, (e.g., shoes, clothing covering the legs and arms). Those patients recalling traumatic events indicate that inoculation was a result of thorns, wood splinters, or cuts from tools. Sporadic cases also occur in temperate climates, even among urban dwellers.

The Institut Pasteur in Madagascar recorded 1323 cases of chromoblastomycosis between 1955 and 1995 (Esterre et al., 1997). During this 40 year period, histopathologic and, where possible, culture isolation determined there are two distinct ecosystems: in the northern rain forest *F. pedrosoi* predominates and, in the south, where the climate is dry desert, *C. carrionii* prevails, accounting for 41% of the entire sample. Deforestation, to produce wood for construction and charcoal for fuel, is a factor associated with this disease. Eighty-seven percent of the patients are male and 96% are more than 16 years old; most lesions are found on the feet or legs. Madagascar is the most important endemic area for chromoblastomycosis with a prevalence of about 1 in 8500 inhabitants.

Adult males are at increased risk, especially where agricultural workers work barefooted. Immunosuppression is another risk factor. Interestingly, in one series of chromoblastomycosis patients, the human leukocyte antigen

HLA-A29 allele was more prevalent (reviewed by López Martínez and Méndez Tovar, 2007).

18.8 TRANSMISSION

The disease is acquired via a penetrating injury from wood and plant material. Occasionally penetrating wounds from other sharp materials may incite infection.

18.9 DETERMINANTS OF PATHOGENICITY

18.9.1 Pathogenesis

Beginning with a penetrating injury with a thorn or splinter containing conidia of one of the melanized causative molds, a papule develops, slowly enlarging over weeks or months and containing the tissue form of the agent—copper-colored muriform cells, which resist lysis in the host. The primary lesion can evolve into various types of skin lesions, including nodules, tumors, or verrucous, cicatricial, and plaque-like lesions. Frequently, the form taken is a tumoral or cauliflower-like lesion at the site of inoculation, from which satellite lesions gradually arise. PMN-rich purulent abscesses form but do not lead to effective killing of the muriform cells. The lesions then are encircled in granulomas and further encapsulated by fibroblasts resulting in the deforming lesions—pseudoepitheliomatous hyperplasia.

The chronic granulomatous reaction is characterized by extensive and progressive dermal fibrosis, perhaps induced by a continuous antigenic stimulus (Ricard-Blum et al., 1998). Verrucous lesions display a dense fibrosis of the whole thickness of the dermis. Extracellular matrix proteins in the fibrotic zone are type I collagen, laminin, elastin, and fibronectin. Collagen of the fibrotic lesions shows increased cross-linkage by pyridinoline, a marker of the severity of fibrosis (Esterre et al., 1992). Chromoblastomycosis patients treated with terbinafine show a rapid clinical response of skin lesions, even with old verrucous lesions. Biopsies of human chromoblastomycosis patients taken before and after a 1 year treatment with daily terbinafine were analyzed. The mean concentration of collagen in the fibrotic lesions of patients infected with *F. pedrosoi* decreased significantly.

18.9.2 Host Factors

Innate and Humoral Immunity

Patients with chromoblastomycosis produce specific IgG_1, IgM, and IgA antibodies, including anti-melanin antibodies which inhibit fungal growth in vitro (Alviano et al., 2004). Such antibodies may play an incremental role in host defense. PMNs from normal human donors exposed in vitro to melanin show enhanced antifungal activity against a mix of muriform cells and conidia of *F. pedrosoi*.

Murine macrophages were tested for the ability to phagocytose and kill conidia of chromoblastomycosis agents (Hayakawa et al., 2006). The phagocytic index was high with conidia of *F. pedrosoi* and *R. aquaspersa* and this phagocytosis was inhibited by *mannan*. Of all chromoblastomycosis agents tested, macrophages were only able to kill *R. aquaspersa* conidia and showed little or no cytotoxic effect on *F. pedrosoi, C. carrionii*, or *P. verrucosa*. All fungi induced the production of TNF-α by macrophages. IL-1β[2] was induced only by *F. pedrosoi* and *R. aquaspersa*. IL-6[3] production by macrophages was induced by contact with *C. carrionii*.

Cell-Mediated Immunity

Murine macrophages challenged with *F. pedrosoi* conidia, in the presence of melanin, demonstrated a significant increase in the number of attached and phagocytosed conidia (Alviano et al., 2004). $CD4^+$ T-lymphocytes are key cells in the immune response in chromoblastomycosis (reviewed by Santos et al., 2007). Mice infected with *F. pedrosoi* conidia show an influx of $CD4^+$ T cells into draining lymph nodes. These T cells respond in vitro to *F. pedrosoi* soluble antigens by producing IFN-γ. Mice lacking $CD4^+$ T cells but not $CD8^+$ cells display decreased delayed-type hypersensitivity and reduced amounts of IFN-γ. Humans with severe chromoblastomycosis produce high amounts of IL-10[4] and low levels of IFN-γ. T lymphocytes from these individuals fail to proliferate in vitro in response to *F. pedrosoi* antigens, whereas patients with mild disease produce IFN-γ and low levels of IL-10, and efficiently respond to *F. pedrosoi* antigens in lymphocyte blastogenesis assays (Mazo Fávero Gimenes et al., 2005).

Monocytes from patients with severe chromoblastomycosis expressed lower concentrations of HLA-DR and costimulatory molecules than those from patients with a

[2]IL-1ß. *Definition*: IL-1β is produced by macrophages, monocytes, and dendritic cells in the inflammatory response to infectious agents. It and IL-1α increase the expression of adhesion factors on endothelial cells, facilitating the transmigration of leukocytes to the site of infection.

[3]IL-6. *Definition*: IL-6 is a cytokine that induces acute phase reactants in the inflammatory response to infection. Macrophages recognize microbial "pathogen associated molecular patterns" (PAMPs) via Toll-like receptors (TLRs) on the macrophage cell surface. This binding sets off an intracellular signaling cascade that initiates production of IL-6 and other inflammatory cytokines.

[4]IL-10. *Definition*: IL-10 is produced by Th2 cells, certain B cell subsets, and LPS-activated macrophages. Its effect is mostly regulatory, including inhibiting Th1 responses, antigen presentation, and TNF-α production by monocytes and macrophages.

mild form of the mycosis. IL-10 levels were higher and IL-12 levels lower in patients with severe chromoblastomycosis when stimulated in vitro with culture supernatant antigens of *F. pedrosoi*, tipping the balance toward production of anti-inflammatory cytokines (Sousa et al., 2008). Addition of anti-IL-10 antibodies to the in vitro system or addition of recombinant IL-12 restored IFN-γ production and lymphocyte proliferation responses to *F. pedrosoi* antigens.

18.9.3 Microbial Factors

DHN Melanin

DHN melanin is present in the cell walls of melanized fungi. (Please see "*Special Topic*: Fungal Melanin" in the Introduction to Part Five.) Melanin protects the fungus against phagocytosis and the oxidative burst. Interaction of *F. pedrosoi* and activated murine macrophages substantially reduced nitrite levels, and this reduction was shown to be due to melanin extracted from the mold (Cunha et al., 2010). *F. pedrosoi* conidia grown in the presence of a melanin inhibitor, tricyclazole, and incubated with macrophages produced significantly more nitrite than melanized conidia. Melanin acts as a trap for the unpaired electron of NO, protecting *F. pedrosoi* against oxidative damage.

Formation of thick-walled muriform cells by the fungi also may resist host cell destruction and stimulate a granulomatous response. Melanin protects muriform cells from recognition by antibodies directed against surface cerebrosides.[5] Microbial cerebrosides differ from the mammalian counterparts. As such they become a target for antibody production. Whether cerebrosides in *F. pedrosoi* are signaling molecules that in some manner affect the mycelial to muriform cell dimorphism is, as yet, hypothetical (Nimrichter et al., 2005).

Secreted Enzymes

A secreted proteinase and wall-associated phosphatases are present in both conidia and mycelial forms of *F. pedrosoi*. Aspartyl proteinase secreted by conidia of *F. pedrosoi* is blocked by pepstatin A, inhibiting germination and outgrowth into a mycelium (Palmeira et al., 2006).

Cell wall-associated acid phosphatase activity in *F. pedrosoi* muriform cells is much higher than that of conidia and mycelia (Kneipp et al., 2004). Conidia cultured in inorganic phosphate-deficient medium respond with a 130-fold increase in the surface phosphatase activity. The adherence of these conidia to epithelial cells is sevenfold higher than conidia cultivated in regular conditions. Surface acid phosphatase appears to contribute to the adherence of *F. pedrosoi* conidia to host cells. Factors influencing the conversion of hyphae into muriform cells, such as platelet activating factor, correlate with increased wall-associated phosphatase. Surface acid phosphatase appears to be involved in the morphogenesis to the tissue form. Both proteinase and phosphatase activities seem to be involved in the early stages of pathogenesis of chromoblastomycosis.

Factors Affecting Dimorphism

C. carrionii, *F. pedrosoi*, and *P. verrucosa* convert from hyphae to muriform cells at 25°C and at 37°C, in chemically defined medium at pH 2.5 at a Ca^{2+} concentration of 1×10^{-4} M (Mendoza et al., 1993). Increasing the Ca^{2+} concentration reversed this trend, maintaining hyphal growth. Addition of the calcium chelator EGTA to medium buffered at pH 6.5 also induces muriform cells. Wall-associated acid phosphatase concentrations are increased in low phosphate medium, corresponding to the mycelial conversion to muriform cells (Kniepp et al., 2004) The role of micronutrients in altering the cell morphology is not new, and it was shown that specific depletion of Mn^{2+} from cultures of *P. verrucosa* in chemically defined medium induced a change in morphology from hyphae to round cells (Reiss and Nickerson, 1974). Evidently there are correlations between phosphate, calcium, and, perhaps, manganese concentrations in the medium and morphogenesis in chromoblastomycosis agents.

Platelet-Activating Factor (PAF)

PAF[6] treatment of mycelial forms of *F. pedrosoi* induces conversion to the muriform cell form in vitro and at the same time promotes a significant increase in wall-associated phosphatase activity (Alviano et al., 2003).

The curious dimorphism from hyphae to muriform cells is induced in different fungal species causing chromoblastomycosis. These tissue forms resist lysis in the host, resulting in a granulomatous and fibrotic response leading ultimately to the disfiguring clinical form, which, however, remains restricted to the subcutaneous tissues. Morphogenesis and the hyperplasia that the muriform cells

[5]Cerebroside. *Definition*: Cerebrosides are glycosphingolipids consisting of monohexyl-ceramides (CMH). The "monohexyl" moiety is a hexose, either glucose or galactose. Ceramides contain a fatty acid chain attached via an amide linkage to sphingosine (an 18 carbon amino alcohol). They have interest because of their ability to act as signaling molecules regulating cell differentiation and apoptosis.

[6]Platelet-activating factor (PAF) is a phospholipid activator and mediator of leukocyte functions, including platelet aggregation and inflammation, even anaphylaxis. PAF is produced by various cell types including PMNs, platelets, and endothelial cells. Chemically it is acetyl-glyceryl-ether-phosphorylcholine.

evoke in the host remain enigmatic, and more research is needed, coupling genomics of the microbe with immunologic studies of the host–pathogen interaction.

18.10 CLINICAL FORMS

The five types of chromoblastomycotic lesions are illustrated in Fig. 18.2. Lesions typically begin at the site of traumatic implantation, initially solitary and unilateral, presenting as a small, pink, smooth surface, papular skin lesion. After some weeks, papules gradually increase and may develop a scaly surface. With time, the inoculation lesion may evolve to several types of skin lesions leading to the polymorphic clinical appearance. They are seen most often on the legs and feet, although any part of the body may be affected. The most common sites are the foot, ankle, and/or leg. Invasion of bone and muscle are rare. Verrucous lesions frequently are found along the borders of the feet. These lesions are granulomatous. Nodular lesions may enlarge to form tumerous growths, often on pedicles, resembling a cauliflower and may be in a variety of colors. If left untreated, lesions gradually progress slowly over months and even years. In persons without access to primary care and referral for specific therapy the lesions may cover large areas. (Please see Section 18.3, Case Presentation.) Five clinical forms of chromoblastomycosis are recognized (Carrion, 1950; López Martínez and Méndez Tovar, 2007; Queiroz Telles et al., 2009). More than one form may be present in the same patient:

1. Nodular lesions with a surface that is smooth, verrucous, or scaly.
2. Tumor-like masses, prominent, papillomatous, sometimes lobulated or "cauliflower-like."
3. Verrucous. Hyperkeratosis is the major feature of these warty, dry lesions.
4. Cicatricial. Flat lesions that enlarge by peripheral extension with atrophic scarring and healing at the center. Usually with annular, arciform, or serpiginous outline. They tend to cover extensive areas of the body.
5. Plaque type. Slightly elevated lesions of various size and shape, reddish to violaceous in color, with a scaly surface.

The extensive fibrosis and chronic inflammatory infiltrate impedes lymphatic flow, eventually causing lymphedema below the affected sites. Impaired circulation caused by lymphedema causes further deformity and ankylosis of joints leads to permanent disability. These lesions are easily traumatized by hitting and may become secondarily infected. Sequelae common in long standing lesions include scarring and deformity, lymphedema, elephantiasis, recurrent bacterial infection, and malignant changes in the lesion and associated tissue.

Illustrative Example:. Chromoblastomycosis in a renal transplant patient in the United States. A 50–yr-old renal transplant patient developed slow-growing violaceous nodules with a thick overlying verrucous crust extending up the leg from the ankle. The lesions were resistant to antibiotic therapy. A skin biopsy demonstrated granulomatous inflammation with melanized muriform cells. (Fig 18.3).

A system for scoring the severity of chromoblastomycosis consists of (i) *mild*—a solitary plaque or nodule $\leq$5 cm diameter; (ii) *moderate*—solitary or multiple adjacent lesions, nodule, plaque or tumor $\leq$15 cm diameter; and (iii) *severe*—extensive cutaneous lesions of any type including nonadjacent areas (Queiroz-Telles et al., 2003).

18.11 THERAPY

Treatment is difficult and surgical excision is common, although lesions may recur. Surgery alone is useful only when there is a single lesion or a few small isolated lesions. There is no single treatment of choice but there are several treatment options. To date there have been low cure rates with many relapses. The best results have

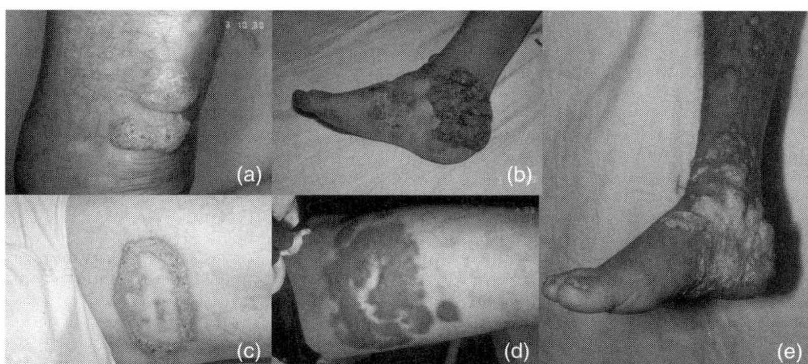

Figure 18.2 The five types of chromoblastomycosis lesions, according to Carrion (1950). (A) Nodular, (B) tumoral, (C) cicatricial, (D) plaque, and (E) verrucous. See insert for color representation of this figure. *Source:* (Queiroz-Telles et al. (2009). Used with permissions from Dr. Flavio de Queiroz-Telles, Universidade Federal do Paraná, Brazil, and from Taylor and Francis Group. Copyright © 2009.

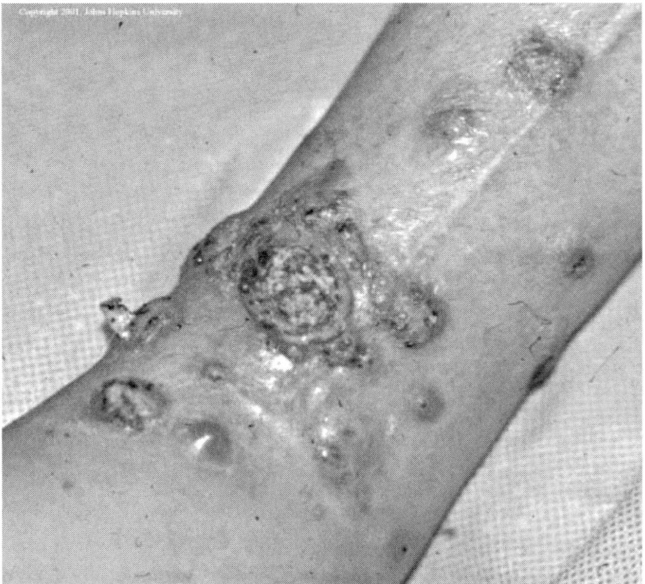

Figure 18.3 Chromoblastomycosis in a 50–yr-old renal transplant patient. Slow-growing violaceous nodules with a thick overlying verrucous crust extending up the leg from the ankle.
Source: Copyright © Adrienne Rencic, M.D., Ph.D. Used with permission from Dermatlas; www.dermatlas.org.

been obtained with itraconazole (ITC) or terbinafine (TRB) monotherapy at high doses for 6–12 months. Oral antifungal therapy is combined with local heat and/or cryotherapy for extensive cases. Early cases may be successfully managed with surgery combined with oral antifungal therapy. Determining the susceptibility of the causal isolate to various antifungal agents is prudent given the refractory nature of the disease and the necessity for long-term therapy.

Clinical and mycologic cure is difficult to achieve. The relapse rate is high when the cases are chronic and extensive. The goal for limited and early cases is clinical and mycologic cure achieved by surgery and oral antifungal therapy combined with either local heat therapy and/or cryosurgery. A considerable reduction of lesions is possible after long-term treatment of very extensive cases. Penetration of drugs into the lesions is hindered because of dense dermal fibrosis. *Fonsecaea pedrosoi* is the causal agent that is a relatively low responder to antifungal therapy compared with *C. carrionii* and *P. verrucosa*. Better outcomes even in extensive cases are seen with the latter two species.

Criteria for assessment of the therapeutic response are:

- Clinical cure—complete healing of all lesions
- Histologic cure—atrophy (resolution) of lesions and replacement of granuloma and abscess by fibrous tissue
- Mycologic cure—failure to grow the fungus from biopsied specimens

A period of at least 2 years of follow-up clinic visits is prudent.

Surgery and Thermal Therapy

Various surgical approaches have been used: standard surgery with appropriate margins, electrodesiccation with electrocautery, and Mohs micrographic surgery in order to reach the deepest areas until the absence of muriform cells, proven histologically. Local thermal therapy for periods of 2–12 months has produced good results in chromoblastomycosis. Local heat to produce a temperature range of 42–45°C will result in growth inhibition as has been used effectively to treat sporotrichosis. In that case *S. schenckii* is very heat sensitive. Cryotherapy by topical application of liquid nitrogen with swabs has had variable results. Cryosurgery using an open-spray method is used to treat small lesions in a single treatment. In extensive cases the areas are divided into sections and treated in separate sessions.

At present the trend is to combine surgical and thermal approaches with oral antifungal therapy to avoid lymphatic spread of the disease. Combined therapy consisting of cryosurgery with ITC is accomplished by first administering oral ITC therapy for 8–12 months and, when maximum reduction of lesions occurs, following with several sessions of cryotherapy. Local heat therapy may be used during the long period of antifungal therapy.

Antifungal Therapy for Chromoblastomycosis

- *Amphotericin B*. Systemic treatment with this agent is accompanied by adverse effects. Intralesional injection causes extreme irritation.
- *Fluconazole*. Fluconazole has little activity against melanized fungi.
- *Flucytosine*. At 100–200 mg/day, oral flucytosine monotherapy may produce rapid reduction of lesions but secondary resistance soon develops and the therapeutic response declines.
- *Ketoconazole*. *Phialophora verrucosa* is very susceptible at doses of 200–600 mg/day, whereas *F. pedrosoi* is less responsive.

The above four drugs are not first-line therapy and have been superseded by itraconazole and terbinafine.

- *Itraconazole (ITC)*. Oral treatment consists of doses of 100–400 mg/day. Higher doses are usually recommended but at 100 mg/day, treatment periods range up to 15 months. In a series of 30 Brazilian patients treated with 200–400 mg/day of oral itraconazole, 8 patients with mild disease (89%) achieved clinical and mycologic cure after 10.9 months (Esterre and Queiroz-Telles, 2006). Patients with moderate disease consisting of a group of 12 showed mycologic and clinical cure in 11 persons after an average of 12.9 months of therapy; whereas of 9 patients with severe lesions, 4 (44%) had a positive clinical and mycologic response after a mean duration of 30 months of therapy. The remaining patients showed significant improvement. Pulse treatment has also been used at 400 mg/day for a week followed by 3 weeks rest (reviewed by Bonifaz et al., 2004).

- *Terbinafine (TRB).* Terbinafine shows considerable in vitro activity against chromoblastomycosis agents with MICs of between 0.03 and 0.125 μg/mL. TRB has shown the best efficacy and tolerability with few drug interactions because its activity does not involve the cytochrome P450 pathway. At a dose of 500 mg TRB/day a series of 42 patients in Madagascar had a cure rate of 74.2% after 12 months of therapy, reaching a cure rate of 81% after a 2 year follow-up (reviewed by Esterre and Queiroz-Telles, 2006). A dosage of 500 mg TRB/day for 12 months resulted in mycologic cure in 82.5% of patients and total cure in 47% of patients with lesions of 10 years' duration (reviewed by Baddley and Dismukes, 2003).

  Type I collagen in skin lesions of chromoblastomycosis was measured before and after a year of TRB therapy (Esterre et al., 1998). Immunostaining of extracellular matrix for mature type 1 collagen and for fibroblasts synthesizing new collagen indicated a significant (30%) reduction in new collagen deposits in the lesions, and a partial reversal of the cutaneous fibrosis. Other studies have shown that TRB interrupts the cell cycle at the G0/G1 transition of various cultured human cancer cell lines (Lee et al., 2003) lending credence to the theory that TRB interferes with exuberant mammalian cell synthesis.
- *Posaconazole (PSC).* Six men with *F. pedrosoi* chromoblastomycosis of 3–29 years' duration had poor responses to standard therapy with TRB and ITC (Negroni et al., 2005). Therapy with PSC consisted of 800 mg/day given orally in divided doses for a maximum of 12 months. Five of six patients responded, with a complete response in four and a partial response in one patient. Four patients had a mycologic cure and in three patients there was complete resolution of skin lesions.

Combination Antifungal Therapy

Combinations of ITC and TRB have been used in chronic cases unresponsive to AmB or to oral antifungal monotherapy. The approach is to treat with ITC at 200–400 mg/day and, in alternate weeks, add 500–750 mg/day of TRB. An alternative regimen consists of concomitant use of ITC at 200–400 mg/day plus TRB at 250–1000 mg/day.

Best Practice

The current best practice therapy is ITC or TRB monotherapy at high doses for 6–12 months to treat small–medium lesions, also including cryosurgery when the lesion size has subsided. Local heat packs also may be used with both drugs. The superficial form of skin plaques responds very well to treatment whereas the verrucous form, which is most frequent, does not easily respond and complications result from fibrosis and lymphostasis, which limit drug penetration. The two drugs of choice are ITC at 200–400 mg/day and TRB at 500 mg/day (monotherapy). The MIC of the causal isolate for each of the drugs should be determined. When monotherapy has failed it is important to consider systemic combination therapy with both drugs given either alternately or concomitantly. The former consists of alternate weeks of ITC, 200–400 mg/day, and TRB, 500–750 mg/day. The latter involves giving ITC, 200–400 mg/day, plus TRB, 250–1000 mg/day, concomitantly (reviewed by Queiroz-Telles et al., 2009).

In extensive cases cure is seldom achieved with monotherapy and the combination of chemotherapy, local heat, and several sessions of open spray cryosurgery seems best. Based on a small series so far, PSC is a promising newer therapy. Invasive surgical methods should always be accompanied by oral antifungal therapy before and after the procedure.

Adverse Effects

ITC is usually well tolerated with side effects being gastric distress or diarrhea. TRB has a favorable adverse effect incidence with most effects being gastric distress, headache, skin rash, and, in exceptional cases, liver damage. Side effects with PSC consisted of headache and gastrointestinal distress.

18.12 LABORATORY DETECTION, RECOVERY, AND IDENTIFICATION

Direct Examination

Specimens for microscopic examination include skin scrapings, crusts, pus, and cyst fluid. Scrapings and biopsy specimens should be taken from surface areas marked by a small dark spot to improve the diagnostic yield.

Biopsied tissue sent to the histopathology laboratory for processing should also be sent for direct microscopy and culture. At the mycology bench tissue specimens are minced, then divided for culture and for microscopy. Diagnosis is facilitated by observation of muriform cells (rounded and irregular at the periphery, thick-walled double septum 5–15 μm in diameter) in biopsied tissue, crusts, and exudates by microscopy in 10% KOH preps or by histologic study of biopsied lesions.

Histopathology

Although other stains may be used, H&E is adequate, as the round-to-oval muriform cells are readily visible in the light microscope, even without special stains due to their natural golden-brown pigmentation ("copper pennies"). This is the tissue form common to all of the agents of chromoblastomycosis. Rarely, hyphal fragments may be seen near the skin surface. Hyperkeratosis and pseudoepitheliomatous hyperplasia are seen in the epidermis, and the muriform cells, 5–15 μm, either singly or in clumps are found within (Fig. 18.4).

Culture

For cultures, crust, exudates, and part of the minced biopsy specimen are planted on SDA with antibiotics and cycloheximide. Small dark green to black colonies usually appear within 10 days to 2 weeks. Species identification is based on conidiogenesis during fungal growth in different media, including PDA.

Cladophialophora carrionii *(Fig. 18.5)*

Colony Morphology Olive-brown to brown-black surface and reverse.

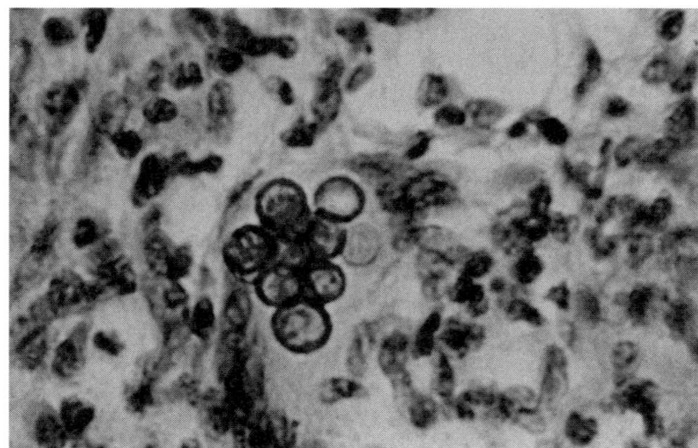

Figure 18.4 Histopathology of muriform cells, the tissue form of chromoblastomycosis. Note the golden-brown melanin pigment, H&E stain. See insert color representation of this figure.
Source: CDC.

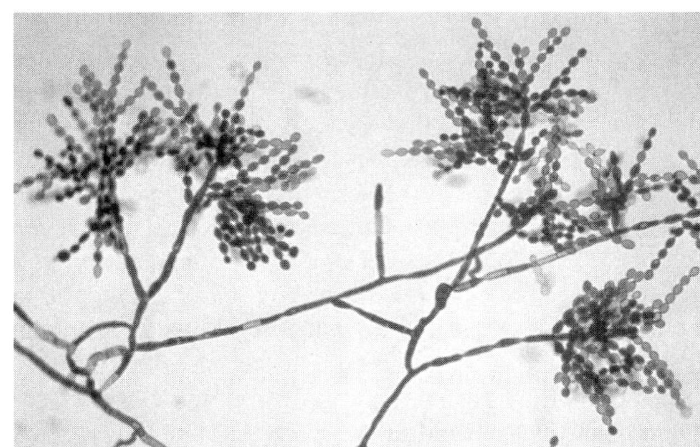

Figure 18.5 *Cladophialophora carrionii*, microscopic morphology, mold form 475×.
Source: CDC PHIL (No. 3061).

Microscopic Morphology Mold form: brown septate hyphae with elongate conidiophores producing long and sometimes branching chains of oval conidia. The conidia are *holoblastic* and formed in an *acropetal* manner. Tissue form: muriform cells.

Fonsecaea pedrosoi *(Fig. 18.6)*

Colony Morphology Slow growing, downy surface, brown-black, olive-brown, gray-black with a black reverse.

Microscopic Morphology Mold form: brown septate hyphae sporulate in a *sympodial* pattern. Two main types of conidiogenesis are seen: branching and nonbranching. The nonbranching sympodial form is called "*Rhinocladiella*"-like. Single conidia are borne on short *denticles* on the upper portion of the conidiophore. The branching form is one where the conidiophore grows around and out from the point where a conidium is produced. The conidia then can produce secondary conidia in an acropetal pattern. In addition, the fungus may at times produce a phialide with phialoconidia and also a "*Cladosporium*" type of conidiophore. The latter two types are *not* used as a basis for defining the genus *Fonsecaea*. The tissue form consists of muriform cells.

Phialophora verrucosa *(Fig. 18.7)*

Colony Morphology Slow growing, gray-black, velvety to wooly surface, with black reverse.

Microscopic Morphology Mold form: brown, septate hyphae with conidiophores consisting of vase-shaped phialides producing a ball of *enteroblastic* conidia which adhere to the distinct, dark-colored and funnel-shaped collarettes. Tissue form: muriform cells.

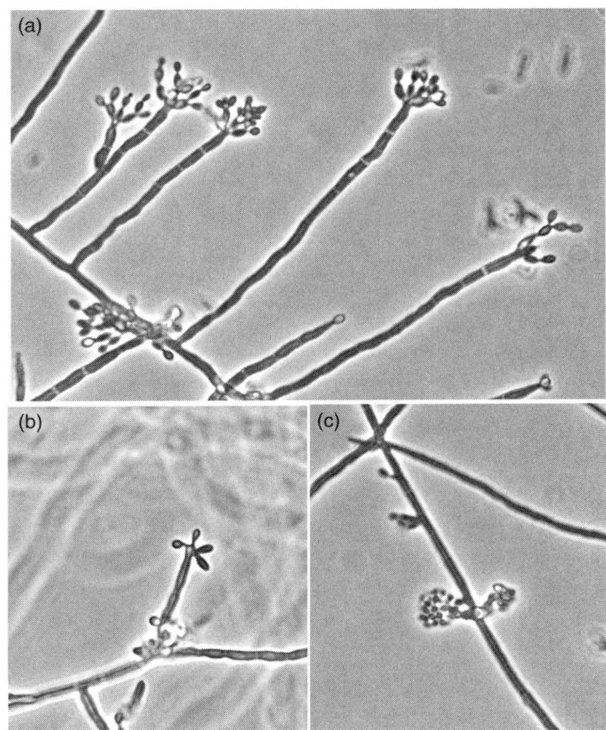

Figure 18.6 *Fonsecaea pedrosoi*, microscopic morphology mold form. This fungus produces three types of asexual sporulation: (a) branching-type, (b) acrothecal type, and (c) phialides. *Source:* E. Reiss, CDC.

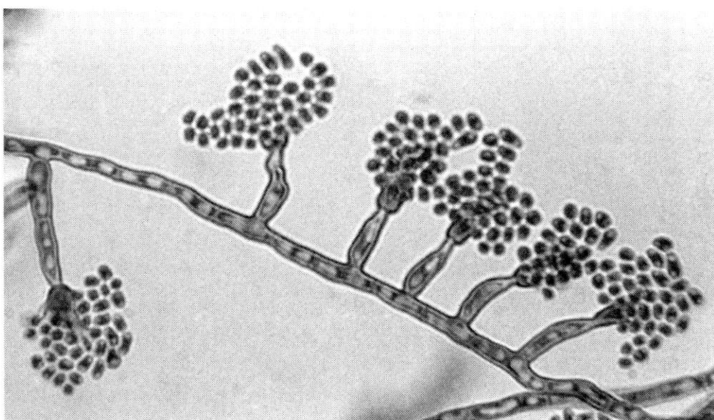

Figure 18.7 *Phialophora verrucosa* microscopic morphology of the mold form: phialides with prominent funnel-shaped collarettes. *Source:* CDC.

Please also see Table 19.1 for some features to differentiate melanized fungi associated with chromoblastomycosis and phaeohyphomycosis.

Rhinocladiella aquaspersa *(Fig. 18.8)*

Colony Morphology Colonies are slow to moderate growers, olive-black, with a velvet textured surface and black reverse.

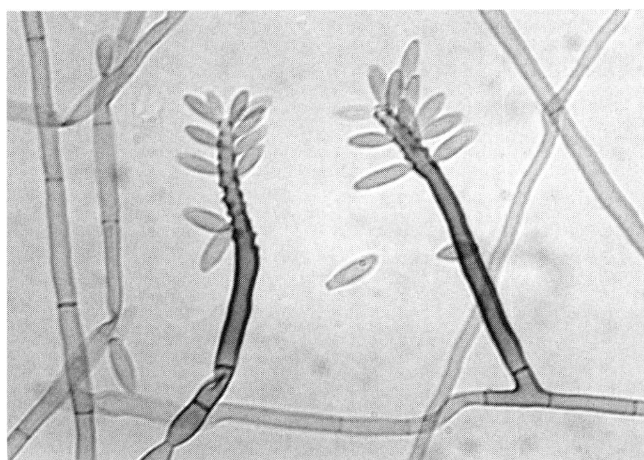

Figure 18.8 *Rhinocladiella aquaspersa* microscopic morphology. *Source:* H. J. Shadomy.

Microscopic Morphology Conidiogenous cells are terminal and intercalary, closely packed, with ellipsoid to clavate pale brown conidia on crowded denticles.

SELECTED REFERENCES FOR CHROMOBLASTOMYCOSIS

Alviano DS, Franzen AJ, Travassos LR, Holandino C, Rozental S, Ejzemberg R, Alviano CS, Rodrigues ML, 2004. Melanin from *Fonsecaea pedrosoi* induces production of human antifungal antibodies and enhances the antimicrobial efficacy of phagocytes. *Infect Immun* 72: 229–237.

Alviano DS, Kneipp LF, Lopes AH, Travassos LR, Meyer-Fernandes JR, Rodrigues ML, Alviano CS, 2003. Differentiation of *Fonsecaea pedrosoi* mycelial forms into sclerotic bodies is induced by platelet-activating factor. *Res Microbiol* 154: 689–695.

Baddley JW, Dismukes WE, 2003. Chromoblastomycosis, pp. 399–404, *in*: Dismukes WE, Pappas PG, Sobel JD (eds.), *Clinical Mycology*. Oxford University Press; New York.

Bonifaz A, Paredes-Solís V, Saúl A, 2004. Treating chromoblastomycosis with systemic antifungals. *Expert Opin Pharmacother* 5: 247–254.

Brown M, Pasvol G, 2005. Images in clinical medicine. Chromoblastomycosis. *N Engl J Med* 352(20): e19.

Butler MJ, Day AW, 1998. Fungal melanins: A review. *Can J Microbiol* 44: 1115–1136.

Caligiorne RB, Licinio P, Dupont J, de Hoog GS, 2005. Internal transcribed spacer rRNA gene-based phylogenetic reconstruction using algorithms with local and global sequence alignment for black yeasts and their relatives. *J Clin Microbiol* 43: 2816–2823.

Carrion AL, 1950. Chromoblastomycosis. *Ann NY Acad Sci* 50: 1255–1282.

Cunha MM, Franzen AJ, Seabra SH, Herbst MH, Vugman NV, Borba LP, de Souza W, Rozental S, 2010. Melanin in *Fonsecaea pedrosoi*: A trap for oxidative radicals. *BMC Microbiol* 10: 80.

de Hoog GS, Attili-Angelis D, Vicente VA, van den Ende AH, Queiroz-Telles F, 2004. Molecular ecology and pathogenic potential of *Fonsecaea* species. *Med Mycol* 42: 405–416.

de Hoog GS, Queiroz-Telles F, Haase G, Fernandez-Zeppenfeldt G, Attili Angelis D, Gerrits Van Den Ende AH, Matos T, Peltroche-Llacsahuanga H, Pizzirani-Kleiner AA, Rainer J,

Richard-Yegres N, Vicente V, Yegres F, 2000. Black fungi: Clinical and pathogenic approaches. *Med Mycol* 38 (Suppl 1): 243–250.

Esterre P, Queiroz-Telles F, 2006. Management of chromoblastomycosis: Novel perspectives. *Curr Opin Infect Dis* 19: 148–152.

Esterre P, Risteli L, Ricard-Blum S, 1998. Immunohistochemical study of type I collagen turn-over and of matrix metalloproteinases in chromoblastomycosis before and after treatment with terbinafine. *Pathol Res Pract* 194: 847–853.

Esterre P, Andriantsimahavandy A, Raharisolo C, 1997. Natural history of chromoblastomycosis in Madagascar and the Indian Ocean. *Bull Soc Pathol Exot* 90: 312–317.

Esterre P, Peyrol S, Guerret S, Sainte-Marie D, Pradinaud R, Grimaud JA, 1992. Cell-matrix patterns in the cutaneous lesion of chromomycosis. *Pathol Res Pract* 188: 894–900.

Hayakawa M, Ghosn EE, da Gloria Teixeria de Sousa M, Ferreira KS, Almeida SR, 2006. Phagocytosis, production of nitric oxide and pro-inflammatory cytokines by macrophages in the presence of dematiaceus (*sic*) fungi that cause chromoblastomycosis. *Scand J Immunol* 64: 382–387.

Kneipp LF, Rodrigues ML, Holandino C, Esteves FF, Souto-Padron T, Alviano CS, Travassos LR, Meyer-Fernandes JR, 2004. Ectophosphatase activity in conidial forms of *Fonsecaea pedrosoi* is modulated by exogenous phosphate and influences fungal adhesion to mammalian cells. *Microbiology* 150: 3355–3362.

Lee WS, Chen RJ, Wang YJ, Tseng H, Jeng JH, Lin SY, Liang YC, Chen CH, Lin CH, Lin JK, Ho PY, Chu JS, Ho WL, Chen LC, Ho YS, 2003. In vitro and in vivo studies of the anticancer action of terbinafine in human cancer cell lines: G0/G1 *p*53 associated cell cycle arrest. *Int J Cancer* 106: 125–137.

López Martínez R, Méndez Tovar LJ, 2007. Chromoblastomycosis. *Clinics Dermatol* 25: 188–194.

Mazo Fávero Gimenes V, da Glória de Souza M, Ferreira KS, Marques SG, Gonçalves AG, Vagner de Castro Lima Santos D, Pedroso e Silva Cde M, Almeida SR, 2005. Cytokines and lymphocyte proliferation in patients with different clinical forms of chromoblastomycosis. *Microbes Infect* 7: 708–713.

Mendoza L, Karuppayil SM, Szaniszlo PJ, 1993. Calcium regulates in vitro dimorphism in chromoblastomycotic fungi. *Mycoses* 36: 157–164.

Negroni R, Tobón A, Bustamante B, Shikanai-Yasuda MA, Patino H, Restrepo A, 2005. Posaconazole treatment of refractory eumycetoma and chromoblastomycosis. *Rev Inst Med Trop São Paulo* 47: 339–346.

Nimrichter L, Cerqueira MD, Leitao EA, Miranda K, Nakayasu ES, Almeida SR, Almeida IC, Alviano CS, Barreto-Bergter E, Rodrigues ML, 2005. Structure, cellular distribution, antigenicity, and biological functions of *Fonsecaea pedrosoi* ceramide monohexosides. *Infect Immun* 73: 7860–7868.

Nosanchuk JD, Casadevall A, 2006. Impact of melanin on microbial virulence and clinical resistance to antimicrobial compounds. *Antimicrob Agents Chemother* 50: 3519–3528.

Palmeira VF, Kneipp LF, Alviano CS, dos Santos AL, 2006. Secretory aspartyl peptidase activity from mycelia of the human fungal pathogen *Fonsecaea pedrosoi*: Effect of HIV aspartyl proteolytic inhibitors. *Res Microbiol* 157: 819–826.

Queiroz-Telles F, Esterre P, Perez-Blanco M, Vitale RG, Salgado CG, Bonifaz A, 2009. Chromoblastomycosis: An overview of clinical manifestations, diagnosis and treatment. *Med Mycol* 47: 3–15.

Queiroz-Telles F, McGinnis MR, Salkin I, Graybill JR, 2003. Subcutaneous mycoses. *Infect Dis Clin North Am* 17: 59–85.

Reiss E, Nickerson WJ, 1974. Control of dimorphism in *Phialophora verrucosa*. *Sabouraudia* 12: 202–213.

Ricard-Blum S, Hartmann DJ, Esterre P, 1998. Monitoring of extracellular matrix metabolism and cross-linking in tissue, serum and urine of patients with chromoblastomycosis, a chronic skin fibrosis. *Eur J Clin Invest* 28: 748–754.

Salgado CG, da Silva JP, Diniz JA, da Silva MB, da Costa PF, Teixeira C, Salgado UI, 2004. Isolation of *Fonsecaea pedrosoi* from thorns of *Mimosa pudica*, a probable natural source of chromoblastomycosis. *Rev Inst Med Trop São Paulo* 46: 33–36.

Santos AL, Palmeira VF, Rozental S, Kneipp LF, Nimrichter L, Alviano DS, Rodrigues ML, Alviano CS, 2007. Biology and pathogenesis of *Fonsecaea pedrosoi* the major etiologic agent of chromoblastomycosis. *FEMS Microbiol Rev* 31: 570–591.

Sousa MG, Azevedo Cde M, Nascimento RC, Ghosn EE, Santiago KL, Noal V, Bomfim GF, Marques SG, Gonçalves AG, Santos DW, Almeida SR, 2008. *Fonsecaea pedrosoi* infection induces differential modulation of costimulatory molecules and cytokines in monocytes from patients with severe and mild forms of chromoblastomycosis. *J Leukoc Biol* 84: 864–870.

Untereiner WA, Angus A, Reblova M, Orr MJ, 2008. Systematics of the *Phialophora verrucosa* complex: New insights from analyses of beta-tubulin, large subunit nuclear rDNA and ITS sequences. *Botany-Botanique* 86: 742–750.

Yan ZH, 1995. Assessment of *Phialophora* species based on ribosomal DNA internal transcribed spacers and morphology. *Mycologia* 87: 72–83.

QUESTIONS

The Answer Key to multiple choice questions may be found at the end of the book.

1. A black mold is growing on SDA from a biopsied skin lesion. What would you do to determine if it is an agent of chromoblastomycosis?
   A. Make a tease mount and observe the microscopic morphology
   B. Set up a temperature tolerance test at 42°C
   C. Transfer it to BHI and incubate at 35°C to see if it demonstrates dimorphism
   D. All of the above
2. If in the previous question you observe moderately long, branched chains of oval acropetal conidia, which agent of chromoblastomycosis would you suspect?
   A. *Phialophora verrucosa*
   B. *Fonsecaea pedrosoi*
   C. *Cladophialophora carrionii*
   D. *Rhinocladiella aquaspersa*
3. How would you distinguish between *Cladophialophora carrionii* and *Cladophialophora bantiana*?
   A. Capable of growth at 42°C
   B. Capable of growth up to 37°C
   C. Moderately long branched chains of oval acropetal conidia
   D. Very long chains of sparsely branched oval acropetal conidia

   Next to each statement indicate if it is true for *C. carrionii* (1) or *C. bantiana* (2).
4. The diagnosis of chromoblastomycosis is made by clinical appearance of the lesions and

A. Essential identification of the causative agent in culture by its microscopic morphology
B. Golden-brown muriform cells in H&E stained tissue section
C. History of travel or residence in the endemic area by the patient
D. Occupational or recreational exposure by the patient
E. All of the above

5. Describe three types of conidiogenesis observed in *Fonsecaea pedrosoi*.

6. Describe the five different lesional types in chromoblastomycosis.

7. Of these three agents of chromoblastomycosis, which is the most refractive to therapy: *C. carrionii, F. pedrosoi*, or *P. verrucosa* ? What therapy is recommended for a single lesion or a few small lesions? What are the relative merits of azole and terbinafine therapy and combinations of the two agents for patients with more extensive lesions?

Chapter 19

Phaeohyphomycosis

19.1 PHAEOHYPHOMYCOSIS-AT-A-GLANCE

- *Introduction/Disease Definition.* Phaeohyphomycosis is a disease category devised for molds with melanin pigment in their cell walls, which, when infecting host tissues, grow as hyphae or as a mix of yeast cells and hyphae.
- *Diagnosis.* Diagnosis is made by the appearance of the fungus in tissue as melanized hyphae, yeast cells, *moniliform* hyphae, or a mix of these morphotypes.
- *Etiologic Agents.* There are many causative agents, including *Exophiala jeanselmei, Wangiella* (*Exophiala*) *dermatitidis, Cladophialophora bantiana, Scedosporium prolificans*, and *Ramichloridium mackenziei*. Less frequent agents are *Alternaria, Aureobasidium, Bipolaris, Curvularia, Exserohilum, Neoscytalidium, Ochraconis*, and *Rhinocladiella* species.
- *Geographic Distribution/Ecologic Niche.* These melanized fungi are found worldwide in soil, thorns, splinters, and decaying plant material. One thermophilic species, *W. dermatitidis*, prefers increased osmolarity and is isolated from fruit juices and walls of sauna and steam baths.
- *Epidemiology.* These are less common cutaneous–subcutaneous and deep-seated mycoses.
- *Risk Groups/Factors.* Immunosuppression is a risk factor but not a requirement.
- *Transmission.* Penetrating injury with conidia-containing soil or plant material. The respiratory route is suspected in cerebral abscess disease.
- *Determinants of Pathogenicity.* Melanin, unidentified neurotropic factors.
- *Clinical Forms.* Cutaneous–subcutaneous cyst, primary cerebral abscess, keratitis, and fungal sinusitis.
- *Therapy.* Surgery and azoles: itraconazole (ITC), voriconazole (VRC), and posaconazole (PSC). Another class of effective agents includes terbinafine (TRB). Cerebral abscess requires surgical excision of the lesion and antifungal therapy—there are different approaches, some emphasizing combination therapy with AmB, flucytosine, and a mold-active azole, but success is difficult to achieve.
- *Laboratory Detection, Recovery, and Identification.* Direct examination of exudates, biopsy; histopathology, culture, microscopic morphology.

19.2 INTRODUCTION

As case reports accumulated of melanized molds causing indolent cutaneous–subcutaneous mycoses it became apparent that the etiologic agents were much broader than the classic causes of chromoblastomycosis: *Fonsecaea pedrosoi, Cladophialophora carrionii*, and *Phialophora verrucosa*. In fact, some of the causative agents were, unlike the chromoblastomycosis agents, molds with the histopathologic appearance of pigmented hyphae in tissue and in culture. Some hyphae in tissue exhibited no obvious pigmentation, requiring special stains to demonstrate cell wall melanin (Fontana–Masson stain). Some grew as a combination of yeast-like forms and hyphae in tissue and in culture.

Fundamental Medical Mycology, First Edition. By Errol Reiss, H. Jean Shadomy and G. Marshall Lyon, III.
© 2012 Wiley-Blackwell. Published 2012 by John Wiley & Sons, Inc.

Examples of the latter include *Cladophialophora bantiana, Wangiella* (*Exophiala*) *dermatitidis*, and other *Exophiala* species. Thus the disease classification of chromoblastomycosis did not fit. Accordingly, a new disease definition, "phaeohyphomycosis," was proposed to include darkly pigmented fungi that grew as molds in culture and in tissue, often with a number of yeast-like cells. The term phaeohyphomycosis (*definition*: *phaios*, Gr.: dun, dusky, gray, or brown) was coined by Dr. Libero Ajello in 1974 to refer to the pigmentation of these fungi in culture and during infection.

Phaeohyphomycosis is admittedly a loose category including cutaneous, subcutaneous, or systemic infections characterized by melanized mycelial elements, which include hyphae, pseudohyphae-like structures, and also yeast-like cells in tissue. The fungi are pigmented but the extent of pigmentation is variable according to the species. *Phaeohyphomycosis is a histopathologic diagnosis* based on the appearance of fungi in tissue, and the spectrum of anatomic sites affected is broad. There are superficial skin infections with melanized fungi but the *major* clinical forms of phaeohyphomycosis are

A. Cutaneous–subcutaneous cyst

B. Cerebral abscess and other extracutaneous sites of infection

C. Sinusitis, including invasive and allergic fungal sinusitis

Other Sites of Involvement

Infections of the eye, lung, heart valves, bones, peritoneal cavity, and gastrointestinal tract also have been caused by melanized fungi. The majority of infections caused by black yeast-like fungi are cutaneous and superficial, but fatal systemic infections are known. One-hundred-eighty-eight *Exophiala* species isolates from the United States were classified according to the body site involved (Zeng et al., 2007). Those from deep-seated mycosis accounted for 40% of the collection, for example, lung, pleural fluid, sputum, digestive organs, heart, brain, spleen, bone marrow, blood, and dialysis fluid. A similar number of *Exophiala* isolates (38%) originated from skin and mucous membranes. The remaining isolates caused subcutaneous disease (12.0%, including sinusitis, eumycetoma, and subcutaneous cyst) or superficial infections (0.5%).

A cluster of *Wangiella* (*Exophiala*) *dermatitidis* meningitis cases in the United States was related to the accidental contamination of injectable methylprednisolone solution by a compounding pharmacy (Engemann et al., 2002).

19A *CUTANEOUS–SUBCUTANEOUS PHAEOHYPHOMYCOSIS*

19A.1 *INTRODUCTION/DISEASE DEFINITION*

Subcutaneous phaeohyphomycosis is an indolent mycosis, usually following traumatic implantation of conidia from contaminated soil, wood splinters, or thorns. Phaeohyphomycotic cyst is a solitary abscess due to infection with a melanized fungus that forms at a traumatized site on an exposed area of the body. Characterized by the presence of melanized hyphae, moniliform pseudohyphae, yeast-like cells, or any combination of these in infected tissue, the disease most commonly presents as a cyst, or a diffuse lesion that usually remains localized (fixed cutaneous) to the site of a penetrating injury, and may be verrucous. In tissue, hyphae may appear regular and uniform in diameter, or may be irregular in shape with many swollen cells. Some fungi may appear hyaline in tissue due to scant production of melanin, although they are melanized in culture. The presence of scant melanin in hyphae may be detected in tissue with the Fontana–Masson stain.

The differential diagnosis includes ganglion cysts, epidermal inclusion cysts, Baker's cysts, or foreign body granulomas (Perfect et al., 2003b). Subcutaneous phaeohyphomycosis may affect normal individuals; in the immunosuppressed patient they are most often seen in solid organ transplant patients or those receiving chronic systemic corticosteroids.

19A.2 *CASE PRESENTATIONS*

Case Presentation 1. Nodule on the Left Hand of a 73-yr-old Man
(Patterson et al., 1999)

History

A 73-yr-old man with no other health problems presented with a 3 cm diameter nodule with pustules on the dorsal surface of his left hand, which developed over a 2–3 month period. He had no memory of trauma.

Diagnostic Procedures

A KOH prep of pus from the lesion yielded lightly pigmented yeast forms. Tissue was obtained for culture and histopathology. The pathology report of the H&E stained specimen of the biopsied skin lesion indicated pseudoepitheliomatous hyperplasia with granulomatous

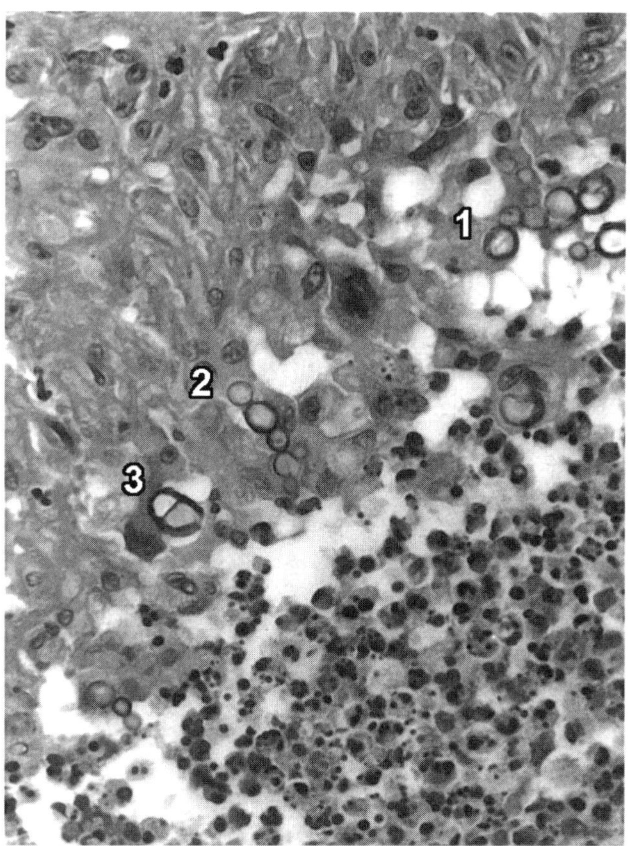

Figure 19.1 Histopathology from a biopsied skin lesion of a phaeohyphomycotic cyst. This high power view of *Cladophialophora bantiana* shows (1) yeast-like cells, (2) short chains of budding cells, and (3) some thick-walled cells with septations dividing in single plane.
Source: Patterson et al. (1999). Used with permission from Elsevier Publishers, Ltd. Copyright © 1999.

dermal inflammation. Numerous pigmented oval yeast-like cells were present, some in short chains and as thick-walled fungal cells with single transverse septa (Fig. 19.1). The culture produced velvety, dark gray to black colonies. In slide culture there were septate, branched, and lightly pigmented hyphae in chains of yeast-like cells produced laterally and terminally from hyphae identified as *Cladophialophora bantiana* (Fig. 19.2).

Diagnosis

The diagnosis was subcutaneous phaeohyphomycosis due to *Cladophialophora bantiana*.

Therapy

The lesion was surgically removed, and treatment was started with ITC. The lesion did not recur.

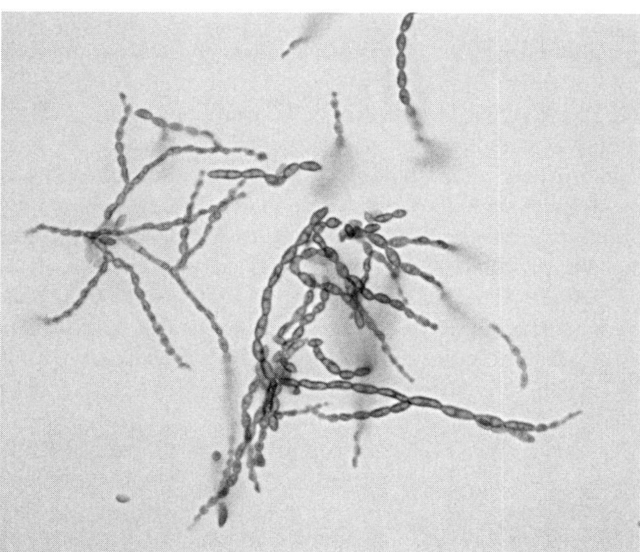

Figure 19.2 Microscopic morphology of *Cladophialophora bantiana* showing brown septate hyphae in long, wavy chains of oval conidia 20–35 cells long, with infrequent branching. These chains remain intact in contrast to *C. carrionii* which readily break apart (400×).
Source: E. Reiss, CDC.

Comment

Cladophialophora bantiana is primarily noted as a neurotropic fungus and an agent of cerebral abscess. Its characteristic appearance in tissue is a mix of budding cells, closely septate hyphae and moniliform hyphae. *C. bantiana* can infect otherwise healthy persons both by a penetrating injury or via the respiratory route. As such, it should be regarded as a primary pathogen and handled under strict BSL-2 level precautions, including personal protective equipment.

Case Presentation 2. Cyst in the Calf of a 60-yr-old Man (Crosby et al., 1989)

History

A 60-yr-old man complained of a mass in his left calf of about 4 months' duration. He had no memory of local trauma. Eighteen years previously he suffered spinal cord trauma and walked with difficulty. Eight years before this clinic visit he had a nephrectomy for renal cell carcinoma with no later recurrence. He was being treated for angina pectoris and peptic ulcer. Palpation revealed a solitary 1 cm × 1.5 cm mass in the subcutaneous soft tissue of his left calf which was mobile and not fixed to deeper tissues.

Diagnostic Procedures

Fine needle aspiration (FNA) of the cyst was performed. Direct smear of aspirates stained with H&E showed yeast forms, aggregated septate hyphae, and tapering pseudohyphae in chains that branched *dichotomously* and formed filaments that passed through giant cells (Fig. 19.3). No muriform cells were observed. Fungal cell walls and cytoplasm varied from colorless to pale red-brown. Histopathologic examination showed a fibrous wall enclosing the cyst surrounded by granulomatous inflammation. Sections stained with H&E appeared light brown. A splinter was observed in the cyst cavity.

After 11 days' incubation at 25°C on SDA, black yeast forms developed (Fig. 19.3, *panel C*). The fungus was thermotolerant and grew at 42°C. Slide cultures on PDA revealed pale brown to hyaline annellides spaced along pale brown septate hyphae. Conidia were single celled and ovoid in shape. Small nonpigmented yeast forms also were observed. The isolate assimilated tyrosine but did not hydrolyze casein. (Please see Fig. 19.7.)

Diagnosis

The diagnosis was subcutaneous phaeohyphomycosis caused by *Wangiella (Exophiala) dermatitidis*. (Please see Section 19.3, Laboratory Detection Recovery, and Identification.)

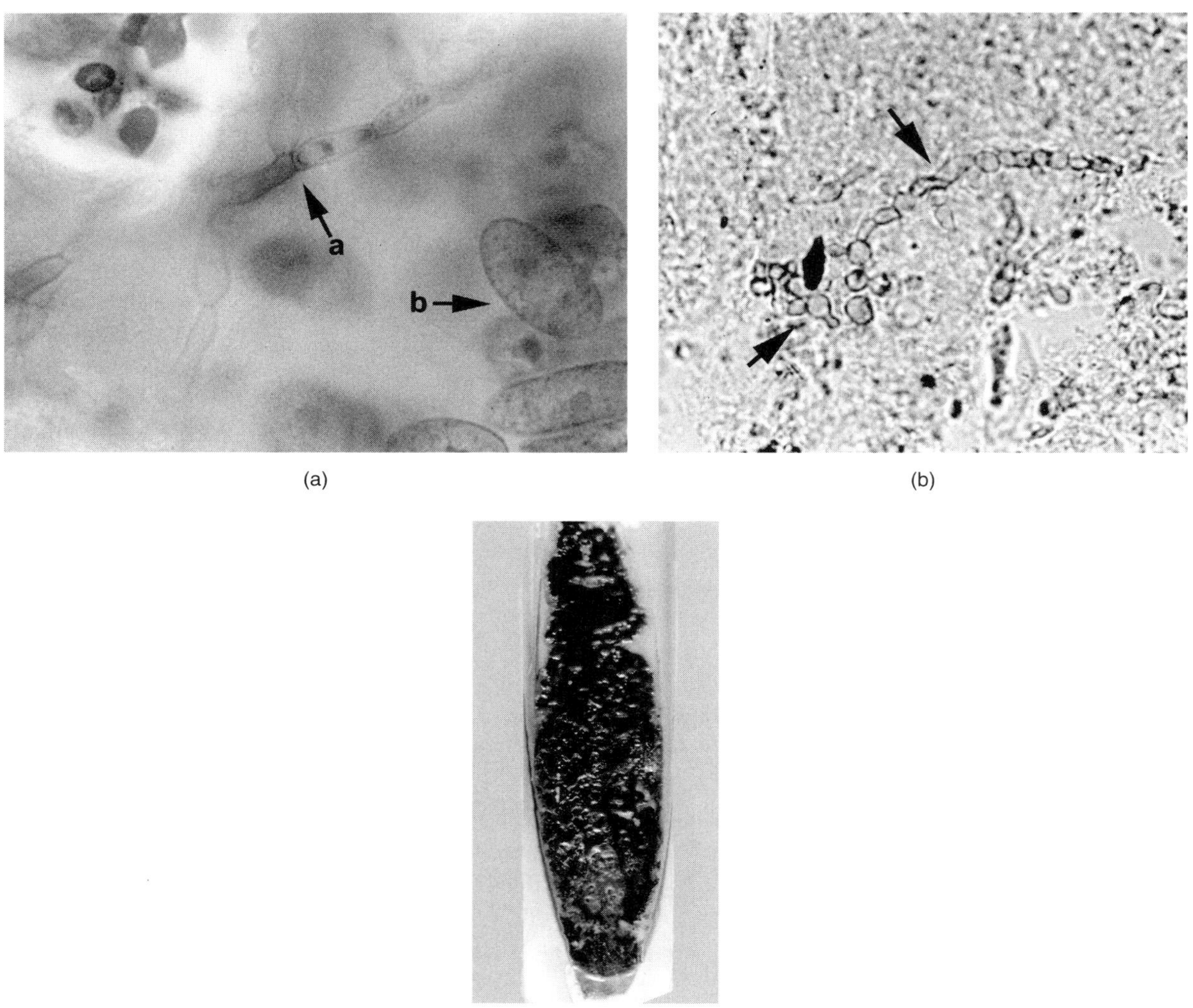

Figure 19.3 *Wangiella* (*Exophiala*) *dermatitidis*. Phaeohyphomycosis and onychomycosis. (*Panel A*.) Cytology of a fine needle aspirate (FNA) from the phaeohyphomycotic cyst on the leg (Case Presentation 2). (a) Naturally pigmented pseudohyphae are present in a branching chain (1000×, H&E stain) (b) *Arrow* points to nuclei of giant cell. *Source:* Used with permission from Dr. John H. Crosby, Medical College of Georgia, Augusta, Georgia. (*Panel B*) *Wangiella dermatitidis* from nail scraping, KOH prep; *arrows* point to naturally pigmented pseudohyphae. *Source:* Used with permission from Dr. Tadahiko Matsumoto, Yamada Institute of Health and Medicine, Tokyo. (*Panel C*) *Wangiella dermatitidis* colony of the culture grown from the specimen in *panel A* (SDA). See insert for color representation of this figure. *Source:* Used with permission from Dr. John H. Crosby, Medical College of Georgia, Augusta.

Therapy

The cystic mass was excised 2 weeks after the FNA procedure. No antifungal therapy was prescribed and there was no evidence of recurrence at follow-up 20 months later.

Comment

This man was a resident of Georgia, U.S.A. The finding of a splinter is consistent with the accepted method of transmission via penetrating injury. Giant cells, some with internal septa, are a response to adverse conditions in the host and may resemble muriform cells, which are the hallmark of a different disease, chromoblastomycosis.

19A.3 *ETIOLOGIC AGENTS*

Agents of subcutaneous phaeohyphomycosis include genera that are numerous and varied, often with only single reported cases. The most commonly reported etiologic agents are *Wangiella* (*Exophiala*) *dermatitidis, Exophiala jeanselmei*, and *Scedosporium prolificans*. Other species causing cutaneous and subcutaneous phaeohyphomycosis are *E. bergeri, E. spinifera, E. mesophila*, and *E. attenuate* (Zeng et al., 2007).

Molecular taxonomy of *Exophiala* species is based on the ITS region of rDNA supplemented by additional genes: *EF 1*-α, β-*TUB*, and *ACT*. The ITS region is a tandem repeat of between 500 and 600 bp. A table of the species-specific fragment ITS region 1 or ITS 2 for eight *Exophiala* species is included in Zeng and de Hoog (2008). Other less frequent agents are *Alternaria* species, *Aureobasidium pullulans, Bipolaris spicifera, Cladophialophora bantiana* (please see Section 19A.2, Case Presentation 1), *Curvularia lunata, Exserohilum rostratum, Rhinocladiella* species, and *Scytalidium* (now *Neoscytalidium*) *dimidiatum*.

19A.4 *GEOGRAPHIC DISTRIBUTION/ ECOLOGIC NICHE*

Fungi producing cutaneous–subcutaneous phaeohyphomycosis are ubiquitous in nature, found worldwide in dusts and soil, wood splinters, decaying vegetation, and forest floor detritus. *Wangiella* (*Exophiala*) *dermatitidis* is isolated from nature on routine cycloheximide-containing mycologic media at elevated temperatures. Its preferred environmental niche is in soil, and in slightly osmotic environments such as fruit juices.

19A.5 *EPIDEMIOLOGY*

Cutaneous–subcutaneous phaeohyphomycosis is less frequent than other subcutaneous mycoses—chromoblastomycosis and sporotrichosis. Cystic lesions occur primarily in adults. Patients who were otherwise healthy have been rural workers with frequent to daily contact with soil, splinters, thorns, and similar natural sources. Trauma at the original inoculation site, however, may not be recalled. At times the substance initiating the trauma has been recovered when pus has been taken from lesions at the inoculation site.

19A.6 *RISK GROUPS/FACTORS*

- Those most at risk are persons working out-of-doors without appropriate clothing (e.g., shoes, long pants, sleeved shirts).
- As indicated in Section 19A.2, Case Presentations, phaeohyphomycosis may occur sporadically in persons in urban areas who have no particular occupational exposure.
- Individuals with devitalized skin are at risk.
- Patients receiving systemic corticosteroid or cytotoxic drug therapy. Immunosuppression is a risk factor, but not a requisite. Subcutaneous phaeohyphomycosis has been reported in persons with cancer, especially hematologic malignancy, and in stem cell and solid organ transplant recipients (reviewed by Silveira and Nucci, 2001).
- Other primary conditions that may predispose patients to infections with melanized molds are diabetes mellitus, renal failure, and peritoneal dialysis (Silveira and Nucci, 2001).

19A.7 *TRANSMISSION*

The route of infection is via traumatic implantation from the environment, but often the patient does not recall any local trauma. Contaminated injectable solutions were identified as the source of an outbreak of *Wangiella* (*Exophiala*) *dermatitidis* infections (Engemann et al., 2002).

19A.8 *DETERMINANTS OF PATHOGENICITY*

The lesions of subcutaneous phaeohyphomycosis are more cystic than granulomatous, such as the lesions of chromoblastomycosis. Rarely do phaeohyphomycotic cysts show the muriform cell form, even though that form can be induced, for example, in *W. dermatitidis*.

19A.8.1 MICROBIAL FACTORS

Wangiella (*Exophiala*) *dermatitidis* serves as a model microbe for delineating pathogenic determinants of black yeast-like fungi. Since the conidia of *W. dermatitidis* are formed by annellides, this mold was assigned to the genus *Exophiala*. Both genus names are in current use and none has chronologic priority. Phylogenetic analysis based on rDNA sequences places *W. dermatitidis* in a cluster with other *Exophiala* species.

Phenotype Variability in Melanized Molds

How do black yeast-like fungal pathogens appear in nature and during infection of the human host? In nature the predominant form is the yeast form. During tissue invasion a mix of yeast forms, hyphae, pseudohyphae, and even muriform cell types are seen. The budding yeast cells are unicellular and haploid, facilitating genetic manipulations. Form development can be altered by culture conditions. Budding yeasts are observed in nutrient-rich media. Changing the culture conditions to a less rich medium stimulates hyphal growth. Another growth form is *isotropic*—where individual cells enlarge in all directions.

Nutritional studies of *W. dermatitidis* have shown that, at low pH in rich media, budding stops and instead isotropic growth occurs, wherein the cells enlarge and divide internally to become multinucleate cells with thickened walls, resembling muriform cells. The same effect is seen in other black molds which are causative agents of chromoblastomycosis. This effect can be reversed by supplying additional calcium ions. At near-neutral pH, withholding of Ca^{+} also induces muriform cell morphology.

Melanin

Black molds and their yeast-like relatives produce *DHN melanin* and deposit it in their cell walls. (Please see Part Five, Introduction—Special Topic: Fungal Melanin.) Hypopigmented mutants of *W. dermatitidis*, arising spontaneously, or induced by UV light mutagenesis, are less pathogenic for mice and more susceptible to killing by PMNs. This supports the idea that melanin is indeed a pathogenic factor. Melanin is laid down in the outer cell wall and may be thought of as armor protection for the fungus. Melanin protects *W. dermatitidis* from killing in the phagolysosome of the human neutrophils (Schnitzler et al., 1999).

Killing rates of melanized saprobes by PMNs are consistently higher than species known as agents of human disease (Peltroche-Llacsahuanga et al., 2003). Melanized fungi are found in both high and low killing rate groups so that melanin is very likely contributive, but pathogenicity is multifactorial and the other determinants await discovery, especially those that are neurotropic. *Wangiella dermatitidis* and *E. jeanselmei* were in the low-intracellular killing group, indicating increased pathogenic potential.

Chitin

A microcrystalline sleeve in the cell wall of many molds, chitin also is present in the black molds, with *W. dermatitidis* serving as a model for the group. The muriform cell form of *W. dermatitidis*, induced in vitro, has 5× more chitin than yeast forms. When chitin synthesis is inhibited in muriform cells by the polyoxin class of antifungal agents, focal ruptures occur in the cell wall and the cells become fragile and burst without osmotic support. These properties prompted investigations into the genetics of chitin synthase in *W. dermatitidis*, which is known to be a series of five distinct enzymes, of which four influence pathogenesis (Szaniszlo, 2006).

1. *WdCHS1.* While not essential for growth, the product of this gene is involved in septum formation and yeast cell separation (cytokinesis).
2. *WdCHS2.* Structurally similar to *CHS1*, this gene may have arisen through gene duplication. Gene disruption results in drastically reduced chitin synthase activity but cells grow normally and have no morphologic defects. Double deletion mutants of *CHS1* and *CHS2* genes indicate their role is in septum formation and cytokinesis. The double deletion mutant does not grow at 37°C without an osmotic stabilizer.
3. *WdCHS3.* Deletion of this gene shows no change in growth at 25°C or 37°C or morphologic change, or rate of chitin synthesis. Double mutants Δ *wdchs2* Δ*wdcghs3* have no morphologic abnormalities but have reduced mouse virulence.
4. *WdCHS4.* Deletion mutants grow more slowly than the wild type but the rate of chitin synthesis is not affected.
5. *WdCHS5.* Deletion mutants fail to grow at 37°C but appear normal at 25°C. At the higher temperature cells enlarge isotropically and blow out in foci of the cell wall, leaking contents. This chitin synthase seems the most important for pathogenesis and is a potential drug target.

19A.9 CLINICAL FORM

The initial presentation is usually a single red nodule at the puncture site, most often located on an extremity. The lesion may remain in the epidermis (cutaneous phaeohyphomycosis) or may develop into a subcutaneous abscess or cyst. Cutaneous lesions may be erythematous crusted plaques with edema which drain a purulent exudate. Fever is not typically present, and only rarely is there spread via the lymphatics or bloodstream. Underlying muscle or bone usually is not affected.

Cystic, pus-filled lesions occur most frequently in adults, occasionally with overlying verrucous lesions, especially in immunocompromised individuals. Common sites of infection are the knee, finger, hand, wrist, or ankle and, less commonly, scalp, face, neck, and buttocks. The majority of patients are adult males, probably because they work more frequently out-of-doors.

19A.10 THERAPY

Localized accessible lesions are surgically excised. It is prudent to combine surgery with antifungal therapy (e.g., oral ITC, 200–600 mg/day). The duration of therapy is not established but 6 months is an approximate length of time. Cases of widespread disease may respond better to extended spectrum azoles (e.g., VRC, PSC). Although in vitro many of the causative agents are susceptible to echinocandins [e.g., caspofungin (CASF), micafungin] there is no knowledge of their in vivo efficacy in treating human disease. *Scedosporium prolificans* is exceptional because it is not susceptible to any known class of antifungal agents.

Phaeohyphomycosis is uncommon and sporadic, making clinical trials difficult to arrange. Murine model infections are instructive in estimating the efficacy of antifungal agents. Mice immunosuppressed with cyclophosphamide were infected intravenously with a human isolate of *Exophiala* species (Rivard et al., 2007). Treatment consisted of daily doses of AmB-deoxycholate, CASF, or PSC, the in vitro MICs of which were 0.25 μg/mL, 0.06 μg/mL, and 0.06 μg/mL, respectively. PSC was the most effective therapy with 80% of mice surviving for 14 days. PSC and AmB-deoxycholate each reduced the fungal burden in brains and kidneys of infected mice.

Five patients with subcutaneous phaeohyphomycosis (due to *Alternaria, Bipolaris*, or *Exophiala* species) were successfully treated with VRC (Perfect et al., 2003a). Recurrent *Exophiala spinifera* phaeohyphomycotic cysts occurred in a woman who suffered five relapses over a 10 year period (Negroni et al., 2004). After initial success with flucytosine and ITC, the patient relapsed and was treated with AmB-deoxycholate and, on another recurrence, with terbinafine. The last recurrence was disseminated with vegetative verrucous skin lesions, endophthalmitis, and osseous lesions. She received oral PSC at 800 mg/day with clinical and mycologic cures achieved within 13 months. Unlike chromoblastomycosis, which remains localized, the agents of phaeohyphomycosis are not restricted, with dissemination to the deep tissues, including bone, eye, and brain a possibility.

Appropriate therapy, and the duration of treatment, depends on identification to species and in vitro susceptibility tests against a panel of standard and newer agents. Therefore, culture and laboratory identification of the pathogen is an important factor in the patient's treatment and leads to a better understanding of the disease. Methods will be found in Section 19.3, Laboratory Detection, Recovery, and Identification.

19B CEREBRAL PHAEOHYPHOMYCOSIS

19B.1 INTRODUCTION/DISEASE DEFINITION

Primary cerebral phaeohyphomycosis is rare and has a poor prognosis. The route of infection is believed to follow inhalation of conidia into the lungs and hematogenous dissemination to the CNS. Dissemination to sites other than the CNS is uncommon. The prognosis is poor primarily because of:

- Inability to diagnose. Hyphal elements may be hyaline early in the infection and, if special melanin stain (Fontana–Masson) is unavailable, may be misdiagnosed as aspergillosis.
- Incomplete abscess encapsulation.
- Presence of multiple lesions.
- Resistance to treatment.
- Complications of surgical and/or medical management.

19B.2 CASE PRESENTATION

Treatment of Fungal Cerebral Abscess (Lyons et al., 2006)

History

A 64-yr-old man presented to the emergency department of a clinic-hospital in Scottsdale, Arizona, with a 6 day history of right upper and lower extremity hemiparesis. He was immune-normal and suffered from mild hypertension and asthma. He had traveled in Asia in the distant past and enjoyed gardening and composting.

Diagnostic Procedures

A CT scan of the brain showed a large area was affected in the left frontal and left parietal lobes. MRI indicated a lobulated mass approximately 2 cm long. Complete blood count, differential platelets, and chemistries were all normal. Coccidioidomycosis and HIV serologies were negative. Stereotactic brain biopsy and intraoperative pathology showed granulomatous inflammation, focal necrosis, and pigmented fungal hyphae. A subsequent culture grew *Cladophialophora bantiana*.

Therapy

The patient received liposomal AmB, 450 mg (5 mg/kg), IV per day and oral VRC, 200 mg twice daily. Liposomal AmB was discontinued after 14 days when he developed renal failure. An MRI at that time showed no improvement in the size of the lesion or intracranial edema. The patient remained neurologically stable. Because the lesion was located in an eloquent area of the brain controlling dominant motor–speech function, surgical resection was deferred in favor of continued VRC treatment.

Clinical Course

One year after initiation of antifungal therapy, preoperative neurologic deficits and resolution of renal failure were achieved. An MRI indicated near-complete resolution of cranial edema and of the abscess. He continued to be maintained on therapy pending complete radiologic resolution.

Comment

Cladophialophora bantiana cerebral abscess is a dire diagnosis and most patients expire despite aggressive surgical debridement and therapy. This report is therefore a milestone in that the successful diagnosis was after stereotactic biopsy, and oral VRC treatment was successful. No visual disturbance was reported by this patient, a common side effect of VRC treatment.

Diagnosis

The diagnosis was fungal cerebral abscess caused by *Cladophialophora bantiana*.

Differential Diagnosis CT and MRI can readily identify a cerebral abscess but cannot differentiate bacterial abscess or primary CNS cystic brain tumors (high grade gliomas or cerebral metastases) from fungal cerebral abscess. Other diseases to be ruled out are toxoplasmosis, listeriosis, and nocardiosis.

19B.3 *ETIOLOGIC AGENTS*

The most frequently isolated melanized neurotropic fungi causing primary cerebral phaeohyphomycosis are *Cladophialophora bantiana* (most frequently isolated), *Wangiella* (*Exophiala*) *dermatitidis*, and *Ramichloridium mackenziei*. These fungi are related to black yeasts. Molecular taxonomic studies classify them as anamorphs of the ascomycete family *Herpothrichiellaceae*, in the order *Chaetothyriales* (Horré and de Hoog, 1999). *Ochroconis gallopava* is another neurotropic fungus unrelated to the first three.

19B.4 *GEOGRAPHIC DISTRIBUTION/ ECOLOGIC NICHE*

Cladophialophora bantiana is isolated worldwide but most reports are from the Americas with a preference for warm climate and high average humidity. It rarely is isolated from the environment and its ecologic niche is unknown. Most isolates of *C. bantiana* are obtained from invasive disease of humans and lower animals—cats and dogs. Isolations from the environment by mouse passage occasionally have been successful: the aqueous phase of a soil suspension is injected intravenously. If or when mice become sick the fungus can be isolated from brain tissue. Analysis of rDNA sequences of many isolates show low strain–strain variation and the universal presence of a specific intron in the 18S rDNA. It is thought that *C. bantiana* underwent rapid worldwide spread during evolutionary time. Its environmental niche, however, remains to be determined.

Ochroconis gallopava is thermophilic and occurs worldwide in heated effluents, hot springs, and also is a known avian pathogen; for example, chicks contract it from broiler house litter. *Ochroconis gallopava* is strongly thermophilic and tolerates temperatures up to 50°C.

Ramichloridium mackenziei has a range restricted to the Middle East (e.g., Saudi Arabia), but there have been no reports of its isolation from the environment.

Wangiella (*Exophiala*) *dermatitidis* can be isolated from environmental sources with selective procedures entailing elevated incubation temperatures (40°C) and media containing cycloheximide. In addition, increased osmolarity favors its growth. *W. dermatitidis* has been isolated from soil and fruit juice and, because it is thermophilic, it is recovered from the walls of sauna and steam baths (Matos et al., 2002). *W.* (*Exophiala*) *dermatitidis* is an agent of cerebral phaeohyphomycosis in East Asia. It is probably worldwide in distribution but most cases of cutaneous–subcutaneous mycosis originate from temperate and subtropical areas of Europe and East

Asia. *W. dermatitidis* is abundant in steam baths which provide an optimal environment for its growth because the fungus has two characteristics that contribute to its survival in this unusual ecologic niche—it is thermotolerant or thermophilic, growing at 42°C, producing quantities of extracellular polysaccharide, which may protect it from heat stress.

Although soil, feces, plants, and fruit have been regarded as environmental sources of *W. dermatitidis*, they do not harbor large quantities of the fungus (Matos et al., 2002). Steam baths (Turkish baths) in The Netherlands and Slovenia were all positive for *W. dermatitidis*. Each swab sample yielded over 1000 colonies within 3 days, indicative of metabolically active cells. The ecologic niche in nature remains to be discovered. Previously, *W. dermatitidis* was isolated from home humidifiers and public bathing facilities in Japan.

19B.5 *EPIDEMIOLOGY*

The majority of infections caused by black yeast-like fungi are cutaneous and superficial, but fatal systemic infections are known.

1. *W. dermatitidis* cerebral infections are rare. Although the fungus is found worldwide, most case reports are from East Asia. Within the East Asian ethnic group, immunocompetent young persons seems to be at risk. The potential for CNS infection with *W. dermatitidis* is not restricted to East Asia, as shown by the accidental infection of five patients in the United States who received contaminated steroid injections for back pain, prepared by a compounding pharmacy that did not adhere to aseptic technique (Engemann et al., 2002). In Europe *W. dermatitidis* is known for subclinical colonization of the lungs in cystic fibrosis patients.

Molecular phylogenetic analysis of black yeast-like fungi assigned to the *Exophiala* genus delineated a number of species of clinical importance, some of which could not be readily identified by morphologic and biochemical criteria alone. In that respect, 188 clinical isolates from the United States identified as *Exophiala* species, were reidentified by sequence analysis of the ITS subregion of rDNA (Zeng et al., 2007).

Wangiella (*Exophiala*) *dermatitidis* comprised 55 or 29.3%. From that point the occurrence of species included ones that are difficult or not possible to identify on morphologic grounds: 37 *E. xenobiotica* (19.7%), 35 *E. oligosperma* (18.6%), 13 *E. lecanii-corni* (6.9%), 12 *E. phaeomuriformis* (6.4%), 7 *E. jeanselmei* (3.7%), 7 *E. bergeri* (3.7%), 6 *E. mesophila* (3.2%), and 5 *E. spinifera* (2.7%). Other even less common species of the total of 15 known *Exophiala* species made up the remaining 66 clinical isolates or they were misidentified and do not belong to *Exophiala* species.

Exophiala species were isolated from sites indicative of deep-seated mycosis in 39.9% of the collection: for example, lung, pleural fluid, sputum, digestive organs, heart, brain, spleen, bone marrow, blood, and dialysis fluid. An almost equal number of *Exophiala* isolates (38.3%) originated from skin and mucous membranes.

The remaining isolates caused superficial infections (0.5%, including hair) or subcutaneous infection (12.0%, including paranasal sinusitis, eumycetoma, and subcutaneous cyst). Species causing cutaneous and subcutaneous phaeohyphomycosis were *E. bergeri, E. spinifera, E. jeanselmei, E. mesophila*, and *E. attenuate*.

2. *Cladophialophora bantiana* is more common in temperate climates in geographic areas other than East Asia: patients have been seen from throughout the temperate and subtropical regions of the world including the southeastern United States. The average age of patients with *C. bantiana* cerebral abscess is 37 years. Most cases occurred in patients whose occupation exposed them to environmental dusts. The cerebral infection is usually the primary disease with neurologic symptoms being the first to appear.

3. Thirteen human cases of phaeohyphomycosis caused by *O. gallopava* were reported from 1996 to 2002: 11 in transplant recipients and 2 in patients with hematologic malignancies (Wang et al., 2003). Of these, 8 patients suffered from cerebral abscess, the remainder had pulmonary mycosis. Seven of the 13 patients died. *O. gallopava* is susceptible in vitro to AmB, ITC, and VRC. A more recent series of three solid organ transplant recipients developed *O. gallopava* disease (Shoham et al., 2008). All three had pulmonary disease and survived with surgical excision and antifungal therapy consisting of ITC, or VRC, or combination therapy with VRC and AmB.

4. *Ramichloridium mackenziei* is geographically restricted to the Middle East: Kuwait, Oman, Qatar, and Saudi Arabia. Lymphoma, leukemia, and renal transplantation were reported as the primary diseases but some patients have had no identified predisposing factors. Patients have ranged in age from 32 to 75 years and there does not seem to be any tendency toward infection of men versus women (Kanj et al., 2001).

19B.6 *RISK GROUPS/FACTORS*

Risk factors include occupational exposure to environmental dusts; treatment with systemic corticosteroids or cytotoxic drugs; and patients with neutropenia or diabetes mellitus. These factors increase risk but otherwise healthy individuals may develop cerebral phaeohyphomycosis; therefore the etiologic agents should be regarded as primary pathogens and handled as such in the clinical laboratory.

19B.7 *TRANSMISSION*

Although the fungus is acquired from the environment, the portal of entry has rarely been identified for certain. Some reports have described infections following traumatic introduction. Most cerebral phaeohyphomycosis cases have been primary with the route of infection proposed to be either by inhalation or ingestion of conidia of a neurotropic fungus with hematogenous spread to the CNS.

19B.8 *DETERMINANTS OF PATHOGENICITY*

Wangiella dermatitidis is a model microbe for this group. Please see Section 19A.7, Determinants of Pathogenicity.

19B.9 *CLINICAL FORM*

The causative agents are neurotropic, but the factors influencing this tropism are at present unknown. Patients present with headache and low-grade or no fever, and eventually develop neurologic signs. Patients usually are immune-normal, they may have no history of remarkable mold exposure, no pulmonary symptoms, and other organs may not be involved. Mycoses by these agents in the CNS are localized to the brain parenchyma. The cerebral abscess is most often an encapsulated cyst with a purulent center containing abundant golden to light brown hyphae visible in H&E stained tissue.

1. Sites of *C. bantiana* CNS infection are single or multifocal lesions of the cerebral parenchyma. Less commonly, the ventricles (CSF-filled spaces) or the cerebellum is affected. CSF specimens have not yielded positive smears or cultures. *C. bantiana* is a rapid grower in vivo, causing acute cerebral disease with edema shortly after the first onset of symptoms, which include severe headache and paralysis of limbs (Horré and de Hoog, 1999). *C. bantiana* cerebral abscess occurs in single cases sporadically and is fatal if untreated. The clinical course in immunocompromised patients is bleak. A review of *C. bantiana* cerebral abscess in solid organ transplant recipients found 11 cases between 1991 and 2008 treated both surgically and with AmB; of these there were 3 survivors (Harrison et al., 2008). Now extended spectrum azoles are available with better CNS penetration (please see Section 19B.10, Therapy). The pathology of *C. bantiana* infection has been compared to that of *Cryptococcus gattii* infection in that both cause cerebral abscess in immunologically intact persons.

2. *Wangiella* (*Exophiala*) *dermatitidis* infection of the CNS occurs as multiple cerebral abscesses. Other sites may be involved, notably the cervical lymph nodes. Cervical adenopathy is among the first symptoms of cerebral abscess. Neurotropic infections by *W. dermatitidis* are not limited to the brain.

3. *Ochroconis gallopava* may infect subcutaneous sites, lungs, and visceral organs. Brain infections are single or multiple well-delimited cerebral abscesses.

4. *Ramichloridium mackenziei* infection may occur as single or multiple cerebral abscesses containing thick-walled cavities.

19B.10 *THERAPY*

When possible, complete surgical removal of the encapsulated abscess combined with antifungal therapy is recommended, but so far success in treating cerebral phaeohyphomycosis is limited. In a review of 101 cases of CNS phaeohyphomycosis the outcomes with AmB therapy were poor with 73% mortality regardless of the immune status of the patient (Revankar et al., 2004). A review of reported cases suggests that combination therapy with AmB, flucytosine, and ITC may improve survival (Revankar, 2006). When there are multiple cerebral abscesses and surgery is not practicable, combination therapy with AmB, flucytosine, and terbinafine, or an extended spectrum triazole has been proposed as a regimen (Perfect et al., 2003a).

Azole drugs have good in vitro activity against melanized molds, as do glucan synthase inhibitors such as CASF. Of the older azoles, ITC has had the most success. Of the extended spectrum triazoles (VRC, PSC, or ravuconazole) PSC has had success in treating cutaneous-disseminated phaeohyphomycosis (Negroni et al., 2004). The role of echinocandins (e.g., CASF or micafungin) in cerebral phaeohyphomycosis remains to be determined. (Please see Section 19B.2, Case Presentation—Treatment of Fungal Cerebral Abscess.) Fungicidal concentrations in the CNS are difficult to achieve because of limited penetration by AmB-deoxycholate and lipid AmB. Of azoles, VRC has better CNS penetration and better bioavailability than itraconazole. PSC has fewer hepatic adverse effects than VRC and is effective against *C. bantiana* cerebral abscess.

Keratitis

Melanized fungi cause an estimated 8–17% of fungal keratitis (Revankar, 2006). Approximately half the cases are associated with trauma. Larger case series have been from India involving *Curvularia* followed by *Bipolaris. Curvularia* keratitis has also been reported in the United States. As therapy, topical natamycin frequently is used, with more severe cases receiving combination therapy with an azole. ITC has good in vitro activity. A significant portion of those treated fail therapy and require keratoplasty, as seen in *Fusarium* keratitis. (Please see Chapter 15).

19C *FUNGAL SINUSITIS*

19C.1 *INTRODUCTION/DISEASE DEFINITION*

Fungal sinusitis is initiated by the inhalation of conidia from the environment. It is often an indolent disease that may remain confined to the nasal sinus cavity (noninvasive), referred to as allergic fungal rhinosinusitis (AFRS) or may spread to contiguous structures over months or years[1]. Melanized fungi are frequently encountered. Rapidly progressing invasive sinusitis is found in patients with immunocompromising conditions or in diabetic ketoacidosis. The causative agents in invasive fungal sinusitis are hyalohyphomycetes: *Aspergillus* species and members of the *Mucorales* and *Fusarium* species.

19C.2 *CLINICAL FORMS*

Four types of fungal sinusitis are recognized by the American Academy of Otorhinolaryngology (http://www.entnet.org/HealthInformation/Fungal-Sinusitis.cfm)

Mycetoma Fungal Sinusitis

This produces a "fungus ball," within a nasal sinus cavity, most often in the maxillary sinuses. The noninvasive nature of this disorder requires treatment consisting of scraping of the infected sinus. Specific antifungal therapy generally is not prescribed.

[1]Paranasal sinuses. *Definition*: Air-filled pockets located within the bones of the face and around the nasal cavity. Each sinus is named for the bone in which it is located:

- Maxillary (one sinus located in each cheek)
- Ethmoid (approximately 6–12 small sinuses per side, located between the eyes)
- Frontal (one sinus per side, located in the forehead)
- Sphenoid (one sinus per side, located behind the ethmoid sinuses, near the middle of the skull) For further information and anatomic images of the paranasal sinuses please see http://www.american-rhinologic.org/patientinfo.sinusnasalanatomy.phtml.

Allergic Fungal Rhinosinusitis (AFRS)

AFRS is an allergic reaction in the paranasal sinuses to environmental fungi inhaled from the ambient air. AFRS is thus *a non-tissue-invasive hypersensitivity reaction to fungi in the sinus cavity*. There often is a long history of allergic rhinitis, recurrent bacterial sinusitis, or occasional sinus pain, unresponsive to antibiotics. Thick fungal debris and allergic mucin are present. Allergic mucin is an extramucosal, cheesy, or viscous, peanut-buttery tan to dark-green amorphous mass. Strongly staining, it contains sheets of eosinophils, necrotic eosinophils, and cell debris surrounded by thin eosinophilic mucin where Charcot–Leyden crystals can often be seen. It is interspersed with nasal polyps. Fungal hyphae are visible in H&E or special stains (GMS). GMS staining shows small areas of sparse, scattered, fungal hyphae within allergic mucin but not within the mucosa. The size and presence or absence of septa in the hyphae, and pigmentation, will help differentiate between *Mucorales* and the melanized fungi (Schubert, 2004).

Allergic mucin accumulates in the sinus cavities and must be surgically removed so that the inciting allergen is no longer present. Histopathology of excised surgical material demonstrates eosinophilic mucin, hyphal fragments, and no fungal invasion of tissue or bone.

Diagnostic criteria for AFRS were described by Bent and Kuhn (1996). They analyzed 15 cases and observed common characteristics:

- IgE-mediated hypersensitivity to fungi.
- Nasal polyposis.
- Characteristic radiographic findings.
- Eosinophilic mucin with hyphae are often present but without fungal invasion into sinus tissue.
- Sinus contents removed at the time of surgery stain positively with fungal stain (GMS).

Many fungi are isolated in AFRS, and melanized fungi are prominent. A partial list of these fungi include *Alternaria* spp., *Bipolaris spicifera, B. australiensis, B. hawaiiensis, Curvularia lunata*, and *Exserohilum rostratum*. [Please see Section 19.3, Laboratory Detection, Recovery and Identification, Table 19.1, and St. Germain and Summerbell (1996) for similarities and differences among these genera.]

If left untreated, pressure from the mucin mass can cause bone resorption in the sinuses. This can result in proptosis, facial asymmetry, loss of vision, frontal lobe compression, and neurologic symptoms. The risk of developing true invasive fungal infection in AFRS seems to be very small (Schubert, 2004).

A widely accepted diagnostic criterion for AFRS is the presence of allergic mucin. Controversy exists on

Table 19.1 Some Special Features to Differentiate Melanized Fungi Associated with Chromoblastomycosis or Phaeohyphomycosis

Melanized fungus	Conidia unicellular or multicelled	Annellides/ phialides	Maximum growth temperature (°C)	Conidial arrangement or specific factors
Alternaria alternata	Multicelled, irregularly shaped with/without apical beak	No	Poorly to 35	Long chains
Aureobasidium pullulans	Arthroconidia, chlamydospores	No	Pref,[a] 25 Max,[b] 35	Hyaline hyphae with blastoconidia; melanized hyphae, arthroconidia
Bipolaris species	Multicelled poroconidia, mostly 4–5 septa, flattened hilum	No	Max, 35	Bipolar germination
Cladophialophora bantiana	Unicellular attached in chains, no apparent hila on conidia	No	42–43	Very long sparsely branched, wavy chains of oval acropetal conidia
Cladophialophora carrionii	Unicelled with hila on conidia	No	35–37	Moderately long, branched chains of oval acropetal conidia
Exophiala jeanselmei	Unicellular with narrow pointed tips	Annellides	Max, 38 (but varies)	Balls, groups of conidia, yeast, elongated cells, annellated black yeast are common. Positive nitrate test = bright blue after 7 days on KNO_3 medium.
Wangiella (Exophiala) dermatitidis	Unicellular yeast cells/conidia; pseudohyphae	Phialides without collarettes	Max, 40–42	Ellipsoid conidia accumulating in groups, many yeast cells
Exophiala spinifera	Unicellular	Annellides, tapered tips	37	Ellipsoid conidia, annellated black yeast abundant
Exserohilum rostratum	Multicelled poroconidia with protruding hilum and 7–9 septa	No	Pref 25–30	Conidiophores long, nonbranched, geniculate with dark basal and distal septa
Fonsecaea pedrosoi	Unicellular	Phialides + *Rhinocladiella* and branching type	37	Strongly branching chains
Hortaea werneckii[c]	Yeast cells and conidia, often two-celled	Annellides	40	Borne singly from annellide, may slide along hyphae
Ochroconis gallopavum	Two-celled constricted at septum, clavate	No	45	Flexuous[d], cylindrical to needle-shaped, usually with 1–2 conidia on denticles; reddish brown diffusible pigment
Phialophora verrucosa	Unicellular	Phialides terminating in collarette	37	Balls of round–oval conidia
Rhinocladiella aquaspersa	Unicellular	*Rhinocladiella* type	35	Conidiogenous cells terminal, intercalary, closely packed hyaline to subhyaline, ellipsoid to clavate conidia on crowded denticles
Scedosporium prolificans	Unicellular	Annellides with inflated base	45	Single conidia on specialized conidiophore along hyphae

[a] Pref, preferred temperature of incubation.

[b] Max, maximum temperature.

[c] This agent of tinea nigra is discussed in Chapter 22.

[d] Conidiophores with smooth bends.

the role that fungi play in chronic rhinosinusitis and the specific immunologic response to fungi in patients with AFRS. Originally, Gell and Coombs types 1 and 3 hypersensitivity[2] responses to inhaled fungal antigens were held responsible for the pathology. In many patients, elevated titers of fungal-specific serum IgE and IgG were found. Serum from patients with *Bipolaris* culture-confirmed AFRS found 82% of AFRS cases had *Bipolaris*-specific IgE detected by RAST[3] and 94% had *Bipolaris*-specific IgG detected by EIA (reviewed by Ryan and Marple 2007).

Fungal-specific serum IgE and IgG levels to multiple fungi were elevated in AFRS patients (diagnosed using Bent and Kuhn criteria) compared to patients with only nasal polyps, nonatopic controls, and allergic rhinitis patients. AFRS therefore would appear to be immunologically mediated via hypersensitivity to fungi.

Thus far there is poor correlation between fungal species present in the eosinophilic mucin of AFRS patients and the species implicated in skin tests and in vitro tests for fungal allergy. Increased fungal-specific IgG_3, however, is a marker that separates AFRS patients from those with other forms of chronic rhinosinusitis (Ryan and Marple, 2007). Immunologic mechanisms other than type 1 hypersensitivity (IgE) may be important in the pathophysiology of AFRS but their relative importance remains to be demonstrated.

Local type 1 hypersensitivity response against fungal antigens may be present even in patients without evidence of systemic allergy, raising the possibility of a local IgE-mediated immune response against fungi. Both IgE- and IgG-dependent mechanisms mediate hypersensitivity responses to fungal antigens in patients with AFRS and are likely responsible for eosinophilic inflammation leading to the eosinophilic mucin that is its hallmark. The clinical criteria used to diagnose AFRS may need to be supplemented with titers of fungal-specific IgE or IgG to delineate patients with AFRS and help direct appropriate treatment.

Therapy

Treatment of AFRS is designed to eliminate the allergen and lessen the allergic response. The first step is nasal sinus surgery to remove all nasal polyps and allergic mucin. The preferred endoscopic tissue-preserving approach is sufficient to remove obstructing polypoid[4] mucosa, evacuate sinus contents, and facilitate continued sinus drainage. Surgical specimens should be sent for pathologic evaluation including special fungal stains and for bacterial and fungal cultures. A treatment plan evaluated in a large series of AFRS patients over 8 years (summarized in Schubert, 2004) consisted of oral corticosteroid therapy, unless concurrent medical conditions rule out the use of oral corticosteroids. Patients received 0.5 mg/kg prednisone postoperatively and then daily as a single morning dose for 2 weeks. Thereafter, therapy was spaced to the same dose every other morning for several weeks, then further tapered from 7.5 mg to 5 mg every other morning. Total duration of therapy was for up to 1 year, or longer in some cases. At least 2 months of postoperative oral corticosteroid therapy resulted in clinical improvement, as compared with no therapy, but 1 year of therapy was clearly better. The reason that systemic therapy is not significantly effective is because AFRS is not a tissue-invasive mycosis. Allergen immunotherapy guided by skin tests for the AFRS-etiologic mold should be considered (Schubert, 2004). Recurrence is not uncommon after the diseased tissue and allergic mucin are removed. Anti-inflammatory therapy with local or systemic corticosteroids are prescribed to reduce AFRS recurrences.

There is controversy about the role of topical or systemic antifungal therapy to treat AFRS because several double-blind, placebo-controlled trials failed to show improvement of symptoms after such treatment (Ebbens and Fokkens, 2008). Their summary and conclusions are summarized:

> *The role of fungi in chronic rhinosinusitis (CRS) remains to be defined because although fungi can be detected in the paranasal sinuses of nearly all CRS patients, they are also present in healthy controls. Further clarity is needed about the role of fungi in CRS, the role of particular fungal species, susceptible individuals, and further characterization of the immune response to fungi that could result in disease. The lack of significant clinical improvement of CRS in placebo-controlled trials of antifungal therapy weaken the case against such therapy in what is a non-invasive disease.*

Chronic Indolent Sinusitis

This is an invasive form of fungal sinusitis in patients without an identifiable immune deficiency. This form is generally found outside the United States, most commonly in the Sudan and northern India. The disease progresses from months to years and presents symptoms that include chronic headache and progressive facial swelling that can cause visual impairment. Microscopically, chronic indolent sinusitis is characterized by a granulomatous inflammatory infiltrate. A subnormal immune system can place patients at risk for this invasive

[2]Readers unfamiliar with this scheme may refer to Janeway et al. (2001), available at the URL http://www.ncbi.nlm.nih.gov/sites/entrez?db<I23>books.

[3]RAST. *Definition*: Radioallergoabsorbent test.

[4]Polypoid. *Definition*: Resembling a polyp in gross features.

disease. Surgical mucosal resection is indicated with or without perioperative oral therapy.

Invasive or Fulminant Sinusitis

This is usually seen in the immunocompromised patient. The disease leads to progressive destruction of the sinuses and can invade the bony cavities containing the eye and brain. The recommended therapy for fulminant sinusitis is aggressive surgical removal of the fungal material and systemic antifungal therapy. Invasive sinusitis usually is seen in the host compromised by immmunosupressive therapy and/or diabetic ketoacidosis. Melanized fungi isolated in cases of invasive fungal sinusitis are *Alternaria, Bipolaris, Curvularia*, and *Exserohilum*. However, this disease is, by far, most often caused by hyaline molds (e.g., *Aspergillus fumigatus, Fusarium* species) or various *Mucorales* (e.g., *Rhizopus*, *Rhizomucor*). Because of the high prevalence of hyaline molds and the acute fulminant clinical course of this disease, please see Part Four, Systemic Mycoses Caused by Opportunistic Hyaline Molds.

19C.3 *TRANSMISSION*

Fungal sinusitis is acquired via inhalation of mold conidia from the environment.

19C.4 *DETERMINANTS OF PATHOGENICITY*

The presence of melanin in fungal cell walls is believed to be a pathogenic factor.

19.3 LABORATORY DETECTION, RECOVERY, AND IDENTIFICATION

Direct Examination

Skin scrapings, wound swabs, biopsy material, crusts, and aspirated pus can be examined directly in a KOH prep. Cysts may be drained or sampled by fine needle aspiration. Although Calcofluor–KOH is not specifically used for melanized identification, it is helpful to detect melanized as well as hyaline molds. In the light microscope one may observe golden brown muriform cells indicative of chromoblastomycosis or, if the tissue form consists of brown, branching, septate hyphae, or moniliform hyphae, phaeohyphomycosis may be suspected. The specimen in AFRS consists primarily of surgically excised polypoid or allergic mucin, fungal debris, or tissue. Laboratory procedures for these specimens includes KOH prep, KOH–Calcofluor, and submission for histopathology with H&E and PAS stains.

Histopathology

Histopathology of cutaneous and subcutaneous phaeohyphomycosis may or may not show granulomatous inflammation. The middle of the subcutaneous abscess consists of necrotic debris and PMNs surrounded by a mixed layer of epithelioid cells, macrophages, giant cells, and neutrophils. Next, is a middle layer of vascularized scar tissue, encircled in a fibrotic capsule. A thick capsule surrounding a phaeomycotic cerebral abscess has also been noted (Sutton et al., 1998). The fungi within are melanized, and the tissue form consists of pale to medium brown irregular hyphae with crosswalls, yeast-like cells, moniliform hyphae, or distorted hyphae. The fungi may show melanin pigmentation in H&E (Figs. 19.1 and 19.3 *panels a, b*) Some isolates produce little or no brown color in the hyphal wall, but latent melanin can be developed with the Fontana–Masson stain. When stained with GMS, fungal elements are visible but the natural brown pigment is masked. *Bipolaris spicifera* appears in tissue as branched, closely septate, brown hyphae.

Morphology of Major Agents of Phaeohyphomycosis and Chromoblastomycosis

Please also see "Etiologic Agents" in Sections 19A.4 and 19B.4. Colonies may be gray, olive-gray, deep green, dark brown, to nearly black. Types of sporulation are summarized in Table 19.1. Depending on the genus and species, phialides and *Rhinocladiella*-like and *Cladosporium*-like sporulation are seen. Colony color also differs among different species and even between strains. Other tests may be helpful such as thermotolerance in the range 37–42°C, and gelatin liquefaction.

Culture

Culturing is essential to provide a definitive diagnosis and to facilitate susceptibility tests. Certain melanized fungi require different cultural procedures, and a few have specific tests to differentiate them from others. They are identified by the development of darkly pigmented colonies when grown on culture medium.

Appropriate material, as in the direct examination (skin scrapings, biopsy material, crusts, and aspirated pus) should be cultured on SDA Emmons and PDA, or CMA (for enhanced conidiation) Petri plates or agar slants, incubated at 25–30°C, and are observed at least weekly. Cultures are maintained for up to 4–6 weeks before discarding them because although some of the fungi grow quickly, others are slow in development. Some melanized fungi begin colonial development as yeasts, including *Hortaea werneckii, Exophiala jeanselmei, E. spinifera*, and *W*. (*Exophiala*) *dermatitidis*.

Mycosel and other culture media containing cycloheximide should not be used, as some of the melanized fungi will not grow in its presence. Rather, penicillin and streptomycin or other antibacterial antibiotics should be used in media, (e.g., SDA).

It also is essential to be sure that the culture is pure. This may require subculturing a small portion of the periphery of a colony that appears to be pure to assure that only a single fungus is present.

Both young and mature colonies must be studied to demonstrate the method of conidial development. Melanized fungi are primarily separated on the basis of type of conidial development, the number of cells and septations within the cells, or in some cases, by PCR-sequencing studies (please see Chapter 2, Section 2.4, Genetic Identification of Fungi. Molecular identification has permitted greater resolution of previously unknown species among clinical isolates of black yeast-like fungi (Zeng et al., 2007). Often it is useful to make slide cultures for study of the conidial development and these are best performed on PDA or CMA (for a discussion of mycologic media, please see Chapter 2).

Keys and Tables for Identification of Melanized Molds

Zeng and de Hoog (2008) have a dichotomous key using morphologic characteristics to identify species of *Exophiala, Fonsecaea, Phialophora*, and *Ramichloridium* which is continued in a separate key to identify "black yeasts" including *W.* (*Exophiala*) *dermatitidis*, and other members of the *Exophiala* complex. A further compilation of assimilation of substrate utilization is included.

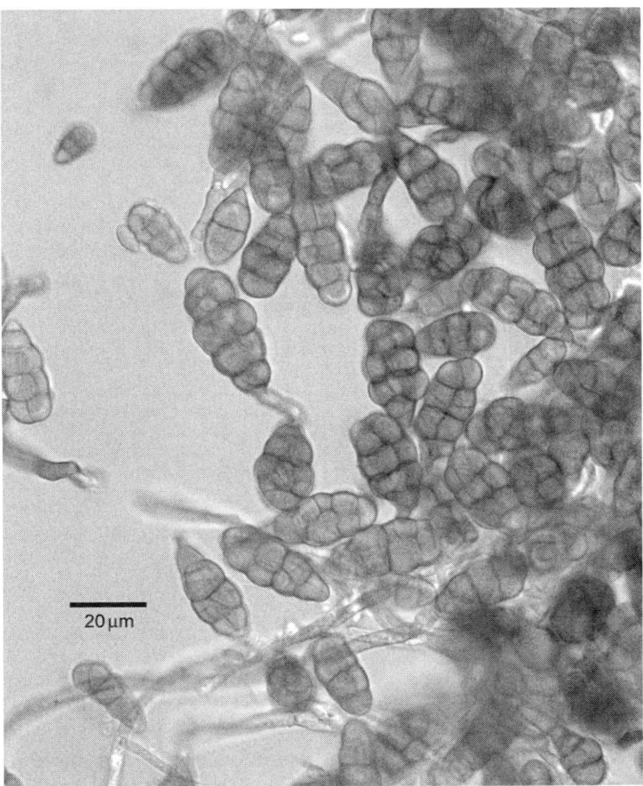

Figure 19.4 *Alternaria alternata*, microscopic morphology. Acropetal chains of multicellular brown conidia (poroconidia) each with a short conical beak. Conidia have both transverse and vertical septa. They arise from simple or branched conidiophores which are less *geniculate* than other similar molds, (e.g., *Ulocladium* species). *Source:* Used with permission from Robert Simmons, Ph.D., Director, Biological Imaging Core Facility, Georgia State University, Atlanta, Georgia.

Alternaria alternata

Alternaria alternata (Fr.) Keissl., *Beih. bot. Zbl.*, Abt. 1 29(2): 434 (1912).

Besides being a common environmental saprobe, *Alternaria* species are implicated as agents of AFRS and are known less common agents of mycotic keratitis. *Alternaria* species are an uncommon cause of phaeohyphomycotic cysts in immunosuppressed patients (Boyce et al., 2010).

Colony Morphology Rapid growing with downy to wooly texture and pale gray to olive brown surface and and brown to black reverse.

Microscopic Morphology See Fig. 19.4 Acropetal chains of multicellular brown poroconidia are produced, each with a short conical beak. Conidia have both transverse and vertical septa. They arise from simple or branched conidiophores which are less *geniculate* than other similar molds (e.g., *Ulocladium* species). (Please also see Chapter 2, Table 6.)

Bipolaris spicifera

Bipolaris spicifera (Bainier) Subram., *Hyphomycetes* (New Delhi): 756 (1971). *Bipolaris spicifera* is an agent of AFRS.

Colony Morphology Rapid growing with a downy texture and gray-brown color. The colony becomes dark olive to black on the surface and dark brown to black on the reverse. Grows up to 40°C.

Microscopic Morphology See Fig. 19.5. Brown septate hyphae produce thick-walled, large (8 μm × 26 μm), ellipsoid, pale brown poroconidia with 3–5 cells. They are borne borne *sympodially*, growing past the pore where the

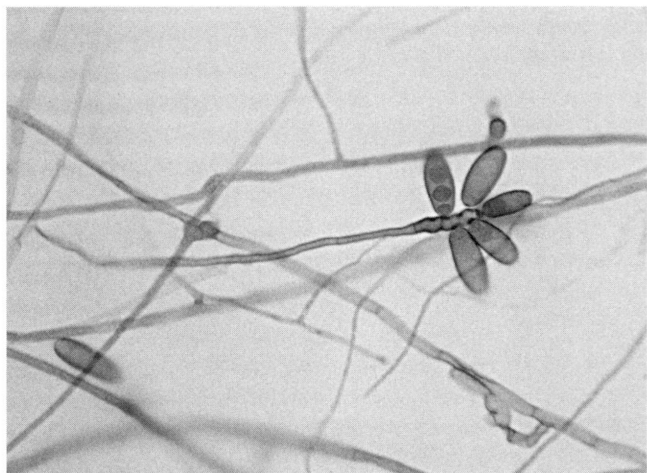

Figure 19.5 *Bipolaris spicifera*, microscopic morphology. Brown septate hyphae with thick-walled, large (8 μm × 26 μm), ellipsoid, pale brown 3–5 celled *poroconidia* borne *sympodially;* 400×, lactofuchsin stain.
Source: scr; E. Reiss, CDC.

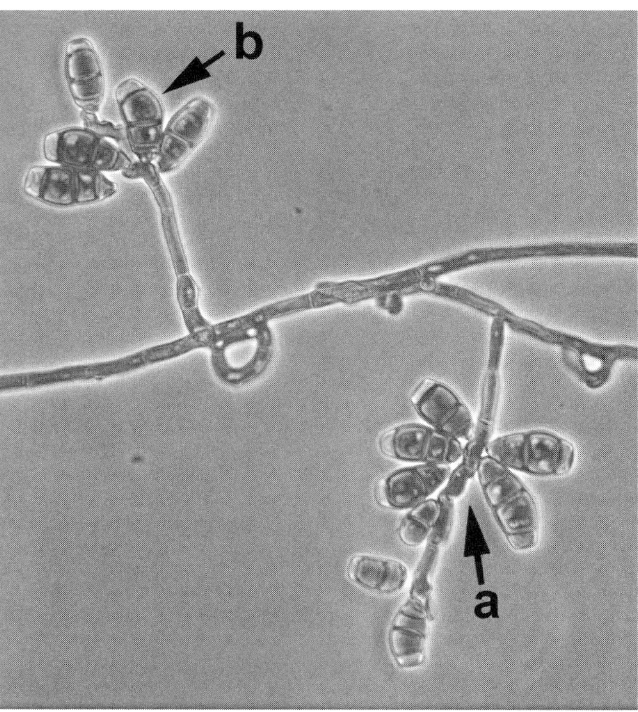

Figure 19.6 *Curvularia lunata*, microscopic morphology. Pale brown multicelled conidia formed apically through a pore (poroconidia) in a *geniculate* conidiophores (a). Conidia are cylindrical or slightly curved, with one of the central cells being larger and darker (b), 400×.
Source: E. Reiss, CDC.

newest conidium is produced, bending around that conidium. This pattern is of a *geniculate* conidiogenous cell. Free conidia have *hila* that protrude slightly beyond the basal contour of the cell.

Cladophialophora bantiana

Cladophialophora bantiana. (Sacc.) de Hoog, Kwon-Chung & McGinnis, in de Hoog, Guého, Masclaux, Gerrits van den Ende, Kwon-Chung et al., *J Med Vet Mycol* 33(5): 343 (1995).

Colony Morphology Slow-growing olive to gray to brown, *floccose*, with a black reverse. *Cladophialophora bantiana* grows at 42–43°C.

Microscopic Morphology See Fig. 19.2. Brown septate hyphae in long, wavy chains of oval conidia (20–35 cells long) with infrequent branching. These chains remain intact in contrast to *C. carrionii*, which readily break apart. *C. bantiana* has characteristic morphology and temperature tolerance. Additional confirmation of its identity is obtained with the ITS region sequence of rDNA.

Curvularia lunata

Curvularia lunata. (Wakker) Boedijn, *Bull Jard bot Buitenz*, 3 Sér 13(1): 127 (1933).

Colony Morphology Rapid growth; the surface is flat, floccose to wooly, dark olive green to brown to gray-black, often with a white edge and a dark brown reverse.

Microscopic Morphology See Fig. 19.6. *Curvularia* species have pale brown multicelled conidia formed apically through a pore (poroconidia) in a geniculate conidiophore. Conidia are slightly curved or cylindrical, with one of the central cells being larger and darker.

Wangiella (Exophiala) dermatitidis

Wangiella (*Exophiala*) *dermatitidis*. (Kano) de Hoog, *Stud Mycol* 15: 118 (1977).

Colony Morphology Beginning as a yeast-like colony that is black, moist, and shiny; as the colony develops hyphae it becomes olive-gray with a velvet texture. Some strains are dark olive-brown-black with a black reverse. Slow growing, grows up to 42°C and is resistant to cycloheximide.

Microscopic Morphology See Fig. 19.7. This melanized species has both yeast-like and mold stages of growth. The mold form consists of brown septate hyphae producing pale brown oval conidia via an *annellidic* conidiophore. The conidiophore becomes narrower,

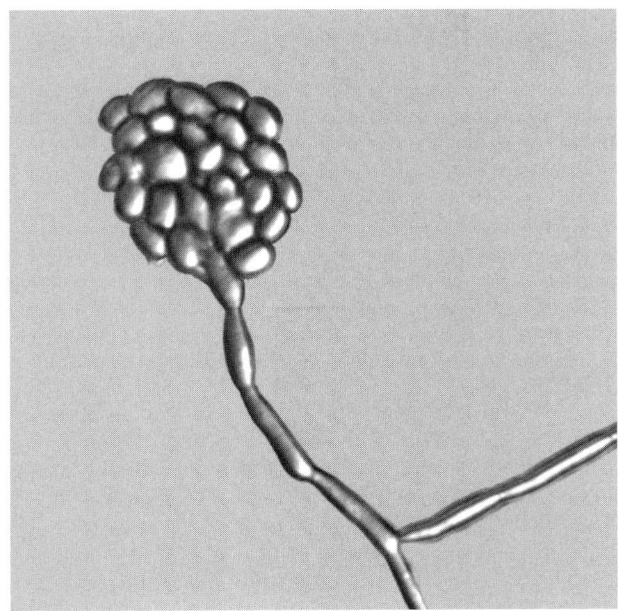

Figure 19.7 *Wangiella (Exophiala) dermatitidis*, microscopic morphology.
Source: Used with permission from Dr. Paul Szaniszlo, University of Texas, Austin.

adding an annellation when a new conidium is produced. Conidia can reproduce by budding or by forming pseudohyphae. Budding yeast may be seen during an infection in host tissue. This fungus is a well-studied model for morphogenesis among the yeast-like melanized molds (Szaniszlo, 2006).

Ochroconis gallopava

Ochroconis gallopava (W.B. Cooke) de Hoog [as "gallopavum"], *in*: Howard DH, *Fungi Pathogenic for Humans and Animals. Part* A, *Biology* (New York): 181 (1983).

Colony Morphology Brown suede-like colony with diffusible red pigment. Grows rapidly and is thermotolerant up to 45°C. Growth is inhibited by cycloheximide.

Microscopic Morphology Pale brown septate hyphae producing two-celled *clavate* conidia attached to the supporting hyphae by *denticles*.

Ramichloridium mackenziei

Ramichloridium mackenziei C.K. Campb. & Al-Hedaithy, *J Med Vet Mycol* 31(4): 330 (1993).

Colony Morphology Slow growing, beginning as shiny dark brown-black, then turning to woolly, domed colonies with a black reverse. Growth occurs at 42°C and in the presence of cycloheximide.

Microscopic Morphology Brown, septate hyphae with conidiophores rising at approximate right angles to the vegetative hyphae. Two smooth-walled pale brown conidia projected from both sides of a thick walled *rachis* presenting a "Mickey Mouse" appearance (Sutton et al., 1998).

Scedosporium prolificans

Please find information about this melanized agent in Chapter 16. The reason for placing it there is that mycologists associate this melanized species with its hyaline relative, *Scedosporium apiospermum*.

Serologic Tests

There are no available serologic diagnostic tests to identify infections due to melanized fungi in the clinical laboratory.

SELECTED REFERENCES FOR PHAEOHYPHOMYCOSIS

Bent JP 3rd, Kuhn FA, 1996. Allergic fungal sinusitis/polyposis. *Allergy Asthma Proc* 17: 259–268.

Boyce RD, Deziel PJ, Otley CC, Wilhelm MP, Eid AJ, Wengenack NL, Razonable RR, 2010. Phaeohyphomycosis due to *Alternaria* species in transplant recipients. *Transpl Infect Dis* 12: 242–250.

Chandler FW, Kaplan W, Ajello L, 1980. *Color Atlas and Text of the Histopathology of Mycotic Diseases*. Year Book Medical Publishers, Chicago.

Crosby JH, O'Quinn MH, Steele JCH Jr, Rao RN, 1989. Fine-needle aspiration of subcutaneous phaeohyphomycosis caused by *Wangiella dermatitidis*. *Diagn Cytopathol* 5: 293–297.

Dismukes WE, Pappas PG, Sobel JD (eds.,) 2003. *Clinical Mycology*. Oxford University Press, New York.

Ebbens FA, Fokkens WJ, 2008. The mold conundrum in chronic rhinosinusitis: Where do we stand today? *Curr Allergy Asthma Rep* 8: 93–101.

Engemann J, Kaye K, Cox G, Perfect J, Schell W, McGarry SA, Patterson K, Edupuganti S, Cook P, et al., 2002. *Exophiala* infection from contaminated injectable steroids prepared by a compounding pharmacy—United States, July–November, 2002. *MMWR Morb Mortal Wkly Rep* 51: 1109–1112.

Harrison DK, Moser S, Palmer CA, 2008. Central nervous system infections in transplant recipients by *Cladophialophora bantiana*. *South Med J* 101: 292–296.

Horré R, de Hoog GS, 1999. Primary cerebral infections by melanized fungi: A review. *Stud Mycol* 43: 176–193.

Janeway Ca JR, Travers P, Walport M, Shlomchik MJ, 2001. *Immunobiology, The Immune System in Health and Disease* (5th ed.). Garland Science, New York.

Kanj SS, Amr SS, Roberts GD, 2001. *Ramichloridium mackenziei* brain abscess: Report of two cases and review of the literature. *Med Mycol* 39: 97–102.

Lyons MK, Blair JE, Leslie KO, 2006. Successful treatment with voriconazole of fungal cerebral abscess due to *Cladophialophora bantiana*. *Clin Neur Neurosurg* 107: 532–534.

Matos T, de Hoog GS, de Boer AG, de Crom I, Haase G, 2002. High prevalence of the neurotrope *Exophiala dermatitidis* and related oligotrophic black yeasts in sauna facilities. *Mycoses* 45: 373–377.

Negroni, R, Helou SH, Petri N, Robles AM, Arechavala A, Bianchi MH, 2004. Case study: posaconazole treatment of disseminated phaeohyphomycosis due to Exophiala spinifera. *Clin Infect Dis* 38: e15–e20.

Patterson, JW, Warren NG, Kelly LW, 1999. Cutaneous phaeohyphomycosis due to *Cladophialophora bantiana*. *J Am Acad Dermatol* 40: 364–366.

Peltroche-Llacsahuanga J, Schnitzler N, Jentsch S, Platz A, de Hoog S, Schweizer KG, Haase G, 2003. Analyses of phagocytosis, evoked oxidative burst, and killing of black yeasts by human neutrophils: A tool for estimating their pathogenicity? *Med Mycol* 41: 7–14.

Perfect JR, Marr KA, Walsh TJ, Greenberg RN, du Pont B, de la Torre-Cisneros J, Just-Nübling G, Schlamm HT, Lutsar I, Espinel-Ingroff A, Johnson E, 2003a. Voriconazole treatment for less-common, emerging, or refractory fungal infections. *Clin Infect Dis* 36: 1122–1131.

Perfect JR, Schell WA, Cox GM, 2003b. Phaeohyphomycoses, pp. 271–282, *in*: Dismukes WE, Pappas PG, Sobel JD (eds.,) *Clinical Mycology*. Oxford University Press, New York.

Queiroz-Telles F, McGinnis MR, Salkin I, Graybill JR, 2003. Subcutaneous mycoses. *Infect Dis Clin North Am* 17: 59–85.

Revankar SG, 2006. Phaeohyphomycosis. *Infect Dis Clin North Am* 20: 609–620.

Revankar SG, Sutton DA, Rinaldi MG, 2004. Primary central nervous system phaeohyphomycosis: A review of 101 cases. *Clin Infect Dis* 38: 208–216.

Rivard RG, McCall S, Griffith ME, Hawley JS, Ressner RA, Borra H, Moon JE, Beckius ML, Murray CK, Hospenthal DR, 2007. Efficacy of caspofungin and posaconazole in a murine model of disseminated *Exophiala* infection. *Med Mycol* 45: 685–689.

Ryan MW, Marple BF, 2007. Allergic fungal rhinosinusitis: Diagnosis and management. *Curr Opin Otolaryngol Head Neck Surg* 15: 18–22.

Schnitzler N, Peltroche-Llacsahuanga H, Bestier N, Zündorf J, Lütticken R, Haase G, 1999. Effect of melanin and carotenoids of *Exophiala (Wangiella) dermatitidis* on phagocytosis, oxidative burst, and killing by human neutrophils. *Infect Immun* 67: 94–101.

Schubert MS, 2004. Allergic fungal sinusitis. *Otolaryngol Clin North Am* 37: 301–326.

Shoham S, Pic-Aluas L, Taylor J, Cortez K, Rinaldi MG, Shea Y, Walsh TJ, 2008. Transplant-associated *Ochroconis gallopava* infections. *Transpl Infect Dis* 10: 442–448.

Silveira F, Nucci M, 2001. Emergence of black moulds in fungal disease: Epidemiology and therapy. *Curr Opin Infect Dis* 14: 679–684.

St. Germain G, Summerbell R, 1996. *Identifying Filamentous Fungi—A Clinical Laboratory Handbook*. Star Publishing Co., Belmont, CA.

Sutton DA, Slifkin M, Yakulis R, Rinaldi MG, 1998. U.S. case report of cerebral phaeohyphomycosis caused by *Ramichloridium obovoideum (R. mackenziei)*: Criteria for identification, therapy, and review of other known dematiaceous neurotropic taxa. *J Clin Microbiol* 36: 708–715.

Szaniszlo PJ, 2006. Virulence Factors in Black Molds with Emphasis on Melanin, Chitin and *Wangiella* as a Molecularly Tractable Model, pp. 407–428, *in*: Heitman J, Filler SG, Edwards JE Jr, Mitchell AP (eds.), *Molecular Principles of Fungal Pathogenesis*. ASM Press, Washington, DC.

Wang TK, Chiu W, Chim S, Chan TM, Wong SS, Ho PL, 2003. Disseminated *Ochroconis gallopavum* infection in a renal transplant recipient: The first reported case and a review of the literature. *Clin Nephrol* 60: 415–423.

Zeng JS, de Hoog GS, 2008. *Exophiala spinifera* and its allies: Diagnostics from morphology to DNA barcoding. *Med Mycol* 46: 193–208.

Zeng JS, Sutton DA, Fothergill AW, Rinaldi MG, Harrak MJ, de Hoog GS, 2007. Spectrum of clinically relevant *Exophiala* species in the United States. *J Clin Microbiol* 45: 3713–3720.

WEBSITES CITED

American Academy of Otorhinolaryngology information on fungal sinusitis at the URL http://www.entnet.org/HealthInformation/Fungal-Sinusitis.cfm.

For further information and anatomic images of the paranasal sinuses please see the URL http://www.american-rhinologic.org/patientinfo.sinusnasalanatomy.phtml.

QUESTIONS

The Answer Key to multiple choice questions may be found at the end of the book.

1. Which melanized molds fit into the disease category "phaeohyphomycosis"?
   A. Grow as hyphae in tissue
   B. Grow as moniliform hyphae in tissue
   C. May also produce yeast forms in tissue
   D. All of the above

2. Which of the following clinical forms of disease are included in phaeohyphomycosis?
   A. Cerebral abscess
   B. Fungal rhinosinusitis
   C. Subcutaneous cyst
   D. All of the above

3. Which of the following is a neurotropic fungus?
   A. *Aureobasidium pullulans*
   B. *Cladophialophora bantiana*
   C. *Exophiala jeanselmei*
   D. *Scedosporium prolificans*

4. What are the characteristics of fungi involved in allergic fungal rhinosinusitis?
   A. Formation of muriform cells in nasal polyps
   B. Known as the "black yeasts"
   C. Production of multicelled poroconidia
   D. Production of long branched chains of oval acropetal conidia

5. Which of the following fungi is isolated in phaeohyphomycotic cyst?
   A. *Hortaea werneckii*
   B. *Madurella mycetomatis*
   C. *Phialophora verrucosa*
   D. *Wangiella* (*Exophiala*) *dermatitidis*

6. The production of annellides is a characteristic of which genus of melanized molds?
   A. *Alternaria*

B. *Bipolaris*

C. *Curvularia*

D. *Exophiala*

7. The genera *Exophiala* and *Wangiella* are referred to as the black yeasts. Do they display the yeast form in culture and during infection? What other forms, if any, occur among the black yeasts? What clinical forms of disease are caused by black yeasts? How would you describe their temperature tolerance? Do these molds have a neurotropism?

8. Phaeohyphomycotic cyst may be due to fungi classed in the *Pleosporaceae* and in the *Herpotrichiellaceae*. Give examples of fungi in both families and describe how they may be identified in the laboratory.

9. Of the two diseases, chromoblastomycosis and phaeohyphomycosis, which remains localized to the skin and subcutaneous tissues and which may disseminate to other organs? What extracutaneous sites of dissemination occur?

10. List and discuss the reasons why primary cerebral phaeohyphomycosis has a poor prognosis.

11. Define the terms nasal polyposis and allergic mucin as they relate to allergic fungal rhinosinusitis.

12. Explain the differences between allergic fungal rhinosinusitis, mycetoma fungal sinusitis, chronic indolent, and invasive fungal sinusitis.

Chapter 20

Eumycetoma (Madura Foot, Maduramycosis)

20.1 EUMYCETOMA AT-A-GLANCE

- *Introduction/Disease Definition.* Eumycetoma is a slow but relentless subcutaneous mycosis, most often affecting the foot and leg accompanied by the triad of (i) tumefaction (an indurated swelling), (ii) draining sinus tracts, and (iii) discharging grains. If untreated, bone erosion (osteomyelitis) may develop.
- *Etiologic Agents.* Eumycetoma (40%). There are many agents but the most prevalent are:
  - *Madurella mycetomatis* accounts for most cases and is present in arid regions.
  - *Madurella grisea* in South America and India.
  - *Pseudallescheria boydii* and *Scedosporium apiospermum* in the United States.
  - Actinomycotic mycetoma (60%); the major etiologic agent is the bacterium *Nocardia brasiliensis*.
- *Geographic Distribution.* Eumycetoma is endemic in the tropics and subtropics between latitudes 15° S and 30° N around the Tropic of Cancer (23.5° N). The major endemic area is in Sudan; other areas are Mexico, Central and South America, India, Indonesia, Pakistan, other African countries, the Middle East, and occasionally in temperate zones including the United States.
- *Pathophysiology.* Initially limited to skin, subcutaneous tissue. Later spreads to muscle, bone, lymphatic vessels, and nerves.
- *Route of Infection.* Penetrating injury.
- *Therapy.* Surgery and long-term oral azole therapy.
- *Laboratory Detection, Recovery, and Identification.* Examination of grains, biopsies, curettage from sinuses, and culture, molecular identification—PCR of rDNA.

20.2 INTRODUCTION/DISEASE DEFINITION

Eumycetoma is a chronic, suppurative, subcutaneous mycosis characterized by fibrotic swelling and tumefaction caused by 30 or more species of molds. Initially limited to the skin and subcutaneous tissue, sinus tracts form to the surface of the skin, discharging serosanguinous pus containing visible grains, which are masses of fungal hyphae. Over months to years tissue destruction spreads to adjoining bone. Lymphatic obstruction and fibrosis can cause lymphedema. Eventually the nerve supply is affected. Secondary bacterial infection complicates clinical management.

The route of infection is by the traumatic intracutaneous inoculation of the causal fungus. Generally it is localized to one area of the body, usually the foot, but occasionally involves the hand, or other body sites (e.g., leg, arm, head, neck, thigh, buttock). Eumycetoma is difficult to treat. Therapy requires surgery and long-term azole antifungal therapy. However, newer therapies with extended spectrum azoles have been effective even without surgery.

There are two categories of mycetoma, *eumycotic*, caused by molds, and *actinomycotic*, due to higher bacteria, the actinomycetes. The discussion of actinomycotic mycetoma is outside the scope of this book. Eumycetoma was known as "Madura foot" because of the high incidence of this disease in Madurai, Tamil Nadu, in India.

Fundamental Medical Mycology, First Edition. By Errol Reiss, H. Jean Shadomy and G. Marshall Lyon, III.
© 2012 Wiley-Blackwell. Published 2012 by John Wiley & Sons, Inc.

20.3 CASE PRESENTATION

Fistulated Infection of the Ankle in a 39-yr-old Man (Ahmed et al., 2003a)

History

A 39-yr-old man, a resident of Mali, sought treatment for a chronic swelling with sinuses draining purulent fluid in his ankle. In December 2000, he attended a clinic in a Paris, France, hospital. His history indicated earlier surgical removal of an abscess near the ankle which was found to contain black grains. At the time he was treated with trimethaprim-sulfamethoxazole with little or no improvement.

Clinical and Laboratory Findings

The patient presented with multiple draining sinuses over the left ankle. Radiography revealed a lytic lesion in the distal fibula. One week after admission the patient was taken to surgery for a massive excision of soft tissue and infected bone. Numerous black grains were seen in surgical specimens. Grains were washed in sterile water, planted on SDA and chocolate agar, and incubated at 30°C and 37°C. Group A *Streptococcus pyogenes* was isolated. After 4–5 days, a slow-growing mold appeared, producing a brown diffusible pigment. No conidia were evident even after 8 weeks incubation, so identification could not be made with morphologic methods. PCR and sequencing of rDNA from the isolate using sequencing primers ITS1 and ITS4 resulted in its identification as *Madurella mycetomatis*.

Diagnosis

The diagnosis was eumycetoma caused by *Madurella mycetomatis* with secondary bacterial infection.

Therapy and Clinical Course

Additional surgical debridements were performed and the patient received amoxicillin and itraconazole (ITC), 400 mg 2×/day, for a total of 20 weeks. No further recurrence was seen at follow-up after 2 months. A relative contacted 2 years later indicated the patient's wound was still draining.

Comment

The initial surgery was not accompanied by antifungal therapy, underlining that surgery alone rarely is successful. The later surgeries also fell short of achieving clinical cure; given that the patient was lost to follow-up and to long duration antifungal therapy. Spread of fungi into adjacent bone is a feature of eumycetoma. Secondary bacterial infection also is a frequent complication. *Madurella mycetomatis* is the major cause of eumycetoma in arid regions of Africa.

20.4 DIAGNOSIS

Diagnosis is made on the presence of painless subcutaneous swellings and, over time, the observation of grains (masses of fungal hyphae) found in the discharge of sinus tracts from these areas to the skin surface. Steps in the mycologic work-up include microscopic examination of the grains, culture of the causative fungal agent, and its identification by morphologic and, if needed, molecular methods. The extent of damage to underlying soft tissue and bone is estimated by radiography (please see Section 20.11, Clinical Forms—Radiography).

Differential Diagnosis

The major diagnostic decision is whether the mycetoma is the result of infection with a fungus or with an actinomycete because that will guide therapy. (Please see Section 20.14, Laboratory Detection, Recovery, and Identification—Histopathology.) Early lesions without draining sinuses may resemble pyogenic granulomas. The larger rubbery lesions may resemble lipomas, but draining sinuses and radiography demonstrating bone lesions assist in the diagnosis. Other diseases to be ruled out include Buruli ulcer, chromoblastomycosis, chronic osteomyelitis, Kaposi sarcoma, melanoma, sporotrichosis, and tuberculosis. Loose clusters of hyphae formed into grains occasionally may be seen in other mycoses (e.g., *Microsporum canis* pseudoeumycetoma). Botryomycosis is a chronic, granulomatous, bacterial infection in which grains are produced. Clinically, it may not be distinguished from eumycetoma.

20.5 ETIOLOGIC AGENT(S)

The number of fungi producing eumycetoma is upwards of 21–30 species (Table 20.1); both hyaline and melanized molds may be involved. *Madurella mycetomatis*, occurring in the arid regions of Africa, accounts for most cases worldwide, causing >70% of cases in Central Africa including Sudan (Lichon and Khachemoune, 2006). *Madurella grisea* is a common etiologic agent in South America. *Leptosphaeria senegalensis* and *Leptosphaeria tompkinsii* are common causes in West Africa. *Pseudallescheria boydii* is the most common cause in United States. (Please see Chapter 16.)

Table 20.1 Fungal Etiologic Agents of Eumycetoma

White-to-yellow grain species	Black grain species
Acremonium species	*Corynespora cassicola*
Aspergillus nidulans	*Curvularia* species
Aspergillus flavus	*Exophiala jeanselmei*
Cylindrocarpon cyanescens	*Leptosphaeria senegalensis*
Cylindrocarpon destructans	*L. tompkinsii*
Fusarium species	*Madurella grisea*
Neotestudina rosatii	*M. mycetomatis*
Polycytella hominis	*Phialophora verrucosa*
Pseudallescheria boydii/Scedosporium species	*Plenodomus auramii*

Pseudallescheria boydii is a teleomorph form related to *Scedosporium* species. *P. boydii* is unusual because the sexual form (cleistothecia containing asci and ascospores) may be demonstrated in the laboratory if grown on cornmeal agar or potato dextrose agar for 2–3 weeks. They also may be seen on SDA-Emmons, but not always on any of these media. The fungus is homothallic and a single culture should be capable of meiosis, unless it has lost this capacity.

Madurella species (de Hoog et al., 2004). The mold *Madurella mycetomatis* was first described as *Madurella mycetomi* (Laveran) Brumpt, *Comptes rendu Seanc Soc Biol* 57: 997–999 (1905). Position in classification: *Sordariales, Sordariomycetidae, Sordariomycetes, Ascomycota*.

Madurella mycetomatis causes eumycetoma in residents of arid climate zones of Africa—from East to West: Somalia, Djibouti, Sudan, Nigeria, Mali, Senegal, occasionally in the Middle East, and definitely in southern India. *M. mycetomatis* is restricted to these countries for climatic reasons. DNA sequences were obtained from the 18S SSU and ITS regions of rDNA from clinical isolates. A distance tree of SSU rDNA sequences found that the nearest neighbor is *Chaetomium*, a member of the ascomycete order *Sordariales*. *M. mycetomatis* is a sterile member of this order. Brumpt erected the genus *Madurella* for eumycetoma agents producing grains in tissue, secreting a brown pigment in vitro into the medium *and remaining sterile*. Strains of nonsporulating black grain mycetoma fungi isolated from other climatic zones differed by 5% from *M. mycetomatis* in their ITS sequences, possibly indicating separate species.

Madurella grisea. J.E. Mackinnon, Ferrada & Montem., *Mycopathologia* 4: 389 (1949). *Madurella grisea* is isolated from eumycetoma patients from India and South America but *not* from dry African climate zones. Alignment of ITS rDNA sequences indicate *M. grisea* is related to *Leptosphaeria thompkinsii*, a rare eumycetoma agent which is a member of the *Pleosporales* and is unrelated to *M. mycetomatis*. Quite likely these two eumycetoma agents are unrelated to each other and do not overlap their endemic areas.

20.6 GEOGRAPHIC DISTRIBUTION/ ECOLOGIC NICHE

The disease is primarily one of the tropics and subtropics. Endemic areas for eumycetoma are found between latitudes 15°S and 30°N, around the Tropic of Cancer (23.5°N). The disease is found in Mexico, Central and South America, India, Indonesia, Pakistan, parts of Africa including Chad, Mali, Nigeria, Senegal, Somalia, Sudan, Yemen, and Zaire, and occasionally in temperate zones, for example, the Appalachians in the United States. The highest incidence of eumycetoma occurs in the Sudan, where it is hot and in the rainy season from June to October there is 50–500 mm of rain. In the dry season of 6–8 months there is little humidity and the daytime high temperature is 42°C. The fungi producing eumycetoma are found on thorny plants, on splinters, and in soil.

Isolation of *M. mycetomatis* from the environment has thus far proved difficult (Ahmed et al., 2002). In Sudanese soil samples, 17 out of 74 (23%) samples were positive for *M. mycetomatis* DNA but cultures did not grow on SDA + chloramphenicol. Only a single thorn of the 22 sampled was positive in PCR. PCR products from clinical and environmental sources had the same PCR-RFLP patterns, indicating identity at the species level. Failure to grow from soil samples underlines that *M. mycetomatis* needs special conditions for growth, and isolation from the environment has not yet been achieved.

As mentioned in Section 20.5, Etiologic Agents, *M. mycetomatis* is localized to dry climate zones whereas *M. grisea* occupies a different ecologic niche in more humid areas.

20.7 EPIDEMIOLOGY

Locally acquired eumycetoma in the United States is due to *Pseudallescheria* or *Scedosporium* species. Cases in the United States due to other fungi are imported. Mycetomata are common in the trans-African belt (Sudan, Somalia, Nigeria) the southern part of Saudi Arabia, in India, and in Central and South America.

Sudan

The prevalence of eumycetoma in highly endemic Sudan was reported as 1231 cases in a 2.5 year period (reviewed in Lichon and Khachemoune, 2006).

Middle East

Eighteen cases of mycetoma were treated at the Riyadh Armed Forces Hospital, in Saudi Arabia between January 1987 and April 1990 (Sharif et al., 1991). Complete or partial amputation was performed in 7 of 14 patients with lower extremity mycetoma. Most patients had actinomycotic mycetoma, only three had eumycetoma—two were cases of *M. mycetomatis* and one isolate could not be identified.

Brazil

Between 1944 and 1978, 147 mycetoma cases were identified in University of São Paulo Mycology Unit. A further 47 mycetoma cases attended this clinic between 1978 and 1989. An additional 27 new mycetoma cases were seen between 1990 and the end of 2000 (Castro et al., 2008). Of the latter, 13 were eumycetomata, and 12 were cases of pedal eumycetoma. Most patients were men between 40 and to 60 years old. Therapy for eumycetoma consisted of trimethoprim-sulfamethoxazole and ITC. Three patients also underwent surgery. Of the 12 patients with eumycetoma, 2 were cured and 4 were clinically improved.

Mexico

In Mexico, actinomycetes are the principal etiologic agent of mycetoma, accounting for about 98% of cases (Welsh et al., 2007).

20.8 RISK GROUPS/FACTORS

Males are infected more frequently than females in ratios from 3:1 to 5:1 depending on the survey. The difference is believed to result from the higher number of outdoor occupations of men over women. Farmers and herdsmen in rural areas are most at risk. Agricultural laborers who work without adequate foot protection are the most likely to suffer from a "Madura foot." There is no known racial differential, and although no age group is exempt, eumycetoma occurs most frequently in males between the ages of 20 and 45 years.

Individuals without shoes or other protective clothing are most likely to become infected via traumatic implantation of the causal fungus. In Sudan, patients with extensive eumycetoma were found to have reduced capacity for DNCB skin-test sensitization when compared to controls. Since other studies in patients with *Madurella mycetomatis* infections did not show defective lymphocyte blastogenesis to specific antigens, it is suggested that host defects are absent or subtle in eumycetoma. Immunosuppression usually is not associated with eumycetoma. In the United Kingdom a study of 27 patients between 1979 and 1985 found that 12 (44%) had diabetes mellitus.

20.9 TRANSMISSION

Intracutaneous inoculation of the causal fungus is the route of transmission. The disease is not communicable from human to human, animal to humans, or humans to animals.

20.10 DETERMINANTS OF PATHOGENICITY

20.10.1 Host Factors

The general health of eumycetoma patients appears to be normal unless decreased activity related to the disease leads to economic malnutrition. Reduced capacity to respond to DNCB skin test may be present. Diabetes mellitus was found in some patients.

Normal, immunocompetent, inbred BALB/c mice injected intraperitoneally with *M. mycetomatis* mycelial fragments develop massive eumycetoma lesions in their peritoneal cavity (Ahmed et al., 2003b). Black grains surrounded by a fibrous capsule, in the mouse model infection, are attached to the peritoneal wall, liver, spleen, and other internal organs. Histopathology of the grains indicates the mass of fungal hyphae is surrounded by inflammatory cells, PMNs in the inner zone and histiocytes in an intermediate zone. Intact growing fungal hyphae are embedded in a hard "cement," which appears to protect the fungus from the host immune response.

20.10.2 Microbial Factors

Melanin is produced by some of the causal fungi. An antigenic protein of *M. mycetomatis* was cloned and sequenced and the recombinant protein was characterized as a translationally controlled tumor protein (TCTP) with a molecular mass of 26 kDa (van de Sande et al., 2006). The TCTP was present in all 38 *M. mycetomatis* strains tested. Immunohistology showed that this protein localizes in the early or growing stage of eumycetoma development when the grain is surrounded by PMNs. Portions of the TCTP protein are crossreactive with other microbial agents, but one peptide is able to discriminate titers to *M. mycetomatis* in patients compared with background levels of antibodies present in healthy controls. Antibody titers with respect to TCTP in eumycetoma patients rise in proportion to disease severity as a marker for tumorous progression in eumycetoma.

TCTP in mammalian cells and simple eukaryotes is upregulated early in the cell cycle, preparatory to growth and cell division (Bommer and Thiele, 2004). TCTP is bound to microtubules including the mitotic spindle apparatus. If the gene is either overexpressed or blocked with RNAi, cells are delayed in growth and cell cycle progression. Site-specific mutagenesis in the phorphorylation sites of the gene disrupts mitosis. TCTP has an anti-apoptotic effect in stabilizing cells from programmed cell death. Studies in parasites have found that secretion of TCTP stimulates inflammatory infiltration of eosinophils and histamine release from basophils. Clearly this molecule has an important role in cell cycle regulation, and when secreted has proinflammatory properties. In addition it is a significant antigenic marker of disease activity in *M. mycetomatis*.

20.11 CLINICAL FORMS (Fahal, 2004)

Pedal Mycetoma

The first written reference to mycetoma is in the ancient Indian religious book *Atharva Veda*, the first Indian text dealing with medicine where it is mentioned as *pada valmikam*, or "anthill foot." The disease rarely is seen in more than one site and the foot is involved ~ 80% of the time. The initial lesion, caused by a small puncture with a thorn or splinter, is a small subcutaneous, painless swelling. Once the fungus begins to grow in the tissue, its mycelium becomes organized into aggregates—these are the "grains." Grain formation is the basic characteristic of eumycetoma. The grains vary in size, color, and hardness, depending on the causative fungus. As the disease progresses fibrosis occurs, with the formation of multiple nodules connected by sinus tracts lined by red granulation tissue. As sinuses form, the overlying skin becomes discolored and scarred. Pus and serosanguinous fluid containing characteristic grains drain from the sinus tracts. Lymphatic spread along a limb with regional lymphadenopathy is very rarely seen. Lymphedema results in tumefaction (Fig 20.1). Hematogenous spread is rare, and eumycetomata typically do not occur at more than one body site. Actinomycotic mycetoma is often more aggressive than eumycetoma.

Secondary bacterial infection with ulceration can occur. Tendons and muscles usually are not affected. There is seldom fever, anemia, or weight loss. Nerve damage can occur late in untreated cases. Bone involvement occurs later in the course of the disease. The net result is a swollen deformed foot with multiple discharging and interconnecting sinus tracts. Pedal eumycetoma due to *M. mycetomatis* is a large tumerous mass.

Bone Involvement

In pedal mycetoma, bone frequently is invaded and eventually destroyed. Eumycotic lesions tend to form fewer but

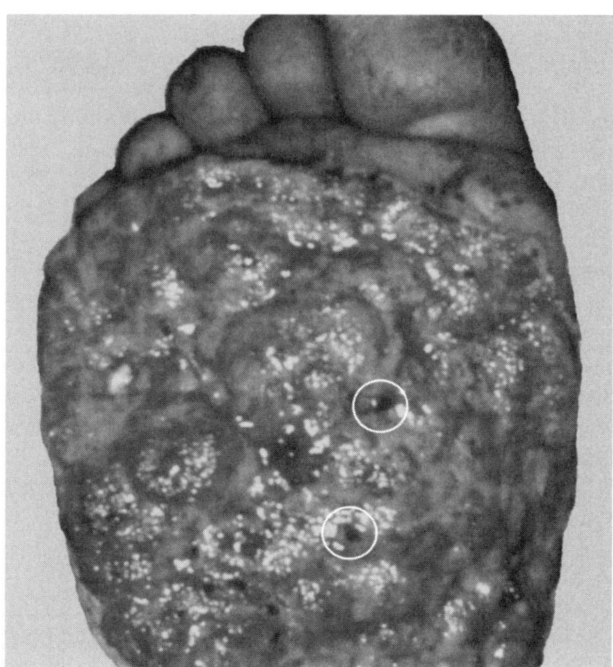

Figure 20.1 Madura foot, pedal eumycetoma. Black grains (encircled) are visible in the plantar surface subcutaneous tissue in a patient with *Madurella mycetomatis* eumycetoma in this postsurgical picture.
Source: Used with permission from Dr. Uma M. Tendolkar, LTM Medical College, Mumbai.

larger cavities in bone, ≥ 1 cm diameter, whereas actinomycotic lesions are smaller but more numerous. The great majority of actinomycotic mycetomata can be treated successfully with antibiotics, whereas eumycetomata often require antifungal therapy and surgery. Recognition of bone involvement is thus very important in designing the treatment plan. Bone is invaded by contiguous spread. In pedal mycetoma multifocal osteomyelitis develops in the foot under the soft tissue swelling, affecting the small bones.

The radiologic appearance in eumycetoma, compared with actinomycetoma, is a mild–moderate periosteal reaction. Without treatment the disease follows a slow relentless course. The periosteal reaction coalesces, forming a "melting snow" appearance on the cortex of the small bones. Eventually the articular cartilage is involved and, due to fungal arthritis or secondary infection, bony fusion develops across the joint. Osseous fusion and multiple infected sinuses produce an amorphous loss of joint definition and a confluent irregular mass develops (Al-Heidous and Munk, 2007).

Radiography

The initial soft tissue nodule can be revealed by ultrasound, radiographs, or MRI before other abnormalities appear. This examination also may reveal an embedded thorn or splinter. MRI helps with the differential diagnoses of the swelling. A diagnostic indicator of eumycetoma is an MRI showing the "dot-in-circle" sign (*definition*: multiple small discrete spherical hyperintense lesions, some with a small hypointense focus—dot—in the center) (Sarris et al., 2003). The small central hypointense foci represent the fungal grains, whereas the surrounding high signal intensity foci are due to granulomatous inflammation.

CT is the method of choice for pretherapy assessment of pedal mycetoma because of its greater sensitivity in detecting abnormalities and early bone involvement (Sharif et al., 1991). CT of pedal mycetoma is most useful in aiding decisions about the need for and extent of surgery (complete versus partial amputation).

Bone involvement (Fig. 20.2). Bone involvement in mycetoma has been radiographically classified:

- *Stage 0* Soft-tissue swelling without bone involvement
- *Stage I* Extrinsic pressure effects on the intact bones in the vicinity of an expanding granuloma
- *Stage II* Irritation of the bone surface without intraosseous invasion
- *Stage III* Cortical erosion and central cavitation
- *Stage IV* Longitudinal spreading along a single ray
- *Stage V* Horizontal spread along a single row
- *Stage VI* Multidirectional spread due to uncontrolled infection

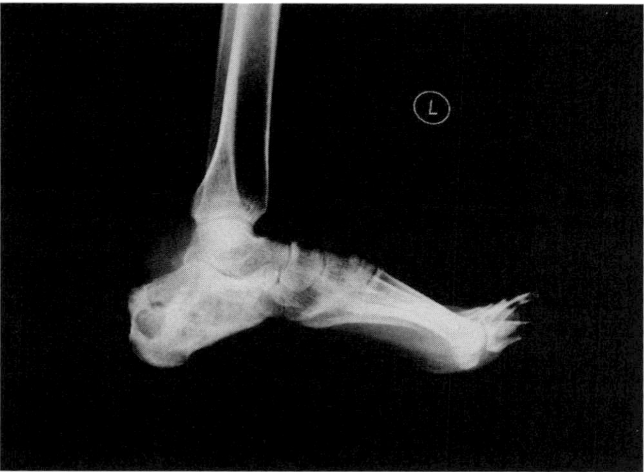

Figure 20.2 X-radiograph, foot, osseous lesions of eumycetoma. Single view of the foot shows several lytic lesions in the calcaneous with no involvement of the talo-calcanear joint. There is also a large soft tissue density in the Achilles tendon near its insertion into the calcaneous. This appearance suggests a chronic infectious process in the calcaneous with extension into the adjacent Achilles tendon. *Source:* Used with permission from Dr. Ahmed Hassan Fahal, University of Khartoum, and Director, Mycetoma Research Centre, Khartoum, Sudan.

Late in the disease, there may be multiple punched out cavities through the bone. The cavities in eumycetoma are larger but fewer in number, with well-defined margins, in contrast to bone cavities in actinomycetoma, which usually are smaller and numerous but without definite margins.

Other Body Sites Involved

Sites other than the foot include the hand, upper or lower arm, and areas of the body exposed by carrying bundles of wood or other plant material—the upper back, neck, and top of the head.

20.12 VETERINARY FORMS

Eumycetoma is seen in dogs, horses, occasionally in cats and cattle, as well as more exotic animals, for example, a squirrel monkey and Grand Eclectus parrot (Clark et al., 1986).

20.13 THERAPY

Eumycetoma is difficult to treat, even when the causal fungus is susceptible in vitro to antifungal agents. Success of chemotherapy depends on the stage of the infection and the causative agent. In advanced lesions the response is poor, especially when bone is involved (Ahmed et al., 2007).

Surgery

Surgery usually is performed when lesions are small, with oral antifungal therapy started well before surgery and continued for a prolonged period afterwards. Ketoconazole or ITC have been used most often. Surgery without antifungal therapy rarely is successful and may facilitate lymphatic spread. Aggressive surgical excision, debulking surgery, or amputation have been used, the latter in advanced disease.

Itraconazole

ITC has achieved good clinical responses over a period of several to many months with low recurrence rates (Lichon and Khachemoune, 2006). Cases refractory to ITC have responded to posaconazole (PSC). *Exophiala jeanselmei* eumycetoma initially responded to ITC but fell short of complete clinical remission even after surgery (reviewed by Ahmed et al., 2007). *Scedosporium apiospermum* eumycetoma was treated successfully with ketoconazole and ITC. More promising than ITC are the extended spectrum azoles and, as an alternative, terbinafine (TRB). Poor responses sometimes are seen even when antifungal agents are given in combination at high doses. *Madurella mycetomatis* has low MICs to a range of polyenes, azoles, and echinocandins.

Voriconazole (VRC)

This extended spectrum azole has proved efficacious in treating eumycetoma, although there are too few cases so treated to generalize (Loulergue et al., 2006). A man with a 6 year history of foot swelling was diagnosed with black grain eumycetoma due to *Madurella mycetomatis*. He was treated with oral VRC at 200 mg, 2×/day. After 3 months the wound closed and swelling dramatically decreased. After a year of therapy, a minor swelling persisted and an MRI showed a major improvement in the subcutaneous lesions.

Posaconazole (PSC)

PSC has had success in treating four of six eumycetoma patients who were refractory to standard therapy with either ketoconazole or ITC (Table 20.2) (Negroni et al., 2005).

Terbinafine (TRB)

Twenty-three eumycetoma patients received 500 mg of TRB/2× day for 24–48 weeks (N'diaye et al., 2006) (Table 20.3). In the 20 patients who completed the study, tumefaction was absent or improved in 80%; sinuses were closed in 50%; and grain emissions were absent in 65% of patients. Of the 16 patients who had repeat mycologic assessment, 4 (25%) were mycologically cured. Ten patients (62.5%) were infected with *M. mycetomatis* at baseline, and 7 (43.7%) were culture-positive at study end. Of 3 patients infected with *L. senegalensis* at baseline, 2 remained culture-positive. Overall, 5 patients (25%) were cured and 11 (55%) were clinically improved. The most common abnormalities were neutropenia, leukopenia, and abnormal liver function tests.

Criteria for Cure

At the Mycetoma Research Centre at the University of Khartoum, Sudan's criteria include disappearance of subcutaneous mass and healing of sinuses and skin; restoration of normal radiologic appearance of bone; absence of hyperreflective echoes and cavities on ultrasonic examination; and absence of grains using FNA. The URL for the Mycetoma Research Centre is http://www.mycetoma.edu.sd/home/staff.htm

The Centre, because of its location in a highly endemic region, also uses serology: monitoring the

Table 20.2 Posaconazole (PSC) Treatment for Mycetoma Refractory to Standard Therapy

Patient number	Months of therapy[a]	PSC (days of therapy)	Six month assessment	Assessment, end of therapy	Outcome
1	KTC (6) ITC (6)	1015	Knee, foot (improved)	Culture negative; foot nodules closed	Success
2 (first round)	ITC (8) KTC (3)	367	Ankle nodules; foot nodules (recurrence)	Negative culture; tumor, nodules resolved	Success
2		398	Improved lesions, mobility	Negative cultures; resolved bone, tumor nodules	Success
3 (second round)	ITC (13)	239	Ankle nodule excised	Negative cultures; resolved bone, tumor, nodules	Success
4	ITC (27)	388	Improved	Cultures grew *P. boydii, S. apiospermum*; symptoms ongoing	Improved
4		445		Skin lesions resolved, improved mobility, osetomyelitis persists	Improved
5	KTC (4) 168	168	Improved symptoms, but eumycetoma persists		Nonresponsive
6	KTC (2) 252	252	No change	Negative cultures; nodule biopsy negative, radiology normal	Success

[a] KTC, ketoconazole; ITC, Itraconazole.

Source: Negroni et al. (2005). Used with permission from the Instituto de Medicina Tropical de São Paulo.

Table 20.3 Efficacy of Terbinafine Therapy for Eumycetoma: Clinical Signs and Symptoms, Mycology, and Overall Clinical Assessment

Clinical signs and symptoms	Baseline, number of patients (%) ($n = 23$)	End of study, number of patients (%) ($n = 20$)[a]
Sinuses		
Open	22 (95.7)	10 (50.0)
Closed	1 (4.3)	10 (50.0)
Tumefaction		
Present	23 (100)	
Absent		6 (30.0)
Improved		10 (50.0)
Unchanged		3 (15.0)
Exacerbated		1 (5.0)
Grains		
Present	22 (95.7)	7 (35.0)
Absent	1 (4.3)	13 (65.0)
Mycology ($n = 16$)[b]		
Madurella mycetomatis	10 (62.5)	7 (43.7)
Leptosphaeria senegalensis	3 (18.7)	2 (12.5)
Unknown fungal species	3 (18.7)	3 (18.7)
Clinical assessment ($n = 20$)		
Cure		5 (25.0)
Improved		11 (55.0)
Unchanged		2 (10.0)
Deterioration		2 (10.0)

[a]End of study = last postbaseline assessment.

[b]Patients with repeat mycological assessment.

Source: N'Diaye et al. (2006). Used with permission from Wiley-Blackwell Publishers.

decline in specific precipitating antibodies as a favorable prognostic sign.

20.14 LABORATORY DETECTION, RECOVERY, AND IDENTIFICATION

Direct Examination

Pus, exudates, or biopsy material with grains are the primary indicators of eumycetoma. The size, shape, color, and consistency of the grains aid in identification, and experienced technologists may make a presumptive identification of the causative fungus (Table 20.4). A common practice is to cover draining sinus openings with sterile gauze pads for 8–12 h to collect the grains. Melanin may be deposited within the grains and extracellularly. Determining the etiologic agent is based on direct examination of the grains, culture of the causative fungus, and genetic identification when sporulation cannot be demonstrated in a reasonable time period.

Eumycetoma agents in grains are composed of packed masses of broad, septate, branching hyphae (2–5 μm diameter), or, if *Exophiala jeanselmei*, as a compact mass of rounded cells. Grains are white or yellow when the agents are hyaline molds, or black when due to melanized fungi. Grains are crushed and a KOH prep is made and examined in the microscope. The structures may be enhanced by staining with lactophenol cotton blue or lactofuchsin.

Histopathology

The histologic picture is one of a granulomatous mass with a purulent center encircled by a thick fibrous capsule. Grains are not always visible in every microscopic section. Infected tissue should be sectioned and stained with H&E, GMS, PAS, or Brown and Brenn stain (tissue Gram stain). The stains are not necessary to visualize the grains, but are useful in demonstrating the large hyphae and chlamydospores within, and to differentiate the grains of fungi from those of actinomycetes.

Differentiation of Actinomycotic Mycetoma from Eumycetoma

This is essential in guiding clinical treatment decisions. Actinomycotic mycetoma responds to antibiotics and may not require surgery, whereas in eumycetoma a combination of antifungal therapy and surgery may be required. Differentiation is based on the appearance of the grains which drain with serosanguinous pus from sinus tracts (Chandler et al., 1980). Actinomycotic mycetoma grains consist of filaments <1 μm in diameter. Eumycetoma grains consist of septate hyphae 2–5 μm in diameter. Often in eumycetoma the hyphae in grains are in distorted shapes. Chlamydospores are frequently present in the periphery of the grain.

The hallmark of mycetoma is the presence of spherical, subspherical, or lobular grains in a tumorous mass of chronically inflamed tissue. The grains may be embedded in a cement-like substance and may be pigmented (Fig. 20.3).

An amorphous, eosinophilic smoothly contoured layer of Splendore–Hoeppli material may surround the grain. Sometimes a foreign body (i.e., thorn or splinter) is seen near the grain, providing a clue to the penetrating injury. The tissue response in eumycetoma is similar regardless of the causative fungus: localized abscesses with grains in the center surrounded by a dense accumulation of PMNs mixed with necrotic debris. The abscess is surrounded by a chronic inflammatory

Table 20.4 Diagnostic Characteristics of Mycetoma Grains Resulting from Selected Species of Actinomycetes and Fungi[a]

Classification	Species	Color	Size range	Cement	Texture	Tinctorial properties
Actinomycotic mycetoma grains	*Actinomadura pelletieri*	Red	0.3–0.5 mm	–	Soft to hard	Non-acid-fast, periphery stained by hematoxylin
	Nocardia brasiliensis	White	<1 mm	–	Soft	Acid-fast, variable H&E staining
	Streptomyces somaliensis	Yellow	0.5–2 mm	+	Hard	Non-acid-fast, very light staining with eosin
Eumycetoma grains	*Acremonium falciforme, A. kiliense*	White	0.5–1 mm	–	Soft	Eosinophilic border darker than interior; network of hyphae and chlamydospores
	Exophiala jeanselmei	Black	0.5–2 mm		Brittle	Compact mass of round thick walled cells; few short melanized hyphae (Severo et al., 1999)
	Leptosphaeria senegalensis and *L. tompkinsii*	Black	0.5–2 mm	+ in periphery	Hard	Lobulated, dark periphery with loose network of hyphae and large chlamydospores in cemented periphery
	Madurella grisea	Black	0.3–0.6 mm	Variable	Hard	Dark peripheral zone, variably shaped; dense network of hyphae and chlamydospores
	Madurella mycetomatis	Black	0.5-4 mm	+	Hard	2 types of grains: 1) Compact—may be lobulated; brown-stained cement throughout grain 2) Vesicular—irregular size and shape; brown stained cement only in periphery
	Pseudallescheria boydii and *Scedosporium* species	White	0.5-1 mm	-	Soft	Eosinophilic border, rest of grain lightly colored consisting of a dense network of hyphae with chlamydospores.

[a]Abridged from Chandler et al., 1980

response consisting of palisaded epithelioid cells and multinucleate giant cells, with smaller numbers of plasma cells and lymphocytes. Between abscesses there is dense granulation tissue, which may be highly vascular, and infiltrated by lymphocytes and macrophages. Fibrosis may be extensive, especially in longstanding lesions, causing the tumefaction characteristic of eumycetoma.

Staining Properties of Mycetoma Grains

Actinomycotic grains are positive in Brown and Brenn Gram and Giemsa stains. Actinomycetes are not reliably stained by PAS and H&E stains. When a pure population of nonfilamentous bacteria (bacilli or cocci) are seen in the grain, a diagnosis of botryomycosis should be considered. Cultures are needed for definitive identification. The stains recommended for fungi within grains are GMS and PAS. Unstained thick sections (15 μm) may be mounted in the usual mounting medium, coverslipped, and examined microscopically with reflected light. The natural color of the grains can be seen by this method, offering a clue as to the causative agent (Winslow, 1971).

Culture

Grains are either white, yellow, pink-red, brown, or black. Grains well washed in sterile saline are examined in a KOH prep and the color, size, shape, and texture are noted (Table 20.4). For examination, press grains between

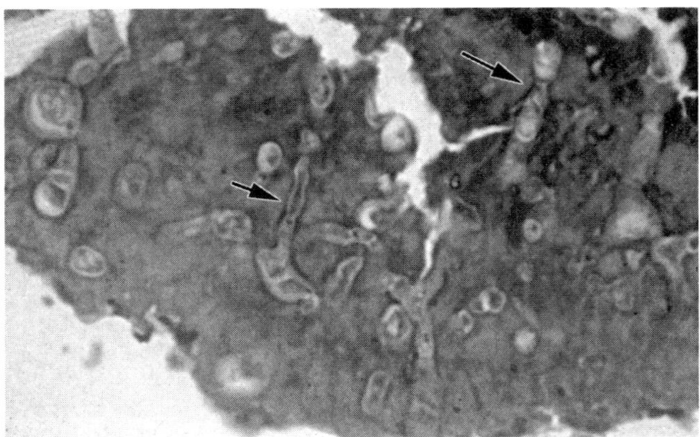

Figure 20.3 Black grain eumycetoma, histopathology. *Madurella mycetomatis* hyphae (*arrows*) embedded in a cemented brown mass. *Source:* Used with permission from Dr. Uma M. Tendolkar, LTM Medical College, Mumbai.

two microscope slides. The texture varies from soft to hard and brittle. Prepare eumycetoma grains for culture by washing them several times in sterile water or saline containing antibacterial antibiotics. Next, they are crushed and cultured in duplicate. Recommended media are SDA and BHI with chloroamphenicol, or penicillin and streptomycin. Cycloheximide is not used because it may inhibit some causative fungi. Incubation temperature should be 25–30°C, with cultures held for 6–8 weeks, as some of the fungi are slow growing.

Pseudallescheria boydii

Pseudallescheria boydii (teleomorph) and anamorphs are in the genus *Scedosporium*. Their colony and microscopic morphology are discussed in Chapter 16.

Madurella mycetomatis

Madurella mycetomatis is slow growing and often produces only sterile hyphae. *M. mycetomatis* is a challenge for the mycology laboratory.

Colony Morphology The surface is white, yellow-brown, olive brown to dark gray with a dark brown reverse. Sometimes a brown diffusible pigment is produced.

Microscopic Morphology Hyphae are septate, sometimes toruloid.[1] Typical isolates are sterile but may sporulate on diluted or weak media, producing phialides with round conidia, or simple or branched conidiophores with pyriform conidia.

[1]Toruloid. *Definition:* Knotted or beaded, like a yeast cell.

Madurella grisea

Differs from *M. mycetomatis* in that it grows poorly or not at all at 37°C.

Molecular Identification of Eumycetoma Agents

Sequence based identification of black granule eumycetoma agents (Desnos-Ollivier et al., 2006) facilitates identification because many black grain producing fungi do not sporulate or require prolonged incubation to do so. Species-specific sequences were obtained by PCR amplification of the ITS region of rDNA, allowing differentiation of *M. mycetomatis* and other eumycetoma agents: *Curvularia lunata, Leptosphaeria senegalensis, Madurella grisea*, and *Pyrenochaeta romeroi. M. mycetomatis* appears to be a homogeneous species. Post-PCR digestion with restriction endonucleases (PCR-RFLP) can differentiate *M. mycetomatis* from *M. grisea*.

SELECTED REFERENCES FOR EUMYCETOMA

AHMED AA, VAN DE SANDE WW, FAHAL A, BAKKER-WOUDENBERG I, VERBRUGH H, VAN BELKUM A, 2007. Management of mycetoma: Major challenge in tropical mycoses with limited international recognition. *Curr Opin Infect Dis* 20: 146–151.

AHMED AO, DESPLACES N, LEONARD P, GOLDSTEIN F, DE HOOG S, VERBRUGH H, VAN BELKUM A, 2003a. Molecular detection and identification of agents of eumycetoma: Detailed report of two cases. *J Clin Microbiol* 41: 5813–5816.

AHMED AO, VAN VIANEN W, TEN KATE MT, VAN DE SANDE WWJ, VAN BELKUM A, FAHAL AH, VERBRUGH HA, BAKKER-WOUDENBERG IAJM, 2003b. A murine model of *Madurella mycetomatis* eumycetoma. *FEMS Immunol Med Microbiol* 37: 29–36.

AHMED A, ADELMANN D, FAHAL A, VERBRUGH H, VAN BELKUM A, DE HOOG S, 2002. Environmental occurrence of *Madurella mycetomatis*, the major agent of human eumycetoma in Sudan. *J Clin Microbiol* 40: 1031–1036.

AL-HEIDOUS M, MUNK PL, 2007. Radiology for the surgeon. Musculoskeletal case 40. *Can J Surg* 50: 467–469.

BOMMER U-A, THIELE B-J, 2004. The translationally controlled tumour protein (TCTP). *Int J Biochem Cell Biol* 36: 379–385.

CASTRO LG, PIQUERO-CASALS J, 2008. Clinical and mycologic findings and therapeutic outcome of 27 mycetoma patients from São Paulo, Brazil. *Int J Dermatol* 47: 160–163.

CHANDLER FW, KAPLAN W, AJELLO L, 1980. *Color Atlas and Text of the Histopathology of Mycotic Diseases*. Yearbook Medical Publishers, Chicago.

CLARK FD, JONES LP, PANIGRAHY B, 1986. Mycetoma in a grand *Eclectus* (*Eclectus roratus roratus*) parrot. *Avian Dis* 30: 441–443.

DE HOOG GS, ADELMANN D, AHMED AO, VAN BELKUM A, 2004. Phylogeny and typification of *Madurella mycetomatis*, with a comparison of other agents of eumycetoma. *Mycoses* 47: 121–130.

DESNOS-OLLIVIER M, BRETAGNE S, DROMER F, LORTHOLARY O, DANNAOUI E, 2006. Molecular identification of black-grain mycetoma agents. *J Clin Microbiol* 44: 3517–3523.

FAHAL AH, 2004. Mycetoma—A thorn in the flesh. *Trans R Soc Trop Med Hyg* 98: 3–11.

LICHON V, KHACHEMOUNE A, 2006. Mycetoma—A review. *Am J Clin Dermatol* 7: 315–332.

LOULERGUE P, HOT A, DANNAOUI E, DALLOT A, POIRÉE S, DUPONT B, LORTHOLARY O, 2006. Successful treatment of black-grain mycetoma with voriconazole. *Am J Trop Med Hyg* 75: 1106–1107.

N'DIAYE B, DIENG MT, PEREZ A, STOCKMEYER M, BAKSHI R, 2006. Clinical efficacy and safety of oral terbinafine in fungal mycetoma. *Int J Dermatol* 45: 154–157.

NEGRONI R, HELOU SH, PETRI N, ROBLES AM, ARECHAVALA A, BIANCHI MH, 2004. Case study: Posaconazole treatment of disseminated phaeohyphomycosis due to *Exophiala spinifera*. *Clin Infect Dis* 38: e15–e20.

NEGRONI R, TOBON A, BUSTAMANTE B, SHIKANAI-YASUDA MA, PATINO H, RESTREPO A, 2005. Posaconazole treatment of refractory eumycetoma and chromoblastomycosis. *Rev Inst Med Trop São Paulo* 47: 339–346.

SARRIS I, BERENDT AR, ATHANSOUS N, OSTLERE SJ, 2003. MRI of mycetoma of the foot: Two cases demonstrating the dot-in-circle sign. *Skeletal Radiol* 32: 179–183.

SEVERO LC, OLIVEIRA FM, VETTORATO G, LONDERO AT, 1999. Mycetoma caused by *Exophiala jeanselmei*. Report of a case successfully treated with itraconazole and review of the literature. *Rev Iberoam Micol* 16: 57–59.

SHARIF HS, CLARK DC, AABED MY, AIDEYAN OA, MATTSSON TA, HADDAD MC, OHMAN SO, JOSHI RK, HASAN HA, HALEEM A, 1991. Mycetoma: Comparison of MR imaging with CT. *Radiology* 178: 865–870.

VAN DE SANDE WWJ, JANSE D-J, HIRA V, GOEDHART H, VAN DER ZEE R, AHMED AOA, OTT A, VERBRUGH H, VAN BELKUM A, 2006. Translationally controlled tumor protein from *Madurella mycetomatis*, a marker for tumorous mycetoma progression. *J Immunol* 177: 1997–2005.

WELSH O, VERA-CABRERA L, SALINAS-CARMONA MC, 2007. Mycetoma. *Clin Dermatol* 25: 195–202.

WINSLOW DJ, 1971. Mycetoma, pp. 589–613, *in*: BAKER RD (ed.), *Human Infection with Fungi, Actinomycetes and Algae*. Springer Verlag, New York.

WEBSITES CITED

Tropical Medicine Central Resource (TMCR Maurice M. Reeder, M.D., FACR, Uniformed Services University of the Health Sciences. Bethesda, MD. http://tmcr.usuhs.edu/tmcr/chapter6/clinical7.htm#fungi3.

The Mycetoma Research Centre in Khartoum, Sudan, is at the URL http://www.mycetoma.edu.sd/home/staff.htm.

QUESTIONS

The Answer Key to multiple choice questions may be found at the end of the book.

1. The signs of mycetoma are
   A. Chain of subcutaneous nodules that trace lymphatic drainage from a lesion on an extremity
   B. Circular lesions with active borders, scaling, and pruritis
   C. Subcutaneous infection of the foot or leg with swelling, draining sinus tracts, and discharging grains
   D. Verrucous, crusted, ulcerated, and dense dermal fibrotic lesions usually on the arms, legs, face, or trunk
2. What type of specimen would you expect to receive at the mycology bench for suspected eumycetoma?
   A. Biopsy tissue
   B. Blood
   C. Grain
   D. Scraping
3. Which of the following characteristics of a grain are used to aid in identifying the causative agent of eumycetoma?
   A. Color
   B. Texture
   C. Size
   D. Microscopic morphology
   E. All of the above
4. A white soft grain 0.5–1.0 mm in diameter, consisting of a dense mass of lightly colored hyphae and chlamydospores is descriptive of
   A. *Exophiala jeanselmei*
   B. *Leptosphaeria senegalensis*
   C. *Madurella mycetomatis*
   D. *Pseudallescheria/Scedosporium* spp.
5. It is found in mycetomata in arid zones of Africa, occasionally in the Middle East, and definitely in southern India. The nearest neighbor is *Chaetomium* in the order *Sordariales*. This mycetoma agent produces black grains in tissue, secretes a brown pigment into the medium, and remains sterile (nonsporulating). It is
   A. *Exophiala jeanselmei*
   B. *Leptosphaeria senegalensis*
   C. *Madurella mycetomatis*
   D. *Pseudallescheria/Scedosporium* spp.
6. Explain how grains are collected and examined in the laboratory. What medium is used to culture eumycetoma agents? When growth occurs but without sporulation, what method is used to identify a eumycetoma agent?
7. With reference to mycetoma grains, how can eumycetoma be differentiated from actinomycotic mycetoma?
8. Discuss the clinical response of eumycetoma patients to extended spectrum azoles and to terbinafine.
9. What is the "dot-in-circle" sign? What is meant by the "melting snow" appearance of bone lesions? Explain how diagnostic imaging is used for clinical assessment and evaluating the response to therapy in eumycetoma.
10. What is the role for surgery in the early phase and late in the development of eumycetomata?

Part Six

Dermatophytosis and Dermatomycoses (Superficial Cutaneous Mycoses)

Fundamental Medical Mycology, First Edition. By Errol Reiss, H. Jean Shadomy and G. Marshall Lyon, III.
© 2012 Wiley-Blackwell. Published 2012 by John Wiley & Sons, Inc.

Chapter 21

Dermatophytosis

21.1 DERMATOPHYTOSIS-AT-A-GLANCE

- *Introduction/Disease Definition.* Dermatophytosis (syn: ringworm, tinea) is a communicable skin disease affecting the outer layer of the epidermis, the stratum corneum,[1] and also may invade the hair and nails, caused by a group of *hyaline* molds. These fungi derive nutrients by breaking down keratin, a protein that is the major component of skin, hair, and nails. These molds, in immune-normal persons, cannot invade living tissue or, in hair, beyond the keratogenous zone. The classic appearance of the skin lesion in dermatophytosis is ringworm: a circular lesion with an active border, inflammation, scaling, and pruritus.
- *Etiologic Agents.* There are two major genera of causative agents: *Trichophyton* and *Microsporum*. Within those genera are five major species: *T. rubrum, T. tonsurans, T. interdigitale, Microsporum canis, and T. violaceum*. A third genus with a single species, *Epidermophyton floccosum*, has declined in importance in the United States but is still found in dermatophytosis in other parts of the world.
- *Geographic/Distribution Ecologic Niche.* Dermatophytes are worldwide with regional differences in species prevalence. The causative species are anthropophilic (humans only), zoophilic (animals and humans), or geophilic (soil dwelling).
- *Epidemiology.* Secular trends show scalp ringworm (tinea capitis) is in decline and is now caused by *T. tonsurans* (United States, United Kingdom, Northern Europe) and *M. canis* (southern and eastern Europe) while *T. violaceum* is the primary cause in Asia. Athlete's foot (tinea pedis) is the major dermatophytosis and is caused by *T. rubrum* and *T. interdigitale*.
- *Risk Groups/Factors.* Dermatophytosis affects all ages, but children are at higher risk for tinea capitis and tinea corporis (ringworm of the smooth skin). Other risk groups include diabetics, those receiving systemic corticosteroid therapy, and persons living with HIV or AIDS. Military troops in the tropics are at risk for tinea cruris (jock itch) and tinea pedis.
- *Transmission.* Person–person transmission is caused by sharing caps and combs, by walking barefoot in public shower rooms, and through exposure to kittens and puppies, which may harbor *M. canis* and cause dermatophytosis, particularly in children. Geophilic dermatophytes such as *M. gypseum* are acquired from soil, causing tinea corporis and tinea capitis.
- *Determinants of Pathogenicity.* Keratinase and lipases cause injury to keratinocytes and inflammation, with IgE production causing pruritus.
- *Clinical Forms.* Depending on the body site affected, the following forms are seen: tinea capitis (scalp ringworm), tinea corporis (ringworm of the smooth skin), tinea cruris (jock itch), tinea pedis (athlete's foot), tinea unguium (nail infections), and other minor clinical forms.
- *Therapy.* The therapy for tinea pedis and tinea cruris is a topical cream containing miconazole, other azoles, or terbinafine (TRB) (Lamisil®, Novartis); for tinea capitis, systemic (oral) therapy with itraconazole (ITC), griseofulvin, or TRB is used; for tinea unguium a regimen of oral TRB or oral ITC with or without topical antifungal nail lacquer,

[1]Stratum corneum. *Definition*: This is the skin's barrier layer, the outermost layer of epidermis, about the thickness of a human hair, and is made up of 25–30 layers of flat, dead cells completely filled with keratin, a waterproof protein.

Fundamental Medical Mycology, First Edition. By Errol Reiss, H. Jean Shadomy and G. Marshall Lyon, III.
© 2012 Wiley-Blackwell. Published 2012 by John Wiley & Sons, Inc.

lasting for weeks to months, is required to clear the infection.

- *Laboratory Detection, Recovery, and Identification.* Direct exam (KOH prep) is performed with or without Calcofluor for skin scrapings, nail clippings, and hair. *Microsporum* scalp and skin lesions fluoresce in *Wood's light*, a useful tool in the dermatologist's office. Cultures are used to determine the cause of an outbreak and to rule out nondermatophyte fungi requiring a different therapy.

21.2 INTRODUCTION/DISEASE DEFINITION

1. *What are dermatophytes?* The dermatophytes are a group of *keratinophilic* molds specialized to grow on the *nonliving keratinized cells* of the outer epidermis (*stratum corneum*), hair, and nails of humans and lower animals. There are five major species: *Trichophyton rubrum, T. interdigitale, T. tonsurans, Microsporum canis*, and *T. violaceum*. A less common genus with a single species, *Epidermophyton floccosum*, is seen most often in Europe. Dermatophytes, in immune-normal persons, do not grow beyond the nonliving stratum corneum, and do not penetrate into the zone of keratin synthesis in the hair follicle. Based on their ecologic niche, dermatophyte species are anthropophilic (humans only), zoophilic (animals and humans), or geophilic (soil dwelling). They are globally distributed with geographic clustering among the five major species (please see Section 21.6, Geographic Distribution).
2. *What is dermatophytosis?* The classic form of this mycosis is ringworm, a circular lesion with an active border, inflammation, pruritus, and scaling caused by mycelial invasion of the stratum corneum of the smooth skin, a condition called tinea corporis. "Tinea" applied to dermatophytosis was coined by ancient Romans because the moth-eaten appearance of scalp ringworm lesions reminded them of the damage to woolens by clothes moths, the *Tineidae*. Dermatophytosis also includes infection of the nails and the hair sheath. The major clinical forms seen worldwide (in addition to tinea corporis) are tinea pedis (foot ringworm), tinea unguium (a form of *onychomycosis* or nail disease), tinea cruris ("jock itch"), and tinea capitis (scalp ringworm). The term "athlete's foot" sometimes considered to be a synonym for dermatophytosis is a more general term that also includes foot infections caused by bacteria and yeast. Dermatophytosis is a communicable disease, spread from person-to-person, or from lower animals to humans. Only very rarely, in severely immunocompromised individuals, do dermatophytes invade living tissue, causing systemic disease.
3. *What are dermatomycoses?* Dermatomycoses are diseases of the skin caused by fungi *other than dermatophytes*. These fungi may produce signs and symptoms mimicking dermatophytosis and should be identified and reported to the physician (please see Chapter 22). Fungi that are not dermatophytes, but resemble them in culture, also may be isolated from skin (please see Chapter 2, Section 2.3.8.11, Saprobic Fungi and Their Pathogenic Look alikes in culture). Fungi causing the dermatomycoses are treated with different antifungal drugs than dermatophytes. In addition to dermatophytoses and dermatomycoses there are cutaneous manifestations of deep-seated mycoses; for example, please see Chapter 4.

Useful Monographs, Texts, and a Web-Based Resource

- Aly R, Maibach HI, 1999. *Atlas of Infections of the Skin*. Churchill-Livingstone, New York. This book provides a practical approach to the diagnosis and management of infectious skin diseases.
- Rebell G, Taplin D, 1970. *Dermatophytes, Their Recognition and Identification*. University of Miami Press, Coral Gables, FL. A classic laboratory manual.
- Kane J, et al., 1997. "Laboratory Handbook of Dermatophytes" is an advanced treatise with numerous illustrations and additional tests to discriminate common and rare dermatophytes.
- Hay RJ, 2005. Dermatophytosis and Other Superficial Mycoses. An authoritative chapter in *Mandell, Bennett, & Dolin: Principles and Practice of Infectious Diseases*.
- Larone DH, 2002. *Medically Important Fungi: A Guide to Identification*, and St. Germain and Summerbell, 1996. *Identifying Filamentous Fungi*, are standard texts for laboratory identification.
- Shadomy HJ Philpot CM, 1980. This is a practical guide to laboratory methods for the dermatophytes.
- "Cutaneous Fungal Infections" is a 50 page illustrated web-based guide to the dermatophytoses and dermatophyte colony and microscopic morphology available at the URL http://www.docstoc.com/docs/526979/Cutaneous-Fungal-Infections.

History of Discovery of Dermatophytosis and the Causative Agents

There is a rich history tracing discoveries in this field. The French school of dermatologists made the most profound discoveries beginning in the 1840s with the pioneering research of the naturalized French citizen, physician, and microscopist, David Gruby, followed by a succession of dermatologists at l'Hôpital Saint-Louis in Paris, culminating with the pioneering work of Raimond Sabouraud, a disciple of the Pasteur Institute (Potter, 2003) who published the treatise *Les Teignes* in 1910 (Sabouraud, 1910) which organized the clinical features, microscopic morphology, and taxonomic classification of the dermatophytoses and their causative agents.

21.3 CASE PRESENTATIONS

Case Presentation 1. Epidemic Ringworm in an Elementary School (Morbidity and Mortality Weekly Report, 1973)

History

Between August 26th and October 10th, 25 cases of dermatophytosis were identified in 109 students attending elementary school in Greenwood, Nebraska. There was an even distribution of children with lesions from kindergarten through grade 6. Lesions were found on the trunk, neck, face, and upper limbs. In 3 of 25 children scalp lesions developed later in the outbreak.

Epidemiologic investigation identified the first 3 cases in children of the same family who formerly had 16 cats; 5 of the 16 of which had clinical ringworm and *alopecia* in the head and neck region. The cats were euthanized before the investigation on the advice of a physician and veterinarian.

Another family whose 3 children had skin lesions had acquired a kitten in late August. This kitten also was euthanized before the investigation. Other children who were culture-positive had two cats and a dog at home. One cat had typical ringworm lesions that developed after the children became infected, suggestive of person-to-cat transmission (Fig. 21.1). All children were given school-based treatment consisting of topical Tinactin® (tolnaftate) that effectively halted both infection and transmission (please see Section 21.13, Therapy).

Comments

Animals with ringworm can be effectively treated with oral itraconazole (ITC). Despite the species name, *M. canis* is more often associated with cats. Preferred therapy for *M. canis* tinea capitis in children is oral griseofulvin or ITC. Both 2% ketoconazole (Nizoral®) and 1% selenium sulfide (Selsun Blue®) shampoos reduce surface colony counts of dermatophytes in infected individuals, and these agents are recommended as adjuvant therapy.

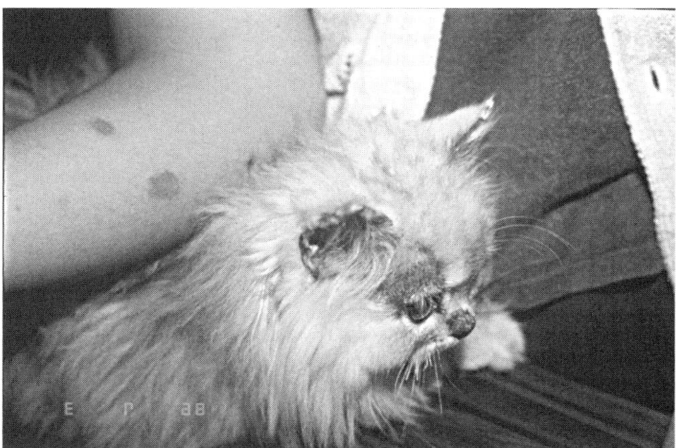

Figure 21.1 Child with tinea corporis and cat.
Source: Dr. Kosuke Takatori, National Institute of Health Sciences, Tokyo.

Microsporum canis tinea capitis and tinea corporis require a longer duration of therapy with either griseofulvin or ITC than other causes of the same dermatophytoses.

Case Presentation 2. Onychomycosis (Tinea unguium): Toenail Infection Following an Injury

History

A 35-yr-old man, wearing sandals on a hot summer day, stubbed his right great toe on a metal coffee table leg. The toe was painful, and the nail was slightly torn. The nail continued growing and the torn portion was epilated. Later, after wearing woolen socks and closed shoes the following winter, a slight discoloration appeared in the same nail and worsened becoming further discolored and thickened. The debris under the distal end of the nail was removed and the undersurface at the junction of the nail and the nail bed was scraped. When viewed microscopically in a KOH prep fungal elements were seen. A culture of nail scrapings grew *Trichophyton rubrum*. The patient received intermittent griseofulvin therapy over the following years, but his nail disease was never completely cured.

Diagnosis

The diagnosis was distal subungual onychomycosis caused by *Trichophyton rubrum*.

Comments

Please see KOH prep in Section 21.14, Laboratory Detection, Recovery, and Identification, and tinea unguium in Section 21.11, Clinical Forms. Tinea unguium can develop even in the absence of a traumatic injury. Untreated tinea pedis can lead to nail disease. *Trichophyton rubrum* is a major cause of distal subungual onychomycosis. Subungual nail scrapings are preferred for microscopic examination and culture. If nail clippings are used they may be subjected to grinding before planting on culture medium or for KOH preps. Therapy of tinea unguium requires oral antifungal agents. Present therapy, unavailable at the time of the "coffee table accident," is more effective and consists of oral ITC or terbinafine (TRB) with or without topical antifungal nail lacquer. Tinea unguium remains difficult to treat, requiring compliance over a period of months. Additionally, there may be genetic risk factors for dermatophytosis.

21.4 DIAGNOSIS

Diagnosis varies with each dermatophyte and syndrome. Scalp ringworm may be diagnosed by fluorescence of infected hair when the scalp is illuminated with Wood's light (a UV source at 365 nm), but only if caused by *Microsporum canis*, other *Microsporum* species, or *Trichophyton schoenleinii*. The major cause of scalp ringworm in the United States, *Trichophyton tonsurans*, does not fluoresce. Specimens consisting of nail clippings, plucked hairs, and skin or scalp scrapings are suitable for direct examination in KOH preps, with or without Calcofluor. Skin scrapings containing dermatophytes demonstrate similar hyphae and must be cultured for correct identification. Brushings or cotton swabs from scalp lesions are more appropriate in children and are suitable for both KOH preps and culture. Slow-growing dermatophytes are usually treated on the basis of positive KOH preps, but culture is still important because:

- Symptoms vary and may be due to either dermatophytosis or another etiology, so that diagnosis cannot be made on clinical appearance alone.
- Periodic culture during therapy is useful in determining mycologic cure.
- Disease caused by nondermatophytes should be identified and reported. (Please see Chapter 22.)
- Identification will provide clues as to the source of infection in individuals and in outbreaks.
- Surveillance studies require species identification.
- Cultures may be positive even in the absence of a negative KOH prep.

Please see Section 21.11, Clinical Forms for more specific diagnostic criteria and differential diagnosis.

21.5 ETIOLOGIC AGENTS

There are two major genera of dermatophytes: *Microsporum* and *Trichophyton*. A third genus, *Epidermophyton*, with the single species, *E. floccosum*, is infrequent in the United States but is seen more often in Europe. Some species are found in specific geographic areas, others are found worldwide. The species are introduced in this section, with important information and illustrations of their clinical presentation explained in Section 21.11, Clinical Forms. Their colony morphology, microscopic morphology, and related topics are explained in Section 21.14, Laboratory Detection, Recovery, and Identification.

General Characteristics of the Genus *Microsporum*

The most common dermatophytosis caused by *Microsporum* species is tinea capitis, although tinea corporis also may occur. Species are zoophilic (*M. canis*), geophilic (*M. gypseum*), or anthropophilic (*M. audouinii*). A key feature of the genus is the *thallic* production of macroconidia, which are large, spindle-shaped, multicellular, and thick walled usually with an *echinulate* surface. This method of conidiogenesis has been described as "a macroconidium develops from a blown-out conidiogenous cell of a hyphal branch from which it secedes with difficulty." Aleuriospores is a term used to describe this type of sporulation (aleuriospores are further described in Chapter 2, Section 2.3.8.8, Microscopic Identification of Molds Based on the Specific Conidia-forming Structures—Aleurioconidia).

Hair Invasion

The genus *Microsporum* displays *ectothrix* sporulation on hair: as the infected hair emerges from the scalp, hyphae grow inside the hair shaft, burst through, and cover the surface with small (2–3 μm) *arthroconidia* in a mosaic pattern. This ectothrix pattern exhibits fluorescence with Wood's light, due to the fungal metabolite, pteridine.

General Characteristics of the Genus *Trichophyton*

The spectrum of dermatophytosis caused by species in the genus *Trichophyton* is broad, including all types of tineas. Four of the five major dermatophyte species are *T. tonsurans, T. rubrum*, *T. interdigitale*, and *T. violaceum*. All are anthropophilic. Other species are zoophilic: *T. verrucosum* (cattle ringworm) and *T. mentagrophytes* (rodent ringworm). A key feature of sporulation in the genus *Trichophyton* is the *thallic* production of pencil-shaped or elongated balloon-shaped *macroconidia* with thin walls, and varying numbers of cells, depending on the species. Although macroconidia, when present, aid in identification, some species do not produce them. In that case, speciation is made by the appearance of *microconidia* and, occasionally, other structures. Fluorescence with Wood's light is not common.

Hair Invasion

When *T. tonsurans* invades hair, hyphae grow *and* sporulate within the hair shaft, filling the hair with arthroconidia, causing it to break off at the hairline leaving a "black dot." *T. violaceum* and *T. soudanense* also invade hair producing the endothrix pattern.

Other *Trichophyton* species when they invade hair (e.g., *T. verrucosum, T. mentagrophytes*) form large ectothrix arthroconidia in chains on the outside of the hair shaft.

Five Major Dermatophyte Species

Microsporum canis E. Bodin ex Guég., *Les Champignons parasites de l'homme et des animaux domestiques*: 137 (1902). Position in classification: *Arthrodermataceae, Onygenales, Ascomycota*.

Microsporum canis is a *zoophilic* dermatophyte causing "cat and dog ringworm," primarily affecting kittens and puppies, causing minimal inflammation. *M. canis* is often asymptomatic in adult animals. In humans, however, considerable inflammation is seen. As an agent of human disease, *M. canis* is worldwide in distribution, occurring in the United States as an uncommon cause of scalp ringworm (<5% prevalence). *M. canis* is the major dermatophyte isolated from human disease in central and Southern Europe with no change in prevalence over the period 1975–2005 (Borman et al., 2007). Children in contact with infected companion animals (please see Section 21.3, Case Presentation 1) can develop tinea capitis or corporis. Adults may develop tinea barbae.

Trichophyton rubrum (Castellani) Sabouraud, *Br J Dermatol* 23: 389 (1911). Position in classification: *Arthrodermataceae, Onygenales, Ascomycota*.

Trichophyton rubrum, worldwide in distribution, is an *anthropophilic* species, which is a major cause of tinea pedis, tinea corporis, and tinea unguium (please see Section 21.3, Case Presentation 2), also causing tinea cruris and tinea manuum. *T. rubrum* does not invade hair. *T. rubrum* causes the most common form of tinea pedis in the United States. *T. rubrum* generally produces a dry, hyperkeratotic, chronic disease; occasionally acute dermatophytosis may be seen. A number of variants of *T. rubrum* are described in Kane et al. (1997).

Trichophyton interdigitale Priestley, *Med J Australia* 4: 475 (1917). Position in classification: *Arthrodermataceae, Onygenales, Ascomycota*.

Molecular taxonomy has clarified that *T. interdigitale* is an anthropophilic species descended in lineage from *A. vanbreuseghem*, and is genetically distinct from the zoophilic species *T. mentagrophytes*, the cause of rodent ringworm. (The term "subspecies *T. mengagrophytes* var. *interdigitale*" persists, but is incorrect.) Worldwide in distribution, *T. interdigitale* primarily causes tinea pedis, less commonly tinea corporis, and, rarely, sporadic tinea capitis and tinea barbae. In 2005 *T. interdigitale* accounted for 21% of all isolates recorded by the Mycology Reference Laboratory, Bristol, United Kingdom (Borman et al., 2007). Cottony and velvety colony types are associated specifically with humans. Velvety cultures retain the microscopic morphology of zoophilic isolates (please see below, *T. mentagrophytes*) whereas cottony cultures produce microconidia *en thyrses* with few or no spiral hyphae. Mating studies and molecular analysis proved that most of the velvety and cottony anthropophilic "*T. mentagrophytes*-like" isolates are correctly identified as *Trichophyton interdigitale*, the principal anamorph name of the *Arthroderma vanbreuseghemii* complex.

Trichophyton mentagrophytes (C.P. Robin) Sabouraud, *Bouchard Traite de Pathologie generale* 2: 912 (1895). Position in classification: *Arthrodermataceae, Onygenales, Ascomycota*. This is the anamorph of *Arthroderma benhamiae*.

Trichophyton mentagrophytes is correctly applied *only* to zoophilic isolates. *T. mentagrophytes* has a granular colony type associated with rodents and small mammals. Epidemiologic data (Borman et al., 2007) tallying cases in the United Kingdom over a 25 year period concluded that the prevalence of tinea pedis caused by the zoophilic species *T. mentagrophytes* is so small as to have no significant impact on overall trends in contrast to *T. interdigitale. T. mentagrophytes* infections are acquired by humans from animal pets or in farm environments. They are inflammatory, acute, and self-limited. There may be lesions on the face, often with a butterfly or lupus distribution, local or widespread inflammatory lesions of the hand and body, ectothrix tinea capitis with an inflammatory kerion component, and multiple lesions of the dorsum of the foot, ankle, and leg under occlusive boots, and in the groin, under the belt; all traced to an animal source (Zaias and Rebell 2003).

Trichophyton tonsurans Malmsten, *Arch Anat Physiol Wiss Med* 1848:14 (1848). Position in classification: *Arthrodermataceae, Onygenales, Ascomycota*. Anamorph only.

Within the past 15 years this *anthropophilic* species has become the leading cause of tinea capitis in the United

States, United Kingdom, and Western Europe. *Trichophyton tonsurans* now accounts for over 90% of all tinea capitis seen in the United States. On the scalp it produces "black dot" tinea capitis in children and adults. (Please see Section 21.11, Clinical Forms.)

Trichophyton violaceum Sabouraud *ex* E. Bodin, *Les Champignons parasites de l'homme et des animaux domestiques* 113 (1902). Position in classification: *Arthrodermataceae, Onygenales, Ascomycota*.

Originally localized in North Africa and the Middle East, *T. violaceum* followed the path of human migration into Europe and the United Kingdom where it has become the second most prevalent cause of tinea capitis. *T. violaceum* has caused outbreaks in families and institutions. (Please see Section 21.11, Clinical Forms.)

In addition to the species described above, several other less common species are seen in different geographic regions (Table 21.1).

The Species Concept in Dermatophytes

The dermatophyte species causing human disease are traced by phylogenetic analysis to a small number of sexually reproducing species (Gräser et al., 2006). The *anamorph Microsporum* and *Trichophyton* species are offshoots of the original *teleomorphs*. These clonal offshoots drifted away from their teleomorph ancestors over evolutionary time. Several molecular markers have been applied to develop independent gene genealogies tracing clonal species back to their sexually reproducing ancestors. This multigene approach, linking clonal species to their phylogenetic history, has enabled the delimitation of species boundaries (Table 21.2).

Molecular markers that have proved useful for multilocus sequence analysis are:

1. Internal transcribed spacer (ITS) region of rDNA
2. Two intergenic transcribed spacer regions in mtDNA: *ATP/CYT II* and N3/N1

Table 21.1 General Features of Less Common Dermatophytes

Genus, species	Host	Geographic distribution	Type of tinea	Comments
Microsporum audouinii	Anthropophilic	Asia, Africa	Capitis in prepuberty children	Ectothrix hair invasion, Wood's light positive, silver fluorescence
M. ferrugineum	Anthropophilic	Asia, Russia, eastern Europe, Africa	Capitis	Ectothrix hair invasion, Wood's light positive
M. gypseum	Geophilic	Worldwide	Inflammatory capitis, corporis, barbae	May resemble impetigo; can result in favus-like crusts on the scalp
M. nanum	Geophilic, zoophilic	Worldwide	Transient inflammatory disease of swine	Occupational disease of swine farmers
Trichophyton concentricum	Anthropophilic	South Pacific, west coast South America	Chronic tinea corporis	Known as tinea imbricata
T. mentagrophytes	Zoophilic	Worldwide	Rodent ringworm, including lab animals; acute inflammatory corporis in humans	
T. schoenleinii	Anthropophilic	Very rare in Europe, People's Republic of China	Favus, favic scutula, alopecia,with scarring	Rare
T. verrucosum	Zoophilic	Worldwide	Cattle ringworm, cattle handlers; fomites; causes capitis, barbae, corporis in adults and children	Ectothrix
Epidermophyton floccosum	Anthropophilic	Worldwide	Cruris, pedis	Does not invade hair

Table 21.2 The Species Concept of Human Pathogenic Dermatophytes Inferred from Multiple Gene Genealogies[a]

Teleomorph	Related anamorphs
Arthroderma vanbreuseghemii	*T. tonsurans* [b]
	T. equinum [c]
	T. interdigitale [b]
Arthroderma simii	*T. schoenleinii* [b]
	T. mentagrophytes [c]
Arthroderma benhamiae	*T. concentricum*[b]
	T. verrucosum [c]
	T. erinacei [c]
Arhroderma otae	*M. audouinii* [b]
	M. ferrugineum [b]
	M. canis [c]
No known teleomorph	*T. rubrum* [b]
	T. violaceum [b]

[a] Data adapted from Gräser et al. (2006).
[b] Anthropophilic.
[c] Zoophilic.

3. *UB/VAR*—3′ noncoding region of ubiquitin gene
4. A metalloproteinase
5. Microsatellite markers (variable number of di- and trinucleotide repeats)
6. Anonymous markers AC4, 9, and 12

Evolutionary Implications of Dermatophyte Taxonomy Determined by Gene Genealogy Studies

The Arthroderma otae *Complex*

This complex contains three related species: *M. canis, M. audouinii*, and *M. ferrugineum. Microsporum audouinii* and *M. ferrugineum* are offshoots of *Arthroderma otae* but *M. ferrugineum* became resident in Asia, and *M. audouinii* in Europe and Africa. This reproductive isolation resulted in differences in morphology, physiology, and ecology. One of the commonest species worldwide is the cat dermatophyte, *M. canis*. Its characteristic macroconidia and lemon pigment make it one of the easiest fungi to identify. Isolates not producing macroconidia on SDA can be stimulated to produce these conidia on PDA or other sporulation medium. The main differential diagnostic is the related species, *M. audouinii*, rarely seen but may be imported from Africa. *M. audouinii* does not produce lemon pigment, but can be confused with pale *M. canis* isolates with few or distorted macroconidia. *M. audouinii* may produce atypical, rough-walled, beaked macroconidia, but many isolates do not sporulate or produce only microconidia. It is distinguished from atypical *M. canis* because it does not grow on polished rice. *M. audouinii* does not perforate hair in vitro in contrast to *M. canis*.

The nearly extinct *M. ferrugineum* is restricted to some rural areas of Asia and Africa. Its colonies resemble atypical *M. canis*, but with an orange colony color, without conidia, and with strongly septate "bamboo hyphae." *M. ferrugineum* does not perforate hair in vitro. *M. canis* strains adapted to horses, referred to previously as *M. equinum*, produce scant or no conidia and do not perforate hair in vitro.

The Arthroderma vanbreuseghemii *Complex*

Arthroderma vanbreuseghemii is the sexual species with the greatest number of asexual offshoots. Three globally distributed anamorph species are in this complex: *T. tonsurans, T. equinum*, and *T. interdigitale*. Anthropophilic *T. interdigitale* is the anamorph of *Arthroderma vanbreuseghemii*. Its colony morphology is velvety and cottony. The host shift of *T. interdigitale* to humans brought few genetic changes so that its host range includes other mammals. Anthropophilic *T. tonsurans* is so closely related to *T. equinum* that genetic markers which separate them are ones used for intraspecific levels of variation.

The natural hosts of zoophilic isolates of *T. mentagrophytes* are rodents, rabbits, and guinea pigs. These isolates have a recognizable phenotype: forming clumps of conidia *en grappe* (in grape-like clusters). Spiral hyphae usually also are present. *T. mentagrophytes* produces urease and perforates hair in vitro. These zoophilic isolates are the anamorphs of *Arthroderma simii*.

The Trichophyton rubrum *Complex*

No teleomorph is known for the *T. rubrum/T. violaceum* clade, which is very distant from the other dermatophyte groups. For this reason the trace back to an ancestral, sexually reproductive interbreeding community has not been determined. These species may be evolutionarily much older than the *M. canis* complex. As a result, there has been much genetic variation in the *T. rubrum/T. violaceum* clade, leading to the formation of several separate species based on morphology and physiology which have been reduced to synonymy by the gene genealogy data (see Gräser et al., 2006, for details). *T. rubrum* is geographically associated with Europe and the United States, whereas *T. violaceum* was imported into Europe and the United Kingdom from Africa and the Middle East.

The T. rubrum complex is the most ancient line of asexual molds. It evolved diverse phenotypes on a single host, the precursors leading to modern *Homo sapiens*. These ancient lineages are associated with tinea capitis and include *T. violaceum* and African populations of *T. rubrum*, referred to as *T. soudanense* or *T. gourvilii*. These fungi are isolated from endothrix tinea capitis in children, or tinea corporis in adult members of affected families. There are three common phenotypes of *T. violaceum*, with different geographic endemic areas (reviewed by Gräser et al., 2008).

Multigene analysis has found that two recognized species, *T. rubrum* and *T. soudanense*, are from the genetic standpoint, the same species. This finding is not, however, comfortable with experience in the clinical laboratory because *T. soudanense* presents with visually distinctive isolates and correlates with the epidemiology of endothrix tinea capitis as well as tinea corporis, whereas *T. rubrum* does not invade hair. The endothrix scalp-infecting African member of the *T. rubrum* complex is the classic *T. soudanense*, with flattened, radially striate, yellow to blood-red colonies. These isolates are most often urease-negative.

The *T. rubrum* complex includes a lineage causing large-spored ectothrix tinea capitis—*T. megninii*. It is more commonly isolated from tinea corporis. *T. megninii* produces cottony colonies with a blood-red reverse, similar to typical *T. rubrum*, but with a requirement for L-histidine, and sometimes is weakly urease-positive. It is the only member of the *T. rubrum* complex that has a (+) mating type; the other varieties are either (−) mating type, *T. rubrum sensu stricto* and *T. violaceum*, or unclassifiable in the case of *T. soudanense*. Most isolates obtained outside Portugal, the endemic area for *T. megninii*, are probably not recognized as distinct from *T. rubrum sensu stricto*.

> *Note on Mating Types in Dermatophytes:* Phyllis Stockdale (1968) discovered that many nonmating dermatophyte species could reveal their mating type in an incomplete mating reaction with tester strains of *Arthroderma simii*. Most recognized asexual species could be tested in this way; for example, *T. rubrum* was shown to be the (−) mating type whereas the highly related or synonymous *T. megninii* was the (+) mating type.

Classic *T. rubrum* isolates infect the lower body and consist of two main groups—members of one group, microsatellite group A (Ohst et al., 2004), are from Africa and Southern Asia. They have a granular colony with macro- and microconidia, a blood-red reverse, are positive for urease, and are involved in tinea corporis and tinea cruris. The most commonly seen dermatophyte is "microsatellite group B," the classic *T. rubrum sensu stricto*, with a cottony colony, a blood-red reverse on SDA, producing clavate microconidia. This group is urease-negative. They cause tinea pedis and tinea unguium, and less commonly, tinea manuum, tinea corporis, or tinea cruris. Scalp infection is exceptional.

Isolates in this complex previously called *T. raubitschekii, T. fischeri*, and *T. kanei* are now unified in the taxon *T. rubrum*.

21.6 GEOGRAPHIC DISTRIBUTION/ECOLOGIC NICHE

Geographic Distribution

The worldwide geographic prevalence of dermatophyte species is linked to these trends:

- Human migration. Patterns of migration brought, and continue to bring, different anthropophilic dermatophytes to countries where the prevalence is low.
- Public health campaigns. When health authorities embark on a campaign to reduce the burden of tinea capitis considerable success has been achieved because of the availability of effective oral antimycotic therapy with griseofulvin, and the utility of Wood's light to diagnose certain mycoses.
- The keeping of companion animals. Dogs and cats can carry zoophilic dermatophytes that are transmissible to humans.
- Geographic isolation. There is a high incidence of certain anthropophilic dermatophyte species in some areas. (Please see Section 21.7, Epidemiology.)

Ecologic Niche

Species are categorized by host preference: anthropophilic, zoophilic, with the term geophilic reserved for soil-dwelling species. These categories are shown for each dermatophyte in Section 21.5, Etiologic Agents. In this section the current prevalence of individual species is summarized in Table 21.3, and the events leading to their incidence and prevalence are discussed in the following section.

21.7 EPIDEMIOLOGY

Trends and Worldwide Incidence of Species

Tinea Capitis

Among children, tinea capitis is the principal dermatophytosis, with symptomatic infection estimated to occur at a rate of 3–8% (i.e., ~1 of every 20 children). Tinea

Table 21.3 Tinea capitis: Geographic Prevalence of Individual Dermatophyte Species[a–c]

Country or region	Prevalent dermatophyte species
Africa	*M. audouinii, T. violaceum, T. soudanense*,
	M. canis, T. gourvilii, T. schoenleinii (rare)
Asia[d,e]	*T. violaceum, T. tonsurans*
Australia/New Zealand[f]	*T. tonsurans, M. canis*
Caribbean	*T. tonsurans*
Eastern Europe	*M. canis, T. violaceum* ; other species include *M. ferrugineum* and *T. verrucosum and T. schoenleinii* (rare)
Mediterranean	*T. violaceum, M. canis*
Middle East	*T. violaceum*
North America	*T. tonsurans, M. canis, T. verrucosum*
People's Republic of China	*M. canis, T. violaceum, T. tonsurans*
India	*T. violaceum, M. audouinii, T. tonsurans*
South America	*T. tonsurans, M. canis, T. violaceum*
South Pacific	*T. concentricum*
United Kingdom	*T. tonsurans, T. violaceum*
Western Europe	*M. canis, T. megninii, T. tonsurans, T. violaceum*
	M. audouinii, T. mentagrophytes, T. schoenleinii (rare)
All areas	*T. soudanense, M. audouinii, T. violaceum*; emergent in immigrants from Africa and parts of Asia

[a] Aly et al. (2000).
[b] Havlickova et al. (2008).
[c] Gupta and Summerbell (2000).
[d] Jha et al. (2006).
[e] Yu et al. (2005)
[f] McPherson et al. (2008).

capitis is a common problem dating from medieval times and continues to afflict children and their families. The epidemiology of tinea capitis is in flux according to geography and patient populations, particularly those underserved by family care practitioners. Although some tinea capitis disease may be subtle, it also may be disfiguring and stigmatizing.

Europe and the United Kingdom Tinea capitis is an Old World mycosis. The dermatophyte species afflicting Europeans were imported into the United States by immigrants. Before 1900 tinea capitis was uncommon in North America and was caused by *M. canis*. Surveys in Europe and England in the late 19th and 20th centuries indicated that in northern Europe *M. audouinii* and *M. canis* were prevalent, whereas *T. tonsurans* and *T. violaceum* were encountered in Mediterranean countries. Beginning in the urban areas of major cities of England, *M. audouinii* was replaced by *T. tonsurans* as the prevalent cause of tinea capitis. As of 2003 in the United Kingdom *T. tonsurans* and *T. violaceum* accounted for approximately 87% of tinea capitis isolations, followed by *T. soudanense* and *M. canis* (Borman et al., 2007).

Zoophilic *M. canis* remains the most common cause of tinea capitis in Europe (Ginter-Hanselmayer et al., 2007). Countries with the highest incidence of *M. canis* infections are mainly in the Mediterranean but also bordering countries such as Austria, Hungary, Germany, and Poland. *M. canis* caused 50–63% of tinea capitis cases in Southwestern Poland during the 1977–2006 period (Jankowska-Konsur et al., 2009) and accounted for 92% of tinea capitis in Bosnia-Herzegovina during 1997–2006 (Prohic, 2008). Most cases of tinea capitis in Western Europe including Italy, Belgium, and Spain are caused by *M. canis* followed by *T. violaceum* and *T. mentagrophytes* (Borman et al., 2007). There is a trend toward anthropophilic tinea capitis in urban areas in Europe.

Eastern Europe saw a very inflammatory dermatophyte species, *Trichophyton schoenleinii*, as the cause of the severe scalp disease favus. *T. schoenleinii* was brought to the United States by immigrants from Europe. Once common among school children, it was eliminated from the United States but can still be found in Asia and Africa, and from persons traveling from those areas. In eastern Europe *T. schoenleinii* was replaced by *T. ferrugineum* from the Far East, and *T. violaceum*, from Mediterranean countries. Another more recently prevalent species causing tinea capitis in eastern Europe is *T. violaceum* (Gupta and Summerbell, 2000).

North America North America has seen a change in the prevalent species causing tinea capitis influenced by trends in immigration and improved diagnosis and treatment. Before the 20th century the most common cause of tinea capitis in the United States was *M. canis*. Tracing the path of immigration from the east coast ports of entry, *M. audouinii* was brought to the United States from the United Kingdom and northern Europe. By 1940 that species had spread in all directions and into Canada, becoming the major cause of tinea capitis. Following the introduction of Wood's light and the availability of griseofulvin, the prevalence of *M. audouinii* subsided to very few cases in the 1970s. *M. audouinii* was replaced by *T. tonsurans* as a cause of tinea capitis beginning in the United States in the 1940s to 1950s after the latter species was introduced by immigrants from Mexico, Puerto Rico, and the Caribbean. *T. tonsurans* is abundant in the Caribbean and South America and now in the United States. *T. tonsurans* is responsible for over 90% of tinea capitis. *T. tonsurans* does not fluoresce in Wood's light. As an example, of 307 cases of tinea capitis diagnosed at the University of California–San Francisco dermatology clinic in a 20 year period, *T. tonsurans* tinea capitis rose from 41.7% of cases between 1974 and 1978, to account for 87.5% of cases in 1986–1993 (Wilmington et al., 1996).

Emergence of different species as causative agents of tinea capitis coincides with recent immigrants from abroad. Between 2000 and 2006, a 24 patients with *T. violaceum* or *T. soudanense* tinea capitis (all in children under 12 years old) were treated in a single hospital in Baltimore, Maryland, coinciding with increased immigration into that city from Africa (Magill et al., 2007).

People's Republic China In the 1950s and 1960s, *T. schoenleinii* and *M. ferrugineum* were the predominant causative agents of tinea capitis (Yu et al., 2005). Most patients were <15 years old. Disease caused by *T. schoenleinii* often presented as *favus*. A successful national public health campaign was begun in the 1960s to identify and treat these cases. *M. ferrugineum* took the place of *T. schoenleinii* by the late 1970s. During the decade 1986–1996 *T. mentagrophytes* and *T. violaceum* became prevalent. After 1996 *M. canis* replaced *T. mentagrophytes* as the major cause of tinea capitis. *M. canis* remains the prevalent species, possibly owing to the increased popularity of keeping companion animals.

India *Trichophyton violaceum* is the most common cause of tinea capitis. As in other countries, it is a disease of children younger than 15 years; most patients are between 5 and 10 years old. In Southern India tinea capitis cases were reviewed in a large public hospital in Madras during a 3 year period (1973–1976) (Kamalam and Thambiah, 1980). Among 357 isolates, *T. violaceum* was most common (74%) followed by *T. tonsurans* (13%). In northern India, of 153 tinea capitis patients, the seborrheic variant was seen in almost half of patients, followed by gray patch, black dot, kerion, and alopecia-areata (Singal et al., 2001). *T. violaceum* was isolated in 38% of patients. *Microsporum audouinii, T. schoenleinii, and T. tonsurans* were isolated in 34%, 10%, and 9% of patients, respectively.

Africa *Trichophyton soudanense, T. violaceum*, and *M. audouinii* are prevalent.

Tinea Pedis

Tinea pedis is the most common dermatophytosis in North America and other developed countries, affecting up to 10% of the population. Those affected probably acquire the infection from an adult family member. The prevalence of tinea pedis in children is from 2.2% to 6.0%. Tinea pedis occurs more often in adult males (20%) than females (5%), and may be acquired during their teens. The dry form is most often caused by *T. rubrum*. (Please see Section 21.11, Clinical Forms—Tinea pedis and Tinea manuum.) Chronic *T. interdigitale* tinea pedis accounts for approximately 15% of the tinea pedis in most urban communities (Zaias and Rebell, 2003).

Tinea Unguium

Fungal disease of the nails affects 6.5–8.7% of the North American population (Gupta et al., 2004). The more general term, onychomycosis, refers to nail disease due to nondermatophytes as well as to dermatophytes. Age and gender are risk factors including the elderly (>60 years old) and men. Primary diseases increasing the risk for fungal nail disease are diabetes mellitus, immunocompromise (e.g., HIV-positive persons), psoriasis, peripheral vascular (arterial) disease, previous tinea pedis, history of nail trauma, or a family history of onychomycosis.

Trends in the Incidence of the Dermatophytoses

Tinea Capitis in the United States

Data from the National Ambulatory Medical Survey in 1996 revealed 172,000 new cases of tinea capitis (Chen and Friedlander, 2001). Most cases were in the 4–7-yr-old demographic, but tinea capitis is not uncommon in adolescents. Children and adolescents 5–18 years old represent approximately 77% of all cases of ringworm in the United States (Howard and Frieden, 1999).

Prevalence of tinea capitis in the grade schools of a representative high-risk city was reported to be 2.5%. The asymptomatic carrier state in the general population is estimated at 5%. In urban school-aged children, the carrier rate appears closer to 15%, and may be as high as 44% for siblings of index cases, and 30% for their adult contacts (reviewed by Chen and Friedlander, 2001). Asymptomatic carriage is characteristic for the anthropophilic species *T. tonsurans*, the predominant agent of tinea capitis. Taking prescriptions for griseofulvin as a marker for the incidence of tinea capitis, during 1984–1993, of children enrolled in the California Medicaid program there was an increased incidence of 84.2%, most dramatic for African-American children (209.7% increase) and also markedly increased for Caucasian children (140.4%), underlining a significant public health issue. African-American children accounted for 80.9% of the 206,000 reported cases of tinea capitis from the 1996 National Ambulatory Medical survey. The relative risk for the African-American population was 29.4 times that of the general population (Lobato et al., 1997).

A 2002 study of 3 million U.S. employees and dependents indicated that 0.1% attended their physician offices or were hospitalized for treatment of tinea capitis, based on health insurance claims data and ICD-9 code 110 (Suh et al., 2006). Fifty-two percent of those treated were <10 years old. By excluding persons without health insurance, a large risk group of uninsured indigent persons were missed. Despite the standard of care indicating the use of oral antifungal agents for treating tinea capitis, only 56% of the patients were so treated, and just 17% used griseofulvin for 6 weeks.

Illustrative Example: To investigate the rates of infection and clinical disease in a high-risk population, 446 preschool-age children attending an urban child care center in Kansas City, Missouri, were visited over a 24 month period (Abdel-Rahman et al., 2006). The children (3.9 ± 1.0 years old) were visited a median of seven times. There was an equal distribution of children by gender; the racial distribution of children was 381 African-American, 47 Caucasian, and 18 other.

Fungal scalp samples were obtained, and a finding of scaling and/or alopecia in addition to other clinical indicators was used to classify the child with symptomatic tinea capitis. Twenty-two percent to 51% of scalp cultures/month were positive, totaling 1390 fungal cultures. More than one-fourth of the children in the child care center harbored a fungal isolate on their scalp, with 10% presenting with symptomatic tinea capitis. Caucasian children were significantly less likely than African-American children to harbor a scalp fungal isolate or to have symptomatic tinea capitis.

Molecular genotyping targeted rDNA and the alkaline proteinase-1 (*ALP-1*) gene locus. Among children with multiple typable isolates, 51% exclusively carried the same strain, 37% demonstrated a single predominant strain with secondary strains transiently acquired, and 12% harbored a different strain of *T. tonsurans* with each typable culture. There was a high probability that the same strain persisted over a period of months. Among children carrying the same strain at every visit the symptomatic disease rate was 18.7%. Symptomatic disease rates were 6.5% in the children where a different strain type was observed with each positive culture. Infection of children in the day care center was endemic and clinical disease occurred as a result of activation of a single strain persisting on the scalp.

21.8 RISK GROUPS/FACTORS

Age, atopy, occupation or recreation, and underlying endocrine or immunosuppressive disorders or treatments are categories of risk groups for dermatophytosis.

Age

Tinea Capitis in Children

The reason for the preponderance of tinea capitis in school-age children is thought to be related to the fatty acid composition of their stratum corneum. Infants and adolescents are less frequently infected. Postpuberty adolescents have medium-chain-length (C_8 to C_{12}) fatty acids in sebum that inhibit the growth of dermatophytes. These fatty acids are absent in prepuberty children (Hay, 2005).

Age Related Prevalence of Onychomycosis

Distal-lateral subungual onychomycosis is the most common clinical presentation. In North America, 90% of toenail tinea unguium is caused by *Trichophyton rubrum* and *T. interdigitale*. The distribution of onychomycosis among 15,000 patients attending dermatology clinics or primary care physicians in Ontario, Canada according to age group is shown Table 21.4. In North America and Europe an estimated 2–13% of the population is affected by onychomycosis (Scher et al., 2007). Another estimate, among middle age and older adults of the prevalence of onychomycosis in North America, was estimated to be >40–50 years old (9.8%), >50–60 years old (13.7%), >60–70 years old (14.1%), and >80 years old (26.9%) (Gupta, 2000). The reasons why onychomycosis occurs more frequently in the elderly are (i) their slower rate of nail growth; (ii) more opportunities for micro- and macrotrauma (please see Section 21.3, Case Presentation 2); (iii) reduced immunocompetence; and (iv) because the increased rate of tinea pedis in the elderly, if left untreated, leads to tinea unguium (Gupta, 2000).

Risk Factors for Onychomycosis

- Tinea pedis, if left untreated.
- Diabetes mellitus. Onychomycosis is prevalent in an estimated one-third of diabetics.

Table 21.4 Age Distribution of Onychomycosis[a]

Age group (years)	0–19	20–39	40–59	60–79
Onychomycosis cases classified by age of patient (%)	0.7	3.1	9.5	18.2

[a]Gupta et al., 2000.

- Genetics. A familial tendency to *T. rubrum* onychomycosis is linked to an autosomal dominant inheritance pattern (reviewed by Faergemann and Baran, 2003).
- Older age (Gupta, 2000). Elderly patients have specific risk factors for poor response to therapy for onychomycosis, including frequent nail dystrophy, slow growth of nails, and increased prevalence of peripheral vascular disease and diabetes mellitus (Loo, 2007).
- Dermatophytosis is common in HIV-positive persons and may be a general sign of an immunodeficiency. The prevalence of onychomycosis in HIV-positive individuals in a sample of 500 Canadian and Brazilian HIV-positive persons was 23.2% (Gupta et al., 2000).
- Smoking and peripheral artery disease are independent predictors.
- Swimmers.

Athletes

Prevalence of tinea pedis is significantly higher in male professional and collegiate athletes than in nonathletes; tinea cruris, tinea corporis, and tinea pedis (Pickup and Adams, 2007). This increased risk arises from opportunities for exposure in locker rooms and showers in gyms and health clubs.

Atopy

Persons who are predisposed to seasonal allergies and have a faulty regulation of IgE are atopic, accounting for ~15% of the human population. Atopic persons suffer from a significantly higher incidence of dermatophytosis.

Diabetes Mellitus

Diabetics are 2.8 times as likely to have onychomycosis as normal individuals, affecting one-third of persons living with diabetes (Gupta et al., 1998). Patients with diabetes are susceptible because they have impaired sensation, making them less aware of foot trauma, including nail changes in onychomycosis. Thickened mycotic nails can cause pressure necrosis of the nail bed, and sharp infected nails can pierce the skin, increasing the risk of serious diabetic foot infections (Scher et al., 2003).

Immunosuppression

Dermatophytosis is common in persons living with AIDS including, in approximate order of frequency, tinea pedis, tinea unguium, tinea cruris, tinea corporis, and tinea capitis (Aly and Bergter, 1996). Widespread *T. rubrum* tinea corporis figures prominently in AIDS–dermatophytosis. The disease may be locally invasive, presenting as subcutaneous nodules around knees, thighs, and buttocks. AIDS–tinea unguium may affect all fingernails and toenails. Dermatophytosis occurred in 13% of a cohort of 1161 HIV-positive persons in Spain followed for 36 months (Muñoz-Pérez et al., 1998). The prevalence of clinical forms was tinea cruris (5%), pedis or manuum (4%), and onychomycosis (4%). Tinea corporis was less common (2%), but was usually diffuse and sometimes was associated with deep dermal involvement.

Persons receiving immunosuppressive therapy are at risk for more serious dermatophytosis. Agents associated with dermatophytosis include azathioprine, cyclosporine, infliximab (TNF-α inhibitor), oral corticosteroids, and prednisolone and its precursor, prednisone. Topical corticosteroids and topical nonsteroidal medications including pimecrolimus and tacrolimus are associated with "tinea incognito," in which the typical ringworm appearance of tinea corporis is altered, but the disease continues to spread and may cover large areas (Serarslan, 2007).

Illustrative Example: *Trichophyton tonsurans* tinea circinata on the right leg of a woman living with AIDS (Rovansek and Papadopolous, 2005). A 46-yr-old woman with a history of HIV infection presented to the emergency department with a painful rash. The $CD4^+$ T- lymphocyte count was 365/ μL, the HIV load was undetectable, and the patient had not received antiretroviral therapy. A violaceous plaque measuring 9 cm × 9 cm and consisting of four concentric rings with intervening areas of normal skin and numerous yellow-white pustules was found on the pretibial surface of the patient's right leg. Around the periphery of the rings, there was fine scaling of the skin. The patient reported being scratched by a cat in the same area 3 weeks earlier and was treated with oral amoxicillin–clavulanate at a dose of 500 mg twice daily for 14 days with no improvement. A skin-scraping specimen, KOH prep, contained large numbers of hyphae. A diagnosis of tinea circinata was made (not to be confused with tinea imbricata). Tinea circinata, an uncommon morphologic variant of tinea corporis, is caused by *T. tonsurans*.

Occupations

Several occupations are at higher risk than the general public to contract tinea of various types.

- *Miners.* In investigations of miners in Germany, the prevalence of tinea pedis and tinea unguium is up to

72.9% (reviewed by Havlickova et al., 2008) resulting from wearing occlusive boots, the humid mine environment, and shared shower facilities.

- *Soldiers.* Known among dermatologists as "barrack dermatophytosis," wearing of boots, prolonged training, and deployments of soldiers are associated with a high rate of tinea pedis. For example, the clinical point-prevalence of tinea pedis in a cohort of Israeli soldiers was 60.1% (Cohen et al., 2005).

Exposure to Animals

Pet owners are at risk from *Microsporum canis*, which may be present on the fur of dogs and cats living with them. (Please see Section 21.9, Transmission, for further details.) Veterinarians, zookeepers, stockmen, laboratory animal handlers, and pet shop workers are at increased risk of tinea corporis and tinea cruris. Farmers and members of their families who frequently handle cattle, hogs, chickens, or other animals may acquire specific fungal infections from infected animals, their stalls, cages, or other *fomites* used by these animals. Stockmen are a well-known category at risk to develop *T. verrucosum* tinea barbae from exposure to cattle. Diagnosis may require a specific occupation-related question, such as, "Are you a dairy farmer?" (Rutecki et al., 2000). Occasionally, workers in greenhouses may acquire dermatophytosis from contaminated soil. Laboratory rats and mice infected with *T. mentagrophytes* can infect laboratory workers.

21.9 TRANSMISSION

Athletic and Recreational Facilities

Tinea pedis is associated with the presence of *T. rubrum* and *T. interdigitale* in locker rooms and shower rooms in gyms, fitness centers, and swimming pool facilities.

Caregivers as Asymptomatic Carriers

Among infected adults, women are more likely to be both asymptomatic carriers and active cases, probably because of their role as caretakers of the family. Household contacts of 56 children with tinea capitis in Milwaukee, Wisconsin, were mostly women: 94% of 50 adult contacts (Pomeranz et al., 1999). High rates of asymptomatic carriage are attributed to anthropophilic dermatophytes, usually *T. tonsurans* or *T. violaceum*. Zoophilic dermatophytes such as *M. canis* cause a brisk inflammatory response, which is less likely to lead to asymptomatic carriage (Frieden, 1999).

The prevalence of asymptomatic carriers correlates with the amount of tinea capitis in the community. The highest rate of asymptomatic carriage, with a prevalence of 49%, was described in the Cape Peninsula of South Africa, where *T. violaceum* tinea capitis is endemic. In at-risk populations of children in a school or a community where tinea capitis is common, the point—prevalence rates for asymptomatic carriage is estimated at $\sim$15%, rising in some communities to $\geq$25% (reviewed by Frieden, 1999).

Asymptomatic carriage also occurs in adults and children living with an index case of tinea capitis. Thirty percent of 46 caretakers in Detroit, Michigan, who were mostly women, had *T. tonsurans* in their scalps despite no evidence of disease (Babel and Baughman, 1989). This population may be a reservoir for reinfection of children.

Companion Animals

After parasitic infections, dermatophytosis from contact with cats and dogs is probably the most common pet-associated disease, causing an estimated 2 million or more infections each year (Stehr-Green and Schantz, 1987). A 5 year study of companion animals, 248 cats and 136 dogs, in the homes of persons with and without tinea corporis found that *M. canis* was isolated from 20.5% of dogs and 28.2% of cats (Cafarchia et al., 2006). *Microsporum canis* was isolated from 36.4% of dogs living with guardians diagnosed with tinea corporis but it was never isolated from dogs whose guardians had no lesions. By contrast, *M. canis* was isolated from 53.6% of cats living with guardians diagnosed with tinea corporis and from 14.6% of cats whose guardians had no signs of the disease. Cats and dogs should be considered a major source of dermatophytes for their human family even when the pets are asymptomatic.

Contact Sports

Ringworm is common among high school and college wrestlers. Known as *tinea gladiatorum*, ringworm in wrestlers is an entity distinct from tinea corporis seen in the pediatric population. Tinea gladiatorum outbreaks have been caused by the dermatophyte *Trichophyton tonsurans*. Epidemiology and microbiology point to person-to-person contact as the main source of transmission in wrestlers. A study in the United States found that 60% of college wrestlers and 52% of high school wrestlers contracted the disease during the 1984–1985 season (reviewed by Adams, 2002). This is an alarming rate of infection. Neither disinfection of wrestling mats nor increased emphasis on personal hygiene have proved to be effective preventive measures. Prophylaxis with FLC or ITC during the wrestling season appears to reduce the percentage of clinically positive wrestlers, but the cost versus benefit of prophylaxis has not been determined. This disease among members of a wrestling team is considered sufficient reason to disqualify team members from competition until completion of therapy and remission of symptoms (Beller and Gessner, 1994). The differential diagnosis of tinea gladiatorum includes herpes gladiatorum caused by HSV-1.

Fomites

Viable conidia can be cultured from inanimate objects in homes or institutions of children with tinea capitis (reviewed by Frieden, 1999). The variety of objects contaminated with hair or skin scales from infected humans and animals and found to be culture-positive for dermatophytes includes combs, hairbrushes, and pillow covers; less commonly blankets, clothing, curtains, mattresses, rugs, sheets, towels, and even through air sampling.

Public Places

Schools

The early recognition by school nurses and treatment of tinea capitis in school children is essential to slow the spread of infection and to prevent reinfection. Transmission occurs via sharing of caps and combs. Infected desquamated epithelial cells can be transmitted by direct bodily contact.

Houses of Worship

Public spaces in which persons remove their shoes have been implicated in the spread of tinea pedis and tinea unguium. Pedal dermatophyte infection seems to be a major problem among the adult Muslim male population regularly attending mosques, especially for men in the 5th and 6th decades of life. The prevalence of the foot dermatomycosis is high among those who practice ablution 3 to 5 times/day (Ilkit et al., 2005).

21.10 DETERMINANTS OF PATHOGENICITY

21.10.1 Host Factors

No single mechanism explains all aspects of susceptibility and immunity to dermatophytes. Please see Section 21.8, Risk Groups/Factors, for disease susceptibility which is affected by occupation, recreation, and immune or endocrine status. Prepuberty susceptibility occurs because, beginning at puberty, sebum contains fungistatic unsaturated fatty acids of C_8–C_{12} length.

Temperature Restriction

Except for rare instances of immunosuppression, dermatophytes generally do not grow or produce disease at or above 37°C. Therefore they are restricted to grow on the dead, keratinized outer surface layers of the epidermis (stratum corneum), as well in the hair and nails.

Barrier Function of the Stratum Corneum

The stratum corneum, along with the avascular epidermis composed of stratified squamous epithelium make up a major barrier against infection. The stratum corneum has an unusual lipid composition consisting of a roughly equimolar mixture of ceramides (45–50% by weight), cholesterol (25%), and free fatty acids (10–15%) plus less than 5% each of several other lipids. Ceramides are composed of sphingosine and a fatty acid. Sphingosine and sphinganine have in vitro antimicrobial activity and may protect against dermatophytosis (Bibel et al., 1995).

Role of Atopy

Atopic persons suffer from a significantly higher incidence of dermatophytosis. Stratum corneum lipids such as sphingosines have antimicrobial activity and are reduced in skin from atopic individuals. Asthma has been associated with dermatophytosis (reviewed by Al Hasan et al., 2004). Patients with chronic rhinitis and asthma also demonstrate immediate hypersensitivity to *Trichophyton* species and positive immediate bronchoconstrictive reactions to extracts of *T. tonsurans*. These patients have eosinophilia and chronic rhinosinusitis in addition to features compatible with "intrinsic or late onset asthma."

Regulation of T-Cell-Mediated Immunity and Its Effect on the Outcome of Dermatophytosis

Individual protein antigens from *Trichophyton rubrum* (Tri r 2) and *T. tonsurans* (Tri t 1) can induce high titers of specific IgG and IgE antibodies (Th2 response) associated with immediate hypersensitivity skin test reactions in atopic persons. These same antigens elicit low to intermediate titers of specific IgG and delayed cutaneous hypersensitivity skin responses (Th1 response) in immune-normal persons (Zollner et al., 2002). The Th2 immune response evokes IgE and IgG_4 antibody production and tends to result in a weak host defense against fungi, whereas the Th1 response evokes production of T-cell lymphokines, IL-2, and IFN-γ, resulting in protective immunity. Immediate hypersensitivity skin tests correlate with chronic dermatophyte infection;

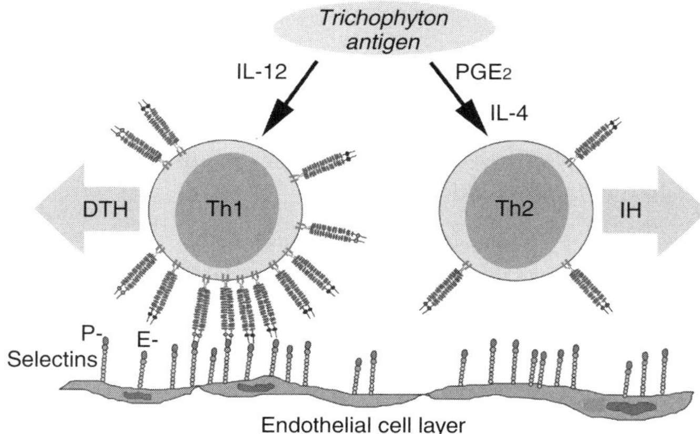

Figure 21.2 A model for differential migration of T lymphocytes into lesions after activation by *Trichophyton rubrum* antigens. Please see text for an explanation.
Source: from Zollner et al. (2002) and used with permission from John Wiley & Sons.

whereas delayed cutaneous hypersensitivity corresponds to more acute disease and a protective immune response causing the lesions to heal spontaneously.

Is there a mechanism to explain how a Th1 immune response results in improved protection in dermatophytosis?

Th1 cells, when induced by IL-12, express a high density of two surface receptors: the cutaneous lymphocyte antigen, CLA, and the second, "post-translational glycosylation of P-selectin glycoprotein 1" (PSGL-1). Figure 21.2 depicts differential regulation of selectin ligands in Th1 versus Th2 lymphocytes. These receptors bind to ligands on endothelial cells in the skin—P-selectin and E-selectin, respectively. Th1 cells rolling along the endothelium of the vascular wall bind to these ligands and are tethered. This enables the Th1 cells to extravasate into the epidermis in the area of the lesion. The result is recirculation of Th1 cells and protective immunity. Th2 cells on the other hand are more often activated in atopic individuals. The Th2 cells have a low expression of the CLA and PSGL-1 ligands and cannot tether to the endothelium and enter the site of the lesions.

21.10.2 Microbial Factors

Early Steps in Pathogenesis

The first stage of attack by dermatophytes is to adhere to the skin surface (Hay, 2006). Adherence involves the binding between a dermatophyte arthroconidium and underlying keratinocytes. The fungal cell wall contains carbohydrate microfibrils. A fine fibrillar layer is seen at the junction of the fungal wall and the underlying human cells. The toe webs and the groin are more frequently attacked, possibly because of increased carbon dioxide tension at the skin surface, which may induce arthroconidia. In invading the hair shaft, the dermatophyte forms a specialized structure, consisting of enlarged cells, called the penetrating organ. At the site where the fungus penetrates hair there is clearing of the keratin, suggesting digestion. After adherence, arthroconidia germinate, and a growing web of hyphae spreads radially from the original arthroconidium. Dermatophyte hyphae migrate toward holes and crevices and grow through them. Dermatophytes can invade skin cells, aided by production of keratinases. Keratinases also are associated with some aspects of virulence; that is, *M. canis* isolates from animals synthesize larger amounts of keratinase producing a more rapid development of symptoms, and quicker resolution, with more inflammation.

Keratin and Keratinases

Keratins are grouped into "hard" or "soft" keratins according to their sulfur amino acid content. Hard keratins, high in cystine, are found in hair and nails, whereas the more pliable soft keratin with a lower cystine content is found in skin. Keratinolysis requires the cooperation of two processes: disulfide reduction and proteolysis. The crosslinking of keratins by disulfide bridges hinders their degradation by proteinases. Most keratinases are proteolytic but none possess disulfide reducing capacity. Disulfide reductase or sulfitolysis[2] works synergistically with keratinases to degrade keratin. There are two steps in keratinolysis: sulfitolysis or reduction of disulfide bridges, and proteolysis. Purified enzymes have difficulty in digesting keratin unless reducing reagents, such as ß-mercaptoethanol, cysteine, or sodium sulfite, are present.

Keratinases are proteinases capable of digesting hair, nails, and skin scales. Proteinases active on keratin are serine or metalloproteinases and are activated by thiols. They

[2]Sulfitolysis. *Definition*: Incubation of proteins or peptides containing disulfide bonds (S–S) with sodium sulfite (Na_2SO_3) cleaves S–S bonds producing equimolar amounts of free thiols (SH) and thiosulfates ($-S-SO_3H$). This process is known as sulfitolysis.

are extracellular enzymes. Partial inhibition of serine proteinases by EDTA underlines the importance of divalent cations as cofactors. Keratinases belong to the subtilisin group characterized by a serine or metallo catalytic center and their pH optima in the neutral to alkaline range. Dermatophytes are capable of reducing disulfide bridges by sulfitolysis. *Microsporum gypseum* secretes sulfites and can mediate sulfitolysis.

Secreted Subtilisin-like and Metalloproteinases

Proteinases have been characterized from *T. rubrum* and other dermatophytes. In tinea unguium, *T. rubrum* hyphae growing in the nail damage keratin-forming channels in the nail plates, which become larger than the hyphae themselves, implying they are produced by secreted proteinases.

Seven genes in *T. rubrum* encode serine proteinases of the subtilisin family (*SUB*) (Jousson et al., 2004b). Sub3 and Sub4 are present in culture supernatants and have high keratinolytic activity. This activity is increased 2–3 times by reducing disulfide bridges in the keratin substrate.

The sequences of three *SUB*s from *T. rubrum* have orthologs in *M. canis*. Four additional *SUB*s are identified in genomic DNA: *SUB*s *4,5* are unique to *T. rubrum. SUB*6 sequence is identical to the previously described allergen, Tri r 2. Subs are synthesized as pre-proenzymes; cleavage of the prosequence results in enzymes with a molecular mass of 28.2–39.7 kDa. Other functions than keratinase activity can be ascribed to the Sub proteinases: Sub1 has a proline rich C-terminal portion reminiscent of parasite surface antigens.

The occurrence of the large assortment of five metalloproteinases and seven subtilisin-like proteinases in dermatophytes makes the case for their involvement in the specialized compartment of the stratum corneum. This is especially relevant for the high keratinolytic activity of Sub3 and Sub4.

A second family of endoproteinases consists of five metalloproteinases (Meps), characterized from *T. rubrum, T. mentagrophytes*, and *M. canis* (Jousson et al., 2004a). The term "fungalysins" is applied to this family of fungal metalloproteinases in the international proteinase database. The five metalloproteinases arose by multiplication of an ancestral gene, reflecting an evolutionary process through which dermatophytes acquired the ability to invade keratinized tissues. Four Meps from *T. rubrum* and *T. mentagrophytes* were estimated to have molecular masses of 40–45 kDa. Mep activity is higher than subtilisin-like serine proteinase activity in culture supernatants of *T. rubrum*.

Allergens of Anthropophilic Trichophyton Species

Well-characterized tools are important for probing the immune response in dermatophytosis. Woodfolk (2005) reviewed the status of the purified protein antigens with allergenic properties.

- *Tri t 1.* This exo-ß-(1→3)-glucanase from *T. tonsurans* with a molecular mass of 30 kDa elicits immediate hypersensitivity in skin tests.
- *Tri t 4.* This 83 kDa *T. tonsurans* protein is a serine proteinase of the prolyl oligopeptidase family. Tri t 4 elicits both immediate and delayed cutaneous hypersensitivity in responders, depending on whether the patient's immune response is directed along the Th2 or Th1 pathway.
- *Tri r 2.* A third allergen, Tri r 2, from *T. rubrum* was cloned and expressed as a recombinant protein. Tri r 2 is a proteinase in the subtilisin family. Tri r 2, 29 kDa, is encoded by the *SUB6* gene and elicits delayed cutaneous hypersensitivity. There is a high prevalence of IgE against Tri r 2 in persons known to have immediate hypersensitivity to whole soluble *T. rubrum* extract.
- *Tri r 4.* A homolog of Tri t 4 was purified from *T. rubrum*, designated Tri r 4. It was cloned, and its deduced amino acid sequence indicates it is a prolyl oligopeptidase. This antigen evokes no skin test reactivity and binds weakly to IgG. It is a weak stimulator in lymphocyte blastogenesis assays.

21.11 CLINICAL FORMS

Tinea: Origin and Definition

The Latin term *tinea* is defined variously as "moth" and "gnawing worm." "Tinea," the term applied to dermatophytosis, was coined by ancient Romans because the moth-eaten appearance of scalp ringworm lesions reminded them of the damage to woolens by clothes moths, the *Tineidae*. Aulus Cornelius Celsus (circa 25 B.C.E.–50 C.E.) in his treatise *De medicina* described tinea as "an acute inflammatory scalp condition that drained purulent fluid." Tinea capitis, or ringworm of the scalp, is one of the earliest human infectious diseases to be documented in medical literature. Other findings of tinea capitis are *alopecia* (hair loss), broken-off hairs, and scabbing. The cardinal signs of ringworm of the smooth skin are circular lesions with active borders, scaling, and pruritus. Infection with a zoophilic or geophilic dermatophyte may produce a strong inflammatory reaction because humans are not the natural host. This may lead

Table 21.5 Tineas Matched with Their Affected Body Sites

Type of tinea	Other term	Body site affected
Tinea barbae	Barber's itch	Beard
Tinea capitis	Scalp ringworm	Scalp
Tinea corporis	Ringworm	Smooth skin
Tinea cruris	Jock itch	Groin
Tinea manuum		Hands
Tinea pedis	Athlete's foot	Feet
Tinea unguium	Onychomycosis	Nails
Tinea versicolor	Pityriasis versicolor (preferred)	Smooth skin

to production of *kerion*s (pustular inflammation). The various types of tinea are defined and identified by the affected body site (Table 21.5).

Tinea Capitis

Tinea capitis (scalp ringworm) (syn: tetter, scald-head, gray patch ringworm, black dot ringworm) is a form of contagious dermatophytosis affecting the scalp, hair follicles, and hair shafts, with scaling, black dot, and hair loss (alopecia). Lesions usually are confined to discrete patches but, in time if left untreated, may cover the entire scalp. Sometimes, the changes can be very subtle with mild scaling and little hair loss, like dandruff, or with patchy loss of hair but no scaling (see alopecia areata type, below), or the black dot form with minimal inflammation. Such types are not always easy to detect. The pathognomonic sign is hair loss. Tinea capitis is a common disease in children, less common in adolescents, and rare in adults. Five types of the disease are classified from the least to most inflammatory:

1. *Alopecia Areata Type* (rare). Hair loss occurring in patches. Smooth alopecia without hyperkeratosis or inflammation (Fig. 21.3.)
2. *Black Dot* (common). Well-defined patches of hair stubs, scaling (noninflammatory). The pattern of hair loss is typical of endothrix hair invasion by *Trichophyton tonsurans* and *T. violaceum*. The fungus, growing and sporulating within the hair shaft (endothrix) weakens it so that it breaks at the scalp, leaving black dots as remnants of the hair shafts. If untreated the disease may last for years. Although kerions are rarely seen, the disease was first noticed because of kerion production. Occasionally, tinea pedis and tinea unguium may develop via autoinoculation. Black dot ringworm is notable for its occurrence in the African-American population. Occipital lymphadenopathy is present in about one-third of affected children (Fig. 21.4).
3. *Seborrheic* (most common). An inflammatory form with scalp erythema, hyperkeratosis, pustules, nodules, oozing and crusting, and hair loss in patchy or matted pattern. The majority of children with this form have occipital lymphadenopathy.
4. *Kerion* (rare). A highly inflammatory form with pustules and a tender, purulent, boggy mass, with lymphadenopathy and caused by *M. canis* and *T. tonsurans*. Kerion may result in scarring alopecia (Fig. 21.5).
5. *Favus* (very rare). A fifth type results from *T. schoenleinii* or *T. violaceum*. Eradicated in the United States, it persists in Africa and the Middle East. The clinical form is one of scutula (*definition*: matted clumps of hyphae and keratin hair debris). Favus, if untreated, can lead to cicatricial alopecia, skin atrophy, and scarring. (Fig. 21.6.) Favus type due to *T. schoenleinii*

Favus: An Illustrative Example

Medical History. A 25-yr-old Arab woman was seen at the medical center. She lived all her life

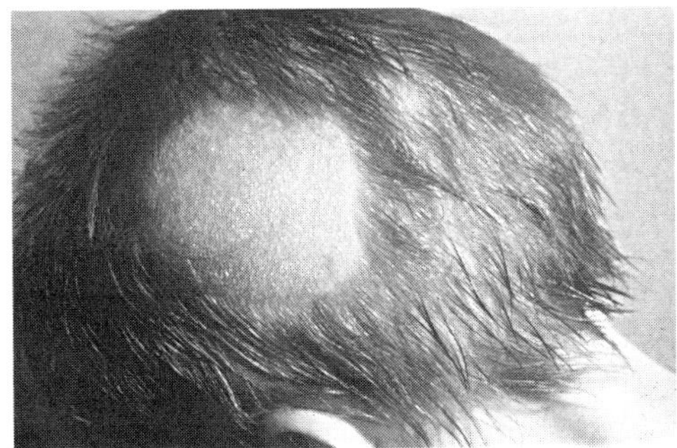
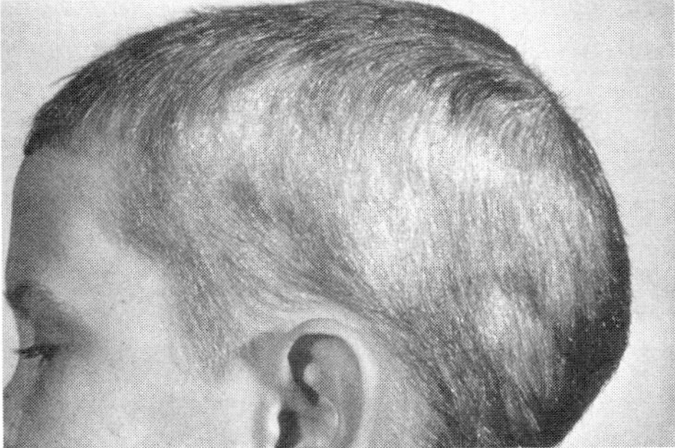

Figure 21.3 Tinea capitis, alopecia areata type due to *Microsporum canis*. Child before and after 6 weeks of treatment with oral griseofulvin. *Source:* CDC.

Figure 21.4 Tinea capitis, black dot ringworm in scalp of a child caused by *T. tonsurans*.
Source: Used with permission from Dr. Tadahiko Matsumoto, Yamada Institute of Health and Medicine, Tokyo.

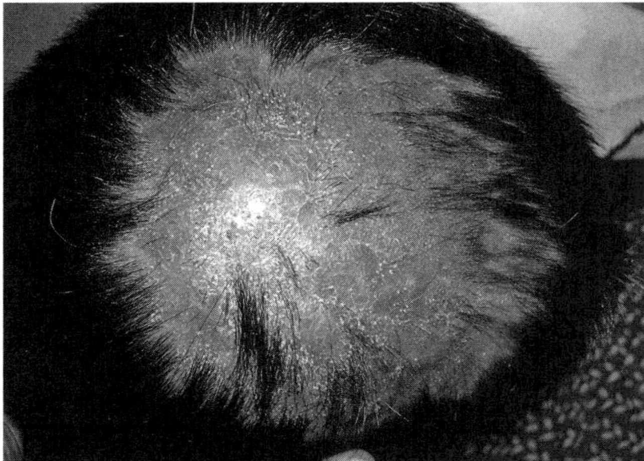

Figure 21.5 Tinea capitis, kerion type. A large boggy swelling, hair loss, erythema and scale in an 8-yr-old boy with 1 month history of regional lymphadenopathy. He was treated with oral griseofulvin (20 mg/kg/day) for 3 months.
Source: © Shahbaz A. Janjua, M.D.; used with permission from Dermatlas; www.dermatlas.org.

in an isolated area of the Sinai Desert. Her scalp lesions were of 5 years' duration, but she had no previous access to medical care (Dvoretzky et al., 1980).

Physical Exam. Extensive lesions on her scalp consisted of crusts and scutula, which had coalesced forming a grayish-white cap from the forehead to the occipit. She was otherwise in good health.

Laboratory Findings. A pierced scutulum specimen mounted in KOH revealed endothrix segmented mycelia in the hair shaft. A slow-growing mold with cerebriform tan-colored colony and waxy folds was isolated on SDA. Microscopic morphology revealed split antler-like filaments, favic chandeliers, and intercalary or terminal chlamydospores. These findings identified the culture as *T. schoenleinii*.

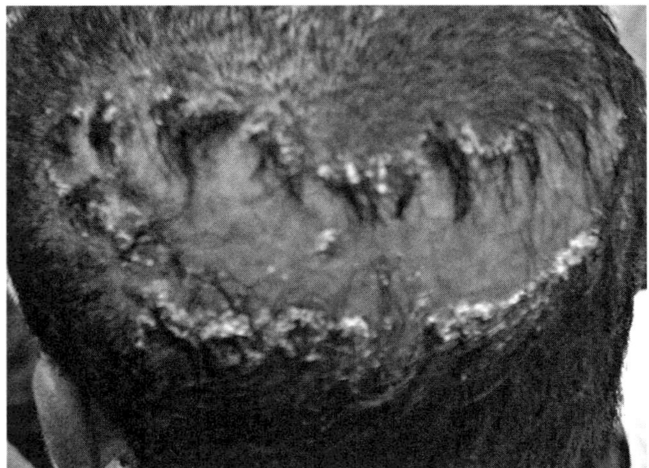

Figure 21.6 Tinea capitis, favus type due to *T. schoenleinii*. A crescent shaped, atrophic, fibrotic plaque with peripheral crusting. A 30-yr-old mentally retarded man was evaluated for a 3 year history of a progressive crusted plaque on the occiput, which healed with scarring and atrophy. Scattered scutulas pierced by hair were still present. A KOH prep from the crusts revealed hyphae and mycelium. *Trichophyton schoenleinii* was isolated from a fungal culture.
Source: © Naeem Alhayani, D.V.D., used with permission from Dermatlas; www.dermatlas.org.

Therapy and Clinical Course. Scutula and crusts were removed by vigorous scrubbing with soap and water. Therapy with oral griseofulvin, 500 mg/day, was initiated. The patient responded quickly, and was seen 2 months later. Tinea capitis was cured but left extensive scarring and almost total baldness.

Hair Invasion Because the cuticle of the hair cannot be penetrated, fungi grow deep into the hair follicle (Habif, 2009), invading the keratinized outer root sheath and the inner cortex. They digest keratin in the hair shaft. Hyphae grow only above the zone of new keratinization. Growing inside the hair shaft, hyphae fragment into short arthrospores (infectious arthroconidia). They remain inside in the endothrix type, but in the ectothrix type, they penetrate the hair surface cuticle to form a sheath of dense packed round arthroconidia, either large (6–10 μm visible in the light microscope) or small (2–3 μm, not visible with low power microscopy).

Endothrix hair invasion: *T. tonsurans, T. soudanense*, and *T. violaceum*.

Ectothrix hair invasion: *M. audouinii, M. canis*, and *T. verrucosum*. Inflammatory tinea related to exposure to a kitten or puppy usually is a fluorescent small-spore ectothrix. Arthroconidia that surround the hair appear as a sheath.

Three main patterns of hair invasion are large-spored endothrix, large-spored ectothrix, and small-spored ectothrix.

- Large-spored endothrix (*T. tonsurans* and *T. violaceum*): chains of large arthroconidia are packed densely in the hair shaft, "like a sack full of marbles."
- Large-spored ectothrix (*T. verrucosum* and *T. mentagrophytes*): chains of large arthroconidia inside and on the surface of the hair shaft are visible with the low-power objective.
- Small-spored ectothrix (*M. canis* and *M. audouinii*): small arthroconidia on the surface of the hair shaft are not visible with the low-power objective.

Tinea Barbae

This dermatophytosis of the bearded area of the face and neck is a variant of tinea capitis, most often affecting men who work with stock animals and associated fomites, providing opportunities for transmission of zoophilic dermatophytes such as *T. verrucosum* and *T. mentagrophytes*. In urban settings, however, anthropophilic fungi may be involved. In either case, the causative fungus invades the hair and hair follicle producing an inflammatory reaction in the host. Infection may be superficial or severe. Superficial tinea barbae lesions resemble those of tinea corporis, with dry central scaling. Hairs may break near the skin surface and fungal masses may occlude hair follicles, causing alopecia (Fig. 21.7).

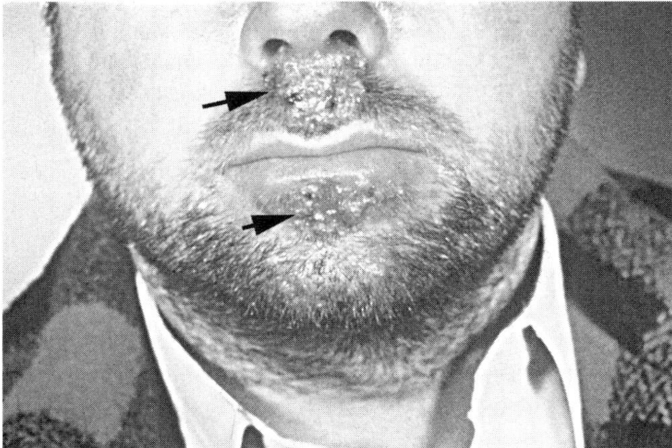

Figure 21.7 Tinea barbae, with lesions (*arrows*) in moustache area and below lips.
Source: H. J. Shadomy.

The deep, pustular type of tinea barbae is accompanied by severe inflammation with pustules, exudates, and crusts. Kerions and draining sinuses undermine adjacent tissue. Hairs become loose and are painlessly removed, revealing a viscous whitish material. Pressure on the epilated areas releases an abundant purulent discharge. Although single lesions are most common, the entire bearded area may become involved. Permanent alopecia may result (Rutecki et al., 2000).

Differential diagnosis includes *Staphylococcus aureus* and other bacterial causes of folliculitis (sycosis barbae and sycosis vulgaris), furunculosis or carbuncles, acne vulgaris, *Candida* folliculitis, rosacea, "pseudofolliculitis," eczema herpeticum, severe contact dermatitis, sporotrichosis, and, rarely, a rapidly growing tumor.

Tinea Corporis

This form is the classic ringworm: well-circumscribed patches or plaques are present on the smooth skin of the trunk, excluding the scalp, bearded area, groin, hands, and feet. Lesions are round to oval, with elevated erythematous scaly borders and central clearing: the annular ("ring") shape of classic ringworm. Pruritus is present (Fig. 21.8). Atypical lesions in immunocompromised patients are flat, with poorly defined borders, and may be erythematous, scaling, large, irregular, and psoriasiform (Fernandes et al., 1998).

Tinea Incognito

This is a form of tinea corporis with the appearance of lesions changed by systemic or topical application of high-potency topical steroids. Lesions present as patches of papules, nodules, or pustules. They may retain a scaly appearance but are more extensive and irregularly shaped. Tinea incognito may affect the face, in which case the presentation resembles rosacea, lupus erythematosus discoid-like- or eczema-like lesions. When present on the trunk or limbs the picture resembles pyoderma or eczema. *T. rubrum* is most often involved but other dermatophytes also are isolated.

Majocchi's Granuloma (Elgart, 1996)

Nodular granulomatous perifolliculitis, commonly known as Majocchi's granuloma, affects immunosuppressed and immunocompetent hosts. The lesions are firm, violaceous nodules and papules accompanying onychomycosis, tinea corporis, or tinea pedis. The histopathologic picture reveals perifolliculitis with granuloma formation, caused by dermal invasion through disrupted hair follicles. Dermatophytes, usually *Trichophyton rubrum*, invade the superficial dermis evoking a granulomatous

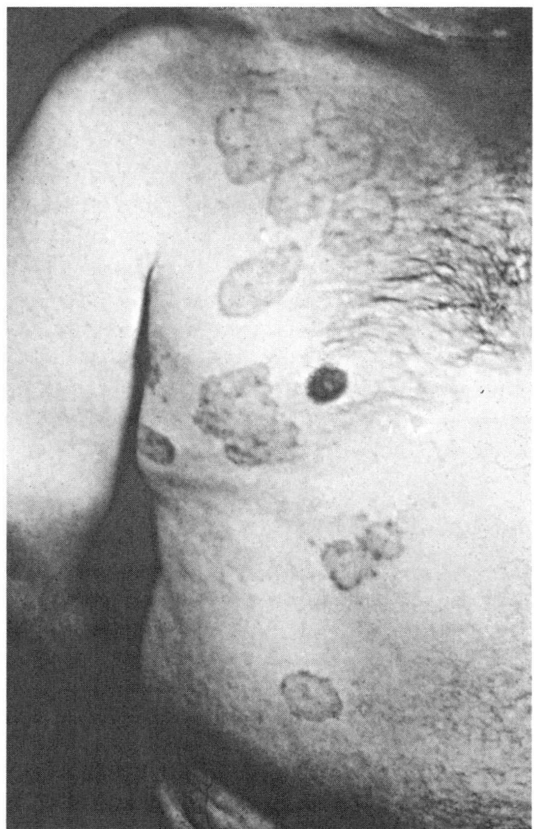

Figure 21.8 Tinea corporis. Numerous circular lesions with active borders.
Source: H. J. Shadomy.

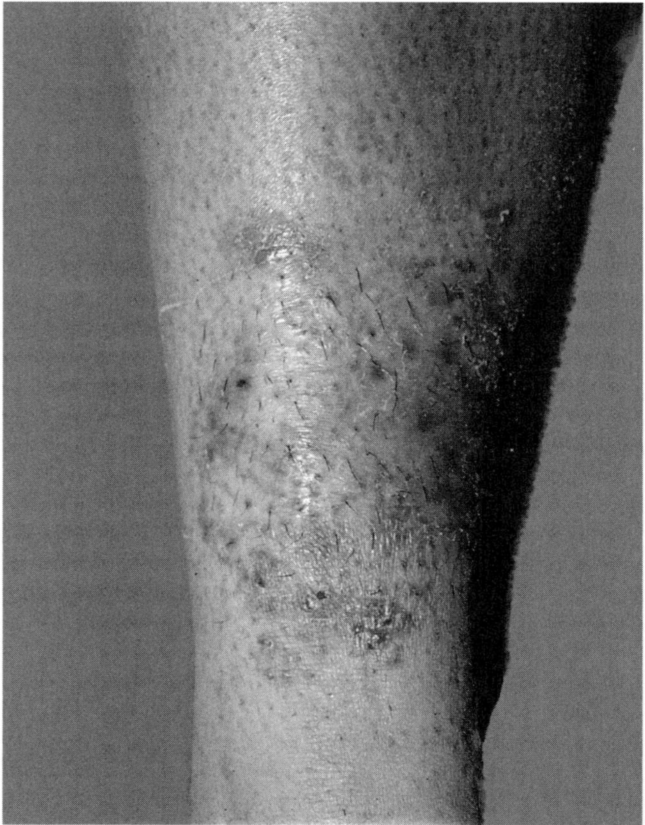

Figure 21.9 Majocchi's granuloma: dermatophyte folliculitis (*T. rubrum*) on lower leg caused by shaving. See insert for color representation of this figure.
Source: Used with permission from the estate of Dr. Orda A. Plunkett.

folliculitis. Biopsy of these lesions discloses fungal elements resembling conidia in chains as well as hyphae, with remnants of the follicle and hair fragments. The inflammatory response attracts neutrophils, histiocytes, and lymphocytes organized in a suppurative granuloma. Splendore–Hoeppli material may be present around the fungal cells.

A precipitating factor in women is shaving of legs (Fig. 21.9). Fungi present on the surface (i.e., from tinea pedis) are deposited deeper, causing fungal folliculitis. Majocchi's granuloma in men may result from razor trauma, implanting fungi beneath the skin of the beard. Biopsy of these lesions reveals fungal elements in and around the hair, with a suppurative granuloma in the surrounding dermis. Other predisposing factors include natural occlusion, as in the groin, or therapeutic occlusion, with topical steroids.

Tinea Imbricata

Tinea imbricata (syn: Tokelau ringworm, Indian or Chinese tinea, scaly tinea, elegant tinea, circinate tinea, lace tinea, chimberé, gogo, grille, cacapash, shishiyotl, roña) (Fig. 21.10) is a chronic variant of tinea corporis in which skin lesions are erythematous concentric rings of scales covering large areas of the trunk or limbs including the palms and soles. The most common form consists of large concentric scaly plaques that overlap in a layered appearance resembling tiles, fish scales, or lace. Pruritus is the most frequent symptom but may be absent. The disease is chronic and does not remit spontaneously. It spares the scalp and, sometimes, flexures. Patients appear to have ineffective immune responses and cannot clear the fungus. Diagnosis is confirmed by identification of *Trichophyton concentricum*, although the clinical appearance is distinctive.

Tinea imbricata is limited to the South Sea Islands, Malay States, central and southern China, India, Sri Lanka, the west coast of Central America, Guatemala, Mexico, and the northwest coast and central regions of South America. It may reach high incidence in the endemic areas affecting up to 30% of the population (Hay, 2006). It is limited to individuals genetically

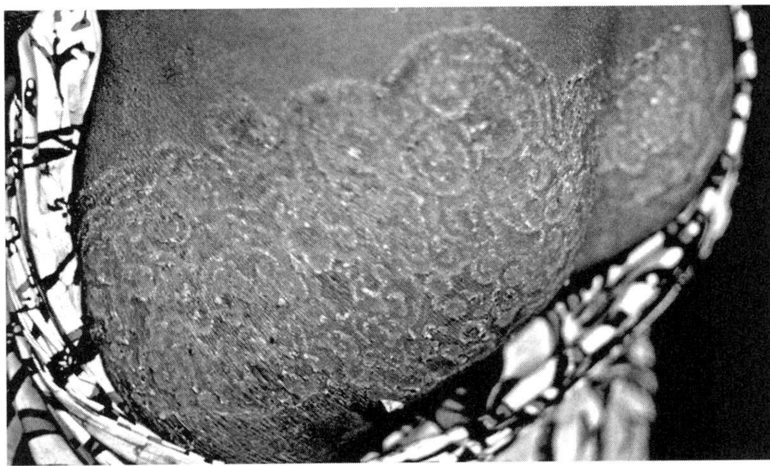

Figure 21.10 Tinea imbricata in an orang asli indigenous person, Malaysia. A chronic non-itchy rash, of several years' duration. Several individuals in his kampong had the same.
Source: Used with permission of Dr. Brian Ward, Director, the J.D. MacLean Centre for Tropical Diseases at McGill University, Montreal, Canada.

related to races in these areas. The disease susceptibility is determined by an autosomal recessive gene and is group-specific. Heterozygotes are not infected, even when other members of the same family are. Infectivity is low within families and there is no sexual preference. The disease can be devastating to individuals and their families, as they often are ostracized from their neighbors and cannot find work.

Tinea Cruris

Tinea Cruris (syn: jock itch) is a dermatophyte infection of the groin, perineum, and perianal areas. Tinea cruris may be chronic or acute, painful, or more often pruritic. Lesions begin as small, round, swollen, inflamed areas that develop into circinate and, later, into serpiginous areas. They are characterized by raised, erythematous margins and dry epidermal scaling. The lesions may be uni- or bilateral on the thighs and may or may not be symmetrical (Fig. 21.11). Tinea cruris in infants may occur in the diaper area and be confused with diaper rash due to *Candida albicans*.

Among adults, men in their 20s and 30s are more likely to become infected than women. Women may have *intertrigo* in the groin area resulting from *C. albicans*. Tinea cruris is found worldwide, more often in the tropics, and other areas where high humidity leads to maceration of the groin. It is common among athletic teams, military troops, persons using locker rooms and dormitories, and persons in prisons. Towels, linens, and clothing are implicated in transmission. *T. rubrum* is the major infecting fungus in the United States and developed countries of Europe.

Tinea Pedis (and Tinea Manuum)

Tinea pedis is dermatophytosis of the feet, especially the interdigital areas (webs) and soles. Tinea manuum is dermatophytosis of the hands. Tinea pedis and tinea manuum

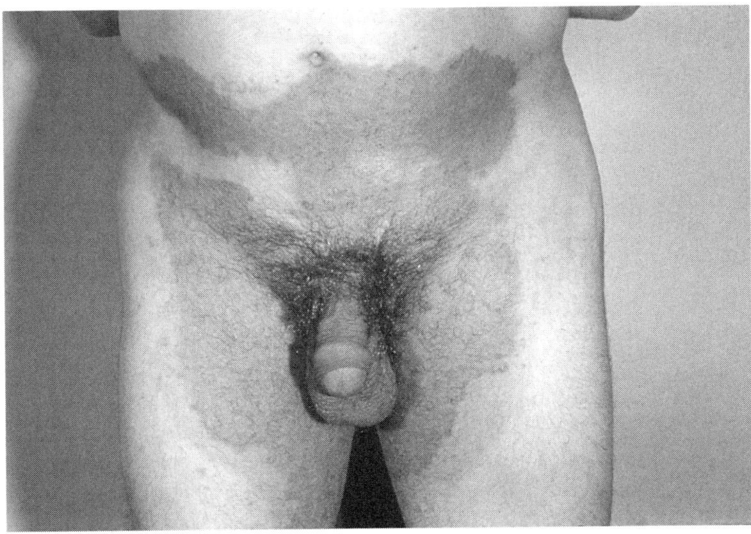

Figure 21.11 Tinea cruris (*Trichophyton rubrum*).
Source: E. Reiss, CDC.

are considered to be the same clinical entity. The disease may be acute or chronic interdigital, chronic hyperkeratotic, or vesicular. The clinical form is influenced by the causative dermatophyte. The moist, acute form is caused primarily by *Trichophyton interdigitale*, whereas the dry form is most often caused by *T. rubrum* (Zaias and Rebell, 2003).

- *Interdigital Type.* Interdigital infection is accompanied by itching, peeling, maceration, and fissuring of the toe webs. If left untreated the infection may spread, including the toenails. *T. interdigitale* infection is also associated with white superficial onychomycosis.
- *Dry Hyperkeratotic Type.* The term "moccasin" or "sandal" foot is used to describe the dry, hyperkeratotic type of infection limited to the heel and sole of the foot, continuing up the side in the area covered by a moccasin (Fig. 21.12). The scaling is very fine and silvery, and the skin underneath is usually pink and tender, sometimes with severe erythema. This dermatophytosis is frequently bilateral and is often accompanied by onychomycosis. *T. rubrum* is the causative agent. Moccasin-type tinea pedis is resistant to topical therapy.
- *Vesicular Type.* Chronic tinea pedis caused by *T. interdigitale* is accompanied by episodes of pruritic, larger than 2 mm in diameter, multilocular vesicles, or bullae, in the thin nonplantar skin of the plantar arch (Fig. 21.13). These pruritic lesions, their residual collarets, and coarse desquamation are a sign of chronic *T. interdigitale* infection, and not a sign of the more prevalent *T. rubrum* infection (Zaias and Rebell, 2003). Initially filled with clear fluid, they may be isolated to coalesce in blisters. After rupturing, they may resolve or recur. Secondary infection results in pustules. Another wave of bullae may involve remote body sites: arms, chest, or the sides of fingers. This is an allergic reaction; please see "Dermatophytid Reaction," below.

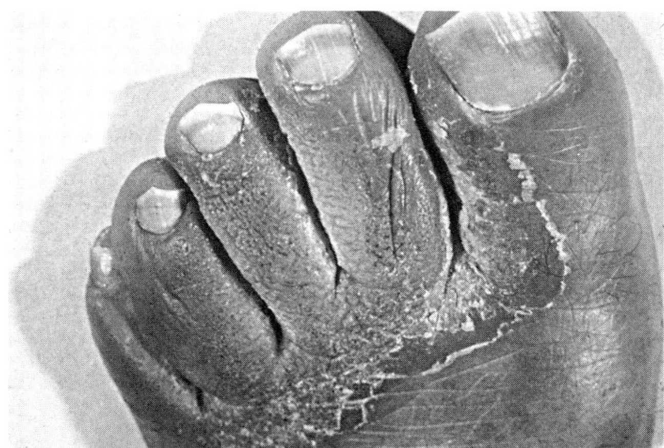

Figure 21.12 Tinea pedis, dry, hyperkeratotic type caused by *Trichophyton rubrum*.
Source: H. J. Shadomy.

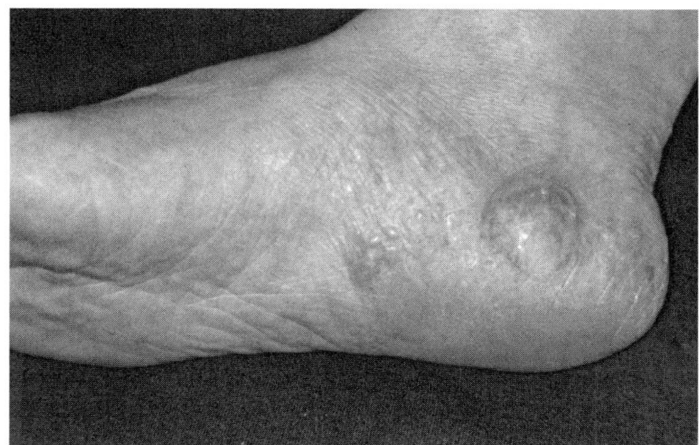

Figure 21.13 Tinea pedis, vesiculobullous type. Vesicles in the plantar arch extend along the left lateral surface to the heel. A 69-yr-old woman presented with this eruption of 2–3 weeks' duration. *Trichophyton interdigitale* was isolated.
Source: Used with permission from Dr. Masataro Hiruma, Department of Dermatology & Allergology, Jutendo University Nerima Hospital, Tokyo.

Tinea pedis caused by *T. interdigitale* is recognized by two signs not characteristic for *T. rubrum* tinea pedis: bullous >2 mm vesicles in the thin skin of the plantar arch and along the sides of the feet and heel adjacent to the thick plantar stratum corneum, and white superficial onychomycosis. This pattern of bullous vesicles is a hallmark of *T. interdigitale* dermatophytosis (Zaias and Rebell, 2003). *T. rubrum* tinea pedis always involves the sole of the foot, producing asymptomatic pinhead-sized vesicles in the thick plantar stratum corneum, and at one end of the spectrum, "moccasin" tinea pedis.

Chronic *T. interdigitale* infection accounts for approximately 15% of the tinea pedis in urban communities and is basically a life-long condition. Episodes of groups of pruritic, larger than 2 mm, vesicles, or bullae, in the thin nonplantar skin of the plantar arch, coincide with periods of foot stress, such as wearing occlusive shoes. Superficial lesions of the surface of the nail plate (white superficial onychomycosis) are usually present. Except in severe cases the thick plantar skin remains free of lesions. The different patterns of foot infection also are reflected in different types of onychomycosis caused by the two species—*T. rubrum* causes mycotic lesions *beneath* the nails in the nail bed (distal subungual onychomycosis), whereas *T. interdigitale* infects the surface of the nail plate. Zaias postulates that the nail surface lesions act as a reservoir to periodically reseed the periplantar skin of the sole and toes and, rarely, of the groin and other body sites.

The differential diagnosis of tinea pedis includes Gram-positive bacterial infection, such as erythrasma caused by *Corynebacterium minutissimum*, or Gram-negative bacterial infection, or a yeast infection, candidiasis. Lesions of erythrasma emit a coral red fluorescence in Wood's light. Cultures are *important* to uncover these causative agents or to realize the presence of a dual infection. *Candida albicans* can cause athlete's foot, but usually is secondary to dermatophyte infection, occurring more often in the tropics. Rarely, nondermatophyte molds are involved (e.g., *Neoscytalidium* species).

"Two Feet One Hand" Syndrome

Occurring predominantly in adult men, the feet are involved before the hands become infected. In over 91% of cases tinea pedis precedes tinea manuum by an average of 8.8 years. Patients relate that tinea manuum developed in the hand that scratched the pruritic soles of the feet or picked the toenails (Daniel et al., 1997).

Dermatophytid Reaction

This reaction (also known as an "id" reaction or autoeczematization) is an acute, local or generalized vesicular eruption that develops in an anatomic location remote from the site of dermatophytosis, for example, vesicles on the hands and tinea capitis or tinea pedis (Brown, 2009). The basis of dermatophytids is sensitization to circulating fungal antigens. The incidence is unknown, but 4–5% of persons with dermatophytosis report having id reactions. Dermatophytids are a confounder in managing tinea capitis and tinea pedis because they often are misinterpreted as a medication allergy by parents and physicians who have never seen an id reaction.

Prevention of Tinea Pedis Preventive steps include avoiding or minimizing the wearing of heavy occlusive shoes or boots; wearing absorbent socks; wearing shower shoes in health spas and gyms; and using of antifungal foot powder.

Tinea Unguium

Tinea unguium (*definition*: invasion of the nail bed by dermatophytes) Onychomycosis refers to nail infections due to dermatophytes *or to other fungi*. Untreated tinea pedis often leads to tinea unguium. A complete visual atlas of clinical types of onychomycosis may be found in Rich and Scher (2003). Nardo Zaias named the *T. rubrum* onychomycosis, a nail bed infection—with no invasion of the nail plate—distal subungual onychomycosis (DSO) (reviewed by Zaias and Rebell, 2003). In contrast, *T. interdigitale* produces onychomycosis of the surface of the nail plate, but no invasion of the nail bed, which he named "white superficial onychomycosis".

Onychomycosis

Five types of onychomycosis are classified (Faergemann and Baran, 2003):

- *Leukonychia.* Also known as white superficial onychomycosis, leukonychia is a superficial white infection with patches or, at times, pits on the nail surface. The causative agent is likely to be *T. interdigitale*. *Black* superficial onychomycosis is caused primarily by melanized molds (e.g., *Neoscytalidium dimidiatum*). Exceptionally, pigmentation affecting a great toenail and simulating longitudinal melanonychia was caused by *T. rubrum* producing a diffusible black pigment (Perrin and Baran, 1994).
- *Distal Lateral Subungal Onychomycosis (DSO).* This is the most common form of tinea unguium (~90%) (Fig. 21.14). *T. rubrum* is the most frequent cause of this nail bed infection, with no invasion of the nail plate. It usually begins as a discolored area at a corner of the big toe, the fungus grows under the nail, slowly spreading toward the cuticle. Eventually the toenail becomes thickened and flaky. This common type of tinea unguium (90%), most often affects the big toenails and is less common on the finger nails. Some *T. rubrum* strains are strongly pigmented, resulting in melanonychia. A study of 13 pedigrees with autosomal dominant patterns of familial *T. rubrum* DSO and tinea pedis implicates genetic susceptibility as well as intrafamiliar contagion in this mycosis (Zaias et al., 1996).
- *Endonyx Subungual Onychomycosis.* The nail plate as well as the nail bed is infected. Causative agents are *T. soudanense* and *T. violaceum*.
- *Proximal Subungual Onychomycosis.* Starting at the cuticle (base of the nail) the infection spreads slowly toward the nail tip, raising the nail up. This is the least common type of tinea unguium in immune-normal persons (3%) but the most common form in persons living with AIDS.
- *Total Dystrophic Onychomycosis.* The whole nail is involved in a combination of all the above types of disease.

The fungi most commonly seen in onychomycosis, depending on climate, migration, and geography, are

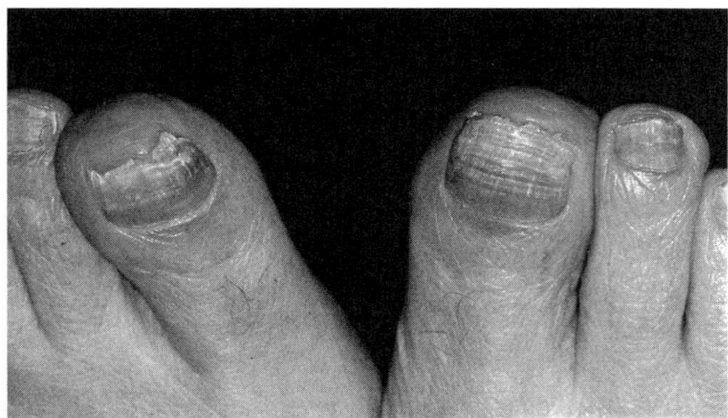

Figure 21.14 Tinea unguium, distal subungual type, great toe. *Trichophyton rubrum* is the most common cause.
Source: Used with permission from Dr. Tadahiko Matsumoto, Yamada Institute of Health and Medicine, Tokyo.

the dermatophytes: *T. rubrum* (71% of all infections) and *T. interdigitale* (20%), *Candida* species (5.6%), other molds including *Scytalidium dimidiatum* (now *Neoscytalidium*) (1.7%), and *Scopulariopsis brevicaulis* (1.7%) (Faergemann and Baran, 2003). In addition to the two major dermatophyte causes of onychomycosis, rarely one of the following may be found: *Epidermophyton floccosum, Microsporum gypseum, M. canis, M. audouinii, M. gourvilii, Trichophyton megninii, T. schoenleinii, T. soudanense, T. concentricum, or T. verrucosum* (Kane et al., 1997).

Differential Diagnosis for Onychomycosis

About half of cases believed to be onychomycosis are other types of disorders, for example, aging, chronic onycholysis, chronic *paronychia*, contact dermatitis to nail polish or shoe dye, hemorrhage or trauma, lichen planus, median canalicular dystrophy, onychogryphosis, psoriasis, subungual malignant melanoma, or subungual squamous cell carcinoma (Gupta and Ricci, 2006).

Yeast onychomycosis caused by a *C. albicans* or other *Candida* species may be the most common cause of fungal fingernails. Yellow, brown, white, or thickened nails may develop and, in addition, there may be *paronychia* (please see Chapter 11).

Laboratory findings necessary to confirm a diagnosis of tinea unguium are discussed in Section 21.14, Laboratory Detection, Recovery, and Identification.

Dermatophytosis in the Immunocompromised Host

Dermatophytosis in the immunocompromised host affects persons living with AIDS and solid organ transplant recipients. Among the latter, an estimated up to 40% of renal transplant recipients may be affected (reviewed by Venkatesan et al., 2005). In addition to the typical appearance of tinea corporis in both of these patient groups, the spectrum of disease also includes deep dermal and nodular granulomatous perifolliculitis. Exceptionally, multisystem disseminated disease may be seen. The deep dermal form presents as multiple fluctuant erythematous, ulcerative nodules on the extremities, near localized chronic superficial lesions. Nodular granulomatous perifolliculitis, commonly known as Majocchi's granuloma, may occur (please see Section 21.11, Clinical Forms—Majocchi's Granuloma). Dermatophytosis is often refractory to treatment in the immunosuppressed patient. Tinea pedis may spread to the dorsum of the foot, the "moccasin" type. *Trichophyton rubrum* onychomycosis, in the proximal subungual form, is a classic nail sign in the HIV-positive patient.

21.12 VETERINARY FORMS

Zoophilic dermatophytes (Chermette et al., 2008; Songer and Post, 2005) are *M. canis* from cats and dogs; *T. mentagrophytes* from rodents and farm animals; *T. verrucosum* from cattle; *M. nanum* from swine; and *T. equinum* from horses. Monkeys, rodents, and horses also are known to be infected with *M. canis*. (Please also see Section 21.8, Risk Groups/Factors—Exposure to Animals, and Section 21.9, Transmission—Companion Animals.)

Cats

Cats are the main reservoir for *Microsporum canis*, the etiologic agent in up to 90% of feline dermatophyte infections. Catteries are a common site for recurrent ringworm, mostly occurring in young animals. Lesions may be small and inconspicuous, particularly in adults, so that the infection may be detected only after the owner or breeder becomes infected. First appearing on the bridge of the cat's nose, circular lesions and hair loss may spread to the external side of the pinnae, distal part of the legs, and the tail. Carriage of dermatophytes is frequent in cats and Wood's lamp can help reveal the presence of *M. canis*.

The disease spectrum extends from mild scaling to severe folliculitis and general hair loss. Persian and Angora cats are more susceptible to ringworm, which may present as severe mycetoma-like lesions. These appear as cutaneous nodules on the back and neck, which may ulcerate or, if opened, discharge an orange viscous fluid containing fungal elements. *M. canis* also causes feline otitis characterized by a waxy discharge from the ear canals.

Dogs

Single or multiple, patchy, hairless, circular lesions, 1–4 cm in diameter, are most often seen on the dog's face and are caused by *M. canis, M. gypseum*, and *T. mentagrophytes*. Other sites affected are the neck, back, haunches, elbow, and paws. Hairs become brittle and the skin is dry and scaly, with crusts and scabs present. Widespread chronic dermatophytosis is accompanied by severe inflammation, pruritus, and, in extreme cases, total alopecia. Kerions may develop when *T. mentagrophytes* is involved.

Horses

Equine dermatophytosis is caused by *T. equinum*, in young and adult horses, and *M. canis*, seen mostly in young horses. Other species involved in are *M. gypseum* and *T. mentagrophytes*. The lesions are initially dry with powdery scales and broken hairs, on any body part, especially under the saddle and girth areas. Inflammation and exudates cause hair to become matted, forming large lesions with a moth-eaten appearance. *T. equinum* infection may spread rapidly to the entire body and to nearby animals. Disease may be spread by means of infectious hairs and scales present in grooming brushes, saddle pads, other tack, and on attendants' clothing.

Cattle

Ringworm in cattle caused by *T. verrucosum* is highly contagious and may spread to humans. Calves are more susceptible than adults. Lesions occurring on the head and neck are circular, scattered, with scaling and alopecia. Scabs and crusts may develop. Cattle ringworm is more frequent when animals are confined in winter and may resolve after turn-out in the spring.

Poultry

Termed "favus" or "white comb," poultry ringworm is caused by *M. gallinae*. The comb may be encrusted with a thick white covering, causing feather loss, emaciation, and even death.

Swine

The classic cause of ringworm in pigs is *M. nanum*, but other dermatophytes may be isolated including *T. verrucosum, M. canis, M. gypseum*, or *M. persicolor*. Factors influencing the causative species are how the animals are housed and their contact with soil or with other domestic animals. Piglets and fattening pigs are more often infected. Lesions, usually on the head and the neck, are small, circular, roughened, and mildly inflamed. They spread, covering large areas, and may form crusts, but usually without hair loss.

Dermatophytosis in swine is self-limiting and lesions may spontaneously resolve in 2–3 months. Risk factors are high animal density, poor sanitation, and high humidity. The trend toward organic pig farming may increase the prevalence of swine ringworm due to closer contact of animals with the outdoor environment.

Systemic Therapy for Animal Dermatophytosis

Therapy includes clipping and applying topical and systemic antifungal agents. Griseofulvin is the standard systemic treatment for animal dermatophytosis. The dosage in dogs and cats consists of oral micronized griseofulvin: 25 mg/kg, 2x/day. The usual dosage in large animals is 7.5–10 mg/kg/day. Griseofulvin is not recommended in food animals, in pregnancy, and for the asymptomatic carrier state. ITC and TRB are current therapies. An effective regimen for feline dermatophytosis consists of oral ITC pulse therapy: 5 mg/kg/day for one week, every 2 weeks, for 6 weeks. TRB is effective in cat and dog dermatophytosis at a dosage of 20–30 mg/kg/day.

Lufenuron is a chitin synthesis inhibitor used to prevent companion animal flea infestations. The use of lufenuron to treat dermatophytosis is controversial because of conflicting reports of its effectiveness. A single oral dose of lufenuron was reportedly effective in treating cat and dog ringworm (Ben-Ziony and Arzi, 2000). The effectiveness of lufenuron was questionable according to a study of experimental infection of juvenile cats treated with either monthly oral lufenuron or placebo (Moriello et al., 2004). Treated cats did not recover faster than untreated cats. Lufenuron did not prevent dermatophytosis or alter the course of infection after direct topical challenge.

Combined Systemic and Topical Treatment of Dermatophytosis

Combined treatment of dermatophytosis in dogs and cats should be maintained for at least 10 weeks. Ringworm is difficult to eradicate in catteries and animal shelters, posing a health hazard for people in contact with the animals. Control of dermatophytosis in catteries requires topical and systemic treatment of all cats coupled with a strong environmental decontamination program.

Topical Therapy for Animal Dermatophytosis

Choices for topical therapy of animal ringworm may be classed in two categories: older therapies with low selective toxicity for fungi, and more modern therapies using azoles and polyenes. Older therapies are lime-sulfur and sodium hypochorite (bleach). Bleach cannot be recommended because it is an irritant. Lime-sulfur is used in the United States to treat dog and cat ringworm in combination with griseofulvin or ITC. It has a disagreeable odor and stains the coats of some animals. Cats, after licking their treated coats, have developed oral ulcers.

Azoles and Polyenes

Enilconazole (0.2%) solution is used to treat dermatophytosis in dogs, cats, horses, and cattle and has few adverse effects. Another topical regimen for animal ringworm consists of miconazole (2%), and chlorhexidine (2%), 2x/week. Topical natamycin is a therapeutic choice within the polyene class.

The effectiveness of topical therapy depends on compliance by the owner in covering the animal's body, not just spot treatment of the lesions. Twice weekly treatments are recommended. Clipping of the coat facilitates the application and penetration of topical therapy, taking care to disinfect the instruments and surfaces afterwards.

21.13 THERAPY

Therapy for Tinea Unguium

See Table 21.6 (de Berker, 2009).

Topical Therapy

Choices for topical therapy are amorolfine lacquer, 5% (Loceryl®[3]); tioconazole paint, 28% (Trosyl®); and ciclopirox olamine lacquer, 8% (Penlac®). These antifungals are best suited to treat distal subungual onychomycosis. Mycologic cure rates have been reported to be 38–54% with amorolfine (2×/week for 6 months); 20–70% with tioconazole (2×/day for 6–12 months); and 28–36% for ciclopirox olamine (daily use for 48 weeks). Discouraging results were observed in that a normal nail was restored after ciclopirox treatment in only 7% of cases. Improved results are obtained by filing the upper nail surface which reduces the amount of infected nail.

Systemic Therapy

Oral agents to treat tinea unguium that have U.S. FDA approval are terbinafine (TRB) (Lamisil®) and itraconazole (ITC) (Sporanox®). Oral dosing to treat toenail mycosis with TRB is 250 mg/day or ITC (200 mg/day). Clinical cure, defined as negative mycologic cultures and clearance of >87.5% of the nail at 72 weeks, occurred in 60% of patients taking TRB for 16 weeks, as compared with 32% of patients receiving 7 day pulses of ITC each month for 3 or 4 months (de Berker, 2009). Debridement, preferably performed by a podiatrist, reduces the amount of infected nail and may be useful in treating more resistant nondermatophytes and for managing deformed nails. FLC may be used to treat tinea unguium in patients unable to tolerate other oral antifungal agents because of adverse effects or drug interactions. The mycologic cure rate was lower with FLC (31.2%) compared with TRB (75%) and ITC (61.1%) (reviewed by Brown SJ, 2009). FLC shows a long plasma $t_{1/2}$ of 30 h and can be detected in hair and nails for 4–5 months after cessation of oral therapy.

[3]Loceryl and Trosyl are not currently available in the United States.

Therapy for Tinea Capitis

An outline of the courses of therapy for tinea capitis is presented here. A fuller discussion may be found in Pomeranz and Sabnis (2002). Those authors also discuss the types of preparations (liquid suspension, pills, mixed with food, etc.) most suitable for younger children.

Topical Therapy

Ketoconazole (Nizoral) shampoo is effective only against *Trichophyton* tinea capitis. Selenium sulfide (Selsun blue) shampoo also is available as adjunctive therapy. Severe inflammatory tinea capitis may benefit from a short burst of oral corticosteroids or topical corticosteroids. Topical corticosteroids for *smooth skin* dermatophytosis, however, may merely mask symptoms and conceal the infection, which may recur.

Systemic Therapy

Griseofulvin The dosage regimens for adults and children, as well as duration and outcome of treatment, may be found in Chapter 3A, Section 3A.14, Griseofulvin.

Allylamine and Azole Therapy TRB and ITC concentrate in skin, hair, and nails, allowing shorter treatment courses than griseofulvin (≤4 weeks). Duration of treatment for allylamine and azole therapy (Gupta and Cooper, 2008) is determined with respect to *T. tonsurans* tinea capitis. Higher doses and/or longer duration may be needed for *M. canis* tinea capitis. A course of TRB therapy for children is 5 mg/kg/day for 2–4 weeks: that is, for a 31–40 kg child the daily dose would be 125 mg. ITC (capsules) dosing for children is 5 mg/kg/day for 2–4

Table 21.6 Antifungal Agents for Dermatophytosis

Chemical class, generic name	Trade name	Formulations[a]
	Azoles (imidazoles)	
Clotrimazole	Fungold (Pedinol Pharmacal)	C, L, S, T
	Lotrimin (Schering-Plough)	
	Mycelex (Bayer)	
	Gyne-Lotrimin (Schering-Plough)	
Econazole	Spectrazole (Ortho Dermatological)	C
Ketoconazole	Nizoral (Janssen Pharmaceutica)	C, S
Miconazole		C, L, S, P
	Micatin (McNeil Labs)	
	Monistat-Derm (Ortho Dermatological)	
	Monistat-3	
	Monistat-7	
Oxiconazole	Oxistat (Glaxo-Wellcome)	C, S
Sulcoconazole	Exelderm (Westwood-Squibb)	C, S
	Azole (triazole)	
Itraconazole	Sporanox (Janssen Pharmaceutica)	OS
	Allylamines and other nonazole ergosterol synthesis inhibitors:	
Amorolfine	Luceryl (Roche Labs)	NL
Butenafine HCL	Meritax (Penederm)	C
Naftifine	Naftin (Allergan)	C, O, P
Terbinafine	Lamisil (Sandoz)	C, S
	Other agents	
Ciclopirox olamine	Loprox (Hoechst Marion Roussell)	C, L
	Penlac (Dermic)	NL
Haloprogin	Halotex (Westwood Squibb)	C
Tolnaftate		C, S, P
	Aftate (Schering-Plough)	
	NP-27(Thompson Med)	
	Tinactin (Schering Plough)	
	Ting (Fisons)	
Undecylenate		C, P, O, S
	Cruex (Fisons)	
	Desenex (Fisons)	

[a]C, Cream; L, lotion; NL, nail lacquer; O, ointment; OS, oral suspension; P, powder; S, solution/spray; T, troche.

Source: Adapted from "Other Topical Agents" Dr. Fungus.org website.

weeks: that is, for a 31–40 kg child the regimen would consist of 100 mg once/day and 100 mg 2x/day on alternating days. ITC pulse therapy is accomplished by 1–3 pulses using the same dosing regimen for 1 week then 3 weeks off. FLC, oral suspension, is given at a dose of 6 mg/kg/day for 20 days.

Control of transmission in families may include Nizoral or selenium sulfide shampoos to reduce or eliminate the asymptomatic carrier state. In the case of zoophilic dermatophytes, pets may be referred for treatment or adjunctive topical therapy (shampoos). Hats, blankets, and combs can be disinfected by washing with dilute bleach.

Therapy for Smooth Skin Dermatophytosis

For tinea corporis, tinea cruris, tinea pedis, and tinea manuum, topical medications applied once or 2×/day are the primary treatment. Various topical agents are available as creams, gels, lotions, and shampoos (Table 21.6). Cure rates of tinea corporis/tinea cruris/tinea pedis are high, with infections resolving after 2–4 weeks of topical therapy.

Topical Therapy for Tinea Pedis, Tinea Manuum

Creams or lotions are applied 2×/day for 1–4 weeks: for example, TRB, clotrimazole, miconazole, butenafine, econazole, naftifine, oxiconazole, or cyclopirox 0.77% cream or gel 2x/day for 4 weeks. An alternative is ketoconazole (KTC) cream, 2%, once/day for 6 weeks. Antifungal powder is used to prevent tinea pedis. Laundering socks in 60°C water reduces or eliminates fungal elements. Moccasin tinea pedis (*T. rubrum*) is characterized by thick hyperkeratotic scale, which hinders absorption of topical therapy. Topical keratolytics, such as 40% urea cream, dissolve the intercellular matrix, increasing the shedding of scale. A study of 12 moccasin tinea pedis patients found that once daily topical urea cream and 2x/day ciclopirox cream resulted in cure in all patients after 2–3 weeks (Elewski et al., 2004).

Topical and Oral Therapy for Tinea Corporis and Tinea Cruris

Topical therapy follows a dosing regimen similar to that for tinea pedis. Oral therapy for smooth skin dermatophytosis is reserved for instances where larger areas are involved and for chronic or recurrent infections. Oral ITC, TRB, and FLC are successful in treating tinea corporis and tinea cruris. For adults, a course of TRB is 250 mg/day for 2–4 weeks. Continuous ITC therapy is effective at 200 mg/day for 1 week, or 100 mg/day for 2 weeks. Oral FLC dosing for tinea corporis and tinea cruris is suggested at 150–350 mg/week for 2–4 weeks, or longer if needed (Niewerth and Korting, 2000). Oral KTC is not recommended because of hepatic adverse effects. Griseofulvin is not recommended because it does not bind effectively to keratin in the stratum corneum.

21.14 LABORATORY DETECTION, RECOVERY, AND IDENTIFICATION

Collection of Specimens

Prepare to collect specimens for dermatophyte identification by gathering sterile instruments for debridement and paper or pill envelopes for submission of material to the clinical laboratory. Nail scrapings may be placed between two microscope slides. Enough material should be obtained for both direct examination and culture. In the clinic, specimens are viewed microscopically before treatment decisions are made and, for confirmation, placed directly on growth medium.

Clean the site prior to collection with sterile gauze squares wetted with 70% ethyl alcohol. Cleaning with cotton balls may introduce cotton filaments, which confuse microscopic examination, as they resemble fungal elements.

- *Skin.* After cleaning, epidermal material should be scraped, or the top of a lesion removed, for microscopy and culture.
- *Nails.* Two types of specimens are nail clippings and subungual debris. Nails should be cleaned and swabbed with alcohol, and samples taken from the deeper part of the discolored or dystrophic parts of nails, from the nail plate, and nail bed. Specimens for KOH prep and for culture are taken at the same time. Nail clippings are prepared for KOH prep (please see below). For culture, the processing laboratory should preferably grind nail clippings with a nail micronizer before they are planted to culture medium (Scher et al., 2003).
    - Review the types of onychomycosis in Section 21.11, Clinical Forms. For distal subungual onychomycosis remove a large amount of the nail with a nail nipper. The nail tip, though infected, is almost always devoid of live fungi. Subungual debris at the juncture with the nail bed is removed with a sterile 1- or 2 mm curette.
    - For proximal subungual onychomycosis the overlying intact proximal nail plate may be burred thin and then penetrated so the curette can scoop up subungual debris from the nail bed.
    - In case of white superficial onychomycosis the affected areas are scraped.
- *Hair.* Hairs are infected from the scalp to Adamson's fringe, the point of incomplete keratinization of the hair where live tissue begins. Hairs should be epilated, or, in the case of black dot ringworm, remove subscalp hair shaft material.
    - *Toothbrush Technique.* Particularly useful to collect children's hair, a sterile, soft-head toothbrush is massaged over the entire scalp to obtain fungal elements and immediately planted to solid culture medium. (Please see below, "Culture.")

Direct Examination

KOH Prep

This is the basic technique for direct examination of skin, hair, and nail specimens. The method for preparation of the KOH reagent with or without fluorescent brighteners is found in Chapter 2. However, this examination may not be positive, even in the presence of clinical disease.

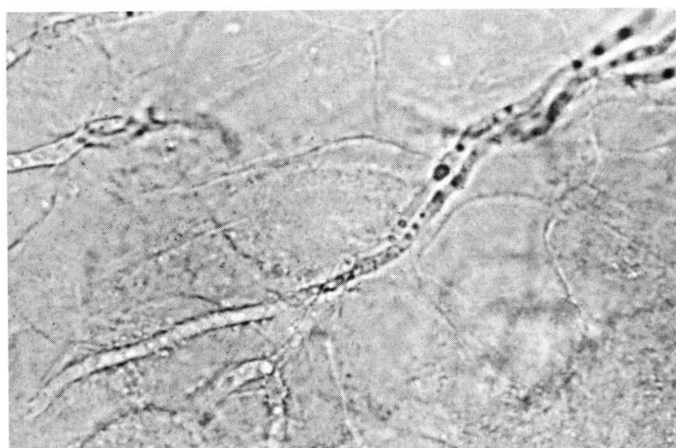

Figure 21.15 KOH prep of skin scrapings showing hyphae of a dermatophyte.
Source: H. J. Shadomy.

Procedure: Place a thin smear of the specimen on a glass slide. Add 1 drop of 10% KOH. Add a coverslip. Wait 10–15 min or gently warm the specimen (51–54°C), taking care not to boil it. When the specimen has cleared, observe it in the microscope at low and high power. It is best to use low light for observation.

Direct Examination of KOH Preps of Skin or Nail Scrapings

Viewed in the microscope, fungal hyphae are translucent, rod-shaped filaments, some with branches, also including beaded chains of swollen cells (arthroconidia) (Fig. 21.15). These are reported as "mycelium present." Other wet preparations (saline, water, or KOH–Calcofluor) of specimens demonstrate the same picture with all dermatophytes: hyaline hyphae and/or arthroconidia. Pigmented filaments indicate a nondermatophyte fungus. Clusters of small yeast with monopolar buds and short filaments indicate *Malassezia* species, the causative agents of pityriasis versicolor. Budding yeast, pseudohyphae indicate a yeast infection with a *Candida* species.

Direct Examination of Hair: Ectothrix Versus Endothrix Hair Invasion

Hairs may be viewed for endothrix or ectothrix arthroconidia and/or air channels. Direct examination of hair from infected individuals may indicate the genus of the fungus involved. Place one or two infected hairs and a drop of 10% KOH on a microscope slide, coverslip it, and observe to see whether arthroconidia are in an ectothrix (outside the hair shaft) or endothrix (within the hair shaft) pattern. The genus *Microsporum* demonstrates fungal arthroconidia in chains in an ectothrix manner (Fig. 21.16.). The most frequent *Trichophyton* species causing tinea capitis, *T. tonsurans*, invades hair in the endothrix manner. Other *Trichophyton* species may produce hyphae and large arthroconidia in either an ectothrix or endothrix pattern, or both.

Laboratory Diagnosis of Onychomycosis

These tests include a positive KOH prep demonstrating hyphae or arthroconidia, followed by a positive culture for a dermatophyte. Filaments of *Neoscytalidium* species, a minor nondermatophyte cause of onychomycosis, resemble dermatophyte hyphae but may have a thick cell wall and may be melanized (*N. dimidiatum*).

Alternative methods include histopathology with PAS stain or confocal microscopy. Sample preparation for histopathology includes an invasive nail biopsy in which the distal free edge of the nail is clipped at its attachment to the nail bed. The nail sample is thinly sectioned and stained. The cell walls of the fungal hyphae appear red. Results are prompt and, in combination with fungal culture, highly sensitive.

In Vivo Confocal Microscopy The nail can be examined with no need for stains or dyes. Fungal hyphae can be detected in the nail plate and the nail bed. The method is more discriminatory than standard microscopy.

Culture

KOH preps of tissue, hair, or nails may show hyaline hyphae and/or arthroconidia, but do not differentiate the genus or species. For this they must be grown. When nail clippings or scrapings are cultured, it is necessary to be sure the entire nail specimen is touching or embedded in the agar medium.

1. *Medium and Temperature for Growth of Dermatophytes.* Material for culture is planted on medium with antibiotics to inhibit bacterial growth. SDA with chloramphenicol and cycloheximide (SDA c+c) selects for dermatophytes and *Candida albicans*. Either original SDA or SDA-Emmons modification may be used. Mycosel™ and Mycobiotic agar have formulations similar to SDA c+c. The contents of these media are found in Table 2.1, Chapter 2. Specimens should also be planted on SDA with chloramphenicol (SDA+c) but without cycloheximide because that medium will allow growth of cycloheximide-sensitive nondermatophyte molds and yeast producing disease.

The temperature for growth is room temperature (approximately 25–30°C), or an incubator set at 30°C. Plated cultures are maintained for 4 weeks. Sufficient

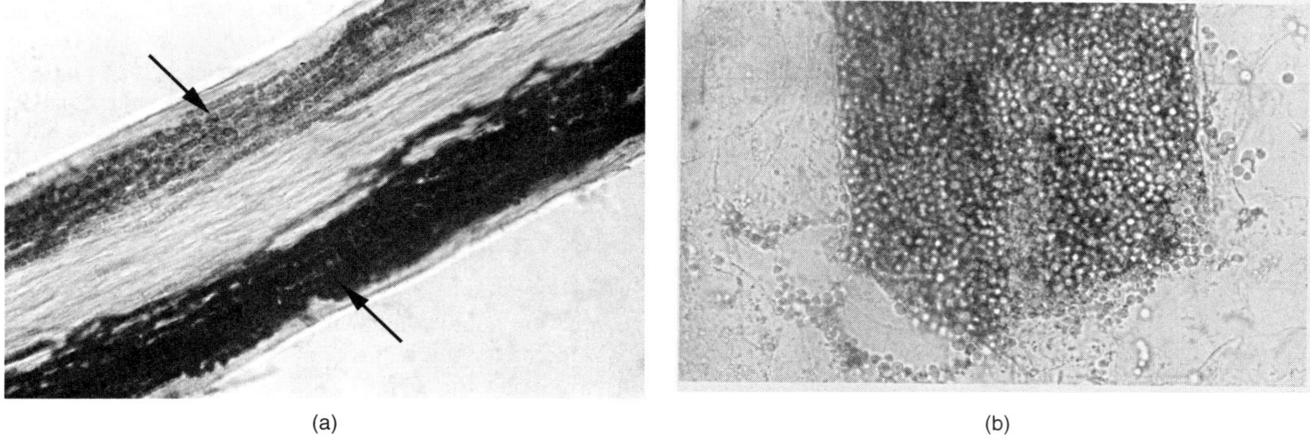

Figure 21.16 (a) Ectothrix sporulation on hair. PAS stain. Hyphae invade hair and sporulate (arthroconidia) on the hair surface (*arrows*). *Source:* Used with permission from Dr. Arvind A. Padhye, CDC. (b) Endothrix hair invasion of hair shaft with sporulation, arthroconidia only inside of hair shaft by *Trichophyton violaceum* (KOH prep, 400×) Wood's light negative.
Source: Used with permission from Dr. Uma M. Tendolkar, LTM Medical College, Mumbai.

material is needed for speciation and, if necessary, DNA isolation and genetic characterization. Subcultures should not be placed in a refrigerator, as some species are killed at refrigerator temperatures. Isolation of fungal DNA from molds and their genetic identification are described in Chapter 2.

2. *Additional Media for Dermatophytes.* Specialty laboratories may wish to learn more about media recommended by Kane et al. (1997) and known as the "Kane/Fischer System" for identification of dermatophytes. Some additional media are described at the end of this section.

3. *Growth characteristics.* Most dermatophytes can be recovered within 7–10 days of incubation at 30°C. Some dermatophytes are slow growers, taking 1–2 weeks incubation before growth is seen, and a further 1–2 weeks or more before sporulation occurs. Pigment production on the surface and reverse also may not form prior to that time. Dermatophyte cultures should not be considered negative before 4 weeks time. Because of this potential delay, treatment is usually initiated on clinical observation alone or in combination with direct examination of hair or skin or nail scrapings. Growth of the culture is monitored every other day, also to note any contaminants on the agar medium. If contaminants are seen, the dermatophyte is transferred to fresh medium for continued growth.

Cultures are worthwhile, even though they may require a lengthy time to grow, in order to trace outbreaks, to determine whether the causative agent is a dermatophyte, another fungus, a mix of yeast and mold, or bacteria and mold, or is another problem entirely. It is advantageous to have a pretreatment culture, should therapy be ineffective and the use of another antifungal agent becomes necessary.

4. *Colony Morphology.* The appearance of cultures is the first line of identification. The colony color, both surface and reverse, surface texture, presence of radiating hyphae in the medium, and any unusual findings are noted.

5. *Microscopic Morphology.* Tease preparations are the first type of microscopic examination, with slide cultures reserved for occasions when tease preps are ambiguous, The presence and appearance of macro- and microconidia and any other specialized structures (e.g., spirals, balloon cells) are used to determine genus and species (Table 21.7). In some cases special media or other tests may be necessary to complete the identification to species.

The Genus *Microsporum*

The genus *Microsporum* is characterized by fusiform or spindle-shaped macroconidia, often with thick, rough, or spiny walls. The rough nature of spines may be difficult to demonstrate in some cultures and may be seen only on the tip of the conidium. Macroconidia and colony morphology are used to identify the genus and species. Although microconidia may be produced, they do not aid in identification. Most *Microsporum* species perforate hair in vitro. (see "In Vitro Hair Perforation Test," below). Examination of tinea capitis or of cultures derived from it, when illuminated by Wood's light, demonstrate a bright yellow-green fluorescence with *M. canis. M. ferrugineum* also has yellow-green fluorescence, while *M. audouinii* produces a silver fluorescence.

Table 21.7 Major Differences Between the *Trichophyton, Microsporum*, and *Epidermophyton* Genera in Culture

Charactieristic	Genus		
	Trichophyton	*Microsporum*	*Epidermophyton*
Macroconidia	Pencil-shaped, thin-walled, often rare	Spindle-shaped, thick-walled with spines; some thin walled	Club-shaped, thin walled
Microconidia	In grape-like clusters (*en grappe*), or disposed linearly along sides and tips of hyphae (*en thyrses*)	Peg-shaped, often rare	None

Microsporum audouinii

- *Colony.* Growth is slow, with glabrous to velvety dense mycelium, wrinkled, with a radiating edge.
- *Surface.* Color is white-gray, tan, or rust-buff.
- *Reverse.* Peach to salmon color.
- *Microscopic.* Macroconidia are rare and poorly developed. When seen they are similar to those of *M. canis* with terminal spine-like projections, *pectinate* hyphae, and chlamydospores. *M. audouinii* is differentiated from other *Microsporum* species by its poor growth on rice grains.
- *Further Tests*
  - In vivo hair invasion: ectothrix.
  - In vitro hair perforation (−).
  - Sterile polished rice grains: poor growth, no conidia, turns grains brown.
  - Urease (−) or (+).

Microsporum canis

- *Colony.* Growth is rapid with a coarse fluffy to wooly or hairy texture and developing cottony areas at maturity. The colony has some radial grooves (Fig. 21.17).
- *Surface.* White to pale yellow/buff.
- *Reverse.* Deep yellow to orange.
- *Microscopic.* Large spindle-shaped, thick-walled macroconidia are produced, often with an asymmetrical knob (15–20 × 60–125 μm) (please see Fig. 2.6). Large warts are seen on mature macroconidia, especially on the knob. Macroconidia are divided by 6–15 thin-walled septa. Microconidia may be present.
- *Further Tests*
  - In vivo hair invasion: ectothrix.
  - In vitro hair perforation (+).
  - Sterile polished rice grains: good growth, yellow pigment, conidia usually produced.
  - Urease (+).

Microsporum gypseum

- *Colony. Microsporum gypseum* rapidly covers the agar surface with floccose to powdery growth that may become *pleomorphic*.
- *Surface.* Cinnamon-buff or fawn.
- *Reverse.* May be colorless, buff, deep mahogany-brown, or variable.
- *Microscopic.* Abundant clusters of macroconidia are produced: clavate or cylindrical to fusiform-shaped, usually thin-walled (7.5–11.5×25–60 μm) with 4–6 septa. The surface of macroconidia may be smooth to

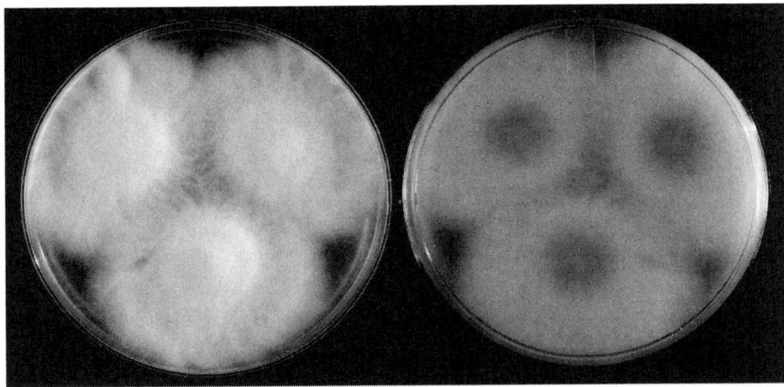

Figure 21.17 *Microsporum canis*, colony. See insert for color representation of this figure.
Source: Mr. Jim Gathany, CDC Creative Arts Branch.

rough. *Racquet hyphae* and *pectinate hyphae* may be present.

- *Further Tests*
  - In vivo hair invasion is large spore ectothrix.
  - In vitro hair perforation (+).
  - Sterile polished rice grains: good growth, no pigmentation, conidia may be produced.
  - Urease (+).

The Genus *Trichophyton*

The genus *Trichophyton* is characterized by pencil-shaped or elongated, balloon-shaped macroconidia with thin walls, and varying numbers of cells, depending on the species. Although macroconidia, when present, may help identify the genus, speciation is made on the appearance of microconidia and, occasionally, other structures. Some species do not produce any macroconidia. Additional tests, nutritional requirements, or biochemical tests may be required. Almost every species has a distinct body site preference.

Only selected species are covered here and are listed alphabetically not in order of importance; for other rare species please see Kane et al. (1997).

Trichophyton concentricum

- *Colony.* Slow growing, densely folded, glabrous, becoming velvety.
- *Surface.* Color is white to brown or orange-brown; resembles *T. schoenleinii* but is more brown and convoluted.
- *Reverse.* Color is similar to the surface.
- *Microscopic.* A tangled mass of branching hyphae, with no conidia or distinctive features.
- *Further tests.* In vitro hair perforation (−). Growth may be stimulated by thiamine.

Trichophyton interdigitale (Anthropophilic)

See Fig. 21.18

- *Colony.* Growth is moderately rapid. Typical colonies are velvety to cottony.
- *Surface.* Color is white, cream-colored, or buff.
- *Reverse.* Reverse is tan to brown.
- *Microscopic.* See Fig. 21.19. The numbers of microconidia produced differ greatly among strains. Some strains sporulate profusely, others only sparsely. Velvety cultures retain the microscopic morphology of zoophilic isolates (please see below, *T. mentagrophytes*), whereas cottony cultures produce small round microconidia disposed along sides and ends of branched hyphae (*en thyrses*) with few or no spiral hyphae. Mating and molecular analysis proved that most of the cottony and velvety anthropophilic "*T. mentagrophytes*-like" isolates are correctly identified as *Trichophyton interdigitale*, which is the principal anamorph name of the *Arthroderma vanbreuseghemii* complex.
- Microconidia are produced both *en thyrses* (cottony colony type) and *en grappe* (velvety colony type). Occasionally thin-walled macroconidia may be seen.

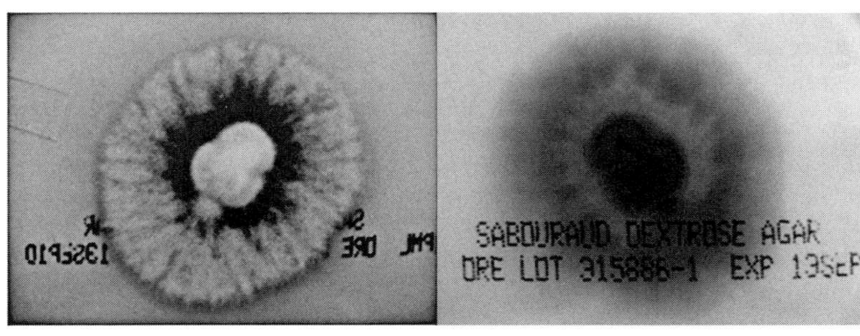

Figure 21.18 *Trichophyton interdigitale* culture on SDA after 12 days' incubation at 30°C. Colonies thin, floccose (cottony), white surface and deep orange-yellowish reverse. See insert for color representation of this figure.
Source: Used with permission from Mr. Jorge Musa, Sr. MLT, Life Laboratories, Toronto, Canada.

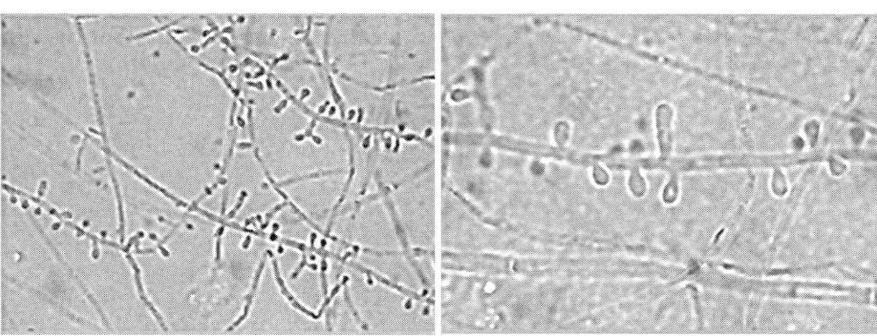

Figure 21.19 *Trichophyton interdigitale* microscopic morphology. Fertile hyphae with microconidia, spherical, teardrop-shaped to clavate, sessile along the sides and ends of branched hyphae. (Left) 50×objective; (right) 100×objective.
Source: Used with permission from Mr. Jorge Musa, Sr. MLT, Life Laboratories, Toronto, Canada.

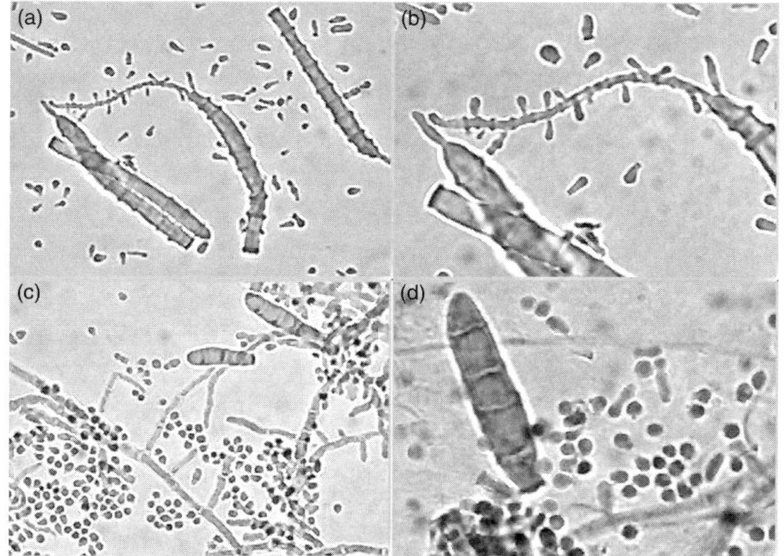

Figure 21.20 Compare *T. mentagrophytes* and *T. rubrum*, microscopic views: *(panel a) T. rubrum*, micro- and macroconidia (50× objective); *(panel b)* (same as *panel a*, but 100× objective; *(panel c) T. mentagrophytes*, micro-and macroconidia (50× objective), the microconidia are in clusters *en grappe*, not distributed along the hyphae; *(panel d)* same as *panel a*, but 50× objective.
Source: Used with permission from Mr. Jorge Musa, Sr. MLT, Life Laboratories, Toronto, Canada.

- *Further Tests*
  - In vivo hair invasion (−).
  - Urease (+). Urea hydrolysis helps to differentiate *T. interdigitale*, urease (+), from similarly appearing colonies of *T. rubrum*.

Trichophyton mentagrophytes (Zoophilic)

The typical *T. mentagrophytes* isolate has larger and more abundant conidia than *T. interdigitale* and sometimes antler-like hyphae suggestive of *Arthroderma* (teleomorph) sexual structures. There are abundant clusters of pigmented microconidia seen as granular or powdery and a more richly colored (yellowish-buff) surface mycelium. These morphologic differences between granular, zoophilic *T. mentagrophytes* and cottony, anthropophilic *T. interdigitale* anamorphs are quantitative, with no specific differences. The description of the colony surface of *T. mentagrophytes* as "granular" (or powdery) and that of *T. interdigitale* isolates as "cottony" can be subtle depending on the appearance of fresh isolates and growth medium. In that situation only genetic identification can definitively identify *T. mentagrophytes*.

- *Colony.* Growth is moderately rapid. The colony is typically flat with a granular or powdery surface due to abundant microconidia, often with radiating hyphae.
- *Surface.* Color is cream-tan or buff.
- *Reverse.* Color is tan to brown, resembling a banana slice in various stages of ripeness. Occasionally red pigment is formed similar to that of *T. rubrum*, or it may be bright yellow to orange.
- *Microscopic.* Granular surface morphology is due to abundant club- to pear-shaped microconidia formed in grape-like clusters "*en grappe*" (Fig. 21.20). Thin macroconidia vary from rare to numerous (7–10 × 20–50 μm). Tightly coiled spiral hyphae are characteristic of the species. Racquet and pectinate hyphae may be seen.
- *Further Tests*
  - In vivo hair invasion: ectothrix.
  - In vitro hair perforation (+).
  - Growth at 37°C (+).
  - Urease test (+).

Trichophyton rubrum

See Figs. 21.21 and 21.22.

- *Colony.* Moderately slow growing, up to 2 weeks, but the growth rate is variable. Typically, dome shaped, the colony is downy to fluffy. Occasional granular forms may be seen growing as a flat white colony.
- *Surface.* Color is white.
- *Reverse.* A deep red color is typical, often described as port wine red. The color does not diffuse into the medium but may be seen through to the surface of flat white colonies. The red pigment may be enhanced by growth on PDA or CMA. Variants are brown, yellow-orange.
- *Microscopic.* Most isolates produce few tear drop microconidia borne sessile on the hyphae ("birds on a wire"). Macroconidia are rare. Pectinate hyphae are rare. An Afro-Asian variant, formerly described as *T. raubitschekii*, produces many club-shaped macroconidia, some with a "rat-tail" extension.

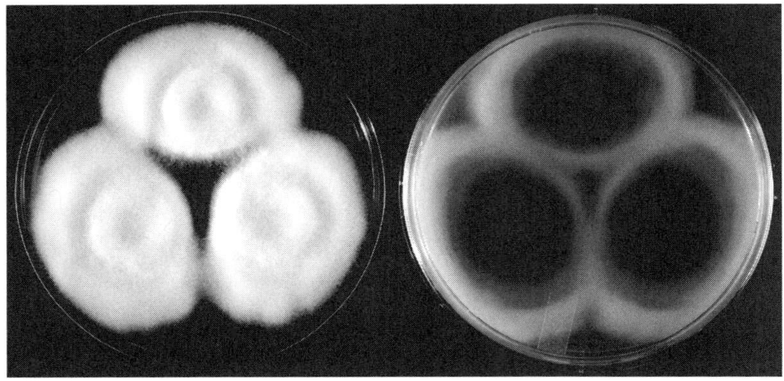

Figure 21.21 *Trichophyton rubrum* colony surface, reverse. See insert for color representation of this figure. *Source:* Mr. Jim Gathany, CDC Creative Arts Branch.

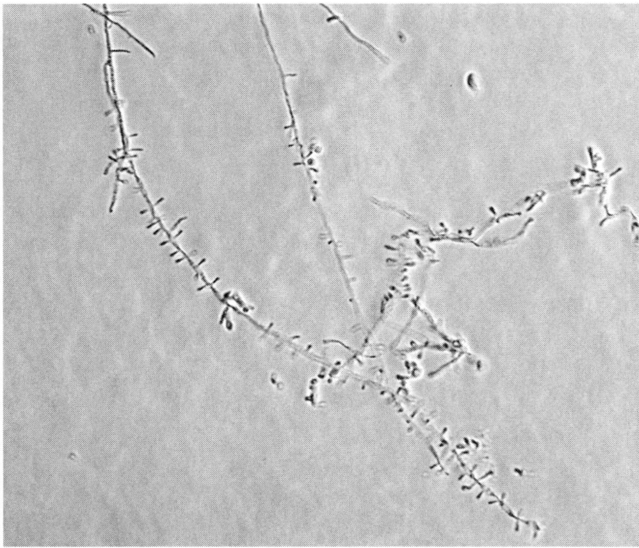

Figure 21.22 *Trichophyton rubrum* microscopic view (400× phase contrast, lactofuchsin stain). *Source:* E. Reiss, CDC.

- *Further Tests*
  - In vivo hair invasion (−).
  - In vitro hair perforation (−).
  - Growth at 37°C (+).
  - *T. rubrum* has no thiamine requirement and is used as a positive-growth control on thiamine-deficient medium. Purple colony variants of *T. rubrum* can be distinguished from *T. violaceum* because the latter has an absolute thiamine requirement.
  - Urease (−) but occasionally variable.

Trichophyton tonsurans

See Figs. 21.23 and 21.24.

- *Colony.* Growth is moderate. There are many colony colors and textures: powdery, suede-like, or granular, flat with a raised center, or folded, often with radial grooves. The edges of young colonies often radiate into the medium.
- *Surface.* Colonies are white, yellow, gray, tan, surface with a mahogany to brown undercoat and reverse. A variant form, *sulfureum*, is yellow on the surface and reverse.
- *Reverse.* It may be pale yellow to red-brown or even resemble the reverse of *T. rubrum*; dark pigment may diffuse into the medium.
- *Microscopic.* Microconidia are usually abundant, variably sized, and shaped (most often tear drop to club-shaped) in open branching clusters, directly on hyphae or at the tips of short lateral "matchstick" stalks. Microconidia may greatly increase in size; these are termed "balloon" cells and may be seen in old cultures. Macroconidia are infrequent, thin-walled, pencil-shaped, 4–6 celled, and often bent rather than straight. Swollen cells, chlamydospores, are common.
- *Further Tests*
  - In vivo hair invasion: endothrix.
  - In vitro hair perforation: variable.
  - Growth at 37°C (+).
  - *T. tonsurans* has a partial requirement for thiamine. Growth is strongly stimulated with a thiamine supplement (*Trichophyton* agar #4). In contrast, *T. rubrum* and *T. mentagrophytes* do not require thiamine.
  - Urease (+).

Trichophyton verrucosum

- *Colony.* Growth is very slow, glabrous, heaped up, and button-like, or mostly subsurface. Growth is faster at 37°C, unlike other dermatophytes, which grow best at room temperature (25–30°C).
- *Surface.* Color is typically white to ochre-brown.
- *Reverse.* Reverse is usually colorless, less often yellow to pale-brown.

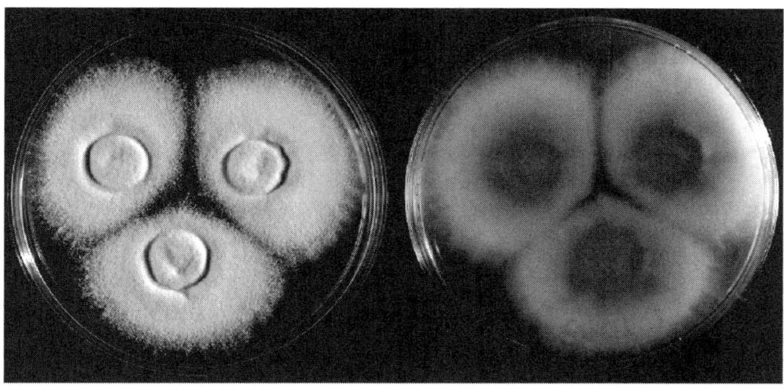

Figure 21.23 *Trichophyton tonsurans* colony: (*left*) surface and (*right*) reverse. See insert for color representation of this figure.
Source: Mr. Jim Gathany, CDC Creative Arts Branch.

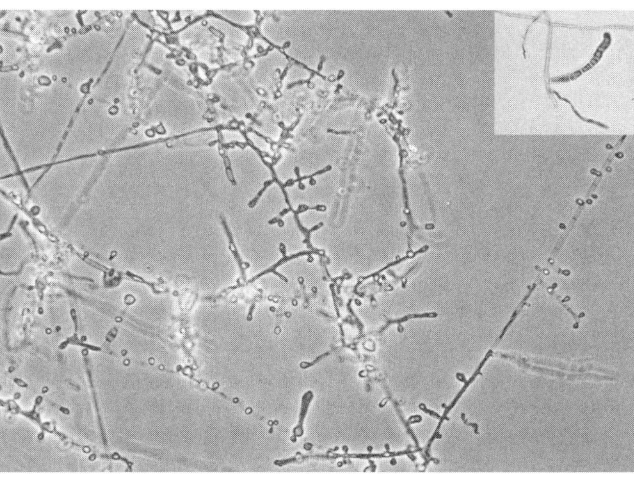

Figure 21.24 *Trichophyton tonsurans*, microscopic morphology (400× phase contrast, lactofuchsin stain). (*Inset*): Macroconidium (rare).
Source: E. Reiss, CDC.

- *Microscopic.* Tortuous hyphae are present, some with antler-like branching. Conidia are not formed on SDA. Addition of thiamine to blood agar base stimulates the production of teardrop-shaped microconidia and rare "rat tail" or string bean-shaped macroconidia.
- *Further Tests*
  - In vivo hair invasion: ectothrix.
  - In vitro hair perforation (−).
  - Growth at 37°C (+); growth at 41–42°C (+).
  - *T. verrucosum* is stimulated by thiamine and inositol (*Trichophyton* agar #3).
  - Urease (−).

Trichophyton violaceum

- *Colony.* Slow growing, heaped or wrinkled, and shiny.
- *Surface.* Color is deep purple. With age or subculture colonies may lose pigmentation or develop white sectors.
- *Reverse.* Reverse is deep purple.
- *Microscopic.* A tangle of branched hyphae, with no conidia produced on primary isolation. Conidia are not formed on SDA. On an enriched medium (*Trichophyton* agar #4) 2–8-celled irregularly shaped macroconidia, and pyriform microconidia are produced.
- **Further Tests**
  - In vivo hair invasion: endothrix.
  - In vitro hair perforation (−).
  - Growth at 37°C (+).
  - Growth is stimulated by thiamine. *T. violaceum* is differentiated from atypical purple *T. rubrum*, which has no thiamine requirement.
  - Urease (−).

Epidermophyton floccosum

- *Colony.* Surface typically suede-like, may have rough surface resembling brown sugar. This grainy surface is similar to that found in *T. tonsurans*. Young colonies may produce subsurface radiating hyphae. Pleomorphic white sterile areas rapidly form.
- *Surface.* Color of primary colonies is a distinctive yellow-mustard green to olive-gray.
- *Reverse.* Reverse of the mature colony is light to dark brown.
- *Microscopic.* Blunt, clavate, macroconidia are formed singly or more commonly in pairs of "rabbit ears" or clusters. The macroconidia are 10 μm × 40 μm with thin, smooth walls, and 2–5 septa. Chlamydospores replace the macroconidia in old cultures.

- *Further Tests*
  - In vivo hair invasion (−).
  - In vitro hair perforation (−).
  - Growth at 37°C (+).
  - Urease (+).

Additional Culture Media and Tests for Dermatophytes

Please also see Chapter 2 for further details.

Additional Media

- *Christensen's Urea Slant or Broth.* Positive control: *T. mentagrophytes* turns medium bright pink. Negative control: *T. rubrum*, medium remains yellow.
- *Cornmeal Agar.* CMA enhances pigment production in the dermatophytes, particularly *T. rubrum* and *M. canis*. When used to stimulate chlamydospores in *Candida albicans*, the surfactant Tween 80 is added at 10 mL/L, prior to autoclaving.
- *Dermatophyte Test Medium (DTM).* DTM (Kane et al., 1997) is used to identify a mould as a dermatophyte, even if it is not sporulating. DTM takes advantage of the property of dermatophytes to produce an alkaline medium during growth. The medium contains phenol red as the pH indicator which turns from pale yellow (acidic pH) to medium red (alkaline pH). Other ingredients include cycloheximide to inhibit many molds, with gentamicin and tetracycline to inhibit bacteria. When growth occurs, the growth must be viewed microscopically, because geophilic keratinophilic molds (nondermatophytes) may grow on this medium. The culture is discarded after 2 weeks, because afterwards false-positive results may occur. Positive control: *T. mentagrophytes;* negative control: *Candida albicans*.
- *Potato Dextrose Agar.* PDA is used to enhance pigment production by *T. rubrum* and *M. canis*.
- *Sterile Polished Rice (Unfortified).* Used primarily to differentiate *M. audouinii* from *M. canis* (as indicated above).
- *Trichophyton Agars (*T-Media*).* Although there are seven T-media, T1 and T4 are most often used. The composition of other T-media and the results expected with each dermatophyte may be found in Kane et al. (1997). T-media contain vitamin-free casamino acids + individual supplements:
  - *T1.* Plain vitamin-free casamino acids.
  - *T2.* Supplemented with inositol.
  - *T3.* Supplemented with thiamine + inositol.
  - *T4.* Supplemented with thiamine.
- *T. tonsurans* (growth stimulated by thiamine, T4) and *T. verrucosum* (growth dependent on thiamine supplement) are used as quality control cultures. Importantly, inoculate T-media with only minimal amounts of hyphae, so that medium constituents from the primary culture is not carried over.

Temperature Tolerance Test

A few dermatophytes (e.g., *T. verrucosum*) are able to grow at temperatures of 40–42°C.

In Vitro Hair Perforation Test

Purpose *Not to be confused with in vivo hair invasion*, in vitro hair perforation is a laboratory test to differentiate *T. mentagrophytes* (perforation-positive) from *T. rubrum* (perforation-negative) (Fig. 21.25).

Procedure A Petri plate is prepared with 25 mL deionized water and 2–3 drops of 10% yeast extract. Sterilized hairs from a prepubescent child, preferably blond, are placed in the plate and inoculated with the culture to be tested. Separate preps are constituted with the positive and negative control cultures. Incubation is carried out at 25°C for up to 28 days and sampled weekly by removing some hair segments, placing them in a drop of water, adding a coverslip, and observing the prep under the microscope. A positive test is when cone-shaped perforations are observed in the hair shafts. Viewed from the top, these perforations appear as round holes. If the control does not become positive, the test is repeated.

Hair Fluorescence with Wood's Light

Tinea capitis caused by members of the *Microsporum* species (*M. canis, M. gypseum, M. audouinii, M.*

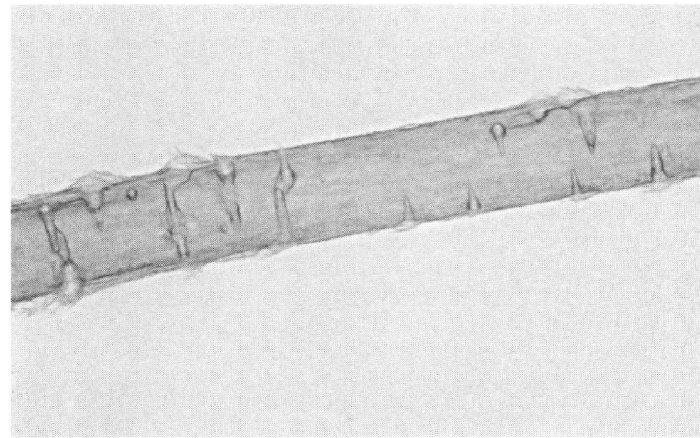

Figure 21.25 Positive hair perforation test.
Source: H. J. Shadomy.

distortum, M. ferrugineum) fluoresce a bright to light yellow-green or silver (*M. audouinii*) due to the presence of tryptophan metabolites. The fluorescence is identified by "black light," that is, a filtered ultraviolet light source at 365 nm.

Genetic Identification of Dermatophytes

Sequencing the ITS region of rDNA is recommended as the "gold standard" for identifying atypical or difficult dermatophytes in reference laboratories. Even for routine laboratories, ITS sequencing is valuable for species identification and may become feasible for direct identification from clinical specimens.

A complete database exists for dermatophyte identification (in the GenBank) and in the commercial SmartGene® Integrated Database Network System, SmartGene Company, Lausanne, Switzerland. Because the GenBank is open access and is not curated, Gräser et al. (2008) developed a table of barcode sequences, that is, verified sequences and their GenBank accession numbers for individual dermatophytes. Only one or two polymorphic nucleotides (signature nucleotides) separate two species, such as *M. ferrugineum* and *M. canis* (in ITS2), *T. equinum* and *T. tonsurans* (in ITS1), and the zoophilic *T. mentagrophytes* and anthropophilic *T. interdigitale*. Another advantage of ITS sequencing is that other fungal species involved in dermatomycoses also can be identified. Sequences for all dermatophyte species are not available for the LSU of rDNA or the topoisomerase gene. The chitin synthase *CHS* gene and SSU of rDNA are too conservative for species recognition.

SELECTED REFERENCES FOR DERMATOPHYTOSIS

Abdel-Rahman SM, Simon S, Wright KJ, Ndjountche L, Gaedigk A, 2006. Tracking *Trichophyton tonsurans* through a large urban child care center: Defining infection prevalence and transmission patterns by molecular strain typing. *Pediatrics* 118: 2365–2373.

Adams BB, 2002. Tinea corporis gladiatorum. *J Am Acad Dermatol* 47: 286–290.

Al Hasan M, Fitzgerald SM, Saoudian M, Krishnaswamy G, 2004. Dermatology for the practicing allergist: Tinea pedis and its complications. *Clin Mol Allergy* 2: 5.

Aly R, Berger T, 1996. Common superficial fungal infections in patients with AIDS. *Clin Infect Dis* 22 (Suppl 2): S128–S132.

Aly R, Hay RJ, Del Palacio A, Galimberti R, 2000. Epidemiology of tinea capitis. Med Mycol 38 (Suppl 1): 183–188.

Babel DE, Baughman SA, 1989. Evaluation of the adult carrier state in juvenile tinea capitis caused by *Trichophyton tonsurans*. *J Am Acad Dermatol* 21: 1209–1212.

Beller M, Gessner BD, 1994. An outbreak of tinea corporis gladiatorum on a high school wrestling team. *J Am Acad Dermatol* 31: 197–201.

Ben-Ziony Y, Arzi B, 2000. Use of lufenuron for treating fungal infections of dogs and cats: 297 cases (1997–1999). *J Am Vet Med Assoc* 217: 1510–1513.

Bibel DJ, Aly R, Shinefield HR, 1995. Topical sphingolipids in antisepsis and antifungal therapy. *Clin Exp Dermatol* 20: 395–400.

Borman AM, Campbell CK, Fraser M, Johnson EM, 2007. Analysis of the dermatophyte species isolated in the British Isles between 1980 and 2005 and review of worldwide dermatophyte trends over the last three decades. *Med Mycol* 45: 131–141.

Brown A, 2009. To itch, perchance to scratch. *Clin Pediatr (Phila)* 48: 334.

Brown SJ, 2009. Efficacy of fluconazole for the treatment of onychomycosis. *Ann Pharmacother* 43: 1684–1691.

Cafarchia C, Romito D, Capelli G, Guillot J, Otranto D, 2006. Isolation of *Microsporum canis* from the hair coat of pet dogs and cats belonging to owners diagnosed with *M. canis* tinea corporis. *Vet Dermatol* 17: 327–331.

Chen BK, Friedlander SF, 2001. Tinea capitis update: A continuing conflict with an old adversary. *Curr Opin Pediatr* 13: 331–335.

Chermette R, Ferreiro L, Guillot J, 2008. Dermatophytoses in animals. *Mycopathologia* 166: 385–405.

Cohen AD, Wolak A, Alkan M, Shalev R, Vardy DA, 2005. Prevalence and risk factors for tinea pedis in Israeli soldiers. *Int J Dermatol* 44: 1002–1005.

Daniel CR 3rd, Gupta AK, Daniel MP, Daniel CM, 1997. Two feet-one hand syndrome: A retrospective multicenter survey. *Int J Dermatol* 36: 658–660.

de Berker D, 2009. Clinical practice. Fungal nail disease. *N Engl J Med* 360: 2108–2116.

Dvoretzky I, Fisher BK, Movshovitz M, Schewach-Millet M, 1980. Favus. *Int J Dermatol* 19: 89–92.

Elewski GE, Haley HR, Robbins CM, 2004. The use of 40% urea cream in the treatment of moccasin tinea pedis. *Cutis* 73: 355–357.

Elgart ML, 1996. Tinea incognito—an update on Majocchi granuloma. *Dermatol Clin* 14: 51–55.

Faergemann J, Baran R, 2003. Epidemiology, clinical presentation and diagnosis of onychomycosis. *Br J Dermatol* 149(Suppl 65): 1–4.

Fernandes NC, Lamy F, Akiti T, da Glória C. Barreiros M, 1998. *Microsporum gypseum* infection in Aids patient: A case report. *Anais Bras Dermatol (Rio de Janeiro)* 73: 39–41.

Frieden IJ, 1999. Tinea capitis: Asymptomatic carriage of infection. *Pediatr Infect Dis* J 18: 186–190.

Ginter-Hanselmayer G, Weger W, Ilkit M, Smolle J, 2007. Epidemiology of tinea capitis in Europe: Current state and changing patterns. *Mycoses* 50(Suppl 2): 6–13.

Gräser Y, de Hoog S, Summerbell RC, 2006. Dermatophytes: Recognizing species of clonal fungi. *Med Mycol* 44: 199–209.

Gräser Y, Scott J, Summerbell R, 2008. The new species concept in dermatophytes—a polyphasic approach. *Mycopathologia* 166: 239–256.

Gupta AK, 2000. Onychomycosis in the elderly. *Drugs Aging* 16: 397–407.

Gupta AK, Cooper EA, 2008. Update in antifungal therapy of dermatophytosis. *Mycopathologia* 166: 353–367.

Gupta AK, Jain HC, Lynde CW, Macdonald P, Cooper EA, Summerbell RC, 2000. Prevalence and epidemiology of onychomycosis in patients visiting physicians' offices: a multicenter Canadian survey of 15,000 patients. *J Am Acad Dermatol* 43: 244–248.

Gupta AK, Summerbell RC, 2000. Epidemiology of tinea capitis. *Med Mycol* 38: 255–287.

Gupta AK, Ricci MJ, 2006. Diagnosing onychomycosis. *Dermatol Clin* 24: 365–369.

Gupta AK, Ryder JE, Summerbell RC, 2004. Onychomycosis: Classification and diagnosis. *J Drugs Dermatol* 3: 51–56.

GUPTA AK, TABORDA P, TABORDA V, GILMOUR J, RACHLIS A, SALIT I, GUPTA MA, MACDONALD P, COOPER EA, SUMMERBELL RC, 2000. Epidemiology and prevalence of onychomycosis in HIV-positive individuals. *Int J Dermatol* 39: 746–753.

GUPTA AK, KONNIKOV N, MACDONALD P, RICH P, RODGER NW, EDMONDS MW, MCMANUS R, SUMMERBELL RC, 1998. Prevalence and epidemiology of toenail onychomycosis in diabetic subjects: A multicentre survey. *Br J Dermatol* 139: 665–671.

HABIF TF, 2009. Chapter 13, Superficial Fungal Infections, *in*: *Clinical Dermatology*, 5th ed., Mosby (an imprint of Elsevier), Oxford, UK.

HAVLICKOVA B, CZAIKA VA, FRIEDRICH M, 2008. Epidemiological trends in skin mycoses worldwide. *Mycoses* 51(Suppl 4): 2–15.

HAY RJ, 2005. Chapter 265, Dermatophytosis and Other Superficial Mycoses, *in*: MANDEL GL, BENNETT JE, DOLIN R (eds.), *Mandell, Bennett, & Dolin: Principles and Practice of Infectious Diseases*, 6th ed. Elsevier-Churchill Livingstone: Philadelphia.

HAY RJ, 2006. How do dermatophytes survive in the epidermis? *Curr Opin Infect Dis* 19: 125–126.

HOWARD RM, FRIEDEN IJ, 1999. Dermatophyte infections in children. *Adv Pediatr Infect Dis* 14: 73–107.

ILKIT M, TANIR F, HAZAR S, GÜMÜŞAY T, AKBAB M, 2005. Epidemiology of tinea pedis and toenail tinea unguium in worshippers in the mosques in Adana, Turkey. *J Dermatol* 32: 698–704.

JANKOWSKA-KONSUR A, DYLAG M, SZEPIETOWSKI JC, 2009. Tinea capitis in southwest Poland. *Mycoses* 52: 193–194.

JHA BN, GARG VK, AGRAWAL S, KHANAL B, AGARWALLA A, 2006. Tinea capitis in eastern Nepal. *Int J Dermatol* 45: 100–102.

JOUSSON O, LECHENNE B, BONTEMS O, CAPOCCIA S, MIGNON B, BARBLAN J, QUADRONI M, MONOD M, 2004a. Multiplication of an ancestral gene encoding secreted fungalysin preceded species differentiation in the dermatophytes *Trichophyton* and *Microsporum*. *Microbiology* 150: 301–310.

JOUSSON O, LECHENNE B, BONTEMS O, MIGNON B, REICHARD U, BARBLAN J, QUADRONI M, MONOD M, 2004b. Secreted subtilisin gene family in *Trichophyton rubrum*. *Gene* 339: 79–88.

KAMALAM A, THAMBIAH AS, 1980. Tinea capitis an endemic disease in Madras. *Mycopathologia* 71: 45–51.

KANE J, SUMMERBELL R, SIGLER L, KRAJDEN S, LAND G, 1997. *Laboratory Handbook of Dermatophytes*. Star Publishing Co, Belmont, CA.

LARONE DH, 2002. *Medically Important Fungi: A Guide to Identification*, 4th ed. ASM Press, Washington, DC.

LOBATO MN, VIRGIA DJ, FRIEDEN IJ, 1997. Tinea capitis in California children. A population based study of a growing epidemic. *Pediatrics* 99: 551–554.

LOO DS, 2007. Onychomycosis in the elderly: Drug treatment options. *Drugs Aging* 24: 293–302.

MAGILL SS, MANFREDI L, SWIDERSKI A, COHEN B, MERZ WG, 2007. Isolation of *Trichophyton violaceum* and *Trichophyton soudanense* in Baltimore, Maryland. *J Clin Microbiol* 45: 461–465.

MCPHERSON ME, WOODGYER AJ, SIMPSON K, CHONG AH, 2008. High prevalence of tinea capitis in newly arrived migrants at an English-language school, Melbourne, 2005. *Med J Aust* 189: 13–16.

Morbidity Mortality Weekly Report (*CDC*), 1973,22 #4.

MORIELLO KA, DEBOER DJ, SCHENKER R, BLUM JL, VOLK LM, 2004. Efficacy of pre-treatment with lufenuron for the prevention of *Microsporum canis* infection in a feline direct topical challenge model. *Vet Dermatol* 15: 357–362.

MUÑOZ-PÉREZ MA, RODRIGUEZ-PICHARDO A, CAMACHO F, COLMENERO MA, 1998. Dermatological findings correlated with CD4 lymphocyte counts in a prospective 3 year study of 1161 patients with human immunodeficiency virus disease predominantly acquired through intravenous drug abuse. *Br J Dermatol* 139: 33–39.

NIEWERTH M, KORTING HC, 2000. The use of systemic antimycotics in dermatotherapy. *Eur J Dermatol* 10: 155–160.

OHST T, DE HOOG GS, PRESBER W, GRÄSER Y, 2004. Origins of microsatellite diversity in the *Trichophyton rubrum*—*T. violaceum* clade (dermatophytes). *J Clin Microbiol* 42: 4444–4448.

PERRIN C, BARAN R, 1994. Longitudinal melanonychia caused by *Trichophyton rubrum*. Histochemical and ultrastructural study of two cases. *J Am Acad Dermatol* 31: 311–316.

PICKUP TL, ADAMS BB, 2007. Prevalence of tinea pedis in professional and college soccer players versus non-athletes. *Clin J Sport Med* 17: 52–54.

POMERANZ AJ, SABNIS SS, 2002. Tinea capitis: Epidemiology, diagnosis and management strategies. *Paediatr Drugs* 4: 779–783.

POMERANZ AJ, SABNIS SS, MCGRATH GJ, ESTERLY NB, 1999. Asymptomatic dermatophyte carriers in the households of children with tinea capitis. *Arch Pediatr Adolesc Med* 153: 483–486.

POTTER BS, 2003. Bibliographic landmarks in the history of dermatology. *J Am Acad Dermatol* 48: 919–932.

PROHIC A, 2008. An epidemiological survey of tinea capitis in Sarajevo, Bosnia and Herzegovina over a 10-year period. *Mycoses* 51: 161–164.

RICH P, SCHER RK, 2003. *An Atlas of Diseases of the Nail*. Parthenon Publishing Group, New York.

ROVANSEK KA, PAPADOPOLOUS A, 2005. Images in clinical medicine: Tinea circinata. *N Engl J Med* 352: e6.

RUTECKI GW, WURTZ R, THOMSON RB, 2000. From animal to man: Tinea barbae. *Curr Infect Dis Rep* 2: 433–437.

SABOURAUD R, 1910. *Maladies du Cuir Chevelu III. Maladies Cryptogamiques. Les Teignes*. Masson et Cie, Paris.

SCHER RK, TAVAKKOL A, SIGURGEIRSSON B, HAY RJ, JOSEPH WS, TOSTI A, FLECKMAN P, GHANNOUM M, ARMSTRONG DG, MARKINSON BC, ELEWSKI BE, 2007. Onychomycosis: Diagnosis and definition of cure. *J Am Acad Dermatol* 56: 939–944.

SCHER RK, WARREN J, ROBBINS J, 2003. Progression and recurrence of onychomycosis. www.Medscape.com Release date: May 30, 2003.

SERARSLAN G, 2007. Pustular psoriasis-like tinea incognito due to *Trichophyton rubrum*. *Mycoses* 50: 523–524.

SHADOMY HJ, PHILPOT CM, 1980. Utilization of standard laboratory methods in the laboratory diagnosis of problem dermatophytes. *Am J Clin Pathol* 74: 192–201.

SINGAL A, RAWAT S, BHATTACHARYA SN, MOHANTY S, BARUAH MC, 2001. Clinico-mycological profile of tinea capitis in North India and response to griseofulvin. *J Dermatol* 28: 22–26.

SONGER JG, POST KW, 2005. *Veterinary Microbiology*: *Bacterial and Fungal Agents of Animal Disease*. Elsevier-Saunders, St. Louis, Mo.

ST GERMAIN G, SUMMERBELL R, 1996. *Identifying Filamentous Fungi*, *A Clinical Laboratory Handbook*. Star Publishing Co, Belmont, CA.

STEHR-GREEN JK, SCHANTZ PM, 1987. The impact of zoonotic diseases transmitted by pets on human health and the economy. *Vet Clin North Am Small Anim Pract* 17: 1–15.

STOCKDALE PM, 1968. Sexual stimulation between *Arthroderma simii* Stockd., Mackenzie and Austwick and related species. *Sabouraudia* 6: 176–181.

SUH D-C, FRIEDLANDER SF, RAUT M, CHANG J, VO L, SHIN H-C, TAVAKKOL A, 2006. Tinea capitis in the United States: Diagnosis, treatment, and costs. *J Am Acad Dermatol* 55: 1111–1112.

VENKATESAN P, PERFECT JR, MYERS SA, 2005. Evaluation and management of fungal infections in immunocompromised patients. *Dermatol Ther* 18: 44–57.

WILMINGTON M, ALY R, FRIEDEN IJ, 1996. *Trichophyton tonsurans* tinea capitis in the San Francisco Bay area: Increased infection demonstrated in a 20-year survey of fungal infections from 1974 to 1994. *J Med Vet Mycol* 34: 285–287.

WOODFOLK JA, 2005. Allergy and dermatophytes. *Clin Microbiol Rev* 18: 30–43.

YU J, LI R, BULMER G, 2005. Current topics of tinea capitis in China. *Jpn J Med Mycol* 46: 61–66.

Zaias N, Rebell G, 2003. Clinical and mycological status of the *Trichophyton mentagrophytes* (*interdigitale*) syndrome of chronic dermatophytosis of the skin and nails. *Int J Dermatol* 42: 779–788.

Zaias N, Tosti A, Rebell G, Morelli R, Bardazzi F, Bieley H, Zaiac M, Glick B, Paley B, Allevato M, Baran R, 1996. Autosomal dominant pattern of distal subungual onychomycosis caused by *Trichophyton rubrum*. *J Am Acad Dermatol* 14: 302–304.

Zollner TM, Podda M, Kaufmann R, Platts-Mills TA, Woodfolk JA, 2002. Increased incidence of skin infections in atopy: Evidence for an antigen-specific homing defect. *Clin Exp Allergy* 32: 180–185.

WEBSITE OF INTEREST ON THE SUBJECT OF DERMATOPHYTOSIS

www.dermatlas.org. A dermatology image atlas authored by Bernard A. Cohen, MD, and Christoph U. Lehmann, MD, DermAtlas, Johns Hopkins University; 2000–2009.

QUESTIONS

The Answer Key to multiple choice questions may be found at the end of the book.

1. The most common causative agent of tinea capitis in the United States is
   A. *Microsporum canis*
   B. *Microsporum interdigitale*
   C. *Trichophyton rubrum*
   D. *Trichophyton violaceum*
   E. *Trichophyton tonsurans*

2. The term ringworm is used to describe lesions of dermatophytosis because
   A. It is derived from the German word *ringelflechte*.
   B. Often seen in children, the name is derived from the children's game, "Ring Around the Rosie."
   C. The circular lesions have active borders.
   D. Their moth-eaten appearance is similar to damage caused to woolens by clothes moths, the *Tineidae*.

3. A 10-yr-old girl developed itchy scalp lesions with active borders and scaling. The family cat also showed skin lesions. Wood's light illumination of both child and cat indicated yellow-green fluorescence. A hair plucked from the child's scalp revealed sporulation (arthroconidia) on the outside of the shaft. Material for culture was obtained from the child's scalp with a toothbrush. A slide culture showed spindle-shaped macroconidia. This is an example of infection caused by
   A. *Microsporum canis*
   B. *Microsporum gypseum*
   C. *Trichophyton violaceum*
   D. *Trichophyton tonsurans*

4. Select the item that is true about medium and culture conditions for dermatophytes.
   A. Cultures are incubated at 37°C to inhibit saprobes.
   B. Cultures should not be considered negative before 4 weeks time.
   C. Do not embed nail specimens in the agar.
   D. Medium containing cycloheximide is inhibitory.

5. What circumstances justify attempts to culture dermatophytes from skin and hair specimens?
   A. To trace outbreaks
   B. To determine if the cause is a dermatophyte, another fungus, a mix of yeast and mold, bacteria and mold, or is another problem entirely
   C. Have a pretreatment culture, in case another antifungal agent becomes necessary
   D. All of the above

6. How long does it usually take for a *Trichophyton* species culture to grow from a clinical specimen?
   A. 3–5 days
   B. 8–10 days
   C. 2–4 weeks
   D. 4–6 weeks
   E. Usually in 8–10 days but may require up to 4 weeks

7. How can *T. mentagrophytes* and *T. interdigitale* be differentiated?
   A. Sequence analysis of the ITS region of rDNA.
   B. *Trichophyton interdigitale* does not invade hair and has a cottony colony texture.
   C. *Trichophyton mentagrophytes* has a granular colony texture because the microconidia are formed *en grappe*.
   D. *Trichophyton mentagrophytes* invades hair in an ectothrix pattern and gives a positive in vitro hair perforation test.
   E. All of the above.

8. What is the proper way to collect specimens of suspected tinea capitis?
   A. Cut the infected hair to avoid disturbing the hair bulb.
   B. In the case of black dot ringworm, do not scrape the affected area.
   C. Rubbing the scalp with a toothbrush to collect hair and skin scales is not useful.
   D. All of the above are false.
   E. All of the above are true.

9. A medium that enhances pigment production by dermatophytes is
   A. DTM
   B. Mycosel
   C. PDA
   D. Urea agar

10. This dermatophyte colony color is reddish brown–mahogany. Abundant microconidia form at the tips of short lateral "matchstick" stalks. Microconidia enlarge forming balloon cells. *Trichophyton* agar #1 stimulates growth. Macroconidia are less common, 4–6 cells long, and bent. Variants may have a white to yellow colony surface. It is
   A. *Microsporum audouinii*
   B. *Trichophyton rubrum*
   C. *Trichophyton tonsurans*
   D. *Trichophyton violaceum*

11. If a pet cat has ringworm that is transmitted to a child in the family, it is best to

A. Have the cat euthanized.

B. Shave the hair of the affected area as that will be sufficient to stop the infection.

C. Treat the cat according to a veterinarian's instructions.

D. Treat the cat with athlete's foot spray.

12. The change in fatty acid composition and related changes at the onset of puberty deter infections in adults with

A. *Microsporum*

B. *Trichophyton*

C. *Epidermophyton*

D. All of the above

13. The difference between favus (1) and kerion (2) types of tinea capitis is

A. Inflammatory with pustules and a tender, purulent, boggy mass

B. Scutula, matted clumps of hyphae and keratin hair debris, with alopecia, skin atrophy, and scarring

14. Discuss the rise and subsidence of *Microsporum audouinii* as a cause of scalp ringworm in North America. What factors led to its subsidence and to the rise of *Trichophyton tonsurans*? Is *T. violaceum* an emerging anthropophilic dermatophyte? Explain.

15. Three unusual forms of tinea are tinea gladiatorum, tinea imbricata, and tinea incognito. Describe the clinical form of each, the risk factors, and causative agent(s).

16. List and define five clinical forms of (a) tinea capitis and (b) tinea unguium.

17. What degree of success is obtained with topical therapy for tinea unguium? What are the therapy options for systemic therapy of that form of dermatophytosis?

18. What is the role of griseofulvin in treating tinea capitis? What are the other therapy options for this dermatophytosis?

Chapter 22

Dermatomycoses

22A MAJOR NONDERMATOPHYTIC FUNGI FROM SKIN AND NAILS

Onychomycosis

Nail infections also may be caused by a nondermatophyte. Although uncommon, the fungi most often isolated are *Candida albicans* and other yeasts which cause *onychia* and *paronychia*, as discussed in Chapter 11. Among the molds, the genera and species most frequently identified in clinical laboratories in the United States are *Acremonium* species, *Aspergillus flavus, Fusarium moniliforme, F. solani, Onychocola canadensis, Neoscytalidium dimidiatum, N. hyalinum*, and *Scopulariopsis brevicaulis*. Unlike the dermatophytes, many of these fungi are inhibited by cycloheximide, and media containing this agent should not be used for their isolation. Other species also may cause dermatomycosis, but are even less common, and are related to persons who have lived or traveled abroad. Fuller description of these molds and the less common dermatophytes may be found in Kane et al. (1997), Larone (2002), and St. Germain and Summerbell (1996).

Aspergillus *and* Fusarium *Species Onychomycosis*

Growth and morphology of these fungi are discussed in Chapters 13 and 15. *Aspergillus* and *Fusarium* species each cause onychia, distal subungual and proximal subungal onychomycosis. In the latter type, periungual inflammation may be present. (Please see Chapter 21, Section 21.11, Clinical Forms—Tinea Unguium, for details of the types of onychomycosis.)

Determining if an *Aspergillus* species, *Fusarium* species, or another hyaline mold is a contaminant or is implicated in colonizing or infecting the patient rests on several criteria. Sterile site isolates are important to report without delay. Confidence that a nonsterile site isolate is infecting the patient depends on isolation of several colonies of the mold, or from sequential specimens from the same patient, and assurance that the agar medium itself is not contaminated. The mold should be capable of growth at 37°C. The significance of *Aspergillus* or *Fusarium* species isolated from nail cultures depends on signs and symptoms of onychomycosis: positive KOH preps showing hyphae in the nail; failure to isolate a dermatophyte in culture; and growth of several colonies of the same mold in at least two consecutive nail specimens. Onychomycosis due to *Aspergillus* species responds well to systemic therapy with terbinafine (TRB) or itraconazole (ITC). Most cases of *Fusarium* species onychomycosis do not improve even after long-term local and systemic treatment (Tosti et al., 2003).

Other molds causing onychomycosis are described below, in alphabetic order, not in order of importance (Table 22.1).

Acremonium

Acremonium species are environmental saprobes and agents of white superficial and distal subungual onychomycosis. *Acremonium strictum* is illustrated in Fig. 22.1.

Mycologic Characteristics *Acremonium* species are hyaline molds developing as flat, slimy colonies which becomes floccose, convoluted, and finely ridged on SDA. The colony is white to pink, with a colorless or pink reverse. The conidiophore is a long slender hyaline phialide tapering toward the tip, bearing ellipsoid or cylindrical phialoconidia held together in a ball or in easily disturbed chains (Fig. 22.1). Conidia may be uni- or multicellular. Differences between *A. strictum,*

Fundamental Medical Mycology, First Edition. By Errol Reiss, H. Jean Shadomy and G. Marshall Lyon, III.
© 2012 Wiley-Blackwell. Published 2012 by John Wiley & Sons, Inc.

Table 22.1 Causative Agents of Onychomycosis Which Are Environmental Molds (but Not Dermatophytes)

Genus, species	Key features of their role in infection	Appearance in nail tissue	Microscopic morphology in vitro
Acremonium species (Fig. 22.1)	Superficial white onychomycosis	Septate hyphae	Long slender hyaline phialides, with a basal septum; oblong to ellipsoid conidia are clustered at the phialide tip
Onychocola canadensis (Fig. 22.2)	Not a typical saprobe	Filaments and arthroconidia	Swollen, one or two-celled arthroconidia in chains
Scopulariopsis brevicaulis (Fig. 22.3)	Common saprobe, often laboratory contaminant	Golden brown, lemon-shaped conidia, short filaments	Simple or branched penicillate-like conidiophores end in annellides producing chains of rough, spiny, pyriform annelloconidia
Neoscytalidium dimidiatum (Fig. 22.4)	Not a typical saprobe; imported from subtropics and tropics	Chains of dark and pale arthroconidia (melanized)	Hyaline to brown septate hyphae produce chains of thick-walled one- or two-celled ellipsoid arthroconidia

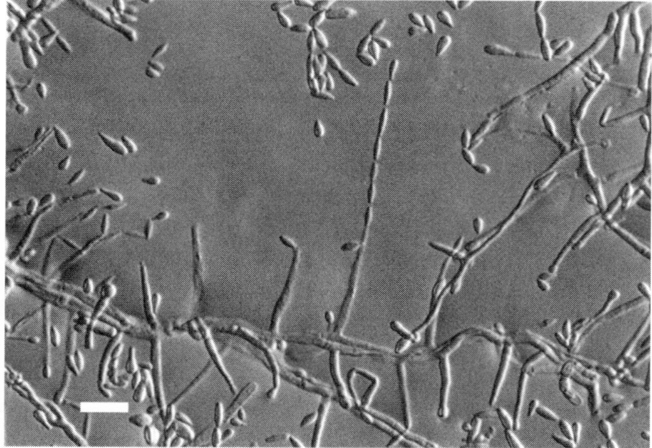

Figure 22.1 *Acremonium strictum*, microscopic morphology.
Source: Used with permission from Dr. Robert Simmons, Director, Biological Imaging Core Facility, Department of Biology, Georgia State University, Atlanta.

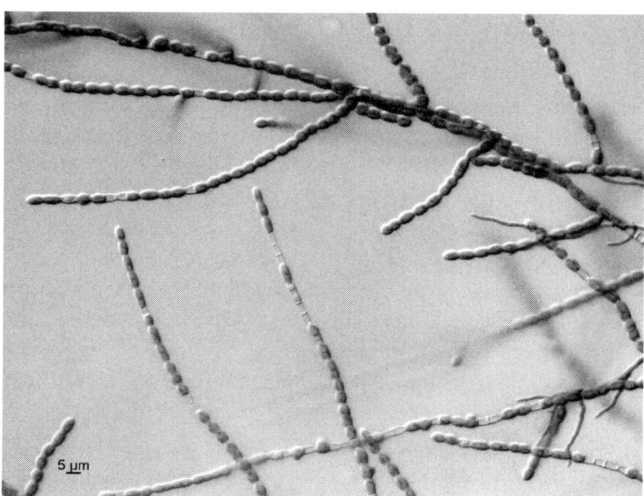

Figure 22.2 *Onychocola canadensis*, DIC microscopy. (Culture reference no. UAMH 10925 sc.)
Source: Used with permission from Professor Lynne Sigler, University of Alberta, Edmonton, Canada.

A. potronii, and *A. kiliense* are best determined by a reference laboratory.

Onychocola canadensis

Onychocola canadensis (teleomorph: *Arachnomyces nodosetosus*) (Fig. 22.2) is not a common saprobe; isolation of *O. canadensis* is strong evidence for its involvement in distal subungual onychomycosis. Filaments and arthroconidia are seen in infected nail tissue. Most cases of *O. canadensis* onychomycosis are not cured by any treatment.

Mycologic Characteristics Growth is very slow, explaining why this hyaline mold may be overlooked. The colony is floccose or velvety, at first flat, and becoming heaped and rounded with age. It is pale gray with a deep brownish-gray reverse. Hyphae sporulate as swollen, one- or two-celled arthroconidia in chains. Arthroconidia may be smooth or with finely rough walls. *O. canadensis* is resistant to cycloheximide.

Scopulariopsis brevicaulis

Scopulariopsis brevicaulis is a common saprobe, often a laboratory contaminant (Fig. 22.3). It is the most common cause of nondermatophyte onychomycosis in temperate climates, and dermatologists are well acquainted with it. *S. brevicaulis* is involved in various types of onychomycosis and, occasionally, in tinea pedis. The microscopic morphology in infected nails is distinctive—with golden

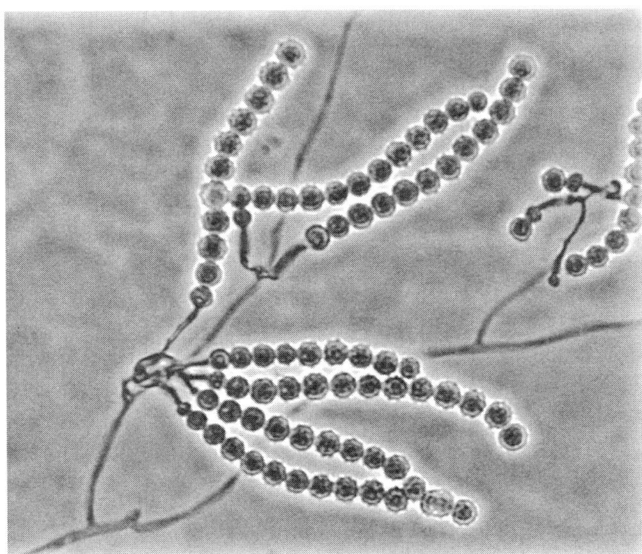

Figure 22.3 *Scopulariopsis brevicaulis*.
Source: E. Reiss, CDC.

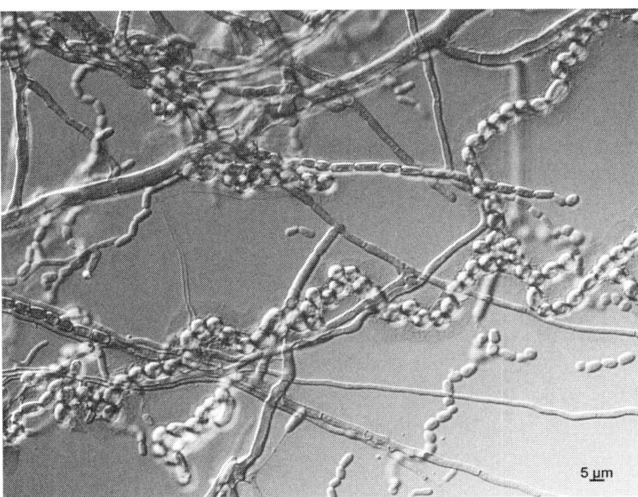

Figure 22.4 *Neoscytalidium dimidiatum*, 7 days' incubation at 35°C, DIC microscopy. (Culture reference no. UAMH 10614 sc.)
Source: Used with permission from Professor Lynne Sigler, University of Alberta, Edmonton, Canada.

brown, lemon-shaped conidia. Systemic antifungals (ITC and TRB) fail to produce mycologic cure in half of the patients. The addition of topical antifungals in nail lacquer increases the percentage of cure. About one-fourth of patients remain mycologically positive even after prolonged therapy (Tosti et al., 2003).

Mycologic Characteristics A rapidly growing flat and velvety to powdery colony develops which is tan to beige to yellow-brown. The reverse is brown in the center becoming paler at the edges. Microscopically the morphology is very distinctive. Hyphae may be dark or hyaline with simple or branched *penicillate*-like conidiophores ending in annellides. Chains of generally rough, spiny, thick-walled pyriform *annelloconidia* with truncate bases are produced and may be either hyaline or brown. The chains of annelloconidia are readily identified by the short truncate base between cells.

Scytalidium dimidiatum

Scytalidium dimidiatum (formerly *Hendersonula toruloidea*) is now classed as *Neoscytalidium* (Fig. 22.4). This melanized mold is a common agent of onychomycosis and dermatophytosis-like foot infections and causes all three types of onychomycosis. Nail and periungual dark pigmentation can occur in *Neoscytalidium* onychomycosis. *Neoscytalidium* spp. onychomycosis do not respond to systemic treatment. Nail avulsion and topical treatment usually are ineffective (Tosti et al., 2003). *Neoscytalidium dimidiatum* is not a typical saprobe and when seen in temperate zones it is imported from subtropics and tropics as an immigrant-related mycosis. In infected nail tissue, chains of dark (melanized) and pale arthroconidia are seen.

Mycologic Characteristics The mold grows rapidly, with a wooly surface and aerial mycelium that reaches the lid of the Petri plate. The surface color is gray-black with a dark brown to black reverse. Hyaline to brownish septate hyphae produce chains of thick-walled one- or two-celled ellipsoid arthroconidia and also narrow pale arthroconidia.

Neoscytalidium hyalinum differs from *N. dimidiatum* in producing hyaline, rather than brown, arthroconidia.

Response to Therapy

Distal and proximal subungual onychomycosis caused by *Acremonium* spp., *Scopulariopsis brevicaulis, Fusarium* spp., *Onychocola canadensis*, and *Neoscytalidium* spp. are very difficult to cure.

22B *SUPERFICIAL MYCOSIS OF THE HAIR CAUSED BY A NONDERMATOPHYTE MOLD: BLACK PIEDRA*

22B.1 *INTRODUCTION/DISEASE DEFINITION*

Black piedra is a superficial fungal infection of hairs of the scalp, beard, and/or moustache where the fungus invades beneath the cuticle,[1] spreads around the hair shaft, forming hard, gritty, discrete dark-brown to black nodules which adhere firmly to the hair. The causative agent is *Piedraia hortae*, a melanized mold. Often multiple nodules occur on a single shaft. Very rarely, nodules may occur on pubic or axillary hair. An unusual characteristic of *P. hortae* is its production of ascoma and ascospores within the nodules.

22B.2 DIAGNOSIS

Diagnosis is made by finding nodules on hair shafts. Viewed under the microscope, squash preps of nodules reveal darkly pigmented hyphae and fusiform ascospores. (Please see Section 22B.10, Laboratory Detection, Recovery, and Identification of *Piedraia hortae*.)

22B.3 ETIOLOGIC AGENTS

Piedraia hortae is a mold in the order *Dothideales, Dothidiomycetidae, Ascomycota*. Black piedra caused by *P. quintanilhae* occurs in chimpanzees in Central Africa and differs from *P. hortae* by lacking terminal appendages on its ascospores (van Uden et al., 1963).

22B.4 GEOGRAPHIC DISTRIBUTION/ECOLOGIC NICHE/EPIDEMIOLOGY

Natural habitats of *P. hortae* (Schwartz, 2004) apart from the hair of humans and nonhuman primates are unknown. It is seen in humid, tropical areas worldwide. The disease is common among natives living in forested areas of Southeast Asia, Africa, and Central and South America (Fischman, 1973). In a study in Amazonia, Brazil, 57% of Zoró Indians older than 11 years were found to have black piedra (Coimbra Júnior and Santos, 1989). Black piedra in temperate zones is an international travel-related mycosis.

22B.5 RISK GROUPS/ TRANSMISSION

The source of infection is unknown but is thought to be environmental exposure. Black piedra is believed by some native tribes to be a sign of beauty and is encouraged. An albino Moon-Child therapeutic chanter on Mulatuppu Island, Panama, was observed by American physicians attached to the Gorgas Memorial Institute. He had blackish color hair mixed with yellow. The black pigment was due to nodules of black piedra. The chanter was careful not to put plant oil on his hair, as is the custom in the Kuna tribe, because it would remove the fungus (Moyer and Keeler, 1964).

Person-to-person transmission occurs—epidemics in families have resulted from sharing of combs and hairbrushes. Reports of outbreaks of black piedra are exceptional. One occurred in a government orphanage in San Juan, Puerto Rico, in 1938 among 45 boys sharing wet combs (Carion, 1965).

22B.6 DETERMINANTS OF PATHOGENICITY

Hairs affected by *Piedraia hortae* from Brazilian indigenous persons of the Xingu tribe were analyzed by light and electron microscopy and revealed destruction of the cuticle and cortex of the hairs, indicating keratinolytic activity (Figueras et al., 1996). Lytic areas in the hair were filled with an electron-dense fibrillar material. The cementing extracellular material holding the nodule together protects it from desiccation and damage, so that unless affected hairs are removed, the infection will remain indefinitely.

22B.7 CLINICAL FORMS

Piedraia hortae infects hair shafts of the scalp and beard. Nodules range in size from microscopic to 3 mm in diameter, and multiple nodules may be seen on a single hair. The fungus grows only on the outside of the hair, but penetrates and disrupts the cuticle. Complete disruption of the hair occurs in lower animals but not in humans.

22B.8 VETERINARY FORMS

P. hortae is seen in mammals including nonhuman primates (even primate pelts in museums; Kaplan, 1959). Development of the nodules in lower animals, unlike those in humans, can lead to damage and destruction of the hair, breaking it at the site of the nodules (van Uden et al., 1963).

22B.9 THERAPY

The condition persists for years if untreated. Treatment consists of shaving or clipping nodule-infested hairs, followed by 2% Nizoral® (ketoconazole, KTC) shampoo. Oral therapy with TRB has also been successful.

[1]Cuticle. *Definition*: The cuticle, or the outermost layer of the hair, is a hard, shingle-like layer of overlapping cells, 5–12 deep. It acts as a protective barrier for the softer inner structure.

22B.10 *LABORATORY DETECTION, RECOVERY, AND IDENTIFICATION OF* PIEDRAIA HORTAE

Direct Examination

See Fig. 22.5. Hairs containing nodules are softened by placing them on a microscope slide. Then

- Add a drop of 25% KOH.[2]
- Cover with a coverslip.
- Warm gently (on a heat block at 51–54°C, not in an open flame). This will accomplish digestion of the nodule.
- Apply pressure to the coverslip to squash the nodule.

Microscopic examination of this squash prep will reveal melanized, thickened, highly septate hyphae with many swollen intercalary cells resembling chlamydospores. No conidia are present, but special cavities, *locules*, are seen at low power. At high power each cavity has an inconspicuous *ostiole* and contains an ascus with up to 8 ascospores. The ascospores are fusiform, curved, and hyaline and possess a single whip-like appendage at each end.

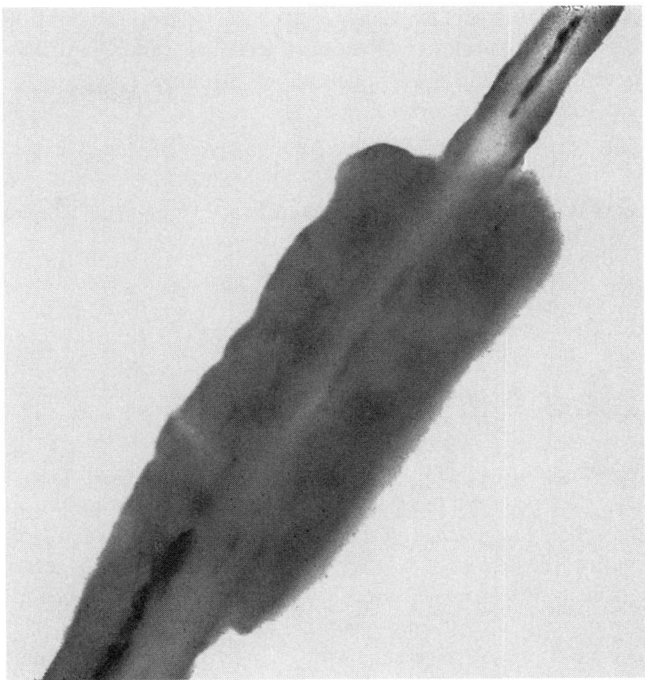

Figure 22.5 Nodule of black piedra on a hair shaft, caused by *Piedraia hortae*.
Source: H. J. Shadomy.

Culture

Although rarely done, infected hairs may be cut and planted on SDA with cycloheximide and incubated at room temperature. *Piedraia hortae* grows slowly as a dark-brown to black colony, either glabrous or covered with short aerial hyphae. A red-brown pigment is occasionally produced. Mixed infection of hair shafts has occurred, with the occasional appearance of a whitish periphery which is due to one of the *Trichosporon* spp. of white yeasts.

Microscopic Morphology

In vitro growth demonstrates closely spaced septate hyphae and intercalary chlamydospore-like cells. Asci with ascospores are less likely to be produced in culture than in vivo.

22C *SUPERFICIAL MYCOSES CAUSED BY YEASTS AND YEAST-LIKE FUNGI*

These fungal infections are defined as those in which a pathogen is restricted to the stratum corneum, with little or no tissue reaction. *Trichosporon* species, however, have increased pathogenic potential to cause invasive disease in the immunocompromised host.

[2]*Safety Note*: This is very strong alkali; **protect eyes and skin**; do not store in a glass stoppered bottle or the stopper will fuse; do not store with a rubber stopper because in removing it the alkali may spatter; only use a screw cap bottle.

22C.1 CANDIDA ALBICANS *ONYCHIA AND PARONYCHIA*

Please see Chapter 11.

22C.2 *WHITE PIEDRA*

22C.2.1 INTRODUCTION/DISEASE DEFINITION

This is a superficial infection of hair of the beard and moustache, occasionally of the axillary or groin hair, caused by basidiomycetous white yeasts of the genus *Trichosporon*. The mycosis is characterized by white to light brown nodules on hair shafts. Nodules are soft and are not firmly adherent to the hair. Usually there is only one concretion per shaft.

22C.2.2 DIAGNOSIS

White piedra is diagnosed by direct microscopic examination of squash preps of the nodules, and confirmed by culture if the direct exam is equivocal. (Please see Section 22C.2.11, Laboratory Detection, Recovery and Identification of *Trichosporon* Species and also for diagnosis of systemic *Trichosporon* disease.)

Differential Diagnosis

Pediculosis capitis (head lice) and pediculosis pubis (pubic lice) are possible diagnoses.

22C.2.3 ETIOLOGIC AGENTS

White piedra is due to yeasts in the genus *Trichosporon* and is discussed here in contrast to hair shaft infection caused by *Piedraia hortae* (causative agent of black piedra). White piedra often is misdiagnosed. The genus *Trichosporon* is classed in the *Trichosporonaceae, Tremellales, Hymenomycetes, Basidiomycota*. These yeasts have a polysaccharide capsule and are related to *Cryptococcus* species. The taxonomy of the genus was revised and the former species *T. beigelii* was succeeded by other pathogenic species (Gueho et al., 1992). Thirty-four *Trichosporon* species were delineated by phylogenetic analysis of the D1/D2 region of 28S (LSU) subunit of rDNA into four clades of which two contain the six medically important species: the *Ovoides* clade contains *T. ovoides, T. asahii, T. asteroides*, and *T. inkin* (Diaz and Fell, 2004). The *Cutaneum* clade contains *T. cutaneum* and *T. mucoides*. They are listed here with the most frequent clinical form (Chagas-Neto et al., 2008; Larone, 2002):

- *Trichosporon asahii* Akagi ex Sugita, A. Nishikawa & Shinoda, *J Gen Appl Microbiol, Tokyo* 40(5): 405 (1994). *Trichosporon asahii* causes systemic disease with hematogenous dissemination in immunocompromised hosts. *T. asahii* is the most clinically important pathogenic yeast in the genus *Trichosporon* because it also causes onychomycosis, visceral infection, and summer-type hypersensitivity pneumonitis.
- *Trichosporon mucoides*—systemic disease, with CNS involvement.
- *Trichosporon cutaneum*—rare skin lesions and white piedra of underarm hair.
- *Trichosporon inkin*—white piedra of pubic hair, occasionally causes systemic disease.
- *Trichosporon ovoides*—white piedra of head hairs, occasionally causes skin lesions.
- *Trichosporon asteroides*—skin lesions and fungemia.

22C.2.4 GEOGRAPHIC DISTRIBUTION/ECOLOGIC NICHE/EPIDEMIOLOGY

White piedra occurs worldwide and is most prevalent in temperate and semitropical climates, for example, South America, the Middle East, India, Southeast Asia, Africa, Europe, Japan, and parts of the southern United States.

Diverse ecologic niches are known for *Trichosporon* species:

- Soil, air, water, vegetation.
- Predominant yeast species from temperate climate forest soils (Sláviková and Vadkertiová, 2000).
- Japanese household tatami mats, pillows, and beds in the homes of summer hypersensitivity patients, a disease that is localized to Japan (Sugita et al., 2004).
- Milk and cheese in Italy (Corbo et al., 2001).
- Minor members of the gastrointestinal microbiota of humans and transient colonizers of the skin and respiratory tract.

Brazil

Most of the 23 Brazilian patients studied with scalp white piedra were female (87%; age 2–42 years), and 74% were children (age 2–6 years) (reviewed by Schwartz, 2004).

Japan

Trichosporon asahii can cause allergic pneumonia and *Trichosporon* species are responsible for summer-type hypersensitivity pneumonitis incited by inhalation of arthroconidia contaminating homes during hot, humid, and rainy summers in western and southern Japan (Sugita et al., 2004; reviewed by Chagas-Neto et al., 2008)

United States

Reports of white piedra in the United States are few. One example is a series of eight children with white piedra of the scalp who resided in Connecticut and New York City (Kiken et al., 2006). There were seven girls, aged 4–16 years and one boy, 4 years old. Five children were immigrants: three from Mexico, one from Britain, and one from Yemen. The girls all had very long hair. They were successfully treated with oral antifungal agents without the need to shave the affected scalp areas. An oral azole (ITC or FLC) for 3 weeks to 1 month combined with topical KTC (Nizoral®) shampoo for 2–3 months was effective in achieving clinical cure in seven patients.

22C.2.5 TRANSMISSION

Sources of *Trichosporon* species are the soil, and from interpersonal contact since these species colonize human skin, nails, and mouth. Human white piedra also can result from contact with infected horses, monkeys, dogs, or other animals.

22C.2.6 DETERMINANTS OF PATHOGENICITY

Trichosporon species possess a glucuronoxylomannan (GXM) capsule which shares properties with the capsular polysaccharide of *Cryptococcus neoformans* (Fonseca et al., 2009). Cryptococcal GXM has antiphagocytic properties and, by analogy, the capsular GXM of *Trichosporon* species should be regarded as a pathogenic factor.

22C.2.7 CLINICAL FORMS

Superficial

The usual form of white piedra is that of single concretions on hair shafts caused primarily by *T. inkin* (on pubic hair) and *T. ovoides* (on head hair) (Chagas-Neto et al., 2008). This clinical form produces no systemic disease, tissue reaction, or discomfort. Genital white piedra can indicate a synergistic coinfection with corynebacteria. Patients with chronic intertrigo unresponsive to usual treatments should be examined for evidence of white piedra. Diagnosis of hair infection is made microscopically in order to differentiate *Trichosporon* species "nits" from those of *pediculosis capitis*. *T. cutaneum* is a causative agent of onychomycosis.

Hypersensitivity Pneumonitis

During summers in southern and western Japan, *Trichosporon* species multiply in homes where repeated inhalation of their arthroconidia causes summer-type hypersensitivity pneumonitis. Symptoms appearing in patients at home during the summer are an onset of cough, sputum, and fatigue (Ando et al., 1991). Cough, fever (38–40°C), and exertional dyspnea gradually increase. The pathologic features are a combination of bronchiolitis and alveolitis with granuloma formation, similar to other forms of hypersensitivity pneumonitis. Pathogenesis of summer hypersensitivity is an immune complex disease evoked by the causative antigens. High titers of *Trichosporon*-specific IgG, IgA, and complement components were detected in patient serum and BAL fluid (Ando et al., 1991). Complement components C1q and C3 correlated with decreased oxygen diffusing capacity in patients. A genetic tendency may be involved because there was an increased frequency of HLA-DQw3 in patients. An environmental survey of 35 homes of persons diagnosed with summer-type hypersensitivity found that *T. dermatis, T. asahii* genotype 3, and *T. montevideense* were the most prevalent species (Sugita et al., 2004).

Invasive Disease

Severe systemic *Trichosporon* species disease occasionally occurs in immunocompromised individuals. When hematogenous dissemination occurs, purpuric or necrotic cutaneous papules and nodules may be seen. *Trichosporon* species infections disseminate mainly in patients with neutropenia and, rarely, in persons living with AIDS. Mortality rates for this form of the disease are high, 60–70%. Fifteen hematology–oncology departments in several regions of Italy cooperated in a retrospective study of *Trichosporon* species and *Geotrichum capitatum* infections treated over a 20 year period, 1983 through 2002 (Girmenia et al., 2005). There were 15 cases of culture confirmed *Trichosporon* spp. infections, 14 with fungemia and, of these, 7 had evidence of tissue invasion. Thirteen cases were treated with AmB, either the deoxycholate formulation in 11 patients or liposomal AmB in 2 patients. The all cause mortality was 64.7%. Neutropenic cancer patients, in particular, those with acute myelogenous leukemia, are at greatest risk of invasive *Trichosporon* disease.

DNA fingerprinting suggests that strains producing superficial infection are distinctly different from those producing invasive disease. Two genotypes representing superficial and invasive disease, respectively, had the identical phenotype when tested in the API 20C® assimilation test panel system (Kemker et al., 1991).

22C.2.8 VETERINARY FORMS

Infections in horses, dogs, and cats have been reported.

22C.2.9 THERAPY FOR WHITE PIEDRA

The first step in therapy consists of clipping the infected hairs containing nodules and shaving the surrounding area, followed by the use of 2% Nizoral® (KTZ) shampoo. Alternative topical antifungal treatment for white piedra includes ciclopirox, selenium sulfide 1–2.5% lotion, 6% precipitated sulfur in petrolatum, chlorhexidine solution,

Castellani paint, pyrithione zinc, 2–10% glutaraldehyde, azoles, and AmB lotion (Schwartz, 2004). Oral therapy with ITC is recommended if topical treatments are ineffective for uncomplicated white piedra of the scalp hair. Genital white piedra remains a therapeutic challenge. The recommendation is to shave pubic hair and to use a topical antifungal. Nodules of white piedra can be detected in cotton fibers, so underwear should be disinfected or discarded.

22C.2.10 THERAPY FOR SYSTEMIC OR DISSEMINATED *TRICHOSPORON* SPECIES INFECTION IN THE IMMUNOCOMPROMISED HOST

Mortality rates for disseminated *Trichosporon* species infection in the immunocompromised host are between 60% and 70%, despite AmB therapy. In vitro MICs for *T. asahii* indicate it is susceptible to AmB and to ITC, but clinical experience in treating systemic *Trichosporon* species infection suggests a poor prognosis. Recovery from neutropenia may be essential for better clinical outcomes. Echinocandins are not considered effective against *Trichosporon* species since breakthrough *Trichosporon* infections have been seen in hematologic malignancy patients on micafungin prophylaxis (Akagi et al., 2006; Matsue et al., 2006). Possibly, voriconazole and other extended spectrum azoles may prove effective, based on a small series of patients thus far treated (reviewed by Chagas-Neto et al., 2008).

22C.2.11 LABORATORY DETECTION, RECOVERY, AND IDENTIFICATION OF *TRICHOSPORON* SPECIES

Direct Examination

Hairs are studied in KOH squash preps for the presence of relatively soft, white to tan nodules. Fungal structures stain easily with Parker blue-black ink. Nodules, 0.5 mm in diameter, can easily be detached from the hair shaft because they affect only the outer lipid layers. Hyphae in the nodules may develop completely within the hair including the cortex, producing a raised cuticle. Remnants of hyphae generally lie perpendicular to the hair shaft. Within the crushed nodule, hyaline, septate hyphae that segment into oval or rectangular, encapsulated arthroconidia, 2–4 × 3.9 μm in diameter, are seen, often along with budding yeast. Coexistent bacteria may be present, typically at the periphery of the nodules. No locules or asci are present.

Histopathology

Diagnosis of systemic disease is made primarily by biopsy study, and culture from sterile body sites, using the same methods for diagnosis as for other yeast species (please see Chapter 2, and Hazen and Howell, 2007). The picture seen in tissue sections is one of branched, septate hyphae and yeasts. Arthroconidia, the hallmark of the growth in vitro, also may be seen in tissue sections, but culture confirmation may be necessary (Chandler and Watts, 1987).

Culture

Trichosporon species grow rapidly on SDA at 25°C (without cycloheximide). Initially, the colony is cream colored, of buttery texture, becoming membranous, wrinkled, with radial ridges and folded with age. When mature, within 2–3 weeks, the colony becomes elevated, yellowish-gray, and consists of arthroconidia and budding cells.

Microscopic Morphology

All of the following forms are produced: budding yeast cells, arthroconidia, pseudohyphae, and hyphae (Fig. 22.6). *Trichosporon* species do not ferment carbohydrates. Speciation can be made on the basis of carbohydrate assimilation tests and temperature tolerance (Hazen and Howell, 2007). All *Trichosporon* species are urease-positive, whereas the morphologically similar nonpathogen, *Geotrichum candidum*, is urease-negative.

Phenotypic methods are difficult for the accurate speciation of members of the genus *Trichosporon*. Genetic identification is now recommended. A useful target within

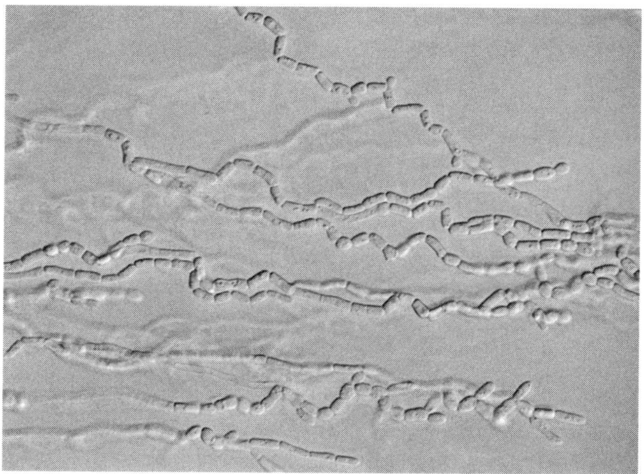

Figure 22.6 *Trichosporon asahii* microscopic morphology, 2 days' incubation on CMA at 27°C.
Source: Used with permission from Dr. Takashi Sugita, Meiji Pharmaceutical University, Tokyo.

rDNA is the IGS1 region between the 26S and 5S rDNA (Sugita et al., 2002). The size of the IGS1 region varies from 195 bp to 704 bp and the sequence differences are sufficient to identify members of the genus.

A rapid Luminex® nucleotide hybridization assay was devised that was capable of identifying 35 different *Trichosporon species* (Diaz and Fell, 2004). Species-specific and group-specific capture probes were designed from the D1/D2, ITS, and IGS1 regions of rDNA. Species-specific amplicons were generated with three sets of primers to yield fragments from the three regions. IGS1 region sequences identified clinical blood isolates as *T. asahii, T. asteroides*, as well as rare species, *T. coremiiforme* and *T. dermatis* (Chagas-Neto et al., 2009).

22C.3 *TINEA NIGRA (TINEA NIGRA PALMARIS)*

22C.3.1 INTRODUCTION/DISEASE DEFINITION/CLINICAL FORM

Tinea nigra is a superficial mycosis of the stratum corneum implanted by traumatic inoculation from soil, sewage, wood, or compost. The causative agent is the melanized ascomycete, and yeast-like fungus, *Hortaea werneckii*. Tinea nigra arises on the palm of the hand and, less commonly, the sole of the foot, body areas with an increased concentration of eccrine sweat glands (please see Section 22C/3.5, Ecologic Niche). Tinea nigra affects only the stratum corneum, appearing as flat, marginated, gray, pale brown, or black *macule*s, due to an accumulation of melanin pigment produced by the fungus. The skin is usually mottled with a deeper coloration at the advancing edges. Tinea nigra is so superficial that it can be scraped off with vigorous effort (Schwartz, 2004). Occasionally, the interdigital web of the hand, fingers, wrist, forearm, trunk, or neck and, rarely, the penis may be affected.

The macules enlarge slowly and are asymptomatic and of cosmetic importance only, except for the fact they have to be differentiated from melanoma. Incubation is between 2 and 7 weeks, although a case was reported 20 years following an experimental inoculation (Blank, 1979; Ritchie and Taylor 1964). Patients generally are older than 19 years of age with a 3:1 ratio of females to males (Fischman et al., 1983; van Velsor and Singletary, 1964).

22C.3.2 DIAGNOSIS

Diagnosis is made by KOH prep of skin scrapings or by histopathology. Culture confirmation is necessary to differentiate *H. werneckii* from other black yeast-like fungi, for example, *Aureobasidium pullulans*.

Differential Diagnosis

The differential diagnosis for black piedra includes a variety of conditions: Addison's disease, melanocytic nevus, and atypical nevus (dysplastic melanocytic nevus), chromhidrosis, contact dermatitis, drug eruption, lentigo, malignant melanoma, pityriasis versicolor, pina, syphilis, or staining from chemicals or dyes (Gupta et al., 2003). The dermatoscopic pattern of homogeneous nonmelanocytic pigment can resemble a melanoma (Schwartz, 2004). Establishing the correct diagnosis is important to avoid excisional surgery and potential scarring.

22C.3.3 ETIOLOGIC AGENT

Hortaea werneckii (Horta) Nishimura and Miyaji, *Jpn J Med Mycol* 26: 145 (1984) (formerly known as *Phaeoannellomyces werneckii*).

Position in classification: *Dothideales, Ascomycota*. This order consists of fungi with thick, melanized cell walls.

22C.3.4 GEOGRAPHIC DISTRIBUTION

Tinea nigra was previously believed to be found only in tropical areas of Central and South America, Africa, and Asia, but is now known to be present in North America and Europe as well.

22C.3.5 ECOLOGIC NICHE

Hortaea werneckii (halophilic) and *Aureobasidium pullulans* (halotolerant) are both black yeast-like fungi isolated from hypersaline waters (3–30% NaCl) of salterns (salt evaporation ponds used to produce edible salt from brine). Halotolerant and halophilic fungi from salterns isolated on saline media were almost exclusively melanized yeast-like fungi. *H. werneckii* is particularly abundant during the halite crystallization period and is the only halophilic species growing on the whole range of NaCl concentrations from 0% to saturation (Gunde-Cimerman et al., 2000).

22C.3.6 EPIDEMIOLOGY

Tinea nigra is common in tropical regions of Central America, South America, Africa, and Asia, and is infrequent in the United States and Europe (Schwartz,

2004). In the United States, it typically affects residents of coastal states, for example, Florida, Texas, Alabama, Louisiana, Virginia, Georgia, and North Carolina. Tinea nigra is an international travel mycosis so that cases appear in other states and cities in patients who relate a history of foreign travel, frequently to the Caribbean islands. Twenty-two patients who resided in humid tropical regions of Mexico were diagnosed with tinea nigra during the period 1997–2007 (Bonifaz et al., 2008). About half of the patients lived in coastal areas where contact with this halophilic yeast-like fungus could occur.

22C.3.7 THERAPY

A topical approach is successful in treating tinea nigra (reviewed by Schwartz, 2004). Topical terbinafine (TRB) (Lamisil®), undecylenic acid, Whitfield's ointment, retinoic acid, 4% aspirin, and epidermal tape stripping are all effective. A short course of cyclopiroxolamine gel applied 2x/day for 3 days was effective in curing tinea nigra (Rosen and Lingappan, 2006). Curative treatments with azoles include topical KTC twice daily for 2 weeks, or oral ITC, 200 mg/day for 3 weeks. Tinea nigra does not tend to recur.

22C.3.8 LABORATORY DETECTION, RECOVERY, AND IDENTIFICATION OF *HORTAEA WERNECKII*

Direct Examination

KOH prep of skin scrapings studied in the microscope reveal one- to two-celled light brown yeasts and masses of thick, septate, frequently branched hyphae (1.5–5 μm diameter) with dark pigment in their walls.

Histopathology

A biopsy specimen can be obtained painlessly from a patient with tinea nigra by carefully scraping off the stratum corneum with a scalpel blade (Schwartz, 2004). Biopsy specimens show hyperkeratosis and mild acanthosis. Because of their natural pigment *H. werneckii* are evident on routine H&E-stained sections. The distinctive morphology of *H. werneckii* is its typically two-celled yeast form. PAS stain reveals red-stained septate hyphae in the stratum corneum.

Culture

Culture is performed to identify the fungus and to differentiate it from noninfectious causes. Scrapings planted on SDA with and without antibiotics at 25°C yields growth in about 1 week.

Colony

Growth progresses from shiny black to moist yeast colonies. After 2 or more weeks velvety gray-black aerial hyphae form.

Microscopic Morphology

Both pseudo- and true hyphae develop which often are thick-walled, multiseptate, and olive colored. The hyphae have intercalary annellides which are the conidiogenous cells. They have annellations at their tips. Hyaline annelloconidia are produced and released to become pale olive-colored 1–2-celled yeast forms. They then bud or form clumps of chlamydospores.

22C.4 *PITYRIASIS VERSICOLOR AND OTHER SUPERFICIAL AND DEEP MYCOSES CAUSED BY LIPOPHILIC YEASTS:* MALASSEZIA *SPECIES*

22C.4.1 INTRODUCTION/DISEASE DEFINITION

Pityriasis versicolor is a chronic, superficial mycosis affecting the stratum corneum of the smooth skin, usually on the upper chest, back, and arms presenting as discolored spots which slowly enlarge and can become confluent patches. The causative agents are lipophilic yeasts of the genus *Malassezia*. They are dimorphic, growing in both yeast and mold forms. These yeasts are basidiomycetes and members of the microbiota of normal skin of humans and lower animals. Pityriasis versicolor is more common in the tropics; in temperate zones it occurs more frequently during the summer.

Malassezia species are implicated in other clinical forms, for example, dandruff, seborrheic dermatitis, folliculitis, and atopic eczema/dermatitis syndrome (AEDS). Patients, especially neonates, receiving intralipid supplement in parenteral nutrition have developed *Malassezia* species fungemia (Dankner et al., 1987; Powell et al., 1984).

In 1847 Eichstedt and Sluyter described pityriasis versicolor; they named the disease but did not propose the name of the fungus. Robin (1853) named the fungus "*Microsporum furfur*" and called the disease tinea versicolor because he thought the fungus was similar to *Microsporium audouinii*. Pityriasis refers to dry, flaking skin; "versicolor" = many-colored, referring to the lesions. In 1889 Baillon named the yeast *Malassezia*. Afterwards, yeasts with short hyphae and monopolar budding were observed widely on healthy skin, in the versicolor condition on smooth skin, in seborrheic dermatitis, and in dandruff. In 1904 Sabouraud placed the lipophilic yeasts into the genus *Pityrosporum*. Additional species were identified and, in 1951, Gordon described *P. orbiculare*, and associated it with *M. furfur*. Before 1996 only 3 species were known, but the subsequent discovery of a total of 14 species by phylogenetic analysis indicated they are a monophyletic lineage, resulting in the erection of the order *Malasseziales* to accommodate them (reviewed by Guillot et al., 2008; Cabañes et al., 2011).

22C.4.2 CASE PRESENTATION

Skin Discoloration in Members of a Swim Team

History

The college swim team practiced daily and, afterwards, applied various creams to avoid dry skin. Several members noticed lightly pigmented areas on their backs, chests, and arms. During the summer season the students became tanned from the sun, which accentuated the spots and increased concern among the students for cosmetic reasons. The coach noticed more swimmers who developed these areas, and that the "spots" enlarged over time. The school infirmary referred the students to a dermatologist.

Clinical and Laboratory Findings

A presumptive diagnosis was made and confirmed when skin scrapings in a KOH prep revealed short irregular hyphae and budding yeast, the classic "meatballs and spaghetti" appearance of *Malassezia* species.

Diagnosis

The diagnosis was pityriasis versicolor caused by a yeast in the genus *Malassezia* (Fig. 22.7).

Therapy and Clinical Course

Therapy was successful with selenium sulfide lotion (2.5%) and Nizoral® (KTC) shampoo. Lesions cleared but there were recurrences.

Comments

In tanned or darkly pigmented persons, pigmentation of lesions is absent or slight, and lesions appear light. Patients with light complexion have skin lesions that vary from pale to salmon colored, to brown. Disease in members of the swim team may have been triggered by the irritating effect of chlorinated water combined with the stimulatory effect of applying oil-based lotion after showering, bearing in mind these yeasts are lipophilic. Recurrences are common.

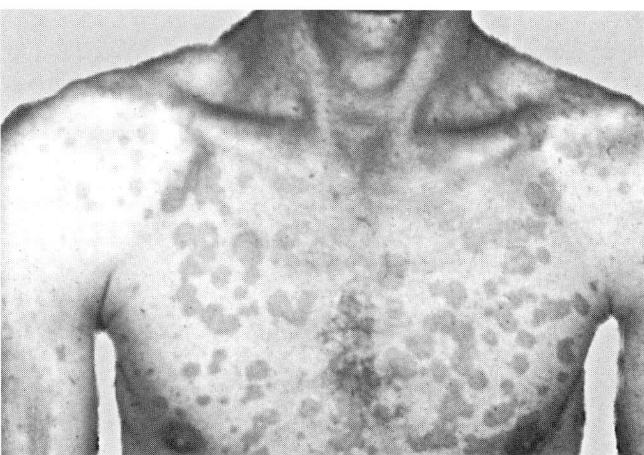

Figure 22.7 Pityriasis versicolor.
Source: Dr. L. Ajello, CDC.

22C.4.3 DIAGNOSIS

The clinical diagnosis of pityriasis versicolor is easy to make for an experienced dermatologist but should be confirmed by direct examination of skin scrapings in a KOH prep: demonstrating monopolar budding yeast, "footprint shaped," and short hyphae. The typical appearance is described as a "spaghetti and meatballs" pattern. Lesions of pityriasis versicolor emit a yellow-green fluorescence when illuminated with Wood's light.

Differential Diagnosis

This includes pigmentation disorders—vitiligo and pityriasis alba (Crespo-Erchiga and Florencio, 2006). Vitiligo affects the hands, feet, and face, whereas pityriasis versicolor affects the trunk. The patches of vitiligo are not scaly and are white under Wood's light. Pityriasis alba, related to atopy, produces diffuse patches with slight scaling, lacks fluorescence, and affects the face and limbs. Less common skin diseases in the differential are pityriasis rosea of Gibert, tinea corporis, secondary syphilis, seborrheic dermatitis, and the rare dermatosis: confluent and reticulated papillomatosis. The last two clinical forms are related to *Malassezia* species inasmuch as yeasts are present in skin scales.

22C.4.4 ETIOLOGIC AGENTS

The causative agents are lipophilic yeasts of the genus *Malassezia*, members of the microbiota of normal human skin and that of lower animals (Batra et al., 2005). They are dimorphic, growing as both monopolar budding yeasts and short hyphae. *Malassezia* species are classed in the order *Malasseziales* in the *Basidiomycota*. The taxonomy of *Malassezia* was rationalized by Guillot and Guého (1995). They assembled 104 isolates of *Malassezia* species representing the various names accorded to them by different mycologists. They used 18S rDNA and DNA–DNA homology studies to establish a monophyletic grouping of seven species: *M. furfur, M. globosa, M. obtusa, M. pachydermatis, M. restricta, M. slooffiae*, and *M. sympodialis*.

Of the seven species of *Malassezia* now known to affect humans, *M. globosa, M. sympodialis*, and *M. restricta* cause >90% of pityriasis versicolor (Guillot et al., 2008) (Table 22.2).

22C.4.5 GEOGRAPHIC DISTRIBUTION/ECOLOGIC NICHE

Malassezia species are commensal members of the normal microbiota of the skin of humans and lower animals, and are found worldwide. *M. globosa* was the predominant isolate from patients with pityriasis versicolor in Greece, Japan, India, Iran, Mexico, Spain, and Tunisia (reviewed by Crespo-Erchiga and Florencio, 2006). *M. sympodialis* was predominant in Canada, whereas *M. furfur* was most often isolated in Panama and Madagascar.

22C.4.6 EPIDEMIOLOGY

- *Pityriasis versicolor* is more common in the tropics; in temperate zones it occurs more frequently during the summer. Prevalence is high in hot and humid climates, where 30–40% of people can be affected (Crespo-Erchiga and Florencio, 2006). Prevalence is much lower in temperate climates (1–4%). The prevalence in the United States and northern Europe is ≤1% of the population, slightly higher in Southern Europe.
- *Seborrheic dermatitis* is one of the most frequent skin diseases. Cradle cap affects up to 70% of newborns in the first 3 months of life and subsides by 1 year of age. The prevalence in adults, assessed by dermatologists, is 11.6%, with 2.8% considered clinically significant enough to prompt an office visit. The condition is rare in African-Americans. Persons living with AIDS are commonly and more severely affected when the $CD4^+$ T-lymphocyte count falls below 400/ μL.

22C.4.7 RISK FACTORS/ TRANSMISSION

Pityriasis versicolor is more common in young adults and equally distributed between men and women. Use of oily tanning lotions and creams and corticosteroids are risk factors (Bonifaz et al., 2010). The only significant associated risk factor is a clinical history of pityriasis versicolor because the disease is recurrent. *Malassezia* species cause catheter-related fungemia in premature neonates. Surprisingly, humans are colonized in the perinatal period at rates in hospitalized neonates ranging from 37% to 100% (reviewed by Ashbee and Evans 2002). *M. pachydermatis* is transmitted to neonates receiving intralipid supplemented parenteral nutrition by healthcare workers carrying the yeast on their hands, likely contracted from their dogs, which harbor the yeast in their ears (Chang et al., 1998; Welbel et al., 1994).

Table 22.2 Selected *Malassezia* Species that Are Most Often Isolated in Pityriasis Versicolor and Systemic Infections (in Addition to Being Commensal Members of the Human or Animal Skin Microbiota)

Malassezia species[a]	Human isolations	Animal isolations	Microscopic features of cells[b]
M. globosa	Pityriasis versicolor, seborrheic dermatitis	Skin lesions; otitis in cats	Cylindrical, 2.5–4.0 μm in length; broad bud base, pronounced bud scar
M. sympodialis	Pityriasis versicolor, atopic dermatitis	Skin lesions; otitis in cats	Elongate, oval or spherical, 6 μm; broad bud base
M. restricta	Seborrheic dermatitis	Not determined	Cylindrical, 1.5–3.5 μm in length; broad bud base
M. furfur	Pityriasis versicolor, fungemia[c]	Not determined	Cylindrical, 4–6 μm; broad bud base
M. pachydermatis[d]	Fungemia	Dogs: otitis, seborrheic dermatitis	Ovoid, globose, 2.5–5 μm long; some sympodial budding

[a]Guilliot et al. (2008).

[b]Ashbee and Evans (2002).

[c]Confirmation of this species involvement in fungemia awaits molecular studies of *Malassezia*.

[d]Zoophilic species.

22C.4.8 CLINICAL FORMS OF SUPERFICIAL MYCOSES CAUSED BY *MALASSEZIA* SPECIES

Pityriasis Versicolor

See Fig. 22.7. This chronic, superficial mycosis affects the stratum corneum of the smooth skin, usually on the upper chest, back, shoulders, and arms, presenting as discolored spots that slowly enlarge and can become confluent patches. The spots may be white, salmon, or brown and are covered with thin scales which contain characteristic yeast cells and short hyphal strands. Hence, the term "versicolor" (*definition*: many colored). The spots are asymptomatic and only a few patients complain of itching (Bonifaz et al., 2010). Two varieties occur; a hypopigmented variety generally presents in dark-skinned individuals as hypochromic patches or macules covered with fine scales. The hyperpigmented variety consists of light brown patches with scales on the surface.

These patches emit a yellow green fluorescence under Wood's light. Pityriasis versicolor is differentiated from *erythrasma* caused by *Corynebacterium minutissimum*, which emits a coral red fluorescence. Pityriasis versicolor is usually asymptomatic, although some patients develop moderate or even severe pruritus.

Other clinical forms in which *Malassezia* species have been implicated are dandruff, seborrheic dermatitis, folliculitis, and atopic eczema/dermatitis syndrome (AEDS) (Batra et al., 2005). Patients, especially neonates, receiving intralipid parenteral nutrition have developed *Malassezia* species fungemia.

Seborrheic Dermatitis and Dandruff

What are they and what evidence is there for a role for Malassezia species as the causative agent? Seborrheic dermatitis (*definition*: a common, chronic, relapsing, inflammatory skin disorder affecting the scalp and other oily areas of the body) presents with scaly, itchy, red skin and stubborn dandruff. Other areas of the body that may be affected are the face, upper chest, and back (Naldi and Rebora, 2009). This disorder is socially embarrassing because of the scaling scalp, creating an impression of poor hygiene.

Seborrheic dermatitis in infants is known as "cradle cap." Evidence linking *Malassezia* species with the condition is that antifungal therapy both causes the symptoms to subside and lowers levels of yeast in the scalp. Patients with AIDS are at greater risk with up to one-third being affected. One view is that the symptoms are an allergic response by the host to the yeast and to fatty acids released by the action of *Malassezia* lipases. *M. restricta* and *M. obtusa* are implicated in this disease. Common clinical variants of the condition are:

- *Cradle Cap.* Red-yellow plaques covered by scales on the scalp of infants, developing after a few weeks of age.
- *Dandruff.* Mild seborrheic dermatitis of the scalp with scaling as the common feature; this is a lay term for any scalp condition causing fine scales.

Differential diagnoses for seborrheic dermatitis include psoriasis, tinea capitis in children, atopic dermatitis, contact dermatitis, rosacea, and erythrasma. For other rarer differential conditions please see Naldi and Rebora (2009).

Folliculitis

Pruritic erythematous, follicular papules, or pustules occur in sites rich in sebaceous glands, often occurring in immunocompromised hosts. *Malassezia* yeast are found in plugged hair follicles.

Atopic Eczema/Dermatitis Syndrome (AEDS)

AEDS is a chronic inflammatory skin condition that is atopic, hereditary, and noncommunicable. Most commonly affected are the flexures of the knees and elbows. Young children are particularly affected; the incidence in young children in developed industrial countries is 10–20%, compared with 1–3% of adults. The disease also is common in dogs. *Malassezia* species frequently are isolated from AEDS patient skin, but as they also are common skin microbiota in normal persons, research has pursued the line that it is allergens of *Malassezia* species, rather than a frank infectious process, that triggers the inflammatory response in the susceptible population (Ashbee and Evans, 2002).

Despite their being mild pathogens, often dismissed as of merely cosmetic importance, *Malassezi*a species evoke various immune responses in the human (and the canine) host response. They stimulate the mononuclear phagocytic system, activate complement, suppress cytokine release, and downregulate phagocytosis and intracellular killing (Ashbee and Evans, 2002).

Psoriasis

Definition: Psoriasis vulgaris is a genetic, systemic, inflammatory, chronic disorder, which can be altered by environmental factors. It afflicts approximately 2% of the U.S. population. It may be associated with other inflammatory disorders such as psoriatic arthritis, inflammatory bowel disease, and coronary artery disease. It is characterized by scaly, erythematous patches, papules, and plaques that often are pruritic. *Malassezia* species do not cause psoriasis, but given the immunologic basis of that disease, and the ubiquitous presence of lipophilic yeasts in the same anatomic compartment as the psoriatic lesions, it is reasonable that *Malassezia* can act as an external trigger, aggravating inflammation. Clinical improvement of scalp lesions of psoriasis after KTC therapy is suggestive of a role for these lipophilic yeasts as capable of evoking or aggravating lesions in psoriasis patients (reviewed by Ashbee and Evans, 2002).

22C.4.9 CLINICAL FORM OF DEEP-SEATED *MALASSEZIA* MYCOSIS: FUNGEMIA

Risk factors for *Malassezia* fungemia include prolonged hospitalization, central venous catheters, and the use of intralipid emulsions for nutritional support. The lipophilic nature of *Malassezia* and the intralipid emulsion therapy explain how these skin-dwelling yeasts become involved in the pathogenesis of sepsis. Often the cases are complex and the specifics of fungemia are difficult to identify because of the severe underlying diseases (Dankner et al., 1987). Neonates often present with the signs and symptoms of sepsis and thrombocytopenia, whereas fever may be the only symptom in adults. When blood cultures are positive, they resolve upon removal of the colonized catheter, usually accompanied by antifungal therapy with AmB + flucytosine, or FLC (Chryssanthou et al., 2001).

> *Illustrative Example: Malassezia* was isolated from Broviac catheter blood cultures in five sick infants who were receiving IV lipid emulsions (Powell et al., 1984). All infants recovered without antifungal therapy after removal of the catheters. Early onset of fungemia after catheter placement in these infants and recovery of *Malassezia* from the skin of nearly 33% of hospitalized premature neonates indicate that contamination of the catheter at the time of placement is the most likely origin of infection.

A prospective study in a French hospital determined that the prevalence of *Malassezia* colonization of central venous catheters in the neonatal intensive care unit was low, but not negligible (Sizun et al., 1994). Only 2.7% of catheters from newborns receiving parenteral nutrition including intralipid emulsion were positive for *Malassezia* when cultured on Dixon medium.

Link to Companion Animals: Dogs

Outbreaks of *M. pachydermatis* fungemia in neonatal intensive care units have been linked by genotyping of the bloodstream cultures of neonates and hand cultures of healthcare workers (Chang et al., 1998; Welbel et al., 1994). *M. pachydermatis* is a zoophilic yeast associated with canines. In one outbreak the link was made to a healthcare worker who acquired the yeast from pet dogs at home (Chang et al., 1998).

22C.4.10 VETERINARY FORMS

Among lower animals, lipophilic yeasts were first described in dogs, causing otitis externa and pruritic, erythematous lesions, primarily on the stomach. Isolation also is documented from cats, ferrets, pigs, sea lions, cattle, and horses (Ashbee, 2007). Weidman in 1925 isolated a yeast he named *Pitrosporum pachydermatis* from skin lesions of an Indian rhinoceros. Morris Gordon reassigned that yeast to *Malassezia pachydermatis* (Gordon, 1979).

Otitis Externa in Dogs and Cats

The yeasty odor of poorly maintained canine ears is familiar to veterinarians. *M. pachydermatis* can be isolated from 50% to 60% of dogs without or with otitis externa. Cats, too, develop this disease—41.2% of symptomatic cats and 17.6% of cats without otitis externa carry this yeast. A small subset of these companion animals also harbor lipid-dependent species: *M. sympodialis* and *M. furfur* in cats, and *M. furfur* and *M. obtusa* in dogs (Crespo et al., 2002). Diagnosis is made by Diff-Quik® stain of a smear made from ear swabs of ceruminous material or skin scrapings showing yeast with monopolar budding ("foot print" shape) mixed with short hyphae.

Therapy of otitis externa in dogs and cats depends on four possible scenarios: without infection, triggered by bacteria, yeast infection, or mixed bacteria and yeast infections. Medication is designed to address the microbial spectrum. Treatment is guided by the clinical appearance and viewing the Diff-Quik stained smear. Ear cleaning products for dogs often are used in conjunction with topical medications to decrease exudates; for example, Epi-Otic®, contains 2.5% lactic acid and 0.1% salicylic acid. Various antibacterial otic preparations are available. Topical antifungal therapy consists of 2% KTC cream or lotion. Topical glucocorticoids also are helpful because they are antipruritic and anti-inflammatory, and decrease glandular secretions. Animals with *Malassezia* otitis externa begin to improve 7–14 days after antifungal treatment.

Animals who fail to improve after topical treatment may benefit from systemic therapy with KTC, ITC, or FLC, given with food. Dosing schedules for antifungal agents for otitis are given in Matousek and Campbell (2002). Dose-related adverse effects include anorexia, diarrhea, and vomiting. Oral TRB is an alternative to azoles with good tolerability and low side effects (Rosales et al., 2005).

Atopic Dermatitis in Dogs

Dogs and humans with atopic dermatitis frequently exhibit concurrent skin infections with *Staphylococcus* bacteria or *Malassezia* yeast, and their treatment is an important part of managing these patients (de Boer and Marsella, 2001). Skin colonization by *Malassezia* yeast contributes to clinical signs of atopic dermatitis; yeast components induce inflammation (please see Section 22C.4.11.2, Microbial Factors). Skin lesions associated with *Malassezia* dermatitis include alopecia, erythema, skin thickening, hyperpigmentation, scaling, and greasy exudates. Affected dogs are pruritic and have an unpleasant odor. Anatomic sites affected are ventral neck, axilla, inguinal region, face, feet, and perineum.

Malassezia dermatitis is overrepresented in certain breeds, for example, basset hounds, cocker spaniels, and West Highland white terriers. Pruritus is a clinical feature and sometimes can be severe. Up to half of dogs with *Malassezia* dermatitis are atopic or are affected by food or flea allergies. Underlying genetic skin and endocrine defects also are common. Thus, not every dog with *Malassezia* dermatitis is atopic, and not every dog with atopic dermatitis will develop *Malassezia* overgrowth.

General principles of therapy for canine atopic dermatitis include a combination of allergen avoidance, anti-inflammatory agents, allergen-specific immunotherapy, and antimicrobial drugs (Olivry and Sousa, 2001). Antimicrobial therapy is one of the most important parts of clinical management. Client dogs with suspected *Malassezia* dermatitis were treated in three groups: oral cephalexin alone, or combined with either oral TRB, or oral KTC. Only the treatment with TRB caused a reduction in pruritus (Rosales et al., 2005).

22C.4.11 DETERMINANTS OF PATHOGENICITY

22C.4.11.1 Host Factors

Normal human lymphocytes and mononuclear cells stimulated with *Malassezia* antigens result in significant levels of lymphocyte blastogenesis and leukocyte migration inhibition (reviewed by Ashbee and Evans, 2002). Thus, although these commensal yeasts are colonizers of the most superficial layers of the epidermis, their antigens stimulate immunologic memory. Neutrophils are capable of phagocytosing *Malassezia* in a complement-dependent process. After 2 h, only 5% of the cells are killed; in contrast, 30–50% of *C. albicans* yeast cells are killed by PMNs. *Malassezia* cells extracted with solvents are phagocytosed efficiently by PMNs, with a parallel release of reactive oxygen intermediates. The lipid-rich layer in the *Malassezia* cell envelope may serve an antiphagocytic function but this inference requires further investigation. Monocyte interaction in vitro with *Malassezia* stimulates IL-8, whereas granulocyte interactions with the yeast produce IL-8 and IL-1α. These two proinflammatory cytokines activate lymphocytes and neutrophils, and increase their chemotaxis, contributing to inflammation.

22C.4.11.2 Microbial Factors

Lipases

Three separate lipases secreted by *Malassezia* species are essential for cell growth (reviewed by Ashbee and Evans, 2002). A phospholipase also is produced which acts on mammalian cells to release arachidonic acid, a precursor of inflammatory prostaglandins. If this mechanism is operative in the skin environment it could contribute to local inflammation.

Factors Affecting Depigmentation in Pityriasis Versicolor

Among metabolites produced by *Malassezia* species are the following:

- *Azelaic Acid and Lipoperoxidase.* When cultures of *Pityrosporum* (now *Malssezia* species) were supplemented with unsaturated fatty acids with double bonds in the 6–12 positions, dicarboxylic acids of chain length C_6 to C_{12} were formed (Passi et al., 1991; Mendez-Tovar, 2010). Azelaic acid is such a product, HOOC–$(CH_2)_7$–COOH, a medium chain length saturated dicarboxylic acid. Azelaic acid is a competitive inhibitor of tyrosinase, a key enzyme in melanogenesis. Exposure to sunlight stimulates the production of azelaic acid, which contributes to the formation of hypopigmented skin spots. Since the connection of azelaic acid to depigmentation was discovered, it has been applied therapeutically as a cream or oral therapy to aid in reversing hyperpigmentary disorders related to increased activity of melanocytes, such as melasma, and for other skin conditions: rosacea and acne vulgaris.

  Linoleic acid, a diunsaturated fatty acid, induces lipoperoxides in a *Malassezia* species (de Luca et al., 1996). *Trans–trans* farnesol and squalene epoxides were detected when an acetone powder of *Malassezia* species was incubated with linoleic acid and squalene. Peroxisomes were induced in the yeast grown with linoleic acid, with resultant production of hydrogen peroxide and hydroxyl radicals. These reactive oxygen species may contribute to damage to epithelial cells and the products of lipid oxidation may contribute to depigmentation.

- *Pityriacitrine and Pityrialactone.* Tryptophan metabolites of *M. furfur* exhibit yellow-green fluorescence under UV light (Mayser et al., 2002). Among tryptophan metabolites, lipophilic substances—pityriacitrine and pityrialactone—exert a potent broad-spectrum UV effect. These compounds may cause hypopigmentation by filtering out UV light in exposed skin of patients.

Morphogenesis to Hyphal Forms

Lipid dependence of *Malassezia* species is attributed to the apparent absence or silencing of a fatty acid synthase gene. Adding glycine, cholesterol, or squalene to *Malassezia* cultures induces conversion of the yeast form to the hyphal form, a change that is a step in the pathogenesis of disease (Ashbee, 2007).

22C.4.12 THERAPY FOR SUPERFICIAL AND DEEP-SEATED MYCOSES CAUSED BY *MALASSEZIA* SPECIES

Therapy for Pityriasis Versicolor

Patients should be cautioned that hypopigmented lesions take some time to resolve after treatment. Effective topical and systemic options are available (Crespo-Erchiga and Florencio, 2006).

- *Topical.* Among topical agents is Selsun blue® shampoo containing selenium sulfide, 2.5%. Recurrences can be prevented by periodic use of the shampoo. Topical KTC gel also is effective. Other topical treatments include sodium thiosulfate, 25%, with salicylic acid, 1%; 50% aqueous propylene glycol; zinc pyrithione shampoo, ciclopirox, 0.1% solution; and topical azoles such as clotrimazole, 1%, or KTC, 2%; as well as topical TRB, 1% solution.
- *Systemic.* Effective systemic medications include:
  - KTC, a single dose of 400 mg
  - ITC, at 200 mg/day, for 5–7 days
  - FLC, a single oral dose of 400 mg, or 300 mg/week for 2 weeks

  To prevent common recurrences, estimated at 60% the first year, suppressive therapy is recommended with KTC, 400 mg once/month; or 200 mg for 3 consecutive days per month, or ITC at 400 mg once/month for 6 months. Selenium sulfide shampoo is safer and also is an effective preventive.

Therapy for Seborrheic Dermatitis

Scalp lesions are treated with shampoo containing KTC 2x/week for 1 month to bring about remission, followed by weekly or bimonthly use of this shampoo. Alternative therapy choices are shampoo with 2.5% selenium sulfide or with ciclopirox. Facial lesions may be treated with KTC foam or gel, ciclopirox cream, or ointment containing lithium salt (Naldi and Rebora, 2009). Methyl prednisolone aceponate cream, 0.1% is another effective choice.

Therapy for Fungemia

Catheter related fungemia usually can be eradicated by removal of the catheter and discontinuation of the parenteral lipid without further treatment (Tragiannidis et al., 2010). Susceptibility testing is complicated for most *Malassezia* species because of a growth requirement for oleic acid. Clinical *Malassezia* isolates, however, demonstrate in vitro susceptibility to AmB and triazoles but there is a lack of comparative clinical experience because of the uncommon occurrence of *Malassezia* fungemia. FLC or VRC are considered rational primary therapy options with AmB reserved for refractory or life-threatening infections.

A treatment regimen for *Malassezia* fungemia with FLC is 400 mg/day (oral or IV). A dosage regimen for AmB is 0.7 mg/kg for AmB-deoxycholate or 3 mg/kg for lipid-AmB preparations (Morrison and Weisdorf, 2000). A course of 14 days of therapy beyond the last positive blood culture and catheter removal is recommended (Tragiannidis et al., 2010). With appropriate clinical management, catheter-associated fungemia in premature neonates and immunocompromised patients reduces attributable mortality.

22C.4.13 LABORATORY DETECTION, RECOVERY, AND IDENTIFICATION OF *MALASSEZIA* SPECIES

Direct Examination

Pityriasis versicolor is diagnosed by clinical appearance of lesions, yellow-green fluorescence under Wood's light, and confirmed by microscopy of a KOH prep of skin scrapings which reveals monopolar budding yeast in a "footprint shape" and with a distinctive scar (Fig. 22.8). Irregular fragments of hyphae round out the picture, often described as "spaghetti and meatballs." Adding Parker ink (blue or blue-black) aids in their detection as does the use of the fluorescent brightener, Calcofluor (please see Chapter 2 for details). Culture of specimens is neither necessary nor recommended, except in cases of systemic infection (fungemia) or for epidemiologic studies.

Histopathology

Skin scrapings or skin biopsy specimens submitted for histopathology are stained with PAS or H&E (Chandler and Watts, 1987). PAS-stained skin scrapings show clusters of spherical to oval budding yeast with short, narrow, and sometimes curved hyphal fragments oriented end-to-end (Fig. 22.8). H&E-stained biopsy specimens show clusters of hematoxylinophilic yeasts and short hyphae within the hyperkeratotic layer.

Culture

Malassezia species can be identified in the clinical laboratory but this is usually unnecessary. Speciation of *Malassezia* based on phenotype is probably best carried out in reference laboratories. Lipid-requiring *Malassezia* species can be isolated on SDA with olive oil incorporated in it, or with an olive oil overlay. Media specialized for recovery and speciation of *Malassezia* are Dixon's medium containing Tween 40 and glycerol-monooleate, and Leeming and Notman medium containing Tween 60, glycerol, and full-fat cow's milk (Ashbee, 2007). No single growth medium, however, can reliably recover and maintain all the various *Malassezia* species.

Blood Cultures

Growth of lipid-dependent *Malassezia* species in blood culture bottles is enhanced by the addition of palmitic acid (3%, wt/vol) (Nelson et al., 1995). Detection of

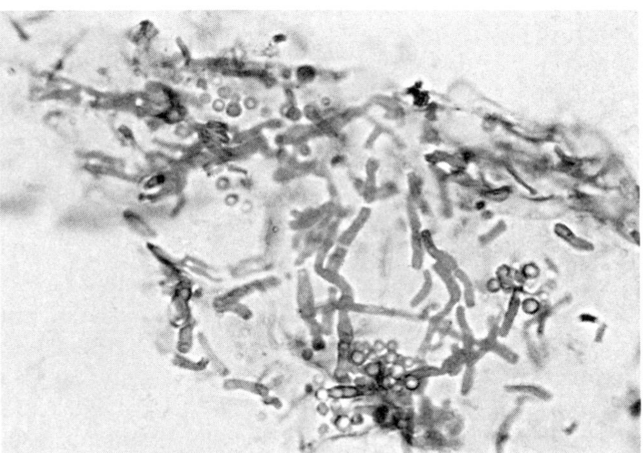

Figure 22.8 Skin scrapings of pityriasis versicolor showing hyphae of *Malassezia* species. Round cells are hyphae cut in cross-section (PAS stain).
Source: E. Reiss.

Malassezia furfur in BACTEC Peds Plus® bottles with nonradioactive detection *required* lipid supplementation. Palmitic acid, a fatty acid with a carbon chain length of 16, was superior to olive oil, oleic acid (C_{18}), and Tween 40 as a supplement, in terms of both speed of detection and growth enhancement. The product is now BD BACTEC Peds Plus®/F Medium, the "F" suffix refers to fluorescent detection in the machine which automatically samples the bottles for CO_2 generation.

Growth and Colony Morphology

Lipid requiring *Malassezia* species grow slowly and poorly at 25°C on SDA, but at 32°C with olive oil supplementation, growth is rapid, producing smooth cream- to yellow-brown colonies which become dry, dull, and wrinkled. *M. pachydermatis* is not lipid requiring but growth is stimulated by an oil supplement in the medium. Cultures should be prepared in Dixon medium if the intent is to differentiate among the *Malassezia* species (Guillot et al., 1998). Incubate the cultures at 30–35°C and protect the Petri plates from desiccation by placing them in plastic bags.

The identification of the different *Malassezia* species can be on the basis of their characteristics, including the colony and yeast morphology, catalase activity, ß-glucosidase (splitting of aesculin), 37°C tolerance, and the ability to utilize Tween 20, 40, 80 and cremophor EL (castor oil) as unique lipid supplements to support growth. A table summarizing these reactions is in Cabañes et al. (2011). *Malassezia* species do not ferment carbohydrates, and can derive all their energy needs from lipids. These yeast cannot form long-chain fatty acids and require lipid supplementation due to a block in synthesis of myristic acid.

Microscopic Morphology

See Fig. 22.9. Yeast cells, $1.5-4.5 \times 2-6$ μm, are actually rounded phialides with small collarettes on the blunt end. They are formed by monopolar *enteroblastic* budding. The mother and daughter cells are separated by a septum; the daughter cell separates by fission, leaving a bud scar or collarette. From this site on the cell single bud-like cells are extruded. Yeast forms predominate in cultures. Hyphae occasionally may be produced.

Genetic Identification

Molecular methods have the advantage of a common method of DNA isolation, PCR, and sequencing. A further molecular identification method adapted for *Malassezia* does not require sequencing. A scheme of DNA probes is capable of identifying an unknown yeast as one of 16

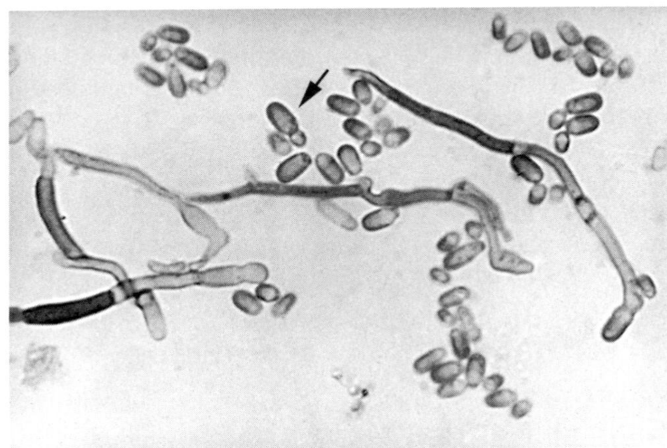

Figure 22.9 *Malassezia* species—microscopic morphology of yeast and hyphal forms; *arrow*—"footprint" shaped yeast. *Source:* H. J. Shadomy.

Malassezia species in 1 h after PCR amplification (Diaz et al., 2006). The target regions are the D1/D2 (~600 bp) region of 28S and ITS (~600–800 bp) regions of rDNA.

Principle: Species-specific probes are affixed to different color beads in the Luminex® detection system. This technology of the Luminex Corp. (Austin, TX) is illustrated at the URL http://www.luminexcorp.com/TechnologiesScience/xTAGTechnology/index.htm.

The biotinylated amplicon generated from the unknown yeast DNA binds to its complementary bead-probe. A fluorescent reporter molecule binds to biotin and the bead-probe–amplicon-reporter complex is sorted by flow cytometry. Two lasers are used to identify the bead–amplicon complex: one detects the bead color, and hence the *Malassezia* species, and the second detects the signal strength of the amplicon, indicating a positive hybridization to the bead.

The Luminex system has appeal as a common platform applicable to a wide variety of bacterial and fungal pathogens where accuracy and high-throughput are needed.

22D CHRYSOSPORIUM *AND OTHER NONPATHOGENIC OR OPPORTUNISTIC FUNGI ISOLATED FROM SKIN AND RESEMBLING DERMATOPHYTES IN CULTURE*

(Please also see Chapter 2 for common saprobic fungi isolated in the clinical laboratory.) Clinical laboratorians may be faced with saprobic look-alikes resembling dermatophytes in culture. The major genus in this differential is *Chrysosporium*. Other look-alike saprobes include *Geomyces* species, *Myceliophthora* species, *Myriodontium keratinophilum*, and *Sporotrichum pruinosum*. These molds must be recognized as nondermatophytes, because their disease producing potential is very exceptional.

- *What is the source of these saprobes?* They grow as contaminants from respiratory and cutaneous specimens, including dermatophytic and psoriatic lesions and nails.
- *What is the common theme that causes them to appear like dermatophytes?*

  The common theme is the production of microconidia resembling those of dermatophytes.

Trichophyton terrestre

Trichophyton terrestre is a soil-dwelling keratinophilic mold that does not cause infection in humans and lower animals. The conidial forms of this species are intermediate between unicellular microconidia and multicellular macroconidia. *T. terrestre* is able to grow in medium containing cycloheximide but, based on microscopic morphology, it can be differentiated from pathogenic *Trichophyton* species (Campbell et al., 2006). *T. terrestre* is isolated as a contaminant in dermatologic and veterinary skin specimens, but with no evidence of infection. The strong keratinolytic abilities of the geophilic *Trichophyton* spp. suggest that *T. terrestre* may be specialized to decompose hair and corpses of animals.

Chrysosporium

Chrysosporium species are keratinophilic environmental molds, decomposers in nature of hair and feathers. It is a large *polyphyletic* genus. Eight species are listed in de Hoog et al. (2000) as being isolated from human specimens (e.g., *C. keratinophilum*.) Atypical dermatophyte cultures, *ones that do not produce macroconidia*, cannot readily be differentiated from *Chrysosporium* species. Characteristics of the microconidia of *Chrysosporium* species are illustrated by de Hoog et al. (2000). Some *Chrysosporium* species have teleomorph states that are classed with the dermatophytes in the genus *Arthroderma* in the *Onygenales*.

Colony Morphology Growth is slow to moderate and colonies are yellow-white in color. Growth is resistant to cycloheximide.

Microscopic Morphology Conidia are single celled, smooth to roughened *aleurioconidia* borne sessile or at the tips or sides of short stalks or long stalks that may be branched. Conidia are *pyriform*, *clavate*, or *obovoid*. Intercalary arthroconidia in a barrel shape or bulged to one side also may be formed. The definitive identification of these fungi in relation to dermatophytes is found in Sigler (1997).

Medical Importance Disease in humans caused by *Chrysosporium* is exceptional. Cutaneous infection caused by *Aphanoascus fulvescens*, an ascomycete with a *Chrysosporium* anamorph, is documented. The medical importance is due to the resemblance that they have to dermatophytes and to the microconidia of *B. dermatitidis*.

Aphanascus fluvescens *(teleomorph of a* Chrysosporium *spp.)*

Ecology Cosmopolitan in soil and plant material.

Colony Morphology Growth is moderate to rapid, with a powdery or felt-like texture with a white to tan surface and colorless reverse.

Microscopic Morphology Pear- or teardrop-shaped conidia resembling very large dermatophyte microconidia are abundant. At maturity the colony produces round, thin-walled cleistothecia containing egg-shaped, rough-walled ascospores in clusters of 8 per ascus.

Geomyces pannorum

Short chains of intercalary arthroconidia are produced which are barrel-shaped and echinulate. Conidiophores are erect and carry short branches with conidia formed at the tip, sides, and intercalary positions. They have been described as "clusters of radiate branches originating in a ring around a central perpendicular main branch, like spokes of a wagon wheel" (Sigler, 1997).

Myceliophthora thermophila

This is a thermophilic species found in composts. It is classed in the *Sordariales* and has a teleomorph: *Corynascus heterothallicus*.

Myriodontium keratinophilum

This is a soil-dwelling mold with a natural role in recycling hair and feathers. It is classed in the *Onygenales* and has a teleomorph: *Neoarachnotheca keratinophilia*.

Sporotrichum pruinosum *(syn.* Chrysosporium pruinosum*)*

Ecology Cosmopolitan on decaying wood, soil.

Colony Morphology Powdery to cottony texture; white, rosy-to-orange surface.

Microscopic Morphology Conidiophores may be short or long, branching at right angles from the supporting hypha, each bearing a single aleurioconidium. The conidiophores may occur in clusters. There is considerable strain-to-strain variation. Conidia are ovoid to ellipsoid with a truncate base, smooth and thick walled. Arthroconidia and spherical chlamydospores may be present.

Cutanous Manifestations

Cutaneous manifestations of the systemic mycoses caused by *Blastomyces dermatitidis, Histoplasma capsulatum*, and *Paracoccidioides brasiliensis* may be confused with dermatophyte infections, and their cultural characteristics (the microconidia) also may be confusing. Please see Part Two, Systemic Mycoses Caused by Dimorphic Environmental Molds (Endemic Mycoses).

SELECTED REFERENCES FOR DERMATOMYCOSES

Akagi T, Yamaguti K, Kawamura T, Nakumura T, Kubo K, Takemori H, 2006. Breatkthrough trichosporonosis in patients with acute myeloid leukemia receiving micafungin. *Leuk Lymphoma* 47: 1182–1183.

Ando M, Arima K, Yoneda R, Tamura M, 1991. Japanese summer-type hypersensitivity pneumonitis. Geographic distribution, home environment, and clinical characteristics of 621 cases. *Am Rev Respir Dis* 144: 765–769.

Ashbee HR, 2007. Update on the genus *Malassezia*. *Med Mycol* 45: 287–303.

Ashbee HR, Evans EG, 2002. Immunology of diseases associated with *Malassezia* species. *Clin Microbiol Rev* 15: 21–57.

Batra R, Boekhout T, Guého E, Cabañes FJ, Dawson TL Jr, Gupta AK, 2005. *Malassezia* Baillon, emerging clinical yeasts. *FEMS Yeast Res* 5: 1101–1113.

Blank H, 1979. Tinea nigra: A twenty-year incubation period? *J Am Acad Dermatol* 1: 49–51.

Bonifaz A, Gómez-Daza F, Paredes V, Ponce RM, 2010. Tinea versicolor, tinea nigra, white piedra, and black piedra. *Clin Dermatol* 28: 140–145.

Bonifaz A, Badali H, de Hoog GS, Cruz M, Araiza J, Cruz MA, Fierro L, Ponce RM, 2008. Tinea nigra by *Hortaea werneckii*, a report of 22 cases from Mexico. *Stud Mycol* 61: 77–82.

Cabañes FJ, Vega S, Castellá G, 2011. *Malassezia cuniculi* sp. nov., a novel yeast species isolated from rabbit skin. *Med Mycol* 49: 40–48.

Campbell CK, Borman AM, Linton CJ, Bridge PD, Johnson EM, 2006. *Arthroderma olidum*, sp. nov. A new addition to the *Trichophyton terrestre* complex. *Med Mycol* 44: 451–459.

Carion AL, 1965. Dermatomycoses in Puerto Rico. *Arch Dermatol* 91: 431–438. (The correct name of the author is Arturo L. Carrion, but was misspelled in the article.)

Chagas-Neto TC, Chaves GM, Melo AS, Colombo AL, 2009. Bloodstream infections due to *Trichosporon* spp.: Species distribution, *Trichosporon asahii* genotypes determined on the basis of ribosomal DNA intergenic spacer 1 sequencing, and antifungal susceptibility testing. *J Clin Microbiol* 47: 1074–1081.

Chagas-Neto TC, Chaves GM, Colombo AL, 2008. Update on the genus *Trichosporon*. *Mycopathologia* 166: 121–132.

Chandler FW, Watts JC, 1987. Figures 163–168 on pp. 114–115; Figs. 311 and 312 on p. 216, *in*: *Pathologic Diagnosis of Fungal Infections*. ASCP Press, Chicago.

Chang HJ, Miller HL, Watkins N, Arduino MJ, Ashford DA, Midgley G, Aguero SM, Pinto-Powell R, von Reyn CF, Edwards W, McNeil MM, Jarvis WR, 1998. An epidemic of *Malassezia pachydermatis* in an intensive care nursery associated with colonization of health care workers' pet dogs. *N Engl J Med* 338: 706–111.

Chryssanthou E, Broberger U, Petrini B, 2001. *Malassezia pachydermatis* fungaemia in a neonatal intensive care unit. *Acta Paediatrica* 90: 323–327.

Coimbra Júnior CE, Santos RV, 1989. Black piedra among the Zoró Indians from Amazônia (Brazil). *Mycopathologia* 107: 57–60.

Corbo MR, Lanciotti R, Albenzio M, Sinigaglia M, 2001. Occurrence and characterization of yeasts isolated from milks and dairy products of Apulia region. *Int J Food Microbiol* 69: 147–152.

Crespo MJ, Abarca ML, Cabañes FJ, 2002. Occurrence of *Malassezia* spp. in the external ear canals of dogs and cats with and without otitis externa. *Med Mycol* 40: 115–121.

Crespo-Erchiga V, Florencio VD, 2006. *Malassezia* yeasts and pityriasis versicolor. *Curr Opin Infect Dis* 19: 139–147.

Dankner WM, Spector SA, Fierer J, Davis CE, 1987. *Malassezia* fungemia in neonates and adults: Complication of hyperalimentation. *Rev Infect Dis* 9: 743–753.

de Boer DJ, Marsella R, 2001. The ACVD task force on canine atopic dermatitis (XII): The relationship of cutaneous infections to the pathogenesis and clinical course of canine atopic dermatitis. *Vet Immunol Immunopathol* 81: 239–249.

de Luca C, Picardo M, Breathnach A, Passi S, 1996. Lipoperoxidase activity of *Pityrosporum*: Characterisation of byproducts and possible role in pityriasis versicolor. *Exp Dermatol* 5: 49–56.

de Hoog GS, Guarro J, Gené J, Figueras MJ, 2000. *Atlas of Clinical Fungi*, 2nd ed. Centraalbureau voor Schimmelcultures, Baarn, The Netherlands (CD version 2004.11).

Diaz MR, Boekhout T, Theelen B, Bovers M, Cabañes FJ, Fell JW, 2006. Microcoding and flow cytometry as a high-throughput fungal identification system for *Malassezia* species. *J Med Microbiol* 55: 1197–1209.

Diaz MR, Fell JW, 2004. High-throughput detection of pathogenic yeasts of the genus trichosporon. *J Clin Microbiol* 42: 3696–3706.

Figueras M J, Guarro J, Zaror L, 1996. New findings in black piedra infection. *Br J Dermatol* 135: 157–158.

Fischman O, 1973. Black piedra among Brazilian Indians. *Rev Inst Med Trop São Paulo* 15: 103–106.

Fischman O, Soares EC, Alchorne MMA, Batista G, Camargo ZP, 1983. Tinea nigra contracted in Spain. *Bol Micol* 1: 68–70.

Fonseca FL, Frases S, Casadevall A, Fischman-Gompertz O, Nimrichter L, Rodrigues ML, 2009. Structural and functional properties of the *Trichosporon asahii* glucuronoxylomannan. *Fungal Genet Biol* 46: 496–505.

Girmenia C, Pagano L, Martino B, D'Antonio D, Fanci R, Specchia G, Melillo L, Buelli M, Pizzarelli G, Venditti M, Martino P, 2005. Invasive infections caused by *Trichosporon* species and *Geotrichum capitatum* in patients with hematological malignancies: A retrospective multicenter study from Italy and review of the literature. *J Clin Microbiol* 43: 1818–1828.

Gordon MA, 1979. *Malassezia pityrosporum pachydermatis* (Weidman) Dodge 1935. *Sabouraudia* 3: 305–309.

Gordon MA, 1951. The lipophilic mycoflora of human skin.1. In vitro culture of *Pityrosporum obiculare* n. sp. *Mycologia* 43: 524–535.

Gueho E, Smith MT, de Hoog GS, Billon-Grand G, Christen R, Batenburg-van der Vegte WH, 1992. Contributions to a revision of the genus *Trichosporon*. *Anton van Leeuwenhoek* 61: 289–316.

Guillot J, Guého E, 1995. The diversity of *Malassezia* yeasts confirmed by rRNA sequence and nuclear DNA comparisons. *Antonie van Leeuwenhoek* 67: 297–314.

Guillot J, Hadina S, Guého E, 2008. The genus *Malassezia*: Old facts and new concepts. *Parassitologia* 50: 77–79.

Guillot J, Breugnot C, de Barros M, Chermette R, 1998. Usefulness of modified Dixon's medium for quantitative culture of *Malassezia* species from canine skin. *J Vet Diagn Invest* 10: 384–386.

Gunde-Cimerman N, Zalar P, de Hoog S, Plemenitas A, 2000. Hypersaline waters in salterns—natural ecological niches for halophilic black yeasts. *FEMS Microbiol Ecol* 32: 235–240.

Gupta AK, Chaudhry M, Elewski B, 2003. Tinea corporis, tinea cruris, tinea nigra, and piedra. *Dermatol Clin* 21: 395–400.

Hazen KC, Howell SA, 2007. Chapter 119, *Candida, Cryptococcus*, and Other Yeasts of Medical Importance, pp. 1742–1788, *in:* Murray PR (ed-in-chief), *Manual of Clinical Microbiology*, 9th ed., (rev) vol 2, Pfaller MA (vol ed.), Section VIII, Mycology. ASM Press, Washington, DC.

Kane J, Summerbell R, Sigler L, Krajden S, Land G, 1997. *Laboratory Handbook of Dermatophytes*. Star Publishing Co, Belmont, CA.

Kaplan W, 1959. The occurrence of black piedra in primate pelts. *Trop Geographic Med* 11: 115–126

Kemker BJ, Lehmann PF, Lee JW, Walsh TJ, 1991. Distinction of deep vs. superficial clinical and non-clinical environmental isolates of *Trichosporon beigelii* by isoenzyme and restriction fragmentation length polymorphism of rDNA generated by the polymerase chain reaction. *J Clin Microbiol* 29: 1677–1683.

Kiken DA, Sekaran A, Antaya RJ, Davis A, Imaeda S, Silverberg NB, 2006. White piedra in children. *J Am Acad Dermatol* 55: 956–961.

Larone DH, 2002. *Medically Important Fungi*: *A Guide to Identification*, 4th ed. ASM Press, Washington, DC.

Mayser P, Schäfer U, Krämer HJ, Irlinger B, Steglich W, 2002. Pityriacitrin—a ultraviolet-absorbing indole alkaloid from the yeast *Malassezia furfur*. *Arch Dermatol Res* 94: 131–134.

Matousek JL, Campbell KL, 2002. *Malassezia* dermatitis. *Compendium* 24: 224–232.

Matsue K, Uryu H, Koseki M, Asada N, Takeuchi M, 2006. Breakthrough trichosporonosis in patients with hematologic malignancies receiving micafungin. *Clin Infect Dis* 42: 753–757.

Mendez-Tovar LJ, 2010. Pathogenesis of dermatophytosis and tinea versicolor. *Clin Dermatol* 28: 185–189.

Morrison VA, Weisdorf DJ, 2000. The spectrum of *Malassezia* infections in the bone marrow transplant population. *Bone Marrow Transplant* 26: 645–648.

Moyer DG, Keeler C, 1964. Note on culture of black piedra for cosmetic reasons. *Arch Dermatol* 89: 436.

Naldi L, Rebora A, 2009. Seborrheic dermatitis. *N Engl J Med* 360: 387–396.

Nelson SC, Yau YC, Richardson SE, Matlow AG, 1995. Improved detection of *Malassezia* species in lipid-supplemented Peds Plus blood culture bottles. *J Clin Microbiol* 33: 1005–1007.

Olivry T, Sousa CA, 2001. The ACVD task force on canine atopic dermatitis (XIX): General principles of therapy. *Vet Immunol Immunopathol* 81: 311–316.

Passi S, Picardo M, de Luca C, Breathnach AS, Nazzaro-Porro M, 1991. Scavenging activity of azelaic acid on hydroxyl radicals "in vitro." *Free Radic Res Commun* 11: 329–338.

Powell DA, Aungst J, Snedden S, Hansen N, Brady M, 1984. Broviac catheter-related *Malassezia furfur* sepsis in five infants receiving intravenous fat emulsions. *J Pediatr* 105: 987–990.

Ritchie EB, Taylor TE, 1964. A study of tinea nigra palmaris: Report of a case and inoculation experiments. *Arch Dermatol* 89: 601–603.

Robin C, 1853. *Histoire naturelle des végétaux parasites qui croissent sur l'homme et sur les animaux vivants*. Bailliere, Paris.

Rosales MS, Marsella R, Kunkle G, Harris BL, Nicklin CF, Lopez J, 2005. Comparison of the clinical efficacy of oral terbinafine and ketoconazole combined with cephalexin in the treatment of *Malassezia* dermatitis in dogs—a pilot study. *Vet Dermatol* 16: 171–176.

Rosen T, Lingappan A, 2006. Rapid treatment of tinea nigra palmaris with ciclopirox olamine gel, 0.77%. *Skinmed* 5: 201–203.

Schwartz RA, 2004. Superficial fungal infections. *Lancet* 364 (9440): 1173–1182.

Sigler L, 1997. Chapter 9, *Chrysosporium* and Molds Resembling Dermatophytes, pp. 261–311, *in*: Kane J, Summerbell R, Sigler L, Krajden S, Land G (eds.), *Laboratory Handbook of Dermatophytes*. Star Publishing Co, Belmont, CA.

Sizun J, Karangwa A, Giroux JD, Masure O, Simitzis AM, Alix D, de Parscau L, 1994. *Malassezia furfur*-related colonization and infection of central venous catheters. A prospective study in a pediatric intensive care unit. *Intensive Care Med* 20: 496–499.

Sláviková E, Vadkertiová R, 2000. The occurrence of yeasts in the forest soils. *Basic Microbiol* 40: 207–212.

St. Germain G, Summerbell R, 1996. *Identifying Filamentous Fungi: A Clinical Laboratory Handbook*. Star Publishing Co, Belmont, CA.

Sugita T, Ikeda R, Nishikawa A, 2004. Analysis of *Trichosporon* isolates obtained from the houses of patients with summer-type hypersensitivity pneumonitis. *J Clin Microbiol* 42: 5467–5471.

Sugita T, Nakajima M, Ikeda R, Matsushima T, Shinoda T, 2002. Sequence analysis of the ribosomal DNA intergenic spacer 1 regions of *Trichosporon* species. *J Clin Microbiol* 40: 1826–1830.

Tosti A, Piraccini BM, Lorenzi S, Iorizzo M, 2003. Treatment of nondermatophyte mold and *Candida* onychomycosis. *Dermatol Clin* 21: 491–497.

Tragiannidis A, Bisping G, Koehler G, Groll AH, 2010. Minireview: *Malassezia* infections in immunocompromised patients. *Mycoses* 53: 187–195.

van Uden N, de Barros-Machado A, Castelo-Branco R, 1963. On black piedra in central African mammals caused by the Ascomycete, *Piedraia quintanilhae* nov. sp. *Rev Bras Port Biol Geral* 3: 271–296.

van Velsor H, Singletary H, 1964. Tinea nigra palmaris: A report of 15 cases from coastal North Carolina. *Arch Dermatol* 90: 59–61.

Welbel SF, McNeil MM, Pramanik A, Silberman R, Oberle AD, Midgley G, Crow S, Jarvis WR, 1994. Nosocomial *Malassezia pachydermatis* bloodstream infections in a neonatal intensive care unit. *Pediatr Infect Dis J* 13: 104–108.

WEBSITE CITED

Luminex bead systems of the Luminex Corp., Austin, Texas, are illustrated at the URL http://www.luminexcorp.com/TechnologiesScience/xTAGTechnology/index.htm.

QUESTIONS

The Answer Key to multiple choice questions may be found at the end of the book.

1. The most common nondermatophyte cause of tinea unguium in temperate climates is
A. *Acremonium potronii*
B. *Fusarium* species
C. *Neoscytalidium dimidiatum*
D. *Scopulariopsis brevicaulis*

2. Nondermatophyte agents of tinea unguium will not grow on medium containing
A. Chloramphenicol
B. Chlortetracycline
C. Colchicine
D. Cycloheximide

3. Hard, gritty, discrete dark-brown to black nodules that adhere firmly to scalp hair or beard describes the superficial mycosis called
A. Black piedra
B. Tinea nigra
C. Pityriasis versicolor
D. White piedra

4. White to light brown nodules occur on hair shafts of the scalp, beard, or mustache, less commonly in the axilla or groin. Nodules are soft and are not firmly adherent to the hair shaft. This is the superficial mycosis called
A. Black piedra
B. Tinea nigra
C. Pityriasis versicolor
D. White piedra

5. Discolored spots slowly enlarge and may become confluent patches on the skin of the upper chest, back, and shoulders. The spots may be white, salmon, or brown and are covered with thin scales. This describes the superficial mycosis called
A. Black piedra
B. Tinea nigra
C. Pityriasis versicolor
D. White piedra

6. Atypical dermatophyte cultures producing microconidia, but which *do not produce macroconidia*, cannot readily be differentiated from other non-disease-causing saprobes isolated from skin specimens such as
A. *Aureobasidium pullulans*
B. *Chrysosporium* species
C. *Cladosporium* species
D. *Penicillium* species

7. List five major nondermatophyte molds causing onychomycosis and for each discuss the success obtained with antifungal therapy.

8. How is *Piedraia hortae* identified in the laboratory?

9. In addition to superficial dermatomycosis, what are the other clinical forms of *Trichosporon* mycosis? What antifungal agents may be effective in their treatment?

10. What nonmycotic diseases are in the differential diagnosis of tinea nigra palmaris? How is the etiologic agent identified in the laboratory? Indicate the antifungal therapy for tinea nigra and the clinical experience with recurrences.

11. Which three species of *Malassezia* are responsible for 90% of the cases of pityriasis versicolor?
12. Describe how examination of skin scrapings aids in the diagnosis of pityriasis versicolor. Include in your description how to prepare the material for microscopic examination, including the use of tinctorial and fluorescent methods. What is meant by "footprint shape" and "meatballs and spaghetti" in describing the microscopic morphology?
13. Explain the major difference in culture conditions for recovery of *M. pachydermatis* and other *Malassezia* species. What role do canines play as a reservoir of this species?
14. Discuss the involvement of *Malasezzia* species in psoriasis.
15. How is the frequent recovery of *Malassezia* species from the skin of neonates linked to sepsis during the use of intralipid nutritional support therapy? What success is obtained with specific antifungal therapy and/or catheter maintenance?
16. How can the various *Malassezia* species be identified in the laboratory?

Glossary

Where definitions were borrowed from specific reference sources, those sources are cited and listed at the end of the Glossary. Terms defined without specific sources were based on the accumulated records of the authors. Terms unrelated to medical mycology are listed here when they have been used in various sections of the text.

ABPA Allergic bronchopulmonary aspergillosis.

Acid Fast A property of microbial cell walls in which the primary stain, carbol fuchsin, resists decoloration after treatment with acid-alcohol, as in the Ziehl–Neelson method for acid fast bacteria.

Acrogenous (Gr., *acros* = highest, outermost) Of conidia, developing at the apex, borne at the tip.

Acropetal (Gr., *acros* = highest, outermost, L., *petre* = to seek) A chain of conidia growing at the tip: having the youngest conidium at the apex of a chain.

Adamson's Fringe That portion of the lower hair shaft at the junction of the living (mitotically active) hair bulb and the keratinized, nonliving hair shaft. Dermatophytes can only infect hair down to the level of Adamson's fringe.

Adhesin A surface molecule that promotes adhesive reactions with other cells or with extracellular matrix molecules (EMMs). Adhesins are microbial surface antigens that bind to specific receptors on epithelial cell membranes, or to the EMMs. Adherence is a prelude to invasion of the host cell; for example, BAD-1 is a surface adhesin of *Blastomyces dermatitidis*.

Adiaspore (Gr., *a* = not + *dia* = across) Conidium of *Emmonsia crescens* or *E. parva*, which, when inhaled and deposited in the lung, undergoes isotropic enlargement without cell division.

Adventitious Sporulation Sporulation arising in an unusual location, for example, in tissue and from there into the bloodstream.

Aerial Hyphae/Mycelium Hyphae growing above the colony surface, for example, on an agar base.

Aerospora The fungal conidia ecology in the ambient air. The burden of fungal conidia and spores in the outdoor ambient air, measured via air sampling and expressed as conidia/m^3/unit time. The identification of individual species is useful for surveillance purposes.

Aflatoxins Toxic metabolites produced by certain fungi in or on food and feed, including peanuts, corn, and cottonseed. The toxin was named aflatoxin because of its origin in *Aspergillus flavus*. Strains of *A. parasiticus, A. nomius*, and *A. niger* also may produce aflatoxins. Please see the Appendix of Chapter 14.

AFRS Allergic fungal rhinosinusitis. Formerly known as AFS—allergic fungal sinusitis.

AFS Antifungal susceptibility, usually referring to an in vitro test to determine the minimum inhibitory concentration of an antifungal agent tested against a fungal isolate.

Air Crescent Sign A radiologic sign describing an area caused by necrosis surrounding a nodule with air filling the space between the devitalized tissue and the surrounding lung parenchyma.

Aleurioconidium (Aleuriospore) A simple terminal or lateral conidium that develops as an expanded end of an undifferentiated hypha or on a short stalk (pedicel). It is the product of thallic conidiogenesis. The aleurioconidium is detached by lysis or fracture of the wall of the supporting cell. It usually is recognized by its truncate base, for example, macroconidia of the dermatophytes.

ALL Acute lymphocytic leukemia.

Allogeneic Pertaining to a graft from an individual of different human leukocyte antigen (HLA) haplotype; or the equivalent genetic region in lower animals.

Alopecia (Gr., *alopex* = fox) Balding; may be temporary or permanent.

ALS A family of *Candida albicans* surface adhesins, which may have analogs in other *Candida* species.

AmB Amphotericin B, a polyene macrolide antifungal agent marketed as Fungizone, an emulsion of AmB and deoxycholate. There are lipid derivatives, please see Chapter 3A.

Anamorph (L., *ana* = away, through, again + *morph* = form) The asexual state in the life cycle of a fungus.

Fundamental Medical Mycology, First Edition. By Errol Reiss, H. Jean Shadomy and G. Marshall Lyon, III.
© 2012 Wiley-Blackwell. Published 2012 by John Wiley & Sons, Inc.

Reproduction via mitosis not involving union of two nuclei as in karyogamy and meiosis.

Anergy A state of unresponsiveness to antigenic stimulation, indicative of downregulation of T-cell-mediated immunity. In a patient with an infectious disease, anergy is manifest as a negative skin test response to an antigen derived from that infectious agent and, at times, a more general lack of response to common recall antigens. Cutaneous anergy is mentioned in connection with coccidioidomycosis, histoplasmosis, and paracoccidioidomycosis.

ANF Anidulafungin.

Angioinvasive Tending to invade the walls of blood vessels.

Ann Arbor Stage IVB A staging system for lymphomas where stage IV indicates involvement of one or more extralymphatic organs, including the liver, bone marrow, or nodular involvement of the lungs. "B" refers to night sweats, significant weight loss, or fever.

Annellide (L., *annellus* = a little ring) A vase- or tube-shaped conidiogenous cell that gives rise to successive conidia in a basipetal manner. With the release of each annelloconidium the annellide becomes longer and narrower. An apical ring composed of outer cell wall material remains as each conidium is released, giving an elongate and tapered appearance to the annellide tip.

Annelloconidium (pl., -ia) A conidium arising from an annellide via enteroblastic conidiogenesis.

Annellophore The conidiophore or stalk supporting an annellide.

Annular Frill (i) The remnant of hyphal wall that remains after detachment of a conidium or (ii) the remnant of a disjunctor cell clinging to an arthroconidium released by rhexolytic cleavage.

Anthropophilic In dermatophytosis, a dermatophyte that preferentially or exclusively infects humans; that is, humans are their natural hosts.

Antler Hyphae Repeatedly branching hyphal tip resembling an antler.

APACHE II Score Acute Physiology and Chronic Health Evaluation II score is a severity of disease classification system. After admission of a patient to an intensive care unit, a score from 0 to 71 is computed based on 13 physiologic variables. Higher scores imply a more severe disease and a higher risk of death.

Apophysis (Gr., *apo* = off + *phyein* = to grow) Depending on the species, it is a bell-shaped or funnel-shaped swelling at the tip of a sporangiophore, below the sporangium, in the *Mucorales*.

ARDS Adult respiratory distress syndrome.

ART Antiretroviral therapy.

Arthroconidium (pl., -ia) (Gr., *arthron* + *konidion*—dim. of *konis* = dust) A conidium formed by the modification of a hyphal cell(s) by forming a double septum, followed by disarticulation at the septation points of the hyphal strand to release the arthroconidium. The formation is a type of thallic conidiogenesis.

Ascoma (pl. -ata) (syn., ascocarp). The entire fruiting body of an ascomycete, containing the asci, characteristic of the phylum *Ascomycota*.

Ascospore A haploid sexual spore formed within an ascus following karyogamy and meiosis. Characteristic of the phylum *Ascomycota*.

Ascus (pl., asci) (Gr., *askos* = wine sac, leather bag) The sexual, sac-like, multicellular structure that gives rise to haploid ascospores within the ascus after karyogamy and meiosis. This meiosporangium is characteristic of the phylum *Ascomycota*.

Aseptate (syn., coenocytic) Lacking septa (cross-walls) between cells.

Aspergilloma A fungus ball generally found in a preformed cavity of the lung. An aspergilloma is composed of a mass of *Aspergillus* species hyphae in a matrix of cell debris and fibrin and is suspended in an air space. Aspergilloma also may occur in a nasal sinus.

Asteroid Body An eosinophilic antigen–antibody complex deposited in vivo surrounding a bacterial, fungal, or parasitic focus of infection. In mycology, it is most often associated with *Sporothrix schenckii* yeast forms in tissue but also may be found surrounding another yeast cell or hypha. Ray-like projections radiate from a yeast form cell at the center. (Please see related term: Splendore–Hoeppli Reaction.)

Atopic See atopy.

Atopy Describes a genetic predisposition to IgE-mediated allergy which may manifest as hypersensitivity to inhaled, injected, or ingested allergens: for example, hay fever, eczema, and asthma.

AUC Area-under-the-curve. The change over time in the amount of drug available in the plasma. The graphic representation of the plasma concentration of the drug over time.

BA Blood agar.

Bactrim (syn., cotrimoxazole) Trimethoprim sulfamethoxazole.

BAL Bronchoalveolar lavage.

Ballistospore A spore that, at maturity, is actively ejected, as in the *Entomophthorales*.

Balloon Cell An enlarged globose microconidium formed by some dermatophytes, and characteristic of *Trichophyton tonsurans* in culture.

Basal Ascomycetes (Archiascomycetes) The Archiascomycetes originated near the evolutionary point of divergence between ascomycetes and basidiomycetes.

Basidiospore A haploid sexual spore formed on the upper outside surface of a basidium following karyogamy and meiosis. Basidiospores are characteristic of the phylum *Basidiomycota*.

Basidium (pl., -ia) (Gr., *bas* = base + L., *-idium*, diminutive suff.) A specialized, usually club-shaped, sexual cell that gives rise to external basidiospores on its upper surface. Characteristic of the phylum *Basidiomycota*.

Basipetal (Gr., *bas* = base + L., *petere* = to seek) A form of conidiogenesis describing the succession of conidia, with the youngest formed at the base of the chain: for example, all conidia formed from phialides and annellides are basipetal.

BGC Agar Blood–glucose–cysteine agar. An alternative medium to maintain yeast forms of dimorphic fungi.

BHI Brain–heart infusion, used as an agar medium, with or without supplementation with 5% defibrinated sheep blood, "BHI+blood."

Biofilm Fungal biofilms are yeast or mycelial communities adherent to an environmental surface. The biofilm is usually encased in an extracellular polysaccharide or slime matrix. A biofilm is a nidus of infection formed on indwelling biomaterials. This is a sessile growth pattern as opposed to fungal cells suspended in blood or other body fluid, which is planktonic growth.

Biomaterial Catheters, implants, or other medical devices made of synthetic materials and implanted temporarily or permanently in the human host; for example, artificial voice prostheses, intravascular catheters, shunts, and heart valves.

Birdseed Agar A selective and differential medium for the isolation of *Cryptococcus neoformans*, which forms tan to nearly black colonies due to the production of melanin pigment. Colonies of other *Cryptococcus* species may produce cream to pale tan colonies. The birdseed referred to is *Guizotia abyssinica*, or Niger-seed.

Biseriate Arranged in two rows. The arrangement of conidiogenous cells in which a row of phialides is supported on a row of metulae attached to the surface of a vesicle (e.g., as in *Aspergillus terreus*) (please see Fig. 14.7).

Biverticillate (L., *bi* = two + *verticillus* = whorl) In *Penicillium* species, having two or, rarely, three levels of branching (whorls) directly below the phialides (please see Fig. 14.8).

BLAST The Basic Local Alignment Search Tool finds regions of local similarity between sequences. The program compares nucleotide or protein sequences to sequence databases and calculates the statistical significance of matches. BLAST can be used to infer functional and evolutionary relationships between sequences and helps identify members of gene families. See the URL http://blast.ncbi.nlm.nih.gov/Blast.cgi.

Blastic (Gr., *blastos* = bud) A mode of conidiogenesis best described as "budding." The hypha or yeast cell bulges, then enlarges, followed by nuclear and cytoplasmic migration, resulting in a blastoconidium. When this formation is complete a septum is formed delineating it from the mother cell. (Please also see Enteroblastic, Holoblastic).

Blastoconidium (pl. -ia) A yeast cell. An asexual (anamorphic) reproductive cell formed by the blowing out of part of the parent cell wall. After mitosis, nuclei form in the mother and daughter cell, cytokinesis occurs, and a thickened basal septum is laid down. Cell separation then occurs. See the animation of mitosis and cytokinesis at the URL http://highered.mcgraw-hill.com/sites/0072495855/student_view0/chapter2/animation__mitosis_and_cytokinesis.html.

BMBL5 *Biosafety in Microbiological and Biomedical Laboratories* (BMBL), 5th ed. Online at the URL http://www.cdc.gov/OD/ohs/biosfty/bmbl5/bmbl5toc.htm.

BMT Bone marrow transplant, now referred to as a hematopoietic stem cell transplant (SCT).

Botryomycosis Botryomycosis, or bacterial pseudomycosis, is an uncommon bacterial infection of domestic animals and humans, erroneously thought to have a fungal etiology. It is characterized by superficial vascular granulomatous masses, associated with wounds, and sometimes with visceral involvement. *Staphylococcus aureus* causes the majority of infections, followed by *Pseudomonas aeruginosa*, and a number of other bacteria.

Breakpoint Breakpoints are the in vitro drug concentrations considered by committees such as the CLSI to correspond to performance of the drug in a population of patients: for example, breakpoints for *Candida* species were determined after consideration of >10,000 AFS tests of *Candida* species and 1295 patient-episode-isolate events.

Brown and Brenn Stain Tissue Gram stain.

BSC Biological safety cabinet.

Bubo Inflammatory swelling of a lymph gland, particularly in the axilla or groin.

Budding A method of asexual reproduction typically in yeasts in which small rounded projections from the parent cell are produced during mitosis resulting in a daughter cell termed a bud or blastoconidium. The parent cell may produce single or multiple buds.

Bullous Adjective for "bulla," a blister more than 5 mm in diameter with thin walls and full of fluid.

Calcofluor White A fluorescent brightener that binds to fungal, bacterial, and plant cell walls (syn., Brightener I, Calcofluor, Calcofluor White M2R New, Cellofluor, Fluorescent brightener 28, Fluostain I, Tinopal LPW). Please see Chapter 2, Section 2.3.3, Direct Examination—Fluorescent Brighteners Used with or without KOH.

Candidemia *Candida* species bloodstream infection.

CAPD Continuous ambulatory peritoneal dialysis.

Capilloconidia (L., *capillus* = hair) A minor type of conidia in *Basidiobolus ranarum* (the major type are ballistoconidia). Zygospore germination may develop into capilloconidia borne on the end of long narrow conidiophores. These conidia lack a propelling apparatus and are passively released. Lacking a means of forcible discharge, capilloconidia produce a sticky adhesive which aids their dispersal via passing insects and arthropods.

Capsule A hyaline, viscous, and usually acidic, halo-like polysaccharide sheath surrounding a yeast or bacterium: for example, the capsule of *Cryptococcus neoformans*.

Carryover Substances stored within cells of an inoculum, nutrients of the original culture medium, or both. Some of this material may be "carried over" to new medium or assimilation studies.

CASF Caspofungin.

Catenulate (L., *catenuia*, dim. of *catena* = chain) An arrangement of conidia in chains, end-to-end.

CBP Calcium binding protein.

CDC U.S. Centers for Disease Control and Prevention.

Cellulitis An infection of skin, subcutaneous, and connective tissues.

CFU Colony forming unit.

CGB Medium Canavanine–glycine–bromothymol blue medium. A specialized agar used to separate *Cryptococcus neoformans* (no growth, medium remains yellow) from *Cryptococcus gattii* (growth, blue color). Please see Chapter 2 for details.

CGD Chronic granulomatous disease.

Charcot–Leyden Crystals Eosinophilic needle-shaped crystalline structures that are breakdown products of eosinophils. Seen in asthma and eosinophilic pneumonia.

Chlamydospore (syn., chlamydoconidium) (Gr., *chlamys* = cloak) An enlarged terminal or intercalary vegetative hyphal cell, usually with a thickened cell wall, and believed to be a resting conidium that functions as a survival propagule. Its actual function, however, is unknown.

CHROMagar *Candida*® A specialized agar for primary isolation and differentiation of some *Candida* species based on their ability to split enzyme substrates causing a change in the colony color: for example, *Candida albicans* produces a green colony color.

Circinate Coiled in a ring or partly so.

Clade A taxonomic group comprising a single common ancestor and all the descendants of that ancestor.

Cladistic Analysis The practice of classifying organisms based on their phylogenetic relationships and evolutionary history, representing these in a branching diagram (a cladogram). Organisms are classified exclusively on the basis of joint descent from a single ancestral species.

Clamp Connection A backwardly directed half-dome shaped structure along the hyphae joining two adjacent cells. The presence of clamp connections on hyphae indicates the fungus belongs in the phylum *Basidiomycota*.

Clavate Club shaped.

Cleistothecium (pl., cleistothecia) (Gr., *kleistos* = closed + *theke* = case) An enclosed, typically rounded, sexual fruiting body (ascoma) of the ascomycetes that contains randomly dispersed asci within. Lacking an opening or ostiole, the cleistothecium must break open before the asci are liberated.

CLIA Clinical Laboratory Improvement Amendments. An act of the U.S. Congress in 1988 establishing quality standards for laboratory testing designed to ensure accuracy, reliability, and timeliness of patient test results regardless of where the test was performed.

CLSI Clinical Laboratory Standards Institute.

CLT Clotrimazole.

CMA Please see Cornmeal Agar.

Cmax Peak plasma concentration of a drug.

CMC Chronic mucocutaneous candidiasis.

CMI Cell (T-cell)-mediated immunity.

CMV Cytomegalovirus.

CNS Central nervous system.

Coccidioidin A toluene-induced lysate of the mycelial form of *Coccidioides* species. The extract is filtered to remove particulates and used as a skin-test antigen.

Coccidioidoma A localized residual granulomatous lesion or scar in a lung following primary coccidioidomycosis containing spherules of *Coccidioides* species.

Coenocytic A multinucleate mass of hyphae without septa (cross-walls); aseptate.

Collagen Vascular Disease Collagen shapes the structure of tendons, bones, and connective tissues. Problems with the immune system can affect these structures causing collagen vascular disease. The types are dermatomyositis, polyarteritis nodosa, rheumatoid arthritis, scleroderma, and systemic lupus erythematosus. Collagen vascular disease is a risk factor for fungal infections.

Collarette (i) A small ring of cell wall remnant material at the tip of a phialide or (ii) the base of a columella left by the wall of a sporangium when it dissolves or ruptures.

Columella (L., dim of *columna* = column) A sterile, dome-like expansion at the apex of a sporangiophore and at the base of, and extending into, a sporangium. This structure is characteristic of the anamorph state of the subphylum *Mucoromycotina*.

Commensalism (L., *com* = together + *mensa* = table). A relationship in which two or more organisms live in close association, where one may derive benefit, but in which, under normal circumstances, neither harms or parasitizes the other(s). Commensal: one of the organisms in this relationship; for example, *Candida albicans* is a commensal living on the oral, intestinal, and vaginal mucosae of humans.

Communicable A disease in which the infectious agent may be transmitted from one person to another (or between lower animals and humans) either directly or indirectly through fomites.

Conidiogenous Describes a cell specialized for asexual reproduction: producing conidia.

Conidiophore (Gr., *konidion* dim of *konis* = dust + *phoros* = bearing) (syn: stipe). A specialized hyphal stalk at the tip or side of a hypha upon which a conidiogenous cell develops.

Conidium (pl.,-ia) A unicellular or multicellular, nonmotile, specialized asexual reproductive propagule of a fungus. Conidia are released by detachment from the conidiogenous cell. There are many types of conidia, please see Chapter 2, Section 2.3.8.8, Microscopic Identification of Molds Based on the Specific Conidia-Forming Structures.

COPD Chronic obstructive pulmonary disease.

"Copper Pennies" In chromoblastomycosis, used to describe the characteristic muriform cells produced in vivo during infection. They are so-named because of their natural pigmentation and shape.

Cording Effect In sporotrichosis, the linear chain of nodules tracing the path of lymphatic drainage from a primary lesion on an extremity.

Coremium (pl., -ia) = synnema. A column of combined filaments; an erect fascicle of hyphae.

Cornmeal Agar (CMA) Cornmeal agar is used for microscopic morphology to promote sporulation of molds. In yeasts, CMA induces pseudohyphae and chlamydospores by *Candida* species. CMA with 1% dextrose stimulates red pigment production in *Trichophyton rubrum*.

Cornmeal–Tween 80 Agar With the addition of polysorbate (Tween) 80, CMA is used primarily to test *Candida* isolates for their ability to produce chlamydospores (*C. albicans* and *C. dubliniensis* are producers).

Cryptococcoma Cyst-like granuloma containing cryptococci, particularly common in the cerebral cortex but also found in the lung and elsewhere.

CSF Cerebrospinal fluid.

CT Computed tomography; also HRCT, high-resolution computed tomography.

Cuticle The outermost layer of the hair shaft is a hard, shingle-like layer of overlapping cells, 5–12 deep, acting as a protective barrier for the softer inner structure.

CXR Chest X-radiograph.

Cyst (i) A sac enclosed in a capsule that is not a normal part of the tissue where it is found. (ii) In pneumocystosis the term cyst was applied by parasitologists to the ascus, before it was known that *Pneumocystis jirovecii* (formerly *P. carinii*) was an ascomycetous fungus.

Czapek–Dox Medium/Agar A completely chemically defined medium containing sucrose as the carbon source and sodium nitrate as the nitrogen source. It is recommended for growth and identification of *Aspergillus* and *Penicillium* species and other molds. It is designed to cultivate bacteria and fungi capable of utilizing inorganic nitrogen.

Dalmau Plate A pattern of inoculation on nutritionally deficient medium. Cornmeal agar is seeded with a suspected colony of *Candida albicans* or other yeast. A light inoculum is cut into, but not through, the agar forming a #-shape in the center of the agar plate. Please see Chapter 2, Section 2.3.7.3, Additional and Confirmatory Tests for Yeast Identification.

Deep-seated (syn: systemic) Referring to an infection that has developed in an internal organ of the body.

Dehiscent (L., *dehiscere* = to split open) Opening of a structure to permit the escape of spores; the separation of spores from the structure that produced them. *Source:* mycolog.com the website of Bryce Kendrick, The Fifth Kingdom.

Deliquescence (L., *deliquescere* = to melt, dissolve) Dissolution of the fungal cell wall: for example, deliquescence of the sporangium of *Rhizopus oryzae* releases the sporangiospores within.

Dematiaceae *Archaic*: A form-family of mitosporic fungi with melanin pigment in their conidia and/or hyphal cell walls. Please see Melanized Fungi.

Dematiaceous Melanized, having brown to black conidia and/or hyphal cell walls.

Denticle Tooth-like projections or thin filaments on which conidia are formed.

Denture Stomatitis An inflammation of the gums beneath dentures occurring as erythema without an adherent white membrane.

Dermatomycosis A fungal disease of the skin, hair, or nails caused by nondermatophytic molds and yeasts.

Dermatophyte (Gr., *derma* = skin + *phyton* = plant) A pathogenic fungus of the genera *Microsporum, Trichophyton*, or *Epidermophyton* causing a cutaneous infection of the stratum corneum, the hair, or nails.

Dermatophyte Test Medium (DTM) A medium used to identify a mold as a dermatophyte, even if it is not sporulating. Upon growth, the medium (containing the pH indicator, phenol red) turns from pale yellow (acidic pH) to medium red (alkaline pH). Occasionally geophilic keratinophilic molds (nondermatophytes) may grow on this medium. Color change also may be induced by growth of *Histoplasma capsulatum, Aspergillus*, and other fungi, which may be neither geophilic nor keratinophilic. Colonies producing such a color change must be studied microscopically to identify the causal organism.

Dermatophytosis A fungal disease of the stratum corneum, hair, or nails caused by members of the genera *Microsporum, Trichophyton*, or *Epidermophyton*.

Desert Rheumatism Joint pain caused by the inflammatory response to infection with *Coccidioides* species.

DHN Melanin Dihydroxynaphthalene melanin. Synthesized by the polyketide pathway using the enzyme laccase, this type of melanin is common in the cell walls of melanized fungi. *Cryptococcus neoformans* utilizes a different pathway. Please see Chapter 12, Section 12.9, Determinants of Pathogenicity—Melanin.

Diaper Rash The most common skin problem in infants resulting from an irritation of the skin caused by damp conditions created by urine or feces. Yeast diaper rash is caused by *Candida albicans*, which also causes oral thrush in infants. The rash is very red, usually with small red bumps on the outer edges. Treatment with topical antifungal drugs is required.

DIC Differential interference contrast. A type of microscopy that results in a pseudo-three-dimensional appearance to the specimen. A polarized light source is passaged through a prism, splitting the incident light into two beams which penetrate the specimen. The objective (eyepiece) has a second prism that recombines the two beams. Having passed through the specimen separately, the two beams have slightly different optical paths. This combination creates the three-dimensional image. Please also see http://www.olympusmicro.com/primer/microscopy.pdf.

Dichotomous Branching Forked with two symmetrical branches.

Dikaryon Pairs of genetically dissimilar unfused nuclei coexist and divide synchronously within hyphae before nuclear fusion (karyogamy) occurs. This pattern of nuclear division is typical of the *Ascomycota* and *Basdiomycota* and explains why these two phyla are classed in the subkingdom *Dikarya*.

Dimorphic Possessing two different morphologies, for example, yeast form and filamentous or mold form. Dimorphic fungi grow as mycelia at 25–30°C and convert to yeast forms or spherules at 37°C in vivo. The structures seen in vivo also may be induced at 37°C on specialized laboratory media. An exception is dimorphism in *Candida albicans*, which produces yeast and hyphal forms at both temperatures.

Dimorphism Please see Dimorphic.

Disjunctor Cell A connective cell that binds together two conidia of a chain. The disjunctor cell fragments, or lyses, releasing the conidium to which it was attached. The liberated conidium may retain a wall remnant called an annular frill. Used especially to describe the intercalary arthroconidia of *Coccidioides* species.

Disk Diffusion A type of antifungal susceptibility test. Paper disks impregnated with a known concentration of an antifungal agent are placed on a lawn freshly seeded with a standard suspension of yeast or mold on standard medium (Mueller–Hinton–glucose agar). After suitable incubation, the diameter of the zone of inhibition is measured. A determination of relative susceptibility is made according to breakpoints established by the CLSI. Please see Chapter 3B, Section 3B.10.6, Disk Diffusion AFS Tests.

Disseminated (i) An infection distributed in multiple organ systems; (ii) scattering of conidia/spores upon release.

DNCB Dinitrochlorobenzene.

Dolipore The septum between cells of the *Basidiomycota* has a septal pore. Each end of the septum, adjoining the pore, is swollen (the dolipore). Behind the septum, and on each side, is a dome-shaped septal pore cap. This pore cap structure shown in Fig. 1.14 is perforated, also called a perforated parenthesome. The dolipore septum of basidiomycetes allows cytoplasm but not organelles to pass through it from one cell to another.

Downy Colonies covered with short, fine hyphae.

Dysgonic Growing with difficulty on artificial media.

Echinulate (L., *echinus* = hedgehog) Covered with small delicate, or rough spines.

ECM The extracellular matrix consists of macromolecules secreted by mammalian cells into their immediate environment. These macromolecules form a region of noncellular material in the interstices between the cells. Extracellular matrices are made up of collagen, proteoglycans, and a variety of specialized glycoprotein molecules, fibronectin, laminin, and vitronectin. These large glycoproteins are responsible for organizing the matrix and cells into an ordered structure. *Source:* Gilbert (2000).

Ecology The relationships between organisms and their environment.

Ectothrix A pattern of sporulation after hair invasion in dermatophytosis, in which hyphae grow in the hair shaft and arthroconidia are formed as a sheath on the outside of the hair shaft.

Eczematous Severe pathologic changes in the skin, accompanied by redness, oozing, crusting, and loss of pigmentation.

Efflux Pumps They pump azole antifungal drugs out of the fungal cell, decreasing intracellular drug concentrations. Two families of transporters function as efflux pumps in *Candida albicans*. They are encoded by CDR genes of the ATP-binding cassette (ABC) superfamily, and the MDR genes of the major facilitators class. The CDR efflux pumps transport multiple azole drugs out of *Candida*, whereas the MDR efflux pump is specific for fluconazole.

EIA Enzyme immunoassay.

EMM Extracellular matrix material produced by a fungus during the formation of a biofilm: for example, ß-(1→3)–D-glucan in *Candida* biofilms. Please also see ECM, extracellular matrix material.

Emmons Modified SDA Please see SDA-Emmons.

Empiric Therapy Treatment of high-risk patients with signs and symptoms of disease in the absence of positive cultures or other evidence of disease.

Encapsulated Having an extracellular capsule, for example, the acidic, polysaccharide capsule of *Cryptocococcus neoformans*.

Endemic (i) Used to describe mycoses restricted to specific geographic areas, as in "endemic mycoses" (e.g., coccidioidomycosis is a New World mycosis); (ii) a mycosis that may have a widespread geographic distribution but with locales of high prevalence (e.g., sporotrichosis in the high plateau of the Andes).

Endogenous From within.

Endophagocytosis Please see Phagocytosis.

Endospore An asexual propagule produced within a spherule, as in *Coccidioides* species.

Endothrix In dermatophytosis, the pattern of sporulation after hair invasion. After the fungal hyphae become well-established in the hair, hyphae fragment into arthroconidia within the hair shaft. There is no sporulation on the external sheath.

En grappe (Fr., grappe = cluster) Microconidia in a grape-like cluster, used to describe sporulation in some dermatophytes, for example, zoophilic *Trichophyton mentagrophytes*.

Enriched Medium A basal medium containing a supplement, for example, 5% defibrinated sheep blood.

Enteroblastic (Gr., *entos* = within + *blast* = bud) A type of conidiogenesis in which the outer wall of a conidiogenous cell ruptures, allowing the inner layer to grow through and become the cell wall of the new conidium. The new conidium is released from an opening in the outer wall of the conidiogenous cell. In this way a succession of conidia are produced in a basipetal manner: for example, the phialides of *Aspergillus* or *Phialophora* spp.

En thyrses Microconidia attached singly along the hypha, used to describe sporulation in some dermatophytes, for example, *Trichophyton rubrum*.

Epidemiology The study of patterns and causes of disease affecting populations, not individuals. This includes the incidence and prevalence of disease, environmental factors, and geographic distribution. The major tool used is the collection and analysis of statistical data. With respect to epidemics, it is the study of the transmission and means of control of disease. The goal is to develop interventions made in the interest of public health and preventive medicine.

Epitrochlear Above the trochlea—the notch within the hook-like proximal end of the ulna that slides in and out of the olecranon fossa of the humerus; for example, epitrochlear bursitis is tennis elbow.

Epizootic A disease attacking many animals in a region at the same time; a disease of high morbidity which is only occasionally present in an animal community.

Erythema Multiforme Target-like lesions, oral involvement, pruritus, and palmar desquamation occurring early in the acute stage of pulmonary coccidioidomycosis.

Erythema Nodosum A rash consisting of tender subcutaneous red nodules, typically on the anterior part of legs (shins). The rash develops in a subset of patients 1–3 weeks into acute pulmonary coccidioidomycosis.

Erythrasma A chronic bacterial infection that usually appears in the area between overlapping skin (skin folds). The main symptoms are reddish-brown slightly scaly patches with sharp borders. Erythrasma is caused by the bacterium *Corynebacterium minutissimum*. It is differentiated from superficial mycoses because lesions of erythrasma fluoresce red in Wood's light.

ESR Erythrocyte sedimentation rate.

Etest A plastic strip is precoated at the factory with a concentration gradient of an antimicrobial compound. The strip is then placed on a freshly seeded lawn of a fungus or bacterium. After incubation, an ellipical zone of inhibition appears if the said microbe is susceptibile to the drug. The point at which the zone touches the strip is read as the endpoint titer expressed in μg/mL.

Eukaryotic Microbe Having cells with a well-defined nucleus containing chromosomes and bounded by a nuclear membrane.

Evanescent Short-lived, as in early evanescent asci, wherein asci rupture soon after they are formed.

Exoantigen Soluble antigens secreted by fungi early in their development. This property is used in the exoantigen test for immunoidentification of the mold form of dimorphic fungal pathogens. The test is conducted as an agar gel immunodiffusion test with reference antiserum in the center well and extracts of the unknown fungus and positive and negative control antigens in outer wells. An immunoprecipitate forms when the test is positive, giving a line of identity with the positive control antigen.

Exogenous From without, for example, the source of infection from outside the body.

Fascicle (L., *fasciculus* = a small fiber bundle) A small, irregular compact bundle of hyphae.

Favus (L., *favus* = honeycomb) A form of dermatophytosis characterized by crusts, or "scutula," that form along the hair shaft. They are composed of pus from the infected follicle combined with hyphae, spores, and cell debris. May result in permanent balding (alopecia). Favus is associated with *Trichophyton schoenleinii* infection.

5FC Flucytosine.

FDA Food and Drug Administration (U.S.A.).

Filament A hyphal strand of a fungus.

FLC Fluconazole.

Floccose Fluffy, loose, cottony in tufts, wooly.

FNA Fine needle aspirate.

Fomite An inanimate object or substance that may harbor and transmit infectious microbes.

Fontana–Masson Stain This "melanin" stain for histology relies on melanin granules to reduce ammoniacal silver nitrate. Examples of fungi that give a positive reaction are *Cryptococcus neoformans* and lightly pigmented causative agents of phaeohyphomycosis, for example, *Bipolaris* and *Curvularia*. Melanin is not the only substance giving a positive reaction: argentaffin, chromaffin, and some lipochrome pigments also will stain black.

Foot Cell (i) In *Aspergillus*, a hyphal cell at the base of a conidiophore; (ii) in *Fusarium*, the sharply angled, heel-like end at the base of the macroconidium where it attaches to the hypha.

Fungi Imperfecti The archaic term for mitosporic fungi with no known sexual stage.

Fungus (pl., fungi) (L., *fungus* = mushroom) Fungi are simple eukaryotes with chitin-containing rigid cell walls which, taken together, constitute the kingdom Fungi. Microfungi are characterized by two growth forms: molds and yeasts. Medical mycology is mostly concerned with microfungi, specifically the zoopathogenic fungi. Their major characteristics are:

- A nonmotile thallus constructed of apically elongating walled filaments (hyphae). This is the mold lifeform. Certain members of the *Ascomycota* and *Basidiomycota* may grow as either branching hyphae or budding yeast for part or all of their life cycles.
- A life cycle with sexual and/or asexual reproduction.
- Heterotrophic nutrition, in which energy is derived from the absorption of organic substances from the environment.

Fungi are not plants, lack chlorophyll, and are not photosynthetic.

Fungus Ball (syn., eumycetoma) A compact mass of fungal mycelium and cellular debris, 1–5 cm in diameter, residing within a lung cavity, paranasal sinus, or urinary tract. Aspergilloma is a type of fungus ball of the lung. Sporulation may occur in hyphae adjacent to an air space.

Fusiform (L., *fusus* = spindle) Spindle shaped; narrowing toward both ends.

G-CSF Granulocyte colony stimulating factor.

Geniculate (L., *genu* = knee) Bent, knee-like. A conidiophore that is sharply bent like a knee at one or more places and is typical of sympodial development.

Geophilic In dermatophytes, those that are terrestrial or soil-inhabiting.

Germ Tube A hyphal initial developing from a conidium, spore, or yeast. Please see germ tube test for *Candida albicans* in Chapter 11.

Germling A germinated conidium or sporangiospore.

Glabrous (L., *glaber* = without hair) Smooth, lacking hairs. A colony producing no aerial hyphae.

Globose Round.

GM Galactomannan; GM-EIA, galactomannan antigenemia enzyme immunoassay.

GM-CSF Granulocyte-macrophage colony stimulating factor

GMS Gomori–methenamine silver, a histologic stain for demonstrating fungi in tissue. The cell walls of fungi stain a dark brown to black color, outlining the fungal elements. It is a good stain to screen for fungi in tissue, providing contrast between fungal elements and the background. Please see Chapter 2.

Gram Stain A differential staining procedure used with bacteria and fungi. It is based on the retention of a primary dye (crystal violet) after treatment with a decolorizing solution (ethyl alcohol). Microbes are identified as Gram+ or Gram−. Young fungus cells are Gram+.

Granuloma A nodule of inflammatory tissue composed of clusters of activated macrophages and T lymphocytes often with associated necrosis and fibrosis. *Source:* Abbas and Lichtman (2009).

Granulomatous Composed of granulomas.

***Graphium* State** Fascicles of conidiophores bound together as in the *Graphium* synanamorph of *Scedosporium apiospermum*.

GVHD Graft-versus-host disease.

Gymnothecium (Gr., *gymnos* = uncovered) A cleistothecium-like ascoma in which the peridial hyphae are loosely organized and in which the asci are randomly dispersed.

HAART Highly active antiretroviral therapy.

Halophilic Requiring a salty environment.

Halo Sign The radiographic halo sign of tissue injury is seen as a halo around a pulmonary nodule. The halo denotes hemorrhage or edema and the nodule is tissue infarcted by a fungus. The fungus is likely to be *Aspergillus* or another mold pathogen.

Halotolerant Tolerant of a salty environment.

"h" Antigen A glucosidase enzyme of *Histoplasma capsulatum* with antigenic properties. Antibodies (h-precipitins) arise 4–6 weeks after the first appearance of symptoms and are associated with pulmonary and extrapulmonary histoplasmosis.

HCW Healthcare worker.

H&E Hematoxylin and eosin. A histologic stain for mammalian cells in which material is stained with hematoxylin, then with an aqueous solution of eosin. H&E is useful in distinguishing melanized and some hyaline fungi, but is often inadequate for detecting fungal elements. H&E will demonstrate Splendore–Hoeppli material.

HEPA High efficiency particulate air filtration.

Heterodimer In biochemistry, a dimer in which the two subunits are different.

Heterothallic (Gr., *heteros* = other + *thallos* = a young shoot) A fungus that is only capable of sexual reproduction after the fusion of two genetically dissimilar nuclei from two different thalli or strains of compatible mating types.

Heterotrophic Requiring preformed organic compounds as an energy source.

Hilum (pl., -a) (L., a little thing, a trifle) A conspicuous scar remaining at the base of a conidium, at the point of separation from the conidiogenous cell.

Holoblastic (Gr., *holo* = whole, entire). Describes a type of conidiogenesis wherein the conidium is produced when both inner and outer walls of the conidiogenous cell swell to form the conidial cell wall. Each conidium may then produce another conidium in an acropetal chain, for example, *Cladosporium* species.

Homothallic A fungus that does not require two distinct thalli or strains for sexual reproduction.

Host A living animal or plant that harbors a parasite; the larger organism in a symbiotic or parasitic relationship.

HPLC High pressure liquid chromatography.

HRCT Please see CT.

Hülle Cell A thick-walled sterile cell produced by some *Aspergillus* species. Although produced as intercalary or terminal cells, they generally are seen surrounding a cleistothecium.

HVAC Heat, ventilation and air-conditioning ducted system.

Hyaline (Gr., *hyalo* = glassy, transparent) Hyphae and conidia without color (transparent) when viewed under the microscope.

Hyalohyphomycosis Disease caused by a heterogeneous group of hyaline molds.

Hyperkeratosis Hypertrophy of the stratum corneum.

Hyperplastic An increase in the normal number of cells in normal tissue.

Hypha (pl., hyphae) (Gr., *hyphe* = web or thread) The microscopic, thread-like branching, tubular body of a filamentous fungus. It may be septate or aseptate; hyphae, a group of such filaments joined together. Large amounts of hyphae comprise a mycelium.

Hyphomycetes An older classification referring to mitosporic fungi—fungi that have no known sexual stage—which produce their conidia on simple conidiophores arising directly from vegetative hyphae.

Iatrogenic (Gr., *iatros* = physician) Any condition in a patient occurring as the result of treatment by a physician or surgeon.

ICD-9 Code International Classification of Diseases—9th revision. A standard classification of disease, injuries, and causes of death, by etiology and anatomic site(s). They are coded by six-digit numbers, allowing physicians, healthcare workers, statisticians, and so on to speak a common language, worldwide.

Id (Dermatophytid) A cutaneous vesicular eruption, usually distant from the site of a dermatophyte infection, but related to it. Ids are thought to result from interaction between circulating antigen and antibodies. Although usually vesicular, they may be papular or eczematous. They are sterile lesions and are most often seen in inflammatory dermatophytosis.

Idiopathic A disease for which there is no known cause.

IDSA Infectious Disease Society of America.

Impetigo An erythematous inflammatory bacterial skin disease caused by *Staphylococcus* or *Streptococcus* species and characterized by isolated pustules and crusty yellow sores.

Incidence In disease, the number of cases developing in a given period of time in a specific population. The incidence rate is the ratio of the number of new cases of a disease over a given period of time (typically per year) divided by the total population.

India Ink Prep India ink is a suspension of fine carbon black particles for use as a negative stain, for example, to visualize the cryptococcal capsule. (Please see Chapter 2, Section 2.3.3.3, Cerebrospinal Fluid.)

Infection A disease caused by microbes (may or may not be contagious).

Infectious Propagules Particles capable of communicating infection; in mycology, spores, conidia or, exceptionally, hyphal fragments.

Inhibitory Mold Agar A complex medium with chloramphenicol to inhibit bacteria. The formula is given in Table 2.1 in Chapter 2.

Integrins Cell surface proteins that mediate adhesion of leukocytes to other leukocytes, to endothelial cells, and to extracellular matrix proteins. They are heterodimers.

Intercalary (L., *intercalere* = to insert) A cell or spore situated in a hyphal strand between two cells.

Interstitium (Gr., *interstitalis*; *inter* = between + *sistere* = set) Gap or space situated between a tissue or structure. The supporting matrix tissue of the lungs, as opposed to the airways or air sacs. May be the site of specific diseases.

Intertrigo Dermatitis in the area between two touching skin surfaces, often chafing, where sweat can accumulate causing a warm, moist site suitable for fungal growth.

Intracystic Bodies These are the ascospores of *Pneumocystis jirovecii*.

IV Intravenous.

In Vitro Outside the host, as on artificial media, in test tube studie and so on.

In Vivo Within the living host.

IRIS Immune reconstitution inflammatory syndrome. Please see Chapter 12.

Isotropic (Gk., *iso* = same; *tropic* = turning) Growth by enlargement in all directions, for example, an adiaspore, spherule, or muriform cell.

Isthmus A narrow neck or pore between mother and daughter yeast cells.

ITC Itraconazole.

ITS Region Intergenic transcribed spacer region of ribosomal RNA genes. Within the rDNA complex is a tandem array of multiple copies of the repeat unit consisting, from the $5' \rightarrow 3'$ direction of 18S ITS1 5.8S ITS2 28S IGS1 5S IGS2 gene segments. Please see Chapter 2, Section 2.4, Genetic Identification of Fungi.

Karyogamy Fusion of two nuclei.

Keratin The insoluble protein and major constituent of hair, nails, stratum corneum, and feathers.

Keratinases Proteinases capable of digesting hair, nails, and skin scales.

Keratinophilic Capable of digesting and using keratin, such as fungi causing cutaneous mycoses, the dermatophytes.

Kerion (Gr., *kerion* = honeycomb). In tinea capitis, or scalp ringworm, a large boggy swelling on the scalp, with ulcerated areas exuding pus, accompanied by hair loss. May become secondarily infected with bacteria.

KOH Potassium hydroxide, a strong alkali used as a 10% aqueous solution in the mycology laboratory as a clearing agent to dissolve organic material to improve the visibility of fungal elements. Caution: KOH is extremely caustic, and great care must be taken to protect eyes and skin when using it.

KOH Prep A method of direct microscopic examination of clinical specimens, for example, wound exudates, sputum, skin, nail scrapings, or hair. KOH is a clearing agent allowing fungal elements to be more apparent. Please see Chapter 2, Section 2.3.3, Direct Examination.

KTC Ketoconazole.

Lactofuchsin A mounting and preservation medium for fungi in tease mounts, and slide cultures.

Lactophenol Cotton Blue A mounting and preservation medium for fungi in tease mounts, and slide cultures. Cotton blue is also known as Aniline blue, Poirrier's blue, or Methyl blue. Formulas for mounting media are found in Chapter 2, Section 2.3.8, Methods Useful for Mold Identification.

Laminar Airflow Undirectional airflow through the room which is turbulence-free and steady.

Lanose A fungal colony having a wooly texture.

Leukoplakia A precancerous lesion that develops on the tongue or the inside of the cheek as a response to chronic irritation.

Locule (L., *loculus* = little spot, little box) A cavity within a stroma (specifically ascostroma) within which asci develop.

LPCB Please see Lactophenol Cotton Blue.

mAb Monoclonal antibody.

Macroconidium (pl., -ia) The larger of two propagules of different sizes that are produced in the same manner in the same colony by a single fungus. They may be single or more often multicellular, thick or thin walled, smooth or rough, single or in clusters, and club, spindle, or oval shaped.

Macule A small circumscribed change in skin color that is neither raised (elevated) nor depressed, and <10 mm in diameter.

Mannan The readily soluble outer layer of the cell wall of *Candida* species is an antigenic peptidopolysaccharide.

"m" Antigen A catalase enzyme produced by *Histoplasma capsulatum*. Anti-m-antibodies are specific indicators of acute pulmonary histoplasmosis. The m-precipitins are the first to be detected, 4–6 weeks after the appearance of clinical symptoms.

MCF Micafungin.

MEC Minimun effective concentration, in μg/mL, applied to the antifungal effect of echinocandin class of drugs, observed microscopically as clubbing at the hyphal tips.

Meiosporangium A sporangium in which meiosis occurs.

Meiosporic With respect to meiosis—sexual reproduction.

Melanized Fungi Replaces the term *Dematiaceae* to describe fungi with melanin in their cell walls.

Merosporangium A small cylindrical sporangium having with it sporangiospores in a single row and found in certain *Mucoromycotina* (e.g., *Syncephalastrum*).

Metula (L., *metula* = obelisk, small pyramid) A specialized sterile branch from which phialides are produced in biseriate aspergilli and pluriverticillate penicillia.

MIC Minimum inhibitory concentration. The minimum concentration of an antifungal agent (μg/mL) required to prevent the in vitro growth of a yeast or mold. The MIC is determined using an in vitro AFS test.

Microbiome The total population of microbes, in a defined environment; usually referring to those colonizing a human in one or more nonsterile organ systems.

Microbiota The microbial population colonizing a normal healthy individual.

Microconidium (pl., -ia) A small asexually produced propagule (conidium) formed on aerial hyphae.

Middlebrook's 7H10 Agar A chemically defined medium supplemented with bovine serum albumin and oleic acid devised by Middlebrook and Cohn to cultivate mycobacteria. This medium stimulates conversion of *Blastomyces dermatitidis* to the yeast form.

Mitosporangium A sporangium in which mitosis occurs.

Mitosporic Asexual reproduction.

MLST Multilocus sequence typing. A method of genotyping that surveys partial sequences (450 bp) of seven gene loci that encode housekeeping proteins. The evolutionary distance between strains is quantified as the number of loci that differ by either a single nucleotide substitution or by a track of nucleotides. Strains identical to a particular strain at five or more loci are members of a clonal complex. The "Burst" algorithm assigns strains to clonal complexes with criteria set by the user (website: www.mlst.net). *Source:* Cohan (2005).

"Moccasin" (or "Sandal") Tinea Pedis Ringworm of the sole of the foot extending up the sides of the foot in the area covered by a moccasin. Most often seen with *Trichophyton rubrum* dermatophytosis.

Mold (alternate spelling: mould) A filamentous fungus. They may be meiosporic (known sexual stage) or mitosporic (only asexual reproduction observed).

Molluscum-like Lesions (L., *molluscus* = soft) Soft pulpy nodules as seen in molluscum contagiosum, a mild chronic skin disease caused by a poxvirus and characterized by small nodules with a central opening and contents resembling curd. They are molluscum bodies or condyloma subcutaneum. Molluscum contagiosum is mainly seen in children. It is often sexually transmitted in teenagers and adults. (It is a benign disorder that usually clears up by itself.)

Moniliform (L., *monilia* = necklace) Hyphae with swellings at regular intervals. Beaded and joined together like a string of beads or pearls; that is, regularly constricted and composed of globose cells.

Monomorphic Growth exclusively in a single form, either as a mold or a yeast.

Monophyletic A group or clade comprised of a single common ancestor and all the descendants of that ancestor.

Morphogenesis Development or change in the form of an organism, for example, mold to yeast or yeast to mold dimorphism.

MRI Magnetic resonance imaging.

MSM Men who have sex with men.

Mucicarmine (Mayer's) A histologic stain that produces bright red color on the partially retracted *Cryptococcus neoformans* capsule.

Mucoromycotina The subphylum of molds with vegetative hyphae which are aseptate or pauciseptate and which undergo asexual reproduction via sporangia that release sporangiospores, and sexual reproduction via the production of a single large zygospore on suspensor cells.

MUD Matched-unrelated donor, for example, in stem cell transplantation.

Muriform (L., *murus* = wall) An enlarged fungal cell with cross-walls produced both vertically and horizontally.

Muriform Cells (syn., sclerotic bodies, "copper pennies") In chromoblastomycosis, a cluster of melanized, thick-walled, usually rounded cells that divide by internal cleavage planes, not by budding.

Mycelium (pl., -ia) (L., *mykēs* = mushroom) A mat of intertwined branched hyphae constituting the body of a fungus.

Mycetoma (i) Eukaryotic mycetoma (eumycetoma): a subcutaneous mycosis, typically on a lower extremity, characterized by tumefaction (swelling) and draining sinus tracts expressing grains, which are masses of fungal hyphae. (ii) Actinomycotic mycetoma: a similar clinical form caused by aerobic actinomycetes.

Mycosel (Mycobiotic) Agar (BD BioSciences, Inc.) Trade names for media similar to SDA- Emmons used to isolate pathogenic fungi from contaminated specimens. These media contain cycloheximide and chloramphenicol. Please see Table 2.1 for differences in their formulation. Cycloheximide inhibits faster growing saprobes. It also is inhibitory for *Cryptococcus* and *Aspergillus* species as well as for the *Mucoromycotina*. Some, but not all, *Candida* species are inhibited by cycloheximide. (Please see Table 2.7, Fungi Inhibited by Cycloheximide.)

Neutropenia A decreased number of PMNs in the blood. Prolonged neutropenia: when the concentration of PMNs is <1000/ μL in the blood for 1 week or more. Profound neutropenia: when PMNs are <100/ μL in the blood.

NICU Neonatal intensive care unit.

Nosocomial Pertaining to a hospital or infirmary. A nososcomial infection is acquired in a hospital setting.

NYT Nystatin.

Obclavate Club shaped and thickened at the outer end.

Obovoid (L, *ob* = reversed; *ovoid* = egg shaped) An inverted egg-shaped conidium with the broad end up and narrow end down.

OD Optical density.

Onychia (Onychomycosis) A chronic fungal infection of the nails of the fingers and toes in which the nails become opaque, thickened, soft or brittle. The nail may separate from the underlying bed.

Onygenales A monophyletic lineage within the *Ascomycota* which produce ascospores in gymnothecia or cleistothecia. Mitosporic stage members of this order include the dermatophytes, *Blastomyces dermatitidis, Histoplasma capsulatum, Paracoccidioides brasiliensis*, and *Coccidioides* species. Please see Chapter 1, for the higher level classification of the kingdom Fungi.

OPC Oropharyngeal candidiasis.

Opportunistic infection An infection seen in debilitated or immunocompromised individuals.

Ostiole A pore, or opening in a sexual reproductive structure, a perithecium, or in an asexual structure, a pycnidium, through which spores or conidia may escape.

Ovoid Egg-shaped conidium with the broad end down, and narrow end up.

PAMP Pathogen associated molecular pattern. Structures shared by pathogens but absent on host cells. They are recognized by the innate immune system early in the immune response and trigger inflammation (e.g., fungal mannans).

PaO_2 The partial pressure of oxygen in arterial blood.

Papanicolaou A cytologic stain used for smears of various body secretions from the respiratory, digestive, or genitourinary tract. Devised for optimal visualization of cancer cells exfoliated from the epithelial surface of the body, this stain is also suitable for fungi, which acquire a pink coloration.

Papule A small, solid, rounded bump rising from the skin, usually <10 mm in diameter.

Paraphysis Sterile hypha growing up between asci in the hymenium (fertile layer in fungi) of many ascomycetes.

Parasexual Cycle A method of recombining genes in which haploid nuclei of a fungal heterokaryon fuse, enabling mitotic recombination without meiosis, and usually involving reduction of a diploid to a haploid through loss of whole chromosomes.

Parenteral Adminstration of medication or nutrition, not through the alimentary tract, but via subcutaneous, intramuscular, or intravenous injection.

Paronychia A fungal infection, either acute or chronic, involving the tissue surrounding the nail.

PAS (Periodic Acid-Schiff) A widely used stain in fungal histopathology, PAS stains the fungal cell wall pink-red. The tissue reaction is also visible when hematoxylin is used as a counterstain.

Pathognomonic A sign or symptom that is so characteristic of a disease that it makes the diagnosis.

Pauciseptate Sparse septations in the hyphae.

PCP *Pneumocytis* pneumonia.

PDA Potato dextrose agar. This medium is used to stimulate conidia production, in slide cultures. Please see Table 2.1 in Chapter 2 for the formula.

Pectinate (L., *pectin* = comb) Hyphal ends with protuberances resembling the teeth of a comb, used to describe specialized hyphae in some dermatophytes.

Penicillate Having a tuft of fine hairs resembling a brush, for example, as in the apices of *Penicillium* conidiophores.

Penicillus (L., = pencil) A brush-like asexual sporing head characteristic of the genus *Penicillium*.

Percurrent Developing through a previous apex.

Perfect (Teleomorph) State The sexual or meiosporic state of a fungus.

Peridium The outer wall of a sporangium or an ascoma, made of hyphae—the peridial hyphae.

Perithecium (Gr., *peri* = around + *theke* = a case) A usually flask-shaped ascoma containing asci as a layer on a fertile base (hymenium) within the ascoma. The asci or ascospores are expelled through an ostiole at the apex of the perithecium.

Perleche Angular cheilitis, an inflammatory lesion at the corners of the mouth caused by *Candida albicans*, other pathogens, or a nutritional deficiency.

Phaeo- (G., *phaios* = dark) Prefix meaning dark-colored and associated with melanized fungi.

Phaeohyphomycosis Disease caused by melanized fungi in which the hyphal or yeast-like tissue forms appear darkly pigmented. Excluded are agents of chromoblastomycosis which have a distinct tissue form.

Phagocytosis The process of engulfment of large particles or infectious agents (>0.5 μm) by cells of the immune system: neutrophils and macrophages. *Source:* Abbas and Lichtman (2009).

Phialide (Gr., *phiale* = a broad, flat vessel) A tube- or vase-shaped conidiogenous cell that gives rise to successive conidia from a fixed site without increasing in length as each conidium is produced (versus annelloconidium). A collarette may surround the phialide opening.

Phialoconidium A conidium extruded from a phialide. Phialoconidia are released in a basipetal manner from within the phialide (enteroblastic conidiogenesis). Conidia may adhere to the phialide as balls held together by polysaccharide (*Phialophora verrucosa*), or they may form easily dispersed dry chains (*Aspergillus fumigatus*).

Pityriasis Versicolor (L., *versicolor* = variously colored) A superficial cutaneous mycosis caused by lipophilic yeasts of the *Malasseziales* and pertaining to the hypo- or hyperpigmentation of the superficial skin lesions. "Pityriasis" is a term based on the former, and now obsolete, genus name: *Pityrosporum*.

Planktonic Please see Biofilm.

Pleocytosis The presence of a greater than normal number of cells in a body fluid such as the cerebrospinal fluid.

Pleomorphic (Gr., *pleion* = more) (i) Displaying more than one form of asexual sporulation. (ii) In dermatophyte mycology it refers to the change from a wild-type mycelial growth to a fluffy white sterile mycelial form.

PMN (syn., neutrophil) Polymorphonuclear neutrophilic granulocyte.

Polyphyletic A taxonomic group of fungi based on morphologic similarity which, on cladistic analysis, is found to consist of individuals that do not share a common ancestry.

Polyploid Having more than two sets of homologous chromosomes.

Poroconidium A type of holoblastic conidiogenesis in which a multicelled (3–6 celled) conidium grows out of a pore in the hyphal cell. After conidium secession, a pore can be observed in the wall of the conidiogenous cell at the point where the conidium was formed. This type of sporulation is typical of the *Pleosporaceae*.

Potassium hydroxide See KOH.

PPD The purified protein derivative of tuberculin used in the tuberculosis skin test.

PPE Personal protective equipment.

Preemptive Therapy Early treatment of an infection with the use of clinical, laboratory, or radiologic surrogate markers of disease in a high-risk patient before clinical signs and symptoms appear.

Prevalence The number of cases of a disease existing in a given population at a specific period of time (period prevalence) or at a particular moment in time (point prevalence). Prevalence can be thought of as a snapshot of all existing cases at a specified time. *Source: Stedman's Medical Dictionary*.

Propagule The reproductive unit.

Proteinase (Extracellular) Microbial secreted enzymes, acting as pathogenic factors, which digest host proteins, for example, aspartyl proteinases of *Candida albicans*.

PRRs Pattern recognition receptors. These receptors recognize PAMPs and facilitate innate immune responses against microbes: for example, Toll-like receptors and mannose receptors. The PRRs are linked to signal transduction pathways, activating genes that promote inflammation. *Source:* Abbas and Lichtman (2009).

Pruritus Severe itching.

PSC Posaconazole.

Pseudoepitheliomatous Hyperplasia A type of epithelial hyperplasia surrounding a chronic inflammatory process. Pronounced thickening is due to proliferation of all layers of the epidermis. Resembling squamous cell carcinoma, but due to nonmalignant skin growth. Pseudoepitheliomatous hyperplasia never has atypical mitotic figures, rarely has atypical nuclei, and does not invade blood vessels, lymphatic, tissue or nervous tissue.

Pseudohypha A distinct form of fungal growth consisting of a chain of elongated cells which grow by unipolar budding from the apex and remain attached, forming a filament resembling true hyphae. Pseudohyphae also may grow subapically to form clusters of buds along the pseudohyphal junctions: whorls or verticils. Pseudohyphae are differentiated from hyphae by the strictures at each attachment in the former. (Please see Chapter 1, Section 1.6.7, Pseudohyphae.)

Pycnidium (pl., -ia) (Gr., *pyknon* = concentrated + *idion* = dimin suffix) A globose or flask-like asexual fruiting body within which conidia are produced that are released through an ostiole. (Compare with Perithecium.)

Pyriform (L., *pyrum* = pear) Pear shaped.

QA and QC Quality assurance and quality control.

Rachis A conidiophore elongating to one side of a terminally produced conidium, often resulting in a "zigzag" or sympodial structure. In botany, a rachis is the primary axis on which leaflets are arranged.

Racquet Hyphae Hyphal cells being swollen at one end resembling tennis rackets with long handles, often in linear series.

RAPD Randomly amplified polymorphic DNA. A method of DNA genotyping using PCR to differentiate strains within a species. A panel of random primers, usually 10-mers, generate a series of PCR products (amplicons), which, when separated by agarose gel electrophoresis, produce a pattern distinctive for the strain.

RFLP Restriction fragment length polymorphism. The action of restriction endonucleases on genomic DNA releases fragments that are sized by agarose gel electrophoresis.

Rhexolytic Cleavage Conidial liberation in which an intermediate cell is disrupted and lysed leaving behind wall remnants, for example, arthroconidium of *Coccidioides* species.

Rhizoid (Gr., *rhiza* = root + *oeides* = like). Branched hyphae, root-like in appearance, which grow into the substrate, anchoring the surface mycelium. In the *Mucorales*, the position of rhizoids in relation to the sporangiophores is a characteristic used for fungal identification. (Please see illustrations in Chapter 17.)

Ringworm (syn., tinea) A communicable skin disease affecting the outer layer of the epidermis, the stratum corneum, which may invade the hair and nails, caused by dermatophytes. The classic appearance of the skin lesion is circular with an active border, inflammation, scaling, and pruritus.

RPMI Medium Developed at the Roswell Park Memorial Institute, it is a chemically defined tissue culture medium used in some in vitro antifungal susceptibility tests.

Rugose Ridged, wrinkled, or roughened surface.

SABHI A combined agar medium consisting of SDA and BHI. The formula is provided in Table 2.2 of Chapter 2. The purpose is to stimulate recovery and growth of dimorphic fungal pathogens. The medium may be supplemented with cycloheximide, chloramphenicol, and defibrinated sheep blood.

Sabouraud Dextrose Agar Please see SDA.

Sandwich EIA An antigen capture enzyme immunoassay in which a capture antibody is affixed to a solid surface (i.e., a dipstick or well of a microtitration plate); next, the source of circulating antigen is added and incubated. Following a buffer wash, the indicator antibody is added. The indicator antibody also will bind to the captured antigen. Color is then developed with a chromogenic substrate. The reaction is observed directly or with a spectrophotometer.

Saprobe (Gr., *sapros* = rotten) A microbe that obtains its nourishment from dead and decaying organic matter.

Saprophyte See Saprobe.

Scissiparity (L., *scissio* = cleavage, + *pario* = to bring forth) Reproduction by fission.

Sclerotic Bodies Please see Muriform Cells.

Sclerotium (Gr., *sklēros* = hard) A resting resistant body composed of a hardened mass of hyphae with a dark rind. When conditions are favorable, hyphae germinate and grow.

SCT A hematopoietic stem cell transplant.

SDA Sabouraud dextrose agar for cultivation of yeasts and molds. Dextrose is glucose (syn., SGA, Sabouraud glucose agar). A widely used medium for growth of fungi with peptone as the nitrogen source, and glucose as the carbon source. The final pH is 5.5–5.6, favoring fungi and suppressing many bacteria. Many fungi do not sporulate well on this medium due to the high concentration of dextrose, 40 g/L, and sporulate better on SDA-Emmons. The addition of antibiotics may be used to inhibit bacterial growth. (Please see Mycosel, Mycobiotic Agar.) Please see Table 2.1 in Chapter 2 for formulas of common mycologic media.

SDA-Emmons Emmons' modification of SDA. This medium is similar to SDA except that glucose is used at 20 g/L (half the glucose content of the original medium) and with a final pH of 6.8–7.0, to produce better sporulation.

Sepsis A bloodstream infection.

Septate Containing cross-walls in hyphae, conidiophore, conidia, or spores. A dividing wall or partition.

Septum (pl., septa) A cross-wall.

Sequela (pl., -ae) Something that follows, especially a pathologic condition (e.g., lesion or affection) resulting from a disease.

Sessile (L., *sessilis* = sitting) (i) A conidium arising directly from the surface of a hypha, without an intervening conidiophore; (ii) attached directly to the substrate.

Settle Plate A Petri plate containing an agar medium which is exposed to the environment for a specified time period. Fungal conidia alighting on the agar surface germinate, grow, and are enumerated after a suitable incubation time.

Shield Cell A conidium having a shield or "V" shape, usually seen in melanized fungi, particularly *Cladosporium* and related organisms. It is formed at the point where a conidial chain branches to form two chains.

Slide Culture A miniature culture method used to study the microscopic morphology and conidiogenesis in order to identify an unknown fungus. A square of agar medium is placed on a microscope slide. Then the middle of each of the four sides is seeded with the mold to be identified by means of an inoculating needle. A coverslip is added and the slide is placed on a bent glass rod in a Petri plate. Sterile water is then added to cover the bottom of the plate, creating a moist chamber. At intervals the prep is examined microscopically for the pattern of sporulation.

"Snowstorm" Appearance A radiographic image of multiple nodules in both lung fields, as seen in acute pulmonary histoplasmosis.

SNP Single nucleotide polymorphism.

SOT Solid organ transplant.

Spathulate Spoon shaped.

Spherule A closed, thick-walled spherical structure, approximately 10–80 μm in diameter, containing endospores and produced by *Coccidioides* species during growth in vivo. Each endospore is capable of developing into a spherule.

Spinose Covered with small spines.

Spinous Cell Layer (syn., prickle cell layer) In the epidermis, the layer between the stratum granulosum and stratum basale. Cells in the stratum spinosum change from being columnar to polygonal and start to synthesize keratin.

Splendore–Hoeppli Reaction Eosinophilic, pseudomycotic structures formed in infected tissues composed of necrotic debris and immunoglobulin. In the United States it is most frequently found in association with botryomycosis. In the tropics, it can be found surrounding schistosome eggs, microfilariae, and a variety of fungi, including *S. schenckii*: for example, in the asteroid body surrounding *Sporothrix schenckii* yeast forms in tissue.

Sporangiole A small, usually spherical sporangium, which contains only one or two sporangiospores characteristic of the genus *Cunninghamella* of the subphylum *Mucoromycotina*, order *Mucorales*.

Sporangiophore The specialized hypha giving rise to a sporangium in the *Mucoromycotina*. It may be branched or unbranched.

Sporangiospore An asexual spore produced within a specialized sac, the sporangium, and freed by deliquescence of the sporangium. Sporangiospores are characteristic of the subphylum *Mucoromycotina*.

Sporangium (pl., sporangia) (Gr., *sporos* = seed, spores + *angeion* = vessel) (i) A sac-like structure, the entire contents of which become converted into a number of spores. (ii) An asexual closed sac structure that develops at the apex of a sporangiophore. Sporangia are the asexual spore-forming structures of the *Mucorales*. A sporangium is delimited by a columella located at the base of and protruding into the sporangium. Haploid nuclei divide in the sporangium, spore walls form around the nuclei and cytoplasmic organelles, and the mature sporangiospores are liberated when the sporangial wall deliquesces. (iii) The spherule of *Coccidioides* species is a specialized type of sporangium found in tissues during infection (please see Chapter 5).

Spore A sexual reproductive propagule as seen in an ascus or on a basidium, or in an asexual propagule produced in a sporangium and produced by a process of cytoplasmic cleavage yielding sporangiospores. (Compare with Conidium.)

Sporodochium (pl., -ia) (sporo- + Gr., *doche* = receptacle) A cushion-shaped stroma or mat-like structure covered with conidiophores bearing conidia.

Sporophore Specialized hyphae in the *Entomophthorales* which produces a single spore.

SSKI A saturated solution of potassium iodide; taken orally in gradually increased doses to treat lymphocutaneous sporotrichosis and other mycoses.

SSU The 18S or "small" subunit of ribosomal DNA (rDNA).

Sterigma (pl., -ata) A thin, spine-like outgrowth of a cell bearing a conidium, usually larger than a denticle.

Stipe (L., trunk of a tree) The technical term for the stem of agarics, boletes, polypores, and other mushrooms. In medical mycology the term usually refers to the stalk or conidiophore of *Aspergillus* species.

Stolon (L., *stolō* = branch) A hypha that grows horizontally along the surface of the substrate from which rhizoids and sporangiophores arise; used to identify some *Mucorales*.

Stratum Corneum The skin's barrier layer; this is the outermost layer of epidermis, the thickness of a human hair, and made of 25–30 layers of flat, dead cells filled with keratin, a waterproof protein.

Stroma (Gr., *stroma* = mattress) A compact mass in which or on which fruiting bodies develop.

Subclavate Somewhat club shaped.

Superficial Mycosis Fungal disease of hair or stratum corneum, with little or no associated pathology. Categorized as dermatomycoses, they are caused by fungi other than dermatophytes: for example, pityriasis versicolor caused by the *Malasseziales*. (Please see Chapter 22.)

Susceptible (Dose-Dependent) A breakpoint determined for isolates tested in vitro and found to have an intermediate level of susceptibility the antifungal agent in question. Clinical experience found that patients infected with such an isolate would have an improved outcome if they received an elevated or even maximum daily dose.

Sympodial The mode of conidiogenesis in which a conidium is formed at the hyphal apex. A new growing point is formed just beneath the apex, growing around the conidum as it elongates. This process is repeated with conidia being formed opposite from each other, giving the conidiophore a zigzag, or geniculate, appearance.

Synanamorph Any one of the two or more asexual reproductive structures formed by the same fungus: for example, the *Graphium* synanamorph of *Scedosporium apiospermum*.

Synnema (pl., -ata) (syn., coremium). A bundle of compact conidiophores; an erect fascicle of hyphae producing conidia at the apex.

Systemic Mycosis A fungal disease involving one of the deep tissues or organs of the body. (Compare with Disseminated.)

Teleomorph (Gr., *teleos* = finished + *morphe* = form) The sexual reproductive state of a fungus; also termed the "perfect" state.

Thallic A mode of conidiogenesis in which a conidium is formed by the transformation of an entire preexisting hyphal cell, previously delineated by one or more cross-walls (e.g., arthroconidia), or when the existing cell develops a thick wall (chlamydospore), or when a conidium forms directly from the side or tip of a hypha (macroconidia of dermatophytes).

Thallus The entire body of a fungus consisting of both vegetative and reproductive structures.

Thrush Mucosal candidiasis affecting the oral (oral thrush) or vulvovaginal (vaginal thrush) mucosae.

Tinea (L., *Tineidae* = clothes moths) The general name for ringworm or dermatophytosis. The term is used as a prefix to designate the various types of dermatophytoses, for example, tinea *corporis* = dermatophytosis of the smooth skin (please see Table 21.5).

Tinea Gladiatorum Dermatophytosis among high school and college wrestlers caused by *Trichophyton tonsurans*. Usually affecting the smooth skin or, less often, the scalp.

Tinea Versicolor Please see Pityriasis Versicolor.

TLR Toll-like receptor. Surface receptors on many cell types which are pattern recognition receptors for different pathogen associated molecular patterns. TLRs are linked to signal transduction pathways to activate genes responsible for inflammation. *Source:* Abbas and Lichtman (2009).

TNF-α Tumor necrosis factor-α. Produced by activated mononuclear phagocytes, this cytokine stimulates recruitment of PMNs and monocytes to sites of infection, activating them for increased killing of target cells. TNF-α stimulates vascular endothelial cells to express adhesion molecules and to secrete chemokines, increasing adherence and extravasation of phagocytes. When produced in large amounts, TNF-α has systemic effects including fever, cachexia, and production of acute phase reactants by the liver. *Source:* Abbas and Lichtman (2009).

TPN Total parenteral nutrition.

TRB Terbinafine.

***Trichophyton* Agar Media** Vitamin test agars to detect dermatophyte vitamin requirements, commercially available from BD Diagnostics Systems Division. The medium is vitamin-free casamino acids (*Trichophyton* agar #1), supplemented with either inositol (#2), thiamine and inositol (#3), thiamine (#4), or nicotinic acid (#5).

Trophic Form (syn., trophozooite) In pneumocystosis, the asexual haploid cell that may divide by fission.

Truncate Used to describe the base of a conidiogenous cell that appears flat, as if cut abruptly, the scar being as wide as the conidium separated from it, as in an aleurioconidium.

Tuberculate Wart-like, spiny, or finger-like projections on the surface of conidia: for example, tuberculate macroconidia of *Histoplasma capsulatum*.

Uniseriate In *Aspergillus*, used to describe a single layer of phialides formed directly on the vesicle (with no intervening metulae).

Urea Agar Please see Urease Test.

Urease Test A test for the enzyme urease using Christensen's urea agar or other modifications. Urea is split by urease releasing ammonia, causing a pH indicator in the agar to turn from colorless to pale–bright pink. This medium is used to differentiate *Candida* species, which do not produce urease (no color change), from *Cryptococcus* species, which turn the medium to a pale–bright pink (magenta) color. Urease testing also is used to differentiate some dermatophytes.

URL Uniform resource locator.

Valley Fever Referring to the disease, coccidioidomycosis, and the valley, which is the Central Valley of California.

Vd ss The volume of distribution at steady state.

Velutinous Velvety.

Verrucous (Verrucose) Having many warts; cauliflower-like.

Verticil (i) A group of conidiogenous cells whose bases are attached in a whorl around a conidiophore (e.g., in *Verticillium*); (ii) verticils of blastoconidia may be found at the junction between *Candida* pseudohyphae.

Vesicle (i) Swollen apex of a conidiophore (e.g., *Aspergillus* spp.); (ii) swollen end of a sporangiophore of the *Mucorales* bearing sporangioles (e.g., *Cunninghamella bertholletiae*); (iii) in cell biology, a small membrane-bound organelle in the cytoplasm; (iv) a circular fluid-containing, epidermal elevation 1–10 mm in diameter.

VF Titer In Valley Fever or coccidioidomycosis, the endpoint serum dilution in the complement fixation test with coccidioidin.

Villous (Villose) Covered with soft long hair.

Volume of Distribution (Vd) The amount of drug in the body divided by the concentration in the blood. (Please see Chapter 3A, Section 3A.1.1, Major Antifungal Agents Approved for Clinical Use.)

VRC Voriconazole.

VVC Vulvovaginal candidiasis.

WBC White blood cell—leukocyte.

Western Blot An enzyme-linked immunoelectrotransfer blot to demonstrate protein antigens, sized with respect to their molecular mass or isoelectric pH.

Wood's Light or Wood's Lamp A specialized lamp producing ultraviolet radiation that excites fungal fluorophores at wavelength 365 nm, emitting fluorescence. Wood's lamp is used by dermatologists to inspect for ectothrix tinea capitis or pityriasis versicolor.

Woronin Body A round, membrane-bound, granular vesicle, comprised of HEX-1 protein. It is located adjacent to the septal pore in the hyphae of the *Ascomycota*. Functionally, the Woronin body blocks the pore in response to cell damage.

Yeast (syn., blastoconidium) A unicellular fungus that reproduces by budding. Small, round, holoblastic projections from the parent cell are produced during mitosis followed by migration of the nucleus and cytoplasm into the bud. Finally, cytokinesis occurs to form a new daughter cell. The parent cell may produce single or multiple buds. Buds may be either solitary or in acrogenous chains. Some yeasts multiply by fission rather than budding. Events in the yeast cell cycle are finely orchestrated. Please see yeast cell cycle in Chapter 1, Section 1.12, General Composition of the Fungal Cell.

Zoophilic Describing a dermatophyte growing preferentially on lower animals rather than on humans or in the soil.

Zygomycetes Fungi that undergo sexual reproduction in a single enlarged cell (zygospore) after plasmogamy of compatible mating types. These fungi are classed in the subphyla *Mucoromycotina* and *Entomophthoromycotina.*

Zygospore (Gr., *zygo* = yoke) Zygospores are characteristic of the subphylum *Mucoromycotina*. A resting sexual spore is formed from the fusion of two genetically compatible cells, the gametangia. A multinucleate dikaryotic zygosporangium forms, surrounded by a thick cell wall ornamented with ridges, and often held by two suspensor cells. A single zygospore develops within the zygosporangium. The zygospore persists in a dormant state for weeks or months. Next, nuclear fusion (karyogamy) produces a diploid nucleus, the zygote. Meiosis occurs only during germination into a haploid form, which quickly produces a short germosporangium and then releases recombined haploid spores.

REFERENCES FOR GLOSSARY

Abbas AK, Lichtman AH, 2009. *Basic Immunology*, 3rd ed. Saunders-Elsevier, Philadelphia.

Cohan FM, 2005. Periodic Selection and Ecological Diversity in Bacteria, *in*: Nurminsky D (ed.), *Selective Sweep*. Landes Bioscience Publishers, Austin, TX (ISBN: 0306482355).

de Hoog GS, Guarro J, Gene J, Figueras MJ (eds.), 2000. A*tlas of Clinical Fungi*, 2nd ed. ASM Press, Washington, DC.

Difco and BBL Manual of formulae of microbiologic media on line at BD Diagnostics Systems Division. http://www.bd.com/ds/productCenter/DCM.asp

Dorland's Medical Dictionary. Available at http://www.mercksource.com/.

Gilbert SF, 2000. *Developmental Biology*, 6th ed. Sinauer Associates, Sunderland, MA.

Kane J, Summerbell R, Sigler L, Krajden S, Land G, 1997. *Laboratory Handbook of Dermatophytes*. Star Publishing Co., Belmont, CA.

Kendrick B. The Fifth Kingdom. Available at http://www.mycolog.com/GLOSSARY.htm#R.

MedlinePlus Medical Encyclopedia. Available at http://www.nlm.nih.gov/medlineplus/ency/article/001223.htm.

Merck Manuals On-line Medical Library. Available at http://www.merck.com/mmpe/index.html.

Rowley Biochemical, Inc., Danvers, MA. http://www.rowleybio.com/procedures.html. The website contains useful information about biologic stains.

Stedman's Medical Dictionary.

Untereiner WA, Scott JA, Naveau FA, Sigler L, Bachewich J, Angus A, 2004. The *Ajellomycetaceae*, a new family of vertebrate-associated *Onygenales*. *Mycologia* 96: 812–821.

Answer Key

Chapter 1. Introduction to Fundamental Medical Mycology

1. B **2**. D **3**. B
4. B **5**. D

Chapter 2. Laboratory Diagnostic Methods in Medical Mycology

1. E **2**. E **3**. E
4. C **5**. A **6**. B
7. B **8**. A with 4; B with 3; C with 2; D with 1

Chapter 3A. Antifungal Agents and Therapy

1. B **2**. B **3**. E
4. D **5**. A **6**. D

Chapter 3B. Antifungal Susceptibility Testing

1. B. Explanation. The precoated drug panels and Etest provide an easy learning curve and are recommended. Commercial disk diffusion tests are not yet available in the United States.
2. A
3. D
4. A
5. A
6. D
7. A

Chapter 4. Blastomycosis

1. A. Explanation. Thick blood sputum specimens should be digested with dithiothreitol (Mucolyse®). Sputum should be tested on more than one occasion. KOH prep is necessary for direct examination. Sputum should be cultured before digestion or KOH treatment. Calcofluor aids in visualizing the yeast forms.
2. B
3. D
4. A
5. D
6. B
7. A

Chapter 5. Coccidioidomycosis

1. C **2**. E **3**. A
4. D **5**. E **6**. B

Chapter 6. Histoplasmosis

1. D **2**. E **3**. D
4. D **5**. B **6**. B
7. B

Chapter 7. Paracoccidioidomycosis

1. A **2**. D **3**. C
4. A (1); B (2); C (2) **5**. D

Chapter 8. Penicilliosis

1. A **2**. D **3**. B
4. C **5**. B

Chapter 9. Sporotrichosis

1. B **2**. B **3**. B
4. C **5**. C

Chapter 10A. Adiaspiromycosis

1. E **2**. A **3**. A
4. A **5**. A **6**. C

Chapter 10B. Lobomycosis (Jorge Lôbo's Disease)

1. A **2**. B **3**. D
4. B **5**. D **6**. A

Chapter 11. Candidiasis and Less Common Yeast Genera

1. D
2. B
3. A
4. B, A, D, C
5. B
6. C. Explanation. 63,000 cases of invasive candidiasis/year. Over the period 1991–2003 the incidence of invasive candidiasis remained steady.

Fundamental Medical Mycology, First Edition. By Errol Reiss, H. Jean Shadomy and G. Marshall Lyon, III.
© 2012 Wiley-Blackwell. Published 2012 by John Wiley & Sons, Inc.

9. Explanation.

Lipid form of AmB: The patient is neutropenic, clinically unstable. As an alternative to echinocandins in the non-neutropenic patient.

Voriconazole: An alternative in both the neutropenic and non-neutropenic patient groups, when additional coverage for molds is desired.

Fluconazole: The patient is not neutropenic, infection is not severe, and the species involved is *C. albicans, C. tropicalis*, or *C. parapsilosis*. As an alternative in the neutropenic patient without recent azole exposure who is not critically ill.

Micafungin and caspofungin: In the neutropenic patient (and, if the infection is moderate to severe, in the non-neutropenic patient if the patient was receiving azole prophylaxis). These echinocandins are specifically indicated if the species involved is *C. glabrata* or *C. krusei*.

Chapter 12. Cryptococcosis

1. E

2. C, D, B, A

3. C

4. A

5. A

6. D

9. Explanation. *Cryptococcus neoformans* acquired the ability to grow on pigeon droppings, and because of that spread world wide along bird migratory routes and, due to the domestication of the pigeon, along trade routes. *Cryptococcus gattii* developed another ecological-niche: an association with the decaying hollows of mature trees (*Eucalypts* and other tree species), or life in the soil detritus. Climate change may explain survival of *C. gattii* in a temperate environment (Pacific Northwest of North America), allowing further spread of this species.

Chapter 13. Pneumocystosis

1. A (2); B (3); C (4); D (1)

2. D

3. B. Explanation. Sulfamethoxazole inhibits dihydropteroate synthetase (DHPS). Trimethoprim inhibits dihydrofolate reductase (DHFR). There is little role for inhibition of DHFR in antifungal action.

4. C. Explanation. A is not correct because other fungal pathogens are unculturable: *Lacazia loboi*. B and D are not correct because the environmental source of other fungal pathogens is unknown or their isolation from the environment is exceptional: *Paracoccidioides brasiliensis, Penicllium marneffei, Lacazia loboi, Blastomyces dermatitidis*.

5. D

Chapter 14. Aspergillosis

1. A (2) these cultures are more significant
B (3) highly significant
C (3) highly significant
D (1) not significant; but repeated isolation could be significant and should be reported

2. D

3. A

4. D. Explanation. Enriched medium is not needed to isolate *Aspergillus* species; Mycosel and Mycobiotic agar contain cycloheximide, which inhibits aspergilli.

5. B

6. A (2); B (2); C (2); D (1); E (1); F (1)

7. D

8. D

9. C. Explanation. *Aspergillus* species are found worldwide as saprobes. They are active decomposers of organic matter. *Aspergillus fumigatus* grows well at 45°C, making it a common microbe in compost piles. *Aspergillus fumigatus* also grows and sporulates at colder than ambient temperatures as a blue-green mold on bread and fruit in the refrigerator. Hospital air often contains conidia of *Aspergillus* species. The density of fungal spores/m^3 is variable, usually much lower than outside air. Spore count is influenced by construction near patients at risk.

10. D. Explanation. Laminar airflow is not recommended routinely in the protective environment.

11. Explanation. Overall: 58%, reaching up to 90% in allogeneic stem cell transplant recipients with GVHD and in those with extrapulmonary dissemination and cerebral abscess.

12. Explanation. The timeline for invasive pulmonary aspergillosis (IPA) follows a bimodal distribution, with a peak in the first month following a SCT, associated with neutropenia. The first peak is currently less significant, because the routine use of stem cells instead of bone marrow for transplantation, nonmyeloablative regimens, the use of colony-stimulating factors during neutropenia, and the widespread use of antifungal agents have all significantly decreased the incidence of IPA during this period. After engraftment takes place, late onset invasive aspergillosis (41–180 days post-transplant) may occur if/when there is GVHD, corticosteroid therapy, secondary neutropenia, and cytomegalovirus (CMV) disease. Very late invasive aspergillosis (>6 months after transplantation) is associated with chronic GVHD and CMV disease.

Chapter 15. Fusarium *Mycosis*

1. D **2.** A **3.** C
4. A **5.** B **6.** A

Chapter 16. Pseudallescheria/Scedosporium *Mycosis*

1. E **2.** D **3.** C
4. C

Chapter 17 A. Mucormycosis

1. B. Rapid growth will fill the Petri plate.
2. D. See Chapter 17A, Table 17.2 for detailed comparison of morphologic features of the *Mucorales*.
3. A
4. B
5. D
6. A
7. A
8. B
10. Explanation. The sparsely septate *Rhizopus* species and other *Mucorales* are disrupted by tissue grinds. The lack of septa causes the cell contents to be lost during grinding. Mincing the tissues or using the Seward Stomacher is recommended.

Chapters 17B and 17C. Entomophthoramycosis Caused by **Basidiobolus ranarum** *and by* **Conidiobolus** *Species*

1. A (1); B (4); C (3); D (2).
2. D. Explanation. Grinding reduces the yield because *Basidiobolus*, although more regularly septate than the *Mucorales*, may still be inactivated in tissue grinds. This fungus is sensitive to cycloheximide and will not grow on Mycosel agar.
3. D. *Basidiobolus* is homothallic and undergoes sexual reproduction on laboratory media.
4. C. Ballistoconidia are ejected in the direction of the light source.
5. C. Explanation. SSKI is ineffective as a topical treatment. Surgery is an important adjunctive therapy. There is no indication for AmB therapy in subcutaneous entomophthoramycosis. Second-line therapy consists of ketoconazole or itraconazole.
6. D
7. B. "A" is *Basidiobolus* entomophthoramycosis; "C" is adiaspiromycosis; "D"is lobomycosis
8. B
9. C
10. B. Explanation. The fungus often fails to grow from biopsy material; *C. coronatus*, unlike *C. incongruus*, is not homothallic and cultures will not demonstrate zygospores; sporangiophores are scarcely differentiated from vegetative hyphae.
11. E

Chapter 18. Chromoblastomycosis

1. A. There is no practical in vitro test to demonstrate the tissue form of chromblastomycosis agents. These agents are not thermophilic.
2. C
3. A. 2 (*C. bantiana*), B. 1 (*C. carrionii*), C. 1 (*C. carrionii*), D. 2. (*C. bantiana*)
4. B. Explanation. The presence of muriform cells is diagnostic. Culture confirmation is useful but not essential to make the diagnosis but is helpful in guiding therapy. History of travel or residence is a normal part of a questionnaire but chromoblastomycosis is worldwide in distribution. Occupation or recreation is useful information but is not essential for the diagnosis.

Chapter 19. Phaeohyphomycosis

1. D
2. D
3. B
4. C. Examples include *Alternaria, Bipolaris*, and *Curvularia* species.
5. D
6. D

Chapter 20. Eumycetoma (Madura Foot, Maduramycosis)

1. C	**2.** C	**3.** E
4. D	**5.** C	

Chapter 21. Dermatophytosis

1. E
2. C. Explanation. "A," Ringelflechte is the German translation of ringworm; "B", the child's game is not related to the mycosis; "D," the term tinea is also applied to ringworm and that term is derived from the *Tineidae*.
3. A
4. B. Explanation. "A," Dermatophytes are incubated at 30°C; "C," importantly, nails should be embedded, but not submersed, in the agar; "D," normally, dermatophytes grow in the presence of cycloheximide.
5. D
6. E
7. E
8. D
9. C
10. C
11. C
12. D
13. **1.** B **2.** A

Chapter 22. Dermatomycoses

1. D	**2.** D	**3.** A
4. D	**5.** C	**6.** B

Index

Page numbers with *t* and *f* indicates table and figure

Fundamental Medical Mycology, First Edition. By Errol Reiss, H. Jean Shadomy and G. Marshall Lyon, III.
© 2012 Wiley-Blackwell. Published 2012 by John Wiley & Sons, Inc.